D1163726

PATTY'S INDUSTRIAL HYGIENE AND TOXICOLOGY

Fourth Edition

Volume I, Parts A and B
GENERAL PRINCIPLES

Volume II, Parts A, B, and C
TOXICOLOGY

Volume III, Parts A and B
THEORY AND RATIONALE
OF INDUSTRIAL HYGIENE
PRACTICE

PATTY'S INDUSTRIAL HYGIENE AND TOXICOLOGY

Fourth Edition

Volume I, Part A
General Principles

GEORGE D. CLAYTON
FLORENCE E. CLAYTON
Editors

CONTRIBUTORS

R. E. Allan
E. J. Baier
D. J. Birmingham
G. M. Breuer
W. A. Burgess
G. D. Clayton
B. D. Dinman
K. J. Donham
D. D. Douglas
A. C. Farrar
M. A. Golembiewski
W. J. Hausler
C. A. Heideman
G. X. Kortsha
R. G. Lieckfield
M. Lippmann
F. A. Madsen
P. L. Michael
P. R. Morey
J. E. Mutchler
C. A. Piantadosi
J. O. Pierce
W. J. Popendorf
H. J. Sawyer
L. K. Simkins
J. Singh
R. G. Smith
G. W. Wright

A Wiley-Interscience Publication
JOHN WILEY & SONS, INC.
New York / Chichester / Brisbane / Toronto / Singapore

In recognition of the importance of preserving what has been written, it is a policy of John Wiley & Sons, Inc., to have books of enduring value published in the United States printed on acid-free paper, and we exert our best efforts to that end.

Library of Congress Cataloging in Publication Data:

Patty, F. A. (Frank Arthur), 1897–1981
[Industrial hygiene and toxicology]
Patty's industrial hygiene and toxicology / George D. Clayton, Florence E. Clayton, editors; contributors, R. E. Allan . . . [et al.].—4th ed.
p. cm.

"A Wiley-Interscience publication."
Includes bibliographical references.
Contents: v. 1. General principles.

ISBN 0-471-50197-2 (v. 1A)
1. Industrial hygiene. 2. Industrial toxicology. I. Clayton, George D. II. Clayton, Florence E. III. Allan, R. E. (Ralph E.) IV. Title.
RC967.P37 1991
613.6′2—dc20 90-13080
CIP

Printed in the United States of America

10 9 8 7 6 5 4 3 2 1

Contributors

Ralph E. Allan, J.D., C.I.H., University of California at Irvine, Irvine, California

Edward J. Baier, C.I.H., Consultant, formerly Director of Technical Support, OSHA, Washington, D.C.

Donald J. Birmingham, M.D., Professor Emeritus, Department of Dermatology and Syphilology, Wayne State University School of Medicine, Detroit, Michigan

George M. Breuer, Ph.D., Chief, Organic Analysis, State Hygienic Laboratory, University of Iowa, Iowa City, Iowa

William A. Burgess, C.I.H., Polaroid, Marion, Massachusetts

George D. Clayton, C.I.H., retired, formerly Chairman of the Board Clayton Environmental Consultants, Inc., San Luis Rey, California

Bertram D. Dinman, M.D., Sc.D., Clinical Professor, Occupational Medicine, Graduate School of Public Health, University of Pittsburgh, Pennsylvania, retired Vice President for Health and Safety, Aluminum Company of America

Kelley J. Donham, D.V.M., Professor, Institute of Agricultural Medicine and Occupational Health, University of Iowa, Iowa City, Iowa

Darrel D. Douglas, C.I.H., Consultant, 2090 Eola Drive, Salem, Oregon

Alice C. Farrar, C.I.H., Assistant Vice President, Clayton Environmental Consultants, Inc., Novi, Michigan

Mark A. Golembiewski, C.I.H., ENSR Health Sciences, Alameda, California

William J. Hausler, Jr., Ph.D., Director, State Hygienic Laboratory University of Iowa, Iowa City, Iowa

Charlotte A. Heideman, C.I.H., Assistant Vice President, Manager of Industrial Hygiene Services, Midwestern Operations, Clayton Environmental Consultants, Inc., Novi, Michigan

Gene X. Kortsha, C.I.H., retired, formerly Director, Industrial Hygiene Department, General Motors Corporation, Detroit, Michigan

Robert G. Lieckfield, Jr., C.I.H., Vice President, Director of Laboratories, Midwestern Operations, Clayton Environmental Consultants, Inc., Novi, Michigan

Morton Lippmann, Ph.D., C.I.H., Professor, Institute of Environmental Medicine, New York University Medical Center, Tuxedo, New York

Floyd A. Madsen, C.I.H., Director of Laboratories, ISDL, OSHA, Salt Lake City, Utah

Paul L. Michael, Ph.D., Paul L. Michael & Associates, Inc., State College, Pennsylvania

Philip R. Morey, Ph.D., C.I.H., Clayton Environmental Consultants, Inc., Wayne, Pennsylvania

John E. Mutchler, C.I.H., Director, Health, Safety and Environmental Programs, The Quaker Oats Company, Chicago, Illinois

Claude A. Piantadosi, M.D., Department of Medicine and The F.G. Hall Hypo-Hyperbaric Center, Duke University Medical Center, Durham, North Carolina

James O. Pierce, Sc.D., Professor and Director, Southern California Research Center, Institute of Safety and Systems Management, Safety Sciences Department, University of Southern California, Los Angeles, California

William Popendorf, Ph.D., C.I.H., Professor of Industrial Hygiene, Institute of Agricultural Medicine and Occupational Health, Department of Preventive Medicine, University of Iowa, Iowa City, Iowa

Howard J. Sawyer, M.D., Director, Occupational, Environmental and Preventive Medicine, Henry Ford Healthcare System, Detroit, Michigan

Lisa K. Simkins, C.I.H., Vice President, Director, Western Operations, Clayton Environmental Consultants, Inc., Pleasanton, California

Jaswant Singh, Ph.D., C.I.H., Senior Vice President, Director, Pacific Operations, Clayton Environmental Consultants, Inc., Cypress, California

Ralph G. Smith, Ph.D., C.I.H., Consultant, Professor Emeritus, University of Michigan, Franklin, Michigan

George W. Wright, M.D., Consultant, 3795 South Hibiscus Way, Denver, Colorado

Preface

In Greek mythology, Hygeia is celebrated as the goddess of *health*. Thus it was most appropriate that in selecting a name, the founders of this profession chose the word "hygiene," a derivative of the Greek word for health. The word has also been used in connection with dental health, and although it is more popularly recognized when associated with the latter, during the past 10 to 15 years there has been more name recognition for "industrial hygiene," due in some measure to the Occupational Safety and Health Act of 1970. *Industrial* was included in the name because at that time the founders were devoting their talents to protecting the health of the *worker*.

Industrial hygiene is a unique profession, in that it utilizes a number of basic disciplines, embracing the domains of the chemist, engineer, physician, physicist and the toxicologist. The expertise of these disciplines is represented in the chapters that follow.

The *Industrial Hygiene and Toxicology* series was the brainchild of Frank A. Patty, who was instrumental in bringing to fruition the publication of Volumes 1 and 2 in 1948 and 1963, respectively. A third volume entitled "Industrial Environmental Analysis" was envisioned by Frank Patty, but did not materialize under that name.

Who was this man who had the vision to recognize an emerging new profession, who saw the need for a tool to guide those who would follow in this profession? And just what is this profession?

It is defined by the American Industrial Hygiene Association as "that science and art devoted to the recognition, evaluation and control of those environmental factors or stresses, arising in or from the workplace, which may cause sickness, impaired health and well-being, or significant discomfort and inefficiency among workers, or among the citizens of the community."

Frank Patty was one of the pioneers of this profession, who emerged on the scene when there was minimal training for persons entering the field and very little guidance obtainable in the literature. He was first a teacher at the University of Washington; his inquisitive mind led him into research and into affiliation with the

U.S. Bureau of Mines. Eventually he became an administrator, author, and editor. His peers likewise were men who were leaders in developing the industrial hygiene profession in their various areas of expertise, such men as Don Cummings, Warren Cook, Phil Drinker, Leonard Greenburg, Ted Hatch, Don Irish, Henry Smyth, Jr., Jim Sterner, George Wright, and Bill Yant. Frank was a past president of the American Industrial Hygiene Association (AIHA), and at the peak of his career was Director of the Division of Industrial Hygiene at General Motors Corporation. He realized the tremendous advantages of communication and the exchange of ideas the members of AIHA enjoyed through their discussions of mutual problems at meetings. He appreciated the power of the printed word, and the value of providing information for the benefit of people who could not be reached through personal discussions. He convinced some of his colleagues to collaborate in preparing a "primer" covering the various facets of industrial hygiene, which would assist newcomers to the field but also would document progress already made. In writing Frank Patty's obituary in 1981, I said that countless lives have been improved and even saved, owing to the knowledge made available to researchers and others in the *Industrial Hygiene and Toxicology* series. All because of the perceptiveness and sagacity of this man.

The first volume of the series was printed in 1948; a revision was published in 1958, adding chapters on air pollution, noise, and heat stress. I (George) had worked with Frank closely after my assignment by the U.S. Public Health Service to head up the International Joint Commission on Air Pollution, based in Detroit. We became good friends, sharing the same philosophy, ideals, and ethics. Thus I was added as a contributor of the chapter on air pollution in the 1958 revision. The friendship with the Pattys continued after his retirement from General Motors. When the publisher approached Frank Patty to revise the series during the mid-1970s, following the passage of the OSH Act, when demand for books on the subject of health and safety accelerated, Frank came to us with the suggestion that we edit the third revision. We agreed to this, and the third revision of Volume 1, "General Principles," was published in 1978. Revision of Volume 2, "Toxicology," was begun immediately after completion of the first volume, and because of the complexity and expansion of chemical hazards, the data were so voluminous that three books were required to contain the material. They were published in 1981 and 1982.

A third volume, which had never before been published, was added at my suggestion, but not with the original title envisioned by Frank Patty. The third volume was entitled "Theory and Rationale of Industrial Hygiene Practice" and is edited by Lewis and Lester Cralley.

The second revision of Volume 1 in 1958 had added chapters on air pollution, noise, and heat stress; the third revision in 1978 was broadened to include epidemiology, ergonomics, odors, agricultural hazards, quality control, and more. This *fourth* revision has been expanded further, with the addition of 18 subjects, including such topics as visual display terminal safety, biological agents in the workplace, indoor air pollution, and design of analytical laboratories.

The addition of these chapters illustrates how the scope of industrial hygiene has changed and broadened over the past 15 years. In its infancy the profession looked only for serious injury to the worker. Now we are concerned with any adverse reaction, however slight. Historically, the primary concern of industrial hygiene was to protect the health and well-being of the worker in the *workplace*. Since the 1970s we have recognized that a person's health must be protected 24 hours a day, not just the hours spent on the job. We recognize that the environment the worker is in, and the stresses encountered the 16 hours when not at work, influence the ability to handle stresses during the work period. Originally it was the factory worker, the laborers, who were deemed to be in need of protection. Now our concern is not only for them, but also for office workers, farmers, hospital workers . . . essentially, *anyone* in any kind of activity.

In the 1950s the working force was not as greatly concerned with healthful working conditions as it is now, as we begin the 1990s. Fifty years ago workers wanted "hazardous pay" to compensate for hazardous occupations. With the 1970s this approach became obsolete as the unions, the workers, and concerned citizens joined scientists employed by industry and government to work in improving health and safety conditions.

In the early 1900s the fundamental role of the federal government was to provide scientific information to permit formulation of sound legislation. In 1970 the Williams–Steiger Occupational Safety and Health Act gave enforcement powers to agencies of the federal government. Thus the role of all concerned has changed.

Although industrial hygiene has been recognized and practiced from the time of Pliny down through the ages, the greatest accomplishments, in most areas, have been made during the past 50 years. This accelerated progress is due primarily to increased public awareness of excess air contaminants and other stresses. As stated earlier, industrial hygiene is both a science and an art, devoted to the recognition, evaluation, and control of environmental factors and stresses. The recognition and evaluation facets have dominated earlier efforts in this field, and it is only in recent years that greater effort has been applied to the area of controls. During the past 10 years, and as yet on a limited scale, the expertise of this profession is being consulted in the design of equipment, design of processes, and in building design. Intelligent use of the industrial hygiene professionals in the above capacities would have eliminated many costly mistakes that have occurred, such as in so-called "sick buildings," and unnecessary exposure to hazardous chemicals and noise in the industrial and community environment.

The future of industrial hygiene is limited only by the narrowness of vision of its practitioners. It is essential that we open up the horizons of the mind. With the advent of space and ocean explorations, problems encountered in these endeavors are, and will continue to be, very challenging. Adequate control of present and new industries will require ever greater sophistication. The safe use of chemicals yet to be developed, machinery yet to be invented, along with those currently in use, will require increasing vigilance by industrial hygienists. As technology progresses, and there is an escalation of available leisure time, the health problems associated with leisure time will call for greater study. As population increases, community environmental problems will also increase, calling for solution. These are some examples of what the industrial hygienists will encounter, and this will require greater numbers of well trained personnel.

The authors of the chapters in this revised volume were selected because of their expertise in their respective fields. Like their predecessors in this complex and challenging profession, they agreed to devote the time to record their knowledge and experiences to benefit society. It is hoped that our united efforts to provide, at the time of writing, the most recent data available in this complex field will be of value to the readers who face the challenges of today and of the future.

George D. Clayton
Florence E. Clayton

San Luis Rey, California
February 1991

Contents

1 **Industrial Hygiene: Retrospect and Prospect** 1

George D. Clayton, C.I.H.

2 **Legislation and Legislative Trends** 13

Edward J. Baier, C.I.H.

3 **Industrial Hygiene Abroad** 29

Gene X. Kortsha, C.I.H.

4 **The Industrial Hygiene Survey and Personnel** 73

James O. Pierce, Sc.D.

5 **Role of the Industrial Hygiene Consultant** 91

Ralph G. Smith, PH.D., C.I.H., and Jaswant Singh, Ph.D., C.I.H.

6 **Hazard Communication and Worker Right-to-Know Programs** 123

Charlotte A. Heideman, C.I.H., and Lisa K. Simkins, C.I.H.

7 **Industrial Hygiene and the Law** 179

Ralph E. Allan, J.D., C.I.H.

8 **Determination of Biological Agents in the Workplace** 193

William J. Hausler, Jr., Ph.D., and George M. Breuer, Ph.D.

9 **The Mode of Absorption, Distribution, and Elimination of Toxic Materials** 205

Bertram D. Dinman, M.D., Sc.D.

10 Occupational Dermatoses 253

Donald J. Birmingham, M.D., F.A.C.P.

11 The Pulmonary Effects of Inhaled Inorganic Dust 289

George W. Wright, M.D.

12 Physiological Effects of Altered Barometric Pressure 329

Claude A. Piantadosi, M.D.

13 Occupational Health Concerns in the Health Care Field 361

Howard J. Sawyer, M.D.

14 Health and Safety Factors in Designing an Industrial Hygiene Laboratory 383

Robert G. Lieckfield, Jr., C.I.H., and Alice C. Farrar, C.I.H.

15 Quality Control 423

Floyd A. Madsen, C.I.H.

16 Calibration 461

Morton Lippmann, Ph.D., C.I.H.

17 Indoor Air Quality in Nonindustrial Occupational Environments 531

Philip R. Morey, Ph.D., C.I.H., and Jaswant Singh, Ph.D., C.I.H.

18 Potential Exposures in the Manufacturing Industry—Their Recognition and Control 595

William A. Burgess, C.I.H.

19 Respiratory Protective Devices 675

Darrel D. Douglas, C.I.H.

20 Agricultural Hygiene 721

William Popendorf, Ph.D., C.I.H., and Kelley J. Donham, D.V.M.

21 Heat Stress: Its Effects, Measurement, and Control 763

John E. Mutchler, C.I.H.

22 Air Pollution Controls 839

John E. Mutchler, C.I.H., and Mark A. Golembiewski, C.I.H.

23 Industrial Noise and Conservation of Hearing 937

Paul L. Michael, Ph.D.

Index 1041

USEFUL EQUIVALENTS AND CONVERSION FACTORS

1 kilometer = 0.6214 mile
1 meter = 3.281 feet
1 centimeter = 0.3937 inch
1 micrometer = 1/25,4000 inch = 40 microinches = 10,000 Angstrom units
1 foot = 30.48 centimeters
1 inch = 25.40 millimeters
1 square kilometer = 0.3861 square mile (U.S.)
1 square foot = 0.0929 square meter
1 square inch = 6.452 square centimeters
1 square mile (U.S.) = 2,589,998 square meters = 640 acres
1 acre = 43,560 square feet = 4047 square meters
1 cubic meter = 35.315 cubic feet
1 cubic centimeter = 0.0610 cubic inch
1 cubic foot = 28.32 liters = 0.0283 cubic meter = 7.481 gallons (U.S.)
1 cubic inch = 16.39 cubic centimeters
1 U.S. gallon = 3.7853 liters = 231 cubic inches = 0.13368 cubic foot
1 liter = 0.9081 quart (dry), 1.057 quarts (U.S., liquid)
1 cubic foot of water = 62.43 pounds (4°C)
1 U.S. gallon of water = 8.345 pounds (4°C)
1 kilogram = 2.205 pounds
1 gram = 15.43 grains
1 pound = 453.59 grams
1 ounce (avoir.) = 28.35 grams
1 gram mole of a perfect gas ≈ 24.45 liters (at 25°C and 760 mm Hg barometric pressure)
1 atmosphere = 14.7 pounds per square inch
1 foot of water pressure = 0.4335 pound per square inch
1 inch of mercury pressure = 0.4912 pound per square inch
1 dyne per square centimeter = 0.0021 pound per square foot
1 gram-calorie = 0.00397 Btu
1 Btu = 778 foot-pounds
1 Btu per minute = 12.96 foot-pounds per second
1 hp = 0.707 Btu per second = 550 foot-pounds per second
1 centimeter per second = 1.97 feet per minute = 0.0224 mile per hour
1 footcandle = 1 lumen incident per square foot = 10.764 lumens incident per square meter
1 grain per cubic foot = 2.29 grams per cubic meter
1 milligram per cubic meter = 0.000437 grain per cubic foot

To convert degrees Celsius to degrees Fahrenheit: °C (9/5) + 32 = °F
To convert degrees Fahrenheit to degrees Celsius: (5/9) (°F − 32) = °C
For solutes in water: 1 mg/liter ≈ 1 ppm (by weight)
Atmospheric contamination: 1 mg/liter ≈ 1 oz/1000 cu ft (approx)
For gases or vapors in air at 25°C and 760 mm Hg pressure:
To convert mg/liter to ppm (by volume): mg/liter (24,450/mol. wt.) = ppm
To convert ppm to mg/liter: ppm (mol. wt./24,450) = mg/liter

CONVERSION TABLE FOR GASES AND VAPORS[a]

(Milligrams per liter to parts per million, and vice versa; 25°C and 760 mm Hg barometric pressure)

Molecular Weight	1 mg/liter ppm	1 ppm mg/liter	Molecular Weight	1 mg/liter ppm	1 ppm mg/liter	Molecular Weight	1 mg/liter ppm	1 ppm mg/liter
1	24,450	0.0000409	39	627	0.001595	77	318	0.00315
2	12,230	0.0000818	40	611	0.001636	78	313	0.00319
3	8,150	0.0001227	41	596	0.001677	79	309	0.00323
4	6,113	0.0001636	42	582	0.001718	80	306	0.00327
5	4,890	0.0002045	43	569	0.001759	81	302	0.00331
6	4,075	0.0002454	44	556	0.001800	82	298	0.00335
7	3,493	0.0002863	45	543	0.001840	83	295	0.00339
8	3,056	0.000327	46	532	0.001881	84	291	0.00344
9	2,717	0.000368	47	520	0.001922	85	288	0.00348
10	2,445	0.000409	48	509	0.001963	86	284	0.00352
11	2,223	0.000450	49	499	0.002004	87	281	0.00356
12	2,038	0.000491	50	489	0.002045	88	278	0.00360
13	1,881	0.000532	51	479	0.002086	89	275	0.00364
14	1,746	0.000573	52	470	0.002127	90	272	0.00368
15	1,630	0.000614	53	461	0.002168	91	269	0.00372
16	1,528	0.000654	54	453	0.002209	92	266	0.00376
17	1,438	0.000695	55	445	0.002250	93	263	0.00380
18	1,358	0.000736	56	437	0.002290	94	260	0.00384
19	1,287	0.000777	57	429	0.002331	95	257	0.00389
20	1,223	0.000818	58	422	0.002372	96	255	0.00393
21	1,164	0.000859	59	414	0.002413	97	252	0.00397
22	1,111	0.000900	60	408	0.002554	98	249.5	0.00401
23	1,063	0.000941	61	401	0.002495	99	247.0	0.00405
24	1,019	0.000982	62	394	0.00254	100	244.5	0.00409
25	978	0.001022	63	388	0.00258	101	242.1	0.00413
26	940	0.001063	64	382	0.00262	102	239.7	0.00417
27	906	0.001104	65	376	0.00266	103	237.4	0.00421
28	873	0.001145	66	370	0.00270	104	235.1	0.00425
29	843	0.001186	67	365	0.00274	105	232.9	0.00429
30	815	0.001227	68	360	0.00278	106	230.7	0.00434
31	789	0.001268	69	354	0.00282	107	228.5	0.00438
32	764	0.001309	70	349	0.00286	108	226.4	0.00442
33	741	0.001350	71	344	0.00290	109	224.3	0.00446
34	719	0.001391	72	340	0.00294	110	222.3	0.00450
35	699	0.001432	73	335	0.00299	111	220.3	0.00454
36	679	0.001472	74	330	0.00303	112	218.3	0.00458
37	661	0.001513	75	326	0.00307	113	216.4	0.00462
38	643	0.001554	76	322	0.00311	114	214.5	0.00466

CONVERSION TABLE FOR GASES AND VAPORS (*Continued*)
(*Milligrams per liter to parts per million, and vice versa; 25°C and 760 mm Hg barometric pressure*)

Molecular Weight	1 mg/liter ppm	1 ppm mg/liter	Molecular Weight	1 mg/liter ppm	1 ppm mg/liter	Molecular Weight	1 mg/liter ppm	1 ppm mg/liter
115	212.6	0.00470	153	159.8	0.00626	191	128.0	0.00781
116	210.8	0.00474	154	158.8	0.00630	192	127.3	0.00785
117	209.0	0.00479	155	157.7	0.00634	193	126.7	0.00789
118	207.2	0.00483	156	156.7	0.00638	194	126.0	0.00793
119	205.5	0.00487	157	155.7	0.00642	195	125.4	0.00798
120	203.8	0.00491	158	154.7	0.00646	196	124.7	0.00802
121	202.1	0.00495	159	153.7	0.00650	197	124.1	0.00806
122	200.4	0.00499	160	152.8	0.00654	198	123.5	0.00810
123	198.8	0.00503	161	151.9	0.00658	199	122.9	0.00814
124	197.2	0.00507	162	150.9	0.00663	200	122.3	0.00818
125	195.6	0.00511	163	150.0	0.00667	201	121.6	0.00822
126	194.0	0.00515	164	149.1	0.00671	202	121.0	0.00826
127	192.5	0.00519	165	148.2	0.00675	203	120.4	0.00830
128	191.0	0.00524	166	147.3	0.00679	204	119.9	0.00834
129	189.5	0.00528	167	146.4	0.00683	205	119.3	0.00838
130	188.1	0.00532	168	145.5	0.00687	206	118.7	0.00843
131	186.6	0.00536	169	144.7	0.00691	207	118.1	0.00847
132	185.2	0.00540	170	143.8	0.00695	208	117.5	0.00851
133	183.8	0.00544	171	143.0	0.00699	209	117.0	0.00855
134	182.5	0.00548	172	142.2	0.00703	210	116.4	0.00859
135	181.1	0.00552	173	141.3	0.00708	211	115.9	0.00863
136	179.8	0.00556	174	140.5	0.00712	212	115.3	0.00867
137	178.5	0.00560	175	139.7	0.00716	213	114.8	0.00871
138	177.2	0.00564	176	138.9	0.00720	214	114.3	0.00875
139	175.9	0.00569	177	138.1	0.00724	215	113.7	0.00879
140	174.6	0.00573	178	137.4	0.00728	216	113.2	0.00883
141	173.4	0.00577	179	136.6	0.00732	217	112.7	0.00888
142	172.2	0.00581	180	135.8	0.00736	218	112.2	0.00892
143	171.0	0.00585	181	135.1	0.00740	219	111.6	0.00896
144	169.8	0.00589	182	134.3	0.00744	220	111.1	0.00900
145	168.6	0.00593	183	133.6	0.00748	221	110.6	0.00904
146	167.5	0.00597	184	132.9	0.00753	222	110.1	0.00908
147	166.3	0.00601	185	132.2	0.00757	223	109.6	0.00912
148	165.2	0.00605	186	131.5	0.00761	224	109.2	0.00916
149	164.1	0.00609	187	130.7	0.00765	225	108.7	0.00920
150	163.0	0.00613	188	130.1	0.00769	226	108.2	0.00924
151	161.9	0.00618	189	129.4	0.00773	227	107.7	0.00928
152	160.9	0.00622	190	128.7	0.00777	228	107.2	0.00933

CONVERSION TABLE FOR GASES AND VAPORS (*Continued*)
(*Milligrams per liter to parts per million, and vice versa; 25°C and 760 mm Hg barometric pressure*)

Molecular Weight	1 mg/liter ppm	1 ppm mg/liter	Molecular Weight	1 mg/liter ppm	1 ppm mg/liter	Molecular Weight	1 mg/liter ppm	1 ppm mg/liter
229	106.8	0.00937	253	96.6	0.01035	277	88.3	0.01133
230	106.3	0.00941	254	96.3	0.01039	278	87.9	0.01137
231	105.8	0.00945	255	95.9	0.01043	279	87.6	0.01141
232	105.4	0.00949	256	95.5	0.01047	280	87.3	0.01145
233	104.9	0.00953	257	95.1	0.01051	281	87.0	0.01149
234	104.5	0.00957	258	94.8	0.01055	282	86.7	0.01153
235	104.0	0.00961	259	94.4	0.01059	283	86.4	0.01157
236	103.6	0.00965	260	94.0	0.01063	284	86.1	0.01162
237	103.2	0.00969	261	93.7	0.01067	285	85.8	0.01166
238	102.7	0.00973	262	93.3	0.01072	286	85.5	0.01170
239	102.3	0.00978	263	93.0	0.01076	287	85.2	0.01174
240	101.9	0.00982	264	92.6	0.01080	288	84.9	0.01178
241	101.5	0.00986	265	92.3	0.01084	289	84.6	0.01182
242	101.0	0.00990	266	91.9	0.01088	290	84.3	0.01186
243	100.6	0.00994	267	91.6	0.01092	291	84.0	0.01190
244	100.2	0.00998	268	91.2	0.01096	292	83.7	0.01194
245	99.8	0.01002	269	90.9	0.01100	293	83.4	0.01198
246	99.4	0.01006	270	90.6	0.01104	294	83.2	0.01202
247	99.0	0.01010	271	90.2	0.01108	295	82.9	0.01207
248	98.6	0.01014	272	89.9	0.01112	296	82.6	0.01211
249	98.2	0.01018	273	89.6	0.01117	297	82.3	0.01215
250	97.8	0.01022	274	89.2	0.01121	298	82.0	0.01219
251	97.4	0.01027	275	88.9	0.01125	299	81.8	0.01223
252	97.0	0.01031	276	88.6	0.01129	300	81.5	0.01227

[a] A. C. Fieldner, S. H. Katz, and S. P. Kinney, "Gas Masks for Gases Met in Fighting Fires," U.S. Bureau of Mines, Technical Paper No. 248, 1921.

PATTY'S INDUSTRIAL HYGIENE AND TOXICOLOGY

Fourth Edition

Volume I, Part A
GENERAL PRINCIPLES

CHAPTER ONE

Industrial Hygiene: Retrospect and Prospect

George D. Clayton, C.I.H.

1 INTRODUCTION

Those of us who have practiced industrial hygiene for many years can recall that a frequently asked question is, "What is an industrial hygienist?" Therefore it seems appropriate to begin this volume of *Industrial Hygiene and Toxicology* with the definition of industrial hygiene (1):

> Industrial hygiene is that science and art devoted to the recognition, evaluation, and control of those environmental factors or stresses, arising in or from the work place, which may cause sickness, impaired health and well-being, or significant discomfort and inefficiency among workers or among the citizens of the community.

A major goal of the early industrial hygienist was to educate the public, the unions, and the workers regarding the benefits of industrial hygiene to both the workers and the nation as a whole. Much has been accomplished in the past 50 years, as is illustrated in the various chapters of this volume. Initially, progress was made at a creeping pace, although it began to accelerate at the end of World War II.

Some people mistakenly believe that occupational diseases began with the introduction of modern production of chemicals; some are of the opinion that prior to 1950 there was minimal activity in the field of industrial hygiene. Among other fallacious beliefs was that unhealthful conditions are inherent in certain trades.

Patty's Industrial Hygiene and Toxicology, Fourth Edition, Volume 1, Part A, Edited by George D. Clayton and Florence E. Clayton
ISBN 0-471-50197-2

Time has proven that with the proper design of facilities and equipment, along with correct ventilation practice and design, no industry need be considered "unhealthy."

2 RETROSPECT

2.1 Early History

It is true that prior to 1900 there was very little concern expressed for the health of the worker. At the beginning of civilization, humans struggled for existence, and survival itself was an occupational disease. As stratification of social classes evolved, common labor was performed by slaves. This practice continued until the nineteenth century. The victories of war provided a steady supply of slaves. Manual labor by others than slaves was scorned. At one period in their culture Egyptians were prohibited by law from performing manual labor. With such an attitude toward the working man, it is not surprising that no efforts were made to control the work environment or to provide a healthful, comfortable workplace.

Early in the fourth century B.C., lead toxicity in the mining industry was recognized and recorded by Hippocrates, although no effort was made to provide protection for the workers. Some 500 years later, Pliny the Elder, a Roman scholar, referred to the dangers imminent in dealing with zinc and sulfur. He described a bladder-derived protective mask to be used by laborers subjected to large amounts of dust or lead fumes. However, the Romans were more concerned with engineering and military achievements than with any type of occupational medicine.

The writings of the Greek physician Galen, who resided in Rome in the second century, present many theories on anatomy and pathology. Galen was authoritative and assertive in his writings, but even though he recognized the dangers of acid mists to copper miners, his writings gave no incentive to the solution of the problem.

During the Middle Ages feudalism made its appearance, and little improvement was made in work standards. However, one advancement of this period was the provision of assistance to ill members and their families by the feudal guilds. During the twelfth and thirteenth centuries, observation and experimentation flourished in the great universities; however, the study of occupational disease was virtually ignored. Thus little was achieved in the field of industrial hygiene until 1473, with the publication of a pamphlet on occupational disease by Ulrich Ellenbog, which included notable hygiene instructions. This was followed in 1556 by the writing of a German scholar, Georgius Agricola, who effectively described hazards associated with the mining industry. His *De Re Metallica* was translated into English in 1912 by Herbert and Lou Henry Hoover. Agricola's 12-section treatment included suggestions for mine ventilation and protective masks for miners, a discussion of mining accidents, descriptions of what is referred to today as "trench foot" (effects on the extremities caused by lengthy exposure to the cold water of damp mines), and silicosis (disease of the lungs caused by inhalation of silica or quartz dust).

In the sixteenth century industrial hygiene still abounded with the mystical. Many believed that demons lived in the mines and could be controlled by fasting and prayer. The published observations of Paracelsus, the alchemist son of a Swiss doctor, are based on the 10 years he worked in a smelting plant and as a laborer in the mines of Tyrol. The book makes many erroneous conclusions: for example, he attributes miners' "lung sickness' to a vapor of mercury, sulfur, and salt. However, his warnings about the toxicity of certain metals and outline of mercury poisoning were quite advanced. He pointed out the fallacies of many medical theories then current, taught the use of specific remedies instead of indiscriminate bleeding and purging, and introduced new medicines.

It is generally agreed that the first comprehensive treatise on occupational disease, *De Morbis Artificum Diatriba* by Bernardo Ramazzini, an Italian physician, was published in 1700. The book described silicosis in pathological terms, as observed by autopsies on miners' bodies. He presented cautions that he felt would alleviate many industrial hazards. Unfortunately, his warnings for vigilance with these hazards were ignored for centuries. Nevertheless, this book had a prodigious effect on the future of public hygiene. Ramazzini, believing the work environment affected health, asked of his patients, "Of what trade are you?" Most physicians to this day include this same question in recording case histories of their patients.

In the eighteenth century more industrial hygiene problems were being recognized and uncovered. Sir George Baker correctly attributed "Devonshire colic" to lead in the cider industry, and was instrumental in its removal from use. Percival Pott recognized soot as one of the causes of scrotal cancer and was a major force in the passage of the Chimney-Sweepers Act of 1788. Charles Thackrah, both a political and a medical influence, wrote a 200-page treatise dealing with occupational medicine. A scientist with vision, Thackrah stated "Each master . . . has in great measure the health and happiness of his workpeople in his power. . . . Let benevolence be directed to the prevention, rather than to the relief of the evils." Others who were beacons of this age were Thomas Beddoes and Sir Humphry Davy, who collaborated in describing occupations that were prone to cause "phthisis" (tuberculosis). Sir Davy also aided in the development of the miner's safety lamp.

It has been reported that health problems suffered by the artists Rubens, Renoir, Dufy, and Klee may have been caused by toxic heavy metals in bright paints they used. These four artists might have been heavily exposed to such poisonous metals as lead and mercury when they used yellow, red, white, green, blue, and violet paints. Rubens (1577–1640), Renoir (1841–1919), and Dufy (1877–1953) had rheumatoid arthritis, and Klee (1879–1940) suffered from sclerosis. It is believed that they were more heavily exposed to pigments containing salts of heavy metals because they used significantly brighter and clearer colors than their contemporaries who did not suffer from rheumatic disease.

Another early example of toxicity of lead has been reported in the *National Geographic* (September, 1990). The episode occurred in 1845, when 129 members of Sir John Franklin's fateful voyage, which was searching for the Northwest Passage, perished. It had been assumed they died because of a combination of starvation, scurvy, and bad luck. It is now suspected they died of lead poisoning. A recent study of the bones of Franklin's sailors on Canada's King William Island,

and of the bodies of three others from Beechey Island, shows that their remains contained higher than normal levels of lead, and that the lead in the skeletons came from a single source, which matched lead in solder used to seal food tins found in a Beechey Island cache. An anthropologist who participated in the study stated that if the lead did not kill the men, it probably affected their judgment, leading to poor decisions that contributed to the death of the entire crew. Interestingly, it was the invention of tinned foods in 1810 that made possible such long voyages. The degree of the toxicity of lead was not recognized until the 1880s; and as a result, new methods of sealing food tins were devised.

Although developments of the eighteenth century surpassed any of the previous centuries, safeguards for workers' safety appeared to be in a suspended state until the English factory acts of 1833 were passed, which indicated an interest of the government in the health of the working man. These acts are considered the first effective legislative acts in the field of industry; they required that some concern be shown to the working population. This concern, however, was in practice directed more toward providing compensation for accidents than controlling the *causes* of these accidents. Various European nations followed the lead of England and developed workmen's compensation acts. These laws were instrumental in stimulating the adoption of increased factory safety precautions and the inauguration of medical service in industrial plants. *Community responsibility* was developing, which was illustrated by the interest of newspapers and magazines, in efforts to control the environment. One of the most popular publications of the nineteenth century, the *London Illustrated News*, affixed the blame in a mine explosion to negligence in proper gas-testing methods. The same article made a point of the fact that no safety lamps had been provided the workers. In 1878 the last of the English "factory acts" centralized the inspection of factories by creating a post for this purpose in London.

2.2 Development of Industrial Hygiene in the United States

There were probably fewer than 50 industrial hygienists in the United States during the early 1930s, devoting their efforts to protecting the health of workers. The general public had little interest in industrial hygiene. Some managers in industry believed it was not only a source of trouble, but an economic waste. A few physicians viewed it as an intolerable invasion of the doctor's domain, and insisted that only medical doctors were qualified to express opinions regarding the effects of any material or any stress on the human body. Unions were more interested in getting "hazardous pay" than in controlling the environment. Workers were kept in ignorance of the actual hazards.

In the early twentieth century, a champion of social responsibility for workers' health and welfare in the United States was Alice Hamilton, a physician. She not only presented substantiated evidence of a relationship between illness and exposure to toxins, but proposed concrete solutions to these problems. This was the start of an "occupational medicine renaissance." The public was becoming increasingly aware of problems that could be encountered in certain industrial environments, and of the need for legislation. In 1908 the federal government passed a compensation act for certain civil service employees, and in 1911 the first state compensation laws were passed. By 1948 all the states had passed such legislation. These workmen's compensation laws significantly influenced the development of industrial hygiene in the United States, for management began to recognize and appreciate that controlling the environment was less costly than paying huge sums in compensation.

In June of 1939 a number of professionals involved with providing health care for workers through the control of their environment formed the American Industrial Hygiene Association (AIHA) to provide a means of evaluating mutual problems. The members of this new organization consisted of all persons interested in industrial hygiene, whether from industry, academia, or government. These founders came from a variety of backgrounds, such as chemistry, engineering, and medicine. Today the typical industrial hygienist, although specializing in one of these sciences, will have a "working knowledge" of the others.

Through the association's efforts, management began to realize that a *healthy* worker is a *productive* worker. Unions began to understand that the health of their members should be of uppermost concern to them, supplanting hazardous pay. Government passed a series of workmen's compensation laws, culminating in 1970 with the passage of the Williams–Steiger Occupational Safety and Health Act (see Chapter 2). Today laws require industry to provide a healthful environment; unions assist in supplying information to their members regarding the hazards of the environment; and employees must be informed of the hazards of the products and materials with which they work (see Chapter 6). As these developments occurred instrumentation was developed, from crude beginnings to the present sophisticated, automated instrumentation.

The founders of the AIHA recognized the importance of promulgating technical information, and early in the association's history (January, 1940) began publishing articles on instrumentation, toxicity, and so on in *Industrial Medicine* (see Section 2.2.2), as a section of the medical journal. This effort was expanded to include a variety of publications authored by committees of the membership; these were influential in the development of the profession.

At this time the number of professionals whose efforts are devoted to protecting the health of others is in the thousands. The AIHA today is the leading technical association in this field; its members number approximately 9000, and there are satellite groups (local sections) in the United States as well as other countries.

A year prior to the formation of the AIHA, a group of industrial hygienists working under the aegis of the U.S. Public Health Service in Washington, DC organized the Conference of Governmental Industrial Hygienists under the guidance of Jack Bloomfield, the Director of State Programs. Membership was limited to professional personnel in governmental agencies or educational institutions who were engaged in occupational safety and health programs and who desired a medium for the free exchange of ideas and experiences and the promotion of standards and techniques in industrial health. This was not an official government agency. One of the outstanding contributions of this group to the field of industrial hygiene has been the formation of a committee that annually evaluates and establishes threshold limit values and biological exposure indexes for various chemicals used in the workplace. The resulting compilation of standards is utilized throughout the world. This conference (ACGIH) now has a worldwide membership of over 3000 persons.

2.2.1 Significant Events

In 1910 the first national conference on industrial diseases was called in Chicago by the American Association for Labor Legislation. A commission consisting of representatives of medicine, engineering, and chemistry was assigned the task of investigating the magnitude of the problem, and of proposing a method of attack in waging war against industrial disease. About this time several other groups began the study of occupational diseases. The U.S. Bureau of Mines was created in 1919. The U.S. Bureau of Labor, established in 1885, became the federal Department of Labor in 1913, with the charge of collecting "information upon the subject of labor, its relation to capital, the hours of labor, and the earnings of laboring men and women, and also upon the means of promoting their material, social, intellectual, and moral prosperity." The Department of Labor had the responsibility of collecting and disseminating information in the field of industrial hygiene, and following the establishment of the Occupational Safety and Health Administration in 1970, which is under the jurisdiction of the Labor Department, its role was greatly expanded.

The American Museum of Safety was created in New York in 1911 and later became known as the Safety Institute of America. The National Safety Council was organized in 1913. The American Public Health Association organized a section on industrial hygiene in 1914. The U.S. Public Health Service organized a Division of Industrial Hygiene and Sanitation in 1915. The American Association of Industrial Physicians and Surgeons was organized in 1916. Accelerated production of munitions and other war materials for World War I resulted in increased mortality and ill health, making many persons conscious of the necessity for technical guidance in the recognition and control of occupational diseases.

2.2.2 Significant Publications

In 1919 the *Journal of Industrial Hygiene* was established; for some years it was the leading publication in the field. In 1949 it was acquired by the American Medical Association, underwent editorial policy changes, and was published as the *AMA Archives of Industrial Health*.

One of the first endeavors of the newly organized AIHA was to make arrangements for publishing papers, and they were successful in having a section in *Industrial Medicine* devoted to industrial hygiene. This section made its debut in the January 1940 issue. In 1945 this section was published as a supplement to the magazine, and in June of 1946, AIHA began publishing the magazine as the *American Industrial Hygiene Association Quarterly*, in the months of March, June, September, and December. From a quarterly publication it grew to a bimonthly in 1958, with the name changed to *American Industrial Hygiene Association Journal*.

January of 1971 saw its emergence as a monthly publication. It has been for many years, and still is today, the leading source of information in the industrial hygiene field.

In 1986 the American Conference of Governmental Industrial Hygienists (ACGIH) began publishing a journal called *Applied Industrial Hygiene*, later changed to *Applied Occupational and Environmental Hygiene*.

3 PROFESSIONAL CERTIFICATION

During the mid-1950s the AIHA Board of Directors explored the feasibility of certification and registration as a means of improving and maintaining professionalism. Two separate committees were established with these charges. Unfortunately the Committee on Registration was not successful in demonstrating the imperativeness of registration at that time. However, at the 1989 May meeting of the AIHA Law Committee round table (2), the committee cited a critical need to determine what action is necessary to prevent industrial hygienists from being displaced as the professionals within their area of expertise. The problem has been intensified because of the increasing number of new and proposed laws and regulations that limit who can practice in areas traditionally believed to be within the scope of industrial hygiene. The committee had distributed a questionnaire in 1988 to determine how the membership felt about this subject. Over 2000 replies were received. In summary, they indicated that if the association does not act now, we will be legislated out of existence, and further, that the survey conducted was about five years late. Regrettably the efforts of the 1957 ad hoc committee did not come to fruition, and in 1990 registration still is not a reality.

The efforts of the ad hoc committee on certification were more successful. The committee recommended to the AIHA board that the association endorse and initiate establishment of a voluntary certification program for qualified industrial hygienists. The Board approved this recommendation and appointed an ad hoc committee on certification standards, which in 1958 recommended to AIHA that the ACGIH should be invited to join with AIHA in initiating a certification program, and should be invited to delegate six members to join with the ad hoc committee in planning. In March 1959 the joint committee recommended to the two associations that voluntary certification should be conducted by an *independent* incorporated board, and that the two associations should sponsor this board and advance funds for initial expenses. Both AIHA and ACGIH accepted these recommendations. Each organization thereupon delegated six of its members to join in organizing the board. A charter as a nonprofit corporation under the laws of the Commonwealth of Pennsylvania was approved in September 1960, and the first annual meeting of the American Board of Industrial Hygiene was held in Pittsburgh on October 28, 1960.

All diplomates of the American Board of Industrial Hygiene become members of the American Academy of Industrial Hygiene. They remain members in good standing so long as they pay the annual dues assessed by the board within the time specified by the board, and maintain their certification as required by the board. During 1966 the diplomates activated the American Academy of Industrial Hygiene as a voluntary professional society. Since 1977 the academy has conducted a meeting each autumn in various locations in the United States and Canada.

At present (1990) there are 3315 diplomates and 485 industrial hygienists in training (3).

4 ACADEMIC PROGRAMS

In 1918 the Harvard Medical School established a Department of Applied Physiology, which in 1922 became a part of the present Harvard School of Public Health as the Department of Physiology and the Department of Industrial Hygiene. This was the first time that instruction and research in industrial hygiene leading to advanced degrees had been offered anywhere in the world. This school cooperates with the graduate school of engineering: any of the courses offered in the School of Public Health may be elected by students working for a degree of master or doctor of science in engineering. The School of Public Health, which is open to graduates of schools of medicine and graduates in arts and sciences with training in basic medical sciences or specialized training and experience in an important phase of public health work, offers the degree of master of public health and, to especially qualified persons, the degree of doctor of public health. The Harvard School of Public Health was the first place in the world where a qualified person could obtain scheduled, broad instruction in industrial hygiene, regardless of whether his undergraduate training had been in medicine or in the sciences. The number of universities in the United States that provide formal training in the occupational health field has increased substantially since 1970.

Passage of the Occupational Safety and Health Act in 1970 ignited the demand for professionals needed to carry out the mandate of the act; this included the disciplines of occupational medicine, occupational health nursing, occupational safety, and industrial hygiene. Although career counselors recommended the field of occupational health to incoming students, and employers sought personnel to fill positions, the sources for training in this burgeoning field and the people to fill the many vacancies were lacking. Attempts were made to publish information on academic sources, first by the National Institute for Occupational Safety and Health (NIOSH), and later by several other groups. Some of these sources are more complete than others, and some contain contradictory information. Directories of available academic programs in occupational safety and health were obtainable from the following sources.

1. NIOSH published the *Directory of Academic Programs in Occupational Safety and Health (DHEW* [*NIOSH*] Pub. No. 70-126) in 1979. This was a comprehensive report that described occupational safety and health academic programs in the United States. Two hundred and one programs, from associate degrees to postdoctoral studies, were identified within 153 institutions.

This report was revised in 1987 by NIOSH, listing only 106 institutions offering 233 academic programs.

2. In 1985 the Research Triangle Institute released its report, *Study of the Impact of Occupational Safety and Health Training and Education Programs on the Supply and Demand for Occupational Safety and Health Professionals* (RTI/2374-01 FR). This document was prepared for NIOSH and described and assessed many facets of occupational safety and health education. Included in this study were the nature and number of occupational safety and health academic programs in the United States. The data, which covered the period of 1984, stated there were 136 academic institutions offering 241 occupational safety and health degree programs. Institutions were not identified.
3. The National Safety Council in August 1985 published a *Directory of College and University Safety Courses*. It listed 405 schools offering safety education. A total of 145 degrees were listed, including 38 entitled "environmental health" or "environmental safety and health."
4. The American Society of Safety Engineers in 1988 published a survey that listed 129 offered degrees, including 30 in the "environmental" categories.
5. In 1989 the AIHA published a report *Education Opportunities in Industrial Hygiene*, which lists 82 institutions offering degrees or certificates in industrial hygiene. There is no indication as to what degrees are available, with the exception of 32 institutions offering "technician programs."*

In summary, the directories listed above contained a diversity of categories and degree designations, as well as inconsistencies in designations. In order to be of maximum value to those desiring such information a directory should be available from a central source and should list *all* types of programs available, all institutions offering programs, and the resulting degrees.

The need for qualified safety and health professionals will continue to escalate, and it would be prudent to provide an organized, centralized system for such information. Those wishing current information on industrial hygiene programs are referred to the American Industrial Hygiene Association, 345 White Pond Drive, Akron, OH 44320.

5 PROSPECT

Although industrial hygiene made great strides in the latter part of the twentieth century, much remains to be achieved in order to provide optimum health care for the working population. We must persevere in our efforts so that the momentum reached in the twentieth century not only will continue into the twenty-first century, but will intensify.

*Data on occupational safety and health academic programs provided through the courtesy of Judith Erickson, Senior Program Administrator, Southern California Educational Resource Center for Occupational Safety and Health, University of Southern California, Los Angeles, California.

The occupational bionomic system is a consequential portion of the total bionomic system. Because it can be measured, we are able to place constraints on it and thereby make contributions to the health and well-being of those in the occupational bionomic system. These contributions can advantageously influence the effect of the total system on the population in general because on the average, workers spend one-fourth of their time in the occupational area; moreover, workers constitute a significant portion of the total population.

Therefore it is essential that the profession expand its area of influence. A basic goal of industrial hygiene should be to "design out" the health problems before they are incorporated into the workplace. For example, the "sick building syndrome" that is frequently a problem is primarily caused by the lack of input from industrial hygienists in the design of the building. Industrial hygienists should be included in the design of machinery to eliminate the source of noise and emission of atmospheric contaminants. Architects should consult, or have on their staffs, industrial hygienists who would participate in the basic design of buildings and factories, to ensure that environmental concerns have been properly addressed. As discussed in Chapter 20, agricultural hazards are a major problem in our country and require the counseling of our professionals. These are illustrations of areas where our profession should expand its influence, breaking the bonds of the "core" of industrial hygiene.

Our objective is the elimination of all physical and mental hazards in the workplace and its environment. To accomplish this, steps should be taken to establish standards for noise and emission rates for all machinery in the workplace. No machinery should be allowed in the workplace that creates more than 80 dBA. The gaseous and particulate emission rates should be essentially zero. Standards for lighting and color should be established to provide an aesthetically pleasing workplace that also adheres to the fundamentals of ergonomics. Accomplishing these objectives would require considerable financial outlay, because it would necessitate redesigning some of the basic equipment in use today. To obtain the necessary funding the two associations, AIHA and ACGIH, should recruit the cooperation of management of major industries, labor unions, academic, and public policy makers, to work toward the basic goal of providing a hazard-free workplace.

Among the problems facing the industrial hygiene profession even today is lack of "popular" recognition of their role in ecology and their name. For example, *Earth Day 1990* was heavily promoted by a series of events throughout the United States. Of the promotional materials that came to my attention, not one included mention of industrial hygiene. It is interesting to note that both AIHA and ACGIH have been in existence more than 50 years; Earth Day was first observed in 1970, just 20 years ago. It is a sad commentary that a profession that for over 50 years has not only been concerned with the environment and health but through its efforts has ameliorated the environment and health is completely ignored in the activities of Earth Day. This lack of awareness on the part of the general public has severely handicapped funding for our profession at both the educational and governmental levels. This lack of recognition has been used by some within the profession as a reason for changing the name of the profession. This shortsighted approach will accomplish nothing. An aggressive effort is needed to bring the profession to the attention of policy makers and various publications. The profession has a long history of accomplishments and deserves public recognition.

The major problem that continues to face industrial hygienists today is the development of maximum allowable concentrations of contaminants. Threshold limit values (TLVs), when first formulated and compiled by ACGIH, were for the use of *professional* industrial hygienists. They were a useful tool in the hands of a professional. Today, however, the TLVs have been incorporated into laws by the federal government and by many of the states, as well as by some foreign countries. They are then used by the legal profession, who have little if any understanding of the background of TLVs, which were intended to be used as *guides* by professionals, rather than to dictate the safety of the environment according to values located above or below the TLV. Some activists recommend a zero maximum allowable concentration as the only safe level. A study of the TLVs established over recent years shows a marked trend toward lower and lower concentrations.

In the development of air pollution standards, in addition to technical considerations, public opinion plays a very important role in their development. Since the beginning of 1989, the Environmental Protection Agency (EPA) at the urging of President Bush, has been formulating a new clean air act. Recently this new clean air act was submitted to Congress for approval. The president's plan would reduce ozone emissions, a component of urban smog, by 2.5 percent a year, instead of the 3 percent a year specified in an earlier version of the bill. Sulfur dioxide emissions would be reduced by 9 million tons by 2001, instead of 10 million tons. In an editorial on August 3, 1989, the *Wall Street Journal* asks:

> Is this really to suggest that people will get sick if sulfur dioxide is reduced by nine million rather than 10 million tons? In short, what do these numbers have to do with the public's health?
>
> In a sense, environmentalism today is primarily a political phenomenon. Sympathetic news stories fan public fears and the public fans the politicians. Thus, to produce "bold" initiatives the Bush administration and Congress decided to disregard the current state of scientific knowledge, whose uncertain status is seen as inhibiting action of any sort. This tactic was expressed succinctly in *Newsweek*: "The most effective environmental standards are based on what's technologically feasible, not on arcane estimates of potential health hazards." In other words, do what you can and don't worry about why you are doing it.
>
> Accordingly, Mr. Bush's clean-air bill, credited to EPA head William Reilly, guarantees no specific public health benefits. Indeed, the President himself acknowledged the initiative's liberation from the constraints of real science by declaring that the "time for study has passed."

As the foregoing quotation illustrates, public opinion is a very important factor in the development of air pollution contamination levels. The industrial hygiene profession should be vigilant that political influence does not contaminate the establishment of TLVs. We must make strenuous efforts to protect the purity of the TLVs.

Because the trend for TLVs is toward establishing the lowest possible atmospheric concentrations, the astute industrial hygienist will prepare management accordingly, urging that management give industrial hygiene a priority status in the planning of new or expanded facilities. The industrial hygienist should be part of the team planning any new facility, and should have the following responsibilities in such projects:

1. Establishing the permissible discharge of chemical emissions.
2. Approving the design of conveyors and other transporting equipment, with special emphasis on emissions and noise.
3. Specifying the noise emissions from each piece of machinery.
4. Approving the general layout of the facility with special regard to calculated emissions and noise.
5. Approving all local and general ventilation systems, including dust collectors.
6. Determining the accuracy of the plant layout to prevent air pollution and noise pollution from becoming a community problem.
7. Studying and recommending proper controls for solid and liquid waste.
8. Recommending insulation of sensing devices for physical and chemical contaminants, and selecting locations within the plant with appropriate computer data processing capabilities.
9. Recommending robots at work stations where machinery design cannot meet the environmental standards.
10. Recommending an adequate budget to maintain the facilities in first-class condition. Included in the budget should be funds for sufficient personnel to provide routine physical inspection of the plant, and sufficient personnel to interpret the computer data for distribution to appropriate individuals.

Future accomplishments of the industrial hygiene profession are without bounds; the only limitations are those we permit ourselves. The founders of the profession had foresight, energy, and integrity. The profession grew during a period when it was not "fashionable" to be concerned with the environment. Now that the public and political policy makers are more knowledgeable as well as concerned, the horizon for expanding our influence is unlimited. The opportunity is here; we have only to take advantage of it.

REFERENCES

1. *Am. Ind. Hyg. J.*, **20**, 428–430 (1959).
2. *Am. Ind. Hyg. J.*, **51**(4), 278–284 (1990).
3. American Board of Industrial Hygiene roster, 1989.

CHAPTER TWO

Legislation and Legislative Trends

Edward J. Baier, C.I.H.

1 INTRODUCTION

It is most difficult to assess historically when the first health standards for workers were developed. Perhaps there were royal decrees but, because territorial boundary battles were frequent, there were always losers who were forced to toil. Little attention was paid to these workers because there was always a fresh supply.

The Industrial Revolution changed the course of history. During this period a number of physicians and scientists made personal observations that linked illnesses and deaths to specific working conditions, (see Chapter 1). There was no systematic approach to preventing occupational illness, however. It was merely an accounting.

The first occupational disease prevention system that could be directly related to the concept of an exposure limit was a study by a German scientist who, in 1833, exposed himself and some animals to carbon monoxide. He achieved a correlation between absorbed dose and physiological effect.

Episodic studies relating effects found in workers with their trades have been reported for centuries, but very little action was taken legislatively from a public policy perspective.

The English Parliament enacted landmark labor legislation in 1833, known as the Factory Act. This act limited hours of work for children and provided inspection of certain workplaces. In 1864 and 1867 revisions broadened coverage to all enterprises employing more than 50 persons, prohibited eating meals in noxious

Patty's Industrial Hygiene and Toxicology, Fourth Edition, Volume 1, Part A, Edited by George D. Clayton and Florence E. Clayton
ISBN 0-471-50197-2

environments, and required mechanical ventilation to control certain dusts. In 1897 further revisions included medical inspection of factories.

The first list of what may be called occupational health standards was published in Germany in 1895. Lehmann of the Munich Department of Hygiene published a List of Tolerable Concentrations in that year.

Labor legislation in the United States lagged behind Europe. It was not until 1896 that Massachusetts created the first Bureau of Labor Statistics. Factory inspection there did not begin until 1877. Similar legislative action was taken soon after by New Jersey, New York, Connecticut, Michigan, Missouri, and Minnesota.

2 EARLY PREVENTION OF OCCUPATIONAL DISEASE IN THE UNITED STATES

There was considerable activity in the United States just prior to, during, and following World War I.

In 1910 the American Association for Labor Legislation convened the first national conference on industrial diseases. A commission made up of representatives of medicine, chemistry, and engineering was assigned the task of defining the magnitude of industrial diseases and proposing solutions.

The first major public act to control occupational disease in the United States was passed in 1912 when a prohibitive federal tax was levied on white phosphorus used in the manufacture of matches. This tax literally eliminated "phossy jaw," an occupational disease of exposed workers.

The Bureau of Mines was created in 1910 and the federal Department of Labor was created in 1913. The National Safety Council was organized in 1913. In 1912 the U.S. Public Health Service created an industrial hygiene program which became the Division of Industrial Hygiene and Sanitation in 1915. The American Public Health Association organized a section on industrial hygiene in 1914. The American Association of Industrial Physicians and Surgeons was organized in 1916. The *Journal of Industrial Hygiene* was first published in 1919. In 1922 the Department of Industrial Hygiene became part of the Harvard School of Public Health.

On the federal scene, the Division of Labor Standards was created in 1934. The Social Security Act was passed in 1935, followed the next year by the Walsh–Healey Public Contracts Act.

Early activities of the Public Health Service, often working with the U.S. Bureau of Mines, were in-depth epidemiologic studies to define those exposures that led to occupational diseases. In today's terms the studies were crude efforts of risk assessment. They were marvels, however, considering the industrial hygiene equipment and medical instrumentation available at the time.

The Social Security Act altered the concept of job safety and health in that it was more logical to keep a worker safe and healthy and paying into the fund than it was to have an injured or ill worker drawing from the fund. To accomplish this, federal funds were made available to state programs, and most existing industrial hygiene programs grew and others were created with these resources.

The Walsh–Healey Public Contracts Act made it mandatory for industries supplying goods in excess of $10,000 to the federal government to maintain a safe and healthy workplace. This act was not a truly effective piece of legislation at the time, because its coverage was limited to those industries with federal contracts. The war effort during the 1940s gave it real meaning, however, because more and more goods came under its purview.

As state programs proliferated, so did state regulations. Most states promulgated standards for many air contaminants in the workplace, but the standards varied from state to state. To obtain some uniformity, the Surgeon General of the Public Health Service convened a meeting of state representatives in 1936 to review each state's standards. An attempt was made for uniformity. Great progress resulted, and the group decided to meet annually. In 1938 the group became the National Conference of Governmental Industrial Hygienists, with full membership limited to one representative from each state. Soon thereafter it changed its name to the American Conference of Governmental Industrial Hygienists (ACGIH), with membership limited to industrial hygienists in government and academia. At its fifth meeting the concept of a formal committee to establish standards for air contaminants in the workplace was considered. This later became the Threshold Limits Committee.

In 1939 the American Industrial Hygiene Association (AIHA) was formed, because industrial hygienists from industry, organized labor, academia, and other employers were not accepted into the National Conference of Governmental Industrial Hygienists. It is of interest that even today, industrial hygienists who leave government service or academia, including past chairs and officers of the conference, are excluded from full membership.

To complement threshold limit values (TLVs), the AIHA sponsored the Z-37 Committee of the American Standards Association (later American National Standards Institute) for several years. Originally, the Z-37 Committee published "allowable concentrations," then "maximal acceptable concentrations," and finally "acceptable concentrations." The term acceptable concentration was not defined as a single concept but was related to the duration and pattern of the exposure.

The Toxicology Committee of the AIHA developed Emergency Exposure Limits for a short time. These were defined as "concentrations of contaminants that can be tolerated without adversely affecting health but not necessarily without acute discomfort or other evidence of irritation or intoxication." They were intended to "give guidance in the management of single, brief exposures to airborne contaminants in the working environment." Three such limits—nitrogen dioxide, 1,1-dimethylhydrazine, and 1,1,1-trichloroethylene—were published in 1964, but no other such limits were published after that.

3 DECADE PRECEDING FEDERAL OCCUPATIONAL SAFETY AND HEALTH LEGISLATION

Promulgation and enforcement of health standards was primarily a function of state, county, and city governments during the 1960s. Forty-two states and 32 city and/or county governments had identifiable occupational health programs. Many occupational health units, however, had small programs, and several employed only one person. Practically all units adopted TLVs carte blanche, but some, notably Michigan, New York, Oregon, and Pennsylvania, developed specific standards for air recirculation, confined space entry, dry cleaning, and ionizing radiation. Pennsylvania developed and adopted specific short-term limits for exposure to airborne air contaminants in addition to TLVs.

Under the Walsh–Healey Act, the Federal Bureau of Labor Standards enforced TLVs for those companies providing goods to the federal government. These limits were used to define "health" and were updated by the bureau annually until 1968. Because congressional hearings on federal job safety and health regulations were active in 1968 and it appeared that the enactment of a law was imminent, no further updates were made. It was this 1968 list of TLVs that was used by the Occupational Safety and Health Administration (OSHA) as start-up standards under the 1970 Occupational Safety and Health Act (OSHAct), Section 6(a).

During this period, expansion of the Atomic Energy Commission and environmental program issues began to have an impact on the growth and, for some, the very existence of state, county, and city occupational health programs. These shifts of program emphasis, together with the drying up of federal funding, cut deeply into occupational health programming. Although the number of staff members involved directly in state programs remained fairly stable during the 1960s, the number of staff involved directly in industrial hygiene activities diminished owing to increased commitments to radiation and community environmental activities.

By the time of the Congressional hearings on federal occupational safety and health legislation in the late 1960s, only three states—Michigan, New York, and Pennsylvania—had full-time occupational health staffs in excess of 25 professional personnel. In these three states, this constituted about 36 percent of those employed by all of the states. These three states, together with New Jersey and California, employed about half of all occupational health professionals employed by all of the states. Seven states—Alabama, Arkansas, Arizona, Delaware, Nebraska, North Dakota, and South Carolina—had no identifiable occupational health program.

4 FEDERAL OCCUPATIONAL SAFETY AND HEALTH LEGISLATIVE HEARINGS

When bills were introduced into the Congress to consider a national policy on occupational safety and health, it became obvious that the task would be complex.

The Metal and Nonmetallic Mine Safety and Health Act of 1966 was undergoing contest. This act set health standards and safety standards and required mandatory reporting of accidents, injuries, and occupational diseases among miners.

Black lung was a concurrent problem that needed to be addressed. Bills patterned after the 1966 legislation and targeted at coal mines were undergoing debate. The magnitude of the cost and definition of the disease were major concerns.

There was a patchwork of occupational safety and health activity at the federal level. The Atomic Energy Act addressed the needs of workers, as did the Walsh–

Healey Public Contracts Act. Job safety and health matters were also implied in Armed Services Procurement Regulations, specifications of the Department of Transportation, the U.S. Air Force safety regulations and technical orders, the U.S. Coast Guard regulations, and regulations concerning toxic chemicals promulgated by the U.S. Environmental Protection Agency.

A particular critical problem for resolution of the many bills introduced had to do with a single industry approach, such as a mine agency or a nuclear energy agency which, at the time, was administered by the Bureau of Mines or the Atomic Energy Commission (AEC), versus the multi-industry approach envisioned by this occupational safety and health legislation. The net effect on employer groups was that it divided the testimony presented by industry. Industries regulated by a single agency, such as the AEC, were familiar with AEC methods of operation and did not want a change. The bills introduced represented a change and a big question of unknown and uncharted territory. At times during the debate on the bills, it appeared that this one conceptual issue could scuttle all efforts to develop any type of job safety and health legislation at all.

Industry was generally fearful of the secretary of labor having virtual dictatorial powers over plant closures for alleged violations. One person testified that "The Secretary would sit as judge, jury, and executioner." Concern was expressed that an unqualified inspection group would be armed with the authority to take drastic action.

There was an effort to remove the myriad of federal regulations that applied to the workplace as well as a thrust to remove applicable state, county, and local government laws and regulations. Federal–state relationships were favored, but a double set of standards was frowned upon. Industry favored the inclusion of government workers for coverage under the law.

Many employers and trade association representatives who testified emphasized differences between major industries and small industries, and pointed out that major industries had job safety and health programs in place. Many showed charts and graphs indicating downward trends in injury experience when they testified. There was strong support for adoption of nationally recognized consensus standards. It was suggested that these be used as guides to educate smaller businesses and not be used as enforcement standards.

At one point the creation of a National Occupational Safety and Health Board of five members to prepare and promulgate standards was proposed. This was widely supported as a means for reducing the powers of the secretary of labor in that a board would set standards for enforcement. The board would represent a separation between standard-setting and enforcement.

A prophetic statement was that the secretary of labor would not be placed in the intolerable position of being called to task for both rule-making and enforcement responsibilities. It was argued that a separate board would achieve a far greater degree of public confidence than a single person who combines the roles of rule maker, inspector, prosecutor, and adjudicator of violations.

Some supported limiting civil penalties to only willful violation of standards. There was strong sentiment in the business community that there be no penalty for violation of the "general duty clause." It was to be used solely as an advisory. If, after it was called to the attention of the employer, the employer failed to abate the conditions, then a penalty could be imposed. This was, basically, the way all state programs operated at the time; there was no first instance sanction.

The thrust of organized labor testimony centered around statistics enumerating fatalities, disabling, injuries, and existing occupational illness statistics. The testimony was generally interspersed with anecdotal case histories and disasters, such as a group of workers made ill by a chemical exposure or a coal mine or plant explosion. It was an effective way of getting the congressional committee's and the media's attention.

The small amount of money each state spent on each worker was highlighted. With practically all of the states having small programs, this was a major point. Also, because much of the tax revenues came from working men and women, it had a dramatic effect politically in that the work force wasn't getting much of a return on its investment.

Organized labor, as opposed to industry, did not support a National Occupational Safety and Health Board. They viewed such a board as an extra layer of bureaucracy. In the final analysis, Congress eliminated further consideration of this board. Congress gave the secretary of labor the broad powers delineated in the final House and Senate bills, and created the Occupational Safety and Health Review Commission to adjudicate disputes arising from enforcement actions of the Secretary. Further, Congress stressed use of the Administrative Procedures Act of 1946 for the development of standards. These two actions provided checks and balances satisfactory to the conflicting constituencies.

Organized labor supported coverage of government workers in the legislation. Government worker coverage was one of the few issues on which there was agreement between organized labor and industry.

Although there was agreement that government workers be covered, there was strong disagreement on employee rights and responsibilities. Industry wanted duties placed on employees. Organized labor wanted more involvement in the conduct of compliance inspections. The compromise on this issue was the clarification of Section 5 of the final bills. Section 5(a) requires the employer to maintain a safe and healthy place of employment, and Section 5(b) requires the employee to cooperate. In addition, employees were given the right to participate during inspections and provide input to the inspection in private.

There were two major concerns expressed by organized labor. The first was how passage of legislation, which they freely supported, might interfere with collective bargaining. There was considerable testimony, directly and indirectly, on limiting interference with this prerogative in any legislation. The other concern was how federal standards would be interpreted and enforced by the many individual states. This presented a dilemma in that the federal enforcement staff was very small and probably could not provide adequate worker coverage; at the same time, organized labor opposed dealing individually with possibly 50 states that might tool up and administer occupational safety and health programs.

Professional organizations that provided testimony were also not of a single voice during the hearing procedure. All viewed the technical aspects of the bills and commented on specific ways to change the draft language to improve, as each organization believed, the achievement of the bills' objectives.

One group suggested that the national board that had been proposed adopt standards and provide them to the individual states for promulgation and enforcement. During this testimony it was suggested that the board's standards be tentative for a year or two, to permit employers to tool up and then submit the standards to the states for enforcement action.

Several professional organization spokespersons pointed out the differences between safety and health and between health standards and safety standards. Knowledge was imprecise, particularly regarding currently available air sampling instruments and chemical analytical procedures.

Practically all groups who testified fostered relationships between the Department of Labor and the Department of Health, Education, and Welfare. That both play major roles regarding job safety and health was highlighted by several persons.

All who testified, either in direct statements or during questioning, also stressed that the federal government was the largest employer in the nation and, as such, should set an example by providing coverage for federal employees.

Technical experts stated that the national board was limited to only five members and the experience of those represented on it was too general. A requirement to specify experts in industrial hygiene or occupational medicine was recommended to Congress. One who spoke recommended the term occupational health scientist, because the language in one of the bills was vague, stating "a background either by reason of training, education, or experience in the field of occupational safety or health" for three of the members of the board. Further, although the board could use consultants, no specific provision was made for advisory groups. Following a parade of witnesses, the board would have grown from five to as many as 12 persons if all recommendations had been adopted.

In universal points not limited to professional groups, but to others as well, it was obvious that no action of Congress would eliminate the occupational safety and health problems of the nation. Any action that Congress would take would complement or supplement a mutual objective. The most significant federal contribution would be funding for research and dissemination of information to both employer and employee.

After several years of congressional committee activity including hearings, debate, discussion, and questions and answers, the marked-up bills went to the floors of the Senate and the House of Representatives. On October 13 and November 16 and 17, 1970, the Senate considered the bill and finally passed the so-called "Senate version." The House considerd the bill on November 23 and 24, 1970, with amendments, and passed the "House version." There were significant differences between the two versions, and a conference committee co-chaired by Senator Harrison Williams of New Jersey and Congressman William Steiger of Wisconsin met several times to resolve the differences. On December 17, 1970, the Senate agreed to the conference report. The House agreed on December 17, 1970, and forwarded the bill to the White House. President Nixon signed the Occupational

Safety and Health Act on December 29, 1970 to take effect 120 days after its enactment.

5 THE OCCUPATIONAL SAFETY AND HEALTH ACT OF 1970

During the first year of operation under OSHAct, four congressional committees held hearings on its various aspects. Over 100 bills were introduced in Congress to amend the act. On two separate occasions amendments to prevent OSHA from inspecting small employers was passed by the Congress. Both were vetoed by the president. Small business exclusions from enforcement by the agency were effected through riders to appropriation bills. These exclusions continue with each appropriation to the present.

OSHAct is one of the most simple, yet one of the most complex and controversial pieces of legislation ever devised. Its basic premise is that each employer must provide a place of employment free from recognized hazards and comply with standards [Section 5(a)1], the "General Duty Clause." Each employee is required to comply with rules, regulations, and orders applicable to his or her actions and conduct [Section 5(b)]. The problem is definition of hazard, definition of compliance and the methods used to develop standards, rules, regulations, and orders.

OSHAct created several distinct governmental entities: The broad powers provided to promulgate, modify, or revoke occupational safety and health standards; provide variances from standards, rules, and regulations; issue citations; assess and settle penalties; approve state plans; conduct training; collect, compile, and analyze statistics; employ experts, consultants, or organizations; and carry out the purposes of the act that reside with the secretary of labor. The act did not specifically create an Occupational Safety and Health Administration (OSHA), but it authorized the creation of an "Assistant Secretary of Labor for Occupational Safety and Health." To administer OSHAct, the secretary delegated much of the Labor Department's authority to this assistant secretary, who must be confirmed by the Senate.

Research, professional and technical training, and workplace investigations, along with related duties, are vested in the secretary of health, education, and welfare (now secretary of health and human services). The act, differing from authorities given to the secretary of labor, specifically established a National Institute for Occupational Safety and Health (NIOSH) to "perform the functions of the Secretary" and gives the institute director specific additional functions. These include developing regulations to carry out prescribed duties, to receive and dispose of property, to appoint and fix the salary of staff, to contract with experts, to enter into contracts, and to authorize grants. The institute director, however, is appointed by the secretary of health and human services and serves for a term of six years unless removed by the secretary.

Section 12 of OSHAct establishes an Occupational Safety and Health Review Commission composed of three members appointed by the president and confirmed by the Senate. The terms of the three members are for six years. The commission is the final administrative authority to rule on a particular case brought to its attention by a person concerning the appropriateness of any action of the secretary of labor. Primarily these actions include citations, proposed penalties, and enforcement actions. The review commission may conduct its own investigations and modify the secretary of labor's findings. The commission's findings and orders are subject, however, to further review by federal courts.

Section 27 of OSHAct established a "National Commission on State Workmen's Compensation Laws" composed of 15 members appointed by the president to study and evaluate state programs and to recommend improvement. The Commission functioned until July 31, 1972, when it submitted its final report to the president and to Congress. It was never reactivated.

Section 7 of the act established a National Advisory Committee on Occupational Safety and Health (NACOSH). NACOSH consists of 12 members appointed by the secretary of labor, four of whom are designated by the secretary of health and human services. The committee has representatives from management, labor, the occupational safety profession, the occupational health profession, and the public. OSHAct requires that the chairperson of NACOSH be a public member. Section 7 further authorizes the appointment of advisory committees to assist the secretary of labor in carrying out functions required by various parts of the act. From time to time advisory committees have been appointed for a specific purpose, most notably in the standard-setting or rule-making process.

6 RELATED LEGISLATION AND EXECUTIVE AND COURT ORDERS

The Occupational Safety and Health Act of 1970 (OSHAct) was one of many pieces of environmental legislation developed by Congress in a relatively short period of time. Each was designed for a specific purpose because, at the time, the public was becoming more and more conscious of the fragile nature of the environment. The impact each would have on the other was not immediately apparent when enacted, and Congress moved to fine tune many of these laws over time.

Several of these laws that relate, directly or indirectly, to OSHAct are the Federal Hazardous Substances Act of 1966, the National Environmental Policy Act of 1969, the Coal Mine Health and Safety Act of 1969, the Clean Air Act of 1970, the Clean Water Act of 1972, the Consumer Product Safety Act of 1972, the Environmental Pollution Control Act of 1972, the Safe Drinking Water Act of 1974, and the Toxic Substance Control Act of 1976. More recent legislation affecting OSHAct includes the Comprehensive Environmental Response, Compensation, and Liability Act of 1980 (CERCLA), the Superfund Amendments and Reauthorization Act of 1986 (SARA), and the Asbestos Hazard Emergency Response Act of 1986 (AHERA). Each of these has components and elements that duplicate some components and elements of OSHAct.

In OSHAct, Congress gave the secretary of labor the authority to promulgate existing federal standards and standards developed by national consensus standards organizations. This authority was granted for a period of two years following the effective date of OSHAct, April 28, 1971. In addition, occupational safety and health standards developed specifically under the Walsh–Healey Act, the Service Contract Act, the Construction Safety Act, and the National Foundation on Arts and Humanities Act were superseded and became standards. All of these are known as "start-up" standards.

To get the legislation enacted, Congress exempted government workers at all levels from direct coverage by the U.S. Department of Labor. It also exempted some private sector workers from coverage when another federal agency exercises its statutory authority to prescribe and/or enforce that agency's existing occupational safety and health standards or regulations. Although Congress exempted employees of the federal government and states and political subdivisions of states from coverage, it required (Section 19) that the head of each federal agency maintain a comprehensive occupational safety and health program consistent with standards promulgated for private sector employers. Enforcement provisions were suspended, however. Further, if a state asserts jurisdiction under state law (Section 18), it must establish and maintain an effective and comprehensive program applicable to all employees of public agencies of the state and its political subdivisions.

Several presidents have affirmed Section 19 of the Act by requiring federal and agency department heads to maintain safety and health programs in accordance with OSHAct requirements for the private sector. This has been accomplished through executive orders, notably Executive Orders 12044 (1978), 12196 (1980), and 12291 (1981). OSHA is required to monitor these federal programs. OSHA has also developed a comprehensive program to review state plans and to monitor state programs.

In addition to requiring federal compliance with OSHA regulations, executive orders require an economic and regulatory analysis, but not a cost/benefit analysis, of each proposed regulation. The analysis includes the evaluation of the economic consequence of regulatory options, the impact on an individual type of industry within a class of industries with similar hazards to be regulated, the impact on industries within a particular geographic region, and the levels of government impact when two or more federal agencies.are regulating the same hazard.

This regulatory concept was also expressed by Congress in the Regulatory Flexibility Act of 1980. This act requires OSHA to determine, in addition to feasibility, whether a regulation will have a significant impact on a substantial number of small workplace entities. Further, the National Environmental Policy Act requires a full environmental impact statement for all proposed regulations.

A detailed analysis of regulatory action is also required by the Paperwork Reduction Act of 1980. This act limits the amount of time required of employers to create and maintain records specified by regulations. Each regulatory agency is assigned a certain number of "burden hours" and, as regulations are developed, the "burden hours" must be kept within allotments and divided appropriately.

The most important legislation for regulatory affairs in the federal government is the Administrative Procedures Act. This act was designed to get the greatest public input into the regulatory process. Basically, this legislation charts exactly how a regulation flows through the agency, how the public is informed of the agency's proposal, and how the public can provide its input.

Although the legislative branch created OSHAct and the executive branch administers it, in the final analysis it is the judicial branch that dictates how the act must be carried out. Court action can result for a number of reasons. Anyone who does not agree with a decision of the Occupational Safety and Health Review Commission can appeal to a federal court. Anyone can request that a federal court take action against OSHA to begin a regulatory action. Anyone can bring any final regulatory action of OSHA to a federal court if that individual believes the action was not properly carried out.

Federal district or circuit courts hear and decide on these cases frequently. Generally an agreement is reached. When agreement is not achieved, and the Supreme Court decides to consider a case, a final decision is made.

A landmark Supreme Court decision was made in 1980 which changed the ground rules and set the limits for OSHA's standard-setting authority. A permanent standard for benzene was published in the *Federal Register* reducing the permissible exposure limit from 10 to 1 part per million parts of air (ppm).

Section 6(b)(5) of OSHAct requires that a standard be set at a level "which most adequately assures, to the extent feasible, on the basis of the best available evidence, that no employee will suffer material impairment of health or functional capacity even if such employee has regular exposure to the hazard dealt with by such standard for the period of his working life. . . . Whenever practicable, the standard promulgated shall be expressed in terms of objective criteria and of the performance desired."

The published standard for benzene was taken to the Fifth Circuit Appeals Court. This court invalidated the standard on the basis that the secretary of labor had failed to satisfy OSHACT's criteria for standard setting. The case was accepted by the Supreme Court, and an opinion was rendered on July 2, 1980. The Supreme Court upheld the decision of the Fifth Circuit Appeals Court. The Supreme Court held that before issuing any new permanent standard the secretary of labor must determine that it is necessary and appropriate to remedy a "significant" risk of material health impairment. The secretary presented no proof that the risk of cancer from benzene existed at 10 ppm but *not* at 1 ppm. The Secretary relied on the belief that some leukemia might result from exposure to 10 ppm and that a reduction to 1 ppm would result in a reduction of leukemia cases. The critical element in the decision was that much of the evidence demonstrated adverse effects from benzene exposures well above 10 ppm. The danger of exposures between 10 and 1 ppm had not be adequately documented. The court further stated that OSHAct is not concerned with absolute safety but with the elimination of "significant" harm. A "safe" workplace is not the equivalent of a "risk-free" workplace. The secretary relied too heavily on OSHA's "no risk" carcinogen policy and had failed to use the "best available evidence" and the "latest scientific data" specified in OSHAct.

The court established the ground rule that OSHA has reasonable leeway in making its findings during the rule-making process. OSHA must determine that a "significant" risk exists and that such risk can be reduced or eliminated by the proposed standard. With this information, OSHA must then set a standard that is both technologically and economically feasible. The aim is to set the lowest feasible level necessary to eliminate "significant" risk.

This supreme court decision altered the process for the manner in which regulations are promulgated. The process begins with the initiation of the need for a standard or regulation. This may be generated by the agency, by a request from NIOSH, by referral from the Environmental Protection Agency (EPA), by a request from a state or local government, by referral from a court as the result of a petition, or by a request from an employer or an employer group, a labor organization or its representatives, a standards-producing organization, or any interested person. Rule making frequently begins as a result of congressional hearings.

Analysis papers are generated that consider various options. An Advanced Notice of Proposed Rulemaking (ANPR) may be prepared for publication in the *Federal Register*. This is to solicit data required for the rule-making activity at an early stage. If data appear to be adequate to proceed without an ANPR, a Notice of Proposed Rulemaking (NPRM) is prepared for publication in the *Federal Register*.

All comments received in response to the ANPR and/or NPRM are stored in the docket office. Comments almost always include a request for a public hearing, which is always granted. The verbatim testimony presented at the hearing as well as documentation to support the position presented in testimony are also stored in the docket office. These records are available for review and perusal during normal work hours.

Concurrent with the development of the standards package are studies, either in-house or by contract, to satisfy the legal specifications required by the related legislation discussed above. When all of the requirements are met, the docket is reviewed and a final rule is developed and published in the *Federal Register*. A reasonable time is granted for all affected parties to comply.

Because of the time frame required to develop standards under OSHAct and related legislation, congressional proposals have been introduced from time to time to increase efficiency. Further, bills have been introduced that would amend existing legislation to address concerns that have been expressed since OSHAct came into being. Over the years, however, no action has been taken. There has always been a concern that opening OSHAct would produce legislation that could create more problems than the changes would remedy.

In 1989, however, it appears that there may be consensus developing that could open up OSHAct for amendments. There is some agreement by practically all sides of issues that there are serious concerns. Activities in occupational safety and health legislation in other countries have demonstrated the value in tighter enforcement, more joint labor–management involvement, more training programs, and more authority for higher penalties and criminal sanction. A number of advisory groups, task forces, and study groups have been established to review the issues. It is and will be an interesting time as well as an opportunity to have a voice in shaping the future of occupational safety and health.

7 FUTURE OF THE OCCUPATIONAL SAFETY AND HEALTH ACT

It is impossible to predict the future with any degree of certainty. Unless one is familiar with all the related factors and the impact of each at various times, one is simply guessing. There are, however, historical events and experiences that may be reviewed in order to indicate trends for the future. Analyzing some of these events and experiences as of January 1, 1990 and using a crystal cube instead of a crystal ball, it may be possible to determine the likelihood for change in OSHAct.

Immediately following passage of OSHAct and on several occasions during its first year, several congressional bills were introduced that were essentially designed to gut it. These bills were met with very strong opposition, and none ever advanced beyond congressional committee. Those persons who were initially involved in passage of the legislation took a paternal interest in protecting it from change.

Administration of the act has been closely monitored by Congress through oversight hearings involving OSHA, and NIOSH, and other witnesses. Congress has also given specific direction to OSHA in appropriation bills. Although the membership of Congress and its committees has changed over time, Congress has been tenacious in maintaining OSHAct in its current form, without amendment.

On the other hand, the day-to-day departmental administration of OSHAct has changed dramatically over time. The OSHA direction during the Carter Administration was very different from the OSHA direction during the Reagan Administration. One of President Reagan's first orders of business when he assumed office was to recall all federal regulations in process and to establish a regulatory review panel chaired by Vice-President Bush. The fact that the well-protected and well-defined OSHAct could be administered with such flexibility became a primary concern to many observers. Several trade associations prepared proposed changes to the act and some of these proposals were introduced by senators and representatives, but none of their bills ever advanced beyond congressional committee.

Also, during the Carter Administration, the president increased the authority of the Executive Office of Management and Budget (OMB) to review regulations from all regulatory agencies in order to try to control inflation and to limit the rising unemployment statistics. President Reagan expanded this authority to OMB to include coordination of regulation between agencies and to monitor the impact of regulation on the economy and on world trade. During the 1980s a number of events took place that prompted Congress to attempt to use OSHAct in a manner that the framers of the act had not anticipated.

A building under construction in Connecticut using a method known as "lift-slab" collapsed. A number of fatalities and serious injuries occurred among the workers. OSHA, following a thorough investigation of the incident, issued citations and proposed several million dollars in penalties. The senators from Connecticut introduced a bill that would require that the monies collected as a result of the fines, which would go into the government's general fund, be used to underwrite coverage for the injured workers and the families of workers killed in the accident. During review of the bill, it was determined that financial penalties assessed under OSHAct were for violation of the regulation and not for the gravity of the consequence of the violation. In some instances, a violation of a regulation could result in only a minor monetary penalty, but the effect of the violation could be very serious. Because the seriousness of the effect of a violation was often unrelated to the proposed penalty, the bill was never reported out of congressional committee.

In the late 1980s a new word, *egregious*, entered the OSHA vocabulary. This term was used to describe a violation of a regulation beyond a *willful* violation. It was used in those instances in which the transgression was determined to be deliberate, premeditated, and/or intentional. Such proposed citations carried many millions of dollars in proposed penalties. Congress at this time saw an opportunity to use OSHAct, with OSHA's new use of *egregious* citations, as a solution to an emerging problem.

Corporate buyouts, mergers, and take-overs became a way of life for many businesses during the 1980s. Many corporations began to use all available sources of financial reserves to accomplish the takeover or to prevent such action and preserve the enterprise. In both situations, this often meant borrowing capital from company retirement funds. Because the federal government operates a retirement insurance program and a number of retirees without valid "company retirement programs" were leaving their jobs, Congress began to search for a source of funds to maintain the federal program. Bills were introduced in Congress to multiply OSHA citation penalties by some mathematical factor and use the monies collected for the retirement insurance program. During review, it was pointed out that there is most often a huge difference between a proposed penalty and the actual amount of money collected into the general fund as a result of a settlement agreement. As with all other proposed congressional actions with regard to OSHAct, this proposal did not progress beyond committee.

Many recent congressional actions, although not involving OSHAct directly, have had a significant impact on job safety and health.

In strengthening legislation for cleanup of hazardous waste sites, Congress specified that OSHA develop and enforce standards to protect workers engaged in the activity. Although the authority for such sites remains with the EPA, OSHA has been given a specific congressional mandate.

The presence of asbestos in schools and public buildings has received much attention from Congress. Again, the removal and cleanup functions related to asbestos require worker protection, which is within the scope of OSHA's responsibilities. Because joint responsibilities exist between EPA and OSHA, jurisdictional enforcement problems arose.

A similar joint responsibility exists with regard to radon in public buildings. Congress has given EPA the authority and funding to ameliorate the health threat from radon exposure. There are, however, many persons employed in public buildings who are covered by OSHAct, and questions of joint jurisdiction have emerged.

Perhaps one of the more substantial occupational safety and health concerns is in the area of indoor air quality. Congress has expressed great concern by formally recognizing that the air quality within buildings is frequently inferior to the air quality in the general community. EPA has been granted extended congressional authority to remedy the situation. Again, EPA and OSHA jurisdictions are in conflict because many persons are employed in indoor activity, including those in most workplaces, where OSHA has been given specific responsibility by Congress in OSHAct.

Under OSHAct Congress has prohibited OSHA from health and safety coverage of employees where another federal agency exercises its mandated statutory authority. This prohibition applies in several government agencies. In 1989, there was a rising public concern with the manner in which contractors in the nuclear energy field were carrying out their job health and safety functions. These contractors are under the jurisdiction of the federal Department of Energy (DOE). Congress held hearings to consider the transfer of job safety and health functions from DOE to OSHA. In an attempt to work by interagency agreement, as opposed to congressional mandate, teams of inspectors from DOE and from OSHA began to conduct inspections of nuclear energy facilities.

From the foregoing it appears that public policy has been shifting. Many persons, including health scientists, safety professionals, employers, trade unions, trade associations, and others involved in matters of health and safety, off the job as well as on the job, believe that the time may be ripe for changes in the OSHAct of 1970. Whether or not Congress will be lobbied to execute change and whether Congress will react is open to debate.

As we enter the 1990s there are many forces, not related to job health and safety, that will impinge on Congress. The dramatic changes in Western Europe will definitely exert an influence. The balancing of the U.S. national budget and the significance of world trade, with the United States becoming a debtor nation, will be priority agenda items. Problems with housing and the homeless, illegal substance abuse, immigration, AIDS, education, the "ozone layer," "acid rain," and "global warming" will need to be addressed.

Because Congress has acted in recent years on several job safety and health matters peripheral to OSHAct, it could conceivably continue this course without dealing with OSHAct directly. Congress considered job safety and health legislation in the late 1960s only because there was a strong lobby to effect its passage. Over the years those involved have left the scene, retired, or become complacent. There does not appear to be a strong lobby to effect change in OSHAct.

On the other hand, it is conceivable that Congress may rethink its more recent activity in the unrelated manner in which it has legislated environmental matters. There appears to be increasing worldwide general public concern for the environment. These public pressures may very well direct Congress to review all existing environmental laws and create a comprehensive environmental program. This could result in a single federal department to consolidate all programs with an environmental mission, including the federal job safety and health program. Such a concept is not without precedent. Prior to passage of OSHAct, many states created environmental resources or environmental services departments that housed occupational and environmental programs. With this state model, such an action may receive favorable reaction from Congress at the federal level.

It is of professional interest that the fields of industrial hygiene and industrial safety spawned many, if not most, of the current environmental programs. Such programs include air pollution, radiation safety, water quality, noise control, hazard abatement, and related programs that have come to be known as "environmental." It may well be, if Congress decides to enact comprehensive legislation, that the progenitor of environmental programs will become just a bit player in the final drama because of lack of expressed concern or because of "tunnel vision"!

CHAPTER THREE

Industrial Hygiene Abroad

Gene X. Kortsha, C.I.H.

This chapter describes the history and state of the industrial hygiene profession in foreign countries. Some countries are treated in more detail because of the advanced level of the profession, because of readily available information, and often because of the generous help of foreign colleagues. Other countries are mentioned in less detail or not at all. Some include highly advanced nations where industrial hygiene may find it hard to grow in the shadow of other established health professions. In other instances, industrial hygiene is still in the early stages and must grow before it can leave its mark on society. Finally, some countries are not mentioned despite the maturity and strength of the profession, because requests for information may have never reached the right office or persons. Whatever the circumstances, this chapter strives to provide information on the status of industrial hygiene abroad. It is a beginning, a first step, for others to build on and expand in the future.

1 AUSTRALIA*

1.1 Historical Development

Australia has a long history of occupational hygiene related activities. In 1923, concerns about occupational illnesses, such as silicosis, lead intoxication, and coal pneumoconiosis, led to the establishment of the first Division of Industrial Hygiene

*Information on Australia in this chapter is based primarily on material provided by A. T. Jones, MAIOH. His courtesy is much appreciated.

Patty's Industrial Hygiene and Toxicology, Fourth Edition, Volume 1, Part A, Edited by George D. Clayton and Florence E. Clayton
ISBN 0-471-50197-2

in the state of New South Wales. Next came the state of Victoria. Eventually, the remaining four states also established occupational hygiene organizations either within their state health departments or as part of governmental analytical laboratories.

Australia is a federation consisting of six states and two territories. In occupational health and safety matters, the federal government has direct responsibility for its employees throughout the country as well as for the population of the territories. The six states have their own autonomous governments but depend on the federal government for federally financed projects. The states have legislated occupational health and safety matters since the late nineteenth and early twentieth centuries through the promulgation of various acts, such as the Factories, Construction Safety, and Mining Acts. More recently, the federal government has taken the lead in regulating health and safety matters.

In 1949 the federal government established the Commonwealth Institute of Health at Sydney University to carry out nationwide projects, conduct research, and teach in areas primarily related to occupational medicine. In the late 1950s, the federal government established the National Health and Medical Research Council of Australia (NHMRC). Through its Occupational Health Committee and the Occupational Hygiene Subcommittee, the council has played an advisory role vis-à-vis the states. This development has led to a degree of uniformity in regulating substances such as lead and asbestos. In 1983, the NHMRC adopted the 1983–1984 threshold limit value (TLVs) of the American Conference of Governmental Industrial Hygienists (ACGIH) as permissible exposure limits without binding legal status except where specified by law (1).

The Robens Report in Great Britain also had a major impact on the approach to occupational health and safety in Australia. Several states have passed acts based on the tripartite approach recommended in the report and adopted by the British Health and Safety Commission. Thus employers, employee organizations, and governmental bodies are brought together to develop and draft occupational health and safety regulations for places of work. Employers and employees are jointly responsible for implementing such regulations. The ultimate responsibility for providing a safe and healthful workplace, however, rests with the employer. Governmental agencies are charged with enforcing the regulations. Jones (2) adds:

> To ensure that there was "expert" contribution to these deliberations, which was not universally the case in the initial stages, a peak council of professional organisations was formed to lobby for the most appropriate person, nominated by it, to be appointed to the committees. This peak council consists of the presidents of the associations of occupational physicians, hygienists, nurses, safety engineers, ergonomists, and it has brought a balance to the development of regulations and standards at the national level.

Acts promulgated by the states call for tripartite councils or commissions responsible for occupational health and safety in mining and industry. Occupational hygienists have participated and have contributed to such activities.

In 1985 the federal government established the tripartite National Occupational Health and Safety Commission known as Worksafe Australia. According to Jones (3),

> In principle, this body has the responsibility to produce legislation, standards, guides and codes of practice as a basis of uniformity. It is anticipated that these will be implemented by the States, although there is no legal requirement for this to take place. Indeed States have, and are, doing their own regulating and standard setting, which is lessening the intent for uniformity in occupational health and safety administration. Endeavors are being made to correct this situation.

In recent years, large employers have established in-house medical and, in some cases, occupational hygiene services. Because of the increasing need for such services among smaller work establishments, a number of consultants offering services-for-fee have entered the field.

1.2 Education and Training

Estimates regarding the number of practicing occupational hygienists in Australia are hard to come by. Findley (4) estimates that about 200 individuals practice occupational hygiene at a professional level, including specialized areas such as noise and radiation protection. Over half work for federal or state departments of health or labor. The remainder are employed in large industries, with a few working in academia. Small companies either rely on governmental occupational hygiene services or purchase such services from consultants.

Most occupational hygiene practitioners have an undergraduate degree in the sciences or in engineering. For years, such individuals received their occupational hygiene training on the job, working for experienced hygienists. Most of the latter either worked for governmental agencies or had worked for such agencies at one time. Thus governmental occupational hygienists in Australia have played a major role in sharing their knowledge and experience with many, both inside and outside of government.

In 1977 the Commonwealth Institute of Health began offering a course in occupational hygiene consisting of 90 days of practical experience in analytical and field work. The course continues to be offered annually and has recently become a part of Worksafe Australia. The course has only a small staff and limited facilities. Even though only a small number of students can attend at any one time, most senior occupational hygienists in Australia have taken it.

Australian universities presently offer neither bachelor's nor master's degree curricula in occupational hygiene. The University of Sydney offers an MPH (Occupational Health) degree within its faculty of medicine. Course work includes occupational hygiene topics. In addition to the University of Sydney, Ballarat College of Advanced Education in Victoria, the South Australia Institute of Technology, and the Western Australian Institute of Technology offer graduate courses with occupational hygiene content as part of their diploma programs in occupational health and/or safety. The Footscary Institute of Technology in Victoria also offers such courses as part of a program leading to an Associate Diploma in Occupational Health and Safety.

Since about 1980, Deakin University in the state of Victoria has offered a postgraduate diploma in occupational hygiene on a part-time attendance basis.

A small number of Australian occupational hygienists have earned postgraduate degrees at universities abroad.

1.3 Professional Organizations

Occupational hygienists established the Australian Institute of Occupational Hygiene (AIOH) in 1978. Conceived as a professional organization, the institute's early concerns focused on membership qualifications and training and a conscious effort on the part of the institute to expect and maintain a high level of education among its members. According to Jones (2),

> There are continuing developments in education, with a number of tertiary institutions wishing to establish courses in occupational hygiene—not generally favored by AIOH, because of content and teaching limitations and the basic premise that training in occupational hygiene should be at the postgraduate level.

In line with such efforts, a number of Australian hygienists have earned certification through the British Examination and Registration Board in Occupational Hygiene (BERBOH) or the American Board of Industrial Hygiene (ABIH). AIOH reached an agreement with ABIH for the latter to offer core and specialty examinations in Australia. Initially, ABIH-certified U.S. hygienists (CIHs) proctored such examinations. Today, examinations take place under the direction of Australian CIHs. As a result, more and more Australian hygienists are submitting to such examinations and becoming certified in various aspects of the profession.

2 CANADA

2.1 Historical Development

Ontario passed the first Act for the Protection of Persons Employed in Factories in 1884. The act was intended to protect children, young girls, and women working in industrial workplaces. Among others, the act stated (5):

> Every factory shall be ventilated in such a manner as to render harmless, so far as reasonably practicable, all gases, vapours, dust or other impurities generated in the course of the manufacturing process or handicraft carried on therein that may be injurious to health.

It established penalties if workers suffered permanent loss of health.

The Workmen's Compensation Act of 1914 set guidelines for worker health education and for control measures intended to prevent accidents and work-related diseases.

The Factory, Shop and Office Buildings Act, promulgated in World War I, was expanded in the late 1920s to require compulsory medical supervision of workers handling hazardous substances. Amendments to the act in 1932 required employers to label containers of benzene and to report any cases of industrial diseases directly to the Director of Industrial Hygiene.

In 1926 a study among workers at the Porcupine Mining Camp indicated that only 98 of the 236 miners employed did not suffer from tuberculosis, silicosis, or both diseases. Thus in 1929, the Workmen's Compensation Board set up clinics for periodic health examinations of underground miners and established the Silicosis Referee Board.

Efforts outlined so far dealt mostly with work-related medical problems. The history of occupational hygiene as such can be traced back to the 1930s, to the era of industrial expansion and of growing recognition within industrialized nations of the need to prevent occupational diseases. Canadian occupational health scientists in industry and government established professional links with their colleagues abroad, particularly in the United States and in Great Britain. Such collaboration originated from their common cultural, political, and industrial heritage. Over the years, links between Canadian occupational hygienists and their American counterparts led to many Canadians becoming members of, and working with, the American Industrial Hygiene Association (AIHA), ACGIH, the American Academy of Industrial Hygiene (AAIH), and ABIH. Similar links were established in Great Britain with the London School of Hygiene and Tropical Medicine, the British Occupational Hygiene Society (BOHS), the Institute of Occupational Hygiene (IOH), and BERBOH. Only twice in its 50-year history did the American Industrial Hygiene Conference meet outside the United States. Both times the event took place in Canada, in Toronto, in 1971 and in Montreal in 1987.

In recognition of his contributions to the broad field of industrial hygiene and occupational medicine, AIHA conferred its 1986 Yant Memorial Award to Dr. Ernest Mastromatteo, a distinguished Canadian scientist and researcher of the health effects of heavy metals and other chemical hazards.

Among Canadian Provinces, Ontario has been one of the leaders in occupational health. In 1976, in addition to safety, the Ontario Ministry of Labour also assumed the responsibility for occupational health. In the ensuing administrative reorganization, existing safety branches in the Ministry of Labour combined with the Occupational Health Protection Branch, formerly with the Ministry of Health, and the Mines Engineering Branch, previously under the Ministry of Mines and Natural Resources, thus giving birth to the Occupational Health and Safety Division.

This reorganization changed the status of occupational hygienists. Up to this time, they had inspected workplaces as technical advisers to safety inspectors of the Ministry of Labour who were responsible for enforcing health regulations. Once occupational hygienists were brought under the Ministry of Labour, they assumed the full responsibilities of inspectors. The Occupational Health and Safety Act of 1978 further changed the way occupational hygiene was practiced in Ontario.

It made industrial hygienists responsible for periodic inspections of workplaces in addition to answering requests for occupational hygiene investigations made by other inspectorate branches and by the Worker Compensation Board.

The act requires that health and safety committees be formed in places of employment with 20 or more workers. Such committees consist of equal numbers of representatives of management and of labor, the latter being selected by the workers. This concept of a cooperative effort between management and labor is based on the recognition that the two parties are best qualified to anticipate and control work-related hazards. The minister of labour may order that such joint management–labor committees be formed in smaller places of employment depending on the use of hazardous materials.

An important feature of the Occupational Health and Safety Act of Ontario is the Internal Responsibility System (IRS) that places joint responsibility for health and safety on both employers and employees. It creates an interlocking set of rights and obligations on the part of management as well as of workers. Both parties are expected to work in concert rather than in conflict, to help and support each other in important health and safety functions in the workplace, with government standing by to see that these responsibilities are fulfilled. The IRS thus represents an effort to promote in-plant problem solving and decrease reliance on government inspectors.

The act also established the Canadian Centre for Occupational Health and Safety (CCOSH) to promote the right of Canadians to a healthful and safe work environment. The center is chiefly funded by the federal government and operates under the direction of a council of governors representing federal, provincial, and territorial agencies, industry, and labor. The center reports to Parliament through the minister of labour.

The CCOSH provides occupational health and safety information to anyone requesting it. It publishes technical information in easy to read language and provides computerized information through access to data banks and data bases.

Other developments have taken place in the broad area of occupational medicine including occupational hygiene. Canada promulgated the Workplace Hazardous Materials Information System (WHMIS) on June 30, 1987. It stipulates the formation of joint management–labor health and safety committees and the right for workers to refuse unsafe work. These concepts are now part of federal legislation and have been adopted by all provinces.

Several provinces have instituted new ways of providing occupational health services to workers and, in some instances, to communities.

The Province of Quebec has granted workers the right to participate in choosing occupational health personnel through joint worker–management committees. Ontario, Manitoba, and Alberta have fostered the establishment of worker-controlled occupational health centers. Such centers differ in several respects from one another.

Yassee (6) reports that the Ontario Workers' Health Clinic has a board of directors composed of union officials, community leaders, representatives of Canadian occupational safety and health groups, and others. It has established clinics in Hamilton, Toronto, Sudbury, and Windsor. All funding comes from union donations and from the health clinic's outside services and contracts.

The Manitoba Federation of Labor Occupational Health Centre was founded with seed money provided by organized labor (6). All members of the board of directors are from the Manitoba Federation of Labor. Presently, the Social-Democratic government of Manitoba provides the operating funds for the center. Continued financial support by organized labor has enabled the center to acquire an extensive occupational health library and a computer link to international data bases. The center provides occupational health information as well as medical and occupational hygiene services to employers and employees alike, irrespective of their union affiliation or lack thereof.

2.2 Professional Training

The University of Toronto in Ontario first offered physicians a diploma in industrial hygiene in 1942. Canadian universities began offering courses leading to master of science degrees in occupational hygiene in the 1950s. The first two institutions to offer such courses were the School of Hygiene at the University of Toronto and McGill University in Montreal, Quebec. Subsequently, both universities also established doctoral programs in occupational hygiene. In 1978, the Ontario Ministry of Labour established manpower training and made funds available for the establishment of a Diploma course in occupational hygiene at McMaster University in Hamilton, Ontario. More recently, Laval University in Quebec has offered courses leading to a certificate in occupational health. Presently, there are no Canadian undergraduate degree programs in occupational hygiene.

In the opinion of Smith (7), "The large number of occupational health and safety technology programs offered at community colleges in Ontario suggests that there is probably an oversupply of safety and hygiene trained technologists in Ontario, and hence in Canada." Also, "It would appear . . . that the present production rate of 30 to 40 hygienists per year is adequate to meet the national need."

2.3 Occupational Hygienists in Canada

Smith (7) estimates the number of Canadian occupational hygienists at between 300 and 350. About 60 are certified, most of them by ABIH. Recently, Canadian occupational hygiene groups, such as the Occupational Hygiene Society of Ontario and others, founded the Canadian Registration Board of Occupational Hygiene (CRBOH). Hygienists certified by CRBOH are called Registered Occupational Hygienists (8).

Over half the Canadian occupational hygienists practice their profession in large industries such as the petrochemical, automotive, and foundry industries. Smaller industries handling highly toxic materials also have occupational hygienists on their staffs. About one-fourth of hygienists work for regulatory agencies with the remainder employed in academia or in other capacities. Occupational health regulations and the need of small industries for occupational hygiene assistance have led to the establishment of a number of consulting services. Such services are available through manpower training programs, universities, and community colleges. Recently, private consulting firms have become available that offer both field and analytical services in occupational hygiene.

Many small establishments continue to assign occupational health responsibilities to safety personnel, chemists, metallurgists, or other individuals familiar with the chemicals and processes used at the plant.

2.4 Professional Organizations

Presently there is no national association for occupational hygienists in Canada. There are, however, occupational hygiene societies in several provinces. The two largest societies are the Occupational Hygiene Association of Ontario (OHAO) and l'Association pour l'hygiene industrielle au Quebec (AHIQ). There are local AIHA Sections in Alberta, British Columbia-Yukon, and Manitoba.

2.5 Permissible Exposure Limits

The federal Department of Labour in Ottawa has jurisdiction over matters affecting the occupational safety and health of workers employed in federal establishments across Canada, such as transportation and civil service workers. The department exercises its authority through the Canada Dangerous Substances Regulations that recognize the most recent ACGIH TLVs as permissible exposure limits.

With the exception of the national fire code, which is generally adopted by all provinces, Canada has no federal codes that apply nationwide. Thus all workers, other than federal workers, come under the jurisdiction of the individual provinces. Because Canada has 10 provinces, applicable rules and exposure limits may vary from province to province and substance to substance. Cook (9), in his book *Occupational Exposure Limits—Worldwide*, provides detailed information on exposure limits adopted by Canadian provinces.

3 EGYPT*

3.1 Historical Development

The International Bank for Reconstruction and Development (IBRD, 1983) categorizes Egypt as a middle-income oil exporter nation with a 1981 GNP per person of more than $410 (10). Like some other developing nations, Egypt has made great efforts and achieved considerable success in modernizing its agriculture and its industry. Ancient tools and advanced technologies are thus coming together in a society known for the splendor of its ancient culture and for its rich documentation reaching back thousands of years. Temples and tombs in Upper Egypt display

*Information presented in this chapter is based on material provided by Professor M. H. Noweir. His courtesy is much appreciated.

drawings showing farmers using plows similar to those still used in some parts of the world. In addition, those murals provide graphic proof that over 2000 years ago, Egyptians processed flax, tanned hides, made pottery, and mined, refined, and cast metals.

Rapid modernization of Egyptian industry began after World War I. It has brought about a higher standard of living, better education, and improved health care for certain segments of the population. Such improvements, however, have also exacted a price. According to Noweir (10), recipient of the 1979 Yant Memorial Award,

> The process of industrialization in most developing countries has not been accompanied by a parallel process in establishing occupational health services to counteract the hazards associated with industrialization, which are usually manifested in the form of work-related diseases. In many instances, both management and workers are not even aware of the risks in their industries.

In that same article Professor Noweir also notes that "the rapid introduction of complex work methods in developing countries has been associated with considerably higher rates of industrial accidents and occupational diseases than in industrial countries." In a presentation made at the International Symposium on the Biomedical Impact of Technology Transfer (11), Professor Noweir identified several contributing factors such as the hot climate, malnutrition, and unsanitary conditions in crowded urban areas.

The minister of manpower first set minimum health standards for workers in industry in 1967 by means of Decree No. 48. This decree was based on Law No. 91 promulgated in 1959. This decree was superseded by Decree No. 55 in 1983, issued on the basis of the new Labour Law of 1981. This latter decree set standards for heat exposure, noise, lighting, radiation, and ventilation in an effort to reduce or possibly even prevent occupational hazards (12).

3.2 Education and Training

To be able to train technical cadres locally, as a first step, Egypt sent a number of prospective scientific educators to the United States to pursue doctoral studies. The University of Alexandria in Egypt established in 1955 the Department of Occupational Health as part of the High Institute of Public Health. Today the department offers courses leading to a Diploma of Science (D.Sc.), a Master of Science (M.Sc.), and a Doctor of Public Health Sciences (Dr.P.H.Sc.) (12).

> The present safety/occupational hygienist practitioners employed in Egypt include five occupational hygienists in academic establishments, about 30 in regulatory agencies, about 100 in industry (of whom only about 20 had actual training), and about ten in other agencies (12).

Developing countries face a number of problems even when they are able to modernize industry as rapidly as local resources permit. Thus, even though Egypt has been able to train a number of capable occupational hygiene professionals, it has lost many to prosperous Middle Eastern countries or to the West because of higher salaries and/or better research opportunities.

Despite considerable difficulties, Egypt has been able to lay the foundation of occupational hygiene. The Egyptian government has promulgated laws and decrees to protect the health and safety of its workers. Egyptian institutions of higher learning have graduated over the years a core of competent occupational hygiene technicians and professional hygienists active in government, industry, and academia. Members of the Labour Inspectorate enforce applicable health and safety standards and permissible exposure limits. In 1987, Cook (9) listed 40 permissible exposure limits for Egypt in his Table V, "Occupational Exposure Limits of Countries other than USA and Canada." There is little question that Egypt is a leader in occupational hygiene among neighboring countries and can be expected to make further progress in the years ahead.

4 FINLAND

4.1 Historical Development

According to the Institute of Occupational Health of Helsinki, Finland (13):

> The provision of occupational health services has a long history in Finland. In the 18th century early industrial enterprises with the most hazardous working conditions provided health services for their employees. Occupational health services in the modern sense were arranged after World War II. During the post-war period, the services were initially provided in order to compensate for defects in the primary health care system; as time passed, occupational health services to an increasing extent have been directed at the prevention of work-related health hazards.

The Labor Safety Act (299/58), enacted by the Finnish parliament in 1958, is the cornerstone of labor protection in Finland. This act is a general document and contains no specific regulations or detailed instructions. The Finnish government, however, has the right to pass decrees under the act and thus becomes the highest labor protection authority.

Initially, the Ministry of Social Affairs and Health was responsible for implementing the Labor Safety Act throughout the country. In 1973, following passage of the Labor Protection Act (131/73), the actual implementation was given to the National Board of Labor Protection, under the supervision of the ministry.

The board supervises the implementation of applicable regulatory requirements in 14 districts, with three being devoted exclusively to agriculture and forestry. These district offices are responsible for workplaces with more than 10 workers. Smaller workplaces come under the jurisdiction of municipal inspectors.

Under the act, the employer is ultimately responsible for protecting employees against accidents and health hazards. Employees, in turn, are required to observe all safety regulations.

Presently the act also covers regulations concerning illumination, ventilation, noise, and vibration.

Since 1974, the Labor Protection Act mandates cooperation between employers and employees on matters of health and safety. At a workplace with more than 10 employees, the employees elect one part-time labor delegate to represent them. If the number of employees exceeds 20, management and labor elect a labor protection committee with equal representation. Such committees are elected for a period of three years. Where the work force exceeds 500 employees, worker safety representatives work full time as ombudsmen with broad safety responsibilities. Occupational health personnel, mainly occupational physicians and nurses, can attend committee meetings as advisors.

The employer appoints the labor protection manager, who is responsible for labor protection at the workplace.

Occupational safety delegates in Finland do not have the authority to stop production for safety reasons. Such responsibilities rest with the labor protection authorities that inspect workplaces to ensure that both the letter and spirit of the law are implemented.

> Labour protection at workplaces in Finland involves the efforts of thousands of people. The number of labour protection supervisors, occupational safety delegates, standby delegates, occupational safety spokesmen, and labour protection committee members is 100,000 people—5% of the working population—a considerable number in a country with a population under five million (14).

4.2 Occupational Health Organizations

The Institute of Occupational Health was founded in 1951 as a private foundation with only a small staff. But the Institute's creators were far-sighted and convinced of its importance. With reason, Leo Noro, the Institute's long-term director, can be considered its father (14).

In recognition of his contributions in the field of occupational health, the American Industrial Hygiene Association honored Leo Noro, M.D., with the 1968 AIHA Yant Memorial Award. He accepted the award at the annual American Industrial Hygiene Conference in St. Louis, Missouri, where he lectured on *Occupational and "Non-Occupational" Asbestosis in Finland*.

Today the Institute of Occupational Health is Finland's most important research center devoted exclusively to labor protection and the creation of more healthful, safer, and better working conditions. Its basic functions involve research, training, and the dissemination of information. Today, in addition to its facilities in Helsinki, it has five regional offices in major cities that pursue research and provide industrial hygiene services within their respective regions.

According to the Finnish publication *World Health Safety* of 1982 (p. 12),

> Official statistics on occupational diseases diagnosed in Finland have been compiled since 1926. The Finnish Occupational Disease Register was set up by the Institute of Occupational Health in 1964. Initially occupational skin diseases were not included, but these have also been recorded since 1 January 1975.

The same publication continues on that same page: "The Register presently contains about 30,000 cases of occupational diseases. Occupational skin diseases account for roughly a quarter of the occupational diseases."

The institute was nationalized in 1978. Since then, most of its budget has been funded by the government. In addition to research on occupational illnesses, since 1979 the institute has also kept a registry of employees exposed to carcinogenic substances. The institute acts as an expert adviser but has no enforcement authority.

4.3 Definition of Occupational Hygiene

According to Kauppinen (15),

> Occupational hygiene is comprised of activities aimed at the identification, evaluation and control of physical, chemical, and biological hazards at workplaces. This is the broad definition of occupational hygiene as understood and practiced in Finland.
>
> Occupational hygiene is accepted as a discipline based on several sciences the most important of which are chemistry, physics, and the engineering sciences. It has established its position in Finland as a profession practiced by occupational hygienists and other persons with related job titles.

Finnish legislation has no official definition of industrial hygiene nor does it specify requirements for those claiming the title of industrial hygienist.

4.3.1 Occupational Health Training

Most industrial hygienists in Finland have a master of science degree in chemistry, physics, or a branch of engineering. Few industrial hygienists have an advanced degree in industrial hygiene.

The Institute of Occupational Health and the University of Kuopio offer full-time training in occupational hygiene.

The Institute of Occupational Health in Helsinki and the regional institutes carry out important training functions for their own personnel and for others desiring occupational health and safety training. Such individuals include occupational health physicians and nurses, industrial hygienists, occupational psychologists and physiotherapists, and labor protection managers and delegates, as well as various kinds of planners and designers.

Employees working at the institute usually have a master of science degree. Those wishing to become industrial hygienists learn the fundamentals of sample collection and analysis and, depending on their background, specialize in the evaluation and control of either physical or chemical hazards. The curriculum takes about three years to complete and provides unofficial certification by the institute.

The University of Kuopio prepares students for advanced degrees in occupational hygiene ranging from a master's degree to Ph.Lic. and Ph.D. degrees

Prospective occupational hygiene technicians attend several occupational safety and health courses as part of their two-year training at technical schools. Many take additional short courses offered by the institute.

Occupational health physicians can attend a one-week training course in industrial hygiene at the Institute of Occupational Health. Occupational health nurses usually receive basic training in occupational hygiene as part of their obligatory four-week course. Both physicians and nurses can receive further training at the institute.

Small companies with fewer than 20 employees retain occupational health physicians and nurses as consultants for a fee. Such consultants, in addition to treating patients, may check noise levels and measure air contaminants with detector tubes.

Companies with up to 1000 employees also use occupational health consultants on a fee-for-service basis. In addition, they have in-house safety personnel and safety committees, the latter being required by law. As in Sweden, management and workers have equal representation on these committees.

Large companies with over 1000 employees have full-time occupational physicians, nurses, safety engineers, and sometimes industrial hygienists on their staffs. In the absence of industrial hygienists, the other members of the occupational health and safety team carry out industrial hygiene functions.

All employers needing in-depth industrial hygiene services can request assistance from the Institute of Occupational Health or from one of its regional offices. Regional labor protection inspectors visit such plants periodically and issue orders and recommendations, depending on the nature and degree of the hazard.

The general population of Finland is 4.9 million, of whom 2.4 million work in industry. According to a 1985 survey by the Finnish Industrial Hygiene Society, there are about 180 practicing industrial hygienists distributed as follows:

Academic establishments	10
Regulatory agencies	15
Industry	60
Institute of Occupational Health	70
Other research institutions	10
Insurance companies	5
Equipment manufacturers, consultants, etc.	10

"In addition, about 1400 occupational health physicians, 1800 occupational health nurses, 300 labor protection inspectors, and nearly 100,000 persons working part- or full-time in the safety organizations of the plants deal with occupational hygiene problems as a minor responsibility" (15).

According to two manpower assessments in 1972 and 1981, the number of practicing industrial hygienists in Finland satisfies existing needs.

4.4 Professional Societies

About 150 industrial hygienists are members of the Finnish Occupational Hygiene Society. A few are also members of the AIHA and about 20 belong to the ACGIH.

Finland has no official certification in occupational hygiene. Individuals with certain prerequisites can apply to take a written examination offered by the Institute of Occupational Health. Candidates must have a master's degrees in chemistry, physics, industrial hygiene, or engineering. They must have worked as industrial hygienists full time for a minimum of three years and must have attended industrial hygiene courses offered by the Institute of Occupational Health. Upon successful completion of the written examination they are issued an unofficial certificate as qualified industrial hygienists. About 40 individuals have earned such certificates.

4.5 Permissible Exposure Limits

The National Board of Labor Protection is responsible for developing and publishing occupational exposure limits for airborne toxic substances. The board is the central administrative agency for occupational safety and health under the Ministry of Social Affairs and Health.

By definition, long-term exposures to toxic substances at levels above Finnish occupational exposure limits may cause deleterious health effects. Such limits are not meant to protect the hypersensitive but should adequately protect healthy workers as long as the permissible exposure limits are not exceeded.

"The Finnish exposure levels are presented for use as a guide only, with no enforcement requirement under the Labor Protection Act. Actually, the occupational safety inspectors and the courts use these values in making decisions as to whether working conditions are hazardous" (9). Detailed information on occupational exposure limits is included in Table VII of the work *Occupational Exposure Limits—Worldwide* (9).

Few Finnish exposure levels are legally binding. The following TWA 8-hour limits are binding only for the viscose rayon industry:

Carbon disulfide	20 ppm*
Hydrogen sulfide	20 ppm
Sulfur dioxide	10 ppm

The following TWA 8-hour limits are binding for all industries:

Asbestos	2 fibers/cm^3 (>5 μm)
Benzene	5 ppm
Lead	0.1 mg/m^3

As to future national commitments, the Institute of Occupational Health states (13) that

*Million parts per cubic foot of air, by volume.

Finland has ratified the ILO Convention no. 161 on occupational health services which encourages member countries to develop further their service systems both at the national and particularly at the level of undertakings. On the other hand, in 1986 the Finnish Government adopted a national level programme of "Health for All by the Year 2000" in which the development of occupational health services plays an important role. These international challenges will certainly guide the development of the Finnish health service systems and help in the achievement of the ambitious goal: health for all by the year 2000.

5 ITALY*

5.1 Historical Development

Before World War II, Italy was among the highly industrialized nations of Europe, known for both its light and its heavy industries. Such industries ranged from agricultural products to footwear and textiles, from passenger vehicles and trucks to aircraft and heavy weapons. In June 1940 Italy entered World War II as a partner of the Rome–Berlin Axis. Five years later, by the end of the war, Italy lay in ruins, from the tip of Sicily to its northern borders.

Italy, with the rest of war-torn Europe, used the immediate postwar years to catch her breath and tend her wounds. The 1950s saw the beginning of an economic rebirth that led, within a few years, to an Italian economy that reached and surpassed prewar levels.

The development of modern industrial hygiene in Italy coincides with the massive effort of industrial reconstruction and expansion. In the words of Dr. Danilo Sordelli (16), President of the Italian Association of Industrial Hygienists, "The early 1950s were characterized by post-war reconstruction and industrial recovery; the available products and their use were not yet a problem for the environment. Social awareness was directed towards problems connected with in-plant conditions."

The rapid growth of industry, an awakening social awareness toward occupational and environmental problems, and the desire to control and prevent such problems as early as feasible are reflected by regulatory efforts by central and local governmental agencies. Beginning in 1955, safety and industrial hygiene regulations followed one another in steady succession (17). Although the early regulations were somewhat generic in nature, they were important because they opened the way for more stringent and specific regulatory efforts. Major industries began to review their operations and seek better workplace controls while occupational physicians were striving to offer their employees better health care. Organized labor and government inspectors participated in such efforts adding to the momentum.

The first Italian regulation limiting air contaminants in the workplace appeared in 1966 (18). Industry came to recognize the need to limit employee exposures to

*I am indebted to Dr. Danilo Sordelli for generously sharing with me his thoughts and informative materials included in this chapter.

air contaminants in factories and began to incorporate such controls into the initial process design. At first, most enterprises emitted the captured air contaminants directly outdoors. Some went beyond indoor controls and installed the first dust collectors and water treatment plants at about this time.

Italian environmental experts soon recognized that early in-plant controls were still inadequate and jointly with organized labor continued their search for better ways to control pollutants. As a result of such efforts, the chemical industry and organized labor incorporated the ACGIH TLVs into their 1969 national contract. The ACGIH TLV booklet was translated into Italian. Updated translations of the booklet have appeared regularly ever since.

On the topic of Italian exposure limits, Cook (9) writes:

> In 1975 a list of occupational exposure limits, termed 'Valori Limite Ponderati (VLP)' was prepared jointly by the Italian Society of Occupational Medicine and the Italian Society of Industrial Hygienists. At about the same time a more nearly complete list with a few differences was prepared by the Technical Committee for Maximum Allowable Concentrations instituted by the ENPI for the Minister of Labor.

By 1969 the ranks of Italian industrial hygienists had swelled to the point where its practitioners felt the need to found the Italian Association of Industrial Hygienists (AIDII). It had a rich historic tradition as it followed in the footsteps of Bernardino Ramazzini (1633–1714) who, more than 200 years before, had delivered a brilliant series of lectures at the University of Padova linking occupational diseases to various trades.

Coincidentally, also in 1969, AIHA awarded Enrico Vigliani, M.D. its prestigious Yant Memorial Award in recognition of his work regarding the toxicity of chemicals and particularly of heavy metals.

By 1970, as the overall pace of industrial growth leveled off, another major environmental problem emerged. Untreated industrial waste was contaminating the soil and many aquifers. Industrial and household detergents were causing eutrophication of rivers and lakes. Chemical fertilizers, pesticides, herbicides, and raw sewage were polluting the waters, including the Adriatic Sea and the Mediterranean.

A law protecting the health and well-being of workers was passed in 1970 (19). This law assumed particular significance. It required, among other things, the formation of plant environmental committees and made them responsible for implementing this law. A law limiting industrial air pollution was passed in 1971 (20). It was followed in 1976 by one addressing water pollution (21) and, in 1978, by another establishing the National Health Service (22). Through this law, the Italian Parliament made the Ministry of Public Health responsible for industrial hygiene and specifically charged the Prime Minister's office with setting and periodically updating occupational exposure limits for places of employment.

Large industries led the way in controlling in-plant and outdoor exposures to contaminants such as vinyl chloride, acrylonitrile, and benzene. These industries also sought to identify and produce biodegradable detergents, selective photobiodegradable pesticides, and other environmentally acceptable products.

At this stage, Italian scientists increasingly recognized the need for interdisciplinary research and cooperation. Large companies added industrial hygienists to their staffs in an effort better to capture and control air contaminants at various processes and operations.

At this juncture, several major industrial accidents happened that threatened the health of workers and of segments of the population living near such plants. In addition, a growing concern for environmental issues further affected the increasingly tense relations between various industries and the public at large. Overall, public opinion came to see the chemical industry as the major polluter and culprit behind the broad environmental threat.

Environmental regulations were passed with increasing frequency, often in reaction to specific events or social pressures rather than because of new hazards or developments. Italy, like other members of the European Economic Community, began to adopt EEC health standards and regulations in an effort to update its own and to achieve conformity within the EEC. Thus Italy replaced its own regulations on packaging and labeling promulgated in the 1970s with corresponding EEC regulations (23).

The first European law for the protection of workers against chemical, physical, and biological hazards in the workplace appeared in 1980 (24). Shortly thereafter, Italy adopted the EEC directive regulating workplace exposures to vinyl chloride (25). Laws regulating air quality (26) and air pollution from industrial sources saw the light in the 1980s (27), followed by laws on waste disposal (28, 29), on prohibiting smoking in public places (30), and on abatement of asbestos in schools, hospitals, and other public buildings (31).

5.2 Professional Training

Although Italy has over 1000 practicing industrial hygienists and various universities and other institutions of higher learning conduct extensive and sophisticated industrial hygiene research, presently there are no university programs offering degrees in industrial hygiene. About 20 medical colleges and occupational health services offer industrial hygiene courses, primarily as part of their curricula for occupational physicians. About half offer occupational physicians advanced industrial hygiene training beyond their regular course work.

Relatively few Italian industrial hygienists have industrial hygiene degrees from universities abroad. Most Italian industrial hygienists begin their professional careers as chemists, physicists, occupational physicians, or graduates of other physical sciences. Once they move toward industrial hygiene, they receive their training on the job under the direction of experienced industrial hygienists and by attending courses offered by AIDII, by universities, and by other institutions. Such courses are available to university graduates, be they in industry or in government. They may cover basic as well as advanced topics and they last from 70 to 80 hours.

The universities of Pavia and Padova in northern Italy, and one at the southern seaport of Bari on the Adriatic, offer two-year courses for industrial hygiene technicians.

According to D. Drown of Utah State University (32), "The time could not be better to establish some formality in industrial hygiene education in Italy. The regulatory requirement, professional interest and backing, and pool of expertise are in place to sustain such an undertaking."

5.3 Professional Organizations

Italian industrial hygienists founded the Italian Association of Industrial Hygienists (Associazione Italiana degli Igienisti Industriali, AIDII) in 1969. Presently, the Association has over 600 members. The number of individuals practicing industrial hygiene in Italy is conservatively estimated at over 1000. The association defines industrial hygiene in a manner similar to its sister societies (33):

> Industrial hygiene is the discipline concerned with the recognition, evaluation, and control of work-related chemical and physical environmental factors both inside and outside the workplace for the purpose of preventing and controlling factors capable of affecting the health and well-being of workers and of the general population.

AIDII has held annual conferences for several years. Such annual conferences also offer technical refresher courses and exhibits of industrial hygiene instruments.

Recently, the AIDII has taken steps to carry out the board of directors' decision to establish professional certification in Italy.

Organizational members of the AIDII include several associations devoted to safety, health physics, occupational medicine, ergonomics, and chemistry, among others. Such associations plan and coordinate future activities with AIDII through regularly scheduled joint meetings.

Italy has a good aggregate of occupational and public health laws. It has a large number of experienced industrial hygienists and of good industrial hygiene laboratories. The level at which industrial hygiene is practiced enables Italy to be counted among industrial countries with advanced industrial hygiene capabilities.

6 THE NETHERLANDS

6.1 Historical Development

The first legislative attempt to regulate working conditions in Dutch factories took place in 1874. As in other countries, such early legislation focused on improving the lot of children working in industry. The law limited the number of hours children were permitted to work and addressed some recognized hazards linked to working conditions characteristic of the times.

Traditionally, such laws were intended to protect persons unable to protect themselves. The law of 1874 covered children and in 1889 the law was broadened to include minors and women. Six years later, in 1895, the so-called Labor Act extended its protection to cover all workers, including mine workers. The act also enabled the Dutch government to set standards for the proper use of hazardous materials and of industrial equipment.

A regulation covering working hours and break periods for adult males was issued in 1919. Next, in 1934, the entire legislation was revised and broadened to regulate safety and health issues in agriculture and in the shipping industry, among others.

In the 1960s, the Occupational Safety Act was broadened to address occupational health care. Initially, medical examinations sought to determine the physical fitness of job seekers and their ability to work in factories. In addition, physicians examined workers to determine whether they were able to return to work after an absence due to illness or an accident.

Over the years, the existing patchwork of safety and health laws and regulations had grown inadequate. Traditionally, legislators had focused exclusively on employers and their responsibilities overlooking the workers' interest in matters of health and safety and their first-hand knowledge of work-related stresses. To correct this situation, Her Majesty's Government passed the Working Environment Act on November 8, 1980, which has been partially in force since January 1, 1982 (34). It consists of a number of nationally and internationally recognized regulations on occupational health, safety, and welfare. It enables legislators further to strengthen such regulations through subsequent amendments. The act provides for the creation of worker Health–Safety–Welfare Committees as partners to management on such matters. It specifies management and worker responsibilities in matters of health and safety and enables employees to participate in the formulation of joint policies.

The act focuses on improving working conditions and promotes cooperation among occupational physicians, safety engineers, ergonomists, and others. Parenthetically, occupational hygienists are not specifically mentioned, because the law was passed before occupational hygiene reached the present stage of activity and development in the Netherlands. Today, governmental agencies recognize occupational hygiene as a discipline side by side with safety and occupational medicine.

The act also requires the establishment of health services for places of employment with more than 500 employees and for certain categories of factories such as those producing lead batteries and lead pigments.

Such health services are charged with promoting and protecting the health and welfare of workers at their place of employment. These services either may exist within larger companies or may be established as regional centers. The latter provide their customers with comprehensive occupational health services for pay and include assistance in occupational hygiene. Private health services may act as consultants to their customers but cannot enforce the law. They may, however, notify official agencies of the existence of occupational hazards in the workplace.

The authority to enforce health and safety laws and regulations rests with the Labor Inspectorate under the Director General of Labor.

6.2 Definition of Occupational Hygiene

Traditionally, occupational physicians have provided medical services to workers and safety engineers have focused on eliminating physical hazards and preventing accidents. The concern of occupational hygienists with chemical, physical, and biological stresses has made them valuable members of the occupational health team. What occupational hygiene may have lacked in tradition in the Netherlands, it has made up with the concerted and well coordinated efforts of Dutch occupational hygienists to protect the work force better and its active role in promoting health and safety matters within the European Economic Community.

In the Netherlands (35),

> Occupational hygiene has been defined as the applied science concerned with recognition, evaluation and control of environmental stresses especially chemical, physical and biological stresses arising from work which might adversely affect the health and/or well-being of people at work and/or their posterity.

The definition is interesting because of both its wording and its omissions. According to the authors of the definition, the word "especially" was inserted to emphasize the preeminent nature of chemical, physical, and biological stresses in the workplace while leaving room for physiological, psychological, and other factors. The definition specifically mentions adverse reproductive effects and excludes effects beyond the workplace because these are considered the domain of environmental health professionals. It is not the authors' intent, however, to absolve occupational hygienists from having to consider the effects of plant effluents and adopted control measures on the environment.

6.3 Professional Training

The Agricultural University of Wageningen is the first and, so far, only institution of higher learning conferring M.Sc. and Ph.D. degrees in occupational hygiene. In the early 1980s, the program was established at this institution because of its long-standing study of the effects of air and water pollution on soil contamination as well as on flora, fauna, and human life. Because of the existing expertise and the need to measure and control the effects of industrial stresses on the working population, it was natural that this university should expand its interests into the area of occupational hygiene. According to Drown (36),

> The program began in 1978 within the Department of Air Pollution and soon branched out to include the Departments of Public Health and Toxicology.
>
> The program is a full-time course of study modelled after the Harvard University Industrial Hygiene Program in the United States. Students are well-schooled in the sciences and public health and receive a substantial amount of practical experience as a part of ongoing industrial research projects carried on by the three Departments involved.

Graduates of the master's program usually go on to positions of responsibility in occupational health services or in academia.

The Post-HBO Hogeschool West-Brabant offers a two-year training program in occupational hygiene. Courses are taught at night to fit the needs of working students, many of whom are employed in occupational health services as technicians, occupational health nurses, and safety specialists. Graduates receive a certificate of completion and find such training helpful, for some are called upon to assist with occupational hygiene-related problems even though fully trained occupational hygienists are becoming more numerous.

6.4 Permissible Exposure Limits

The Labor Inspectorate, under the Director General of Labour, use the *Nationale MAC-list 1985 Arbeidsinspectie P no 145* as guides in enforcing regulatory requirements. These MAC values are taken largely from the ACGIH TLV list. Some are based on limits listed by the Senate Commission for the Investigation of Health Hazards of Work Materials of the Federal Republic of Germany. Others draw from recommendations published by NIOSH in the United States (37).

Substances on the MAC list are also identified by their Chemical Abstracts Service (CAS) numbers. Permissible exposure levels fall into two categories:

1. Maximal accepted concentrations (Maximale Aanvaarde Concentratie—Tijdgewogen Gemiddelde, 'MAC-TGG' for short), averaged over a period of up to 8 hours per day, 40 hours per week.
2. Ceiling concentrations (Maximale Aanvaarde Concentratie—Ceiling, or MAC-C), which may not be exceeded at any time because of their acute toxic effects.

For detailed information on permissible exposure limits, see Table VII, in *Occupational Exposure Limits—Worldwide* (37).

6.5 Professional Employment

It is estimated that there are about 200 occupational hygienists in the Netherlands. About 65 percent work within occupational health services. Another 20 percent work for governmental agencies or research institutions, or hold teaching positions. Five percent or so are with the Labor Inspectorate, and the remainder work as consultants either in occupational hygiene or in technical services such as safety and environmental departments.

6.6 Professional Organizations

In the early 1980s, Dutch occupational hygienists felt the need for a forum of their own that would allow them to discuss and reach agreement on technical matters and to promote and protect the interests of the profession and its practitioners. On May 27, 1983, 33 individuals active in occupational health founded The Dutch Occupational Hygiene Society (DOHS) (Nederlandse Vereniging voor Arbeidshygiene—NVvA). Presently, the Society has about 180 members or 90–95 percent of occupational hygienists working in the Netherlands (38). These members come from the ranks of industry, academia, the occupational health services, and government. Applicants must be actively engaged in the field of occupational hygiene. They must have a degree from the University of Wageningen, the two-year West-Brabant certificate, or equivalent qualifications.

The DOHS is very active in national and international affairs. In addition to sponsoring professional conferences at home and participating in conferences abroad such as the American Industrial Hygiene Conference, it has proposed a limit for hand–arm vibrations (Voorstel voor een Grenswaarde voor Hand-Arm Trillingen—January 1988) as a guide for practicing occupational hygienists and researchers. It maintains close liaison with BOHS, ACGIH, AIHA, and other societies. It has a number of active technical committees. Among the most active is the Education Committee, which has compiled recommendations for training programs in occupational hygiene. The committee is presently considering the need for professional refresher courses and for certification in line with the needs and conditions prevailing in the Netherlands.

DOHS is also considering ways to influence EEC occupational health regulations to represent specific Dutch interests in cooperation with BOHS and other European sister societies.

7 SWEDEN

7.1 Historical Development

Sweden passed its first industrial safety law in 1889 intended mainly to protect workers against fire hazards. A more comprehensive act, the Workers' Protection Act, was promulgated in 1913, then amended in 1949 and again in 1974.

The 1949 version of the act compares in many ways to the Occupational Safety and Health Act of 1970 (OSHAct) in the United States. Like OSHAct, the amended Swedish Workers' Protection Act applies nationwide. It establishes a central enforcement agency and makes employers responsible for providing safe and healthful working conditions in their places of employment. The governing board is known as the National Board of Occupational Safety and Health (NBOSH). It consists of 11 members: a director-general, an assistant to the director-general, four representatives from the Confederation of Trade Unions (LO), three from the Swedish Employers' Confederation (SAF), and two members of the Swedish Parliament.

The board is responsible for issuing "directions" or "codes of practice," comparable to OSHA standards. Employer and labor representatives on the board participate directly in drafting such "directions" and "codes of practice." Other employers, equipment manufacturers, labor representatives, and any other interested parties also have the opportunity to comment on such documents before promulgation.

Another important function of the board and its five technical departments is to explain how such standards apply to specific workplaces and to review plans of new processes and industrial plants for adequacy of health and safety controls. The 1974 version of the Workers' Protection Act makes it mandatory that experts employed by the board review such plans and blueprints before construction begins. The act further requires that local joint safety committees also approve such plans and drawings, unless they participated in the earlier stages.

The board also supervises the Labor Inspectorate, which is responsible for enforcing board rules in 19 districts.

Over half the annual health and safety inspections are conducted by local municipal inspectors. These inspectors visit small firms with simpler operations and fewer than 10 employees.

Swedish inspectors function as technical consultants rather than as enforcers. They do have the authority, however, to issue "warnings." Employers who ignore such "warnings" may be fined or even sent to prison for up to one year. Employers can appeal such "warnings" directly to the NBOSH, for Swedish law does not provide for a review commission such as stipulated by OSHAct in the United States.

The Swedish approach to occupational health and safety stresses cooperation over contention between industry and labor unions. Traditionally, industrial firms participate in and act through SAF. On the labor side, over 85 percent of workers are represented by unions that are members of LO. As early as 1938, SAF and LO agreed to settle important issues through negotiations. Over the years, as the strength of SAF and LO remained fairly evenly matched, most local issues have been resolved locally. Swedish industry and unions both want to maintain and further the economic health of the country. According to Clack (39), Birger Viklund, former Secretary of the Swedish Metal Workers' Union and Labor Attaché at a Swedish Embassy, explains: "Swedish industry is geared to export, and the workers realize that strikes could destroy foreign trade built up over the years."

Swedish firms with more than 100 employees have two or more union representatives on their boards of directors. Swedish management recognizes the benefits of worker involvement and cooperation, and gives its employees much latitude and responsibility in how they carry out their jobs.

Management members of joint safety committees are principally responsible for providing safe working conditions in their plants. They must train workers to work safely with equipment and chemical substances and must also provide health services for workers in establishments with over 1000 employees. Such health services are on-site and include occupational physicians and industrial hygienists. Smaller plants have access to health services provided by over 100 regional health centers built by the SAF. Temporary work sites, such as construction projects, are serviced by healthmobiles staffed with nurses and health technicians.

Factories with more than five employees are required by law to have a safety steward. Factories with more than 50 workers must have a joint labor–management safety committee.

Traditionally, elected safety stewards have carried much responsibility in Swedish plants. The amended Swedish Workers' Protection Act of 1974 gives safety stewards new and broader powers. They may stop production in case of immediate and serious danger, pending a decision by the Labor Inspectorate or in case of refusal by the employer to comply with a labor inspector's warning. Safety stewards have access to all pertinent health and safety documents and are paid regular wages while performing safety related duties.

Workers receive extensive health and safety training. They are expected to observe applicable regulations, use chemical substances properly, and operate mechanical equipment as intended. They are required to use proper personal protective equipment and must work safely using proper precautions to protect themselves and the health and safety of their co-workers.

7.2 Occupational Hygiene: Definition and Training

The Swedish definition of industrial hygiene parallels that adopted by American industrial hygiene organizations. Scandinavian countries, including Sweden, have not defined what qualifications an individual must meet to be a professional industrial hygienist. In the early 1950s the Swedish Institute of Occupational Health proposed that governmental occupational hygienists have a degree in the physical sciences or in engineering, plus a minimum of five years of appropriate experience at the Swedish National Institute of Occupational Health, at qualified Occupational Health Service Centers, or in industries with occupational hygiene services.

Early on, some Swedish governmental occupational hygienists received scholarships from the Rockefeller Foundation and earned master's degrees in industrial hygiene at universities in the United States. For a number of years, no formal university courses leading to degrees in occupational hygiene were available in Sweden. The Swedish Institute of Occupational Health offered class work and practical training in occupational hygiene patterned after courses offered at American universities and in particular at Harvard. Such courses taught students how to measure and evaluate chemical and physical stresses in the plant environment and emphasized control methods to reduce or eliminate such stresses.

Since 1952, SAF has offered occupational hygiene courses and on-the-job training to members of their occupational health staffs, including physicians and other technical personnel. Beginning in 1959, SAF offered a one-year course in occupational health for occupational hygienists and safety engineers with particular emphasis on prevention and control of hazards.

In 1966, the Swedish Parliament combined four major organizations devoted to occupational health and safety research into the National Institute of Occupational Health in Stockholm. Six years later, in 1972, it placed the Institute under NBOSH. At this point, the board assumed the important role of providing technical and practical training for occupational safety and health (OSH) engineers. Such individuals have an engineering degree from a university or from an equivalent institution and several years of field experience. According to Dahlner (40), "Sweden has about 1400 trained OSH engineers (specialists within occupational safety and health services) and 1200 of them have been trained at the National Board of Occupational Safety and Health (NBOSH)." Practicing OSH engineers attend 15 weeks of theoretical training and are required to complete a special project. Prospective OSH engineers receive 20 weeks of classroom instruction and work for another 20 weeks at an OSH service unit under the supervision of an experienced OSH engineer.

Dahlner (40) reports the following distribution of occupational hygienists in Sweden:

1. Academia and occupational clinics in hospitals	40
2. Regulatory agencies	60
3. Industry (OSH engineers)	1400
4. Other	20

7.3 Academic Training

In 1984, the University of Lund began offering academic training for individuals wishing to become occupational hygienists, the first and only Swedish university to do so. Candidates must have a master's degree in chemistry or be a professional safety engineer with a strong chemistry background. The program is geared mainly toward control of chemical hazards.

Full-time students can complete the program in three semesters. Part-time students can attend the program with a modified attendance schedule.

7.4 Occupational Hygiene in Industry

In attempting to describe Swedish industry to American readers, Clack (39) writes: "Many Americans assume Sweden is a 'socialist' country with state-operated industries, but, actually, 91 percent of Swedish industry is privately owned."

Yet because of extensive cooperation between unions and employers in health and safety, duties of occupational hygienists in industry are determined mostly by agreements among SAF, LO, and the Negotiation Cartel for Salaried Employees in the Private Business Sector (PTK). According to Gerhardsson (41), three such agreements are of fundamental importance:

1. The Development Agreement of 1982, which strives to foster industrial efficiency and preserve employment as areas of common interest among the three signatories.
2. The Equal Employment Agreement, also signed in 1982, providing for equal employment opportunities for men and women with regard to promotions and equal pay for equal work. This agreement applies to salaried and hourly employees alike.

   Both agreements foster a spirit of cooperation between management and labor so necessary to promote occupational health and safety.

3. The Work Environment Agreement, reached in 1983, and its impact on occupational health. It sets health and safety rules for the work environment. It establishes guidelines for industrial health programs and defines details of occupational health training for employees. The Agreement specifies that corporate health programs must include occupational hygiene and medical services with special emphasis on psychosocial factors. It also calls for a joint industrial safety committee or a company health program committee to supervise the occupational health staff and thus ensure that its activities are based on sound science and experience. This approach is intended to maintain strict impartiality and fairness on the part of company health programs and strengthen their commitment to prevent potential health problems. Disagreements over scientific or other important issues are brought before a national scientific committee for review and resolution.

The three signatories of these agreements contribute funding and staff in support of joint activities. All three parties are also free to pursue individual initiatives within the framework of the agreed-upon programs and guidelines.

Under such tripartite agreements, occupational hygienists are responsible for a number of activities. Such activities include preparing for and working toward ever more stringent permissible exposure levels, considering indoor as well as outdoor pollution problems, and maintaining effective industrial emission controls. Swedish occupational hygienists are also expected to strive for energy and cost savings while reducing and controlling environmental problems.

7.5 Future Needs

In 1981, a governmental advisory committee estimated that by the year 2010 the number of occupational hygienists with governmental and municipal affiliations would more than double, and that their numbers in industry would increase fourfold. Gerhardsson (41) disagrees:

> From the present terms of reference—which are too narrow—this assessment has been shown to be an over-estimation of the numbers needed. In industry managerial skills are upgraded; enforcement tasks are increasingly transferred to line functions. In governmental and municipal agencies, systems strategies require fewer occupational hygienists but more specialized ones.

7.6 Professional Associations

In Sweden there is no official certification or registration mechanism for occupational hygienists.

Occupational hygienists working at occupational medicine clinics belong to the Swedish Occupational Hygiene Association, which, in addition to pursuing technical goals, represents its members in collective bargaining for wages and in other contractual matters. All other occupational hygienists may join the Swedish Occupational Hygiene Society. The Society requires that all its members also become members of AIHA.

7.7 Permissible Exposure Limits

The Swedish Institute of Occupational Medicine published the first list of permissible exposure limits in 1969. NBOSH has had the responsibility for issuing lists of limit values (TWA) for chemical substances since 1974. According to Edling and Lundberg (42),

> That list, which included values for about 70 substances, was based on the ACGIH-list and Swedish experiences but did not recognize carcinogens specifically. In the second list, published in 1974, background data from NIOSH, OSHA, ANSI, BRD [the German Federal Republic], Czechoslovakia and ILO were considered. That list included an appendix of recognized carcinogens. The list has been revised several times (1978, 1981, 1984) and the latest is in effect since July 1, 1988, including about 300 substances.
>
> . . . These lists have gradually been expanded and legally strengthened. From 1981 the list of TWA is a regulation and not a recommendation only.
>
> . . . The Work Environment Act, 1977:1166, effective as of July 1, 1978, provides in its Chapter 2, Section 6, that substances liable to cause ill health or accidents may only be used in conditions affording adequate security. Authorized by this Act, the National Swedish Board of Occupational Safety and Health issued its Hygienic Limits Values Ordinance on May 6, 1981, effective January 1, 1982.

The ordinance established two types of permissible exposure limits for airborne contaminants. "Level limit values" apply to daily 8-hour exposures and "ceiling limit values" represent maximum permissible exposure levels measured over 15-minute periods, unless otherwise specified in Appendix I of the Ordinance. For details see Reference 9.

In recent years, Swedish exposure limits have become increasingly stringent to avoid or minimize biological effects. Some limits were lowered based on worker complaints rather than on demonstrable dose–effect relationships. Others were made more stringent because of demonstrable effects on the central and/or peripheral nervous system(s). Thus Swedish permissible exposure limits for many chemicals, and particularly for some industrial solvents, are lower than corresponding limits in several industrialized countries.

7.8 Conclusions

Occupational hygiene, as practiced in Sweden, is characterized by a spirit of cooperation between management and labor. Such cooperation is not limited to health and safety but extends into other important areas of labor relations. Swedish occupational hygiene is characterized by strong research and the broad implementation of occupational hygiene principles and practices in industry, particularly with regard to control of chemical hazards and the implementation of ergonomic tenets.

The American Industrial Hygiene Association has conferred the Yant Award on three distinguished Swedish scientists:

		Yant Memorial Lecture
1967	Sven Forssman, M.D., Stockholm	Occupational Health Institutes: An International Survey
1978	Harry G. Ohman, Vasteras	Prevention of Silica Exposure and Elimination of Silicosis
1985	Lars T. Friberg, M.D., Stockholm	The Rationale of Biological Monitoring of Chemicals—With Special reference to Metals

8 UNITED KINGDOM

8.1 Historical Development

In 1833 the United Kingdom established the first Factory Inspectorate. Its charge was to deal, among others, with the problem of the long hours that children and men spent working in factories. These "factory children" worked from the early hours of the morning till late at night performing essential work for the adult work force in Britain's expanding industries. The thinking that led to the establishment of the Factory Inspectorate was that if the children's hours were shortened, those of the men they helped would be also. Things turned out differently. Eventually, more children were assigned to work shorter shifts and the men continued to keep the same hours as before.

The Factory Inspectorate broke new ground in several respects. The four original inspectors divided the country into four districts. They recognized early on that they could not visit all establishments in their districts and thus could not play a true policing role. They could, however, promote good will among employers and elicit their cooperation. Inspectors had the opportunity to see problems in the factories they visited and to seek solutions that would benefit employers as well as those working for them. As the inspectors traveled from workplace to workplace, they appealed to common sense and to the good will of factory owners and left enforcement as a matter of last resort.

Despite the Factory Inspectorate's successes, serious risks continued to exist in British factories as industry continued to grow and expand. Such risks were serious enough to claim the attention of the British public. Luxon (43) writes,

> The risks of the new industrial processes were highlighted by contemporary writers including Charles Dickens who drew attention forcibly to the dangers. In the latter part of this period there were several disastrous fires which paved the way for additional legislation and industrial hygiene, as we know it today, was born. The British Factories

Act of 1864 contained the first requirements. Every factory was to be ventilated to render harmless any gases, dusts or impurities that may be damaging to health, i.e., what we would call now dilution ventilation. An Act of 1878 took a further step and required exhaust ventilation by means of fans for the removal of dust likely to be injurious to health. It is interesting to note that wording almost identical to that setting out these two concepts appears in present day UK legislation.

One-hundred and fifty years after its establishment, the Factory Inspectorate is still active. It has become even more effective but is not without its problems. Simpson (44), Chairman of the Health and Safety Commission, writes:

> From the 1833 quartet has developed the Factory Inspectorate of today: more professional, more scientific and better serviced, but still facing enforcement problems, still giving advice and information, still being abused on odd occasions but still with the instincts of service and caring which have distinguished the organisation throughout its history.

The Factories Act of 1901 marked another major step forward for it also regulated hazardous trades. The act led to in-depth investigations and control of work-related hazards. Early surveys focused on the hazards of silicosis in the pottery industry. The public of the time had read about the shakes of the Mad Hatter in *Alice in Wonderland*. A study revealed that Lewis Carroll had given a good description of occupational poisoning among hatters owing to their using mercury to change fur into felt. Another study linked phossy jaw to the use of yellow phosphorus in the production of matches. Based on this study, Great Britain was the first country to outlaw the use of yellow phosphorus.

By 1914, World War I had engulfed Europe. Once again, long working hours became a matter of concern in Great Britain. In 1915 the British government appointed the Health of Munitions Workers' Committee under the chairmanship of the Right Honorable David Lloyd George, later Prime Minister of Great Britain from 1916 to 1922. The Committee was asked to investigate and advise on questions regarding working hours and matters that may affect the health and efficiency of munitions workers. As in other wars, women had entered the work force in large numbers. Warner (45), in his Yant Memorial Lecture delivered at the American Industrial Hygiene Conference in 1974, reported that

> . . . for women engaged on turning shell cases, a reduction in the weekly hours worked from 68 to 60—a reduction of 12%—resulted in a quick increase in total output of 11%. Further reduction to a 55-hour week showed yet an additional improvement in output of 5%. Sickness rates and involuntary absenteeism were also substantially reduced with the modified hours of work.

As in the case of the Factory Inspectorate of 1833, efforts to investigate productivity versus hours worked and the related health effects seemed to link once more economic issues with health and safety matters. July 1918 saw the establishment of the Industrial Fatigue Research Board with a mandate similar to that given the Munitions Workers' Committee under Lloyd George. The Industrial Fatigue Research Committee under the direction of the Medical Research Council and the Department of Scientific and Industrial Research would have suffered an early demise barely three years after its founding had it not been for the vigorous protests of the Medical Research Council. This could have inflicted a severe blow to the industrial hygiene efforts carried out by this group. In 1928 the Industrial Fatigue Research Committee was appropriately renamed the Industrial Health Research Board, thus reflecting its charge to investigate all aspects of the workers' occupational health and well-being.

The need for people trained in industrial physiology and psychology prompted the London School of Hygiene and Tropical Medicine to offer several related courses in 1938.

By then, Europe was once again approaching the abyss of war. Hitler unleashed his armed forces on Poland on September 1, 1939. Two days later, Great Britain and France entered the war against the aggressor. Fortunately for Great Britain, the results of previous research into human fatigue and ill health caused by unfavorable working conditions paid off and allowed British industries to avoid past mistakes.

By 1942 the responsibilities of the Industrial Health Research Board increased to include responsibilities for studying occupational and environmental factors that could cause ill-health and disease. Following the end of World War II, a number of research centers, mostly at universities, began to specialize in various areas of knowledge based mainly on staff interest and existing facilities. Thus Cambridge specialized in applied psychology. The Research Unit at Penarth in South Wales conducted studies on pneumoconioses whereas groups at Oxford and London concentrated on the effects of heat stress. An atmospheric pollution research unit was set up in London, and the Atomic Research Establishment at Harwell conducted investigations into the hazards of ionizing radiation.

Like many other fields of knowledge, industrial hygiene had made important advances because of the armed conflict that had convulsed the nations in World War II. Major employers had come to realize that protecting the health of their workers paid handsome dividends. Thus about 20 percent of British workers employed by large industries came to enjoy the benefits of in-house occupational hygiene and medical services. The Slough Industrial Health Service was set up in 1947 and was intended to provide first medical services to small industrial establishments with a total work force of some 20,000 workers. Two years later the service was expanded to include industrial hygiene support, the sole provider of such services in Great Britain at the time. Unfortunately, the intended beneficiaries lacked the financial resources and may have failed to grasp the need for such services. The Slough Industrial Health Service was available to its customers for about 18 years. By 1964 it had to close its doors for lack of financial support.

Financial justification of the need for any service and financial support have always played a role in determining whether a product or service could withstand the test of the market place. Warner (45) reports an instance where economic considerations may have received unusual emphasis. At a conference held by the British Occupational Hygiene Society in April 1965,

> A leading industrial medical officer pleaded that industry must take a broad view of the role of occupational hygiene. It was needed not only to control occupational disease but also the wide range of adverse environmental conditions which led to sub-standard work performance. Occupational hygiene should be developed primarily to improve efficiency and its human and moral values would follow.

By the mid-1960s, occupational hygiene was receiving strong support and increased recognition in Great Britain. The Factory Inspectorate added an industrial hygiene unit to advise industry on occupational hygiene problems and provide field as well as analytical services. The Trades Union Congress expressed strong interest in occupational hygiene and what it could do for British workers. Although the congress did not add hygienists to its staff, on the occasion of the one-hundredth aniversary of its founding, it did endow an Institute of Occupational Health at the London School of Hygiene and Tropical Medicine to provide consultation and field and laboratory services to various organizations, industries, and individuals.

Concern about the in-plant environment soon broadened to include outdoor pollution. The year 1970 was declared International Conservation Year. Great Britain established a Department of the Environment with a minister in Parliament. Environmental problems including air, water, and industrial pollution of neighborhoods received more attention than ever before.

Also in 1970, the Secretary of State for Employment and Productivity set up a Committee of Inquiry into Safety and Health at Work to review and report on problems that hampered the progress of occupational health and safety in Great Britain. Willis (46) writes:

> Until the early 70s, British industry was subject to many safety statutes, mostly under some earlier parent statute and usually specifying duties of factory occupiers (typically the employers). The inspectorate was fragmented by industry, and there was no national body to oversee the systematic development of safety legislation. A government committee, under Lord Robens, was appointed in 1970 to review this situation.

The committee published its report two years later and identified a series of shortcomings that existed in industry. It recommended increased involvement and self-regulation on the part of industry. It also stressed the need for greater cooperation in matters of health and safety between management and workers, and for a more effective role on the part of the government.

Two major events, unrelated to each other, followed and had a major impact on occupational medicine and hygiene in the United Kingdom. In 1973 Great Britain joined the European Economic Community (EEC). The original signatories, Belgium, France, the Federal Republic of Germany, Holland, Italy, and Luxembourg had signed the treaty of Rome committing themselves mainly to promoting economic cooperation among the members. In 1973, in addition to Great

Britain, Ireland and Denmark also joined the EEC. Three more countries, Spain, Portugal, and Greece, joined later, bringing the total number of member countries to 12.

Luxon (47) notes that

> The treaty requires the removal of barriers impeding trade between member states and provides for the harmonisation of their Safety and Health policies. It also lays down regulatory procedures for achieving these objectives. Thus far, action has been taken to implement a common classification and labelling system and to prohibit the use of certain dangerous substances both of which directly affect cross boundary trade. More general regulatory requirements have been promulgated, setting out good industrial hygiene practices and laying the basis for more detailed action in particular areas. Very recent proposals have been discussed which could harmonise occupational exposure limits and monitoring procedures in the 12 member states. These developments on a European front are now tending to overshadow national initiatives which will undoubtably in the future follow rather than lead.

The other major event resulted from the findings of the Robens report published in 1972. Two years later, in 1974, the British Parliament passed the Health and Safety at Work Act that placed all occupational health and safety agencies under one umbrella. The act covers areas such as occupational health and safety, explosive and flammable materials, noxious contaminants, radioactive materials, and nuclear installations. It imposes a number of obligations on manufacturers, designers, importers, and suppliers of substances and articles. It requires such individuals to test items covered by the act. They either have to carry out, or have others carry out, research to ensure that their products can be used safely or with minimum risk. They also have to provide customers with adequate information for the proper and safe use of such products. If, however, customers use these substances or articles improperly and in violation of instructions provided, the act places the responsibility for any harm on the user.

The act also imposes duties on employees. Employees are expected to use due care and caution to protect their own health and safety and that of their co-workers. Otherwise, they can be held accountable if they knowingly violate health and safety instructions.

The act makes industry responsible for the health and safety of the public if work-related hazards threaten the well-being of the latter. It expects regulatory bodies to replace overly complex regulations with more flexible codes of practices. Such codes carry weight in certain legal proceedings but allow industry to satisfy the standards by equivalent means.

The act lists mandatory requirements and identifies others as being applicable "so far as is reasonably practicable." Although the act does not clearly define this expression, the courts have interpreted it to mean that the risks of a given process or activity must be weighed against the cost and effort involved in controlling such risks. Though cost is clearly a factor, British courts have not required that such costs be compared against or linked to an employer's ability to pay. It is equally clear, however, that without the moderating expression "so far as reasonably practicable," agencies of the government might tend to enforce provisions of the act more strictly.

Parliament created the Health and Safety Commission (HSC) and the Health and Safety Executive (HSE) to implement the provisions of the act. The HSC must develop strategies, formulate policies, and act as an advisor on occupational health and safety to other governmental bodies. It consists of a chairman and eight members, three appointed by the National Employers Organisation, three by the National Trade Union Organisation, and two representing the public at large. The HSC operates through the HSE, which consists of three members supported by six inspectorates, a research laboratory, and an Employment Medical Advisory Service. Cullen (48), who served as Chairman of the Health and Safety Executive, has defined its purpose as follows:

> On general safety issues and on specific standards, the aim of the legislators is clear: To move away from 'Fiat' and towards a system which improves by involving and encouraging those directly concerned. There is a new system of improvement and prohibition notices and while prosecution powers are retained, they are very much a last resort for the flagrant offender or the very serious breach.

By its very membership, the HSC is the principal arena where technical and economic issues are raised and major interests voiced. It is subject to political debate and exposed to public scrutiny. Interested parties have the opportunity to review and scrutinize proposed regulations and codes of practice. Only after all parties have had the opportunity to be heard does the HSE present a proposal to the HSC for ratification. By the time a regulation or code of practice is promulgated, all concerned parties have been heard. Although the process may be slow and laborious, it is time-tested and works to the satisfaction of the participants for the purpose and within the framework for which it was created.

A few years ago, the HSC identified a further need to regulate the use of hazardous substances in industry. The commission set up an Advisory Committee on Toxic Substances, or ACTS for short. Members of ACTS represent employers, employees, and professional organizations. They were charged with drawing up a document assessing occupational risks, recommending control measures, and establishing requirements for health surveillance. They provided information to affected parties and training to minimize or avoid ill health among those exposed in the work place or among the public. After efforts which extended over several years, on October 12, 1988 A. J. Lord, Secretary to the Health and Safety Commission, published the following *Notice of Approval* as a preamble to the new regulation:

> By virtue of section 16(1) of the Health and Safety at Work etc. Act 1974 and with the consent of the Secretary of State for Employment, the Health and Safety Commission has on October 12, 1988 approved the Code of Practice entitled Control of Substances Hazardous to Health.

> The Code of Practice gives practical guidance with respect to the Control of Substances Hazardous to Health Regulations 1988 (SI 1988 No 1657).
>
> The Code of Practice comes into effect on 1 October, 1989 which is the date when the Regulations come into force."

This document is bound to have a major impact on the extent and depth to which industrial hygiene hazards will be controlled in Great Britain over the next few years.

8.2 Professional Training

There are presently five universities in Great Britain that offer degrees in occupational hygiene:

- London School of Hygiene and Tropical Medicine
  (M.Sc., Ph.D.)
- University of Newcastle-upon-Tyne
  (M.Sc., Ph.D.)
- University of Manchester
  (M.Sc.)
- University of Bradford
  (B.Sc.)
- South Bank Polytechnic of London
  (B.Sc. in Environmental Science with emphasis on occupational hygiene)

Three other institutions of higher learning are considering whether to offer courses leading to degrees in occupational hygiene, two at the master's level (University of Birmingham, University of Surrey) and one at the baccalaureate level (University of Bristol).

The wealth of experience, the number of academic institutions offering professional training in occupational hygiene, and its professional certification program, may enable Great Britain to play a prominent role within the EEC. Drown (49) has characterized Great Britain's potential role as follows:

> From this accounting of British occupational hygiene degree granting institutions and those additional degrees proposed, it appears the profession of occupational hygiene is well served by academia. Perhaps those established and proposed programs, coupled with the vast amount of expertise in the U.K., could be put to good use to accommodate students from member states of the EEC where occupational hygiene degree programs are not yet in place or planned for. Certainly there has been a good deal of cooperation in the past by British degree granting programs to serve the needs of overseas students, but surely there is opportunity to increase offerings if demand so dictates and financial support is forthcoming.

8.3 Permissible Exposure Limits

As some of their colleagues in other countries, industrial hygienists in Great Britain followed for many years the TLVs published by ACGIH. The need for limits that reflected conditions prevailing in the United Kingdom and the need to proceed in step with initiatives of the EEC finally prompted Great Britain to enact its own mechanism for setting permissible exposure limits.

The establishment of the HSC gave a major impetus to the initiative of developing British limits. In 1979, the commission adopted a new system of setting control limits intended to protect industrial workers from airborne substances at work. Subsequently the HSE published a list of "control limits" and of "recommended limits" in a publication entitled *Guidance Note Environmental Hygiene (EH) 40* (April 1984).

Control limits were based on Approved Codes of Practice of the European Community Directives and on recommendations of ACTS. Exposures in excess of time-weighted control limits were viewed as endangering the health of the unprotected individual. The 1984 list of control limits can be found in Tables IV.r and IV.s of Cook's *Occupational Exposure Limits—Worldwide* (9).

Although control limits apply to about a dozen substances, published recommended limits cover over 460 chemicals. The latter (9) are listed in Cook's Table VII. HSE inspectors determine whether an industrial establishment complies with applicable regulations based on its implementation of recommended limits and on overall working conditions.

8.4 Professional Organizations

British professionals from diverse occupational health-related areas founded the British Occupational Hygiene Society (BOHS) as a Learned Society in 1953. Over the years, the society grew rapidly both in the field of comprehensive practice as well as in specialized areas. In 1958 the Society began publishing a technical journal entitled *Annals of Occupational Hygiene*.

By 1975 the ranks of British occupational hygienists had grown to the point where they decided that they needed their own professional organization, distinct and separate from BOHS. Thus they founded the Institute of Occupational Hygienists (IOH) but continued to work closely with BOHS. In fact, many members hold dual membership.

8.5 Professional Certification

By 1967, to strengthen the profession through expanded training and after the example of other learned professions, BOHS formed the British Examining Board in Occupational Hygiene (BEBOH) with members nominated by the society. The examining board set professional criteria and granted diplomas to those qualified and able to pass appropriate oral and written examinations.

Over the years, BEBOH grew in both strength and influence. To serve its award holders better, under a constitution agreed to jointly by BOHS and IOH, the functions of the board were expanded in 1978 to include registration of qualified individuals. Thus 11 years after its founding, BEBOH became the British Examining and Registration Board in Industrial Hygiene (BERBOH). Initially, BOHS and IOH each appointed four board members. By 1985 BERBOH amended its constitution. Henceforth, award holders, rather than BOHS and IOH members, would elect candidates to BERBOH.

The 1978 official BERBOH booklet explains the board's aims and functions as follows: "The aims of the Board are to certify to the attainment of recognised standards of competence in the practice of Occupational Hygiene, and to establish and maintain a public Register of those people who have achieved such recognised standards."

As stated in the BERBOH publication, the board recognizes three levels of certification and four of registration:

Levels of Certification

- "Preliminary Certificates in Occupational Hygiene are awarded to those who satisfy the Board on their knowledge of the principles and practice of occupational hygiene relative to specified aspects of the profession."

  A number of institutions of higher learning offer preparatory courses covering ten areas of occupational hygiene. Owners of six Preliminary Certificates are exempted from taking the written portion of the Occupational Competence Examination.
- "The Certificate of Operational Competence in Comprehensive Occupational Hygiene is awarded to those who satisfy the Board on their knowledge of the broad principles of occupational hygiene and demonstrate by their education, training and experience that they are competent to practice in the comprehensive field."

  Such individuals must have a minimum of three years of comprehensive field experience but are not required to make the interpretations and judgments expected of a diplomate.
- "The Diploma of Professional Competence in Comprehensive Occupational Hygiene is awarded to those who satisfy the Board of their knowledge of occupational hygiene and demonstrate that by their education, training and professional experience they are competent to practice in the comprehensive field at an advanced level.

  "The Diploma is available to those who demonstrate that they have held a post requiring them to accept professional responsibility for all aspects of an occupational hygiene programme for at least 5 years and who show on examination that they have the necessary knowledge to make competent assessment of the probable extent of hazard in various circumstances, and to advise on suitable control procedures."

Levels of Registration

- Professional Hygienist: "A person who is recognised as being capable of practicing at a professional level in all aspects of occupational hygiene."
- Limited Professional Hygienist: "A person who is recognised as being capable of practising at a professional level in a specified area of occupational hygiene practice, as indicated in the Register."
- Operational Hygienist: "A person who is recognised as being capable of making environmental surveys in the comprehensive field of occupational hygiene."
- Competent Person: "A person recognised as being competent to carry out a specified range of environmental surveys, as indicated in the Register, to the same standard as an Operational Hygienist."

In Great Britain as in many other countries, including the United States, recognition of the industrial hygiene profession has been slow. Thus there are no universally recognized requirements an individual must fulfill before calling himself or herself an industrial hygienist. Equally slow in coming have been regulations or standards reserving professional responsibilities for professional industrial hygienist. Anderson (50) describes the situation in Great Britain as follows:

> Despite the modest growth of occupational hygiene as an accepted science, there is no statutory requirement for certification or registration prior to practice. Nevertheless, it is increasingly common to find industrial employers requiring evidence of competence on the basis of academic or professional certification, and hence there is much greater reliance on the British Examining and Registration Board's standards than was the case only a few years ago.

Also addressing conditions in Great Britain, Hickish (51) said at the Workshop and Conference held in Luxembourg in 1986: "The requirement for the possession of qualifications has not yet been established. However, we have nearly 20 years experience in assessing professional competence. Over this period our detailed regulations have undergone changes, but the basic concepts have continued unaltered."

8.6 Conclusions

Great Britain can lay claim to having been among the first to recognize the need for occupational hygiene. It has offered the rest of the world outstanding examples of scientific advances and legislative controls in this field. It is rich in occupational health organizations, academic institutions, and publications. Five British scientists have been honored with the AIHA Yant Memorial Award. The themes of their Yant Memorial Lectures show the depth and diversity of their scientific interests and contributions:

		Yant Memorial Lecture
1965	Henry L. Green, Salisbury, Wiltshire	Respiratory Protection Against Particulates—Problems Solved and Unsolved
1971	W. H. Walton, Edinburgh, Scotland	Mining—Problems and Progress in Industrial Hygiene
1974	C. G. Warner, South Wales, England	Fifty Years of British Occupational Hygiene
1984	S. G. Luxon, Buckinghamshire, England	A History of Industrial Hygiene
1987	R. J. Sherwood, Abingdon, England	Occupational Hygiene: An Appropriate Potpourri

As a member of the EEC, Great Britain will have the opportunity to play a role in shaping the future of occupational health within and beyond the borders of member countries. Its impact is likely to contribute to the health and well-being of industrial workers, the public at large, and the progress of the profession of occupational hygiene worldwide.

9 Closing Comments

A number of countries not mentioned in this chapter have developed important elements in industrial/occupational hygiene but may lack others that are needed to promote systematic growth of the profession.

France and the Federal Republic of Germany, for example, have strong underpinnings in occupational health law. Both rely heavily on occupational physicians to protect the health and well-being of workers. They have a body of permissible exposure limits, have recognized research centers, and have demonstrated strength in certain areas of industrial hygiene. France has developed expertise in engineering controls for in-plant air contaminants. West Germany is a leader in toxicologic research, detector instruments, and a well-coordinated infrastructure in occupational medicine and safety. Both countries have some industries with traditional industrial hygiene setups. But, although they have practicing industrial hygienists, they have not yet developed academic curricula in industrial hygiene. They also lack professional industrial hygiene societies and certifying bodies for field and laboratory activities.

A number of countries not listed in this chapter are recognized for their contributions to the field of industrial hygiene and occupational health. Aside from their technical literature, one need only read the list of Yant awardees to see how scientists from Switzerland and Czechoslovakia, from West Germany, France, and Israel, from India, China, and Japan have enriched the profession. Future chapters on industrial hygiene worldwide will have the opportunity to present a more complete listing of accomplishments and contributions by occupational health professionals from these and other countries.

In reviewing the status and the likely development of our profession in the United States and abroad, one is struck by emerging forces in Europe and elsewhere that will have a major impact on the future of the profession. One such factor is the European Community (EC) that will hasten and integrate the development and growth of occupational health and safety in Europe and beyond. Today the EC includes 12 member countries with more than 100 million workers. These countries are:

Belgium	Italy
Denmark	Luxembourg
France	Netherlands
Federal Republic of Germany	Portugal
Greece	Spain
Ireland	United Kingdom

As reported by Cook (52), in July of 1978 the Council of Ministers of the EC adopted the first Action Program on Safety and Health at Work. The Council Resolution emphasized the need to protect workers from dangerous chemicals and to distribute pertinent information on occupational health and industrial hygiene. The Division of Employment, Social Affairs and Education has formulated directives regarding provisions intended to promote better protection of workers. Such directives cover worker training, control measures, permissible exposure limits, and monitoring procedures.

Presently, the EEC has established permissible exposure limits for vinyl chloride, lead, asbestos, and ionizing radiation. An Environmental Chemicals Data and Information Network (ECDIN) has been established and is being maintained as part of the ISPRA Establishment in Copenhagen. The network can be accessed through EURONET and through the American data transmission networks from the "DC Host Center" in Copenhagen.

The Commission of the EC has also published common methodologies for assessing occupational hazards from chemical, physical, and biological agents, for setting occupational exposure limits, and for collecting occupational morbidity and mortality data (53). The European Council of Federations of the Chemical Industry (CEFIC) (54) has issued a report that describes procedures on permissible exposure limits in six member countries (Germany, United Kingdom, France, Italy, Netherlands, and Belgium). It discusses monitoring and control procedures in detail.

In addition to the thrust of the European Council and the European Parliament, there are other forces in pursuit of those same goals, according to Ripley (55). Among international organizations, well known are the efforts of the International Labor Organization (ILO), the World Health Organization (WHO), and the International Agency for Research on Cancer (IARC). Less known, but quite significant, are the contributions by the World Bank in the area of health and safety. The United Nations Organization participates actively through its Environment

Program (UNEP), the Industrial Development Organization (UNIDO), and the United Nations Center on Trans-National Corporations. At the United Nations, the ILO often speaks for the International Conference of Free Trade Unions (ICFTU) and the International Federation of Chemical, Energy, and General Workers' Unions (ICEF). Among others, the International Chamber of Commerce (ICC) and the International Organization of Employers (IOE) represent management interests.

Similar organizations represent worker and management interests in Europe. Thus European employers participate through a number of organizations, including CEFIC and the European Federation of Pharmaceutical Industries' Association (EFPIA), whereas the European Trade Union Confederation (ETUC) and others represent the interests of organized labor.

Although the primary purpose of the EEC is to avoid trade barriers and to streamline and promote investments and the free flow of industries, there is also a strong desire to protect and promote occupational health and hygiene among member countries. It is the clear desire of the EC to reach this goal by adopting uniform standards and practices and thus to create a well-trained work force able to find work in any or all member countries under comparable conditions and equal levels of protection. This upgrading and harmonizing of occupational health concepts and procedures will take several years to formulate, adopt, and implement. Furthermore, if past history is an indication, there will be differences among the various members with regard to promptness and level of implementation of such directives and standards. Nonetheless, the impact of such strategies will be considerable and is bound to spread to other countries and continents.

A number of occupational hygiene organizations from Great Britain, Italy, and Holland are striving to lend their expertise to various policy-making groups of the EC in an effort to provide technical input to directives and regulations under consideration.

Clearly, such regulatory efforts will affect all member countries of the EC and anyone outside interested in cultivating commercial links with member countries. More than likely, it is also only a matter of time before such regulatory efforts have an impact on similar legislation in nonmember countries in Europe and elsewhere. Industrial hygienists everywhere will do well to keep abreast of and, whenever possible, provide direct or indirect input into the regulatory effort in occupational health taking place within the United Nations Organization, major international health agencies, and, of course, the EC.

Another initiative likely to have an impact on the growth and development of industrial hygiene is the creation of the International Occupational Hygiene Association (IOHA). A multinational effort to increase contacts among practitioners of the profession and share the benefits of industrial hygiene on an international scale began in the mid-1980s in the United States and abroad. Formal contacts in Luxembourg and in Rome in 1986 led to the founding of IOHA at the Annual American Industrial Hygiene Conference in Montreal, Canada in 1987. Representatives of the following nations signed the founding charter: Australia, Canada, Great Britain, Holland, Italy, Spain, Switzerland, and the United States.

IOHA represents the efforts and the collective good will of a relatively small group of individuals compared to the total number of practitioners represented by charter members and those active in other countries. Administratively, the office of the president for 1989–1990 is in Australia. Next it will be in Switzerland, and the year after in Spain. The secretariat has been in England since its inception and will remain there over the next few years for reasons of continuity. For the time being, the association draws its financial support from charter member associations. Individual memberships are under consideration once the association is firmly established and has proven its value to the membership at large.

The challenges facing IOHA are many but so are the opportunities. The next few years will tell whether the association will be able to serve the interests of working men and women everywhere and of the profession at large. At that point it will become clear to all whether IOHA has measured up to the dreams and hopes of its founders.

REFERENCES

1. W. A. Cook, *Occupational Exposure Limits—Worldwide*, American Industrial Hygiene Association, 1987, p. 92.
2. A. T. Jones, *Am. Ind. Hyg. Assoc. J.*, **49**, 593–599 (Dec. 1988).
3. A. T. Jones, private communication.
4. A. Findley, *Training and Education in Industrial Hygiene: An International Perspective*, American Conference of Governmental Industrial Hygienists, 1988, pp. 125–127.
5. The Ontario Factories Act. 1884, c.39, s.11(3), as quoted in *Occup. Health Ontario*, **9**(1), 20 (Winter 1988).
6. A. Yasse, *Am. J. Public Health*, **78**(6), 689–693 (1988).
7. J. W. Smith, *Training and Education in Industrial Hygiene: An International Perspective*, Vol. 15, American Conference of Governmental Industrial Hygienists, 1988, p. 135.
8. G. S. Rajhans, private communication.
9. W. A. Cook, *Occupational Exposure Limits—Worldwide*, American Industrial Hygiene Association, 1987.
10. M. H. Noweir, "Occupational Health in Developing Countries with Special Reference to Egypt," *Am. J. Ind. Med.*, **9**, 125–141 (1986).
11. M. H. Noweir, "Overview of Occupational Health Research in Egypt," *International Symposium on the Biomedical Impact of Technology Transfer, Faculty of Medicine Ain Shams University Cairo-Egypt*," National Institute of Environmental Sciences, North Carolina, 1986.
12. M. H. Noweir, *Training and Education in Occupational Hygiene: An International Perspective*, Vol. 15, American Conference on Governmental Industrial Hygienists, 1988, p. 143.
13. *Occupational Heatlh Services in Finland*, 2nd rev. ed., Institute of Occupational Health, Helsinki, Finland, 1988.
14. Jukka Hakko, "Reforms Bring Labor Protection Closer to Workers," *Work Health Safety*, 4–7 (1982).

15. T. Kauppinen, *Training and Education in Occupational Hygiene: An International Perspective*, Vol. 15, American Conference of Governmental Industrial Hygienists, 1987, pp. 145–147.
16. From a speech by Dr. Danilo Sordelli, President AIDII, at the II Conferencia Nacional de Higiene Industrial, on Nov. 17, 1988 in Valencia, Spain.
17. DPR April 27, 1955, No. 547, Norme per la prevenzione degli infortuni sul lavoro e successive integrazioni. DPR of March 19, 1956, No. 303, Norme generali per l'igiene del lavoro, followed by 320/1956, 321/1956, 322/1956, and 128/1959.
18. Legge No. 615, July 13, 1966, Provvedimenti control l'inquinamento atmosferico.
19. Legge No. 300, Articolo 9, Tutela della salute e dell'integrita' fisica.
20. DPR of April 15, 1971, No. 322, Regolamento per l'esecuzione della legge 13.7.1966 No.615, provvedimenti contro l'inquinamento atmosferico, limitatamente al settore delle industrie.
21. Legge 319/76 sulla tutela delle acque e successive modifiche e integrazioni.
22. Legge 23.12.1978, No. 833, Istituzione del Servizio Sanitario Nazionale.
23. Direttiva 79/831 EEC, Sesta modifica della direttiva 67/548 CEE, classificazione, imballaggio, etichettatura sostanze e preparati pericolosi, recepiti dalla Legge No. 256, 29.5.74 e DRP 927, 24.11.1981.
24. Direttiva 80/1007/CEE, 27.11.1980, Sulla protezione dei lavoratori contro i rischi derivanti da una esposizione ad agenti chimici, fisici e biologici durante il lavoro.
25. DPR September 10, 1982, No. 962, Attuazione della direttiva CEE No. 78/610 relativa alla protezione sanitaria dei lavoratori esposti a cloruro di vinile monomero.
26. DPCM March 23, 1983, Limiti massimi di accettabilita' delle concentrazioni e di esposizione relativi agli inquinanti dell'aria nell'ambiente esterno.
27. DPR May 24, 1988, No. 203, Attuazione delle direttive CEE numeri 80/779, 82/884, 84/360 e 85/203 concernenti norme in materia di qualita' dell'aria, relativamente a specifici agenti inquinanti, e di inquinamento prodotto dagli impianti industriali, ai sensi dell'art. 15 della Legge 6.4.1987 No. 183.
28. DPR September 10, 1982, No. 915, Attuazione delle direttive CEE No. 75/442 relative ai rifiuti, No. 76/403 relativa allo smaltimento dei policlorotrifenili e No. 78/319 relative ai rifiuti tossici e nocivi.
29. D.L. September 9, 1988, No. 397, Disposizioni urgenti in materia di smaltimento di rifiuti industriali, e decreto applicativo 22.9.1988.
30. Legge 11.11.1975, No. 584, Art.7, Divieto di fumare in locali pubblici.
31. Circolare Ministeriale No. 45, 10.7.1986, Min. Sanita': Ordinanza Ministero Sanita' 26.6.1986 e Circolare Ministeriale 1.7.1986, No. 42, Min. Sanita'
32. D. Drown, "Occupational Hygiene Education in the EEC, a Survey of Existing Programs," draft report, Wageningen Agricultural University, The Netherlands, 1988.
33. Statuto dell'AIDII, Art.2, 6.10.87.
34. P. B. Mayer, *Training and Education in Occupational Hygiene: An International Perspective*, Vol. 15, American Conference of Governmental Industrial Hygienists, 1988, pp. 165–166.
35. From "Note 'Occupational Hygiene' NVvA," English translation, Utrecht, The Netherlands, May 1987.
36. D. B. Drown, "Occupational Hygiene Education in the EEC: A Survey of Existing

Programs," draft report, Wageningen Agricultural University, The Netherlands, 1988, pp. 105–110.
37. W. A. Cook, *Occupational Exposure Limits—Worldwide*, American Industrial Hygiene Association, 1987, pp. 99–100 and Table VII.
38. S. H. S. van der Meulen, personal communication, 1988.
39. G. Clack, *Job Safety Health*, **2**, 11–16 (June 1974).
40. B. Dahlner, *Training and Education in Occupational Hygiene: An International Perspective*, Vol. 15, American Conference of Governmental Industrial Hygienists, 1988, pp. 177–181.
41. G. Gerhardsson, *Training and Education in Occupational Hygiene: An International Perspective*, Vol. 15, American Conference of Governmental Industrial Hygienists, 1988, pp. 69–78.
42. C. Edling and P. Lundberg, "The Procedure of Occupational Standard Setting in Sweden," *Proceedings of the Seventh Annual Conference*, Australian Institute of Occupational Hygiene, December 1988.
43. S. G. Luxon, *Am. Ind. Hyg. Assoc. J.*, **45**, 731–739 (Nov. 1984).
44. W. Simpson, *Her Majesty's Inspectors of Factories, 1833–1983*, Directorate of Information and Advisory Services (IAS2), 1983, p. 4.
45. C. G. Warner, *Am. Ind. Hyg. Assoc. J.*, **35**, 381–391 (July 1974).
46. T. Willis, "Safety in the Workplace—The British Approach," *Prof. Safety*, 40–44 (Jan. 1985).
47. S. G. Luxon, *Am. Ind. Hyg. Assoc. J.*, **48**, pp. A787–790 (Dec. 1987).
48. J. J. Cullen, Presentation at Opening Session, American Industrial Hygiene Conference, May 1986.
49. D. B. Drown, "Occupational Hygiene Education in the EEC, a Survey of Existing Programs," draft report, Wageninen Agricultural University, The Netherlands, 1988, p. 122.
50. J. T. Anderson, *Training and Education in Occupational Hygiene: An International Perspective*, Vol. 15, American Conference of Governmental Industrial Hygienists, 1988, pp. 189–193.
51. D. E. Hickish, *Training and Education in Occupational Hygiene: An International Perspective*, Vol. 15, American Conference of Governmental Industrial Hygienists, 1988, p. 49.
52. W. A. Cook, *Occupational Exposure Limits—Worldwide*, American Industrial Hygiene Association, 1987, pp. 115–116.
53. See Hunter et al., *Activities at the European Community Level Regarding Toxic Chemicals and the Protection of Workers Exposed to Them*, Vol. 12, American Conference of Governmental Industrial Hygienists, 1985, pp. 59–72.
54. "CEFIC Report on Occupational Exposure Limits and Monitoring Strategy," European Council of Chemical Manufacturers' Federation, Brussels, Belgium, July 1983.
55. K. Ripley, "Role of International Agencies, Governmental and Commercial, in Europe," from a paper presented before the American Academy of Industrial Hygiene, San Antonio, Texas, September 22, 1989.

CHAPTER FOUR

The Industrial Hygiene Survey and Personnel

James O. Pierce, Sc.D.

1 INTRODUCTION

In today's industrial society, with the advent of numerous health and safety regulations, many of which are contradictory and overlapping, the industrial hygienist finds that what was once considered to be the industrial hygiene survey has now taken on new ramifications. In the past, these surveys were regarded as scientific investigations using all of the skill, experience, and art of the professional industrial hygienist to recognize, to evaluate, and eventually to devise methods of controlling exposure to potentially harmful substances in the workplace. The results of these efforts were intended to provide a safe and healthy work environment, to prevent undue stress on the worker, and to prevent injury or illness resulting from the workplace environment.

Prior to the passage of the Occupational Safety and Health Act of 1970 (OSHAct), the accomplishment of these aims was considered to be a fundamental moral obligation on the part of the employer because few health and safety regulations existed and there was little or no enforcement at the federal level, although enforcement authority existed under statute authority such as the Walsh–Healy Act (see Chapter 2).

One of the primary industrial health professional groups in the country, the U.S. Public Health Service, was instrumental in establishing policy and procedures that would eventually result in the formation of the Bureau of Occupational Health

Patty's Industrial Hygiene and Toxicology, Fourth Edition, Volume 1, Part A, Edited by George D. Clayton and Florence E. Clayton
ISBN 0-471-50197-2

(BOSH) and, finally, the National Institute of Occupational Health and Safety (NIOSH), which is the premier research and training arm of the federal government in occupational health and safety.

The changes brought about by the act and the resulting emphasis, especially in later years, on industrial hygiene practices and procedures, have been significant. Abetted by increased concern of employer, employee, and society in general, along with a changing definition of societal well-being, today's industrial hygiene practice has taken on a new meaning and direction, and what was once considered the industrial hygiene survey has now been upgraded to include many new skills and requirements. No longer is the practice of industrial hygiene viewed as one based upon the investigation of acute episodes and events with the long-term goal of reducing exposures to potentially toxic substances, but one based upon the new regulatory mode of having to comply not only with complex regulations promulgated by the Occupational Safety and Health Administration (OSHA), the Environmental Protection Agency (EPA), and the Consumer Product Safety Commission, but also with a myriad of regulations from federal, state, and community agencies.

One highly recognized and respected expert in the field (1) has recently described the progression of industrial hygiene as characterized by six distinct periods of time:

1950–1960	Benign neglect of occupational safety and health
1960–1970	The environmental counterpart and stimulus
1970–1975	OSHA—start-up: early childhood patterns
1975–1980	Adolescence
1980–1988	Deregulation and "down-scaling"
1989–1990	Adulthood

To this impressive analysis up to the 1990s must be added a prediction for 1990–1995, and to the year 2000 and beyond, which will be characterized by a different kind of work force and by more concern for limiting the risks associated not only with the workplace, but with society in general.

In 1994, when there no longer exists a mandatory retirement age for most employees, we will see a much more aging work force, and statistics prove that industrial accidents increase with the age of the work force. Thus the industrial work site will be the focus of increased vigilance to provide for a safe and totally healthful environment. In addition, in future years more intensive automation and emphasis on robotics will continue to develop. The major employers of the future will be in the service-related industries. Furthermore, the continued trends of stress-related health effects and disorders will rise and cause us further to evaluate our approach to the field of preventive health. All of these factors, including our continuing dependency upon computers and computer-controlled work sites, will change our approach to the industrial hygiene survey. The field of industrial hygiene must change to meet and accommodate these future challenges.

However, we must be practical and realize that change comes about slowly and, although we must look toward the future and try to prepare adequately for its eventualities, we must still face the realities of the 1990s. We must look toward the leaders of today to cope with the realities of what we must accomplish.

The industrial hygiene era of 1980–1988, Corn (1) points out, is one of deregulation accompanied by the shock of early OSHA childhood. He cites the following parameters as the primary factors responsible for this era:

1. Reduced effort of federal OSHA
2. State programs facing budgetary constraints
3. Improved cost effectiveness of control methods
4. Pressure to abandon "hierarchy of controls"
5. Trimming of corporate staffs ("decentralization") and increased utilization of consultants
6. Employee and community "right-to-know" laws
7. Comparative risks utilized for priority setting
8. International coordination at the professional level
9. University programs in occupational safety and health

Corn views the 1990s and beyond as having reached "adulthood" and as characterized by increased regulations, increased employee and community knowledge, increased worker/employer joint safety and health programs, increased presumption of risk in absence of data, increased concern with stress and ergonomics, and increased litigation.

These factors among others directly affect today's industrial hygienists and the duties and responsibilities they face in a modern society. These factors play an important role in the industrial hygiene survey because no longer is such a survey the personal province of the industrial hygienist and management; the results and all data thus obtained can now be construed to be in the public domain.

However, the industrial hygiene survey is affected not only by OSHA regulations, but also by the proliferation of health-related regulations being promulgated by a wide variety of federal and state programs. These include those regulations that require complicated record-keeping systems ranging from SARA Title III to the emerging legislative activity at the state level, requiring risk management and prevention programs (RMPPs). These regulations are a direct result of tragic events such as Bhopal, the Union Carbide disaster, and the asbestos episodes.

In addition, we now are faced with overlapping legislations such as AHERA (Asbestos Hazard Emergency Response Act, 1986); SARA (Superfund Amendments and Reauthorization Act, 1986); TSCA (Toxic Substances Control Act, 1976); CERCLA (Comprehensive Environmental Response Compensation and Liability Act, 1980); and numerous other state and federal regulations related to health and safety. All these affect the field of industrial hygiene and further complicate the importance and significance of the industrial hygiene survey. Today's hygienist must have thorough knowledge of all such related regulations in order to function in our complex society of the 1990s and beyond.

If I were to attempt to predict the future of the industrial hygiene survey, I would look toward a model of risk assessment/computer enhanced procedures. A recent article on risk assessment by Renshaw (2) states:

> Techniques for assessing risks to the health of employees and property in the occupational setting are evolving at a rapid pace. These techniques embrace a wide variety of risks including chronic disease from repeated exposures to toxic substances and fatalities from vapor cloud releases. An understanding of these techniques by the industrial hygiene manager is necessary to provide insight into the effective allocation of resources and to make sound decisions on control measures which minimize risks. An understanding of basic terms related to risk assessment is necessary in order to apply these techniques and, more importantly, to determine where such techniques are worthy of application.

In defining the future role of the industrial hygiene survey, one needs to address the emerging proliferation of computer software programs that are coming out with increasing frequency. Separate reference books are being published on computer systems for occupational safety and health management. Record-keeping programs for tabulating industrial hygiene exposures and records are readily available. It is undeniable that computers and advanced software programs will play an integral part in the future of industrial hygiene and will have a direct impact on the future performance of industrial hygiene surveys. Also, owing to the complexities involved, the "team" personnel will change, requiring new disciplines to be integrated into this already complex field. Industrial hygiene is fast becoming less of an art and more of a science.

In summary, I believe the industrial hygiene survey and the type of personnel required to meet the needs of the year 2000 and beyond will be dramatically different from today's, requiring changes in the traditional education of industrial hygienists and occupational health and safety personnel to equip them with the advanced techniques and knowledge necessary to meet the challenges of the future. Of increasing importance will be innovative educational curricula that will address these needs, including new approaches that will include risk assessment, risk management, and management techniques.

The role of the industrial hygienist will continue to increase in the industrial management structure of the future; it will also require more skills. The role of the industrial hygiene technician or associate industrial hygienist, discussed below, will also change dramatically and will eventually assume the role of what has in the past been the assumed role of the primary industrial hygienist. In the interim, we must still rely upon a modified approach to the traditional industrial hygiene survey, a basic tool that all such professionals must still be adept at using.

2 TYPES OF SURVEY

In the current practice of industrial hygiene there are three types of survey:

1. The reconnaissance or observational survey
2. The investigational or appraisal survey
3. The combined industrial hygiene and medical survey

The second type has as its chief purpose the evaluation and control of potentially harmful situations, and the third type integrates the investigational survey with medical examinations of the work force. It must be noted that this approach is becoming much rarer because the reporting structures of many current and future corporate entities separate the occupational health and safety component from the traditional medical department and have put the latter group into a direct line reporting organization. A number of corporations still follow the older, traditional approach, but the future is clear: industrial hygienists have become corporate managers and, as such, must be prepared to enter this more rarefied atmosphere. New computer software programs automatically correlate medical and toxicology data and nowadays even the results of genetic screening into highly sophisticated programs that will individually characterize work stations with exposure data.

2.1 The Preliminary Industrial Hygiene Survey

The preliminary or observational industrial hygiene survey is usually the immediate forerunner of a survey employing technical instruments. In this survey, experience and familiarity with industrial processes are a *must*.

This survey is of paramount importance, especially when familiarizing oneself with a new plant or workplace; it is done for the purpose of selecting locations in a plant where exposures or hazards are later to be evaluated by analytical studies, to determine whether additional control is necessary. During this survey pertinent data should be collected, such as the number of male and female workers employed at various operations or processes; safety and systems management practices, emergency response programs, and medical services; availability of accident and illness records; the different types of operations conducted; raw materials, processing aids, products, and recognized by-products; measures employed for dust, fume, and vapor or gas control; and methods of solid, liquid, or gas waste disposal.

Detailed notes, preferably a tape-recording followed by a typewritten transcription, may prove invaluable, as will also discussions with workers involved with the actual performance of the work in progress.

If a plant layout showing the location of the operations or process equipment in the industrial establishment is immediately available or can be obtained in a short time, it will be invaluable in helping to orient the industrial hygienist to the complexities of an industrial situation. The preliminary survey will also determine the field survey equipment necessary for the investigational survey.

Depending on the size and complexity of the operations, the time required for the preliminary survey may vary from one day in a small shop to a week or more in large industrial plants. In a small plant the preliminary survey may immediately precede the investigational survey, and both may be accomplished on the same day by the same individual, although care must be taken if the necessary goals are to be achieved.

2.1.1 The Industrial Hygiene Survey and Personnel

The preliminary survey is usually made with no equipment for measurement purposes other than those portable pieces of equipment that can be conveniently carried on the person, such as a sound level meter. This survey relies heavily upon experience and expertise in the field.

This survey should be made by an individual who is familiar with the type of industry involved, especially the chemistry of its products and by-products, and one who is well grounded in the field of industrial hygiene. This individual may or may not be the person who is to make the final technical analysis, but he or she should be at least equally familiar with the problems involved in recognizing and evaluating exposures to potentially harmful materials. Also, a complete familiarity with local, state, and federal regulations pertaining to the workplace and to the outside environment should be taken into account and notes appropriately made, along with references to pertinent regulations.

This survey is best accomplished by following the industrial process through the plant from raw materials to finished products. The industrial hygienist should be accompanied by the production superintendent, or some other qualified plant employee, to explain any process or steps in manufacture that are not evident to the surveyor. Among the plant personnel who are best suited to the role of guide for the investigator are the production superintendent, the chief process engineer, the plant chemist, the foreman of the department under investigation, and the safety director. Consultation with the medical department may also be of help in attempting to identify potential exposures with physiological effects. Familiarity with industrial processes is a *must*.

As a part of this type of survey, which draws upon the expertise of the trained and experienced industrial hygienist, he or she must rely upon a well-developed and trained sense of smell along with other senses that tend to detect and recognize potentially hazardous exposures. The person should be able to recognize and identify all the common gases and vapors that possess characteristic odors, tastes, or irritant effects, as well as to judge qualitatively whether the concentration of a substance may be exceeding permissible levels or recommended levels of exposure. This highly unobjective evaluation must of course then be followed by a full evaluation employing "state-of-the-art techniques" for quantifying atmospheric workplace exposures.

The industrial hygienist should also be alert to any signs in the personal appearance of the workers that may indicate adverse occupational exposures. Of course, interviews with individual workers are a primary source of information for the industrial hygienist evaluating potential problems in the workplace. In this way it may be possible to identify a serious exposure that might otherwise be overlooked because of the intermittent or obscure nature of its cause. Interviews with such workers can often reveal health problems resulting from workplace exposures that can easily be overlooked by any qualitative or even quantitative evaluation of the workplace.

Frequently there are contributory or intermittent operations not in evidence at the time of the survey. These may include preliminary treatment of raw materials, disposal of by-products or waste products, "turn-around" or periodic maintenance and rebuilding, and warehouse operations. The industrial hygienist may learn of these through experience in the industry, from a knowledge of similar operations, or by an extended discussion with potentially affected employees.

One should always determine the presence of control measures and provide an opinion about:

1. The probable need for, or effectiveness of, control
2. The type of personnel, in terms of training, skill, or knowledge of the potential hazards in the workplace
3. The attitude of management, supervising staff, and the personnel employed at the work site toward health and safety practices, along with the control measures currently in effect and proper maintenance procedures

At the end of the preliminary survey it may be desirable to give the management a report of findings and plans.

Having completed the preliminary survey, and having decided where all investigations are to be made, the industrial hygienist now selects the necessary equipment and undertakes the technical analysis. However, all findings and observations made during the preliminary survey should be fully documented in writing and forwarded to the appropriate supervisor for the record.

2.2 The Investigational Industrial Hygiene Survey

2.2.1 Surveying the Plant

The investigational survey, or the technical industrial hygiene analytical survey, involves the evaluation of all exposures to potentially harmful situations and the development of control measures. This is what is ordinarily referred to in the profession as an industrial hygiene survey. If no previous survey of the plant has been made, it is best to leave any equipment in a safe place and conduct a preliminary survey before undertaking the detailed investigation.

It is now necessary to evaluate exposures by exact measurements owing to the necessity for legal documentation. When control is the only objective, and it is obvious to the investigator that contaminants are excessive, it is not prudent to devise satisfactory controls without the benefit of quantitative measurements. Nowadays one must document all such situations to protect against future liability claims.

When there is any question about the necessity of control, however, or when facts are desired for the record because of medical, legal, or other needs, samples of the atmosphere must be collected and evaluated by chemical or instrumental methods.

In surveys to determine the merits of a claim for compensation for injury allegedly arising from harmful exposures in industry, it is essential to obtain quantitative information about exposures, in order to ensure the establishment of facts, rather than the recording of opinions, even though it may be obvious to the experienced industrial hygienist that the exposures are either insignificant or excessive. In a system handling a potentially harmful chemical, even though the investigator is completely satisfied that control is ample, sampling may be desirable for its psychological effect; it eliminates personal factors and gives the uneasy employees assurance of a safe environment.

Under the requirements of OSHAct, records of periodic analyses of atmospheric contamination are required when the concentration of a chemical agent is borderline with respect to the threshold limit value for an 8-hour daily exposure. More frequent analyses are required when control measures are to be installed to reduce the air contamination to acceptable levels. The detailed regulatory requirements of the act with respect to chemical air contaminants are still being developed, and some of the procedures specified for air sampling, work practices, and other administrative matters are subject to dispute. However, there is still room for the knowledgeable hygienist to exercise judgment in evaluating exposures and creating a safe working environment, as long as this judgment decision is fully documented. Again, caution is advised.

2.2.2 The Industrial Hygiene Survey and Personnel

The analytical survey can be accomplished with the cooperation of the production superintendent, the safety director, or the foremen of the particular departments in which investigations are to be made. If the results are to be representative, air sampling or other quantitative measurements should be made during normal, as well as the most unfavorable, operating conditions, and over a period of time long enough to yield the complete exposure picture. The foremen can assist in maintaining the normal conditions as well as in simulating abnormal conditions if such are required for a true evaluation. It is good policy to visit the physician or nurse on duty, if there is one, to establish cordial cooperation and to review any records of ill health. The plant physician may wish to accompany the hygienist during the survey of the plant, and this may prove advantageous to both; but it is an exceptional physician who can supply needed information about plant processes.

In the present social climate, it is a rare employee indeed who has no inkling of the possible occupational health hazards of a particular industry or a specific workplace.

Both worker and management are constantly bombarded by articles on industrial health hazards that appear in the daily newspapers, union periodicals, trade journals, and scientific literature. Some of the information is factual and straightforward, whereas some articles have an inflammatory tone. In addition to such educational efforts, some of the regulations promulgated by OSHA require that employee representatives be informed of the nature and purpose of an industrial hygiene survey and be granted the opportunity to observe air sampling, instrumental measurements, and other procedures carried out during the survey by the employer, the employer's agent, or a regulatory officer. Furthermore, during the inspection or survey made by an OSHA compliance officer, the inspector has the legal right to interrogate employees in private concerning their knowledge of adverse working conditions and complaints of ill health due to alleged occupational exposure.

In view of these unsettling influences, how should professional industrial hygienists carry out their responsibilities without creating any friction between the workers and their employer or without contributing to any further apprehension among the workers concerning the healthfulness of their occupation?

It is advisable for management to explain the purpose of the survey to the employees and/or the union, thereby creating a feeling of partnership with the workers in their health protection efforts. At the same time, the element of apprehension due to the presence of strangers or relatively unfamiliar persons making unfamiliar measurements will have been removed. A request can be made for management to assign a temporary guide to answer workers' questions at the different locations surveyed in the establishment. The normal exchange of greetings and courtesies between the fellow employees and the hygienist should be maintained. The plant escort can be useful in answering questions regarding the survey. Although all possible sources of friction between humans cannot be prevented, a friendly attitude will have a calming influence on many hostile situations.

Air samples may be taken at one or several locations to determine individual exposures, general room air conditions, and sources of contamination. Many variable factors should be considered at the time of sampling, in relation to normal, adverse, and optimal conditions. Single samples, though not to be scorned, should not be accepted as true indications of existing conditions because the concentration of contaminants vary from minute to minute, day to day, and season to season.

The more samples taken, the more evident will be the reasons for any variance in results, trends, peak concentrations, and other factors, versus the single extended-time sample. Here again, the trained observer, from the senses of smell, taste, or feeling, may be able to evaluate an exposure accurately enough to establish either the necessity for its control or its insignificance. In large measure, this depends on previous observations of known amounts and similar magnitudes of the material to be evaluated, or a calibration of individual sensitivity, so to speak, because there is some variation among persons in sense of smell.

Three is the minimum number of samples to be taken to evaluate any one operation in a series of successive steps of a process or job assigned to a worker. If the job consists of four different tasks successively repeated, 12 air samples should be collected if confidence in the results is to be had.

All shifts of a similar process should be observed, for a night shift may present an exposure during some operation that the day shift conducts safely, or vice versa. A survey was made many years ago in which an unusual night operation created a very high overexposure to toxic substances. The process involved the smelting of zinc sulfide ore in gas-fired retorts, and the molten zinc metal was collected in cylindrical clay condensing units. A certain number of condensers were removed and replaced each night, successively. This was a hot and dirty job that was rotated among all smelter employees. Tests for employee exposure to airborne heavy metal contaminants, such as lead, antimony, arsenic, and cadmium, were below acceptable levels during the day and "graveyard" shifts. However, the night shift workers who performed this special operation for 4 to 5 hours had very high exposures to these agents, which offered an explanation for the subjective symptoms and clinical signs of illness among the employees.

The opportunity for employee exposure to airborne chemical contaminants is low in the chemical and petroleum processing industries, because the processing is carried out in closed systems. When such systems are shut down for periodic safety inspections, or cleaning and replacement of worn equipment, gross exposure to chemicals may occur unless the work is planned and safety precautions followed. During these periods, exposures can occur to the process chemicals, cleaning agents, and air contaminants arising from such construction activities as welding in confined spaces.

During the course of the technical survey it may be desirable to trace the air contaminants from their origin to the point or points of dispersion.

Because of the possible synergistic effect of exposures to combined aerial contaminants or the enhancing influence of abnormal temperatures on chemical exposures, the industrial hygienist should also search for the unusual circumstances or combinations of factors that may give rise to adverse physiological responses.

Up to this point this discussion has centered on a survey designed to quantify aerial contaminants and evaluate employee exposure to these agents. However, associated matters such as the types of emergency respiratory protective device issued for use against a specific air contaminant also should be examined.

The survey should determine that the device conforms with the NIOSH approval schedules, making sure that it will provide protection against the concentration of contaminant that may be encountered. No respiratory protective device is presently permitted to be used for continuous protection against air contaminants in lieu of engineering control, although regulatory enforcement may differ on this aspect. Nevertheless, certain devices—namely, mechanical filter-type respirators and chemical cartridge-type respirators—may be used temporarily over protracted periods while engineering controls are being installed.

2.2.3 Developing Control Procedures

If the industrial hygienist or associates have knowledge and skills in the design of mechanical ventilation systems that may be suitable for controlling or reducing the dispersion of a chemical agent from a certain operation, advice and assistance in the form of elementary design specifications can be offered to the plant management.

To be more specific, the hygienist should at least choose between general ventilation and exhaust at the source of contamination, the size, shape, and face velocity requirements of hoods, type of ventilating fan, necessity of a dust collector, suitable types of dust collector, location of dust collector inside or outside, and possibility of recirculation of the air. The judicious use of a smoke tube or Velometer to locate and measure air currents and cross-drafts may demonstrate to the plant engineer the ineffective control of a present exhaust system. In today's environment, the effective industrial hygienist should rely upon competent outside consultants to provide the expertise in designing such advanced systems.

At the end of this investigational survey, the hygienist possesses many accumulated facts that must be correctly applied; otherwise they become of academic interest only and serve little purpose in the promotion or control of health. The hygienist should develop methods or procedures for controlling overexposures and should have such ideas sufficiently crystallized to present to the plant management, albeit they must also be able to survive outside and regulatory scrutiny.

Any unusual hazards found during the plant survey should be pointed out and discussed with the physician or nurse, to ensure that by cooperation and knowledge of the complete picture, both the industrial hygienist and the plant medical department are aided in their work. It is important that the medical advisors have knowledge of all information developed as a result of the survey.

At the end of the survey, if not before, the plant manager, personnel director, production superintendent, safety engineer, and maintenance engineer should be acquainted with all significant findings, along with recommendations or suggestions for control. Objections may be raised: some of them can be overcome, whereas others may indicate that alternatives to some of the original proposals are more practical. It is obviously desirable that any major control program be discussed in detail and its merits carefully explained. Because federal law requires preventing excessive exposure to potentially harmful agents, it is not a matter of "selling" the need for a control program but rather "selling" a system that is most effective and most economic for the specific situation. The professional industrial hygienist should not attempt to coerce management to undertake any corrective action nor allow any negative management attitudes to discourage reporting adverse findings or to color the conclusions and recommendations that are made.

2.2.4 Records and Reports

A written confirmation report of the survey, complete with findings and recommendations, should be delivered to plant management and other appropriate levels of a corporate establishment. The report should contain the results of all measurements in tabulated form. Interpretation of the findings in each table should be presented in an adjacent narrative, along with the surveyor's conclusions and any necessary recommendations for improving the working environment. A summary of the findings, conclusions, and recommendations should be located in the beginning of the report to draw the reader's attention to information of special importance.

The report may vary from a brief, unbound set of written pages to a voluminous manuscript in an attractive and expensive binding. The report is more likely to be read when brief, but it should always be sufficiently detailed to establish the necessity for the accompanying recommendations.

The industrial hygienist may wish to use prepared forms to accommodate personal needs to record each sample collected and all environmental and work information pertinent to the sample. Calculated air concentration of the chemical agent derived from the laboratory analysis of the sample can also be recorded on such a form. Many new and emerging computer software programs are available for recording and permanently storing such data. Many software programs are specifically tailored to meet regulatory requirements and can save much time and effort.

A permanent record of observations made during the course of the survey should be entered into a field notebook on the job for reference during preparation of the report; these notes will also serve as admissible evidence in any possible court action. The surveyor should not jot down random bits of information on loose pieces of paper that may be lost. Details that are not recorded on the spot are often forgotten. If management permits, snapshots of specific operations taken with a camera that uses self-developing film are extremely helpful for refreshing memory in report preparation. A decision to include photographs in a report should be made after crucial consideration for their need, but only with prior approval of management.

It is important that the report of a survey reach the plant management while the situation is fresh in mind, never later than 30 days after the survey and preferably within 2 weeks.

The value of the report is enhanced if it is delivered in person by the surveyor, or this effect may be achieved by a personal call as soon as the management has read the report, in time to explain or support any findings or recommendations that may be questioned. Even though a recommendation was agreed on at the time of the survey, if it arrives at an inopportune time—or so late that it has been forgotten or interest has waned—management may assume an antagonistic attitude, necessitating another personal contact. Reports or copies of reports should never be presented to anyone other than the person requesting the survey without the knowledge and permission of that person.

The written report should be delivered personally by the chief industrial hygienist, to ensure that the findings are discussed at the appropriate management level. A report sent by mail, or delivered by someone with little experience, detracts from the importance of the survey and can also lead to misinterpretation of the recommendations.

2.3 The Combined Industrial Hygiene and Medical Survey

The combined industrial hygiene and medical survey is a comprehensive study of the environment and of the health status of the workers. Its purpose is to determine any relationship between exposure to atmospheric contaminants and the workers' health. Depending on the size of the plant and the number of different kinds of agent to which the employees may be exposed, this type of survey may require a week to several months or even longer to complete. In addition to in-depth studies of the various environmental factors, this kind of investigation includes examination of individual workers by physicians and medical technicians. It includes tests to observe the various organs of the body and to measure their physiological functions by microscopic and chemical examination of the blood and urine as well as X-ray and vital capacity tests of the pulmonary tract. Other special examinations such as the electrocardiogram or an electroencephalogram may be required.

The interpretation and evaluation of medical information belong in the field of medicine and call for the services of physicians qualified by training to evaluate the tests just mentioned. Likewise, the environmental findings should be handled by persons technically trained in the scientific principles of industrial hygiene, preferably with long experience in this field.

Because the purpose of the combined survey is to determine any possible relation of exposure to the environmental factors and any detectable adverse effects among the workers, another important professional, namely, the statistician or epidemiologist, completes the survey team.

In years past such studies evolved from the discovery of unexplained illnesses in certain occupations, but at present they are more likely to originate from a desire to learn whether the known presence of air contaminants with ill-defined or low toxic properties may exert little known or obscure adverse effects among occupational exposed workers. Many of the maximum permissible concentrations of exposure for atmospheric contaminants, notably free-silica-bearing dusts, mercury vapor, lead dust and fume, and carbon monoxide, were derived from such classical, comprehensive industrial hygiene studies (5–9) carried out in the 1920s and 1930s by the Division of Industrial Hygiene of the U.S. Public Health Service. The standards or "safe limits" established by these studies have been modified in recent years because of new evidence based on advances in analytical techniques and more sensitive clinical test procedures.

For that reason, in all reports proposing safe working standards, the environmental and clinical methods of analysis should be explained and documented. At some later time, more accurate or sensitive analytical test methods may reveal new findings that show an accepted figure to be erroneous.

3 THE INDUSTRIAL HYGIENE UNIT OR ORGANIZATION

3.1 Recent Pattern of Growth

For a span of about 30 years, beginning in the middle 1930s, the centers of industrial hygiene activity were concentrated in the governmental units of about 25 states. The federal industrial hygiene activity was located in the Public Health Service in its Division of Industrial Hygiene, which provided leadership in research, grants in aid, and loans of personnel to the various states. The size of these industrial hygiene units in the states varied from one or two persons up to a dozen or more. At the zenith of its field investigational activity, from about 1942 to about 1955, the Industrial Hygiene Division of the Public Health Service had a staff of several hundred technical persons.

Obviously the smaller organizational units could not accomplish a great deal in the way of technical surveys, and much of their time was devoted to educational efforts and the investigation of complaints.

During the same period there was a slow but steady growth of industrial hygiene activity in private industry, primarily among larger corporations. These units ranged from one to three or four persons and, in a few instances, as many as 10 hygienists. Most of the units were staffed by experienced industrial hygienists who had "won their spurs" in state or federal employment. Many of these companies made notable advances in researching areas of potential hazard to their employees and initiating corrective action to protect workers against adverse effects to their health. All industrial hygiene units in private industry have environmental evaluation and control as their primary objective, but their activities vary greatly depending on the views of management and the size of the establishment. Several types of industrial hygiene consulting organizations were also created during this period. The Industrial Health Foundation (formerly the Industrial Hygiene Foundation) is an organization having a general, educational, and investigational interest, whereas the industrial hygiene units of insurance companies were principally engaged in investigational surveys and individual control programs for their insurance clients.

OSHAct has exerted great influence on the field of industrial hygiene and its centers of activity. At present, governmental activity in industrial hygiene is primarily centered in the organizational structure of OSHA, in the U.S. Department of Labor, and in NIOSH, in the U.S. Department of Health and Human Services. The continued existence of most state units is problematic for many reasons, including lack of financing, personnel, and legislative interest. The federal industrial hygiene activities are divided into two broad objectives: the inspection and compliance activities of OSHA and the research and investigational activities of NIOSH. Each of these organizations employs about the same number of industrial hygienists, with a total of about 150 technically trained personnel.

Unquestionably, the largest current employer of industrial hygienists is private industry, and its need for workers will continue. The number of industrial hygienists required for an individual plant or company depends on a number of factors: the number of employees, the number of different chemical agents and physical factors providing opportunity for exposure, the number of such physical or chemical agents that may require periodic evaluation during the year, and the physical dimensions of the plant.

Plants with 350 to 500 employees where there may be opportunity for exposure to three or four chemical agents or factors requiring semiannual evaluation may be able to utilize one professional industrial hygienist on a full-time basis. Larger plants may require a proportionate number of professional industrial hygiene personnel, including one experienced in the analysis of industrial hygiene air samples. Multiplant corporations may have an industrial hygiene staff of six or more professional persons providing services from a central location, or they may be located in selected company regions from which they can best serve their employers' needs.

Insurance carriers each employ at least one industrial hygienist and usually several, located in regional areas, to advise their clients on the hazards of various chemical agents and to provide minimum services. A client with 50 to 150 workers usually finds it necessary to employ a private consultant to obtain adequate evaluation of exposures in his establishment.

A number of competent, private consulting organizations devoted to industrial hygiene and air pollution activities have been in existence for 30 or more years.

These organizations are generally staffed with professional industrial hygienists of long experience who carry out industrial hygiene evaluations and may supervise the work of younger, less experienced technical people.

Consultants in this category provide complete in-house services and are prepared to undertake a job of any magnitude. Newer organizations, equally competent in industrial hygiene, have been started in more recent times and may not be as well known in their field. Since 1970, the number of industrial hygiene consultants has mushroomed. Many are retired professional industrial hygienists who operate privately and obtain analytical services from special consulting laboratories. Among this large group of capable consultants are a few organizations with little knowledge of industrial hygiene and no professional abilities or judgment. Employers seeking to utilize the services of consultants for industrial hygiene purposes should inquire into their qualifications just as they would ask about the qualifications of a product research organization or other types of management service.

3.2 Administrative Location in Organizational Structure

The administrative location of an industrial hygiene unit in the organizational structure of an establishment can exert a marked influence on its effectiveness and accomplishments. Private firms, entirely engaged in providing consulting services in industrial hygiene, air and water pollution control, and similar environmental matters, integrate such functions into a cohesive body, to utilize all available talents.

In governmental units, industrial hygiene has attained equal status with other preventive health and safety functions that are necessary in today's society. The activity and effectiveness of such units depends on the needs of the public and the support of the legislative entity. At this writing, governmental activity in industrial hygiene is rapidly becoming centralized in the federal sphere, with laboratory and field research located in NIOSH, whereas authority for the framing of regulatory standards and enforcement activities resides in OSHA.

The industrial hygiene functions in industry show the widest variation in their location within the organizational structure and thus are reflected in their functioning.

Some units may report directly to the head of manufacturing, and may or may not be combined with safety and fire protection. Frequently such a unit serves no function beyond monitoring employee exposures for in-company compliance purposes. The industrial hygiene function may be located in the central engineering department of a large establishment, and its obvious focus of attention is on measuring the effectiveness of control systems. In large corporations, especially those with multiplant activities, industrial hygiene generally is structured within the corporate headquarters.

However, branch units may be found in individual large plants or may be located to serve a number of smaller plants in regional areas of the corporation. Even the research departments of several large corporations have successfully accommodated the centralized industrial hygiene functions of their company. Among the corporate industrial hygiene units that have been operating for 15 years and longer, units tend to be located in the medical department of the establishment. Although there are honest differences of opinion regarding the most desirable administrative location of the industrial hygiene unit, the success of the large number of such units in medical departments appears to favor this location.

It provides flexibility in functions, helps bridge normal departmental barriers, and provides access for communicating with top management. Functionally, this location permits the unit to engage in comprehensive medical–industrial hygiene studies, environmental surveys, training of plant personnel in monitoring techniques, research in sampling and measurement of exposures, and other activities. Traditionally, the physician who is trained and experienced in occupational medicine considers industrial hygiene and occupational health and safety to be an important and equal partner in the integrated occupational health team.

4 QUALIFICATIONS AND TRAINING OF PERSONNEL

4.1 Director of Industrial Hygiene

The qualifications specified here for the administrator of an industrial hygiene unit represent opinions molded from the observations and experience of numerous persons who have served in this capacity in the field of industrial hygiene during the past 25 years. The director should have at least a master's degree from an accredited college or university and a basic education in science, preferably in physical chemistry or chemical engineering. The graduate training should include courses in physiology, toxicology, biochemistry, ventilation and air conditioning, and the practical application of statistical parameters to numbers obtained in scientific measurements.

Certification as a diplomate by the American Board of Industrial Hygiene should be a consideration for the administrator's position. A prerequisite to certification is five years of practical field experience followed by passing a comprehensive written examination.

Although these are the basic requirements, the administrator should have at least 10 additional years of broad experience in industry, including involvement with the type of manufacturing over which he will maintain environmental supervision.

In the process industries, the absence of such experience would be a serious obstacle to successful performance of the administrative function.

Directors should be able to understand and discuss all phases of industrial hygiene and toxicology and should be interested in public health and industrial relations. They should be able to set up systems for keeping records of results achieved and for planning future activities. They should be avid readers and should devise a systematic bibliography of references covering the field of their interests and activities. Directors should be able to locate the cause of cases of industrial ill health presented by a physician, to make comprehensive analyses of health conditions in industries, to draw adequate conclusions, and to prepare clear and informative reports for publication.

Tact, good judgment, and courage of personal convictions are also important assets. The director must do sufficient fieldwork to keep the viewpoint of the worker in the field, and it is necessary to maintain a wide personal acquaintance with the men and women active in promoting industrial health.

The administrator should support and attend public meetings where current problems in the field are discussed, and should develop the ability to address a group of peers as well as the public.

4.2 Associate Industrial Hygienists

The industrial hygienists reporting to the director should have received basic education in the scientific disciplines similar to that of the administrator. A degree in chemistry or chemical engineering is desirable, and a degree in mechanical engineering provides greater knowledge in the disciplines required for designing control measures. A graduate degree that includes training in illumination, ventilation and air conditioning, energy conservation, principles of air sampling, and the toxicology of the commonly used industrial chemicals is essential. As indicated earlier, the role of the associate industrial hygienist is rapidly changing as newer and more administrative responsibilities fall upon the industrial hygiene administrator.

The position of associate has become more technical and field-oriented and thus will require more involvement with the design of sampling schemes and quantitative evaluation of the workplace environment.

If the unit is so large that one of the industrial hygienists acts in the capacity of a supervisor, this person should be certified by the American Board of Industrial Hygiene or should possess sufficient experience to command respect of the junior members.

Depending on their individual experience, the associates must be able to recognize and evaluate a potentially harmful situation, to judge the necessity for carrying out a specific study, and, of course, conduct a survey.

The knowledge required for developing a control measure varies directly with experience and the type of basic education. However, one of the most important abilities required of industrial hygienists is that of preparing a clear report of the survey findings, bearing a definitive conclusion. Because so much of their time is spent in studying plant situations, industrial hygienists must have the ability to communicate their ideas and thoughts to production personnel, safety engineers, maintenance supervisors, and plant physicians.

4.3 Other Unit Personnel

To operate effectively an industrial hygiene unit may find it desirable to add such specialists to its staff as an industrial toxicologist, an industrial hygiene analytical chemist, a ventilation engineer, a statistician, a health physicist, and perhaps survey monitors; each one will have a personal set of qualifications, and each will make special contributions to the output of the unit. Occasionally, a person may start out as a specialist and by close association broaden the scope of his or her activities and interests, to be able to claim properly the title of industrial hygienist. With today's utilization of increasingly sophisticated computer hardware and software for both record-keeping and reporting purposes, an additional resource would be a technician skilled in the use and application of computers.

Many young persons are now being graduated from recognized colleges with bachelor's degrees in environmental health, air and water conservation, ecology, and other general fields relating to the biosphere in which humans lives.

On examining numerous employment applications of persons with such degrees, one finds that many of their studies have emphasized the social aspects of environmental conservation with relatively brief exposure to the basic sciences.

These young persons can be trained as technicians or scientific aides to carry out routine procedures such as exposure monitoring, sample collection, and other uncomplicated assignments.

It could be a serious error for an employer to place the responsibility of a new industrial hygiene program in the hands of an inexperienced person on the basis of a degree with an attractive sounding title. Helpful advice and assistance in seeking competent industrial hygienists can be obtained from the office of the manager of the AIHA (10).

REFERENCES

1. M. Corn, "The Progression of Industrial Hygiene" *Appl. Ind. Hyg.* **4**(6), 153–157 (June, 1989).
2. F.M. Renshaw, "Risk Assessment, Industrial Hygiene Management, XIII," *Appl. Ind. Hyg.*, F23–F26 (July 1989).
3. Public Law 91-596, 91st Congress, S.2193, December 29, 1970.
4. B. Ramazzini, *De Morbis Artificum Diatriba*, 1700 and 1713.
5. "Health of Workers in a Portland Cement Plant," Public Health Service Bulletin No. 176, U.S. Government Printing Office, Washington, DC, 1926.
6. "Exposure to Siliceous Dust in the Granite Industry," Public Health Service Bulletin No. 187, U.S. Government Printing Office, Washington, DC, 1929.
7. "Lead Poisoning in a Storage Battery Plant," Public Health Service Bulletin No. 205, U.S. Government Printing Office, Washington, DC, 1933.
8. "Mercurialism and Its Control in the Felt-Hat Industry," Public Health Bulletin No. 263, U.S. Government Printing Office, Washington, DC, 1941.
9. "Effect of Exposure to Known Concentrations of Carbon Monoxide," *J. Am. Med. Assoc.*, 118, 585–588 (1942).
10. American Industrial Hygiene Association, P.O. Box 8390, 345 White Pond Drive, Akron, OH 44320.

CHAPTER FIVE

Role of the Industrial Hygiene Consultant

Editorial Note: Concurrent with the development of the industrial hygiene profession has been the increase in the number of consultants specializing in industrial hygiene services. Although the functions of the *individual* consultant and that of the *organizational* consultant may differ because the latter's capacity may be broader in scope of available specialties, they are similar in their objectives: that of helping the client and that of achieving, in their modus operandi, the highest degree of ethics. Therefore the reader will find some duplication and overlapping in the two sections that follow. This is deliberate.

A THE NONORGANIZATIONAL INDUSTRIAL HYGIENE CONSULTANT

Ralph G. Smith, Ph.D., C.I.H.

1 CONSULTATION BY INDIVIDUALS

1.1 History: Growth of Industrial Hygiene Consultation

Every professional endeavor and, indeed, many activities that might not qualify as professional endeavors appear to have individuals and organizations who offer their specialized services to society as consultants. The profession of industrial hygiene is no exception. The reasons for the proliferation of consultants in this field are not difficult to identify. Industrial hygiene is a specialty requiring expertise not

Patty's Industrial Hygiene and Toxicology, Fourth Edition, Volume 1, Part A, Edited by George D. Clayton and Florence E. Clayton
ISBN 0-471-50197-2

immediately available within many organizations; in addition, specialized equipment for monitoring the industrial environment is also required.

There is no record of the number of industrial hygiene consultants or organizations that have existed at any time in the past or, for that matter, even today. In all probability, though, the number of consultants has increased in proportion to the growth of industrial hygiene as a profession, although it seems likely that the percentage of professional industrial hygienists engaged in consulting activities today is greater than in past years. There has always existed an enormous need for consulting industrial hygienists to provide services to the many hundreds of thousands of small industries in the United States that may require professional assistance, but do not have professional industrial hygienists on their staffs.

When the American Industrial Hygiene Association (AIHA) was founded in 1939, its list of founding members numbered about 160 individuals. It is probable that not more than 10 percent of these individuals offered industrial hygiene consultation services, so although no records exist to support any estimates, it is likely that in the years between World War I and World War II, only a handful of individuals were engaged in industrial hygiene consulting, and very few organizations existed for this purpose. Subsequent to World War II, steady, sustained growth of industrial hygiene as a profession began, until today the AIHA membership numbers in excess of 8000 persons, and the American Academy of Industrial Hygiene (AAIH) consists of almost 4000 certified individuals. It might be estimated that the number of industrial hygiene consultants and individuals employed by consulting organizations today is conceivably as high as 25 percent of those practicing the profession. During the post-World War II period, industrial hygiene consulting organizations began to form, and such organizations have grown greatly both in number and in size during the past 10 or 20 years. The many reasons for this rapid growth, as well as the kinds of industrial hygiene services presently available, are detailed in Part B of this chapter. The remainder of this discussion addresses matters related to consultation offered by individuals, or small groups of individuals, operating with little or no laboratory or field analysis equipment.

1.2 Backgrounds of Industrial Hygiene Consultants

There are many individuals offering services related to industrial hygiene matters who are not actually considered industrial hygiene consultants, although at times the distinction may be difficult to define. There are, for example, individuals whose principal specialty is measuring or controlling noise, or ionizing radiation, or perhaps they consider themselves to be industrial ventilation engineers. This discussion is limited, however, to individuals offering full-service industrial hygiene assistance, which is to say that they are prepared to consider or study any workplace, and make recommendations related to potential or actual health hazards arising from work operations. They may, of course, call upon other specialists, such as noted, to be of further assistance, but in general the industrial hygiene consultant is capable of taking appropriate actions related to any risk of adverse health effects.

All industrial hygiene consultants, as just defined, should have a common back-

Table 5.1. Professional Backgrounds of Practicing Industrial Hygiene Consultants

1. Industrial hygienists presently employed by industry, government, insurance companies, etc.
2. Industrial hygienists formerly employed by industry, government, insurance companies, consulting organizations, etc.
3. University faculty and research associates, primarily from undergraduate and graduate programs in industrial hygiene.
4. Retired industrial hygienists, all sectors.

ground, namely, proficiency in industrial hygiene, and they should be professionals who are primarily dedicated to the practice of industrial hygiene. It is reasonable to expect such individuals to be members of the AIHA, possibly the American Conference of Governmental Industrial Hygienists, and also to be Certified Industrial Hygienists (CIHs). There are and will be exceptions to this latter qualification, because some of the more senior industrial hygienists may not have elected to become CIHs, even though well qualified to be considered competent professionals. Today, with the rapid growth in the number of CIHs, and the growing requirement by many employers that job candidates and consultant be certified, it is likely that few, if any, future industrial hygiene consultants will lack this qualification.

Although it is true that all consultants are or should be industrial hygienists, their individual backgrounds may differ greatly, and it is useful to list some of the many backgrounds of individuals actively consulting today. Table 5.1 lists the most probable prior employment of individuals who choose to become consultants. Prior to becoming industrial hygienists, of course, their backgrounds differed to an even greater extent, with the more common educational backgrounds of industrial hygienists listed in Table 5.2. To some extent, the capabilities and preferences of individual consultants vary with the particular educational background that preceded their employment in industrial hygiene, even though they may be qualified to offer broad-spectrum industrial hygiene consultation. Thus the analytical chemist who became an industrial hygienist will in all likelihood tend to emphasize sampling and analytical activities, whereas a physicist may feel more comfortable dealing with physical phenomena such as radiation and noise, and seek to become more knowledgeable in these areas.

It is apparent that industrial hygienists who have resigned their positions with industry, government, labor, and so on are perhaps the most logical prospects for becoming consultants, but it is not so apparent that individuals presently employed by the private sector, government, and so forth may also be consultants. In fact, a limited number of corporations, government agencies, and other organizations

Table 5.2. Educational Backgrounds of Industrial Hygiene Consultants

1. Undergraduate and graduate degrees in one of the Physical Sciences, e.g. Chemistry, Physics, Geology, etc.
2. Undergraduate degree in one of the Physical Sciences, graduate degree in Industrial Hygiene.
3. Undergraduate and graduate degrees in one of the Biological Sciences, e.g. Biology, Biochemistry, etc.
4. Undergraduate degrees in one of the Biological Sciences, graduate degree in Industrial Hygiene.
5. Undergraduate and graduate degrees in Engineering (Chemical, Industrial, other).
6. Undergraduate degree in Engineering, graduate degree in Industrial Hygiene.
7. Undergraduate degree in a variety of specialties not included above, graduate degree in Industrial Hygiene.

do permit their employees to engage in consulting, provided generally that they do it on their own time and meet certain other requirements. Clearly they must be careful to avoid conflicts of interest with their employers, and the amount of time they can devote to consulting will be limited. It is important that consultants who do have corporate or other employers inform their clients of this relationship so that there can be no misunderstandings and possible embarrassments owing to potential or actual conflicts of interest.

Industrial hygienists from every sector tend to consider becoming consultants once they retire, on either a full- or part-time basis. Such individuals are usually very well qualified, and frequently have assurance from previous employers and closely related organizations that they will become clients of the new consultant. There are pitfalls to be considered, however, and not infrequently they are unanticipated by prospective retirees, who may be disappointed at their lack of success as consultants. More often than not, individual consultants tend to rely on word-of-mouth advertising among their professional associates and clients, and in this regard, some individuals who retire tend to be somewhat disadvantaged. Unlike retired university staff, for example, who may have been consulting on a part-time basis for many years, the corporate industrial hygienist who retires may suddenly learn that although well known within his ordinary circle of professional acquaintances, he is not well known to the larger community that may require consulting assistance. Direct advertising is of limited usefulness, because the universe of persons or organizations who may need assistance is very large, and essentially unknown to most would-be consultants.

Most, if not all, industrial hygienists on university staffs do some limited amount of consulting, and it is logical and even essential that they do so. In a dynamic field such as industrial hygiene, a university professor who does no consulting runs the risk of being poorly prepared to familiarize the students with important current problems and the challenges of the ever-changing workplace. Consulting activities, particularly when they involve on-site visits and workplace studies, are extremely valuable to university professors, and spawn many research projects both large and small for the professor and the students. The benefits of consulting by university staff are twofold, of course, including in addition to the benefits just noted, a reverse benefit to the client in retaining an individual who may possess unique skills in the area of expertise required. It may be argued that the more practical reason for encouraging university staff to consult is the suppplementation of income that is often required so that university persons will not succumb to the inevitable economic pressure to accept more lucrative positions elsewhere. Given the inevitability of inadequate university budgets, this factor is an important one that helps ensure the maintenance of competent staff in our universities.

1.3 Ideal Qualifications of an Industrial Hygiene Consultant

In view of the varied backgrounds of industrial hygiene consultants, it is reasonable to ask what would be considered the ideal or optimal qualifications for such an individual? There is probably no single answer to which all professionals would agree, but certainly the single qualification that would be of the greatest importance is proficiency in the field of industrial hygiene. Today this goal is best realized by taking formal graduate training in a good university program. As yet most university industrial hygiene programs are not accredited, although an effort in this direction has been commenced by the AAIH. Nevertheless it is not difficult to obtain information concerning the best of the universities currently active in this field, as well as those that serve the particular needs of students with special interests.

In years past, the probable answer to the question concerning the ideal educational background of an industrial hygienist would have been an engineering degree, preferably chemical engineering, or perhaps a degree in chemistry. These backgrounds presuppose primary concern with the effects of substances in the environment rather than physical phenomena, whereas today it would be difficult to argue that the general subject areas of noise, radiation, or ergonomics are not equally deserving of special educational backgrounds in the appropriate engineering or physical sciences. Also, given the close relationship between the effects of substances and various forms of energy in the workplace, and the importance of toxicologic studies, a case can certainly be made for the desirability of a strong background in the biological sciences, and, in fact, a substantial percentage of incoming students to graduate industrial hygiene programs have been graduates of biological science programs.

In the final analysis it can be argued that both undergraduate and graduate training are secondary to the field experience of the individual who contemplates industrial hygiene consulting as a career. There is probably little disagreement that inexperienced persons who have just completed graduate studies should not begin their professional careers as consultants. The reasons are self-evident, of course, but perhaps in industrial hygiene more than many professions, a period of learning on the job and solving real-life problems is of the greatest importance in developing the degree of proficiency required of a consultant. No minimum time of employment prior to becoming a consultant can be defended, but certainly someone entering the field at present should be a CIH prior to becoming a consultant, and for this and other reasons it is not recommended that anyone with less than five years experience attempt to become established as a consultant.

Brief mention may also be made of personal qualifications, and here, of course, it is difficult to defend any particular set of attributes. As just noted, proficiency and technical competence are of the utmost importance, but second only to these attributes are those personal qualities that enable an individual to interact with other persons in a natural and easy manner. Consultants, by definition, are constantly meeting new persons, and must always attempt to sell both their own personal capabilities and the importance of carrying out their recommendations. Individuals who find it difficult to interact with strangers on short notice will probably not fare well as consultants. Consultants are also required to live a somewhat "unstructured" life, which is to say that they may travel considerably and on short notice, and must be prepared to offer their services when needed, and not necessarily at their own convenience. Those individuals who are most comfortable with a five-day week, nine-to-five schedule are probably not suited to the irregular hours and working habits frequently required of consultants.

1.4 Relationships Between Consultants and Client's Professional Staff

Consultants are frequently identified and retained by industrial hygienists, safely engineers, or other health professionals if in fact the clients have such persons on their staffs. Not infrequently, however, a consultant may be retained by someone else in the client organization who is either unaware of the industrial hygiene staff, or has not felt it necessary to consult with it. Regardless of by whom retained, it is important that the industrial hygiene consultant make every effort to meet with the other staff members concerned with employee health and safety. These persons are obviously familiar with the principal problems within their own organization, and in all probability have ongoing programs that may be attempting to address problems that have resulted in calling in a consultant in the first place. Generally their insights will be invaluable and, in addition, they may be expected to have records that can be of value to the consultant. If the organizational industrial hygienist has not been involved in the particular problem that necessitated the services of a consultant, then it is incumbent upon the consultant to attempt to see that that person is properly informed and, if possible, becomes part of the effort to address the problem. Transcending these considerations, of course, is the propriety of exercising ordinary professional courtesy toward one's professional associates.

Not infrequently a given organization may request the services of a consultant, while unknown to the consultant that organization may be part of a larger corporation or entity that does have professional industrial hygiene staff, who may be located at a central headquarters elsewhere. In such instances, the consultant should make a special effort at least to discuss the problems with the staff industrial hygienist, if permitted to do so by the organization retaining him or her.

Ideally, an industrial hygiene consultant would report to a counterpart in the client's organization, but frequently this will not be possible, and instead a consultant may report to some other individual, in all probability the person who retained the consultant initially. Such individuals may be legal counsel, department heads, or other administrative persons with special interests in the matter that prompted the need for consultation assistance. The consultant may be unable to insist that the report be submitted to the corporate industrial hygienist, but certainly every effort should be made to ensure that that person is fully informed and is given a copy of any reports that may be generated.

1.5 Advantages and Limitations of Being a Nonorganizational Industrial Hygiene Consultant

An individual may choose to become an industrial hygiene consultant, and may eventually be sufficiently successful that pressure builds to form an organization to handle the increasing work load. This expansion may take several forms, ranging from a loose federation with other professionals, available on a part-time basis, to the formation of a consulting organization requiring larger quarters, secretarial staff, more equipment, and perhaps a laboratory. Each individual consultant must weigh the advantages and shortcomings of expanding and make whatever decision seems best suited to his or her needs. There are readily definable advantages and shortcomings to any course of action selected. For example, it is obviously attractive to form an organization and begin an expansion that may lead to becoming a full-service organization. The opportunity for substantially greater income is probably the greatest motivation, but the opportunity to offer services not otherwise possible may also be an important factor. Additionally, acquiring field sampling equipment and other devices certainly is desirable, and the convenience of a laboratory serving not only the in-house consulting activity but also offering such services to others is readily apparent. Offsetting these advantages are all of the problems related to the formation of any organization, such as finding and keeping qualified personnel, greatly increased overhead costs, concern with attracting sufficient professional activity to ensure the payment of staff salaries, the large capital investment required to acquire modern equipment and to form a modern laboratory, and other matters.

If an individual consultant chooses, on the other hand, to remain a one-person organization or at most to form a loose relationship with others, to be available on a part-time basis, most of the problems associated with large organizations will be avoided, but the consultant must continue to deal with the pressures of trying to accommodate clients' needs within limited time, and will in all likelihood find it difficult or impossible to acquire as much equipment as might be needed. It will also be necessary to rely upon one or more existing consulting laboratories if analysis of collected samples is required.

Table 5.3. Principal Industrial Hygiene Consulting Activities

- Problem Solving - Industry, Labor, Government
- Performing Industrial Hygiene Surveys of Plants, Industries, Specific Operations, etc.
- Industrial Hygiene or Occupational Health Program Auditing
- Industrial Hygiene Training, Workers, Management
- Assisting Industry or Other Organizations Required to Comply with Regulatory Orders
- Activities Related to Litigation, Including Serving as Expert Witness
- Designing and Conducting, or Assisting in the Conduct of Scientific - Studies Related to Health Problems in Industry
- Appearing at Hearings, or Meeting with Regulatory Officials re: Specific Proposed Legislation, Problems Unique to Client, re: New Standards, etc.
- Serving as Advisor to Trade Organizations, Labor Unions, Professional Organizations and Similar Groups.

There is obviously no single best choice, and it remains for each individual consultant to practice the profession in the manner that seems best suited to his or her needs. There are many examples today of individual consultants continuing to practice more or less on their own, and there are many examples of individuals who have started out as one-person organizations but have built large and successful consulting organizations.

2 PRINCIPAL INDUSTRIAL HYGIENE CONSULTING ACTIVITIES

All industrial hygiene consulting activities will in all probability bear some clear relationship to industrial hygiene-related problems, but there are a number of different kinds of activities that deserve classification and discussion. Table 5.3 lists the principal kinds of industrial hygiene consulting activities that may be anticipated, although occasional consulting assignments will not fit into any of the suggested categories.

2.1 Problem Solving—Industry, Labor, Government

Consulting for the purpose of problem solving, or problem-stimulated consultation, is probably the most common consulting activity performed. Typical problems may arise because of management concerns, or concerns expressed by individual workers, often identified by labor organizations. Citations by the Occupational Safety and Health Administration (OSHA) or local regulatory agencies are also a common basis for identifying problems requiring solutions, but no matter what the source of the original concern may have been, it has subsequently been determined that outside assistance by a qualified consultant is required. There are several reasons why such a decision may have been made, and it is well for the consultant to be aware of the real reasons for seeking outside assistance. The more common reasons for requesting consultation assistance are genuine lack of adequate expertise within the requesting organization; "political" considerations, wherein the opinions of a disinterested expert are required; insufficient time and staff of the requesting organization to address the problem adequately, even though they possess the expertise; and perhaps the need for someone to act as an arbitrator in an attempt to reconcile opposing points of view. The consultant will normally approach the problem to be solved in the same manner, regardless of the reasons that stimulated the request for services, but he or she may stand a better chance of satisfying the client if fully informed as to the circumstances prompting the investigation. It goes without saying that no matter by whom retained, or why, the consultant must make every effort to be completely neutral with respect to the political aspects of the problems involved, and totally objective, regardless of the opinions of the particular individual who retained him or her initially.

2.2 Performing Industrial Hygiene Surveys of Plants, Industries, and Specific Operations

For a variety of reasons, industrial hygiene consultants may be requested to perform "wall-to-wall" industrial hygiene surveys of entire plants, or specific buildings, units, or operations. Such surveys may be related to problems that initially required some problem-solving activity, but after which it became apparent that a complete survey was needed. In some instances, complete surveys may be completed in one or two days, whereas others may require weeks of effort involving several persons. Regardless of the size or complexity of the area to be surveyed, each survey should be based on the classic industrial hygiene principles of recognition, evaluation, and control.

The recognition phase of the survey is perhaps the most important, and should be accomplished by extensive discussions with management, labor, and other concerned parties; detailed studies of all of the processes, raw materials, by-products, and finished products involved; and careful examination of all existing records, including plant layout prints, ventilation diagrams, and of course, all existing industrial hygiene records. One or more thorough walk-through inspections should also be made, accompanied by knowledgeable persons who can explain everything that is of possible interest. Consultations with the plant medical department, including interviews with both the physician and nurse, are also of great value and may provide insights into the existence of problem areas or operations.

The evaluation phase of the survey may consume the greatest amount of time, and may require the acquisition of additional personnel and equipment. Basic industrial hygiene principles should be followed, of course, with particular attention directed both to obvious problem areas and to jobs or operations that are subject to regulation by specific OSHA standards or regulations. Whenever applicable, efforts should be made to supplement an air sampling program with biospecimen analyses, whenever these analyses are recognized as being useful in evaluating the extent of exposures.

Recommendations for controlling proven overexposures, or potential overexposures, must be made with the realization that a consultant is not a compliance officer for a regulatory agency, and must instead present options in a clear and lucid manner, emphasizing those actions that are most urgently required but also presenting those that may be optional. Individual clients may possess the required expertise and may be capable of providing their own controls, but in many instances it will be the responsibility of the consultant to make specific recommendations, or to recommend expert organizations capable of designing and installing good industrial ventilation, for example.

2.3 Industrial Hygiene or Occupational Health Program Auditing

Most good corporate industrial hygiene programs periodically audit their programs to ensure the level of performance that has been deemed necessary. Many corporations have internal auditing procedures and units for just this purpose, and are quite capable of doing an excellent job. Other organizations may not elect to have such self-auditing capability, and may choose instead to retain an industrial hygiene consultant for this purpose. Such audits are by their nature very demanding, for the consultant is required to evaluate the performance of peers, who may be well informed and quite capable of evaluating their own performance level. The requested audits may be quite limited in scope, or may be so broad as to include not only the performance of the industrial hygiene group but also its interrelationships with other members of the occupational health team. In the latter event, it is necessary to meet with and interview members of the other groups, in order to make an appraisal of the relationships that are essential to the successful operation of the entire program. Typically, audits require extensive interviews with all staff personnel, meeting with key individuals in private and being sensitive to the need for respecting confidences. Additionally, interviews should be held with the administrative head of the department of which industrial hygiene is a part and with other administrative personnel, up to and including the corporate president if possible. Ideally, interviews should also be held with selected members of the work force, including union representatives, and perhaps a random sampling of workers who are most concerned with the quality of industrial hygiene programs. Not infrequently, however, corporate policy will not permit the consultant to interview workers, unlike OSHA compliance officers, who may do so.

In addition to extensive discussions and interviews with staff persons, all relevant records should be made available to the consultant, as well as all documents such as the current industrial hygiene manual, safety publications, and standard procedures. The consultant should note whether these are current and complete and whether they are consistent with good practices in industrial hygiene and, if applicable, in industrial hygiene laboratories. Samples of reports generated by the staff should be reviewed, as well as records indicating the extent of follow-up activities.

It is important that the consultant draw up a comprehensive outline of the audit to be performed, or else acquire such an outline from organizations that have developed them over a number of years. As noted earlier, many corporations perform their own audits and generally have formalized and very complete outlines and operating procedures. Similarly, other consultants who have performed such audits in the past may likewise have prepared reliable operating procedures, and the consultant who is performing an audit for the first time would do well to seek out such preexisting outlines, if possible.

Prior to performing the audit there should be agreement concerning the form in which the results will be summarized. Some formalized audit systems have devised numerical rating systems, whereas others rely simply on word descriptions and listings of deficiencies, strengths, and so on. Either approach may be satisfactory, but there should be agreement prior to conducting the audit concerning how the results will be expressed.

2.4 Industrial Hygiene Training, Workers, Management

Specialized training of groups of workers or management personnel is frequently required, sometimes in relation to special projects, or else as a part of an ongoing routine training program. Companies or other organizations may elect to request that such training be given by an industrial hygiene consultant, or that the consultant organize a training session or sessions that assemble a staff of qualified plant personnel and others with expertise in selected areas. Consultants who are associated with a university are frequently asked to organize such activities and are generally well qualified and prepared to do so. Subject areas to be addressed may be as general as a survey of the principles of industrial hygiene, or may be very specific, treating relatively advanced subjects. Training activities are sometimes requested in order to satisfy contractual agreements between management and labor, and in such instances, the required subject matter is ordinarily a matter of contractual agreement.

In preparing a training activity of any kind, it is extremely important that the consultant design the activity for the particular students expected to attend the course. There is little value in presenting complex subjects using language that is unfamiliar to the students, and care must be taken to avoid, whenever possible, technical jargon that may impress but not educate the students. It is equally wasteful of time to underestimate the level of knowledge of the class and explain fundamentals that may be well known to them.

Whenever possible, training sessions should use visual aids to the maximum extent, and take advantage of modern computer programs and interactive techniques, as well as hands-on laboratory sessions whenever these are appropriate.

For example, if a research program requires that air samples be taken in various locations by persons not familiar with air sampling techniques, it is useful to have a training session at some central location, in order to explain carefully the purpose of the entire effort, its goals, and the techniques to be used. The actual sampling equipment to be used should be available, and there should be enough units to permit each student to use the unit in the manner required for the study. In addition, printed material should be prepared and distributed. This material should describe the procedures in as much detail as required, discussing the importance of calibration procedures, methods of handling the samples, and instructions concerning recording the required sampling data.

Occasionally, a consultant may decide that for various reasons it may not be possible to provide the depth of coverage required by the client and instead may elect to make the client aware of training courses offered by the AIHA, National Institute for Occupational Safety and Health (NIOSH), or some other well established training organization. In the long run, a client's needs may better be met by such a decision, and the consultant will be spared the trauma of conducting a training activity that did not work.

2.5 Activities Related to Litigation, Including Serving as Expert Witness

All industrial hygiene consultants may anticipate being asked to become involved in matters related to litigation, including offering their expertise to attorneys, and ultimately appearing as an expert witness. There are a number of individual consultants for whom litigation-related activities are their most important single source of income, and there are some who do nothing else. Typically, such litigation involves matters related to workmen's compensation cases, toxic torts, or even environmentally related problems. Much has been written about this kind of litigation, and it is beyond the scope of this chapter to address in detail the complexities of actual lawsuits and the particular roles of the industrial hygiene consultant. A few general observations may be useful, however.

There is no single activity of an industrial hygiene consultant that requires more adherence to a strict code of ethics than does litigation (see Chapter 7). All industrial hygiene activities, of course, must be conducted in accordance with ethical principles, but it is more likely that ethical dilemmas may arise in matters related to litigation than in most other consulting activities. By definition, litigation always involves two legal teams with opposing views, and each seeks to do the best possible job for its client. It is important that the industrial hygienist avoid taking an adversary position, if at all possible. The role of the industrial hygiene consultant is to provide information based upon expertise, and to offer opinions that, to the best of his or her knowledge, arise from an objective consideration of all of the variables involved. An industrial hygiene consultant should be able to state that no matter whether retained by counsel for the plaintiff or counsel for the defense, the opinions, given the same body of facts, will be the same. It may well be that in some instances the opinions will not be those that the retaining attorney finds useful, in which case the relationship in all probability will terminate.

It is extremely important that the industrial hygiene consultant recognize the boundaries of his or her expertise, and not be persuaded to go beyond them. Obviously, an industrial hygienist is knowledgeable concerning toxicologic, medical, epidemiologic, and environmental matters, and it is quite proper that such knowledge be factored into the formation of his or her opinions. The industrial hygienist must never, however, assume the role of a physician, for example, and offer opinions that he or she is unqualified to make. Conversely, other experts should not assume the role of an industrial hygienist if they are not qualified to do so, but unfortunately in practice it is difficult to determine whether an individual is qualified as an industrial hygienist, and not infrequently poorly qualified individuals are in fact rendering opinions requiring industrial hygiene expertise.

A consultant may be approached by counsel for the plaintiff or the defense, and there is no reason why any consultant should refuse to offer assistance to one or the other. Ideally, in fact, it could be argued that all consultants should be equally involved with litigation for the plaintiff, as well as for the defense, but in practice such is frequently not the case. Certain individuals apparently choose, or are chosen, repeatedly to assist counsel for the plaintiff, while others routinely offer their services to counsel for the defense. The single, most important principle that should guide the activities of any consultant in this regard is the avoidance of any semblance of conflict of interest. Obviously, the consultant cannot offer advice to both plaintiff and defense in the same case, but perhaps not so obvious are the potential conflicts of interest that may arise with other cases, which may have some connection with either the organizations involved in previous litigation or the central issues. Common sense should serve to prevent most conflicts of interest from developing, but the consultant is well advised to give considerable thought to each request for assistance in relation to previous activities for the same clients or other clients.

The code of ethics (Fig. 5.1) developed and used by the AAIH and adopted subsequently by the AIHA is totally adequate both for its intended purpose of defining ethical practices for the professional practice of industrial hygiene and more particularly for the practice of industrial hygiene, in relation to litigation.

Note particularly the statement in the code, "Avoid circumstances where compromise of professional judgment or conflict may arise." Also of particular concern in litigation is the admonition to "be objective in the application of recognized scientific methods and in the interpretation of findings," and finally the warning to "state professional opinions founded on adequate knowledge and clearly identified as such."

2.6 Scientific Studies Related to Health Problems in Industry

On occasion, a consultant may be fortunate enough to be asked to conduct or to participate in a study designed to provide required information regarding a health problem in industry. These studies may be of several kinds, perhaps the most common being limited to specific work operations or perceived problems in a given department or building. Such studies will generally require that the industrial hygiene consultant perform all the industrial hygiene activities normally associated

THE AMERICAN ACADEMY OF INDUSTRIAL HYGIENE CODE OF ETHICS

FOR THE PROFESSIONAL PRACTICE OF INDUSTRIAL HYGIENE

PURPOSE

This code provides standards of ethical conduct to be followed by Industrial Hygienists as they strive for the goals of protecting employees' health, improving the work environment, and advancing the quality of the profession. Industrial Hygienists have the responsibility to practice their profession in an objective manner following recognized principles of industrial hygiene, realizing that the lives, health, and welfare of individuals may be dependent upon their professional judgment.

PROFESSIONAL RESPONSIBILITY

1. Maintain the highest level of integrity and professional competence.
2. Be objective in the application of recognized scientific methods and in the interpretation of findings.
3. Promote industrial hygiene as a professional discipline.
4. Disseminate scientific knowledge for the benefit of employees, society, and the profession.
5. Protect confidential information.
6. Avoid circumstances where compromise of professional judgment or conflict of interest may arise.

RESPONSIBILITY TO EMPLOYEES

1. Recognize that the primary responsibility of the Industrial Hygienist is to protect the health of employees.
2. Maintain an objective attitude toward the recognition, evaluation, and control of health hazards regardless of external influences, realizing that the health and welfare of workers and others may depend upon the Industrial Hygienist's professional judgment.
3. Counsel employees regarding health hazards and the necessary precautions to avoid adverse health effects.

RESPONSIBILITY TO EMPLOYERS AND CLIENTS

1. Act responsibly in the application of industrial hygiene principles toward the attainment of healthful working environments.
2. Respect confidences, advise honestly, and report findings and recommendations accurately.
3. Manage and administer professional services to ensure maintenance of accurate records to provide documentation and accountability in support of findings and conclusions.
4. Hold responsibilities to the employer or client subservient to the ultimate responsibility to protect the health of employees.

RESPONSIBILITY TO THE PUBLIC

1. Report factually on industrial hygiene matters of public concern.
2. State professional opinions founded on adequate knowledge and clearly identified as such.

Figure 5.1.

with plant surveys, problem solving, and so on, but with a different focus. It is essential that such studies, or any research endeavors for that matter, be carefully designed so that they are capable of answering the questions that initiated them. The investigators should also anticipate preparing the comprehensive report that will ultimately be required, as well as a publication, preferably in a refereed journal in which findings of such studies are ordinarily published.

The most comprehensive studies likely to be carried out in the workplace are epidemiologic studies of large working populations believed to be at risk for one or more environmental factors. Such studies generally require the cooperative efforts of a team consisting of an epidemiologist, a physician, who may also be the epidemiologist, an industrial hygienist or industrial hygiene consultant, and other persons technically proficient in areas of importance to the study. All such studies are generally expensive, and require a substantial period of time for completion. The industrial hygiene consultant will generally not be the principal investigator of a study, but may share responsibility with the epidemiologist who will be responsible for the overall design of the study.

The principal concern of the industrial hygiene consultant will ordinarily be the collection of required quantities of all environmental data. Under ideal conditions the air sampling study, if one is required, will be carried out under the direction of the consultant utilizing an appropriate number of industrial hygienists or industrial hygiene technicians, probably from a cooperating industrial hygiene consulting organization. Frequently, however, the realities are that budgetary limitations and other concerns require that sampling be carried out by plant personnel, thus entailing training activities, quality control procedures, and other necessary steps to ensure the objectivity and validity of the data.

One of the most common motivations for conducting such large-scale studies is the pressure exerted by pending or existing governmental standards or regulations. Thus a study may be intended to address specifically the question of an appropriate standard for a substance, and the investigators must take care to ensure that the findings will have the required degree of credibility with the regulatory agencies concerned. In all such instances, it is strongly recommended that the study be planned very carefully, and that prior to commencing the study, meetings be held with the regulatory agency personnel or perhaps with the Threshold Limit Values Committee if appropriate, with their inputs welcomed and used to modify the studies where required. These same principles, of course, apply to any studies conducted for the purpose of providing information to regulatory agencies relative to proposed or actual standards. As noted earlier, all large-scale epidemiologic studies can be very time consuming and demanding, but can also be very rewarding professionally to the industrial hygiene consultant who finds himself engaged in a research activity that may result in significant information affecting the health of the working populations under study.

2.7 Activities Related to Proposed Legislation and Open Hearings

Given the fact of ever-increasing regulatory pressures and activities by federal agencies such as OSHA and the Environmental Protection Agency, as well as numerous state and local government agencies, it is almost a certainty that an industrial hygiene consultant will be asked to assist some organization having a direct interest in such matters. Most commonly, given a new proposed standard, for example, special-interest groups such as industrial establishments, industrial associations, labor groups or public interest groups may have sharply different opinions regarding the standards being considered, and will request the opportunity to make statements at the public hearings that are usually held in connection with such standard-setting activities. Any or all of these special-interest groups may send persons to testify or give opinions at hearings, and on occasion the most suitable person is an independent industrial hygiene consultant. The consultant selected will ordinarily have had some prior familiarity with the problem area, and may have consulted for the company or other organization requesting the consultant's services. Regardless of the extent of prior involvement or lack of it, the consultant must become thoroughly familiar with the issues involved and make every effort to acquire all available information. Then opinions must be formulated with which he or she is completely comfortable, and which he or she strongly believes. Should these opinions be consistent with those of the client, the consultant will logically then express them in open hearings and ordinarily will prepare a written document for submission. As in activities related to litigation, the consultant should be guided by the code of ethics that govern all of his or her activities, for there may well be suspicions of bias favoring the group represented. Such charges are frequently leveled at consultants who have been retained by industry, and who thereafter make appearances at hearings on behalf of industry. Even though the consultant may have been selected because of not being an employee of the concerned industry, the simple fact of the consulting relationship with the organization will not infrequently be used as a basis for charges of industry-slanted bias. Similar charges, of course, may be made against consultants representing labor, public interest groups, or other organizations, but they are not as probable as those made against industry-retained consultants. It is unfortunate when such charges are made without cause, but in anticipation of this possibility, the conduct of the consultant should always be consistent with the highest principles required by the AAIH code of ethics. It is obviously very important that the consultant not be tempted to make statements favorable to the client's position that cannot be supported by objective consideration of all available information.

2.8 Advisory Activities

Industrial hygiene consultants are frequently asked to become members of advisory groups established by industry, labor, professional organizations, and others. Providing such services should be considered a particularly desirable activity, for it allows the consultant to express opinions without the pressures associated with litigation, regulatory activities, and so on, while being presented with the opportunity to make significant contributions to the organization that formed the advisory group. Typical advisory groups include those established by specific industrial or trade organizations sharing common problems, governmental agencies, and somewhat less frequently, combined labor union and management groups. For example, the union–management contracts of the tire and rubber industry have for some time provided for jointly funded research programs requiring advisory groups, and more recently the United Auto Workers union and the plants of the General Motors Corporation, Chrysler Corporation, and Ford Motor Company that employ UAW workers have signed contracts involving the expenditure of substantial sums on employee health and safety related matters. The advisory committees formed are charged with the very important task of assisting the joint union–management committees in training efforts, research directed at potential problems, and related matters.

In contrast to the usual "problem-solving" activities of a consultant, service on such advisory committees provides unique opportunities to help identify present and future problems and design research programs or other measures that can significantly affect the health and safety of many thousands of employees. As might be anticipated, it is frequently the practice in forming such committees to tend to select consultants who are affiliated with university programs.

3 ETHICAL CONSIDERATIONS

The importance of ethical considerations in relation to ligitation related activities has been stressed. In fact, however, strict adherence to a code of ethics is an essential activity for anyone practicing industrial hygiene, whether the person be a consultant or otherwise employed. Because of the great variety of situations in which an industrial hygiene consultant may be involved, it is perhaps even more important, or at least potentially difficult, to maintain the highest degree of ethical conduct. The AAIH code of ethics discussed earlier is a well conceived document that has withstood the test of time. Every industrial hygiene consultant should review it carefully, and make constant effort to comply with its various provisions. All the clients of a consultant should be made aware that this code of ethics exists and that the consultant does indeed conduct himself or herself accordingly. Perhaps the central theme of the code is that all industrial hygienists must constantly be aware that their primary mission is to protect the health of people who work. It does not matter whether the employer or the client, in the case of a consultant, is a corporate entity, a labor organization, or an insurance company; the industrial hygienist is committed to recommending whatever actions are best for the working population involved. It will be recognized, of course, that industrial hygiene consultants cannot implement all of the recommendations they may make, but they should never fail to make such recommendations because of the costs involved, or because of other practical considerations if they believe that in doing so adverse health consequences will follow. There may be many reasons why management will not implement recommendations made by a consultant, and may select alternative strategies that are better suited to their overall needs, at least in their own opinion. Nevertheless, the consultant should make such recommendations as are believed necessary to whatever extent possible and take all such steps as may be feasible to see that they are implemented. The code of ethics summarizes this major responsibility very succinctly. It states, "Hold responsibilities to the employer or client subservient to the ultimate responsibility to protect the health of employees."

3.1 Confidentiality Concerns

Another provision of the code of ethics is perhaps uniquely important to the industrial hygiene consultant. The code calls for the industrial hygienist to "protect confidential information." Virtually all industrial hygienists may have access to confidential information, but in the specific instance of consultants, it is probably true that not only will the consultant have access to considerable confidential information, but the opportunity also exists to violate this simple rule more frequently. A consultant must always exercise the greatest caution in transmitting to one client information obtained from another. Certainly there are instances when some kinds of information may be freely exchanged, but whenever trade secrets or proprietary matters are involved, exchanging such information with others is a clear breach of trust. It is generally wise for the consultant to assume the latter circumstances and avoid discussing information concerning one client with another.

Practical difficulties may arise within an organization when information of a confidential nature is provided by an individual who does not wish it to be known to others in the organization, or when employees confide in a consultant with the understanding that such information will be kept confidential. In such instances, common sense and the exercise of good judgment are obviously indicated, but a consultant is urged to attempt to avoid such situations whenever possible. It may even be advisable for the consultant to advise everyone with whom he deals that all information provided to him or her will not be kept confidential, at least insofar as the client is concerned.

4 BUSINESS ASPECTS OF CONSULTING

Consulting is ordinarily conducted on a fee-for-service basis, and the client has no obligation to pay the consultant or provide fringe benefits except when requested services are being rendered. On occasion, however, it may suit the needs of both the consultant and the client to establish a continuing relationship to provide services throughout a stated period of time, frequently one year. In such instances, contracts may be drawn up by the client and will probably be found acceptable by the consultant. Unfortunately for consultants, there does not exist a document entitled "Code of Business Practices" comparable to the code of ethics to use as a reliable guideline. There does not exist a generally accepted fee schedule for services rendered by industrial hygiene consultants, for example, nor are there universally accepted practices concerning such matters as charging for travel time, or differential fee rates for certain activities such as serving as an expert witness. Nonetheless, there does exist at any given time a general awareness of the fees being charged by consultants, and the individual consultant is advised to be aware of such information. All of the larger consulting organizations have fee schedules, of course, and individuals acting as consultants are well advised to obtain a sufficient number of such fee schedules to enable them to arrive at an appropriate fee for their services. Certain generalities concerning the fee schedules are possible: a young, relatively inexperienced, and noncertified industrial hygienist is logically going to command the lowest fee, whereas individuals with increasing amounts of experience, who are certified, will generally earn more. Other things being equal, an individual with a Ph.D. will in all probability command a higher fee than one without such a degree. In general, however, an individual considering consulting as a profession will in all probability find it necessary to resort to the time-honored practice of asking friends, associates, and those already familiar with consulting practices for guidance in arriving at a reasonable basis for charging for services rendered. It is perhaps self-evident also that the individual contemplating consultation as a career should have adequate resources to be able to weather periods of decreased activity and, more commonly, periods during which payments are slow. A consultant cannot count on receiving payments in a regular and predictable manner, as do corporate employees, for example, and in addition, a consultant frequently must make expenditures of substantial size for travel, secretarial assistance, or assistance by other professional associates prior to billing his client.

The basic principles of successfully running a business are no different for a consultant than for others, but many industrial hygienists who consult may not have strong business backgrounds and are well advised to seek advice and guidance from a certified public accountant or other persons qualified to give such advice. Keeping good records is extremely important, and particularly so in relation to income taxes. A consultant receives payments from a large number of sources, and may spend a great deal on travel, entertainment, and so on, all of which are of special interest to the Internal Revenue Service.

5 AIHA INDUSTRIAL HYGIENE CONSULTANT LISTING

For a number of years, the AIHA has assembled and published a list of industrial hygiene consultants. Presently, this listing is updated twice annually, and printed in the January and July issues of the *American Industrial Hygiene Association Journal*. The listing has become a very useful source of information concerning industrial hygiene consultants to those seeking assistance, but it is important to recognize its limitations. The AIHA makes the following statement concerning this listing: "The American Industrial Hygiene Association provides these listings as an informational service, accepting no responsibility for the performance of services by the consultants, their claims of specialization and their competence as consultants." The only requirements for being placed on the list are that the consultants must be full members of the AIHA, and must pay the required fees. The July 1989 listing contains the names of approximately 225 individuals and organizations who offer a spectrum of industrial hygiene consulting services. Obviously, there are many more industrial hygiene consultants than choose to be listed, for a simple inspection of the list will reveal the absence of a number of well-known consultants, and many, if not most, part-time consultants. The potential user of consulting services is well advised to keep in mind that the listing, as stated by AIHA, is noncritical, and there is no assurance of competence. At the same time, it is probable that most of the individuals and organizations listed are competent, and the listing provides a very useful service to those seeking assistance. For convenience, the listing is presented first according to the state in which a consultant does business, then all names are presented alphabetically, and finally the consultants are listed according to the specialty codes in which they have indicated competence. Although as noted, many consultants have apparently not found it necessary to be included in this listing, it is potentially useful both to consultants and to those seeking assistance.

6 RELATIONSHIP WITH CONSULTING ORGANIZATIONS

Individual industrial hygiene consultants will rarely be capable of providing all requested services without the assistance of others to help perform field studies, for example, or to prepare and analyze samples collected in the field. Such assistance is readily found today, and simple reference to the consulting listing just described should be ample to identify consulting organizations who are geographically convenient to the consultant. In some cases, geographical convenience may not be important, and the consultant should then try to identify the organization best suited to meet his or her needs. It is reasonable to require that the consulting laboratory be accredited by the AIHA, and in all probability most, if not all, of the consulting laboratories in the AIHA consultant listing are also accredited. It is recommended that consultants identify consulting laboratories and organizations upon whom they plan to rely, and learn in a general way the spectrum of services available and the prevailing fee schedules. Given the number of competent laboratories and organizations that presently exist, the individual consultant should have little difficulty in satisfying every need which may arise.

7 CONCLUSIONS

The profession of industrial hygiene is relatively young, and the process of maturing requires that attitudes and practices gradually change to meet contemporary needs. The substantial growth in the number of industrial hygiene consultants and their use by those requiring such assistance is part of the maturing process, and these trends are likely to continue indefinitely. Industrial hygienists who retain consultants should not be viewed as professionals who lack basic skills, but rather as enlightened professionals who choose to supplement their skills as required to meet their employer's needs in the best manner possible. In this regard, industrial hygiene follows traditions established many years ago by such long established professions as engineering, medicine, and accounting, whose practitioners routinely retain required consultative assistance as soon as the need is perceived.

REFERENCES

1. American Academy of Industrial Hygiene, inside cover page, 1989, roster of diplomates and members.

B THE CONSULTING ORGANIZATION

Jaswant Singh, Ph.D., C.I.H.

1 INDUSTRIAL HYGIENE CONSULTATION FIRMS

1.1 Introduction

Emergence of the industrial hygiene consulting firm in the United States is a relatively recent phenomenon. With few exceptions most of the industrial hygiene firms in business today have their beginnings in the passage of the Occupational Safety and Health Act (OSH) of 1970. A 1966–1967 survey of the industry conducted by the American Industrial Hygiene Association (AIHA) showed (1) that as recently as 1967, only about 10 to 15 percent of the workforce ever came in contact with an industrial hygienist. Dr. Morton Corn, in his keynote address at the opening session of the 1976 American Industrial Hygiene Conference, said, "In the pre-OSHA (Occupational Safety and Health Administration) days, larger, enlightened firms can be pointed to as the main consumers of our knowledge and our primary employers."

1.2 Growth of Industrial Hygiene Consulting Firms

The industrial hygiene consultants in the "pre-OSHA" days were to a large extent a few individuals, many of whom were well-known pioneers in the field. The OSHAct imposed considerable health and safety requirements on employers, many of whom had no previous exposure to the field of industrial hygiene or its practitioners (2). Suddenly, there was a need for consultants who could provide diversified industrial hygiene services to thousands of employers. This growth in industrial hygiene consulting paralleled the need for industrial hygiene laboratory services leading to the emergence of a number of small, independent industrial hygiene analytical laboratories. The NIOSH-sponsored AIHA laboratory accreditation program provided further impetus to the growth of commercial industrial hygiene laboratories. Some consulting firms integrated analytical services with their professional consulting practice.

In the 1970s, industrial hygiene certification programs offered by the American Board of Industrial Hygiene (ABIH) also became popular as more and more employers became aware of the certification process and its value. At the time OSHA came into being, certified industrial hygienists were few and far between (Figure 5.2). For many employers, their only access to a certified industrial hygienist was through a consulting firm.

The following factors lead to the growth of industrial hygiene consulting firms:

- OSHAct imposed detailed requirements on employers to comply with the permissible exposure limits (PELs) and a variety of other health and safety measures.

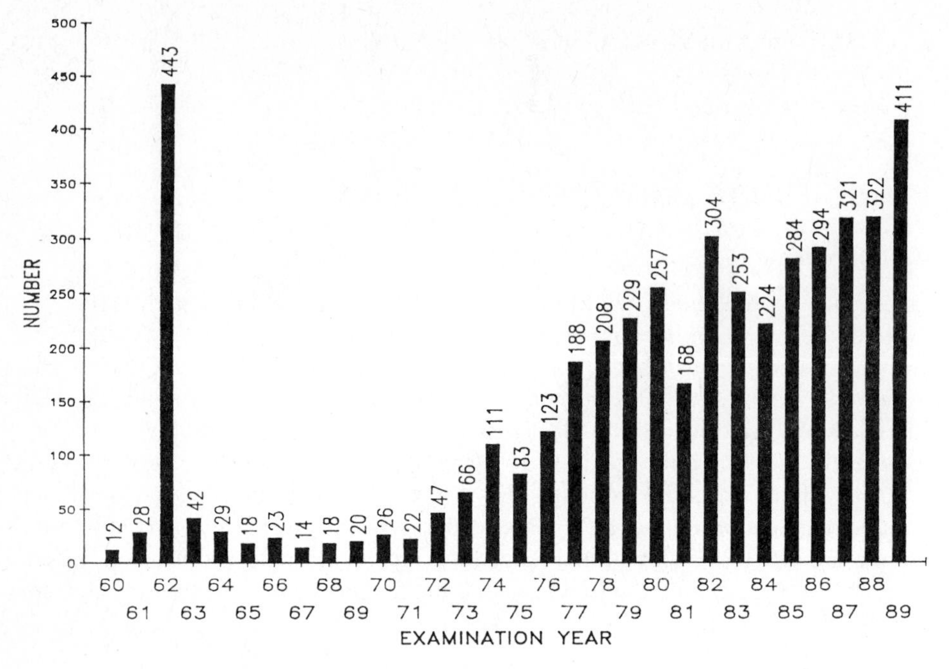

Figure 5.2. CIH certificates issued by year (1960–1989).

- The OSHAct mandated that OSHA develop health standards. NIOSH had the task of producing criteria documents to support the OSHA rule-making process. To do this, NIOSH had to rely heavily upon outside sources; several industrial hygiene consulting firms emerged to fulfill this need.
- Labor unions, in their contract negotiations, often specified industrial hygiene surveys of their workplaces to be conducted by independent third-party industrial hygienists.
- Trade and professional organizations also recognized the need to conduct industry-wide studies to characterize workplace, safety hazards, and to develop control measures and safe work practices for their industries. These organizations also perceived the need to determine the technological and economic feasibility of complying with the proposed and increasingly stringent governmental standards.
- Beginning in the mid-1980s, public concern about asbestos generated a flurry of industrial hygiene activity. Many consulting firms emerged to fulfill this growing need to assess and remediate asbestos hazards. Building owners worried about legal liabilities, resulting from perceived or actual health risks of occupants from exposure to asbestos, sought assistance from industrial hygiene professionals to assess and remediate asbestos problems. Many of the new consulting firms almost exclusively provided asbestos services. Asbestos consulting firms continued to grow rapidly as commercial building owners, school

districts, and manufacturers increasingly sought to manage asbestos hazards at their properties.

- Threat of liabilities resulting from claims related to exposures of workers and the community to toxic materials also generated demand for industrial hygiene consulting services. Recent trends in the U.S. courts have been to allow employees to file criminal charges against employers for the alleged willful and negligent conduct in exposing their employees to toxic substances. The threat of employee lawsuits and employer liabilities beyond worker compensation laws has driven many companies to seek industrial hygiene consultants.
- In addition to federal regulations, state and local regulatory requirements have created a demand for industrial hygiene consulting services. Notable among these regulations are requirements by state agencies such as California (Cal-OSHA). Other state laws include California's Proposition 65, which requires risk assessment and public notification by employers who may discharge carcinogens and teratogens into the environment.
- Worker and community right-to-know laws have contributed significantly toward public awareness of workplace hazards. Businesses have become increasingly sensitive to employee and public concerns about worker safety and health, thus placing greater emphasis on industrial hygiene services.

2 WHAT NEEDS DO THE CONSULTANTS SERVE?

Industrial hygiene consulting firms fill a critical need. For companies that are small or medium size, independent consultants are the main source of industrial hygiene expertise because many smaller firms cannot justify hiring a full-time professional. In addition, larger companies such as service organizations and academic and religious institutions may not always need a full-time in-house industrial hygiene staff. For these organizations, engaging consulting firms provides a cost-effective means of procuring industrial hygiene services because consultants are used only on an as-needed basis.

Companies with in-house industrial hygiene capabilities may also retain consultants from time to time in areas these companies may lack in-house. These areas may include toxicological evaluation, developing material safety data sheets (MSDS) and industrial hygiene programs, and performing exhaust ventilation engineering or epidemiologic studies.

Consultants are also called upon in situations where independent third-party investigations are required because of legal considerations or because of certain labor contract agreements.

3 SERVICES OFFERED

Services provided by consulting firms range from asbestos management only to comprehensive industrial hygiene-related services. Increasingly, larger consulting firms combine industrial hygiene services with related environmental activities in the areas of air pollution, water pollution, hazardous waste management, and subsurface environmental investigations (to assess and remediate contaminated soil and groundwater problems). Although the depth and range of services offered by industrial hygiene consulting firms vary, consulting firms offer the following typical services.

3.1 Field Surveys

Field surveys are a major activity for many industrial hygiene consulting firms; they include sampling to develop baseline exposure data and/or conduct routine monitoring to measure worker exposures. Such measurements are often made to satisfy mandatory periodic monitoring required by federal or state regulatory agencies. For example, OSHA mandates periodic monitoring for substances such as asbestos, lead, arsenic, acrylonitrile, and benzene. Periodic monitoring is also required to measure noise exposures using personal dosimeters and/or sound level meters. Consulting firms often perform sampling surveys for companies that have in-house industrial hygiene functions but may lack sufficient manpower resources and field equipment.

3.2 Program Development and Program Auditing

Motivated by concerns for worker health and safety and/or as a result of requirements imposed by law, many firms in the United States are confronting industrial hygiene issues for the first time and turning to consulting firms for assistance. Because of the passage of the OSHA's hazard communication regulations (1910.1200), firms in both the manufacturing and the nonmanufacturing segments of U.S. industry have had to develop specific written hazard communication programs. Such programs require generating chemical inventories, preparing MSDS, training employees, and instituting personal protection maintenance programs. Many companies have found it advantageous to engage consultants to develop computerized data bases for chemical inventories, MSDS, employee exposure data, and medical records.

3.3 Asbestos Management

More and more, consulting firms are being called upon to provide asbestos management services. Assessment of asbestos hazards has become particularly important during property transfers, lease negotiations, mergers and acquisitions, and financing of commercial properties. Asbestos consulting services are also in high demand by educational institutions and owners of commercial and industrial buildings concerned about employee exposure to asbestos and resultant legal liabilities. As a result, a major asbestos consulting industry has developed in the United States and overseas. In the United States, a large number of consulting firms are almost exclusively engaged in asbestos consulting services including:

- Building inspections to identify suspect asbestos-containing materials (ACM)
- Laboratory analyses of ACM for asbestos content
- Risk assessments
- Identification of areas needing abatement
- Development of plans and specifications for the safe removal and disposal of asbestos
- Surveillance of the asbestos abatement projects to provide safe work practices, air sampling, and on-site microscopic analysis of samples to determine exposure of abatement workers and building occupants as well as to verify the integrity of fiber-containment barriers
- Final inspections and "clearance" monitoring to determine if the abated area has been adequately cleaned up and whether airborne fiber concentrations are within the specified "acceptable" limits

3.4 Legal Testimony

In our increasingly litigious society, employers are facing mounting environmental liabilities. In the United States, litigation related to environmental and safety and health matters has proliferated at an accelerated pace. Such litigation is often complex, and the prosecution or defense of such cases sometimes requires extensive literature searches and gathering of field data. Consulting firms are often called upon to generate such data in support of litigation. More experienced consultants, particularly those who specialize in certain aspects of industrial hygiene and have attained recognition in their field, are generally called upon as expert witnesses.

3.5 Indoor Air Quality

Indoor air quality (IAQ) is a fast developing area of endeavor for industrial hygiene consultants. Energy conservation measures along with the rising use of synthetic building materials and interior furnishing have resulted in poor air quality in many buildings. Moreover, employers and building occupants have become more cognizant of IAQ problems and their resultant effects on employee health, productivity, and employee relations. In the United States, legislation has been proposed to address IAQ issues. This legislation is likely to result in further demand for industrial hygiene consulting services.

Indoor air quality projects can be complex and often require a team of experts including an industrial hygienist, a heating, ventilating, and air-conditioning (HVAC) engineer, and a microbiologist. In addition, an epidemiologist may be needed, unless the industrial hygiene professional is well versed in epidemiologic techniques. Very few consulting firms have such in-house multidisciplinary expertise and are therefore not equipped to perform comprehensive IAQ investigations. In some IAQ projects, an occupational physician may be needed to interview building occupants or examine them physically. Services of an occupational physician are either subcontracted by the consulting firm or engaged directly by the employer.

3.6 Industrial Hygiene Engineering

Full-service industrial hygiene consulting firms provide local exhaust ventilation engineering services including conceptual design as well as detailed drawings and specifications, field supervision during installation, and system performance testing. To a lesser extent, industrial hygiene consulting firms provide general building ventilation design services. Such services, however, are generally limited to a review of the ventilation design performed by mechanical engineering firms specializing in HVAC system design.

Industrial hygiene engineering services also include noise control engineering, including machine and process enclosure design and, occasionally, process modifications. Unfortunately, fewer industrial hygiene firms provide engineering services today, for the number of engineering graduates entering the practice of industrial hygiene has gradually dwindled.

3.7 Laboratory Services

Several industrial hygiene consulting firms maintain analytical laboratories. Some firms specialize only in laboratory services. The number of consulting industrial hygiene laboratories has steadily grown since the AIHA laboratory accreditation program began (3).

Capabilities and resources of accredited industrial hygiene laboratories vary considerably. Services offered range from limited microscopic analyses of asbestos samples to comprehensive analyses covering the entire range of contaminants encountered in the workplace. Most industrial hygiene laboratories now use modern and automated analytical instrumentation such as gas and liquid chromatography, mass spectrometry, atomic absorption spectrometry, and optical and electron microscopy in addition to conventional wet chemical analyses.

Consulting industrial hygiene laboratories serve a vital function. Even large industrial firms with in-house industrial hygiene programs rely heavily on consulting laboratories for chemical analyses because commercial laboratories, owing to the large volume of analyses they may perform, conduct certain tests more cost effectively than in-house laboratories that perform such specialized tests only occasionally.

3.8 Other Services

An increasing trend among larger consulting firms has been to provide multidisciplinary environmental services including industrial hygiene, hazardous waste management, and air pollution and water pollution control. In particular, environmental engineering firms specializing in assessment and remediation of hazardous waste sites have added industrial hygiene expertise to their capabilities because safety and health issues are prominent in hazardous waste site cleanup programs.

The broad range of industrial hygiene related services offered by consulting firms is reflected in the directory of industrial hygiene consultants (4) published in the AIHA journal. The recent listing included the following 20 areas of specialization:

Air pollution
Asbestos
Audiometry
Biological monitoring
Ergonomics
Expert witness testimony
Industrial hygiene comprehensive plant studies and/or analyses
Industrial hygiene chemistry
Meteorology
Noise control
Occupational medicine
Product safety labeling
Radiological control
Respiratory protection
Safety specialist
Toxicology
Training instruction
Ventilation
Waste disposal
Water pollution

4 INDUSTRIAL HYGIENE CONSULTING AS A CAREER

Consulting firms now employ significant numbers of industrial hygienists. Moreover, the industrial hygiene consulting business appears to be poised for considerable growth as governmental regulations become more complex and stringent, and workers and employers become more aware of and concerned about industrial hygiene issues. Another reason for the growth of the industrial hygiene consulting business is a rising trend among large U.S. corporations to contract out work that is not directly related to the manufacture of their basic product. Similar trends are also apparent for environmental, legal, accounting, medical, and design engineering services.

Industrial hygiene consulting can be a rewarding career for professionals who strive for diversity, desire to solve problems, enjoy interaction with others, and derive satisfaction and pride in helping clients to maintain safer workplaces. If the consulting firm is multidisciplinary, there is opportunity to interact with other specialists and develop skills in related areas. Because industrial hygiene service is the end product of a consulting firm, an industrial hygiene professional is the most valuable commodity in that business, for the success and growth of the business depends directly on the industrial hygienist's productivity and competence.

To succeed in a consulting carrier, an industrial hygienist should:

- Possess good communication (oral and written) skills
- Have a relentless drive to please clients
- Be adaptable to highly variable situations
- Possess appropriate certifications (such as certification by the ABIH)
- Be service-motivated, flexible, and patient
- Be willing and able to travel

A typical industrial hygiene consultant handles several projects simultaneously, faces critical deadlines, travels on short notice, and sometimes works long and odd hours.

Frequently consulting services are requested at a moment's notice due to emergency situations. Report deadlines and frequent emergencies can sometimes strain certain individuals. Consulting careers may therefore prove stressful to those whose life styles are not amenable to such conditions. Another discouraging factor for some consultants sometimes can be their inability or helplessness to implement recommendations. If a client chooses not to act on a consultant's recommendations, the industrial hygienist generally has little influence or recourse.

Successful industrial hygiene consultants, however, derive considerable satisfaction from helping clients solve problems, making workplaces safer, and preventing liability for the client. Professional and financial awards for the consultant can be substantial. A competent and energetic industrial hygienist can advance quickly in a growing consulting firm and can attain a managerial rank. In some firms, an aspiring industrial hygienist can attain the status of a partner.

5 ETHICS AND RESPONSIBILITIES

The primary responsibility of every industrial hygienist is to protect the health of the worker (5). Industrial hygiene consultants are responsible for fulfilling the requirements of the client who engages their services to identify problems and develop cost-effective solutions. Consultants are also responsible for providing documentation and accountability to the client in support of their findings.

Industrial hygiene consultants of course subscribe to the code of ethics imposed by their profession and as defined by the AIHA (6). The code of ethics has been described earlier in this chapter. For the industrial hygiene consultant, one of the most important ethical issues is to avoid conflict of interest. For example, a consultant who is affiliated with an equipment manufacturer or a contractor who installs the recommended control system has an obvious conflict of interest. In the asbestos abatement field, conflict of interest also occurs when the abatement contractor acts as the industrial hygiene investigator and project consultant, thus specifying procedures and then certifying his or her own work.

6 LIABILITIES

Failure to perform adequately entails certain legal risks for all industrial hygienists. An industrial hygiene consultant could face additional potential liability from an unhappy client not satisfied with the consultant's performance. Some legal experts feel (7, 8) that developments in the area of toxic tort liability can engulf industrial hygiene consultants for negligence or breach of duty if the conduct of the industrial hygienist is below the standard normally accepted in the profession. (See also Chapter 7.) To avoid liability, industrial hygiene consultants should use procedures and test methods generally recognized in the profession as reliable and accurate.

To guard against such liabilities, most industrial hygiene consulting firms carry insurance coverage, generally in the form of errors and omissions. Recently, such coverage has been difficult to obtain. Moreover, when available, insurance coverage can be costly and may carry exclusions for certain types of work involving relatively higher risks.

7 SELECTING INDUSTRIAL HYGIENE FIRMS

7.1 Consulting Firms

Listings of qualified industrial hygiene consulting firms can be obtained from several sources. Many firms advertise their services through professional journals. The *AIHA Journal* publishes an annual list of consultants. The July 1989 listing of the journal (4) includes practitioners representing individual consultants as well as those affiliated with larger firms. The *Air and Waste Management Association Journal* (8) also publishes a guide to consultants that includes practitioners in industrial hygiene.

Selection of a qualified industrial hygiene firm is a crucial task and can prove to be difficult and frustrating because, currently, any person can claim to be an industrial hygiene expert and legally offer services as an industrial hygiene consultant. To avoid consultants who are inexperienced, incompetent, or lacking suitable training and equipment, the following criteria should be used in the selection:

- Certification by the ABIH
- References from past clients
- Adequacy of staff, equipment, and laboratory facilities
- Insurance coverage for professional liabilities, including errors and omissions
- Successful completion of AHERA training programs (for asbestos projects)
- AIHA-accredited laboratory facilities (preferably in-house)
- Financial stability
- Competitive fees
- No conflict of interest

7.2 Analytical Laboratories

Equal care should to be exercised in selecting an industrial hygiene laboratory. Prospective clients should take the following steps before selecting a laboratory:

- Talk with the key laboratory personnel about their capabilities and credentials.
- Visit the laboratory and examine sample handling facilities, chain-of-custody procedures, instrumentation, written procedures, and quality control logs.
- Look for evidence of competence such as laboratory accreditation and proficiency in external testing programs.
- Ask for references from colleagues and professionals.

The best source for finding a qualified laboratory is the listing of AIHA-accredited laboratories published bimonthly in the *AIHA Journal*. The AIHA accreditation program requires participating laboratories to demonstrate proficiency in analyzing the following seven categories of analyses:

- Lead
- Cadmium
- Zinc
- Asbestos
- Silica
- Organic solvents
- All proficiency analytical testing materials

The AIHA listing indicates the proficiency of the listed laboratories based on the laboratories' successful participation in the AIHA's Proficiency Analytical Testing (PAT) program. In the 1989 listing (3), 323 laboratories were listed as accredited; a number of these laboratories were determined to be proficient in all of the above categories.

8 CONCLUSIONS

Industrial hygiene consulting firms serve a vital function by providing services to employers who lack in-house expertise or need backup from time to time. As employers and public become increasingly aware of workplace hazards, and regulatory requirements become more complex and stringent, demand for services of consulting firms will continue to rise. Consulting firms will provide challenging careers to many who strive for diversity, who desire to help others, and who seek rapid career advancement.

REFERENCES

1. M. Corn, "Role of the American Conference of Governmental Industrial Hygienists and the American Industrial Hygiene Association in OSHA Affairs," *Am. Ind. Hyg. Assoc. J.*, 391–394 (1976).
2. M. Corn, "Influence of Legal Standards on the Practice of Industrial Hygiene," *Am. Ind. Hyg. Assoc. J.*, (6), 353–356 (1976).
3. "Accredited Laboratories," *Am. Ind. Hyg. Assoc. J.*, **50**, A473–A482 (1989).
4. "Industrial Hygiene Consultants," *Am. Ind. Hyg. Assoc. J.*, **50**, A562–A577 (1989).
5. P. D. Halley, "Industrial Hygiene—Responsibility and Accountability (Cummings Memorial lecture—1980)," *Am. Ind. Hyg. Assoc. J.*, **41**, 609–615 (1980).
6. R. L. Harris, "Information, Risk and Professional Ethics," *Am. Ind. Hyg. Assoc. J.*, **47**, 67–71 (1986).
7. M. E. Alexander, "Professional Hazards for the Industrial Hygienist," *Econ. Environ. Contractor*, 85–87 (1989).
8. Consultant Guide, *J. Air Pollut. Control Assoc.*, **38**, 1605–1627 (1989).
9. M. Nash, "A Question of Balance," *OH & S Canada*, **4**, 92–98 (1988).

CHAPTER SIX

Hazard Communication and Worker Right-to-Know Programs

Charlotte A. Heideman, C.I.H. and Lisa K. Simkins, C.I.H.

1 INTRODUCTION

Consumer demand for new products and conveniences has resulted in the introduction of vast numbers of new chemicals. Recent estimates show that more than two million chemical compounds are known to exist, with at least 25,000 new chemicals introduced every year, many of which find significant commercial application. Most of these chemicals may not be dangerous, but many chemicals could cause harm to human health and the environment, if handled improperly.

Until recently, the concern has been with the intended product, not with the unintended effects of the chemicals used in the process. Much of the earlier concern about chemical safety was with the flammable and explosive nature of chemicals and not with their toxicity and carcinogenicity.

Now, however, concern has focused on health and environmental issues. Since the early 1970s, there has been growing support for the right of employees to know the nature of the hazards associated with chemicals they encounter on the job. The obligation to inform employees of the nature and potential harmful effects of chemicals is now regulated on a federal level under the Hazard Communication Standard. This standard applies to all employees covered under Occupational Safety and Health Administration (OSHA) regulations.

Initially, most of the emphasis was in the workplace. However, the tragic release of methyl isocyanate in Bhopal, India in 1984 and the ensuing public outrage

Patty's Industrial Hygiene and Toxicology, Fourth Edition, Volume 1, Part A, Edited by George D. Clayton and Florence E. Clayton
ISBN 0-471-50197-2

expanded the demand for information from the workplace to the community. Citizens throughout the United States began to question whether safety precautions and emergency plans would be effective in preventing a similar incident in their own communities.

The Emergency Planning and Community Right-to-Know Act of 1986, also known as SARA (Superfund Amendments and Reauthorization Act), Title III, was passed by the U.S. Congress in response to public pressure for information and emergency planning arising in part from the Bhopal incident.

In addition to these two federal regulations, state and community right-to-know laws have proliferated, creating a confusing patchwork of overlapping and, in some cases, contradictory regulations. The common thread running through all of these regulations is the employer's duty to provide information regarding the hazardous chemicals used and stored at the workplace.

The industrial hygienist is in a unique position to assist in this endeavor, interpreting the complex issues of chemical exposure and explaining them in terms that can be understood by the layperson. More and more, the practice of industrial hygiene is extending from the workplace out into the community.

The effectiveness of the industrial hygienist in helping to prevent disease and promote good health depends not only on the practitioner's technical skill but, more and more, on communication skills. For this reason, communication should be added to the triad of industrial hygiene: recognition, evaluation, and control.

This chapter focuses primarily on the requirements of the Hazard Communication Standard, providing some interpretation of the standard and tools for the implementation of a program. Section 2 provides insight into the evolution of right-to-know regulations. The essential elements of a hazard communication program and suggested application strategies are described in Section 3. Related regulations are summarized in Section 4. Sections 5 through 7 discuss management of a program, the materials and systems used to implement a program, and finally, recommendations for auditing a program.

2 HISTORICAL PERSPECTIVE

Community and worker right-to-know laws have been evolving over a period of more than 10 years. During this period, legislation has been promulgated by federal agencies, state governments, and local authorities. Concern regarding the environment and chemicals in the workplace continue to be prominent in the news. The Hazard Communication Standard was created in an atmosphere of widely diverse interests, changing political administrations and consequent positions, and a proliferation of state and local laws. As a result, its history is a confusing patchwork of court challenges, rule making, and revision.

Understanding and interpreting the current standard depends, to some extent, on an appreciation for its history and the influences that shaped it. This section is intended to provide that perspective. A summary of the major events in the development of the federal Hazard Communication Standard is presented in Table 6.1.

2.1 Background

The Occupational Safety and Health Act of 1970 (OSHAct), Section 6(b)(7) states that "Any standard promulgated under this subsection shall prescribe the use of labels or other appropriate forms of warning as are necessary to insure that employees are apprised of all hazards to which they are exposed, relevant symptoms and appropriate emergency treatment, and proper conditions and precautions of safe use or exposure" (1). However, only 25 chemicals were regulated by a specific standard. Although warning signs and labels were used by some chemical manufacturers, there was no general requirement for consistent labeling, nor was it specifically required that employees be informed of the potential hazards associated with materials in their workplace. When it became obvious that it would be impossible for OSHA to regulate every chemical by a specific standard, the need for a generic chemical information or labeling standard was clear.

In 1974, the National Institute for Occupational Safety and Health (NIOSH) published a criteria document titled *A Recommended Standard . . . An Identification System for Occupationally Hazardous Materials*, "in an effort to minimize risk by alerting all concerned to the hazards inherent in chemicals found in the workplace" (2).

The objective of this document was to provide a uniform and comprehensive information system utilizing placards, labels, and material safety data sheets (MSDSs). The recommendations in this document were intended to apply to materials stored, distributed, made available, furnished, or supplied for use by employees. Further, it recommended labeling work areas that contained hazardous materials, and was intended to apply to all workers (3).

Also in 1974, the Standards Advisory Committee of Hazardous Materials Labeling was established under Section 7(b) of OSHAct to develop guidelines for implementing Section 6(b)(7). This committee published its final report in 1975 recommending guidelines for categorizing and ranking chemical hazards. The report further recommended that hazards be communicated by means of labeling, MSDSs, and training (4).

At the request of the Subcommittee on Labor, Committee on Human Resources and the U.S. Senate, NIOSH prepared a document in 1977 entitled *The Right to Know: Practical Problems and Policy Issues Arising from Exposures to Hazardous Chemicals and Physical Agents in the Workplace*. Although the primary thrust of this document was notification of workers regarding past and present exposures, a "major stumbling block" to this notification process was identified as the lack of information on labels regarding the chemical composition of materials. A uniform labeling and warning system was recommended for consideration. Material safety data sheets were suggested as an interim step to alleviate the difficulty anticipated because of the reluctance of manufacturers to divulge chemical compositions of proprietary formulations (5, 6).

Table 6.1. Chronology of the Hazard Communication Standard

DATE	EVENT
1970	Occupational Safety and Health Act becomes law. Section 6(b)(7) states that standards promulgated shall prescribe warning labels.
1974	NIOSH publishes "A Recommended Standard...An Identification system for Occupationally Hazardous Materials."
	OSHA establishes the Standards Advisory Committee on Hazardous Materials under Section 7(b) of the Occupational Safety and Health Act (OSH Act).
1975	Standards Advisory Committee on Hazardous Materials Labeling established in 1974, publishes a report recommending labels, MSDSs, and training.
January 28, 1977	OSHA publishes ANPR on chemical labeling.
1979	Assistant Secretary of Labor Eula Bingham announces OSHA and EPA will work on a labeling standard.
1980	EPA withdraws from the joint labeling project.
January 16, 1981	OSHA publishes NPRM for "Hazard Identification" standard.
February 12, 1981	OSHA withdraws NPRM for "Hazard Identification" standard
July 1981	House Education and Labor Subcommittee on Health and Safety holds hearings regarding labeling chemicals in the workplace.
March 19, 1982	OSHA publishes NPRM for "Hazard Communication" Standard.
1983	ANSI publishes revised labeling standard (ANSI 129.1-1982).
November 22, 1983	United Steel Workers of America, AFL-CIO-CLC, and Public Citizen, Inc. petition for judicial review in the U.S. Court of Appeals for the Third Circuit.
November 25, 1983	OSHA publishes final Hazard Communication Standard limited to manufacturing industries.
May 24, 1985	U.S. Court of Appeals for the Third Circuit issues a decision in United Steel Workers of America v. Auchter. Trade secret issues and scope of standard remanded to the secretary of labor.
November 27, 1985	OSHA announces ANPR on the expanded standard.
September 30, 1986	OSHA revises sections addressing trade secrets and issues final standard.

2.1.1 The Need for a Uniform Standard

The need for a uniform standard requiring labeling and communication of hazards was further supported by the National Occupational Hazard Survey, which concluded that one in every four American workers was exposed to hazardous chemicals (as identified by NIOSH). This survey, as well as other data on occupational illness, was used by OSHA to justify the need for a standard (4). In 1979, Assistant Secretary of Labor, Eula Bingham identified drafting a proposal for a labeling

Table 6.1. (Continued)

DATE	EVENT
January 27, 1987	United Steel Workers of America and Public Citizen file a motion in the Third Circuit Court to force OSHA to issue the expanded standard immediately.
May 29, 1987	The Third Circuit Court orders OSHA to issue a Hazard Communication Standard applicable to all workers covered by OSH Act within 60 days.
August 24, 1987	OSHA publishes final Hazard Communication Standard (expanded); public comment is invited for 60 days.
October 28, 1987	Office of Management and Budget issues a letter to OSHA disapproving certain sections of the expanded Hazard Communication Standard.
April 13, 1988	OMB issues limited approval of expanded Hazard Communication Standard.
May 20, 1988	Associated Builders and Contractors et al. challenge the final rule in the U.S. Court of Appeals for the Third Circuit.
June 24, 1988	The Third Circuit Court grants stay of the Hazard Communication Standard as it applies to the construction industry.
August 19, 1988	The U.S. Court of Appeals for the Third Circuit invalidates OMB's disapproval and orders OSHA to announce that all parts of the Hazard Communication Standard are effective (United Steelworkers of America v. Pendergrass).
September 2, 1988	U.S. Department of Justice files a petition in the Third Circuit Court requesting a rehearing.
November 25, 1988	The Third Circuit Court denies the petition for review; the challenge to the Hazard Communication Standard by the Associated Builders and Contractors et al is without merit. The stay issued on June 24 is vacated.
November 29, 1988	Request for a rehearing by U.S. Department of Justice is denied by the Third Circuit Court.
January 24, 1989	A motion is entered by industry representatives for a stay of the Third Circuit Court decision.
February 15, 1989	OSHA publishes notice that all provisions of the Hazard Communication Standard are in effect for all industries.

standard as a primary goal for OSHA and indicated that OSHA was working on such a standard in close coordination with the Environmental Protection Agency (EPA) (7).

The need to communicate chemical hazards was supported by chemical manufacturers. Speaking at a symposium in San Francisco in 1979, officials of the chemical industry stated that informing employees of workplace hazards was the most critical issue facing the industry, and that more information was being disseminated than ever before. A spokesperson for the chemical manufacturers commented that employees are entitled to information, and that withholding it only leads to mistrust and concern among employees. Management's role was described as making the information useful and understandable to employees. MSDSs were cited as the primary vehicle for this communication (8). OSHA subsequently announced its intention of publishing a proposed standard in the fall of 1980 (9).

In the summer of 1980, the Consumer Product Safety Commission (CPSC) issued a request for proposals to establish a coordinated Interagency Regulatory Liaison Group (IRLG) strategy for the labeling of toxic substances (9). At this point, labeling of chemicals was regulated by CPSC, EPA, OSHA, and the Food Safety and Quality Service. The potential for overlapping and conflicting labels was perceived as a serious barrier to the effective communication of the hazards of toxic substances. The mission of the IRLG was to develop and present options for a uniform system to member agencies.

The EPA withdrew from this joint labeling standard project in 1980, leaving OSHA with the responsibility of developing a labeling standard (10). A draft standard prepared by OSHA in September of 1980 drew statements of concern from the chemical industry primarily on the issues of trade secret protection, the contents of labels, the extent of research needed to determine a hazard, and the specificity of the language in the standard (11).

2.1.2 The Federal Hazard Communication Standard

On January 16, 1981, the last day of the Carter administration, OSHA published a notice of proposed rule making (NPRM) entitled "Hazard Identification" (12). The proposal focused on labeling and did not contain provisions for MSDSs and training as called for in the Standard Advisory Committee recommendations and the NIOSH criteria document. The initial cost to industry was estimated at $870 million with annual costs ranging from $360 to $374 million. The proposal drew sharp criticism from industry as being burdensome to industry because of the detailed labeling requirements for all containers and piping systems (13).

On February 12, 1981, OSHA, under the Reagan administration, withdrew its proposed standard, citing a need to consider regulatory alternatives fully (14). Although the proposal had been criticized by labor for its limited scope (the manufacturing sector, Standard Industrial Classification Codes 20–39 only), labor union leaders protested its withdrawal (15).

Subsequently, the Chemical Manufacturer's Assocation proposed a voluntary hazard communication program to the EPA that stressed training and communication, as an alternative to the proposed rule recently withdrawn by OSHA. The definitions used in the American National Standards Institute precautionary labeling standard were suggested for acute hazards (16).

In May of 1981, the House Education and Labor Subcommittee on Health and Safety held hearings regarding workplace labeling of chemicals. Chemical industry representatives testified that flexibility in the means used to communicate hazards was essential, and that labels listing all ingredients of a chemical mixture would be unnecessarily confusing (17).

Also in these hearings, labor representatives voiced strong support of complete labeling and criticized recommendations to substitute hazard warnings, MSDSs, and symbols. Further, they encouraged OSHA to work swiftly to generate a generic standard covering a broad range of industries and requiring complete chemical identification, including warnings regarding chronic health hazards. Reserving such information to protect trade secrets was to be treated as an exception (18).

In October of 1981, OSHA announced that a draft proposal would be issued in the fall (19). At informal rule-making hearings, controversy focused on key issues:

- Protection of trade secrets
- Determination of hazard
- The definition of an article
- Warnings to "downstream users"
- The scope of the standard (manufacturing only)
- Treatment of mixtures (the "1 percent rule")
- The definition of carcinogens
- The enforceability of the standard
- The contents of labels
- Chronic hazard warnings
- Preemption of state standards/regulations
- Blanks on material safety data sheets
- The performance-oriented nature of the standard
- Duplicate federal labeling requirements

In the midst of this controversy, a NPRM was published on March 19, 1982 for the Hazard Communication Standard (20). Unlike the previous proposal, piping systems were exempted from strict labeling requirements; labels would carry simple warnings, rather than a full list of ingredients; and information regarding composition would be detailed on MSDSs. Training on the dangers associated with hazardous materials would also be required. The proposal was termed "performance oriented" in that employers were allowed flexibility to tailor plans to different industries and environments rather than be held to detailed compliance requirements. Staggered compliance dates were designated, hazard assessments and recordkeeping requirements were limited, and the issue of trade secrets was addressed by means of generic names and confidentiality agreements. Public hearings were scheduled for June of 1982.

Criticism of the proposal again focused mainly on the areas of trade secret protection, the "1 percent rule," difficulty in enforcement because of the lack of specific requirements, its limited scope (manufacturing industries), and the exclusion of chemical importers.

Unions continued to support the concept of a strong federal regulation, but encouraged the passage of state and local regulations as well. Chemical manufacturers hoped that the anticipated federal regulation would reduce the confusion resulting from a proliferation of state laws (21).

The final Hazard Communication Rule was published in the *Federal Register* on November 25, 1983 (22). The way in which the more controversial issues have been addressed are discussed in Section 2.2.

In the preamble to the final rule published in the *Federal Register*, OSHA estimated that the rule would affect 318,000 chemical manufacturers and 10,000 importers and distributors. The estimated annual cost was $160.9 million (22).

The issuance of the federal rule, however, did not stem the tide of local and state right-to-know regulations. Although OSHA stated in the preamble that the rule would reduce the regulatory burden of multiple state and local laws by preempting those laws with respect to hazard communication requirements for employees in the manufacturing industry, the issue was still the subject of diverging legal opinion and was far from settled. Estimates of the number of right-to-know bills pending at the state and local level ranged as high as 28 when the final rule was published.

2.1.3 Legal Challenges

Soon after the final rule was published, suits were filed against OSHA by the Flavor and Extract Manufacturers Association (who later withdrew the suit) and the United Steelworkers of America. The states of New Jersey, New York, and Connecticut filed as intervenors in the steelworkers' suit; they were joined by Massachusetts, Illinois, and Public Citizen, Inc. of Wisconsin, New Mexico, and West Virginia. The AFL–CIO filed as amicus in the suit (23).

The steelworkers' suit focused on three areas of the rule: (1) the exclusion of nonmanufacturing employees from the rule, (2) the nonaccess to trade secret information for workers with a demonstrated need to know, and (3) the too-liberal mechanism for chemical manufacturers to refuse to provide information by claiming a trade secret (24).

In its response to the suit described above, OSHA defended limiting the rule to the manufacturing sector as consistent with its priority-setting discretion, addressing the sector with the clearest need for a standard based on illness and injury documentation. On the issue of trade secrets, OSHA reiterated its position that the balance struck in the standard was reasonable and necessary to protect the interests of both employees and chemical manufacturers.

Late in the summer of 1984 four industry associations, joined by nine companies having chemical plants in New Jersey, filed suit contending that the New Jersey Worker and Community Right-to-Know Act was unconstitutional because it was preempted by the Federal Hazard Communication Standard, which it allegedly exceeded and with which it was in conflict (25).

In November, the New Jersey Department of Health proposed amendments to the state's act that would expand the exceptions to the labeling requirements and allow employers to protect trade secrets (26).

In January of 1985, the U.S. District Court for the District of New Jersey ruled that OSHA's Hazard Communication Standard preempted that portion of the state's regulation that covered manufacturing industries. In response to this decision, James J. Florio (D–NJ) introduced a set of bills in the U.S. House of Representatives intended to prevent OSHA standards from preempting state laws, and to create comprehensive national community right-to-know legislation. In addition, the package addressed recourse for victims of chemical disasters, reform of environmental regulations to include the regulation of underground storage tanks and certain pesticide processes, and the training of emergency response personnel dealing with truck and train wrecks (27).

A resolution was introduced by Representative Robert Edgar (D–PA) in February calling on the assistant secretary of labor to expand the Hazard Communication Standard to all employees. The resolution also urged the revision of the labeling and trade secret portion of the standard, and criticized the preemption of state and local laws (28).

2.1.4 The Expanded Rule

With the first deadline of November 25, 1985 (by which chemical manufacturers were to label containers and provide MSDSs) approaching, OSHA indicated in March of 1985 that it was considering extending the Hazard Communication Standard to all employers. The extension of the standard to all employers was likely to affect the construction industry most, according to OSHA. The National Advisory Committee on Occupational Safety and Health was asked to make recommendations regarding this issue and on OSHA's policies regarding the chemical industry, in light of the catastrophic release of methyl isocyanate at a Union Carbide plant in Bhopal, India (29).

In April another resolution was introduced in the U.S. House of Representatives that would broaden the scope of the Hazard Communication standard to include all employers (30).

In May of 1985, the U.S. Court of Appeals for the Third Circuit ruling on the *United Steelworkers et al.* v. *Auchter et al.* found that the secretary of labor must explain why the Hazard Communication Standard should apply only to manufacturing industries. The court also found that OSHA had not justified the protection of trade secrets provided in the standard, and had failed to support its position that employees should not have access to trade secrets. The court, however, upheld OSHA's contention that Hazard Communication was a standard rather than a regulation, thus allowing preemption of state regulations (31, 32).

In light of this decision, coverage under the Hazard Communication Standard was extended to federal agency workers. OSHA did not appeal the ruling by the Third District Court (33).

In November of 1985, OSHA published an interim final rule, initiated a study to evaluate the economic impact of expanding the standard to nonmanufacturing industries, and published an advance notice of proposed rule making (ANPR) for the expansion to include all employees (34).

In this ANPR, OSHA sought information particularly on the extent to which nonmanufacturing employers had implemented Hazard Communication programs, and the applicability of the standard to other industries. OSHA received over 200 responses, and was preparing a proposed rule based on this new evidence and on the previous rule-making record (35).

On January 27, 1987, the United Steelworkers of America, the AFL–CIO–CLC, and Public Citizen, Inc. (petitioners in the 1985 court challenge) filed a motion claiming that OSHA should have made a feasibility determination based on the 1985 rule-making record without additional fact gathering, and that OSHA's pace was unduly slow (36).

On May 29, 1987, the court found in favor of the petitioners and ordered the agency to issue a Hazard Communication Standard applicable to all workers covered by the OSHAct, or to state why such a standard was not feasible. This was to be accomplished within 60 days of the order (37).

The NPRM for the expanded rule was published in the *Federal Register* on August 24, 1987. In the preamble, OSHA restated its position that a more current feasibility determination was needed, and invited comment for a period of 60 days after publication of the standard. This version of the Hazard Communication Standard differed from the previous rule only in scope (38).

OSHA received 136 comments, and found that most did not contain data or evidence concerning feasibility. However, the Office of Management and Budget (OMB) had disapproved certain information collection requirements of the expanded standard under the Paperwork Reduction Act, specifically:

- The requirement that MSDSs be provided on multi-employer worksites
- Coverage of any consumer products falling within the exemption covering these materials in Section 311(e)(3) of the Superfund Amendments and Reauthorization Act of 1986
- Coverage of any drugs regulated by the FDA (39, 40)

OSHA, abiding by the OMB decision, did not enforce these provisions. In addition, the OMB determined that OSHA should reopen rule making to redefine articles in the Hazard Communication Standard and made approval of the paperwork conditional on OSHA consulting with the U.S. Small Business Administration and the Department of Commerce to develop a plan to provide assistance to regulated industries in alleviating the paperwork burden and expense (41).

In the year following its publication, the final rule was challenged in the U.S. Court of Appeals by the Associated Builders and Contractors, National Grain and Feed Association, Associated General Contractors of Virginia, Associated General Contractors of America, and United Technologies Corporation (42, 43).

Pending the outcome of this case, the Third Circuit Court granted a stay of the standard as it applied to the construction industry on June 24, 1988 (44).

The United Steelworkers of America, AFL–CIO–CLC, and Public Citizen filed a motion with the Third Circuit Court requesting that the court order OSHA to enforce the provisions disapproved by the OMB (45).

On August 8, 1988 OSHA issued a notice of proposed rule making to address issues raised by the OMB, solicit additional public comment, and modify certain sections in response to meetings with the Advisory Committee on Construction Safety and Health (39).

On August 19, 1988 the Third District Court ordered the secretary of labor to publish a notice that the parts of the Hazard Communication Standard disapproved by the OMB were in effect (46). On September 2, 1988 the Department of Justice filed a petition with the Third Circuit Court requesting a rehearing of the case which was subsequently denied (47). In January of 1989, the Supreme Court lifted a stay of the Third Circuit Court's decision regarding the application of the Hazard Communication Standard to the construction industry (48).

On February 15, 1989, OSHA published notice in the *Federal Register* that all provisions of the Hazard Communication Standard are in effect for all segments of industry (49). Other motions brought before the Supreme Court are pending at this time.

2.2 Controversial Issues

2.2.1 Scope

The rule published in 1983 was limited to the manufacturing sector, Standard Industrial Classification (SIC) codes 20 through 39. Although requests were made for exemption of certain industries within these codes, no exemptions were granted in the rule. The scope of the rule was extended to importers or distributors, and to laboratories within the manufacturing sector.

OSHA's rationale for limiting application of the standard to the manufacturing sector was that the need for information was the greatest in this sector, that requiring manufacturers to comply with the standard would result in the dissemination of the information to other industrial sectors, and that labels on containers would already provide information on hazards outside the regulated sectors. OSHA based its decision on Section 6(g), which gives the secretary of labor permission to set priorities and promulgate standards sequentially.

This limitation brought criticism primarily from labor unions (see Section 2.1.3) which culminated in a court decision on May 24, 1985 that directed OSHA to reconsider its decision. OSHA reopened the rule-making procedure, but in 1987 the Third Circuit Court ordered OSHA to issue an expanded rule, which it did on August 24, 1987.

Many spokesmen for small businesses and those outside the manufacturing sector protested the expansion of the rule based on the high cost of compliance versus little perceived impact on accidents and injury rates. Building contractors concerned with the difficulties involved with providing information to employees on multiple work sites and on sites with various contractors using materials brought an unsuccessful suit against OSHA.

However, in February of 1989, OSHA published a notice in the *Federal Register* that all provisions of the standard were in effect for all industries. At this writing, the Hazard Communication Standard applies to all employees covered by the OSHAct.

2.2.2 Mixtures

The definition of health hazard is broad in the standard. The health hazards of any components present in the material at concentrations of 1 percent (0.1 percent if the component is a carcinogen) or more must be presented on the label and MSDS if the results of any study designed and conducted according to established scientific principles report statistically significant evidence of a health hazard.

The difficulty of dealing with mixtures was the subject of concern for both formulators and users. The mixture must be treated as if it represents the same hazards as the pure components unless the mixture has been tested to determine its specific hazards.

Many formulators believed that this would result in a dismaying number of warnings and health hazard statements on labels that would overstate the hazards of some mixtures. The alternative of testing a mixture would be expensive and would therefore not be done routinely by smaller manufacturers. This provision remains unchanged at this writing.

2.2.3 Intrinsic Hazard versus Risk

The Hazard Communication Standard requires warning regarding potential adverse health effects or physical hazards based on the intrinsic properties of the material. All of the provisions of the standard are applicable in most cases, even if air sampling or other means of investigation demonstrates that there is virtually no risk of exposure. This application has been modified somewhat by OSHA in the 1987 final rule for those handling sealed containers, office employees, and those using consumer goods in essentially the same manner as an ordinary consumer. OSHA views the probable risk of exposure to be minimal in these circumstances, and has, as a result, waived certain provisions (50, 51).

OSHA's rationale for this position is that exposure may occur during nonroutine use of a material, accidental spills, or misuse; therefore, even though ordinary use may not result in overexposure, the employees using materials must be warned of intrinsic properties so that they will be able to protect themselves during a foreseeable emergency.

2.2.4 Articles

The Hazard Communication Standard does not apply to the following kinds of articles (38):

> A manufactured item: (i) which is formed to a specific shape or design during manufacture; (ii) which has end-use function(s) dependent in whole or in part upon its shape or design during end-use; and (iii) which does not release, or otherwise result in exposure to, a hazardous chemical under normal conditions of use.

Although this exclusion appeared fairly straightforward at first reading, it was not as simple to apply. Certain articles might be expected to emit small amounts of their original "components," for example, varnish solvents from new furniture.

Articles such as mercury switches would not emit hazardous materials under ordinary conditions of use, but could during installation if broken. And finally, certain materials, for example, piping, would not emit hazardous chemicals in use but could when being cut, as during repair or remodeling, or as scrap. Because vehicles would fit the first two conditions, but not necessarily the third, would they require labels? Could they be considered "containers?"

OSHA retained the definition of an article unchanged, but clarified the intent of this exclusion in the preamble to the 1987 final rule. The following specific points helped to clarify OSHA's position on this issue:

- Releases of trace amounts of materials are not covered by the rule.
- Chemical analysis and testing are not required to prove that hazardous chemicals are *not* present.
- The article exemption applies to the end use. Intermediate users, such as installers, are covered when hazardous materials could be released as a result of their actions; for example, the cutting of lead-containing tile for installation on a ship's hull.
- Information regarding potential hazards must be passed "downstream" until the item is installed, if there is potential for exposure.
- Information need not be passed downstream solely because of the possibility of exposure during repair.
- Solid metal (for example, steel beams) are *not* exempted because of the possibility that downstream use will result in potential exposure to hazardous materials.

OSHA later revised the definition of "container" specifically to exempt engines, fuel tanks, or other operating systems in a vehicle.

2.2.5 Trade Secrets

Much of the controversy surrounding the Hazard Communication Standard focused on the "trade secrets" provisions. The primary issues involved were:

- The chemical manufacturers' concern about guarding formulations they viewed as "secret," which provided an advantage in the marketplace
- The concern of labor unions, employees, and others that too broad a definition of trade secrets would provide an easy mechanism for manufactures to avoid giving specific information about materials
- The broadening definition of health professionals to include nurses

The 1987 final rule defined a trade secret with a more precise description of the nature of the trade secret and the extent of the secrecy which is necessary to support the exclusion. These descriptions are contained in Appendix D of the 1987 final rule.

The final rule allows a chemical manufacturer, importer, or employer to withhold the specific chemical identity of a hazardous chemical from the MSDS if:

- The claim that the information being withheld as a trade secret can be supported.
- The properties and health effects of the hazardous chemical are provided on the MSDS.
- The MSDS indicates that the information being withheld is a trade secret.
- The specific information being withheld is made available to health professionals, employees, and designated representatives in accordance with the provisions described below.

Conditions under which such information must be released include:

- Medical emergency determined by a treating physician or nurse in which the identity of the chemical is necessary for treatment.
- In a nonemergency situation, a health professional requests the information in writing describing in reasonable detail the occupational health needs for the information as listed below:

  To assess the hazards of the chemicals to which employees will be exposed

  To conduct or assess sampling of workplace atmosphere to determine employee exposure

  To conduct preassignment or periodic medical surveillance of exposed employees

  To provide medical treatment to exposed employees

  To select or assess appropriate personal protective equipment for exposed employees

  To design or assess engineering controls or protective measures for exposed employees

  To conduct studies to determine the health effects of exposure

As a further requirement, the rule stipulates that requests for trade secret information detail why the disclosure of the specific chemical was necessary for the purpose stated, and why the request would not be satisfied by the information provided without this disclosure. The request must also include a description of the methods to be used to maintain the confidentiality of the information provided, and an assurance that the health professional would not use the information for other than the stated purpose.

Provisions covering the content of the confidentiality agreement, release of the information to OSHA, denial of the request by the manufacturer, importer, or employer, and the recourse available in the event of denial or the request were also added to the rule.

A complete description of OSHA's opinion and reasoning regarding these and other issues surrounding the rule are contained in the preamble which was published in the *Federal Register* on Friday, November 25, 1983, and in the preamble to the expanded final rule published in the *Federal Register* on Monday, August 24, 1987.

3 SCOPE OF HAZARD COMMUNICATION PROGRAM

3.1 Application of Hazard Communication

The Hazard Communication Standard applies to chemical manufacturers, importers, and distributors of hazardous chemicals, as well as all employees who use hazardous chemicals. A "hazardous chemical" is broadly defined in the Hazard Communication Standard as any chemical that is a physical hazard or a health hazard. However, there are a few exceptions to the standard including:

- Any hazardous waste subject to the Resource Conservation and Recovery Act (RCRA)
- Tobacco or tobacco products
- Wood or wood products (except wood dust)
- Articles (defined as "a manufactured item which (a) is formed to a specific shape or design during manufacture, (b) has an end use function dependent in whole or in part upon its shape or design during end use, and (c) does not release, or otherwise result in exposure to, a hazardous chemical during normal conditions of use") (52)
- Food, drugs, or cosmetics intended for personal use or consumption by employees while in the workplace

Some chemicals are subject to labeling requirements of other laws or regulations and are, therefore, exempt from hazard communication labeling requirements. Chemicals exempt from labeling requirements include:

- Pesticides subject to the labeling requirements of the Federal Insecticide, Fungicide, and Rodenticide Act or pursuant regulations
- Foods, drugs, and cosmetics subject to the labeling requirements of the Federal Food, Drug and Cosmetic Act or pursuant regulations
- Distilled spirits (beverage alcohols), wine, or malt beverage intended for non-industrial use that are subject to the labeling requirements of the Federal Alcohol Administration Act or pursuant regulations
- Consumer products or hazardous substances as defined in the Consumer Product Safety Act and Federal Hazardous Substances Act, respectively, when subject to a consumer product safety standard or labeling requirement of these Acts or pursuant regulations

In the final standard issued on August 24, 1987, OSHA added a provision addressing work situations in which employees handle only sealed containers (53), such as retail sales, warehousing, and cargo handling. Requirements under the regulation are limited for these operations to:

- Ensuring that labels on incoming containers are not removed or defaced
- Maintaining copies of MSDSs received with incoming shipments of sealed containers
- Requesting MSDSs from the manufacturer if an employee asks for one
- Ensuring that MSDSs (as described above) are accessible
- Providing training for employees that must include the following subjects to the extent necessary to protect them in the event of a leak or spill (54):
  - The training requirements under the regulation
  - Operations in the work area in which hazardous materials are present
  - Methods and observations that may be used to detect the presence or release of a hazardous chemical in the workplace
  - The physical and health hazards related to the general classes of chemicals present in the workplace
  - Measures employees can use to protect themselves from chemical hazards during spill and leak episodes
  - An explanation of the labels and MSDSs, and how an employee may obtain appropriate hazard information

The training must be provided to new employees and again whenever a new hazard is introduced into the work area.

In the preamble to the final rule referenced above, OSHA recognized the unique characteristics of other businesses as well, for example, offices, banks, and similar operations. Employees in operations of this type "who encounter hazardous materials only in nonroutine isolated instances are not covered" (55) by the regulation.

There may, however, be employees within such organizations who *would* be covered by the regulation. Examples of employees who could be exposed to hazardous chemicals in companies in which the bulk of employees are exempted include:

- Copy room employees whose routine duties would include handling copier fluids, toners, or similar materials
- Employees in the graphics or art departments
- Maintenance and/or custodial employees
- Employees in a photography or blueprint reproduction shop

Laboratories are exempt from some of the provisions of the Hazard Communication Standard. The provisions of the standard that do apply to laboratories include the following:

- Labels on incoming containers of hazardous chemicals must not be removed or defaced.

- Employers must maintain MSDSs that are received with incoming shipments of hazardous chemicals and ensure that they are readily accessible to laboratory employees.
- Employers must ensure that laboratory employees are apprised of the hazards of chemicals in their workplaces in accordance with the information and training provisions of the standard.

3.2 Program Elements—Chemical Manufacturer, Supplier, and Importer Responsibilities

The basic responsibilities of chemical manufacturers, suppliers, or importers of hazardous substances in a hazard communication program are described as follows:

3.2.1 Hazard Determination

Hazard determination is the responsibility of chemical manufacturers and importers. When evaluating chemicals, the chemical manufacturer or importer must consider the available scientific evidence concerning such hazards. For health hazards, hazard determination criteria are specified in Appendix B of the Hazard Communication Standard (38). This appendix specifies criteria for (*a*) carcinogenicity, (*b*) human data, (*c*) animal data, and (*d*) adequacy and reporting of data.

Determination of whether a chemical is a carcinogen or a potential carcinogen is based on findings by the National Toxicology Program, the International Agency for Research on Cancer, and/or OSHA. Positive determinations on carcinogenicity by any of these groups is considered conclusive.

When human data such as epidemiologic studies and case reports of adverse health effects are available, they must be considered in a hazard evaluation. Because human evidence of health effects is not generally available for many chemicals, animal data must often be relied upon. The results of toxicology testing in animals must often be used to predict the health effects that may occur in exposed workers. This reliance is particularly evident in the definition of some acute hazards that refer to specific animal data, such as lethal dose (LD_{50}) or lethal concentration (LC_{50}).

The results of any positive studies that are designed and conducted according to established scientific principles, and have statistically significant conclusions regarding health effects of a chemical, are considered sufficient basis for hazard determination and reporting on a MSDS.

The Hazard Communication Standard specifies the following categories for hazard determination in its mandatory Appendix A (38).

- Carcinogen
- Corrosive
- Highly toxic
- Irritant
- Sensitizer

- Toxic
- Target organ effects

Examples of target organ effect categories include (*a*) hepatotoxins, (*b*) nephrotoxins, (*c*) neurotoxins, (*d*) agents that act on the blood or hematopoietic system, (*e*) agents that damage the lung, (*f*) reproductive toxins, (*g*) cutaneous hazards, and (*h*) eye hazards.

3.2.2 Material Safety Data Sheets

Chemical manufacturers and importers are required to obtain or develop an MSDS for each hazardous chemical they produce or import. Chemical manufacturers and importers must then ensure that distributors and purchasers of hazardous chemicals are provided with an MSDS with the initial shipment of each hazardous chemical and with the first shipment following an MSDS update. Distributors must ensure that MSDSs and updated information are provided to other distributors and purchasers of hazardous chemicals.

The preparer of the MSDS must ensure that the information reported is accurate and reflects scientific evidence used in making the hazard determination. If the MSDS preparer becomes newly aware of significant information regarding the hazards of a chemical or ways to protect against the hazards, the new information must be added to the MSDS within three months.

The MSDS may be in any format, but must provide specific information regarding the hazardous chemical. Required information on an MSDS includes:

(a) *Identity of the Material.* The identity on the MSDS must match the identity on the label. The MSDS must also provide the chemical and common name(s) of the hazardous chemical. In the case of a mixture, the chemical and common name(s) of the ingredients must be listed. If the mixture has been tested as a whole to determine its hazards, the chemical and common name(s) of any ingredients contributing to the known hazards must be listed. If the hazardous chemical has not been tested as a whole, the chemical and common name(s) of all hazardous ingredients that comprise 1 percent or greater of the composition must be listed, except for carcinogens, which must be listed if they comprise 0.1 percent or greater of the composition. In all cases, the chemical and common name(s) of ingredients that present a physical hazard when present in the mixture must be listed. In some states, the Chemical Abstract Services (CAS) number is also required for each component. The standard does include provisions for protecting trade secrets of chemical manufacturers. The specific identity of hazardous chemicals can be withheld from the MSDS if this information is a trade secret; however, all other information regarding the properties and effects of the chemical must be provided.

(b) *Physical Hazards.* This includes information regarding potential for fire, explosion, and reactivity.

(c) *Health Hazards.* Health hazard information must include signs and symptoms of exposure and any medical conditions that are generally aggravated by exposure to the chemical.

(d) *Primary R Route(s) of Entry into the Body*. Typical routes of entry are inhalation, ingestion, skin contact, and skin absorption.

(e) *Exposure Limits*. These limits include the OSHA permissible exposure limit (PEL), the ACGIH Threshold Limit Value (TLV), and any other exposure limit used or recommended.

(f) *Carcinogenicity*. The MSDS must state whether the chemical has been determined to be a carcinogen or potential carcinogen by the National Toxicology Program (NTP) Annual Report on Carcinogens, the International Agency for Research on Cancer (IARC) Monographs, or OSHA.

(g) *Precautions for Safe Handling and Use*. This includes precautions that are generally applicable including hygienic practices, protective measures during repair and maintenance of contaminated equipment, and procedures for cleanup of spills and leaks.

(h) *Control Measures*. These include engineering controls, work practices, or personal protective equipment.

(i) *Emergency and First Aid Procedures*.

(j) *Date of Preparation or Latest Change*.

(k) *Name, Address, and Telephone Number*. This is required for the chemical manufacturer, importer, employer, or other responsible party preparing or distributing the MSDS. This is provided in case the user needs additional information on the hazardous chemical and emergency procedures.

If the MSDS preparer cannot find any information for one of these categories to include on the MSDS, then that section should be marked as not applicable or a notation that no information was found should be made. There must be no blanks on an MSDS.

OSHA has an optional form that may be used to prepare an MSDS. This OSHA form is shown in Figure 6.1. It is important to note that the old OSHA Form 20 is no longer adequate, because it does not include all the information required by the Hazard Communication Standard.

3.2.3 **Labels**

The chemical manufacturers, importers, or distributors must ensure that each container of hazardous chemicals leaving their workplaces is labeled. The label on a "shipped" container must include:

- Identity of the hazardous chemical
- Appropriate hazard warning
- Name and address of the chemical manufacturer, importer, or other responsible party

Material Safety Data Sheet
May be used to comply with
OSHA's Hazard Communication Standard,
29 CFR 1910.1200. Standard must be
consulted for specific requirements.

U.S. Department of Labor
Occupational Safety and Health Administration
(Non-Mandatory Form)
Form Approved
OMB No. 1218-0072

IDENTITY *(As Used on Label and List)*	*Note: Blank spaces are not permitted. If any item is not applicable, or no information is available, the space must be marked to indicate that.*

Section I

Manufacturer's Name	Emergency Telephone Number
Address *(Number, Street, City, State, and ZIP Code)*	Telephone Number for Information
	Date Prepared
	Signature of Preparer *(optional)*

Section II — Hazardous Ingredients/Identity Information

Hazardous Components (Specific Chemical Identity; Common Name(s))	OSHA PEL	ACGIH TLV	Other Limits Recommended	% *(optional)*

Section III — Physical/Chemical Characteristics

Boiling Point		Specific Gravity (H_2O = 1)	
Vapor Pressure (mm Hg.)		Melting Point	
Vapor Density (AIR = 1)		Evaporation Rate (Butyl Acetate = 1)	

Solubility in Water

Appearance and Odor

Section IV — Fire and Explosion Hazard Data

Flash Point (Method Used)	Flammable Limits	LEL	UEL

Extinguishing Media

Special Fire Fighting Procedures

Unusual Fire and Explosion Hazards

(Reproduce locally) OSHA 174, Sept. 1985

Figure 6.1. Material Safety Data Sheet

Section V — Reactivity Data

Stability	Unstable		Conditions to Avoid
	Stable		

Incompatibility (*Materials to Avoid*)

Hazardous Decomposition or Byproducts

Hazardous Polymerization	May Occur		Conditions to Avoid
	Will Not Occur		

Section VI — Health Hazard Data

Route(s) of Entry:	Inhalation?	Skin?	Ingestion?

Health Hazards (*Acute and Chronic*)

Carcinogenicity:	NTP?	IARC Monographs?	OSHA Regulated?

Signs and Symptoms of Exposure

Medical Conditions
Generally Aggravated by Exposure

Emergency and First Aid Procedures

Section VII — Precautions for Safe Handling and Use

Steps to Be Taken in Case Material Is Released or Spilled

Waste Disposal Method

Precautions to Be Taken in Handling and Storing

Other Precautions

Section VIII — Control Measures

Respiratory Protection (*Specify Type*)

Ventilation	Local Exhaust	Special
	Mechanical (*General*)	Other

Protective Gloves	Eye Protection

Other Protective Clothing or Equipment

Work/Hygienic Practices

Page 2

☆ USGPO 1986-491-529/45775

Figure 6.1. (Continued)

The label is meant to be an immediate warning. The appropriate hazard warning should cover the major effects of exposure; however, it will not include all of the detailed information provided by the MSDS.

Information required on secondary in-plant containers is discussed in Section 3.3.3.

3.3 Program Elements—Employer Responsibilities

An employer that uses hazardous substances must have a hazard communication program that includes the components described as follows.

3.3.1 Written Hazard Communication Program

The employer must prepare and make available to employees a written hazard communication program that explains how compliance with the standard is achieved. The written program must include a description of how the employer complies with labeling, MSDS, and employee information and training requirements. The written program must also include a list of hazardous chemicals known to be present at the workplace. The identity on the list must match the identity used on the MSDS and label. The list can be for the entire facility or for individual work areas.

The written program must also include an explanation of how the employer informs employees of the hazards of nonroutine tasks, such as cleaning reactor vessels, and the hazards associated with chemicals contained in unlabeled pipes in their work areas.

The written program also must include the methods used to inform contractors working in the employer's workplace of the hazardous chemicals their employees may be exposed to while performing their work; suggestions for protective measures for contractor's employees must be included.

3.3.2 Material Safety Data Sheets

The employer may rely on the chemical manufacturer, importer, or distributor for the hazard determination and MSDS preparation. The employer is responsible for maintaining copies of the MSDSs in the workplace and ensuring that MSDSs are readily accessible during each work shift to employees when they are in their work areas. Material Safety Data Sheets may be kept in any form, including operating procedures, and may be designed to cover groups of hazardous chemicals in a work area where it may be more appropriate to address the hazards of a process rather than individual hazardous chemicals. However, the employer shall ensure that in all cases the required information is provided for each hazardous chemical, and is readily accessible during each work shift to employees when they are in their work area(s) (38).

3.3.3 Labels

The employer must ensure that each container of hazardous chemicals in the workplace is labeled, tagged, or marked with (*a*) the identity of the hazardous chemical and (*b*) appropriate hazard warning.

The labels provided by suppliers on original containers are typically adequate for those containers. The employer must label in-plant containers into which hazardous chemicals are transferred from original containers. Labels similar to those used on original containers can be used, or an alternate method can be employed. Signs, placards, process sheets, batch tickets, operating procedures, or other such written materials may be used in lieu of actual labels on individual stationary process containers provided the method conveys the required label information.

Employers are not required to label portable containers into which hazardous chemicals are transferred from labeled containers, and which are intended for immediate use by the employee who performs the transfer. Immediate use is typically interpreted to mean within the same work shift that the transfer was performed.

Existing labels on original containers of hazardous chemicals must not be defaced or removed, unless the required information is immediately marked on the container again.

Labels must always be present or available in English. It is acceptable and often advantageous to supplement the English labels with hazard information in other languages, where employees speak other languages.

3.3.4 Employee Information and Training

Employers must provide specific information and training to employees to comply with the Hazard Communication Standard. Employees must be trained on hazardous chemicals in their work area at the time of their initial assignment and whenever a new hazard is introduced into the work area.

Information that must be provided to employees includes:

- Requirements of the Hazard Communication Standard
- Operations in the work area where hazardous chemicals are present
- Location and availability of the written hazard communication program, including the list(s) of hazardous chemicals and MSDSs

Employee training must include, at a minimum, the following:

- Methods and observations that may be used to detect the presence or release of a hazardous chemical in the work area
- Physical and health hazards of chemicals in the work area
- Measures employees can take to protect themselves from hazards
- Procedures the employer has implemented to protect employees from exposure to hazardous chemicals, such as work practices, emergency procedures, and personal protective equipment

Product Name: ______________________________

MSDS contains the following:

[ ] Name, address, and telephone number of manufacturer, or other responsible party

[ ] Date of MSDS preparation

[ ] Chemical name(s) and CAS number(s)

[ ] ACGIH TLV and OSHA PEL

[ ] Physical properties (e.g., vapor pressure, boiling range, appearance, odor, etc.)

[ ] Fire and explosion (e.g., flash point, auto ignition temperature, flammable limits, fire extinguishing media, etc.)

[ ] Potential routes of exposure

[ ] Symptoms of overexposure

[ ] Health effects (acute and chronic)

[ ] Emergency and first aid procedures

[ ] Carcinogen (cancer-causing) determination

[ ] Medical conditions aggregated by exposure

[ ] Reactivity data (stability, incompatibility, hazardous decomposition products, hazardous polymerization)

[ ] Spills/disposal procedures

[ ] Ventilation and engineering controls

[ ] Respiratory protection

[ ] Protective clothing

[ ] Work practices

Figure 6.2. Material safety data sheets (MSDSs) checklist.

- Details of the hazard communication program developed for the facility, including an explanation of the labeling system, the MSDS, and how employees can obtain and use appropriate hazard information

3.4 Good Practice

In addition to the legal requirements of the program, there are elements of hazard communication that are considered "good practice."

An employer can improve an MSDS program by performing a review of each MSDS provided by a supplier. A review checklist is shown in Figure 6.2. After review, the employer can request additional information from suppliers if MSDSs are found to be inadequate. The employer will then help ensure that the essential information regarding hazards provided to the workers is adequate.

Some employers prefer to produce "in-house" MSDSs from information provided on manufacturers' MSDSs. This approach requires some interpretation during transfer of manufacturers' information, and the employer therefore may assume liability for incorrect information on "in-house" MSDSs. The advantage of this approach is a uniform MSDS form for employee use. Uniform MSDSs are sometimes automated for retrieval at computer terminals, as discussed in Section 6.1.2.

Training is required; however, its form is left to the discretion of the employer. Research has shown that learning by doing is the most effective. Therefore, the more the workers can be actively involved in the training program, the more likely it is to be effective. Testing for comprehension is not required; however, it provides valuable feedback on how much of the information was actually received and retained.

Supplemental labeling systems such as NFPA and HMIS are not required, but may be useful in providing a single uniform warning system to supplement written hazard warnings. When such labeling systems are used, they must be included in training to have any value. They also must be viewed as supplemental, because they don't provide specific target organ hazard warnings.

A good industrial hygiene program, including air sampling when potential overexposures exist, is essential as a "good practice" to supplement a hazard communication program. An employee will not know when to use respiratory protection specified on an MSDS or whether a PEL or TLV shown on an MSDS is exceeded without representative air monitoring data.

Hazard communication requirements are clearly only a starting point for providing a safe work environment when hazardous substances are used.

3.5 Application of Elements

3.5.1 Involvement of Professionals

Implementation of a hazard communication program in an employer's workplace often requires an multidisciplinary approach. A typical implementation team may include:

- Industrial hygienists
- Safety professionals
- Communications specialists
- Trainers
- Lawyers
- Physicians/nurses

Professional support for a hazard communication program can come from within the company (local or corporate staff) or from outside consultants. Professionals can provide initial program guidance followed up by periodic audits. Involvement of other individuals within a company is also needed for a program to be truly effective. Examples of functions these professionals might perform in the program are discussed in Section 5.2.

3.5.2 Involvement of Management

Top management support for a hazard communication program is absolutely essential to its success. As discussed in Section 5.1, a company should have a specific policy regarding hazard communication, ideally issued and signed by the president.

An effective program typically requires that time and effort be spent by many different individuals within the organization. Many of the costs are hidden; for example, training programs for production workers require time away from the production process. Administrative support is required to receive, distribute, and maintain MSDSs. The management must provide adequate resources to support the program and allow key individuals time to implement it.

3.5.3 Input from Employees

Employees and employee representatives can provide valuable information regarding hazard communication. During training sessions, employees can alert the program coordinator, or supervisory representative of potential hazards in the workplace. Additionally, employees may provide feedback after reviewing MSDSs and labels.

Employee involvement and interest in hazard communication is important to the success of the program. If an employee is trained on hazards, yet does not heed precautions, the program is ineffective.

Employee involvement can be stimulated by:

- Direct input from employees via quality circles or employee participation groups
- Involving employees in the preparation of one or more training modules
- Team teaching with selected employees

Employee input before, during, and after training is particularly important to determine how effective this aspect of the program has been. Input can be solicited directly through questionnaire/comment forms following training classes. A more indirect form of measurement of the effectiveness of a program is through testing of participants following training.

Employee representative groups often can be active participants in hazard communication programs.

If there is no mechanism for addressing questions from employees that arise as a result of training, the entire program will suffer from a lack of credibility. If nontechnical trainers are used, a mechanism for addressing technical issues is essential. One such mechanism is to forward questions to the appropriate "in-house expert" for response, but it must be rapid and accurate to maintain employee confidence.

4 RELATED REGULATIONS

The complex issue of managing hazardous chemicals safely from manufacture to disposal has spawned a variety of federal regulations. Although each regulation deals with a slightly different aspect of this problem, they are related and, in some cases, overlap. This section provides a summary of some of these regulations and discusses how they relate to the Hazard Communication Standard.

4.1 Other OSHA Regulations

4.1.1 Access to Employee Exposure and Medical Records (29 CFR 1910.20)

The purpose of this regulation is to provide employees and their designated representatives access to relevant exposure and medical records. MSDSs are identified in the regulation as one of the types of information that constitute an employee exposure record. As such, they must be available to employees, their designated representatives, and OSHA.

Exposure records must be maintained for 30 years; however, MSDSs need not be maintained for materials no longer in use, if there is some record of the identity of the chemical, where it was used, and when it was used in the exposure record.

4.1.2 Air Contaminants (20 CFR 1910.1000)

When the permissible exposure limits (PELs) were revised in 1989, limits for 376 chemicals were added or changed. These changes will require amending the information on virtually every data sheet written prior to the revision of this regulation.

4.2 Superfund Amendments and Reauthorization Act

On October 17, 1986, the Superfund Amendments and Reauthorization Act of 1986 (SARA) was enacted into law. Title III of SARA: the Emergency Planning and Community Right-to-Know Act of 1986 (56) establishes requirements for federal, state, and local governments and industry regarding emergency planning and community right-to-know reporting on hazardous and toxic chemicals.

When SARA Title III was enacted, the Hazard Communication Standard applied only to the manufacturing sector. However, in 1987 the Hazard Communication Standard was expanded to cover all employers. As a result, SARA Title III requirements were also extended to all employers.

Under Section 311, Community Right-to-Know Reporting Requirements, facilities that must prepare or have MSDSs available under the OSHA Hazard Communication Standard must submit copies of MSDSs or a list of the MSDS chemicals to:

- The local emergency planning committee
- The state emergency response commission

• The local fire department

If a list of chemicals is submitted, it must include the chemical name or common name of each substance and any hazardous components as provided on the MSDS. The list must be organized in categories of health and physical hazards as set forth in OSHA regulations unless modified by EPA.

Until October 17, 1989 facilities having less than 500 pounds of extremely hazardous chemicals or 10,000 pounds of hazardous chemicals were exempted from this requirements. However, after this date the threshold is zero unless revised by the EPA. If the threshold is zero, all employers covered by the Hazard Communication Standard will be obligated to comply with Section 311 (57).

Under Section 312, all facilities required to have MSDSs must also prepare a hazardous chemical inventory (Tier I or Tier II forms) for the above emergency response groups. These requirements are subject to threshold quantity exclusions described under Section 311.

4.3 Resource Conservation and Recovery Act

The Resource Conservation and Recovery Act (RCRA) enacted in 1980 regulates the management of hazardous wastes. One of the requirements of this regulation is that employees involved in the handling of hazardous wastes receive training. Although materials covered under this regulation are specifically exempt from the provisions of the Hazard Communication Standard, the requirements for training are similar. Because employees handling hazardous materials must receive training under the Hazard Communication Standard, time can be saved by combining this training in many cases.

4.4 Regulatory Trends

In 1977 NIOSH presented a report to Congress entitled *The Right-to-Know, Practical Problems and Policy Issues Arising from Exposures to Hazardous Chemicals and Physical Agents in the Workplace* (5). In it, NIOSH explored the issue of notifying workers of exposure to potentially harmful chemicals. Although provisions for notifying workers of exposures to chemicals are contained in some OSHA standards, for example, those covering specific materials, and in some state regulations, there continues to be concern regarding the completeness of this notification.

In 1985, a bill was introduced in the U.S. House of Representatives (H.R.162) by Representative Gaydos (D–PA) calling for NIOSH to identify and notify workers at risk of contracting occupational diseases. A similar bill (S.79) was introduced by Senator Metzenbaum (D–OH) in the Senate. The purpose of both of these bills is to provide for a comprehensive notification plan for workers who are currently being exposed or, at some time, have been exposed to a hazardous chemical. At this writing neither bill has been enacted into law.

5 PROGRAM MANAGEMENT

The objective of hazard communication is to provide information to employees on the chemicals in the workplace so that they understand the hazards involved, thus reducing the risk of injury or illness. It is essential to keep this objective in mind when designing a hazard communication program. Assembling the essential elements, such as MSDSs, labels, or a written program, although vital to the program, will not in itself accomplish this objective.

There is a danger that once a compliant program is developed it will become stagnant with ever-growing collections of MSDSs clogging file cabinets, and canned training programs presented over and over again. Clearly, the real objective can be lost in the flood of paper.

To be effective, a hazard communication program must be a dynamic process that includes continuous updating of information and training, program audits, quality control, and continued management commitment.

5.1 Policies and Procedures

Support for an effective program must be reflected in company policies. For example, the responsibility for obtaining a MSDS might rest with purchasing agents. Company policy might state that purchases would be made only from companies supplying complete and reliable data sheets. The manner in which these sheets would be requested, received, recorded, and forwarded to the user within the company would be reflected in the operating procedures.

A company policy statement should establish full management support of the hazard communication program. Individual policies should address:

- Developing and updating MSDSs, labels, product information (chemical manufacturer)
- Acquiring, reviewing, distributing MSDSs
- Verification of labeling on containers, in-house labeling systems, and labeling of pipes and temporary containers
- Content and frequency of training
- Review of new chemicals
- Contractor compliance with the Hazard Communication Standard
- Control of purchase and use of chemicals within the company
- Selection of personal protection equipment
- Audit of the program

Each company's policies should clearly define (1) responsibility for the functions described and (2) the means used to ensure that policies are implemented, for example, quality control and auditing. Accompanying procedures would then explain how each policy is to be implemented.

5.2 Program Responsibilities

The departments within a company most likely involved with the hazard communication process are indicated below with a list of potential responsibilities. This list is not complete, because the exact functions performed will depend on a specific company's organizational structure, size, and other factors.

Purchasing

- Request MSDS.
- Make sure that chemical orders have appropriate approvals before orders are placed.
- Affirm documentation of contractors' compliance with the company hazard communication program.

Receiving

- Check that MSDSs are with shipment of chemicals, or that an appropriate MSDS is on file.
- Check name on MSDS against name on container.
- Check containers for labels.
- Check labels for readability.
- Apply internal company labels, if required by company program.

Operating Department, Maintenance, and Construction Groups

- Order chemicals from approved list.
- Follow company procedure for obtaining approval for new or replacement materials.
- Check labels on chemicals received.
- Check for appropriate MSDSs for materials.
- Update list of chemicals used in department.
- Apply additional labeling (in-house labeling system, if used).
- If material is new, train employees on hazards and safe use.
- Maintain training records.
- Make sure new employees receive hazard communication training before beginning to work with chemicals.
- Respond to employee concerns regarding usage or exposure.
- Make sure all employees have appropriate safety equipment.
- Conduct periodic self-audits of work areas to make sure that containers are labeled, MSDSs are available for materials in use, employees are using personal protective equipment, and training records are complete.
- Conduct periodic audits of employee's knowledge of the chemicals used.
- Participate in continuing hazard communication training.

Health and Safety Departments

- Participate in approval process for new chemicals.
- Participate in the review and selection of alternative chemicals.
- Conduct audits of the workplace to ensure compliance with the Hazard Communication Standard.
- Review incoming MSDSs for completeness and accuracy.
- Establish labeling practices and select labeling system.
- Make sure that appropriate MSDSs are provided to emergency responders in cases of chemical exposure incidents (58).
- Conduct assessments, including air sampling to determine potential employee exposure.
- Review content of or develop training programs.
- Assess/develop procedures for safe handling of chemicals in routine and non-routine tasks.
- Provide guidance in the selection of personal protective equipment.
- Develop and update existing MSDSs for products.
- Develop and update existing in-house MSDSs.

Laboratories (59)

- Conduct physical, chemical, and toxicologic tests to provide data for MSDSs.
- Review test data from outside laboratories.
- Research potential substitute chemicals.
- Participate in development or review of product MSDSs.

Engineering

- Make sure that MSDSs for new chemicals are received and reviewed with users before processes are changed or new designs are implemented.
- Make sure that new chemicals receive appropriate review.
- Consider pipe system and vessel labeling in major redesign and new projects.
- Make sure contractors have fulfilled hazard communication program requirements before beginning projects.
- Consider hazard communication requirements when inspecting project work sites.

Legal Department

- Review initial hazard communication written program.
- Review program annually.
- Interpret regulations.
- Participate in review of product MSDSs.

- Assist in the negotiation of confidentiality agreement in cases of trade secret disclosure.

Medical Department

- Participate in preparation or review of in-house or product MSDSs.
- Review manufacturer's MSDSs for products in use in the plant.
- Make sure that MSDSs are provided to emergency care providers and hospitals, in cases of chemical exposure incidents (58).
- Contact manufacturers for additional information, when required, in cases of trade secret formulations.

Training Department

- Help select/develop training programs.
- Support training effort with equipment/technical advice.
- Conduct training.

This list is not comprehensive, nor is it meant to imply that certain departments or functional groups within a company must perform the duties indicated. It is intended as a guide for delineating responsibility for specific aspects of a hazard communication program within a middle-to-large size company that buys and uses chemicals.

5.3 Flow of Communication

For a hazard communication program to function efficiently, information regarding chemicals must flow smoothly from the source (the chemical manufacturer, importer, or distributor) to the employee using the chemical. This transfer of information utilizing MSDSs and labels as the source of information and training as the means of communication has already been discussed. However, to be complete, there must be provisions for addressing specific employee questions regarding hazards or specific uses, and direct input from employees, health and safety, operating group management, and engineering, regarding the effectiveness of controls.

A mechanism must be in place to "field" employee questions regarding chemical hazards, and to respond to them efficiently and reliably. Such questions may be addressed by referring the question to in-house expert staff, obtaining additional information from the manufacturer, making additional information sources available to the employee, or a combination of all of these. Failure to address employee's questions will result in loss of confidence in the program and more serious employee relations problems.

This response mechanism must also address employee concerns regarding the level of exposure to chemicals and the efficiency of personal protective equipment or engineering controls. Such concerns may be expressed through routine safety committees and employee participation groups or directly to area supervisors or safety/health representatives. Regardless of the route, once management becomes aware of the concern, quick action should be taken to investigate the concern, and evaluate the best means of addressing the issue. The employee(s) should then be informed of the planned action. Ideally, an approach involving the employees, their representatives, and management is the most effective in obtaining resolution.

In addition, the effective management of the chemicals in the workplace depends on a smooth flow of communication between many departments within a company. For example, as a matter of convenience and cost savings, chemicals are frequently purchased in large quantities. The perceived economy of purchasing chemicals in large quantities or in highly concentrated form may be deceptive, owing to the rising storage and handling costs. Another major factor is the liability associated with the potential health and safety risks from exposure to stored chemicals and the potential for environmental damage from leaks and accidental spills. It is obvious that a company must maintain a sufficient chemical inventory to avoid business interruptions. However, a balance must be maintained between the amount of inventory maintained and the ability to deal with the associated risks.

Purchasing decisions should not be made solely on the basis of direct cost, but require input from a wide variety of knowledgeable people within the company to arrive at a truly economical purchase. Factors and associated costs to be considered in the example cited above would include:

- Adequate storage facilities with spill containment
- The effect on air, soil, and water of a release or spill
- The effect of a spill or release on the surrounding community, as well as employees
- Capabilities to respond to a leak or spill of the stored quantity
- Distribution of the chemical to users
- Health and safety considerations of distribution and handling (including dilution, if appropriate)
- Waste disposal (dilute and concentrate)

The efficiently functioning program requires feedback from all active participants. Examples of such feedback are listed below:

- Industrial hygienists
  - Employee exposure evaluations
  - Evaluation of the effectiveness of personal protective equipment
  - Exposure control options
  - Level of employee knowledge of chemicals
- Safety professionals
  - Use of personal protective equipment or control devices in the workplace
  - Employee concerns
  - Accidents/incidents involving chemicals, including fire and explosion

- Communications/training specialists
  - –Level of employee comprehension of training
  - –Problems and needs in training programs
- Lawyers
  - –Modifications to written program
  - –Updates on court cases and decisions involving hazard communication
  - –Updates on legislation trends
- Physicians/nurses
  - –Medical cases involving chemicals
  - –Effectiveness of personal protective equipment
  - –Level of employee knowledge regarding chemicals
- Line supervisors/managers
  - –Changes in processes
  - –Practical problems with controls and personal protective equipment
  - –Employee concerns
  - –Effectiveness of in-house labeling system (if used)
  - –Status/difficulties with training program
- Engineers
  - –Changes in processes or materials being planned
  - –New control devices being planned (e.g., upgrade of ventilation, fume hoods)

5.4 Planning

Early in the program development process, planning must focus on key elements of the program and the tasks related to each element:

- Responsibility for each task
- Interaction of responsible parties
- Systems that will be used to maintain data and records
- Policies and procedures necessary to support those functions
- Mechanisms to evaluate, upgrade, and measure the effectiveness of the program
- Cost analysis and budgeting

Table 6.2 provides guidelines for isolating the steps associated with each element of the program, as well as program upgrading functions. These continuing activities form the basis for the quality control of the program.

5.4.1 Program Costs

The costs associated with a hazard communication program will be reflected in the budgets of many departments of a company. Realistic planning for these costs will ensure the continued effectiveness of the program. Actual costs will, of course, depend on the size of the company and means chosen to implement the program. Both staff time and material costs should be considered for the tasks listed below:

Program Start-up Costs

- Developing the program
- Reviewing existing policies and writing new ones
- Developing procedures
- Meeting with key persons to plan strategy

Collecting MSDSs

- Compiling a list of chemicals used
- Compiling and organizing MSDSs
- Contacting manufacturers for missing MSDSs
- Copying and distributing MSDSs
- Developing data management systems for data sheet collection
- Reviewing MSDSs for completeness

Labeling

- Developing in-house labeling system (option)
- Checking and/or applying in-house labels
- Developing labeling systems for piping systems, process vessels, and temporary containers

Training

- Training materials (audio/visuals, printed materials)
- Training the trainers
- Training current employees initially
- Training new employees
- Training current employees on new chemicals
- Indirect cost of employee training (hours off the job)

Recordkeeping

- Written program
- List of chemicals
- MSDSs

Table 6.2. Key Program Elements

KEY ELEMENTS	STEPS	PROGRAM UPGRADING
Written Program	▪ Write initial program ▪ Review Compliance	▪ Audit Annually - Comment by operating department - Annual compliance audit - Program revision, if needed
Chemical List	▪ Assemble initial list ▪ Input into data management system ▪ Check against chemical purchasing record ▪ Establish review procedure for chemical purchases	▪ Check list periodically against chemical purchase list ▪ Update list ▪ Establish committee(s) - Review chemical usage - Review purchase quantities - Substitute less hazardous materials ▪ Limit off-the-shelf purchase ▪ Eliminate low usage/infrequent usage chemicals
MSDS	▪ Prepare MSDSs for all products* ▪ Match MSDSs with chemical list ▪ Request MSDSs for new chemicals ▪ Negotiate confidentiality agreements ▪ Document requests for MSDSs ▪ Check MSDSs for completeness ▪ Contact supplier if inadequate ▪ Disseminate MSDSs to user areas	▪ Review MSDS annually ▪ Purge MSDS files ▪ Audit use area - MSDS versus containers - MSDS versus chemical list

- –Documentation of source of information for chemical manufacturers' or in-house MSDSs
- –Written requests for missing or inadequate MSDSs and follow up

- Training records
  - –Content of training, date, person conducting training, materials used
  - –Names of employees trained, department or work area, job classifications or position
  - –Follow-up training: chemicals or hazards reviewed, persons conducting training, date, materials used, persons in attendance
  - –Refresher training
- Contractors' acknowledgments that hazard communication information was provided

Table 6.2. (Continued)

KEY ELEMENTS	STEPS	PROGRAM UPGRADING
Labeling	▪ Develop labels for all products* ▪ Check incoming containers for labels ▪ Apply in-house labels ▪ Label vessels, portable containers ▪ Design system for pipes, process flow diagrams, batch tickets, etc. ▪ Isolate unlabeled containers at receiving stations ▪ Contact supplier if labels unreadable, missing	▪ Establish procedures to make pipe/vessel content identification part of redesign procedure ▪ Spot check containers routinely for readable labels ▪ Audit all piping systems/vessels annually
Training	▪ Develop training materials ▪ Train the trainers ▪ Train employees ▪ Develop system to train new employees ▪ Develop training for new materials/systems ▪ Document training	▪ Audit employee training ▪ Compare contents of training with chemical list ▪ Interview employees for knowledge of chemicals ▪ Review training materials annually - Upgrade audio visuals - Insert input from operating departments - Insert real-life examples ▪ Provide skills development for trainers
Contractors	▪ Provide information on hazardous materials at worksite ▪ Request information on contractor's materials ▪ Conduct pre-project meetings ▪ Obtain written agreement to abide by all safety and health rules	▪ Audit worksite - Contractor information matches containers on site - Containers are labeled - Unused materials/empty containers removed

*Chemical manufacturers, distributors, and importers only.

On-Going Costs

- Maintaining chemical lists
- Procuring replacement labels
- Auditing program
- Reviewing MSDSs
- Copying and distributing MSDSs

- Training

5.5 Record Keeping

One of the most difficult aspects of hazard communication program management is record keeping. Early in the planning stages, the questions of how and where the records will be kept, and how files will be updated and purged must be addressed. The major recordkeeping needs are outlined in Section 5.4.

At a minimum, the following documentation must be kept for MSDSs:

- Sources for each item of information on MSDS (in-house MSDS or chemical manufacturers)
- Collections of MSDSs at the workplace
- Documentation of attempts to obtain MSDSs

The record-keeping system used may be hard copy or electronic; however, the plan must include a mechanism for updating and purging files.

6 MATERIALS/SYSTEMS/EQUIPMENT

6.1 Hazardous Material Information Systems

Hazard communication programs for medium to large manufacturers and employers require gathering and communication of large volumes of data. Because the Hazard Communication Standard is performance oriented, the system used is left to the discretion of the individual company and a multitude of options are available.

6.1.1 MSDS Preparation

The federal standard provides manufacturers with flexibility in MSDS format and presentation. The only strict requirement for MSDS preparation is that all of the required elements as discussed in Section 3.1.2 be included on the sheet. This flexibility has prompted most manufacturers of chemicals and many suppliers and distributors to develop computer-generated MSDSs for their products. Computer generation allows for easier updating of MSDSs as new technical information becomes available and provides a convenient method for storing and retrieving MSDSs for a wide variety of chemicals.

Some individual publishers have developed data bases of thousands of MSDSs for sale to manufacturers and users of chemicals. These data bases typically include MSDSs for pure chemicals only. Therefore, they are not a complete answer to MSDS preparation, but they can provide a starting point. Because the manufacturers, suppliers, distributors, or importers will have the name of their companies on the MSDS and are ultimately reponsible for hazard determination and other contents, they should verify all information prior to distribution of MSDSs.

6.1.2 Hazardous Material Tracking

Employers who use hazardous chemicals are faced with the task of managing very large volumes of paper, particularly MSDSs. Once MSDSs are received and stored for existing hazardous chemicals, the employer is still confronted with the task of receiving, tracking, and storing MSDSs for new chemicals and updating MSDSs for existing chemicals. The employer will also typically track "approved" hazardous materials that may be purchased for use in the facility.

Many approaches are available for tracking MSDSs, ranging from a completely manual system to a completely automated system. Except in cases where very few chemicals are used, some form of automated tracking is typically needed. Tracking of MSDSs by a data base program allows for (*a*) cross-referencing and updating of hazardous chemical list, (*b*) tracking the date of the latest MSDS revision, (*c*) checking chemicals and suppliers for preapproval, and (*d*) tracking departments using each chemical for training.

A data base system facilitates communication between users and purchasing and health/safety professionals regarding chemicals used and approved. Materials approved by either a health/safety professional or review committee can be entered into the data base as "approved" so that purchasers and users of the chemical know that the material is acceptable for use. This can help avoid duplication of effort in reviewing substances and speed the purchasing process for commonly used materials. The hazardous materials tracking system will also allow tracking of departments that use each chemical. This information can be used to determine training requirements within the department. An enhancement to the system may include tracking the training (initial and update) of employees within each department. When updated MSDSs are received, the departments using the chemical can be altered to provide updated information to their employees on that chemical.

A data base tracking system can be simple or elaborate, depending on the needs of the company. Some companies have opted for almost total automation, with all MSDS information put into a computer data base system in a uniform format. The MSDS can then be retrieved at any terminal throughout the system at any location company-wide. This form of total automation is very costly, because MSDSs received from manufacturers vary in format and must be reformatted for computer input. However, over time, the cost may be offset by savings in staff time, especially in large companies with many plant locations. Employers should do a careful cost analysis to determine what benefit, if any, would be gained from a fully automated system. The major advantage of this type of system is the ease in use. The MSDS user need only understand one MSDS format to access the information easily through a computer terminal. A disadvantage is the increased potential liability the end user incurs by changing the format of, and interpreting information from, the original MSDS.

6.1.3 Related Programs

An information system set up for hazard communication compliance should be flexible enough to use in complying with other related regulatory requirements, such as those discussed in Section 4. Additionally, other state or local regulations may require information tracking that could be easily incorporated into a hazardous material tracking system set up for hazard communication.

An example of another use of the hazard communication tracking information is the requirement of Section 311 of SARA Title III to produce copies of MSDS chemicals. In addition, Section 312 of SARA Title III requires the submission of an inventory of hazardous chemicals with additional information regarding quantities present at a facility, daily use, locations, and storage information. This information can be submitted either as an aggregate by OSHA categories of health and physical hazards (Tier I) or by individual chemicals (Tier II). Including the information required for producing SARA Title III reports with the hazard communication data base helps avoid duplication of efforts. The hazard communication data base may also be useful as a starting point for preparing a Toxic Chemical Release Form as required by Section 313 of SARA Title III.

If the information tracking system includes a provision for tracking training, other training programs may be added to the system for tracking purposes. This may be particularly useful when training programs are combined. For example, a hazard communication training program may be expanded to include training of employees who handle hazardous waste, as required by RCRA regulations. The training may be combined, but records must be kept to show compliance with both sets of regulations. An automated system for tracking this is helpful, particularly when many employees require training.

6.2 Training Tools

The performance-oriented Hazard Communication Standard allows the employer to determine "how" training will be conducted and specifies only "who," "when," and/or "what" topics training must be conducted. This provides employers with a great deal of flexibility to implement a training program that is both effective and practical for their organizations. Obviously, different approaches may be applicable depending on the number of employees, number of worksites, number of chemicals, projected turnover of personnel, and the need for refresher or update training.

There are a variety of training techniques and tools available. Typically, a combination of techniques will enhance the learning process and allow for more effective training. The following are some of the most common techniques used for hazard communication training:

- Group lectures (This method is used when all members of a large group need to learn the same information and the information is conducive to mass display.)
- Group and individual role plays (Group members play roles and make decisions based on real-life situations. This method would be used to simulate emergency evacuations and other emergency procedures that must be performed within a critical time frame.)
- Emergency equipment demonstrations (People learn by watching demonstrations and practicing behavior with fellow class participants.)

- Self-paced programmed instruction (The worker responds to written questions or situations at his/her own pace and receives immediate feedback.)
- Charts and/or diagrams (Pictures and diagrams reinforce learning.)
- Audiotape/slides (This media allows flexibility of site-specific photographs that are readily available for playback and review.)
- Film/videotape (An appealing alternative to lectures, graphics and animation can be used to hold participants' interest and provide valuable information for new behaviors or performance.) (60)

Training effectiveness can be evaluated by a number of different methods. As a minimum, training should be documented using a sign-in roster, both to document training and to track each employee's need for retraining. Additionally, the instructor may test effectiveness of the training by administering either oral or written tests at the end of each session. An auditor or supervisor may later verbally quiz employees during field visits to test for retention of information. This method is often used by OSHA compliance officers. Additionally, the employee's work practices may be observed to determine whether principles taught during training are being implemented in the workplace.

6.3 Material Labeling

The standard allows flexibility in the format and content of labels, provided they include the required information.

6.3.1 Manufacturer's Responsibility

As discussed in Section 3.2, manufacturers must ensure that each container of hazardous chemical leaving the workplace is labeled with (1) identity of the hazardous chemical, (2) appropriate hazard warning, and (3) name and address of the chemical manufacturer, importer, or other responsible party. For substances shipped in a tank truck or rail car, the appropriate label or label information may be either posted on the vehicle or attached to the shipping papers (61).

What constitutes an "appropriate hazard warning" is the portion of the labeling requirement that is most open to interpretation. It is not necessarily appropriate to include every hazard listed on the MSDS on the label (61).

The identity of the chemical must match the MSDS chemical identity so that an employee seeking additional information can link the material in a container with an MSDS. The label must provide an immediate warning of specific acute and chronic health hazards as well as physical hazards. A precautionary statement such as "caution," "harmful," or "harmful if inhaled" does not provide sufficient information on the actual hazard (61). The hazard warning should include target organ effects information, such as "causes lung damage when inhaled," when a specific target organ effect is known. Some manufacturers include additional information, such as emergency first aid procedures, that can prove useful. These, however, are not mandatory for compliance.

There are numerous labeling systems in use in industry. Labeling systems that rely on numerical or alphabetic codes to define hazards do not provide target organ effect information and are not appropriate for shipped containers unless additional narrative information is also included on the container's label (61). In some cases of chemicals regulated by OSHA in a substance-specific health standard, the warning label must meet the requirement of that standard.

The manufacturer's label on shipped containers must not conflict with the requirements of the Hazardous Materials Transportation Act and Department of Transportation regulations (38).

6.3.2 Employers' Responsibilities

As discussed in Section 3.3.3, employers are responsible for ensuring that all containers in their facility are properly labeled with (*a*) the identity of the hazardous chemical and (*b*) appropriate hazard warning.

The employer has even more flexibility than the manufacturer regarding labeling containers. In the case of individual stationary process containers, employers have the option of using signs, placards, process sheets, batch tickets, operating procedures, or other written materials in lieu of affixing labels directly on the container. The alternative method used must identify the container to which it is applicable and convey the required label information. The written materials must be accessible to employees throughout their work shift (38).

Portable containers need not be labeled if the material is transferred to the portable container from a labeled container and is intended for immediate use by the employee who performed the transfer.

"The employer shall not remove or deface labels on incoming containers of hazardous chemicals, unless the container is immediately marked with the required information" (38).

Labeling systems that include numerical or alphabetic codes to convey hazards may be permissible for in-plant labeling provided the entire hazard communication program is effective. The target organ effects must be communicated to employees in some manner. Proper training on a specific labeling system is essential to its effectiveness in promoting safe handling and use of chemicals.

6.3.3 Carcinogen Labeling

General guidance for assessing labels for substances that are potentially carcinogenic are provided in the OSHA Instruction Standard 2.1 (61).

"In general, those chemicals identified as being 'known to be carcinogenic' and those substances that may 'reasonably be anticipated to be carcinogenic' by National Toxicology Program must have carcinogen warnings on the label" (61).

Substances regulated by OSHA as carcinogens must be labeled in accordance with the substance-specific standard.

If chemicals studied by IARC have one or more specific study that includes positive human evidence indicating carcinogenic potential, its label must contain hazard warnings.

6.3.4 American National Standards Institute (ANSI) Labeling

The American National Standards Institute (ANSI) published a voluntary labeling standard in 1988 as guidance for precautionary information on chemical container labels (62). ANSI recommended that labels provide much more detailed precautionary information than that required by OSHA in the Hazard Communication Standard. Additional information included by ANSI includes:

- Signal word
- Precautionary measures
- Instructions in case of contact or exposure
- Antidotes
- Notes to physicians
- Instructions in case of fire and spill or leak
- Instructions for container handling

ANSI also requires the identity of the material and a statement of the hazard to be included, as does the Hazard Communication Standard. The ANSI standard does not include the supplier's name and address, and therefore would not meet the OSHA requirement for shipped containers.

The additional information provided by the ANSI labels is generally available on the MSDS. The possible dilution of the immediate warning must be considered before deciding to use the ANSI labeling system.

7 PROGRAM AUDIT

7.1 Program Appraisal

Every hazard communication program should be periodically reviewed to determine its effectiveness and degree of compliance with the OSHA standard. The best form of review is a formal audit, performed either by an independent outside party, such as a consultant, or an internal knowledgeable individual, such as an industrial hygienist.

The audit should include a review of records to determine the adequacy of the documentation of the program, as well as a determination of how well the program is being performed. These items are discussed in Sections 7.2 and 7.3. A checklist for use during the audit is provided in Section 7.4.

7.2 Record Review

A key part of every OSHA compliance officer's visit is a review of an employer's written hazard communication program. As discussed earlier, the written hazard communication program provides an explanation of all the elements of an employer's hazard communication program and how they are implemented. Therefore, a logical first step of any audit is a review of the written hazard communication program. The program must include, at a minimum: (*a*) a list of hazardous chemicals in the workplace, (*b*) an explanation of how labeling requirements will be met, (*c*) an explanation of how MSDS requirements will be met, (*d*) an explanation of how employee information and training requirements will be met, (*e*) methods of informing employees of the hazards of nonroutine tasks and chemicals contained in unlabeled pipes, and (*f*) methods for informing contractors of hazardous chemicals and protective measures. The MSDS is an essential ingredient of every hazard communication program. For manufacturers, backup information should be available for MSDS preparation. The source of hazard determination and protective measure information should be available. In some cases each individual MSDS may warrant auditing.

For an employer using chemicals, the availability of MSDSs to employees should be audited. Documentation to be reviewed includes (1) MSDS copies in the workplace, (2) written requests to suppliers for missing or inadequate MSDSs, and (3) tracking method for new and updated MSDSs.

Hazard communication training should be well documented and therefore a review of training records is an important part of the records review. The records should include a description of (1) what was covered at each training session, (2) who was in attendance, and (3) when the training occurred. Training of new employees and employees transferred to a department with different chemicals should also be audited. Documentation of follow-up training for new hazards in the workplace should also be reviewed. Refresher training is required at specific time intervals in some states. The documentation should be reviewed for compliance with state as well as federal OSHA requirements.

Specific labeling systems should be audited for their compliance with OSHA's requirements. Manufacturers must include (1) identity of the material, (2) hazard warnings, and (3) the name and address of the manufacturer, supplier, importer, or other responsible party. Containers used at a specific facility, such as bulk tanks or small containers used at work stations, must include the first two items. In both cases, the labels, particularly the hazard warning, should contain specific information regarding target organ effects, when applicable.

7.3 Performance

The Hazard Communication Standard is a "performance standard." Thus employers have much latitude in determining how to comply with specific provisions of the standard.

The most important test of its effectiveness is whether a program actually results in a more aware worker. The OSHA compliance officer tests this by actually interviewing employees at random and asking them about the chemicals in their workplace (61). A good audit will also include this spot check. It is not feasible to interview every worker, but a few employees in several different work settings should be interviewed. For example, maintenance, process control, in-house construction workers, and outside contractors should be included in the interview (when possible), in addition to production employees with routine exposure. Employees should be asked key questions about hazard communication such as:

- What are the potential hazards of that chemical?
- What protective equipment do you use when working with that chemical?
- Where would you find the MSDS?
- What would you do if some of that chemical spilled?
- Did you receive any training regarding chemicals?

Other elements of the program, such as MSDS and labeling, should be checked in the workplace for (1) compliance with OSHA standards, (2) implementation in the workplace, and (3) actual performance. This part of the audit should be through a random check, as with training. A number of chemicals should be selected in the workplace. The adequacy of the labels on the containers should be checked at the work station. The MSDS should then be located and checked for completeness. The MSDS must be easily accessible to the worker, not locked in a supervisor's office. The protective measures and air monitoring data should also be reviewed as appropriate.

7.4 Checklist

The checklist shown in Figure 6.3 may be used a guideline for a hazard communication audit. This checklist may be augmented to include state and local regulations. It also can be supplemented to include related regulations such as SARA Title III. The checklist is prepared in two parts, one for employers and one for suppliers. The appropriate portions may be used for specific audits.

8 CONCLUSION

The Hazard Communication Standard has been called a performance-oriented standard, in that the specific means of accomplishing the requirements of the standard are left to the employer. The measure of an effective hazard communication program, therefore, is whether employees recognize the nature and extent of the hazards posed by chemical substances and understand what steps must be taken to protect themselves.

All components of an effective safety and health program must be integrated into the overall management structure within a company. The level of commitment, the implementation strategies, the measurement of effectiveness, funding, and responsibility for program implementation must all be considered when evaluating the effectiveness of a hazard communication program.

Employers with effective hazard communication programs will reap the benefits of enhanced employee and community relations. Overall, OSHA expects the standard to benefit the American workforce in the ways indicated below:

PART I
HAZARD COMMUNICATION AUDIT - EMPLOYERS

WRITTEN PROGRAM

A. Statement of Purpose *(optional)*

[ ] Purpose of Hazard Communication Program

Comments: ______________________________

B. Location: ______________________________

[ ] Available for review upon request to employees and OSHA

Comments: ______________________________

C. Labeling

[ ] Who is responsible for checking and maintaining original labels?
[ ] Who is responsible for making and maintaining inplant labels?
[ ] Description of inplant labeling system for secondary containers.
[ ] Description of labeling system for bulk storage tanks *(if applicable)*.
[ ] Description of alternative systems used for batch processes *(if applicable)*.
[ ] Description of fixed labels *(if applicable)*.

Comments: ______________________________

D. MSDS

[ ] Who is responsible for requesting MSDS and keeping files up to date?
[ ] How are MSDS obtained for every Hazardous Substance?
[ ] Where are MSDS kept?
[ ] Accessibility to employees.
[ ] Review of MSDS for omissions (checklist).
[ ] How are missing MSDS requested?
[ ] MSDS alternatives *(if used)*.

Comments: ______________________________

E. Training

[ ] Who is responsible?
[ ] Initial training plan.
[ ] New employee training.
[ ] Refresher training plan.
[ ] Update of training for new hazards.
[ ] Documentation of training.

Figure 6.3. Hazard Communication Audit—Employers and Suppliers

Comments: ______________________________

F. List of Hazardous Substances

[ ] Included in written program.
[ ] Identities match MSDS and labels.
[ ] Procedure to update list.
[ ] Locations listed by work area.

Comments: ______________________________

G. Non-Routine Tasks and Unlabeled Pipes

[ ] How employees will be informed.
[ ] Who is responsible for informing employees?
[ ] Incorporation of existing procedures.

Comments: ______________________________

H. Contractors

[ ] How they will be informed of hazards and protective measures.
[ ] How they will provide MSDS for hazardous substances they bring onsite.
[ ] Who is responsible for informing contractors and obtaining MSDS?

Comments: ______________________________

I. Hazard Determination *(optional)*

[ ] How chemicals are evaluated (e.g. rely on supplier or do own evaluation).
[ ] Who is responsible for evaluating chemicals?

Comments: ______________________________

LABELING

A. Labeling System

[ ] Include product identity and hazard warning (health and physical hazards).
[ ] Labels displayed and legible.
[ ] Method for replacing damaged labels.
[ ] Checking labels on incoming containers.

Comments: ______________________________

Figure 6.3. Continued

MATERIAL SAFETY DATA SHEETS

[ ] MSDS available for each hazardous substance.

[ ] Accessible location to employees on all shifts.

[ ] Requests for missing MSDS documented.

[ ] Requests for update of incomplete MSDS documented.

[ ] Review of MSDS for completeness.

[ ] MSDS in all files up-to-date.

[ ] Method for updating MSDS files.

[ ] Provisions for providing MSDS to employees, physicians, and/or representative upon request.

Comments: ______________________________

____

____

INFORMATION AND TRAINING

A. Training Curriculum Include:

[ ] Explanation of MSDS and information it contains.

[ ] MSDS contents for each substance or class of substances used.

[ ] Explanation of requirements of Hazard Communication Standard.

[ ] Location and availability of written program.

[ ] Location of operations where hazardous substances are used.

[ ] Observation and detection methods for hazardous substance presence or release.

[ ] Physical and health hazards of substances used in work area.

[ ] Protective measures for work area.

[ ] Details of employer's hazardous communication program, including labeling and MSDS program.

Comments: ______________________________

____

____

B. Training Update

[ ] Periodic refresher training.

[ ] Provision for informing employees within 30 days of receiving a new or revised MSDS which indicates significantly increased risks or protective measures.

Comments: ______________________________

____

____

C. Training Documentation

[ ] Attendance lists.

[ ] Tests of comprehension (Pre/Post)

Comments: ______________________________

____

____

Figure 6.3. Continued

PART II
HAZARD COMMUNICATION AUDIT - SUPPLIERS

HAZARD DETERMINATION

[ ] Who is responsible for hazard determination?

[ ] List of information sources used.

[ ] Written procedure for evaluating all products.

[ ] Plan for getting updated information.

Comments: ______________________________

____

____

MATERIAL SAFETY DATA SHEETS

[ ] Who is responsible for:

  [ ] Preparing and updating MSDS.

  [ ] Distributing MSDS.

  [ ] Replying to MSDS requests from customers.

[ ] Complete and accurate MSDS written for each product.

[ ] Copies of MSDS provided to each purchaser (even if not requested)

[ ] Method of providing MSDS: (e.g., with bill of lading, mailed to purchaser).

____

____

____

[ ] Procedure for updating MSDS when new information becomes available.

Comments: ______________________________

____

____

LABELS

[ ] Who is responsible for:

  [ ] Determining label working.

  [ ] Updating label wording.

  [ ] Making and maintaining labels.

[ ] Product containers labeled with:

  [ ] Name and address of manufacturer or supplier.

  [ ] Product identity (same as on MSDS).

  [ ] Hazard Warning (specific physical and health hazards).

TRADE SECRETS *(if applicable)*

[ ] Who is responsible for making trade secret determination?

[ ] Justification documentation.

[ ] Procedure for applying to OSHA for trade secret status.

[ ] Procedure for providing trade secret chemical identities in emergencies.

[ ] Procedure for providing trade secret chemical identities in non-emergencies under specified conditions.

[ ] Standard Confidentiality Agreement?

Figure 6.3. Continued

- Reduction of accidents, injuries, and illness related to exposure to hazardous materials
- Reduction of workers' compensation cost
- Reduction of lost time accidents
- Improved productivity
- Improved employee morale

These benefits notwithstanding, communicating technical information is not without risk. Communication on any subject can be plagued with misinterpretation. It is doubly so for issues that have become emotionally charged.

However, the consequences of noncompliance with the Hazard Communication Standard include OSHA citations and penalties (63), adverse publicity, and increased liability for failure to inform employees of hazards that, in extreme cases, can involve criminal prosecution.

To the employer, the right-to-know laws are really "duty-to-warn" requirements. Failure to warn employees of a potential health hazard may be interpreted as a negligent violation of the workers' rights. This may allow workers to pursue civil actions against the employer outside the protection of the workers' compensation system.

In other situations, employers have been held criminally negligent when information was withheld from the employee who subsequently sustained injuries. In a court case in Illinois in 1985, top officers of a company were convicted in the death of a worker by chemical poisoning. Failure to warn the employee of the hazards of the chemical figured prominently in the case.

REFERENCES AND NOTES

1. The Occupational Safety and Health Act of 1970 (P.L. 91-596, 91st Cong., S. 2193, December 29, 1970) is codefined at 29 U.S.C. 651 et seq.
2. U.S. Department of Health, Education, and Welfare, U.S. Public Health Service Center for Disease Control, National Institute for Occupational Safety and Health, *A Recommended Standard . . . An Identification System for Occupationally Hazardous Materials*, P.B., 246 698, Cincinnati, OH, 1974, p. 1.
3. A hazardous material for the purposes of this document was defined as ". . . a substance or mixture of substances having intrinsic properties capable of producing adverse effects on the health or safety of a worker" (p. 1). Risk, defined as ". . . the probability of adverse effects occurring in a defined set of circumstances" (p. 1), was not a factor in the definition of a hazardous material. The philosophy behind these definitions has been central to some of the major controversy surrounding the Hazard Communication Standard.
4. See the *Federal Register*, **48**(228), November 25, 1983, p. 53280 (40FR 53280). Subsequent references will use the notation in parentheses.
5. U.S. Department of Health, Education, and Welfare, U.S. Public Health Service Center for Disease Control, National Institute for Occupational Safety and Health, *The Right*

to Know: Practical Problems and Policy Issues Arising from Exposures to Hazardous Chemical and Physical Agents in the Workplace, Cincinnati, OH, 1977.

6. The material safety data sheet was originally Form OSHA-20, which was required in Section 57, paragraphs (b) and (c), Parts 1915, 1916, and 1917 of the *OSHA Safety and Health Regulations*, promulgated under Public Law 85-742 and adopted under Public Law 91-596 for Ship Repairing, Shipbuilding, and Shipbreaking.
7. *BNA Occupational Safety and Health Reporter, Current Report*, July 25, 1979, pp. 100–101.
8. *Ibid.*, September 20, 1979, pp. 361–362.
9. *Ibid.*, July 17, 1980.
10. *Ibid.*, November 27, 1980, p. 665.
11. *Ibid.*, various articles, September to December 1980.
12. Published in the *Federal Register*, 46FR 4412, January 16, 1981, reprinted in the *BNA Current Report*, January 16, 1981, reprinted in the *BNA Current Report*, January 22, 1981, pp. 875–915.
13. *BNA Occupational Safety and Health Reporter*, February 5, 1981, p. 1230.
14. *Ibid.*, February 19, 1981, p. 1265.
15. *Ibid.*, February 26, 1981, p. 1291. Also see April 9, 1981; April 30, 1981; and May 28, 1981.
16. *Ibid.*, March 26, 1981, pp. 1365.
17. *Ibid.*, June 11, 1981, pp. 38–39.
18. *Ibid.*, July 23, 1981, pp. 150–1151; and July 30, 1981, pp. 173–174.
19. *Ibid.*, November 5, 1981, p. 443.
20. See the *Federal Register*, 47FR 12092.
21. *BNA Occupational Safety and Health Reporter, Current Report*, various articles, March 1982 to November 1983.
22. See the *Federal Register*, 48FR 53280-53348.
23. *BNA Occupational Safety and Health Reporter, Current Report*, various articles, December 1983 to June 1984. A summary of these suits is provided on pages 19–20, June 14, 1984.
24. *Ibid.*, various articles, June 1984 to December 1984.
25. *Ibid.*, September 27, 1984, pp. 356–357.
26. *Ibid.*, November 15, 1984, pp. 463–464.
27. *Ibid.*, January 10, 1985, p. 579, and February 7, 1985, pp. 701–702.
28. *Ibid.*, February 14, 1985, pp. 718–719.
29. *Ibid.*, March 7, 1985, pp. 763–764.
30. *Ibid.*, April 11, 1985, pp. 879–880.
31. *U.S. Steel Workers* v. *Auchter*, 763 F.2d 728, 12 OSHC 1337 (3rd Cir 1985). The decision is summarized in *BNA Occupational Safety and Health Reporter*, May 30, 1985, p. 1020 and June 6, 1985, pp. 3–4.
32. Central to the issue of preemption is the question of whether the Hazard Communication Rule was a standard issued under Section 6 of the Occupational Safety and Health Act, or a regulation issued under Section 8. OSHA contended that, as a standard, the Hazard Communication Rule would preempt state laws in nonstate plan states; however, if it

was determined to be a regulation, it would not preempt. The pending litigation was expected to make this distinction and thus help to clarify the issue.

33. *BNA Occupational Safety and Health Report, Current Report*, August 22, 1985, pp. 251–252.
34. See the *Federal Register*, 50FR 48750 and 50FR 48794, November 27, 1985. Full text reprinted in *BNA Occupational Safety and Health Reporter, Current Report*, Vol. 15, pp. 572–582.
35. *BNA Occupational Safety and Health Reporter, Special Report*, April 3, 1986, pp. 1108–1110.
36. U.S. Court of Appeals for the Third Circuit, January 21, 1987, Steelworkers v. Pendergrass (Nos. 83-3554 et al.). This motion was summarized in *BNA Occupational Safety and Health Reporter, Current Report*, Vol. 16, January 28, 1987, pp. 939.
37. See United Steelworkers v. Pendergrass, 819 f. 2nd 1263, 1268, 13 OSHC 1305, 1308-1310 (3rd Cir 1987). Summary of the decision and comments are found in *BNA Occupational Safety and Health Reporter, Current Report*, Vol 17, June 3, 1987, p. 3.
38. See the *Federal Register*, 52FR 31852, August 24, 1987, OSHA Hazard Communication; Final Rule. Full text of preamble reprinted in *BNA Occupational Safety and Health Reporter*, Vol. 17, August 26, 1987, pp. 520–544.
39. See Notice of Proposed Rulemaking, published on August 8, 1988, in the *Federal Register*, Vol. 53, No. 152 for a summary of these events and actions.
40. *BNA Occupational Safety and Health Reporter, Current Report*, various articles from September 1987 to August 1988.
41. *Ibid.*, Vol. 17, December 2, 1987, pp. 995–996; December 9, 1987, pp. 1045–1046; January 6, 1988, p. 1360.
42. See *Associated General Contractors of Virginia* v. *OSHA* (No. 87-1185), Associated Builders and Contractors, Inc. v. Brock (No. 87-1582), *United Technologies Corporation* v. *OSHA* (No. 87-4143), and *National Grain and Feed Association* v. *OSHA* (No. 87-1603).
43. These cases were consolidated in the U.S. Court of Appeals for the District of Columbia Circuit. They were transferred to the U.S. Court of Appeals for the Third Circuit Court on May 20, 1988, because the Hazard Communication Standard was promulgated in response to orders of the Third Circuit Court.
44. *BNA Occupational Safety and Health Report, Current Report*, June 22, 1988; June 29, 1988, pp. 459.
45. *Ibid.*, Vol. 18, July 27, 1988, p. 567.
46. See OSHA Instruction CPL 2-2.38B. The full text of this instruction is reprinted in *BNA Occupational Safety and Health Reporter*, Vol. 18, August 24, 1988, pp. 742–766. A summary of the court decision is on p. 723 of the same issue.
47. *BNA Occupational Safety and Health Reporter, Current Report*, Vol. 18, September 7, 1988, pp. 788–789, September 21, 1988, p. 852.
48. *Ibid.*, Vol. 18, January 25, 1989, pp. 1515–1516.
49. See the *Federal Register*, 54FR 6886. Notice summarized in *BNA Occupational Safety and Health Reporter, Current Report*, Vol. 18, February 22, 1989, p. 1651.
50. Employers whose workers handle only sealed containers must keep labels intact; make MSDSs available to employees, if an MSDS is received. Obtain an MSDS if requested

by an employee; train employee on spill and leak response, and the hazards associated with the general classes of chemicals in the workplace.

51. See the Preamble to OSHA's Final Rule to Expand Scope of Hazard Communication Standard, 52FR 31852, August 24, 1987, and the standard itself, for a detailed explanation of these limitations and others.
52. A proposed change was published in the August 8, 1988 *Federal Register* to exclude fluids and particles from this definition.
53. 52FR 31852, August 24, 1987, OSHA Hazard Communication; Final Rule, 29CFR 1910, 1915, 1917, 1918, 1926, and 1928, paragraph (b)(4). The rationale is explained in the preamble, see p. 31861.
54. *Ibid.*, paragraph (h).
55. *Ibid.*, paragraph (c). The rationale for this statement is explained in the preamble, p. 31864.
56. Enacted by Public Law 99-499, October 17, 1986.
57. On October 12, 1989, the EPA published an interim final rule extending the limitations for another year. During this period comments will be received.
58. The exact procedures to be followed should be reflected in the emergency procedures.
59. Because of the nature of their work, laboratories have been exempted from some of the requirements of the hazard communication regulations for the chemicals used in the laboratory. This is a list of ways in which a laboratory may function as a participating member of an overall company hazard communication program.
60. *Hazard Communication Training Programs and Their Evaluations*, Golle & Holmes Companies, March, 1986.
61. OSHA Instruction Std. 2.1, January 20, 1987.
62. ANSI Z129.1-1988, American National Standard for Hazardous Industrial Chemicals Precautionary Labeling.
63. From January, 1985 until May, 1989, OSHA issued more than 49,000 citations for violations of the Hazard Communications Standard. Of these, no written program and inadequate training and information were the most frequently cited, followed by inadequate labeling and missing or inappropriate MSDSs.

BIBLIOGRAPHY

Several of the books or journals listed below are also cited in the References and Notes; other books are listed because they provide excellent background information on the topic.

1. *Occupational Safety and Health Reporter*, The Bureau of National Affairs, Inc., Vol. 6 (June 3, 1976 to May 26, 1977) through Vol. 19 (June 7, 1989 to August 30, 1989).
2. R. Worobec, Ed. *Chemical Right-to-Know Requirements: Federal and State Laws and Regulations: A Status Report*, The Bureau of National Affairs, Inc., Washington, DC 1984.
3. American National Standards Institute, Inc., American National Standard for Hazard-

ous Industrial Chemicals—Precautionary Labeling, ANSI z129.1-1988, ANSI, New York, 1988.

4. S. G. Hadden, "Providing Citizens with Information About Health Effects of Hazardous Chemicals," *J. Occup. Med.* **31**(6) (June 1989).
5. The Bureau of National Affairs, Inc., "Hazard Communications Training Programs and Their Evaluation," Washington, DC, 1986.
6. OSHA, "Occupational Safety and Health Administration Directives Pertaining to the Hazard Communication Standard," reprinted by the Bureau of National Affairs, Inc., Washington, DC, 1988.
7. American Chemical Society, *Chemical Risk Communication: Preparing for Community Interest in Chemical Release Data*, ACS, Department of Government Relations and Science Policy, Washington, DC 1988.
8. J. C. Silk, "Hazard Communication: Where Do We Go From Here?," *Appl. Ind. Hyg.*, **3**(1), F27–28.
9. Occupational Safety and Health Standards, Subpart C—General Safety and Health Provisions, 1910.20, Access to employee exposure and medical records, 53FR 38162, September 29, 1988.
10. Occupational Safety and Health Administration Rules Concerning OSHA Access to Employee Medical Records, 1913.10, Rules of Agency Practice and Procedure Concerning OSHA Access to Employee Medical Records, 51FR 24527, July 7, 1986.
11. B. M. Eisenhower, T. W. Oaks, and H. M. Braunstein, "Hazardous Materials Management and Control Program at Oak Ridge National Laboratory—Environmental Protection," *Am. Ind. Hyg. Assoc. J.*, **45**, 212–221 (April 1984).
12. L. W. Keller, K. L. Schaper, and C. D. Johnson, "A Hazardous Materials Identification System for the Coatings and Resins Industry," *Am. Ind. Hyg. Assoc. J.*, **41**, 901–907 (December 1980).
13. W. E. Porter, C. L. Hunt, Jr., N. E. Bolton, "A System for Labeling and Control of Toxic Materials in a Large Research Facility," *Am. Ind. Hyg. Assoc. J.*, **38**, 51–56 (January 1977).
14. Emergency Planning and Community Right-to-Know Act of 1986, enacted by Public Law 99-499, October 17, 1986.
15. A. B. Waldo, "Guide to SARA Title III Deadline: March 1, 1988," Thompson Publishing Group, Washington, DC, 1988.
16. Comprehensive Environmental Response Compensation and Liability Act of 1980, enacted by Public Law 96-510, December 11, 1980, 94 Stat. 2767; 42 USC 9601 et seq.
17. S. A. Bokat, and H. A. Thompson, III, *Occupational Safety and Health Law*, The Bureau of National Affairs, Washington, DC, 1988.
18. Commerce Clearing House, Inc., *Occupational Safety and Health Hazard Communication Federal/State Right-to-Know Laws*, Chicago, IL, 1985.
19. U.S. Department of Health, Education, and Welfare, U.S. Public Health Service Center for Disease Control, National Institute for Occupational Safety and Health, *A Recommended Standard . . . An Identification System for Occupationally Hazardous Materials*, P.B. 246 698, Cincinnati, OH, 1974.

20. U.S. Department of Health, Education, and Welfare, U.S. Public Health Service Center for Disease Control, National Institute for Occupational Safety and Health, *The Right to Know: Practical Problems and Policy Issues Arising from Exposures to Hazardous Chemical and Physical Agents in the Workplace*, Cincinnati, OH, 1977.
21. G. G. Lowry, and R. C. Lowry, *Handbook of Hazard Communication and OSHA Requirements*, Chelsea, MI, 1985.

CHAPTER SEVEN

Industrial Hygiene and the Law

Ralph E. Allan, J.D., C.I.H.

1 INTRODUCTION

The complex area of toxic substances law and regulation is one of the fastest growing areas of the law and industrial hygiene activity is an inherent component of this exploding legal area. Society is mandating that irresponsibility in reference to toxic substance management be dealt with forthwith in order to protect the current and future health of all of the earth's population. The occupational and environmental health issues are complex and in many instances rest largely upon inferences derived at the very frontiers of science and technology. Lawyers and courts assume their roles in adjudicating complex social issues not easily resolvable whether dealing in the areas of administrative, civil, or criminal law. It is therefore important that the industrial hygienist in the process of anticipation, recognition, evaluation, and control of health hazards in the workplace and community understand the legal system in order to assist more favorably in unraveling some of the most complicated legal issues that have ever needed resolution in our society. It is extremely important to analyze the role of law and its interface with the science and art of industrial hygiene.

Lawyers and industrial hygienists do share a common goal of problem solving. The initial aspect of problem solving is understanding the problem in order to provide the optimum in decision making. Therefore, understanding the legal system, the vehicle for legal problem solving, becomes especially important for the industrial hygienist as a precursor in order to clear the way for efficient attorney–industrial hygiene team effort for toxic substance issue resolution.

Patty's Industrial Hygiene and Toxicology, Fourth Edition, Volume 1, Part A, Edited by George D. Clayton and Florence E. Clayton
ISBN 0-471-50197-2

It is the purpose of the following material to provide more awareness of the legal system and its interaction with industrial hygiene activity.

2 DEFINITION OF INDUSTRIAL HYGIENIST

The title of industrial hygienist has been used by many throughout the years, however, The Board of Directors of the American Board of Industrial Hygiene has defined the industrial hygienist as a person having a college or university degree or degrees in engineering, chemistry, physics, or medicine or related biological sciences who, by virtue of special studies and training, has acquired competence in industrial hygiene (1). The board has thus defined what an industrial hygienist is and their definition emphasizes a strong science education. The board further defines a Certified Industrial Hygienist (C.I.H.) using the aforementioned definition with the amended clause "and has successfully completed the CORE examination in combination with either the comprehensive practice or as Aspect Examination" (2). Thus use of the C.I.H. designation requires a strong science education plus successful completion of an examination.

The Occupational Safety and Health Administration (OSHA) safety and health standards provide that "any equipment and/or technical measures used for this purpose must be approved for each particular use by a competent industrial hygienist or other technically qualified person" (3). OSHA has recognized the importance of professional practice as a major component of hazard control and worker protection; however, OSHA has not elaborated any further concerning this matter.

3 THE CRITERIA OF A PROFESSIONAL

There are certain criteria that can be used to judge whether or not a field of endeavor qualifies as a profession. The term profession originally referred only to theology, law, and medicine, but as applications of science and learning were extended, other vocations are now considered professions. The name "professional" thus implies professed attainments in certain knowledge, for example, law or medicine. Thus the term profession refers to a vocation or occupation, as distinguished from a trade, with unusually advanced education and skill. The skill involved in a profession is predominately mental or intellectual, rather than physical. Using this definition, one can easily defend industrial hygienists as professionals (4).

The above definition pertaining to professions requires individual members of a specific field to become experts in an identifiable body of knowledge. As experts, the individuals must convey information to others and constantly improve their expertise. A professional code of ethics for each profession is necessary to establish a code of conduct for each practitioner.

4 ETHICAL RESPONSIBILITIES

The question of professional liability cannot be dealt with effectively without a consideration of ethics. It seems that ethical considerations raise difficult questions; however, they are tied to legal issues, including that of proof. The American Academy of Industrial Hygiene has provided standards of ethical conduct to be followed by industrial hygienists: All industrial hygienists have an ethical responsibility to protect employee health, improve the work environment, and advance the quality of the profession. These words were taken from "The American Academy of Industrial Hygiene Code of Ethics for the Professional Practice of Industrial Hygiene" (5). The code is divided into four categories: professional responsibility, responsibility to employees, responsibility to employers and clients, and responsibility to the public.

Perhaps the most important issues are listed under the section entitled "Professional Responsibility." Number 1 under this section states that the industrial hygienist must "maintain the highest level of integrity and professional competence." Number 2 states that the industrial hygienist must "be objective in the application of recognized scientific methods and the interpretation of findings." Number 6 states that the industrial hygienist must "avoid circumstances where compromise of professional judgement or conflict of interest may arise." Underlying all of these statements is the responsibility of the industrial hygienist to employees, namely, that the industrial hygienist must "recognize that the primary responsibility is to protect the health of employees."

The establishment of guidelines that define both what constitutes an industrial hygienist and the corresponding ethical and moral requirements of the profession can help toward establishing the professional nature of the practice of industrial hygiene.

With the responsibility of acting as a professional comes the potential of being subjected to toxic tort or environmental legal liability. An industrial hygienist is constantly asked to make expert decisions when identifying the hazards of toxic chemicals to workers within occupational setting. As the field of industrial hygiene has expanded, the industrial hygienist now renders expert advice in other environmental areas, including those prescribed by statutory regulations. Because of the expert nature of the profession, legal liability can extend into both civil and criminal arenas, as well.

5 THEORIES OF POTENTIAL LIABILITY

5.1 Torts

The 1980s could well be classified as the decade of torts. Simply speaking, a tort is defined as a private or civil wrong or injury (6). Because of the increase in occupational health tort litigation, more and more lawyers are becoming educated in environmental and occupational health issues. The complexity of the issues and subsequent effects of litigation regarding environmental and occupational health matters are nearly overwhelming.

Tort litigation concerning an environmental or occupational health matter can involve the issue of malpractice. Although the issue of malpractice has been used in litigation, its exact definition is somewhat unclear. The dictionary defines malpractice as professional misconduct or unreasonable lack of skill. The term has been employed in a broad sense; an Ohio court has defined malpractice as any professional misconduct, unreasonable lack of skill or fidelity in professional or fiduciary duties, evil practice, or illegal or immoral conduct (7). In professional malpractice, the injured party generally files suit under the theory of negligence.

5.1.1 Negligence

The Restatement (second) of Torts (8) defines negligence as "conduct which falls below the standard established by law for the protection of others against unreasonable risk of harm." The elements of a cause of action in negligence are:

- A duty or obligation recognized by the law requiring the actor to conform to a certain standard of conduct for the protection of others against unreasonable risks or harm
- A failure to conform to the standard required
- A reasonably close causal connection between the conduct and the resulting injury, commonly known as "legal cause," or "proximate cause"
- Actual loss or damage to the injured party

Before a plaintiff can get to a jury on the issue of negligence he or she must establish all elements of negligence. As a matter of law, the first element to be proven relates to the establishment of a duty or standard of conduct. Some negligent conduct involves unintentional harm resulting from lack of care. The duty or conduct is judged on average carefulness, what the ordinary reasonable prudent person (industrial hygienist) would have done under the same or similar circumstances. This duty or standard of care is then compared with what was actually done by the individual (industrial hygienist).

For professionals, what standard of conduct needs to be followed can only be clearly defined by other professionals in the field. Therefore associate professionals become the resource for determining the standard of care. The standard of care for all professions would seem to be basically the same, that is, the general average of professionally acceptable conduct or the learning and skill ordinarily possessed and experienced by the profession.

For example, an industrial hygienist has contracted to perform an industrial hygiene review of a degreasing operation including identification of any toxic substances, evaluation of exposures, and control of the exposures, if necessary. The contract or agreement establishes the duty of the industrial hygienist to perform the indicated services. This represents the first element involved in the cause of action for negligence—duty. If the industrial hygienist did not perform the duty indicated with a standard of care expected as a reasonable and prudent industrial hygienist would, under the same or similar circumstances, there would be a breach of the duty and therefore the second element of the cause of action for negligence would be fulfilled.

Another element related to the negligence cause of action is sufficient causal connection between the defendant's conduct and the injured party's damages. Establishing the causal relationship between the defendant's conduct and the plaintiff's damages can be difficult in toxic tort litigation. For instance, the long latent periods, the unforeseen effect of conduct undertaken years ago by persons unknown, the exposure of the plaintiff to many harmful elements over the years, and loss or spoliation of evidence tend to obscure causation.

In the example presented above, if the industrial hygienist failed to identify excessive exposure to the toxic substance present at the degreasing operation, and thereby breaches the recognized duty, there may be no clear causation element because the adverse health effects of the excessive exposure may occur due to other causes, that is, other environmental exposure not related to the degreasing operation or other disease consequences. Particularly in toxic tort litigation, there is often little proof of cause and effect, and courts tend to be skeptical of statistical correlations. In time, these problems of proof may be solved by science. The test of causation does not require absolute medical certainty.

Thus the newly developing field of toxic torts may leave the legal system no option but to be unfair on the causation issue. For example, cancers caused by a toxic chemical or agent (like ionizing radiation) may not be scientifically distinguished from the many background cases of naturally occurring cancers.

Sometimes the deleterious exposure is easy to prove; the real challenge comes in connecting the exposure to the illness. Where the disease itself is poorly understood or is of unknown etiology, the likelihood of establishing causation is slim, even with strong evidence of chronic chemical exposure. For example, what caused the plaintiff's shortness of breath? Asbestos from long-term exposure to asbestos fibers, or chronic obstructive pulmonary disease, from his two-pack-a-day history of smoking? Or what about exposure to formaldehyde? It may have caused the plantiff's running nose and tearing eyes, but did it cause gastritis or asthma?

Many treating doctors are not trained in occupational medicine and epidemiology; therefore, a treating doctor may not be able to identify toxic substances as factors in a disease process with sufficient particularity to substantiate the injured party's complaint. In the absence of many similar cases exhibiting the same symptoms, establishing causation can be difficult. If the evidence is compelling and the disease is serious, however, decisions are easier resolved.

The absence of readily identifiable acute symptoms in a person with a chronic illness after long-term exposure may not rule out a link between the original exposure and the resulting illness. If the chemical is a carcinogen to which there is no known safe level of exposure, for example, the worker may have been exposed at a level long enough to produce noticeable acute warning signs of exposure, like a rash or blurred vision. Yet that same low-level exposure may ultimately contribute to illness years later. Even where chronic disease has not appeared, a number of jurisdictions now recognize the damage element of the cause of action based upon fear of future illnesses. If such a claim is to be pursued, the lawyer must establish that there was a deleterious exposure to a substance capable of inducing chronic illness.

The following scenario is typical in trying to prove negligence in a toxic tort setting. First, the defendant had a duty and usually did something or failed to do something that has caused harm to the plaintiff. Second, there has been at least some recognition by medical authorities that exposure to the alleged substance causes some physical harm. For example, chest doctors generally agree that exposure to asbestos can cause asbestosis, mesothelioma, and lung cancer. Third, the defendant usually has reason to know of the contaminant's harmful effects. The plaintiff proves this through discovery based upon the complaint or through articles in the medical literature or in trade journals. Fourth, the plaintiff has been exposed to the contaminant, and the plaintiff claims that the exposure caused some injury or disease.

The etiology of the plaintiff's physical harm becomes a major battleground. Etiology can become the plaintiff's nemesis, however, because the plaintiff who has the burden of proof on this issue must also fight the strategy of a resourceful defense tactician.

When dealing with causation it is important to review the medical history of the plaintiff, the family background, and when the plaintiff realized that health problems might be related to the chemical exposure. It is important to amass a complete work history, ascertain whether the plaintiff is involved with certain hobbies where chemicals are used, and find out if there was more than one job involved.

The fourth and last element of a negligence cause of action is that the plaintiff must suffer actual loss or damage. Damages can include medical expenses, loss of income and other economic loss, and pain and suffering.

With regard to the negligence cause of action, the duty of an industrial hygienist is to recognize, evaluate, and control environmental and occupational stresses by virtue of special studies and training in a competent manner. The duties include keeping abreast of current statutes and regulations, using state-of-the-art equipment when measuring environmental and workplace operations, and evaluating data in a competent manner. After evaluating the situation with or without data, the industrial hygienist would have to act appropriately to identify potential hazards, including the use of proper personal protection, if applicable. The duty of the industrial hygienist could extend toward anyone whose interests may be harmed by the failure of the industrial hygienist to exercise the standard of care in a competent manner.

5.1.2 Intentional and Other Tort Actions

Intentional or fraudulent misrepresentation has been identified with the common law action of deceit. In the typical case, one will find that the plaintiff has parted with money or property of value in reliance upon the defendant's representations.

The court relies upon the following elements when dealing with intentional misrepresentation (9):

- A false representation ordinarily of fact made by the defendant
- Knowledge or belief on the part of the defendant that the representation is false
- An intention to induce the plaintiff to act or refrain from acting in reliance upon the misrepresentation
- Justifiable reliance upon the representation on the part of the plaintiff in taking action or refraining from it
- Damage to the plaintiff resulting from such reliance

If a professional is liable for negligent performances of his or her duties, liability may also be found for fraudulent misrepresentation. The issue of fraudulent misrepresentation concerning information exchange goes to the very essence of ethical and moral responsibilities of industrial hygienists. Failure to inform adequately concerning industrial hygiene issues can result in allegation of fraudulent misrepresentation.

An example of a situation where fraudulent misrepresentation may apply is as follows: One of the problems of many industrial hygiene projects is that multiple parties are involved with the project. The fact that an industrial hygienist is in a position to supervise the project creates a possibility for a conflict of interest with the client, for example, a building owner. A client building owner may employ an industrial hygiene firm without knowing that the industrial hygiene firm was employed by another contractor on the project; the court may find liability if the contractor did a poor job under the industrial hygienist's supervision, causing the building owner to pay additional money to remedy the situation.

At times, industrial hygienists will be asked to recommend another industrial hygienist or contractor for a particular project. The industrial hygienist should be careful not to make defamatory statements to a third person either by libel (written words) or by slander (usually oral).

Generally, the theory of strict liability, or liability without fault, has been limited to litigation involving products and the abnormally dangerous commercial activities of industry. A court will probably be reluctant to apply strict liability in a toxic tort case involving professionals. Strict liability will generally be found in cases involving product liability. In *LaRossa* v. *Scientific Design* (10), the court stated that "those who hire (experts) are not justified in expecting infallibility, but can expect only reasonable care and competence. They purchase service, not insurance."

5.1.3 Contracts

A contract is defined as a promissory agreement between two or more persons that creates, modifies, or destroys a legal relation. The same act by an individual can be, and very often is, both negligence and a breach of contract.

It is important to know under what theory of law the underlying suit is being filed. For instance, the statute of limitations can be different for contract versus tort law; the defenses the defendant may raise are generally altogether different in contract than in tort law, and the damages are different as well.

An industrial hygienist who occupies a specific position by contract could be held liable to anyone injured as a result of his or her failure to discharge that duty with due care. In cases where an individual is injured, the courts may look to the contract to find the existence of a duty on the part of the industrial hygienist to protect the plaintiff from injury. Thus, industrial hygienists who voluntarily, and for a fee, make it part of their business to supervise work and enforce safety regulations may be responsible for failure to discharge those duties carefully.

5.2 Workers' Compensation and Individual Liability

Historically, workers' compensation cases have been filed against an employer when an employee sustained an injury. Regardless of how the injury was sustained, whether the fault of the employer or employee, workers' compensation was the exclusive remedy. Because many states provide inadequate benefits, plaintiffs in an occupational disease case may try to find a theory that will circumvent the exclusivity provisions of the Workers' Compensation Act.

The doctrine of exclusivity limits the common law remedy that the injured employee would have had if the Workers' Compensation Act did not apply. In return for agreeing to forego other legal remedies, workers are guaranteed a swift and sure payment, although circumscribed, which covers lost wages and medical payments. To avoid the limitation of remedies, workers may seek to avoid the exclusivity provisions.

Most large companies employ full-time industrial hygienists. The industrial hygienists may be employed by the employer, the employer's parent companies, or an industrial hygiene consulting firm. Plaintiffs can circumvent the workers' compensation ban on suing employers of co-employees for negligence by asserting that the person causing the injury was not a co-employee but an independent contractor.

The injured worker may attempt to sue a co-employee or employer by reclassifying the employer as another entity or "dual capacity." The claim presents the theory that the employer is functioning toward the worker in a manner no different from the way the employer acts toward the general public. It is important that a company employing industrial hygienists have a carefully worded job description establishing employment status; an established industrial hygiene protocol and a limitation on performance of services to others outside the company; or a carefully worded description of policy where such actions are performed so as to limit dual capacity exceptions.

A contract-based negligence action against an industrial hygienist also might be raised where workers' compensation insurance companies provide safety inspections to their clients. The action might occur where the insurance company renders loss control prevention as a service to its client and not for the protection of the insurance company.

6 CRIMINAL SANCTIONS

Senior level officials at the U.S. Environmental Protection Agency (EPA) and the U.S. Department of Justice have assigned high priority to the criminal enforcement of environmental laws. The number of cases referred to the Department of Justice by the EPA has increased dramatically. Many of the indictments have included management-level employees, directors, presidents, vice presidents, or owner-operators.

Environmental protection statutes have been created to address widespread public recognition of the need to control pollution strictly. The environmental statutes seek to establish a balance among business, protecting the public health and welfare, and preserving natural resources. In the United States, Congress recognized the public's concern for violations of environmental laws by providing criminal sanctions and incarceration to ensure that the goals of the legislature are achieved (12).

6.1 History

Federal programs to prosecute environmental crimes first began in 1982. Although the programs have been in existence for a relatively short period of time, an enforcement presence has been established in nearly every geographic region of the country. Investigators and prosecutors hired to enforce environmental laws have been trained in both the complex technical and scientific areas of law.

Before initiation of the criminal program, most of the cases referred to the Department of Justice for prosecution were declined usually because they lacked merit, were insufficiently investigated, or did not receive the staff support necessary to bring the violators to trial. The initiation of the criminal program has changed the situation. The two organizations largely responsible for the success of the program are the Land and Natural Resources Division of the U.S. Department of Justice and the EPA.

The EPA began to hire its first criminal investigators in 1982. Many of the investigators had little environmental experience but were experienced criminal investigators. The Land and Natural Resources Division of the Department of Justice ultimately organized an Environmental Crimes Unit staffed by attorneys with both criminal and environmental law experience. One of the unit's main purposes has been to prosecute cases and set substantial penalties.

The majority of convictions have been against managerial level officials acting illegally in their capacity. The conduct of the typical defendant in the cases under prosecution is no different and no less serious than the conduct of one who has been convicted of traditional "white collar" or "street crime" felonies. The acts of the defendant have shown some degree of intent whether it be willful, deliberate, or premeditated, over a period of time.

The prosecutor's objective is clearly to deter illegal conduct by implementing strong penalties including incarceration. Initially, financial profit may motivate an illegal act; thus fines must exceed by a substantial amount the illegal gain of the violator. If not, noncompliance will be viewed as simply a cost of doing business (13).

6.2 The Statutes

A number of federal statutes create specific sanctions for violators of their provisions and the regulations promulgated pursuant to them. Prosecutions may be brought under statutes such as: The Resource Conservation and Recovery Act (RCRA) (14); The Comprehensive Environmental Response, Compensation, and Liability Act (CERCLA) (15); The Hazardous Substance Act (16); Occupational Safety and Health Acts (OSHA) (17); The Clean Air Act (18); and the Water Pollution Control Act (Clean Water Act) (19).

The criminal enforcement provisions of the aforementioned statutes define a liable person to include an individual and generally require that the liable person have an awareness of wrongful conduct.

Local district attorney's offices are also filing criminal charges against individuals violating health and safety statutes. The County of Los Angeles, Office of the District Attorney, Occupational Safety and Health Section was the first of its kind in the country created specifically to protect workers' rights to a safe and healthy workplace. This objective is pursued in a number of ways: aggressive prosecution of cases referred by the California Division of Occupational Safety and Health; identification and investigation of other possible cases involving safety and health violations through the assistance of employees, organized labor, and health professionals; and strengthening regulations, legislation, and the practices and policies of the relevant regulatory agencies relating to worker health and safety issues.

Violations of laws relating to occupational safety and health are prosecuted either criminally or civilly. Most of the cases are multiple defendant, generally with the corporation and corporate officers and managers charged. Complex factual issues and a substantial amount of scientific or medical testimony are involved (20).

6.3 The U.S. v. Park

The general rule concerning officers, directors, or agents of a corporation is that they may be criminally liable individually for their acts on behalf of the corporation. They cannot, in the absence of a statute, be held liable for acts either in which they have not actively participated or which they have not directed or permitted. But the Supreme Court has broadened the traditional theory of individual criminal liability under welfare statutes by holding corporate officers personally culpable even when they do not have knowledge of the unlawful acts.

In *U.S.* v. *Park* (21), the Supreme Court upheld the chief executive officer's conviction for violation of the Food, Drug and Cosmetics Act (FDCA) for allowing unsanitary conditions to exist in a company warehouse. The court stated that the "government establishes a prima facie case when it introduces evidence sufficient to prove that the defendant had, by reason of his position in the corporation, responsibility and authority either to prevent the unsanitary conditions in the first instance, or promptly to correct the violation complained of, and that he failed to do so." The failure to fulfill the duty imposed by the interaction of the corporate agent's authority, and the statute, furnished a sufficient causal link.

Although the application of the Park holding is only beginning to see its effect in environmental statutes, the potential liability for officers and employees of firms involved in the handling of toxic materials is clear: the Public Welfare statutes may be the vehicle for convicting employers using the responsible corporate officer principle.

7 EXPERT WITNESS

7.1 Introduction

The biological system responds in very complex ways to the numerous and varied toxic exposures. The puzzle is even more difficult to unravel because toxic-related disease may be indistinguishable from nontoxic origins sharing the same symptoms, indications, and biological test findings. Therefore toxic torts are distinguished from nontoxic tort counterparts by the medical and scientific questions that must be analyzed when toxic torts are alleged. Whether there were any breach of duty and any causality related to toxic substances exposure is a question that must be addressed by experts. Although other experts may be needed, the primary team for case analysis includes epidemiology, medicine, toxicology, and industrial hygiene. Specifically, the industrial hygienist can assist the case analysis team by establishing the facts of exposure, or history, and depending upon experience, advise as to the need for additional specific experts.

After a complete work and personal medical history is obtained, a thorough review of the pertinent literature is required in order to determine any causal relationship between the exposure and the disease or injury. Questions to be answered include:

1. Is the suspect substance capable of producing this type of toxic effect?
2. What is the target organ?
3. Has there been sufficient exposure to produce the indicated effect?

As part of a team, the industrial hygienist can assist in assessing the exposure. The toxicologist and occupational health physician can provide assistance in validating the cause–effect relationship between the specific substance and the harm done to the individual, and the epidemiologist can help assess the relationship between the exposure of interest and the medical facts presented.

Although current terminology includes industrial hygienists as those who practice the recognition, evaluation, anticipation, and control of health hazards in industry and the community, the practice of industrial hygiene as indicated in the restated definition is not currently confined to industry. Early studies in the area of radiological health and air pollution were conducted by industrial hygienists, just as currently industrial hygienists with their broad areas of expertise have become involved in the health aspects of such issues as product safety and indoor air pollution. Studies related to asbestos issues within industry and in the general environment, as well as formaldehyde studies in manufactured homes, are specific issues that industrial hygienists have become deeply involved in in recent history.

Because of the complexities involved in the applied and basic science knowledge, the central issue involved in toxic tort litigation very frequently comes down to risk. Risk is the likelihood or probability that an adverse health effect will occur in a given set of circumstances (22).

Because many factors such as basic toxicity, health of the individual, the frequency and duration of exposure, and the environmental conditions under which exposure occurs are important in the assessment of risk, it is helpful to determine the roles of experts by the following equation (23): toxic substance + susceptible individual + excessive exposure = adverse health effect. The element of excessive exposure is the responsibility of the industrial hygienist.

7.2 Participation of the Industrial Hygienist

The specific role of industrial hygiene to assist in unraveling the mysteries of a toxic substance case relate directly to the basic question of exposure. The specific evaluation techniques of industrial hygiene, the general knowledge of potential sources, experts associated with interrelating the toxicological consideration with the historical information, and general and specific understanding of control considerations all contribute to the industrial hygienist's invaluable specific and general assistance in assessing the merits of a toxic substance controversy. The specific controversy in question will dictate the extent of any individual expert's value and participation. Knowledge of the specific issue and the associated credential to support the indicated knowledge will dictate the extent that any one expert can provide assistance in resolving a particular controversy. The selection of the experts becomes a matching of education and experience with the specific facts associated with the case undergoing review.

Because expert testimony usually requires advocacy as related to the specific investigation of the facts, it is a normal consideration to avoid participation in a court proceeding. However, from a practical standpoint, it is exceedingly important that industrial hygienists participate in such proceedings even though ideally it would be preferable to develop opinions based upon the evidence reviewed and argue that position regardless of the benefit to either plaintiff or defendant. The selection of experts by the attorney is based upon benefit for the respective attorney's client. The industrial hygienist who is chosen and agrees to testify therefore should be comfortable with the position undertaken and provide an objective viewpoint of the facts despite the advocacy situation. It is important to remember that an attorney never testifies. The attorney has the responsibility to represent a client zealously and therefore will present the case through witnesses (including experts) who are carefully selected in order to provide the best position possible for the client. The testimony presented therefore is a key element of the ultimate decision by the judge or jury. Use common sense, therefore, as to the worthiness of supporting a position and always keep in mind the code of ethics of the profession when testifying in an advocacy situation.

Personal aspects such as honesty, appearance, and demeanor are important factors, as well as qualification and credibility. The ability to remain calm under intensive cross-examination, to appear modest, yet convincing by articulation and explanation, and regardless of position, to indicate empathy for the plaintiff's plight are notable characteristics for the successful expert. Regardless of the background of the expert, whether it is labor or industry orientation or government or academic neutrality, an appreciation of all sides of the issue is also a critical element of effective testimony; therefore thorough and complete discussion of the issues with the attorney is necessary. The objective of participating as an expert witness is to provide a team effort with the attorney for the scientific–legal resolution of the issue in question. The expert should be supported by the best available current science and its logical application to the issue at hand for effective and efficient problem resolution.

8 CONCLUSION

It is clear that professional industrial hygienists are faced with more complex legal issues today than in past years, and the future holds more of the same. The steady growth of applicable statutes plus the expansion in the area of tort law requires industrial hygienists to conduct their activities completely and competently. Industrial hygienists must perform their jobs with the utmost of care, be ethical, keep abreast of the current statutes and regulations, and learn to work with the legal system as an ally rather than as an enemy.

REFERENCES

1. *Bull. Am. Board Ind. Hyg.*, 3 (1984).
2. *Ibid.*, p. 5.
3. 29CFR, Section 1910. 1000(e).
4. *Black's Law Dictionary*, 4th ed., p. 1375.
5. The American Academy of Industrial Hygiene Code of Ethics.
6. *Black's Law Dictionary*, 4th ed., p. 1660.
7. *Mathews* v. *Walker*, 34 Ohio App ed. 128., 1996 N.E. ed. 569, 630 0 ed 208.
8. Restatement (2nd) of Torts.
9. Prosser, *Law of Torts* 1982, p. 31.
10. *LaCossa* v. *Scientific Design*, 402 F 2nd 937 (3d Cir 1968).
11. *Black's Law Dictionary*, 4th ed., p. 394.
12. American Bar Association, National Institute on Environmental Compliance, May 7 and 8, 1987, Los Angeles, California.

13. J. W. Starr, "Countering Environmental Crimes," *Boston College Environ. Affairs Law Rev.*, **13**(3) 103 (1986).
14. 42 U.S.C. S 6901 et. seq.
15. 42 U.S.C. S 9601 et. seq.
16. 15 U.S.C. S 1261 et. seq.
17. 29 U.S.C. S 651 et. seq.
18. 42 U.S.C. S 7401 et. seq.
19. 33 U.S.C. S 1251 et. seq.
20. From Department Mission Statement, Office of the District Attorney, County of Los Angeles, Los Angeles, California, 1985.
21. 421 U.S. 658, 95 S.Ct, 9903, 1975.
22. National Institute of Occupational Safety and Health, *A Recommended Standard. An Identification System for Occupational Hazardous Materials, NIOSH Publication No. 75-126, Cincinnati, OH, 1975.*
23. G. Z. Northstein, *Toxic Torts*, Shepard/McGraw Hill, Colorado Springs, CO, 1984.

CHAPTER EIGHT

Determination of Biological Agents in the Workplace

William J. Hausler, Jr., Ph.D. and George M. Breuer, Ph.D., C.I.H.

1 INTRODUCTION

The primary purpose of this chapter is to assist the industrial hygienist in developing an increased awareness of biological agents and their interactions in the workplace. There will be no attempt to provide a treatise on infectious diseases in the workplace; the reader will be referred to excellent reference texts. We want simply to call attention to the potential for workplace exposures to biological agents and subsequent interactions that may produce illness.

Biological agents are those bacterial, fungal, viral, rickettsial, and parasitic microorganisms that cause an infectious or pathogenic process when provided with a susceptible host. This definition excludes microbial fragments and allergens.

The workplace is not limited to the industrial setting but is defined to be wherever compensated effort occurs. In a number of instances this may be beyond the area of responsibility of the industrial hygienist and more in the purview of the public health disease investigator or epidemiologist. Regardless of who has responsibility, it is important to possess adequate information in order to provide a useful and complete evaluation. These workplace environments can range from the operatory of the solo practicing dentist through day-care centers to very busy and large abattoirs. In addition, a host may be a healthy and immunocompetent person or one who is immunocompromised through medical treatment or has underlying disorders.

Patty's Industrial Hygiene and Toxicology, Fourth Edition, Volume 1, Part A, Edited by George D. Clayton and Florence E. Clayton
ISBN 0-471-50197-2

2 ROUTES OF TRANSMISSION

It is necessary to understand that common routes of disease transmission should not be overlooked in any workplace evaluation. The usual routes are air, water, fecal–oral transmission, and fomites.

2.1 Air

Airborne transmission of infectious particles is probably the most important and the most frequent mode of transmission. Sneezing, coughing, and spitting all produce infectious aerosols that can be transmitted to a susceptible host depending upon the mass of the particle and air currents. It is important to remember that infectious particles can travel considerable distances, complicating the evaluation of an exposure. It must also be borne in mind that investigations take place after the actual incident when original conditions may have changed and the offending organism may no longer be present in the air or cannot be viably collected. The collection and examination of air for infectious materials is covered elsewhere in this work.

2.2 Water

Water can serve as a very effective dispersant of infectious agents through contamination of the potable water source by cross-connections, back-siphonage, or eroded pipes and weak connections. Water can also be the vehicle for transporting an infectious agent from a contaminated article or surface to the exposed skin, face, or mouth of the worker.

2.3 Fecal–Oral

The fecal–oral route is very common in most infectious processes and can be the principal route for spread of infections in workplace settings such as day-care or nursery-care facilities, hospitals, nursing homes, and other types of institutional care giving. This route can be coupled with the following identified important transmission mechanism.

2.4 Fomites

Inanimate objects that in themselves are not harmful but that may serve as a vehicle for transmission of disease are termed fomites. Examples are a contaminated eating utensil or an object that has been in an infected worker's mouth, such as a respirator mouthpiece, and is now placed in a susceptible worker's mouth.

3 CAUSAL DETERMINATION

Perhaps one of the most perplexing questions faced by anyone when evaluating a workplace problem is whether or not an infectious agent is involved. Many times this is not readily discernible and most often is further complicated by multiple anecdotal comments. Similar complaints by co-workers and examination of sick leave records can usually reveal a commonality of exposure or experience. An episode is made extremely complicated by protracted or delayed incubation periods or by the health status of what can be called the index case should one be present. Oftentimes it is necessary to examine records over several weeks to determine if there has been a series of amplifying waves or if it is simply a single point exposure. Quite often it still is not clear whether or not the episode is a community-acquired illness (i.e., outside the workplace) or something other than an infectious agent. Support by medical and public health practitioners is a primary necessity in these investigations.

4 POTENTIAL WORKPLACE BIOLOGICAL AGENTS

4.1 Anthrax

A common name for anthrax is "wool sorter's disease," connoting workplace acquisition. It is primarily an occupational hazard of those who process animal hair and hides, wool, and bone and bone products, as well as of agricultural workers and veterinarians. The primary site of infection is usually the skin but inhalation anthrax also occurs and is frequently fatal.

4.2 Brucellosis

Brucellosis is an occupational disease of veterinarians, farm workers, and abattoir workers. Infected animals or their tissues are the source and the route of infection can be aerosol, ingestion, or direct contact.

4.3 Campylobacteriosis

Campylobacteriosis is a zoonotic disease of veterinarians, packing plant workers, poultry processors, and livestock producers, as well as day-care and nursery workers. It is a gastroenteritis transmitted by the fecal–oral route and also can be foodborne or waterborne.

4.4 Chlamydia

Chlamydia is an occupational disease of those in the poultry industry. Psittacosis is the term used for human infection, which is transmitted usually by inhaled dust particles; cases present as respiratory infections.

4.5 Cytomegalovirus

Cytomegalovirus is a member of the herpes virus family that may cause inapparent infection in workers or may cause a syndrome clinically and hematologically like mononucleosis but distinguishable from Epstein-Barr Virus mononucleosis. Day-care and nursery workers may manifest this illness or serve as an inadvertent carrier to immediate family.

4.6 Erysipelothrix Infections

An occupational disease of veterinarians, abattoir workers, and fish handlers, erysipelothrix produces cutaneous lesions (especially on the hands) of those workers with domestic animal contact.

4.7 Giardiasis

Giardiasis is a protozoan infection of the upper small intestine associated with a variety of intestinal symptoms. The parasite is transmitted by the fecal–oral route and the disease is prevalent in day-care centers and in institutions.

4.8 Histoplasmosis and Coccidioidomycosis

Both histoplasmosis and coccidioidomycosis are systemic mycoses with primary pulmonary involvement. Construction workers in the semiarid southwestern United States may be exposed to *Coccidioides immitis*, the etiologic agent of coccidioidomycosis, while moving dry or friable soil. Construction tradesmen from the Midwest to the East Coast may be exposed to the biological agent of histoplasmosis as they move topsoil or rehabilitate buildings containing bird or bat droppings.

4.9 Legionellosis and Pontiac Fever

Although reports of workplace acquisition of the agent causing these pulmonary illnesses are few, they do occur and have been documented. The primary route of infection is airborne from cooling water in cooling towers or in power generation.

4.10 Leptospirosis

Some of the more common names for leptospirosis are cane-cutter's or swineherd's disease, noting workplace acquisition. Those involved in animal husbandry and abattoir workers are commonly affected. Clinical manifestations of the disease are quite variable, from headache and muscle pain to meningitis and jaundice.

4.11 Listeriosis

Listeriosis is an occupational disease of abattoir workers and veterinarians that ranges from asymptomatic through mild flu-like illness to severe manifestations such as meningitis.

4.12 Q Fever

Q fever is an acute febrile illness with considerable variation in severity and duration. Clinical manifestations range from inapparent infection through pneumonitis to acute pericarditis and chronic endocarditis. The disease affects veterinarians, farmers, dairy workers, and abattoir workers. Transmission is commonly by airborne contaminated dust that may travel considerable distances from the original site.

4.13 Shigellosis

Shigellosis is an acute bacterial enteritis transmitted by the fecal–oral route. Workers in institutions, mental hospitals, and day-care centers may be exposed.

4.14 Sporotrichosis

Sporotrichosis is a fungal disease most commonly of the skin, initiating by a nodule that grows and extends to lymphatics draining the involved area. It is often a disease of farmers, gardeners, horticulturists, and golf-course maintenance workers.

There are many other communicable diseases not noted above, such as influenza and viral gastroenteritis (particularly among wastewater treatment plant workers and sewer workers), that may be acquired at the workplace but usually do not involve industrial hygienists or occupational health specialists. Because there is an element of community involvement, these outbreaks are commonly investigated by local or state health department epidemiologists. An exception would be nosocomial infections that may or may not involve workers. Nosocomial infections are those that involve or originate in a hospital or other health-care facility. Most institutions in this category have established infection control programs under the leadership of a physician or nurse epidemiologist.

5 PROBLEM EVALUATION

Providing specific guidance for the evaluation of any workplace disease problem is difficult because multiple factors and many individuals are involved. For instance, an outbreak of disease in an abattoir in all likelihood will involve physicians, veterinarians, public health or clinical microbiologists, infectious disease specialists, and industrial hygienists. Knowledge of infectious diseases is essential but each of these specialists provides specific insight from their own professional perspective. Utilizing these combined efforts in a cooperative effort greatly reduces the time required from problem discovery to conclusion as well as providing the most economical assistance.

Excellent texts and handbooks on infectious diseases are readily available in most medical libraries and in many personal libraries. A few of the more useful texts are:

- *Control of Communicable Diseases in Man*, 14th ed., 1985. A. S. Benenson. Ed. American Public Health Association, Washington, DC
- *Infectious Diseases*, 2nd ed., 1977. P. D. Hoeprich. Harper and Row, Hagerstown, MD
- *Laboratory Diagnosis of Infectious Diseases: Principles and Practice*, Vols. I and II, 1988. A. Balows, W. J. Hausler, Jr., and E. H. Lennette. Springer-Verlag, New York
- *Occupational Mycoses*, 1983. A. F. DiSalvo. Lea and Febiger, Philadelphia, PA
- *Principles and Practice of Infectious Diseases*, Vols. I and II, 1979. G. L. Mandell, R. G. Douglas, Jr., and J. E. Bennett. John Wiley and Sons, New York

Of greater difficulty than finding suitable reference texts is locating qualified microbiologists knowledgeable of biological agent peculiarities, specifics, and interactions with humans and the environment. Although small, there is a growing number of exceptionally qualified clinical and public health microbiologists establishing consultative practices and laboratories. Contact may be made with these board-certified doctoral level professionals by contacting the American Board of Medical Microbiology at the American Academy of Microbiology, 1325 Massachusetts Avenue NW, Washington, DC 20005.

6 PREVENTION

The old adage "an ounce of prevention is worth a pound a cure" applies perhaps more strongly to the control of biological agents in the workplace than to any other area of industrial hygiene. Not only are the consequences of disease serious, as with other occupational illnesses, but the identification and evaluation process is more difficult because of the ubiquitous presence of other organisms (usually in much greater numbers than the pathogenic organism of interest) and the wide variation in culture requirements of microorganisms.

It is often apparent that a workplace is subject to transmission of biological agents. For example, the day-care center is often the focus of investigation of outbreaks of giardiasis among children and parents as well as workers; the abattoir may be the source of infection for several of the agents listed previously; blood-borne diseases may be spread by needle sticks in hospitals; and diseases may be transmitted via aerosol generation in the dental practice. In these situations it is common that the precautions necessary to prevent spread of disease are widely known or are prescribed by state health department requirements or guidelines, recommended practices of professional or industrial associations, or guidance from other agencies such as the Centers for Disease Control. The industrial hygienist's role in such cases may simply be to promote the "eternal vigilance" necessary for conscientious application of the guidance. In other cases more education and motivation may be required using such guidance for the particular industry or a closely related one.

Specific guidelines for each industry and especially each situation are obviously beyond the scope of this chapter. Reference should be made to the appropriate organization for that industry. Nearly all situations involving biological agents, however, share certain facets and thus certain practices that represent "common sense":

- Personal hygiene is always a good work practice for prevention of the spread of disease, whether occupational or incidental.
- Avoidance of any disruption—cut, abrasion, needle-stick, or dermatitis—of the integrity of the skin and its barrier to infection is always appropriate.
- Proper provision and use of personal protective equipment such as clothing, gloves, and face masks or respirators must be encouraged.
- Workplace cleanliness conforming to an appropriate standard (e.g., from refuse removal and hosing down in abattoirs to disinfection of surfaces and articles in day-care centers and medical practices) is always indicated.
- Proper disposal of wastes—especially those suspected of pathogen contamination—and proper cleaning of contaminated articles is always appropriate.
- Proper maintenance of equipment and facilities is usually cost effective and helpful. Because of their common neglect, special mention should be made of air handling equipment (1–3); even redesign of the system may be necessary in some cases to direct air flow away from sources of contamination to the outside rather than to occupied areas in both summer and winter, to enable drain pans to drain properly and be properly cleaned, to prevent or limit humidifier contamination, and so on.
- Consideration also should be given to other aspects of environmental control in the workplace, including such things as maintenance of relative humidity at levels less than 70 percent to minimize fungal growth and restriction of unnecessary access to areas of possible contamination.
- Special care may be necessary in the case of immunocompromised or particularly susceptible employees.

Attention to these items and adherence to up-to-date, industry-specific guidance may reduce significantly the number of cases of illness that must be addressed after the damage is done.

7 MONITORING

It can be categorically stated that routine monitoring for microbiological agents in the absence of identification of the agent to be sought is improper except possibly in a research setting. Medical diagnosis and isolation of a particular organism is a necessary first step in the monitoring effort. Selective media and guidance on appropriate sampling techniques and locations can then be provided by a microbiologist. In general, the laboratory chosen for the monitoring should be one familiar with environmental monitoring for microbiological agents as well as one familiar with infectious diseases; most clinical laboratories have a substantially different role and orientation from what is required and may not be familiar with procedures used for samples other than human clinical specimens. The laboratory must be able to supply collection and culture media of appropriate type, quantity, and quality for monitoring as well as the appropriate equipment.

A variety of techniques are available for sampling solid, aqueous, and surface contamination by microbiological agents. The emphasis must be on obtaining a representative and viable sample from which the organism can then be isolated and identified, if present, without overgrowth by ubiquitous organisms. The media and techniques specified by the laboratory must be used conscientiously; attention to detail is required.

For air sampling, if appropriate, a variety of more specialized techniques and instrumentation is available. Reference should be made to standard texts covering this instrumentation, in particular the latest edition of the American Conference of Governmental Industrial Hygienists' (ACGIH) text, Air Sampling Instruments for Evaluation of Atmospheric Contaminants (4), which contains an article on "Sampling Airborne Microorganisms" by Mark Chatigny. The ACGIH Committee on Bioaerosols has also provided guidance for sampling and analysis of airborne microorganisms, especially in office environments (5). Air monitoring for fungal and other microbiological agents using various instruments has also been reviewed in the recent industrial hygiene literature (6–22). These articles and their references may be consulted for additional guidance. Prior to any sampling, however, consideration must be given to how the data obtained will or can be interpreted. Exposure limits for biological agents have not been set and are unlikely to be set in the near future, although occasional suggestions may be made. Indoor/outdoor microbial ratios may be useful, as is certainly the identification of a pathogenic organism determined from a medical evaluation.

8 SELECTED CASE HISTORIES

Case histories are often helpful in focusing one's attention on a possible problem solution. We have selected a few such cases that emphasize the interaction of biological agents, workplace environment, and the worker in the causation of illness.

8.1 Case 1

Two women worked in a small plastics factory and both were admitted to different hospitals within a day of each other with primary complaints of fever and shortness of breath. Subsequently both women were diagnosed with pneumonia and one of them ultimately died.

Not only did both receive complete clinical evaluation, but the homes of both and their workplace received a complete environmental investigation. *Legionella pneumophila* was recovered from cooling water used to cool hydraulic injection mold presses as well as nine other sites in the plant. A variety of serogroups were found including *L. pneumophila* serogroup I, which was also recovered from one of the patients (23).

Procedures used for control are described in the authors' reports (23, 24). Related control measures have been described elsewhere (25). Protocols for assessing environmental sites for Legionellae also have been discussed elsewhere as well (e.g., see 26, 27).

8.2 Case 2

Ten of 12 men cleaning an electric power steam turbine condenser clogged with freshwater sponge contracted an acute febrile illness. Similar previous outbreaks had been attributed to *Legionella*, but etiology could not be established in this case. Sampling was not carried out until after completion of cleaning. Delayed onset and nature of symptoms suggested an infectious agent. In the absence of definitive medical diagnosis and identification of the source in the workplace, precautions consisting of use of supplied air respirators with full facepiece and initial thorough flushing of condensers prior to entry were recommended (28).

8.3 Case 3

Two of 14 workers in a small clerical office reported recurrent episodes of respiratory illness characterized by cough, wheezing, dyspnea, malaise, and fever. Symptoms were associated with presence in the workplace. One was eventually diagnosed as having bronchial asthma, perhaps exacerbated by levels of *Penicillium* substantially elevated compared to an asymptomatic office; the other was diagnosed as having alveolitis and probable hypersensitivity pneumonitis (HP) based on extensive clinical examination, although more definitive (for HP) lung biopsy was not performed and there was an absence of infiltrates on chest roentgenograms. Extensive cleaning and subsequent maintenance of the ventilation system reduced *Penicillium* to background levels, although the patient diagnosed as having HP continued to show symptoms, possibly due to sensitization to even background levels (29).

8.4 Case 4

A review of the health status of 50 workers of a predominantly stable, older population at a municipal sewage treatment plant revealed complaints including dermatitis, frequent "influenza-like" illness, and intermittent acute illness characterized by cough, fever, and sore throat. A variety of techniques were used in an attempt to identify causes; these included clinical diagnosis of cutaneous disorder, lung function testing, and serum levels of polychlorinated biphenyls (the last was a concern specifically raised by workers). Several suggestive factors were indicated, such as respiratory impairment due to fly ash generated during sludge incineration, but no clear cause was indicated for many of the symptoms (30). In a different plant using spray irrigation, coliphages and bacteria were isolated from aerosol samples (31); obviously, the opportunity exists for transmission by other routes as well.

REFERENCES

1. R. T. Hughes and D. M. O'Brien, *Am. Ind. Hyg. Assoc. J.*, **47**; 207–213 (1986).
2. P. R. Morey and J. E. Woods, *State Art Rev. Occup. Med.*, **2**, 547–563 (1987).
3. B. P. Ager and J. A. Tickner, *Ann. Occup. Health*, **27**; 341–359 (1983).
4. P. J. Lioy and M. J. Y. Lioy, eds., *Air Sampling Instruments for Evaluation of Atmospheric Contaminants*, 6th ed., American Conference of Governmental Industrial Hygienists, Cincinnati, OH, 1983.
5. P. Morey, J. Otten, H. Burge, M. Chatigny, J. Feeley, F. M. LaForce, and K. Peterson, *Appl. Ind. Hyg.*, **1**, R19–23 (1986).
6. I. M. Lundholm, *Appl. Environ. Microbiol.*, **44**, 179–183 (1982).
7. A. I. Donaldson, N. P. Ferris, and J. Gloster, *Res. Vet. Sci.*, **33**, 384–385 (1982).
8. K. L. Morring, W. G. Sorenson, and M. D. Attfield, *Am. Ind. Hyg. Assoc. J.*, **44**, 662–664 (1983).
9. J. M. Macher and M. W. First, *Am. Ind. Hyg. Assoc. J.*, **45**, 76–83 (1984).
10. G. Blomquist, G. Strom, and L. H. Blomquist, *Scand. J. Work Environ. Health*, **10**, 109–113 (1984).
11. G. Blomquist, U. Palmgren, and G. Strom, *Scand. J. Work Environ. Health*, **10**, 253–258 (1984).
12. V. Lach, *J. Hosp. Infect.*, **6**, 102–107 (1985).
13. W. Jones, K. Morring, P. Morey, and W. Sorenson, *Am. Ind. Hyg. Assoc. J.*, **46**, 294–298 (1985).
14. B. K. Lewis and W. W. Herbert, *Am. Ind. Hyg. Assoc. J.*, **46**, B10 (1985).
15. H. A. Burge, *Clin. Rev. Allergy*, **3**, 319–329 (1985).
16. K. L. Melvaer and H. Ringstd, *Acta Pathol. Microbiol. Immunol. Scand. Sect. B*, **94**, 325–328 (1986).
17. T. M. Madelin, *J. Appl. Bacteriol.*, **63**, 47–52 (1987).
18. N. J. Zimmerman, P. C. Reist, and A. G. Turner, *Appl. Environ. Microbiol.*, **53**, 99–104 (1987).
19. J. M. Macher and H. C. Hansson, *Am. Ind. Hyg. Assoc. J.*, **48**, 652–655 (1987).
20. P. Heikkila, M. Kotimaa, T. Tuomi, T. Salmi, and K. Louhelainen, *Ann. Occup. Hyg.*, **32**, 241–248 (1988).
21. J. Lacey and B. Crook, *Ann. Occup. Hyg.*, **32**, 515–533 (1988).
22. T. Smid, E. Schokkin, J. S. M. Boleij, and D. Heederik, *Am. Ind. Hyg. Assoc. J.*, **50**, 235–239 (1989).
23. P. W. Muraca, J. E. Stout, V. L. Yu, and Y. C. Lee, *Am. Ind. Hyg. Assoc. J.*, **49**, 584–590 (1988).

24. P. W. Muraca, V. L. Yu, and J. E. Stout, *J. Am. Water Works Assoc.*, **81**, 78–85 (1988).

25. J. S. Colbourne, P. J. Dennis, J. V. Lee, and M. R. Bailey, *Lancet*, **1**, 684 (1987).

26. J. M. Sykes and A. M. Brazier, *Ann. Occup. Hyg.*, **32**, 63–67 (1988).

27. J. M. Barbaree, G. W. Gorman, W. T. Martin, and B. S. Fields, *Appl. Environ. Microbiol.*, **53**, 1454–1458 (1987).

28. J. F. Lauderdale and C. C. Johnson, *Am. Ind. Hyg. Assoc. J.*, **44**, 156–160 (1983).

29. R. S. Bernstein, W. G. Sorenson, D. Garabrant, C. Reaux, and R. D. Treitman, *Am. Ind. Hyg. Assoc. J.*, **44**, 161–169 (1983).

30. J. R. Nethercott and D. L. Holness, *Am. Ind. Hyg. Assoc. J.*, **49**, 346–350 (1988).

31. K. P. Brenner, P. V. Scarpino, and C. S. Clark, *Appl. Environ. Microbiol.*, **54**, 409–415 (1988).

CHAPTER NINE

The Mode of Absorption, Distribution, and Elimination of Toxic Materials

Bertram D. Dinman, M.D., Sc.D.

1 INTRODUCTION

The toxicity of a material and the hygienic standard applying to its use are of obvious importance in the practice of health protection; however, they must not be confused with the *hazard* of using the material. As this chapter will attempt to demonstrate, multiple physical and biological factors will, under conditions of exposure, determine the risk associated with the use of workplace chemicals.

Multiple physical factors will determine whether chemical agents ever reach target tissues to produce toxic sequelae. Gases may replace all the air in the atmosphere, but the amount of vapor in the atmosphere at any time is limited by the vapor pressure of the liquid or solid from which it arises. However, the amount of suspended particulate matter in the atmosphere (e.g., dust, mist, or fume) is not limited by vapor pressure. It is well established that particulates may adsorb gases and vapors, thereby altering—and in some instances increasing—their physiological action. This follows because gas or vapors sorbed on particulates must follow the same physical principles governing the deposition of such particles in the pulmonary tree. Because particulates are more likely than gases to impinge on various portions of respiratory tract surfaces, the physiological consequences of particulate inhalation can be more significant than would be the case for gas or

Patty's Industrial Hygiene and Toxicology, Fourth Edition, Volume 1, Part A, Edited by George D. Clayton and Florence E. Clayton
ISBN 0-471-50197-2

vapor molecules that do not deposit and are exhaled. In addition to these purely physical considerations outside the body, other biophysical factors in the body, for example, pH, pK_a, lipid/water partition coefficient, and dissociation characteristics, discussed in this chapter also determine whether the xenobiotic and tissue target ever have the opportunity to interact.

Similarly, a multiplicity of biological variables determine whether a material will ever reach a target site. In the course of its being absorbed into the body, transported to, and entering a specific target organ or being eliminated, numerous conditions must be met before the cellular target site may be reached. The blood supply, conditions of vascular and cell permeability, integrity of the muco-ciliary escalator system, actions of neurohumoral mediators, and so on all determine whether cellular target molecules ever are at risk.

To plan the prevention of injury from toxic materials in industry, it is essential to have a clear understanding of the dynamics of how materials enter the body, how they are handled therein, and how they are eliminated. To comprehend these processes more clearly, we need to understand the mechanics of respiration and circulation and their role in the kinetics of absorption, transport, body cellular uptake, and body elimination. Many of these processes require comprehension of the gas laws and the ability to apply them to the solution of gases in liquids and specifically in the body fluids.

On the basis of such understanding it will become apparent why the occurrence of exposure to a toxic material is not tantamount to the risk of an untoward effect. Indeed, after review of this chapter, it would seem that, given all the conditions required before toxicity will occur, such an outcome is less probable than no event at all.

2 CLASSIFICATION OF CONTAMINANTS

The earth is surrounded by a gaseous atmosphere of rather fixed composition: 78.09 percent nitrogen, 20.95 percent oxygen, 0.93 percent argon, 0.03 percent carbon dioxide, insignificant amounts of neon, helium, and krypton, and traces of hydrogen, xenon, radioactive emanations, oxides of nitrogen, and ozone, with which may be mixed up to 5.0 percent water vapor. Any of these gases in greater proportions than usual, or any other substance present in the atmosphere, may be regarded as a contaminant or as atmospheric pollution. The possibilities of contamination are legion, but we may classify them according to their physical state, their chemical composition, or their physiological action.

2.1 Physical Classifications

2.1.1 Gases and Vapors

Although strictly speaking a gas is defined as a substance above its critical temperature and a vapor as the gaseous phase of a substance below its critical temperature, the term "gas" is usually applied to any material that is in the gaseous state at 25°C and 760 mm Hg pressure; "vapor" designates the gaseous phase of a substance that is ordinarily liquid or solid at 25°C and 760 mm Hg pressure. The usage distinction between gas and vapor is not sharp, however. For example, hydrogen cyanide, which boils at 26°C, is always referred to as a gas, but hydrogen chloride, which boils at −83.7°C, is sometimes referred to as an acid vapor.

2.1.2 Particulate Matter

There are at least seven forms of particulate matter:

1. *Aerosol.* A dispersion of solid or liquid particles of microscopic size in a gaseous medium—for instance, smoke, fog, and mist.
2. *Dust.* A term loosely applied to solid particles predominantly larger than colloidal and capable of temporary suspension in air or other gases. Derivation from larger masses through the application of physical force is usually implied.
3. *Fog.* A term loosely applied to visible aerosols in which the dispersed phase is liquid. Formation by condensation is implied.
4. *Fume.* Solid particles generated at condensation from the gaseous state, generally after volatilization from melted substances and often accompanied by a chemical reaction, such as oxidation. Popular usage sometimes loosely includes any type of contaminant.
5. *Mist.* A term loosely applied to dispersion of liquid particles, many of which are large enough to be individually visible without visual aid.
6. *Smog.* A term derived from "smoke" and "fog" and applied to extensive atmospheric contamination by aerosols arising from a combination of natural and man-made sources.
7. *Smoke.* Small gas-borne particles resulting from incomplete combustion and consisting predominantly of carbon and other combustible materials.

2.2 Chemical Classifications

Chemical classifications are variously based on the chemical composition of the air contaminants and may vary widely depending on the aspect of the composition to be emphasized. The classification used in Volume 2 of *Industrial Hygiene and Toxicology* is an example of chemical classification.

2.3 Pathological Classifications

The physiological classification of air contaminants is not entirely satisfactory because with many gases and vapors, the type of physiological action depends on concentration. For instance, a vapor at one concentration may exert its principal action as an anesthetic, whereas a lower concentration of the same vapor may, with no anesthetic effect, injure the nervous system, the hematopoietic system, or some visceral organ. Although it is frequently impossible to place a material in a single class correctly, a pathophysiological classification may be suggested.

2.3.1 Irritants

Irritant materials are corrosive or vesicant in their action. They inflame moist or mucous surfaces. They have essentially the same effect on animals as on humans, and the concentration factor is of far greater significance than the time (duration of exposure) factor. Some representative irritants are as follows:

1. Irritants affecting chiefly the upper respiratory tract: aldehydes (acetaldehyde, acrolein, formaldehyde), alkaline dusts and mists, ammonia, chromic acid, ethylene oxide, hydrogen chloride, hydrogen fluoride, sulfur dioxide, sulfur trioxide.
2. Irritants affecting both the upper respiratory tract and the lung tissues: bromine, chlorine, chlorine oxides, cyanogen bromide, cyanogen chloride, dimethyl sulfate, diethyl sulfate, iodine, ozone, sulfur chlorides, phosphorus trichloride, phosphorus pentachloride, and toluene diisocyanate.
3. Irritants affecting primarily the terminal respiratory passages and air sacs: arsenic trichloride, nitrogen dioxide and nitrogen tetroxide, and phosgene. (To the extent that their action frequently terminates in asphyxial death, lung irritants are related to the chemical asphyxiants.)

2.3.2 Asphyxiants (Anoxia-Producing Agents)

Strictly speaking, "asphyxia" should be restricted to descriptions of agents that produce oxygen lack and concomitant increases of carbon dioxide tension in the blood and tissues. Our concern with effects of chemicals on oxygen availability to the body requires that we refer to agents producing this effect as producing anoxia (i.e., lack of oxygen). Many of the chemical agents noted below produced effects at multiple loci (e.g., carbon monoxide affects hemoglobin as well as various tissue respiratory catalysts such as cytochrome P-450). However, the classification to follow considers the major sites of action of such chemical agents.

This group can be subdivided into three classes, depending on how they cause anoxia within the body:

2.3.2.1 Anoxic Anoxia. Lack of oxygen availability to the lungs and blood stems from simple displacement or dilution of atmospheric oxygen; this may occur even in the case of physiologically inert gases. This, in turn, results in reduction of the partial pressure of oxygen required to maintain oxygen saturation of the blood sufficient for normal cellular respiration. Agents producing anoxic anoxia include ethane, helium, hydrogen, nitrogen, and nitrous oxide.

2.3.2.2 Anemic Anoxia. Anemic anoxia implies a total or partial lack of availability of the blood pigment hemoglobin for oxygen carriage by the red blood cell. Whereas simple hemorrhage reduces the total loading of blood oxygen in proportion to red blood cell loss, arsine has an effect similar to hemorrhage by producing red blood cell breakdown, making it unavailable for oxygen carriage. In addition, numerous chemical agents can produce a similar effect by impairing or blocking the oxygen uptake and carrying capacity of hemoglobin in the lungs. Examples of this group are carbon monoxide, which combines with hemoglobin to form carboxyhemoglobin, and aniline, dimethylaniline, and toluidine, which form methemoglobin.

2.3.2.3 Histotoxic Anoxia. Histotoxic anoxia results from the action of agents that impair or block the action of cellular catalysts necessary for tissue oxidative metabolism. Because the ability of hemoglobin to take up oxygen is not necessarily altered, and because the tissues cannot avail themselves of this oxygen, the capillary and venous oxygen saturation of the blood usually is higher than normal. Examples in this group are cyanogen, hydrogen cyanide, and nitrites. Hydrogen sulfide blocks cellular oxidation at the respiratory center controlling respiration and directly stops pulmonary air moving action. The ultimate effect of this agent is to cause anoxic anoxia secondarily.

2.3.3 Anesthetics and Narcotics

The anesthetics and narcotics group exerts its principal action as simple anesthesia without serious systemic effects, and the members have a depressant action on the central nervous system governed by their partial pressure on the blood supply to the brain. The following examples are arranged in the order of their decreasing anesthetic action compared with other actions: (*a*) acetylene hydrocarbons (acetylene, allylene, crotonylene), (*b*) olefin hydrocarbons (ethylene to heptylene), (*c*) ethyl ether and isopropyl ether, (*d*) paraffin hydrocarbons (propane to decane), (*e*) aliphatic ketones (acetone to octanone), (*f*) aliphatic alcohols (ethyl, propyl, butyl, and amyl), and (*g*) esters. (Although the last are not particularly anesthetic, they are placed here for want of a better classification.)

2.3.4 Systemic Poisons

The following substances are classified as systemic poisons:

1. Materials that cause injury to one or more of the visceral organs: the majority of the halogenated hydrocarbons.
2. Materials damaging the hematopoietic system: benzene, phenols, inorganic lead, warfarin.
3. Nerve poisons: carbon disulfide, methyl alcohol, methyl *n*-butyl ketone.
4. Toxic metals: lead, mercury, cadmium, antimony, manganese, beryllium.
5. Toxic nonmetal inorganic materials: compounds of arsenic, phosphorus, selenium, and sulfur; fluorides.

2.3.5 Sensitizers

Strictly speaking, materials producing an antigen–antibody complex resulting in allergic type reaction are referred to as sensitizers. Such sensitizing materials, that is, antigens, capable of inducing the immune response usually are of large molecular size, for example, proteins of bacterial, plant, or animal origin. Most industrial compounds are of insufficient molecular weight to act as antigens and to elicit the immune response.

However, an exception to this generalization must be noted. It has become particularly clear that relatively small reactive molecules, for example, notably toluene diisocyanate, are indeed capable of inducing classical evidence of the immune response. Such small molecules demonstrating this capacity have been referred to as haptens. In these cases it has been found that relatively low molecular weight haptens are capable of conjugating by covalent binding with a host protein or other macromolecules. The chemical structure of the foreign compound may play a role in determining its bioreactivity toward appropriate macromolecules. The ability to act as a hapten is largely determined by the chemical reactivity of the compound (electrophilicity) and the molecular configuration; slight changes in shape can markedly alter antigenicity. The net result of this combination of a low molecular weight compound with a macromolecule is formation of a complete antigen capable of initiating the first step in the typical sensitization sequence.

The next step in the process is the combination of the antigen with formed elements of the reticuloendothelial system, for example, macrophages and various types of lymphocytes. Upon contact with an antigen the lymphoid cells are stimulated to divide and mature. The B lymphocytes elaborate antibodies. The antibody is a large, complex protein molecule with a configuration specific for the eliciting antigen. Owing to the size and complexity of the antigen more than one antibody may result from this interaction with a single antigen. On subsequent exposures to the specific antigen such antibodies combine or complex with this antigen. Such complexes are responsible for many of the clinical immunologic response syndromes.

Antibodies produced by the immune system are specific proteins of the immunoglobulin (Ig) series, that is, IgA, IgG, IgM, and IgE. These proteins in common all consist of heavy and light polypeptide chains. Within these chains are regions common to the entire Ig series; in addition, all Ig of the same class demonstrate the same number of amino acid residues. However, the amino acid content varies with the specificity of the antibody, that is, its ability to recognize and combine with antigen. Some antibodies are found in plasma, namely, IgG and IgM, whereas others, notably IgE, frequently are associated with cells fixed in specific tissues.

The complexing of an antigen and an antibody may involve complement, a complex cascade-type system of serum proteins. The complement proteins released by the antigen–antibody complex may increase phagocytosis by leukocytes, cause smooth muscle contractions (see below), or damage cell surfaces. Also, IgE on cell surfaces combines with antigen with the release of histamine from mast cells and platelets. Such complexing of antigen and antibody results in cell damage and release of heparin, histamine, or serotonin. Another destructive interaction results when the antibody complexes with antigens that have become bound to the surface of specific cells, for example, red blood cells. This complexing leads to the agglutination of such elements; if complement is released here the cells will lyse. In a third type of reaction circulating antigen–antibody complexes may initiate blood cell damage; complement mediates this response. In other cases, tissue injury may result when antigen combines with a lymphocyte that has been sensitized and as a result has specificity for the antigen.

The clinical result of the immune or allergic response may be localized or widespread, immediate or variably delayed reactions. The severity of response may be so localized as to affect only a few square millimeters of skin, or may extend to involve the whole body as in severe anaphylactic type response. The dose–response relationship governs induction of the magnitude of elicited response. However, there are individual variations in responsiveness and complex modulating mechanisms regulating production of immunologic cells and antibodies.

The most severe immunologic reaction, anaphylaxis, has infrequently been associated with industrial exposures. By contrast, less life threatening but serious reactions are seen in industry. These most commonly affect the skin and less frequently the respiratory tract. Anaphylaxis may reflect topical direct contact with sensitizer or, less commonly, result from absorption via other portals of entry. The effect may be local or widespread, independent of how much dermal contact occurs.

3 PORTALS OF ENTRY AND ABSORPTION OF XENOBIOTICS

If a foreign material is to produce a biological effect, it must breach some structural and/or biochemical barrier(s) interposed between the external environment and a susceptible locus within the body. In the course of transit of these barriers exogenous materials may be (1) chemically altered (e.g., hydrolyzed by gastric acids), (2) biotransformed (i.e., altered by enzymatic action into more or less toxic moieties), (3) stored in depots where they usually produce little effect, or (4) ultimately excreted, primarily by the kidney and biliary tree. But regardless of such alterations, before any toxic event can occur the material—in whatever form or transformate—must usually pass through multiple barriers. The nature of these barriers and how they may be breached are the focus of this section.

3.1 Principles of Cell Membrane Transit

3.1.1 Cell Membrane Structure

On a gross level cells may be arranged in a variety of fashions; they may consist of layers of multiple cell units, for example, the skin, or a single layer of cells, for example, alveolar walls. However, all individual cell membranes are strikingly similar at the molecular level.

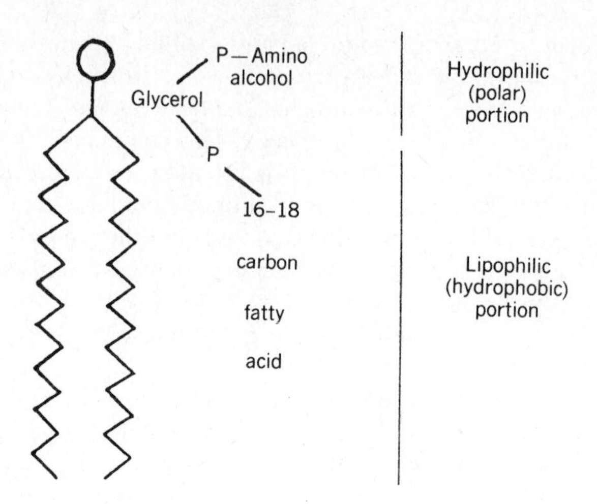

Figure 9.1. Structure of phospholipid molecular component of the bimolecular cell membrane model.

The cell membrane at the molecular level of organization consists of phospho(glyco) lipid (Figure 9.1) as well as protein and cholesterol molecules organized in a bilayer arrangement (Figure 9.2).

These major components of the plasma membrane consist of phosphoglycerides and fatty acids (Figure 9.1.a). The phosphoglycerides consist of glycerol esterified to phosphoric acid further esterified to an amino alcohol, namely, choline or ethanolamine. The amino alcohol, glycerol, and phosphoric acid components are hydrophilic; this assures membrane stability within the hydrophilic environment with which they are in contact, that is, the intracellular cytosol or the extracellular water. The cholesterol molecules present in the membrane are also believed to impart a degree of stiffness to this fluid-like lipid bilayer. This "head" oriented arrangement (Figure 9.1) contrasts with the twin "tails" of the two 16- to 18-carbon fatty acids. Their lipophilic nature is consistent with their orientation within the mass of the membrane. The membrane is thus made up in greater part by a lipid bimolecular layer with "heads" oriented toward both the inside and outside of the cell (Figure 9.2, 9.3). The fluidity of the cell membrane is in large part a reflection of the proportion and structure of unsaturated fatty acids. As unsaturated fatty acid content increases the membranes become more fluid, resulting in more rapid active transport. It is the predominance of the lipid cell wall components that contributes to the relative ease with which lipid xenobiotics penetrate into the cell.

In addition, crystalline water containing various ions (e.g., Ca^{2+}, Mg^{2+}) is located within this two-molecule-thick layer; these also influence the permeability and the gel state of the membrane.

Immersed in this bimolecular structure approximately 6 to 10 nm thick are holo-, glyco- and lipoproteins (Figure 9.3). The proteins may penetrate the entire bimolecular structure, that is, the intrinsic proteins, or may occur in either half of

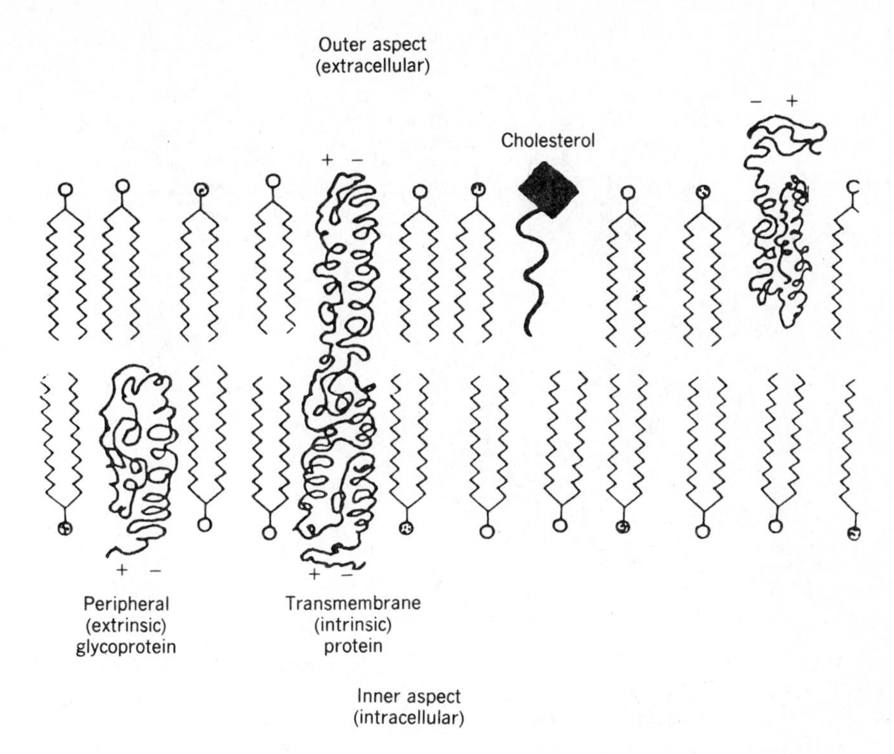

Figure 9.2. Schematic diagram of lipid bimolecular cell membrane model.

the bilayer (Figures 9.2, 9.3). These extrinsically oriented proteins are in contact with both aqueous and lipid media. Accordingly, they tend to demonstrate amphipathic properties. The intrinsic proteins, found largely in a lipid region, consist largely of hydrophobic amino acids; that is, they contain hydrophilic amino acids where they are in contact with water and hydrophobic amino acids within the bimolecular phospholipid components of the membrane. These latter proteins have been proposed as an explanation for the presence of pores in the cell membrane in connection with the simple diffusion of ions and large molecules (see below). Because it is the nature of proteins to (1) immobilize adjacent lipid molecules and (2) perform translational or rotational motion in the lipid bilayer, they play important roles in the maintenance of rigidity or fluidity of membrane structures. Within constraints of ambient temperature and "tail" fatty acid saturation, the number of extrinsic and intrinsic protein molecules markedly affect the physical structural properties of cell walls.

The physical chemical activities of these structures are highly dynamic. The relative mix of the chemical components varies both within an individual cell wall and from cell to cell. The membrane is constantly renewing these components; thus the lipid molecules have a 3- to 5-day half-life, the large proteins a ½- to 5-day half-life, and small proteins a 7- to 13-day half-life. Some membrane protein molecules drift laterally in the plane of the membrane whereas others are anchored

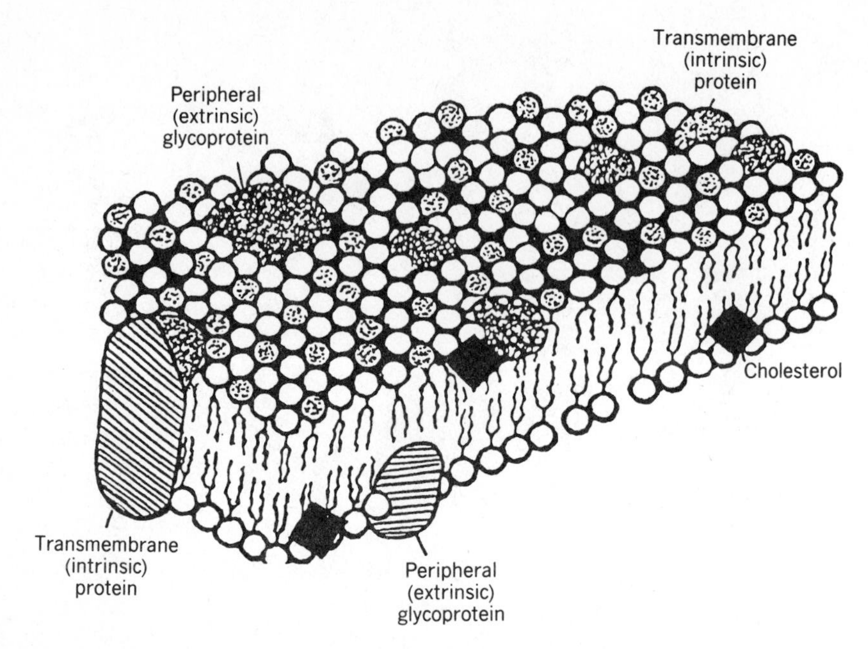

Figure 9.3. Three-dimensional proposed model of a typical cell membrane. White or stippled spheres indicate two different types of polar head groups of the membrane phospholipid with associated fatty acid chains. Both intrinsic and extrinsic membrane proteins are indicated by large hatched and mosaic topped bodies.

to the cyto-skeleton. It should thus be apparent that these cell-limiting membranes are dynamic and can be degraded and rebuilt rapidly in keeping with needs of the organism.

3.1.2 Passive Transport

Passive transport depends upon three primary determinants, namely, (1) concentration gradients across membranes, (2) liposolubility of a compound, and (3) ionization of the compound.

3.1.2.1 Determinants of Passive Transport. The first of these largely reflects the difference in concentration of the compound in question on both sides of the membrane. Diffusion rates also will reflect the thickness and area of the membrane as well as the diffusion constant for the material in keeping with Fick's law. This phenomenon is essentially a first-order rate process.

The interplay of these variables represents a dynamic process, because their rates of occurrence change as concentrations vary on either side of the membrane. Such variations may reflect blood flow rates, concentration changes exterior to the cell, or rates of metabolic transformation occurring within the cell. The rate of simple diffusion also will reflect the rate of dissolution of a compound in the largely lipid domains of the cell membrane. Accordingly, the rate of transfer by this process will in turn depend upon the lipid–water partition coefficient. Thus small hydrophobic molecules, such as carbon tetrachloride, rapidly pass through cell walls of the gastric mucosa and pulmonary alveoli, respectively.

Although lipid solubility is an important determinant of the degree of passive diffusion, the extent of ionization also plays a large role in this type of transmembrane movement. The pH partition theory in effect states that only lipid-soluble, nonionized chemical species will passively diffuse over plasma membranes.

Depending upon the pH of its environment a compound may exist in a ionized or nonionized state. In addition, each compound may be characterized by its own intrinsic dissociation constant (i.e., pK_a), the pH at which an acid or basic compound is 50 percent associated. Knowledge of the pH of the environment and the pK_a of a compound permits use of the Henderson–Hasselbach equation for calculation of the percentage in the nonionized form and hence availability for passive diffusion across the largely lipoprotein cell membrane.

By contrast polar materials, for example, sugars, transit passively at a slow rate—if at all—at these sites. However, it should not be implied that polar compounds do not transit across cell barriers by simple diffusion. Such a process does occur via aqueous channels, that is, the transmembrane protein pores of the cell wall. However, the *relative paucity* of such channels in what is mainly a lipid structure makes such occurrence less frequent, that is, transitting at a lower rate relative to lipophilic compounds.

Nevertheless, movement of hydrophilic and large lipophilic molecules as well as ions readily occurs through these pores. The membrane proteins can gain both positive and negative charges; thus small-diameter pores will permit passage only to ions whose charge is opposite to that of the membrane protein. This selectivity decreases as pore diameter increases so that both cations and anions can pass; in addition, penetration is increased with ion-pair formation. The potential gradient, that is, the electrochemical potential between outer and inner membrane surfaces, also induces diffusion of ions through pores.

The number and size of pores vary in any one membrane, because these attributes can be altered by numerous extrinsic factors, for example, hormones. Thus, for example, if the number and diameter increase, hydrodynamic flow (governed by Poiseiulle's law) increases as the relative proportion of diffusion decreases. As the pore diameter approaches 20 nm, hydrodynamic flow should account for 90 to 95 percent of total transmembrane movement. In view of this high degree of biophysical plasticity, it is apparent that the cell membrane can exhibit considerable dynamics and a remarkable potential for responsiveness to changing cell and body needs.

A dynamic reality applies in vivo; ionization continuously occurs as within the cell nonionized moieties are altered or removed through metabolic events. On the other hand, changes in concentration of a nonionized compound outside the cell determine how much of it is available for passive transport, as a reflection of associated changes in transmembrane concentration gradient.

3.1.2.2 Filtration. Filtration, or the "solvent drag" effect reflects the fact that water flowing through a porous membrane may carry solutes. Obviously, this form of transport involves uniquely hydrophilic compounds. If such solutes are of sufficiently small molecular weight they will pass through pores in the cell membrane as determined by osmotic or hydrostatic forces. There is a wide range in pore sizes in various cell membranes of the body, that is, from 4 nm in most cells to as large as 40 to 70 nm in blood capillaries. Thus filtration only allows passage of small molecules (100 to 200 Da) from the usual capillary; however, molecules as large as 60,000 Da can be passed through the glomerular capillaries in the kidney (1).

3.1.3 Carrier-Mediated Modes of Transport

3.1.3.1 Active Transport. Active transport occurs across concentration or electrochemical gradients. Polar compounds, electrolytes (e.g., Na^+, K^+), nonelectrolytes (e.g., sugars), and zwitterions (e.g., amino acids), which do not diffuse passively, require this mode of transport. This process of transport requires metabolic work. It postulates macromolecular carriers that form a complex at one side of the membrane with the substances to be carried. This complex moves to the other side of the cell membrane at which place the substance carried is released and the carrier returns to the surface of origin, repeating once more the cycle of transport. The carriers are specific insofar as certain basic structural characteristics must exist before complex formation and subsequent transport occur. Thus among compounds with similar critical structural characteristics competitive inhibition of transport may occur for one of the several compounds present at the membrane's exterior. In addition, such systems are subject to saturation at high substrate concentrations; at this point zero-order transfer rates obtain regardless of concentration gradients. Finally, because this process requires expenditure of energy, metabolic inhibitors have the potential for blocking this form of transport. These active transport systems are frequently specific for certain forms of exogenous compounds, for example, amino acids and sugars. Active transport is also relevant to toxicologic processes because it represents an important method for transport of xenobiotics from the system. This process is also particularly important in the removal of organic acids and bases from, for example, the central nervous system, liver, and kidney.

3.1.3.2 Facilitated or Exchange Diffusion. In contrast to the type of active transport just described, this type of transport is not against an electrochemical or concentration gradient and does not require expenditure of metabolic energy. However, this type of carrier-mediated transport occurring down a concentration gradient plays a significant role in the transport of important nutrients such as glucose. It occurs at the gastrointestinal epithelial surface, the red cell membrane, and the capillaries of the blood–brain barrier.

A protein molecule acting as the carrier is formed from two subunits whose new configuration facilitates binding and transport across the plasma membrane. This carrier–substance complexation process resembles the interaction between enzyme and substrate. As with active transport, similar compounds may compete for a

Table 9.1. Membrane Function/Structure Alterations Associated with Toxic Agents

Membrane Alteration	Causal Agent
Structural change	
Extraction of lipid elements	Alcohols
Altered protein configuration	Anesthetics
Alteration of ligands maintaining integrity, e.g., SH	Hg, X-rays, ozone
Transport interference	
Diffusion impairment, secondary to altered steric relationships	Detergents
Change of terminal charge on membrane protein ligands	Mercury, lipid solvents
Competition for carrier substrate	5-Fluorouracil

carrier site; irreversible binding by and of such molecules can block this form of transport.

3.1.4 Special Transport Processes, Endocytosis

The processes referred to as pinocytosis and phagocytosis represent primitive forms of transport across cell membranes. In these processes the cell wall invaginates or flows around a foreign material in order to engulf it and bring it into the cell interior. When the engulfed foreign material is a liquid, the process is referred to as pinocytosis, and when particulates are involved, phagocytosis. The former process occurs in the transport of a variety of compounds, for example, proteins and glycoproteins, hormones, and lipids. It occurs in a variety of cells, for example, leukocytes, hepatocytes, and gastrointestinal cells. Phagocytosis is an important process for the removal of foreign particulates by cells involved in the immune system, e.g., alveolar macrophages in the lung and hepatic Kuipfer. The importance and frequency with which this process occurs in other parts of the body is not presently clear.

3.1.5 A General Overview of Membranes and Toxic Interactions

Xenobiotics first contact living organisms at their interface with the environment, namely, the cell membrane. Given the delicately balanced, dynamic state of this structure it would be expected that its interaction with toxic compounds would almost inevitably produce deleterious consequences. But this inherent dynamic also suggests—within dose limits at the cell membrane—that a finite degree of interaction and alteration can be tolerated with subsequent restitution to a previous functional and structural state. Within such limits, Table 9.1 summarizes a few known relationships between toxic agents and the specific membrane effects they may precipitate.

3.2 Penetration and Absorption Via the Lungs

Although most xenobiotics reach the interior milieu through the gastrointestinal tract, in occupational settings the pulmonary portal of entry is the more usual pathway for uptake of environmental agents. Because of the (1) very large surface area of the lungs (namely, varying between 30 and 100 m^2), (2) extremely large blood flow to the organ carried by a capillary bed approximately 2000 km in length, and (3) the extremely thin physical chemical barrier between air and blood (approximately 0.8 μm), the potential for the ready access of ambient air and its contents to the internal environment is self-evident. Ultimately how body penetration by toxicants occurs in the lungs largely depends on the physical state of such xenobiotics, namely, whether particulate, aerosol, gas, or vapor.

3.2.1 Lung Structure and Function

These factors determine how xenobiotic penetration and absorption occur. As a general principle, the basic purpose of the respiratory system is to bring oxygen laden air into intimate contact with capillary blood. All other structures in the lung are ancillary to this end.

Examination of the cellular components of the lung (Figure 9.4) reveals major differences among cells over the course of the airways that are in contact with the ambient environment. The site of the primary oxygen–blood interface is the pulmonary alveoli found at the terminus of the air-conducting pathways. The alveoli consist of very thin-walled sacs about which is intimately juxtaposed a rich capillary network. The alveoli are made up of mainly flattened epithelial cells (type I) and fewer granular cuboid cells (type II) with a very few brush cells (type III). Thus the capillary network, the more common flattened cells (type I)—approximately 0.2 μm thick—and a basement membrane about 300 to 350 Å thick constitute this basic unit of respiration, the alveolus. Upon the air surface of these alveoli is found a fluid-containing surfactant that is produced by these cells. It is at this very thin alveolar sac wall site that gases are exchanged between ambient air and the blood; here the internal environment makes its closest approximation to the ambient world. Blind terminal vessels containing lymph surround the terminal bronchioles. This lymphatic system drains the secretions of the alveoli, preventing fluid accumulation and thus facilitating gas exchange. This especially protein-rich lymph drains into the lymph nodes at the lung hila or root. Thus the lymphatic system not only serves as a recovery system for protein, but also provides a channel for fluid-media based movement (active and passive) of phagocytes carrying exogenous and endogenous materials out of the lung.

3.2.2 Penetration and Absorption of Gases and Vapors into the Lungs

Gases and vapors come into equilibrium with blood passing through the alveolar capillaries practically instantaneously. This reflects Dalton's law of partial pressures, Henry's law (solubility of gases over a liquid), and the gases' lipid–aqueous partition coefficient.

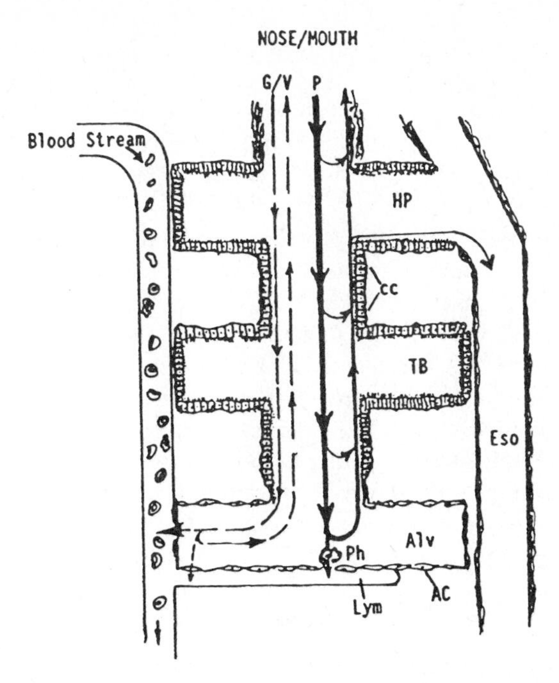

Figure 9.4. Schematic diagram of gas, vapor, and particulate movement in the respiratory system. Gases and vapors generally are exhaled as readily as they diffuse into the bloodstream at the alveolar compartment. To an insignificant extent they may also diffuse into the lymphatics. Particulates (depending upon particle aerodynamics or mass median diameter) deposit at various locations on surfaces of the respiratory tract. Those sufficiently small (<5–10 μm) to be deposited upon the alveolar membrane are generally phagocytosed. Phagocytes may transport particulates into the lung interstitium via lymphatics to lymph nodes or may carry them back up to the mucociliary escalator (see text). Physical penetration by particulates occurs less frequently. HP = hypopharynx; Eso = esophagus; TB = tracheobronchial compartment; Alv = alveolar compartment; Lym = lymphatics; AC = alveolar cells (flattened type I); cc = ciliated columnar epithelium; Ph = phagocyte; G/V = gas/vapor; P = particulate.

Because the alveolar surface is covered with an aqueous film, the water solubility of a gas or vapor (as well as of a particulate or aerosol) assists in determining its rate of alveolar penetration and diffusion. Absorption also varies directly as a function of temperature, ventilation, and blood flow rates.

After passage through the alveolar membrane and capillary wall xenobiotics reach and enter the blood. This process can occur as the result of simple diffusion–dissolution in the blood or by chemical binding or absorption with blood solid or liquid components. Water constitutes 75 percent of the blood volume; however, although water-soluble compounds, for example, methanol, are readily taken up by the blood, other nonaqueous soluble materials can be equally rapidly taken up. Recalling the highly lipid nature of the cell membrane, simple passive diffusion of lipophilic substances should be expected. Indeed, because the blood also contains lipoidal elements, fat-soluble gases or vapors may be rapidly absorbed by the blood. In addition, because nonpolar compounds are rapidly removed from the blood to fat storage depots, the blood returning to the lung will contain low lipid concentrations and/or vapor pressures so that a gradient from alveolar air to blood is maintained.

In addition, the presence of salts, proteins, lipoproteins, and formed membrane elements may variably take up and bind a wide variety of chemical classes independent of lipid–water partitioning.

It should be clear that the uptake of gases and vapors represents the interaction of many variables, physical and chemical. Ultimately, pulmonary gas absorption will largely depend upon the concentration or partial pressure gradient between alveolar gas and blood. Because this gradient depends as much upon the partial pressure of a compound in the *venous* blood returned to the lung, it should be evident that penetration and absorption rates in the lung will indirectly reflect blood saturations at tissue sites resulting from various extrapulmonary biological, physical, and chemical processes.

3.2.3 Absorption of Aerosols by the Lung

Consistent with general principles of membrane passage discussed previously, the absorption of xenobiotics through the lung can be predicted in part. Brown and Schanker (1) indicated that for nonlipid soluble compounds, in the course of a nonsaturable passive diffusion process, the rate of absorption varies inversely with molecular size. Such hydrophilic chemicals pass over as expected through aqueous paths or pores between alveolar cells. Lipophilic aerosols cross the pulmonary epithelium even more rapidly by a similar passive, nonsaturable transfer phenomenon. The higher the lipid–water partition coefficient, the more rapid the passage. In addition, there are energy-dependent, high-affinity active transport systems in type I and II alveolar cells. The active transport mechanisms appear to be cytochrome P-450 dependent. Paraquat appears to be absorbed through alveolar epithelia by this method. By contrast, alveolar phagocytes do not appear to be as active in the uptake of aerosols by such processes.

3.2.4 Particulate Penetration and Absorption in the Lung

3.2.4.1 Principles of Particulate Movement and Deposition in the Lung. The general aerodynamic properties of particulates and aerosols determine if and where they may be deposited in the lungs. Thus, depending upon diameter, mass, and shape, particulates may either preferentially deposit in various regions in the airways or not deposit at all. Accordingly, particles may be sedimented (depending largely upon gravity), impacted (mainly reflecting inertial forces acting with flow-direction change), or subject to diffusion (largely caused by Brownian movement). In addition, as regards irregularly shaped particles (especially fibers) their deposition depends almost entirely upon aerodynamic properties.

Sedimentation is determined by density and diameter (density $\times$ diameter^2), the density of the medium (in accordance with Stokes' law), and the shape of the particle. Particles demonstrating deposition as a result of this essentially gravitational force usually have aerodynamic diameters of 2 μm or more. They usually are deposited in the nasopharynx and the upper reaches of the tracheobronchial tree. Because settling of fibrous particles occurs mainly as a function of the square of their diameter, their gravitational deposition occurs mainly in large airways. Fibers capable of reaching the smaller airways have aerodynamic diameters of less than 3 μm. Thus length and shape rather than falling speed determine their deposition.

Inertial impaction occurs as a fluid stream changes direction and entrained particulates are likewise forced to change their flow direction. In effect such particulates continue their direction of flow because of inertial energy imparted by movement. Accordingly, depending upon the size of the particle, its mass, diameter, and the change of direction of the airway, as well as the velocity of flow, particles either (1) change their direction or (2) do not change and consequently impact upon the wall. This mode of deposition is particularly relevant to large particulates, that is, larger than 5 μm; such particles generally are impacted in the nasopharynx and tracheobronchial region of the lungs. By contrast, particles less than this size tend to reach the smallest airways and alveoli.

The diffusional behavior of very small particles is independent of density; rather it reflects their collision with vibrating gas molecules, that is, Brownian movement. Thus in regions of laminar air flow (alveoli, terminal bronchioles), particles less than 0.1 to 0.5 μm will tend to contact surfaces by simple diffusion.

Another set of determinants governs the deposition of fibrous particulates; that is, their length and shape are more important than falling speed. As the length of a particle becomes greater than its diameter (i.e., more fibrous in configuration), impaction and sedimentation are less relevant to the probability of tracheobronchial contact. Rather, such impaction occurs at bifurcations of terminal and respiratory ducts. The curled fibers of chrysotile tend to exhibit this behavior, whereas the straight fibers of amphibole asbestos follow the aerodynamic flow lines parallel to the airway axis and tend to avoid impact at bifurcations. Thus these latter fibers penetrate more deeply into the lung than those of chrysotile asbestos.

Other physical factors may also govern pulmonary deposition, particularly because these can affect aerodynamic size. Thus particulates that can act as condensation centers may grow in size owing to agglomeration with collisions in humid air. Electrical charge, radius, shape, and retention time can also modify the agglomeration process.

Finally, physiological events must be considered in assessing deposition potentials. With increases in respiratory rate, the volume, air stream velocity, and inertial forces are increased. Airflow rate changes can affect deposition by alteration of the various combination of laminar and turbulent flows; these will have important effects upon deposition probabilities. Conversely, increased respiratory rates will decrease retention time of inhaled air streams, thus potentially decreasing diffusional deposition in the finer radicles of the respiratory tract.

The net effect of these many physical and physiological considerations as they

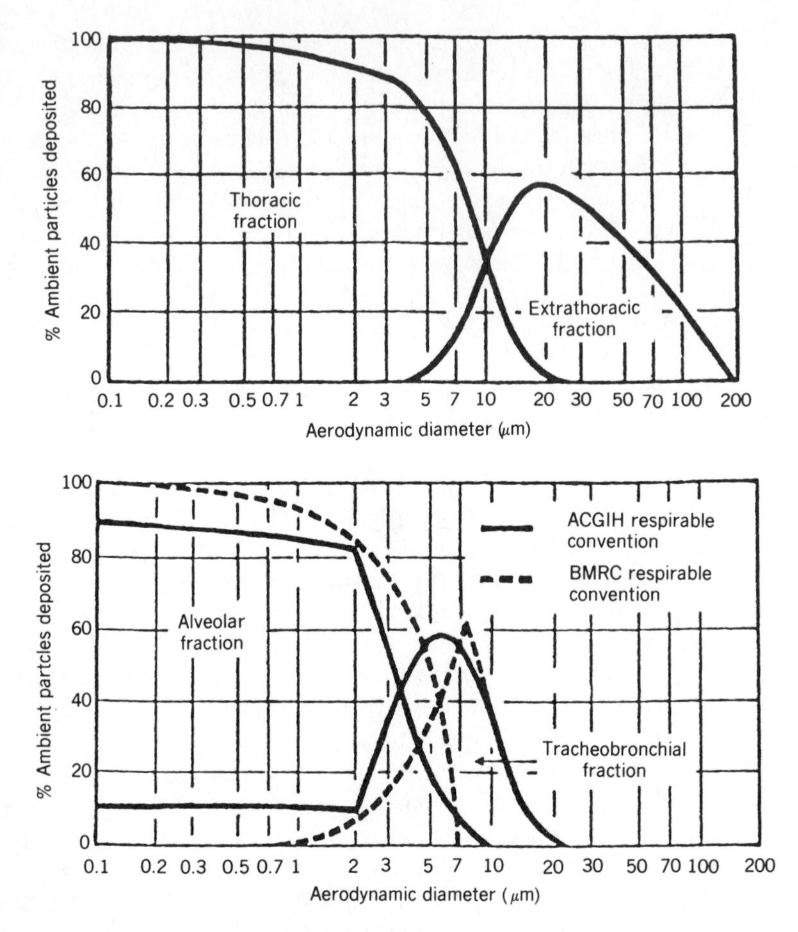

Figure 9.5. Deposition of inspirable aerosols in the extrathoracic (head) and thoracic segments of the respiratory system according to the ISO Model (1983) (2). Top: division of inspirable fraction into extrathoracic (head) and thoracic subfractions, based primarily on oral inhalation (3). Bottom: further division of thoracic subfraction into tracheobronchial and alveolar inspirable particulate fractions. Both the American Conference of Governmental Industrial Hygienists (ACGIH) and British Medical Research Council (BMRC) definitions of respirable fractions are provided in the lower figure.

determine loci of respiratory tract deposition of particles is shown in Figure 9.5. The curves are specific for a 1450-ml tidal volume (moderate to heavy activity) at the rate of 15 respirations per minute. As the respiratory tidal volume increases, the predicted gravimetric deposition shifts upward; conversely, at lower tidal volumes, predicted deposition decreases. Caution must be exercised in using these predictions, for the following factors militate against too-literal application of the data: variations in the mode of breathing (e.g., mouth, nose), the need for idealized models of lung anatomy and airflow patterns in making these estimates, and the occurrence of nonuniform ventilatory distribution. Nevertheless, the use of these relationships to provide approximations of pulmonary deposition patterns is warranted insofar as their approximate nature is appreciated.

3.2.4.2 Physical Penetration of the Alveolar Wall. The smaller of the two main cell types making up the alveolar epithelium normally extend multiple cytoplasmic extensions into the alveolar space. Electron microscopic studies have revealed withdrawal of such extensions with alveolar wall irritation by particulate matter. Penetration of the alveolar wall into its interstitium and lymphatics may be actively facilitated by these villous withdrawals.

3.2.5 Clearance from the Respiratory Tract

3.2.5.1 From the Nasopharynx. Particles having an aerodynamic diameter of 10 μm or more are usually deposited in the nasopharynx; the probability of deposition here increases directly with particle diameter. Nasal hairs and changes in flow direction over the nasal turbinates enhance entrapment and impaction. Ciliated epithelium present here directs the mucus flow downward to the pharynx so that entrapped particles are swallowed, making them eventually available for possible gastrointestinal tract absorption. Thus even though a particle may be too large to reach airways below the level of the nasopharynx, it is carried in this mucus to the lower throat (i.e., the hypopharynx) and swallowed.

3.2.5.2 From the Tracheobronchial Tract. Ciliated columnar epithelial cells lining the air-conducting tubes are found from the level of the seventeenth branching upward to the nose. These cilia, that is, fine, 7 to 10 μm whiplike extensions of the cell membrane, simultaneously beat upward during their rapid phase and then are restituted relatively slowly to their previous state. These projections are immersed in and covered by thin, sticky mucus formed by specific mucus secreting cells of the airways. Thus, by such repeated cycles of coordinated, upward beating of the cilia this overlying mucus film is moved headward up the tracheobronchial tree, eventually reaching the hypopharynx. Particles landing upon this sticky mucus film covering the interior of the airways are gradually carried in this mucus much as they would be transported by an escalator. This effective particulate transport system, that is, the mucociliary escalator, moves this mucus blanket headward at a rate of 1 to 15 mm/min. Clearly it provides an effective mode of particle removal from this major portion of the respiratory tract. It has been found that clearance is effected in two phases: a rapid phase taking 2 to 4 hr, and a slow phase requiring 24 to 30 hr. With large particles the rate of transport is more rapid. However, irritant gas and vapors decrease the ciliary beat rate, which in turn lessens mucociliary escalator flow.

It should be apparent that particulate compounds deemed too large to reach the alveolus for possible absorption eventually may be made available, as a consequence of the escalator, for body uptake via the gastrointestinal portal of entry. Thus in considering total body burden with respiratory tract exposure, particulates larger than 10 μm should not be disregarded as contributing to potential total body burdens.

3.2.5.3 Clearance by Phagocytosis. Pulmonary phagocytes are similar to mononuclear phagocytes found elsewhere in the body, for example, the liver and peritoneal cavity. There are at least three major types of pulmonary phagocytes: (1) alveolar phagocytes found in the seromucous, surfactant-rich fluid on the surface of the alveolar cells, (2) airway phagocytes found either on the surface or beneath the mucus layer of the conducting airways, and (3) macrophages found in various interstitial tissue compartments, for example, alveolar wall, peribronchial, and perivascular spaces and lymph vessels and nodes. Although they demonstrate variability in their specific structure and function, they all appear to play an important role in defending the airway and alveolar surfaces from toxic particles, gases, and pathogens, keeping these surfaces clean and sterile.

In the event of injury by particle or pathogen deposition, pools of macrophages are tapped by (1) an active cell removal system in the interstitium, (2) increased cell precursor production, (3) release from lung reservoirs, or (4) increased influx of blood monocytes. Although smaller particles seem to be more effective in eliciting phagocytic mobilization, it is not clear if particle number or surface area is the relevant factor precipitating this response (4).

Within each phagocyte are found subcellular vesicles (diameter 0.5 μm or less) rich in lytic enzymes, for example, ribonuclease, glucuronidase, phosphatases, and phospholipases. After a phagocyte engulfs a particulate and incorporates it into the cell, lysosomes attach themselves to the phagosomal membrane. These membranes become continuous and the lytic enzymes kill the pathogens.

For particulates it has been found *in vivo* that among dusts not known to be reactive that 50 to 75 percent of particles were taken up in 2 hr, more than 95 percent by 10 hr, and essentially all by 24 hr. By contrast, when alpha quartz was tested in vivo, clearance was cut by 50 percent in the early phases of exposure (5).

However, these protective tissue elements may also play a deleterious role in the development of lung disease, for example, emphysema and fibrosis. In addition, certain agents may be inappropriately handled by phagocytes [e.g., silica, lead (6)] or may lead to local segregation and concentration of toxic particles (7).

In general, as in most cases when phagocytosis is effective, it provides a useful removal system for transport of toxic compounds or particles from lungs. The phagocytic route of egress is mainly via the mucociliary escalator and toward the pharynx or by direct carriage in the pulmonary lymphatic vessels and system. Ultimately these cells disintegrate with the particulates (and cell debris) reaching the pharynx and eventual gastrointestinal ingestion. A smaller proportion of the phagocytes move through alveolar walls to the lymphatic drainage system.

3.2.5.4 Transport of Particulates in the Lymphatic System. Because lymphatic vessel walls are permeable to macromolecules, large protein molecules in body fluids easily move out of these structures. Small particles that succeed by direct penetration—or phagocytosis—in reaching the lymphatics slowly make their way to the pulmonary lymph nodes where they can accumulate. After protracted periods such particulates may ultimately find their way to the reticuloendothelial system elements found in the various organs, namely, liver, bone marrow, and spleen.

3.3 Gastrointestinal Absorption

The gastrointestinal tract is usually considered a less important portal of entry than the lungs for the toxic agents encountered in the occupational setting. Although in general this is true, it is too frequently forgotten that ambient particles or aerosols too large to reach the alveoli eventually are available for absorption via the mucociliary escalator. Thus it should be remembered that those particulates considered irrespirable may nonetheless be presented for absorption in the gastrointestinal tract. Actually these particles are of a size range (i.e., 10 to 100 μm) that is quite readily absorbed from the gut. This consideration is especially important in industrial hygiene because sampling strategies designed only to measure respirable particle sizes are appropriate only for those materials that produce only pulmonary pathology. By contrast, measurement of only respirable airborne chemicals that are: (1) capable of producing systemic toxicity and, (2) that can be absorbed both via alveoli and the gastrointestinal tract *will underestimate potential occupational dose*, for example, for particulate lead and fluorides.

In addition, for work environments where personal hygiene is poor, hand to mouth ingestion and absorption via the gastrointestinal tract is important. For the nonoccupational physician—and especially the pediatric clinician—this portal of entry is undoubtedly highly significant.

3.3.1 Structure and Absorptive Function of the Gastrointestinal Tract

Basically the gastrointestinal tract (GIT) can be considered as a tube going *through* the body; essentially, until xenobiotics transit this tube's wall, they remain outside the body's economy or its internal environment. Accordingly, poisons orally ingested and swallowed may produce no systemic effect until they traverse this tube, that is, are absorbed *across* the GIT wall.

This process of transepithelial absorption may occur from mouth (e.g., nitroglycerin) to anus (e.g., chlorpromazines); however, most absorption takes place in the small intestine. Until recently, the stomach was considered to be a significant site of absorption only for water and ethanol. In the past this was believed to reflect a short passage time in the stomach as well as its low pH. However, in keeping with the Henderson–Hasselbach equation, strong acids with a low pK_a would be expected to exist in a nondissociated, nonpolar form in the stomach and thus will readily be absorbed. Nevertheless, in a larger perspective, gastric absorption is relatively less important in GIT absorptive processes.

Digestion is accomplished by enzymes excreted in a mucous medium by the GIT and its associated structures (e.g., liver bile acids, pancreatic enzymes); these catalysts accomplish the breakdown of dietary protein, carbohydrate, and lipid. Mechanical mixing is achieved by autonomic system controlled pulsatile contraction of the digestive tube effected by successive contractions of muscle layers beneath the epithelial lining of the GIT lumen. Interposed between these two elements (i.e., the epithelia and muscle layers) is an extremely rich capillary network. Thus absorption of GIT contents proceeds through the epithelial cells lining the inner surface of this tube, across a cell basement membrane and subepithelial layer into the capillary network. The surface of the GIT is characterized by numerous upfoldings providing a markedly increased surface area for absorption. In contrast to the intestine, the stomach's increased surface area provides for more cells of a secretory type that produce enzymes, mucus, and hydrochloric acid.

Microscopic examination of the intestinal wall reveals that its surface consists of numerous, fingerlike projections covered largely by a single layer of columnar cells. On the surface of each cell in turn are microvillous projections that actively "pump in" materials for presentation to a bed of capillaries associated with each cell. These capillaries all drain into the portal veins leading to the liver where multiple biotransformation systems are available for further chemical alteration of absorbed compounds. By virtue of these systems of villi and microvilli, which increase surface area by about 600 times, that is, to about 2000 F^2, the absorptive capacity of the small intestine is markedly enhanced. Almost one-half of the total GIT mucosal area is found in the upper quarter of the small intestine.

Absorption of ingested material proceeds by passive transfer in the proximal small intestine; in the distal portion both passive and active transport occur. Most essential nutrient moieties (e.g., amino acids, sugars) depend upon active transit through the gut wall. However, active transport and absorption of xenobiotics appear to be relatively uncommon in the GIT. But for toxic materials that are structurally or electrochemically similar to such nutrients, active transport can facilitate absorption of what normally would not be absorbed, for example, 5-fluorouracil by the pyrimidine transport system, or cobalt, manganese, and thallium by the iron transport system.

The microvilli are covered by a coat of mucopolysaccharide; this probably acts as a barrier against (1) bacteria, (2) macromolecules, and (3) ionized compounds. At the intestinal surface is found a thin acid layer at pH 5. Although this layer is found in what is generally a basic environment of pH 7.3, such an acidic layer affects the extent of ionization of compounds in the GIT. This layer is immersed in a relatively thicker second layer of water that will restrict passage of nonpolar compounds that might otherwise penetrate into the lipophilic components of the walls of the microvilli.

As previously noted the pH of the gut contents plays a critical but not limiting role in absorption. The Henderson–Hasselbach equation predicts that nonionizing weak acids would be minimally absorbed in the small intestine. However, the small amount that is absorbed into the pH 7.4 environment of the plasma is transported away, maintaining the concentration gradient across the membranes. In addition, because of the huge surface area of the small intestines, the capacity for net absorption is great although local absorption rates are low.

Physical factors further determine absorption in the GIT. Although nonpolar, nonionizable compounds (e.g., lipophilic organic solvents) most readily penetrate cell membranes, such dissolve poorly in the largely aqueous GIT fluids with consequent minimal absorption. Similarly, solid particulates of a low water solubility in such an essentially aqueous medium dissolve poorly and thus do not effectively contact the absorptive membranes. If such particulates are large, less is absorbed by diffusion, because the rate of dissolution is directly proportional to the surface area; it is upon this basis that ingested metallics demonstrate little if any toxicity potential, for example, mercury metal.

The presence of multiple metal salts may alter the absorption of other individual metallic compounds. For example, cadmium decreases zinc and copper absorption (8). The presence of phosphate and/or calcium may inhibit absorption of aluminum by formation of less soluble metallic-salt complexes. Although lead is not readily absorbed from the GIT, EDTA chelation in the gut increases its absorbability and bioavailability.

Multiple physiological factors directly affect GIT absorption. Enzymes, microflora, and biliary emulsifiers all may alter or degrade toxic compounds, minimizing or enhancing their toxicity. For example, although oral ingestion of snake venoms is innocuous, the pH of the stomach promotes formation of carcinogenic amines when native secondary amines present in food interact with nitrites used as food additives. Intestinal flora readily convert aromatic nitro compounds to carcinogenic aromatic amines, and other flora can degrade DDT to DDE.

Other physiological factors may alter GIT absorptive capacity. In general decreased peristalsis will increase absorption; however, for a compound that may have an inherently low absorption rate, decreased GIT motility can markedly increase its residency time and thus its net absorption. Variations in local blood flow may alter absorptive rates, especially for lipids. Because these readily penetrate cell membranes, decreased blood flow will decrease local lipid uptake. Finally, age itself markedly affects GIT absorption; thus newborn rats may absorb cadmium 24 times as rapidly as adults (9).

Although the foregoing has considered xenobiotics occurring either as solutes or otherwise in liquid form, it has also been found that particulates may move across the epithelial cell walls of the GIT. Thus colloids and small particles can move through aqueous cell membrane "pores." In addition, such particles as large as 23 μm have been observed to be taken up by pinocytosis into cellular vesicles that are subsequently discharged in the connective tissues beneath these cells. Ultimately such particles are either ingested by phagocytic cells or directly taken up in lymphatic fluids.

As should now be readily apparent, the wide variety of food components, their hydrolysis and other breakdown products, and the potentials for binding to such variegated diet-derived components provides a highly variable and rapidly changing GIT mileau. Accordingly, prediction of absorption rates under these complex conditions all too readily approaches conjecture.

Regardless of the effects of this multiplicity of variables, once a xenobiotic traverses the epithelial wall of the GIT it is subject to one of two fates. For most materials absorbed, carriage in the portal circulation to the liver is the rule; here multiple biotransformations take place. Such metabolic transformation may produce a less toxic product so that little of the parent toxic compound reaches the systemic organs. If hepatic metabolism over the specifically required pathway were saturated or detoxifying rate limits were approached, extrahepatic target tissues may not be protected. Conversely, compounds of high cytotoxicity may selectively damage the liver by exposing it to high portal vein concentrations whereas other organs are spared such relatively higher, toxic concentrations.

For some small percentage of GIT-absorbed materials lymphatic drainage bypassing the liver is the rule. In such manner ingestion particularly of lipophilic material might avoid a first pass through the liver's detoxifying systems, thus allowing toxic consequences to occur elsewhere in the body.

3.4 Absorption and Penetration Through the Skin

The skin, a very important portal of entry in the occupational setting, is specifically considered in Chapter 10.

4 TRANSPORT AND DISTRIBUTION OF XENOBIOTICS

4.1 Introduction

After foreign materials have penetrated the epithelial barriers, the next medium encountered is the blood or the lymphatic or interstitial fluid. The liquid phase of the blood, in which are suspended various formed elements (i.e., red blood cells, platelets, white blood cells, and other reticuloendothelial derived cells in various forms) is referred to as the plasma. Humans possess from 45 to 68 ml blood/kg body weight, or from 5 to 5.5 liters in a 70-kg man. Materials carried in this medium may leave blood vessels via physical "pores" in the capillary wall, by diffusion or by special transport mechanisms previously discussed. In effect, the major determinant of capillary bed egress is molecular size. With capillary wall pore sizes as large as 30 Å, compounds of up to 6000 Da—regardless of electrical charge—may directly exit from capillaries into the extracellular compartment. Larger molecules may directly exit via pinocytosis but at a very much lower rate.

After leaving the vascular system, compounds enter the extracellular compartment whose constituents differ slightly from the plasma because of variations among plasma components' ability to leave the blood system. This extracellular compartment serves as the final liquid medium before a xenobiotic encounters a tissue cell membrane. The processes governing transit of that membrane have been described in Section 3.1.

4.2 Blood Elements and Their Transport of Xenobiotics

4.2.1 Transport by Formed Elements

Because only red blood cells play a role in transport, this discussion does not consider leukocytes or platelets in this context. Mature erythrocytes as they are found in the blood possess no nucleus. The carriage of the essential element oxygen occurs because of the presence of a divalent iron contained in the hemoglobin molecule; this particular molecule's configuration results in an extremely strong affinity for oxygen. If this ferrous iron is oxidized to the trivalent state, methemoglobin is formed and oxygen transport is severely restricted.

Each 100 ml of arterial blood contains about 15 g of hemoglobin, which combines with 19 or 20 ml of oxygen. In blood that is in equilibrium with air at normal atmospheric pressure, about 1 percent of the total oxygen is in solution in the plasma. Normal venous blood carries 55 to 60 volume percent carbon dioxide; it drops off about 10 percent of this in the lungs where the blood is once more oxygenated. The amount of carbon dioxide in simple solution is about 3 volume percent in the venous blood and 2.5 volume percent in the arterial. Blood flows through the capillaries of the lungs in approximately 1 sec (10), and through active tissue in a similar time. At such a flow rate it is evident that mere solution of oxygen and carbon dioxide cannot account for the exchange of these gases, even though the equilibration in the lungs is highly efficient. The speed of exchange is due to the reversible combination of these gases with hemoglobin in the oxyhemoglobin and carbamino reactions, respectively, as well as reflecting the presence of catalysts in the red cells that play an important part in the oxygen and carbon dioxide exchange. Carbonic anhydrase greatly increases in either direction the reversible chemical reaction $H_2CO_3 \rightleftharpoons CO_2 + H_2O$, and oxidation and reduction in the tissues are greatly accelerated by oxygen-activating catalysts or by dehydrogenases.

It must be recognized that the transport of oxygen or other gases is strictly a passive, partial pressure gradient-dependent phenomenon. Transport by diffusion, either from alveolar air sacs to blood or from tissue cells to blood, involves passage through cell membranes. Because oxygen and gases other than carbon dioxide never move counter to partial pressure gradients, it may be said that such exchanges are governed simply by physical laws of diffusion and solubility.

Although the oxygen binding affinity is about 1000 times greater than its dissociation constant, carbon monoxide's binding affinity is 242,000 times greater than its dissociation constant. Accordingly, this 242-fold greater affinity favors not only a greater binding capacity for CO, but also produces a considerable reluctance for CO dissociation in the capillary bed.

The aggregate surface area of red blood cells (namely, 3000 to 4000 m^2) favors transport by adsorption (van der Waals forces) and/or stromal binding of physiologically necessary as well as xenobiotic materials. Most such molecules are in equilibrium with their fraction bound to plasma. Because the stromal density of the erythrocyte is low and readily permeable, xenobiotics can have ready access at multiple sites to the constituents of the cell, namely, the protein fibrils (albuminoids) surrounded by lipids (cholesterol and esters, neutral fats, cephalin, lecithin, and cerebrosides) as well as its other contents, for example, hemoglobin. Thus, whereas O_2, HCN, CO, and Se are bound to heme, As and Sb bind with the sulfhydryl ligands of the globin protein. Other toxic or essential compounds are bound to the red cell although their locus is not well defined. This group includes organic mercury salts, Cr, Pb, Zn, Th, Cs, and steroids.

Although the erythrocyte has no nucleus it must be considered a metabolically active entity. Accordingly, this mature, aneucleated cell contains the enzymes lipase, catalase, and anhydrase, as well as the others required for porphyrin synthesis. It is this latter activity and changes in δ-aminolevulinic acid dehydrase activity that has proved to be such a highly sensitive indicator of lead exposure.

4.2.2 Transport of Xenobiotics in Plasma

4.2.2.1 General. Numerous electrolytes found in plasma as ions are in equilibrium with nondissociated molecules. The dissociated ions readily move from blood to interstitial fluid and into cells by diffusion. For example, Be and the alkaline earths Ca and Sr exist in this form; others, for example, Yt and Pu, exist as microcolloids in association with plasma proteins. Gases and vapors are essentially physically dissolved in the plasma.

It should be noted that very few substances are transported in the blood in association with only *one* blood component or element. For example, the uranyl ion is approximately 40 percent bound to plasma protein and the other 60 percent occurs as a diffusible bicarbonate. In such fashion, most elements exist in a state of dynamic equilibrium between various formed and/or liquid plasma components.

4.2.2.2 Transport by Plasma Proteins. Notwithstanding the foregoing, transport by these proteins represents a most important vehicle for carriage of exogenous materials through the body. This reflects in part the finding that most plasma proteins possess surface areas of 600,000 to 800,000 m^2 for physical interactions, for example, electrostatic attraction, van der Waal forces, and ion radius, all of which affect binding behavior. Thus divalent cations, for example, Ca^{2+} and Ba^{2+}, form unstable complexes with proteins, whereas heavy element polyvalent cations form more tightly bound and less dissociable complexes.

Protein binding sites may be saturated; competition for binding sites may result in the displacement or inavailability to another compound. Such an inavailability of binding sites result in marked increases in plasma concentrations of the nonbound compound, with consequent cellular availability and effect potentials.

Albumin, with a molecular weight of 6800, is the most abundant and important plasma protein involved in transport. This molecule possesses 109 cationic and 120 anionic ligands for binding, distributed over a relatively large surface area. The most common of the ligands are carboxylic binding groups associated with the asparaginic and glutaminic amino acids. There appear to be six binding regions on albumin (11) where protein–ligand interactions occur. This binding results mainly from van der Waals and hydrophobic forces, the latter reinforced by hydrogen bonds. Many physiologically important compounds are carried in the albumin fraction, for example, bilirubin, porphyrins, ascorbic acid, fatty acids, and cholinesterase. Many cations (e.g., Cu and Cd) and anions (e.g., Br, I, and CNS) are partially bound to albumin, the former via imidazole or carboxyl ligands.

Generally, metal ions are bound by sulfhydryl, amino, carboxyl, and imidazole groups of albumin amino acids; compounds with an *o*-carboxylic or *o*-hydroxylic group on a benzene ring (e.g., *o*-cresols, nitro- and halo-substituted aromatic hydrocarbons, and phenols) demonstrate considerable binding affinities for albumin.

By contrast with albumin, the globulin fractions are believed to participate in relatively fewer transport-related interactions with other compounds. The transport of mono- and bivalent copper is accomplished by ceruloplasmin found in the α-2 globulin fraction, whereas divalent iron is carried by transferrin found in the β-1 globulin fraction. Steroid hormones are transported by interactions with α-1 globulins, as are vitamin B_{12} and thyroxine. The γ-globulin fraction is believed to be essentially largely associated with the immune mechanisms.

The binding of multiple toxicants with plasma protein has been extensively reported; however, specific binding site locations are less clear. Phosgene and nitrous oxide fumes react with amino groups of proteins, whereas methyl bromide reacts with sulfhydryl binding groups on amino acid proteins. Isocyanate, aromatic amines, and carbon disulfide react with proteins, the latter binding to peptides forming dithiocarbanates and also cyclic mercaptothioazolinone.

Because lipoproteins are found in the globulin fraction, many lipophilic compounds may be transported by these proteins, for example, benzopyrene.

Variations in sites and strengths of binding in the blood affect the bioavailability and hence metabolism and elimination from the body of xenobiotics, for increase in binding diminishes diffusibility. For example, lead bound to red blood cells has a biological half-life of 30 hr, in contrast to lead binding to plasma proteins, with a half-life at that site of only 30 to 40 min. Thus while the protein-bound fraction has a greater bioavailability, lead bound to erythrocytes will tend to spare target organs from exposure to peaking dosages. If a substance binds at various sites more strongly than at others, its bioavailability to other target loci is likewise diminished. Protein binding by a toxic substance may, by such successful competition for such sites, cause the release of other protein-bound physiological substances. This in turn can produce other untoward effects, for example, as in the release of histamine or serotonin. Finally, binding by toxic substances may pathologically alter protein structure, as is believed to be the case for ozone, isocyanates, aromatic amines, and fresh zinc fume.

4.2.2.4 Transport by Blood Organic Acids. The organic acids found in plasma are potent complexing agents and hence important transport media for xenobiotics. The complexing of xenobiotics by blood organic acids depends upon the physicochemical properties of both reactants (e.g., valence, ionic radius) as well as the metabolism of these acids.

Although citric acid is the most effective of this class of compounds, lactic acid—though less active—is present at four- to six-fold greater concentrations in blood. Glutaminic acid and the amino acid α-glutamic acid are also effective complexing agents. The anions of organic oxy- and amino acids readily complex with alkaline earths and some heavy elements present in blood as cations.

The consequences of such bindings are variable, depending upon diffusibility, stability, and metabolism of these complexes. Thus although citric acid may complex cations of Bi, Po, and Y, such complexes may ultimately enhance the bioavailability of such toxicants as the organic acid is metabolically broken down within the cell. In the case of bone-seeking elements, for example, Pu, which complex with citric acid, the breakdown of such complexes in the region of bone minerals effectively makes such osteotrophic elements readily available for bone incorporation. Conversely, complexing of Yt by glutamic acid to form a diffusible soluble form allows excretion of this element in the urine.

4.3 Transport by the Lymphatic Vascular System

In contrast to the high-pressure blood circulatory system, the lymphatics are a low-pressure aspirating system for fluid lost from capillaries and tissues. Because such vessels do not possess basement membranes, they are extremely permeable, permitting ingress of relatively large molecules. The role of alveolar macrophages and their access to the lymphatics are discussed in Section 3.2.5.3. As the lymphatics ultimately drain macrophages into the circulatory system at the thoracic duct, such particulates are thus made available for metabolism, particularly by the reticuloendothelial system.

5 XENOBIOTIC DISTRIBUTION TO AND DEPOSITION IN ORGANS AND TISSUES

5.1 Introduction

In terms of its constituents, the body is largely water, namely, 70 percent by weight. For this reason, in considering where xenobiotics are distributed within the body, it is convenient first to divide the body into three fluid compartments, that is, the vascular or plasma, the interstitial or extracellular, and the intracellular compartments.

It must be emphasized, however, that this construct simply represents a convenient model for analyzing the distribution and movement of xenobiotics throughout the body. As discussed in Section 4, these transfers operate under constraints imposed by the physicochemical nature of the compound and the bodily environment in which it is found.

The vascular system that brings each organ its supply of blood provides a unifying channel from the external environment, across the interstitial compartment, and finally to each cell membrane. The priority the body places upon the needs of each organ is suggested by comparisons between the mass of each organ and their relative blood flow rates (Table 9.2). The concept that considers the body as a liquid continuum "separated" into permeable compartments allows for kinetic analyses of body xenobiotic distributions.

Further breakdown of the intracellular compartment into organs or regions of interest, each supplied and drained by blood flow, permits more specific multi-

Table 9.2. Relative Body Mass and Blood Flow for a 70-kg Man

Tissue or Organ	% Total Body Mass[a,b]	% Total Cardiac Output[c]
Brain	2	15
Liver	2.6	30
Kidneys	0.4	20
Heart	0.4	20
Skin, muscle	43.7	10
Bone, connective tissue	17	
Fat	18	3

[a]Tissues here constitute 70% of total body weight.
[b]Data from Reference 12.
[c]Data from Best and Taylor (13).

compartmental modeling and toxicokinetic analysis of xenobiotic distribution and movement.

5.2 General Factors Affecting Distribution

As noted previously, physicochemical factors as well as the state in which a material exists in blood or plasma (e.g., ion, molecule, binding affinity for blood elements, molecular weight) affect transfer potentials and thus distribution. Variations in blood supply and flow rates may affect how a toxic compound is distributed. Such variations may result from the relative blood flow to specific organs (see Table 9.2), neurohumoral-induced flow alterations, and so on. In addition, concentration gradients between compartments, the presence of other substances in the cell, affecting cellular permeability or intracellular distribution all may have an impact upon how a xenobiotic distributes itself within the body.

When a xenobiotic is ultimately conveyed to a tissue cell membrane, numerous factors determine its transmembrane movement. Thus nonionized compounds that are lipid soluble readily diffuse through cell membranes. In the event the cell or tissue has a high fat content, such nonpolar compounds readily dissolve in such tissue, often remaining there for long time periods. Such is the case for a large group of compounds of environmental interest, for example, PBB and DDT, in which case they are sequestered in fat depots for months and years. Other materials accumulate in association with specific macromolecules, for example, binding of CO with hemoglobin. In still other cases physical dimensions of xenobiotics may be similar to those of physiological components found in specific tissues. Such similarities may result in exchange with these specific cellular elements, for example, alkaline earths and fluoride incorporated by ion exchange with calcium of bone.

In the process of biotransformation multiple events occurring within the cell will affect body distribution. Reactive intermediates of biotransformation may combine with cell macromolecules, impeding their excretion from that site. Any reaction within the cell resulting in altered permeability of cellular and intracellular membranes, for example, endoplasmic reticulum, or which competes for binding sites within the cell, or alters pH, pK_a, or blood flow can affect permeability, transfer rates, and ultimately distribution. Depending upon whether such actions take place at a target cell or at a storage site, a toxic outcome or attenuation of damage potentials, respectively, may result.

5.3 Specific Structures Limiting Distribution

5.3.1 The Blood–Brain Barrier

This structural concept is of major importance among the several structural barriers to transfer of substances from blood to tissue. The blood–brain barrier, found at the capillary–glial cell junction, is relatively less permeable than all other such membranes. (However, because of the usual high lipid content of membranes the transit of fat-soluble compounds are *relatively* less impeded than ionized, dissociated compounds.) This lesser degree of permeability results in part from the capillary–endothelial cell anatomical junctions being in unusually close apposition, which leaves fewer intercellular pores. In addition, brain capillaries are closely surrounded by the processes of glial brain cells; accordingly, at least two cells must be traversed if a blood-borne chemical is to flow from blood to brain cells. Another impediment to movement reflects the relatively lower protein content of the brain interstitial fluid. Although this barrier is highly effective in controlling extravascular transfer, it is not uniformly distributed throughout this organ. Finally, this barrier is not as effective in infants as in adults.

5.3.2 The Placental Barrier

The so-called placental barrier has been considered to represent an effective impediment to transfer of toxins from the maternal bloodstream to that of the fetus. However, it has become clear that many biopathogens (e.g., viruses, spirochetes), large molecules (e.g., antibodies), and even whole cells (e.g., erythrocytes) can cross this barrier.

Although toxic chemicals can readily cross this barrier, because biotransformation systems are present in this organ, some chemical substances can be detoxified in situ here. Further, if some toxicants have a high affinity for specific maternal tissues (e.g., PCB in fat depots), little may be available for transfer to the fetus. Furthermore, the behavior of various organs of the fetus differs from that of the adult. Thus the immature blood–brain barrier more readily permits passage and deposition of lead or methylmercury into the fetal brain. By contrast, xenobiotics concentrated in the human liver may not be accumulated in that fetal organ.

Accordingly, although the placental barrier may alter the potential for transfer from maternal to fetal blood, this movement is carried out in a highly variable and compound specific manner. Any toxicant in question cannot be presumed to place the fetus at more or less hazard; each compound presented to the placenta can be considered only on the basis of its individual chemical, biological, or physical properties.

5.4 Redistribution of Xenobiotics

Xenobiotic concentrations in most organs will change with the passage of time. This may occur in order to maintain intercompartmental equilibrium as tissue/blood concentration gradients change. Blood flow rates to various organs (see Table 9.2) initially will determine tissue extracellular concentrations; the permeability of the tissue and relative availability of cell binding sites will determine subsequent deposition. The latter two factors will change with tissue uptake and subsequent biotransformation. Redistribution of lipophilic substances occurs regardless of biotransformation, with such xenobiotics transferring readily to and from adipose tissue and brain.

With the passage of time and uptake by specific tissues, blood carriage becomes relatively less important. At this point tissue-specific binding affinities are more important in determining tissue concentration of xenobiotics. Thus although the liver takes up 50 percent of a dose of lead within 2 hr of administration, 30 days later 90 percent of the remaining lead is found in bone, having substituted for calcium in the hydroxyapatite crystal lattice of that organ.

5.5 Specific Body Compartments of Deposition and Accumulation

5.5.1 Liver and Kidneys

Although these two organs have a high binding capacity for chemicals and metals, the mechanisms responsible for this behavior have not been clearly elucidated. On the basis of more recent studies the role of active transport and site of specific cell component binding has become clearer. Thus cadmium, mercury, and bismuth have been found to bind preferentially with the cytoplasmic soluble fraction of hepatocytes; the latter two elements are also bound to nuclear soluble protein and to a lesser extent with mitochondrial fractions of liver cells. Upon further investigation of hepatocellular protein binding, it has been found that a number of heavy metals are bound to a specific protein, metallothionine (14). Biosynthesis of this protein is induced by increased metal loading of the liver and kidney. This low molecular weight protein (i.e., 10,000 to 12,000 Da) consists of a dimer of equal size; being rich in cysteine it possesses many sulfhydryl groups that readily bind certain metals. It was found that a metallothionine molecule can bind to metal ions, 3 SH groups being required to bind each ion of such metals as Zn, Cd, Hg, Bi, Co, and Cu. This binding of free metal ions represents a form of detoxification of xenobiotics. However, any benefit may be transient, because mobilization of metal ions can occur owing to their displacement as a consequence of unsuccessful competition for binding sites.

Another hepatocellular cytoplasmic protein, ligandin (15), one of the forms of glutathione reductase, readily binds a number of organic ions. This protein has been shown to bind azo dye carcinogens, steroids, and other xenobiotic organic acids. Such binding suggests a role for this protein in facilitating transfer from plasma to liver, thus effectively removing such compounds from further circulation.

5.5.2 Deposition and Accumulation in Lipid-Rich Organs and Tissues

In general terms, nonpolar, nonelectrolyte substances that minimally dissociate and that demonstrate a high Overton–Meierhof partition coefficient for lipids will readily permeate lipid components of the cell membrane. Also, as previously noted, changes in pH and pK_a will also alter dissociation. In addition, chemical structural characteristics can change liposolubility; for example, substitution of an alkyl group greatly increase the liposolubility of hexobarbital.

With these caveats in mind one would thus expect that such substances, which would dissolve in neutral body fats, would also readily accumulate in large quantities in the body, because such fats constitute from 20 to as high as 50 percent of total body weight. It should therefore be predicted that compounds of industrial importance such as industrial solvents (e.g., alcohols, ketones, glycoesters) and chlorinated hydrocarbons (e.g., chlorinated aliphatics, pesticides, PCBs) all would tend to accumulate in the large lipid compartment.

Of particular importance in this regard is the central nervous system, which is rich in various lipids. As previously noted, the blood–brain barrier does not provide a completely effective blockade to nonpolar, nondissociated compounds. Thus compounds such as organic solvents (e.g., halogenated aliphatics, alcohols) and organometallics (e.g., alkyl mercury and leads) readily pass through this barrier.

Once this constraint to their movement is overcome, the liposolubility of this class of compound can further affect brain structure and nervous system functional integrity. Neurofibers acting as electrical conductors require an intact, electrically functional "insulation." The lipid myelin sheath, or a layer of Schwann cells rich in lipid that serve such function, can be disrupted by uptake of liposoluble xenobiotics. For yet other reasons, compounds such as CS_2, Pb, Hg, and Mn appear specially to have an effect on the brain's cellular elements. Yet other phenomena may precipitate neurotoxicity. For example, tertiary configuration and consequent successful competitive inhibition of acetylcholine esterase activity at the neural synapses appear to be responsible for the toxicity of the organophosphate pesticides.

In addition to the relatively high lipid concentrations of the nervous system, body fat depots that act as body stores of readily mobilized energy-rich triglycerides make up from 10 percent to as much as 50 percent of total body mass. Thus a large compartment for uptake and storage of xenobiotics is represented by such fat depots. Accordingly, accumulation at such sites can temporarily remove liposoluble xenobiotics from circulation. It may thus minimize the available substrate for biotransformations occurring elsewhere and the toxicity that might follow. Further, because of the relatively smaller blood supply (cf. Table 9.2) in such storage sites, compounds once deposited here tend to turn over less rapidly from these depots. For these reasons, biopsy of fat stores can provide an indication of previous exposures long after exposures have ceased. However, although such deposition sites are considered capacious and turnover is less rapid, these loci do not necessarily imply sequestration or protection of the body from the effects of their xenobiotic contents. For example, fat—like all other body components—undergoes continuous turnover of its chemical elements. In the event of sudden needs for energy-rich substrates (e.g., as with starvation), fat stores—and their associated xenobiotics—can be rapidly mobilized to the general circulation. This process under neurohumoral control is mediated via multiple enzyme systems found in adipose tissue, for example, lipase, diastase, phosphatase, and lecithinase, as well as catalysts involved in intermediate carbohydrate metabolism, for example, hexokinases and dehydrogenases. In addition, this tissue, which is now understood to be metabolically active, is known to carry out biotransformations not directly related to these energy mobilizing activities. Experimental evidence indicates that oxidation of benzene to phenol or hydrolysis of esters may take place in fat storage sites.

Finally, deposition in other tissues that contain lipids in somewhat higher concentrations than in general may produce direct effects at such places, for example, the hematopoietic disruptions of the bone marrow caused by benzene.

5.5.3 Deposition in Bone

Because bone constitutes 10 to 15 percent of total body mass, these structures represent a significant potential site of deposition for osteotrophic compounds or elements. Minerals are the major components of bone, representing 72 percent by weight of this tissue. The primary bone mineral is constituted by hydroxyapatite, $Ca_{10}(PO_4)_6(OH)_2$, which occurs as a crystal 25 to 30 Å thick and approximately 400 Å long and wide. The surface area of this crystalline structure averages 100 m^2/g of bone, totaling 7×10^6 m^2 for a standard 70-kg person. This mineral portion of bone contains primarily (in descending order of concentration) calcium, phosphate, hydroxyl, carbonate, and citrate plus lesser amounts of sodium, magnesium, and fluoride. Structurally, bone consists of a variably dense matrix, made up of (1) active bone cells continually breaking down and reforming bone and (2) a stable, relatively inert, dense mineral condensation. Depending upon systemic (e.g., calcium regulatory, growth, and other hormones) and local factors (e.g., mechanical stress or loadings) there is variation in the percentage of bone metabolically active. The extremes of activity are presented by age, that is, almost 100% of bone is active in the young, whereas only 30% is active in adults.

Xenobiotics that demonstrate osteotrophic proclivities, though occurring as ions, molecules, or colloids, usually are transported as complexes, for example, with organic acids. Ultimately, such complexes dissociate, so that newly freed ions can exchange with calcium or phosphate ions in bone mineral. Colloids react with bone by chemisorption or adsorption. All these reactions take place only in metabolically active areas of bone.

Bone mineral structurally and functionally can be considered to consist of three layers: (1) a deep layer consisting of the hexagonal hydroxyapatite crystal, (2) an intermediate hydration layer or "shell," and (3) an outer shell consisting of nonspecific cations and anions in solution. All these three layers are in dynamic equilibrium with each other, the outermost ultimately in equilibrium with body interstitial fluid. Ions must diffuse through each layer or shell; diffusion through the hydration shell is restricted to specific ions. Ultimately ions reach the crystal, which can act as an anion or cation exchanger. In such fashion calcium cations or phosphate anions are exchanged for ions of similar size. Calcium ions can be exchanged for other calcium ions (i.e., by iso-ionic exchange) or with other ions of a similar ionic radius (i.e., by hetero-ionic exchange). In such a fashion phosphate ion can be exchanged for phosphate, or citrate and carbonate can be exchanged. The hydroxyl anion of the hydroxyapatite crystal can be exchanged for fluoride ion. Ultimately, as ions are incorporated at the surface of the hydroxyapatite crystal, they are repeated, overlaid by another layer of crystal. Thus with time and repeated crystallization such incorporated ions are deposited "deeper" into the crystalline bone structure and thus less readily accessible for subsequent exchange with body fluids.

The rate of diffusion and ultimate crystal deposition is complex, determined by multiple factors. The relative ionic concentration (Guldberg–Waage law) in interstitial fluid at the two outer shells and the crystal surface dynamically determine the net direction and velocity of flow. The chemical characteristics and valency of ions, pH, fluid volume, temperature, and so forth also determine velocity of flow and direction. Although all these principles are applied to ionic transfer, the rate of transfer seems to diminish as an ion moves from the outer shell toward the inner crystalline layer. Within the limits of the number of exchange sites at the outermost shell, exchange takes place more rapidly with interstitial fluid than with the immediately deeper hydration shell. Similarly, exchange between the hydration shell and the outermost layer takes place more rapidly than does exchange between hydration shell and the deeper hydroxyapatite crystal surface. About 30 percent of bone ion is loosely bound; exchange takes place with a biological half-life of 15 days. Because the other 70 percent is firmly bound by crystallization, mobilization of this fraction has a biological half-life of about 2.5 years.

In view of these dynamics, the behavior of small, large, or repeated doses of osteotrophic elements is predictable. For example, a single large dose of fluoride mainly appears in the urine and is not deposited in bone as the finite number of exchange sites on the outer shell are saturated. By contrast, repeated smaller doses of fluoride are more readily accumulated deeper within this three-shell system. With cessation of fluoride exposure this ion diffuses slowly out of the bone. This diffusion out of bone proceeds over scores of months and is manifested by a slight but persistent increase in urinary fluoride concentrations. Similar behavior is generally applicable to a large range of osteotrophic elements. Indeed, calcium, citrate, fluoride, radium, and strontium being deposited in bone by ion exchange should follow similar dynamic transfer mechanisms. However, there are wide variations in the kinetics of such movements; for example, the biological half-life of barium in bone is 35 days, whereas that of radium is approximately 60 years.

Colloidal absorption represents another important mechanism by which osteotrophic elements become incorporated into bone structure. After being carried by organic acids in plasma, these complexes dissociate so that the resulting free ions form colloidal particles that ultimately reach the bone surface. Because of the enormous surface area per gram of bone (i.e., 100 m^2/mg), adsorption occurs there due to van der Waals forces and chemisorption reflecting electrostatic or covalent binding. The irreversible absorption as a mono- or multilayer film on the crystal surface is covered here by successive layers of bone crystal. By cyclic repetition these colloids are progressively buried deeper in the bone. Such colloid deposition appears to be the mechanism whereby transuranics and lanthanides accumulate in bone.

It should be apparent that bone is not a metabolically inert tissue. The dynamics of these exchange reactions makes it self-evident that ionic flow may move either toward the crystalline depths of the bone mineral or to the outer shell for ready exchange and possible excretion. Accordingly, bone deposition cannot be regarded either as permanent or irrevocable, although the dynamics of such movements is usually relatively slow and highly variable.

6 THE BIOTRANSFORMATION OR METABOLISM OF XENOBIOTICS

There is a broad diversity of reactions whereby xenobiotics are transformed in order to facilitate their elimination from the body. Substances that exist in the plasma as polar, dissociated, and hydrophilic forms are readily eliminated by the kidneys. By contrast, most lipophilic, nonpolar compounds are not easily excreted via the renal route. Although some of this class of compounds can be eliminated either into the bile or by volatilization from the lungs, most such nonpolar materials must undergo transformation to water-soluble metabolites, which are then readily excreted into the urine.

In the larger perspective, a two-stage process affects the alteration of foreign materials required for their excretion. The first phase, that is, a "nonsynthetic" stage, largely involves lipophilic, nonpolar substances and consists largely of enzymatically mediated oxidation, reduction, or hydrolysis. In most cases this step alters molecular structure so that a "conjugation" or linkage center (i.e., a carboxyl, amino, sulfhydryl, or halogen group) is available for subsequent reaction. In most instances this process succeeds in making possible subsequent steps leading to detoxification and elimination. Occasionally this first phase leads to the opposite situation, "lethal synthesis," that is, the formation of a more toxic product.

The second basic step in the detoxification process consists of a synthesis phase. At this juncture an endogenous substance is linked at the conjugation center of the previously altered xenobiotic. This step almost always results in detoxification, largely because the newly conjugated molecule is more water soluble and thus more easily excreted. (In an evolutionary context, these processes, which all lead to water solubilization, reflect the aquatic origin of most terrestrial animal life.)

However, conjugation may also result in the formation of a larger molecule that displaces or blocks access of other toxic molecules at receptor sites. Alternatively, a newly conjugated—or altered—molecular structure may not sterically "fit" at a target cell's receptor site. In both cases the end result is prevention of toxic sequelae, that is, detoxification.

Multiple thermodynamically driven chemical processes, participant tissues, and biological mediators are involved in these complex activities referred to as detoxification and biotransformation. These are more extensively discussed in Volume III, Chapter 7, as well as in each chapter of Volume II, which considers various classes of chemicals and their individual members.

7 ELIMINATION AND EXCRETION OF XENOBIOTICS

7.1 Introduction

The rate and effectiveness of the elimination of xenobiotics or toxics from the body is a major determinant of the severity and duration of their possible actions. Thus if they are rapidly eliminated any effect tends to be minimized; if residence in the body is protracted they may act for longer periods of time and possibly with greater severity.

Xenobiotics are eliminated from the body via several routes. The onset of elimination reflects in part the route of absorption, so that inhaled vapors and gases may be immediately desorbed from plasma and exhaled almost simultaneously with exposure. By contrast, material absorbed from other portals of entry will first be transported, distributed, and transferred to organs, where they can be transformed and ultimately excreted.

Because excretion is the primary function of the kidneys, the majority of absorbed chemicals are eliminated via that organ. Because compounds must be converted to water-soluble materials for kidney excretion, this process is usually less rapid than is pulmonary elimination. The bile and gastrointestinal tract are also important modes of elimination for a number of xenobiotics of industrial significance, for example, lead.

Although several other organs are capable of excretion, namely, the skin via sweat and the salivary, lachrymal, and mammary glands, these are of lesser significance. In addition, via deposition and shedding of body parts (e.g., hair, fingernails), several xenobiotics of industrial significance can exit the body, for example, mercury and arsenic.

The factors that determine a material's "mobility" in the body have been discussed previously. Such factors also play a role in the elimination process by influencing transferability and hence accessibility to the organs of excretion.

7.2 The Kinetics of Elimination and Excretion

The study of the kinetics of elimination of xenobiotics requires quantitative data describing their concentrations in blood and/or urine. Such information collected over a time course permits study of the rate and character of xenobiotic elimination, in addition to providing valuable insights into their body uptake, distribution, and mobilization.

After a single dose of a compound is rapidly administered intravenously, a logarithmic plot of its concentration in plasma over time will generally demonstrate one of at least two patterns, namely, (1) a linear decrease or (2) a nonlinear decrease. Derivation of a linear excretion curve (Figure 9.6*a*) suggests rapid and relatively uniform distribution of the compound throughout the body, that is, as if the body was acting as one homogeneous vessel or compartment. The shape of this curve also indicates that a compound is eliminated using all transformation and excretion routes by first-order kinetics.

The time necessary for the compound to be eliminated from the body is known as the biological half-life, or $t_{1/2}$. It can be determined visually by inspection of the derived curve in Figure 9.6*a* or by calculation using the compound's elimination rate over time. Among chemicals showing first-order elimination kinetics the half-life is independent of dose *up to individual limits*. In a single open-compartment model tissue concentrations decrease at similar half-lives as in the plasma. Because the ratio of tissue to plasma concentration is constant within limits, a determination of such a ratio can permit subsequent tissue concentration to be calculated from an assay of plasma levels (See Section 8.2, Equation 3)

The slope of the derived curve (Figure 9.6*a*) represents a first-order constant of elimination; it can be derived from the relationship:

$$t_{1/2} = \frac{\ln 2}{k_{el}} \qquad k_{el} = \frac{0.093}{t_{1/2}}$$

so that the elimination rate constant demonstrates units of reciprocal time. Thus the half-life of the compound in Figure 9.6*a* is expressed by $k_{el} = 0.093/t_{1/2}$ because 20 percent of the element is excreted each hour. This open model and its first-order kinetics assume (1) the compound is excreted by essentially one route of elimination, (2) the substance is not biotransformed, and (3) it is distributed to tissues that are all similar in their exchange rates.

These relationships and the elimination rate hold constant, regardless of dosages, except when the dose exceeds certain limits. But as these limits are exceeded, the initial rates of elimination do not hold constant, indicating that biotransformation, protein binding, active transport, and so on all operate within finite limits. This indicates that some essential pathway(s) has become saturated, in keeping with Michaelis–Menten kinetics. Thus below the Michaelis–Menten constant (K_m) for the process first-order kinetics apply, but as the K_m for the process is exceeded, nonlinear kinetics apply. Such an event indicates that a change from first-order to saturation kinetics is occurring; that is, the body's processing mechanisms are operating in a different fashion than seen with low doses. Such points of departure may also indicate the attainment of a level of dosage compatible with toxicity.

Elimination curves that are not linear on a logarithmic plot but rather are exponential (Figure 9.6*b*) suggest the need for a multicompartment analysis. By the analysis and determination of residuals the observed curve can be distinguished as consisting of two different slopes, that is, one having a rapid and the other a slow rate of excretion, each curve having its own elimination rate constant. Such suggests

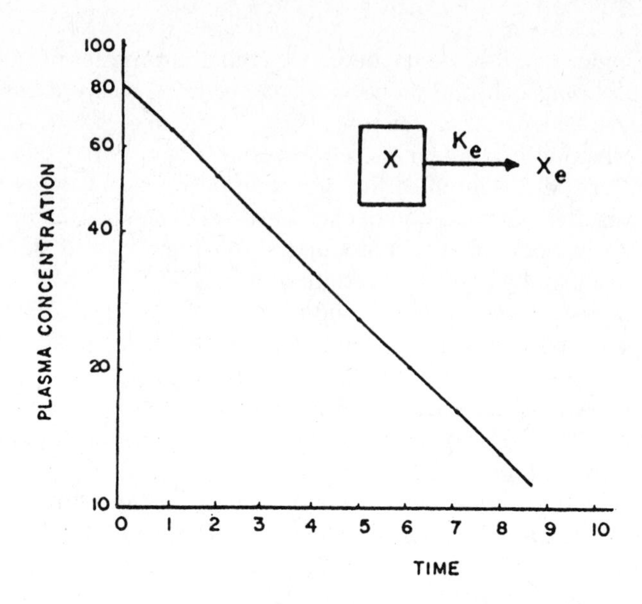

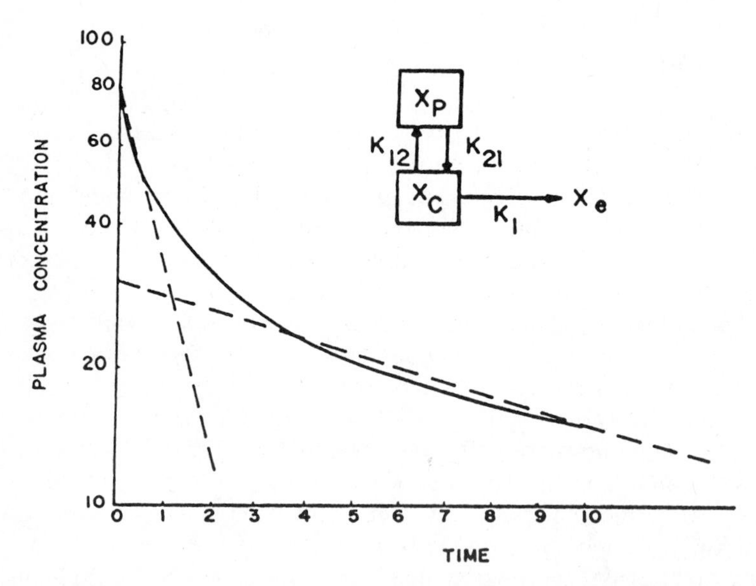

Figure 9.6. Plasma concentration of an injected xenobiotic. (*a*) Experimentally derived first-order excretion in a one-compartment model system, where X and X_e are the amount of the compound, and k_e = first-order rate constant for excretion. (*b*) Biexponential excretion curves for a two-compartment system (1 and 2), where C and P are central and peripheral compartments, respectively, and k = individual first-order rate constants characteristic of compartment (1) and (2).

that the body consists of a two-compartment system. One compartment manifests rapid equilibration, biotransformation, and excretion components, whereas a second compartment performs these activities less rapidly.

Regarding the first of these compartments, this could represent the liver and kidneys, which have a high perfusion rate and thus rapidly encounter the effect of xenobiotic dosing. In toxicokinetic modeling these may be referred to as a "central compartment." Conversely, given the relatively low blood perfusion rates and long-term deposition characteristic of fat or bone, the slow excretion component of the biphasic excretion curve may reasonably represent the essentially similar behavior of these organs. These organs with slower uptake and release in contrast to central compartment organs, for example, kidney, are referred to in multicompartment models as "peripheral compartments."

The comparison of such a biexponential system with the linear pharmacokinetics of a single compartment system reveals characteristic differences. Noting that the former has a relatively slow component, one would expect it to demonstrate (1) increased half-life with increasing dose, (2) that the composition of excretion products may quantitatively and qualitatively change, (3) that saturation of processes in the slow compartment may show a marked change in dose–response relationships as dose increases, and (4) that when operating at or near rate limits, competitive inhibition of metabolic or transport processes by other chemicals can more likely occur. Thus the elements of Michaelis–Menton saturation kinetics can reveal valuable information. Chronic dosing with compounds that produce small cumulative insults will eventually reveal indications of excretory organ damage by increases in biological half-life. Likewise with such small dosings, if half-life is less than the dosing interval, such schedule can be predicted as not leading to bioaccumulation.

In the case of chemicals whose metabolic transformation mechanisms are readily saturated, zero-order kinetics are followed. In these cases an *arithmetic* plot of plasma concentration versus time will produce a straight line. In effect, this indicates that a rate-limiting step is occurring at some point in the metabolic process so that a constant amount is being transformed per unit time, independent of the amount found in the organ or the body. For such compounds a true half-life does not exist, because complete excretion rather than logarithmic kinetics apply. A prime example of such a compound is ethyl alcohol.

7.3 Specific Routes of Excretion

7.3.1 The Kidneys

The kidneys have as their prime function the removal of nonessential metabolites from the body; by these same mechanisms xenobiotics are also removed. By receiving 20 to 25 percent of the total cardiac output, large quantities of blood are brought to the kidneys for elimination of materials contained therein. This is accomplished by three basic processes: (1) glomerular filtration, (2) passive tubular diffusion, and (3) active tubular secretion.

Despite such large blood volumes brought to the kidneys, the efficiency of the

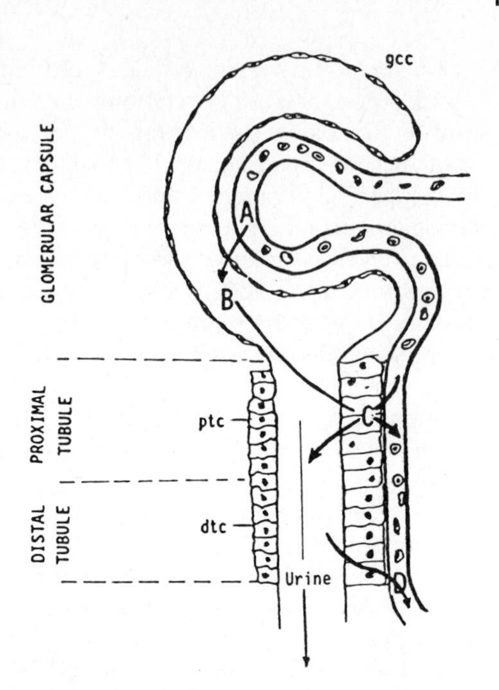

Figure 9.7. Schematic diagram, structural and functional elements of the renal glomerulus and tubules. (A) Formed elements and plasma enter glomerular capillary. (B) Ionic or free small molecules (< 60,000 Da) are filtered through capillary and glomerular walls into the uriniferous lumen of the glomerular capsule. As ionic or free form concentrations decrease in plasma, the equilibrium with the bound components is altered, causing further dissociation of these components and thus an increase in their availability for filtration. (C) Energy-dependent active transport at the proximal tubules allows either (1) excretion from the blood to the glomerular filtrate, or (2) extraction from the glomerular filtrate for recovery to the blood. These shifts may occur in the presence or absence of protein binding. (D) Passive diffusion allows transfer of nonionized, nonpolar substances, consistent with basic membrane diffusion theory. gcc = glomerular capsule membrane; ptc = proximal tubule cells: dtc = distal tubular cells.

organ is such that only 1 percent of the fluid passing through the glomerular filtration segment reaches the urinary bladder. Passage across the single-cell glomerular capsule (see Figure 9.7) by molecules of less than 60,000 Da is readily accomplished. This transfer is enhanced by the presence of pores of up to 70 μm diameter in the glomerular capillary walls. In this filtration process, as the unbound forms of plasma elements (e.g., metal ions) decrease in concentration, the equilibrium between free and bound forms is altered. This leads to dissociation of the protein-bound forms and their subsequent availability for passive filtration.

As the filtrate and postglomerular capsule plasma stream pass to the area of the proximal tubules, both are subjected to two transfer forces, that is, passive diffusion and active tubular transport (Figure 9.5). Both these processes may act to move compounds in either direction, that is, back across the tubular cell from the glomerular filtrate toward the tubular blood supply (i.e., reabsorption), and vice versa. Passive filtration of xenobiotics through the tubules to the urine is of relatively minor significance. Nevertheless, reabsorption and excretion follow basic principles of cell membrane transfer. Thus in the acidic environment of the glomerular filtrate, weak organic acids move toward a nonionized state, which leads to consequent facilitated passage across the tubular cell wall, that is, passive reabsorption. Such a principle also applies to the fate of hydrophilic, dissociated ions, which will readily be filtered through the glomerulus. That is, materials in this state are not passively transferred across cell membranes and thus are not recovered by passive transport across renal tubules, subject to variations in pK_a and pH.

Active transport is accomplished largely in the proximal tubules. As with any active process it is an energy-dependent process. As has been previously seen, similar configurations or binding constants may compete at binding sites involved in the tubular transport system involved in the secretory process. In the region of the proximal tubules there is active reabsorption of vital elements that are passively filtered through the glomerular capsule, for example, amino acids, ions, and small proteins. In contrast to filtration at the glomerulus, protein-bound compounds are accessible for active transport and tubular excretion.

7.3.3 Biliary Excretion

Study of the gross and microscopic anatomy of the liver reveals its central role in xenobiotic metabolism. Because it receives a large portion of its blood supply from the GIT, it has the first opportunity to sequester, bind, or excrete xenobiotics brought into the internal milieu from the oral portal of entry. Microscopic examination of the liver demonstrates that its individual cells on one side are in direct contact with mixed portal and systemic blood and on its opposite side with the biliary vascular system. Thus materials are presented from either blood source to hepatocytes for biotransformation and excretion to either the blood or biliary stream. Transfer from either stream to within the hepatocyte occurs by active transport; a number of specific transport mechanisms have been identified. A xenobiotic must have either a strongly polar group or a large molecular weight (>300 Da) for biliary excretion to occur. Organic xenobiotics excreted in the bile are conjugated, for example, as glucuronic acids or with glutathione. Multiple factors that are poorly understood appear to modulate hepatocytic excretory capacity. Some of these factors appear to be related to increases in intracellular binding by proteins, increased conjugating activities, and stimulation of microsomal biotransformation. Additionally, specific transport systems for organic acids, bases, and neutral organics have been identified.

Xenobiotics absorbed via the GIT may be eliminated after hepatic biotransformation via the blood—and ultimately the kidney—or in the bile to the GIT. However, nonpolar or lipophilic metabolites excreted in bile are most likely to be reabsorbed and carried back via the portal system for re-presentation to the liver. By such recirculation by the enterohepatic circulatory cycle (i.e., bile to GIT with reabsorption and transport via the portal vein back to the liver), compounds that are hepatotoxic may be concentrated in the liver with resultant damage to that organ. On the other hand, many organic compounds are biotransformed to polar metabolites or conjugates before excretion in the bile. Such forms are less susceptible to reabsorption and entry into the enterohepatic circulation.

7.3.4 The Lungs as an Organ of Elimination

Although some investigators have attempted to ascribe to the lung an active excretory capability for gases, vapors, and volatile metabolites, it is now universally agreed these are eliminated by simple diffusion. In addition, volatile liquids in equilibrium with their gas phase may be similarly excreted regardless of their route of uptake. Thus volatile organic solvents absorbed via the GIT may be eliminated through the lungs. Gases are eliminated at a rate that is inversely related to their solubilities; thus chloroform, which is highly soluble in blood, is more slowly excreted than ethylene. However, part of the prolonged retention of the highly soluble gases (e.g., chloroform) reflects lipid depot deposition, that is, representing the expression of a two compartment system.

Other factors determining the rate of pulmonary elimination are related to pulmonary physiological factors, for example, ventilation and blood perfusion rates in the lungs. In addition, it has become clear that the lung contains—although at generally lower concentrations—almost all of the hepatic pathways required for biotransformation of xenobiotics. However, it is generally considered that this represents simply a metabolic capacity directed toward ends unrelated to excretory functions.

Previous discussions (cf. Section 3.2.5) have considered the elimination of particulate matter by the lung.

7.3.5 The Gastrointestinal Tract as an Organ of Excretion

Because the gastrointestinal tract actively secretes about 3 liters/day in humans, most xenobiotics are excreted here by passive diffusion. Thus transfer from blood via the GIT could and does occur for highly lipophilic, nonpolar substances, for example, polychlorinated biphenyls (PCBs) (16) and organochloro pesticides.

However, although other xenobiotics may be found in the feces, such does not necessarily represent or imply GIT excretion. Rather their presence can be the result of (1) excretion in the bile, (2) secretion in the saliva, stomach, pancreatic, or intestinal secretory fluids, (3) failure of absorption following oral intake, or (4) clearance from the respiratory tract (cf. Section 3.2.5) followed by swallowing.

7.3.6 Excretion via Perspiration and Saliva

Both these routes are of minor significance; they largely are associated with simple diffusion of nonionized lipophilic forms. Thus xenobiotics excreted via sweat may accumulate at sudorific and pilocarpal glands sufficient to produce dermatitis; the chloracne associated with PCB exposure may be a case in point. Evidence exists that fluoride is excreted in sweat, as are Hg, Bi, Pb, and As. The secretion of lead in saliva in chronic lead absorption may be associated with the gingival lead line. Some other diffusible substances found in saliva are bromides, iodides, alkaloids, and ethanol.

7.3.7 Excretion by Milk

Because xenobiotics may be conveyed by cows to humans, and from maternal milk to infants, such secretions are of some significance as an excretory medium. Because of its important lipid content, many foreign compounds may be concentrated in milk. Thus lipophilic organics such as DDT, PCB, and PBB concentrated in milk can be an interesting source of dietary contamination. In addition, metals whose physicochemical characteristics are similar to calcium, for example, lead, can be found in this medium.

8 A SYNTHESIS OF ABSORPTION, DISTRIBUTION, METABOLISM, AND EXCRETION OF XENOBIOTICS BY KINETIC ANALYSIS

8.1 An Introduction to Pharmaco(toxico)kinetic Modeling

With consideration of the previous discussions it should be almost overwhelmingly evident that a multiplicity of interacting variables determine absorption, transfer, distribution, and excretion of xenobiotics. Indeed such a plethora of information is associated with the development of a large body of quantitative data describing such phenomena. In the past 20 years these accrued data have inevitably stimulated efforts to arrange them in a fashion sufficient to provide a coherent, dynamic, integrated representation of these simultaneous events. With the availability of electronic computing technology the possibility arose that unifying models of these phenomena could be constructed and tested. Bringing the data together in this manner would serve, at the minimum, possibly to bring order to it. Further, testing of these models by insertion of experimentally derived data would validate their correctness. Upon validation, such models could be used to predict the behavior of materials not yet tested, providing the possibility of economies of effort and resources, for example, for prediction of trans-species differences. Finally, it is possible that such models might provide unifying insights into basic biological phenomena as yet unperceived among a mass of otherwise disparate information.

Reflecting these potentials there has developed over the past 15–20 years the field of pharmaco- and toxicokinetics. Models developed by this discipline are operational means of expressing working hypotheses for mechanisms of absorption, distribution, transfer, and disposition. However, it must always be held in mind that kinetic modeling is a simplified representation of inherently complex biological systems. These models must be recognized as incorporating a number of assumptions and approximations because many subtleties inherent in these systems are just barely perceived. Accordingly, despite the apparent sophistication of such models these reservations must not be forgotten. But within the limits of such caveats, models have proved useful in predicting movements of xenobiotics and

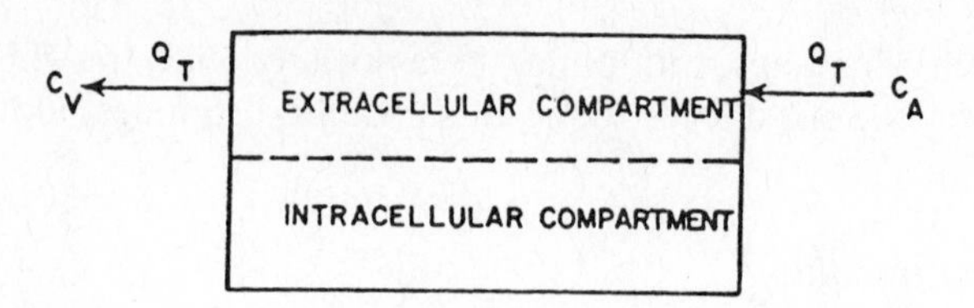

Figure 9.8. A "lumped compartment" model, which may represent an organ or regions in which concentrations are assumed to be uniform. Q = blood flow rates per unit, T = time, C = concentration of compound in arterial (A) and venous (V) blood.

their trans-species variations, as well as suggesting testable hypotheses and fruitful directions for future investigation.

8.2 A Toxicokinetic Model for Chlorinated Biphenyls

The study of Lutz and co-workers (17) of PCB pharmacokinetics in the rat can serve as a useful example of modeling techniques. In this case relevant physical, biochemical, and physiological parameters such as blood flow rates, organ volumes, tissue binding ratios, and metabolic distributions have been previously determined or available. The kinetic model provides insights into modes of distribution because postulated systems can be tested by comparison of experimental data with simulations. The model helps verify the magnitude of important parameters such as clearance and metabolic rates otherwise difficult to determine in a complicated in vivo system. The specific purpose of the study was to demonstrate the effects of varying degrees of chlorination and different positions of chlorine atoms upon the disposition of the PCBs.

A generalized representation of their pharmacokinetic model is shown in Figure 9.8. In this concept arterial blood carrying a PCB at a concentration C flows at a rate Q into an organ or compartment, and leaves at a concentration C in venous blood at a determined rate Q. The xenobiotic (i.e., PCB) partitions into various compartments via the extracellular space, often reaching higher concentrations than those found in blood.

The principle underlying the flow-limited model used in this study can be expressed by the general statement

Rate of accumulation in a tissue

$$= \text{Rate of arterial blood influx} + \text{Rate of venous blood efflux} \quad (1)$$

or as a differential equation:

$$V\frac{dC_T}{dt} = Q_T C_A - Q_T C_V \quad (2)$$

where C = concentration in blood, subscript A = arterial blood, subscript V =

venous blood, subscript T = tissue, Q = blood flow rate, t = time, and V = compartment volume (or size).

The authors chose a flow-limited model, in which plasma and cell membrane permeabilities are large as compared to the tissue or organ blood perfusion rate; that is, a blood-borne compound readily diffuses into tissues at a considerably larger rate than blood influx. The compound enters into a number of physical interactions, for example, physical bindings with blood and tissue macromolecules. As a consequence, the partitioning between blood (B) and tissue (T) occurs as a function of the xenobiotic's specific affinities. Thus when equilibrium is reached, the partitioning can be expressed by the equation:

$$R_T = \left.\frac{C_T}{C_B}\right]_{eq} \tag{3}$$

where C = concentration and R = partition ratio, tissue to blood. Based upon the assumption that exchange rates between blood and tissue region are rapid, the concentration of a chemical leaving a tissue in venous blood *will essentially be in equilibrium* with the tissue concentration. Thus Equation 3 can be written:

$$C_V = \frac{C_T}{R_T} \tag{4}$$

Accordingly, the term C in Equation 2 can be substituted for by Equation 4, which yields the basic statement of a tissue mass balance equation for a tissue T as

$$V_T \frac{dC_T}{dt} = Q_T\left(C_B - \frac{C_T}{R_T}\right) \tag{5}$$

Because blood-borne input and output and tissue storage were the major effects taking place in skin, muscle, and fat, this model could be used to represent these compartments.

Based upon this generalized model, which represents a simple mass balance statement for any compartment, it becomes possible next to develop a multiple compartment model. Such a model represents the flow (Figure 9.9) of a PCB and its metabolites in various tissues and organs.

However, PCBs are biotransformed to their metabolites in the liver; this effectively decreases the mass of the parent PCB molecule in that compartment. Accordingly, such metabolically produced decrements in liver PCB mass can be accounted for by insertion of a negative term in the generalized model, Equation 5. Thus the mass balance equation for the liver compartment can be represented as

$$V_L \frac{dC_L}{dt} = Q_L\left(C_B - \frac{C_L}{R_L}\right) - K_M \frac{C_L}{R_L} \tag{6}$$

where K represents a metabolism rate constant.

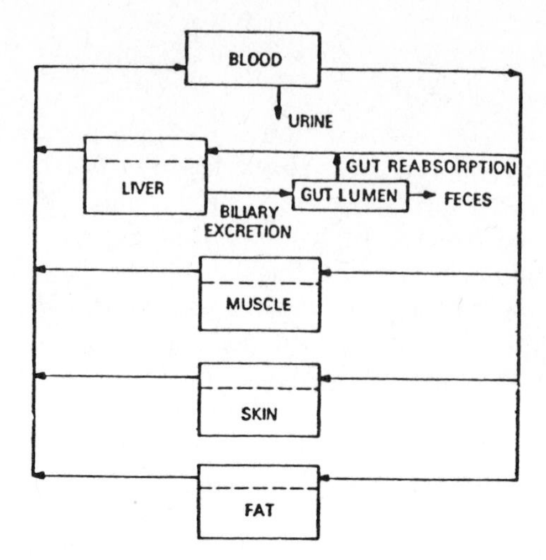

Figure 9.9. Flow diagram for pharmacokinetic model of chlorinated biphenyl in the rat, from Lutz et al. (17).

For the other tissues and organs, that is, muscle, skin, and fat, where there is no significant metabolic or excretory function, the model in Equation 5 was used to define those organs' participation in uptake and storage.

In this multicompartment model there still remained the not insignificant blood compartment for which a model was required. The basic statement of Equation 1 was expanded in order to account for other sources of influx and efflux to and from blood. Thus each of the organs emptying PCB in their venous drainage add to the blood compartment's PCB loading, whereas all those organs containing PCB-laden blood are considered as diminishing PCB in the overall blood compartment. Thus the mass balance model for the blood compartment can be written:

$$V_B \frac{dC_B}{dt} = Q_L \frac{C_L}{R_L} + Q_M \frac{C_M}{R_M} + Q_S \frac{C_S}{R_S} + Q_F \frac{C_F}{R_F} - [Q_L + Q_M + Q_S + Q_F]C_B + Mg(t) \quad (7)$$

where M = dose, $g(t)$ = an injection function, and the subscripts B, L, M, S, and F represent blood, liver, muscle, skin, and fat, respectively.

It is noteworthy that renal PCB is not considered in any of these equations. It was found that *unmetabolized* PCBs were not excreted by the kidneys; only metabolites were eliminated by this route. Similarly, the liver excreted PCB metabolites into the bile where they were either disposed of in the feces or else taken up by the enterohepatic circulation, reentering the liver; accordingly, they were considered as an influx to the total blood supply to the liver.

Because of the significance of the total fate of PCB and its metabolites in the body, it was necessary to construct similarly a series of models describing the distribution and disposition of these biotransformation products. Similar principles applied, so each set of models for the unmetabolized PCBs was modified to express movements and masses of PCB metabolites. Thus in each organ (See Figure 9.9) where simply uptake and clearance of PCB metabolites occurred, influx and efflux terms were required to express the PCB metabolite mass balance. Thus the model for metabolites took essentially the same form as given in the parent PCB mass balance statement for these organs. For the blood contained in the cardiovascular compartment, a decremental term was required to reflect efflux from that compartment owing to renal excretion of PCB metabolites. For the gut lumen compartment model constructed to analyze metabolic disposition, the model added the increment due to biliary clearance to that organ, minus losses due to reabsorption from the gut and fecal losses with elimination. For the liver, not only did losses due to excretion of metabolite require an efflux expression, but an additional term to account for influx of newly formed metabolites also had to be added to the incremental portion of the basic equation. Finally, the same equation could be used to express the mass balance for fat, skin, and muscle.

To test these models' validity, basic data were needed; such could then be substituted in each equation for computer-assisted solution. Many of these data had been previously reported. Other data, in particular the predominance of a particular PCB metabolite, required modification of the model by specific assumptions; each of these was reported. Other data deficiencies required experimental study of PCB and its metabolic fate, both in vitro and in vivo. Inserting each of the specific parameters in the appropriate term in the model, a computer-assisted solution of the 13 simultaneous differential equations was carried out.

A test of the model's validity was required to determine if it could be useful in predicting tissue concentrations and excretion of PCBs and their metabolites as a function of time. The validity of this model's predictions as reasonable approximations were confirmed following animal experiments that studied PCB concentrations in blood, tissues, and excreta. In addition, initial attempts were made to use this model for extrapolating from one species to another, that is, from rat to mouse. Scaled-down simulations were performed and compared with experimental data obtained using this smaller species. Again the general validity of this model was confirmed by determination of blood, tissue, and excreta concentration in mice; these also demonstrated that the model reasonably simulated metabolic events in mice by predicting various compartment concentrations and kinetics for PCB and their metabolites.

Although it is prudent to recall the caveats stated earlier, nevertheless such models may be of considerable utility. As more basic data is accumulated (e.g., cell membrane transport determinants), more precise modeling of these inherently complex systems will become feasible. In any event, the present value—and limitations—and future promise of these unifying approaches to the understanding of toxicity phenomena should be self-evident.

REFERENCES

1. R. A. Brown and L. S. Schanker, *Drug Metab. Disp.*, **11**, 355–360 (1983).
2. ISO Technical Committee 146—Air Quality, "Particle Size Fraction Definitions for Health-Related Sampling," International Standards Org. ISO/TR 7708-1983(E), 1983.

3. T. L. Chan and M. Lippmann, *Am. Ind. Hyg. Assoc. J.*, **41**, 399–409 (1980).
4. J. D. Brain, *Inhaled Particles III*, W. H. Walton, Ed., Unwin, London, 1971, pp. 205–225.
5. B. D. Beck, J. D. Brain, and D. Bohannon, *Toxicol. Appl. Pharmacol.*, **66**, 9–29 (1982).
6. C. R. DeVries, P. Ingram, S. R. Walker, R. W. Linton, W. F. Gutknecht, and J. D. Shelbourne, *Lab. Invest.* **48**, 35–44 (1983).
7. J. D. Brain, *Ann. Rev. Pharmacol. Toxicol.*, **26**, 547–655 (1986).
8. C. J. Pfeiffer, *Handbook of Physiology*, Section 9, "Reactions to Environmental Agents," American Physiological Society, Bethesda, MD, 1977, pp. 349–374.
9. L. B. Sasser and G. E. Jarboe, *Toxicol. Appl. Pharmacol.*, **41**, 423–431 (1977).
10. C. N. Davies, Ed., *Inhaled Particles and Vapors*, Vol. 11, Pergamon Press, Oxford, 1967, pp. 121–131.
11. U. Kragh-Hansen, *Pharmacol. Rev.*, **33**, 17–53 (1981).
12. International Commission on Radiological Protection, *Report on the Task Group on Reference Man*, ICRP No. 23, Pergamon Press, Oxford, 1975, pp. 325–327.
13. Best and Taylor's *Physiological Basis of Medical Practice*, 11th ed., J. B. West, Ed., Williams and Wilkins, Baltimore, MD, 1985, p. 136.
14. R. A. Goyer, in *Chemical Toxicology and Clinical Chemistry of Metals*, S. S. Brown, Ed., Academic Press, London and New York, 1984, pp. 199–209.
15. A. J. Levi, Z. Gatmaitan, and I. M. Aria, *J. Clin. Invest.*, **48**, 2156–2167 (1969).
16. T. Rozman, L. Ballhorn, K. Rozman, C. Klaassen, and H. Greim, *J. Toxicol. Environ. Health*, **10**, 277–283 (1982).
17. R. J. Lutz, R. L. Dedrick, H. B. Mathews, T. E. Eling, and M. W. Anderson, *Drug. Metab. Disp.*, **5**, 386–396 (1977).

CHAPTER TEN

Occupational Dermatoses

Donald J. Birmingham, M.D., F.A.C.P.

Diseases of the skin caused by agents or conditions at work continue to outnumber all other work-associated illnesses. They occur in practically all work pursuits and can vary from a trivial localized redness or dryness to complicated widespread lesions or tumors of benign or malignant nature. Most texts identify them as occupational skin disease, occupational dermatitis, and less often as professional eczema. Among the work force, a number of more descriptive titles associated with cause are commonly used, for example, asbestos wart, cement burn, chrome holes, fiber glass itch, hog itch, oil acne, rubber rash, and tar smarts. In view of the variety of skin lesions known to result from contactants within the workplace, the term "occupational dermatoses' is preferred because it includes any abnormality of the skin resulting directly from or aggravated by the work environment (1).

1 HISTORICAL

Just when and how occupational affections of the skin first occurred is a matter of conjecture; however, if we apply our past and present knowledge of diseases associated with work, it can be reasonably suspected that skin disorders in one or another form were expressed soon after humans began to perform various types of work. Archaeology has shown that ancient inhabitants invented and used primitive tools and weapons made of stone, flint, bone, and wood (2). It is quite likely that abrasions, blisters, bruises, lacerations, punctures, and probably more serious traumas were incurred as part of daily living associated with hunting for food, building shelter, making clothing, and gaining protection. Poisonous plants and

Patty's Industrial Hygiene and Toxicology, Fourth Edition, Volume 1, Part A, Edited by George D. Clayton and Florence E. Clayton
ISBN 0-471-50197-2

biological agents, including various parasites, probably took their toll; however, such harmful effects remain suspect rather than documented in medical history. Perhaps the earliest reference occurred in the writings of Celsus about 100 A.D. (3), when he described ulcers of the skin caused by corrosive metals. During later centuries, several authors enhanced the knowledge of certain occupational diseases, but cutaneous ulcerations seem to have been the major occupational skin disease of record. An explanation may reside in the fact that ulcerations of the skin were easily recognized, especially among those handling metal salts in mining, smelting, tool and weapon making, creating objects of art, glassmaking, gold and silver coinage, casting, and similar metallics. It would be strange indeed if none of these tradesmen incurred skin problems caused by a substance or condition met with at work. Nonetheless, little was recorded about occupational skin disease until Ramazzini's historic treatise on diseases of tradesmen in 1700 (4). In this tome he described skin disorders experienced by bath attendants, bakers, gilders, midwives, millers, and miners, among other tradespeople. Seventy-five years later Sir Percival Pott published the first account of occupational skin cancer when he described scrotal cancer among chimneysweeps (5).

The Industrial Revolution of the eighteenth century brought change from an agricultural and guild economy to one dominated by machines and industrial expansion. As cities and industries grew, so also occurred the growth of science and the eventual discovery and use of new materials such as chromium, mercury, and petroleum, among many others. The chemical age brought enormous numbers of materials, natural and synthetic, into industrial and household use (2). As a result physicians began to recognize occupational dermatoses and publish their observations in England, Germany, Italy, and France. Similarly, industrialization within the United States led to the recognition of old and new causes of occupational skin disease. Numerous dermatologists, industrial physicians, practitioners, and allied scientists have added to the information bank dealing with clinical investigations, clinical manifestations, causal factors, diagnostic procedures, and treatment and prevention of these disorders. A number of updated texts and related publications are available (6–15). Today we live in an age of technical ingenuity brought about by atomic power, electronics, computers, great speed, precision tools, machinery, and, of course, space (2).

1.1 Incidence

In the 1978 edition of Patty's *Industrial Hygiene and Toxicology* text, dermatologic diseases were shown to account for about 40 percent of all occupational diseases reported to the U.S. Department of Labor (15). In the 1984 Bureau of Labor Statistics, dermatologic disease had decreased to 34 percent of all reported occupational disease. Most of the cases were associated with manufacturing, services, wholesale–retail trade, construction, agriculture–forestry–fishing, and transportation. In this Bureau of Labor Statistics breakdown of occupational diseases by type, lung disease due to dust accounted for 1.4 percent, disorders due to repeated trauma 27.8 percent, disorders due to physical agents 7.2 percent, poisonings 3.6 percent, respiratory disease (toxic) 8.5 percent, and skin disease 34.1 percent (17).

Since 1940 there has been a 35 to 40 percent decrease in the number of reported cases of skin disease, but reasons for the impressive decrease are not well defined. It is reasonable to assume that the Toxic Substances Act and the Occupational Safety and Health Act improved hygiene controls in industry, better management surveillance, and the "right-to-know" (see Chapter 6) laws have had significant impact on the lowered incidence. Yet there are sound reasons for questioning the number of cases that go unrecognized, unreported, and misclassified. Consequently, it is estimated that a 10- to 50-fold increase could escape the annual survey figures (18). No one really knows the actual cost involved.

1.2 Skin as an Organ of Defense

Work of any type can cause an occupational skin disorder, yet the majority of the millions employed remain free of disabling work-related skin disease. This seeming exemption is due in large part to the natural defense provided by the anatomic, physiological, and chemical characteristics of the skin. It is a protection subserved by the skin's role in controlling body heat, secreting and excreting body sweat, receiving and reacting to various sensory stimuli, manufacturing pigment, and replacing its own cell layers.

Anatomically, the skin is composed of two main levels, epidermis and dermis. The outer or epidermal layer varies in thickness, being most protective on the palms and soles. Because it is contiguous with the dermis, it also acts as the outer cover of the cushion of connective and elastic tissue that guards the blood and lymph vessels, nerves, secretory glands, hair shafts, and muscles. Its resiliency provides protection within limits against blunt trauma and its flexibility accounts for the return of stretched skin to its normal location (1–6, 7–15, 19).

The epi or outer dermis has two layers, the stratum corneum, or keratin layer, and the epidermal cell layer. The keratin cells make up a strong, tightly packed fibrous protein shield of dead cells that once were living epidermal cells. This outer dead cell stratum is highly important in resisting the mass entrance of water and electrolytes. Friction and pressure cause it to thicken and in certain areas form callous. Natural and artificial ultraviolet light also causes it to become thicker. Besides acting as a water barrier and a physical shield, it provides modest protection against acids and acidic substances. In contrast, it is quite vulnerable to the action of organic and inorganic alkaline materials. Such chemical substances attack the keratin cell by dissolving the cellular content, thus altering its cohesiveness and its capacity to retain water, which is essential in the maintenance of the barrier layer. In short, any physical or chemical force such as lowered temperature and humidity or repetitive action of soaps, detergents, and organic solvents generally leads to impairment of the barrier efficiency because of water loss and dryness (19, 20).

The living epidermal cells are produced by the basal or germinative cells that reside side by side in the lowermost portion of the epidermal layer. These are the cells that divide and constantly supply the living epidermal layer (19).

Located also in the basal layer are the melanocytes that manufacture the melanin pigment, which in turn is a major defense against ultraviolet light. The pigment granules that arise from complex enzymatic reactions within the melanocytes are picked up by the epidermal cells and eventually are shed by way of the keratin exfoliation (21). Besides the natural production of melanin, certain agents such as coal tar, pitch, selected aromatic chlorinated hydrocarbons, petroleum products, and trauma can cause excess melanin production, leading to hyperpigmentation (15, 16). In contrast, members of the quinone family and selected phenolics can inhibit pigment formation following percutaneous absorption by direct action upon the melanin enzymatic system (1, 6, 9, 13, 22–24).

Immediately below the epidermal region lies the dermis, which is much thicker than its outer covering. It is composed of connective tissue made up of collagen, elastic tissue, and ground substance that constitutes an encasement for the sweat glands and ducts, the hair follicles, sebaceous glands that secrete natural sebum (fatty substance), the blood and lymph vessels, and the nerve endings (6, 9, 15, 19).

Heat control of the body is regulated by the excretion of sweat, the circulation of the blood, and the nerve centers in the brain. Thus body temperature and circulating blood are physiologically stabilized at a constant temperature despite climatic variations. Sweat gland function and the consequent delivery of sweat, which evaporates from the skin surface, are important in the control of body heat. Simultaneously heat loss is facilitated by the dilation of the cutaneous blood vessels. The opposite occurs when cold weather causes the blood vessels of the skin to constrict and preserve body heat (6, 9, 15, 19).

Secretory functions within the skin are relegated to sweat gland and sebaceous gland activity. Sweat is composed largely of water but also contains several metabolic products. Sweat gland function and sweat delivery is essential for physiological normalcy. Too much or too little sweat delivery can have deleterious effects on the whole physiological behavior. Sebaceous glands reside within the dermis adjacent to the hair follicles with which they connect. Their function product, sebum, is excreted through the hair follicle and the orifice on the skin surface. Overfunction of this gland (acne) is a frequent problem of the adolescent, but the sebaceous glands are also target sites for occupational acne resulting from working with coal tar, heavy oils, greases, and certain aromatic chlorinated hydrocarbons (6, 7, 9, 15, 19).

Special receptors within the skin are part of a network of nerve endings and fibers that receive and conduct various stimuli, later recognized as heat, cold, pain, and other perceptions such as wet, dry, sharp, dull, smooth, and rough (6, 9, 15, 16, 19).

From experience in laboratory testing and what happens at the work site, we know that skin provides a fairly efficient defense against percutaneous absorption of most materials encountered in the work environment. Nonetheless, absorption by way of the skin can occur particularly when the outer barrier has been compromised by laceration or abrasion or massive harmful exposure. Skin can absorb materials by way of the epidermal route, the hair follicles, and less so through sweat openings and ducts. It is generally accepted that the appendageal routes provide entry only during the early stages of absorption. The skin itself is the major avenue of entry for a number of toxic agents present in the agricultural and industrial areas (6, 9, 15, 19, 20). Best known for absorption through the skin are the organophosphate and certain chlorinated hydrocarbon pesticides, the cyanides, aromatic and amino nitro compounds, mercury, and tetraethyllead, among a few other substances. Fortunately, systemic toxicity by way of the skin is not a common happening, but that does not lessen the importance of skin as a route of entry. It is well to remember that skin is not a perfect barrier (9, 15, 19, 25).

2 CAUSAL FACTORS

To detect the cause of an occupational dermatosis it is fundamental to consider a spectrum of factors that can have indirect and direct relationship to the disease.

2.1 Indirect Factors

The importance of recognizing the direct cause or an occupational dermatosis is obvious; however, there can be one or more elements that greatly influence the induction of these diseases, for example, the race, type of skin, age, perspiration, seasonal influence, presence of other skin disease, and level of personal hygiene (1, 3, 6, 7, 9, 13, 15, 16).

2.1.1 Race

It is often stated that black skin fares better in the industrial environment, but this is true only in part. Without doubt, black skin is well endowed with built-in protection against sunlight, and thus it is not commonly associated with malignant lesions induced by sunlight. However, black skin is not endowed with special immunity against the action of photoreactive contactants or ingested medications (1, 6, 7, 9, 13, 26).

It is well documented that light-complexioned and redheaded individuals, particularly those of Celtic extraction, are readily affected by overexposure to sunlight and outdoor work, particularly in Sun Belt areas. People with this type of skin also seem to be less resistant to skin damage when working with numerous chemicals, for example, plastic systems and various irritant dusts and solvents (1, 3, 6, 9, 13, 15, 27).

2.1.2 Skin Type

Caucasians differ in their reactivity to cutaneous irritants. Those with dark swarthy skin appear to be less vulnerable (at least within limits) to the action of soaps, detergents, soluble metalworking fluids, and solvents. Yet these same individuals with hairy arms and legs are prey to embedment in the hair follicles while working with insoluble oils, greases, heavy dusts, tars, and waxes.

Workers with inherently dry skin (xerosis and ichthyosis) are at increased risk when exposed to alkaline agents, acids, detergents, and most solvents (1, 3, 6, 7, 9, 13, 15, 27).

2.1.3 Age

Young workers, particularly those in the adolescent group, often incur acute contact dermatitis. Increased vulnerability because of age is not the reason. More often, young people are placed in service jobs, for example, fast food, janitorial, or car wash, where wet work prevails and protection is difficult. At times disregard for safety and hygiene measures may be the reason.

Older workers usually are more careful through experience, but aging skin is often dry and when work is largely outdoors, sunlight can cause skin cancer (1, 3, 6, 9, 13, 15, 16).

2.1.4 Perspiration

It is normal and physiologically desirable to sweat. However, excessive delivery of sweat can cause sponginess of the skin in the groins, the armpits, and other sites where skin surfaces are opposed to each other.

Sweat can partially solubilize nickel, cobalt, and chromium in small amounts, a situation troublesome to those individuals with cutaneous allergy to these metals. Sweat also can put dry chemical agents into solution which can be irritating. Beneficially, sweat can act as lavage to keep the skin relatively free of irritant contact (1, 3, 6, 9, 11, 12, 15, 16).

Work in poorly ventilated surroundings and during hot weather can induce prickly heat because of excess perspiration.

2.1.5 Season

Occupational skin diseases are generally more frequent in the warmer months for two reasons. First, hot weather discourages the use of protective clothing gear, thus allowing more unprotected skin for exposure to environmental contactants. Secondly, hot environments induce excess sweating (see Chapter 21).

Cold weather is associated with dry skin because of lowered temperatures and humidity. Further, during cold weather some workers simply do not like to take a shower and then go into the cold after work (1, 3, 6, 7, 9, 13, 15, 19, 27).

2.1.6 Presence of Other Skin Diseases

New and old employees with preexisting skin disease are prime candidates for a supervening occupational dermatitis or an aggravation of a preexisting disease of the skin. Hirees with adolescent acne should not be placed in jobs in which they will be exposed to insoluble oils, tar products, greases, and certain polychlorinated aromatic hydrocarbons known to cause chloracne (1, 3, 6, 7, 9, 13, 15).

Atopics (those with allergic tendency) are at greater risk when exposed to any number of irritant chemical agents. The capricious skin behavior of these people does not tolerate exposure to dusty or oil-laden jobs (9, 13, 15, 16, 27).

Other skin diseases known to be worsened by physical or chemical trauma, even though mild, are psoriasis, lichen planus, chronic recurrent eczema of the hands, and those skin conditions to which light exposure can be detrimental. The need of careful skin evaluation and job placement is clear (6, 7, 9, 13, 15, 27).

2.1.7 Personal Hygiene

Poor washing habits breed prolonged occupational contact with agents that harm the skin. Personal cleanliness is a sound preventive measure, but it depends upon the presence of readily accessible washing facilities, quality hand cleansers, and the recognition by the workers of the need to use them (1, 3, 6, 7, 9, 15, 16).

3 DIRECT CAUSES OF OCCUPATIONAL SKIN DISEASE

Chemical agents are unquestionably the major cutaneous hazards; however, there are multiple additional agents that are categorized as mechanical, physical, and biological causalities.

3.1 Chemicals

Chemical agents have always been and will most likely continue to be a major cause of work-incurred skin disease. Organic and inorganic chemicals are used throughout modern industrial processes and increasingly on the farm. They act as primary skin irritants, as allergic sensitizers, or as photosensitizers to induce acute and chronic contact eczematous dermatitis, which accounts for the majority of cases of occupational skin disease, probably no less than 75 to 80 percent (1, 3, 6, 7–16).

3.1.1 Primary Irritants

Most diseases of the skin caused by work result from contact with primary irritant chemicals. These materials cause a dermatitis by direct action on normal skin at the site of contact if allowed to act in sufficient quantity and intensity for a sufficient time. In other words, any normal skin can be injured by a primary irritant. Certain irritants, such as sulfuric, nitric, or hydrofluoric acid, can be exceedingly powerful in damaging the skin within moments. Similarly, sodium hydroxide, chloride of lime, or ethylene oxide gas can produce rapid damage. These are absolute or strong irritants and they can produce necrosis and ulceration resulting in severe scarring. More commonly encountered are the low-grade or marginal irritants that through repetitious contact produce a slowly evolving contact dermatitis. Marginal irritation is often associated with contact with soluble metalworking fluids, soap and water, and solvents such as acetone, ketone, and alcohol. Wet work in general is associated with repetitive contact with marginal irritants (1, 3, 6–16).

3.1.2 Primary Irritant Action on Skin

Clinical manifestations produced by contact with primary irritant agents are readily recognized but not well understood. General behavior of many chemicals in the laboratory or in industrial processes is fairly well known and at times their chemical action can be applied theoretically, and in some instances actually, to chemical action on human skin. For example, we know that organic and inorganic alkalies damage keratin; that organic solvents dissolve surface lipids and remove lipid components from keratin cells; that heavy metal salts, notably arsenic and chromium, precipitate protein and cause it to denature; that salicylic acid, oxalic acid, and urea, among other substances, can chemically and physically reduce keratin; that arsenic, tar, methylcholanthrene, and other known carcinogens stimulate skin to take on abnormal growth patterns. Just how these interactions take place in a molecular biological sequence remains to be explained (1, 3, 6, 7, 9, 11–16).

Irritant chemicals are commonly present in agriculture, manufacturing, and service pursuits. Hundreds of these agents classed as acids, alkalies, gases, organic materials, metal salts, solvents, resins, soaps including synthetic detergents, can cause absolute or marginal irritation (Table 10.1) (15).

Table 10.1. Typical Primary Irritants

Acids	
Inorganic	*Organic*
Arsenious	Acetic
Chromic	Acrylic
Hydrobromic	Carbolic
Hydrochloric	Chloroacetic
Hydrofluoric	Cresylic
Nitric	Formic
Phosphoric	Lactic
Sulfuric	Oxalic
	Salicylic
Alkalis	
Inorganic	*Organic*
Ammonium	Butylamines
Carbonate	Ethylamines
Hydroxide	Ethanolamines
Calcium	Methylamines
Carbonate	Propylamines
Cyanamide	Triethanolamine
Hydroxide	
Oxide	
Potassium	
Carbonate	
Hydroxide	

Table 10.1. (continued)

Alkalis (cont.)	
Sodium	
Carbonate (soda ash)	
Hydroxide (caustic soda)	
Silicate	
Trisodium phosphate	
Cement	
Soaps	
Metal Salts	
Antimony trioxide	
Arsenic trioxide	
Chromium and alkaline chromates	
Cobalt sulfate	
Nickel sulfate	
Mercuric chloride	
Zinc chloride	
Solvents	
Alcohols	*Ketones*
Allyl	Acetone
Amyl	Methyl ethyl
Butyl	Methyl cyclohexanone
Ethyl	
Methyl	
Propyl	
Chlorinated	*Petroleum*
Carbon tetrachloride	Benzene
Chloroform	Ether
Dichloroethylene	Gasoline
Epichlorohydrin	Kerosene
Ethylene chlorohydrin	Varsol
Perchloroethylene	White spirit
Trichloroethylene	
Coal Tar	*Turpentine*
Benzene	Pure oil
Naphtha	Turpentine
Toluene	Turpineol
Xylene	Rosin spirit

Source: L. Schwartz, L. Tulipan, and D. J. Birmingham, *Occupational Diseases of the Skin*, 3rd ed., Lea & Febiger, Philadelphia, 1957 (with revisions).

3.2 Cutaneous Sensitizers

Any chemical can act as a cutaneous sensitizer in a given individual. Fortunately, not every chemical substance behaves in this manner. For many years it has been accepted that allergic dermatitis accounted for about 20 percent of the occupational dermatoses. It now appears from experience here and abroad that the figure of 20 percent is too conservative because more chemical agents (potential allergens) are being encountered in industrial processes and more dermatitis cases are being associated with specific cause by diagnostic tests. Allergic dermatitis (occupational or otherwise) represents a specifically acquired alteration in the capacity to react brought about by an immunologic mechanism. The allergic substance (antigen) is usually a simple chemical that does not cause visible change upon the skin during the first contact and frequently after many contacts. However, after a period of incubation (usually five or many more days) subsequent contact with the same or a chemically similar substance induces a dermatitic reaction. [Some primary irritants can also act as sensitizers (1, 6, 7, 9–12, 14, 15).]

For the reaction to occur, the antigenic material must traverse the keratin layer and gain access to epidermal cells and possibly other sites within the skin. The antigenic material then combines or conjugates with protein carrier to become a complete antigen. After conjugation has taken place the antigen is taken up by Langerhans cells, which in turn present processed antigen to T lymphocytes in the outer (cortical) area of the lymph nodes, and more cells with the ability to recognize antigen (sensitized cells) enter the circulation. At this time the individual is said to be specifically sensitized. Upon reexposure to the antigen, sensitized T lymphocytes release mediating materials (lymphokines) that bring about an inflammatory reaction expressed as an acute eczematous contact dermatitis (10–12, 28). To distinguish between a primary irritant and allergic contact dermatitis it is necessary to recognize that (*a*) allergic reactions usually require a longer induction period than occurs with primary irritation effects; (*b*) cutaneous sensitizers generally do not affect large numbers of workers except when dealing with epoxy resin systems, phenol–formaldehyde plastics, poison ivy, and poison oak. Some other well-known sensitizers associated with occupation are potassium dichromate by itself or contained in cement, nickel sulfate, hexamethylenetetramine, mercaptobenzothiazole, and tetramethylthiuram disulfide, among several other agents (1, 6, 7, 9–13, 15).

3.3 Plants and Woods

Many plants and woods cause injury to the skin through direct irritation or allergic sensitization by their chemical nature. Additionally, irritation can result from contact with sharp edges of leaves, spines, thorns, and so on, which are appendages of the plants. Photosensitivity may also be a factor.

Although the chemical identity of many plant toxins remains undetermined, it is well known that the irritant or allergic principal can be present in the leaves, stems, roots, flower, and bark (29, 30).

High-risk jobs include agricultural workers, construction workers, electric and telephone linemen, florists, gardeners, lumberjacks, pipeline installers, road builders, and others who work outdoors (31).

Poison ivy and poison oak are major offenders. In California several thousand cases of poison oak occupational dermatitis are reported each year. Poison ivy and oak and sumac are members of the Anacardiaceae, which also includes a number of chemically related allergens as cashew nut shell oil, Indian marking nut oil, and mango. The chemical toxicant common to this family is a phenolic (catechol), and sensitization to one family member generally confers sensitivity or cross-reactivity to the others (1, 9, 11–14, 29–31).

Plants known to cause dermatitis are carrots, castor beans, celery, chrysanthemum, hyacinth and tulip bulbs, oleander, primrose, ragweed, and wild parsnip. Other plants including vegetables have been reported as causal in contact dermatitis (9, 11–15, 29–31).

A number of woods are known to provoke skin disease. Woods do not cause as many cases as are reported from plants, but carpenters, cabinetmakers, furniture builders, lumberjacks, lumberyard workers, and model makers (patterns) can incur primary irritant, allergic dermatitis or traumatic effects from the wood being handled. Sawdust, wood spicules, and chemical impregnants in the wood may cause irritation, whereas most allergic dermatitis is caused by oleoresins, the natural oil, or chemical additives. Woods best known for their dermatitis-producing potential are acacia, ash, beech, birch, cedar, mahogany, maple, pine, and spruce. Other agents capable of causing cutaneous injury are the chemicals used for wood preservation purposes as arsenicals, chlorophenols, creosote, and copper compounds (9, 15, 30, 32).

3.4 Photosensitivity

Dermatitis resulting from photoreactivity is an untoward cutaneous reaction to ultraviolet light. The effect may be phototoxic, which is similar to primary irritation, or it may be allergic. The simplest example of phototoxicity is sunburn. Thus thousands of outdoor workers in construction, road building, fishing, forestry, gardening, farming, and electric and phone line erection are exposed to sunlight. Additionally, exposure to artificial ultraviolet light is experienced by electric furnace and foundry operators, glassblowers, photoengravers, steelworkers, welders, and printers in contact with photocure inks. Phototoxic reactions due to certain plants, a number of medications, and some fragrances have been well documented (1, 9, 11–13, 15, 26, 33–36).

In the coal tar industry, distillation can offer exposure to anthracene, phenanthrene, and acridine—all well-known photoreceptive chemical agents. Related products, as creosote, pitch roof paint, road tar, and pipeline coatings are well demonstrated causes of hyperpigmentation resulting from interaction of tar vapors or dusts with sunlight (see Table 10.2) (1, 3, 9, 15, 26, 33, 36).

Occupational photosensitivity is further complicated by a number of topically applied and ingested drugs that can interact with specific wavelengths of light to

Table 10.2. Photosensitizers

Antibacterials	Coal Tar
Dibromosalicylanilide	Acridine
Hexachlorophene	Anthracene
Sulfonamides	Certain chlorinated hydrocarbons
Tribromosalicylanilide	Creosote
	Phenanthrene
	Pitch

produce a phototoxic or photoallergic reaction. Among such agents known to produce these effects are drugs related to sulfonamides, certain antibiotics, tranquilizers of the phenothiazine group, and a number of phototoxic oils that are used in fragrances (11–13, 15, 26, 33, 36).

Among the plants known to cause photosensitivity reaction are members of the Umbelliferae. They include celery that has been infected with pink rot fungus, cow parsnip, dill, fennel, wild carrot, and wild parsnip. The chemical photoreceptors in these plants are psoralens or furocoumarins (see Table 10.3) (15, 26, 33, 36).

3.5 Mechanical

Work-incurred cutaneous injury may be mild, moderate, or severe. The injuries include cuts, lacerations, punctures, abrasions, and burns, and these account for about 35 percent of occupational injuries for which worker's compensation claims are filed (National Electric Injury Surveillance System. U.S. Consumer Product Safety Commission, 1984; U.S. Bureau of Labor Statistics Supplementary Data Systems [SDS], 1983 data) (17).

Table 10.3. Photosensitizing Plants

Moraceae
Ficus carica (fig)
Rutaceae
Citrus aurantifolia (lime)
Dictamnus albus (gas plant)
Ruta graveolens (rue)
Umbelliferae
Anthriscus sylvestris (cow parsley)
Apium graveolens (celery, pink rot)
Daucus carota var. *sativa* (carrot)
Heracleum spp. (cow parsnip)

Contact with spicules of fiber glass, copra, hemp, and so on induce irritation and stimulate itching and scratching. Skin can react to friction by forming a blister or a callus; to pressure by changing color or becoming thickened; and to shearing or sharp force by denudation or a puncture wound. Any break in the skin may become the site of a secondary infection (1, 3, 6, 9, 13–15).

Thousands of workmen use air-powered and electric tools that operate at variable frequencies. Exposure to vibration in a certain frequency range can produce painful fingers, a Raynaud-like disorder resulting from spasm of the blood vessels in the tool-holding hand. Slower-frequency tools such as jackhammers can cause bony, muscular, and tendon injury (37, 38).

3.6 Physical

Heat, cold, electricity, ultraviolet light (natural and artificial), and various radiation sources can induce cutaneous injury and sometimes systemic effects.

3.6.1 Heat

Thermal burns are common among welders, lead burners, metal cutters, roofers, molten metalworkers, and glass blowers.

Miliaria (prickly heat) often follows overexposure to increased temperatures and humidities. Increase in sweating causes waterlogging of the keratin layer with blockading of the sweat ducts and their skin exits.

Excess heat can lead to heat cramps, heat exhaustion, and heat stroke (1, 6, 7, 9, 39, 40).

3.6.2 Cold

Frostbite is the common injury from too much exposure to cold. Fingers, toes, ears, and nose are the usual sites of injury (policemen, firemen, postal workers, farmers, construction workers, military personnel, and frozen food storage employees are at risk) (1, 6, 7, 9, 13, 40).

3.6.3 Electricity

Severe cutaneous burns of local or widespread proportions can result from electrical injury (1, 9, 13).

3.6.4 Ultraviolet Light

Ultraviolet wavelengths are divided into three ranges: UVA 320–400 nm; UVB 290–320 nm, and UVC 200–290 nm.

Sunlight (range 290–320 nm) is a formidable hazard for those who work outdoors, notably in the Sun Belt states.

Artificial ultraviolet light arising from electric furnaces, foundry operations, glassblowing, steel work, and welding, though generally less hazardous than natural light, can produce thermal burns and chronic skin changes (11, 13, 15, 26).

3.6.5 Microwave

Thermal burn is the major hazardous potential in contact with this radiation source (39).

3.6.6 Laser

Radiation effects from laser beams include burns, loss of pigment, and possibly damage to deeper organs (39).

3.6.7 Ionizing Radiation

Modern industry and technology have many applications of this radiation type. It is important in the production and use of fissionable materials, radioisotopes, X-ray diffraction machines, electron beam operations, industrial X-ray for detecting metal flaws, and various uses in diagnostic and therapeutic radiology. Accidental exposures may result in severe cutaneous and systemic injury depending upon the level of radiation received (1, 13, 39).

3.7 Biological

Primary or secondary infection can happen in any occupation following exposure to bacteria, viruses, fungi, or parasites. Simple lacerations or embedment of a thorn or a wood splinter or metal slug can lead to infection. Certain occupations are associated with greater risk of bacterial infection, for example, anthrax among sheepherders, hide processers, and wool handlers; erysipeloid infection among meat, fish, and fowl dressers; and folliculitis among machinists, garage workers, candymakers, sanitation and sewage employees, and those exposed to coal tar (1, 6, 7, 9, 13, 15, 43).

Fungi can produce localized or systemic disease. Yeast infections (*Candida*) occur among those employees engaged in wet work, for example, bartenders, cannery workers, fruit processors, or anyone who works in a wet environment. Sporotrichosis is seen among garden and landscape workers, florists, farmers, and miners. Ringworm infection of animals can be transmitted to farmers, veterinarians, laboratory personnel, and anyone in frequent contact with infected animals (1, 9, 13, 15, 43).

Certain parasitic mites inhabit cheese, grain, and other foods and will attack bakers, grain harvesters, grocers, and longshoremen. Mites that live on animals and fowl similarly are known to attack humans. In the southeastern states animal hookworm larvae from dogs and cats are deposited in sandy soils and lead to infection among construction workers, farmers, plumbers, and of course, anyone who works in the infected soils. Ticks, fleas, and insects can produce troublesome skin reactions and in certain instances, systemic disease such as spotted fever, Lyme disease, yellow fever, and malaria, among other vector-borne diseases (1, 13, 43).

Several occupational diseases are associated with virus infections, for example, Q fever, Newcastle disease, and ornithosis. However, the occupational dermatoses best known as caused by virus infections are eczema contagiosum (orf) contracted from infected sheep, milker's nodules from infected cows, chicken pox from infected children, and herpes infections among dentists, nurses, physicians, and others whose work occasions contact with open lesions. Rare cases of herpes zoster (shingles) have been associated with trauma (1, 15, 16).

Rare but real dermatologic effects can follow animal bites, both domestic and wild (e.g., snakes, spider, shark, various animal bites, and bee stings) (1, 13, 15, 16).

4 CLINICAL APPEARANCE OF OCCUPATIONAL SKIN DISEASE

The hazardous potential of the work environment is unlimited. Animate and inanimate agents can produce a wide variety of clinical displays that differ in appearance and in histopathological pattern. The nature of the lesions and the sites of involvement may provide a clue as to a certain class of materials involved, but only in rare instances does clinical appearance indicate the precise cause. Except for a few strange and unusual effects, the majority of occupational dermatoses can be placed in one of the following reaction patterns. Several materials known to be causal for each clinical type are included (1, 3, 6, 7, 9, 13, 15, 16).

4.1 Acute Eczematous Contact Dermatitis

Most of the occupational dermatoses can be classified as acute eczematous contact dermatitis. Heat, redness, swelling, vesiculation, and oozing are the clinical signs; itch, burning, and general discomfort are the major symptoms experienced. The backs of the hands, the inner wrists, and the forearms are the usual sites of attack, but acute contact dermatitis can occur anywhere on the skin. When forehead, eyelids, ears, face, and neck are involved, dust and vapors are suspected. Generalized contact dermatitis comes from massive exposure, the wearing of contaminated clothing, or autosensitization from a preexisting dermatitis.

Usually a contact dermatitis is recognizable as such, but whether the eruption has resulted from contact with a primary irritant or a cutaneous sensitizer can be ascertained only through a detailed history, a working knowledge of the materials being handled, their behavior on the skin, and a proper application and evaluation of diagnostic tests. Severe blistering or destruction of tissue generally indicates the action of an absolute or strong irritant; however, the history is what reveals the precise agent.

Acute contact eczematous dermatitis can be caused by hundreds of irritant and sensitizing chemicals, plants, and photoreactive agents. Some examples are (1, 3, 6, 7, 9, 11–15, 41, 42):

Acids, dilute	Herbicides	Resin systems
Alkalies, dilute	Insecticides	Rubber accelerators
Anhydrides	Liquid fuels	Rubber antioxidants
Detergents	Metal salts	Soluble emulsions
Germicides	Plants and woods	Solvents

4.2 Chronic Eczematous Contact Dermatitis

Hands, fingers, wrists, and forearms are the favored sites affected by chronic eczematous lesions. The skin is dry, thickened, and scaly with cracking and fissuring of the digits and palms. Chronic nail dystrophy is a common accompaniment. Periodically, acute weeping lesions appear because of reexposure or imprudent treatment. A large number of materials sustain the marked dryness that accompanies this chronic recurrent skin problem. Among these are the following (1, 3, 6, 7, 9, 11–15, 44):

Abrasive dusts (pumice, sand, fiber glass)
Alkalies
Cement
Cleansers (industrial)
Cutting fluids (soluble)
Chronic fungal infections
Oils
Resin systems
Solvents
Wet work

4.3 Folliculitis, Acne, and Chloracne

Hair follicles on the face, neck, forearms, backs of hands, fingers, lower abdomen, buttocks, and thighs can be affected in any kind of work entailing heavy soilage. Comedones (blackhead) and follicular infection are common among garage mechanics, certain machine tool operators, oil drillers, tar workers, roofers, and tradesmen engaged in generally dusty and dirty work.

Acne caused by industrial agents usually is seen on the face, arms, upper back, and chest; however, when exposure is severe lesions may be seen on the abdominal wall, buttocks, and thighs. Machinists, mechanics, oil field and oil refinery workers, road builders, and roofers exposed to tar are at risk. Such effects are far less prevalent than was noted in the past.

Chloracne is hallmarked by the presence of follicular lesions, blackheads, acneform lesions, cysts, hyperpigmentations, and sometimes accompanying scars. The disease may be mild or severe, localized or widespread, but under any circumstance it is a disorder with prolonged course. Severe exposure may lead to toxic liver injury, including porphyria cutanea tarda. Early lesions (blackheads and cysts) of chloracne first appear on the sides of the forehead and around the lateral aspects of the eyelids. Cystic lesions may be seen behind the ears. As exposure persists, lesions can be seen in widespread areas except for the palms and soles.

The following agents are known to cause these conditions (1, 6, 7, 9, 13, 44–52, 55):

Folliculitis	Acne	Chloracne
Asphalt	Creosote	Chloronaphthalenes
Creosote	Crude oil	Chlorinated diphenyls
Greases	Insoluble cutting oil	Chlorinated triphenyls
Lubes	Pitch	Hexachlorodibenzo-*p*-dioxin
Oils	Tar	Tetrachloroazoxybenzene
Pitch		Tetrachlorodibenzodioxin

4.4 Sweat-Induced Reactions

Miliaria (prickly heat) results from waterlogged keratin that eventually blocks the sweat duct and surface opening.

Intertrigo occurs at sites where the skin opposition allows sweat and warmth to macerate the tissue. Favored locations are the armpits, the groins, between the buttocks, and under the breasts (1, 6, 7, 9, 13–15).

4.5 Pigmentary Abnormalities

Color changes in skin can result from percutaneous absorption, inhalation, or a combination of both entry routes. The color change may represent chemical fixation of a dye to keratin or an increase or decrease in epidermal pigment (melanin).

Hypermelanosis results from stimulation of the melanocytes to produce pigment. It may follow an inflammatory dermatosis, exposure to sunlight alone, or the combined action of sunlight plus a number of photoreceptive chemicals or plants. The opposite (loss of pigment or leukoderma) results from direct injury to the epidermis and melanin-producing cells by burns, chronic dermatitis, trauma, or chemical interference with the enzyme system that produces melanin.

Antioxidant chemicals used in adhesives, cutting fluids, sanitizing agents, and rubber have caused loss of pigment (leukoderma).

Inhalation or percutaneous absorption of certain toxicants as aniline or other aromatic nitro and amino compounds causes methemoglobinemia. Jaundice can result from hepatic injury by carbon tetrachloride or trinitrotoluene, among other hepatotoxins.

Pigmentary abnormalities are caused by the following agents (1, 6, 7, 9, 10, 13, 15, 22–24):

	Melanin Abnormalities	
Discolorations	Plus	Minus
Arsenic	Chloroacnegens	Antioxidants
Certain organic amines	Coal tar products	Hydroquinone
Carbon	Petroleum oils	Monobenzyl ether of hydroquinone
Dyes	Photoreactive chemicals	Tertiary amyl phenol
Mercury	Photoreactive plants	Tertiary butyl catechol
Picric acid	Radiation (sunlight)	Tertiary butyl phenol
Silver	Radiation (ionizing)	Burns
Tetryl		Chronic dermatitis
Trinitrotoluene		Trauma

4.6 Neoplasms

New growths on the skin can be benign or malignant. Benign lesions such as asbestos warts and the warty growths associated with petroleum and tar exposures are ready examples. Malignant lesions include basal cell epithelioma and squamous cell carcinoma. Whether certain cases of melanoma are related to certain occupations remains to be seen (15, 54).

Several chemical and physical agents are classified as industrial carcinogens, but only a few are frequent causes of skin cancer. Admittedly, more cancers appear on the skin than at any other site; however, the number of these that are of occupational origin is not known. Sunlight is probably the major cause of skin cancer, particularly among those engaged in agriculture, construction, fishing, forestry, gardening and landscaping, oil drilling, road building, roofing, and telephone and electric line installations (15, 26, 54, 55).

The following are known carcinogens (1, 3, 15, 54–60):

Actinic rays
Anthracene
Arsenic
Burns
Coal tar
Coal tar pitch
Creosote oil
Mineral oils containing various additives and impurities
Crude oils
Radium and roentgen rays
Shale oil
Soot
Ultraviolet light

In European countries, mule spinners exposed to shale oil and pressmen exposed to paraffin experienced a high frequency of carcinomatous lesions of the scrotum and lower extremities. Similar experiences with paraffin happened in the United States, but improved industrial practices and hygienic controls have all but eliminated the problem. As of 1984, the International Agency for Research on Cancer determined that mineral oils containing various additives and impurities used in mule spinning, metal machining, and jute processing were carcinogenic to humans. Oils formerly in use that were responsible for cutaneous cancers including those affecting the scrotum were not as well refined as the lubricating oils being used today (60).

It is important to understand that skin cancers noted on workers in contact with carcinogenic agents are not necessarily of occupational origin. A certain number of people develop skin cancer irrespective of their jobs. For instance, residents in the southwestern United States or in Australia are associated with a high frequency of skin cancer because of the exposure to sunlight. Ascertaining whether a skin cancer is truly of occupational origin in individuals residing in Sun Belt regions can be controversial (1, 3, 13, 15, 26, 54–60).

4.7 Ulcerations

Cutaneous ulcers were the earliest documented skin changes observed among miners and allied craftsmen. In 1827, Cumin reported on skin ulcers produced by chromium (61). Today the chrome ulcer (hole) caused by chromic acid or concentrated alkaline dichromate is a familiar lesion among chrome platers and chrome reduction plant operators. Perforation of the nasal septum also occurs among these employees, though in smaller numbers than occurred 20 years ago because many of the operations are now well enclosed (62). Punched out ulcers on the skin can result from contact with arsenic trioxide, calcium arsenate, calcium nitrate, and slaked lime (63). Nonchemical ulcerations may be associated with trauma and ulcers of the lower extremities in diabetics, pyogenic infections, vascular insufficiency, and sickle cell anemia (1, 3, 13, 15, 64).

4.8 Granulomas

Cutaneous granulomas are caused by many agents of animate and inanimate nature. Such lesions are characterized by chronic, indolent inflammatory reactions that can be localized or systematized and result in severe scar formation. Granulomatous lesions can be the result of bacterial, fungal, viral, or parasitic elements such as atypical mycobacterium, sporotrichosis, milker's nodules, and tick bite, respectively. Additionally, minerals such as silica, zirconium, and beryllium and substances such as bone, chitin, coral, thorns, and grease have produced chronic granulomatous change in the skin (1, 15, 16, 65).

4.9 Other Clinical Patterns

The clinical patterns described above represent well-known forms of occupational skin diseases. However, there are a number of other disorders affecting skin, hair, and nails that do not fit into these patterns. Some examples follow.

4.9.1 Contact Urticaria

This form of urticarial reaction can be allergic or nonallergic, occupational or nonoccupational. Several foods such as apple, carrot, egg, fish, beef, chicken, lamb, pork, and turkey have been reported as causal. Other sources have arisen from animal viscera and products handled by veterinarians and food dressers; from contact with formaldehyde, rat tail, guinea pig, and streptomycin; and several cases from latex rubber gloves. It is likely that many of these cases go unreported and unrecognized as such (10, 11, 66, 67).

4.9.2 Nail Discoloration and Dystrophy

Chemicals such as alkaline bichromate induce an ochre color in nails; tetryl and trinitrotoluene induce yellow coloring; dyes of various colors may change the nail color; carpenters may have wood stains on their nails. Dystrophy can follow chronic contact from acids and corrosive salts, alkaline agents, moisture exposures, sugars, trauma, and infectious agents such as bacteria and fungi (1, 9, 16, 68–70).

4.9.3 Facial Flush

This peculiar phenomenon has been reported from the combination of certain chemicals such as tetramethylthiuram disulfide, trichloroethylene, or butyraldoxime following the ingestion of alcohol. Because trichloroethylene is known to cause liver damage, the facial flush may be related to the intolerance associated with alcohol ingestion (1, 15, 71, 72).

4.9.4 Acroosteolysis

Several years ago, a number of workers involved in cleaning vinyl chloride polymerization reaction tanks incurred a peculiar vascular and bony abnormality involving the digits, hands, and forearms. Bone resorption of the digital tufts was accompanied by Raynaud's symptoms and sclerodermatous-like changes of the hands and forearms. Removal from the tank cleaning duties led to vascular and bone improvement (73, 74).

5 SOME SIGNS OF SYSTEMIC INTOXICATION FOLLOWING PERCUTANEOUS ABSORPTION

A number of chemicals with or without direct toxic effect on the skin can cause systemic intoxication following percutaneous entry. To do so the toxicant must traverse the keratin layer and the epidermal cell layers, and then pass through the epidermal–dermal membrane. At this point the toxicant has ready access to the vascular system and conveyance to certain target organs prior to being excreted as such or being identified by way of a metabolite. The following examples represent target organs and demonstrable effects (16, 75, 76):

Aniline → RBC → methemoglobinemia

Benzidene → urinary bladder → carcinoma

Carbon disulfide → CNS → psychopathological symptoms and signs as peripheral neuritis + cardiac disease

Carbon tetrachloride → liver, kidney, and CNS → liver and kidney damage, depression

Chlorinated naphthalenes, diphenyls, and dioxins → skin → chloracne, hepatitis, peripheral neuritis

Ethylene glycol ethers → CNS → lungs, liver and kidney damage

Methyl butyl ketone → CNS and peripheral nerves→ polyneuritis, depression

Organophosphate pesticides → inhibition of cholinesterase → cardiovascular, gastrointestinal, neuromuscular, and pulmonary disturbances

Tetrachlorethane → CNS → depression and liver damage

Toluene → CNS and liver → confusion, dizziness, headache, paresthesias

Over a period of years, industrial and contract laboratories have developed and used tests for predicting the toxicologic behavior of a new product or one in use whose toxic action becomes questioned. Such tests on animals and humans are designed to demonstrate signs of systemic and cutaneous and toxic effects, for example, primary irritation, allergic hypersensitivity, phototoxicity, photoallergenicity, interference with pigment formation, sweat and sebaceous gland activity, dermal absorption routes, metabolic markers, and cellular aberrations indicating malformation or frank carcinoma. Conducting these tests leads to prediction of toxicologic potential and diminishment thereby of the number of untoward reactions that might appear if such tests had not been done (16, 77).

6 DIAGNOSIS

It is a common assumption among employees that any skin disease they incur has something to do with their work. At times the supposition is correct, but often there is no true relationship to the work situation. Arriving at the correct diagnosis may be quite easy, but such is not a routine occurrence. The industrial physician has a distinct advantage in being familiar with agents within the work environment and the conditions associated with contacting them. The dermatologist may find the diagnosis difficult if unfamiliar with contact agents in the work environment. The practitioner with little or no interest in dermatologic problems associated with work and also lacking dermatologic skill will find it most difficult to make a correct diagnosis. At any rate the attending physician, specialist or otherwise, should satisfy certain basic tenets in establishing a diagnosis of occupational skin disease.

6.1 History

Only through detailed questioning can the proper relationship between cause and effect be established. Taking a thorough history is time consuming because it should cover the past and present health and work status of the employee, for example, family history, particularly of allergies, personal illness in childhood and the past, the title of the job, the nature of the work performed, the materials handled, how long the job has been done, when and where on the skin the rash appeared, the behavior of the rash away from work, whether other employees were affected, what was used to cleanse and protect the skin, and what has been used for self or prescribed treatment. It is important to ascertain whether the employee has had dry skin or chronic hand eczema or psoriasis or other skin problems; what drugs, if any, have been used for any particular disease; and finally, which materials have been used in home hobbies such as the garden or woodworking or painting and moonlighting (1, 6, 7, 9, 10, 13, 15, 16).

6.2 Appearance of the Lesions

The eruption should fit into one of the clinical types in its appearance. Although the majority of the occupational dermatoses fall into acute or chronic eczematous contact dermatitis classification, other clinical types such as follicular, acneform, pigmentary, neoplastic, ulcerative, and granulomatous can occur. Further, one must be on the lookout for the oddities that show up in unpredictable fashion, for example, Raynaud's disease and contact urticaria (1, 6, 7, 9, 13, 15, 16).

6.3 Sites Affected

Most cases of occupational skin disease affect the hands, the digits, the wrists, and the forearms, for the upper extremities are truly the instruments for work. However, the forehead, face, V of the neck, and ears may also display active lesions, particularly when the employee is exposed to dusts, vapors or fumes. Although most of the work dermatoses are usually seen in the above sites, generalization can occur from massive exposure, contaminated work clothing, and also from autosensitization (spread) of an already existent rash (1, 6, 7, 9, 13, 15, 16).

6.4 Diagnostic Tests

Laboratory tests should be employed when necessary for the detection of bacteria, fungi, and parasites. Such tests include direct microscopic examination of surface specimens; culture of bacterial or fungal elements and biopsy of one or more lesions for histophatological definition. When allergic reactions are suspect, diagnostic patch tests can be used to ascertain occupational, as well as nonoccupational allergies, including photosensitization. At times, useful information can be obtained through the use of analytical chemical examination of blood, urine, or tissue (skin, hair, nails) (1, 6, 7, 9, 10–16).

6.5 The Patch Test

Diagnostic patch testing, properly performed and interpreted, is a highly useful procedure. The test is based on the theory that when an acute or chronic eczematous dermatitis is caused by a given sensitizing agent, application of the suspected material to an area of unaffected skin for 24 to 48 hr will cause an inflammatory reaction at the application skin. A positive test usually indicates that the individual has an allergic sensitivity to the test material. When the employee is working with primary irritants and fellow employees also are affected with dermatitis, the cause is self-evident and patch testing is neither necessary nor indicated. Exception to this rule occurs when the employees are working with irritant agents that can also sensitize (epoxy and acrylic systems, resin hardeners of the amine group, formalin, chromates). When an employee incurs a dermatitis suspected to be due to work and fellow employees remain free of skin problems, the diagnostic patch test is then indicated. The number of industrial, agricultural, and service materials capable of causing allergic dermatitis is limitless and because allergic dermatitis is believed to account for not less than 20 percent and probably more of all cases of occupational skin disease, the patch test is being used with increasing frequency, notably in the European clinics (1, 6, 9, 11, 12, 14, 15, 78).

If the test is to have relevance and reliability, it must be performed by one with a clear understanding of the difference between a primary irritant and a sensitizer.

When patch tests are conducted with strong or even marginal irritants, a skin reaction is inevitable. However, this does not mean that a patch test cannot be performed with a diluted primary irritant. There is an abundance of published material pointing out proper patch test concentrations and appropriate vehicles considered as safe for skin tests (9–12). Further, the one performing the tests should have a working knowledge of environmental contactants, particularly those well known as potential cutaneous allergens (1, 6, 9, 11, 12, 14, 15, 78–81).

The technique of the test is simple. Liquids, powders, or solids can be applied as open or closed tests. The test agents (usually 12 or more) are applied to the back. The North American Contact Dermatitis Group and the International Contact Dermatitis Group advocate standardized test concentrations applied in vertical rows on the back and covered by hypoallergenic tape (Al-Test). Some clinicians prefer the test material to be contained within the Finn chamber attached to Scanpore tape. Contact with the test material is maintained for 48 hr and readings are made 30 min or later after removal and again at 72 and 96 hr.

Reading the tests and interpreting them for the degree of reaction requires experience. The levels of reaction currently in use are:

? = doubtful: faint, macular redness only

+ = weak (nonvesicular) positive reaction: redness, infiltration, possibly papules

++ = strong (vesicular) positive reaction: redness, infiltration, papules, vesicles

+++ = extreme positive reaction: bullous reaction

− = negative reaction

IR = irritant reaction of different type

NT = not tested

If a reaction is not present at the time of patch removal and reading, but appears at a later time (less than 5 days after initial application), the reaction is considered as being of the delayed type. True allergic reactions tend toward increased intensity for 24 to 48 hr after test removal, whereas irritant reactions usually subside within 24 to 48 hr after removal period (9, 11, 15, 78–81).

Interpreting the significance of the test reactions is all important. A positive test can result from exposure to an irritant or a sensitizer. When specific sensitization is the case, it means that the patient was reactive to the allergen at the time of the test. When the positive test coincides with a positive history of contact, it is considered strong evidence of allergic reactivity. Conversely, the examiner must be aware that false positive tests can occur if the patient is tested (1) during an active dermatitic phase leading to one or several nonspecific reactions; (2) with a marginal irritant; or (3) with a sensitizer to which the patient had developed an early sensitization, for example, nickel, but which is not relevant to the present occupational dermatitis.

A negative test, if correct, indicates the absence of an irritant or an allergic reaction. However, a negative reaction can also mean (1) testing with the wrong allergen; (2) insufficient strength and quantity of the test allergen; (3) failure to duplicate physical or mechanical factors as poor adherence, lack of friction, and occlusion; or (4) hyporeactivity by the patient at the time of the test.

Performing the patch test with unknown substances the employee has brought to the physician's office can be most misleading. The material could be a caustic and thus produce a strongly positive chemical burn or perhaps no reaction will occur. In either case, the test has been misused and provides no help in diagnosing the cause. Useful information concerning unknown materials can be obtained by contacting the plant manager, physician, nurse, industrial hygienist, or safety supervisor (9, 11, 12, 79–81).

A patch test being employed more frequently has to do with suspected photoallergy. Most of the photodermatoses incurred in the work arena are phototoxic and thus do not require photo patch tests for diagnosis. Some dermatologists perform photo patch tests in their offices; however, many of these patients suffering with suspected photodermatoses are referred to an office or center where photo patch tests are performed routinely (9, 13, 26, 33, 36).

The importance associated with the patch test as a diagnostic tool is well represented in dermatologic case reports and textbooks that provide updated information concerned with the demonstration of contact allergens in the work place. Some well-known examples are (9, 11, 12, 47, 79, 80):

a. Biocides and germicides: formaldehyde and formaldehyde releasers, glutaraldehyde, quarternium-15, isothiazolone-3-1 derivatives
b. Dyes: paraphenylenediamine, color developers, azo dyes
c. Metal salts: chromates, cobalt, mercury, nickel
d. Plastics and resins: acrylates, epoxies, epoxy hardeners, phenolic resins
e. Rubber accelerators and antioxidants: thiurams, mercapto compounds, dithiocarbamates, paraphenylenediamine derivatives

7 TREATMENT

Immediate treatment of an occupational dermatosis does not differ essentially from that used for a similar eruption of nonoccupational nature. In either case, treatment should be directed toward providing fairly rapid relief of symptoms. The choice of treatment agents depends upon the nature and severity of the dermatitis. Most of the cases are either an acute or a chronic contact eczematous dermatitis, and most of these can be managed with ambulatory care. However, hospitalization is indicated when the severity of the eruption warrants in-patient care.

Acute eczematous dermatoses caused by a contactant generally respond promptly to wet dressings and topical steroid preparations, but systemic therapy with corticosteroids should be used when deemed necessary. Corticosteroids have definitely lessened the morbidity in the acute and chronic eczematous dermatoses caused by work. Once the dermatitis is under good control, clinical management must be directed toward:

1. Ascertaining the cause
2. Returning the patient to the job when the skin condition warrants, but not before
3. Instructing the patient in the means necessary to minimize or prevent contact at work with the offending material

In any contact dermatitis it is essential to establish the causal agents or situations that contributed to the induction of the disease. Follicular or acneform skin lesions, notably chloracne, are notoriously slow in responding to treatment. Pigmentary change similarly may resist the run of therapeutic agents and remain active for months. New growths can be removed by an appropriate method and studied histopathologically. Ulcerations inevitably lead to the formation of scar tissue. Similarly, granulomatous lesions generally scar.

Almost all cases of occupational skin disease respond to appropriate therapy; however, when chloracne or pigmentary changes or chronic dermatitis due to chrome or nickel are the problem, therapeutic response may take months or years. It cannot be overemphasized that contact with the causative agents must be minimized, if not eliminated; otherwise return to work is accompanied by the return of the rash (1, 6–9, 13, 15).

7.1 Prolonged and Recurrent Dermatoses

As a rule, an occupational dermatosis can be expected to disappear or to be considerably improved within a period of 4 to 8 weeks after initiating treatment. Yet there are cases that refuse to respond to appropriate treatment and continue to plague the patient with chronic recurrent episodes. This situation is commonly noted when the dermatosis was caused by cement, chromium, nickel, mercury, or certain plastics. However, all cases of recurrent disease are not necessarily associated with the above materials. The following situations may be operable in prolonged and recurrent disease (82, 83):

1. Incorrect clinical diagnosis
2. Failure to establish cause
3. Failure to eliminate the cause even when direct cause has been established
4. Improper treatment, often self-directed
5. Poor hygiene habits at work
6. Supervening secondary infections
7. Cross-reactions with related chemicals
8. Self-perpetuation for gain

8 PREVENTION

The key to preventing occupational skin disease is to eliminate or at least minimize skin contact with potential irritants and sensitizers present in the workplace. To do so requires:

a. Recognition of the hazardous exposure potentials
b. Assessment of the workplace exposures
c. Establishment of necessary controls

Achieving these steps is more likely to occur in large industrial establishments with trained personnel responsible for the maintenance of health and safety practices. In contrast, many small plants or workplaces that employ the largest percentage of the work force have neither the money nor personnel to initiate and monitor effective preventive programs. Nonetheless, any work establishment has the responsibility of providing those preventive measures that at least minimize, if not entirely eliminate, contact with hazardous exposures.

8.1 Direct Measures

Time-tested control measures known to prevent occupational diseases are classed as primary (immediate) and indirect. The primary categories are

a. Substitution
b. Process change
c. Isolation/enclosure
d. Ventilation
e. Good housekeeping
f. Personal protection

8.2

The indirect measures include:

a. Education and training of management, supervisory force, and employees
b. Medical programs
c. Environmental monitoring

Although small plants generally lack in-house medical and industrial hygiene services, they do have access to such services through state health departments or through private consultants knowledgeable in health and safety measures.

8.2.1 Substitution and Process Change

When a particular agent or process is recognized as a trouble source, substituting a less hazardous agent or process can minimize or eliminate the problem. This has been done with a number of toxic agents, for example, substituting toluene for benzene or tetrachloroethylene for carbon tetrachloride. Substitution has potential value in allergen replacement of known offenders such as with chromium, nickel, certain antioxidants and accelerators in rubber manufacture, and certain biocides in metalworking fluids. When feasible, substituting a nonallergen for a hazardous agent is a recommended procedure.

8.2.2 Isolation and Enclosure

Isolation of an agent or a process can be used to minimize hours of exposure or the number of people exposed. Isolation can mean creation of a barrier, or distance, or time, as the means of isolation to lessen exposure. Enclosure of processes provides a high level of safety when hazardous agents are involved. Local enclosures against oil spray and splash from metalworking fluid lessen the amount of exposure to the machine operators.

Radiation exposures can be shielded with proper barriers and remote control systems. Bagging operations can be enclosed to lessen, if not entirely eliminate, exposure to the operators involved.

8.2.3 Ventilation

Movement of air can mean general dilution and/or local exhaust ventilation used to reduce exposures to harmful airborne agents. Local exhaust ventilation is effective in controlling vapors of degreasing tanks and in mixing, lay-up, curing, and tooling of epoxy, polyester, phenol–formaldehyde resin systems.

8.2.4 Good Housekeeping

A clean shop or plant is essential in controlling exposure to hazardous materials. This means keeping the workplace ceilings, windows, walls, floors, workbenches, and tools clean. It means providing adequate storage space, properly placed warning signs, sanitary facilities adequate in number, cleanup of spills, and emergency showers for use after accidental heavy exposure to harmful chemicals.

8.2.5 Personal Protection

8.2.5.1 Clothing. It is not necessary for all workers to wear protective clothing, but for those jobs in which it is required, good-quality clothing should be issued as a plant responsibility. Protective clothing against cold, heat, and biological and chemical injury to the skin is available. Depending upon the need, equipment such as hairnets, caps, helmets, shirts, trousers, coveralls, aprons, gloves, boots, safety glasses, and face shields are available. Similarly, clothing to protect against ultraviolet light and ionizing, microwave, and laser radiation is readily available.

Once protective clothing has been issued, its laundering and maintenance should be the responsibility of the plant. When work clothing is laundered at home it becomes a ready means of contaminating family wearing apparel with chemicals, fiber glass, or harmful dusts.

Specific information concerned with protective equipment can be obtained from the National Institute for Occupational Safety and Health and from any of the manufacturers listed in the safey and hygiene journals.

8.2.5.2 ***Gloves.*** Gloves are an important part of protective gear because the hands are valuable instruments at work. Leather gloves, though expensive, offer fairly good protection against mechanical trauma (friction, abrasion, etc.). Cotton gloves suffice for light work, but they wear out sometimes in a matter of hours. Neoprene and vinyl-dipped cotton gloves are useful in protecting against mechanical trauma, chemicals, solvents, and dusts. Unlined rubber gloves and plastic gloves can cause maceration and sometimes contact dermatitis from the chemical accelerator or antioxidant leach-out by sweat in the wearer's glove. Of no small importance in choosing gloves is the reason for use. Much time and money can be saved by reviewing the catalogs and tables provided by the manufacturers.

8.2.5.3 ***Hand Cleansers.*** Of all the measures advocated for preventing occupational skin disease, personal cleanliness is paramount. Although ventilating systems and monitoring are important in controlling the workplace exposures, there remains no substitute for washing the hands, forearms, and face and keeping clean. To do this the plant must provide conveniently located wash stations with hot and cold running water, good-quality cleansers and disposable towels.

Several varieties of acceptable cleansers are available on the market and these include conventional soaps of liquid, cake, or powdered variety. Conventional soaps are used each day by millions of people and are generally considered as safe. Liquid varieties including "cream" soaps are satisfactory for light soil removal. Powdered soaps are designed for light frictional removal of soil and may contain pumice, wood fiber, or corn meal.

Waterless cleansers are popular among those who contact heavy tenacious soilage such as tar, grease, and paint. They should not serve as a substitute for conventional removal of soilage. Daily use of waterless cleansers leads to dryness of the skin and, at times, eczematous dermatitis from the solvent action of the cleanser.

In choosing an industrial cleanser the following dicta are suggested:

1. It should have good cleansing quality.
2. It should not dry out the skin through normal usage.
3. It should not harmfully abrade the skin.
4. It should not contain known sensitizers.
5. It should flow readily through dispensers.
6. It should resist insect invasion.
7. It should not clog the plumbing.

8.2.5.4 ***Protective Creams.*** Covering the skin with a barrier cream, lotion, or ointment is a common practice in and out of industry. Easy application and removal plus the psychological aspect of protection account for the popularity of these materials. Obviously, a thin layer of barrier cream is not the same as good environmental control or an appropriate protective sleeve or glove. Yet there are circumstances in which a barrier cream can be used, for example, when wearing a glove inhibits dexterity or feel, or when a glove can be caught in machinery. Some benefit is derived by washing the barrier cream off the skin along with the entrapped soil three or four times a day. There is no controversy about the benefit creams provide in protecting against sunlight and the combined effects of sunlight and coal tar exposures. Similarly, certain lotions offer protection against insects. Preparations that protect against poisonous plants still remain a most useful goal (1, 6, 7, 9, 13, 15, 79, 84–88).

8.3 Indirect Measures

8.3.1 Education

An effective prevention and control program against occupational disease in general, including diseases of the skin, must begin with education. A joint commitment by management, supervisory personnel, workers, and worker representatives is required. The purpose is to acquaint managerial personnel and the workers with the hazards inherent in the workplace and the measures available to control the hazards. The training should be in the hands of well qualified instructors capable of instructing the involved people with:

a. Identification of the agents involved in the plant
b. Potential risks
c. Symptoms and signs of unwanted effects
d. Results of environmental and biological monitoring in the plant
e. Management plans for hazard control
f. Instructions for emergencies
g. Safe job procedures

Worker education cannot be static. It must be periodic through the medical and hygiene personnel, during job training, and periodically thereafter through health and safety meetings.

Special training courses are available at several universities with departments specializing in occupational and environmental health and hygiene.

8.3.2 Environmental Monitoring

Periodic sampling of the work environment detects the nature and extent of potential difficulties and also the effectiveness of the control measures being used. Monitoring for skin hazards can include wipe samples from the skin as well as the work sites and use of a black light for detecting the presence of tar product fluorescence on the skin before and after washing. Monitoring is particularly required when new compounds are introduced into plant processes.

8.3.3 Medical Controls

Sound medical programs contribute greatly to preventing illness and injury among the plant employees. Large establishments have used in-house medical and hygiene personnel quite effectively. Daily surveillance of this type is not generally available to small plants; however, small plants do have access to well trained occupational health and hygiene specialists through contractual agreements. At any rate, medical programs are designed to prevent occupational illness and injury, and this begins with a thorough preplacement physical examination, including the condition of the skin. When the preplacement examination detects the presence of or personal history of chronic eczema (atopic), psoriasis, hyperhidrosis, acne vulgaris, discoid lupus erythematosus, chronic fungal disease, dry skin, or other skin diseases, extreme care must be exerted in placement to avoid worsening a preexisting disease.

Plant medical personnel, full time or otherwise, should make periodic inspections of the plant operations to note the presence of skin disease, the use or misuse of protective gear, and hygiene breaches that predispose to skin injury.

When toxic agents are being handled, periodic biological monitoring of urine and blood for specific indicators or metabolites should be regularly performed.

Plant medical and industrial hygiene personnel should have constant surveillance over the introduction of new materials into the operations within the plant. Failure to do so can lead to the unwitting use of toxic agents capable of producing serious problems.

Of great importance in the medical control program is the maintenance of good medical records indicating occupational and nonoccupational conditions affecting the skin, as well as other organ systems. Medical records are vital in compensation cases, particularly those of litigious character (1, 6, 7, 9, 13, 15, 88).

A well detailed coverage of prevention of occupational skin diseases is present in the publications, "Proposed National Strategies for the Prevention of Leading Work-Related Diseases and Injuries, Part 2" (88) and the "Report of the Advisory Committee on Cutaneous Hazards to the Assistant Secretary of Labor, OSHA, 1978" (16).

REFERENCES

1. L. Schwartz, L. Tulipan, and D. J. Birmingham, *Occupational Diseases of the Skin*, 3rd ed., Lea & Febiger, Philadelphia, 1957.
2. "History of Technology," in *The New Encyclopedia Brittanica Macropedia*, 15th ed., Vol. 18, 1978.
3. R. P. White, *The Dermatergoses or Occupational Affections of the Skin*, 4th ed., H. K. Lewis & Company, London, 1934.
4. B. Ramazzini, *Diseases of Workers* (translated from the Latin text, *De Morbis Artificum*, 1713, by W. C. Wright), Hafner, New York-London, 1964.
5. P. Pott, *Cancer Scroti*, Chirurgical Works, London, 1775, pp. 734; 1790 ed., pp. 257–261.

6. R. R. Suskind, "Occupational Skin Problems. I. Mechanisms of Dermatologic Response. II. Methods of Evaluation for Cutaneous Hazards. III. Case Study and Diagnostic Appraisal," *J. Occup. Med.*, 1 (1959).
7. M. H. Samitz and S. R. Cohen, "Occupational Skin Disease," in *Dermatology*, Vol. II, S. L. Moschella and H. J. Hurley, Eds., W. B. Saunders Company, Philadelphia, 1985.
8. K. E. Malten and R. L. Zielhius, "Industrial Toxicology and Dermatology," in *Production and Processing of Plastics*, Elsevier, New York, 1964.
9. R. M. Adams, *Occupational Dermatology*, Grune & Stratton, New York, 1983.
10. H. I. Maibach, *Occupational and Industrial Dermatology*, 2nd ed., Year Book Medical Publishers, Chicago, 1987.
11. A. A. Fisher, *Contact Dermatitis*, 3rd ed., Lea & Febiger, Philadelphia, 1986.
12. E. Cronin, *Contact Dermatitis*, Churchill Livingston, London, 1980.
13. G. A. Gellin, *Occupational Dermatoses*, Department of Environmental, Public and Occupational Health, American Medical Association, 1972.
14. R. J. G. Rycroft, "Occupational Dermatoses," Chapter 16 in *Textbook of Dermatology*, 4th ed., Vol. I, A. Rook, D. S. Wilkinson, F. J. G. Ebling, and J. L. Burton, Eds., Blackwell Scientific Publications, Oxford, 1986.
15. D. J. Birmingham, "Occupational Dermatoses," in *Patty's Industrial Hygiene and Toxicology*, 3rd ed., Vol. I, G. D. Clayton and F. E. Clayton, Eds., John Wiley, New York, 1978.
16. Report of Advisory Committee on Cutaneous Hazards to Assistant Secretary of Labor, OSHA, U.S. Department of Labor, 1978.
17. "Occupational Illness by Type," Bureau of Labor Standards Annual Survey 1984, Fig. 1 in *A Proposed National Strategy for the Prevention of Dermatological Conditions*, Part 2, Association of Schools of Public Health Under a Cooperative Agreement with NIOSH, U.S. Government Printing Office, Washington, DC, 1988.
18. National Institute for Occupational Safety and Health, *Pilot Study for Development of an Occupational Disease Surveillance Method*, NIOSH Publication No. 75-162, U.S. Government Printing Office, Washington, DC, 1975.
19. I. H. Blank, "The Skin as an Organ of Protection Against the External Environment," in *Dermatology in General Medicine*, 2nd ed., T. B. Fitzpatrick et al, Eds., McGraw-Hill, New York, 1979.
20. A. M. Kligman, "The Biology of the Stratum Corneum," in *The Epidermis*, W. Montagna and W. C. Lobitz, Jr., Eds., Academic Press, New York, 1964, pp. 387–433.
21. T. B. Fitzpatrick, "Biology of the Melanin Pigmentary System," in *Dermatology in General Medicine*, T. B. Fitzpatrick et al, Eds. McGraw-Hill, New York, 1979.
22. G. Kahn, "Depigmentation Caused by Phenolic Detergent Germicides," *Arch. Dermatol.*, **102**, 177–187 (1979).
23. G. A. Gellin, P. A. Possick, and V. B. Perone, "Depigmentation from 4-Tertiary Butyl Catechol—An Experimental Study," *J. Invest. Dermatol.*, **55**, 190–197 (1970).
24. K. Malten, E. Seutter, and I. Hara, "Occupational Vitiligo due to *p*-Tertiary Butyl Phenol and Homologues," *Trans. St. John Hosp. Dermatolol. Soc.*, **57**, 115–134 (1971).
25. R. Scheuplein, "Permeability of the Skin," in *Handbook of Physiology-Reactions to Environmental Agents*, American Physiology Soc., Bethesda, MD, 1974, pp. 299–322.
26. J. Epstein, "Adverse Cutaneous Reactions to the Sun," in *Year Book of Dermatology*,

F. D. Malkinson and R. W. Pearson, Eds., Year book Medical Publishers, Chicago, 1971.

27. E. Schmunes, "The Role of Atopy in Occupational Skin Disease," in *Occupational Medicine—State of Art Review*, Vol. I, Hanley & Belfus, Philadelphia, 1986, pp. 219–28#.
28. R. L. Baer, "The Mechanism of Allergic Contact Hypersensitivity," in *Contact Dermatitis*, 3rd ed., A. A. Fisher, Ed., Lea & Febiger, Philadelphia, 1986.
29. K. F. Lampke and R. Fagerstrom, *Plant Toxicity and Dermatitis (A Manual for Physicians)*, Williams & Wilkins, Baltimore, 1968.
30. T. Barber and E. Husting, "Plant and Wood Hazards," in *Occupational Diseases—A Guide to Their Recognition*, rev. Ed., M. M. Key et al, eds., U.S. Department of Health, Education and Welfare, PHS, CDC, NIOSH, DHEW-NIOSH Publication 77-181, U.S. Government Printing Office, Washington, DC, 1977.
31. G. A. Gellin, C. R. Wolf, and T. H. Milby, "Poison Ivy, Poison Oak and Poison Sumac—Common Causes of Occupational Dermatitis," *Arch. Environ. Health*, **22**, 280 (1971).
32. *Wood Preservation Around the Home and Farm*, Forest Products Laboratory Publication No. 1117, Canadian Department of Forestry, Ottawa, 1966.
33. D. J. Birmingham, "Photosensitizing Drugs, Plants and Chemicals," *Michigan Med.*, **67**, 39–43 (1968).
34. K. E. Malten and W. J. M. Beude, "2-Hydroxy-Alkyl Methacrylate and Di and Tetra-Ethylene Glycol Di-Methacrylate, Contact Photo-Sensitizers in Photo Polymer Printing Plate Procedure," *Contact Dermatitis*, **5**, 214 (1976).
35. E. Emmett and J. R. Kaminski, "Allergic Contact Dermatitis from Acrylates in Ultra Violet Cured Inks," *J. Occup. Med.*, **19**, 113 (1977).
36. V. De Leo and L. C. Harber, "Contact Photodermatitis," Chapter 25 in *Contact Dermatitis*, A. A. Fisher, Ed., Lea & Febiger, Philadelphia, 1986.
37. N. Williams, "Biological Effects of Segmental Vibration," *J. Occup. Med.*, **17**, 37–39 (1975).
38. G. A. Suvorov and I. K. Razumov, "Vibration," in *Encyclopedia of Occupational Safety and Health*, 3rd ed., Vol. II, International Labor Office, Geneva, 1983.
39. E. Meso, W. Murray, W. Parr, and J. Conover, "Physical Hazards: Radiation," in *Occupational Diseases—A Guide to Their Recognition*, rev. ed., U.S. Department of Health, Education and Welfare, NIOSH, U.S. Government Printing Office, Washington, DC, 1977.
40. F. N. Dukes-Dobos and D. W. Badger, "Physical Hazards—Atmospheric Variations," in *Occupational Diseases—A Guide to Their Recognition*, rev. ed., U.S. Department of HEW, NIOSH, U.S. Government Printing Office, Washington, DC, 1977.
41. J. V. Klauder and M. K. Hardy, "Actual Causes of Certain Occupational Dermatoses: Further Study of 532 Cases with Special Reference to Dermatitis Caused by Certain Petroleum Solvents," *Occup. Med.*, **1**, 168–181 (1946).
42. J. V. Klauder and B. A. Gross, Actual Causes of Certain Occupational Dermatoses: A Further Study with Special Reference to Effect of Alkali on the Skin, Effect of Soap on pH of Skin, Modern Cutaneous Detergents, *Arch. Dermatol. Syphilol.*, **63**, 1–23 (1951).
43. D. S. Wilkinson, "Biological Causes of Occupational Dermatoses," in *Occupational*

and Industrial Dermatology, H. I. Maibach and G. A. Gellin, Eds., Year Book Medical Publishers, Chicago, 1982.

44. M. M. Key, E. J. Ritter, and K. A. Arndt, "Cutting and Grinding Fluids and Their Effects on Skin," *Am. Ind. Hyg. Assoc. J.*, **27**, 423–427 (1966).
45. M. H. Samitz, Effects of Metal Working Fluids on the Skin, *Prog. Dermatol.*, **8**, 11 (1974).
46. G. Hodgson, "Eczemas Associated with Lubricants and Metal Working Fluids," *Dermatol. Digest*, 11–15 (Oct. 1976).
47. R. J. G. Rycroft, "Cutting Fluids, Oils and Lubricants," in *Occupational and Industrial Dermatology*, H. I. Maibach and G. A. Gellin, Eds., Year Book Medical Publishers, Chicago, 1982.
48. J. Kimmig and K. H. Schultz, "Occupational Chloracne Caused by Aromatic Cyclic Ethers," *Dermatologica*, **115**, 540 (1970).
49. J. Bleiberg, M. Wallen, R. Brodkin, and I. L. Applebaum, "Industrially Acquired Porphyria," *Arch. Dermatol.*, **89**, 793–797 (1964).
50. K. D. Crow, "Chloracne," *Br. J. Dermatol.*, **83**, 899 (1970).
51. N. Z. Jensen, I. B. Sneeden, and A. F. Walker, "Tetrachlorobenzodioxin and Chloracne," *Trans. St. John's Hosp. Dermatol Soc.*, **58**, 172 (1972).
52. J. S. Taylor, "Environmental Chloracne—Update and Overview," *Ann. N.Y. Acad. Science*, **320**, 295–307 (1979).
53. R. R. Suskind and V. Hartzburg, "Human Health Effects of 2,4,5,T and its Toxic Contaminants," *J. Am. Med. Assoc.*, **251**, 2370–2380 (1984).
54. F. C. Combs, *Coal Tar and Cutaneous Carcinogenesis in Industry*, Charles C Thomas, Springfield, IL, 1954.
55. E. A. Emmett, "Occupational Skin Cancer—A Review," *J. Occup. Med.*, **17**, 44–49 (1975).
56. R. E. Eckhard, *Industrial Carcinogens*, Grune & Stratton, New York, 1959.
57. W. D. Buchanan, *Toxicity of Arsenic Compounds*, Elsevier, Amsterdam, 1962.
58. I. Berenblum and R. Schoental, "Carcinogenic Constituents of Shale Oil," *Br. J. Exp. Pathol.*, **24**, 232–239 (1943).
59. E. Bingham, A. V. Horton, and R. Tye, "The Carcinogenic Potential of Certain Oils," *Arch. Environ. Health*, **10**, 449–451 (1965).
60. N. Rothman and E. A. Emmett, "The Carcinogenic Potential of Selected Petroleum Derived Products," Chapter 7, in *Occupational Medicine—State of the Art Review*, Vol. 3, Hanley & Belfus, Philadelphia, July–Sept., 1988, pp. 475–494.
61. W. Cumin, Remarks on the Medicinal Properties of Madar and on the Effects of Bichromate of Potassium on the Human Body, *Edinburgh Med. & Surg. J.*, **28**, 295–302 (1827).
62. S. Cohen, D. Davis, and R. Kozamkowski, "Clinical Manifestations of Chromic Acid Toxicity Nasal Lesions in Electoplate Workers," *Cutis*, **13**, 558–568 (1974).
63. D. J. Birmingham, M. M. Key, D. A. Holaday, and V. B. Perone, "An Outbreak of Arsenical Dermatoses in a Mining Community," *Arch. Dermatol.*, **91**, 457–465 (1964).
64. M. H. Samitz and A. S. Dana, *Cutaneous Lesions of the Lower Extremities*, J. B. Lippincott, Philadelphia, 1971.
65. H. Pinkus and A. H. Mehregan, "Granulomatous Inflammation and Proliferation,"

Section IV in *A Guide to Dermatopathology*, 2nd ed., H. Pinkus and A. H. Mehregan, Eds., Appleton-Century-Crofts, New York, 1976.

66. R. B. Odom and H. I. Maibach, "Contact Urticaria: A Different Contact Dermatitis in Dermatotoxicology and Pharmacology," in *Advances in Modern Toxicology*, Vol. 4, F. N. Marzulli and H. I. Maibach, Eds., Hemisphere Publishing Corp., Washington, DC, 1977.
67. A. A. Fisher, "Contact Urticaria," in *Contact Dermatitis*, 3rd ed., Lea & Febiger, Philadelphia, 1986, Chapter 39.
68. F. Ronchese, "Occupational Nails," *Cutis*, **5**, 164 (1965).
69. F. Ronchese, *Occupational Marks and Other Physical Signs*, Grune & Stratton, New York, 1948.
70. A. A. Fisher, "Regional Contact Dermatitis," Chapter 6 in *Contact Dermatitis*, 3rd ed., 1986, pp. 92–96.
71. W. Lewis and L. Schwartz, "An Occupational Agent (*N*-Butyraldoxime) Causing Reaction to Alcohol, *Med. Ann. D.C.*, **25**, 485–490 (1956).
72. R. D. Stewart, C. L. Hake, and I. E. Peterson, "Degreaser's Flush, Dermal Response to Trichloroethylene and Ethanol," *Arch. Environ. Health*, **29**, 1–5 (1974).
73. R. H. Wilson, W. G. McCormick, C. F. Tatum, and J. L. Creech, "Occupational Acroosteolysis: Report of 31 Cases," *J. Am. Med. Assoc.*, **201**, 577–581 (1967).
74. D. K. Harris and W. G. Adams, "Acroosteolysis Occurring in Men Engaged in the Polymerization of Vinyl Chloride," *Br. Med. J.*, **3**, 712–714 (1967).
75. I. R. Tabershaw, H. M. D. Utidjian, and B. L. Kawahara, "Chemical Hazards," Section VII in *Occupational Diseases—A Guide to Their Recognition*, rev. ed., U.S. Department of HEW-NIOSH Publication No. 77-181, 1977.
76. N. H. Proctor and J. P. Hughes, *Chemical Hazards in the Workplace*, J. B. Lippincott, Philadelphia, 1978.
77. D. Hood, "Practical and Theoretical Considerations in Evaluating Dermal Safety," in *Cutaneous Toxicity*, V. Drill and P. Lazar, Eds., Academic Press, Inc., New York, 1977.
78. M. B. Sulzberger and F. Wise, "The Patch Test in Contact Dermatitis," *Arch. Dermatol. Syphilol.*, **23**, 519 (1931).
79. S. Fregert, *Manual of Contact Dermatitis*, Munksgaard, Copenhagen, 1974.
80. K. E. Malten, J. P. Nater, and W. G. Von Ketel, *Patch Test Guidelines*, Dekker & Van de Vegt, Nigmegen, 1976.
81. H. I. Maibach, "Patch Testing—An Objective Tool," *Cutis*, **13**, 4 (1974).
82. G. E. Morris, "Why Doesn't the Worker's Skin Clear Up? An Analysis of Factors Complicating Industrial Dermatoses," *Arch. Ind. Hyg. Occup. Med.*, **10**, 43–49 (1954).
83. D. J. Birmingham, *Prolonged and Recurrent Occupational Dermatitis. Some Whys and Wherefores, Occupational Medicine—State of the Art Reviews*, Vol. 1, No. 2, Hanley & Belfus, Philadelphia, April–June, 1986.
84. "Methods for Prevention and Control of Occupational Skin Disease," Chapter IV in *Report of the Advisory Committee on Cutaneous Hazards to the Assistant Secretary of Labor*, OSHA, 1978.
85. C. D. Calnan, Studies in Contact Dermatitis XXIII Allergen Replacement, *Trans. St. John's Hosp. Dermatol. Soc.*, **56**, 131–138 (1970).

86. D. J. Birmingham, *The Prevention of Occupational Skin Disease*, Soap & Detergent Association, New York, 1975.
87. C. G. T. Mathias, "Contact Dermatitis from Use and Misuse of Soaps, Detergents and Cleansers in the Work Place," *Occupational Medicine—State of the Art Reviews*, Vol. 1, Hanley & Belfus, Philadelphia, 1986, pp. 205–218.
88. Prevention Planning, Implementation, Evaluation and Recommendations in Proposed National Strategy for the Prevention of Dermatological Conditions in Proposed National Strategies for the Prevention of Leading Work-Related Diseases and Injuries, Part 2, The Association of Schools of Public Health under a Cooperative Agreement with NIOSH, 1988.

CHAPTER ELEVEN

The Pulmonary Effects of Inhaled Inorganic Dust

George W. Wright, M.D.

1 INTRODUCTION

Nature has provided man with a remarkable physiological apparatus for living in dust-laden air. This is fortunate, for dust-laden air is the normal environment of man. There is continuous exposure to inhaled particles that vary from time to time in number and in chemical and physical properties. It seems likely that the remarkable ability of the respiratory apparatus to cope with inhaled particles accounts in part for the survival and evolution of humans, because we developed through our various stages in an environment that over long periods was in all probability even dustier than now as the result of volcanic activity and the scouring effect of wind flowing across the land. Fortunately the respiratory apparatus intercepts a large portion of inhaled particles as they move through the nose or mouth and along the course of air-conducting channels before they reach the more delicate distal parts of the lung. Moreover, the lung has a large capacity to remove deposited dust by way of the mucociliary escalator and macrophage system. It also is important that the lung cells constituting the surfaces in contact with air normally have a rapid turnover or replacement rate; hence partially damaged surface cells are quickly replaced by new and normal cells. In addition, mechanisms are available for the repair of injured tissue in the lung. As is true of all other organ systems, however, the capacity for self-protection and repair of injury can be exceeded, and excessive dust deposition can cause adverse effects within the breathing apparatus.

Patty's Industrial Hygiene and Toxicology, Fourth Edition, Volume 1, Part A, Edited by George D. Clayton and Florence E. Clayton
ISBN 0-471-50197-2

Depending on the durability, the intrinsic chemical and physical nature of the inhaled particles, and also the chemicals adsorbed onto their surfaces, the biological response may be noninjurious, slight or serious, or even fatal. Of equal importance governing the biological outcome are the quantity of particles deposited and the amount retained. The weight of evidence thus far overwhelmingly supports the conviction that all biological responses to nonliving agents are dose related.

Dust particles in contact with tissue cells evoke a response specific to each particular kind of cell. This response may be fleeting or temporary with no persistent or serious cell injury. On the other hand, persistent injury or death of the cell may occur, leading to secondary tissue alterations of varying degrees of gravity. Because of variations in the number of cells responding, the capacity or normality of cell replacement and repair and immunologic cell surveillance, and the hyperreactivity or sensitivity of the reacting cells, the overall tissue response can be expected to differ considerably in magnitude among individuals exposed in quite similar ways.

There are many kinds of respirable particles in the total human environment. Some are essentially peculiar to the occupational environment, but others occur commonly in both the general and occupational environment. Thus biological effects of the occupational environment must always be related to the sum of the particles in the total environment, including those of nonoccupational origin. There is also the possibility that coexisting particles of different kinds and from various sources may act not only additively but also in a synergistic or even an inhibiting manner.

Those interested in industrial toxicology and hygiene or in occupational medicine quite properly are concerned with the biological reactions of humans exposed to dust composed of various agents acting alone or in combination and in varying concentrations. One must be aware not only of the biological effects but also of the mechanisms underlying and controlling their occurrence. Such toxicologic knowledge provides the *raison d'être* for industrial hygiene programs aimed at prevention of injury.

This chapter discusses principles relating the lung response to inorganic dust particles occurring in occupational environments. Specific agents are used to exemplify various kinds of reaction, but the biological effects of all the kinds of particle coexisting with other gases and fumes throughout all sorts of industrial environments cannot be covered. The reader is urged to consult more comprehensive texts for greater medical detail or to rectify omissions of the effect of specific agents. Among such texts, the recent one by W. Raymond Parkes, *Occupational Lung Disorders*, is recommended.

2 PERTINENT ANATOMY OF THE LUNGS

The primary purpose of the respiratory apparatus is to act as a gas exchange mechanism. A very thin layer of venous blood, pumped into the lungs by the right ventricle via the pulmonary artery system, is circulated through the pulmonary capillaries, which are arranged in a network over the surface of approximately 300 million air-containing alveoli. The total effective gas exchange surface of the alveoli has been estimated at $\pm 70\ m^2$. The tissue barrier separating the blood from the gas phase is the alveolocapillary wall, which averages 0.55 μm in thickness. During passage of venous blood over the surface of this barrier, the diffusion gradient favors the rapid movement of molecular oxygen from the air into the blood and of carbon dioxide out of the blood into the air phase. Thus the blood becomes arterialized in its transit over the alveolocapillary surface.

As can readily be appreciated, the partial pressure of these gases in the blood and gas phase would quickly become equal unless the air phase is continuously maintained at partial pressures of O_2 and CO_2 favoring the gradient for diffusion. External respiration brings ambient air low in CO_2 and high in O_2 content to the gas phase overlying the alveolocapillary membrane by mass movement and continuous replacement of air within the lungs. Mass movement of air does not go all the way to the surface of the alveolar wall, and there is an interface, probably within the alveolar sac, across which the final pathway for molecular delivery to the alveolar membrane occurs by diffusion.

After moving through the nose or mouth, inspired air enters the larynx and then the trachea, which divides into two main bronchi, each leading into a lung. Each main bronchus divides into 16 further generations of branches that serve solely the purpose of air conduction (Figure 11.1). There are seven subsequent generations of branching having alveoli in their walls. Counting limbs and twigs, there are approximately 116,300 solely air-conducting bronchi including the terminal bronchioles. Additional air-conducting branches that are also alveolus bearing, the so-called transitional and respiratory zone, bring the total number of branched limbs and twigs through which air moves to about 82 million. The main bronchi measure 1.2 cm in diameter. By the eleventh generation the lumen is reduced to 1 mm, and by the sixteenth and subsequent generations to 0.5 mm (2). These dimensions and numbers have major implications with respect to deposition and clearance of particles entering the lung while entrained in the respired air. The opportunity for impingement of air-entrained particles at the myriad of bifurcations before reaching any alveoli or gas exchange area is enormous. This means that particles of aerodynamic size favoring deposition by impingement, sedimentation, or interception will fall on and be removed by the mucociliary escalator, a structure that extends through the first 16 generations of branching. Hence the probability that a particle suspended in the air bolus of a single breath will actually reach and be deposited in a specific alveolus-bearing portion of the lung is small. Moreover, because of the extremely large number of branched units and their small individual volume, the probability that the moiety of air entering a single unit during a single breath will contain an entrained particle is very low by the time the bolus of air has passed beyond the seventh or eighth generation of branching and is small indeed by the time it has reached or passed the eighteenth generation. This type of distribution reduces the likelihood of overloading any single small conduit during prolonged exposure and helps maintain a high efficiency of dust removal.

Equal in importance to an understanding of the morphometric aspects of the respiratory apparatus is knowledge of the histological or cellular structure of the

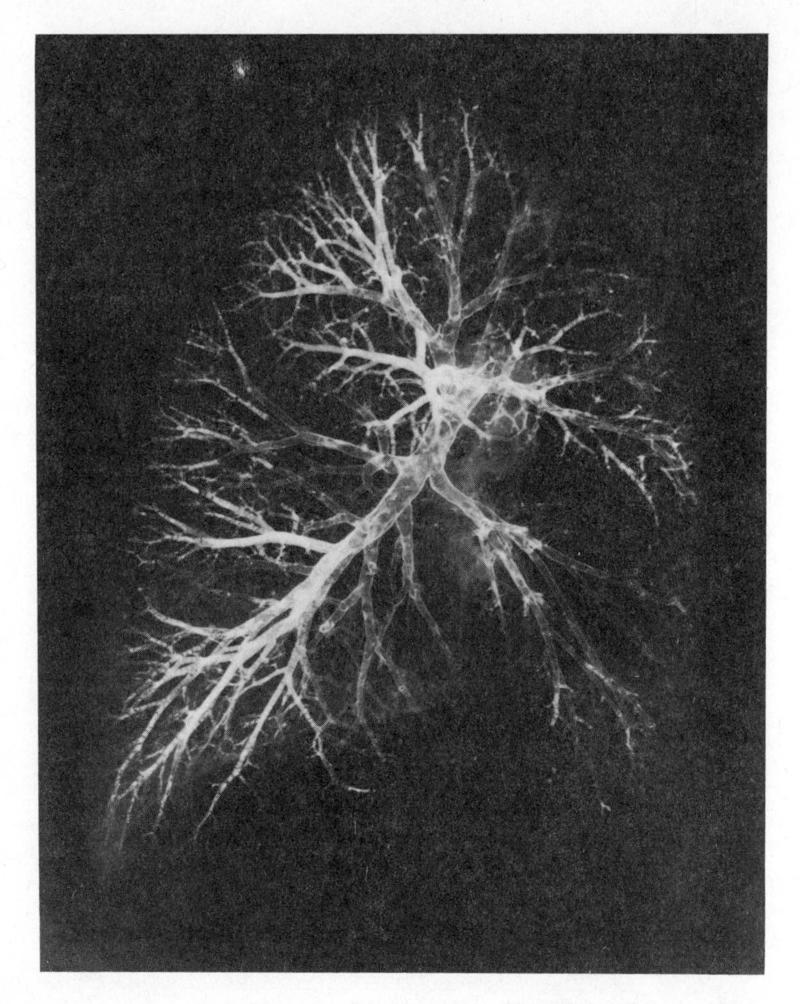

Figure 11.1. Radiograph of airway of distended human lung; right main bronchus with lobar and a few subsequent branchings (probably not beyond eighth generation) as outlined by contrast media.

lung. These components determine the manner in which dust particles entering the lung are disposed of and also what kind of biological reaction does or does not ensue. For details of the normal lung histology, the reader is referred to standard texts of histology and to the special texts on the lung (3, 4).

The first few generations of the air-conducting system, the trachea and bronchi, have rather thick connective tissue walls containing cartilaginous plates that support the wall except in the posterior aspect where the encircling plates are incomplete. There is also a thin layer of smooth muscle located just beneath the mucous membrane lining these conduits. The muscle bundles are arranged in a spiral fashion encircling the conducting tube. As the air tubes further divide, the cartilaginous elements disappear and the walls become thinner, but the spiraling smooth muscle becomes proportionately larger. As the last few generations of conducting airways are reached, the wall becomes still thinner, and by the time the terminal bronchiole, the sixteenth generation, is reached it is very thin. In the seventeenth generation of branches an occasional small outpouching appears in the wall. This cup-shaped outpouching is the alveolus, the site of molecular gas exchange between air and blood. The airways having only a few alveoli protruding from the walls are termed respiratory bronchioles, but when the alveoli make up most of the wall they are called alveolar ducts. When the final division is reached and the walls are entirely formed by alveoli, the passage is termed an alveolar sac or atrium. Elastic fibers, coursing roughly parallel to the air tubes, exist from the trachea all the way to the alveolus, where they form a network within the alveolar wall.

Lining the surface of the air tubes is a membrane composed of an epithelial layer supported on a base of interlacing reticulin, collagen, and elastic fibers embedded in a proteinaceous matrix, the ground substance, containing cells such as fibroblasts and macrophages. The epithelial surface of the trachea and bronchi as far out as the terminal bronchioles is composed of columnar and cuboidal ciliated cells; interspersed between these are mucus-secreting goblet cells, discharging onto the surface of the air-ways. In addition, there are clusters of large mucus-secreting glands scattered throughout the subepithelial region of the bronchi out to the ninth or tenth generation. These glands discharge their contents by way of ducts opening onto the surface of the airway lining. Their discharge plus that from the goblet cells form a carpet of mucus that is distributed rather evenly over the surface of the ciliated cells, which extend out to the terminal bronchiole or sixteenth generation of airways. This carpet of mucus is continuously moved by the action of the underlying cilia from the places of manufacture up through the trachea, where the mucus is then swallowed or expectorated. This system, known as the mucociliary escalator, is the mechanism whereby particles deposited on the airway surface are removed.

Beyond the terminal bronchiole the surface lining of the airways is no longer ciliated and becomes thinner as the cuboidal cells grow shorter and assume the shape of progressively flatter cells. The alveolus is the terminal unit, whether it arises from the wall of the respiratory bronchiole, from the alveolar duct, or from the lumen of the alveolar sac. The alveoli are clustered in a tightly packed manner, abutting upon themselves and the walls of conducting airways of their own and adjacent units. Each alveolus is lined by large, flat, very thin cells, the type I pneumocytes. These lie on a thin layer of reticulin plus elastic fibers that cover a network of capillaries. The capillaries in turn are separated from the reticulin and elastic fibers of the interstitium by large flat cells making up the endothelium that lines the capillaries. A scant amount of ground substance plus occasional connective tissue cells and macrophages lie in the interstices of the reticulin network. The alveoli are so closely packed that a single pulmonary capillary usually serves two or more alveoli abutting upon the pulmonary capillary network. The net pressure of the blood within the capillaries is below that in the interstitial space; thus liquid normally does not pass into the alveolar spaces.

Other cells of importance are found in the alveolar unit. The type II pneumocyte is found scattered at various places in the alveolocapillary wall as part of the surface lining cells. It apparently secretes surfactant, a material that affects the surface tension of the ultrathin liquid layer located at the actual air–tissue interface of the alveolus. Alveolar macrophages (4), large free cells of the reticuloendothelial system measuring 10 to 50 μm in diameter, are located both in the interstitial tissue and within the alveolar spaces lying free on the alveolar surface. These are large mobile cells whose most spectacular function is phagocytosis. They also are involved in immunologic reactions. The number of these cells can be greatly augmented by the presence of dust particles within the alveoli. Particles not exceeding 10 μm in greatest dimension are readily ingested by the macrophage. The fate of such particles is discussed later.

The major branches of the pulmonary artery and veins follow closely the branching of the airways. In addition to these there are very thin walled vessels, the lymphatics, which begin within the interstitial spaces at the level of the respiratory bronchiole and alveolar ducts and ascend in a merging fashion with lymphatic channels from other regions of the lung along the course of the arteries and veins. Similar vessels underlie the pleura, the membrane that covers the lungs and lines the chest cage. These channels are for the purpose of removing excess interstitial fluid. Along their course are collections of lymphoid tissue, the smallest being arranged in cuffs around the respiratory bronchiole and adjacent arterioles. The lymph fluid flows toward the lung root, and larger aggregates of lymphoid tissue are found along the course of these lymphatics as the channels coalesce and become larger. The nodular clusters of lymphoid tissue in the root of each lung are quite large. The various regional aggregates of lymphoid cells serve as collecting points where particles penetrating to the interstitial fluid can be sieved or filtered out of the liquid phase and accumulated in the lymphoid tissue, thus minimizing introduction of such particles into the circulating blood by way of the main lymph vessels that spill their contents into the innominate veins.

Each lung is covered by a thin membrane except at its root, where the main bronchi, blood vessels, and lymph channels pass into or out of the lung. This membrane, the pleura, reflects off the lung root and passes onto the surface of the inner aspect of the thoracic cage, thus forming a closed sac bordered by the lung, the chest wall, and the mediastinum. It contains a small amount of constantly exchanged fluid. Covering the surface of the pleura are large flat cells of mesothelial origin lying on a thin layer of loose areolar connective tissue containing blood and lymph vessels, nerves, and a well-developed layer of elastic tissue along with its complement of macrophage and fibroblasts. Of particular importance for this discussion is the fact that alveoli abut directly onto the inner aspect of the elastic layer and supporting structures of the pleura. Hence there is the possibility that particles reaching these alveoli might pass directly into the pleura. In this way particles in addition to those that move by way of the lymphatics might ultimately come into contact with cells of the pleura.

3 BEHAVIOR AND FATE OF PARTICLES THAT ENTER THE RESPIRATORY SYSTEM

Dust deposition to a large degree depends on the concentration and the physical nature of the specific dust particles in the air. A substantial body of information exists with respect to the respirability of particles of differing dimensions and shapes, their deposition in various compartments of the air-conducting and distal spaces of the lung, and the speed and effectiveness with which rapid and slow clearances occur (6–9).

If, by suitably gentle technique, one digests the lungs of older individuals who have worked in the "dusty trades," a residue is obtained that can be assumed to have come from exogenous sources by way of the airborne route over the years (10). These tiny particles have a most interesting size range. Approximately half are smaller than 0.5 μm in diameter. Of those that are larger, almost all are between 0.5 and 5.0 μm in diameter. Fewer than 0.2 percent of the total are larger than 5.0 μm in diameter, and less than 0.002 percent are larger than 10 μm in diameter. If a fiber is defined as a particle having parallel sides and an aspect ratio of 3 or more, fibers for the most part are observed to be less than 50.0 μm long, although some may be as long as 200 μm. The diameters of these fibers are less than 3 μm. In contrast, in samples of the ambient air to which the general public or those who work in the dusty trades are exposed, one finds particles of these dimensions, but in addition, many of much larger diameter and length. Why is the long-term retention of particles limited to the sizes just described, even though millions of particles of greater diameter or length become airborne and therefore have the potential for entry into the respiratory system? The explanation arises from our knowledge of the behavior of particles suspended in air (aerosols) and the anatomic structure of the lung as described in the preceding paragraphs.

Particles can vary markedly in shape and density, and both these factors play a role in the behavior of particles in air suspension. For our purposes we consider all particles as being spheres of unit density, with the understanding that there could be some variation between particles with respect to speed of settling, depending on their shape and density.

Several physical forces are conducive to the removal of particles from an air suspension and their deposition on surfaces of the respiratory system (6). Particles suspended in a moving airstream possess inertial forces tending to maintain the direction of motion of the particle. When the air column changes its direction, as at a branching point of the conducting system, or in the tortuous passages of the nose, the entrained particle tends to continue in its previous direction and to be cast on the surface. This effect is directly proportional to the size of the particle and the speed of the airstream, thus of the particle, and inversely proportional to the radius of the tube. Gravitational forces also remove particles from the airstream and precipitate them on the surface of the respiratory system. The terminal settling velocity of a particle is directly related to its density, the gravitational constant, and the square of the particle diameter. It is inversely related to air viscosity. Because the gravitational constant and air viscosity are the same at all times, the terminal settling velocity is predominantly related to the other two factors, plus the distance through which the particle must fall and the time permitted for the event to occur. Deposition of particles by diffusion is limited to those of a diameter smaller than 0.5 μm and predominantly to those smaller than 0.1 μm. The smaller the particle, the more rapid the diffusion movement that can occur. The electron microscope size particles are relatively uninfluenced by any deposition force other than that of diffusion, and that fact that such large numbers of particles of this size are found in the lung residue indicates that diffusion can play a major role in their deposition. It has been suggested that electrostatic and thermal forces may play a role in deposition of particles in the lungs, but this is still uncertain.

On the basis of known physical behavior of particles in air suspension and the anatomic arrangement of the conducting tubes, it can be predicted that particles larger than 10 μm in diameter would be removed completely in the passage of the airstream through the nose and upper airways and that the particles between 5 and 10 μm would be deposited primarily in the upper airways of the mucociliary escalator. Only the particles in the range of 1 to 2 μm would be likely to penetrate into the deeper portions of the lung, where deposition in the alveoli could occur by gravity and diffusion. On the basis of these calculations, particle deposition by sedimentation would be least for particles having a diameter of 0.5 μm, but deposition of particles smaller than this might be increased by diffusion, particularly in the most distal portions of the air system. Numerous actual experimental determinations have confirmed this general distribution for location of deposition. Particles larger than 10 μm in diameter are almost completely removed in the nose, and few, if any, reach the smaller conducting tubes of the lungs. Some smaller particles also are deposited in the nose, but the majority pass through and then are deposited, depending primarily on their diameter, along the upper or lower airways. Particles greater than 3 μm in diameter have very little opportunity to penetrate beyond the mucociliary apparatus and to be deposited in the most distal alveolated portions of the conducting tubes. Because almost all the particles larger than 3.0 μm in diameter fall on the mucociliary escalator and are removed, there is a reasonable explanation for the very small number of particles of larger size that are found in the lung residue after a lifetime of exposure to aerosols of ambient air that undoubtedly contains particles of larger size.

A fiber represents a special case in terms of deposition. As is true of other particles, the settling velocity of a fiber depends primarily on its diameter. In a moving airstream, fibers tend to align their length parallel to the direction of airflow. The fibers that are straight and rigid therefore present an end-on aspect essentially that of their actual diameter. Fibers that are curved, curled, or bent in a U-shape have an end-on aspect equal to the width of the curl or curvature. Insofar as interception is concerned, there is thus a much greater chance for deposition of the nonstraight fibers, a factor of considerable importance in the narrow airways and along the boundaries of airflow close to the surface. It has been demonstrated that curly fibers penetrate to the deeper portions of the lung much less readily than do straight fibers of equivalent diameter (11). Length becomes important also to the degree that the fibers are distributed in a random way in the moving airstream.

Thus a fiber 100 μm long oriented at right angles to the direction of flow will have a much greater chance of impacting on the surface than will fibers oriented parallel to the direction of flow. Although occasionally a fiber 200 μm long is observed in the lung dust residues, by far the majority are shorter than 50 μm.

The particles deposited beyond the terminal bronchiole and mucociliary escalator experience one of five fates. It has been suggested that the mucus sheet of the terminal bronchiole is contiguous with a more distal, slow-moving liquid layer manufactured by the Clara cell and the type II pneumocyte. Particles falling on this sheet could slowly be pulled up onto the mucociliary system and thus be removed. As a second fate, there is abundant evidence that particles deposited in the alveolus-bearing part of the lung can quickly be ingested by macrophages. The small size of the particles deposited in this area favors ingestion by the macrophage except for the long fibers thin enough to penetrate to this area but too long to be completely engulfed. The limiting length has not been proved, but fibers up to 10 μm long are readily ingested, and even longer fibers can become surrounded by the spreading membrane of the macrophage. By the mechanism of ingestion, particles of various shapes are taken into a cell whose function is to house and process foreign material. In addition, many of the mobile macrophages with their ingested particles move up the mucociliary escalator and appear in the sputum or are swallowed. Other macrophages filled with particles simply live out their life and die, discharging the particles, which are reingested by new macrophages. This process is repeated indefinitely. The life of the macrophage under ordinary circumstances is measured in terms of weeks and perhaps even a month or more. Its life can, of course, be shortened if the ingested particles are especially toxic, as is the case with free crystalline silica, which kills the macrophage in a period of hours or days. Materials such as coal, iron, asbestos, and glass do not appear to alter the longevity of the macrophage substantially.

A third fate of particles involves their movement, either naked or inside the macrophage, across the alveolar surface and into the interstitial substance. Once having moved there, some particles appear to remain in this space either free or within macrophages, and others enter the lymphatics and are sieved out in the regional lymphoid tissue, where they may remain indefinitely. A few particles pass through by way of the lymph into the systemic blood, which then carries them to other organs of the body. The precise mechanism by which particles move through the alveolar surface may involve more than one method, and there is little agreement on this question. A fourth fate of particles is simply to remain free on the surface of the alveolus.

The fifth fate of particles is to be dissolved partially or completely or to disintegrate into smaller particles over a period of time. The small diameter of the respirable particles, hence the large ratio of surface area to weight, promotes the rate at which these reactions can occur. Some inorganic particles such as those comprised of $CaSO_4$ (gypsum) or $CaCO_3$ (limestone, marble, dolomite) are sufficiently soluble in body fluids to be more or less rapidly removed. Fibers of asbestos or glass, thin enough to be respirable, commonly are thought of as being impervious to the influence of body fluids and enzymes such as those contained within macrophages. Evidence is accumulating, however, to show that fibers of this nature fracture transversely into shorter fragments. Moreover, there now is reason to believe that both glass and asbestos fibers of diameters below 0.5 μm actually may, over a substantial period of time, dissolve within the lung tissue.

The self-cleaning mechanism of the lung appears to be efficient. Although there is some variation of this capacity between individuals, it has been reported that less than 1 percent of the total calculated amount of dust inhaled remains more or less permanently in the lung (6, 12). The paucity of studies bearing on this effect in humans warrants some caution in accepting the figure of 1 percent as applying under all conditions. Nevertheless, the mechanism for removing deposited particles is remarkably effective. It should be borne in mind that the process of lung cleansing adds a portion to the total gastrointestinal tract exposure to exogenous particulate, for a substantial part of the dust brought up on the mucociliary apparatus is swallowed.

4 THE RESPIRATORY FUNCTION OF THE LUNGS

A continuous supply of oxygen to and removal of carbon dioxide from the environment of tissue cells is an essential prerequisite for life. The demand varies from the resting state to the 10 times or more higher metabolic rates encountered during heavy exercise. The respiratory and cardiovascular systems are intimately associated in this accomplishment. The mass movement of air into and out of the lungs required to satisfy this wide range of demands by the tissue cells is accomplished by the rhythmic application of muscle force to the thoracic boundaries, causing the thorax to enlarge and diminish its volume in the familiar act of breathing. The lungs are attached to the thoracic boundaries by a thin layer of liquid that permits them to follow the excursions of the thoracic walls and thus change their volume in a spatially even manner. The lungs not only are stretchable but in addition are elastic and can recoil by virtue of the abundant elastic tissue in the walls of the airways, alveoli, and blood vessels, and also because of the tension in the ultrathin layer of liquid overlying the surface of the alveoli. This overlying liquid layer is elastic by reason of its simulation of a "bubble," with consequent surface tension properties. In fact, this force would collapse the lung if not counteracted by phospholipid particles, the "surfactant" manufactured by the type II pneumocyte and discharged into the thin liquid layer (13).

The rate of air exchange within the lungs reflects the depth and frequency of respiration, which in turn is controlled by the respiratory center. Under normal circumstances the ratio between alveolar ventilation and capillary blood flow is essentially the same throughout the lungs, and the arterialized blood is maintained at a remarkably constant oxygen and hydrogen ion concentration over a wide range of physical activity. There is, however, an inequality of distribution of air and blood flow to the individual units of the lung, the lower or more dependent units receiving a larger portion of the total air and blood flow than do the units higher on the vertical axis.

Because both the respiratory and cardiovascular systems are intimately involved in the movement of oxygen and carbon dioxide between the ambient air and the body cells, it is important to know which of these systems is the limiting factor. Without going into the details, the evidence supports the view that the cardiovascular system limits maximum ability for physical performance in normal persons. Because of this, in an overall sense the respiratory apparatus has a larger reserve capacity for function than does the cardiovascular system. Therefore a substantial impairment of pulmonary function can occur before there is a recognizable loss of capacity for physical effort, especially because the lungs are made up of literally millions of identical units and most types of pulmonary injury involve only some of the units while sparing the others.

5 REACTION OF THE LUNG TO INHALED PARTICLES OF DUST

Utilizing knowledge of the cellular components and organization of the lungs plus the manner in which cells respond to stimulation or injury, one might anticipate the various reactions this organ would display to the deposition of dust. As stated earlier, whether such reactions develop depends on the nature and number of specific particles deposited and retained, as well as the influence of coexisting inhaled agents and the reactivity of the host or individual.

That dust particles can stimulate the smooth muscle distributed in a circular fashion within the thin-walled portions of the airways, thus narrowing the lumen and raising the resistance to airflow, has been demonstrated to occur as a reversible reaction following the inhalation of high concentrations of dust (14). This response in most persons requires high concentrations of dust and is not recognized to occur commonly in the workplace. In persons who for one or another reason are hyperreactive, however, lower concentrations of dust may evoke a recognizable response.

The deposition of dust on the mucociliary apparatus normally stimulates a flow of mucus. If the production of mucus is excessive, or if it is not removed adequately, it can accumulate in the airways, thus reducing the lumen of the conducting tubes and elevating the resistance to airflow. Furthermore, prolonged stimulation of the mucus-secreting glands and cells can lead to hypertrophy or enlargement of these structures. The enlarged subepithelial glands can encroach upon the lumen and cause persistent narrowing of the airways and elevation of resistance to airflow.

Particles of dust lodging on and beneath the surface of the alveoli stimulate the recruitment and accumulation of macrophages in this area. There can be enough of these cells to fill some alveoli partially or completely, but because of the cells' ability to move onto the mucociliary escalator or into the lymphatics, this is unusual. Some particles are always found lying free on the surface of the alveolus, but whether these have never been phagocytized or whether they are waiting for reingestion cannot be determined. Presumably particles could stimulate any of the lung cells they come into contact with directly or indirectly, or they may act through the intermediary of cells such as the macrophage. Some particles lodged on or just beneath the alveolar surface appear to induce a proliferative reaction and become overgrown by the surface cells—at least they are found embedded in a mass of cells involving the surface and the immediate substructure of the alveoli. This reaction can evolve to include cells such as lymphocytes, macrophages, and polymorphonuclear leukocytes, as well as other components of the interstitial or connective tissue and the regional lymphoid collections of the lung. At times such reactions are transitory, but they may become chronic or persistent.

Because fibroblasts are present in the interstitium of the lung, these cells may be stimulated to form excessive amounts of reticulin or collagen. Excessive collagen formation is likely to accompany prolonged or chronic inflammation in most organs of the body. This is a part of the familiar formation of scar tissue, whether it be in the skin or in deeper portions of the body or in the lung. Pulmonary fibrosis is a common sequela of chronic pulmonary inflammation. The pathogenesis of pulmonary fibrosis caused by deposition of some kinds of dust is not entirely clear. It has been postulated in recent years that because the macrophage contains small vesicles holding potent enzymes, these substances may be released by certain ingested particles, but not all kinds, thus altering the cell milieu and causing premature death of the macrophage. The products released by the macrophage are then believed to stimulate the fibroblasts to lay down excessive amounts of collagen, thus promoting the development of tissue fibrosis (15). It has also been suggested that fibers not completely engulfed within the macrophage but protruding through the wall permit leakage of the macrophage content, and this material might also stimulate the fibroblasts to increased activity (16). These concepts are interesting because some dust particles are known to be relatively inert and not capable of causing an excess of fibrous tissue production. Such particles, though readily ingested by the macrophage, do not cause the premature death of the macrophage. In contrast, free crystalline silica, a known fibrogenic dust, has been demonstrated to exert this effect.

Malignant transformation can be the fate of any cell that can divide, and the cells of the lung are no exception. The cells lining the surface of all the airways and alveoli have a particularly fast turnover rate and hence are probably more vulnerable to carcinogenic alteration by a dust or other agents. In addition, particles lodged on the surface of the air-conducting tubes might promote an even higher turnover rate of the surface cells, thus making them unusually vulnerable to carcinogenic substances. Toxic agents adhering to particles could leach off onto the cells with a similar effect. A combination of these effects could lead to metaplasia or could play a role directly or indirectly in malignant transformation of the cells lining the conducting airways.

On the basis of the foregoing considerations, one could anticipate that dust deposited in the lungs might induce:

1. Little or no reaction of any kind
2. Hyperproduction of mucus secretion
3. Hypertrophy of mucus-secreting glands

4. Macrophage recruitment and ingestion of particles
5. Chronic proliferative or inflammatory reaction
6. Reticulinosis
7. Fibrosis
8. Cell metaplasia or malignant transformation

6 INFLUENCE OF ALTERED LUNG STRUCTURE ON PULMONARY FUNCTION

Narrowing of the airways, whether caused by muscle contraction, accumulation of mucus, or hypertrophy of the mucous glands, increases the resistance to airflow through the involved tubes. It should be noted that resistance to airflow through a tube is a function not only of the length of the tube but of the fourth power of the radius (Poiseuille's law). For this reason, relatively small changes in the radius over a substantial length of the tube have a marked influence on the resistance to airflow. If the increased resistance to airflow occurs more or less uniformly throughout all the lung, the work of breathing is commensurately increased, and the volume of air moved per unit of effort is reduced. This not only decreases the maximum available capacity for mass movement of air but causes the expenditure of unusual respiratory effort at all volumes of air movement. Breathlessness or "shortness of breath" is closely related to the awareness of respiratory effort; hence the increased use of respiratory force to move each unit of air leads to the development of breathlessness at lower than normal levels of exercise and to a reduction of the maximum capacity for physical effort. If, as is usually the case, there is focal or regional variation of airway resistance, there will be a shunting of airflow away from the alveoli supplied by the airways experiencing the higher resistance. This leads to regional alveolar underventilation and produces a mismatch between ventilation and capillary blood perfusion in the lung region involved, causing the blood leaving the underventilated area to have a subnormal oxygen and an elevated carbon dioxide content. As a consequence, the arterial blood going to the tissues is below normal in oxygen content. To some degree, coexisting overventilated alveoli can compensate by increased carbon dioxide removal; but if the regional hypoventilation is extensive, there can also be an accumulation of carbon dioxide in the arterial blood returning to the tissues.

The body attempts, with partial success, to correct this mismatch between ventilation and perfusion in the following manner. The low alveolar oxygen tension developing in the hypoventilated regions produces an effect on the alveolar blood capillaries causing partial closure of the capillaries and thus an increase in capillary blood flow resistance and a shunting of blood from the hypoventilated alveoli to those that are more normally ventilated. If the number of underventilated alveoli involved is relatively small, this is an effective mechanism; but if many alveoli are involved, the narrowing of the alveolar capillaries produces an elevated resistance to blood flow in the entire pulmonary circuit, with consequent pulmonary artery hypertension and the bad effects that flow from it. From these comments it can be seen that airway narrowing can lead to rather serious malfunction of the cardiorespiratory system and can be the cause of severe physical impairment, ultimately leading to death.

Chronic intrapulmonary inflammatory and fibrotic changes cause distortions of pulmonary architecture leading to functional derangements. In regions where the cellular infiltrate or fibrosis destroys functioning lung tissue *en bloc*, the effect is similar to that of simply removing this amount of lung. If the volume of tissue involved is relatively small, there is little or no measurable alteration of pulmonary function. If the volume of destroyed tissue is large, the remaining normal lung is much smaller and behaves in an overall less compliant fashion than would a larger volume of lung. In addition, there is a measurable loss of pulmonary vascular bed or diffusing surface. Moreover, because of the loss of arterioles and capillaries, more force is needed to drive the normally required total amounts of blood flow through the pulmonary circulation. This in turn leads to pulmonary artery hypertension and the abnormalities consequent to that condition. If the fibrotic or inflammatory reaction is arranged as ribbons running diffusely throughout the lung, the entrapped normal lung units are less compliant and the work required to change lung volume and produce a unit of mass movement of air is increased. The work of breathing is augmented, and the maximum ability to move air into and out of the lungs is curtailed. Some portions of the entrapped normal lung between the ribbons of diseased tissue are underventilated, and a mismatch between ventilation and perfusion will ensue.

It is of interest to observe that although an elevated resistance to airflow and a decrease in pulmonary compliance affect lung mechanics differently, the end results in terms of overall physiological impairment of the respiratory apparatus are quite similar. Both cause an increase in work of breathing, a loss of maximum ability to move air into and out of the lungs, a mismatch of the ventilation perfusion relationship, with consequent arterial hypoxia, and an elevation in the pulmonary artery pressure. One difference is that in airway obstruction there is carbon dioxide retention and an elevation of arterial pCO_2, whereas in the abnormalities associated with loss of compliance caused by chronic inflammation or fibrosis, carbon dioxide transfer is interfered with relatively slightly or not at all and an elevated pCO_2 usually does not develop.

7 CRITIQUE OF THE METHODS USED TO STUDY THE PULMONARY EFFECTS OF INHALED INORGANIC DUST

Knowledge about the way in which humans respond to inhaled dust should be derived from the demonstrable effects of inhaled particles as observed in humans. Extrapolation from other modes of study such as animal experimentation or cell culture is speculation, although this may be warranted or even necessary under some circumstances. Unfortunately the technical conditions imposed on observations of humans, especially when retrospective observations are made or new substances are being considered, at times limit the availability of the human study modality. Therefore experimental animals or isolated organ or cell systems are often utilized to explore the possibilities of inducing tissue reaction in the lung, the factors that control the reaction, and the biological mechanisms involved. The animal and cell modalities are used most appropriately for examining aspects of the overall question that cannot be approached by human study alone. This does not, however, assure the validity of extrapolating from animal studies to human experience.

The advantages provided by animal or cell experiments are several. The exposure can be controlled with respect to varying its concentration, its duration, and whether single agents or combinations are presented. The opportunity to follow the course of events by serial sampling of the tissue is available only with animal or cell experiments. Also it is possible to repeat experiments under controlled conditions in animals. The shorter life-span of small animals is an advantage in that the process of carcinogenesis is telescoped into a shorter period of time, thus shortening the duration of the experiment. Although the methods used for examining tissues anatomically or biochemically are essentially the same whether animals or humans are being examined, there are some advantages to the use of animal tissue. Because dust particles are distributed in a nonhomogeneous way, the effects have a patchy distribution also. There is an advantage then to being able to examine whole lung slices and to do biochemical analysis, such as for hydroxyprolene on whole lungs of animals.

These manifest advantages explain why the animal or cell models are so widely used. There are, however, some very important and virtually insurmountable disadvantages associated with the use of these models if one intends to extrapolate to human experience. With respect to isolated cell systems, there are two important disadvantages. First, the systems are not under the normal influence of humoral and cellular contols provided by neighboring cells or emanating from other organs and circulating cells that exist in the intact animal or man. Second, long-term experiments cannot be done, because cell survival is rather brief in any event.

Animal experiments avoid these two difficulties but are limited in several other ways. There is always the influence of species difference with respect to carcinogenesis. Humans are but a single species of animal and do not invariably mimic the tissue reactions of other species. Also, experimental animals usually are kept under environmental conditions quite different from those of humans. For example, animals other than humans do not smoke tobacco, a common biologically active coexisting agent in human respiratory exposure patterns. Although a 2-year-old rat may be equivalent to a 60-year-old human with respect to cell line aging, it will have had 58 fewer years of exposure to the coexisting agents usually being studied. Moreover, physical and emotional stress induced by caging conditions can influence the biological response of animals. How similar are the environmental conditions of humans throughout their lifetime to those of the experimental animal in this respect?

The exposure concentrations used in animal experiments are usually at a level very much higher than those occurring in human experience. Understandably, this is done deliberately to guarantee an effect in animals if it is at all possible to produce one. Unfortunately, when a biological effect is produced, there is all too often little effort to determine or take into account the response at lower concentrations—the dose-response curve.

There is also a difficulty posed by the number of animals available for use in experimental protocols. Contrary to popular belief, an unlimited number of animals—or, in many experiments, the number actually needed for the kinds of extrapolation commonly made—is not realistically available for most animal experiments. This is especially true of the experiments seeking a no-effect level, as well as when the biological response of any type is meager. As a matter of fact, the number of individuals at risk and as controls is usually superior in human epidemiologic studies to the number available for most animal experiments. Few research laboratories can afford to buy or breed, feed, and house the number of animals needed to establish epidemiologic validity in their studies. Because of this fundamental defect, the conclusions reached by some animal studies are of questionable validity.

Observation of the effects of inhaled dust in humans has been the conventional way for establishing whether a biological reaction occurs. Clinical and autopsy or histological examination of the lungs of people working in the presence of the agent under study is the mainstay of this approach. A difficulty arises, however, because the cellular components of the lung react in the same manner to a number of different kinds of stimulating agent. For example, the development of a biological abnormality in a person employed in a particular way might suggest the possibility of a relationship between an agent in the occupational environment and the biological disturbance observed. On the other hand, because the same biological abnormality of the lung can develop from various causes, the single isolated case may in fact be caused by something other than the agent present in the specific dust exposure. Therefore, although the clinical and histological approach is of great importance in determining whether an abnormality is present, other modalities to *prove its relationship to a specific agent must be used*. For this purpose the techniques of epidemiology are useful, and the reader is referred to Chapter 26 for discussion of this discipline.

To return to the advantages and disadvantages inherent in observation of human populations, it cannot be stressed too strongly that the greatest advantage of this approach is that no extrapolations are needed. Moreover, the exposure is that of a real-life circumstance with respect to the multiple other factors that may be playing a role. There is an advantage also in the great possibility for observation of larger number of persons both at risk and in the control population. These advantages sometimes are offset by other problems that pose large difficulties. The first of these is the difficulty of making retrospective estimates of the total exposure of the individual to the specific agent in question, expressed in quantitative terms over long periods of time. At times the only available expression is that of duration of employment. Undocumented changes in the environment of the workplace over the many years of employment and the lack of records or measurements of the various agents present during the same period cannot be corrected for. Under such circumstances one can only compare the overall clinical or mortality experience of the observed group to that of some other group, usually the population at large. From such observations it is impossible to derive information usable for establishing the dose–effect relationship. Under such circumstances one also cannot be sure of just which of several coexisting agents is the cause of the biological reaction. Nor can one be sure that cofactors are not operating as the ultimate biological stimulus. To attribute the biological reactions observed to specific agents under such circumstances, though at times a temptation, is a practice of questionable validity.

A second disadvantage associated with human epidemiologic studies is that whereas in animals the lungs may be examined by microscope or chemical methods at any point in time relative to the exposure, these modalities are available in human epidemiologic studies only at death. Because only a portion of the deaths from any population are autopsied, there is a strong possibility of bias posed by autopsy selection, and epidemiologic data derived from autopsy studies must be accepted with some caution. In addition, autopsies rarely are done with the epidemiologic study in mind, and important observations may be lacking, whereas the opposite is true in animal experiments. To the extent that epidemiologic studies depend on death certificates, they are subject to faults such as erroneous diagnosis, "styles" or "fashions" of diagnosis, incomplete diagnosis, and very commonly, unavailability of some death certificates. These faults become particularly grievous when small numbers of a highly localized population are compared to the population of a state or of the entire country.

Because of the difficulties attending the use of mortality data, an alternative epidemiologic approach is to relate the data derived by medical or physiological examinations during life to exposure up to that point in time. This approach does not take into account the results that might occur over a full life-span, and it is particularly limited with respect to the search for relationships between dust exposure and carcinogenesis. Nevertheless, with respect to ill health or impairment of function, a properly designed epidemiologic study during life can in a very useful way serve to compare the frequency of types of impairment or disease in the study group versus an internal control population ranked by the severity of exposure or to an external control population of persons not exposed to the agent in question. This approach has a high degree of validity.

It is certain that biological reactions too minor in extent or intensity to be discovered occur in lung tissue before evolving to an extent recognizable by our methods of searching for these changes during life. How should one view the importance of these subclinical alterations of tissue? The goal is to *prevent disease*, with its associated impairment of function leading to a reduced capacity for normal activity and in some instances to shortened life expectancy. Therefore it is important to distinguish between a small and limited area of tissue injury and what we mean by disease. The differentiation hinges on whether impairment of vital function will occur or has occurred. Cell death and orderly replacement by replication of cells characterizes the human lung from birth to death. Physical factors and external agents such as viruses and bacteria active in small areas of the lung commonly influence the orderly process of cell death and replacement during the state we call "good health." When one examines the body of an adult who has given no prior evidence of lung disease, some pulmonary fibrosis representing the end stage of injury is invariably found. Because these effects are minor in extent and occur in an organ possessing more reserves than can be used by normal persons, such minor scars do not in any way interfere with pulmonary function or life expectancy. There is therefore a quantitative as well as a qualitative aspect to body impairment and disease. In the light of these comments, it is imperative to realize that there is no sharp line of demarcation between being well or ill, normal or diseased, or injured or uninjured. We can speak in rather broad terms of the way in which gases and particles may or may not affect the lung, but insofar as impairment or disease is concerned, the quantitative aspects are probably more important under most circumstances than are the qualitative.

With the exception of sputum cytology, microscopic examination of the lungs of humans is not available for the purpose of early recognition of change in the individual. Epidemiologic studies during life are limited, therefore, to clinical, radiologic, pulmonary function, and sputum cytology modalities. Advantageous as these are, their inherent weaknesses must be recognized if the validity of conclusions derived from their use is to be evaluated.

The clinical history is subjective and difficult to confirm. Physical examination, chest radiography, and pulmonary function measurement, although more objective, exhibit a rather large variation of values among normal persons. Therefore, the recognition of "change" in an individual is difficult unless the observation can be made before as well as after the onset of exposure. This difficulty can be lessened by comparing the average of observations of one group to that of another group serving as a control. The various techniques for these examination modalities require skilled personnel and equipment not always available. In spite of these difficulties, virtually without exception these modalities, when properly used, will permit recognition of deviation from normal before there is interference with vital body functions; thus they are of great use for studying the tissue reaction to inhaled dust during life.

The chest radiogram exhibits shadows cast because of the greater absorption or deflection of X-ray photons by tissues of greater density. Highly cellular or fibrotic tissue is more dense than normal tissue, and when the difference is great enough or when the shadow is located where it should not exist, the normal shadows are altered. A difficulty arises because there is considerable variation between normal persons in the shadows cast by normal lung structure such as blood vessel and supporting lung tissue. Grossly abnormal lungs are readily recognized by alteration of the normal pattern of shadows, but minor tissue alterations make either no or very slight change in the pattern of densities. Therefore there is always a gray area of differing opinions among readers of the x-ray films of persons experiencing slight tissue alteration: are the shadows present in the radiogram those of normal or abnormal structure? For epidemiologic studies, these differences can be lessened by the use of multiple readers using the UICC system of notation and standard reference films (17).

Measurements of pulmonary function are aimed primarily at recognition and quantification of changes in pulmonary airway resistance, pulmonary compliance, the alveolar ventilation–perfusion relationship, and the total effective area of the alveolocapillary membrane. Using rather sophisticated methods, these aspects of pulmonary mechanics can be measured rather accurately. Recognition that an alteration of any of these functions has occurred is hampered by the wide range of values existing between normal persons. One of the less difficult, thus more widely used measurements, is the fast vital capacity (FVC), or volume of air expelled by prolonged, explosively applied maximum expiratory force beginning at the position of full inspiration. This not only indicates the stroke volume of the respiratory pump but also, because maximum expiratory effort is applied rapidly, reflects the speed with which the air can be pushed through the distribution system, and this in turn reflects airway resistance. For this purpose, the volume expelled in the first second (FEV_1), expressed either as a percentage of the predicted volume or as a percentage of the total volume expelled, serves as the reflection of airway resistance. Other portions of the expiratory curve can be used for the same purpose.

With few exceptions, such as the presence of large space occupying masses within the thorax or extreme obesity, pulmonary compliance is in part reflected by the total lung volume at maximum inspiration. Accurate measurement of total volume requires the use of a tracer gas or a plethysmograph, and it is therefore seldom used. An indirect expression of compliance can be obtained in the following manner. Both an elevated airway resistance and a lessened compliance will cause a reduction in FVC; thus it alone will not differentiate between the two conditions. When airway resistance is the cause of the subnormal FVC, however, the FEV_1/FVC ratio is substantially lower than normal. In contrast, when the subnormal FVC is caused by a lessened compliance, this ratio is either normal or slightly reduced, or it may actually be increased.

The foregoing methods of measurement are useful in epidemiologic studies when a contrast between normal and moderately severe alteration of pulmonary function is expected. They will not suffice for recognition of minor changes in airway resistance or pulmonary compliance. This calls for more sophisticated measurement, utilizing body plethysmography.

In most circumstances the ventilation–perfusion relationship is reflected by the arterial blood gas measurements. Alteration in the size and character of the alveolocapillary membrane is best measured by the diffusion coefficient for carbon monoxide. Because these measurements require carefully controlled technical procedures that are not always available, they are rarely used in epidemiologic studies. Nevertheless, when one searches for biological reactions to dust in the lungs, these modalities should be used at some point in the overall study.

8 BIOLOGICAL EFFECTS OF SPECIFIC INHALED INORGANIC DUSTS

The variety of inorganic dusts to which humans are exposed in either pure or mixed form is large. A prototype of the effects of several will demonstrate the biological effects produced.

8.1 Nonfibrogenic or "Inert" Dust

Although some dusts are classified as biologically inert, the designation can be misleading. A more correct implication would be that such dusts do not cause pulmonary fibrosis, physical impairment, or disease. In that context, the classification is useful.

All durable inorganic dusts evoke mobilization of macrophages. This response is a daily occurrence in all walks of life. Depending on the number of particles entering the alveoli over a period of time and the resistance of such particles to destruction by the macrophage, there is an accumulation of these particles either free or inside macrophages within and around the alveoli. Particles classified as being inert (1) appear to be slightly if at all toxic to the macrophage in which they come to reside. The host macrophage dies after a normal or near-normal cell life-span, and the ingested particles are released but then undergo reingestion by another macrophage. Repetition of this cycle appears to have no limit.

After a long period of exposure to durable nonfibrogenic particles of respirable size, the particles are found distributed within alveoli; within the interstitial tissue supporting the alveoli, airways, and blood vessels; and within the small and large collections of lymphoid tissues scattered throughout the lung. The particles at any point in time may be found free or within macrophages at any one of these locations. The particles attached to the alveolar surface at times appear to be overgrown by the type I alveolar pneumocyte, forming a plaque within the alveolar wall. In locations where the particles accumulate in small collections, within either the supporting connective or the lymphoid tissue, there is a slight proliferation of monocytic cells accompanied by a slight increase of interlacing reticulin fibers. Collagen fibers, thus genuine fibrosis, do not develop either in these small dust foci or within the lymphoid tissue more centrally located along the course of the lymphatics. The size of the focal collection of dust particles is directly related to the intensity and duration of the exposure.

Insofar as an effect on pulmonary function or longevity of life is concerned, it is of paramount importance that these particles evoke neither fibrosis nor alteration of the basic lung structure. The tiny particles are present, but no destruction or replacement of normal tissue occurs. Because the fundamental mechanical quality of the lung is that of distensibility and elasticity based on the arrangement of tissue fibers having elastic qualities combined with an intact surface lining of the alveoli, one can understand why the mere physical presence of small foci of particles scattered sparsely in the interstices or on the surface of the alveolus would not affect the physical quality of the lung. It is only when the normal structure of the lung is destroyed or immobilized by replacement with nondistensible fibrous tissue that measurable impairment of pulmonary function ensues.

When the nonfibrogenic dust has a low atomic number, the foci of accumulated particles do not interrupt or deflect sufficient X-ray photons to cast a shadow, and the radiogram exhibits no unusual pattern of densities. In contrast, foci of nonfibrogenic particles having a higher atomic number do interrupt and deflect X-ray photons sufficiently to cast shadows, even though they do not evoke an inflammatory or fibrotic reaction within the tissues. They cause the radiogram to have a striking appearance when sufficient quantities of the particles have accumulated. Among the inert dusts having a high atomic number are iron, tin, barium, and antimony.

Workers may be exposed to oxides of iron either as a dust or as a fume, as during welding, and the biological result is referred to as siderosis even though no fibrosis is induced (18). When the lungs are examined, they are rust-brown as a result of the focal depositions of iron, and microscopic examination reveals particles of iron oxide distributed in the manner common to all respirable particles. Characteristically, the particles are collected in discrete foci or macules associated with lymphoid tissue and around the smaller blood vessels. There is the usual minor accumulation of reticulin fibers and monocytes in the focal deposits of particles, but no true fibrotic reaction occurs. The dust foci of this material of higher atomic number cast a distinct, small, rounded nodular shadow on the film, tending to be no greater than 1 to 2 mm in diameter. When materials of an even higher atomic number such as tin or barium are inhaled, the shadows cast by the focal deposits may have a very sharp margin, and the large collections of particles in the lymph nodes at the lung root may cause these structures to be unusually radiopaque because of their metal content, but they are not enlarged as is so often the case in silicosis. There is no evidence of any impairment of cardiopulmonary function or any reduction of life expectancy in individuals with siderosis.

An excess of malignant transformation manifested as bronchogenic cancer has been reported among iron ore miners, but unusually high levels of radon have been reported in the same mines, and this is considered to be the more likely cause (19). The evidence to incriminate iron oxide or any of the inert dusts as a carcinogen is not convincing. As is the case with other of the inert dusts, exposure to iron oxide, especially in mining, is often combined with exposure to free crystalline silica, and siderosilicosis is the resulting entity. It should be borne in mind that silica is the etiologic or active agent in any impairment that ensues.

The possibility must be recognized that deposition of inert particles on the surface of the upper airways where the mucus-secreting apparatus is located might stimulate that mechanism to an inordinant degree. Such stimulation might lead to excessive production of mucus and to hypertrophy of the mucous glands or to impairment of the ciliary system. Thus although inert dusts do not produce tissue fibrosis, they could lead to excessive production of mucus and mucous gland hypertrophy and to increased airway resistance. The question of whether chronic bronchitis is an outcome of heavy exposure to dusts of various kinds over a long period of time is discussed below.

8.2 Fibrogenic Dusts

8.2.1 Free Crystalline Silica

In contrast to the inert dusts, particles of free crystalline silica (FCS) have a strikingly different biological end effect (20). Silicon, the element, does not exist free in nature. It occurs most commonly as silica in the combined form of crystalline or amorphous silica (SiO_2) or as SiO_2 combined with one or another cation as a simple or complex crystalline or amorphous silicate.

FCS exists as quartz, tridymite, cristobalite, coesite, and stishovite. Of these, stishovite is essentially nonfibrogenic, whereas cristobalite and tridymite are more fibrogenic than quartz. In nature, quartz occurs as the most common of the three forms and is the one most frequently encountered during occupational exposures. It is converted to tridymite when heated to between 860 and 1470°C, and above 1470°C it becomes cristobalite.

When the intensity and duration of exposure are too high to be defended against effectively, respirable sized particles of FCS accumulate in the alveolated regions of the lung. As is the case with inert dust particles, at any point in time the FCS particles are distributed throughout the lung tissue in the manner described earlier for all respirable particles. Thus focal deposits of FCS, either free or within macrophages, form in a discrete fashion throughout the lungs and lymphoid tissue. It has been suggested that FCS particles move more easily, therefore in larger numbers, through the alveolar wall to the interstitium and the lymph nodes than is the case with inert particles. This may be because particles of FCS are toxic to the macrophage. The life of the host macrophage is thus greatly shortened, their number is diminished, and the defense against movement of particles into the interstitium and lymphoid tissue is thereby reduced.

The subsequent course of events within the foci of FCS particles is strikingly different from that characterizing similar foci of inert particles. At first, but with greater speed of development, fibroblasts increase in number and reticulin fibers are laid down in an interlacing fashion throughout the focal deposit. Instead of stopping at the point of reticulin development, which is the case with inert particles, the foci of FCS go on to develop masses of interlacing collagen fibers, and mature fibrosis characterized by the deposition of hyalin ensues.

With the passage of time the focus or nodule undergoing fibrosis enlarges, the active process going on at the periphery leaving behind a progressively larger central mass of mature fibrous tissue. The mature nodule has a characteristic whorled fibrous tissue appearance almost devoid of cells at its center. This mass is surrounded by a zone of reticulin and collagen fibers containing a few fibroblasts, and surrounding this layer is a peripheral zone of cells in which can be seen the process of laying down reticulin and collagen by fibroblasts, mixed in with macrophages and other mononucleated cells, in the process of forming fibrous tissue. At first the nodules are spaced rather far apart, the intervening lung or lymphoid tissue being free of abnormality. As the number of nodules increases, there is a tendency for them to coalesce and form a larger multicentric nodule, which may reach several millimeters in diameter. At all times large amounts of normal lung tissue persist between the individual or clustered nodules. In many individuals the reaction does not progress beyond this point, and the lung can be described as being the site of simple discrete nodular fibrosis. In other cases, however, there develops a massive conglomeration of nodules, merging in a cement of dense fibrosis. These masses can become quite large, occupying in some cases as much as half of each lung. On the immediate periphery of the larger conglomerate masses and even some of the large single discrete nodules, the air spaces of the lung are larger than normal. This type of perifocal emphysema should not be confused with diffuse generalized emphysema of the lung, which is not a characteristic part of the anatomic alterations accompanying the interaction between FCS and lung tissue.

In most cases the fibrotic tissue development in the advanced nodular stage and especially in the conglomerate stage extends to involve the pleura. As a result, the pleural space commonly is bridged by adhesions or is obliterated over large parts of the lung surface. The pleura itself may be unusually thickened, sometimes markedly so, over a large portion of the lung.

There is no precise location in the lung structure at which a silicotic nodule can be said to begin. A cluster of alveoli and adjacent subalveolar tissue, including small aggregations of lymphoid tissue, become involved virtually simultaneously. The air passages and pulmonary capillary bed caught up in this reaction are destroyed in the process of formation of the nodule. The lung tissue intervening between the nodules remains free of any discernible change. This is true even when large conglomerate masses also are present. Although there is undoubtedly some migration of the particles of FCS from their locus of deposition into aggregates or foci, the distribution of the nodules and the freedom from abnormality in the intervening tissues strongly suggests that the FCS particles taken into the lung are not dispersed in a completely even manner. One must assume that some regions of the lung receive far fewer particles than do others, and the lung cleansing mechanism keeps the total particles in such areas below a critical level.

In a similar manner, although the small regional lymph collections are usually destroyed in the fibrotic mass of the nodule, in the lymph nodes at the lung root the development of silicotic nodulation with sparing of intervening lymphoid tissue is the rule. The involved lymph nodes become quite large and firm or hard as a result of the process. In some cases the nodules, whether located in lung or lymph node, may become calcified, producing a striking radiograph.

A number of explanations for the mechanism whereby FCS stimulates fibrogenesis have been put forth over the years. The most compelling, and the one having the greatest experimental support, is that developed by Heppleston (15). Leaving aside the question of just how the silica surface disrupts cell membranes, it can be stated that the ingested particles of silica rather quickly kill the host macrophage. In cell culture this happens within hours, but in the normal milieu of the body it probably takes considerably longer, perhaps several days. Heppleston deduces from his studies that "on present evidence it may be concluded that silica in some way reacts with cell constituents, probably lysosomal, to produce or activate a relatively soluble substance capable of stimulating collagen formation, possibly acting in low concentrations." Because the fibrogenic action of FCS depends on a continuous supply of macrophages in the foci where the particles have accumulated, a mechanism to promote this recruitment is needed. On the basis of his own work and that of others, Heppleston further suggests that the lipid material abundantly released from the dying macrophage killed by FCS acts as a stimulant for the accelerated production and delivery of macrophages to the foci containing FCS.

There is evidence that interaction between the membranes of the macrophage and the surface of the silica particles is the key to the death of the host macrophage. Alteration of the surface of the FCS particles by polyvinylpyridine *N*-oxide and other agents, including dusts such as clay or iron, which commonly coexist with silica in occupational exposures, considerably lessens or may even prevent the fibrogenic response.

On the basis of both human epidemiologic study and animal experiments, as is true of other toxicologic reactions, the fibrogenic activity of FCS appears to be a dose-related phenomenon. One can envision the following train of events. A critical amount of FCS must accumulate and be ingested by a sufficient number of macrophages to supply the necessary amount of stimulation to fibroblasts before the process is pushed beyond the stage of reticulin formation into that of collagen formation. The development of hyalin, to reach the mature stage of the nodules, is more complex and appears in part to involve immunologic processes. The rate at which the entire process goes on, once it has reached the critical concentration of FCS and macrophages, is related to the speed with which silica particles accumulate and the rate at which the macrophages die, plus a still ill-defined immune response. The rate of FCS accumulation in the various foci depends on the balance between deposition and removal from the lung. The entire process is in constant change, even after all external exposure ceases. Some particles are removed from contact with the macrophages within the focal deposit by being sequestered within the central mass of acellular fibrotic tissue, and in addition, there is a continuous migration of particles away from the pulmonary foci to the lymph nodes and even to the mucociliary escalator.

According to the concept just expressed there is a level of silica accumulation below which the fibrogenic train of events is not set in motion. This is the tolerable dose. Moreover, once the critical level is reached, acceleration of the whole process occurs if the effective factor, probably the number of free bonds on the surface of the crystals, is rapidly augmented. This is consistent with the facts because if FCS accumulation occurs rapidly, the biological response, though fundamentally the same, has a distribution different from the one just described. In these circumstances, it appears that not all the particles have time to be transported to foci, and they accumulate along the way with a more diffuse macrophage mobilization and consequent fibrogenesis distributed in a more widespread, nonnodular manner. This produces a linear, mixed with a nodular, fibrotic reaction that has been called acute silicosis but is probably better termed accelerated silicosis, for it requires several years to develop. The accelerated process appears to be further aggravated if the particles are not only abundant but also especially small, less than 1 μm in diameter, thus providing an unusually large surface area per unit of weight.

A third variety of reaction to FCS has been reported under circumstances of extremely severe exposure, for example, in the process of sandblasting. This reaction shows combined interstitial fibrosis plus lipoproteinosis-like lesions in which the alveoli are filled with a pink-staining, relatively acellular material. This reaction may develop within less than a year, proceeding rapidly to a fatal outcome, and it merits the term "acute silicosis" (21).

As might be anticipated from knowledge of the effects of FCS on lung tissue, the evidences observable during life that these reactions have occurred vary. At slight stages of development there are no observable evidences of tissue alteration except by direct examination of the tissue itself. The earliest sign during life is observed in the chest roentgenograms as an increase in the size and visibility of the shadows cast by the bronchovascular structures located in the outer reaches of the lungs. Presumably these changes are caused by accumulation of FCS and the slight cellular reaction produced. When the nodules within the lung develop sufficient size and density as a result of the mature cellular and fibrotic reaction, they cast abnormal shadows ranging from barely perceptible, multiple, small, rounded densities to larger and conglomerate densities plus enlargement of the lung root shadows. The number of densities varies from scant to myriads, and the nodules are predominantly located in the upper half of each lung. The very large conglomerate shadows are almost invariably in this region. Shadows cast by thickened pleura can also be seen.

Even at the stage of easily recognized discrete nodular silicosis, there usually are no accompanying symptoms and no detectable evidence of pulmonary or overall physical impairment unless the volume of lung involved is unusually large or other nonrelated causes of impairment coexist. In part, this reflects the fact that the functions of the lung are not the limiting factor for physical performance. Of greatest importance, however, is that in pure simple nodular silicosis, the intervening tissue is normal. Physically the diseased areas have destroyed the lung in a focal or nodular fashion, and they constitute a relatively minor part of the total lung. The remaining normal portion of the lung tissue can readily accommodate the cyclic changes in thoracic volume during respiration; in addition, it can accept an increase in blood perfusion without an elevation of arterial pressure. In this fashion, cardiopulmonary function in the range that is needed during physical exercise is maintained. It must be said that if measurements of pulmonary function could be made in each individual before occupational exposure begins and after the nodulations appear, the person could serve as his own control, and alterations of pulmonary function might be recognized at an earlier stage than is possible when the first examination occurs after the nodules are already present.

Individuals showing overt radiographic evidence of nodular silicosis who also have a lessened capacity for physical effort because of unusual breathlessness on exertion are found in most instances to have a restrictive type of functional impairment. In others, however, there is an obstructive type or a combination of the two. The restrictive type of impairment can readily be accounted for by loss of lung tissue consequent to the extensive fibrosis, and it is most readily demonstrated in those who exhibit a combined massive conglomerate and linear type of disease. When severe, this kind of impairment can lead to pulmonary artery hypertension and cardiac failure. In some persons, though by no means all, impairment of oxygen transfer with associated arterial hypoxia develops.

Restrictive impairment is consistent with the specific pathology of silicosis. An obstructive type of impairment, though sometimes observed in individuals with simple discrete nodular silicosis, is difficult to reconcile with the specific pathology attributable to FCS per se. Chronic bronchitis or obstructive emphysema or a combination of the two are the common causes of airway obstruction in the general population. In view of the great frequency with which an obstructive impairment occurs in men over age 50 among the general population, when an obstructive impairment is observed in persons who have silicosis, it would seem to be more consistent with the facts to attribute it to causes other than FCS per se. Cigarette smoking and the nonspecific effects of FCS or other coexisting dusts and fumes in the workplace play an etiologic role in the production of chronic bronchitis and ultimately may be shown to influence the development of obstructive emphysema. Cognizance of this probability directs our attention to the need for controlling the total dust, fume, and irritant gas exposure, not just that of FCS in the workplace.

It is rather remarkable in view of the multifocal fibrosis characteristic of silicosis that there is no convincing evidence of malignant transformation or carcinogenic effect in humans. At the same time it must be said that up to the present a properly conducted epidemiologic study designed to explore the question of carcinogenesis in those exposed to FCS has not been made.

From what has been presented thus far, it could be concluded that silicosis in and of itself is not a severe crippling or life-shortening disease unless the fibrosis is very extensive. This in fact is the case. Why, then, has silicosis been an important cause of morbidity and death? The answer lies not in the discrete nodules characteristic of silicosis but in the very adverse influence of the tissue reaction to FCS on the tissue defenses against the tubercle bacillus and similar mycobacteria (22, 23).

In years past there was little chance that a person with coexisting silicosis and tuberculosis would recover. The infection slowly or rapidly but relentlessly progressed to a fatal outcome. Moreover, in the presence of quiescent or inactive tuberculosis, the development of silicosis often converted the dormant tuberculosis to an active, spreading disease. Because until recently almost all members of the adult population harbored one or more inactive tuberculosis foci, the development of silicosis fell on fertile ground and the likelihood of reactivating the tuberculosis followed by its relentless progression was high. This is no longer the case in the United States. With the advent of specific drugs to combat the tubercle bacillus and public health measures for tuberculosis control, the proportion of the adult population harboring latent tuberculosis is now very small. Moreover, specific antituberculosis drugs are quite effective, even in the presence of silicosis.

Because the reaction of lung tissue to FCS has a demonstrable dose-response relationship, reducing the severity of occupational exposure offers the possibility of abolishing the disease. Measures for environmental control in the workplace have been effective when instituted properly. New cases in metal mines where the levels of FCS have been kept for the most part below 5 million particles per cubic foot have been few (24). One large mine under my surveillance employs several hundred underground miners, and no new cases of silicosis have developed up to the present (1988) in men initially employed after 1940. Over this period a program of dust measurement shows the exposures to FCS have been kept below 5 million particles per cubic foot with few and temporary exceptions. In the same mine prior to 1938 and before installation of extensive dust control procedures, approximately

25 percent of the underground miners had radiographic evidence of silicosis. If the even more stringent current ACGIH-recommended threshold limit value for FCS is scrupulously adhered to, silicosis as a disease should be abolished. Unfortunately and regrettably, exposure to excessive levels of FCS in some places of work still exists, and silicosis continues to be induced. This is not because the recommended TLV is too high but because it is not complied with.

8.2.2 Carbonaceous Dust

Exposure to respirable particles of carbon may occur in several occupations, the most common being those associated with the mining, processing, and handling of coal. Exposure also occurs in the mining and use of graphite and in the production of carbon black and carbon electrodes. In all these cases the biological effects are similar but not identical, and only the consequences of exposure as a coal worker are discussed. "Coal workers' pneumoconioses" (CWP) is the proper term to apply to the abnormalities caused by the dust inhaled in this class of occupation. "Black lung," which is a socioeconomic term, has no meaning in the scientific or medical context.

Respirable, free crystalline silica in varying concentrations from slight to high commonly coexists with coal particles in the miner's environment. Some is in the coal seam, some is released by roof bolting where sandstone or hard shale is present, and in U.S. mines sand used on the haulageways is ground to respirable size. In the process of mining, the exposure commonly is to a mixed dust, though in subsequent handling of the commercial product the contamination by FCS is much less.

Pure coal or graphite inhaled by experimental animals produces dust foci characterized by reticulin fibers but little or no collagen, and in all respects it can be classed as an inert dust reaction (25). In general, coal particles come to reside in the same localities in the lung as already described under inert and FCS particles. In humans, the number of coal particles retained in the lung usually exceeds by far the number observed as a result of other inert or FCS exposures.

The "coal macule" is different in appearance from the FCS nodule previously described (26). It takes on a more stellate form and is localized more intimately around the respiratory bronchiole, the alveoli of which are filled with dust-laden macrophages. The normal architecture of the alveolus persists, even though it may be shrunken. In other words, there is no actual destruction of lung tissue. Alveoli become effaced by a combination of shrinkage and filling with macrophages. A sleeve of dust-laden tissue interlaced by reticulin fibers and a few collagen fibers forms a cuff, obliterating all structures except the lumen of the respiratory bronchiole. This bronchiole may become abnormally dilated, producing a form of focal or centrilobular emphysema. It is noteworthy that not every macule is associated with the development of focal emphysema. This raises the unanswered question of why focal emphysema develops at all. The mechanism is not clear.

The coal macule is a discrete focus, the intervening tissue remaining normal. The stellate densities may be 1 to 2 mm or more in diameter. Ingestion of coal particles by the macrophage does not prematurely kill the cell; thus the coal macule is fundamentally the same as that described earlier for inert materials. The larger area involved by the coal macule and obliteration of alveolar spaces appear to be the result of overwhelming the cleansing mechanism of the lung by very heavy exposure and retention of large amounts of coal. It has been shown that the number and size of the macules in the lung is closely related to the total exposure to coal dust (27). Although the coal macule may have a slight amount of mature collagen fiber interlaced throughout, characteristically this is a rather minor development, and reticulin fibers constitute the vast majority of the fibrillar tissue. Quite often, however, there is more than the usual scant amount of collagen laid down in an irregular fashion in the structure that ordinarily would be termed a coal macule. The components of the reaction are closer to that ascribed to FCS, but the organization is quite different in that the whorled arrangement is absent. Some have referred to these structures as nodules in contrast to macules. Obviously, when collagen is present in this amount the reaction can no longer be thought of as characteristic of an inert substance. Whether this nodular form is in fact a type of mixed silicotic reaction or is a response solely to coal particles has not been determined. Because FCS exists so commonly in the coal mine workplace, the former is a likely probability.

In some individuals, superimposed on the background of simple discrete coal macules a marked coalescence of macules and proliferation of mature fibrous tissue causes large masses to develop. These are referred to as progressive massive fibrosis (PMF). Because the coal macule in contrast to the silicotic nodule has rather indeterminate borders, the macules lose their identity when they are caught up in this large mass and do not exhibit the agglomerated separate structures that characterize the appearance of conglomerate silicosis. The mechanism of PMF development is not clear. It does not appear to be closely related to the coexisting FCS, although some role for this substance in the development of PMF cannot be excluded. It seems likely that PMF is related in part to the extraordinarily high content of coal dust in the involved areas. On the whole, in pure silicosis the tissue effects are induced by a lung burden of 2 to 5 g of FCS. The lung burden of coal in simple CWP or PMF is 10 or more times greater than this. It is tempting to believe that uncomplicated or simple CWP is basically a moderate to severe example of an inert dust reaction and that PMF would develop in all types of inert dust exposure if it was heavy enough. This probably is too simplistic an approach. It has been suggested that immunologic factors may play a role in PMF.

A quite different kind of massive lesion, Caplan's syndrome, may develop rapidly in some persons having simple CWP (28). A large proportion of these individuals have rheumatoid arthritis, a disease having a large immunologic component. The intrapulmonary mass characteristic of this syndrome does not have the histological appearances of PMF but does have some of the characteristics associated with immunologic reactions. This entity clearly raises the question of whether an immunologic response plays some role in the development of PMF, even though PMF is definitely not the same lesion as that characterizing Caplan's syndrome.

In contrast to the adverse effect on tissue resistance to the tubercle bacillus caused by the silicotic reaction, the coal macule does not appear to influence the body defense against this organism. When tuberculosis and simple CWP coexist, the infection may play a role in causing PMF, but rapid acceleration of tuberculosis does not occur.

For the most part, the radiographic evidences of CWP including PMF are similar to those of silicosis and need not be repeated. The shadows cast by CWP are a mixture of round and irregular shapes measuring 1 to 5 mm in diameter. The attack rate and severity of this stage of the disease is related closely to the amount of dust in the lung. Many individuals never progress beyond this stage to that of PMF, which is characterized by large, irregular shadows, usually developing in the upper lung areas.

Measurement of pulmonary function in persons having CWP shows essentially the same range of alteration with respect to type and degree as discussed for simple and conglomerate silicosis. There is less evidence of a restrictive impairment in CWP except when severe PMF exists. In simple CWP, impairment of function is so minor that it does not interfere with physical effort or shorten the life-span (29, 30). In some individuals having severe PMF, however, there is marked restrictive impairment of pulmonary function, reduction in capacity for physical effort, and shortened life-span.

Several studies searching for evidence that CWP is related to an excess risk of bronchogenic cancer have failed to reveal such an association (1).

It should be emphasized that some occupations in coal mines (e.g., a motorman, a roof bolter, or a development man) are characterized by a predominance of FCS exposure. For this reason a coal miner can develop pure nodular silicosis. If he changes jobs during his total employment, his lungs may develop coexisting coal macules, coal nodules, and pure nodular silicosis. These combinations of abnormality led, particularly in the United States, to the term anthracosilicosis, before it was demonstrated that coal in the virtual absence of FCS could produce the coal macule and PMF. Because simple CWP is strongly related to total exposure or the dose of coal, and because PMF virtually always develops on a background of simple CWP, control of the dust exposure on the mine premises prevents CWP.

8.3 Chronic Bronchitis and Emphysema

Examination of large numbers of persons who have been exposed to inert, fibrogenic, or mixed dusts invariably discloses a paradox. One is faced with the observation that among a group of persons demonstrating mild radiological or histological evidences of tissue alteration—for example, simple CWP or nodular silicosis—most have no physical complaint and demonstrate no impairment of function. Nevertheless, a few complain of moderate to severe symptoms of cough and sputum production or breathlessness on exertion and have moderate to severe measurable alteration of pulmonary function. Almost without exception such individuals have either chronic bronchitis or destructive emphysema or both, at a stage associated with increased airway resistance. We then ask whether these two diseases are caused or influenced in their natural course by the exposure to the specific dust.

Emphysema, defined as pathological enlargement of the air spaces distal to the terminal bronchioles, has been found at autopsy to involve more than 5 percent of the lung in approximately 50 percent or more of all males over age 50. The individuals come from all walks of life, and there is no preponderance of men from the dusty trades. Possibly 10 to 20 percent of these individuals from the population at large have severe enough involvement to exhibit moderate to severe increase of airway resistance and to have experienced abnormal breathlessness on exertion (dyspnea).

Chronic bronchitis (1) defined as periodic or persistent bouts of coughing caused by the production of excessive bronchial mucus, is a common condition in males after the age of 30 and is one of the rather frequent causes of death. The prevalence among adult men may be as high as 15 to 20 percent. By far the most potent factor, though not the only one causing and aggravating this disease, is cigarette smoking. In view of the high prevalence of these two conditions in the adult male population at large, it is not surprising to meet the paradox just described. Efforts to disclose the role of occupational exposure to dust as a causative factor in the development of chronic bronchitis or emphysema have been made. There is surprinsingly modest support for the concept that industrial dust or the irritating fumes commonly also present play a role in the development of aggravation of chronic bronchitis or emphysema. It is difficult to accept the belief that no role is played by these agents, particularly with respect to chronic bronchitis. What seems likely is that the powerful effect of cigarette smoking, the influence of social class with respect to place of residence, medical care, and so on, and general air pollution overshadow any demonstrable strong influence of industrial dust, fumes, or gases (31). In studies in which these other factors are corrected for, there is some suggestion that occupational dust per se may contribute at least to chronic bronchitis, even though it is not the most potent factor.

9 ASBESTOS

"Asbestos" is a term customarily reserved for commercially mined and milled, naturally occurring crystalline fibers of the serpentine and amphibole family of minerals. Chrysotile, the fibrous form of serpentine, is by far the most abundantly produced. The amphiboles crocidolite, amosite, and anthophyllite are produced and used as such. On a lower scale, tremolite is found primarily as a contaminant of talc and chrysotile. Actinolite is rarely encountered in industry.

In their natural state, the fibers are found massed together, forming a layer of cross or slip fiber within parent rock of the same chemical composition. In the process of separation from their parent rock or of "opening" during milling and manufacturing uses, the airborne fibers of asbestos vary enormously in length and diameter, but some ultimately become thin and short enough to be of respirable size. Because the aerodynamic behavior of fibers is determined by their diameter rather than length, fibers more than 3 μm thick to not penetrate into the alveolated regions of the lung.

As discussed earlier, length and shape as well as diameter play a role in governing deposition. Although fibers as long as 200 μm may occasionally be found in the lung, the length of the majority is less than 50 μm.

As observed in humans or experimental animals, all the ultimate fates described earlier for particles entering the upper airways occur also for fibrous particles but with some important differences. Fiber deposited on the mucociliary apparatus is removed and expectorated or swallowed. Those less than 10 μm long can be ingested readily by the macrophage. Longer fibers, though thin enough to be respirable, are not readily ingested by single macrophages; yet occasional fibers 20 to 50 μm or longer may be seen with several macrophages attached and spread out over the fiber surface. At times two or more macrophages clump together, forming a giant cell totally surrounding one or more fibers. In contrast to respirable-sized nonfibrous particles, any one of which can be ingested by a macrophage, fibers longer than 20 μm are not readily ingested, and macrophage defense against this length of fiber is impaired or not available. In spite of this, asbestos fibers free on the surface of the alveoli are relatively scant. Almost always, fibers remaining in the lung either are within macrophages or appear to be embedded within the thickened wall of the alveolus or within the interstitial tissue, which then becomes the site of fibroblast proliferation and collagen fiber deposition. Fibers longer than 10 μm within the regional lymph nodes are unusual, and even shorter ones are not found in large numbers.

Some of the fibers within both the lungs and lymph tissue become coated with a proteinaceous iron-containing substance and are referred to as "ferruginous" or "asbestos" bodies. It is believed that the coating is laid down in some way by contact with the macrophage. This coating persists virtually indefinitely, but it is of interest that these bodies can undergo fragmentation. Curiously, even in persons who experience moderate to heavy exposure, many and perhaps most fibers are free of this coating even after long periods of residence within the lung. There is some evidence to indicate that the coated fibers are nonfibrogenic and that the coating process is a mechanism whereby fibers entering the lung are neutralized. It should also be pointed out that ferruginous bodies may be formed by any kind of durable fiber that enters the lung, including such things as glass and vegetable fibers of silicious composition.

It has been shown experimentally that asbestos fibers shorter than 5 to 10 μm whether coated or naked, behave essentially like inert particles (32, 33). This length of fiber is readily ingested by the macrophage, and in contrast to the effect of FCS, the host macrophage is not materially damaged by the ingested fiber. Its life cycle is not shortened. Experimental animals exposed solely to asbestos fibers shorter than 5 to 10 μm do not develop pulmonary fibrosis. If longer fibers in sufficient numbers are introduced into the lungs of animals, they produce a diffuse, non-nodular type of fibrosis of the mature variety. The fibrotic reaction spreads out along the supporting structures of the lung, obliterating the alveoli and associated vascular units. The diffuse spreading nature of this involvement stands in striking contrast to the discrete nodular fibrosis that is characteristic of silicosis. Although some areas of normal-appearing lung tissue persist between the areas of fibrosis, the volume of residual normal lung tissue is less, and the amount of lung tissue entrapped between strands of fibrosis is greater than is the case in silicosis (34).

The mechanism of fibrogenesis induced by asbestos is not clear. In animals, the early stages suggest an organizing alveolitis closely associated with the presence of long fibers. Eventually the fibrosis completely destroys the alveolar architecture and its intimately related vascular structures. Commonly the extensive mature fibrotic tissue contains long asbestos fibers, singly or in clusters, distributed throughout, but they are sparse and far apart. The fibrotic reaction seems to be excessive for the few fibers visible. Nevertheless, fibrogenesis must have something to do with the fibers and specifically with their shape because nonfibrous particles having identical chemical and crystalline structures do not produce fibrosis when introduced into experimental animals. The relative paucity of long fibers in mature fibrotic tissue may be the result of fragmentation and destruction of long fibers after they have induced the fibrotic reaction. The concept that asbestos fibers are biologically indestructible cannot be supported.

It has been suggested that long fibers induce alveolitis and subsequent fibrosis, whereas short fibers or particles of the same chemical nature do not, because the fibers are too long to be taken entirely into the macrophage. As a result, the macrophage leaks at the point of fiber entry, and the discharge of some of the macrophage contents leads to fibrogenesis (33). It is of interest that little if any fibrogenesis is associated with the relatively few asbestos fibers that reach the lymph nodes. It would appear that the lymphoid tissue treats even the long fibers as inert material.

That immunologic processes may play a role in fibrogenesis consequent to inhalation of asbestos or FCS has been suspected not only because of the potential role of this mechanism for governing the actual tissue response but also because there appears to be some variation of individual reactivity to these dusts. Measurement of antinuclear antibodies, rheumatoid factor, and the W27 antigen of the HL-A system indicates higher values in those having asbestos-induced pulmonary fibrosis than among controls or those exposed but without fibrosis (35, 36). The importance of these studies, especially with regard to recognizing prior to exposure those who would be hyperreactors, remains to be determined.

In contrast to particles of FCS, asbestos fibers tend to remain on the surface of the alveoli; the cellular reaction to them develops inside the alveolus, and this alveolitis ultimately goes on to organization, with obliteration of the alveoli and the diffuse fibrotic reaction flowing from that. Perhaps it is this difference in particle transport that explains the diffuse rather than the nodular tissue reaction to asbestos. Also, in contrast to the long-term effects of free crystalline silica, the fibrosis caused by asbestos develops more extensively in the lower half of the lungs, and it is unusual for large conglomerate masses surrounded by normal tissue to develop.

In addition to the diffuse intrapulmonary tissue reaction to asbestos, in some exposed persons there is also a fibrotic reaction of two kinds involving the pleura. First, there may be slight to widespread fibrosis of the visceral pleura associated with the underlying pulmonary fibrosis, especially in severely involved cases. Obliteration of the pleural space may occur in these areas. Second, there may be small to large plaque formation involving virtually only the parietal pleura. The plaques are thickened tissue having quite sharp, raised margins, and they are composed of fibrous hyalinized tissue that at times becomes calcified. The parietal pleura over these plaques usually is not adherent to the visceral pleura, and the space is not obliterated. When visible in the radiogram, the pleural plaques are striking, especially when they are calcified. Plaque formation can develop without pulmonary fibrosis; the mechanism, especially the reason for its predilection for the parietal pleura, is unknown.

The pathology as seen in experimental animals and in humans is essentially the same and explains reasonably well the impairment of function and health consequent to the reaction. The diffuse fibrous bands radiating throughout the lung, especially in the lower regions, reduce the distensibility of the normal lung tissue entrapped between the strands of fibrous tissue. This lowered distensibility plus actual loss of lung substance by replacement with fibrous tissue cause a restrictive type of pulmonary impairment. As a result, the total lung volume, vital capacity, and residual volume are reduced without an increase of resistance to airflow in the proximal conducting system. If the restrictive impairment is severe, it causes an increase of the work of breathing and a lessened maximum capacity for ventilation. Regional hypoventilation of entrapped lung plus loss of total vascular surface may lead to impaired oxygen transfer and arterial hypoxia. The volume of air breathed in response to exercise may become abnormally high, and this plus the increased work of breathing leads to abnormal breathlessness and dyspnea. If the vascular bed is sufficiently reduced, the pulmonary artery pressure rises, especially during exercise, and right ventricle hypertrophy and heart failure may ensue.

With mild asbestosis, the radiogram is not as distinctly abnormal as is the case with silicosis because the tissue changes are diffuse and the well-demarcated densities so characteristic of silicosis are not present. The usual appearance of asbestosis in the radiogram is that of poorly defined, small, irregular densities that obscure the normally sharp margins of the bronchovascular structures, particularly in the lower parts of both lungs. Thickened pleura may cast a diffuse haze over both lungs which often obscures the intrapulmonary densities. When the disease is severe the radiographic abnormalities are readily recognized; but in the slight stages, the question of whether the radiogram is in fact abnormal is commonly disputed and may be answered differently by the same reader on different occasions.

Physical examination in the early and subsequent stages often reveals persistent dry crackling sounds (rales), audible over the lung bases posteriorly. Clubbed fingers with or without cyanosis may or may not be present.

In any given case all the abnormalities thus far described may exist, but commonly only one or two are present. Moreover, all the physical, radiographic, and altered pulmonary function measurements are characteristic of chronic interstitial lung disease, no matter what its etiology may be. Although idiopathic chronic interstitial lung disease is rather uncommon, it occurs frequently enough to pose a diagnostic problem, especially in persons exposed very slightly to asbestos fiber—that is, for the intermittent user of this material or the general public. The diagnosis of asbestosis therefore requires exclusion of other causes of observable abnormalities plus establishment of the occurrence of adequate exposure to respirable asbestos.

In contrast to the experience with exposure to inert particles, to free crystalline silica, and to coal dust, exposure to respirable asbestos fiber is associated with an excess risk of cancer (37, 38). The excess risk of bronchogenic cancer is virtually confined to those having a long history of cigarette smoking plus heavy exposure to asbestos. The greatest risk appears to be in those exposed in the insulation working trade, where the "total" exposure to multiple factors by reason of coexisting agents in their variable working conditions may play an important role. On balance, asbestos per se is a low-order carcinogen for the bronchial cells, and its role is almost certainly that of a promoter, not an initiator.

There is evidence to suggest that an excess of bronchogenic cancer occurs only among those having diffuse pulmonary fibrosis or asbestosis (39, 40). Thus because fibers shorter than 5 to 10 μm do not cause diffuse pulmonary fibrosis, it seems likely they also do not play a role in the causation of an excess bronchogenic cancer risk in persons exposed to asbestos fiber in the workplace. Early studies (37) suggested there might be a slight excess risk of gastrointestinal cancer in those heavily exposed. More recent studies (41) do not disclose any such risk.

An excess risk of developing either pleura or peritoneal mesothelioma has been observed among workers exposed to chrysotile, crocidolite, and amosite, but not to anthophyllite or tremolite (40). The dose at which the excess risk occurs appears to be lower than for the other kinds of asbestos-related cancers. Some believe that exposure to crocidolite carries a greater risk than that to chrysotile or amosite, but this attitude is not universal (42). As is true for fibrogenesis, the risk of carcinogenesis related to respirable asbestos shows a dose–response relationship including that for mesothelioma. Reconstruction of dose data for crocidolite exposure and associated mesothelioma risk has never been presented in a fashion to exclude dose as the factor responsible for what in some circumstances appears as a higher risk of mesothelioma among those exposed to crocidolite.

Mesothelial tumors can be produced in experimental animals by placing asbestos fibers directly into the pleural or peritoneal space (43–45). Using this technique, it has been shown that long, very thin fibers induce the greatest tumor response. If fibers of thicker or shorter dimensions are used, tumor induction decreases. When the fibers used do not exceed 5 μm in length, tumor production, if it does occur, is not statistically significant.

Because all the manifestations of impaired health and function related to respirable asbestos are dose related, it should be possible to continue the use of this material under adequate controls.

10 FIBROUS GLASS

Glass, an amorphous silicate containing various other ingredients, can be produced in fiber form from the molten material by drawing, as in textile glass, or by blowing or centrifugation, as in glass wool manufacture. In view of the biological effects of abestos fibers, concern for possible effects of fibrous glass has often been voiced.

Fibrous glass began to be produced in large amounts only after 1935–1940. Because the fibers of drawn glass characteristically were thicker than 5 μm and the same was true of most of the early-produced blown glass fibers, it was believed that these would be nonrespirable. In addition, mineral wool, a blown glass, has been in use since early in this century; and although those exposed to it have not been studied by modern epidemiologic methods, their clinical experience did not lead observers to believe that there was an unusual health risk. Early animal experiments using glass fibers ground to respirable lengths were not entirely satisfactory because their large diameter interfered with carrying out adequate inhalation experiments. When these were supplemented by intratracheal injection, however, the results were those of an inert dust reaction, no mature fibrosis being produced. On these grounds, fibrous glass was considered to be an inert or nuisance dust. Radiographic, pulmonary function, and clinical plus autopsy studies performed since 1965 confirmed the absence of recognizable pulmonary alteration other than those of an inert nature in persons having experienced 25 to 30 years of exposure to fibrous glass in the workplace (46–48).

More recently, technological developments and requirements have led to the production for special purposes of blown glass fibers having diameters as thin as 1 μm or less. In addition, electron microscope studies of the workplace where conventional glass fiber is being produced show that a small number of fibers of this dimension are present in the air where nominally thicker glass fibers are being manufactured. It is apparent that fibers thin enough to be respired have been in the workplace for some time and are now being manufactured for specific purposes.

In contrast to the earlier animal experiments showing only an inert dust reaction, contemporary animal experiments using repeated intratracheal instillations of dilute suspensions of glass fibers thinner than 2 μm and longer than 25 μm have revealed the development of fibrosis in the interstitial tissues around the respiratory bronchioles (33). This reaction, though qualitatively similar, is much less severe in comparison to that resulting from asbestos fiber, thus strongly suggesting that something in addition to fiber shape and dimension plays a role in the induction of fibrosis. The same experiments also indicate that glass fibers of identical composition and diameter but shorter than 10 μm produced only an inert dust reaction (33).

Experiments over the past several years have shown that introduction of long glass fibers thinner than 3 μm directly into the pleura and peritoneum of experimental animals has produced mesothelial tumors (49). Though such experiments are important in the study of mechanisms of carcinogenesis, it should be pointed out that such implantation of large numbers of fibers of this size directly into the pleura or peritoneum completely bypasses the defense mechanisms available to minimize the movement of fibers to the pleura by the usual inhalation route in the intact respiratory system. Furthermore, the number of fibers introduced in these experiments is doubtless several orders of magnitude higher than might conceivably result by way of the inhalation route. As was the case with asbestos, these experiments show that as the length of the fiber is reduced, the yield of tumors is lessened, and when all the fibers are shorter than 8 μm, no statistically significant excess of tumors is induced (49). One would anticipate that by the normal inhalation route, the long fibers would be less likely than short ones to reach the subpleural regions, and this would tend to lessen further the number of long thin fibers ultimately reaching the pleura by the inhalation route. Although these animal experiments suggest the possibility that glass fibers thinner than 2 μm and longer than 10 μm might cause pulmonary fibrosis or induce mesothelioma in humans if sufficient numbers reached the alveolar areas of the lung and the pleura, they are so artificial in their manner of placement of the fibers within the animal that they have questionable relevance to human exposures. Two more recent animal inhalation experiments using long, thin glass fibers did not show either pulmonary fibrosis or malignant tumor development (50, 51).

Human experience with exposure to these unusually thin, long fibers is rather limited, but some data are available to place animal experiments such as those just referred to in perspective. As indicated earlier, fibers of these long, thin dimensions have been present in small numbers in the air of glass wool manufacturing plants for the past 40 years. In fact, for several years they were deliberately manufactured for the production of flotation materials during World War II and for a short time thereafter. Recent epidemiologic studies of workers in manufacture of fibrous glass where long, very thin fibers were present 40 or more years ago have shown an excess neither of cancer, including mesothelioma, nor of pulmonary fibrosis in workers exposed over this period (52).

11 PNEUMOCONIOSIS?

The reader must wonder why the term "pneumoconiosis" has been used only with regard to coal workers' pneumoconiosis (CWP). I have used it in that connection because in this particular circumstance it has a rather specific, well-entrenched meaning. As generally used, "pneumoconiosis" is defined as the circumstance of deposition of dust in lung tissue and the tissue reaction thereto. In this sense it has no specific implications, because all adults, no matter what their work experience, fall within this category. Moreover, some of the pneumoconioses have serious biological implications whereas others do not, and all too often there is a false inference that all pneumoconioses are harmful. The term is so ambiguous that its use should be discouraged, and more specific designations should be used for each of the circumstances.

REFERENCES

1. W. R. Parkes, *Occupational Lung Disorders*, Butterworths, London, 1982.
2. E. R. Weibel, *Morphometry of the Human Lung*, Academic Press-Springer-Verlag, New York, 1963.

3. H. von Hayek, *The Human Lung* (translated by V. E. Krahl), Hafner, New York, 1960.
4. H. Spencer, *Pathology of the Lung; Excluding Pulmonary Tuberculosis*, 3rd ed., Pergamon Press, Oxford, 1976.
5. B. Vernon-Roberts, *The Macrophage*, Cambridge University Press, Cambridge, 1972.
6. T. F. Hatch and P. Gross, *Pulmonary Deposition and Retention of Inhaled Aerosols*, Academic Press, New York, 1964.
7. W. H. Walton, Ed., *Inhaled Particles*, Vol. 3, Unwin Bros., Ltd., Gresham Press, Old Woking, Surrey, England, 1971.
8. M. Newhouse, J. Sanchis, and J. Bienenstock, *Lung Defense Mechanisms* (2 parts), *New Engl. J. Med*, **295**, 990, 1045 (1976).
9. M. Lippmann, "Review: Asbestos Exposure Indices," *Environm. Res.*, **46**(1), 86 (1988).
10. J. Cartwright and G. Nagelschmidt, "The Size and Shape of Dust from Human Lungs and Its Relation to Relative Sampling," in *Inhaled Particles and Vapours*, C. N. Davies, Ed., Pergamon Press, New York, 1961.
11. V. Timbrell, "Inhalation and Biological Effects of Asbestos," in *Assessment of Airborne Particles: Fundamentals, Applications and Implication to Inhalation Toxicity*, T. Mercer, P. Morrow, and W. Stäber, Eds., Thomas, Springfield, IL, 1972, pp. 429–445.
12. C. N. Davies, "Deposition of Dust in the Lungs: A Physical Process," in *Industrial Pulmonary Diseases*, King and Fletcher, Eds., Little Brown, Boston, 1960.
13. E. M. Scarpelli, *The Surfactant System of the Lung*, Lea & Febiger, Philadelphia, 1968.
14. J. Kaufman and G. W. Wright, "The Effect of Nasal and Nasopharyngeal Irritation on Airway Resistance in Man," *Am. Rev. Resp. Dis.*, **100**, 626 (1969).
15. A. G. Heppleston, "The Fibrogenic Action of Silica," *Br. Med. Bull.*, **25**(3), 282 (1969).
16. J. Bruch, "Response of Cell Cultures to Asbestos Fibers," *Environ. Health Perspect.* **9**, 253 (1974).
17. "Guidelines for the Use of ILO International Classification of Radiographs of Pneumoconioses," Revised ed., 1980, International Labour Office, Geneva.
18. M. Kleinfeld et al., "Welders Siderosis," *Arch. Environ. Health*, **19**, 70 (1969).
19. J. T. Boyd et al., "Cancer of the Lung in Iron Ore (Haematite) Miners," *Br. J. Ind. Med.*, **27**, 97 (1970).
20. M. Ziskind, R. N. Jones, and H. Weill, "Silicosis," *Am. Rev. Resp. Dis.*, **113**, 643 (1976).
21. H. A. Buechner and A. Ansari, "Acute Silico-Proteinosis," *Dis. Chest*, **55**, 274 (1969).
22. L. U. Gardner, "Studies in Experimental Pneumoconiosis: Reactivation of Healing Primary Tubercles in Lung by Inhalation of Quartz, Granite and Carborundum," *Am. Rev. Tuberc.*, **20**, 833 (1929).
23. L. U. Gardner, "Etiology of Pneumoconiosis," *J. Am. Med. Assoc.*, **111**, 1925 (1938).
24. "Silicosis in the Metal Mining Industry, A Revolution," U.S. Public Health Service Publication No. 1076, Government Printing Office, Washington, DC, 1963.
25. S. H. Zaidi, *Experimental Pneumoconiosis*, Johns Hopkins Press, Baltimore, 1969, pp. 187–198.
26. A. G. Heppleston, "The Pathological Anatomy of Simple Pneumoconiosis in Coal Miners," *J. Pathol. Bacteriol.*, **66**, 235 (1953).
27. S. Rae, "Pneumoconiosis and Coal Dust Exposure," *Br. Med. Bull.*, **27**(1), 53 (1971).

28. A. Caplan, R. Payne, and J. Withey, "A Broader Concept of Caplan's Syndrome Related to Rheumatoid Factors," *Thorax*, **17**, 205 (1962).
29. W. K. C. Morgan et al., "Respiratory Impairment in Simple Coal Workers' Pneumoconiosis," *J. Occup. Med.*, **14**, 839 (1972).
30. C. E. Ortmeyer et al., "The Mortality of Appalachian Coal Miners," *Arch. Environ. Health*, **29**, 67 (1974).
31. I. T. T. Higgins, "The Epidemiology of Chronic Respiratory Disease," *Prevent. Med.*, **2**, 14 (1973).
32. P. Gross, "Is Short-Fibered Asbestos Dust a Biological Hazard?" *Arch. Environ. Health*, **29**, 115 (1974).
33. G. W. Wright and M. Kuschner, *Proceedings of the Fourth International Symposium on Inhaled Particles and Vapours*, British Industrial Hygiene Society, 1975.
34. A. J. Vorwald, T. M. Durkan, and P. C. Pratt, "Experimental Studies of Asbestosis," *Arch. Ind. Hyg. Occup. Med.*, **3**, 1 (1951).
35. M. Turner-Warwick and W. R. Parkes, "Circulating Rheumatoid and Antinuclear Factors in Asbestos Workers," *Br. Med. J.*, **3**, 492 (1970).
36. J. A. Merchant et al., "The HL-A System in Asbestos Workers," *Br. Med. J.*, **1**, 189 (1975).
37. J. C. McDonald et al., "The Health of Chrysotile Asbestos Mine and Mill Workers of Quebec," *Arch. Environ. Health*, **28**, 61 (1974).
38. I. J. Selikoff, E. C. Hammond, and J. Churg, "Asbestos Exposure, Smoking and Neoplasia," *J. Am. Med. Assoc.*, **204**, 106 (1968).
39. H. Bohlig, G. Jacob, and H. Müller, *Die Asbestose der Lungen*, Georg Thieme, Stuttgart, 1960, p. 60.
40. J. C. Wagner, J. C. Gilson, G. Berry, and V. Timbrell, "Epidemiology of Asbestos Cancers," *Br. Med. Bull.*, **27**, 71 (1971).
41. D. A. Edelman, "Exposure to Asbestos and the Risk of Gastrointestinal Cancer: A Reassessment," *Br. J. Ind. Med.*, **45**(2), 75–82 (1988).
42. I. Webster, "Asbestos and Malignancy," *S. Afr. Med. J.*, **47**, 165 (1973).
43. M. F. Stanton, "Fiber Carcinogenesis: Is Asbestos the Only Hazard?" *J. Nat. Cancer Inst.*, **52**, 633 (1974).
44. M. F. Stanton and C. Wrench, "Mechanisms of Mesothelioma Induction with Asbestos and Fibrous Glass," *J. Nat. Cancer Inst.*, **48**, 797 (1972).
45. J. C. Wagner and G. Berry, "Mesothelioma in Rats Following Inoculation with Asbestos," *Br. J. Cancer*, **23**, 567 (1969).
46. G. W. Wright, "Airborne Fibrous Glass Particles: Chest Roentgengrams of Persons with Prolonged Exposure," *Arch. Environ. Health*, **21**, 175 (1968).
47. R. T. de Treville and H. M. Utidgian, "Fibrous Glass Manufacturing and Health. Part I. Report of an Epidemiological Study. Part II. Results of a Comprehensive Physiological Study," *Trans. Bull.*, Industrial Health Foundation, Pittsburgh, 1970, pp. 98–111.
48. J. W. Hill et al., "Glass Fibers: Absence of Pulmonary Hazard in Production Workers," *Br. J. Ind. Med.*, **30**, 174 (1973).
49. M. F. Stanton et al., "The Carcinogenicity of Fibrous Glass: Pleural Response in the Rat in Relation to Fiber Dimension," *J. Nat. Cancer Inst.*, **58**, 1977.

50. R. T. Drew et al., "The Chronic Effects of Exposure of Rats to Sized Glass Fibres," *Ann. Occup. Hyg.*, **31**, 711–729 (1987).
51. D. M. Smith et al., "Long-Term Health Effects in Hamsters and Rats Exposed Chronically to Man-Made Vitreous Fibres," *Ann. Occup. Hyg.*, **31**, 731–749 (1987).
52. *Man-Made Mineral Fibres: Environmental Health Criteria, 77*, World Health Organization, Geneva, 1988.

CHAPTER TWELVE

Physiological Effects of Altered Barometric Pressure

Claude A. Piantadosi, M.D.

1 INTRODUCTION

Scientists of the last half of this century have been fortunate to participate in one of the most exciting periods of discovery in human environmental physiology. The rich growth of biomedical knowledge in this field has been stimulated by a technological revolution that has given human beings the opportunity to visit and explore the depths of the ocean and the frontiers of near space. These extreme physical environments present the applied physiologist with the challenge of defining reasonable limits for physiological tolerance so that the engineer can fashion adequate life support systems that maximize our opportunity to interact with the environment. Among the most troublesome environmental variables are extremes of heat transfer, exposure to radiation, untoward effects of acceleration and microgravity, hypoxia and hyperoxia, and drastic changes in barometric pressure. Nonetheless humans in capsule environments can survive for many weeks or months in diverse settings ranging from high altitude and the near vacuum of space to simulated undersea depths approaching 70 atm of pressure absolute (ATA).

Abrupt or great changes in barometric pressure have precipitated many of the injuries and deaths connected with the practice of aviation and diving medicine. Physiological problems arise most commonly during and after decompression from high pressure or hyperbaric conditions; these problems relate to our still rudimentary understanding of inert gas transport and elimination from body tissues and

Patty's Industrial Hygiene and Toxicology, Fourth Edition, Volume 1, Part A, Edited by George D. Clayton and Florence E. Clayton
ISBN 0-471-50197-2

their consequences. The most prominent physiological responses to changes in barometric pressure and in particular the physiological consequences of exposure to hyperbaric pressure, are the focus of this chapter.

The hyperbaric environment has assumed an ever-expanding role in modern society for many reasons. The earliest engineering application of high pressure, to maintain dry working environments in caissons and tunnels, continues unabated today and extracts its toll in decompression sickness and aseptic bone necrosis from many workers who undergo daily decompression after long shifts in compressed air. As the demand to exploit undersea resources such as oil has increased, so have the working depth and time, and hence the medical vulnerability, of the diver. In relatively shallow diving using air and mixed gases, many commercial and military divers rely on surface-supplied apparatus to provide an uninterrupted supply of respirable gas and continuous communication with surface tenders. These diving rigs provide an unlimited supply of breathing gas to the working diver in open, semi-closed, or closed (rebreathing) circuits. In addition, hundreds of commercial, military and scientific divers also have been trained to use self-contained underwater breathing apparatus (SCUBA) to work in shallow water. More recently, the development of lightweight, open-circuit SCUBA and low-resistance regulators has resulted in unprecedented growth in the popularity of recreational diving with compressed air. Finally, renewed interest in hyperbaric medicine and the need to provide recompression therapy to a plethora of injured divers and patients results in the routine exposure of physicians, nurses, and paramedical personnel to hyperbaric environments.

2 PRESSURIZED ENVIRONMENTS

2.1 General Principles

In order to understand the physiological responses to pressurized environments, the fundamental principles of the physical behavior of gases and the mechanical effects of pressure must be made clear. The units of pressure measurement, force per unit area, can be expressed in a variety of ways. Some examples of equivalent units of pressure are provided in Table 12.1. Normal atmospheric pressure is approximately 760 torr or 14.7 pounds per square inch (psi). The pressure exerted by a column of water varies linearly with its height and must be added to the normal atmospheric pressure to obtain the absolute pressure. As shown in Table 12.1, a column of seawater 33.1 ft deep (fsw) exerts the same pressure as the normal atmosphere at sea level. Thus a diver immersed in seawater at 33.1 ft is exposed to a total pressure of 2 atm (ATA).

In diving, the pressure of the diver's breathing gas must be increased in proportion to the absolute pressure in order to allow the diver to inhale against the pressure of the water column above him. As a result, more gas molecules must occupy the lungs and other gas-containing cavities of the body to maintain constant volume at a given temperature. The relationship between pressure (P), volume

Table 12.1. Pressure Equivalents

Condition	Altitude or Depth	Pressure torr	Pressure psig[a]	Atmospheres (ATA)[b]
Altitude	+18,000 ft	380	—	0.5
Sea level	0	760	0	1.0
Seawater	33 fsw	1,520	14.7	2.0
Seawater	66 fsw	2,280	29.4	3.0
Seawater	330 fsw	8,360	147.0	11.0
Seawater	660 fsw	15,960	294.0	21.0

[a]psig = pounds per square inch gauge. Note that a pressure gauge at sea level reads zero.
[b]ATA = atmospheres absolute. Note that depth in ATA = (fsw + 33)/33.

(V), temperature (T), and the number of moles of gas (n) can be described by the ideal gas law:

$$PV = nRT$$

where R is the universal gas constant (0.082 g·liter·ATA·mole^{-1}·°K). Several limited gas laws relevant to diving can be derived from the ideal gas law. For a constant number of moles of gas under condition 1,

$$\frac{P_1V_1}{T_1} = nR$$

After changing one of the variables on the left side of the equation to create condition 2, we have

$$\frac{P_2V_2}{T_2} = nR$$

and

$$\frac{P_1V_1}{T_1} = \frac{P_2V_2}{T_2}$$

Three special cases of the ideal gas law, where one of the three variables is held constant, are relevant to diving and are shown in Table 12.2.

Air and other respirable gases are mixtures of oxygen and various other molecules. In a mixture of gases, the total pressure is equal to the sum of the partial pressures of the individual gases (Dalton's law). This means that each gas in the mixture behaves as though it alone occupies the entire space available. The concentration of gas available to the body tissues is determined primarily by the

Table 12.2. Special Gas Laws

Boyle's law	$P_1V_1 = P_2V_2$
Charles's law	$V_1T_1 = V_2T_2$
Gay-Lussac's law	$T_1P_1 = T_2P_2$

diffusion of gas into or out of blood from the alveolar spaces. The amount of a gas dissolved in liquid at any temperature, for example, blood or tissues of the body at 37°C, is also directly proportional to its partial pressure (Henry's law). The gas concentration is related to the partial pressure times the solubility coefficient. The solubility of a gas in a liquid also increases inversely with temperature. In general, the physiological effects of diving gases, such as inert gas narcosis, correlate directly with the partial pressure of the gas in body tissues.

2.2 Immersion and Breath-hold Diving

Immersion in water produces profound adjustment in several physiological parameters including responses by the respiratory, cardiovascular, renal, and endocrine systems (1, 2). The immediate physiological consequences of immersion are pulmonary restriction, a sustained increase in cardiac output, and dehydration. These responses are related to the effects of hydrostatic pressure and to the high density of water relative to air. During head-out immersion, the extrathoracic blood vessels are supported by the water and the body is exposed to a hydrostatic pressure gradient proportional to the vertical distance from the water surface. The hydrostatic pressure surrounding the body compresses the abdomen and thorax causing negative pressure breathing of approximately −20 cm H_2O. By displacing the diaphragm upward, these hydrostatic effects decrease the thoracic gas volume and expiratory reserve volume. The pressure gradient across the diaphragm created by immersion, together with hydrostatically decreased venous capacitance in the legs, increases the volume of blood in the intrathoracic vasculature including the heart. The increase in intrathoracic blood volume during immersion is also facilitated by the high density of water, which largely abolishes venous pooling. Central blood volume also may be augmented by peripheral arterial vasoconstriction if the water temperature is not thermoneutral (34°C).

Cardiovascular distension during immersion activates cardiac merchanoreceptors that normally respond to hypervolemia. Although there is no immediate change in the circulating blood volume, the apparent hypervolemia is detected at the hypothalamus via vagal afferents and an immersion response ensues. There are two components to this response, which include diuresis and natriuresis, respectively. The diuresis and natriuresis after immersion have different time profiles, suggesting that they operate by different mechanisms. Peak diuresis occurs in the first 1 to 2 hr, whereas peak natriuresis occurs after 4 to 5 hr of immersion. Fluid restriction and the administration of aqueous vasopressin before immersion prevent diuresis but do not affect natriuresis.

There is a correlation between the amount of distension of the central circulatory organs and the excretion of urinary sodium (2). Studies have shown that seated immersion increases central blood volume by about 700 ml. The mechanism of the response to increased water loss during immersion is suppression of antidiuretic hormone (ADH) release. This is also known as the Gauer–Henry response. The mechanism of the natriuresis is related to decreased tubular reabsorption of sodium, not to an increase in filtered sodium. The most important factors in the natriuresis appear to be (1) aldosterone suppression via decreased renin–angiotensin activity, (2) increased release of atrial natriuretic factor, (3) increased release of renal prostaglandins, and (4) decreased sympathetic activity.

The increase in central blood volume during immersion distends the cardiac chambers and enhances ventricular diastolic filling, which increases cardiac output (3). The increase in cardiac output is attributable almost entirely to an increase in stroke volume because resting heart rate usually remains unchanged or decreases slightly with immersion. The stroke volume and cardiac output may double. The increase in cardiac output is persistent and occurs with no measurable increase in systemic oxygen uptake. The mechanism of the increase in stroke volume appears to be primarily an increase in cardiac preload, although changes in cardiac contractility have not been ruled out.

During immersion, the physiological responses to breath-hold diving have special significance (4). With apnea, the lungs act as a reservoir for exchange of O_2 for CO_2 in pulmonary capillary blood. During a breath-hold in air, the mean alveolar pO_2 decreases linearly with time and its rate of change is a function of the decline in the mixed venous pO_2. During the first seconds of apnea, the rate of O_2 uptake by the body is relatively constant. Shortly thereafter, O_2 uptake from the lungs can no longer support the needs of the body, the O_2 uptake falls, and anaerobic metabolism increases. Carbon dioxide enters the lungs during apnea in proportion to both the pulmonary blood flow and the diffusion gradient for CO_2, that is, the difference between mixed venous pCO_2 and alveolar pCO_2. The rate of CO_2 transfer is high initially but decreases rapidly because the CO_2 diffusion gradient decreases as alveolar pCO_2 approaches the mixed venous value. As CO_2 production continues, mixed venous pCO_2 and alveolar pCO_2 rise. The point at which high pCO_2 causes breathing to resume is called the breakpoint. The time to the breakpoint can be extended by maneuvers that lower pCO_2 or raise PO_2 such as hyperventilation, oxygen breathing, or both. Notably, the body's O_2 store is not increased appreciably by hyperventilation because the increase in alveolar pO_2 made possible by the decrease in alveolar pCO_2-only increases blood O_2 content slightly. Therefore hyperventilation, associated with a greatly extended breath-hold time, can produce hypoxia-induced unconsciousness before the CO_2 rises to the breakpoint.

Alveolar gas exchange during breath-holding is altered considerably by an actual descent when lung volume decreases from thoracic compression and the partial pressures of O_2, CO_2, and N_2 increase in the lungs. As the descent progresses, alveolar O_2 and CO_2 concentrations decrease owing to a greater transfer of those gases to pulmonary capillary blood compared to N_2, which is metabolically inert and not very soluble in blood. The alveolar pO_2 is also higher during breath-hold diving than during simple breath-holding and the transfer of CO_2 during early descent is opposite normal; that is, CO_2 moves from alveoli to pulmonary capillary blood. During ascent, the lung re-expands and the alveolar pO_2 and pCO_2 decline. Near the surface, the alveolar pO_2 may be almost as low as the mixed venous pO_2. If the dive has been of sufficient magnitude, reverse O_2 transfer from mixed venous blood to alveoli may occur during ascent. Carbon dioxide retained in the blood during the dive also leaves the blood on ascent as the alveolar pCO_2 decreases. Carbon dioxide elimination usually continues post-dive, as does elimination of the small amount of N_2 that entered the blood during the dive. This amount of N_2 is so small that it is not a significant decompression risk; however, frequent repetitive breath-hold diving theoretically may allow enough N_2 retention to produce decompression sickness.

During breath-hold diving, a dangerous situation may arise if hyperventilation is used to extend the dive. During the dive, the alveolar pO_2 increases due to compression of the lung. Thus the primary signal to return to the surface is the pCO_2. When the diver hyperventilates before the dive, the arterial pCO_2 begins at a lower level and the time to the breath-hold break point is prolonged. The O_2 reserve is more severely depleted during a longer dive, and life-threatening hypoxia may occur when the diver approaches the surface. This phenomenon is known as shallow water blackout and probably causes many drownings.

The human body's responses to breath-hold diving can be modified by a rudimentary diving response induced by apnea and facial immersion, particularly in cold water. This diving response, manifested primarily as bradycardia, is most pronounced in young children. The diving response has been interpreted as an O_2-conserving response that extends breath-holding time by redistributing blood flow from organs that are relatively resistant to hypoxia to organs with continuous requirements for oxygen such as heart and brain. The validity of this interpretation in humans is uncertain, although a diving response may contribute to survival of children resuscitated up to an hour after submersion in cold water.

2.3 Diving with Compressed Air and Mixed Gas

The limitations of breath-hold diving have been circumvented by diving systems that allow the diver access to a continuous supply of breathable gas at any depth. Diving in shallow water, for example, 130–150 fsw, is usually carried out with compressed air as the breathing gas because air is inexpensive and can be obtained readily, even at remote dive sites. Air divers may be free swimming, as with SCUBA, for example, or supplied by an umbilical to the surface. The tethered diver has a number of advantages over the SCUBA diver, namely, access to continuous air supply, access to surface communication, the possibility of being retrieved, and supply of power for thermal protection. The umbilical, however, can be an encumbrance and has occasionally caused a diver to become trapped by an underwater obstruction.

The maximum safe depth for SCUBA is about 140 fsw and for surface-supplied air diving, approximately 200 fsw (5). These limitations are brought about by three factors, which are (1) nitrogen narcosis, (2) decompression time, and (3) oxygen toxicity. In order to dive beyond the practical range of compressed air, special gas mixtures must be supplied. Usually helium is substituted for nitrogen as the diluent gas, although other non-narcotic inert gases, such as hydrogen, have been used successfully. In early helium dives, decompression sickness was a serious obstacle. In 1937 the first safe helium decompression schedules were developed and helium–oxygen (heliox) mixtures were proven to work effectively for deeper diving. Heliox was first used successfully in 1939 by the U.S. Navy to salvage the sunken submarine *Squalus* from 243 fsw (6). In current heliox operations, the oxygen pressure is usually maintained at a constant 0.4 to 1.0 ATA and the diver often operates out of a diving bell. The helium may be recycled by gas reconditioning equipment located on the bell. Surface-supplied heliox diving is expensive, but effective and relatively safe to depths to 400 fsw.

2.4 Saturation Diving

The ability to dive deeper with helium–oxygen mixtures led inevitably to longer decompression time relative to effective work time for the diver. In an effort to solve this problem, U.S. Navy medical officers led by George Bond, in 1957 explored the feasibility of saturation diving (6). The physiological principle was that after prolonged exposure (12 to 24 hr) at any depth, a finite limit of gas uptake by tissues would occur (saturation). Then no further decompression penalty would be incurred. The diver residing at that depth could work for long intervals and would require only a single, slow decompression. Using these principles divers have lived and worked for extended periods at ocean depths of 1500 fsw and at simulated depths of 2250 fsw in dry chambers.

Saturation diving is routine practice for many military and commercial diving applications when extended periods of time are needed to complete a task at 400 fsw or greater. Saturation techniques are widely employed in the offshore oil drilling industry in the North Sea and Gulf of Mexico. The divers generally live in a deck decompression system mounted on board the ship or the drilling platform. The divers work in shifts and are transported to the work site in a pressurized transfer capsule or bell that mates with the deck system. At the working depth, the hatch of the capsule is opened and the divers in their rigs are free to leave the bell to work. The divers are tethered to the bell by umbilicals containing safety cables, communications, and water lines for hot water suits. The bell or capsule systems are managed by an inside tender with communications to the mother ship or platform. After each shift, the divers return to the bell, the hatch is closed, and the bell is returned to the ship. This procedure allows shipboard personnel to tend the divers continuously and obviates the need to supply large amounts of power to great depths 24 hours a day.

2.5 Hyperbaric Chambers

Hyperbaric chambers are pressure vessels designed for controlled compression and decompression of gas. Hyperbaric chambers for human occupancy may be of either multiplace or monoplace design. These chambers are used primarily to administer hyperbaric oxygen as a therapeutic modality. Multiplace chambers are large and expensive and require considerable expertise to operate. Multiplace facilities, however, offer significant advantages over monoplace chambers. The monoplace chamber, by definition, holds only one person, the patient, in a 100 percent oxygen environment. Therefore procedures that require an attendant or tender cannot be accomplished in the monoplace setting. Multiplace chambers also are generally more suitable for critically ill patients. Multiplace chamber complexes may also be used to simulate the diving environment and therefore have been invaluable to scientists conducting diving-related research.

The safe operation of hyperbaric chambers requires a thorough understanding of the equipment, procedures, and potential hazards of the hyperbaric environment. This requires specialized training and extensive knowledge of principles of pressure integrity, gas handling, built-in-breathing systems (BIBS), electrical systems, and fire safety. Hazards of chamber operation include abrupt changes in gas pressure, gas contamination, hypoxia, thermal problems, oxygen toxicity, and fire. Safe chamber operation requires careful, full-time supervision by a qualified operator. Operational and emergency procedures must be established in advance and be readily available on site at all times. Careful operations and maintenance records should be maintained for each facility. The U.S. Navy and ASME have standardized requirements for certification of hyperbaric chambers for human occupancy.

2.6 Tunnels and Caissons

The use of compressed air to keep water out of shafts and tunnels evolved during the middle of the nineteenth century. The engineering problems associated with providing a continuous supply of compressed air were solved first in France in the 1830s by the engineer Triger. Successful caissons containing compressed air were used for bridge construction in the 1850s and for tunnels in the 1880s. Compressed air workers were exposed daily to pressures as high as 4.5 ATA. Similar air pressures are still used today in the construction of tunnels, bridges, and mine shafts (7).

Traditional practice for the construction of underwater shafts is to force compressed air into the entire working chamber at a pressure high enough to keep water out of the face of the shaft. The workers usually work one shift per day. The length of the shift is governed by the maximum working pressure, the decompression profile, and local regulations. For example, in the United Kingdom, the maximum working pressure permitted by law is 50 psig (4.5 ATA) and the maximum time at pressure cannot exceed 10 hr in any 12-hr period. Thus a compressed air worker at 50 psig is allowed just over 4 hr at depth because the decompression time approaches 6 hr. Work shifts in the United States have been kept relatively short to limit the incidence of aseptic bone necrosis; however, this practice has been only partly effective.

The working chamber at each site is fitted with a number of air locks that serve different purposes. Separate air locks for divers are required for timed decompressions. Equipment and supplies pass through muck-locks, and medical locks for the treatment of decompression sickness are required when the working pressure is 2 ATA or greater. In some places, decant locks are used for decompression; in these cases workers are rapidly decompressed in the man-lock and then recompressed in the decant lock for completion of the appropriate decompression schedule.

Compressed air workers often face difficult working conditions ranging from cool, damp environments to hot, humid atmospheres containing a number of potentially toxic gases and fumes. Although very large quantities of clean air are supplied to tunnels and caissons by on-site compressors, welding fumes and other contaminants such as CO tend to accumulate in pockets where ventilation is inadequate. Current guidelines require that at least 10 ft^3 of air at working pressure per minute be supplied per person in the tunnel or caisson. Other immediate hazards encountered in caisson or tunnel environments are flooding, heat exhaustion, decompression sickness, and hypoxia. Hypoxia may occur with influx of oxygen-depleted air from soil beds upon reduction of the pressure inside the chamber. The risk of fire is enhanced greatly in compressed air environments because of the elevated partial pressure of oxygen. The most troublesome long-term occupational health problem for tunnel workers has been aseptic necrosis of bone (see Section 4.4).

2.7 Submarines and Submarine Escape

The ambient environment of the nuclear-powered submarine is sealed but not pressurized. The most significant problem in submarine atmosphere control remains the metabolic production of CO_2 by the crew (8). Atmospheric carbon dioxide concentrations are allowed to approach 1 percent for the duration of a submarine patrol. Continuous exposure to increased ambient CO_2 levels are associated with small increases in the respiratory minute ventilation and cyclic changes in blood bicarbonate and pH. The acid–base disturbance correlates with cycles in calcium homeostasis probably related to the uptake and release of CO_2 from bone. Other minor atmospheric contaminants include CO, freons, and various aerosols. The only detectable health effect of these contaminants has been a slight increase in hematocrit attributed to low-level CO exposure.

Though modern submarines do not operate at pressures above 1 ATA routinely, they can be pressurized by virtue of their pressure-tight internal hulls and gas storage capacity. This capability is restricted in U.S. submarines to approximately 6 ATA, but it provides a measure of protection against accidental flooding, which in peacetime is most likely to result from collisions in shallow coastal waters. The occurrence of submarine disasters in relatively shallow waters initiated the development of an apparatus to facilitate escape from U.S. submarines in 1929 by Momsen, Tibbels, and Hobson. The Momsen "lung" consists of a rebreathing bag containing a canister for CO_2 absorption. The device is charged with air or oxygen and excess pressure in the lungs and apparatus during ascent is relieved by exhaust of gas through a ventral flutter valve. Simulated submarine escapes were practiced by submariners for many years from 100-ft training towers constructed at New London, Connecticut and Pearl Harbor, Hawaii. Many thousands of safe ascents have been made, but occasionally pulmonary overinflation accidents occurred, resulting in air embolism. The routine "free ascent" practice for submariners has been abandoned by the U.S. Navy, and the Momsen "lung" has been superseded by a hooded pressure suit developed by the British Royal Navy. Royal Navy divers have made escapes safely from depths of 600 ft with this type of equipment.

The current operational plan for rescue of submarine crews in the U.S. Navy is centered around the Deep Submergence Rescue Vessel or DSRV (6). This maneuverable submersible is designed to interface with the hatches of a sunken submarine. The DSRV can operate to 5000 fsw and can be pressurized in accord with the capacity to pressurize the inner hull of the submarine. There are a number of potential operational and physiological problems associated with the use of DSRVs and the system has never been put to actual operational use.

3 PHYSIOLOGICAL EFFECTS OF COMPRESSION AND PRESSURIZED GASES

3.1 Compression Arthralgias

Compression of the body in air or mixed gas during diving can produce a variety of joint symptoms (9). These symptoms are most prominent during rapid compression to great depth and include both pain on movement and joint popping or cracking. The former is more common in helium–oxygen diving and the latter in nitrogen–oxygen diving, although both symptoms occur during compression with helium. Compression arthralgias most often involve the shoulders, knees, hips, and back, and the symptoms resolve spontaneously after a few hours on the bottom.

The etiology of compression arthralgias is unknown. The two leading hypotheses about their origin relate to (1) direct effects of hydrostatic pressure and (2) osmotic shifts of fluid out of the synovium during the uptake of inert gas by well-perfused tissues near the joint space. Whether or not the joints are damaged by compression is unknown; however, the presence of symptoms raises the concern that divers may be predisposed to degenerative joint disease. Compression arthralgia can be ameliorated by using slow rates of compression.

3.2 Barotrauma of Descent

The physical pressure–volume relationships of gases have important consequences for how gas-containing body cavities equalize during diving. Body structures that normally contain gas in equilibrium with ambient air include the paranasal sinuses, lungs, middle ear space, and hollow viscera of the gastrointestinal tract. Any gas-filled cavity that is not fully collapsible must equilibrate with ambient pressure during both descent and ascent. If the pressure within the space does not equalize, barotrauma may occur to the tissues. Common examples of barotrauma of descent include middle ear squeeze, sinus squeeze, and lung squeeze. Ear squeeze is the most common form of barotrauma of descent and may result in a range of injuries including tympanic membrane perforation and rupture of the round and oval windows (10). Lung squeeze occurs in either deep breath-hold diving or when the pressurized gas supply of the diver cannot be maintained during rapid descent. Mask squeeze may also occur in SCUBA diving if the diver fails to equalize the space in the face mask on descent. This sometimes produces dramatic periorbital hemorrhage, which is of little clinical consequence and generally does not require medical intervention. Barotrauma also may occur on ascent, usually in compressed air or gas environments when an equilibrated space fails to vent during the ascent. Common examples of barotrauma of ascent include reversed middle ear and reversed sinus, alternobaric vertigo of ascent (rarely on descent), and pulmonary barotrauma syndromes. Pulmonary barotrauma is discussed in Section 4.3. One important but rare consequence of middle ear overpressurization during ascent is facial baroparesis (11). Facial baroparesis is a transient, ischemic paresis of the seventh cranial nerve in divers with defects in the wall of the facial canal. Similar palsies of the fifth cranial nerve due to its compression in the maxillary sinus also have been reported. These cranial nerve findings are important because they may be confused with type II decompression sickness (see Section 4.2) and/or arterial gas embolization (see Section 4.3).

3.3 Inert Gas Narcosis and High-Pressure Nervous Syndrome

For more than a century, divers have known that compressed air has narcotic properties when breathed at pressures greater than 3 ATA. Compressed air narcosis is related directly to the partial pressure of N_2; therefore manifestations increase in direct proportion to the ambient pressure. Symptoms of nitrogen narcosis such as euphoria and impaired cognitive function are similar to those produced by alcohol and the early stages of anesthesia. At depths beyond 10 ATA, N_2 may result in unconsciousness, thus limiting the useful depth of compressed air as a breathing gas. Narcosis is a property of nitrogen shared with other noble gases, such as helium, neon, argon, and xenon. The narcotic properties of these inert gases correlate well with their solubility in lipids. For example, helium, which is the least narcotic diluent gas, has a lipid solubility and narcotic potency only one-fourth that of nitrogen whereas xenon is about 25 times more lipid soluble and narcotic than N_2. Carbon dioxide is also a narcotic gas and at one time was proposed erroneously as the cause of compressed air narcosis.

The precise biological mechanisms of inert gas narcosis have not yet been delineated (12, 13). Most investigators favor mechanisms related to the critical volume hypothesis of general anesthesia. According to this hypothesis, anesthetic gas molecules are absorbed by lipid layers in nerve cell membranes, thereby causing an increase in volume. This increased volume is thought to alter the electrical and ionic permeability characteristics of neuronal membranes, particularly at sites of synaptic transmission. Volatile gases with high lipid solubility, for example, nitrous oxide (N_2O), achieve the critical volume effect at lower partial pressures than narcotic gases such as N_2. The critical volume hypothesis is supported indirectly by pressure reversal of anesthesia, wherein increased hydrostatic pressure or compression with a gas of low narcotic potency, such as helium, ablates the narcotic effect.

Pressure reversal of anesthesia is probably a consequence of the broader phenomenon of increased excitability in the CNS produced directly by high pressure (12). In practical terms, this neuronal hyperexcitability has been seen with (1) rapid compression and (2) very deep diving. These two effects of pressure have become known collectively as the high-pressure neurological syndrome (HPNS), although they may well have separate biological mechanisms. HPNS is experienced by both humans and animals at pressures above 15 to 20 ATA. The neurophysiological basis of the syndrome is poorly understood, but may involve decreased membrane fluidity and altered presynaptic neurotransmitter release and/or decreased postsynaptic binding or responsivity at high pressure. The hallmarks of HPNS in humans are rapid tremor (6 to 12 Hz), incoordination, myoclonus, microsleep, and various electroencephalographic abnormalities. At extremely high pressures in animals, convulsions and death occur, depending on species and individual susceptibility. In accordance with the critical volume hypothesis, some of the neurophysiological manifestations of HPNS can be ameliorated by increasing the concentration of a narcotic diluent gas in the breathing gas mixture. This approach has resulted in the use of trimix (helium, nitrogen, and oxygen) for many deep manned dives. At this time, however, HPNS, more than any other single physiological factor limits the depth at which humans can dive.

3.4 Pulmonary Function and Gas Exchange

Although the physical work capacity of humans in hyperbaric conditions is normal at moderate depths, oxygen uptake ultimately becomes limited as a function of depth, primarily because of limitations in pulmonary mechanical function (14, 15). These mechanical limitations are related most closely to the increased density of the breathing gas rather than to either increased hydrostatic pressure or increased gas viscosity. The density (ρ) of the gas that a diver breathes increases in proportion to both the molecular weight of the gas and ambient pressure. Thus N_2, which has a density of 1.1g/l at Body Temperature and Pressure (BTP), is about seven times as dense as helium under the same conditions. Helium–oxygen breathing gases at 50 ATA, however, are approximately seven times as dense as air at 1 ATA.

Three important respiratory consequences of high gas density are an increase in airflow resistance, an increase in the functional residual capacity (FRC), and a decrease in the ventilatory response to CO_2. The first two factors translate to a greater work of breathing needed to sustain the high levels of ventilation accompanying heavy exercise. The third effect, a depressed ventilatory response to CO_2, is probably secondary to the increased resistance to airflow. High flow resistance decreases the amount of ventilation produced by the muscular response to a particular CO_2 level (16). Ultimately, then, the capacity to exercise at high gas densities becomes limited by ventilation, unlike the normal environment, where maximum oxygen uptake is usually limited by cardiovascular factors. Gas density also has three other potentially important, although less well evaluated effects on respiratory function. These are increased respiratory heat loss, increased inertance of flowing gas, and decreased gas-phase diffusivity. The latter effect may impair alveolar gas mixing and decrease effective alveolar ventilation in divers by increasing the physiological dead space and therefore contribute to CO_2 retention. The potential ventilatory limitation to exercise is exacerbated by the additional external resistance intrinsic to all underwater breathing apparatus.

As gas density increases, respiratory resistance increases linearly at moderate flow rates for densities up to 25 g/l. The increase in airways resistance produced by breathing dense gases is distinct from resistance loading produced by chronic obstructive pulmonary disease or breathing through an external resistance; however, the mechanical effects of dense gas may be understood in much the same way as the other processes (17). During a forced expiration, the maximum rate of airflow is limited by lung volume and dynamic compression of airways. Dynamic airways compression during expiration is a direct result of the high pressure generated by the expiratory muscles, which adds force to the normal elastic recoil pressure of the lung. As the pressure inside the airways drops enroute to the larger airways, the sum of the expiratory plus elastic recoil pressures imposed outside the airways equals (equal pressure point) and eventually overcomes the pressure inside the airways and any resistance to deformation by the tissue. At the equal pressure point, the airways collapse, creating a flow-limiting segment. Beyond this point, airflow is independent of effort; that is, it will not increase further with additional expiratory force. In the human respiratory system, flow-limiting segments are of sufficient radius that the airflow is primarily turbulent. Therefore airflow is inversely related to gas density. Simple calculations yield the useful relationship that airflow through flow-limiting segments is proportional to the reciprocal of the square root of gas density. This square-root relationship describes how high gas density decreases maximum respiratory flow rates; it also approximates the strong inverse correlation between gas density and maximum voluntary ventilation (MVV). If the MVV is limited by dynamic airways compression, and the MVV at depth determines maximum exercise ventilation, then gas density becomes the limiting factor for the amount of aerobic exercise that a diver can perform.

During exercise, the minute ventilation ($\dot{V}_E$) increases in proportion to the amount of oxygen consumed (V_{O_2}). The ventilatory equivalent, $\dot{V}_E$ [Body Temperature Pressure Saturated (BTPS)]/V_{O_2}, of approximately 30 generally maintains the arterial pCO_2 near 40 torr. If the maximum exercise ventilation is decreased by dense breathing gas, then additional O_2 uptake can occur only at a smaller ventilatory equivalent and thus a higher pCO_2. Hypercapnia is exaggerated further by the depressed ventilatory response to CO_2. Respiratory acidosis and high sustained levels of respiratory work may produce respiratory muscle fatigue limiting exercise at high gas densities. In fact, p_aCO_2 values above 60 torr have been reported in working divers just before they stop exercise (18).

As noted previously, breathing dense gas can increase the FRC during exercise. This change in FRC, however, has not been documented by all investigators, particularly during immersed exercise. When the FRC is larger, exercise ventilation is done at higher lung volumes where airflow resistance is less. The pressure–

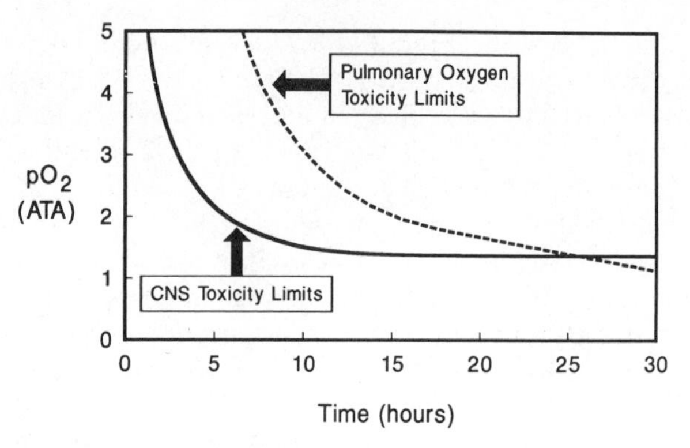

Figure 12.1. Effects of elevated oxygen partial pressure.

volume characteristics of the respiratory system, however, require a large increment in the inspiratory work needed to overcome elastic recoil forces at high lung volumes. Ventilation at high lung volume is also inefficient because the angle of apposition between the diaphragm and chest wall may widen and decrease inspiratory efficiency, thereby paradoxically resulting in inspiratory muscle fatigue.

3.5 Oxygen Toxicity

Toxic effects of hyperoxia have been known since 1878 when Paul Bert reported convulsions in animals exposed to hyperbaric pressures of oxygen (19). Pulmonary oxygen toxicity was discovered more than 20 years later by Lorrain-Smith. Oxygen concentrations sufficient to cause manifestations of pulmonary and CNS toxicity are encountered frequently in hyperbaric environments. High concentrations of oxygen are used to accelerate inert gas elimination during decompression in mixed gas diving. Pure oxygen is also used in closed-circuit diving with a CO_2 scrubber to prevent detection of exhaust bubbles in shallow water military operations. During clinical hyperbaric oxygen exposures, patients routinely breathe pure oxygen at treatment depths of 2 to 3 ATA.

Oxygen is probably toxic to all organs in the body at high enough concentrations, although oxygen poisoning is expressed primarily in three organ systems including the lungs, CNS, and eyes. The toxic effects of oxygen can be avoided if the hazards of oxygen are appreciated and efforts are made to limit exposure times at potentially toxic doses of oxygen. As a general rule, the rate at which oxygen toxicity develops increases as a hyperbolic function of the inspired partial pressure of the gas (20, 21). This principle is illustrated in Figure 12.1 for both pulmonary and CNS oxygen toxicity. The most effective practice for avoiding oxygen toxicity has been the use of empirical tables that limit exposure time in inverse proportion to oxygen partial pressure. There are several ways to estimate the physiological affects of a known dose of oxygen (22); however, these calculations are used infrequently for practical

Table 12.3. Common Signs and Symptoms of CNS Oxygen Toxicity

Visual symptoms
Loss of acuity
Scotomata
Constriction of visual fields
Acoustic symptoms
Tinnitus
Music
Unpleasant odors and tastes
Twitching of lips and face
Behavioral changes
Hiccups
Nausea and vomiting
Syncope
Convulsions

reasons. When high levels of inspired oxygen are needed, the oxygen tolerance time can be extended greatly be interrupting the exposure briefly.

The most dangerous form of oxygen toxicity for the diver is CNS O_2 toxicity and its most serious manifestation is O_2-induced convulsions. Although oxygen convulsions are transient, they may occur at pO_2 above 1.7 ATA and sometimes happen without warning. This may have disastrous consequences, particularly for free-swimming divers. The common signs and symptoms of CNS oxygen toxicity are given in Table 12.3 (23).

The mechanisms of O_2 toxicity have been clarified in recent years by elegant experimental work. Hyperoxia accelerates the production of reactive oxygen intermediates that serve as mediators of the toxic effects of O_2 produced in brain and lung tissues (24). The exact relationship between these reactive O_2 metabolites and symptoms of O_2 poisoning, however, remains to be defined. Many intracellular enzymes and electron transport systems catalyze single electron transfers to dioxygen as by-products of normal metabolic processes. In these metabolic reactions, the elaborated superoxide anion ($O_2^{-}\cdot$) is highly reactive toward biological molecules. The risk of cellular injury is reduced by dismutation of superoxide to H_2O_2 spontaneously or via a reaction catalyzed by superoxide dismutase (SOD). Hydrogen peroxide, which also may be toxic to cells, is converted catalytically to water plus O_2 by the enzyme catalase. There is also the potential for generating hydroxyl radical ($\dot{O}H$) by redox interactions between $O_2^{-}\cdot$, H_2O_2, and iron (Haber–Weiss reaction). Under normal conditions, SOD, catalase and low molecular weight nonenzymatic free radical scavengers, such as reduced glutathione, provide sufficient antioxidant defenses to limit free radical-mediated damage to intracellular organic molecules. During hyperoxia, however, production of reactive oxygen metabolites greatly increases and may exceed the ability of radical scavengers to contain them. These radicals may oxidize a variety of intracellular components including proteins, nucleic acids, and membrane lipids (lipid peroxidation).

In the brain, the free radical hypothesis appears to be linked closely to the effects of hyperoxia upon cellular energy metabolism (25). During hyperbaric oxygenation, brain tissue pO_2 increases despite compensatory vasoconstriction of cerebral blood vessels. Tissue hyperoxia is accompanied by oxidation of mitochondrial pyridine nucleotides (NADH) and cytochrome c oxidase. These intramitochondrial events precede O_2-induced convulsions, and together with evidence of enhanced glycolysis, have been interpreted to indicate activation of cerebral energy metabolism and impairment of ATP availability by hyperbaric oxygen. These events also may disturb the normal balance between excitatory and inhibitory neurotransmitters, for example, γ-aminobutyric acid depletion, in specific regions of the brain and produce convulsions.

Lung injury from reactive O_2 species has also been associated with activation of lung metabolism. The resulting pathophysiological process has been divided into two distinct phases, the early or exudative phase and the late or proliferative phase (26). During lethal O_2 exposures the early injury is not associated with early structural changes in the lung. Later, inflammation leads to destruction of lung parenchyma. If the O_2 exposure is less severe, repair processes supervene, and a proliferative phase follows. The proliferative response diminishes the destruction of lung parenchyma, but it may produce pulmonary fibrosis.

There are a number of ways to increase or decrease tolerance to oxygen. These include a host of biochemical, pharmacological, and physiological factors. Several physiological factors are important amplifiers of CNS oxygen toxicity in humans including hypercarbia, immersion, exercise, fever, and sympathetic stimulation. The only practical approach to the extension of human oxygen tolerance has been to decrease the inspired oxygen concentration to nontoxic levels for brief (5 to 10 min) intervals of time during exposure. Pharmacological measures to prevent CNS O_2 toxicity, such as the administration of anticonvulsants, are often recommended in high-risk clinical settings, but have not been proven to be beneficial in this setting.

3.6 Thermal Problems and Energy Balance in Diving

Most diving activities are conducted in water that is considered cold relative to the thermoneutral temperature of about 34°C for the immersed human at rest. This rather high thermoneutral point in water is a result of its high density and thermal conductivity relative to air. Surface water temperatures range from 4°C near the poles to about 30°C near the equator; at 600 fsw water temperatures are approximately 4 to 10°C. Thus the physiological responses to body cooling and the principles of thermal protection are important factors in most diving operations. Human thermoregulation has been studied extensively; however, only those physiological responses to cold stress with relevance to diving are summarized here. A more comprehensive discussion of this topic has been provided by Webb (27).

Heat transfer from or to the human body obeys the second law of thermodynamics; that is, transfer of heat occurs from a system of higher temperature to one of lower temperature in direct relation to the temperature difference between the two systems. There are four modes of heat transfer including conduction, convection, radiation, and evaporation. Conduction, or transfer of heat from two physical systems in direct contact, and convection, or heat transfer in moving fluids, account for the bulk of the heat exchange from the skin during underwater immersion and in hyperbaric environments. Radiant heat transfer, or transmission of heat between two bodies at a distance, is less important in diving; and evaporative heat transfer, which is the heat required to change a liquid into a vapor at constant temperature, is usually limited to insensible respiratory heat losses by the diver. Respiratory heat loss can be quite significant in deep diving, and convective heat loss, which varies as a direct function of gas density and temperature, supplants evaporative heat loss as the major component of respiratory heat loss at great depth.

The exchange of heat between the body and its environment can be expressed in the form of a body heat balance equation:

$$M \pm C \pm R - E = S$$

where M = metabolic heat production, C = convective (and conductive) heat loss or gain, R = radiant heat loss or gain, E = evaporative heat loss, and S = heat loss or gain (storage). The respiratory heat losses contribute to both the evaporative and convective terms in the heat balance equation. In general, however, convective losses from the skin (C_{sk}) account for most of the heat loss from the body during immersion in both cold water and cold dense gas, for example, dry hyperbaric habitats. Convective heat loss (C_{sk}) from the body surface can be expressed as

$$C_{sk} = h_c\,(\overline{T}_{sk} - T_a)$$

where h_c is the convective heat transfer coefficient (W/m^2·°C), $\overline{T}_{sk}$ = mean skin temperature (°C), and T_a = ambient temperature (°C). The convective heat transfer coefficient (h_c) is affected by the velocity, density (ρ), and the specific heat of the fluid (liquid or gas) surrounding the body. In gas-filled hyperbaric habitats, the density of the gas is proportional to the ambient pressure; thus as the depth increases, the convective heat loss increases linearly for any temperature difference ($\overline{T}_{sk} - T_a$). This heat loss can be compensated for partly by vasoconstriction, which decreases $\overline{T}_{sk}$, and by decreased evaporative heat loss. The practical approach to this difficulty, however, is to increase T_a as a function of depth to maintain the diver's habitat within the thermal comfort zone. This principle is illustrated by the list of optimal temperatures for thermal comfort in hyperbaric helium–oxygen in Table 12.4 (28). Even this practice, however, does not appear to be entirely successful in preventing energy expenditure in excess of energy intake in saturation divers. Many saturation divers do not appear to maintain proper energy balance despite high caloric intakes during a dive (29). This has resulted in small but significant decreases in body weight in some individuals in long saturation dives independent of hyperbaric diuresis. Hyperbaric diuresis also occurs, and its pathogenesis is complex. It has been variably attributed to convective cooling, suppres-

Table 12.4. Thermal Comfort Zone for Hyperbaric Helium

Depth (ATA)	Comfort Zone (°C)
1	24–26
10	26–28
20	28–30
30	30–31
40	31–32
50	32–33

sion of evaporative heat loss, osmotic gas gradients, and direct effects of hyperbaric pressure.

Convective heat loss from the body is the major thermal problem in both wet and dry hyperbaric environments, and it can be defended against to some extent by thermoregulatory responses such as vasoconstriction of the skin, decreased evaporative heat loss, and increased metabolic heat production (shivering). The physiological responses to cold stress also provide a warning to the diver about his thermal balance. In addition, passively insulated garments, for example, wet suits and dry suits, offer reasonably good thermal protection in shallow water. Thermal protection of the hands and feet, however, remains a serious problem in passive garments at water temperatures below approximately 10°C. Active thermal protection systems, for example, hot water suits, are used for deep diving because of compression and loss of insulation by passive garments such as wet suits. These actively heated systems rely on a continuous flow of hot water to the diver through an umbilical line.

There are three other notable thermal concerns in deep diving operations; these are slow cooling, respiratory heat loss, and the lost bell situation (30). These thermal problems are significant because each may lead to lethal consequences for the diver. The problem of slow cooling has been recognized primarily through the work of Webb, who has reported that humans may lose up to 300 kcal without significant shivering responses (27). Shivering is one of the diver's early defenses against hypothermia, and its absence may allow the diver to cool to the point of serious fatigue or decrement in performance or judgment. A similar problem may occur with high convective respiratory heat losses, which are not readily detected by thermoregulatory sensory mechanisms. At 600 fsw in 4°C water, the entire metabolic heat production of the body may be lost directly from the body core by respiratory convection. Such massive respiratory heat losses cannot be made up by shivering or exercise because of the accompanying increase in ventilation. Hazardous cooling of the respiratory tract can be averted by appropriate heating of the inspired gas. The accidental loss of a diving bell results in loss of power to heat the dense gas in the diver's environment. Hence the gas in the bell cools rapidly to ambient temperature, and trapped divers will survive only as long as their passive thermal protection will prevent lethal hypothermia. The survival time is directly related to the depth (gas density), the temperature of the water, and respiratory minute ventilation.

4 DECOMPRESSION FROM PRESSURIZED ENVIRONMENTS

4.1 Principles of Decompression

The use of compressed air in diving during the last century led inevitably to physiological problems caused by release of excess gas from tissues after diving. The process of inert gas elimination from the body after a decrease in pressure during ascent from diving or to altitude is known as decompression. Although the basic concept that rapid decompression could produce decompression sickness was formulated by Pol and Watelle in 1854, it was not until 1878 when Paul Bert reported that bubbles consisting mainly of nitrogen were formed in tissue during rapid decompression (19). Guidelines for safe decompression did not exist, however, until 1908 when Boycott, Damant, and Haldane published empirical decompression tables based on exponential half times for elimination of nitrogen from body tissues (31). This seminal paper has provided a scientific basis for much of modern decompression theory.

The rate of uptake or elimination of nitrogen or other inert gas from the body after a change in pressure is exponential with respect to time. Inert gas exchange is determined primarily by gas solubility in blood and tissue, blood flow, and the volume of tissue. This assumes that the diffusion resistance between blood and tissue is inconsequential for nitrogen and other inert gases. Tissues that behave this way are perfusion-limited, and the characteristics of their inert gas exchange can be defined by a half time. Because the tissues of the body are perfused differently and nitrogen is more soluble in fat than other tissues, the half times for different tissues may vary considerably. This principle, recognized by Boycott, Damant, and Haldane, was used to calculate their original tables for safe decompression. They also knew that gas bubbles would form and decompression sickness would occur only if the tissue was supersaturated, that is, if the nitrogen partial pressure was about twice as great as the absolute pressure. The Haldane approach has been modified over the years, particularly with respect to the magnitude and duration of allowable supersaturation. Most current decompression tables are still based on multiple parallel exponential models, although a variety of other modeling approaches also have been used successfully (32). Even though decompression models are conceptually useful, two great unknowns still remain in decompression research, namely, the degree and the duration of supersaturation with inert gas in actual tissues.

Although perfusion is the primary variable affecting uptake and elimination of inert gas, diffusion may limit the rate of inert gas exchange under some conditions. Diffusion limitation may occur if the rate of exchange of inert gas by perfusion is extremely rapid, if the tissue perfusion is very low, or if part of the tissue such as the capillary endothelium has a high diffusion resistance. Diffusion may also be important when two adjacent tissues have very different rates of perfusion. In such a situation, the elimination of inert gas from the faster tissue may allow additional inert gas to enter by diffusion from the slower tissue. Thus the faster tissue may maintain supersaturation for a longer time. Diffusion is also important after gas bubbles have formed in a tissue during decompression. Gas bubbles contain large amounts of inert (N_2) gas that can be removed by tissue perfusion only after the N_2 diffuses back into the tissue. The rate at which N_2 diffuses away from a gas bubble is determined by the bubble surface area, the intrabubble pressure, and the partial pressure difference between the bubble and the tissue. It is noteworthy that decompression also can be modeled successfully using slab diffusion and the resulting decompression profiles are quite satisfactory.

During decompression, bubble formation appears to occur at specific nucleation sites in the body. The number and location of nucleation sites vary according to physiological circumstances. For example, exercise may increase the number of bubbles formed by tribonucleation, a mechanism that creates bubbles in areas such as joint capsules, where large negative pressures can be generated by traction between surfaces lubricated by a liquid. Bubbles created by this so-called vacuum phenomenon may arise either from preexisting gas nuclei or by de novo formation. Tiny unstable bubbles, also called gas nuclei, may have long lifetimes in the body if they stabilize at hydrophobic sites. Such gas nuclei may grow into bubbles during decompression. There is little doubt that experimentally gas bubbles cause decompression sickness, although clinically silent bubbles can be detected in the circulation by ultrasonic devices during normal decompression. There is still uncertainty, however, about the mechanisms of bubble formation in the body and how they are related to the diverse manifestations of decompression sickness.

Bubble formation occurs most commonly after a decrease in ambient pressure, but bubbles also may form whenever the sum of the dissolved gas pressures in tissue exceeds ambient pressure (33). Supersaturation does not require a decrease in hydrostatic pressure if a diver, saturated with one inert gas, suddenly changes breathing gas at constant pressure, or if a diver breathes a soluble inert gas while surrounded by a less soluble gas. These forms of isobaric inert gas exchange may lead to supersaturation and bubble formation by various mechanisms. For example, body tissues do not exchange all inert gases at the same rate. If a diver saturated with a slowly exchanging inert gas switches to a more rapidly exchanging gas, the new gas enters the tissue faster than the slow gas leaves and transient supersaturation of the tissue occurs. Other mechanisms of isobaric inert gas exchange are counterperfusion and counterdiffusion. These forms of isobaric inert gas exchange require that two adjacent layers of tissue with dissimilar gas exchange characteristics be exposed to different gases on the unopposed surfaces. This situation may arise when a diver breathes one inert gas while surrounded by another less soluble gas. Each gas diffuses across the two tissues independently at rates proportional to the diffusion gradients, which differ as a function of the diffusive resistances of the dissimilar tissues. Therefore it is possible for isobaric supersaturation to occur across the tissues, particularly at tissue interfaces.

Table 12.5. Clinical Manifestations of Decompression Sickness

Type I (mild DCS) 75–80%

1. Limbs
   Pain (bends), niggles, lymphatic obstruction, numbness and paresthesias usually involving the large joints, e.g., shoulders, elbows, knees
2. Skin
   Mottling, itching, rash, pallor, urticaria, edema

Type II (serious DCS) 20–25%

1. CNS
   a. Brain
      Headache, seizures, loss of consciousness, visual disturbances, hemiparesis, aphasia, tremor, ataxia (staggers)
   b. Spinal cord
      Low back or pelvic girdle pain, paraparesis, urinary retention, incontinence
   c. Audiovestibular DCS
      Tinnitus, vertigo, nystagmus, decreased hearing, nausea, vomiting
2. Cardiopulmonary
   Substernal pain, cough, tachypnea, asphyxia (chokes)
3. Systemic
   Extreme fatigue, hemoconcentration, shock

4.2 Decompression Sickness

Understanding the factors that govern the formation of bubbles during decompression has helped investigators develop tables for safe decompression. The computation of depth–time profiles for decompression usually assumes that decompression sickness (DCS) is related to the breach of some critical supersaturation threshold. This assumption conflicts with clinical data which indicate that DCS often occurs as a statistical rather than as a threshold event (34). Therefore, even proper use of well-tested decompression tables is associated with a small but definite risk of decompression sickness for dives of approximately 27 fsw or deeper. Most serious DCS, however, is the result of omission of decompression after deep air or mixed gas dives.

The clinical manifestations of DCS are attributed to growth of bubbles in body tissues. Primary sites of bubble growth are the joint spaces, tendon sheaths, and periarticular tissues including peripheral nerve endings. Bubbles in these locations produce pain that is known as type I decompression sickness, or bends. Bubble formation also occurs in the spinal cord and other areas of the nervous system. This is called type II or serious decompression sickness. DCS involving the audiovestibular system also occurs, but it is relatively rare (35). The most common clinical manifestations of DCS are summarized in Table 12.5. Reports of DCS in military and commercial divers indicate a 4:1 predominance of type I DCS. Altitude bends, or dysbarism, is generally accompanied by mild to moderate type I DCS.

Tunnel workers have a reported incidence of DCS of 0.7 to 1.5 percent with a great predominance of type I symptoms that involve the knee and lower leg in two-thirds of the cases (36). In contrast, recent statistics from recreational divers suggest a 3:1 predominance of type II DCS in these individuals (37). The reasons for this difference in type of DCS are uncertain, although under-reporting of type I DCS is likely, and in some cases prolonged delays in the diagnosis and treatment of type I DCS may allow it to evolve into type II disease. Also, recreational divers are more likely to omit part of their decompression obligation than are professional divers, thus increasing the probability of serious DCS.

One of the most troublesome forms of serious DCS is spinal cord injury (38). The precise mechanisms of spinal cord DCS are not yet fully understood. In experimental animals, spinal cord injury may be precipitated by intravascular bubbles that form in the low-pressure, epidural venous plexus of the spinal cord. This venous lake is susceptible to bubble formation because of its slow, bidirectional blood flow. Bubbles therein resolve slowly, and bubble-induced thrombi can easily obstruct venous outflow from the plexus, leading to ischemic injury of the spinal cord. Despite evidence for bubble formation in the spinal venous plexus, intravascular formation of bubbles appears to be otherwise uncommon. Most intravascular bubbles probably originate at the tissue–blood interface and stream into the circulation where they are absorbed by the lung. In the circulation, the surface-active effects of bubbles at the blood-to-blood interface may cause a number of hematologic changes. These include complement activation, platelet aggregation and destruction, mediator release, and procoagulant activity. The precise role of these factors in the pathogenesis of decompression sickness is undetermined. Although the bubble hypothesis of DCS certainly explains many of the physiological problems encountered during decompression, there are diving-related conditions, such as aseptic bone necrosis, which are not well understood in terms of the physical properties and behavior of gas bubbles.

When massive numbers of bubbles are released into the circulation, they may overwhelm the pulmonary circulation and cause chest pain, shortness of breath, and cough, a syndrome called the chokes. Some of these bubbles may also cross the pulmonary capillary bed as micro air emboli. Recently it has been pointed out that intravenous bubbles may pass into the arterial circulation in a significant number of divers with type II DCS by a right-to-left intracardiac shunt such as a patent foramen ovale (39). This cardiac anomaly, however, is present in about 20 percent of the normal population, and its precise role as a risk factor for DCS is uncertain.

4.3 Pulmonary Barotrauma and Arterial Gas Embolism

Pulmonary barotrauma is a potentially serious consequence of the failure of expanding gases in the lung to escape from alveoli or acini during ascent. Overinflation of the lung may lead to alveolar rupture and pulmonary interstitial emphysema. These problems may cause pneumothorax, mediastinal or soft tissue emphysema, pneumopericardium, and/or arterial gas embolization (AGE). Pulmonary overinflation is most likely to occur during ascent with breath-holding loss of consciousness or local obstruction of airways with trapping of gas, or after explosive decompression of aircraft at altitude. Lung rupture during a high-risk ascent also depends upon several physiological factors including pulmonary compliance, transrespiratory pressure, and lung volume. Dahlback and Lundgren also have suggested that airway closure and air trapping induced by immersion in the upright position might increase the risk of lung rupture in free ascent (40).

When the pressure within the lung exceeds that at the body surface by about 100 cm H_2O, the lung will rupture. During breath-holding at total lung capacity (TLC), the pressure difference between the alveoli and ambient is approximately 50 cm H_2O. Thus ambient hydrostatic pressure outside the body during ascent must fall another 50 cm H_2O for the lung to rupture. If the compliance of the lung and chest wall at TLC is 15 ml/cm H_2O, then the lung volume during ascent must increase by 15 ml × 50 cm H_2O or 750 ml before the lung will rupture. If the TLC is 6250 ml, then using Boyle's law, we can determine the approximate ambient pressure from which a diver must ascend for pulmonary rupture during a breath-hold. Thus at the surface, $P_1 = 1.0$ ATA, and $V_1 = 7000$ ml. At P_2, $V_2 = 6250$ ml. Therefore

$$(1.0)(7000 \text{ ml}) = P_2 \, (6250 \text{ ml})$$

and $P_2 = 1.0\ (7000/6250) = 1.12$ ATA or 4.0 ft. This calculation explains how pulmonary overinflation and arterial air embolization may occur in shallow water when an inexperienced diver breathing compressed gas ascends too rapidly. The example also points out that fractional volume changes are greatest at low hydrostatic pressures. For example, 1 liter of gas at 1 ATA decreases in volume to 0.5 liter at 2 ATA. A liter of gas at 2 ATA, however, only decreases to 0.67 liter at 3 ATA. This principle operating in reverse during ascent from depth accounts for the fact that the amount of gas released into the arterial circulation after pulmonary overpressurization can be quite great. The distinction between AGE and DCS is based upon the release of expanding gas from the air spaces of the lung into the pulmonary venous circulation during AGE. In DCS, bubbles are formed from gas that is physically dissolved in tissues and comes out of solution during decompression.

The clinical onset of AGE is quite sudden and generally presents with ischemic symptoms involving the brain. Neurological symptoms and signs usually appear within a few seconds to a few minutes after the ascent. Initial symptoms of AGE rarely appear more than 2 hr after ascent and later onset of symptoms usually indicates DCS. Common clinical findings of acute AGE include headache, pleuritic chest pain, confusion, loss of vision, hemiparesis, seizures, and unconsciousness.

4.4 Aseptic Necrosis of Bone

Aseptic necrosis of bone or dysbaric osteonecrosis is seen frequently in compressed air workers and saturation divers. This poorly understood condition is felt by most

Table 12.6. Common Causes of Aseptic Necrosis of Bone[a]

Hyperbaric exposure	Rheumatoid arthritis
Chronic alcoholism and cirrhosis	Gout
Steroid therapy	Syphilis
Sickle cell anemia	Radiation
Diabetes	Trauma

[a]After Walder, 1984 (41).

investigators to be related to decompression stress, and it may actually represent subclinical DCS due primarily to the delayed occlusive effects of bubbles (41). The uncertainty about the pathogenesis of aseptic necrosis is because of the lack of a definite relationship between the site of the bone lesions and the location of previous episodes of type I DCS. Aseptic necrosis of bone is also associated with a variety of other medical conditions that may confound its appearance in a diver (see Table 12.6). Most other causes of bone necrosis, however, can be excluded by a thorough clinical evaluation.

Aseptic bone necrosis in compressed air workers and divers is usually restricted to the humerus, femur, and tibia. In compressed air workers, the shoulders and hips are affected approximately equally, whereas in divers, the shoulders are affected most often. At present, timely diagnosis of aseptic necrosis is established by radiographic criteria because clinical symptoms appear only in advanced disease. The sites of necrosis are divided into those located next to an articular surface (juxta-articular or JA lesions) and those located away from articular surfaces (head, neck, and shaft or HNS lesions). In 1966, the Decompression Sickness Panel of the Medical Research Council proposed a standard system for classifying the radiographic changes of aseptic necrosis (42). This classification is summarized in Table 12.7. Radiographic diagnosis of aseptic necrosis of bone is the current gold standard, although there is considerable research interest in other diagnostic modalities such as bone scintigraphy and magnetic resonance imaging.

The early diagnosis of JA bone necrosis is of paramount importance to the compressed air worker or diver because these lesions may collapse and give rise to clinical symptoms and various degrees of disability. Approximately 8 to 10 percent of JA lesions progress to the point of collapse and irregularity of the articular surface. At this time, however, there is no way to determine which lesions will progress and which will stabilize. Because further diving may increase the risk of progression, workers with JA lesions should be prohibited from further diving. In contrast, HNS lesions are usually inconsequential and rarely become symptomatic.

5 RECOMPRESSION THERAPY AND HYPERBARIC OXYGEN

Gas bubbles in tissues will resolve spontaneously, but the rate at which they are removed can be enhanced greatly by recompression and oxygen breathing. Re-

Table 12.7. Radiographic Classification of Aseptic Bone Necrosis

Juxta-articular lesions
A1 Dense areas, with intact articular cortex
A2 Spherical segmental opacities
A3 Linear opacities
A4 Structural failure
A4a Transradiant subcortical bands
A4b Collapse of articular cortex
A4c Sequestration of articular cortex
A5 Osteoarthritis
Head, neck, and shaft lesions
B1 Dense areas
B2 Irregular areas of calcification
B3 Transradiant areas

compression and hyperbaric oxygen administration are primary therapy for both DCS and AGE (43). A comprehensive treatise on the clinical management of DCS and AGE is beyond the scope of this chapter; the reader is referred to more thorough discussions of these topics (44–46). A few points about the pathophysiological basis of treatment of gas bubble disease, however, are relevant to this chapter.

The resolution of bubbles is related to their size and the partial pressure difference that exists between the gas cavity and respiring tissue. This partial pressure difference is due to the inherent unsaturation of venous blood, which in turn is a result of the difference in solubility of O_2 and CO_2 in tissue. CO_2 is 20 times more soluble in body tissues than is O_2. As O_2 is consumed, it is replaced by CO_2 from substrate oxidation at a ratio of about 0.8 moles of CO_2 for each mole of O_2 reduced. Thus oxygen entering a tissue at an arterial pO_2 of 100 torr leaves the venous capillary at a pO_2 of 40 torr. In contrast, CO_2 enters at 40 torr and leaves at only 46 torr. The remainder of the gas pressure in the tissues is primarily N_2, which at equilibrium at sea level has the same partial pressure, 573 torr, on both sides of the circulation. Therefore the sum of the partial pressures in the arterial system is 54 torr less than in the venous system. This "oxygen window" provides a potential pressure gradient for elimination of N_2. As a gas pocket or bubble collapses from removal of O_2 by metabolism, the internal pN_2 must rise above tissue pN_2 because the total pressure in the bubble remains in equilibrium with ambient pressure. In this way, the inert gas is resorbed gradually. The oxygen window can be expanded during decompression at the surface and after recompression by administration of high partial pressures of O_2. Oxygen decreases the tissue pN_2 and increases the partial pressure gradient for N_2 between the bubble and tissue. Surface oxygen administration at the scene of a diving accident before recompression therapy can be initiated is clearly beneficial in the management of DCS and AGE.

There are a variety of recompression tables available that are effective for DCS provided the treatment is begun promptly. Treatment tables that list minimal recompression and hyperbaric oxygen have become the treatment of choice in most situations. U.S. Navy Treatment Table 6, which employs intermittent hyperbaric oxygen at a maximum depth of 2.8 ATA (60 fsw), has become a therapeutic standard for primary DCS as well as for recurrent symptoms. Also, Table 6 or its equivalent is probably adequate initial therapy for AGE, although deeper recompression to 165 fsw (e.g. Table 6A) may be appropriate for very recent occurrence of AGE or when the quantity of intravascular gas is very large and maximum reduction of intravascular bubble size is desired. Deep recompression on air can be associated with additional N_2 uptake by the body, thus most specialists recommend recompression with a 50% N_2–50% O_2 gas mixture for Table 6A and other deep treatment tables.

In some instances of DCS, oxygen recompression is not available and air recompression tables must be used. Air treatment tables are longer and less effective than oxygen tables, and they are more likely to produce DCS in the attendants inside the chamber. In compressed air workers, recompression with air is usually carried out to the minimum depth of relief. The use of hyperbaric oxygen for DCS in compressed air workers has not been adopted industry wide because of the increased risk of fire. Hyperbaric oxygen therapy, if administered with adequate fire safety precautions, can be used safely and effectively for relief of DCS occurring at compressed air construction sites (7).

6 FITNESS TO DIVE

6.1 Individual Variation and Acclimatization

The question of fitness to dive represents a complex issue in industrial and occupational health. Few workplaces expose the worker to such an inhospitable environment and such a wide range of physiological stresses. As a result, there is no universally accepted consensus of physical qualifications to dive. Standards for diving fitness undoubtedly should be tailored to each diving situation and to the required level of performance. It is also quite clear that work in hyperbaric environments is associated with a higher risk of injury, for example, DCS, for individuals with certain physical attributes. There are also many preexisting medical conditions that constitute contraindications to exposure to alterations in barometric pressure.

There is considerable individual variation in susceptibility to pressure-related diseases such as aseptic necrosis of bone and DCS (47). For DCS, several factors have been connected routinely to enhanced susceptibility including age, weight, body fat, and lack of acclimatization. Acclimatization or habituation is well-described in compressed air workers who become less susceptible to DCS after daily exposure to the hyperbaric environment. Acclimatization requires approximately 14 days for maximum effect, whereas ceasing to work in compressed air leads to more rapid deacclimatization. The mechanism of such acclimatization, however, has proved quite elusive.

The roles of age, weight, and body fat in susceptibility to DCS also have been evaluated. During World War II, Gray's analysis (48) of thousands of decompressions in altitude chambers revealed that relative susceptibility to dysbarism based on mean age increased about 11 percent per year between ages 18 and 28. Similarly, mean susceptibility to DCS correlated highly with increase in body weight per unit of height over the same age range. The role of body fat as a risk factor for DCS has been well described. Inert gases such as N_2 are highly soluble in lipid, but the total perfusion of a fat diver is approximately the same as for a thin diver. Hence large amounts of inert gas may be stored in body fat during long dives. The inert gas in fatty tissues is eliminated more slowly during decompression, and the risk of critical supersaturation is increased during decompression. This rationale also has been used to suggest that women should be more susceptible to DCS than men because they average about 10 percent more body fat per unit of mass. A male–female difference in susceptibility to DCS has not been proven to be the case, although occasional reports have suggested that dysbarism might be more frequent and more serious in women than in men.

There are a number of other well-known risk factors for DCS. These include poor physical conditioning, heavy exercise before and during the dive, and diving in cold water. The effects of exercise during decompression, however, may lessen the risk of DCS by enhancing inert gas elimination. Recently Ward has proposed that individuals whose blood is sensitive to complement activation by gas bubbles are more susceptible to DCS (49).

6.2 Physical Evaluation of Divers

The physician who examines the diver for fitness to dive must have a thorough familiarity with diving physiology and an understanding of the particular environment and type of work that the individual will perform. There are special requirements and waivers that pertain to military and commercial divers. For the recreational diver, it is important to remember that the diving activity is a sport; hence any medical condition that could predispose to DCS, AGE, or other injury should be disqualifying. Any diving candidate with chronic disease, significant physical disability, or requirement for long-term medication should not be cleared physically for diving until he or she is evaluated by a physician trained in diving medicine. The need for ancillary laboratory studies in routine screening of asymptomatic candidates is arguable, although I suggest a base-line chest radiograph and audiogram. A set of recommended medical standards for unrestricted sports diving was published by the Undersea and Hyperbaric Medical Society (UHMS) in 1987 (50). A summary of the UHMS consensus contraindications to diving is provided in Table 12.8. This table is by no means a comprehensive list of medical contraindications and specific questions should be referred to a specialist in diving medicine.

Finally, important questions arise about returning to diving after a decompres-

Table 12.8. Medical Contraindications to Diving[a]

I. Eye
- Recent ocular surgery
- Uncorrected loss of visual acuity

II. Ear, nose, and throat
- External ear obstruction, e.g., exostoses
- Chronic middle ear disease, e.g., otitis media, otorrhea
- Tympanic membrane perforation or tympanoplasty
- Middle ear or mastoid surgery, e.g., stapedioectomy, mastoidectomy
- Inner ear disease, e.g., deafness, tinnitus or vertigo
- Nasal or sinus obstruction, e.g., infection or nasal polyps

III. Pulmonary
- Upper airway obstruction
- Asthma
- Chronic obstructive pulmonary disease
- Pulmonary restriction
- Spontaneous pneumothorax
- Thoracic surgery involving the lung or pleura

IV. Cardiovascular
- Congestive heart failure
- Post myocardial infarction or coronary artery bypass graft with persistent left ventricular dysfunction
- Angina pectoris

V. Musculoskeletal diseases
- Metabolic or traumatic bone or joint diseases with compromised circulation
- Infectious diseases of the bone

VI. Miscellaneous medical disorders
- Chronic renal failure
- Diabetes mellitus with insulin or oral agent dependence
- Hemoglobinopathies, e.g., HbS disease
- Thrombocytopenia

VII. Nervous system
- Epilepsy
- Severe migraine headaches
- Serious head injury with a history of prolonged unconsciousness
- Stroke or other cerebrovascular disease
- Craniotomy
- Spinal cord diseases
- Spinal surgery
- Type II DCS with residual neurological abnormalities

VIII. Psychiatric disorders
- Alcohol or other drug dependence
- Schizophrenia, severe depression, and personality disorders
- Need for psychotropic medication

[a]Adopted from Reference 50.

sion accident. After type I DCS, a symptom-free interval of 2–6 weeks is generally recommended; after type II DCS, this interval should be longer. There is also justification for recommending permanent disqualification from diving after DCS involing the spinal cord.

REFERENCES

1. M. Epstein, A. De Nunzio and M. Ramachandran, "Characterization of the Renal Response to Prolonged Immersion in Normal Man. Implication for an Understanding of the Circulatory Adaptation to Manned Space Flight, *J. Appl. Physiol: Resp. Environ. Exercise Physiol.*, **49**, 184–188 (1980).
2. M. Epstein, "Water Immersion and the Kidney: Implications for Volume Regulation," *Undersea Biomed. Res.*, **11**, 113–121 (1984).
3. Y. C. Lin, "Circulatory Functions during Immersion and Breath-hold Dives in Humans," *Undersea Biomed. Res.*, **11**, 123–138 (1984).
4. J. C. Mithoefer, "Breath Holding," in *Handbook of Physiology, Respiration*, Section 3, Vol. II, W. Fenn and H. Rahn, Eds., American Physiological Society, Washington, DC, 1965, pp. 1011–1025.
5. *U.S. Navy Diving Manual*, Vol. 1, Air Diving, Revision 1, U.S. Department of the Navy, Washington, DC, 1985 (NAVSEA 0994-LP-001-9010).
6. *U.S. Navy Diving Manual*, Vol. 2, Mixed Gas Diving, U.S. Department of the Navy, Washington, DC, 1985 (NAVSEA 0994-LP-001-9021).
7. E. P. Kindwall, "Caisson Decompression," in *The Physiological Basis of Decompression, Proceedings of the Thirty-eighth Undersea and Hyperbaric Medical Society Workshop*, R. D. Vann, ed., UHMS Publication #75, Undersea and Hyperbaric Medical Society, Bethesda, MD, 1989, pp. 375–395.
8. K. E. Schaefer, "Physiological Stresses Related to Hypercapnia during Patrols on Submarines," *Undersea Biomed. Res.*, **6**(*suppl.*), 515–547 (1979).
9. J. A. Kylstra, I. S. Longmuir, and M. Grace, "Dysbarism: Osmosis Caused by Dissolved Gas?" *Science*, **161**, 289–291 (1968).
10. J. C. Farmer, Jr. and W. G. Thomas, "Ear and Sinus Problems in Diving," in *Diving Medicine*, R. H. Strauss, ed., Grune and Stratton, New York, 1976, pp. 109–133.
11. O. I. Molvaer and S. Eidsvik, "Facial Baroparesis: A Review," *Undersea Biomed. Res.*, **14**, 277–295 (1987).
12. W. L. Hunter and P. B. Bennett, "The Causes, Mechanisms and Prevention of the High Pressure Nervous Syndrome," *Undersea Biomed. Res.*, **1**, 1–28 (1974).
13. R. A. Smith, B. A. Dodson, and K. W. Miller, "The Interactions between Pressure and Anaesthetics," *Phil. Trans. R. Soc. London B*, **304**, 69–84 (1984).
14. E. H. Lanphier and E. M. Camporesi, "Respiration and Exercise," in *The Physiology and Medicine of Diving*, 3rd ed., P. B. Bennett and D. H. Elliott, Eds., Best Publishing, San Pedro, CA, 1982, pp. 99–156.
15. H. D. Van Liew, "Mechanical and Physical Factors in Lung Function during Work in Dense Environments," *Undersea Biomed. Res.*, **10**, 225–226 (1983).
16. J. Milic-Emili and J. M. Tyler, "Relation between Work Output of Respiratory Muscles and End Tidal CO_2 Tension," *J. Appl. Physiol.*, **18**, 497–504 (1963).

17. N. B. Pride, S. Permutt, R. L. Riley, and B. Bromberger-Barnea," Determinants of Maximum Expiratory Flow from the Lungs," *J. Appl. Physiol.*, **23**, 646–662 (1967).
18. J. V. Salzano, B. W. Stolp, R. E. Moon, and E. M. Camporesi, "Exercise at 47 and 66 ATA," in *Proceedings of the Seventh Symposium on Underwater Physiology*, A. J. Bachrach and M. M. Matzen, Eds., Undersea Medical Society, Bethesda, MD, 1981, pp. 181–196.
19. P. Bert, *Barometric Pressure. Researches in Experimental Physiology* (English translation by Hitchcock and Hitchcock), Long College Book Co., Columbus, OH, 1943.
20. J. M. Clark, and C. J. Lambertsen, "Pulmonary Oxygen Toxicity: A Review," *Pharmacol. Rev.*, **23**, 37–133 (1971).
21. J. M. Clark and C. J. Lambertsen, "Rate of Development of Pulmonary O_2 Toxicity in Man during O_2 Breathing at 2.0 ATA," *J. Appl. Physiol.*, **30**, 739–752 (1971).
22. A. L. Harabin, L. D. Homer, P. K. Weathersby, and E. T. Flynn, "An Analysis of Decrements in Vital Capacity as an Index of Pulmonary Oxygen Toxicity," *J. Appl. Physiol.*, **63**, 1130–1135 (1987).
23. K. W. Donald, "Oxygen Poisoning in Man," *Br. Med. J.*, **1**, 712–717 (1947).
24. B. A. Freeman and J. D. Crapo, "Free Radicals and Tissue Injury," *Lab. Invest.*, **47**, 412–426 (1982).
25. A. Mayevsky, "Brain Oxygen Toxicity," in *Proceedings of the Eighth Symposium on Underwater Physiology*, A. J. Bachrach and M. M. Matzen, Eds., Undersea Medical Society, Bethesda, MD, 1984, pp. 69–89.
26. J. D. Crapo, "Morphologic Changes in Pulmonary Oxygen Toxicity," *Ann. Rev. Physiol.*, **48**, 721–731 (1986).
27. P. Webb, "Thermal Problems," in *The Physiology and Medicine of Diving*, 3rd ed., P. B. Bennett and D. H. Elliott, Eds., Best Publishing, San Pedro, CA, 1982, pp. 297–318.
28. L. W. Raymond, E. D. Thalmann, G. Lindgren et al., "Thermal Homeostasis of Resting Man in Helium–Oxygen at 1–50 ATA," *Undersea Biomed. Res.*, **2**, 51–68 (1975).
29. S. K. Hong, R. M. Smith, P. Webb, and M. Matsuda, "Hana Hai II: A 17-day Dry Saturation Dive at 18.6 ATA," *Undersea Biomed. Res.*, **4**, 211–220 (1975).
30. L. A. Kuehn, "Thermal Effects of the Hyperbaric Environment," in *Proceedings of the Eighth Symposium on Underwater Physiology*, A. J. Bachrach and M. M. Matzen, Eds., Undersea Medical Society, Bethesda, MD, 1984, pp. 413–439.
31. A. E. Boycott, G. C. C. Damant, and J. S. Haldane, "The Prevention of Compressed Air Illness," *J. Hyg. Camb.*, **8**, 342–443 (1908).
32. R. D. Vann, "Decompression Theory and Application," in *The Physiology and Medicine of Diving*, 3rd ed., P. B. Bennett and D. H. Elliott, Eds., Best Publishing, San Pedro, CA, 1982, pp. 352–382.
33. B. G. D'Aoust, "Inert Gas Exchange and Counterdiffusion in Decompression Sickness and Diving Medicine," in *Proceedings of the Eighth Symposium on Underwater Physiology*, A. J. Bachrach and M. M. Matzen, Eds., Undersea Medical Society, Bethesda, MD, 1984, pp. 159–170.
34. P. K. Weathersby, L. D. Homer, and E. T. Flynn, "On the Likelihood of Decompression Sickness," *J. Appl. Physiol.*, **57**, 815–825 (1984).
35. J. C. Farmer, "Inner Ear Decompression Sickness," in *The Physician's Guide to Diving*

Medicine, C. W. Shilling, C. B. Carlston, and R. A. Mathias, Eds., Plenum Press, New York, 1984, pp. 312–316.

36. T. H. Lam and K. P. Yau, "Manifestations and Treatment of 793 Cases of Decompression Sickness in a Compressed Air Tunneling Project in Hong Kong," *Undersea Biomed. Res.*, **15**, 377–388 (1988).
37. Divers Alert Network, Report on 1987 Diving Accidents. Duke University Medical Center, Durham, NC.
38. J. M. Hallenbeck, A. A. Bove, and D. H. Elliott, "Mechanism Underlying Spinal Cord Damage in Decompression Sickness," *Neurology*, **25**, 308–316 (1975).
39. R. E. Moon, E. M. Camporesi, and J. A. Kisslo, "Patent Foramen Ovale and Decompression Sickness in Divers," *Lancet*, **1**, 513–514 (1989).
40. G. O. Dahlback and C. E. G. Lundgren, "Pulmonary Air-trapping Induced by Immersion," *Aerospace Med.*, **43**, 768–774 (1972).
41. D. N. Walder, Osteonecrosis, in *The Physician's Guide to Diving Medicine*, C. W. Shilling, C. B. Carlston, and R. A. Mathias, Eds., Plenum Press, New York, 1984, pp. 397–405.
42. R. I. McCollum, D. N. Walder, R. Barnes, et. al., "Bone Lesions in Compressed Air Workers with Special Reference to Men Who Worked on the Clyde Tunnels 1958 to 1963," *J. Bone Joint Surg.* (*London*), **48B**, 207–235 (1966).
43. Committee Report, "Hyperbaric Oxygen Therapy," UMS Publication #30, Undersea and Hyperbaric Medical Society, Bethesda, MD, 1986.
44. D. R. Leitch and J. M. Hallenback, "Neurological Forms of Decompression Sickness," in *The Physician's Guide to Diving Medicine*, C. W. Shilling, C. B. Carlston, and R. A. Mathias, Eds., Plenum Press, New York, 1984, pp. 316–327.
45. R. R. Pearson, "Diagnosis and Treatment of Gas Embolism," in *The Physician's Guide to Diving Medicine*, C. W. Shilling, C. B. Carlston, and R. A. Mathias, Eds., Plenum Press, New York, 1984, pp. 333–367.
46. J. C. Davis and D. H. Elliott, "Treatment of Decompression Disorders," in *The Physiology and Medicine of Diving*, 3rd ed., P. B. Bennett and D. H. Elliott, Eds., Best Publishing, San Pedro, CA, 1982, pp. 473–487.
47. M. L. Dembert, "Individual Factors Affecting Decompression Sickness," in *The Physiological Basis of Decompression. Proceedings of the Thirty-eighth Undersea and Hyperbaric Medical Society Workshop*, R. D. Vann, Ed., Publication #75, Undersea and Hyperbaric Medical Society, Bethesda, MD, 1989, pp. 355–367.
48. J. S. Gray, "Constitutional Factors Affecting Susceptibility to Decompression Sickness," in *Decompression Sickness*, J. Fulton, Ed., W. B. Saunders, Philadelphia, 1951, pp. 182–191.
49. C. A. Ward, P. K. Weathersby, D. McCullough, and W. D. Fraser, "Identification of Individuals Susceptible to Decompression Sickness," in *The Ninth International Symposium on Underwater and Hyperbaric Physiology*, A. A. Bove, A. J. Bachrach, and L. J. Greenbaum, Jr., Eds., Undersea and Hyperbaric Medical Society, Bethesda, MD, 1987, pp. 239–247.
50. P. G. Linaweaver and J. Vorosmarti, Jr., "Fitness to Dive," in *Proceedings of the Thirty-fourth Undersea and Hyperbaric Medical Society Workshop*, UHMS Publication #70, Undersea and Hyperbaric Medical Society, Bethesda, MD, 1987.

CHAPTER THIRTEEN

Occupational Health Concerns in the Health Care Field

Howard J. Sawyer, M.D.

1 HISTORICAL PERSPECTIVE

Since the preceding edition of *Industrial Hygiene and Toxicology*, the health care field (HCF) has become the target of a specific Occupational Safety and Health Administration (OSHA) rule intended to deal with the growing threat of hepatitis, AIDS (acquired immune deficiency syndrome), and other blood-borne infectious diseases among potentially exposed employees (1). The rule and the inclusion of this new chapter respond to a heightened awareness by employees, occupational health professionals, health care administrators, and government (2) of the unusual occupational health challenges in this complex workplace. Reminiscent of the 1970 Occupational Safety and Health Act in which *recommended* standards became enforceable, the CDC (Centers for Disease Control) *guidelines* for health care institutions relative to blood-borne infectious agents are the intended standard to be relied upon in complying with the new requirements (3, 4).

Although the advent of OSHA in 1970 was bemoaned by industry, it became the stimulus not only for compliance with minimum standards but sometimes for even more stringent voluntary practices. Similarly, there were those in the HCF and in OSHA who opposed OSHA's new infectious agent responsibility in 1988, recognizing that OSHA had neither the knowledge nor the experience to police the new requirements effectively. Nevertheless, this new emphasis will stimulate the provision of minimum or better HCF employee health programs, especially by

Patty's Industrial Hygiene and Toxicology, Fourth Edition, Volume 1, Part A, Edited by George D. Clayton and Florence E. Clayton
ISBN 0-471-50197-2

smaller providers of health care, which might otherwise escape industry internal auditing activities, such as those performed by the Joint Commission on Accreditation of Healthcare Organizations (JCAHO). Moreover, the new OSHA attention will likely enhance compliance relative to other exposures peculiar to this industry.

Since its inception, OSHA has always had authority in hospitals over such general industry considerations as electrical safety, noise, and repair shop conditions. Although it has never been involved in infection control practices and is only marginally concerned with such exposures as radiation, lasers, and chemotherapeutics, compliance officers are already being briefed on the relevant CDC guidelines.

Prior to the new OSHA rule, there had been an assumption that each kind of exposure to health care workers had some regulatory agency responsible for health and safety issues, as is the case with radiological hazards, but this is not so. HCF emphasis has primarily been on the health and safety of patients, often ignoring equally relevant concerns of its employees. New chemicals that may have been carefully studied for short-term patient therapeutic contact may have no regulated exposure limits for those who regularly work with these agents. For example, most chronic exposures of hospital pharmacy employees who compound drugs are not specifically regulated by OSHA.

There are many other exposures common to this industry, such as heat and biomechanical stresses, that remain without a specific regulatory standard. This deficiency emphasizes the responsibility of HCF management reasonably to assure, *voluntarily*, an environment free of recognized hazards at levels known or likely to cause significant harm to employees or those who may be secondarily affected, such as family members. HCF management must be as committed to employee health as to the health and safety of patients, a new emphasis for some.

2 DESCRIPTION OF THE HEALTH CARE FIELD

Many call the HCF an industry, an offensive term to those who care for the sick and wounded, but there is a sense in which the term is appropriate. The traditional view of hospitals as almost autonomous charitable institutions has given way to other priorities, along with the nearly sacrosanct positions formerly held by physicians. The public's medical information base and expectations have increased, heightening their anticipation of superior medical results. That, along with cost containment mandates and litigation experience, has contributed to a rapid new wave of change in health care "delivery." A significant part of this change is an increased concern by health care workers about their potentially hazardous exposures and the rapid technological changes that characterize their work and that they often understand poorly. Attempts at educating workers about workplace hazards often increase anxiety because of inappropriate communication efforts that may be too detailed or not placed into the proper perspective.

The HCF is vast and intricate, with exposures as complex as in any other. It employs somewhere between 9 and 11 million people and includes employees in the major health care networks as well as paramedics at an accident scene. Most are employed outside of hospitals (5) in such areas as nursing homes, hospices, physicians' offices, and home health care. Whereas four out of five of such employees are involved in direct patient care, less than half the employees in major institutions actually care for patients (5). Large hospitals employ their own extensive support services, which are usually purchased from outside by smaller clinics and other providers. Large institutions have staffs to support quality assurance, legal, personnel, benefits, laundry, dietary, maintenance, parking services, and other departments involving many different trades and professions.

Each of these employee groups has its own peculiar workplace hazards. The incinerator operator might encounter noise, heat, offensive odors, biomechanical stresses, or a carelessly discarded, contaminated needle. A dietary worker may experience temperature and humidity extremes and repetitive stresses in the production-line rush to feed hundreds of people. The exposures are often numerous and complex.

Although the treatment of the subject in this chapter is primarily from the viewpoint of a large institution, the principles used to manage the various exposures can be applied as appropriate to clinics, dental and medical offices, and other work sites.

3 THE ROLE OF PSYCHOSOCIAL FACTORS

In order to understand the new regulatory emphasis on HCF occupational health concerns, physical, biomechanical, biological, chemical, and psychosocial factors must all be appreciated. Too often, industrial hygiene projects focus on scientific, analytical assessment to the exclusion of equally important human factors.

Among the most difficult challenges facing all occupational and environmental health professionals are unrealistic employee and public concerns about hazardous exposures. These concerns may be heightened by careless, unsubstantiated reporting in communications media when neither time nor expertise is applied to evaluate the information reported. The resultant anxiety, coupled with rapid changes in technology, contribute to the stress reactions cited by futurist Alvin Toffler as the space/information/electronic wave of change. This change is occurring so rapidly that there is often insufficient time for people to adapt (6). Failure to recognize this phenomenon is sure to delay the speedy resolution of problems and increase costly environmental analytical procedures, often without clearly addressing the real concerns of those potentially harmed.

Those not schooled in this expanding technology (older or poorly educated workers and other citizens) experience the greatest anxiety because of their lack of understanding of, and loss of control over, their environments. What they do not understand and cannot control, they tend to distrust and oppose. Because the HCF exhibits some of the most dramatic examples of ultratechnology, it should be no surprise that in 1986 its workers successfully petitioned OSHA for tighter regulatory controls over their workplace. Although their petition was about blood-borne diseases, chiefly AIDS and hepatitis, their concerns are far more inclusive.

If we are to address such problems as tight building syndrome, chemophobia, and fear of new technologies properly, we must understand the sources of these concerns. If we do, we will increase our efforts at directly communicating with workers about their true exposure risks while dispelling baseless fears.

4 SPECIAL FACTORS IN EARLY DETECTION OF OCCUPATIONAL HEALTH PROBLEMS

People are drawn to this industry for many different reasons, but there are two that tend to impede early identification of occupational health problems.

- The first is a strong work ethic that arises out of the drive to serve those who suffer, even at the sacrifice of personal health considerations. A corollary to this is an intolerance of peers who become incapacitated. This can cause those with early symptoms to endure them beyond a reasonable point. One report cites a *lower* incidence of back complaints among nurses as compared with other groups (7), even though one would expect an incidence rate of back problems at least as high or higher in this population which does so much moving of heavy patients. Work ethic could be the reason.
- The industry also attracts many low-wage support staff who often cannot find employment in higher-paying industrial jobs. Some of these employees have no benefits, working as "temporary" help, so that any work-related medical problem is perceived by them as potentially threatening to employment. Even if they receive workers' compensation benefits for a prolonged period, it is common to remove such employees from the payroll after a year, with resultant loss of most fringe benefits. Moreover, workers' compensation benefits generally do not follow the cost-of-living index. All these factors combine as disincentives for the early reporting by this group of adverse workplace conditions. Among this population are some very hard-working people with no financial buffer who are only marginally out of poverty. They will endure much before risking loss of a job for medical reasons. These impediments to early reporting are strong, especially for those who wish to avoid the indignity of a welfare existence.

Although there are many who are not in these two groups, they focus attention on the need for periodic medical evaluations as a means of early detection, rather than depending on employee complaints as the index of workplace-related health problems.

5 EVALUATING THE EMPLOYEE

5.1 Cost Effectiveness

It is unthinkable that a company would buy a machine with a $100,000 price tag without carefully specifying what performance is expected and checking its quality, reliability, suitability, and durability. Yet in this labor-intensive industry, where employees potentially represent a greater investment, there are hospitals, clinics, nursing homes, and other employers where many, if not all, of these precautions are ignored, relative to their work force. Conversely, there are institutions where preemployment health assessments are far too extensive and costly without advancing the legitimate needs of the business. A large, diverse, multi-component health care corporation is also potentially vulnerable to prejudicial discrimination charges if preemployment health assessment requirements and the resulting decisions are not consistent, from component to component, for identical work and workplace exposures.

The content of the preemployment health assessment should be determined by the legitimate needs of the employer. Anything else that is included may be otherwise desirable, but cannot be the basis for determining fitness to work. Expected workplace exposures, along with the likelihood of finding a relevant abnormality at a given age, combine to determine the appropriateness of any component of the evaluation.

5.2 Goal of and Justification for the Preemployment Health Evaluation

At the core of industrial hygiene is the avoidance of adverse interactions between people and their work environment. This mandates a careful evaluation not only of the environment but of those who experience its exposures. The goal of the preemployment, as well as periodic and other preplacement evaluations, is to assure reasonably the placement of employees into positions in which they can work effectively without increased risk of harm to themselves or to others. The objectives required to achieve this goal are:

- To determine the general ability to work, keeping in mind the potential for changes in job assignment.
- To assess the potential for the applicant to endure special environments safely. For example, the asthmatic should be excluded from the animal dander exposures of a research laboratory. The applicant who reports Raynaud's disease-like symptoms should be excluded from working in cold environments such as coolers or freezers.
- To detect abnormalities not presently limiting but with potential to cause future impairment, such as a history of alcoholism or recurrent back symptoms. These would likely require some verification of prior and/or continuing corrective action.
- To establish base lines against which future changes can be measured.

- To establish the signed medical history statement as a document upon which discharge can be based if the applicant willfully has materially falsified statements. Routine *patient* medical history questionnaires do not generally meet this requirement without modification.

If these criteria are strictly applied to the preemployment health assessment, not everyone will get the same evaluation. This creates no "fairness" problem, however, as long as assessment criteria are strictly and consistently followed. There are different preemployment medical evaluation needs for a teenage finance clerk who works in a separate building with minimal exposure risks and a 45-year-old nurse assistant working in a critical care unit where there are lifting, straining, and chemical and psychological stresses. The former is not likely to have an occupationally significant medical impairment that would escape detection by an experienced occupational health nurse. In this case, the nurse would review the medical history statement with the applicant and a full physician-conducted examination would be done only if indicated by that assessment, such as finding marked obesity or hypertension.

5.3 Functional Job Descriptions

Each job in a health care setting should have a hazardous exposure evaluation to include all categories of risk (physical, biomechanical, biological, chemical, and psychosocial) so that the correct mix of evaluation components can be selected. To do this by manual methods alone is confusing and unrealistic. Without computer tracking of job titles, as well as appropriate task/exposure suffixes, the apparently easiest and safest practice often followed is to do the same *basic* evaluation on everyone. This assumes that everyone potentially has the same exposures, resulting in a practice that is costly not only in terms of the actual expense of tests and procedures, but of wasted time, facility, overhead, and salary expenses.

5.4 Recommended Components of a Preemployment Health Evaluation

Everyone hired to work in the HCF should be medically evaluated. The specific components of that evaluation are dictated by three major considerations:

- Infection control requirements
- Potential risks of the work to be performed
- Age, as it relates to the probability of finding occupationally significant impairments

5.4.1 Infection Control Considerations

In the health care field infection control (IC) is of such central concern that each job where communicable diseases are a major consideration *should be viewed independently of whether a full-scale evaluation is otherwise indicated.* Applicable

CDC categories and/or hospital directives relative to infection control immunization practices should be the sole determinant of immunization practices. Whether a full preemployment assessment is otherwise indicated (because of age or exposures to such hazards as heavy lifting, noise, or chemicals) is an issue that will be discussed separately.

There are three employee potential exposure categories, relative to infectious agent and patient contact (8, 9):

1. Those where significant contact is integral to the job (OSHA Cat. I)
2. Those where such contact is only occasionally likely (OSHA Cat. II)
3. Those where contact will not forseeably occur in the performance of the job (OSHA Cat. III)

There appear to be no good reasons to consider employees in the first two OSHA categories differently, so only two categories are considered here:

Category I (OSHA Cat. I and II): those who predictably have infectious agent or patient contact whether routinely or only occasionally

Category II (OSHA Cat. III): those with no foreseeable infectious agent or patient contact as a part of their job

Sometimes it is not clear, by job title alone, which IC category is appropriate for a given individual. This is because persons with identical job *titles* may perform different tasks. Therefore, because IC preventive procedures may be scheduled every 6 or 12 months, this often-overlooked opportunity should be taken to upgrade demographic and medical history information. Also, any change in patient, infectious agent, or other environmental exposures can be recorded and appropriate procedures scheduled. Such information can be obtained by a simple checklist questionnaire that may be retained or discarded after information is electronically or manually recorded in the medical record.

Many years of industrial hygiene experience clearly document the inadequacy of job numbers or titles as reliable sources of exposure information. There could be a job transfer, change in the same job in a given department, or simply a different expectation by a new supervisor about, for example, whether a clerk or secretary will now be expected to have patient or infectious agent exposure not formerly required.

There has also been discussion about whether to immunize *all* employees against those diseases spread by droplet dissemination, regardless of their potential patient contact. The rationale is that those in category II could be in the prodromal (infectious but not necessarily ill) phase of such a disease and infect an employee in category I, who can then infect patients or other employees. The risk would be greatest during the early stage of an epidemic, perhaps not yet recognized as such. However, immunization would not block transmission from the general public.

Although the idea is being challenged by some (8), OSHA will likely continue to use those CDC guidelines that relate to infection control as the basis for potential enforcement action. Hospitals also have their own IC committees, a mandate of the JCAHO, which may have even more restrictive practices, based on local conditions such as endemic organisms, anticipated or actual local epidemics, and the like. Such internal hospital standards may also be enforceable under the OSHA "general duty clause."

5.4.2 Criteria for the General Preemployment Medical Assessment, Based on Factors Other Than Infectious Agent Exposures

Whereas each employee, irrespective of other hazardous exposures, should be evaluated relative to what IC procedures may be required, the extent of that employee's main preemployment evaluation should be based on age and/or activities or hazardous exposures other than to infectious agents. These categories are:

A Age less than 35 and *no* noninfectious exposures (no heavy lifting, no noise or antineoplastic agent exposures, for example) for which a physician's professional assessment is needed to detect a significantly disqualifying condition. Examples include receptionist, typist, accountant, and attorney. Such employees younger than 35 years of age need not *routinely* see the physician but should have a *core* occupational health evaluation before hire. This is in addition to, and unrelated to, any applicable IC practices.

Employees in this category should complete a detailed medical/exposure history form with a signed statement of awareness that willful, material falsification will be grounds for dismissal. The occupational health nurse, or other paramedical so designated by the physician, reviews this information with the applicant. Height, weight, temperature, pulse, respiration, blood pressure, and urine "dipstick" for protein and glucose should be obtained. A microscopic urine check should be done if the specimen gross appearance or dipstick protein warrants it.

Unless this core assessment determines that a physical examination is indicated, the individual is then approved for the assigned work by the occupational health nurse or other paramedical designee of the responsible physician. Those with such evident problems as marked obesity, diabetes mellitus, or an orthopedic limitation would have physician review and a full or limited physical examination as indicated. Only the physician should establish a limited medical work classification.

Anyone already working in Category A who is to enter a new job that requires a Category C assessment would need to be specifically approved by the employee health service physician.

B Persons otherwise in category A but age 35 or older should be screened the same as those in Category A but would also have physician review, a baseline electrocardiogram, and a full physical examination by the health service physician.

	Core Eval.	Physician Review and Exam	Base-line ECG	Exposure Specific Tests	Infection Control Screening
A (no exposure, age < 35)	Yes	Only if OHN screen indicates need	No	Not appl.	IC Cat. I only
B (no exposures, age 35 +)	Yes	Yes	Yes	Not appl.	IC Cat. I only
C (exposures, all ages)	Yes	Yes	Age 35+	Yes—Exposure dictates specific tests	IC Cat. I only

Figure 13.1. Preemployment health assessment in the health care industry.

C An applicant of *any* age with specific *noninfectious* exposures that require a physician's evaluation would also have specific tests based on individually required exposures such as noise, which will require an audiogram, or antineoplastics, which require a complete blood count. If the applicant is aged 35 or older, a resting electrocardiogram should be done, as a base line at least.

These recommendations are summarized in Figure 13.1.

6 WHO HIRES?

Once the preemployment evaluation is completed the information about employability, from an occupational health viewpoint, must be communicated. It is at this point that issues often get cloudy. There are those in the HCF who argue the futility of preemployment health evaluations because "you can't turn anyone down anyway." This is simply untrue. But personnel department decisions to reject an applicant must be based only on medical findings that are clearly relevant to the exposures likely to be encountered.

The *physician* should not reject anyone from employment. The doctor's task is to evaluate job applicants medically to establish a functional *medical work classification* (MWC). The personnel manager considers that classification, along with such other factors as job qualifications, urgency to fill the position, equal opportunity considerations, background checks, and the like, in arriving at a determination.

The evaluating physician can of course postpone a decision pending further medical information or correction of a problem. If there are "automatic" disqualifiers, such as a positive urine drug screen, the results are still reported along with the MWC to Personnel, where the final determination and notification are made. A physician who rejects someone from employment on medical grounds alone assumes a variety of legal and other risks that need not be taken. The physician is truly a consultant to the personnel manager, who must consider medical along with other information in determining whom to hire.

6.1 Avoiding Prejudicial Discrimination

The physician may establish a functional work restriction, described in the MWC, based entirely on the objective health assessment. The MWC is then communicated to Personnel. The MWC may then result in removal of a previous conditional job offer, based on *germane* work restrictions.

In light of this, it may occasionally be appropriate for some job titles to be reassessed, recognizing a broader work responsibility that the job title suggests. For example, if the title is "Pediatric Nurse Assistant," a given impairment might not be considered disqualifying. However, if the title is the broader "Nurse Assistant," the impairment might be disabling if it is understood that a nurse assistant on a pediatric floor may also occasionally have to work on relief in areas where incapacitated adult patients need to be moved about.

Offering another perspective on the notion of having to accept "everyone," consider the hypothetical applicant whose weight is twice normal. Such an individual may be unable to fit physically into areas where work is to be performed without work space modification, might lack the agility to work effectively and/or safely, or could lack the physical stamina to support the 100 percent excess weight load in addition to the physical stresses required of the job. In such a case, germane work restrictions should be written into the MWC. Technically, the individual would not then be *medically* rejected, although the personnel manager would likely advise that neither the job applied for nor other posted jobs meet the stated MWC.

Although the result may be functionally equivalent to a medical rejection, it is far different; it is much more legally defensible and more considerate than for the physician to advise the applicant, in effect, "You're too fat and out of condition to work here." In this way, the individual is not prejudically discriminated against because of physical appearance or personal bias. The ultimate personnel department rejection is based on real *functional* impairments that are verifiably related to the job for which application has been made.

7 FACTORS IN EVALUATING HAZARDOUS EXPOSURES

7.1 Keeping Alert for Previously Unrecognized Problems

There is nothing more potentially dangerous than complacency about environmental health hazards. There is no shortage of examples in industrial hygiene history of apparently inert substances becoming major health concerns, such as asbestos and silica. New technology and equipment or new uses for existing equipment continually present the potential for new adverse human responses.

With escalating technology, medical equipment may be so expensive that few units are in use, resulting in minimal operator exposure experience. Those regularly exposed occupationally to new ultratechnological diagnostic or treatment modalities should be made aware that any unusual symptoms should be medically assessed, even though there is nothing published to implicate etiologically the new technology. For example, the potential long-term employee health effects of high-energy electric or magnetic fields, not presently known to present a problem, must be respected. In the early days of radiology it was not generally appreciated that the peculiar property of X-rays to penetrate tissue and then still expose photographic plates can also ionize that tissue, causing delayed malignant changes.

For reasons mentioned earlier, we know that new technology increases the opportunity for psychogenic illness. Significant new symptoms may therefore be masked by an anxiety state. Although it may seem insultingly simplistic, it should be noted that every valid, new, but unpublished finding was first reported to someone whose literature search was void. What is reported in "the literature" is history! Those responsible for employee health must remain open-minded, on hearing of newly reported symptoms, to the possibility that such exposures could result in impairment in ways not suspected or yet recognized.

7.2 Health and Safety Committee

To assess such matters in an integrated way, each establishment should have one or more Health and Safety Committees. In the past, such committees were more patient/visitor oriented, but they now have responsibility for all aspects of environmental health and the safety of all patient and work areas.

Levels of contact that are safe for short-term patient treatment may not be protective for the chronic exposures of those who administer care. The wearing of protective equipment by a radiology technician who delivers the full dose to the patient is a classic example of an awareness of this concern, which was not recognized during the first years of medical radiology. The aerosolized administration of pentamidine to AIDS patients, an antibiotic normally administered by conventional routes, caused a flurry of concern by therapists doing the administration at our hospital during the field trials of this new administration vehicle. Though there proved to be no health threat to employees, their fears indicate a heightened expectation by HCF employees relative to the safety of new methods, procedures, and technologies (10).

Every major department of the establishment should be represented on the committee, minimally including Environmental safety, Facilities Engineering, Quality Assurance, the Medical Staff, Nursing, Clinical Pathology, Research and Development, Security, Materials Management, Industrial Hygiene, Employee Health, Equipment Standards, and other affiliated committees. Additionally, probably at several levels, those who do the hands-on work such as maintenance personnel, technical support staff, and general services employees should participate actively in health and safety committees because they are often in the best position to observe developing or threatening problems.

7.3 Enhanced Proclivity to Environmental Reactions

There are many possible exposures in this industry that might affect employees because of individual sensitivity reactions. There is a tendency by management in any industry to conclude that if exposure levels are kept at legally "acceptable" concentrations there will be no adverse medical reactions. But persons with an allergic diathesis can respond to sensitizers such as molds, bacteria, and chemicals or to irritants such as sanitizing sprays, whereas others do not. Through planned educational programs, workers and management must understand why this is *not* so, and periodic health assessments must also take into account this possibility.

7.4 Procreative Considerations

Birth defects can result from exposure to teratogenic chemicals and from at least three prevalent viruses: varicella zoster virus (VZV), cytomegalovirus (CMV), and rubella virus. Unfortunately, teratogenic effects are possible before a definitive clinical diagnosis is made. Moreover, with teratogenic chemicals or other agents such as ionizing radiation, there may be no acute illness at all.

This has led to confusion with respect to standard-setting and personnel practices throughout industry. Although many companies have published policies to "protect the fetus," some of these are poorly thought out. They usually require that a woman report her pregnancy as soon as it is discovered, allegedly for the purpose of averting injury to her fetus. However, unless there is a significant likelihood in a work area of accidental exposure, the reporting of pregnancy *for the purpose of avoiding risk from routine exposures* has little merit. There is no change that can be made several weeks into pregnancy that will reverse whatever harm might have already come to the fetus during the prior period of greatest teratogenic vulnerability. Furthermore, consider the reaction of the employee who, after announcing her pregnancy, gets a change of job assignment *to avoid a specific environmental exposure.* She could easily believe that her baby might have already been harmed.

Dow Chemical Company takes an exemplary position by specifically *not* writing a separate policy (11). Their position is that their general statement about employee health and safety is already fully protective of the fetus because it considers procreation to be a natural part of the human condition. Thus male and female germinal tissues, as well as the fetus they may create, are included. The policy is therefore protective in every sense. Moreover, it eliminates the absurdity of after-the-fact reporting of pregnancy and the potential anxiety and legal problems resulting therefrom.

8 SPECIAL PROBLEMS WITH HOSPITALS AND OTHER CLOSED OR TIGHT BUILDINGS

The analytical aspects of "sick building syndrome" are covered in Chapter 17. This section considers how to evaluate those whose complaints suggest an etiology related to closed or tight buildings, such as hospitals.

The expression sick building syndrome overemphasizes the building environment and, by inference, deemphasizes the crucial role of those potentially reacting to the "sick" building. Hospitals are necessarily closed buildings, often tightly separated from the external environment, especially in urban areas. This results from valid infection control considerations as well as those related to proper air handling, filtration, and odor and temperature control.

Large building mechanical systems are seldom properly equipped to compensate fully for accidental spills, such as a formalin leak while sterilizing a dialysis unit, or when paint or adhesive vapors associated with renovations are temporarily excessive.

Moreover, the renovations themselves may have impaired otherwise well designed, installed, and maintained systems. Temperature or humidity may become inadequate and air movement deficient. This could permit excessive exposures to occur now from aerosolized hazards in processes that had been identically performed in the same area previously without offensive or dangerous buildups.

In considering the human side of the formula (those who react to the environment and why they respond as they do), the following approach applies not only to hospitals but to the evaluation of any indoor or outdoor condition where environmental exposures may be of concern to those affected. Failure to assess carefully the human, reactive component of the problem increases the risk of leaving those concerned with anger, frustration, and persistence or even aggravation of their symptoms.

8.1 Recognition of the Problem

The first step in any industrial hygiene activity is recognition of the problem. Here, as always, the reaction by those within an environment is "the problem." Whatever environmental abnormality may be documented, the human reaction is increased by the inability of those within the building to alter their circumstances directly. Additionally, there are individual variations among co-workers. "Some like it hot," and some like it cold. Failure to accommodate these differences reasonably promotes hostility among employees and toward management, which might be more concerned about a modern appearance than employee comfort.

Problems can be precipitated by insensitivity to legitimate needs, such as in the turning off of a building air-handling system for 20 min each hour to "save energy." In a tight hospital building, with windows that could otherwise be opened, variations from an optimal environment become even less tolerable because of the prohibition of opening the windows for some fresh air. The supervisor who is hypercritical, overbearing, or otherwise insensitive will certainly contribute to an adverse reaction to a troublesome physical or chemical environment.

External causes are often precipitators of closed-building environmental concerns. It is not uncommon for one outbreak to occasion domino-like reactions elsewhere. I was called to investigate three school "outbreaks," all within a 3-month period. In the first, there had been wide publicity about two teacher deaths suspected as arising from building exposures. Ill-informed "environmentalists" or offbeat medical practitioners may dote on weak individuals who are anxious to explain their psychosomatic symptoms by environmental exposures.

8.2 The Investigation

The environmental investigative principles and practices, detailed in Chapter 4, may precede, follow, or accompany the human assessment recommended below. The earlier those affected are medically assessed, the more likely is a successful outcome. The following steps must be taken:

1. A physician experienced in interviewing victims of environmental "occurrences" such as closed building syndrome should be engaged whenever and as soon as possible. Serious oversights can result when industrial hygienists or others not medically trained attempt to elicit details of the acute symptom complex, especially where complaints may be strongly associated with anxiety. When people are not feeling well, they *expect* to talk with a physician, preferably one experienced with environmental health concerns.
2. The physician should avoid interviewing more than one person at a time, in order to discourage symptom sharing, promote a caring physician–patient atmosphere of confidentiality, and authenticate a valid symptom list for later analysis.
3. The assessing physician should include a general medical history; precise dates, times, and durations of exposures relative to onset; and patterns of occurrence and resolution of symptoms.
4. Helpful clues may result from soliciting, at the time of the interview, each person's perception about the cause of the problem.
5. The reliability and emotional stability of the person being interviewed should be assessed, recognizing that a *few* symptoms are more likely the result of a significant environmental medical problem than are multiple, vague complaints.
6. Relevant human factors should be noted. References to attitudes of management such as unheeded prior complaints about workplace issues, or disputes between employees and management or other employees, may obscure or intensify the problem.
7. Special observations such as peculiar noises, odors, electrical arcing, and other such clues should be noted.

8. Representative (preferably demographically matched) individuals at potential risk who are unaffected should always be included.
9. Any appropriate medical/biological testing and/or physical examinations should be provided for all those affected or in a selected control group to explain individual symptoms and enhance the epidemiologic effort.
10. A symptom chart and area layout should be prepared showing where individuals are located, what symptoms predominated, the timing of appearances of new cases, and other appropriate factors, paying special attention to the possibility of earlier cases triggering the appearance of subsequent cases, as can happen with infectious disease or psychogenic outbreaks.
11. The outbreak should be studied with appropriate epidemiologic methods no matter how small the population. The services of a medical epidemiologist may be indicated.
12. The physician should also be part of the team that finally concludes the evaluation and recommends control measures that could require follow-up medical evaluations. Correcting factors that may legitimately need correction but whose correction does not satisfy the health complaints may be unnecessary and wasteful.

8.3 Preparation of the Report

The report is relatively easy to prepare when specific, *relevant*, correctable problems are identified. The report identifies them and the proper action to be taken is recommended. Where few or no evident problems are found to account for the complaints, other factors must be considered:

1. One must always appreciate that the cause for the complaints may have eluded detection. Exposures may intermittently enter the site from outside, providing unusual investigative problems. I was involved in one outbreak in which initial, "negative" results failed to detect the source. Careful medical interviewing of affected persons eventually revealed that an adjacent plant was periodically dumping used solvents into its sewer, permitting the vapors to enter the affected building.
2. Although a careful industrial hygiene assessment and medical review usually provide answers to resolve the problem, occasionally no really harmful health hazard is identified. In this case, reporting that the workplace is free of hazards is likely to be rejected by those affected. Moreover, their natural anxiety may be further heightened by the idea that "even the experts can't find the cause!" When only relatively minor problems are found, report language is critical. The "negative" report implies to those who are unaffected that the complaints are imagined or trumped-up for some non-health purpose. This may cause embarrassment and defensiveness by those affected, who may then organize efforts to disprove the "negative" assessment. Such reports rarely reassure those affected and may lead to allegations of management insensitivity.

3. Occasionally one or more individuals with significant allergic reactions may be found among a larger group that includes a few suggestible "secondary responders," whose symptoms are psychogenic. This can be especially troublesome if the allergic responders are credible individuals when it is not clearly understood by others that the reactions are based on *individual sensitivity* and not on general toxic effects.
4. Where the outbreak appears to be entirely inexplicable on the basis of anything other than psychogenic factors, report language is particularly critical. In such a case, the following wording has been effective: "No significant health hazards were presently found to explain the symptoms about which most complained. It appears likely, therefore, that whatever may have caused the problem is no longer of concern. Moreover, the medical assessment reveals no evidence of active disease."

8.4 Summary

All relevant factors, not just environmental exposure measurements, must be addressed before the successful resolution of an indoor or outside environmental outbreak can be *credibly* resolved. The intervention by a physician experienced in such outbreaks, or nurses or other paramedical personnel working under the physician's direction, is likely to identify more quickly the nature of the critically important human response, leading to the most speedy resolution of the outbreak. Psychosocial factors can be intricate and often more difficult to assess than objective measurements of specific environmental hazards.

9 MONITORING RESPONSES TO SPECIFIC EXPOSURES

9.1 Finding the Right Monitoring Parameter

In addition to infectious agents there are hundreds of hazardous exposures in the HCF including every category of toxic chemicals, physical agents, and biomechanical and psychological stresses. *Optimal* assessment of human reactions to these hazards includes the use of both biological monitoring (tissue measurement of the agent or its metabolite) and medical monitoring. A good biological marker is one whose presence or change gives enough information upon which to base some corrective action, examples of which are blood lead levels and audiograms. However, as for carcinogens, most toxic exposures do not have *dependable* indicators of acute risk.

9.2 Resisting Wrong Choices

In spite of the lack of dependable indicators, there is a disturbing trend to use unreliable indicators as a convenient basis for such administrative actions as medical removal from exposure. There are tremendous pressures to establish *something* as a standard, rather than to admit frankly the lack of a reliable basis for accepting a given value as a valid biological marker. An example of poor use of an indicator would be obtaining a serum cobalt or isocyanate level to screen persons who might have a pulmonary sensitivity reaction. In such cases the blood or urine measurement has no bearing on the safety of continued exposure to specific individuals. Rather, *clinical* factors must be the basis for such administrative decisions.

9.3 Prevalent Monitoring Practices

There is wide divergence of monitoring practices among health care institutions. Some practices determine the presence of valid direct indicators of early human reactions. Other substances are checked by tracking a change in *non*specific indicators. An example is the obtaining of blood counts in those exposed to antineoplastics, a common practice.

Where there are allegations of a medical effect, with meager or spurious scientific evidence, routine testing may be indicated. For example, the rare, anecdotally reported bronchospasm from formaldehyde may justify regular, periodic spirometry to provide a data base to verify the nonresponse of an exposed population or, as appears less likely in the case of formaldehyde, to confirm a dose-related reaction.

Detailed discussions for most agents encountered in this industry, such as formaldehyde and ethylene oxide, are found in other chapters or in industry consensus or NIOSH recommendations (12, 13). However, with the exception of those few exposures covered by permanent OSHA standards (noise and benzene, e.g.), none have OSHA-mandated monitoring surveillance requirements. One can argue that specific monitoring standards ought to be prepared, preferably by consensus, but no such common, *reliable*, authoritative recommendations exist, and they may be years in coming. Although there are many NIOSH-recommended medical and biological assessments, they must be considered only advisory, because they have not yet endured full peer review.

10 CONTROLLING ADVERSE HUMAN RESPONSES

We have been considering the recognition and evaluation of HCF environmental hazards so that effective preventive measures can be established. It is clear that the severe limitations of medical and biological markers, mentioned above, mandate environmental and administrative controls, with heavy worker participation, against harming health care workers. These controls must:

- Recognize the possibility that harmful exposures may cause significant subclinical changes or problems not previously associated with the exposure
- Evaluate exposures both by objective measurements and by attentively listening to those at potential risk who may relate new symptoms
- Minimize exposures not only by established but also by innovative engineering and administrative practices

10.1 Infectious Agent Exposures

For many, infection control concerns have been the major role of a hospital employee health service. Little will be said here about this very important area because the subject is complex and important enough to merit full treatment by itself. Indeed, JCAHO requires health care organizations to have infection control committees that must establish, for each institution, appropriate protocols.

Hepatitis, AIDS, and tuberculosis are three diseases that merit at least brief comment. As previously mentioned, Public Law 100-607 (2), which was passed in response to the AIDS threat, specifies that "methods shall be developed to reduce the risk of AIDS transmission in the workplace." These methods have been synthesized into the CDC's "Universal Precautions" protocol (14), which provides a degree of protection not only against AIDS and hepatitis B but against all blood and body fluid infectious agents.

10.1.1 Hepatitis B

Although AIDS was the concern that stimulated this law, hepatitis B virus (HBV) represents as great or a greater threat in this industry. The CDC estimates that 12,000 HCI workers get HBV infections each year, 500 to 600 of whom become hospitalized, 250 of whom will die from the infection. Twelve to 15 of those affected will succumb to acute, fulminant disease, 170 to 200 to subsequent liver cirrhosis, and 40 to 50 to liver cancer (15). This is a serious threat for which a vaccine is now available.

10.1.2 AIDS

As of September 19, 1989, a total of 3182 HCF workers had contracted AIDS. Of these 95 percent reported high-risk behavior. For the remaining 169, the specific cause could not be identified. Of 860 HCI workers whose skin had been pierced by sharps *from AIDS patients*, only three seroconverted with no other explanation than this contact. Of 103 workers who had contamination of mucous membranes or non-intact skin, none seroconverted within 6 months (14).

The purpose of the HBV/AIDS comparison is not to diminish the importance of AIDS but to emphasize the great immediate threat of the more highly communicable hepatitis B virus. There is no reason why workers with significant potential exposure to this dangerous virus should not be immuniologically protected, for the vaccine is readily available.

10.1.3 Tuberculosis

Although tuberculosis had decreased around 6 percent per year since 1953, this decline has recently reversed, especially in large urban communities like Detroit. This development emphasizes the need to maintain an aggressive prevention and early diagnosis program for health care workers. Specifics of such recommended efforts may be obtained from the U.S. Public Health Service (16).

10.2 Other Exposures (noninfectious)

In recommending biological or medical monitoring practices, as previously noted, factors in addition to early detection of abnormal responses or establishment of base lines may be appropriate, such as collection of data to document a *non-response*, as a defense against unfounded future injury or illness claims. Individual HCF institutions may also elect to collect data to aid in establishing a sound clinical basis for OSHA standard setting. But apart from these valid considerations, recommendations for monitoring specific responses to hazardous exposures should be based on:

- Mandates of existing regulations, whether or not they appear to have scientific merit
- The dictates of sound practice, irrespective of regulatory requirements
- The need to establish preexposure base-line data for individuals at risk of adverse effects
- The need to detect actual or imminent impairment

10.2.1 Exposures with Established OSHA Standards

For those few agents with "permanent" standards (established after full OSHA hearings), the monitoring requirements arguably are fully protective.

10.2.2 Exposures That Have Temporary or No OSHA Standards

Because new substances are not automatically regulated, except for the OSHA general requirement to provide a healthy workplace, the bulwark of worker protection must be an informed management with a responsible attitude toward occupational health and safety and an aware work force. This chapter cannot properly list each agent in the HCF, then encyclopedically pontificate on what is to be done for each exposure. To do so would be to include erroneous or insufficient criteria on the basis of which wrong conclusions would certainly result.

A sterile list of blood, urine, or physiological performance levels, *helpful* as it may be, is far from the sentinel of occupational health and safety it is too often taken to be. Yet there is a tendency to ascribe a sanctity to such published values, letting them substitute for the judgment required of a truly effective protective effort. The level that protects most will not protect the sensitized. As seen in the setting of "allowable" workplace concentrations for sensitizers, the standard that will truly protect *all* will be prohibitively expensive and unnecessarily protective of *most* workers.

10.2.3 Specific Recommendations

There already exist semiauthoritative recommendations for exposure limits and for biological and medical monitoring "standards" in the HCF (13). These are laudable attempts to be precise in areas where the supporting science is often lacking. Still, these reasonably enlightened efforts can be a basis for appropriate action plans. However, the following thoughts are offered to ameliorate the rigidity with which these *recommendations* are likely to be viewed.

In considering antineoplastic agents, for example, it is a fairly common practice to get complete blood counts (CBCs) on workers exposed to these agents that of course have the potential to affect production of blood cells. But current CBC changes, which reflect *past* exposures, are of little help in evaluating *current* practices. Moreover, without a truly informed work force and management, the finding of leukemia in the "population at risk" is likely to trigger panic among the others exposed as well as inappropriate administrative responses. This is especially true when the situation becomes amplified and distorted by news releases and the always impossibly brief radio and TV "sound bites." The obtaining of annual CBCs is no substitute for a really effective employee hazard communication effort and a genuinely motivated and informed administrative staff. Employees must have a high regard for the great potential of these agents to cause delayed illness or death. Nothing can substitute for this periodically renewed awareness, without which no list of test results has any meaning or usefulness.

Notwithstanding these precautions, some will prefer a list of "safe" numbers to a mandate for careful, sensitive, and reflective attention to the complex interactions necessary to understanding occupational health concerns in the health care industry. Those who prefer the simple solution, a list of environmental limits, are like Plato's cave dwellers, insisting on shadows, blinded by the sunlight. Such persons will find comfort in the recent NIOSH publication on the subject (12); it is an excellent source of reference material. In addition there is an industry-based reference, edited by Professor Edmund Emmett of Johns Hopkins, that is excellent (13). Also, Dr. Horvath's chapter in Zenz's *Occupational Medicine* will prove helpful (17). Furthermore, the few permanent OSHA standards each have specific requirements.

11 ACHIEVING THE GOAL

In accomplishing the goal of health care industry health and safety, the following areas of primary concern are offered: As already indicated, the cornerstone of an effective program lies in *application* of the knowledge that will result from the genuine concern about and by workers for the health and safety of the workplace. Management must devise ways of reducing the unnecessary ergonomic stresses in this industry. Employees must learn proper body mechanics. All must participate in ways to cope with death and dying by methods other than to become recalcitrant. We must consider the true cost of sleep deprivation among students of the healing arts whose minds must remain particularly alert under some of the most rigorous stresses. The commitment must be to all aspects of the recognition, evaluation, and control of workplace hazards, not merely the identification and achievement of presumed-safe levels, no matter how useful such numbers may or may not be.

ACKNOWLEDGMENTS

Thanks to Drs. Irene Jessick, Jos. Szokolay and Panayotis Pesaros for sharing their experience in caring for health-care workers. Marilyn Kelemen, R.N., M.S., and Mark Dittman offered helpful comments. Darrel Kemble's help in resource gathering and Dan Sawyer's assistance with writing and editing was very much appreciated.

REFERENCES

1. OSHA Field Instruction CPL 2-2.44 & CPL 2-2.44A, August 15, 1988.
2. U.S. Public Law 100-607, Subtitle E, General Provisions Section 253 (a).
3. NIOSHA Program Directive No. 88-3, October 12, 1988.
4. Bureau of National Affairs, *Occup. Safety Health Rep.*, **19**(32) 1372 (Jan. 17, 1990).
5. U.S. Dept. of Labor Statistics, *Occupational Employment in Selected Non-manufacturing Industries*, 1984.
6. A. Toffler, *Future Shock*, Bantam Books, New York, 1970.
7. S. J. Bigos, D. M. Spengler, and N. A. Martin, "Back Injuries in Industry. A Retrospective Study. III. Employee Related Factors," *Spine*, 11 (1986).
8. Bureau of National Affairs, *Occup. Safety Health Rep.*, **19**(32), 1372 (Jan. 17, 1990).
9. Dept. of Labor/Dept. of Health & Human Services Joint Advisory Notice. HBV/HIV. *Fed. Reg.*, **50**(210): 41818–41824 (Oct. 30, 1987).
10. Dept. of Health & Human Services, USPHS, FDA Drug Bulletin 19, 2. *Aerosolized Pentamidine for P. Carinii Pneumonia.* July, 1989.
11. John M. Lanham, M.D., Chief Medical Officer, Dow Chemical Co., personal communication.
12. *Guidelines for Protecting the Safety and Health of Health Care Workers*, USPHS (NIOSH) Publ. No. 88-199, 1988.
13. E. A. Emmett, Ed., "Health Problems of Healthcare Workers," in *Occupational Medicine*, Vol. 2, No. 3, Hanley and Belfus, Philadelphia, 1987.
14. *MMWR*, **37**(S6), 377–382, 387–388 (1988).
15. *MMWR*, **38**(S6), 5–8 (1989).
16. *MMWR*, **37**(43), 663–675 (1988).
17. E. P. Horvath, Jr., "Occupational Health Programs in Clinics and Hospitals" in *Occupational Medicine, Principles and Practical Applications*, 2nd ed., Carl Zenz, Ed., Year Book Medical Publishers, Chicago, 1960.

CHAPTER FOURTEEN

Health and Safety Factors in Designing an Industrial Hygiene Laboratory

Robert G. Lieckfield, Jr., C.I.H., and Alice C. Farrar, C.I.H.

1 INTRODUCTION

Health and safety are important concerns in any laboratory. But because the industrial hygiene laboratory in particular exists to help ensure worker protection, it should be the exemplary safe and healthful workplace. The manner in which a laboratory is designed, including the care, knowledge, and foresight that go into its planning, will do much to ensure the safety and health of its occupants and even the surrounding community. This chapter discusses health and safety factors that must be considered in designing the industrial hygiene laboratory and is intended to help management plan new laboratory construction or remodel existing laboratory spaces. The references listed at the end of this chapter may be consulted for more information on the broad topic of laboratory design.

1.1 Nature and Scope of the Industrial Hygiene Laboratory

An industrial hygiene laboratory analyzes samples that have been collected to assess worker exposures or potential exposures to dusts, fibers, fumes, mists, gases, and vapors in the workplace. Samples may include airborne contaminants collected on various sampling media, such as solvent vapors on activated charcoal tubes; bulk

Patty's Industrial Hygiene and Toxicology, Fourth Edition, Volume 1, Part A, Edited by George D. Clayton and Florence E. Clayton
ISBN 0-471-50197-2

materials, such as asbestos-containing sprayed-on fireproofing or insulation; body fluids for biological monitoring, such as lead in blood; or microbiological samples, such as microorganisms sampled to assess indoor air quality.

In almost all cases, samples are analyzed for trace quantities of materials. The industrial hygiene laboratory is not a laboratory for physical testing or for gross analytical work, such as is done in a quality-control laboratory.

1.2 Laboratory Accidents

This chapter outlines steps to take to help ensure a building design that is safe and healthful to personnel, property, and the community. Poor laboratory building design is the culprit in many laboratory accidents. The following section lists common industrial hygiene laboratory accidents. Some of these may also result from poor organization and management. The reader should consult the references at the end of this and other chapters in this text for help in organizing and managing health and safety plans.

Industrial hygiene laboratory accidents can be classified by injury to personnel, property, and the community.

Injuries to personnel (1) can be caused by:

- Lack of or failure to wear protective equipment, such as safety glasses
- Lack of hazard communication to workers, such as the use of material safety data sheets (MSDS) in providing information on precautions to be taken with specific chemicals
- Lack of proper ventilation, such as inoperative or ineffective chemical fume hoods
- Personal hygiene problems, such as eating in the laboratory or pipetting by mouth
- Electrical hazards, such as wiring that violates the National Electric Code
- Storage problems, such as failure to secure compressed gas cylinders or to store incompatible chemicals in segregated storage locations
- Inadequate emergency procedures and equipment, such as inoperative safety showers or fire extinguishers
- Lack of proper management, such as failure to implement an industrial hygiene and safety plan or a hazard communication program
- Lack of personal responsibility, such as a worker's failure to follow procedures

The most common accidents resulting in property damage are fires and explosions. Water damage from firefighting also damages property. Several of the causes of injury to personnel listed above can also result in fire and explosion.

Fire and explosion can also harm people or property in the community. However, a very common and serious community concern is improper disposal of laboratory wastes. This is discussed further in Section 5.2.

2 GENERAL DESIGN CONSIDERATIONS

Building a laboratory (2) can be a challenge for the laboratory director. But for those who have worked in a less-than-ideal facility, it is exciting to be able to help plan and design a new one.

2.1 Design Elements

2.1.1 Location

Industrial hygiene laboratories exist as part of government agencies, academic institutions, corporations, and consulting groups. Most independent consulting laboratories are housed in one- or two-story buildings specially designed for the consulting firm or laboratory. Therefore geographic location is a matter of choice of the owners of the facility. However, industrial hygiene laboratories with parent organizations are usually located within a major office complex. Location of the laboratory within such a complex is a major consideration because of the special ventilation, plumbing, electrical, storage, and access needs of the laboratory. The following should be considered when choosing a location for a laboratory within an existing building:

- The need to exhaust fume hoods to the external environment
- The need for waterlines and acid-resistant drains
- The need for special electrical installations
- Convenient access for laboratory users
- Convenient access for delivery of gases and supplies
- Fire and safety considerations for laboratory staff and other building occupants

2.1.2 Aesthetics and Environment

The laboratory director and the consulting firm or parent organization must provide a workplace that is not only free of health and safety hazards but that is comfortable and conducive to good employee morale and productivity.

Aesthetics is important. Analysts and laboratory staff often spend their entire workday in the laboratory. The laboratory should be designed to be visually attractive as well as functional. Consideration should be given to modest amenities for staff, including lockers or areas designated for personal belongings; lunch- or breakroom equipped with sink, refrigerator, and microwave; coat closet or rack; and convenient rest rooms.

Proper lighting, temperature, and humidity are necessary to keep both the employees and many of the instruments working well. Instruments produce a good deal of heat and contribute to the temperature of the laboratory environment. A separate thermostat control for areas with analytical instrumentation is desirable. Air exhausted by hoods and air contributed to makeup air also affect the temperature. These factors should be considered in planning for laboratory comfort and temperature control.

Table 14.1 Health and Safety Regulations

Agency	Regulation	Pertains To
OSHA	29CFR1910.132-134, 1910.136, 1910.212	Personal protective equipment (eyes, face, head, extremities, respiratory tract)
	29CFR1910.106, 1910.157-164	Flammable and combustible liquids
	29CFR1910.101-105, 1910.166-167	Compressed gases
	29CFR1910.96-97	Radiation/ionizing and nonionizing
	29CFR1910.137, 1910.301-308	Electrical hazards
	29CFR1910.176, 1910.141	Materials handling and storage, sanitation, and housekeeping
	29CFR1910.35-37	Means of egress
	29CFR1910.1000-1500	Air contaminants
	29CFR1910.20	Access to medical records
	29CFR1910.1200	"Right-to-Know"
	29CFR1910.1450	Occupational exposures to hazardous chemicals in laboratories
EPA	40CFR720.36, 720.78	New chemicals
	40CFR355.10, 355.20, 355.30, 355.40, 355.50	Chemical emergency plans

2.2 Health and Safety Considerations

2.2.1 Regulations

There are no federal regulations that apply to the design of industrial hygiene laboratories. However, laboratory activities related to chemical health and safety are governed by more than 100 separate regulations promulgated by the Occupational Safety and Health Administration (OSHA) and the Environmental Protection Agency (EPA). Table 14.1 lists several of these regulations and the areas of safety and health they address. These regulations must be consulted when constructing or remodeling laboratories.

Of the many regulations that apply to laboratories, the OSHA regulation for Occupational Exposures to Hazardous Chemicals in Laboratories (29CFR1910.1450) is particularly important when planning and managing a safe laboratory facility.

The OSHA laboratory standard applies to all laboratories that use small quantities of hazardous chemicals on a nonproduction basis. This includes laboratories in industrial, clinical, and academic settings. Laboratory operations excluded from the OSHA standard are laboratory operations that produce commercial quantities of materials, are part of the production process, or whose uses of hazardous chemicals provide no potential for employee exposure.

The main emphasis of the OSHA regulation is the implementation of a written Chemical Hygiene Plan (CHP) that documents procedures, equipment, and work practices that are capable of protecting employees from potential health hazards encountered in a particular laboratory. The CHP must outline procedures for keeping chemical exposures below either (1) permissible exposure limits (PELs), as outlined in Subpart Z, 29 CFR 1910.1000 of the OSHA standard, or (2) recommended exposure limits for hazardous chemicals where there is no applicable OSHA standard.

A CHP must include the following eight elements:

- Safety and health standard operating procedures for working with hazardous chemicals
- Criteria to determine and implement control measures to reduce employee exposure to hazardous chemicals
- Procedures to ensure the proper functioning of fume hoods and other protective equipment
- Provisions for employee information and training
- Designation of circumstances in which a particular laboratory operation, procedure, or activity will require prior approval from the employer or supervisor
- Provisions for medical consultation, surveillance, and examination
- Designation of a responsible individual (a chemical hygiene officer or chemical hygiene committee) to oversee implementation of the CHP
- Provisions and procedures for employee protection when hazardous chemicals (e.g., carcinogens, reproductive toxins, teratogens, or acutely toxic chemicals) are used in a laboratory

The OSHA Hazard Communication standard (29CFR1910.1200) should be consulted for guidance in determining the scope of health hazards and determining the hazard clarification of a particular chemical.

2.2.2 Ventilation

Laboratory ventilation is needed to provide a safe and comfortable environment. Balanced amounts of supply and exhaust air are needed as well as temperature and humidity control. Good laboratory exhaust ventilation contains or captures toxic contaminants and transports them out of the building. Exhaust ventilation must be designed to prevent contamination of other areas of the building that can occur by reentrainment of contaminants from discharge points into outside air inlets, or by negative pressure inside the building that can cause downdrafts in fume hoods. The same supply ventilation system may be used to provide makeup air for exhaust air systems as well as a comfortable and safe work environment, or a separate supply system may be used for each function.

The design of a ventilation system for the industrial hygiene laboratory should consider air balance, pressure relationships among areas, supply and exhaust air criteria, fume hoods and local exhaust, and criteria for outside air intakes and hood exhaust discharges. These items are discussed in Sections 3.4.2 and 3.4.3.

2.2.3 Egress and Handicapped Person Access

Safe egress (1), as defined by the National Fire Codes 45 and 101 prepared by the National Fire Protection Association (NFPA) and adopted as the building code in many localities, is an essential part of laboratory planning. Two means of egress are usually required. Exceptions are based on limitations of square footage and in the types and quantities of chemicals housed within the laboratory. The primary exit door should be no more than 75 ft in travel distance from the farthest occupied space of the laboratory. The recommended minimum width of exit doors is 36 in. clear. Exit doors must swing in the direction of egress. Doors opening into corridors should not reduce the required width of the egress passage. Design of building egress outside of individual laboratories is governed by NFPA 101 provisions.

Arrangement of laboratory benches should facilitate egress. Benches arranged in parallel rows form regularly spaced working aisles. The recommended minimum width of working aisles is 5 ft, with a maximum of 7 ft. Working aisles should join directly to at least one egress aisle of equal or greater width. The egress aisle should lead directly to a fire-protected exit.

Hazardous operations affect planning for safe egress. The most hazardous operations, such as those using flammable solvents, should be located farthest from the primary exit. Less hazardous activities such as paperwork, should be next to the primary exit.

State and local regulations specify laboratory design for access by persons in wheelchairs or those otherwise handicapped. The prime concern is safe egress. The regulations specify minimum door widths, minimum clearances on the latch side of in-swinging doors, the directions of egress, elimination of threshold conditions, ramp access and allowable slopes, handrail heights, and appropriately designed personal hygiene facilities. In general, slightly more floor space is required to comply with most of these regulations.

2.2.4 Emergency Facilities

Emergency facilities and supplies include emergency deluge showers, eyewash fountains, spill kits, fire extinguishers, fire blankets, telephones and communications devices, control panels, and building fire protection systems. These are discussed in Sections 5.3 through 5.6.

2.2.5 Laboratory Furnishings

Selection of wood or metal casework is largely a matter of economy or aesthetics, not safety. Wood rots under wet conditions and it can burn; it can be treated to minimize these disadvantages but not eliminate them. Metal corrodes, especially in typical laboratory atmospheres. Although it can burn, this is not likely in promptly extinguished fires. Metal can be coated to resist corrosion, but cannot be totally protected. A good quality finish and construction are primary considerations in selecting laboratory casework.

All wall-hung storage units, shelves, and equipment should be securely attached with fittings of sufficient strength to support units under maximum loading conditions. Walls that support storage units should be reinforced to bear the maximum load. In normal construction of gypsum wallboard on metal stud partitions, lateral bracing is not included. Even lightweight concrete masonry unit wall construction cannot hold the weight of typical laboratory storage. Reinforcing for interior partitions must be directly called for in the specifications and drawings.

2.2.6 Laboratory Features

Lighting should be designed and installed so there is low glare at the laboratory bench height and no shadows. Normal lighting levels range from 50 to 100 footcandles (500 to 1000 lux) at countertop height. The intensity of light depends on the nature of the work. All emergency lighting and exit signs should be provided as required by local building codes.

If compressed gas cylinders are used in the laboratory, sturdy supports must be provided for securing cylinders in use and empty cylinders or those awaiting use.

An area out of the flow of traffic should be provided for lab staff to store personal articles such as coats, hats, raingear, and briefcases. Recessed locations out of the flow of traffic are desirable for waste containers, including receptacles not only for ordinary trash but also for temporary storage of different chemical wastes.

Walls between adjacent laboratories and between laboratories and corridors should have a $\frac{3}{4}$-hr or more fire-resistant construction rating according to local building codes.

3 FACILITY DESIGN

3.1 Introduction

The basic layout of laboratories, with fixed benches and utilities, has changed little since the mid-1800s. The laboratories of this period consisted of fixed benches along two walls with cupboards and drawers underneath and shelves above. Center rows of parallel benches with cupboards underneath were spaced about 6 ft apart. The side benches were 2 ft deep, and the center benches approximately 3 ft wide (3). As is true in today's laboratories, every surface of the laboratory was covered with apparatus. This basic layout has been used in almost every laboratory built in the last 100 years. The concept of parallel benches was indeed practical, because most laboratory analyses have been traditionally performed on the bench surface. However, the industrial hygiene laboratory has evolved from classical wet chemistries, such as titrations and colorimetric and gravimetric methods, to analyses using sophisticated electronic instrumentation with computer-based data reduction.

The widespread use of video display terminals (VDT), where the analyst sits before a VDT conducting the experiment and data calculations, requires much more flexibility in design than that provided by fixed benches.

In any case, the design of the physical space is the biggest challenge in building or renovating a laboratory. The most important and effective tool is a sound master plan. This plan must cover all aspects of the design process and reflect a clear understanding of both aesthetic and functional requirements by the designers and the users of the facility.

The difficulty in laboratory design is predicting the growth of both the individual components of a laboratory and that of the whole organization. Even so, it is imperative that the design team develop a strategy to address future requirements. Benches, services, and partitions must be planned with anticipation of changes in work patterns, scientific instrumentation, staffing, and work demands, which all affect space requirements. If such growth is not planned for in the original design, the result will be a misuse of space or the need for major redesign. Ignoring future laboratory requirements for space, services, and function is extremely costly in terms of both additional capital investment and loss of analyst productivity caused by inefficient space design.

The capital investment in a laboratory expansion is generally not just the cost of constructing new walls or expanding into new areas. Laboratory expansion often requires extensive redesign of the ventilation and heating and cooling systems, as well as other utilities. Because the costs of ventilation and heating and cooling systems are linear, upgrading and expansion is often two to three times the cost of systems that are designed for expansion (3).

With all the traps associated with laboratory design, how can anyone get it right? First, it must be acknowledged that some aspect of the laboratory design will end up insufficient for the work being performed. But this should not deter the designer from covering all aspects of the design in minute detail with a goal of a functional, practical, and efficient work environment. The design will have succeeded if the facility can accommodate change easily at a reasonable cost. Successful laboratory design requires a careful and well-thought-out planning and design process involving the architects, engineers, and users of the laboratory. Please note the term "users" refers to both laboratory management and analysts.

Anderson, deBartolo, Pan, Inc., an architectural engineering firm, offers the following recommendations based on its design of the Sandia National Laboratory in Albuquerque, New Mexico (4):

- A multidisciplinary architectural and engineering team working intimately with the client, particularly the user, is the best combination for ensuring successful architectural and engineering solution to design problems.
- The architect and engineer and the user must all consciously push for innovation, to guarantee that new concepts will be explored.
- Sufficient time must be available to test the proposed innovations thoroughly.
- The client's management and decision-makers should be involved in the details

of the project at the earliest stages of design. This is an excellent way to minimize red tape and complications in interactions between the architect and engineer and laboratory management.

- The architect and engineer and the laboratory must achieve a balance between solution concepts and follow-through on the details.
- The architect and engineer and laboratory must share an absolute commitment to satisfying the analysts' requirements and achieving design goals.

3.2 Space Planning

Space planning is conducted for the entire laboratory facility and includes planning the physical layout, individual laboratory space, and necessary adjacencies and work flow patterns, as well as planning for future expansion. There is no single best design and layout for an industrial hygiene laboratory because each laboratory has unique space requirements and work habits. Therefore space planning must be project-specific as identified by laboratory management and personnel.

3.2.1 Physical Layout

The physical layout of the laboratory depends on the particular constraints imposed by the actual building structure. Although these constraints are certainly minimized in construction of a new building, land use and economics are always a consideration.

There are a number of basic design plans that can be used as a starting point. However, because function and space requirements will certainly change over time, flexibility must be built into the planning process.

The first step is to develop a building design program. The building design program is a written document that provides details on the construction site and the various functions and requirements of the laboratory. Because laboratories require extensive ventilation and safety features, the building design program must address the relationship between function and safety and health aspects.

The building design program should provide a description of all rooms, and include information on the type of work performed, instrumentation and equipment used, and physical activities in each area. There are five general areas within every laboratory: laboratory areas; support areas, such as computer and data calculation facilities; glassware washing and storage; offices; and personnel support facilities, such as lunchroom, study and library areas, meeting rooms, and mechanical rooms.

The building design program must also consider the issue of shared versus single use of the laboratory areas described above. Depending on the size and function of the particular laboratory, areas such as sample and chemical storage, data calculation, breakrooms, and shipping and receiving can have either centralized or decentralized functions. Although it is generally more economical to build laboratories around centralized support facilities rather than to duplicate support areas in each laboratory or department, the long-term costs of centralized areas must be calculated in terms of work flow inefficiencies and loss of productivity.

A major consideration in the physical design of a laboratory is the circulation of staff and work flow. This can be greatly enhanced by a carefully thought-out floor plan. Aside from the more obvious adjacencies, such as locating a sample receipt area next to a sample storage area, and locating the chemical storeroom central to all work areas, there are six principal patterns of use in a laboratory:

- Circulation of staff and samples
- Individual laboratory function
- Distribution of mechanical equipment and services
- Structural system
- Site regulations
- Building enclosure

For each laboratory area, the building design program must detail the placement of the extensive mechanical service areas needed for laboratory ventilation, electricity, plumbing, and compressed gas. Figures 14.1 through 14.4 show various physical layout plans and the advantages and disadvantages of each compiled by DiBerardinis et al. (5). The purpose of illustrating these basic plans is to show the variety of approaches that can be taken. As noted above, the end product must be developed as a consensus of the architect, engineer, and laboratory management and analysts.

3.2.2 Room Size Planning

Determination of space requirements for each laboratory area is critical to the overall design plan. The space used by existing laboratory areas can be used as a starting point. If a new laboratory is being built, area requirements can be approximated from review of similar laboratories.

Starting from the space occupied by the existing laboratory areas, the design plan focuses on area needed for instrumentation, benches, hoods, data reduction, and research, and on the mechanical and storage requirements of each laboratory area, factoring in future needs.

Space requirements can be determined by construction of a model of the laboratory. This is done using scale models (two- or three-dimensional) of laboratory benches, instrumentation, equipment, hoods, sinks, work stations, and computer terminals. The design team then constructs a scale laboratory showing the bench plan, instrument layout, and hoods. Using the existing laboratory as a guide, the design team can physically determine current and future laboratory space needs. Areas for data review and report writing should not be forgotten in this process. Scale models should also be constructed for storage, computer, and library areas.

The results of the scale model study will then define the individual laboratory area space requirements. The total facility space needs are then determined by adding up the individual laboratory area space needs and the space needed for corridors, offices, lunchrooms, and mechanical rooms.

As discussed previously, the mechanical and service requirements of a laboratory

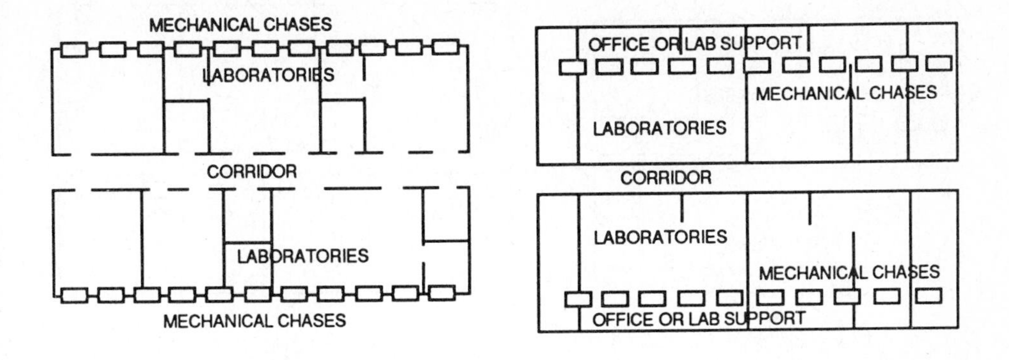

Advantages:

- Short horizontal exhaust duct from fume hood
- Fume hood is at the end of an aisle
- Major inner lab traffic near corridor egress
- Equipment and desk at corridor wall

Disadvantages:

- Pipes and ducts only accessible from laboratory
- Secondary egress near corridor exit
- Restricted window area
- Outswinging doors temporarily obstruct corridor, wide corridors are recommended

Figure 14.1. Service chase at exterior walls.

are extensive and must be carefully planned. Adjacencies for efficient distribution of services should be planned as practical to reduce the overall design and construction costs. The design plan must include access to the mechanical or utility service by both installation and maintenance personnel.

In addition to the mechanical and service design, special design and construction are required for certain instrumentation, that which is sensitive to vibration, such as mass spectrometers, electron microscopes, or inductively coupled argon plasma spectrophotometers.

Local government regulations on land use and construction methods have to be consulted. Local zoning ordinances govern the building use classification, easements, building height, allowable floor area ratio, number of parking spaces required and use of sewer and water utilities. Some state regulations also require permits for exhausting air from laboratory hoods.

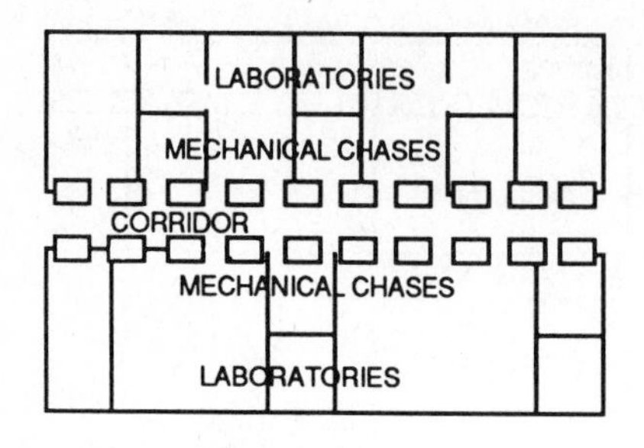

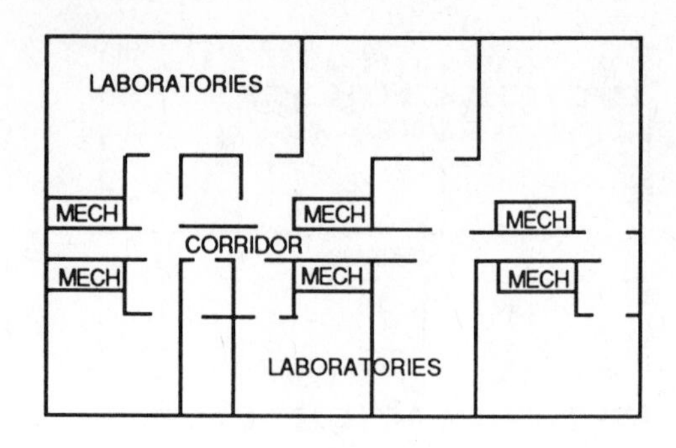

Advantages:

- Pipes and ducts accessible from corridor
- Secondary egress opposite corridor
- Expansion window area possible on perimeter wall
- Outswinging door shielded in alcove
- Equipment and desk at corridor wall

Disadvantages:

- Long horizontal exhaust duct from fume hood
- Persons pass in front of fume hood, causing turbulence

Figure 14.2. Service chase at interior walls.

The actual building enclosure can be determined after all the above criteria are established.

3.2.3 Expansion Planning

Once constructed, a laboratory is difficult to expand. It is extremely important to build in a well thought-out expansion plan to meet the ever-changing requirements of the laboratory. Expansion of laboratory facilities into laboratory support areas such as offices and storage rooms can be planned. Because these areas are usually converted to laboratory use anyway, up-front planning of the inevitable will save considerable renovation expense. Locating mechanical and utility services in expansion areas during initial construction will certainly provide the necessary flexibility. As discussed, expansion planning must consider the capacity of ventilation

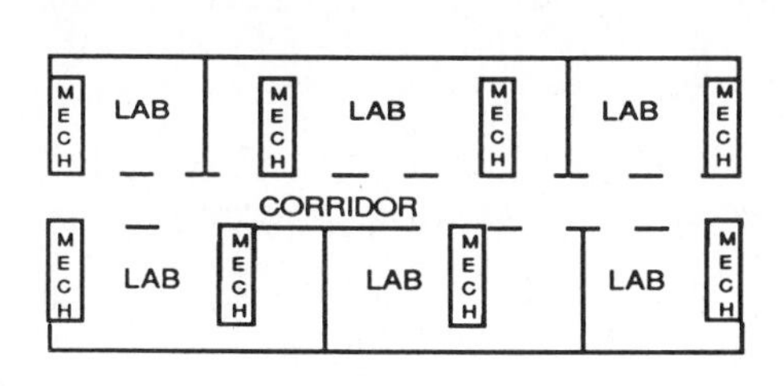

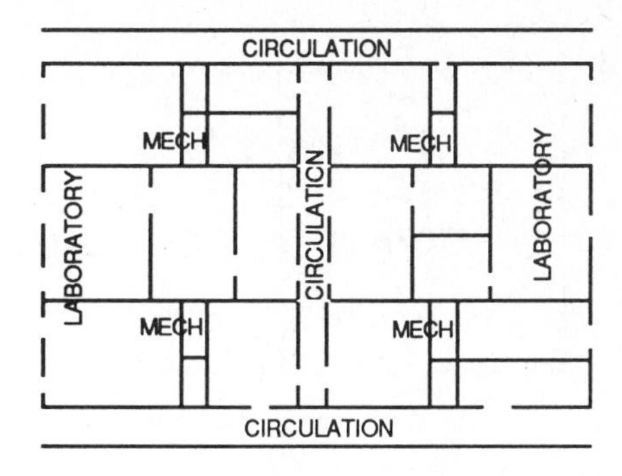

Advantages:

- Short horizontal exhaust duct from fume hood
- Fume hood is at the end of an aisle
- Major inner lab traffic near corridor egress
- Expansive window area possible on perimeter wall
- Applicable to conversion of building to laboratory use

Disadvantages:

- Single module lab has only one wet wall
- Outswinging doors temporarily obstruct corridor, wide corridors are recommended
- Pipes and drains have long horizontal runs to peninsula benches
- Drains require additional venting

Figure 14.3. Service chase between modules.

systems, utilities, and other services. It is more cost effective to oversize these services in initial construction than to upgrade them later.

3.3 Laboratory Planning

Once the physical space is determined, the design should be tested using the layout model of actual bench and instrument placement discussed in Section 3.2.1. The purpose of this portion of the design phase is to test the physical space of the entire facility both for current requirements and for future expansion. In this stage of planning, it is important to establish (if not already done) the location of doorways, corridors, and hoods. In general, the bench and instrument layout must first be planned for easy egress and efficient travel within the laboratory.

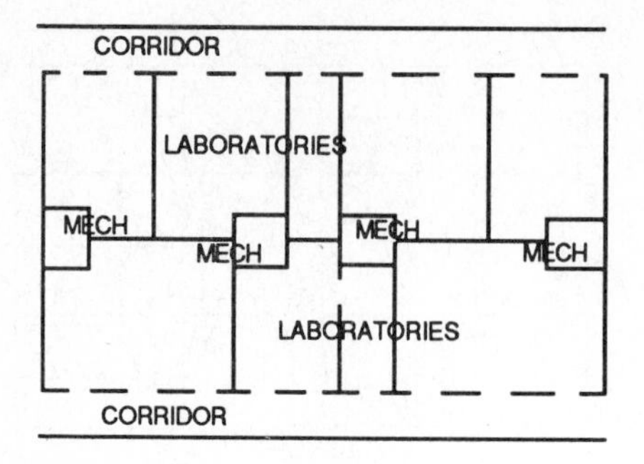

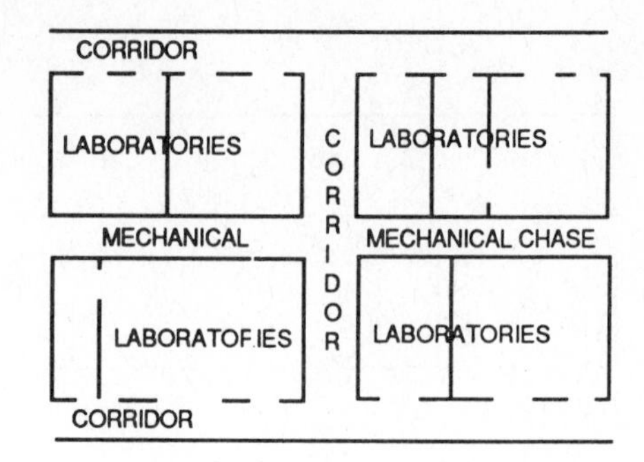

Advantages:

- Minimum mechanical area, efficient for low buildings
- Fume hood is at the end of an aisle
- Short horizontal exhaust duct from fume hood
- Pipes and ducts for cluster of 4 labs accessible from each
- Maximum variation in laboratory size possible

Disadvantages:

- Limited chase area not suitable for multi-storied buildings
- No window area in laboratory
- Outswinging doors temporarily obstruct corridor, wide corridors are recommended

Figure 14.4. Central service chase.

3.3.1 Bench Layout

The laboratory bench layout is unique to a particular laboratory's needs and use. Typical bench configurations are parallel to a partition containing utility hookups, extending from a partition at right angles, or free-standing. Any or all of these design concepts can be incorporated into a laboratory bench layout.

In deciding configuration, the ease of utility hookups (gas, water, electric) should be considered. Services can easily be extended when benches are parallel or perpendicular to a partition. Services to free-standing or island benches are somewhat more difficult, but can be extended from the ceiling or from underground through mechanical chases.

The minimum clearance between benches should be 5 ft. This allows for free traffic flow with chemists working at benches. Bench surfaces and storage units should not block utility outlets. Storage units placed above the benches should be easily accessible. The standard work surface is 24 in. deep. Deeper surfaces may be required depending on the type of instruments being placed on the surface. If wall-mounted cabinets are placed over benches, a minimum clearance of 2 ft should be provided to accommodate taller instruments, allow easy access to top and back of instrument, and minimize heat buildup under the wall cabinet.

The work surface height of the bench depends on the use intended for the bench. If workers are seated, the bench can be designed at a height of 30 to 32 in., typical of normal desk height. If workers stand at the bench, the bench height should be 35 to 37 in. and knee spaces should be provided. A drawback of the standing height knee space is its awkwardness and need for laboratory stools.

The lower-height bench is recommended in laboratories doing microscopy and microbiological analyses.

The remainder of the benches should be the standard 35 to 37 in. high. This height has been found to be ergonomically efficient, allowing easy access to the rear of the bench surface, and a comfortable height at which to perform analytical tasks in a standing position.

Incorporated into the bench plan may be desk surfaces for calculations and related activities. Laboratory desks must be positioned away from potentially dangerous laboratory operations and not impede the aisles when the desk is occupied.

An integral aspect of the bench layout plan is the placement of fume hoods. In general, hoods should be located away from entrances and exits and major corridors. This will reduce the traffic flow in front of the hood, thus minimizing airflow disturbance, which affects hood performance. Hoods handling highly hazardous material should be isolated to minimize safety hazards and potential for explosion and fire damage.

3.3.2 *Instrument Layout*

Once the basic bench and hood layout is complete, the location of each piece of instrumentation and equipment should be planned. A complete list of instrumentation should be drawn up to include dimensions, operating characteristics and specifications, clearances, utilities required, related equipment such as glassware, and ventilation requirements. A checklist is shown in Figure 14.5 to aid in defining these needs.

Considerations in planning for instrumentation placement include:

- A centralized equipment room to service multiple laboratories with utilities and layout designed specifically for instrumentation
- A review of lighting requirements for areas using video display terminals

1. List all major pieces of equipment in each laboratory according to the following characteristics:

   (a) Size and weight
   (b) Whether bench mounted, free standing, or other mounting method
   (c) Fixed to wall or floor with permanent utility connections
   (d) Moveable with quick disconnects to utilities
   (e) Required utilities and services
       (1) Water: city, purified, recirculated process
       (2) Drain: floor, piped
       (3) Gas: piped, manifolded tanks
       (4) Compressed air
       (5) Vacuum
       (6) Electricity: voltage, amperage, phases, location and number of standard and special receptacles
       (7) Local exhaust services: type, volume, filters, other treatment
       (8) Air supply for local exhaust services: tempered, filtered, humidified
       (9) Stream: pressure, volume, flowrate
   (f) Maintenance clearances
   (g) Operator's position and clearances
   (h) Equipment attachments
   (i) Heavy rotary components

2. Determine the location of and area required for supplies associated with each piece of equipment:

   (a) Glassware
   (b) Instruments, manipulators
   (c) Chemicals
   (d) Disposables: paper goods, plastic ware

Figure 14.5. Checklist for selection and location of laboratory furnishings and equipment (5).

- The need for vibration-free and temperature- and humidity-controlled locations (for balances, electron microscopes, mass spectrometers)
- Clearances for instrument delivery and service by maintenance staff

Because many newer instruments are controlled by remote terminals, placement of instrumentation adjacent to offices may be considered for efficient use of analysts' time.

3.4 Services

A good laboratory design requires accurate specifications for ventilation, electrical, and plumbing services. We have concentrated on space planning and physical layout of equipment and benches with little mention of the services required to make the laboratory functional. Ideally, all of these aspects are considered together, with the best possible solutions for each design component addressed at the same time.

Laboratory services can be provided by either a centralized or decentralized laboratory distribution system. Factors and constraints that affect this decision are:

- Ceilings/access space
- Number of floors
- Structural system
- Types of laboratories and services
- Flexibility desired
- Expansion plans
- Budget limitations

3.4.1 Distribution Schemes

Typical distribution schemes include continuous wall service corridors, vertical distribution, and horizontal distribution (6).

3.4.1.1 Continuous Wall Service Corridors. The mechanical and electrical services in continuous wall service corridors design are distributed horizontally along a central service corridor. The services are distributed via a main vertical service feed. In this plan, each laboratory area has direct access to the utilities. The continuous wall service corridor has many advantages, including easy maintenance and the flexibility to rearrange and modify services without affecting the entire operation.

3.4.1.2 Vertical Distribution. The vertical distribution design introduces the necessary services to each laboratory area through a series of vertical utility plenums located to service one or more laboratories. Studies are accessed by tapping into the vertical service plenums and extending services to each laboratory area horizontally either through the ceiling or below the floor. This type of design has a lower initial construction cost. However, it has limited flexibility unless the service plenums are closely spaced.

3.4.1.3. Horizontal Distribution. The horizontal distribution design organizes the laboratory services through ceiling plenums. The individual services are then accessed through vertical service chases at the particular bench area. Horizontal distribution offers flexibility and comparable simplicity of design.

Planning the most effective distribution scheme requires knowledge of present and future design needs. The type of distribution system selected should be determined in conjunction with the laboratory space planning and layout phases of the design process, because layout and services access are mutually dependent.

3.4.2 Heating, Ventilation, and Air Conditioning

Laboratory heating, ventilation, and air conditioning (HVAC) systems must be capable of maintaining a uniform temperature throughout the laboratory. Much of the analytical equipment in use at modern laboratories is extremely temperature sensitive. Wide fluctuation in temperature conditions is detrimental to instrument performance and typically results in higher-than-normal maintenance expense.

In many cases, simultaneous heating and cooling are required for different laboratory areas. Both local cooling units and central heating and cooling systems may be necessary. For example, laboratories equipped with gas chromatograph/mass spectrometers and electron microscopes need individualized cooling units because of the larger heat loads (150,000 to 200,000 BTUs) and extremely temperature-sensitive equipment.

If the laboratory is located within office space, the design team should specify separate heating and cooling units for the laboratory and office areas. In areas where office HVAC allows for 20 to 25 percent recirculation of exhaust air, separate systems may be the only viable alternative because recirculation of laboratory exhaust air is generally discouraged.

Discharge and supply air volumes must be carefully balanced in the laboratory. Laboratories handling hazardous materials must always be maintained at a negative pressure relative to corridors and other access areas such as offices, lunchrooms, and public access areas.

There are three types of laboratory ventilation systems. Comfort ventilation considers the supply and removal of air for breathing and temperature control. Exhaust ventilation, through fume hoods, is designed principally for protection of health and safety. Makeup air supply is designed to replenish the amount of air discharged through the fume hoods.

Laboratory ventilation design is affected by the following factors:

- Fume hood design
- Minimum exhaust air required by fume hoods
- Location of the fume hoods within the laboratory
- Pressure relationships with respect to surrounding areas
- Removal of chemical vapors, gases, and odors

A flow chart of ventilation design is given in Figure 14.6.

3.4.3 Fume Hoods

3.4.3.1 General. The purpose of a fume hood is to remove toxic and harmful fumes, gases, and vapors from the laboratory environment by exhausting air at a sufficient velocity to capture and remove the substances. The fume hood must be designed and located to allow the analyst to perform tasks or observe processes both easily and safely. The laboratory fume hood accomplishes this by providing both a work area and a sufficient face velocity (airflow across the hood opening). Examples of typical laboratory hood configurations are shown in Figures 14.7 through 14.9.

The fume hood face velocity is crucial for protection of the analysts' health and safety. The face velocity and, therefore, hood performance are adversely affected by drafts across the hood face, large temperature differences between the hood and surrounding air, equipment contained within the hood cabinet, and analysts' work habits.

The upper limit on face velocity is approximately 150 feet per minute (fpm).

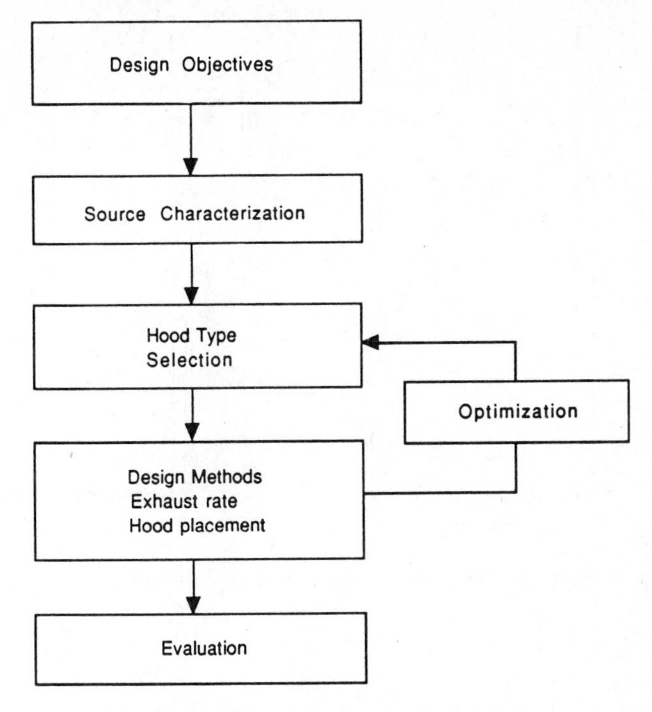

Figure 14.6. Development of hood design.

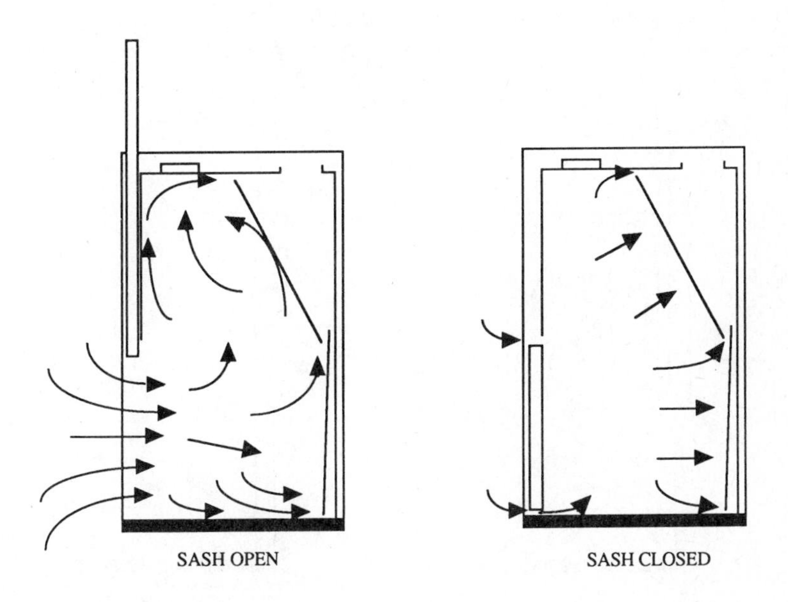

Figure 14.7. Conventional fume hood.

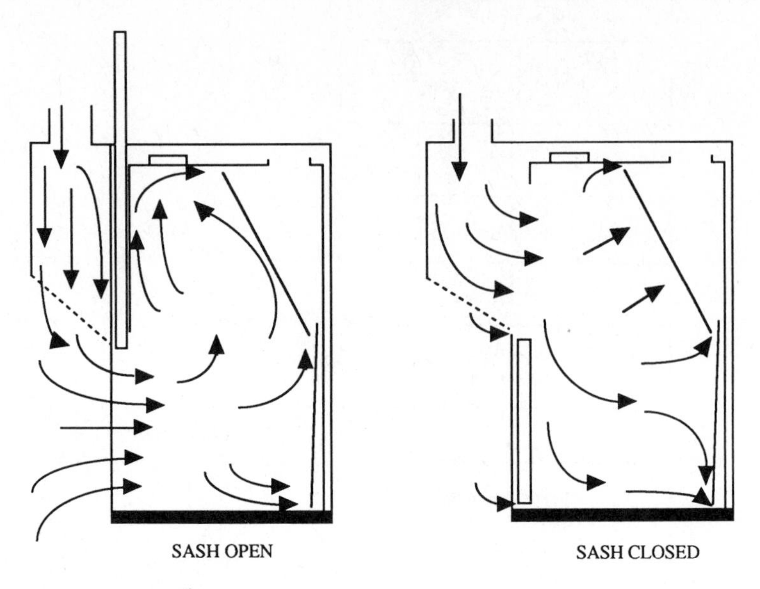

Figure 14.8. Auxiliary air fume hood.

Flow rates above this level develop air disturbances, created by the analysts' position in front of the hood, resulting in a negative pressure in front of the analyst, which actually pulls vapors and gases back toward the analyst.

Until recently, it was generally accepted that high face velocities offered better protection. However, most recent data support a maximum face velocity of 100

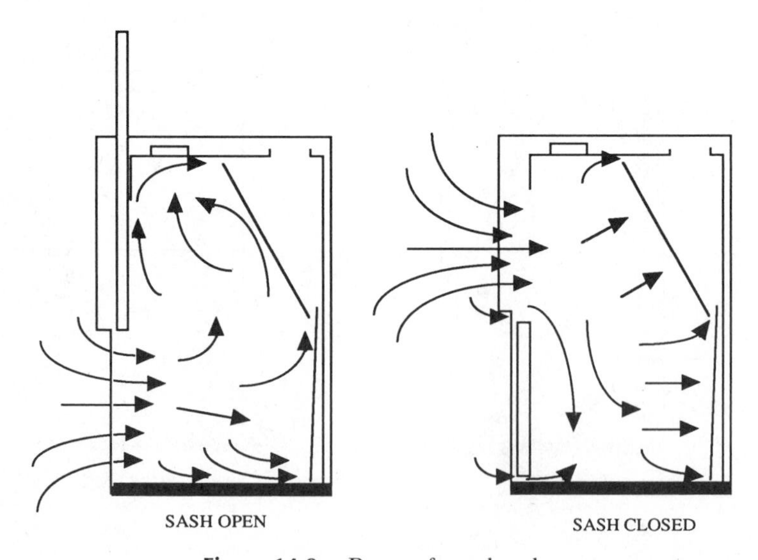

Figure 14.9. Bypass fume hood.

fpm with face velocities of 60 to 80 fpm, provided proper hood safety procedures are followed (8).

Ultimately, hood face velocities should be determined by the specific use and practices in the laboratory. The American Society of Heating, Refrigerating, and Air Conditioning Engineers (ASHRAE) has published guidelines that should be consulted during the design phase (9).

A properly designed laboratory hood should have the following characteristics:

- Uniform air velocity (± 20 percent over the access opening)
- Smooth, rounded, and tapered openings to minimize turbulence at the hood face
- Constant face velocities maintained regardless of sash opening height
- Airflow monitors incorporated into each hood such as liquid-filled draft gauges and magnehelic gauges

Most laboratory furniture supplies have a selection of models that meet these performance standards.

The supply of makeup air is critical to the optimum performance of laboratory hoods. The design of the room air supply system is as important to lab hood performance as face velocity. The volume of air exhausted from fume hoods must be replaced with an equal amount of incoming air. The location of the makeup air outlets and temperature of the supply air are also important design factors.

Supply air should be introduced at low velocity (approximately $\frac{1}{2}$ to $\frac{2}{3}$ the hood face velocity) and in a direction that does not cause disruptive cross-drafts at the hood opening.

Supply air can be introduced through diffusers, wall grilles, or perforated ceiling panels located adjacent or opposite to the fume hood. Introducing supply air through perforated ceiling panels is the most efficient means because these panels are of simple design and easy to apply, and do not require precise adjustment. The ceiling panels should be located at least one meter from the hood face.

Proper location of fume hoods within the laboratory is another important aspect of effective and safe hood performance. Fume hoods should be located away from:

- Doorways, for a fire or explosion in the hood could block an exit
- Major traffic corridors to minimize cross-drafts and turbulence
- Room corners and openable windows

In addition to the standard laboratory fume hood, special hoods may be needed to handle specific chemical hazards. These include perchloric acid hoods, glove boxes for highly toxic materials, high-efficiency particulate air (HEPA)-filtered hoods, and clean bench hoods. Each of these hoods is discussed briefly below.

3.4.3.2 Perchloric Acid Hoods. Perchloric acid hoods must meet the same face velocity and other design criteria noted for general fume hoods. They do, however, have several specific design features. The perchloric acid hood should be constructed of stainless steel and have welded seams. Taped seams, putties, or sealers can be used in the fabrication of the hood or duct system. An internal water wash capability is required to eliminate the buildup of perchlorate material. Extremely explosive organic perchlorate vapors can condense in the hood exhaust system and perchlorate residue can detonate on contact during cleaning and repair. The water system must be capable of rinsing all portions of the hood system including ductwork. A vertical duct system should be designed for ease in cleaning and visual inspection.

3.4.3.3 Glove Boxes. The typical laboratory fume hood is very effective in protecting the analyst from volatile solvent. However, when highly toxic or carcinogenic materials are being handled, it is prudent to use a glove box enclosure. The glove box is the ultimate enclosure. It is a type of hood contained in a sealed box fitted with flexible airtight gloves for performing manual operations inside the box.

The exhaust airflow is calculated to effectively remove contaminants generated from work performed inside the enclosure, maintain the glove box interior at a negative pressure, and prevent contaminant escape if a glove fails. Generally, the airflow exhaust is 50 cubic feet per minute (cfm) per square foot of open door area.

The glove box must be an isolated system to protect the analyst from exposure to toxic and carcinogenic materials. The isolated glove box includes an air lock system. The airlock allows materials to be transferred into and out of the glove box while preventing the escape of contaminants from inside the box.

Air filtering equipment is used in glove box enclosures to remove contaminants within the glove box before they are exhausted to the outside. The typical glove box contains three filters. The first is a particulate filter. The second filter is generally a charcoal bed, which is effective in removing volatile solvents. The third filter is a HEPA filter.

3.4.3.4 HEPA Filters. The HEPA filter is constructed of pleated paper or other filter material bonded to the filter frame, and is at least 99.97 percent efficient in collecting a 0.3-μm aerosol. Because HEPA filters are highly efficient in removing small particles from the air stream, the prefilter must be kept in proper working condition. The effectiveness of the prefilter will extend the life of the HEPA filter.

There are a number of products on the market designed specifically for use in preparation of asbestos samples. The HEPA filter hood is one of two ventilation designs that must be considered when a laboratory is handling asbestos. The second, the clean bench hood, is discussed in the next section.

It is important to prevent analyst exposure and release of asbestos to the outside air during asbestos fiber counts and analysis of asbestos-containing materials. Use of HEPA filters in a standard hood design can be extremely complicated because of pressure differences upstream and downstream of the filter and fan speed required to maintain adequate face velocity.

The advantages of commercial specialty hood systems are their small size and self-contained HEPA filter and fan assemblies. At a relatively low cost of $600 to $800 per hood, the design team should consider this purchase before designing a HEPA ventilation system for a standard fume hood.

3.4.3.5 Clean Bench Hoods. The second specialty unit for asbestos is the clean bench hood which is used in preparing samples for transmission electron microscopy. The purpose of the clean bench is to ensure that samples do not become contaminated with laboratory air. Laboratory air is filtered through a HEPA filter before being exhausted across a work surface. Clean benches are commercially available, and purchase, rather than design, of a separate system is recommended.

3.4.4 Electrical

The electrical power requirements of laboratories are significantly greater than those of the typical office building. Careful study must be done to plan for current and future requirements. The design of electrical systems must allow for growth and change with reasonable ease and without major disruption to operations during renovation. Recommended electrical design features include 50 percent excess capacity in electrical service and reserve for additional equipment and distribution panels.

The first step in designing the electrical system is to assemble information on the use of electrical power by the laboratory. The electrical requirements should be listed by area according to laboratory equipment in use and the voltage and operating amperage of each instrument. All laboratory equipment should be listed, including such items as hot plates, centrifuges, and meters.

Current and anticipated use rates should also be provided to the electrical system design team. Use rates will allow calculation of continuous power required and peak power.

The initial space plan model should be used to place electrical equipment together with electrical requirements. This allows the design team to determine optimum placement and sizing of distribution panels within the laboratory. Noting the growth potential of individual laboratory areas will also allow the design team to plan future growth in areas of probable need, rather than oversizing all electrical systems in all areas of the laboratory.

Because laboratories rely on computers to operate instruments and analyze data, it is important to consider the special electrical requirements of computer operations. Because most computers require minimal line disturbances, an isolated ground should be installed that reduces the electrical fluctuations to the equipment. Isolated grounds should be installed where necessary and as specified by the equipment manufacturer. Installation of line filters, static voltage regulators, or shielded transformers may also be necessary. Electrical service should be located so that it can be easily and quickly disconnected. This is an important safety consideration and should not be overlooked during the design phase.

3.4.5 Plumbing

Laboratories require complete sanitary and storm water drainage systems. Complicating water supply design are special requirements for neutralizing sumps before contents are discharged to the public sanitary sewer. A comprehensive report on each laboratory function should be developed to determine the types of chemicals to be discharged. Waste piping can then be selected to handle the particular chemicals. Piping materials for these systems include high silicon iron, borosilicate glass, polypropylene, and polyvinyl chloride.

Acid waste must be neutralized by chemical reaction. A typical application is a neutralization sump filled with limestone chips to raise the pH of the discharge to suitable levels. A centrally located neutralizing sump can be used to handle wastes from a number of sources.

The water system design for a laboratory is more complicated than that for typical office buildings. Systems usually encountered in laboratories are potable water, both hot and cold; cooling water for mass spectrometers, electron microscopes, and X-ray diffraction instruments; and high-purity distilled water.

All laboratory areas should have access to the water supply. Again, an inventory of needs will help the design team in locating water outlets. Laboratory water supplies should be designed to prevent the drinking water supply from being contaminated. All plumbing fixtures should be equipped with anti-siphon devices or backflow preventers.

Most laboratories require a source of pure water. Pure water systems consist of distilled water, deionized water, or demineralized water. There are four basic methods of producing pure water—distillation, demineralization, reverse osmosis, and filtration. Depending on the type of pure water that is required, one or more of these methods will be needed.

The use of mass spectrometers, electron microscopes, and X-ray diffraction units, for example, requires a continuous cooling water system. Cooling water systems can be designed to use either a cold water supply and discharge or a recirculating cooling unit depending on the frequency of use, economics, and environmental considerations.

3.4.6 Compressed Gas System

A wide variety of compressed gases are needed to operate laboratory instrumentation, including air, nitrogen, hydrogen, helium, acetylene, and argon. The design of compressed gas distribution depends on the overall requirements of the laboratory. In many cases, compressed gas is delivered to the instruments by compressed gas tanks located within the particular laboratory.

In other cases, compressed gases can be delivered via gas manifold systems. When designing a gas manifold system, the design team must pay particular attention to local building and fire protection codes and pressure requirements. Safety features that should be considered are backflow preventers, spark arrestors, and automatic shutoffs.

If the manifold method is chosen, it is imperative that ultra-clean piping (oxygen supply grade) be used. Joints must not be soldered. Fittings such as those marketed by Swagelok must be used.

4 CONSTRUCTION

4.1 The Construction Team

The efforts of the exhaustive design planning and successful building of the laboratory hinge on the quality of the general contractor. The general contractor should be hired early in the design phase. Employing the contractor at this time allows the general contractor to be part of the design process and provide valuable insight into specifics on the actual construction details.

Selection of a general contractor should be approached in much the same manner as choosing a doctor, lawyer, or accountant. All aspects of a contractor's experience and qualifications must be thoroughly investigated. All general contractors will claim to be able to build buildings, and most could do a fine job. However, the building of laboratory facilities requires special expertise and qualifications. In many cases, the general contractor can also arrange services of architects and engineers.

The construction team includes the client, the architect, the mechanical engineer, and the general contractor. To ensure success, each representative of the construction team must be qualified with respect to experience specific to laboratory design.

In addition to reviewing the qualifications of the general contractor, each trade participating in the construction should be investigated. Trades such as HVAC, plumbing, electrical, and carpentry should submit a list of previous laboratory projects completed.

Visits to laboratories designed and constructed by the construction team should be conducted to help assess qualifications. Interviews with laboratory staff at these locations will be helpful. Staff should be asked about satisfaction with design and workmanship. As many visits and interviews should be made as are necessary to satisfy questions concerning qualifications and experience completely. The time spent in qualifying the construction team is easily justified when put in terms of the useful life of the laboratory facility.

The construction of a laboratory is a complicated process subject to many changes during the building process. To aid communication between the design team and the construction team, a construction manager should be appointed. The construction manager should handle questions, changes, and problems that are inevitable in the building process. The construction manager must be intimately involved in the project and knowledgeable in laboratory operation and construction techniques. Above all, the manager must be an effective communicator and must be detail-oriented and diplomatic.

An independent expert in building construction should also be employed to serve as the "tenant's representative." The tenant's representative is in fact the client's building inspector. The principal purpose of this position is to ensure the building owner that the construction blueprints are being followed by the contractor and trades. In other words, the tenant's representative is a project overseer.

Although it may appear to be unnecessary to employ a tenant's representative, because the qualifications of the contractor and trades have been thoroughly investigated, it does provide another means of ensuring the quality of the work. As noted before, the construction of a laboratory is extremely complex, involving extensive HVAC, electrical, and plumbing construction. An outside expert can help prevent, and quickly correct, building errors not uncovered by the general contractors, site supervisors, or local building inspectors.

4.2 Review of Final Blueprints

With the construction team of construction manager, general contractor, architect, and tenant's representative in place, the next phase of the construction process is to perform a critical review of the final blueprints. As noted before, laboratory design and engineering are extremely complicated. Neglecting a final exhaustive review may result in a less-than-optimal working design and can prove costly in future changes. Depending on the construction team's overall confidence in the design and engineering phases, the blueprint review process may need to include outside consultants and engineers to give a third-party unbiased evaluation of the final blueprints.

Each building blueprint should be reviewed in the context of the original design and specifications. The following are questions to ask during the review phase:

- Are the overall plans consistent with the original design concept?
- Will the ventilation and hood systems operate according to desired specifications?
- Are the electrical and plumbing plans consistent with instrument placement and service requirements, such as voltage and amperage necessary, hot and cold water, and sanitary and chemical drains?

During this phase of the construction program, the laboratory staff or at least senior level staff should critically review the final plans. It is important for the construction team to be open-minded, at this point, to small changes in space requirements and bench layout.

Typically, the process from early design to final blueprint is 9 to 12 months. Depending on the degree of change in the laboratory organization over time, it is sometimes necessary to rethink the original design in terms of space allocations and services. This is not to imply that wholesale changes should and can be made, but the design team should not rule out current thinking that would be best addressed before construction. Types of changes could include varying interior room sizes or bench layout or modifying service requirements.

The decision to make changes during final blueprint review should rest with the construction team, after consideration of up-front costs versus after-construction costs. Much depends on the overall flexibility and expansion built into the original designs.

The review phase should also include inspection and definition of the actual construction materials specified in the plan. This does not necessarily mean reviewing the color scheme or wall coverings, although these may be considered. Because laboratories use and store corrosive chemicals, and background contamination must be minimized or eliminated, the construction team must determine that all construction materials meet laboratory specifications. For instance, the acid and solvent resistance of the floor covering and plumbing and ventilation ductwork should be determined. Adhesives used in construction, such as flooring and drywall adhesives, can cause long-term off-gassing problems with background solvents interfering in laboratory analysis of samples. Because of typical solvent off-gassing interfering with future analysis, we recommend the use of latex rather than solvent-based paints.

4.3 Communication During Construction

Once the design and construction teams are satisfied with the plans, specifications, and materials, the physical construction of the facility can begin. The construction manager and tenant's representative must take an active role in overseeing the facility construction. Constant communication and sometimes arbitration and mediation are required to ensure the timely completion of the project. Written communication is highly recommended for all exchanges of information between the contractor, trades, and construction manager. All conversations involving problems or change orders must be summarized in writing and copied to the design and construction team. This will provide some assurance that changes are appropriate and implemented.

4.4 Building Checklist

The design and construction teams must guard against a hasty move-in and start-up of the new facility. When construction is complete, a general excitement at completion of the project can overcome even the most stoic of laboratory managers. However, prior to move-in and start-up, the entire facility must be carefully checked to ensure proper operation of services and utilities.

To facilitate this process, a checklist should be created listing all electrical outlets, hot and cold water supplies, drains, hoods, and compressed gases if manifolds are used. The next step is to test each individual point for proper operation. This entails verifying that electrical outlets are functioning by using a voltmeter, hot and cold water supplies and drains are functioning by allowing sufficient flow to verify the hot water supply, and hood and ventilation system operations are functioning using flowmeters, such as thermoanenometers and Velometers. In addition to the fume hood and ventilation systems check, a complete ventilation balancing audit should be conducted by a certified air-balancing contractor.

The gas lines and vacuum system, if used, should be leak tested with soap solution. If a gas manifold is used for compressed gases such as air, helium, hydrogen, argon, and acetylene, it may be appropriate, depending on the complexity of the manifold, to verify that the correct gas is being delivered to the exit point and the lines are free of contamination. This can be done by connecting the appropriate laboratory instrument and performing an operations check. For instance, on a manifold designed for multiple gas chromatographs, a single unit could be used to verify all exit points, rather than testing the operation after moving all the instrumentation.

Conducting a thorough check of all systems before move-in will help ensure a smooth start-up and minimize costly downtime.

5. SAFETY DESIGN AND PLANNING

5.1 Chemical Storage

Space for storing chemicals in a storeroom and within areas of use in the laboratory must be planned for in the design of the laboratory. Lack of sufficient storage space can create hazards because of overcrowding, storage of incompatible chemicals together, and poor housekeeping. Adequate, properly designed and ventilated storage facilities must be provided to ensure personnel safety and property protection.

5.1.1 Chemicals Used in the Industrial Hygiene Laboratory

The industrial hygiene laboratory analyzes trace amounts of contaminants and typically does not store large volumes of chemicals. However, chemicals used include flammable liquids, high chronic toxicity substances, and incompatible chemicals.

Chemicals that may be stored in quantities of 1 gal or more include:

- Carbon disulfide for desorption of organic compounds from activated charcoal and analysis by gas chromatography (GC)
- Solvents for use in high-pressure liquid chromatography (HPLC), including methanol, acetonitrile, and methylene chloride
- Solvents, including chloroform and dimethyl formamide, for use in preparing air sample filters for analysis of asbestos by transmission electron microscopy (TEM)
- Acids such as nitric and hydrochloric for digestion of air sample filter materials for analysis of metals

Because of the large number of analytes within the laboratory's scope of work, the laboratory may store small quantities of several hundred materials to be used as standards. Some of the chemicals may be known human carcinogens or suspect carcinogens. A small number of reagents may be stored typically in quantities of 500 g or less.

5.1.2 Chemical Storeroom Design

Storeroom size must be large enough to meet current chemical storage needs and to accommodate future growth. Areas must be planned for segregation of incompatible chemicals, storage of high chronic toxicity substances in an identified area, and storage of certain chemicals in an explosion-proof refrigerator. More than one room may be required.

All chemical storerooms should be under negative pressure with respect to the surrounding area. Air supplied to the room should not be returned to the building HVAC system but should be exhausted to the roof.

5.1.2.1 Flammable Liquids. The most effective way to minimize the impact of a hazard such as fire and explosion is to isolate it. A flammable liquid storeroom is best located in a special building separated from the main building. If the room must be located within a main building, the preferred location is a cut-off area on the first floor (at grade) level with at least one exterior wall. In any case, storage rooms for flammable liquids should not be placed on the roof, on a below-grade level, an upper floor, or in the center of the building. All of these locations are undesirable because they are less accessible for firefighting and are potentially dangerous to personnel in the building.

The walls, ceiling, and floor of an inside storage room for flammable liquids should be constructed of materials having at least a 2-hr fire resistance. The room should have self-closing Class B fire doors [see the OSHA standard (29CFR 1910.106 (d) (4) (i)) or the National Fire Protection Association (NFPA) standard (No. 30.4310)]. All storage rooms should have (1) adequate mechanical ventilation controlled by a switch outside the door and (2) explosion-proof lighting and switches. Other potential sources of ignition, such as cigarettes and open flames, should be forbidden.

When space is limited or renovation out of the question, a flammable liquid storage cabinet exhausted to the outside is an alternative for safely storing small quantities of flammable liquids.

5.1.2.2 High Chronic Toxicity Substances Including Carcinogens. The industrial hygiene laboratory stores and uses small quantities of high chronic toxicity substances including known and suspected human carcinogens. Examples are benzo[*a*]pyrene, bis(chloromethyl) ether, and nitrosamines.

High chronic toxicity substances should be segregated from other substances and stored in a well-defined or identified area that is cool, well ventilated, and away from light, heat, acids, oxidizing agents, and moisture. For protection of laboratory and nonlaboratory personnel, such chemicals should be stored in a locked cabinet or in an area accessible only by designated personnel. The area should be identified with warning signs such as: WARNING! HIGH CHRONIC TOXICITY OR CANCER-SUSPECT AGENT. All containers of substances in this category should have labels that identify the contents and include a warning as above.

Procedures for working with substances of known high chronic toxicity should be adopted and followed (see Reference 10).

5.1.2.3 Incompatible Chemicals. Incompatible chemicals should not be stored together. Such contact could result in a serious explosion or the formation of highly toxic and/or flammable substances.

Table 14.2 gives examples of incompatible chemicals. Several of the chemicals listed are not in common use in industrial hygiene laboratories. However, information on these chemicals is provided since many labs change their scope of work over time or inherit chemicals from previous operations.

5.1.2.4 Water-Sensitive Chemicals. If the laboratory uses any water-sensitive chemicals, such as sodium metal, it must store them in an area with no sources of water. The area should not have an automatic sprinkler system. Storage areas for such chemicals should be of fire-resistant construction, and other combustible materials should not be stored in the same area.

5.1.2.5 Chemicals Requiring Refrigeration. Some chemicals that are very volatile or unstable at room temperature must be refrigerated. Chemical storage refrigerators must be explosion-proof and wired to an explosion-proof outlet. Such refrigerators should be located in a designated storeroom that is supplied with single-pass air exhausted to the outside and that is under negative pressure with respect to the surrounding area.

5.1.2.6 Compressed Gases. Storage and use of compressed gases are covered in Sections 3.4.6 and 5.1.3.3.

5.1.3 Chemical Storage in the Laboratory

Every chemical in the laboratory should have a definite storage place and should be returned to that location after each use. Chemicals should not be stored on benchtops because of the potential exposure to fire and the possibility of being knocked over. Chemicals should not be stored in hoods because they will interfere with the airflow, clutter up the working space, and increase the amount of materials that could become involved in a hood fire.

Volatile solvents, concentrated acids, and toxic substances should be handled in a chemical fume hood. Design of the industrial hygiene laboratory should include an adequate number of hoods of appropriate size for staff to conduct their work safely.

5.1.3.1 Beneath-Hood Storage. Because most laboratory workers tend to store chemicals in the cabinet space under the hood, ventilated cabinets designed by the manufacturer for safe storage of acids or solvents should be provided. Cabinets located directly under the hood also allow for the safe practice of making transfers of hazardous materials in the hood. The use of each hood should be planned and the appropriate base cabinet purchased for beneath-hood chemical storage.

5.1.3.2 High Chronic Toxicity Substances. Only minimum quantities of toxic materials should be present in the work area. Such substances should be handled in glove boxes or hoods according to procedures established by the laboratory.

5.1.3.3 Compressed Gases. In the industrial hygiene laboratory, compressed gases are used for atomic absorption, gas chromatography, and other instrumentation. If the laboratory is small, these specialty gas needs may be best handled by the use of gas cylinders stored and tied down at the bench or point of use. However, if the laboratory is large or its needs extensive, a central piping system may be needed. Gases may be provided from banks of high-pressure cylinders located in a convenient and safe place from which gas is piped through reducing valves to desired locations. This system is also an asset to laboratory housekeeping and safety, because gas cylinders are removed from the immediate work area.

5.2 Disposal of Waste

Typical wastes generated daily from the analysis of industrial hygiene samples include:

- Carbon disulfide from analysis of solvents on charcoal tubes
- Dilute nitric acid from analysis of metal fumes on filters
- Bulk samples of building materials containing asbestos
- Aqueous mixtures of reagents from colorimetric analyses

All wastes generated by the laboratory must be disposed of in accordance with applicable federal, state, or local waste disposal regulations. Proper disposal methods vary with the type of waste, the quantities generated, and regulatory requirements.

5.2.1 Regulations

Laboratories (11) have a moral and legal obligation to see that chemical waste is handled and disposed of in ways that pose minimum potential harm, both short-term and long-term, to health and the environment. The legal obligations began on November 19, 1980, when the U.S. Environmental Protection Agency (EPA) put into effect federal regulations on a Hazardous Waste Management System (40 CFR Parts 260 through 266), under the authority of the Resource Conservation and Recovery Act (RCRA) of 1976, as amended. These regulations were designed to establish a "cradle-to-grave" system for the management of hazardous waste from all sources.

EPA's November 19, 1980 RCRA regulations focused on large generators who produce most of the hazardous waste in the country. As a result, the initial regulations exempted small-quantity generators—those producing less than 1000 kg per calendar month (kg/mo) of hazardous waste—from manifesting and record-keeping requirements applicable to large generators. More recently, however, the

Table 14.2 Examples of Incompatible Chemicals (10)

Chemical	Is Incompatible With
Acetic acid	Chromic acid, nitric acid, hydroxyl compounds, ethylene glycol, perchloric acid, peroxides, permanganates
Acetylene	Chlorine, bromine, copper, fluorine, silver, mercury
Acetone	Concentrated nitric and sulfuric acid mixtures
Alkali and alkaline earth metals (such as powdered aluminum, magnesium, calcium, lithium, sodium, potassium)	Water, carbon tetrachloride or other chlorinated hydrocarbons, carbon dioxide, and halogens
Ammonia (anhydrous)	Mercury (in manometers, for example), chlorine, calcium hypochlorite, iodine, bromine, hydrofluoric acid (anhydrous)
Ammonium nitrate	Acids, powdered metals, flammable liquids, chlorates, nitrites, sulfur, finely divided organic or combustible materials
Aniline	Nitric acid, hydrogen peroxide
Arsenical materials	Any reducing agent
Azides	Acids
Bromine	See Chlorine
Calcium oxide	Water
Carbon (activated)	Calcium hypochlorite, all oxidizing agents
Carbon tetrachloride	Sodium
Chlorates	Ammonium salts, acids, powdered metals, sulfur, finely divided organic or combustible materials
Chromic acid and chromium trioxide	Acetic acid, naphthalene, camphor, glycerol, alcohol, flammable liquids in general
Chlorine	Ammonia, acetylene, butadiene, butane, methane, propane (or other petroleum gases), hydrogen, sodium carbide, benzene, finely divided metals, turpentine
Chlorine dioxide	Ammonia, methane, phosphine, hydrogen sulfide
Copper	Acetylene, hydrogen peroxide
Cumene hydroperoxide	Acids (organic or inorganic)
Cyanides	Acids
Flammable liquids	Ammonium nitrate, chromic acid, hydrogen peroxide, nitric acid, sodium peroxide, halogens
Fluorine	Everything
Hydrocarbons (such as butane, propane, benzene)	Fluorine, chlorine, bromine, chromic acid, and sodium peroxide
Hydrocyanic acid	Nitric acid, alkali
Hydrofluoric acid (anhydrous)	Ammonia (aqueous or anhydrous)
Hydrogen peroxide	Copper, chromium, iron, most metals or their salts, alcohols, acetone, organic materials, aniline, nitromethane, combustible materials

Table 14.2 (continued)

Chemical	Is Incompatible With
Perchloric acid	Acetic anhydride, bismuth and its alloys, alcohol, paper, wood, grease, oils
Peroxides, organic	Acids (organic or mineral), avoid friction, store cold
Phosphorus (white)	Air, oxygen, alkalis, reducing agents
Potassium	Carbon tetrachloride, carbon dioxide, water
Potassium chlorate	Sulfuric and other acids
Potassium perchlorate (See also chlorates)	Sulfuric and other mineral acids
Potassium permanganate	Glycerol, ethylene glycol, benzaldehyde, sulfuric acid
Selenides	Reducing agents
Silver	Acetylene, oxalic acid, tartartic acid, ammonium compounds, fulminic acid
Sodium	Carbon tetrachloride, carbon dioxide, water
Sodium nitrite	Ammonium nitrate and other ammonium salts
Sodium peroxide	Ethyl or methyl alcohol, glacial acetic acid, acetic anhydride, benzaldehyde, carbon disulfide, glycerin, ethylene glycol, ethyl acetate, methyl acetate, furfural
Sulfides	Acids
Sulfuric acid	Potassium chlorate, potassium perchlorate, potassium permanganate (similar compounds of light metals, such as sodium, lithium)
Tellurides	Reducing agents
Hydrogen sulfide	Fuming nitric acid, oxidizing gases
Hypochlorites	Acids, activated carbon
Iodine	Acetylene, ammonia (aqueous or anhydrous), hydrogen
Mercury	Acetylene, fulminic acid, ammonia
Nitrates	Sulfuric acid
Nitric acid (concentrated)	Acetic acid, aniline, chromic acid, hydrocyanic acid, hydrogen sulfide, flammable liquids, flammable gases, copper, brass, any heavy metals
Nitrites	Acids
Nitroparaffins	Inorganic bases, amines
Oxalic acid	Silver, mercury
Oxygen	Oils, grease, hydrogen, flammable liquids, solids, or gases

U.S. Congress also has directed EPA to regulate small-quantity generators of hazardous waste. Specifically, Congress lowered from 1000 kg/mo to 100 kg/mo the quantity at which generators of hazardous waste are exempt from compliance with certain RCRA regulations. This means that generators of more than 100 kg/mo of hazardous waste—which include many small laboratories—are now subject to specific RCRA regulations.

Wastes generated in laboratories can differ significantly from wastes generated at industrial production sites. Laboratories, which typically produce small quantities of many different wastes, face problems that set them apart from production facilities that generate large quantities of homogeneous wastes. Proper handling and disposal of laboratory wastes may involve procedures different from those used in a typical manufacturing environment. Therefore, when following the RCRA regulations, the unique nature of a laboratory environment must be taken into consideration.

The EPA can authorize state hazardous waste programs to operate in lieu of the federal program (12). Such state programs must be at least equivalent to and consistent with the federal program. The RCRA regulations do not preclude states from adopting more stringent hazardous waste regulations, and many have done so. All generators should check with state and local authorities to determine their local responsibilities. State contacts for information on local requirements are listed in Table 14.3.

5.2.2 Hazardous Waste Management System

The selection of a hazardous waste manager or coordinator for each laboratory or facility is important because laboratory managers are obligated to understand and to comply with the applicable elements of RCRA and the corresponding U.S. Department of Transportation (DOT) regulations. The laboratory hazardous waste coordinator should be familiar with the regulations and be able to organize a program that is practical to implement and that addresses both environmental and economic concerns.

As mentioned earlier, typical wastes generated by industrial hygiene laboratories include spent solvents, unused chemicals, samples, and various contaminated containers, supplies, and equipment. Although many wastes generated by laboratories are not wastes appearing on EPA lists, many will be regulated because of hazardous characteristics (i.e., ignitability, corrosivity, reactivity, and EP toxicity). A hazardous waste management program should take into account both regulations and a common sense approach to hazardous material safety.

5.2.3 Waste Management Options

Until recently, the majority of laboratory wastes were placed in secure chemical landfills. Because of long-term concerns with liability, landfill space, and regulations that may restrict land disposal, alternative disposal methods are now being considered. Landfilling of wastes still may be the most economical disposal method in some geographic areas, but long-range concerns suggest that the generator should

Table 14.3 State Solid and Hazardous Waste Agencies

The following are telephone numbers of state agencies responsible for the proper disposal of waste materials. The asterisks (*) indicate states having EPA-authorized state hazardous waste programs.

Alabama	(205)	271-7737	Montana*	(406)	444-2821
Alaska	(907)	465-2671	Nebraska*	(402)	471-4217
Arizona*	(602)	257-2211	Nevada*	(702)	885-5872
Arkansas*	(501)	562-7444	New Hampshire*	(603)	271-2900
California	(916)	324-1826	New Jersey*	(609)	292-8341
Colorado*	(303)	331-4830	New Mexico*	(505)	827-2611
Connecticut	(203)	566-5712	New York	(518)	457-3274
Delaware*	(302)	736-3689	North Carolina*	(919)	733-2178
Florida*	(904)	488-0300	North Dakota*	(701)	224-2366
Georgia*	(404)	656-2833	Ohio	(614)	644-2917
Hawaii	(808)	543-8335	Oklahoma*	(405)	324-5033
Idaho	(208)	334-5879	Oregon*	(503)	229-5913
Illinois*	(217)	782-6760	Pennsylvania	(717)	787-7381
Indiana*	(317)	232-3210	Rhode Island*	(401)	277-2797
Iowa	(515)	281-8308	South Carolina*	(803)	734-5213
Kansas*	(913)	296-1607	South Dakota*	(605)	773-3153
Kentucky*	(502)	564-6716	Tennessee*	(615)	741-3424
Louisiana*	(504)	342-1354	Texas*	(512)	463-7830
Maine	(207)	289-2651	Utah*	(801)	538-6170
Maryland*	(301)	631-3000	Vermont*	(802)	244-8702
Massachusetts*	(617)	292-5853	Virginia*	(804)	225-2667
Michigan	(517)	373-2730	Washington*	(206)	459-6301
Minnesota*	(612)	296-6300	West Virginia	(304)	348-5935
Mississippi*	(601)	961-5062	Wisconsin*	(608)	266-2111
Missouri*	(314)	751-3176	Wyoming	(307)	777-7752

give careful consideration to alternative disposal methods. The most common alternative methods are discussed briefly below.

5.2.3.1 ***Waste Reduction.*** The most effective method of waste management is minimization of waste generation. The control of chemical purchasing practices and sharing of chemicals among departments, labs, or facilities offer two ways of doing this. By reducing the volume of chemicals entering the user system, the generator ultimately reduces the waste. Increasing costs of treatment or disposal will provide an incentive for waste reduction.

5.2.3.2 ***Recycling.*** In-house recycling is an environmentally sound and cost-effective way to manage certain waste chemicals, particularly solvents. However, the wastes accumulated for recycling are regulated under RCRA from the time of accumulation until they are recycled. The waste materials must be labeled, segregated, and stored according to the regulations applicable to the particular class of hazardous waste generator performing the recycling. The recycling activities and resulting products are not regulated under RCRA, but the residues from recycling are generally hazardous wastes.

5.2.3.3 ***Thermal Treatment.*** Thermal treatment of hazardous waste involves using high temperatures to change the chemical, physical, or biological character or composition of a waste. The objective of this process is to convert hazardous materials into nonhazardous by-products.

Incineration is the type of thermal treatment of most use to laboratories. Incineration can be used to destroy organic hazardous wastes safely by reducing waste materials into nontoxic gases and small amounts of ash and other residues (note, however, that the ash itself is considered a hazardous waste unless proved otherwise to EPA).

Many materials, however, are not suitable for incineration because of their high toxicity or reactivity. Mercury and mercury compounds, for instance, are not accepted for disposal by most incinerators. Highly reactive or explosive materials should not be incinerated.

Incineration may involve a different choice of outer packaging for off-site transportation from that of other disposal methods. Check with the incineration facility for packaging procedures before preparing a shipment. Because of a shortage of available facilities, lead times prior to shipment may vary considerably. This variation could affect the length of on-site storage. In the short term, incineration may be an expensive option; however, in the long term, if may offer the lowest liability.

5.2.3.4 Chemical Treatment. By altering the character or composition of a material by chemical treatment, a hazardous waste can be rendered less hazardous. Chemical treatment includes solidification, neutralization, oxidation, reduction, hydrolysis, and precipitation. Chemical treatment often can be done in the laboratory by the chemist who generated the material and knows its properties best. Information on reducing or destroying chemical wastes can be found in the National Research Council's *Prudent Practices for Disposal of Chemicals from Laboratories* (11).

5.2.3.5 Recovery. Sometimes it is economically feasible to regenerate chemical wastes so that the original products are reusable, especially large volumes of solvents. To date, laboratory recovery practices have been limited. However, the recovery of laboratory chemicals such as valuable metals (e.g., mercury, silver, platinum) and common water-insoluble solvents (e.g., chloroform) is increasing. As disposal costs rise, other recoveries will become more economical.

5.2.3.6 Sewage System. Under certain circumstances it is possible to flush certain waste materials into a municipal sewage system. Neutralized acids and bases and organic chemicals with 3 percent or greater solubility in water that are not highly toxic, flammable, or foul smelling are possible candidates if local regulations permit this type of disposal. Local regulations must be consulted and complied with.

5.2.4 Sources of Information on Chemical Waste Disposal

The most widely used reference on laboratory waste disposal is *Prudent Practices for Disposal of Chemicals from Laboratories* (11). Other references on waste disposal are also listed at the end of this chapter.

Questions on EPA regulations can be directed to EPA's RCRA "hotline" at (800) 424-9346 (toll-free) and (202) 382-3000 (metropolitan Washington, DC area).

5.3 Safety Showers and Eyewash Stations

Each laboratory should be equipped with at least one safety shower and one eyewash station. Safety showers and eyewash stations should be placed near operations with the greatest risk of splashes, spills, or fire.

The American National Standards Institute (ANSI) standard on emergency deluge showers (13) provides the following design criteria for safety showers and eyewashes:

- Both must be within 30 steps walking distance from the farthest occupied space in the laboratory
- Shower and fountain should be at least 5 ft apart so they could be used by two people simultaneously
- Both should be equipped with tempered water (the necessary 15-min minimum flushing period is difficult to achieve when the water is cold)
- Showers should have approximately 50 gal/min flow
- Eyewash fountains should have approximately 2.5 gal/min flow

An adequately sized floor drain is useful to speed cleanup.

If possible, deluge showers and eyewash fountains should be positioned so that laboratory staff can quickly and reliably locate them in an emergency. Bright distinctive floor markings and signage aid in safety shower and eyewash fountain identification.

5.4 Fire Extinguishers and Fire Suppression Systems

Fire extinguishers should be of the type and size appropriate to the work conducted in the laboratory. They should be mounted on a wall at a comfortable height in a convenient location. They should be placed where coats or equipment cannot impede their use.

The use of fire blankets in laboratories is controversial. While a tightly wrapped blanket may extinguish burning clothing, it may also press molten synthetic fabric deeply into burned skin and tissue. If mounted in a laboratory, a fire blanket should be within 30 steps walking distance from any part of the lab and well marked.

Typically, laboratory fire suppression systems are water sprinklers. Water systems are hazardous in high-voltage locations and in locations where chemicals that are incompatible with water are used or stored. Also, a water system may not be the best choice in rooms where expensive equipment would be ruined by water. Alternative, but more expensive, fire suppression systems are available for these applications. Fire detection devices should be installed in all areas not protected by fire suppression systems. Local fire codes dictate requirements for fire suppression systems, when used, and should be consulted during the design phase. In buildings in which water sprinkler pipes run through unheated plenum space, pipes may freeze and burst during cold weather. The design phase should include consideration of pipe insulation or a compressed-air dry system to prevent water damage from frozen pipes.

5.5 First Aid Planning

Most industrial hygiene laboratories do not have medical personnel on staff or within the building. The lab should plan to have personnel trained in first aid, including cardiopulmonary resuscitation (CPR), available during working hours to respond to emergencies until medical help can be obtained. Emergency phone numbers should be displayed on all laboratory phones.

An emergency room staffed with medical personnel specifically trained in proper treatment of chemical exposures should be identified and readily accessible. Prior arrangement with a nearby hospital or emergency room may be necessary to ensure treatment will be available promptly. The services of an ophthalmologist especially alerted to and familiar with chemical injury treatment should also be available to minimize the damage to eyes that may result from many types of laboratory accidents. Proper and speedy transportation of the injured to the medical treatment facility should be available. In addition to the normal facilities found in an emergency room, there should be specific standing orders for emergency treatment of chemical accidents. Emergencies that should be anticipated include:

- Thermal and chemical burns, including those caused by hydrofluoric acid
- Cuts and puncture wounds from glass or metal, including possible chemical contamination
- Skin irritation by chemicals
- Poisoning by ingestion, inhalation, or skin absorption
- Asphyxiation (chemical or electrical)
- Injuries to the eyes from splashed chemicals

5.6 Emergency Evacuation Design

The laboratory should establish an emergency plan that includes evacuation routes, shelter areas, medical facilities, and procedures for reporting all accidents and emergencies.

Evacuation procedures (10) should be established and communicated to all personnel. Evacuation routes and alternatives may have to be established and, if so, communicated to all personnel. An outside assembly area for evacuated personnel should be designated.

An emergency alarm system should be available to alert personnel in an emergency that may require evacuation. Laboratory personnel should be familiar with the location and operation of this equipment. A system should be established to relay telephone alert messages; the names and telephone numbers of personnel responsible for each laboratory or department should be prominently posted in case of emergencies outside regular working hours.

Isolated areas (e.g., microscopy dark rooms) should be equipped with alarm or telephone systems that can be used to alert outsiders to the presence of a worker trapped inside or to warn workers inside of the existence of an emergency outside that requires their evacuation. Where unusually toxic substances are handled, it may be desirable to have a monitoring and alarm system that is activated if the concentration of the substances in the work environment exceeds a set limit.

Brief guidelines for shutting down operations during an emergency or evacuation should be communicated to all personnel. Return procedures to ensure that personnel do not return to the laboratory until the emergency is ended, as well as start-up procedures that may be required for some operations, should be prominently displayed and regularly reviewed.

All aspects of the emergency procedure should be tested regularly (e.g., every 6 months to a year), and trials of evacuations (if there are such procedures) should be held periodically.

5.7 Facilities Maintenance

5.7.1 Housekeeping

There is a definite relationship between safety performance and orderliness in the laboratory. When housekeeping standards fall, safety performance inevitably deteriorates. The work area should be kept clean, and chemicals and equipment should be properly labeled and stored. The following guidelines should be established.

- Work areas should be kept clean and free from obstructions. Cleanup should follow the completion of any operation or be done at the end of each day.
- Wastes should be deposited in appropriate receptacles.
- Spilled chemicals should be cleaned up immediately and disposed of properly. Disposal procedures should be established, and all laboratory personnel should be informed of them.
- Unlabeled containers and chemical wastes should be disposed of promptly using appropriate procedures. Such materials, as well as chemicals that are no longer needed, should not accumulate in the laboratory.
- Floors should be cleaned regularly; accumulated dust, chromatography adsorbents, and other assorted chemicals pose respiratory hazards.
- Stairways and hallways should not be used as storage areas.
- Access to exits, emergency equipment, and controls should never be blocked.
- Equipment and chemicals should be stored properly; clutter should be minimized.

5.7.2 Equipment Maintenance

Good equipment maintenance is important for safe, efficient operations. Equipment should be inspected and maintained regularly. Servicing schedules will depend on both the possibilities and the consequences of failure. Maintenance plans should include a procedure to ensure that a device that is out of service cannot be restarted.

5.7.3 Security

Laboratory security is important in protecting the safety of employees and those who might enter the laboratory, and in protecting laboratory work and capital assets. The laboratory should be in an area that has limited access. All individuals entering the laboratory should come through a reception area, provide identification, and sign in. During hours when the laboratory is not staffed, areas housing equipment, samples, or laboratory records should be locked. Only laboratory, custodial, and maintenance staff should have access to the laboratory during non-office hours.

REFERENCES

1. J. A. Young, Ed., *Improving Safety in the Chemical Laboratory*, John Wiley, New York, 1987.
2. A. C. Farrar, and H. A. Hurley, "Industrial Hygiene Laboratory," in J. Garrett, L. J. Cralley and L. V. Cralley, Eds., *Industrial Hygiene Management*, John Wiley, New York, 1988.
3. S. Braybrooke, Ed., *Design for Research, Principles of Laboratory Architecture*, John Wiley, New York, 1986.
4. AIA, C. Linn, Ed., "Interactive Approach Produces Innovations," *Lab. Planning Design*, **1**(1) (Spring 1989).
5. L. DiBerardinis et al., *A Guide to Laboratory Design: Health and Safety Considerations*, Wiley-Interscience, New York, 1987.
6. D. Guise, *Design and Technology in Architecture*, John Wiley, New York, 1985.
7. H. D. Goodfellow, *Advanced Design of Ventilation Systems for Contaminant Control*, Elsevier, New York, 1985.
8. American Conference of Governmental Industrial Hygienists, *Industrial Ventilation, A Manual of Recommended Practice*, 20th ed., 1988.
9. American Society of Heating, Refrigerating, and Air-Conditioning Engineers, Inc., "1985 Fundamentals," *ASHRAE Handbook*, Atlanta, GA, 1985.
10. National Research Council, *Prudent Practices for Handling Hazardous Chemicals in Laboratories*, National Academy Press, Washington, DC, 1981.
11. National Research Council, *Prudent Practices for Disposal of Chemicals from Laboratories*, National Academy Press, Washington, DC, 1983.
12. American Chemical Society, Task Force on RCRA, *RCRA and Laboratories*, American Chemical Society, Washington, DC, 1986.
13. American National Standards Institute, ANSI Z358.1, New York, 1985.

CHAPTER FIFTEEN

Quality Control

Floyd A. Madsen, C.I.H.

1 INTRODUCTION

All systems are subject to errors. The purpose of a quality-control (QC) program is to define and control errors to the extent possible. The driving force for quality control in industrial hygiene and related fields is at least twofold. The first is, by nature, primarily ethical. The quality of life and life itself are at stake, requiring the practitioner to exercise the greatest care to avoid misrepresentation of the facts. The second is the force of law that has been placed behind many of the accepted exposure indices in the field.

It is interesting to note that the ethical reasons were not a great driving force for improvement of quality control because QC programs were generally of a "bare bones" nature prior to the passing of the Occupational Safety and Health Act of 1970. Prior to that time quality control in the laboratory consisted mostly of running a standard curve and analyzing blank samples. With the force of law being placed behind industrial hygiene sampling, proof of quality was necessary and more rigorous programs were developed. Quality control is still evolving and its place in existing programs is still being defined. One reason for continued evolution, other than the obvious advances in skill and technique, is the fact that costs and benefits of any program must be continually evaluated and balanced to achieve an optimum mix.

The hoped-for result of a cost–benefit analysis in this area is a program that can be funded reasonably and still meet the goal of identifying and minimizing errors. Quality control then becomes an extension of the role of management and assures effective management of the industrial hygiene program. Incorporation of

Patty's Industrial Hygiene and Toxicology, Fourth Edition, Volume 1, Part A, Edited by George D. Clayton and Florence E. Clayton
ISBN 0-471-50197-2

the QC plan into the management of an industrial program will trigger timely, consistent actions and will provide evidence of the industrial hygiene program's effectiveness. In actuality, most effective QC programs will pay for themselves over the years by reducing liabilities and repetitions due to undiscovered errors.

Quality control in industrial hygiene is analogous to quality control for plant processes. The industrial hygiene program itself is the embodiment of the principles and practices of quality control. The identification, evaluation, and control of health hazards, the triad of industrial hygiene, is the QC approach or philosophy. A plant process produces certain occupational exposures of the workers to the noise, heat, dust, gases, vapors, and so on being produced by the process. These exposures can be considered as defects in the production process. The perfect process would be so engineered that the loss of energy, the formation of by-products or contaminants, and the wear or breakdown of equipment would not occur. We must, however, deal with reality. The industrial hygiene engineer contributes to the design of optimum, if not perfect, plant processes. The monitoring program validates the effectiveness of the process design and control techniques. It detects the impending or recent failure of these control measures. The cause of the "out-of-control" situation is determined, and corrective action is taken to put the plant operation back into control—that is, no unacceptable occupational exposures. Monitoring is resumed. The level of monitoring is proportional to the magnitude and frequency of problems or "out-of-control" situations. According to Hagan, a good ground rule to follow might be: "Let the level of detail be equal to the value of preventing problems" (1).

This chapter provides guidance needed by the formulators of the industrial hygiene program for incorporating efficient QC practices. These practices will assure timely and consistent actions to achieve the program's goal. The material is presented as an overall plan or system. The order of presentation follows the sequence that the industrial hygiene program would use to consider and install the various elements of the system. Initial consideration must be given to the scope and application of quality control in the industrial hygiene program. The program management section includes such elements as objectives and policies, organization, planning, operating procedures, chain of custody, records, corrective actions, training, and costs. The section on equipment, standards, and facilities covers the areas of sampling equipment, direct reading instruments, analytical instruments, calibration, preventive maintenance, reagents and reference standards, control of purchases, and facilities. The laboratory analysis control section is analogous to process quality control and deals with sample identification and control, intralaboratory control, and interlaboratory testing. The area of sample handling, storage, and delivery is considered in light of the logistic needs of the industrial hygiene program. The section on statistical QC presents aspects of control charts, data validation, data analysis techniques, and the use of sampling plans. The final section discusses the third element of management responsibility, that is, evaluation in the context of a QC program audit.

2 SCOPE OF THE QUALITY CONTROL PROGRAM

2.1 Program Elements

The elements of an industrial hygiene QC program include (1) statement of objective, (2) policy statements, (3) organization, (4) quality planning, (5) standard operating procedures, (6) chain-of-custody procedures, (7) record keeping, (8) corrective action, (9) quality training, (10) quality costs, (11) document control, (12) calibration, (13) preventive maintenance, (14) reagent and reference standards, (15) procurement control, (16) sample identification and control, (17) laboratory analysis and control, (18) intra- and interlaboratory testing programs, (19) sample handling, storage, and delivery, (20) statistical quality control, (21) data validation, and (22) system audits.

Most of these program elements are already operational in a more or less adequate and documented system. Evaluation of the degree of conformity of the program operations to program needs is the key to fulfilling the program's responsibility.

2.2 Application of Program Elements

The application of program elements to the industrial hygiene program must be accomplished in consonance with the program's operational needs and resources. A systematic approach to program evaluation and development is an investment whose payoff is the reduction of errors in reported results and recommendations. In a more positive tone, the QC aspects of the industrial hygiene program contribute value in achieving the performance, cost, and schedule commitments of the program. A principal aspect of performance relates to the professional quality rather than quantity of results and decisions of the industrial hygiene program.

3 QUALITY CONTROL PROGRAM MANAGEMENT

Several areas or elements of a QC program are part of the overall management plan. These elements include objectives, policies, organization, planning, operating procedures, chain of custody, record keeping, corrective actions, training, and costs.

3.1 Objective

The objective of the QC function of the industrial hygiene program is to "assure the medical and or scientific reliability of data and subsequent decisions of the program." The subobjectives include the following: (1) the use of rugged methods meeting the program's needs, (2) routine determination of the level of performance of the program, (3) making item 2 compatible with item 1, (4) monitoring routine performance to assure long-term adequacy, and (5) validation of performance by comparison with peer groups. The National Institute for Occupational Safety and Health (NIOSH) published this quantified objective statement for its Industrial Hygiene Analytical Laboratory (2).

3.2 Policies

Quality policies provide the framework of procedures that the industrial hygiene program uses to accomplish the foregoing objective(s). The policies are based on good QC practices and compliance with applicable regulations, and they will assure the implementation of the QC program elements just outlined. They may also specify the manner of implementation and frequency of implementing certain procedures such as preventive maintenance, calibration and checking of field and sampling equipment, laboratory internal QC procedures, and participation in interlaboratory QC or evaluation programs.

3.3 Organization

The QC function of the industrial hygiene program must be carried out by both the line supervision and by nonline monitoring of the total effort. The QC staff function may be a "collateral-duty" or part-time responsibility, but better results are achieved with one or more persons having a specific QC assignment without other interfering responsibilities. The QC supervisor should be a designated individual, who has defined responsibilities (3). These responsibilities require the authority and organizational freedom to identify and evaluate quality problems and to initiate, recommend, or provide solutions. The job description of the QC supervisor includes such areas as developing and carrying out QC programs, monitoring QC activities of the program, and advising management with respect to the quality aspects of industrial hygiene work.

3.4 Planning

Planning for quality control should be an integral part of the general planning process. For every organization or process, quality control must be an integral part of every decision process. The special controls, methods, equipment, and skills necessary for upcoming work should be identified and provided for in timely fashion. The planning allows for necessary research or development work to provide analytical methods or instrumentation or correlation among these. Compatibility of the program's needs with both field and laboratory capabilities must be assured.

3.5 Operating Procedures

The operating procedures necessary to the industrial hygiene program include survey protocols, sampling procedures (4), analytical procedures (4–7) (may include sampling), instrumentation calibration and maintenance procedures, and QC procedures (3) to ensure the validity of the industrial hygiene program's data.

Once such operating procedures are established, they must be communicated to the appropriate staff members. On-line data bases with laptop computers for the field staff are the preferred method for data access. Should it be necessary to use hard copies, then enough sets of operating procedures must be maintained for easy access for all individuals. The industrial hygiene program manager must institute some means of ensuring that all personnel are using the current adopted procedures.

3.6 Chain of Custody

There are two main reasons for maintaining a chain of custody for industrial hygiene samples. First, one can be assured that the sample analyzed has not been confused or exchanged with another. Second, one will have a degree of certainty that the sample has not been tampered with to destroy its integrity.

If the samples or measurements taken may be part of a litigation, the need for chain-of-custody procedures is self-evident. Because such procedures guarantee the integrity of the evidence collected, data developed for standards setting or for regular program work will benefit by such assurances. The manager should consciously assess the "benefits versus risk" of not following a chain-of-custody requirement. The laws or regulations of the state applicable to chain of custody should be complied with. A discussion with legal personnel would provide the basic requirements of that state.

3.6.1 General

A general discussion (9) of the chain-of-custody procedure follows. A sample is in your custody if (*a*) it is in your actual physical possession, or (*b*) it is in your view, after being in your physical possession, or (*c*) it was in your possession and you locked it up in a manner that would prevent any tampering.

Chain-of-custody record tags should be prepared before the work-site work and should contain as much information as possible to minimize clerical work by field personnel. The source of each sample should also be written on the container itself, if feasible, prior to the field work. Field log sheets, if used, should also be completed to the extent practicable before arriving at the site.

If more than one person is involved in the survey, all participants should receive a copy of any study plan and should become acquainted with its contents before the survey. A presurvey briefing and a postsurvey debriefing should be held. The debriefing should determine adherence to chain-of-custody procedures.

3.6.2

Specific points applicable to the industrial hygiene survey include the following:

a. To the extent achievable, as few people as possible should handle the sample.

b. Standard sampling techniques should be used.
c. A chain-of-custody record or seal should be attached to the sample container at the time the sample is collected. The record should be attached to the sample in such a way that the sample container (cassette, tube, etc.) is sealed and cannot be opened without destroying the record. The record should contain the following information: sample number, time taken, date taken, source of sample (includes type of sample and name of person or area sampled), preservative, if applicable (e.g., for biological samples), analyses required, name of person taking sample, and names of witnesses. The record should be signed, timed, and dated by the person doing the sampling. The records must be legibly filled out in ballpoint (waterproof ink). Should the sample size preclude the inclusion of this amount of information in a sample seal then an additional sheet can be used and included with the sample shipment. One must be certain that the seal includes at least the date, sample number, and signature of the industrial hygienist.
d. Blank samples should also be taken. They will be analyzed by the laboratory to exclude the possibility of sample contamination. Each set of samples should have at least one blank included. For larger sample sets blank samples may be limited to 5 to 10 percent of the total number. Blanks should be treated in the same way as the sample, excluding the actual sampling process.
e. A preprinted field data record should be maintained to record field measurements and other pertinent information necessary to refresh the sampler's memory. This may be a part of the chain-of-custody record or a separate record. A separate set of field records should be maintained for each survey and stored in a safe place where they can be protected and accounted for at all times. The entries should be signed by the field sampler. The preparation and conservation of the field records during the survey are the responsibility of the lead industrial hygienist. Once the survey is complete, field records are retained by the lead industrial hygienist, or by a designated representative, as a part of the permanent record. Current OSHA regulations require that personnel exposure records be kept for a period of 30 years following the end of employment (10).
f. The field sampler is responsible for the care and custody of the samples collected until properly dispatched to the receiving laboratory or turned over to an assigned custodian. The field sampler must assure that each sample container is in physical possession or within view at all times or is secured in an area designated specifically for that purpose.
g. If photographs are taken to substantiate any conclusions of the survey, written documentation on the back of the photo should include the signature of the photographer, time, date, and site location. Photographs of this nature, which may be used as evidence, should also be handled in accordance with chain-of-custody procedures to prevent alteration.

3.6.3 Transfer of Custody and Shipment

Some considerations for transfer of custody and shipment are as follows:

a. When turning over the possession of samples, the transferee must sign and date the chain-of-custody record. If a third person takes custody, that person must also sign and date the record. Keeping the number of persons who handle the sample to a minimum limits the chance for error. It also keeps the number of people who may be called to testify as to the sample integrity to a minimum, in case of litigation.
b. The field custodian or sampler has the responsibility of properly packaging and dispatching samples to the laboratory for analysis. The sample identification and sample accountability portions of the chain of custody record must be completed, dated, and signed.
c. Samples must be properly packed in shipment containers to avoid breakage. The shipping containers are sealed for shipment to the laboratory.
d. All packages must be accompanied by the field data record and chain of custody record. A copy of each is mailed directly to the laboratory.
e. If sent by mail, the package is sent registered or certified with return receipt requested. If sent by common carrier, a bill of lading should be obtained. Receipts from post offices and bills of lading are retained as part of the permanent chain of custody documentation.

3.6.4

Laboratory operations must be so organized that the principles and practices of chain-of-custody procedures are observed. Clearly these procedures require the application of the best principles of management and scientific investigations.

a. The laboratory must designate one or more employees as "sample custodian(s)." In addition, the laboratory must set aside a "sample storage security area." This should be a clean, dry, isolated area that can be securely locked from the outside.
b. All samples should be handled by the minimum possible number of persons.
c. All incoming samples are to be received only by the custodian, who indicates receipt by signing the sample transmittal sheets accompanying the samples.
d. Immediately upon receipt, the custodian places the sample in the sample storage area, which is locked at all times except when samples are removed or replaced by the custodian. To the maximum extent possible, only the custodian should be permitted in the sample storage area.
e. The custodian must ensure that heat-sensitive or light-sensitive samples, or other sample materials having unusual physical characteristics or requiring special handling, are properly stored and maintained.
f. Only the custodian can distribute samples to personnel who are to perform

tests. The custodian must enter into a permanent logbook the laboratory sample number, time and date, and the name of the recipient. A computer record with appropriate backup (disk and hard copy) may be used.

g. The analyst records in his laboratory notebook or worksheet the name of the person from whom the sample was received, whether it was sealed, identifying information describing the sample (by origin and sample identification number), the procedures performed, and the results of the testing. The notes should be signed and dated by the person performing the tests and retained as a permanent record in the laboratory. In the event that the person who performed the tests is not available as a witness at time of trial, the laboratory may be able to introduce the notes in evidence under the Federal Business Records Act.

h. To the extent possible, standard methods of laboratory analyses are to be used. If laboratory personnel deviate from standard procedures, they should be prepared to justify their decision during cross-examination.

i. Laboratory personnel are responsible for the care and custody of the sample once it is handed over to them and should be prepared to testify that the sample was in their possession and view, or securely locked up, at all times from the moment it was received from the custodian until the tests were completed.

j. Once the sample testing is completed, the unused portion of the sample (if stable), together with all identifying tags and laboratory records, should be returned to the custodian, who makes the appropriate entries in his log. The returned, tagged sample is retained in the sample room until it is required for trial. Raw data and other documentation of work are also turned over to the custodian.

3.7 Record Keeping

Records are considered to be a principal form of objective evidence of the operation and effectiveness of the QC system. Any laboratory needs a system of records to establish and maintain control over the samples received for analysis. An adequate system would certainly provide confirmation that chain-of-custody requirements have been met. The preceding section indicates the nature of such requisite records. Any laboratory also needs administrative records. Because these are not unique to industrial hygiene or to a laboratory operation, however, they are not discussed further.

Quality-control records are an integral part of an effective and economical QC program. The records provide the assurance that calibrations, "blind sampling," recycled sampling, and similar procedures have been carried out in accordance with the laboratory's plan. The records of the day-to-day system and instrumental QC checks document routine control and the initiation of necessary corrective actions. Control charts for instrument performance and check samples perform an efficient and economical record-keeping function. As each instrumental check is made, for instance, the result should be posted immediately on the control chart near or on the instrument itself. The analyst has access to the record without difficulty and can make an instantaneous decision about the action to be taken. If the chart shows a pattern of expected variability, no action is necessary. If a real trend is developing or an out-of-control point is plotted, corrective action is initiated. This achievement of real-time decision making is hard to match, although current computer-controlled instrumentation can be utilized for nearly automatic updates and monitoring of control charts.

The test of the QC records system is analogous to a certified public accountant's audit of a company's financial statement and its bookkeeping system. A knowledgeable QC person should be able to audit the records system and come to the same conclusion given by the laboratory personnel about the "state of quality control." An affidavit analogous to that appearing with the annual financial report attesting to the validity of the laboratory's claim about its real state of quality control represents the ultimate compliment to the records system.

3.8 Corrective Actions

Problems are bound to occur with any analytical system. For example, components of the system wear out. When a problem or defect begins to affect the performance of the analytical system, corrective action should be initiated. For major problems, management must initiate the correction action plan and assure that all affected individuals cooperate in a well-planned program. In many instances, the analyst can detect erratic behavior or changed response level, either based on experience or from out-of-control points on a control chart. The analyst should check the operational parameters likely to cause such "nonroutine" response. Seeking out the cause of a problem and correcting it requires analytical or technical expertise rather than statistical knowledge. One of the advantages of the analyst or instrument operator being the principal in the bench-level QC program is that he or she on many occasions can initiate corrective actions before problems become major.

When there are significant (out-of-control) differences between the observed and expected results on QC test samples, notification should be sent to the analyst. The analyst should investigate, determine the cause of difference if possible, take necessary corrective action, and report results to the laboratory management. If the required corrective action is beyond the responsibility and authority of the analyst (e.g., in the case of an incompatibility between the sampling and analytical procedures) the appropriate level of laboratory management must decide what type of corrective action is necessary and ensure its implementation.

As the time increases between the "out-of-control" incident and its investigation, the probability of the analyst finding the cause rapidly decreases. In some cases, the original sample may have been discarded already, or the reagents changed, or the instrument reprogrammed for another analysis. In these cases, the QC system rather than the analyst must be faulted for what is usually a futile effort to determine an appropriate corrective action.

3.9 Training

The personnel of the industrial hygiene program must have sufficient training and/or experience to develop the necessary knowledge and skills to perform their job functions proficiently (11,12). In the laboratory setting there are basic levels of training and experience required to obtain and maintain accreditation by the American Industrial Hygiene Association. It would be well to follow these requirements even if the laboratory has no interest in this particular association. Similar levels of training should be required of the industrial hygienist or other persons involved in the sampling and field evaluations.

Beyond these basic levels, a relatively large number of graduate programs and professional improvement seminars are now available to most professionals. Advantage should be taken of these to improve and maintain skill levels. The effectiveness of the training should be evaluated by the program management to ensure that the person trained should be able to perform the new function completely and accurately. Performance measures should relate to the degree of proficiency of the staff already performing the function. Values for precision and recovery for an analytical procedure are relatively easy to develop and provide an objective measure of performance. Other functions are more difficult to quantify. In some cases the performance test is the skill with which the staff person can perform the given function—for example, organizing and conducting a plant survey independently.

The QC function of the industrial hygiene program can perform the function of auditing the effectiveness of the training program. It is self-evident that personnel charged with the responsibility for the QC function must also require the appropriate training to develop and implement the QC system.

3.10 Costs

The industrial hygiene program allocates financial and personnel resources among the various elements of the QC system. These elements can be grouped into four categories. Prevention costs involve elements associated with planning, implementing, and maintaining the QC program itself—for example, objectives, policies, organization, planning, operating procedures, chain-of-custody procedures, record keeping, training, equipment calibration, preventive maintenance, reference standards, control of purchases, and facilities.

Appraisal costs are those entailed in efforts to evaluate and maintain the quality performance levels of the industrial hygiene program. The elements included in appraisal consist of sample identification and control, intralaboratory QC testing, interlaboratory proficiency testing programs (including accreditation program costs), control charts, data analysis, and data validation. The reports generated on the effectiveness of the QC system are also included in appraisal costs.

Internal failure costs are the costs to the industrial hygiene program attributable to defective materials, reagents, instruments, sampling, and/or analytical procedures, which cause data to be discarded and repeated or lost. Corrective actions to determine and correct these problems are included in internal failure costs. The costs of having to resurvey a plant, including the travel and personnel time costs, would also be included if the survey had been done by the industrial hygiene program. If the resurvey is to be done by an outside group, the cost would belong to the next category as an external failure cost.

Also included in the external failure costs category are those associated with investigating complaints from defective data that have already been reported. The repetition of procedures or effort spent in validating or revalidating procedures because of reported defective data also constitutes an external failure cost. The costs to other programs or departments may be considerable when defective data are reported. Should the defective data be discovered during an enforcement proceeding, the indirect costs will greatly outweigh the direct costs to plant or agency management. The loss of credibility to the industrial hygiene program cannot be calculated in terms of dollars but may include the very existence of the industrial hygiene program.

The value of categorizing costs is to allow and encourage a programmed and budgetary apportionment of these costs. Because the costs can be concentrated in the prevention category, the total costs decrease. The cost effectiveness of the QC system and of the industrial hygiene program itself also improve.

4 EQUIPMENT, STANDARDS, AND FACILITIES

4.1 Sampling Equipment

The variability or error associated with sampling equipment is considerable. In a collaborative test on the sampling and analysis of solvents using personal sampling pumps, charcoal adsorption tubes, carbon disulfide elution, and gas chromatographic analysis, the sampling equipment error was of the same order of magnitude as the analytical errors; the relative standard deviation ranged from 5 to 14 percent (13). This error or variability existed when all participating laboratories calibrated their sampling pumps with a single calibration system. The report indicates that the procedure was used to reduce the calibration variability apparent with each laboratory having calibrated its own pumps at its own facility before meeting at the site for the collaborative test. The variability was measured by sampling from a homogeneous atmosphere. No spatial or time variability was involved in sampling from the controlled concentration level test chamber. No real plant sampling situation can possibly approach the homogeneous conditions in a test chamber.

The appropriate frequency of calibration depends on the handling and use (abuse) the sampling pump has undergone. Specifically, pumps should be recalibrated after suspected abuse (such as dropping), when received from the manufacturer, and when repaired. Experience will determine an appropriate frequency of recalibration. A control chart approach can efficiently develop the information necessary for such decisions. All pumps should be calibrated before being used in the field and at intervals during the field work if numerous samples are taken. The calibration should be checked upon return from the field. This is especially important if changes in altitude are significant between work sites.

The accuracy of the calibration depends on the calibration system itself. The choice of calibration system depends on the location at which the calibration is made and the facilities available to the industrial hygiene program (14). In the laboratory, a burette is generally used as a soap bubble flowmeter. Other standard calibrating systems could also be used.

The procedure for calibration with the soap bubble flowmeter follows (15). The calibration setup for personal sampling pumps with a cellulose membrane filter follows standard procedures. Because the flow rate indicated by the flowmeter of the pump depends on the pressure drop across the sampling device, the pump flowmeter must be calibrated while operating with a membrane filter and appropriate backup pad in the line.

1. While the pump is running, the voltage of the pump battery is measured with a voltmeter to assure that the battery is charged adequately for calibration.
2. The cellulose membrane filter with backup pad is placed in the filter cassette.
3. The calibration setup is assembled so that the air flows from room atmosphere through the soap bubble meter to the cassette, through the cassette to the sampling pump, and to room atmosphere. A U-tube water manometer monitors the pressure drop between the cassette and the pump.
4. The pump is turned on, and the inside of the soap bubble meter is moistened by immersing the burette in the soap solution and drawing bubbles up the tube until they are able to travel the entire length of the burette without bursting.
5. The pump is adjusted to provide a flow rate of 2.0 l/min.
6. The water manometer is checked to ensure that the pressure drop across the sampling train does not exceed 13 in. H_2O at 2 l/min.
7. A soap bubble is started up the burette, and the time it takes the bubble to travel a minimum of 1.0 liter is measured with a stopwatch.
8. The procedure in step 7 is repeated at least three times, the results are averaged, and the flow rate is calculated by dividing the volume between the preselected marks by the time required for the soap bubble to travel the distance.
9. Data for the calibration should include the volume of air measured, elapsed time pressure drop, air temperature, atmospheric pressure, serial number of the pump, date, and name of the person performing the calibration.

The specific requirements for the calibration procedure, the frequency of its use, and maintenance and operational notes must be permanently recorded because these records form an integral part of the QC system and are invaluable in any court proceeding. Requirements for the calibration system itself are discussed later.

4.2 Direct-Reading Instruments

The calibration of direct-reading instruments involves not only flow but concentration level response calibration. For guidance on the general technical problems consult Lippmann (14), Saltzman (16), and the air sampling instruments manual (17) of the American Conference of Governmental Industrial Hygienists (ACGIH). In addition, a number of evaluation reports have been developed; Johnson on NO_2 meters (18), McCammon on portable, direct-reading combustible gas meters (19), and Parker and Strong on CO meters (20). These reports provide detailed performance evaluations for those classes of instrument. The calibration of direct-reading instruments for physical agents is handled in a manner equivalent to that for chemical agents. Guidance for specific instruments should be available from the manufacturer or the literature. In some cases a calibration capability, such as for ultraviolet radiation, must be developed (21).

4.3 Analytical Instruments

The calibration system records and approach are equivalent for sampling equipment, direct-reading instruments, and analytical instruments. The following discussion on the calibration system has been adapted from the NIOSH QC manual TR-78 (3).

All equipment to be calibrated should have a permanently affixed record number (serial or property number). A calibration control card or sheet for each instrument should show identification number, description (including manufacturer, model, and serial number), location of use or storage, calibration procedure used, calibration interval, date of last calibration, signature of the person performing the calibration, due date for next calibration, and values obtained during the calibration. Calibration reports and compensation or correction figures should be filed with the calibration control card.

All equipment should be calibrated in a room in which the laboratory has provided controls for environmental conditions to the degree necessary to assure measurements of the specified and required accuracy. The calibration area should be reasonably free of dust, vibration, and radio-frequency interferences and should not be located near equipment that produces acoustic noise or vibration. Isolation of pressure, mass, and acceleration equipment from vibrations is particularly essential; isolation mounts, seismic masses, and other means of protection should be provided.

The laboratory calibration area should have adequate temperature and humidity controls. A temperature from 68 to 73°F and a relative humidity of 35 to 55 percent normally provide a suitable environment. A filtered air supply is a necessity in the calibration area. Dust particles are more than just a nuisance; they can be abrasive, conductive, and damaging to instruments. A measure of dust filtration can be provided in the air conditioning system by the washing action of sprays and atomizers, but this may have to be supplemented by electrostatic and/or mechanical filters of the activated charcoal, oil-coated, or ribbon types.

Recommended requirements for electrical power within the laboratory should include voltage regulation of at least 10 percent (preferably 5 percent), low values of harmonic distortion, minimum line transients as caused by interaction of other users on the main line to the laboratory (separate input power if possible), and a suitable grounding system established to assure equal potentials to ground throughout the laboratory (or isolation transformers may be used to operate individual pieces of equipment). Adequate lighting (suggested values, 80 to 100 footcandles) is necessary for workbench areas. The lighting may be provided by overhead incandescent or fluorescent lights. Fluorescent lights should be shielded properly to reduce electrical noise.

All instruments should be calibrated and checked by qualified personnel. An outside calibration service group may schedule and perform these checks, but this does not relieve the laboratory of the responsibility for controlling, monitoring, and identifying calibration intervals and having the checks made on time.

All measurements or calibrations performed by or for the laboratory for the calibration program should be traced, directly or indirectly, through an unbroken chain of properly conducted calibrations (supported by reports or data sheets) to some ultimate or national reference standard maintained by a national organization such as National Institute of Standards and Technology (NIST); an example of proper calibration procedure for standard quartz cuvettes appears in the literature (22). The ultimate reference standard can also be an independent, reproducible standard (i.e., a standard that depends on accepted values of natural physical constants). A typical example is the cesium beam type of microwave frequency standard.

There should be an up-to-date report for each reference standard (except independent reproducible standards) used in the calibration system, and for any subordinate standards or measuring and test equipment if their accuracy requires supporting data. If calibration work is contracted out to a commercial laboratory or facility, copies of reports issued by the subcontractor should be kept available.

All reports should be kept in the calibration system file and should contain the following information: (1) identification or serial number of standard to which the report pertains, (2) conditions under which the calibration was performed (temperature, relative humidity, etc.), (3) accuracy of standard (expressed in percentage or other suitable terms), (4) deviation or corrections, and (5) report number or designation. In addition (6) reports for the highest level standards of sources other than NIST or a government laboratory should bear a statement that comparison has been made with national standards at periodic intervals using proper procedures and qualified personnel; and (7) corrections that must be applied if standard conditions of temperature, gravity, air buoyancy, and so on, are not met, or if they differ from those at place of calibration, must be noted.

The description of the calibration program indicates that an extensive effort may be required. If the data to be generated by the industrial hygiene program are sufficiently important that related decisions may be challenged in court or elsewhere, the complete calibration program is cost effective. For other situations, the program manager must decide on an appropriate allocation of his resources among this and the other elements of the QC system.

4.4 Preventive Maintenance

The preventive maintenance program complements the calibration program. An effective preventive maintenance program increases the reliability of measurement systems and decreases downtime. An inadequate preventive maintenance program will be responsible for increased unscheduled downtime, a probable increase in total maintenance costs, and possibly a lack of trust in the validity of the data being generated.

A schedule of preventive maintenance should be keyed to the calibration program schedule. The frequency of the preventive maintenance program should be based on the equipment manufacturer's recommendations and cumulative experience. All records from the maintenance program should be kept as part of the records system and used to assure the validity of data generated by the industrial hygiene program.

4.5 Reagent and Reference Standards

A number of reagent standards suitable for laboratory programs are available from commercial sources. Standard reference materials (SRMs) are available from NIST.

Calibration gases may be obtained from commercial suppliers. A cross-check program comparing the old and new tanks or tanks from several suppliers will serve to validate the calibration gases for the program. Such checking may be necessary because the responsibility for valid data lies with the industrial hygiene program itself.

The use of all available commercial analytical standards still leaves some of the industrial hygiene program's analytical needs without standards. The only recourse is the internal preparation of needed working standards for calibration and QC purposes. The development of a diverse range of working standards for many elements in wide concentration ranges has been covered by Hill (23), who uses the working standards in a ratio of 1:10 when the unknowns cover a narrow concentration range and in an even greater ratio for more diverse samples. Hill states: "Having calibrated our laboratory, the standards are still used daily on every set run as a datum and an early warning system to detect any problems that may develop. Any set exhibiting poor results on the included standards is immediately rerun, and the problem is identified and eliminated." Some literature (3) uses the term "control sample" in much the same sense as Hill used the term "working standard." As Hill concludes, "Millions of dollars may be spent upon the results of a large suite of chemical analyses. There is no mystery to obtaining the good analytical results needed to make the right decisions the first time; one must get or make the needed standards and use them continuously."

4.6 Control of Purchases

Common laboratory reagents and chemicals are generally bought to a specified grade or quality such as "ACS Reagent" grade or "Spectroquality." Experience will indicate differences between such materials from various suppliers. The main assurance of the quality of such materials is the routine use of reagent blanks in the analytical procedures. The shelf life should be determined and stocks rotated on a first in, first used basis.

The purchase of equipment may offer the program a greater opportunity to set performance requisites in the purchase specifications. Acceptance testing against the performance specifications protects the program.

In many cases the industrial hygiene program is obliged or desires to purchase analytical measurements from another laboratory. Because the data reported by the program are the program's responsibility regardless of where the analyses were made, the program must obtain valid data from outside sources. It must assure that such data are being generated by a laboratory whose QC system is at least as effective as its own. A routine technique is to submit split samples, spiked samples, and blanks as blind samples.

4.7 Facilities

The role of facilities in assuring the quality of an industrial hygiene program's results is flexible. The best possible building and physical facility does not guarantee "good work"; on the other hand, "good work" may be produced under poor conditions. Good facilities certainly simplify and minimize an additional set of variables that the program does not want to have affecting the quality of the data produced. Downtime and time-consuming corrective action investigations should be minimized with adequate facilities. A number of books and manuals on the design of laboratory facilities can be consulted (see Chapter 14). It is important to minimize occupational safety and health problems in the design and/or development of laboratory space for the industrial hygiene program.

5 LABORATORY ANALYSIS CONTROL

The QC system to provide for precise and accurate industrial hygiene measurements has been discussed in earlier sections for the program management and for its physical elements. Sections 5 and 6 cover the operational elements.

5.1 Sample Identification and Control

All samples taken or received for analysis must be logged in and identified with respect to type and source of the sample. The sample identity must be carefully and clearly maintained throughout the sampling and analysis-reporting cycle of events. The earlier discussions on chain of custody and record keeping (Sections 3.6 and 3.7) apply.

5.2 Intralaboratory Control

The principal operational element of the industrial hygiene program's QC system is its intralaboratory control program.

5.2.1 Method Adaption and Validation

The adaption and validation of reference or other methods to the specific needs of the program constitute one of the most important elements. These specific needs require validation because of collection characteristics, interference, sensitivity, storage problems, or other operational aspects unique to the sampling and analysis problems to be expected. Numerous texts (24–26), articles (27, 28), and other publications (13, 29–34) discuss the approaches and techniques of the evaluation or validation of sampling and analytical methods. The individual laboratory must assure itself that the methods it uses are "rugged" enough for the conditions of use entailed by its problems. Youden and Steiner (24) extensively discuss "ruggedness testing" and its value to the analytical laboratory. Linch (35), Linning et al. (36), and Wilson (37–39) describe the types of error that may afflict a method, as well as programs to detect, quantitate, and eliminate or correct for determinate errors.

When the method is ready for intralaboratory validation, any number of protocols may be used. One protocol is that used by the Occupational Safety and Health Administration (OSHA). The method of choice is evaluated at each of three levels (0.5, 1, and 2 times the level of primary interest, which is usually the OSHA PEL). Contaminant atmospheres are generated, sampled using the prescribed collecting techniques, and analyzed. Experimentally, it has been found desirable to determine sampling pump variability separately. Laboratory air-moving devices with low variability can then be used in the sampling of the contaminant. Figure 15.1 is a sample outline of one of the schemes used by OSHA.

5.2.2 Instrumental Quality Control

The industrial hygiene laboratory many times finds itself using a basic analytical technique for a wide variety of materials. Atomic absorption spectrophotometry, X-ray analysis, and gas chromatography are common examples. In considering the QC aspects of such procedures, emphasis should be directed to the common points, rather than the differences.

As an example, a common sampling and analytical technique for many organic materials involves charcoal tube adsorption, desorption with carbon disulfide (CS_2), and gas chromatography instrumental analysis (39). The sampling and analytical method has been validated by collaborative testing for seven solvents (12). The method has been evaluated for application to numerous materials. There are still materials that have not been checked out. Each new challenge is not unique, with the result that prior developmental and operational information does not apply. By checking the operation of the parts of the procedure and by using control samples of a similar "evaluated" material, the laboratory can, with great confidence, analyze for the new "unknown." The desorption efficiency must be specifically determined for the laboratory's operating condition for all materials. The gas chromatograph must be calibrated for the "unknown" for quantitative analysis. However, these operations can be done routinely rather than on a research project basis (41). The use of control charts for instrumental quality control is most appropriate. The

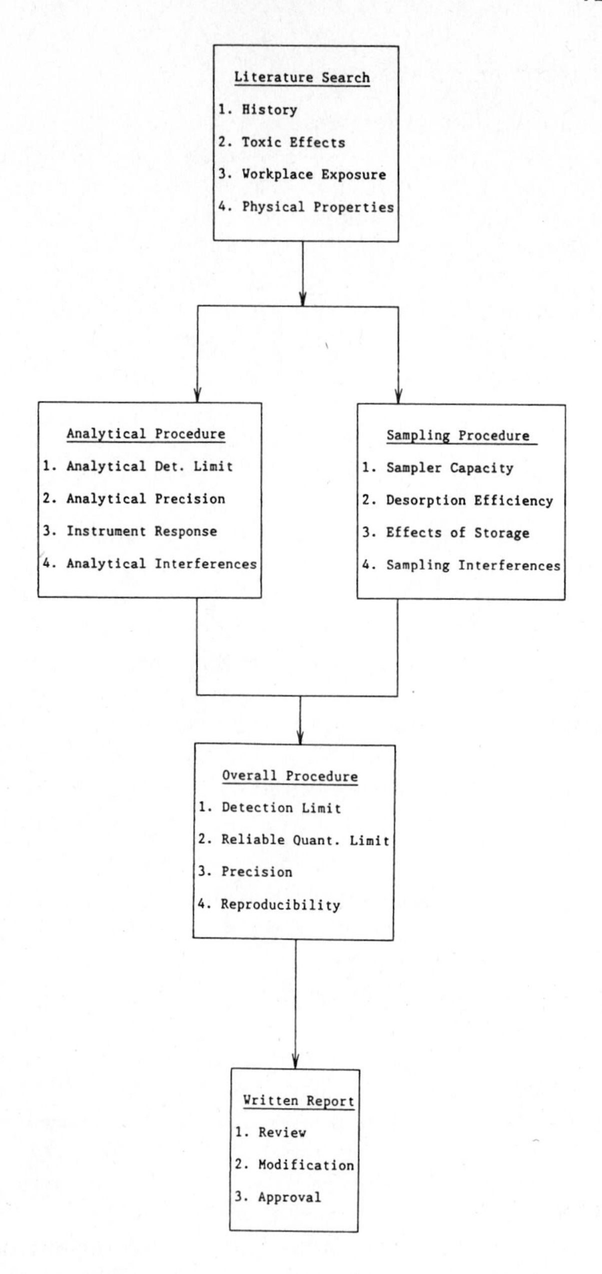

Figure 15.1. Sample evaluation protocol for method validation.

monitoring of a critical performance indicator assures the analyst that the instrument is operating reliably. Control charts are developed in Section 7.1.

5.2.3 Routine Samples

One of the most generally applicable and easily applied statistical quality control techniques is the Shewhart control chart. The Shewhart control chart technique can be applied to almost any area of somewhat repetitive measurements, including monitoring critical instrumental performance characteristics, analytical blanks, instrument calibration standards, and total analysis control samples, as well as monitoring the desorption efficiency determination for consistent technique, and ultimately, plant environmental control monitoring (42). For routine analytical procedures, control charts should be set up for the "control samples" or "working standards" discussed by Hill (23). The control chart for the "control sample" (3) will monitor the recovery of the procedure, thereby detecting errors that cause a shift in the process average. These sources of errors are due to calibration-type problems or changes in concentration of reagents (43).

A control chart for blank determinations will aid in detecting contamination-type problems. Reagent or sample contamination or certain instrumental operational problems will tend to cause a small constant absolute value shift in the determination. Such a shift would be difficult to detect at concentration levels up in the working range, but it is more easily found using blanks, because the absolute value would be lost in the usual variability at higher concentrations. Excellent summaries of general precautions and techniques applicable to such problems are presented by the Intersociety Committee on Air Sampling and Analysis (44) and by Linch (45, 46).

If the "control sample" used in a determination is not a realistic simulation of regular environmental samples, it will be of value in determining the precision for recycled routine samples. The use of a "mercuric chloride in distilled water" (3) control sample because of instability problems associated with a "pooled mercury in urine" control sample is such a case. Such recycled samples should be more affected by interferences and matrix effects therefore will yield a more realistic value, within laboratory precision or repeatability (47) on routine samples. Special considerations and techniques are applicable when automated procedures are used for high-production routine work (48–50).

5.2.4 Single Samples or Infrequently Employed Procedures

The single-time or infrequently employed procedures present the greatest problems to the laboratory analyst. Because of unfamiliarity with the procedure to be used and probable unfamiliarity with the sample matrix and interferences presented to the procedure, the analyst is deservedly wary. Because almost certainly no SRMs or reasonable substitutes will be readily available, the analyst must rely on generally good technique (as validated with other procedures) and a limited assessment of recovery (51) and interferences on the subject sample. OSHA refers to this as a "stop-gap" method. For stop-gap determinations an abbreviated version of the evaluation scheme can be used (52).

Because the occasional sample presents the greatest uncertainty and potential problems to the industrial hygiene analytical laboratory, major efforts of senior staff are required to ensure the quality of such work. The difficulties, the time involved, and the infrequency of such samples also present a special challenge to the person responsible for the reporting of such efforts. Elwell and Lawton (53) have reported on a "relative value structure" to aid in such reporting.

5.3 Interlaboratory Testing

It has been recognized that external proficiency analytical testing on a continuing basis is essential to assure quality (54). The American Industrial Hygiene Association (AIHA) laboratory accreditation program (55) includes satisfactory continued performance in a proficiency analytical testing program as an integral requirement for accreditation. Such a program offers the laboratory director confirmation of the effectiveness of the internal quality control program. Ratliff (56) has reported the development of the National Institute for Occupational Safety and Health Proficiency Analytical Testing (PAT) program used by NIOSH and OSHA, and subsequently AIHA.

The industrial hygiene program will have to supplement such a program by splitting or exchanging samples with one or several other laboratories interested in the same problem. These informal programs should be documented and appropriate credit taken for the value of the data so generated.

6 SAMPLE HANDLING, STORAGE, AND DELIVERY

The logistic aspects of sample handling, storage, and delivery impose a significant constraint on some industrial hygiene programs. When the industrial hygienist must rely on common transportation systems, whether plane, truck, mail, or parcel delivery, the options are limited. The use of bubblers and absorbing solutions become a "method of last resort." The risk of loss of samples, or of instability of the collected samples over a several-day transport time, and the attendant cost of doing a resurvey, make other alternatives attractive despite any inherent limitations or difficulties. The industrial hygienist who can personally, or by courier, transport the samples back to the laboratory in a few hours may be able to choose more specific, more accurate, or more economical sampling and analytical methods.

The industrial hygiene program manager should ensure that operating procedures, chain-of-custody procedures, and records are adequate for the specific needs of the program. Sampling and shipping procedures are available for many industrial hygiene samples (4). Perishable samples, even milk (57), can also be safely and routinely shipped. The shipping procedures must consider Department of Transportation rules and other applicable shipping regulations.

7 STATISTICAL QUALITY CONTROL

Statistical quality control involves the application of statistical techniques to the appropriate areas of the industrial hygiene QC program. For routine samples or for instrumental QC programs, the Shewhart control chart is the simplest and most appropriate technique (58, 59).

7.1 Control Charts

The industrial hygiene program should institute a control chart for an instrument or a procedure (control sample, or blank or recycled samples) when it is more efficient to consider the procedure to be semiroutine than to use the single or occasional sample procedures outlined earlier and advocated by Linch (35, 45, 46).

7.1.1 Purposes

The control chart is both a diagnostic and a reliability tool. Because the control chart differentiates between the usual pattern of random variation (from indeterminate errors) and mistakes or biases (due to determinate errors), it can be used to troubleshoot a procedure or to make it more rugged. In the process, the weak points of the procedure, whether due to interferences, instability of sample, reagents, or equipment, or required operator judgments, are identified (or determined) and eliminated. The resulting procedure has become "ruggedized" by this use of the control chart. The second use of the control chart is to establish the normal operational precision and stability of the sampling and/or analytical method or measurement system. As experience is gained, a reliable, valid value of routine precision will be developed. This value has built into it the "real world" sources of variability that the industrial hygiene program routinely faces. The evaluation of the measuring process stability (accuracy or bias) permits the program to make validated decisions with respect to whether plant processes have actually changed environmental or biological concentration levels or whether an apparent change is due to the inability of the industrial hygiene program to make reliable measurements. Such knowledge may obviously affect management decisions on hazard control equipment investments. The third use of control charts is as a monitoring instrument on the procedure or measurement system itself. The control chart provides immediate objective evidence of reliable operation or detects impending (trends) or actual problems (out-of-control points) with the procedure or measurement system. Corrective action can then be initiated by the industrial hygiene program, making the corrective action costs (Section 3.10) internal and not external (including loss of credibility).

7.1.2 Parameters to be Controlled

The calibration of a direct-reading instrument such as a combustible gas meter can be checked by the use of commercial calibration kits. A calibration check on each day of use, before and after the day's survey, guarantees valid data throughout the day and backup if the measurements are challenged. A number of radiation survey instruments have "check sources" on the instrument. The surveyor performs the check and can even plot the result directly on a small control chart taped to the side of the instrument. These calibration checks are not a substitute for full individual contaminant response calibrations, but they do provide an objective means of determining their appropriate frequency and offer assurance of reliable measurements between such calibrations.

Laboratory instruments, whether used to calibrate sampling equipment or to measure analytical response, such as spectrophotometers, gas chromatographs, atomic absorption instruments, or radiation spectrometers, can be monitored for continued calibration and reliable performance. If an instrument's response can be completely calibrated for a given procedure in a few minutes, it may be more efficient to recalibrate each time a procedure is run. If, however, the instrument can be calibrated only infrequently by some reference or transfer standard, if calibration is quite expensive, involving weeks of effort, if economies can be achieved by checking a couple of points on a calibration curve, or if a critical response parameter can be monitored, a control chart is appropriate. Not only may economies in operational costs be realized, but greater reliability and comparability of data over longer periods of time will be achieved.

Total analysis quality control uses the control chart for routine analytical procedures to monitor the procedure's performance on "control samples," blanks, and "recycle samples" where used.

7.1.3 Type of Control Chart

The Shewhart control chart measures both measurement process variability and calibration stability. The Shewhart or $\overline{X}$ and R chart uses the range R to estimate process variability. The average of these individual measurements provides the $\overline{X}$ for monitoring the stability of the procedure or indicating how well it is maintaining its calibration. There are other types of charts that use the standard deviation to measure variability or a moving average, or individual values, to measure process stability (60). These are special application tools that may be more appropriate after the procedure or instrument is known to be in a "state of control" through use of the $\overline{X}$ and R chart. The cumulative sum or cu-sum chart is more complex to institute and is not as rugged for situations not already in a "state of control." In most cases the basic Shewhart or $\overline{X}$ and R chart can be easily and effectively used.

7.1.4 Trial Control Charts

The industrial hygiene program often has some historical data on which to base "trial control limits." These data are invaluable in getting a start on an individual problem. As experience is gained, the control limits are recalculated, and decisions based on these control limits are better. The advantage of initially using historical data is the gain of several weeks to several or many months in developing effective control limits.

7.1.5 Calculation of Control Limits

The calculation of control limits is a straightforward exercise. Linch (35) and others (3, 58, 60–62) have shown the mechanics of such calculations. Another presentation (8) develops the following format:

Table 15.1. Control Chart Lines Factors (58)

Factor	Two Measurements per Set	Three Measurements per Set
D_4	3.27	2.57
D_5	2.51	2.05
A_2	1.88	1.02

1. Calculate R for each set of measurements.
2. Calculate $\overline{R}$ from the sum of R divided by the number of sets (or subgroups) k.
3. Calculate the upper control limit (UCL, approximately $3s$) on the range by $UCL_R = D_4\overline{R}$. There will be no lower control limit where there are six or fewer values in each set (subgroup) k. The value of D_4 for duplicate analyses from Table 15.1 is $D_4 = 3.27$.
4. Calculate the upper warning limit (UWL, approximately $2s$) on the range by $UWL_R = D_5\overline{R}$. The value of D_5 for duplicate analyses from Table 15.1 is 2.51.
5. Construct the control lines and plot the consecutive analyses on graph paper. Circle or highlight any value outside the control limits. Corrective action to investigate and eliminate the causes will be taken after each out-of-control point as it happens. Less vigorous follow-up should happen for each value outside the warning limit but within the control limit.
6. Calculate $\overline{X}$ for each set of measurements.
7. Calculate $\overline{X}$ or the mean of the $\overline{X}$'s for the set of measurements.
8. Calculate the upper and lower control limits by $UCL_X = \overline{\overline{X}} + A_2\overline{R}$ and $LCL_{\overline{X}} = \overline{X} - A_2\overline{R}$. The value of A_2 for duplicate analyses from Table 15.1 is 1.88.
9. Calculate the upper and lower warning limits by $UWL_{\overline{X}} = \overline{\overline{X}} + \frac{2}{3}A_2\overline{R}$ and $LWL_{\overline{X}} = \overline{\overline{X}} - \frac{2}{3}A_2\overline{R}$.
10. Construct control lines and plot the consecutive analyses on graph paper. See step 5.

The development of trial control limits from the laboratory's own experience is most appropriate and realistic. However, other sources of information can be used. Collaborative test data can be used for developing these trial limits until the laboratory's own experience is developed. The data of Table 15.2 are taken from a NIOSH report.

The calculation of trial control limit values from the data of Table 15.2 is set up in Table 15.3.

The calculation of control chart lines using the information from Table 15.3 and factor values from Table 15.1 is given in Table 15.4.

Table 15.2. Benzene Found in Duplicate Analyses (13)

Laboratory	Benzene (mg/tube)
1	1.50, 1.70
2	1.53, 1.73
3	1.56, 1.57
4	1.67, 1.88
5	1.40, 1.98
6	1.33, 1.42
7	1.45, 1.62
8	1.49, 1.69
9	1.42, 1.61
10	1.12, 1.27
11	1.36, 1.53
12	1.43, 1.57
13	1.59, 1.59
14	1.38, 1.55
15	1.17, 1.44

Table 15.3. Data for Calculation of Control Chart Lines

Laboratory	Measurements		Average, X	Range, R
	First	Second		
1	1.50	1.70	1.60	0.20
2	1.53	1.73	1.63	0.20
3	1.56	1.57	1.57	0.01
4	1.67	1.88	1.78	0.21
5	1.40	1.98	1.69	0.58
6	1.33	1.42	1.38	0.09
7	1.45	1.62	1.54	0.17
8	1.49	1.69	1.59	0.20
9	1.42	1.61	1.52	0.19
10	1.12	1.27	1.20	0.15
11	1.36	1.53	1.45	0.17
12	1.43	1.57	1.50	0.14
13	1.59	1.59	1.59	0.00
14	1.38	1.55	1.47	0.17
15	1.17	1.44	1.31	0.27
Σ	—	—	22.82	2.75
$\bar{\bar{X}}$	—	—	1.52	—
$\bar{R}$	—	—	—	0.18

Table 15.4. Calculation of Control Chart Lines

Formula	Example
1. $\overline{R} = \Sigma R \div k$	1. $\overline{R} = 2.75 \div 15 = 0.18$
2. $\mathrm{UCL}_R = D_4\overline{R}$	2. $\mathrm{UCL}_R = 3.27 \times 0.18 = 0.59$
3. $\mathrm{UWL}_R = D_5\overline{R}$	3. $\mathrm{UWL}_R = 2.51 \times 0.18 = 0.45$
4. $\overline{\overline{X}} = \Sigma\overline{X} \div k$	4. $\overline{\overline{X}} = 22.82 \div 15 = 1.52$
5. $\mathrm{UCL}_{\overline{X}} = \overline{\overline{X}} + A_2\overline{R}$	5. $\mathrm{UCL}_{\overline{X}} = 1.52 + 1.88 \times 0.18 = 1.86$
6. $\mathrm{LCL}_{\overline{X}} = \overline{\overline{X}} - A_2\overline{R}$	6. $\mathrm{LCL}_{\overline{X}} = 1.52 - 1.88 \times 0.18 = 1.18$
7. $\mathrm{UWL}_{\overline{X}} = \overline{\overline{X}} + \frac{2}{3}A_2\overline{R}$	7. $\mathrm{UWL}_{\overline{X}}\ 1.52 + \frac{2}{3} \times 0.34 = 1.75$
8. $\mathrm{LWL}_{\overline{X}} = \overline{\overline{X}} - \frac{2}{3}A_2\overline{R}$	8. $\mathrm{LWL}_{\overline{X}} = 1.52 - \frac{2}{3} \times 0.34 = 1.29$

7.1.6 Interpretation of Control Limits

Standardized control chart calculation and graphing formats are available from many sources (3, 8). The trial control limits were calculated for one concentration range. Other parts of the working range may require different control limits. The NIOSH collaborative test data (13) indicate that, at least for all levels except the level near the detection limit of the method, a single percentage value could be used to express the variability. In such a case it would be easy to convert all ranges into a percentage value and calculate for all values a "%*R*." The accuracy or bias plot then would become useful over the working range by plotting the percentage deviation of the average of each set of control tubes from its known or nominal value. The control tubes would ideally be generated using known concentrations of airborne contaminant but could also be made by spiking a known quantity of the liquid contaminant onto the charcoal in the sampling tube. The latter technique is essentially that used for the determination of desorption efficiency. Analogous techniques for producing control samples can be developed for other measurement systems.

As the industrial hygiene program gains experience with a measurement system, the current data are used to calculate new control limits to replace the trial control limits. Periodically, the validity of the control limits should be checked and recalculated if appropriate.

7.2 Data Analysis Techniques

Numerous data analysis techniques are useful in the industrial hygiene program. Standard textbooks and handbooks such as the NBS handbook by Natrella (63) and QC manuals (3, 70) should be consulted for approaches and sample problems. Consultation with a statistician can be of great help in selecting and applying the most appropriate approach to a specific data analysis problem. Computer programs for QC data analysis are widely available (68, 69).

7.3 Data Validation

Data validation is accomplished by a critical review of a set of data based on previously determined criteria. For large amounts of data from automatic measurement systems, a computer program may be used (8). For more usual industrial hygiene program sets of data, the analyst or surveyor and supervisor should determine whether the set of data conforms to the program's requirements for precision and accuracy. At least a spot check to assure the accuracy of calculations and conversions should be made. The measurement or analytical report (may be a report form) should be checked for transcription errors, omissions, and mistakes. Any outlier values or values widely different from the expected, based on a history of measurements from the area of the survey, should be specifically checked to assure the industrial hygiene program that a real change in plant operations, rather than a measurement error, has taken place. This determination of consistency with past (expected) values may not be possible. An efficient way of automatically alerting the industrial hygiene program to such a situation is to develop a control chart based on operational experience. Trends and changes in plant operations may be more readily detected and interpreted using such an approach. Linch (35) discusses such a case, a lead in urine analysis program.

A method used by OSHA to validate measurements, because historical data are not often available, is the use of a large volume of blind samples included in otherwise normal sample sets (5 to 10 percent of total sample throughput). These samples are prepared by the Laboratory Quality Control Division and are identified by the analyte of interest. They are assigned to analysts along with the routine samples that they are processing. Following analysis and before reporting results of routine field samples in the same set, the blind sample results are checked against control limits set through historical laboratory data. In conjunction with other control procedures, this provides a very good tracking system for precision and accuracy. The system is very easily applied using computers normally available for most analytical equipment so that calculations, personnel time, and costs are minimized.

7.4 Sampling Plans

Sampling plans for acceptance sampling of such items as detector tubes can be an efficient tool for the industrial hygiene program. Their use in developing a sampling program for environmental or biological monitoring is discussed elsewhere in this book (Chapter 17 and 8, respectively).

8 QUALITY-CONTROL PROGRAM AUDIT

8.1 Internal Evaluation

The industrial hygiene program should periodically seek assurance that the elements of the QC system are living up to the program's expectations for the "return on invested resources." Other program elements not directly related to quality control, such as compliance with statutory rules and regulations, and other controlling programmatic standards such as company or agency policies should also be audited.

Auditing of operating procedures can be accomplished by having a supervisor or an operator/analyst other than the person conducting the routine measurements or analyses perform the procedures. A second set of calibration equipment and reference standards can be acquired for checking sampling equipment calibration procedures and operational equipment. A "transfer standard" such as the National Institute of Standards and Technology uses for some basic measurements could be employed by several programs in a cooperative program.

8.2 Self-Appraisal Quality-Control Check List

Ratliff (56) has adapted standard manufacturing quality program audits and vendor qualification procedures to the needs of the analytical laboratory. The following checklist has been modified to cover more broadly the needs of the overall industrial hygiene program laboratory functions. Ratliff used a standard of 3.8 average score as acceptable. An average score of 2.5 to 3.7 indicated a need for improvement. It was felt that an average score of less than 2.5 indicated a risky situation, requiring immediate correction.

INDUSTRIAL HYGIENE PROGRAM SELF-APPRAISAL QUALITY-CONTROL CHECKLIST

1. The acceptance criteria for the level of quality of the industrial hygiene laboratory's routine performance are:

	Score
a. Clearly defined in writing for all key or critical characteristics.	5
b. Defined in writing for some characteristics, and some depend on experience, memory, and/or verbal communication.	3
c. Defined only by experience and verbal communication.	1

2. Acceptance criteria for the level of quality of the industrial hygiene program's routine performance are determined by:

a. Monitoring program performance in a structured program of inter- and intralaboratory evaluations.	5
b. Program determination of what is technically feasible.	3
c. Program determination of what can be done using currently available equipment, techniques, and manpower.	1

3. The QC coordinator has the authority to:

a. Affect the quality of measurements by inserting controls to assure that the methods meet the user's needs for precision, accuracy, sensitivity, and specificity.	5

	Score
b. Reject suspected results and stop any method that produces high levels of discrepancies.	3
c. Submit suspected results to laboratory management for a decision on disposition.	1

4. Accountability for quality is:

a. Clearly defined for all program elements and their chiefs where their actions have an impact on quality.	5
b. Vested with the QC coordinator, who must use whatever means possible to achieve quality goals.	3
c. Not defined.	1

5. "Quality" in the industrial hygiene program long-range planning:

a. Is considered an important factor with regard to changing user demands, new applications, legal considerations, and technical advances in control and methods.	5
b. Is considered part of the technology or service.	3
c. Is not considered a factor for planning purposes.	1

6. Calibration, measuring, gauging, and analytical instruments are:

a. Maintained operative, accurate, and precise by regular checks and calibrations against stable standards that are traceable to the National Institute of Standards and Technology.	5
b. Periodically checked against a zero point or other reference and examined for evidence of physical damage, wear, or inadequate maintenance.	3
c. Checked only when they stop working or when excessive defects are experienced that can be traced to inadequate instrumentation.	1

7. Reagents and chemicals (critical items) and sampling system components such as detector tubes are:

a. Procured from suppliers who must submit samples for test and approval before initial shipment.	5
b. Procured from suppliers who certify that they can meet all applicable specifications.	3
c. Procured from suppliers on the basis of price and delivery only.	1

8. Reagents, chemicals, and sampling systems (or components) are:

a. Checked 100 percent against specification and quantity, and for certification where required, and accepted only if they conform to all specifications.	5

	Score
b. Spot-checked for proper quantity and for shipping damage.	3
c. Released to program personnel by the receiving clerk without being checked as described in a or b.	1

9. Discrepant purchased systems and materials are:

a. Submitted to a review by quality control and chief chemist for disposition.	5
b. Submitted to the operational program elements for determination on acceptability.	3
c. Used because of scheduling requirements.	1

10. Inventories are maintained on:

a. First-in, first-out basis.	5
b. Random selection in stock room.	3
c. Last-in, first-out basis.	1

11. Inventories are:

a. Identified with respect to type, age, and acceptance status.	5
b. Identified with respect to material only.	3
c. Not identified in writing.	1

12. Reagents and chemicals and sampling system components (e.g., detector tubes) that have limited shelf life are:

a. Identified with respect to shelf life expiration date and systematically issued from stock only if they are still within that date.	5
b. Issued on a first-in, first-out basis, expecting that there is enough safety factor that the expiration date is rarely exceeded.	3
c. Issued at random from stock.	1

13. The operating conditions of the methods are:

a. Clearly defined in writing in the method for each significant variable.	5
b. Controlled by supervision based on general guidelines.	3
c. Left up to the field personnel or bench chemist/analyst.	1

14. Operational procedures are checked:

a. During the measurements for conformity to operating conditions and to specifications.	5

	Score
b. After the measurements or analyses to determine acceptability of the results.	3
c. Not at all.	1

15. Revisions to technical operational procedures and sampling/analytical methods are:

a. Clearly spelled out in written form and distributed to all parties affected on a controlled basis, which assures that the change will be implemented and permanent.	5
b. Communicated through memoranda to key people who are responsible for effecting the change through whatever method they choose.	3
c. Communicated verbally to operating personnel, who depend on experience to maintain continuity of the change.	1

16. Changes to methods and other operational procedures are:

a. Analyzed to make sure that any harmful side effects are known and controlled before revision implementation.	5
b. Installed on a trial or gradual basis, monitoring the product to see whether the revision has a net beneficial effect.	3
c. Installed immediately with action for correcting side effects taken if they are evident in the final results.	1

17. Revisions to operational procedures and sampling/analytical methods are:

a. Recorded with respect to date, serial number, and so on, when the revision becomes effective.	5
b. Recorded with respect to the date the revision was made on written specifications.	3
c. Not recorded with any degree of precision.	1

18. The capability of the measurement or sampling/analytical method to produce within specification limit is:

a. Known through method capability analysis (X–R charts) to be able to produce consistently acceptable results.	5
b. Assumed to be able to produce a reasonably acceptable result.	3
c. Unknown.	1

19. Measurement or sampling/analytical method determination discrepancies are:

a. Analyzed immediately to seek out the causes and apply corrective action.	5

	Score
b. Checked out when time permits.	3
c. Not detectable with present controls and procedures.	1

20. Decisions on acceptability of questionable results are made by:

a. A review group consisting of the chief chemist, the chief field industrial hygienist, QC personnel, and others who can render expert judgment.	5
b. An informal assessment by quality control.	3
c. The bench chemist/analyst or field personnel.	1

21. Final acceptance of the results is made:

a. By replicating statistically adequate samples.	5
b. By routine acceptance because of lack of complaints.	3
c. On faith.	1

22. Follow-up action is:

a. Taken to identify assignable causes for "suspect" determinations.	5
b. Taken to make sure that method errors have been corrected.	3
c. Not considered necessary.	1

23. Data reports on quality are distributed to:

a. All levels of management.	5
b. One level of management only.	3
c. Quality control only.	1

24. Quality reports contain:

a. Information on trends, required action, and danger spots.	5
b. Information on suspected analyses and their causes.	3
c. Number of analyses per month.	1

25. Quality control analysis is performed to:

a. Seek out the optimum levels of operation.	5
b. Provide for highlighting trouble spots.	3
c. Fix the blame for substandard results.	1

26. When key personnel changes occur:

a. Specialized knowledge and skills are retained in the form of documented methods and descriptions.	5

	Score
b. Replacement people can acquire the knowledge of their predecessors from co-workers, supervisors, and detailed study of specifications and memoranda.	3
c. Knowledge is lost and must be regained through long experience or trial and error.	1

27. The people who have an impact on quality, bench chemists, industrial hygiene field personnel, supervisors, and so on, are:

a. Trained in the reasons for and the benefits of standards of quality and the methods by which high quality can be achieved.	5
b. Told about quality only when their work falls below acceptable levels.	3
c. Reprimanded when quality deficiencies are directly traceable to an individual's work.	1

28. Training of new employees is accomplished by:

a. A programmed system of training where elements of training, including quality standards, are incorporated in a training checklist; the employee's work is immediately rechecked by supervisors for errors or defects, and the information is fed back instantaneously for corrective action.	5
b. On-the-job training by the supervisor, who gives an overview of quality standards; details of quality standards are learned as normal results are fed back to the employee.	3
c. On-the-job learning, with training on the rudiments of the job by senior co-workers.	1

29. Auditing of the QC program is:

a. Performed on a random but regular basis, to verify that all quality procedures are being implemented and are effective.	5
b. Performed whenever a suspicion arises that there are areas of ineffective performance.	3
c. Never performed.	1

30. If the costs of quality are known, the major portion of the expenditure is in:

a. Prevention.	5
b. Appraisal.	3
c. External or internal failure (duplicate determinations to correct errors; reruns).	1

Score

31. Corrective action to reduce failure costs is:

a. An ongoing program with specific objectives measurements, and target dates. 5
b. Taken when deficient results threaten schedules. 3
c. Taken after considerable losses have occurred in the laboratory or in the field. 1

32. The management, through the NIOSH Proficiency Analytical Testing program or other external evaluation mechanisms, regards the laboratory's performance quality as:

a. Significantly better than peer laboratories. 5
b. About the same as peer laboratories. 3
c. Significantly worse than peer laboratories. 1

33. Support for laboratory quality goals and results is indicated by:

a. A clear statement of quality objectives by the top executive, with continuing visible evidence of sincerity to all levels of the organization. 5
b. Periodic meetings among the section heads of service, field operations, research and development, and quality assurance on quality objectives and progress toward their achievement. 3
c. A "one-shot" statement of the desire for product quality by the top executive, after which the QC staff is on its own. 1

34. The QC system is:

a. Formalized and documented by a set of procedures clearly describing the activities necessary and sufficient to achieve desired quality objectives from initial design through final delivery to the user. 5
b. Contained in operational methods and procedures or is implicit in those procedures; experience with the materials, product, and equipment is needed for continuity of control. 3
c. Undefined in any procedures and left to the current managers or supervisors to determine as the situation dictates. 1

Summary

Strong points:

Weak points:

Improvement goals:

8.3 External Evaluations

External evaluations of environmental programs such as that for water laboratories (64) have been in existence for many years. Clinical laboratories have been involved in such programs at least since the passage of the Clinical Laboratories Improvement Act of 1967. The AIHA program has initiated (55) and validated (65) an accreditation program for industrial hygiene analytical laboratories. The Occupational Health and Safety Program Accreditation Commission (66) (administrative aspects are being handled by the AIHA) was set up by sponsoring occupational health and safety professional societies and has developed and validated (67) program standards and audit criteria for evaluating the overall quality of the occupational health and safety function of an employer (whether a private company, a nonprofit institution, or a governmental agency). Such external evaluation programs will assure industrial hygiene program management and facility management that the safety and health function compares favorably with peer groups, or will objectively identify areas of needed improvement to achieve that status. These programs are continuing, thus ensuring that operational capabilities are maintained over a long period of time.

9 SUMMARY

This chapter covers the philosophy and scope of quality control in the industrial hygiene program. The elements of such a program are grouped and developed for the areas of program management; equipment, standards, and facilities; laboratory analysis control; sample handling, storage, and delivery; statistical quality control techniques; and program audits. The use of computers is briefly discussed. Available software programs can be utilized to track and manage all aspects of a QC program. A high score on the Industrial Hygiene Self-Appraisal Quality-Control Checklist indicates that an industrial hygiene QC program is effective. External evaluation, however, provides corroborative evidence that may carry more weight with the users of the industrial hygiene program's efforts.

REFERENCES

1. John T. Hagen, *A Management Role for Quality Control*, American Management Association, New York, 1968, p. 18.
2. National Institute for Occupational Safety and Health, NIOSH Specification, "Industrial Hygiene Laboratory Quality Control Program Requirements," NIOSH, Cincinnati, OH, 1976.

3. National Institute for Occupational Safety and Health, *Industrial Hygiene Service Laboratory Quality Control Manual*, TR No. 78, NIOSH, Cincinnati, OH, 1974, revised 1976 and 1979.
4. Occupational Safety and Health Administration, *OSHA Manual of Analytical Methods*, ACGIH, Cincinnati, OH, 1990.
5. P. M. Eller, *NIOSH Manual of Analytical Methods*, 3rd ed, U.S. Department of Health and Human Services Publication No. (NIOSH) 84-100, Cincinnati, OH, 1984.
6. Stanford Research Institute, "Laboratory Validation of Air Sampling Methods Used to Determine Environmental Concentrations in Work Places," NIOSH Contract No. CDC-99-7445, NIOSH, Cincinnati, OH, 1976.
7. American Society for Testing and Materials, *1988 Annual Book ASTM Standards*, Part 26, "Gaseous Fuels; Coal and Coke," ASTM, Philadelphia, 1988.
8. Quality Assurance and Environmental Monitoring Laboratory, *Quality Assurance Handbook for Air Pollution Measurement Systems*, Vol. 1, *Principles*, U.S. Environmental Protection Agency, Research Triangle Park, NC, 1975.
9. R. L. Crim, Chairman, Water Monitoring Task Force, Ed., *Model State Water Monitoring Program Environmental Protection Agency*, No. EPA-44019-74002, Washington, DC, 1975.
10. 29 CFR 1910.20(d) Access to Employee Exposure and Medical Records, Preservation of Records (As Revised by 53 FR 38162, September 29, 1989).
11. Division of Training and Manpower Development, NIOSH, "Announcement of Courses," U.S. Department of Health and Human Services, Cincinnati, OH.
12. OSHA Training Institute, *Catalog of Courses*, U.S. Government Printing Office, Washington, DC, 1986.
13. L. R. Rechner and J. Sachdev, "Collaborative Testing of Activated Charcoal Sampling Tubes for Seven Organic Solvents," NIOSH Contract No. HSM 99-72-98, U.S. Department of Health, Education and Welfare Publication No. (NIOSH) 75-184, Cincinnati, OH, 1975.
14. M. Lippmann, "Instruments and Techniques Used in Calibrating Sampling Equipment," in *The Industrial Environment; Its Evaluation and Control*, Stock No. 1701-00396, U.S. Government Printing Office, Washington, DC, 1973, Chapter 11.
15. National Institute for Occupational Safety and Health, "Criteria for a Recommended Standard Oçcupational Exposure to Sodium Hydroxide," U.S. Department of Health, Education and Welfare Publication No. (NIOSH) 76-1005, Rockville, MD, 1975.
16. B. E. Saltzman, "Preparation of Known Concentrations of Air Contaminants," in *The Industrial Environment: Its Evaluation and Control*, Stock No. 1701-00396, U.S. Government Printing Office. Washington, DC, 1973, Chapter 12.
17. American Conference of Governmental Industrial Hygienists, *Air Sampling Instruments*, 7th ed., ACGIH, Cincinnati, OH, 1989.
18. B. A. Johnson, "Evaluation of Portable, Direct Reading NO_2 Meters," U.S. Department of Health, Education and Welfare, Publication No. (NIOSH) 74-108, Cincinnati, OH, 1974.
19. C. S. McCammon, "Evaluation of Portable, Direct-Reading Combustible Gas Meters," U.S. Department of Health, Education and Welfare Publication No. (NIOSH) 74-107, Cincinnati, OH, 1974.
20. C. D. Parker, and R. B. Strong, "Evaluation of Portable Direct-Reading Carbon Mon-

oxide Meters," NIOSH Contract No. HSM-99-73-1 (T.O. No. 1), U.S. Department of Health, Education and Welfare Publication No. (NIOSH) 75-106, Cincinnati, OH, 1974.

21. R. P. Madden, "Ultraviolet Transfer Standard Detectors and Evaluation and Calibration of NIOSH UV Hazard Meter," Interagency Agreement No. NIOSH-IA-73-20, U.S. Department of Health, Education and Welfare Publication No. (NIOSH) 75-131, 1975.
22. R. Mavrodineanu and J. W. Lazor, "Standard Reference Materials: Standard Quartz Cuvettes for High-Accuracy Spectrophotometry," *Clin. Chem.*, **19**(9), 1053–1057 (1973).
23. W. E. Hill, Jr., "Analytical Standards for Quality Controlled Analysis of Ore and Pilot Plant Products," *Am. Lab.*, **8**(2), 65–67 (1976).
24. W. J. Youden and E. H. Steiner, *Statistical Manual of the Association of Official Analytical Chemists*, AOAC, Washington, DC, I 975.
25. American Society for Testing and Materials, *ASTM Manual for Conducting an Inter-Laboratory Study of a Test Method*, ASTM Special Technical Publication No. 335. Available from University Microfilms, Ann Arbor, MI, 1963.
26. H. H. Ku, Ed., "Precision Measurement and Calibration," *NBS Special Publication No. 300*, Vol. 1, U.S. Government Printing Office, Washington, DC, 1969.
27. American Society for Testing and Materials, Committee D-19 on Water, "Practice for Determination of Precision of Methods of Committee D-19 on Water-D-2777," in *1975 Annual Book of ASTM Standards*, Part 31, *Water*, ASTM, Philadelphia, 1975.
28. American Society for Testing and Materials, "Practice for Developing Precision Data on ASTM, Methods for Analysis and Testing of Industrial Chemicals-E1 80," in 1974 *Annual Book of ASTM Standards*, Part 30, *General Methods*, ASTM, Philadelphia, 1974.
29. J. F. Foster and G. H. Beatty, "Interlaboratory Cooperative Study of the Precision and Accuracy of the Atmosphere Using ASTM Method D 2914," Publication No. DS 55-51, ASTM, Philadelphia, 1974.
30. H. F. Hamil and D. E. Camann, "Collaborative Study of Method for the Determination of Nitrogen Oxide Emissions From Stationary Sources," U.S. Environmental Protection Agency Contract No. 68-02-0623, EPA, Research Triangle Park, NC, 1973.
31. H. F. Hamil, "Laboratory and Field Evaluations of EPA Methods 2, 6 and 7," U.S. Environmental Protection Agency Contract No. 68-02-0626, EPA, Research Triangle Park, NC, 1973.
32. H. F. Hamil and D. E. Camann, "Collaborative Study of Method for the Determination of Sulfur Dioxide Emissions Front Stationary Sources," U.S. Environmental Protection Agency, Contract No. 68-02-0623, EPA, Research Triangle Park, NC, 1973.
33. J. A. Winter and H. A. Clements, "Interlaboratory Study of the Cold Vapor Technique for Total Mercury in Water," in *Water Quality Parameters*, American Society for Testing and Materials Special Technical Publication No. 573, ASTM, Philadelphia, 1975, pp. 566–580.
34. J. Mandel and T. W. Lashof, "Interpretation and Generalization of Youdens' Two-sample Diagram," *J. Qual. Technol.*, **6**, 22–36 (1974).
35. A. L. Linch, "Quality Control for Sampling and Laboratory Analysis," in *The Industrial Environment. Its Evaluation and Control*, Stock No. 1701-00396, U.S. Government Printing Office, Washington, DC, 1973, Chapter 22.
36. F. J. Linning, J. Mandel, and J. M. Peterson, "A Plan for Studying the Accuracy and Precision of an Analytical Procedure," *Anal. Chem.* **26**(7), 1102–1110 (1954).

37. A. L. Wilson, "The Performance-Characteristics of Analytical Methods, I," *Talanta*, **17**, 21–29 (1970).
38. A. L. Wilson, "The Performance-Characteristics of Analytical Methods, II," *Talanta*, **17**, 31–44 (1970).
39. A. L. Wilson, "The Performance-Characteristics of Analytical Methods, III," *Talanta*, **20**, 725–732 (1973).
40. National Institute for Occupational Society and Health, "Organic Solvents in Air," P&CAM No. 127, in *NIOSH Manual of Analytical Methods*, U.S. Department of Health, Education and Welfare Publication No. (NIOSH) 75-121, 1975.
41. G. M. Bobba and L. F. Donaghey, "A Microcomputer System for Analysis and Control of Multiple Gas Chromatography," *Am. Lab.* **1**(2), 27–34 (1976).
42. S. A. Roach, "Sampling Air for Particulates," in *The Industrial Environment: Its Evaluation and Control*, Stock No. 1701-00396, U.S. Government Printing Office, Washington, DC, 1973, Chapter 13.
43. B. A. Punghorst, "Methods for Detection and Elimination of Determinate Errors," in *Statistical Method Evaluation and Quality Control for the Laboratory*, Training Course Manual in Computational Analysis, Department of Health, Education and Welfare, Public Health Service, Washington, DC, 1968.
44. Intersociety Committee on Air Sampling and Analysis, "Methods of Air Sampling and Analysis," American Public Health Association, Washington, DC, 1972.
45. A. L. Linch, *Evaluation of Ambient Air Quality by Personnel Monitoring*, CRC Press, Cleveland, OH, 1974.
46. A. L. Linch, *Biological Monitoring for Industrial Chemical Exposure Control*, CRC Press, Cleveland, OH, 1974.
47. J. Mandel, "Repeatability and Reproducibility," *Mater. Res. Stand.*, 8–16 (1971).
48. D. A. B. Lindberg and H. J. Van Peenen, "The Meaning of Quality Control with Multiple Chemical Analyses," paper presented at Technicon Symposium on Automation in Analytical Chemistry, New York, 1965.
49. M. H. Shamos, in *Proceedings of the 1974 Joint Measurement Conference*, Instrument Society of America, Pittsburgh, PA, 1974.
50. R. B. Conn, "Effects of Automation on Clinical Laboratory Operations," in *Proceedings of the 1974 Joint Measurement Conference*, Instrument Society of America, Pittsburgh, PA, 1974.
51. American Industrial Hygiene Association, Analytical Committee, Quality Control for the Industrial Hygiene Laboratory, AIHA, Akron, OH, 1975.
52. Intersociety Committee on Air Sampling and Analysis, Methods of Air Sampling and Analysis, American Public Health Association, Washington, DC, 1972, p. 70.
53. G. R. Elwell and H. E. Lawton, "A Relative Value Structure Helps Laboratory Management Fight the Numbers Racket," *Health Lab. Sci.*, **10**(3), 203–208 (1973).
54. Clinical Laboratories Improvement Act of 1967 Notice of Effective Date, 42 CFR Part 74, Fed. Reg. 33: 253, December 31, 1968; as amended.
55. L. J. Cralley et al., "Guidelines for Accreditation of Industrial Hygiene Analytical Laboratories," *Am. Ind. Hyg. Assoc. J.*, **31**, 335 (1970).
56. T. A. Ratliff, Jr., "Laboratory Quality Program Requirements," paper presented at 30th Annual Technical Conference Transactions, American Society for Quality Control, Milwaukee, 1976.

57. C. B. Donnelly et al., "Containers, Refrigerants and Insulation for Split Milk Samples," *J. Milk Food Technol.*, **21**(5), 131–137 (1958).
58. American Society for Testing and Materials, *ASTM Manual on Quality Control of Materials*, Special Technical Publication 15-C, ASTM, Philadelphia, 1951.
59. C. C. Craig, "The *X*- and *R*-Chart and Its Competitors," *J. Qual. Technol.*, **1**, 2 (1969).
60. W. D. Kelley, Ed., *Statistical Method-Evaluation and Quality Control for the Laboratory*, Training Course Manual in Computational Analysis, Department of Health, Education and Welfare, Public Health Service, Washington, DC, 1968.
61. Analytical Quality Control Laboratory, *Handbook for Analytical Quality Control in Water and Wastewater Laboratories*, U.S. Environmental Protection Agency, Cincinnati, OH, 1972.
62. F. S. Hillier, "*X*- and *R*-Chart Control Limits Based on a Small Number of Subgroups," *J. Qual. Technol.*, **1**(1), 17–26 (1969).
63. M. G. Natrella, *Experimental Statistics, NBS Handbook No. 91*, U.S. Government Printing Office, Washington, DC, 1963.
64. "Evaluation of Water Laboratories Recommended by the U.S. Public Health Service," Public Health Service Publication No. 999-EE-1, Washington, DC, 1966.
65. American Industrial Hygiene Association, "Development of a Laboratory Accreditation Program for Occupational Health Laboratories," NIOSH Contract No. HSM99-72-58, NIOSH, Cincinnati, OH, 1975.
66. Occupational Health Programs Accreditation Commission, Transactions of the 37th Annual Meeting of the American Conference of Governmental Industrial Hygienists, ACGIH, Cincinnati, OH, 1975, pp. 105–108.
67. Occupational Health Institute, "Develop and Validate Criteria for Performance Standards of Occupational Health Programs," NIOSH Contract No. HSM-99-72-109, NIOSH, Washington, DC, 1975.
68. T. F. Hartley, "Computerized Quality Control: Programs for the Analytical Laboratory," ACGIH, ISBN: 0-470-20761-2, 1987.
69. J. A. Burkart, L. M. Eggenberger, J. H. Nelson, and P. R. Nicholson, "A Practical Statistical Quality Control Scheme For The Industrial Hygiene Chemistry Laboratory," *Am. Ind. Hyg. Assoc. J.*, **45**(6), 386–392 (June 1984).
70. James P. Dux, "*Handbook of Quality Assurance for the Analytical Chemistry Laboratory*," ACGIH, ISBN: 0-442-21972-5, 1986.

CHAPTER SIXTEEN

Calibration

MORTON LIPPMANN, Ph.D., C.I.H.

1 INTRODUCTION

Proper interpretation of any environmental measurement depends on an appreciation of its accuracy, precision, and whether it is representative of the condition or exposure of interest. This chapter is concerned primarily with the accuracy and precision of industrial hygiene measurements. Although considerations of the location, duration, and frequency of measurements may be equally or even more important in the evaluation of potential hazards, they require knowledge of the process variables, the kinds of hazard and/or toxic effect that may result from exposures, and their temporal variations. Such considerations, which require the exercise of professional judgment, are beyond the scope of this chapter.

The accuracy of a given measurement depends on a variety of different factors including the sensitivity of the analytical method, its specificity for the agent or energy being measured, the interferences introduced by cocontaminants or other radiant energies, and the changes in response resulting from variations in ambient conditions or instrument power levels. In some cases the influence of these variables can be defined by laboratory calibration, providing a basis for correcting a field sample or instrument reading response. In cases such as the effect of variable line voltage on an instrument's response, they can be avoided by modifications in the circuitry or by the addition of a constant voltage transformer. When the effects of the interferences cannot be controlled or well defined, it may still be desirable to make field measurements, especially in range-finding and exploratory surveys. The interpretation of any such measurements is greatly aided by an appreciation of the

Patty's Industrial Hygiene and Toxicology, Fourth Edition, Volume 1, Part A, Edited by George D. Clayton and Florence E. Clayton
ISBN 0-471-50197-2

extent of the uncertainties. Laboratory calibrations can provide the basis for such an appreciation.

It is important to document the nature and frequency of calibrations and calibration checks to meet legal as well as scientific requirements. Measurements made to document the presence or absence of excessive exposures are only as reliable as the calibrations upon which they are based. Formalized calibration audit procedures established by federal agencies provide a basis for quality assurance where they apply. They can also provide a systematic framework for developing appropriate calibration procedures for situations not governed by specific reporting requirements.

State and local air monitoring networks that are collecting data for compliance purposes are required to have an external performance audit on an annual basis (1). The audit also summarizes the performance of the instruments. In the case of ozone, for example, this would include the records of the weekly multipoint calibrations at 0.1, 0.2, and 0.4 ppm.

The *NIOSH Manual of Analytical Methods* (2) recommends that sampling pumps should be calibrated with each use, and that this calibration be performed with the sampler in line. It also recommends that records of calibration be recorded with each unit.

2 TYPES OF CALIBRATIONS

Occupational health problems can arise from exposure to airborne contaminants, heat stress, excessive noise, vibration, ionizing radiations, and nonionizing electromagnetic radiations. Each of these types of exposure involves a different set of measurement variables and calibration considerations, and they are considered separately. Other types of measurements requiring calibration are associated with the evaluation of ventilation systems used to control exposures to airborne contaminants.

2.1 Air Sampling Instruments

Air samples are collected to determine the concentrations of one or more airborne contaminants. To define a concentration, the quantity of the contaminant of interest per unit volume of air must be ascertained. In some cases the contaminant is not extracted from the air (i.e., it may simply alter the response of a defined physical system). An example is the mercury vapor detector, where mercury atoms absorb the characteristic ultraviolet radiation from a mercury lamp, reducing the intensity incident on a photocell. In this case the response is proportional to the mercury concentration, not to the mass flow rate through the sensing zone; hence concentration is measured directly.

In most cases, however, the contaminant either is recovered from the sampled air for subsequent analysis or is altered by its passage through a sensor within the sampling train, and the sampling flow rate must be known to be able ultimately to determine airborne concentrations. When the contaminant is collected for subsequent analysis, the collection efficiency must also be known, and ideally it should be constant. The measurements of sample mass, of collection efficiency, and of sample volume are usually done independently. Each measurement has its own associated errors, and each contributes to the overall uncertainty in the reported concentration.

The sample volume measurement error often is greater than that of the sample mass measurement. The usual reason is that the volume measurement is made in the field with devices designed more for portability and light weight than for precision and accuracy. Flow-rate measurement errors can further affect the determination if the collection efficiency depends on the flow rate.

Each element of the sampling system should be calibrated accurately before initial field use. Protocols should also be established for periodic recalibration, because the performance of many transducers and meters changes with the accumulation of dirt, as well as with corrosion, leaks, and misalignment due to vibration or shocks in handling, and so on. The frequency of such recalibration checks should be high initially, until experience indicates that it can be reduced safely.

2.1.1 Flow and/or Volume

If the contaminant of interest is removed quantitatively by a sample collector at all flow rates, the sampled volume may be the only airflow parameter that need be recorded. On the other hand, when the detector response depends on both the flow rate and sample mass, as in many length-of-stain detector tubes, both quantities must be determined and controlled. Finally, in many direct-reading instruments, the response depends on flow rate but not on integrated volume.

In most sampling situations the flow rates are, or are assumed to be, constant. When this is so, and the sampling interval is known, it is possible to convert flow rates to integrated volumes, and vice versa. Therefore flow-rate meters, which are usually smaller, more portable, and less expensive than integrated volume meters, are generally used on sampling equipment even when the sample volume is the parameter of primary interest. Normally little additional error is introduced in converting a constant flow rate into an integrated volume, because the measurement and recording of elapsed time generally can be performed with good accuracy and precision.

Flowmeters can be divided into three groups on the basis of the type of measurement made: integrated volume meters, flow-rate meters, and velocity meters. The principles of operation and features of specific instrument types in each group are discussed in succeeding pages. The response of volume meters, such as the spirometer and wet test meter, and flow-rate meters, such as the rotameter and the orifice meter, are determined by the entire sampler flow. In this respect they differ from velocity meters such as the thermoanemometer and the Pitot tube, which measure the velocity at a particular point of the flow cross section. Because the flow profile is rarely uniform across the channel, the measured velocity invariably differs from the average velocity. Furthermore, because the shape of the flow profile usually changes with changes in flow rate, the ratio of point-to-average velocity also changes. Thus when a point velocity is used as an index of flow rate, there is an additional potential source of error, which should be evaluated in laboratory calibrations that simulate the conditions of use. Despite their disadvantages, velocity sensors are sometimes the best indicators available, for example, in some electrostatic precipitators, where the flow resistance of other types of meters cannot be tolerated. Velocity sensors are also used in measurements of ventilation airflow and to measure one of the components in the determination of heat stress.

2.1.2 Calibration of Collection Efficiency

A sample collector need not be 100 percent efficient to be useful, provided its efficiency is known and consistent and is taken into account in the calculation of concentration. In practice, acceptance of a low but known collection efficiency is reasonable procedure for most types of gas and vapor sampling, but it is seldom if ever appropriate for aerosol sampling. All the molecules of a given chemical contaminant in the vapor phase are essentially the same size, and if the temperature, flow rate, and other critical parameters are kept constant, the molecules will all have the same probability of capture. Aerosols, on the other hand, are rarely monodisperse. Because most particle capture mechanisms are size dependent, the collection characteristics of a given sampler are likely to vary with particle size. Furthermore, the efficiency will tend to change with time because of loading; for example; a filter's efficiency increases as dust collects on it, and electrostatic precipitator efficiency may drop if a resistive layer accumulates on the collecting electrode. Thus aerosol samplers should not be used unless their collection is essentially complete for all particle sizes of interest.

2.1.3 Recovery from Sampling Substrate

The collection efficiency of a sampler can be defined by the fraction removed from the air passing through it. However, the material collected cannot always be completely recovered from the sampling substrate for analysis. In addition, the material sometimes is degraded or otherwise lost between the time of collection in the field and recovery in the laboratory. Deterioration of the sample is particularly severe for chemically reactive materials. Sample losses may also be due to high vapor pressures in the sampled material, exposure to elevated temperatures, or reactions between the sample and substrate or between different components in the sample.

Laboratory calibrations using blank and spiked samples should be performed whenever possible to determine the conditions under which such losses are likely to affect the determinations desired. When it is expected that the losses would be excessive, the sampling equipment or procedures should be modified as much as feasible to minimize the losses and the need for calibration corrections.

2.1.4 Sensor Response

When calibrating direct-reading instruments, the objective is to determine the relation between the scale readings and the actual concentration of contaminant present. In such tests the basic response for the contaminant of interest is obtained by operating the instrument in known concentrations of the pure material over an appropriate range of concentrations. In many cases it is also necessary to determine the effect of such environmental cofactors as temperature, pressure, and humidity on the instrument response. Also, many sensors are nonspecific, and atmospheric cocontaminants may either elevate or depress the signal produced by the contaminant of interest. If reliable data on the effect of such interferences are not available, they should be obtained in calibration tests. Procedures for establishing known concentrations for such calibration tests are discussed in detail in Sections 4.6 and 4.7.

2.2 Ventilation System Measurements

2.1.1 Air Velocity Measurements

Most ventilation performance measurements are made with anemometers (i.e., instruments that measure air velocities), and with the exception of the Pitot tube, all require periodic calibration. Instruments based on mechanical or electrical sensors are sensitive to mechanical shocks and/or may be affected by dust accumulations and corrosion. Calibration requirements are indicated in Table 16.1.

Anemometers are usually calibrated in a well-defined flow field that is relatively large in comparison to the size of meter being calibrated. Such flow fields can be produced in wind tunnels, which are discussed in Section 4.5.

2.2.2 Pressure Measurements

Although the standard Pitot tube and a water-filled U-tube manometer may not require calibration, Pitot tubes and other flowmeters may be used with pressure gauges that do. Many direct-reading gauges can give false readings because of the effects of mechanical shocks and/or leakage in connecting tubes or internal diaphragms. For pressures of approximately 1 in. H_2O or greater, a liquid-filled laboratory manometer should be an adequate reference standard. For lower pressures, it may be necessary to use a reference whose calibration is traceable to a National Institute of Standards and Technology (NIST) standard. Calibration requirements for pressure gauges are given in Table 16.2.

2.3 Heat Stress Measurements

The four environmental variables used in heat stress evaluations are air temperature, humidity, radiant temperature, and air velocity. The measurement of air velocity is discussed in detail elsewhere in this chapter. Liquid-in-glass thermometers used to measure dry-bulb, wet-bulb, or globe thermometer temperatures should have calibrations traceable to certified NIST standards, but should not need recalibration. Other temperature sensors will require periodic recalibration.

Table 16.1. Characteristics of Flow Instruments

Instrument	Range (fpm)	Hole Size (for ducts) (in.)	Range Temp.[a]	Dust, Fume Difficulty	Calibration Requirements	Ruggedness	General Usefulness and Comments
Pilot tubes with inclined manometer	600 up	3/8	Wide	Some	None	Good	
Standard	600 up	3/8	Wide	Some	None	Good	Good except at low velocities
Small Size	600 up	3/16	Wide	Yes	Once	Good	Good except at low velocities
Double	500 up	3/4	Wide	Small	Once	Good	Special
Swinging vane anemometers	25–10,000	1/2–1	Medium	Some	Frequent	Fair	Good
Rotating vane anemometers							
Mechanical	30–10,000	Not for duct use	Narrow	Yes	Frequent	Poor	Special; limited use
Electronic	25–200 25–500 25–2,000 25–5,000	Not for duct use	Narrow	Yes	Frequent	Poor	Special; can record; direct reading

[a]Temperature range: Narrow, 20–150°F; medium, 20–300°F; wide, 0–800°F.

Source: *Industrial Ventilation*, 20th ed. Courtesy American Conference of Governmental Industrial Hygiensts.

Table 16.2. Characteristics of Pressure Measuring Instruments: Static Pressure, Velocity Pressure, and Differential Pressure

Instrument	Range (in. H_2O)	Manufacturer's Stated Precision (in. H_2O)	Comments
Liquid manometers			
Vertical U-tube	No limit	0.1	Portable; needs no calibration
Inclined 10:1 slope	Usually up to 10	0.005	Portable; needs no calibration; must be leveled
Hook gauge	0–24	0.001	Not a field instrument; tedious, difficult to read; for calibration only
Micromanometer (Meriam model 34FB2TM)	0–10 0–20	0.001	Heavy; must be located on vibration-free surface; not difficult to read; uses magnifier
Micromanometer (Vernon Hill type C)	0.001–1.2	0.0004	Small; portable; uses magnifier; need experience to read to manufacturer's precision; calibration needed
Micromanometer (Electric Microtonic, F.W. Dwyer, Mfg.)	0–2	0.0003	Portable; needs vibration-free mount; no magnifier, slow to use; no eyestrain; no calibration needed
Diaphragm and mechanical			
Diaphragm-magnehelic gauge	0–0.5 0–1 0–4	0.01 0.02 0.10	Calibration recommended; no leveling, no mounting needed, direct reading
Swinging vane anemometer	0–0.5 0–20	5% scale	Calibration recommended; no leveling, no mounting needed; use manufacturer's exact recommendation for size of SP hole
Pressure transducers	0.05–6	0.3%	Must be calibrated; remote reading responds to rapid change in pressure

Source: *Industrial Ventilation*, 20th ed. Courtesy of the American Conference of Governmental Industrial Hygienists.

2.4 Electromagnetic Radiation Measurements

The electromagnetic spectrum is a continuum of frequencies whose effects on human health are discussed in Chapters 38 and 39. Most of these effects are

Figure 16.1. Sound level calibrator.

frequency dependent, and some are attributable to narrow bands of frequency. Thus many of the instruments used to measure frequency and/or intensity are designed to operate over specific frequency regions. Other instruments, known as bolometers, measure the total incident radiant flux over a wide range of frequencies.

The literature on calibration techniques for the measurement of electromagnetic radiation is too extensive to summarize adequately in the space available. Instead, the reader is directed to Chapters 38 and 39 and to the reference works cited in those chapters.

2.5 Noise-Measuring Instruments

The accuracy of sound-measuring equipment may be checked by using an acoustic calibrator (Figure 16.1) consisting of a small, stable sound source that fits over a microphone and generates a predetermined sound level within a fraction of a decibel. The acoustic calibration provides a check of the performance of the entire instrument, including microphone and electronics. Some sound level meters have internal means for calibration of electronic components only. Sound level calibrators should be used only with the microphones for which they are intended. Manufacturers' instructions should be followed regarding the use of calibrators and indications of malfunction of instruments.

An ANSI standard for acoustic calibrators (ANSI S1.40-1984) stipulates that calibrators "normally include a sound source which generates a known sound pressure level (SPL) in a coupler into which a microphone is inserted. A diaphragm or piston inside the coupler is driven sinusoidally and generates a specified SPL and frequency within the coupler. The calibrator presents to the inserted microphone of a sound level meter or other sound measuring system a reference or known acoustic signal so one can verify the system sensitivity or set the system to indicate the correct SPL at some frequency."

Many acoustic calibrators provide two or more nominal SPLs and operate at two or more frequencies. Multiple levels and multiple frequencies are useful for checking the linearity and, in a limited way, the frequency response of a measuring system, respectively. The latter is useful for gross checks of microphones and weighting filters for failure. An acoustic calibrator that produces multiple frequencies may also be used to determine a single-number composite calibration for a broad-band sound through calibration at several frequencies. Additional signals such as tone bursts may be provided for use in checking some important electroacoustic characteristics of sound level meters (SLMs) and other acoustical instruments. Such signals may also be useful in checking performance characteristics of instruments to measure sound exposure level or time-period average sound level.

The ANSI standard specifies tolerances for SPLs produced by calibrators. These range from ± 0.3 dB for calibration of microphones expected to be used with types 0 and 1 SLM instruments to ± 0.4 dB for calibration of type 2 instruments including dosimeters.

One should not place blind trust in any calibrator or instrument. A simple procedure, when several calibrators and/or SLMs are available, is to perform a cross check among them. This should produce results within the published tolerance limits.

Coupler-type calibrators should be used only with microphones for which they are intended. Instructions supplied by the manufacturer on instrument use and corrections for barometric pressure and temperature should be carefully followed. Use of single-frequency calibrations may result in overlooking damage to microphones that is manifested at frequencies other than the calibrator frequency.

3 CALIBRATION STANDARDS

Calibration procedures generally involve a comparison of instrument response to a standardized atmosphere or to the response of a reference instrument. Hence the calibration can be no better than the standards used. Reliability and proper use of standards are critical to accurate calibrations. Reference materials and instruments available from or calibrated by NIST should be used whenever possible. Information on calibration aids available from NIST is summarized in Table 16.3.

Test atmospheres generated for purposes of calibrating collection efficiency or instrument response should be checked for concentration using reference instruments or sampling and analytical procedures whose reliability and accuracy are well documented. The best procedures to use are those that have been refereed or panel tested, that is, methods that have demonstrably yielded comparable results on blind samples analyzed by different laboratories. Organizations that publish standard calibration methods are listed in Table 16.4; a summary of the procedures is provided in Table 16.5.

Calibrations of flow rate or sampled volume are generally made by comparing the indicated rate or volume passing through the sampler with that passing through or into a calibration standard in series with the sampler. The standards used are classified as primary, intermediate, and secondary.

Primary standards provide direct and unequivocal calibration because they are based on direct and easily measurable linear dimensions, such as the length and diameter of a cylinder of displaced fluid. Secondary standards are instruments or

Table 16.3. National Institute of Standards and Technology Standard Reference Materials (SRMs)* Used for Industrial Hygiene Calibrations

A. Compressed Gases

Contents	Nominal Concentrations
SO_2 in N_2	50, 100, 1500, 2000, 3500 ppm
NO in N_2	5, 10, 20, 50, 100, 250, 500, 1500, 3000 ppm
NO_2 in air	250, 500, 1000, 2500 ppm
CO in N_2	10, 25, 50, 100, 250, 500, 1000, 2500, 5000 ppm
CO in N_2	1, 2, 4, 8 mol%
CO in air	10, 20, 45 ppm
CO_2 in N_2	300, 400, 800 ppm
CO_2 in N_2	0.5, 1.0, 1.5, 2.0, 2.5, 3.0, 3.5, 4.0, 7.0, 14.0 mol%
CO_2 in air	335, 342, 351 ppm
O_2 in N_2	2, 10, 20 mol%
CH_4 in air	1 ppm
C_3H_8 in air	3, 10, 50, 100, 500 ppm
C_3H_8 in N_2	100, 250, 500, 1000, 2500, 5000, 10,000, 20,000 ppm
C_6H_6 in N_2	0.25, 10 ppm
C_2Cl_4 in N_2	0.25, 10 ppm

B. Permeation Devices

Contents	Permeation Rates (μg/min at 25°C)
SO_2	0.56, 1.4, 2.8
NO_2	1.0
C_6H_6	0.4
C_2Cl_4	1.0

meters that have been calibrated against primary standards. Intermediate standards are secondary standards that can provide an accuracy comparable to primary standards, for example, 1 percent. Examples include wet test meters and dry gas meters. Secondary standards must be handled with care and properly stored and maintained when not in use, and they should be periodically recalibrated against a suitable primary standard.

4 INSTRUMENTS AND TECHNIQUES

4.1 Cumulative Air Volume

Many air sampling instruments utilize an integrating volume meter for measurement of sampled volume. Most of them measure displaced volumes that can be determined from linear measurements and geometric formulas. Such measurements usually can be made with a high degree of precision.

Table 16.3. (*Continued*)

C. Reference Particles

Material	Description
Coal fly ash	75 g
Urban particulate matter	2 g
Urban dust/organics	10 g
Diesel particulate matter	Set of 5
Polystyrene spheres	0.3, 1.0, 10, 30 μm
Respirable alpha quartz	5 g
Respirable cristobalite	5 g

D. Analytical Reference Materials

Material	Description
Metals on filter media	Set of 12
Be and As on filter media	Set
Membrane blank filter	Set of 10
Quartz on filter media	Set of 4
Ashless blank filter	Set of 10
Coal	Certified for content of trace elements
Coal fly ash	Certified for content of trace elements
Fuel oil	Certified for content of trace elements
Dried urine	Certified for content of fluorine
Dried urine	Certified for content of mercury
Dried urine	Certified for content of trace metals
Bovine liver	Certified for content of Pb, Hg, and 10 other elements

[a]Available from Office of Standard Reference Materials, Room B311 Chemistry Bldg. National Institute of Standards and Technology, Gaithersburg, MD 20899. Phone: (301) 921-2045.

4.1.1 Water Displacement

Figure 16.2 is a schematic drawing of a Mariotte bottle. When the valve at the bottom of the bottle is opened, water drains out of the bottle by gravity, and air is drawn by way of a sample collector into the bottle to replace it. The volume of air drawn in is equal to the change in water level multiplied by the cross section at the water surface.

4.1.2 Spirometer or Gasometer

The spirometer (Figure 16.3) is a cylindrical bell with its open end under a liquid seal. The weight of the bell is counterbalanced so that the resistance to movement as air moves in or out of the bell is negligible. The device differs from the Mariotte bottle in that it measures displaced air instead of displaced liquid. The volume change is calculated in a similar manner (i.e., change in height times cross section).

Table 16.4. Organizations Publishing Recommended or Standard Methods and/or Test Protocols Applicable to Air Sampling Instrument Calibration

Abbreviation	Full Name and Address
ANSI	American National Standards Institute, Inc. 1430 Broadway New York, NY 10018
ASTM	American Society for Testing and Materials D-22 Committee on Sampling and Analysis of Atmospheres, and E-34 Committee on Occupational Health and Safety 1916 Race Street Philadelphia, PA 19103
AWMA	Air and Waste Management Association P.O. Box 2861 Pittsburgh, PA 15230
EPA/EMSL	U.S. Environmental Protection Agency Environmental Monitoring Systems Laboratory Quality Assurance Division (MD-77) Research Triangle Park, NC 27711
ISC	Intersociety Committee on Methods for Air Sampling and Analysis % Dr. James P. Lodge, Editor *Intersociety Manual*, 3rd ed. 385 Broadway Boulder, CO 80303
NIOSH	National Institute for Occupational Safety and Health NIOSH Manual Coordinator Division of Physical Sciences and Engineering 4676 Columbia Parkway Cincinnati, OH 45226

Spirometers are available in a wide variety of sizes and are frequently used as primary volume standard (1).

4.1.3 "Frictionless" Piston Meters

Cylindrical air displacement meters with nearly frictionless pistons are frequently used for primary flow calibrations. The simplest version is the soap bubble meter illustrated in Figure 16.4. It utilizes a volumetric laboratory burette whose interior surfaces are wetted with a detergent solution. If a soap-film bubble is placed at the left side, and suction is applied at the right, the bubble is drawn from left to right. The volume displacement per unit time (i.e., flow rate) can be determined by measuring the time required for the bubble to pass between two scale markings that enclose a known volume.

Soap-film flowmeters and mercury-sealed piston flowmeters are available com-

Table 16.5. Summary of Recommended and Standard Methods Relating to Air Sampling and Instrument Calibration

Organization	No. of Methods	Type of Methods
ANSI	1	Sampling airborne radioactive materials
ASTM	46	Test methods for sampling and analysis of atmospheres
ASTM	28	Recommended practices for sampling and calibration procedures, nomenclature, guides, etc.
AWMA	3	Recommended standard methods for continuous air monitoring for fine particulate matter
EPA/EMSL	6	Reference methods for air contaminants
ISC	115	Methods of air sampling and analysis
NIOSH	300	Analytic methods for air contaminants

mercially from several sources (3). In the mercury-sealed piston, most of the cylindrical cross section is blocked off by a plate that is perpendicular to the axis of the cylinder. The plate is separated from the cylinder wall by an O-ring of liquid mercury, which retains its toroidal shape because of its strong surface tension. This floating seal has a negligible friction loss as the plate moves up and down.

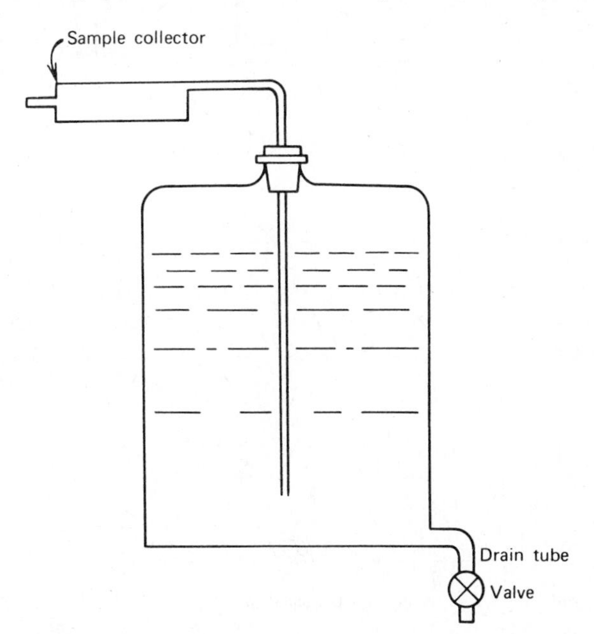

Figure 16.2. Mariotte bottle. From *The Industrial Environment—Its Evaluation and Control*, 2nd ed., C. H. Powell and A. D. Hosey, Eds., Public Health Service Publication No. 614, U.S. Goverment Printing Office, Washington, DC, 1965.

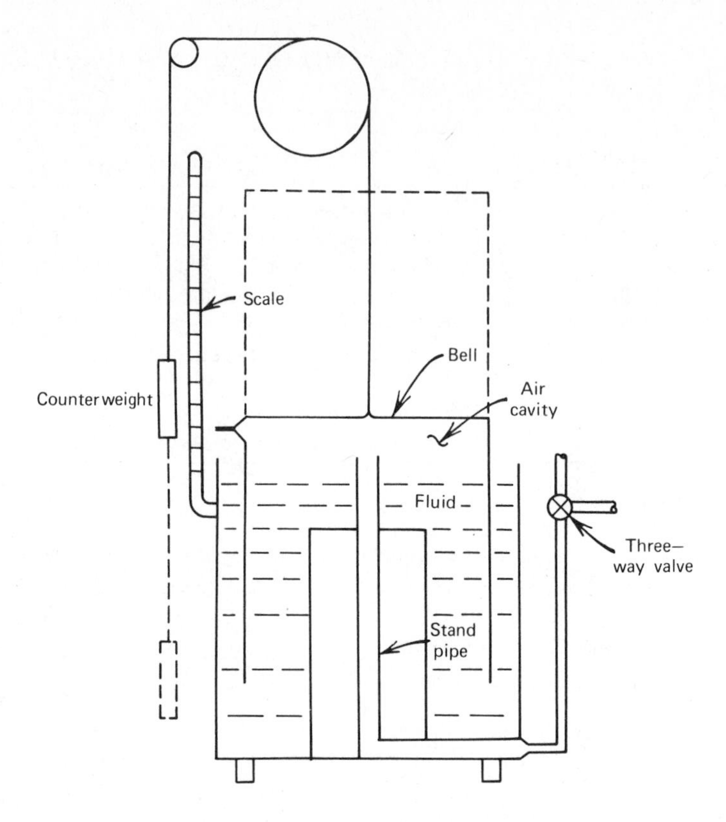

Figure 16.3. Spirometer.

4.1.4 Wet Test Meter

A wet test meter (Figure 16.5) consists of a partitioned drum half-submerged in a liquid (usually water), with openings at the center and periphery of each radial chamber. Air or gas enters at the center and flows into an individual compartment,

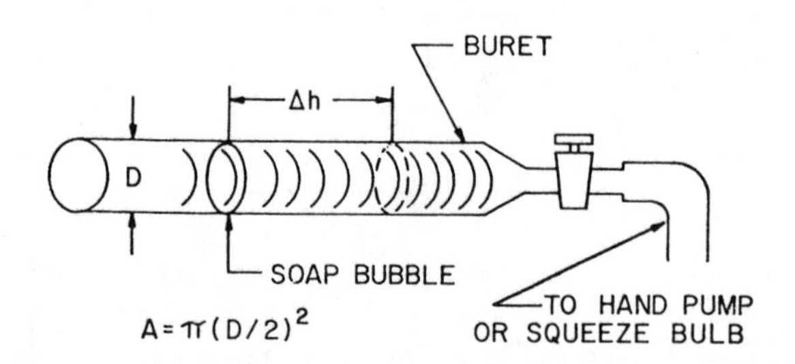

Figure 16.4. Bubble meter. From *The Industrial Environment—Its Evaluation and Control*, 2nd ed., C. H. Powell and A. D. Hosey, Eds., Public Health Service Publication No. 614, U.S. Government Printing Office, Washington, DC, 1965.

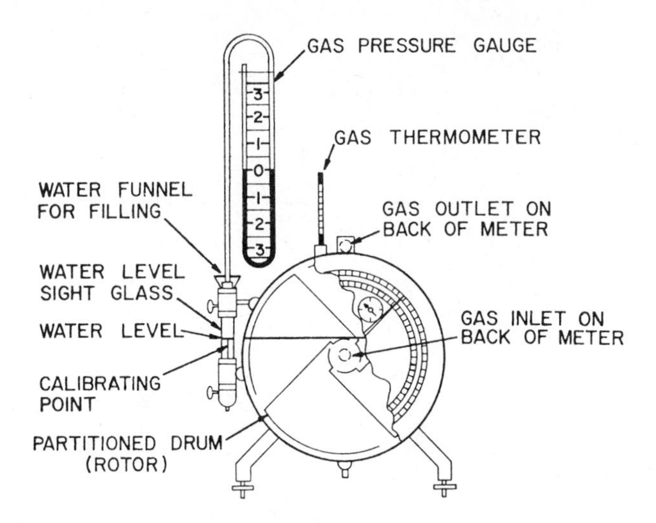

Figure 16.5. Wet test meter. From *The Industrial Environment—Its Evaluation and Control*, 2nd ed., C. H. Powell and A. D. Hosey, Eds., Public Health Service Publication No. 614, Government Printing Office, Washington, D.C., 1965.

causing the drum to rise, thereby producing rotation. This rotation is indicated by a dial on the face of the instrument. The volume measured depends on the fluid level in the meter, because the liquid is displaced by air. There is a sight gauge for determining fluid height, and the meter is leveled by screws and a sight bubble provided for this purpose.

Several potential errors are associated with the use of a wet test meter. The drum and moving parts are subject to corrosion and damage from misuse, there is friction in the bearings and the mechanical counter, and inertia must be overcome at low flows (<1 rpm), whereas at high flows (>3 rpm) the liquid might surge and break the water seal at the inlet or outlet. In spite of these factors, the accuracy of the meter usually is within 1 percent when used as directed by the manufacturer.

4.1.5 Dry Gas Meter

The dry gas meter shown in Figure 16.6 is very similar to the domestic gas meter. It consists of two bags interconnected by mechanical valves and a cycle-counting device. The air or gas fills one bag while the other bag empties itself; when the cycle is completed, the valves are switched, and the second bag fills while the first one empties. Any such device has the disadvantages of mechanical drag, pressure drop, and leakage; however, the advantage of being able to use the meter under rather high pressures and volumes often outweighs the disadvantages created by the errors, which can be determined for a specific set of conditions. The alternate filling of two chambers is also used as the basis for volume measurement in twin-cylinder piston meters. Such meters can also be classified as positive displacement meters.

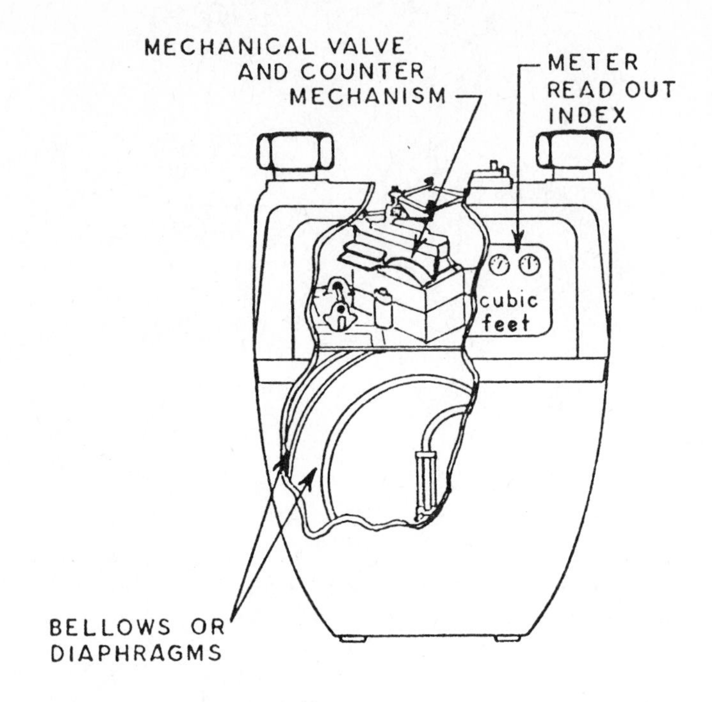

Figure 16.6. Dry gas meter.

4.1.6 Positive Displacement Meters

Positive displacement meters consist of a tight-fitting moving element with individual volume compartments that fill at the inlet and discharge at the outlet parts. Another multicompartment continuous rotary meter uses interlocking gears. When the rotors of such meters are motor driven, these units become positive displacement air movers.

4.2 Volumetric Flow Rate

The volume meters discussed in the preceding paragraphs were all based on the principle of conservation of mass—specifically, the transfer of a fluid volume from one location to another. The flow rate meters in this section all operate on the principle of the conservation of energy; more specifically, they utilize Bernoulli's theorem for the exchange of potential energy for kinetic energy and/or frictional heat. Each consists of a flow restriction within a closed conduit. The restriction causes an increase in the fluid velocity and therefore an increase in kinetic energy, which requires a corresponding decrease in potential energy (i.e., static pressure). The flow rate can be calculated from a knowledge of the pressure drop, the flow cross section at the constriction, the density of the fluid, and the coefficient of discharge, which is the ratio of actual flow to theoretical flow and makes allowance for stream contraction and frictional effects.

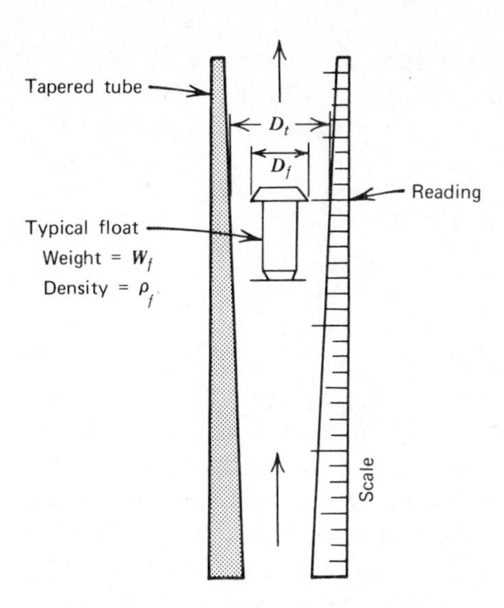

Figure 16.7. Schematic of rotameter.

Flowmeters that operate on this principle can be divided into two groups. The larger group, which include orifice meters, venturi meters, and flow nozzles, have a fixed restriction and are known as variable head meters because the differential pressure head varies with flow. The other group, which include rotameters, are known as variable area meters, because a constant pressure differential is maintained by varying the flow cross section.

4.2.1 Variable Area Meters (Rotameters)

A rotameter consists of a "float" that is free to move up and down within a vertical tapered tube that is larger at the top than the bottom (Figure 16.7). The fluid flows upward, causing the float to rise until the pressure drop across the annular area between the float and the tube wall is just sufficient to support the float. The tapered tube is usually made of glass or clear plastic, and the flow-rate scale is etched directly on it. The height of the float indicates the flow rate. Floats of various configurations are used. They are conventionally read at the highest point of maximum diameter, unless otherwise indicated.

Most rotameters have a range of 10:1 between their maximum and minimum flows. The range of a given tube can be extended by using heavier or lighter floats. Tubes are made in sizes from about ⅛ to 6 in. (3 mm to 15 cm) in diameter, covering ranges from a few cubic centimeters per minute to more than 1000 cfm (28 m^3/min). Some of the shaped floats achieve stability by having slots that make them rotate, but these are less commonly employed than previously. The term "rotameter" was first used to describe such meters with spinning floats but now serves generally for tapered metering tubes of all types.

Rotameters are the most commonly used flowmeters on commercial air samplers, especially on portable samplers. For such sampler flowmeters, the most common material of construction is acrylic plastic, although glass tubes may also be used. Because of space limitations, the scale lengths are generally no more than 4 in. (10 cm), and most commonly nearer 2 in. (5 cm). Unless they are individually calibrated, the accuracy is unlikely to be better than ±25 percent. When individually calibrated, ±5 percent accuracy may be achieved. It should be noted, however, that with the large taper of the bore, the relatively large size of the float, and the relatively few scale markers on these rotameters, the precision of the readings may be a major limiting factor.

Calibrations of rotameters are performed at an appropriate reference pressure, usually atmospheric. Because good practice dictates that the flowmeter be located downstream of the sample collector or sensor, however, the flow is actually measured at a reduced pressure, which may also be a variable pressure if the flow resistance changes with loading. If this resistance is constant, it should be known; if variable, it should be monitored, to permit adjustment of the flow rate as needed, and the making of appropriate pressure corrections for the flowmeter readings.

For rotameters with linear flow-rate scales, the actual sampling flow will approximately equal the indicated flow rate times the square roots of the ratios of absolute temperatures and pressures of the calibration and field conditions (4). The ratios increase when the field pressure is less than the pressure in the calibration laboratory or the field temperature is greater than that in the laboratory. Thus if the flowmeter was accurate at ambient pressure and the flow resistance of the sampling medium was relatively low, for example, 30 mm Hg for a flow rate of 11 l/min, the flow rate indicated on the rotameter would be $11 \times (730/760)^{1/2} = 10.8$ l/min, a difference of only 1.8 percent. On the other hand, for a 25-mm diameter AA Millipore with a 3.9 cm^2 filtering area and a sampling rate of 11 l/min, the flow resistance would be ~190 mm Hg, and the indicated flow rate would be $11 \times (570/760)^{1/2} = 9.5$ l/min, a difference of 14 percent.

A further correction is needed when the sampling is done at atmospheric pressures and/or temperatures that differ substantially from those used for the calibration. For example, at an elevation of 5000 ft above sea level, the atmospheric pressure is only 83 percent of that at sea level. Thus the actual flow rate would be 9.6 percent greater than that indicated on a rotameter scale, based upon the altitude correction alone. If the temperature in the field was 35°C while the meter was calibrated at 20°C, the actual flow rate in the field would be $(308/293)^{1/2} \times 100 =$ 2.5 percent greater than the indicated rate.

In a situation where there were corrections needed for the pressure drop of the sampler, high altitude, and high temperature, the overall correction could be, for the examples cited, $1.14 \times 1.096 \times 1.025 = 1.28$ or 28 percent.

Craig (5) has shown that the change in calibration with air density cannot be made by simple computation, especially for the small-diameter rotameter tubes and floats commonly used on air sampling equipment. Figure 16.8 gives experimental calibration data at various suction pressures for a specific glass rotameter. Clearly it is not practical to generate such a family of empirical calibration curves for each rotameter. Craig recommends (1) that a pressure gauge be used at the

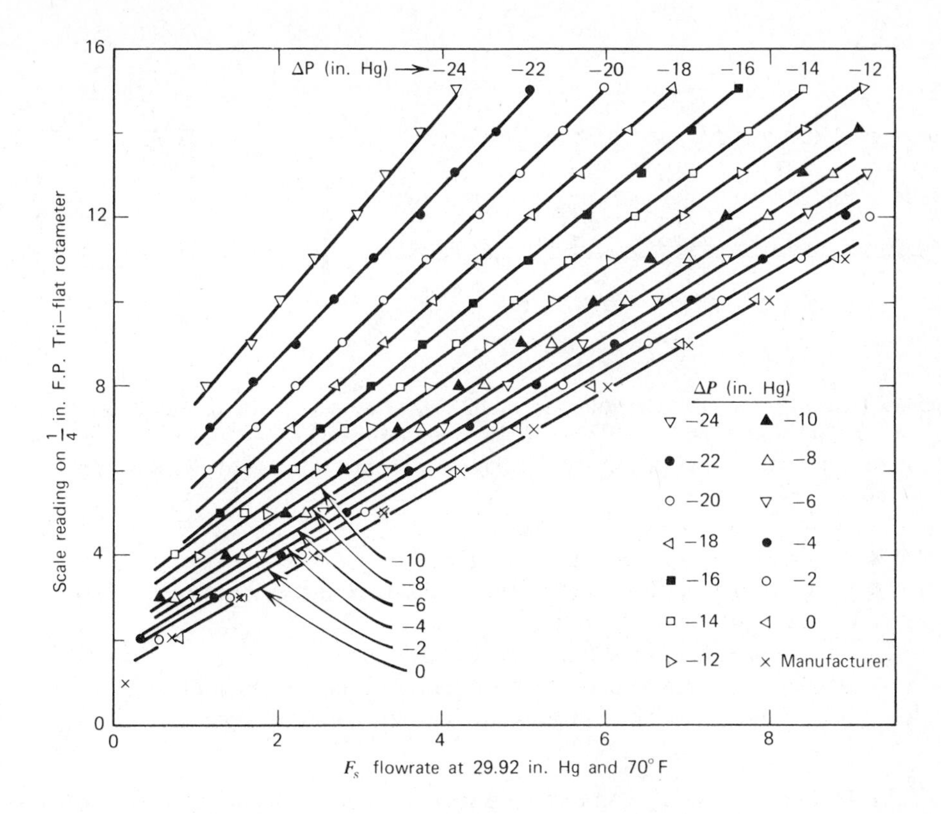

Figure 16.8. Rotameter reading versus airflow rate under standard conditions for various pressure gauge readings at rotameter.

inlet to the rotameter, (2) that the flow rates used be those which give as low a pressure drop as possible, and (3) that the meter size be selected to give readings near the upper end of the scale.

4.2.2 Variable Head Meters

When orifice and venturi meters are made to standardized dimensions, their calibration can be predicted with ~ ±10 percent accuracy using standard equations and published empirical coefficients. The general equation (6) for this type of meter is

$$W = q_1\rho_1 = KYA_2\sqrt{2g_c(P_1 - P_2)\rho_1} \tag{1}$$

where $K = C/\sqrt{1 - \beta^4}$
C = coefficient of discharge (dimensionless)
A_2 = cross-sectional area of throat (ft^2)
g_c = 32.17 ft/sec^2
P_1 = upstream static pressure (lb/ft^2)

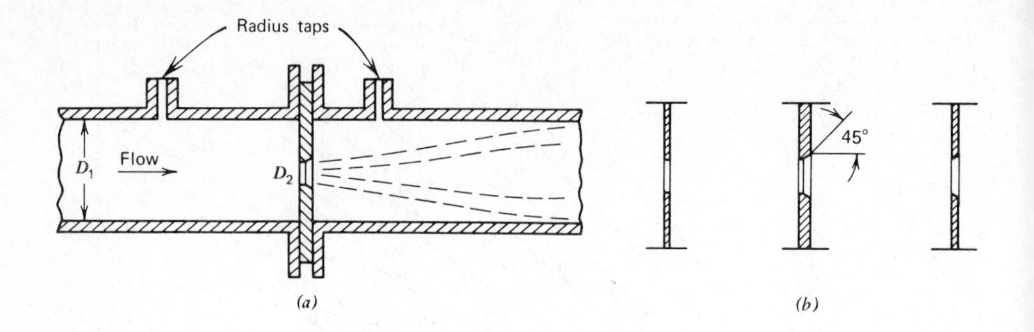

Figure 16.9. Square-edged or sharp-edged orifices. The plate at the orifice opening must not be thicker than 1/30 of pipe diameter, 1/8 of the orifice diameter, or 1/4 of the distance from the pipe wall to the edge of the opening. (*a*) Pipeline orifice. (*b*) Types of plate.

P_2 = downstream static pressure (lb/ft²)
q_1 = volumetric flow at upstream pressure and temperature (ft³/sec)
W = weight rate of flow (lb/sec)
Y = expansion factor (see Figure 16.11)
β = ratio of throat diameter to pipe diameter (dimensionless)
ρ_1 = density at upstream pressure and temperature (lb/ft³)

4.2.2.1 Orifice Meters. The simplest form of variable head meter is the square-edged or sharp-edged orifice illustrated in Figure 16.9. It is also the most widely used because of its ease of installation and low cost. If it is made with properly mounted pressure taps, its calibration can be determined from Equation 1 and Figures 16.10 and 16.11. However, even a nonstandard orifice meter can serve as a secondary standard, provided it is carefully calibrated against a reliable reference instrument.

Although the square-edged orifice can provide accurate flow measurements at low cost, it is inefficient with respect to energy loss. The permanent pressure loss for an orifice meter with radius taps can be approximated by $(1 - \beta^2)$ and often exceeds 80 percent.

4.2.2.2 Venturi Meters. Venturi meters have optimal converging and diverging angles of about 25° and 7°, respectively, which means that they have high pressure recoveries; that is, the potential energy that is converted to kinetic energy at the throat is reconverted to potential energy at the discharge, with an overall loss of only about 10 percent.

For air at 70°F and 1 atm and for $1/4 < \beta < 1/2$, a standard venturi would have a calibration described by

$$Q = 21.2\,\beta^2 D^2 \sqrt{h} \tag{2}$$

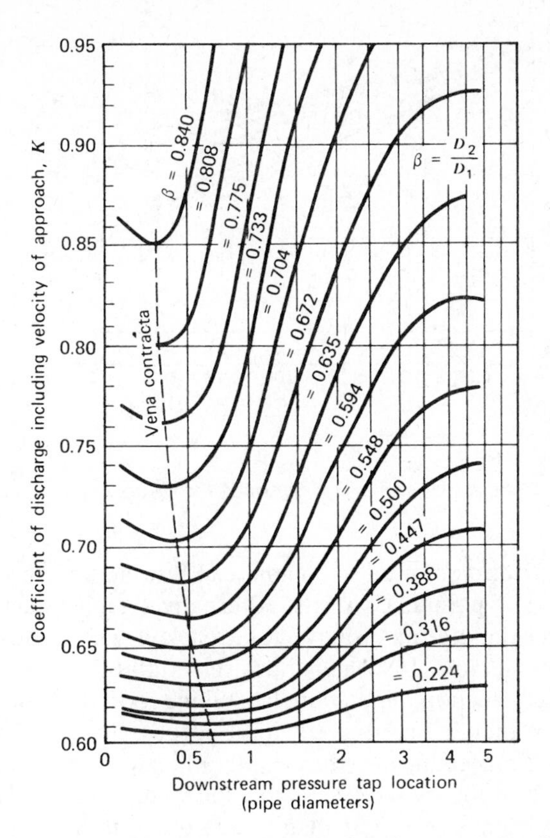

Figure 16.10. Downstream pressure tap location in pipe diameters. Coefficient of discharge for square-edged circular orifices for $N_{Re_2} > 30{,}000$ with the upstream tap location between one and two pipe diameters from the orifice position. From Spitzglass, *Trans. ASME*, **44**, 919 (1922).

where Q = flow (cfm)
β = ratio of throat to duct diameter (dimensionless)
D = duct diameter (in.)
h = differential pressure (in. H_2O)

4.2.2.3 Other Variable Head Meters. The characteristics of various other types of variable head flowmeters (e.g., flow nozzles, Dall tubes, centrifugal flow elements) are described in standard engineering references (6, 7). In most respects they have similar properties to the orifice meter, the venturi meter, or both.

One type of variable head meter that differs significantly from all the foregoing is the laminar flowmeter. These devices are seldom discussed in engineering handbooks because they are used only for very low flow rates. Because the flow is laminar, the pressure drop is directly proportional to the flow rate. In orifice meters, venturi meters, and related devices, the flow is turbulent and flow rate varies with the square root of the pressure differential.

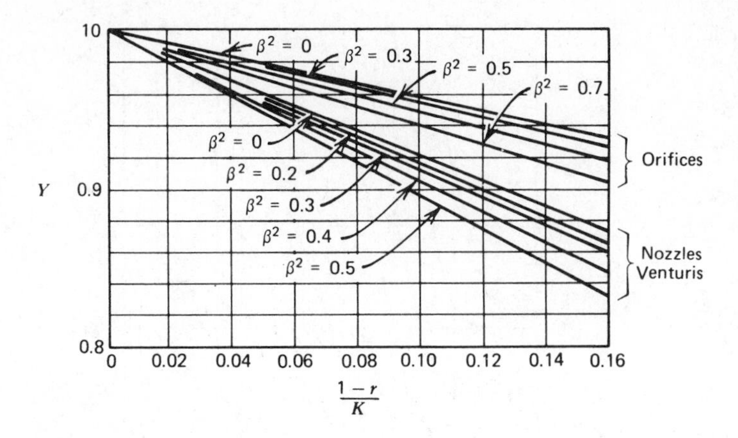

Figure 16.11. Values of expansion factor Y for orifices, nozzles, and venturis.

Laminar flow restrictors used in commercial flowmeters consist of egg-crate or tube bundle arrays of parallel channels. Alternatively, a laminar flowmeter can be constructed in the laboratory using a tube packed with beads or fibers as the resistance element. Figure 16.12 illustrates this kind of homemade flowmeter. It consists of a "T" connection, pipette or glass tubing, cylinder, and packing material. The outlet arm of the "T" is packed with a porous plug, and the leg is attached to a tube or pipette projecting into the cylinder filled with water or oil. A calibration curve of the depth of the tube outlet below the water level versus the rate of flow should produce a linear curve. Saltzman (8) has used tubes filled with mineral fiber to regulate and measure flowrates as low as 0.01 cm^3/min.

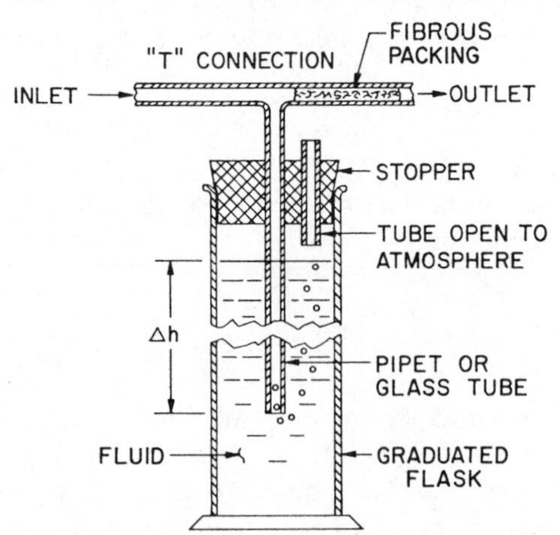

Figure 16.12. Packed plug flowmeter. From *The Industrial Environment—Its Evaluation and Control*, 2nd ed., C. H. Powell and A. D. Hosey, Eds., Public Health Service Publication No. 614, U.S. Government Printing Office, Washington, DC, 1965.

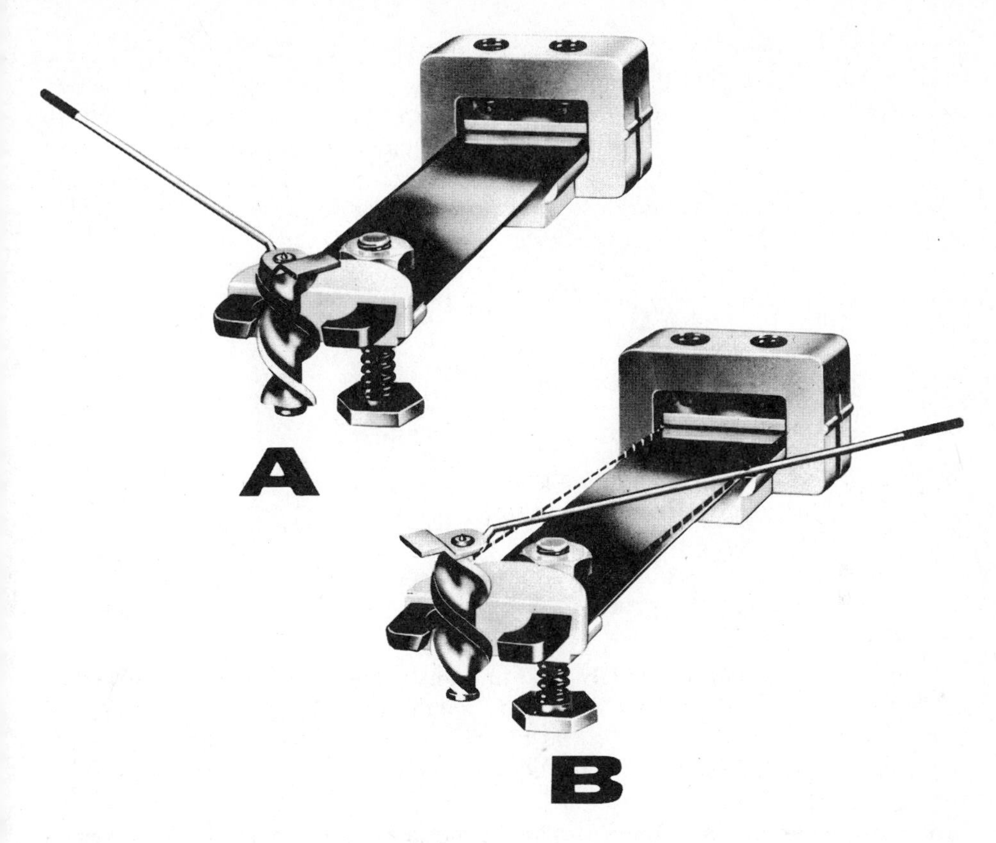

Figure 16.13. Workings of the magnetic linkage. Photograph courtesy of Dwyer Instruments, Inc., Michigan City, Ind., Bulletin No. A-20.

4.2.2.4 Pressure Transducers. All the variable head meters require a pressure sensor, sometimes referred to as the secondary element. Any type of pressure sensor can be used, although high cost and fragility usually rule out many electrical and electromechanical transducers.

Liquid-filled manometer tubes are sometimes used, and if they are properly aligned and the density of the liquid is accurately known, the column differential provides an unequivocal measurement. In most cases, however, it is not feasible to use liquid-filled manometers in the field, and the pressure differentials are measured with mechanical gauges with scale ranges in centimeters or inches of water. For these low-pressure differentials the most commonly used gauge is the magnehelic, illustrated in Figure 16.13. These gauges are accurate to ± 2 percent of full scale and are reliable, provided they and their connecting hoses do not leak and their calibration is periodically rechecked.

4.2.2.5 Critical Flow Orifice. For a given set of upstream conditions, the discharge of a gas from a restricted opening increases with a decrease in the ratio of absolute pressures P_2/P_1, where P_2 is the downstream pressure and P_1 the upstream pressure, until the velocity through the opening reaches the velocity of sound. The value of P_2/P_1 at which the maximum velocity is just attained is known as the critical pressure ratio. The pressure in the throat will not fall below the pressure at the critical point, even if a much lower downstream pressure exists. When the pressure ratio is below the critical value, therefore, the rate of flow through the restricted opening depends only on the upstream pressure.

It can be shown (6) that for air flowing through rounded orifices, nozzles, and venturis, when $P_2 < 0.53\ P_1$ and $S_1/S_2 > 25$, the mass flow rate w is constant.

$$w = 0.533 \frac{C_v S_2 P_1}{T_1} \text{ lb/sec} \tag{3}$$

where C_v = coefficient of discharge (normally ~1)
S_1 = duct or pipe cross section (in.^{2})
S_2 = orifice area (in.^{2})
P_1 = upstream absolute pressure (psi)
T_1 = upstream temperature (°R)

Critical flow orifices are widely used in industrial hygiene instruments such as the midget impinger pump and squeeze bulb indicators. They can also be used to calibrate flowmeters by using a series of critical orifices downstream of the flowmeter under test. The flowmeter readings can be plotted against the critical flows to yield a calibration curve.

The major limitation in their use is that the orifices are extremely small when they are used for flows of 1 cfm (28.3 l/min) or less. They become clogged or eroded in time and therefore require frequency examination and/or calibration against other reference meters.

4.2.2.6 Bypass Flow Indicators. In most high-volume samplers, the flow rate depends strongly on the flow resistance, and flowmeters with a sufficiently low flow resistance are usually too bulky or expensive. A commonly used metering element for such samplers is the bypass rotameter, which actually meters only a small fraction of the total flow—a fraction, however, that is proportional to the total flow. As shown schematically in Figure 16.14, a bypass flowmeter contains both a variable head element and a variable area element. The pressure drop across the fixed orifice or flow restrictor creates a proportionate flow through the parallel path containing the small rotameter. The scale on the rotameter generally reads directly in cubic feet per minute or liters per minute of total flow. In the versions used on portable high-volume samplers, there is usually an adjustable bleed valve at the top of the rotameter that should be set initially, and periodically readjusted in laboratory calibrations so that the scale markings can indicate overall flow. If the rotameter tube accumulates dirt, or the bleed valve adjustment drifts, the scale readings may depart greatly from the true flows.

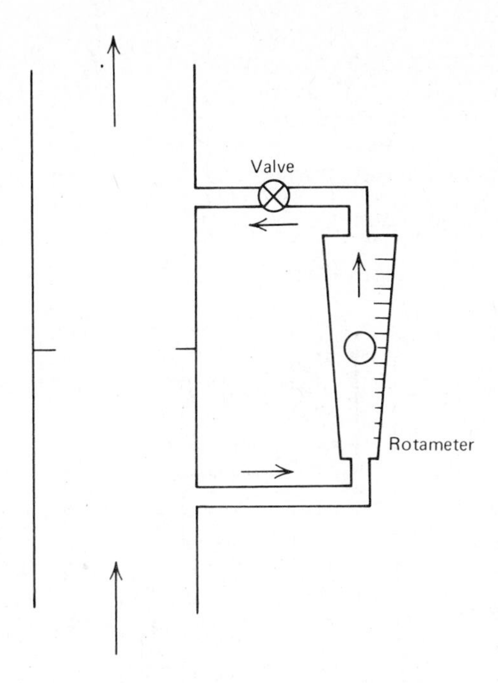

Figure 16.14. Bypass flow indicators.

4.3 Mass Flow and Tracer Techniques

4.3.1 Thermal Meters

A thermal meter measures mass air or gas flow rate with negligible pressure loss. It consists of a heating element in a duct section between two points at which the temperature of the air or gas stream is measured. The temperature difference between the two points depends on the mass rate of flow and the heat input.

4.3.2 Mixture Metering

The principle of mixture metering is similar to that of thermal metering. Instead of adding heat and measuring temperature difference, a contaminant is added and its increase in concentration is measured; or clean air is added and the reduction in concentration is measured. This method is useful for metering corrosive gas streams. The measuring device may react to some physical property such as thermal conductivity or vapor pressure.

4.3.3 Ion-Flow Meters

In the ion-flow meter illustrated in Figure 16.15, ions are generated from the central disk and flow radially toward the collector surface. Airflow through the cylinder causes an axial displacement of the ion stream in direct proportion to the mass

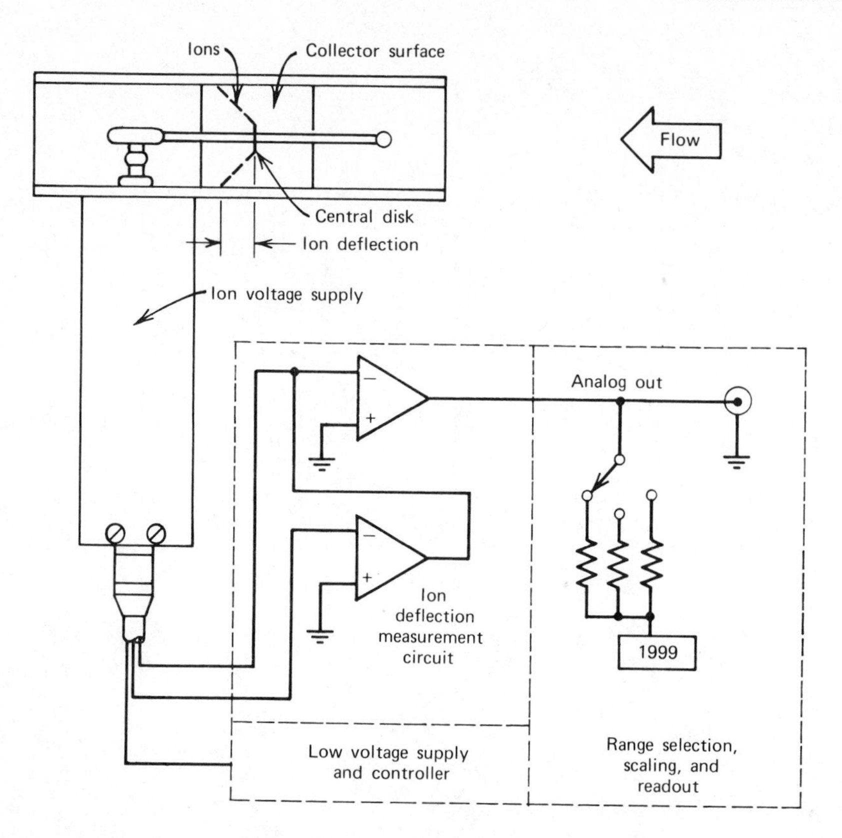

Figure 16.15. Ion-flow mass flowmeter. Schematic courtesy of Thermosystems, Inc., St. Paul, Minn., Leaflet No. TSI-54100671.

flow. The instrument can measure mass flows from 0.1 to 150 scfm (3–4.250 l/min) and velocities from 1 to 12,000 fpm (5 cm/sec to 60 m/sec).

4.4 Air Velocity Meters (Anemometers)

Air velocity is a parameter of direct interest in heat stress evaluations and in some ventilation evaluations. Though it is not the parameter of interest in sampling flow measurements, it may be the only feasible parameter to measure in some circumstances, and it usually can be related to flow rate, provided the sensor is located in an appropriate position and is suitably calibrated against overall flow.

4.4.1 Velocity Pressure Meters

4.4.1.1 Pitot Tube. The Pitot tube is often used as a reference instrument for measuring the velocity of air. A standard Pitot device, carefully made, will need no calibration. It consists of an impact tube whose opening faces axially into the flow, and a concentric static pressure tube with eight holes spaced equally around it in a plane that is eight diameters from the impact opening. The difference between the static and impact pressures is the velocity pressure. Bernoulli's theorem applied to a Pitot tube in an airstream simplifies to the dimensionless formula

$$V = \sqrt{2g_c P_v} \tag{4}$$

where V = linear velocity
g_c = gravitational constant
P_v = pressure head of flowing fluid or velocity pressure

Expressing V in linear feet per minute (fpm), P_v in inches of water, that is, (h_v), and with

$$g_c = 32.17 \frac{(\text{lb} - \text{mass})(\text{ft})}{(\text{lb} - \text{force})(\text{sec}^2)}$$

we get

$$V = 1097 \left(\frac{h_v}{\rho}\right)^{1/2} \tag{5}$$

where ρ is the density of air or gas (lb/ft^3).

If the Pitot tube is to be used with air at standard conditions (70°F and 1 atm), Equation 5 reduces to

$$V = 4005\sqrt{h_v} \tag{6}$$

where V = velocity (fpm)
h_v = velocity pressure (in. H_2O)

There are several serious limitations to Pitot tube measurements in most sampling flow calibrations. It may be difficult to obtain or fabricate a small enough probe, and the velocity pressure may be too low to measure at the velocities encountered. Thus at 1000 fpm (5.1 m/sec), h_v = 0.063 in. H_2O (1.6 mm H_2O), a low value even for an inclined manometer.

4.4.1.2 Other Velocity Pressure Meters. There are several means of utilizing the kinetic energy of a flowing fluid to measure velocity besides the Pitot tube. One way is to align a jeweled-bearing turbine wheel axially in the stream and count the number of rotations per unit time. Such devices are generally known as rotating vane anemometers. Some are very small and are used as velocity probes. Others are sized to fit the whole duct and become indicators of total flow rate; sometimes these are called turbine flowmeters.

The Velometer, or swinging vane anemometer, is widely used for measuring ventilation airflows, but it has few applications in sample flow measurement or calibration. It consists of a spring-loaded vane whose displacement indicates velocity pressure.

4.4.2 *Heated Element Anemometers.*

Any instrument used to measure velocity can be referred to as an anemometer. In a heated element anemometer, the flowing air cools the sensor in proportion to the velocity of the air. Instruments are available with various kinds of heated element, including heated thermometers, thermocouples, films, and wires. They are all essentially nondirectional (i.e., with single element probes); they measure the air speed but not its direction. They all can accurately measure steady state air speed, and those with low mass sensors and appropriate circuits can also accurately measure velocity fluctuations with frequencies above 100,000 Hz. because the signals produced by the basic sensors depend on ambient temperature as well as air velocity, the probes are usually equipped with a reference element that provides an output that can be used to compensate or correct errors due to temperature variations. Some heated element anemometers can measure velocities as low as 10 fpm (50 cm/sec) and as high as 8000 fpm (41 m/sec).

4.5 Procedure for Calibrating Velocity; Flow and Volume Meters

It is not possible to describe here all the techniques available, or to go into great detail on those commonly used. This discussion is limited to selected procedures that should serve to illustrate recommended approaches to some commonly encountered calibration procedures.

4.5.1 *Producing Known Velocity Fields*

Known flow fields can be produced in wind tunnels of the type illustrated in Figure 16.16. The basic components needed have been described by Hama (9) as follows:

1. *A Satisfactory Test Section.* Because this is the location of the probe or sensing element of the device being calibrated, the gas flows must be uniform, both perpendicular and axial to the plane of flow. Streamlined entries and straight runs of duct are essential to eliminate pronounced vena contracta and turbulence.
2. *A Satisfactory Means of Precisely Metering Airflow.* A meter with adequate scale graduations to give readings of ± 1 percent is required. Venturi and orifice meters represent optimal choices, because they require only a single reading.
3. *A Means of Regulating Airflow.* A wide range of flows is required. A suggested range is 50 to 10,000 fpm (2.50 cm/sec to 51 m/sec); therefore the fan must have sufficient capacity to overcome the static pressure of the entire system at the maximum velocity required. A variable drive provides for a means of easily and precisely attaining a desired velocity.

Meters must be calibrated in a manner reflecting accurately their use in the field. Vane-actuated devices should be set on a bracket inside a large test section with a streamlined entrance. Low-velocity, probe-type devices may be tested through appropriate openings in the same type of tunnel. High-velocity ranges of probe-type devices and impact devices should be tested through appropriate openings in a circular duct at least 8.5 diameters downstream from any interference. If straighteners (Figure 16.16) are used, this requirement can be reduced to seven diameters.

NOTE: *Devices must be calibrated at multiple velocities throughout their operating range.*

4.5.2 Comparison of Primary and Secondary Standards

Figure 16.17 presents an experimental setup for checking the calibration of a secondary standard (in this case, a wet test meter) against a primary standard (a spirometer). The first step should be to check all the system elements for integrity, proper functioning, and interconnections. Both the spirometer and the wet test meter require specific internal water levels and leveling. The operating manuals for each should be examined, because they usually outline simple procedures for leakage testing and operational procedures.

After all connections have been made, it is a good policy to recheck the level of all instruments and determine that all connections are clear and have minimum resistance. If compressed air is used in a calibration procedure, it should be cleaned and dried.

Actual calibration of the wet test meter in Figure 16.17 is accomplished by opening the bypass valve and adjusting the vacuum source to obtain the desired flow rate. The optimal range of operation is between 1 and 3 rpm. Before actual calibration is initiated, the wet test meter should be operated for several hours in this setup to stabilize the meter fluid with respect to temperature and absorbed gas, and to work in the bearings and mechanical linkage. After all elements of the system have been adjusted, zeroed, and stabilized, several trial runs should be made. During these runs, the cause of any indicated difference in pressure should be determined and corrected. The actual procedure would be to divert the air instantly to the spirometer for a predetermined volume indicated by the wet test meter (minimum of one revolution), or to near capacity of the spirometer, then return to the bypass arrangement. Readings, both quantity and pressure of the wet test meter, must be taken and recorded while the device is in motion, unless a more elaborate system is set up. In the case of a rate meter, the interval of time that the air is entering the spirometer must be accurately measured. The bell should then be allowed to come to equilibrium before displacement readings are made. Enough different flow rates are taken to establish the shape or slope of the calibration curve, and the procedure being repeated three or more times for each point. For an even more accurate calibration, the setup should be reversed so that air is withdrawn from the spirometer. In this way any imbalance due to pressure differences would be canceled.

A permanent record should be made, consisting of a sketch of the setup and a list of data, conditions, equipment, results, and personnel associated with the

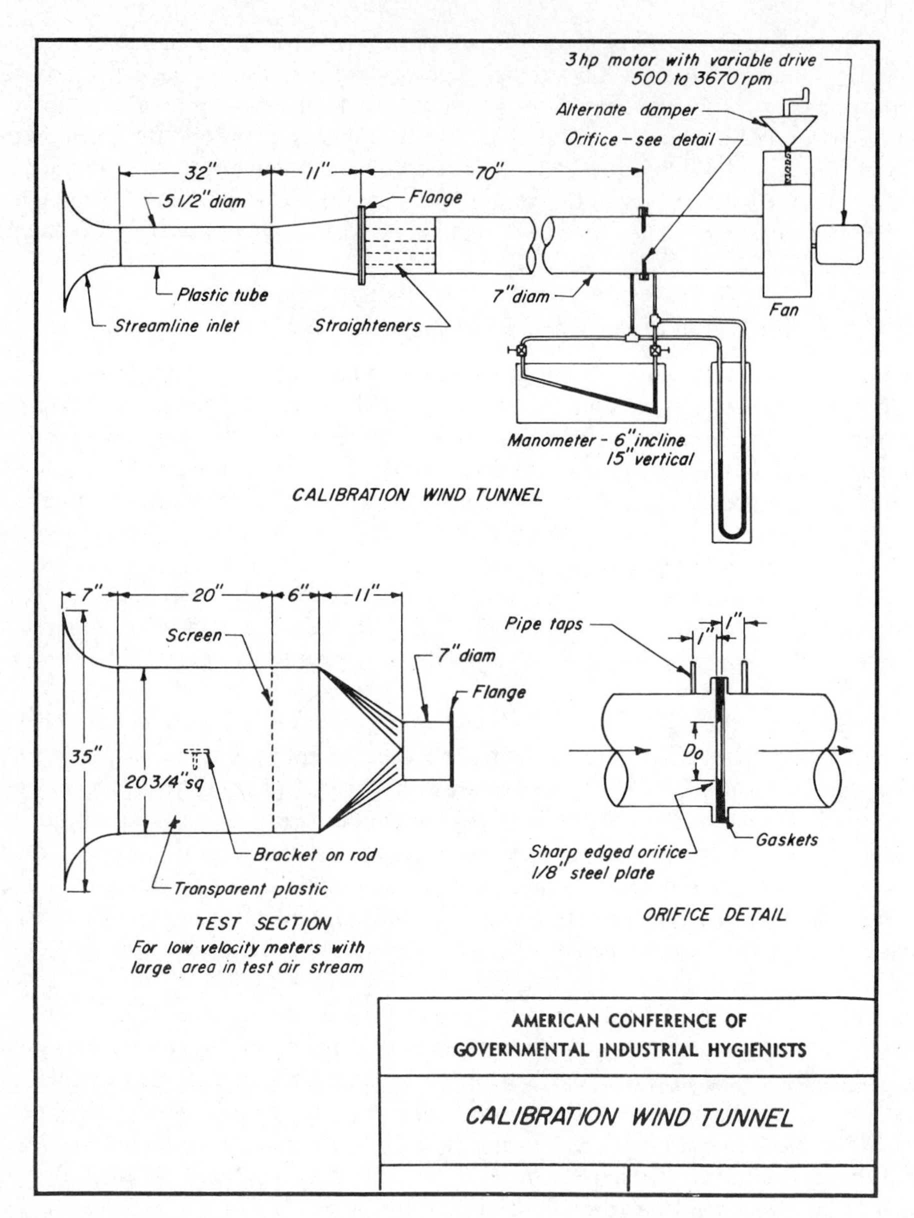

Figure 16.16. Wind tunnel and its use for calibration of anemometers. From *Industrial Ventilation*, 19th ed. Courtesy of American Conference of Governmental Industrial Hygienists.

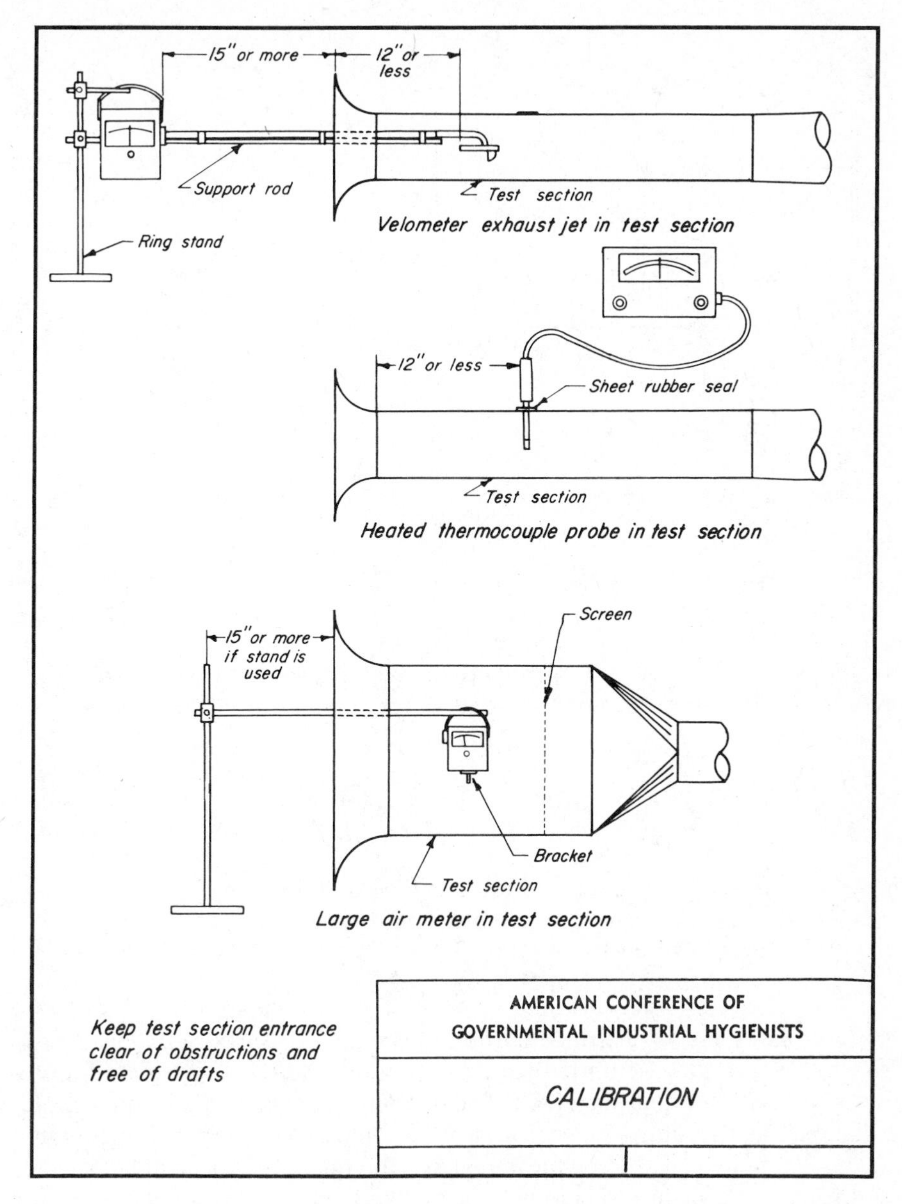

Figure 16.16. (*Continued*)

calibration. All readings (volume, temperatures, pressures, displacements, etc.) should be legibly recorded, including trial runs or known faulty data, with appropriate comments. The identifications of equipment, connections, and conditions should be complete, enabling another person, solely by use of the records, to replicate the same setup, equipment, and connections.

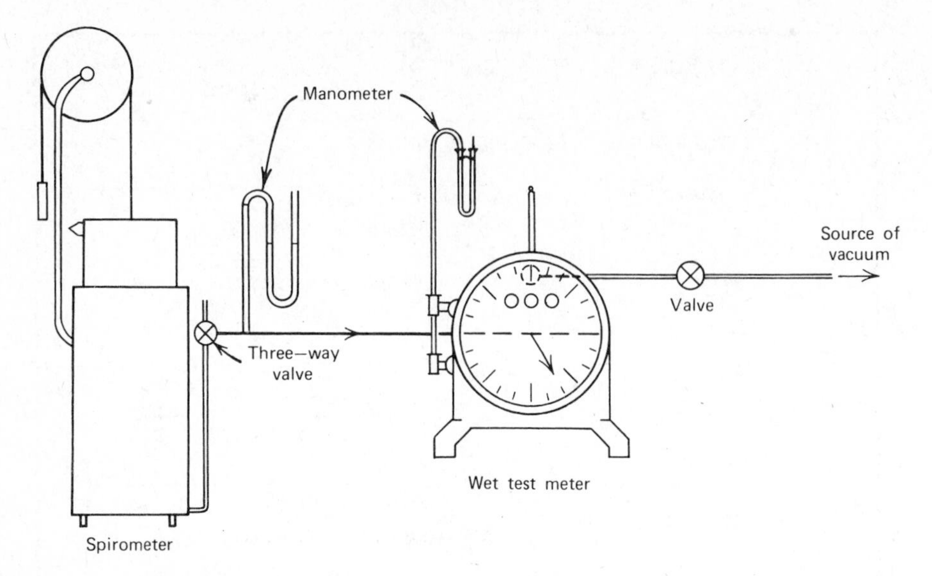

Figure 16.17. Calibration setup for calibrating a wet test meter.

After all the data have been recorded, the calculations (e.g., correction for variations in temperature, pressure, and water vapor) are made, using the ideal gas laws:

$$V_s = V_1 \times \frac{P_1}{760} \times \frac{273}{T_1} \tag{7}$$

where V_s = volume at standard conditions (760 mm and 0°C)
V_1 = volume measured at conditions P_1 and T_1
T_1 = temperature of V_1 (°K)
P_1 = pressure of V_1 (mm Hg)

In most cases the water vapor portion of the ambient pressure is disregarded. Also, the standard temperature of the gas is often referred to normal room temperature (i.e., 21°C rather than 0°C). The instruments, data reading and recording, calculations, and resulting factors or curves should be manipulated with extreme care. If a calibration disagrees with previous calibrations or the supplier's calibration, the entire procedure should be repeated and examined carefully to assure its validity. Upon completion of any calibration, the instrument should be tagged or marked in a semipermanent manner to indicate the calibration factor, where appropriate, the date, and the identity of the calibrater.

4.5.3 Reciprocal Calibration by Balanced Flow System

It is impractical to remove the flow-indicating device for calibration in many commercial instruments. This may be because of physical limitations, characteristics of

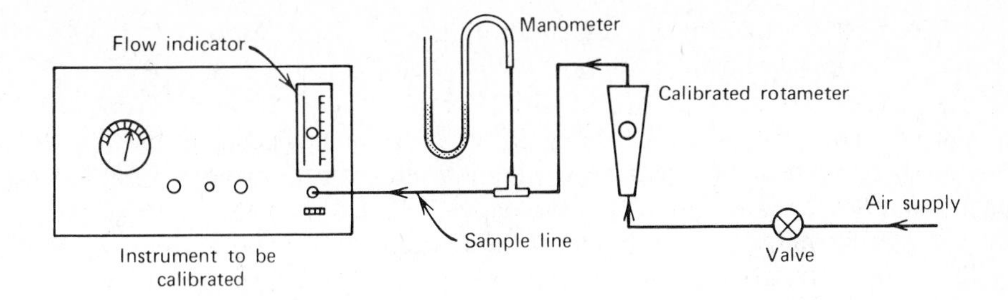

Figure 16.18. Setup for balanced flow calibration.

the pump, unknown resistance in the system, or other limiting factors. In such situations it may be necessary to set up a reciprocal calibration procedure: that is, a controlled flow of air or gas is compared first with the instrument flow, then with a calibration source. Often a further complication is introduced by the static pressure characteristics of the air mover in the instrument. In such instances supplemental pressure or vacuum must be applied to the system to offset the resistance of the calibrating device. An example of such a system appears in Figure 16.18.

The instrument is connected to a calibrated rotameter and a source of compressed air. Between the rotameter and the instrument an open-ended manometer is installed. The connections, as in any other calibration system, should be as short and resistance-free as possible.

In the calibration procedure the flow through the instrument and rotameter is adjusted by means of a valve or restriction at the pump until the manometer indicates "0" pressure difference to the atmosphere. When this condition is achieved, both the instrument and the rotameter are operating at atmospheric pressure. The indicated and calibrated rates of flow are then recorded, and the procedure is repeated for other rates of flow.

4.5.4 Dilution Calibration

Normally gas dilution techniques are employed for instrument response calibrations; however, several procedures (9–11) have been developed whereby sampling rates of flow can be determined. The principle is essentially the same except that different unknowns are involved. In airflow calibration a known concentration of the gas (e.g., carbon dioxide) is contained in a vessel. Uncontaminated air is introduced and mixed thoroughly in the chamber to replace that removed by the instrument to be calibrated. The resulting depletion of the agent in the vessel follows the theoretical dilution formula

$$C_t = C_0 e^{bt} \tag{8}$$

where C_t = concentration of agent in vessel at time t
C_0 = initial concentration at $t = 0$
e = base of natural logarithms

b = air changes in the vessel per unit time
t = elapsed time

The concentration of the gas in the vessel is determined periodically by an independent method. A linear plot should result from plotting concentration of agent against elapsed time on semilog paper. The slope of the line indicates the air changes per minute b, which can be converted to the rate Q of air withdrawn by the instrument from the relationship $Q = bV$, where V is the volume of the vessel.

This technique is advantageous in that virtually no resistance or obstruction is offered to the airflow through the instrument; however, it is limited by the accuracy of determining the concentration of the agents in the air mixture.

4.6 Production of Known Vapor Concentrations

4.6.1 *Introduction and Background*

Methods of producing known concentrations are usually divided into two general classes: (1) static or batch systems and (2) dynamic or continuous flow systems. With static systems, a known amount of gas is mixed with a known amount of air to produce a known concentration, and samples of this mixture are used for calibration. Static systems are limited by two factors, loss of vapor by surface adsorption and the finite volume of the mixture. In dynamic systems, air and gas or vapor are continuously metered in proportions that will produce the final desired concentration. They provide an unlimited supply of the test atmosphere, and wall losses are negligible after equilibration has taken place.

In the field of industrial hygiene, gas or vapor concentrations are usually discussed in terms of parts per million (ppm). In this case, "parts per million" refers to a volume-to-volume relationship (i.e., so many liters of contaminant per liter of air). Thus by definition, both 1 μl of SO_2 per liter of air and 1 ml of SO_2 per cubic meter of air are equal to 1 ppm SO_2. In the field of air pollution these concentrations may also be discussed as parts per 100 million or parts per billion, also based on a volume-to-volume ratio.

Occasionally with direct-reading instruments, and more frequently with chemical analysis of the atmosphere, confusion arises in converting milligrams per cubic meter to parts per million. Dimensional analysis is very useful in avoiding these errors. Thus if one has a concentration in milligrams per cubic meter of air, it must be converted to millimoles per cubic meter, and to milliliters per cubic meter or parts per million:

$$\left(\frac{\text{mg}_x}{\text{m}^3\ \text{air}}\right)\left(\frac{\text{mmole}_x}{\text{mg}_x}\right)\left(\frac{22.4\ \text{ml}_x}{\text{mmole}_x}\right)(F_t)(F_p) = \frac{\text{ml}_x}{\text{m}^3\ \text{air}} = \text{ppm} \tag{9}$$

where F_t and F_p are the pressure and temperature conversion factors from the well-known gas laws, and the subscript x refers to a trace contaminant. Conversely,

$$\text{ppm} = \left(\frac{\text{ml}_x}{\text{m}^3\ \text{air}}\right)\left(\frac{\text{mmole}_x}{22.4\ \text{ml}_x}\right)\left(\frac{\text{mg}_x}{\text{mmole}_x}\right)(F_t)(F_p) = \frac{\text{mg}_x}{\text{m}^3\ \text{air}} \tag{10}$$

Chemical analysis of atmospheric samples is further complicated by procedures that call for some fixed volume of absorbing or reacting solution and sometimes for dilution. In this case it is convenient to determine the concentration of contaminant in solution and, by multiplying by the volume of solution, calculate the total amount of contaminant collected. This is then related to the volume of air sampled and converted to parts per million. For example, after bubbling 5 liters of air at 25°C and 755 mm Hg through 25 ml of an appropriate absorbing solution (100 percent collection efficiency) it was determined that the SO_2 (molecular weight = 64) concentration in solution was 5 μg/ml.

The total amount of SO_2 measured was

$$\frac{5\ \mu\text{g}}{\text{ml}} \times 25\ \text{ml} = 125\ \mu\text{g} \tag{11}$$

The volume of 1 μmole of SO_2 at 25°C and 755 mm Hg is found as follows:

$$1\ \mu\text{mole} \times \frac{22.4\ \mu\text{l}}{\mu\text{mole}} \times \frac{298}{273} \times \frac{760}{755} = 24.6\ \mu\text{l} \tag{12}$$

The concentration in parts per million is

$$\frac{125\ \mu\text{g SO}_2}{5\ \text{liters air}} \times \frac{\text{mole SO}_2}{64\ \mu\text{g SO}_2} \times \frac{24.6\ \mu\text{l SO}_2}{\mu\text{mole SO}_2} = \frac{9.6\ \mu\text{l SO}_2}{\text{liter of air}} = 9.6\ \text{ppm} \tag{13}$$

When producing test atmospheres, many factors may interfere with the contaminant gas or the instrument or analytical procedure. These include (1) specificity of reagents or instrument being used to measure the particular contaminant and (2) loss of the contaminant by reaction with, or adsorption onto, other trace contaminants in the carrier gas or elements of the system. Thus before establishing test atmospheres, the dilution gas should be purified. In addition, the nature and chemistry of the material to be analyzed as well as the detection principle must be thoroughly understood.

4.6.2 Static Systems: Batch Mixtures

In static systems, a known amount of material is introduced into a container, either rigid or flexible, then diluted with an appropriate amount of clean air. If flexible systems are desired, bags of Mylar, aluminized Mylar, Teflon, or polyethylene can be used. They provide the advantage that the entire volume of the bag is usable. Polyethylene is simple to use, but many pollutants either diffuse through it or are adsorbed onto the walls (10). Mylar and aluminized Mylar are less permeable (11),

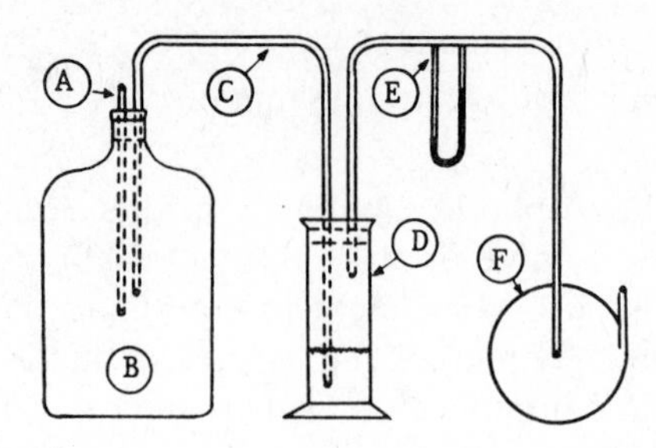

Figure 16.19. Five-gallon bottle for static calibration. *A*, intake tube; *B*, 5-gal. bottle; *C*, withdrawal tube; *D*, collecting device of direct reading instrument; *E*, flowmeter; *F*, suction pump. From *J. Ind. Hyg. Toxicol.*, **29**, 408 (1947).

and because they are inelastic, offer the additional advantage of filling to a constant volume.

Rigid containers such as 5-gal bottles (Figure 16.19) are commonly used for static systems. The bottles are usually equipped with a glass tube, a valved inlet, and a similar outlet. A third inlet or pass-through port for introduction of the contaminant may also be provided. In practice, after the mixture has come to equilibrium, samples are drawn from the outlet side while replacement air is allowed to enter through the inlet tube. Thus the mixture is being diluted while it is being sampled.

Under ideal conditions, the concentration remaining is a known function of the number of air changes in the bottle. If one assumes instantaneous and perfect mixing of the incoming air with the entire sample volume, the concentration change, as a small volume is withdrawn, is equal to the concentration times the percentage of the volume withdrawn:

$$dC = C\frac{dV}{V_0} \tag{14}$$

which integrates to

$$C = C_0 e^{-V/V_0} \quad \text{or} \quad 2.3 \log_{(10)} \frac{C_0}{C} = \frac{V}{V_0} \tag{15}$$

where C = the concentration at any time
V = total volume of sample withdrawn
C_0 = original concentration
V_0 = the volume of the chamber

Thus if we remove one-tenth the volume, we can write

$$2.3 \log_{(10)} \frac{C_0}{C} = 0.1 \tag{16}$$

$$\log_{(10)} \frac{C_0}{C} = \frac{0.1}{2.3} = 0.0435 \tag{17}$$

$$\frac{C_0}{C} = 1.1053 \quad \text{or} \quad \frac{C}{C_0} = 0.9047 \tag{18}$$

The average concentration of the sample withdrawn is 0.9524. If instantaneous mixing does not occur, and the inlet and outlet port are separated, the average concentration may be even higher.

If one were interested in a maximum of 5 percent variation from the average concentration, only about 10 percent of the sample could be used. Setterlind (12) has shown that this limitation can be overcome by using two or more bottles of equal volume (V_0) in series, with the initial concentration in each bottle being the same. When the mixture is withdrawn from the last bottle, it is not displaced by air but by the mixture from the preceding bottle. If, as previously, a maximum of 5 percent variation in concentration can be tolerated, two bottles in series provides a usable sample of $0.6V_0$. With five bottles, the usable sample will increase to about $3V_0$. A table in Setterlind's paper gives both residual concentration and average concentration of the withdrawn sample as a function of the number of volumes withdrawn for each of five bottles in series.

A rigid system can also be modified to give greater usable volumes by attaching a balloon to the intake side inside the bottle. Air from the bottle can then be displaced without any dilution by merely blowing up the balloon.

Introduction of Material into Static Systems. Calibrated syringes provide a simple method for introduction of materials, either gaseous or liquid, into static systems. A wide variety of both gas and liquid syringes are available down to the microliter range (1). A second method is to produce glass ampuls containing a known amount of pure contaminant and break them within the fixed volume of the static system. Setterlind (12) has discussed the preparation of ampuls in detail. Other devices such as gas burettes, displacement manometers, and small pressurized bombs have been used successfully (13). Gaseous concentrations can also be produced by adding stoichiometrically determined amounts of reacting chemicals.

Finally, a standard cylinder can be evacuated, filled with a measured volume of gas or liquid, and repressurized with compressed air or other farrier gas to produce the concentrations required. This mixture can be used with further dilution if necessary. The techniques for filling cylinders have been discussed by Cotabish et al. (14). A number of various gases and vapors are available in different concentrations from several manufacturers (1). Analysis, usually gravimetric, is provided on request. These data should always be checked, for the trace gas may not be adequately mixed or may be partially lost because of wall adsorption.

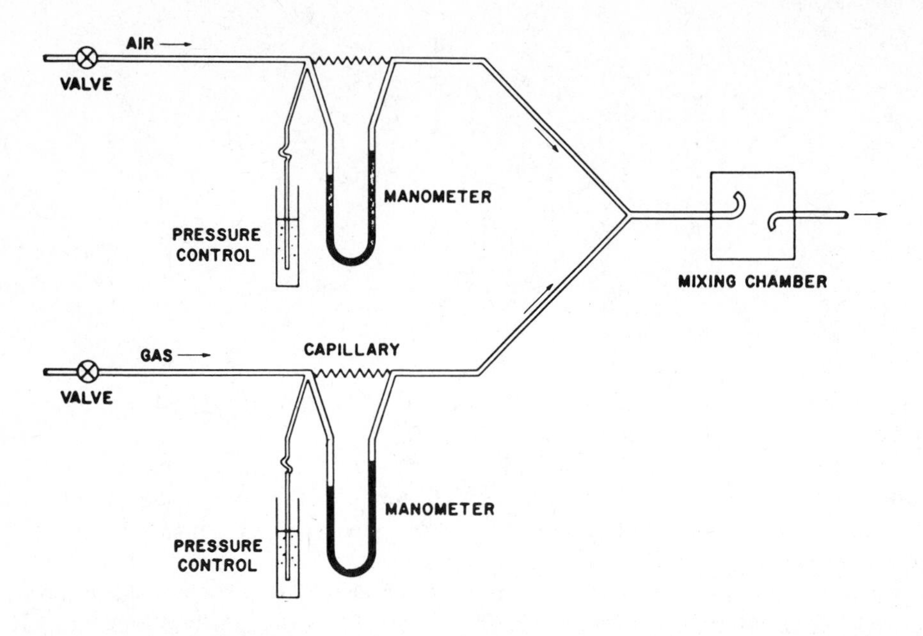

Figure 16.20. Continuous mixer for dynamic gas concentrations. From *Am. Ind. Hyg. Assoc. J.*, 22, 393 (1961); courtesy of Williams & Wilkins Company, Baltimore, and Mine Safety Appliances Company, Pittsburgh.

4.6.3 Dynamic Systems: Continuous Flow

In dynamic systems the rate of airflow and the rate of addition of contaminant to the airstream are carefully controlled to produce a known dilution ratio. Dynamic systems offer a continuous supply of material, allow for rapid and predictable concentration changes, and minimize the effect of wall losses as the contaminant comes to equilibrium with the interior surfaces of the system. Both gases and liquids can be used with dynamic systems. With liquids, however, provision must be available for conversion to the vapor state.

4.6.3.1 Gas Dilution Systems. A simple schematic of a gas dilution system appears in Figure 16.20. Air and the contaminant gas are metered through restrictions and mixed. The output can be used as is or further diluted in a similar system. In theory this process can be repeated until the necessary dilution ratio is obtained. In practice, series dilution systems are subject to a variety of instabilities that make them difficult to control.

Saltzman (3) has described two flow dilution devices; Figure 16.12 shows a porous plug flowmeter, a device that assures that a restricting asbestos-plugged capillary receives a constant pressure of the contaminant gas. The contaminant gas flow, a function of the pressure, is controlled by the height of the column of water or oil. A second device (Figure 16.21) minimizes back pressure and includes a mixing chamber, because the airstream is split. The majority of the gas can be piped to

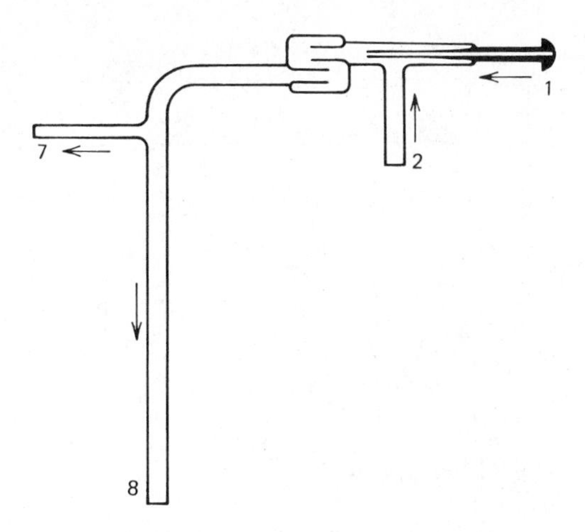

Figure 16.21. All-glass system. From *Anal. Chem.*, **33**, 1100 (1961).

waste through the larger tube. Immersing the end of this tube in water provides a slightly positive pressure at the smaller sidearm delivery tube.

Cotabish et al. (14) have described a system originally patented by Mase for compensation of back pressure (Figure 16.22) in which both the air and contaminant gas flow are regulated by the height of a water column, which in turn is controlled by the back pressure of the calibration system. Thus an increase in back pressure causes an increase in the delivery pressure of both air and contaminant gas.

Calibrated Instruments, Inc., has two instruments available for calibration purposes: the ppm Maker (Figure 16.23) and the Stack Gas Calibrator (Figure 16.24). The ppm Maker consists of a four-output, positive displacement pump and two mechanized four-way stopcocks with single-bore plugs. The bore is normally aligned with the carrier gas flow. When activated, the stopcock is rotated 180°, momentarily aligning the bore with the contaminant gas airflow and delivering a precise volume to the carrier gas. A mixing chamber downstream mixes the carrier gas and the contaminant. The mixture is then pumped through a second identical system. By varying the flow rates of the carrier gas, dilution ratios in the order of $1:10^9$ can be achieved. The stepwise increments of the pumps and the stopcocks provide more than 10,000 different concentration ratios.

The Stack Gas Calibrator traps a fixed amount of gas by a series of valves and tubing and releases this volume into the carrier gas. The number of volumes released can be varied from 1 to 10 per minute. Depending on the size of the volume, three dilution ranges of 10 steps are available, 200 to 2000 ppm, 660 to 6600 ppm, and 1320 to 13,200 ppm.

Another gas dilution system, the Dyna-Blender of Matheson Gas Products, is represented schematically in Figure 16.25. Still another device for constant delivery of a pollutant gas has been described by Goetz and Kallai (15) (Figure 16.26). It consists of a large, gastight syringe with a centrifugal rotor attached to the piston

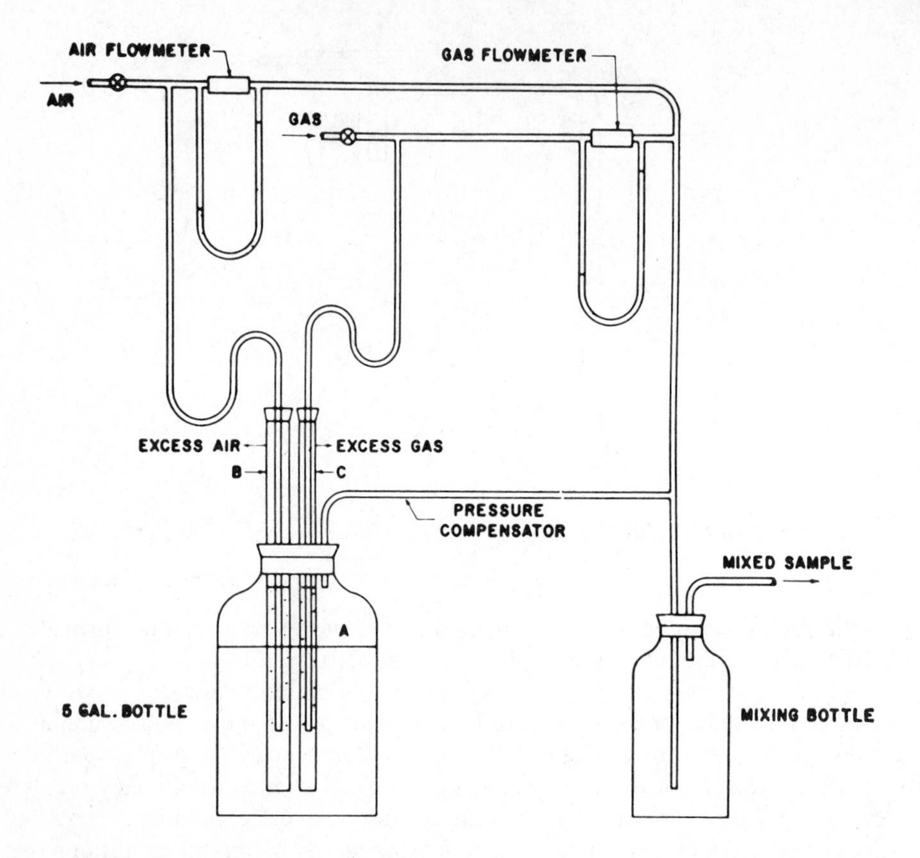

Figure 16.22. Modified Mase gas mixer for compensation of back pressure. From *Am. Ind. Hyg. Assoc. J.*, **22**, 392 (1961); courtesy of Williams & Wilkins, Baltimore, and Mine Safety Appliances Company, Pittsburgh.

so that the piston rotates around its axis. The rotation, caused by a jet of air directed tangentially toward the rotor, is nearly friction-free and induces a constant pressure in the gas. The outlet of the syringe is connected on one side of a glass T-tube. Dilution air is piped into the base of the T, and the mixture exits the T-tube from the outer sidearm.

4.6.3.2 Liquid Dilution Systems. When the contaminant is a liquid at normal temperatures, a vaporization step must be included. One procedure is to use a motor-driven syringe (13, 14) and meter the liquid onto a wick or a heated plate in a calibrated airstream. Nelson and Griggs (16) have described a calibration apparatus that makes use of this principle (Figures 16.27 and 16.28). The system consists of an air cleaner, a solvent injection system, and a combination mixing and cooling chamber. A large range of solvent concentrations can be produced (2 to 2000 ppm). The device permits rapid changes in the concentrations and is accurate to about 1

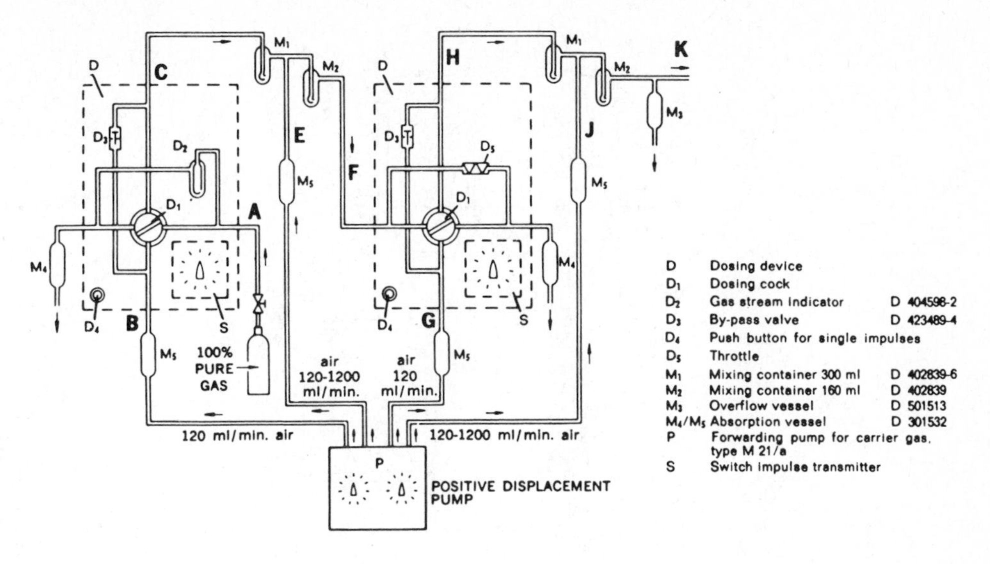

Figure 16.23. The ppm Maker. Schematic courtesy of Calibrated Instruments, Incorporated.

percent. It can also be used to produce gas dilutions with an even wider range of available concentrations (0.05 to 2000 ppm).

A second vapor generation method is to saturate an airstream with vapor and dilute to the desired concentration with makeup air. The amount of vapor in the saturated airstream depends on both the temperature and the vapor pressure of the contaminant and can be precisely calculated. A simple vapor saturator is illustrated in Figure 16.29. The inert carrier gas passes through two gas-washing bottles in series, and these contain the liquid to be volatilized. The first bottle is kept at a higher temperature than the second one, which is immersed in a constant temperature bath. By using the two bottles in this fashion, saturation of the exit gas is assured. A filter is sometimes included to remove any droplets entrained in the airstream as well as any condensation particles. A mercury vapor generator using this principle has recently been described by Nelson (17).

Diffusion cells (Figure 16.30) have also been used to produce known concentrations of gaseous vapors. In this case, the liquid diffuses up a center tube and into a mixing chamber through which air is passed. Devices of this type can be used with dynamic systems; however, they are limited to fairly low rates.

O'Keefe and Ortman (18) found that any material whose critical temperature was above 20 to 25°C could be sealed in Teflon tubing. Three sealing techniques are depicted in Figure 16.31. The material permeates the walls of the tube and diffuses out at a rate dependent on wall thickness and area (fixed parameters) and temperature. At constant temperature, the investigators showed that the rate of weight loss was constant as long as there was liquid in the tube. In use, rigid precautions are necessary to assure fixed temperature (Figure 16.32) because, for

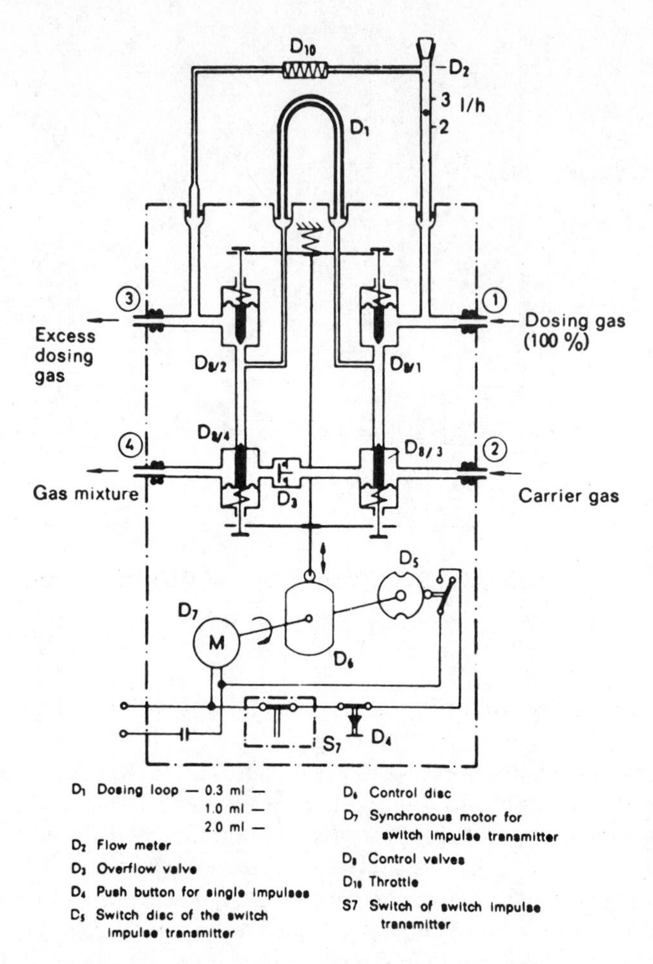

Figure 16.24. Stack Gas Calibrator. Schematic courtesy of Calibrated Instruments, Incorporated.

example, the sulfur dioxide permeation rate more than doubles for every 10°C increase in temperature. Permeation tubes have been successfully used as primary standards (see Table 16.3). Permeation tubes of nitrogen dioxide, hydrogen sulfide, chlorine, ammonia, propene, propane, butane, butene, and methyl mercaptan are also available.

4.7 Production of Calibration Aerosols

4.7.1 Introduction

The generation of a test aerosol having the desired combination of physical and chemical properties is in most cases more difficult than the generation of gas of

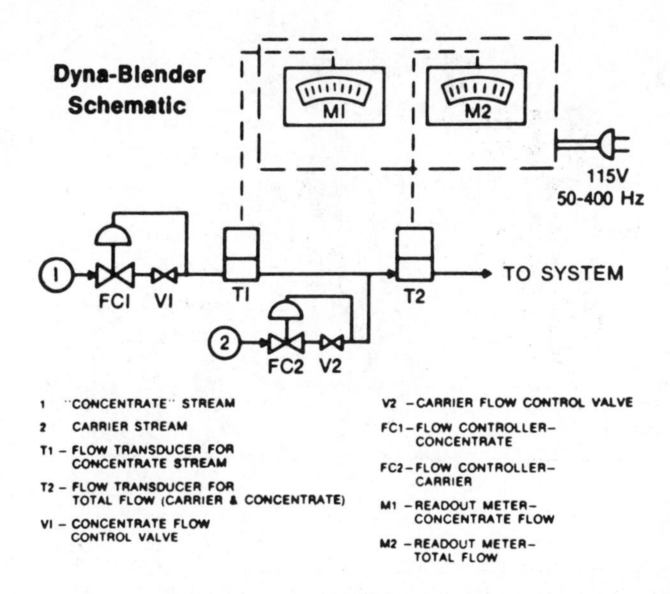

Figure 16.25. Matheson Dyna-Blender. Schematic courtesy of Matheson Gas Products, Division of Will Ross, Inc.

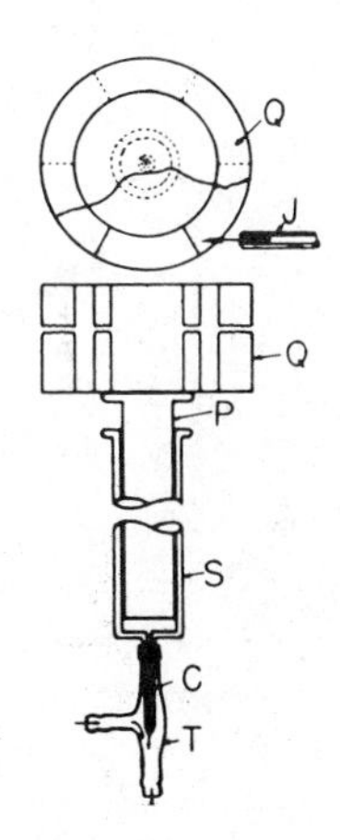

Figure 16.26. Schematic of spinning syringe calibrator assembly: Q, fan vanes; J, air jet; P, glass piston; S, large glass syringe; C, capillary tube; T, T-tube. From *Journal of the Air Pollution Control Association*, **12**, 437 (1962); courtesy of the Air Pollution Control Association.

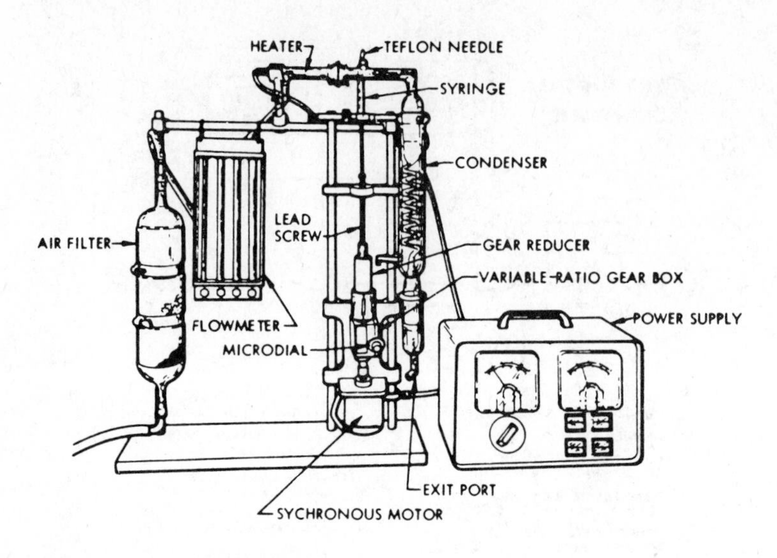

Figure 16.27. Syringe drive calibration assembly. Reprinted from U.C.R.L.-70394; courtesy of Lawrence Radiation Laboratory and the U.S. Atomic Energy Commission.

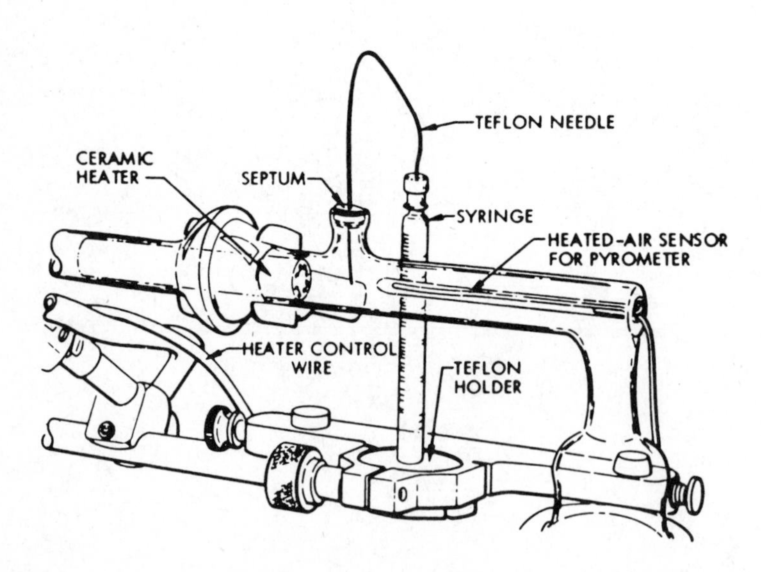

Figure 16.28. Detailed view of heating system and injection port. Reprinted from U.C.R.L.-70394; courtesy of Lawrence Radiation Laboratory and the U.S. Atomic Energy Commission.

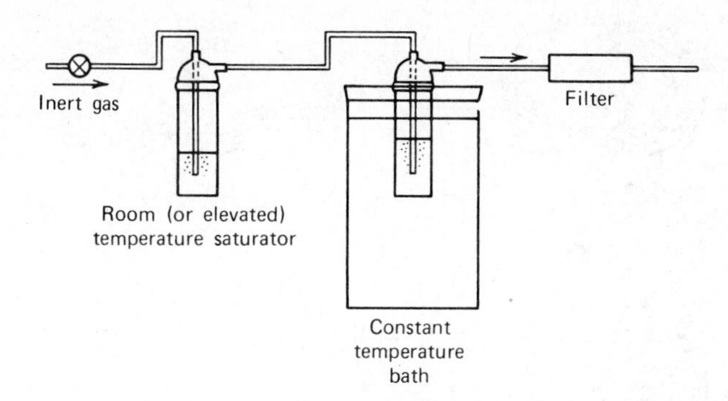

Figure 16.29. Vapor saturator. From *Am. Ind. Hyg. Assoc. J.*, **22**, 392 (1961); courtesy of the Williams & Wilkins Company, Baltimore, and Mine Safety Appliances Company, Pittsburgh.

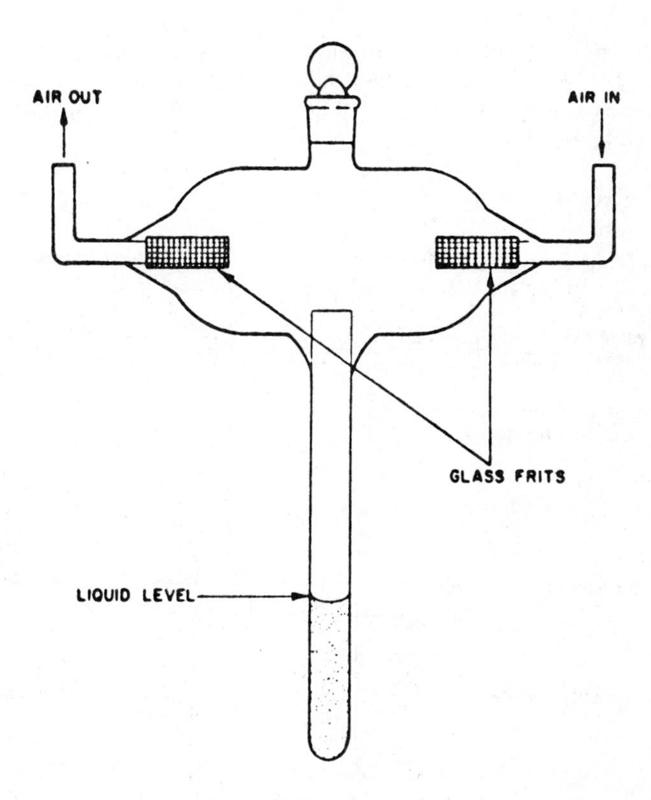

Figure 16.30. Diffusion cell. From *Anal. Chem.* **32**, 802 (1960); courtesy of the American Chemical Society.

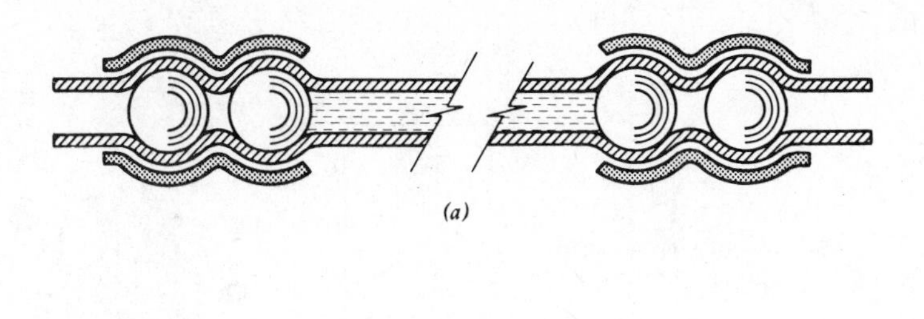

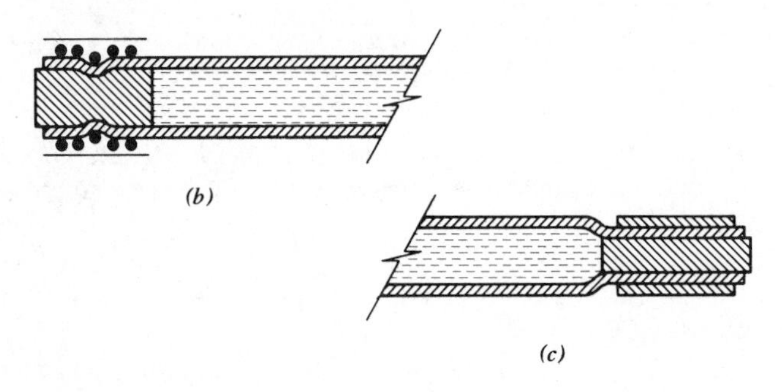

Figure 16.31. Three types of seal. (*a*) Steel on glass balls. (*b*) Teflon plug bound with wire. (*c*) Teflon plug held by a crimped metal band. From International Symposium on Identification and Measurement of Environmental Pollutants, National Research Council of Canada, Ottawa, Ontario, 1971.

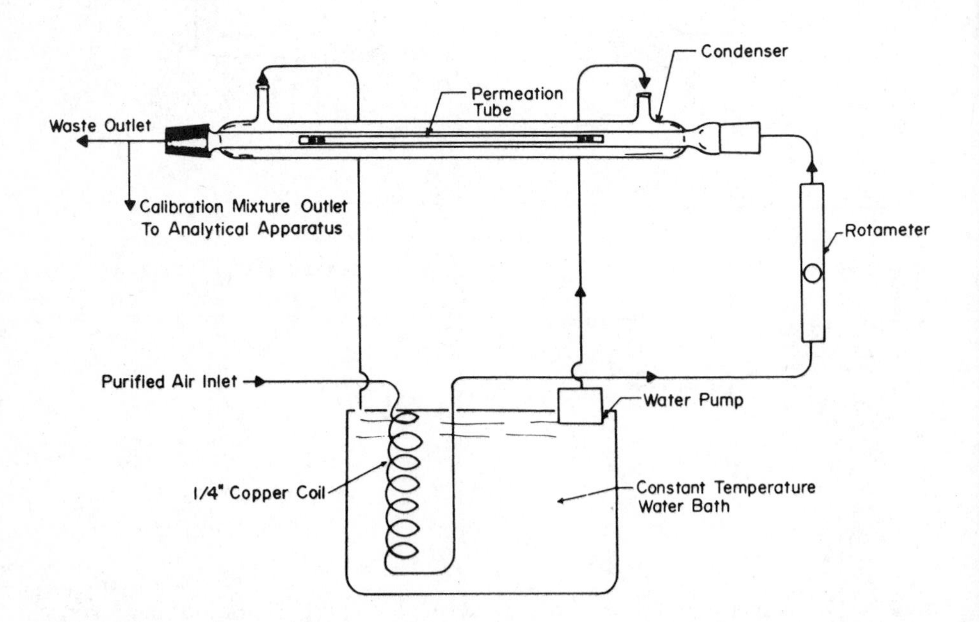

Figure 16.32. Permeation tube apparatus with constant temperature bath.

vapor test atmospheres. There are few commercially available aerosol generators capable of producing a stable, reproducible aerosol over extended intervals. Those that are available and reliable may not be capable of producing the desired airborne concentration, particle size distribution, shape, density, charge, or surface properties. Furthermore, it is difficult to obtain useful generalized data on the operational characteristics of commercial aerosol generators, which were designed for specific limited applications. For example, the common drugstore variety of DeVilbiss No. 40 nebulizer has been widely used as a laboratory aerosol generator for many years, (19). The performance characteristics of this and several other compressor air nebulizers have been discussed recently in a comprehensive review paper by Raabe (20) and are summarized in Table 16.6.

The types of laboratory aerosol generators that are available or have been described in the literature have also been described in detail in a comprehensive review by Kerker (21). Previous reviews were provided by Fraser et al. (22), Silverman (13), and Axelrod and Lodge (10). A detailed review of techniques and equipment for producing monodisperse aerosols has been prepared by Fuchs and Sutugin (23). From these and other sources, a condensed summary of techniques for generating monodisperse test aerosols has been constructed (Table 16.7). Sources of commercially available devices for producing polydisperse test aerosols are tabulated in Reference 3.

Aerosol generators can be divided into two types: those that produce condensation aerosols, and those that produce dispersion aerosols. In the former type, the material to be aerosolized is dispersed in the vapor phase and allowed to condense on airborne nuclei.

4.7.2 Generation of Monodisperse Condensation Aerosols

In an isothermal supersaturated environment, vapor molecules diffuse to and condense on airborne nuclei. Wilson and LaMer (24) demonstrated that the surface area of the resulting droplets increases linearly with time. Thus as the droplets become large in comparison to the nuclei, the size range becomes quite narrow, even when the nuclei on which the droplets grew may have varied widely in size.

The LaMer–Sinclair (25) aerosol generator, illustrated schematically in Figure 16.33, was based on these considerations. Improvements in the basic LaMer–Sinclair design have been described by Muir (26). Further refinements were made by Huang et al. (27). Rapaport and Weinstock (28) have described a condensation aerosol generator that is simpler and less expensive to produce, and requires less critical control of temperature and flow rate for the production of monodisperse aerosol. A more sophisticated version of this generator has been described by Liu and Lee (29). This type of generator is capable of producing high-quality aerosols of high-temperature boiling, low vapor pressure liquids (e.g., dioctyl phthalate, triphenylphosphate, and sulfuric acid in the size range of about 0.03 to 1.3 μm). Prodi (30) has described a modified LaMer–Sinclair generator for monodisperse particles in the 0.2 to 8 μm size range at a concentration of $\sim 100/cm^3$. Apparatus for producing monodisperse condensation aerosols of lead, zinc, cadmium, and

Table 16.6. Representative Characteristics of Selected Compressed Air and Ultrasonic Nebulizers

Compressed Air Nebulizers[a]

Air Pressure (psig)	DeVilbiss, Setting No. 40 (19) (jet = 33 mil; vent closed)			Lovelace[b] (jet = 9.2 mil)			Dautrebande D-30 (19) (jet = 41 mil)			Lauterbach (89) (jet = 13 mil)			Collison (90) (3 jets = mil)			Retec X-70/N (91)		
	Output (evap.) (μl/liter)	Total Air (liters/min)	VMD (σ_g) (μm)	Output (evap.) (μl/liter)	Total Air (liters/min)	VMD (σ_g) (μm)	Output (evap.) (μl/liter)	Total Air (liters/min)	VMD (σ_g) (μm)	Output (evap.) (μl/liter)	Total Air (liters/min)	VMD (σ_g) (μm)	Output (evap.) (μl/liter)	Total Air (liters/min)	VMD (σ_g) (μm)	Output (evap.) (μl/liter)	Total Air (liters/min)	VMD (σ_g) (μm)
5	16.0	7.5	4.6	1.6	0.8		1.0	13.4		2.6	1.2							
			(1.8)				(9.7)											
10	16.0	10.8	4.2	15.3	1.2		1.6	17.9	1.7	3.9	1.7	3.8						
			(1.8)	(11)			(9.6)		(1.7)			(2.0)						
15	15.5	13.5	3.5	19.5	1.4					5.2	2.1							
	(8.6)		(1.8)															
20	14.0	15.8	3.2	30.0	1.7	5.8	2.3	25.4	1.4	5.7	2.4	2.4	7.7	7.1	2.0	53	5.4	5.7
	(7.0)		(1.8)	(10)		(1.8)	(8.6)		(1.7)			(2.0)	(12.7)		(2.0)	(12)		(1.8)
30	12.1	20.5	2.8			4.7	2.4	32.7	1.3	5.9	3.2	2.4	5.9	9.4		54	7.4	3.6
	(7.2)		(1.8)			(1.9)	(8.2)		(1.7)			(2.0)	(12.6)			(11)		(2.0)

Commercial Ultrasonic Nebulizers[a] (88)

Nebulizer	Output (Evap.) (μ/liter)	Total Air (liters/min)	VMD (σ_g) (μm)
DeVilbiss, setting No. 4	150	41.0	6.9
	(33.1)		(1.6)
Mist-O_2-Gen, with reservoir	61.5	24.7	6.5
	(22.2)		(1.4)

[a]Outputs are given in microliters of solution per liter of total aerosols (evaporation losses are in parentheses). Total volume of aerosol is indicated as total air in liters per minute. The droplet distribution of usable aerosol at initial formation is assumed to be log-normal with data given for the volume median diameters VMD and geometric standard deviations σ_g in parentheses. The sources of the values are indicated by reference numbers.

[b]Baffle setting has been optimized for operation at 20 psig. The data on the Lovelace Nebulizer are by Dr. Otto G. Raabe (91), G. J. Newton, and J. E. Bennick of the Lovelace Foundation.

Table 16.7. Techniques for Generating Monodisperse Test Aerosols

Name or Type	Operational Mechanism of Generator	Types of Monodisperse Aerosol Produced	Typical Diameter Range, μm (σ_g)	Approximate Output (no./sec)	Approximate Flow (l/min)	Utilities Required	Techniques for Tagging	Commercial Source or Reference
Uniform spheres	Nebulization	Latex spheres T3 *E. coli* phage (63) Type 3 Poliomyelitis Virus (66)	0.109–1.947 (1.02) 0.035 0.026	10^4	10	10 psig air	Emulsion polymerization (81–83)	Duke[d] IOC[e] Polysciences[f]
Atomizer-impactor	Nebulizer with impactor cut-off	Any liquid or solid residue	0.03[a]–3 (1.4)	10^9	≥57	45 psig air	—[b]	(49)
Spinning disk	Rotary atomizer	Any liquid or solid residue	1[a]–30 (1.1)	10^7	≥283	60 Hz ac	—[b]	(56, 57)
Spinning top	Rotary atomizer	Any liquid or solid residue	0.5[a]–200 (1.1)	10^7	NA[c]	40 psig air		BGI[g]
Other vibrating reed or capillaries (transverse)	Displacement of liquid from reed or capillary in transverse vibration	Any liquid or solid residue	1[a]–200 (1.1)	10^2	NA	60 Hz ac	—[b]	(58, 60)

(Continued on following page)

Table 16.7. Continued

Name or Type	Operational Mechanism of Generator	Types of Monodisperse Aerosol Produced	Typical Diameter Range, μm (σ_g)	Approximate Output (no./sec)	Approximate Flow (l/min)	Utilities Required	Techniques for Tagging	Commercial Source or Reference
Vibrating orifice (axial)	Liquid filament disruption—mechanical instability	Any liquid or solid residue	1[a]–200 (<1.1)	10^5	NA	60 Hz ac	—[b]	TSI[h]
Electrostatic classifier	Mobility stripping	Any liquid or solid residue	0.01[a]–0.3 (<1.1)	10^6	3	60 Hz ac	—[b]	TSI[h]
Electrostatic nebulizer	Liquid filament disruption—electrical instability	Liquids with low electrical conductivity or their solid residue	<0.1–200 (NA)	10^8	NA	10 kV ac or dc	—[b]	(62, 63)
Condensation	Condensation on nuclei	DOP, TPP, other low vapor pressure, high-boiling liquids, subliming solids	0.01–8 (1.2)	10^8	3	ac or dc Line	Use of radioactive nuclei	BGI[g] TSI[h] In-Tox[i]

[a]Lower size limit based on dried residue particles from dilute solutions or suspensions.
[b]Tags can be dissolved or suspended in feed liquid; see text for further discussion.
[c]Not available.
[d]Duke Scientific Co., 11350 San Antonio Rd., Palo Alto, CA 94303.
[e]Interfacial Dynamics Corp., P.O. Box 279, Portland, OR 97207-0279.
[f]Polysciences, Inc., 400 Valley Rd., Warrington, PA 18976-9990.
[g]BGI, Inc., 58 Guinan St., Waltham, Mass. 02154.
[h]TSI, Inc., 500 Cardigan Rd., St. Paul, MN 55164.
[i]In-Tox Products, 1712 Virginia, NE, Albuquerque, NM 87110.

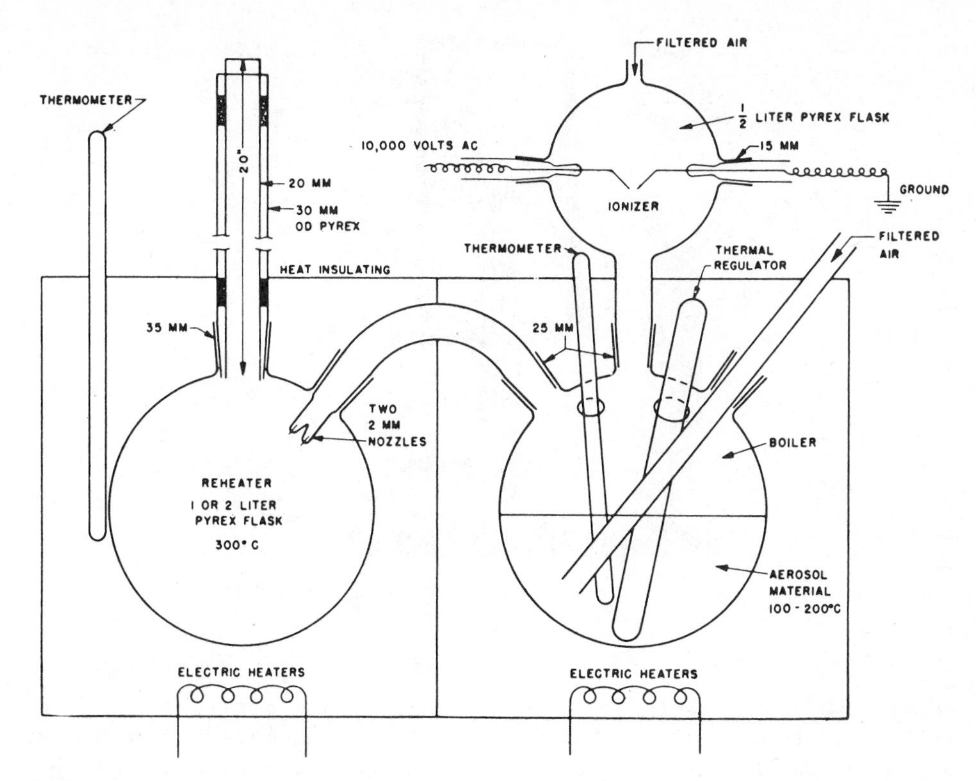

Figure 16.33. LaMer–Sinclair type of condensation aerosol generator. From U.S. Atomic Energy Commission *Handbook on Aerosols*.

antimony using a high-frequency induction furnace has been described by Homma (31) and Movilliat (32). Matijevic et al. (33) and Kitani and Ouchi (34) have produced monodisperse condensation aerosols of sodium chloride.

The particles produced by condensation generators are liquid and spherical unless the material vaporized has a melting point above ambient temperature. In this case the particles solidify and, if crystalline, may form nonspherical shapes. A summary of techniques for producing radioactively labeled monodisperse condensation aerosols with 18 organic compounds and eight inorganic materials has been presented by Spurny and Lodge (35).

Kerker's review (21) provides the most complete summary of the state of knowledge on the factors affecting the performance of condensation generators.

4.7.3 Generation of Dry Dispersion Aerosols

Dispersion aerosol generators may be classified as wet or dry. Dry generators comminute a bulk solid or packed powder by mechanical means, usually with the aid of an air jet. They often include an impaction plate at the outlet for removal of oversize particles and for breaking up aggregates. The aerosol particles produced

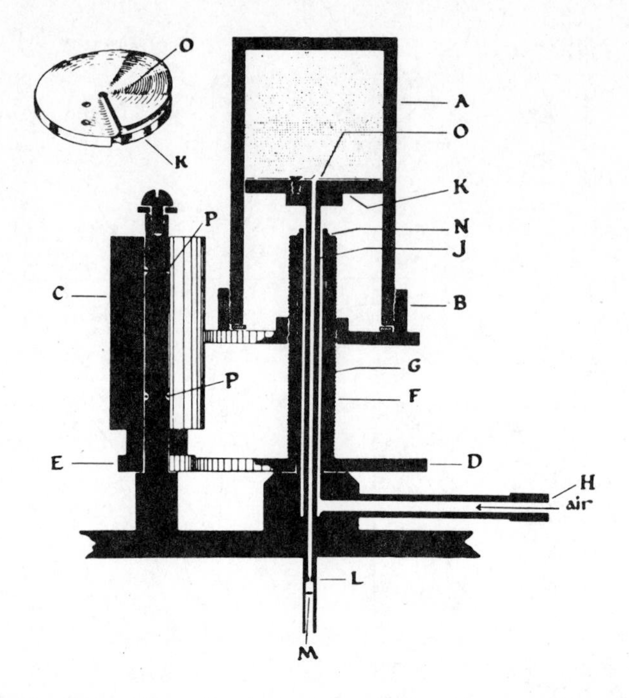

Figure 16.34. The Wright Dust Feed: *A*, dust cylinder; *B*, cap, with peripheral gear; *C*, pinion; (D), wheel; (E), pinion; *F*, threaded tube; *G*, tube, connected to *H*, compressed air line; *J*, small tube, carrying scraper head *K*, which communicates with jet *L*, which is above impaction plate *M* for breaking up aggregates; *O* is a spring disk with cutting edge. From *J. Sci. Inst.*, **27**, 12 (1950).

are typically composed of solid, irregularly shaped particles having a broad range of sizes. Also, the rate of generation is usually not perfectly uniform, because it depends on the uniformity of hardness, or friability, of the bulk material being subdivided, as well as on the uniformity of the feed–drive mechanism and air jet pressure.

The characteristics of a variety of types of dry dust generators have been described by Ebens (36), including the widely used Wright Dust Feed (37) illustrated in Figure 16.34. Among the more difficult kinds of dry dust aerosol to generate are plastics that develop high electrostatic charges. Laskin et al. (38) have described two types of generator for such materials. One uses a high-speed fan to create a stable fluidized bed from which aerosol can be drawn; the second uses a high-speed grinder to comminute a block of solid material.

Other generator designs developed for "problem" dusts include those by Dimmock (39) for viable dusts, by Brown et al. (40) for deliquescent dust, and by Timbrell et al. (41) and Holt and Young (42) for fibrous dust.

Useful aerosols of dry particles of metal and metal oxides are also produced with electrically heated (43, 44) or exploded wires (45, 46). These techniques have some disadvantages because of the very broad size distributions of the resulting particles and because of the tendency of particles to coalesce. There are applications, however, for this type of aerosol, and it is possible with a wire-heating method to produce spherical particles of many different metals or their oxides. Aerosols of very small particles have also been produced by arc vaporization (47).

4.7.4 Wet Dispersion Aerosol Generators

Wet dispersion generators break up bulk liquid into droplets. If the liquid is nonvolatile, the resulting aerosol is a mist or fog. If a volatile liquid is aerosolized, the resulting particles are composed of the nonvolatile residues in the feed liquid and are much smaller than the droplets dispersed from the generator. Solid particles can be produced by nebulizing salt or dye solutions or particle suspensions. Aqueous solutions, of course, produce water-soluble particles that may be hygroscopic. This may be an important factor, because the aerodynamic size for such aerosols varies with ambient humidity.

A variety of techniques can be used to subdivide bulk liquid into airborne droplets. In most cases the liquid is accelerated by the application of mechanical pneumatic, or centrifugal forces and drawn into filaments or films that break up into droplets because of surface tension. Centrifugal pressure nozzles and fan spray nozzles use hydraulic pressure to form a sheet of liquid that breaks up into droplets, but these generally have high liquid feed rates and produce very large droplets. They are seldom employed for producing aerosols for instrument calibration.

A commonly used type of aerosol generator is the two-fluid nozzle, which uses pneumatic energy to break up the liquid. Several laboratory-scale compressed air-driven nebulizers have been described in detail by Mercer et al. (19). Table 16.6 summarizes the operational characteristics of six such nebulizers, including the Lauterbach and Lovelace designs, which are illustrated in Figures 16.35 and 16.36, respectively. The DeVilbiss No. 40 is made of glass, which not only makes it fragile but also limits its precision of manufacture and reproducibility. Ready reproducibility led Whitby to select the British Collison (48) nebulizer for his atomizer-impactor aerosol generator (49). Other commercially available nebulizers, including those of Wright (50) (Figure 16.37) and Dautrebande (51), are machined to close tolerance from plastic materials.

Nebulizers produce droplets of many sizes, and resultant aerosol particles after evaporation are therefore polydisperse, although relatively narrow size dispersions can be obtained with Whitby's atomizer-impactor (49) and Dautrebande's D-30 (51). The droplet distributions described for nebulizers are the initial distributions at the instant of formation; droplet evaporation begins immediately, even at saturation humidity, because the vapor pressure on a curved surface is elevated (52). The rate of evaporation depends on many factors, including solute concentration, the hygroscopicity of the solute (53, 54), the presence of immiscible liquids or evaporation inhibitors (55), and the size of the droplets.

Evaporative losses cause an increase in the concentration of the solution or of the suspended particles, resulting in an increase in the size of the dry particles

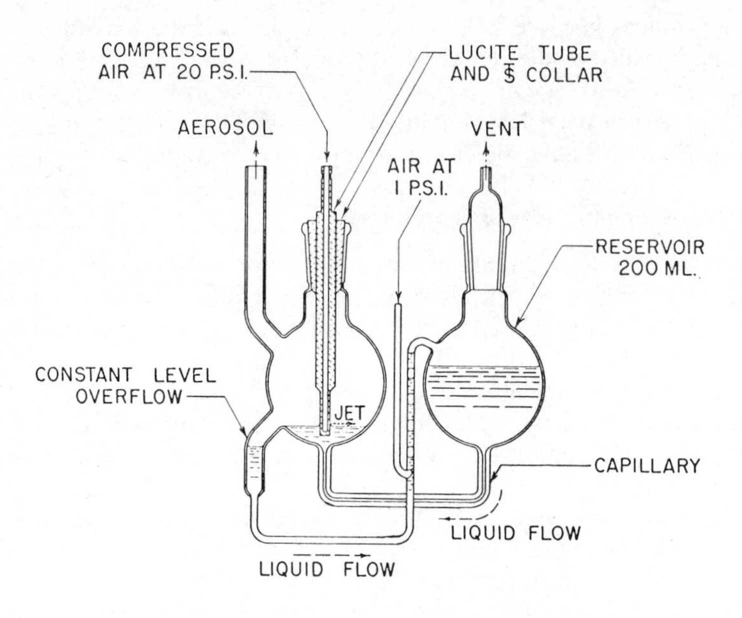

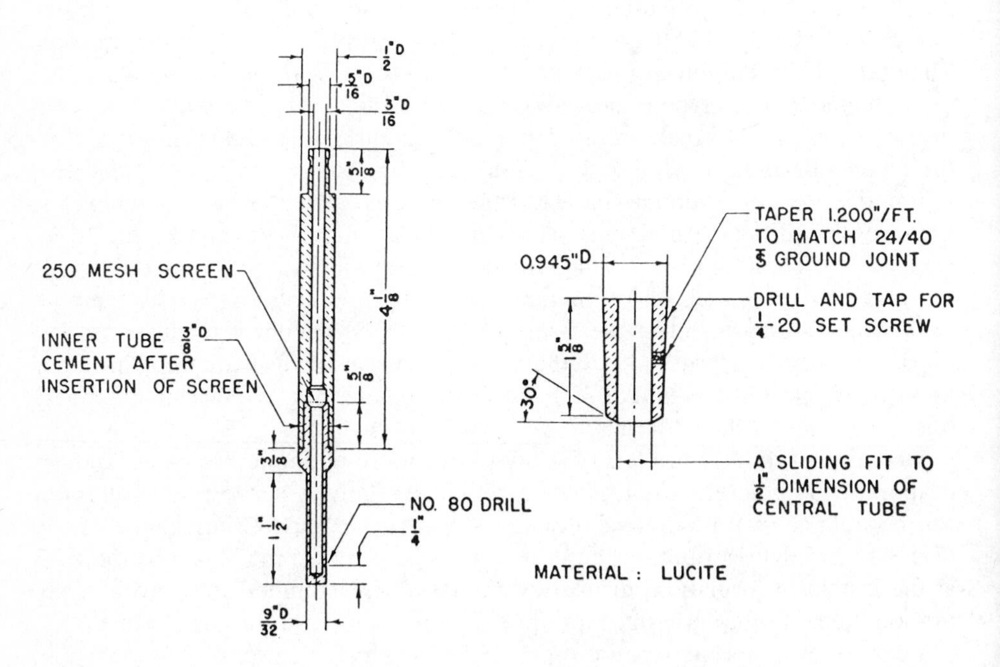

Figure 16.35. The Lauterbach aerosol generator and its jet tube. From *AMA Arch. Ind. Health*, **13**, 156 (1956).

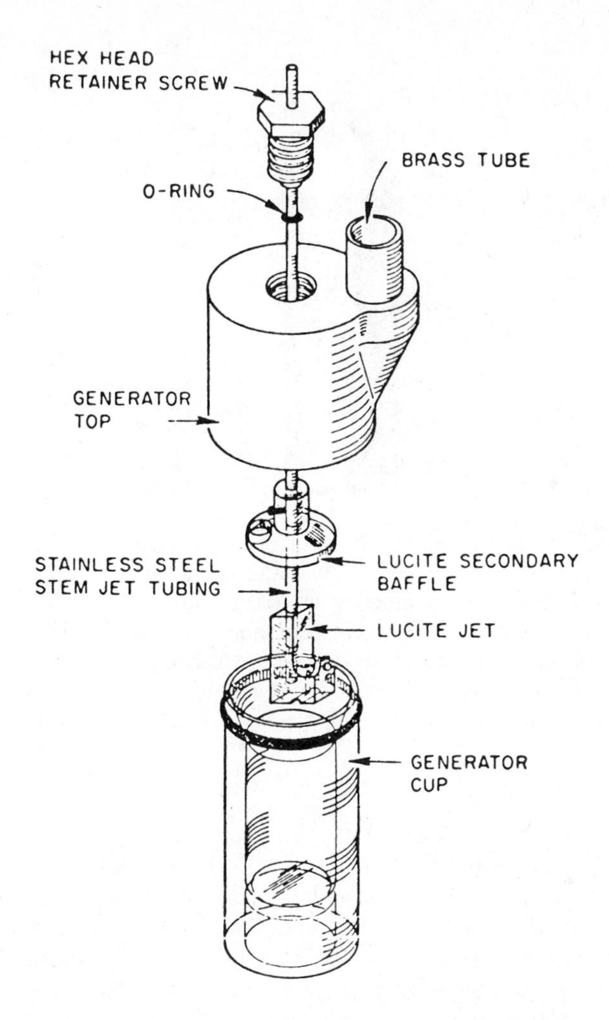

Figure 16.36. The Lovelace nebulizer, which operates with a liquid volume of ~4 ml and incorporates a jet baffle similar to that of Wright (50). Schematic courtesy of Dr. Otto G. Raabe.

formed when the liquid evaporates. Evaporation occurs both from the surface of the liquid and from the droplets, which evaporate slightly and hit the wall of the nebulizer to be returned to the reservoir; it is most important in nebulizers with small reservoirs but large volumetric airflows.

Rotary atomizers, such as the spinning disk, utilize centrifugal force to break up the liquid, which undergoes an acceleration as it spreads from the center to the edge of the disk. The liquid leaves the edge of the disk as individual droplets or as ligaments that disintegrate into droplets. Walton and Prewett (56) demonstrated that these atomizers can produce monodisperse aerosols when operated with low liquid feed rates and high peripheral speeds. A spinning disk generator designed

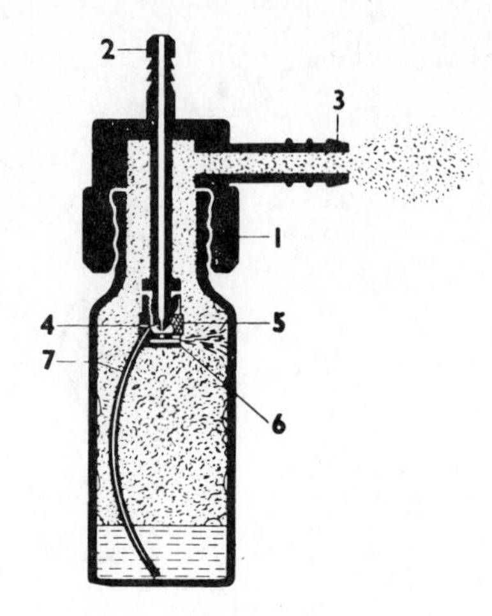

Figure 16.37. The Wright nebulizer. It consists of a solid cap 1, into which can be screwed any suitable bottle; 2, inlet connection; 3, outlet connection. The inlet connection communicates with a fine jet 4 on to which is screwed a knurled nozzle 5. The nozzle carries a circular baffle plate 6 mounted on an eccentric pillar through which passes a flexible feed tube 7. As the air jet passes through the nozzle 5, a vacuum is created that draws liquid up the feed pipe 7. The resulting spray impacts against the baffle plate and the coarser droplets (more than about 8 μm diameter) are trapped, coalesce, and fall back into the liquid. From *Lancet*, 24 (1958).

specifically for the production of monodisperse test aerosols with radioactive tags has been described by Lippmann and Albert (57) and is illustrated in Figure 16.38.

Monodisperse test aerosols can also be produced by a variety of techniques that break up a laminar liquid jet into uniform droplets. Most of them vibrate a capillary at high speed with a variety of transducers and types of motion. Dimmick's (58) generator, for example, uses transverse vibrations, whereas Strom's (59) uses axial vibrations. Wolf (60) uses a vibrating reed, wetted to a constant length by passage through a liquid reservoir, to create the droplet stream. The generator described by Raabe (20) has an air jet above the orifice and uses an ultrasonic transducer to convert a high-frequency power signal into mechanical axial vibrations of the orifice. The vibrating orifice generator of Berglund and Liu (61), which is illustrated in Figure 16.39, uses a cylindrical piezoelectric ceramic to vibrate a thin orifice plate, with holes from 3 to 22 μm in diameter, producing droplet diameters from about 10 to 50 μm. The particles produced by these generators can be made to vary less than 1 percent in volume, less than the variation in size of aerosols generated by a spinning disk device.

Electrostatic atomization can also produce monodisperse aerosols. Electric charges on a liquid surface act to decrease the surface tension. Liquid flowing through a capillary at high voltage is drawn into a narrow thread that breaks up into very small droplets (62–64).

Liu and Pui (65) developed a generator (Figure 16.40) for quite monodisperse

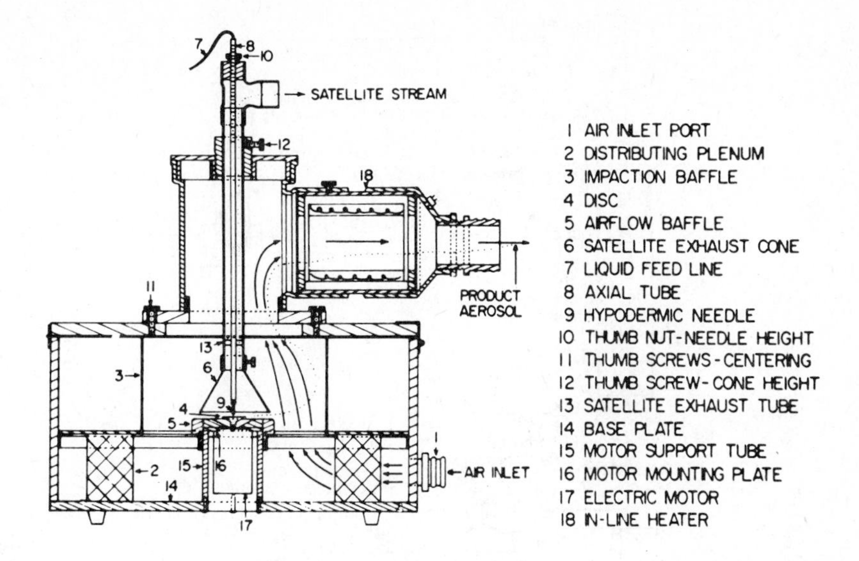

Figure 16.38. Electric motor driven spinning disk aerosol generator of Lippmann and Albert (57).

submicrometer aerosol particles in which the polydisperse output of a compressed air nebulizer is classified electrostatically. A solution or colloid is aerosolized in a Collison atomizer, mixed with dry air to form a solid aerosol, and brought to a state of charge equilibrium with the aid of a ^{85}Kr source. This aerosol, which is polydisperse (geometric standard deviation ≅ 2.7, median particle diameter ranging from 0.009 to 0.65 μm), is introduced into a differential mobility analyzer, which functions as a particle size classifier, based on the electrical mobility of the different

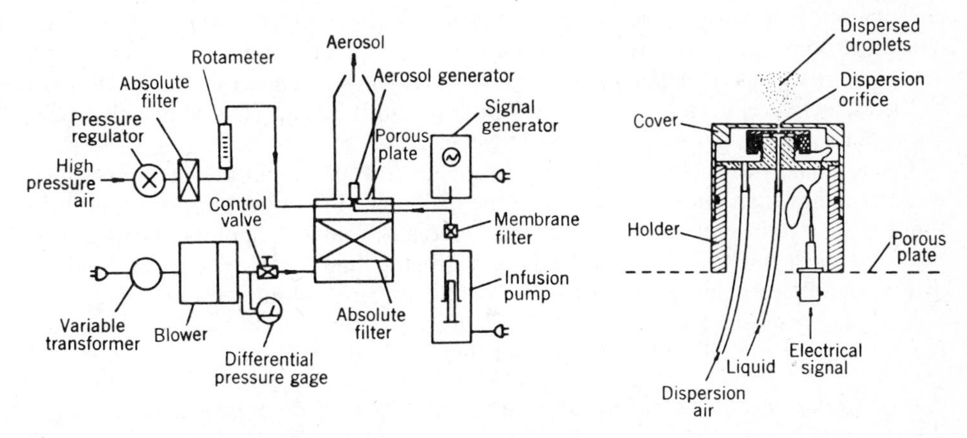

Figure 16.39. Vibrating orifice monodisperse aerosol generator. *Left:* schematic of system. *Right:* generator head. From *Air Pollut. Control Assoc. J.*, **24**, 12 (December 1974).

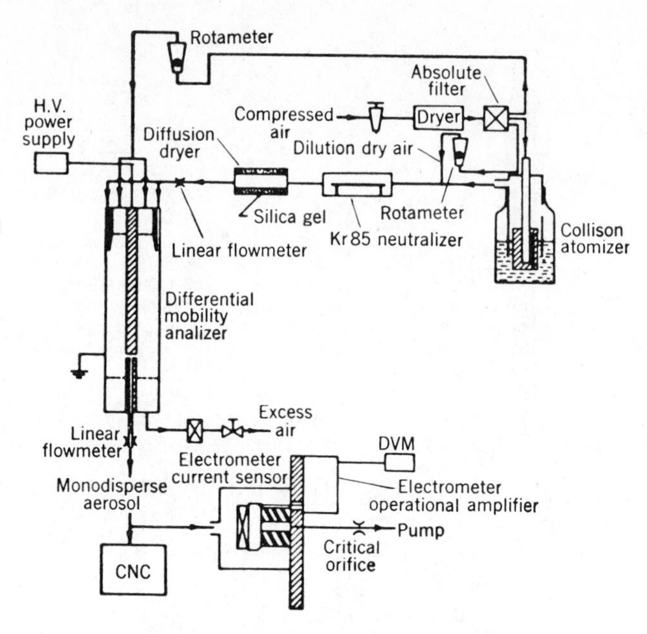

Figure 16.40. Apparatus for generating submicron aerosol standards.

size categories. This apparatus consists of an inner cylindrical electrode along which flows a sheath of clean air surrounded by an outer concentric sheath of the aerosol. Depending on the voltage of the electrode and the flow rate and the geometry, the more mobile particles drift through the clean air sheath to the electrode, where they are discharged and adhere, thereby being removed from the aerosol stream. Under a given set of operating conditions, a particular class of particles drift to a particular position, where they can be vented. These particles make up the monodisperse aerosol. For the particular design described by Liu and Pui, there was a coefficient of variance of .04 to .08 in particle size for singly charged particles. The concentration of aerosol is measured by collection on a filter, where the electrostatic charge is discharged through an electrometer. The presence of doubly charged particles increases greatly for particles larger than 0.3 μm. Thus above this size, the mobility no longer defines the particle size.

Commercially available ultrasonic aerosol generators can vibrate a liquid surface at a frequency high enough to result in the disintegration of the surface liquid into a polydisperse droplet aerosol. For mass median droplet diameters below 5 μm, the transducer must vibrate at a frequency greater than 1 MHz. The output characteristics of two commercial ultrasonic nebulizers are summarized in Table 16.6. An experimental ultrasonic generator (Figure 16.41) designed by G. J. Newton of the Lovelace Foundation has been described by Raabe (20).

4.7.5 Generation of Solid Insoluble Aerosols with Wet Dispersion Generators

Solid insoluble aerosols can be produced by nebulizing particle suspensions. One technique for producing monodisperse test aerosols is to prepare a uniform sus-

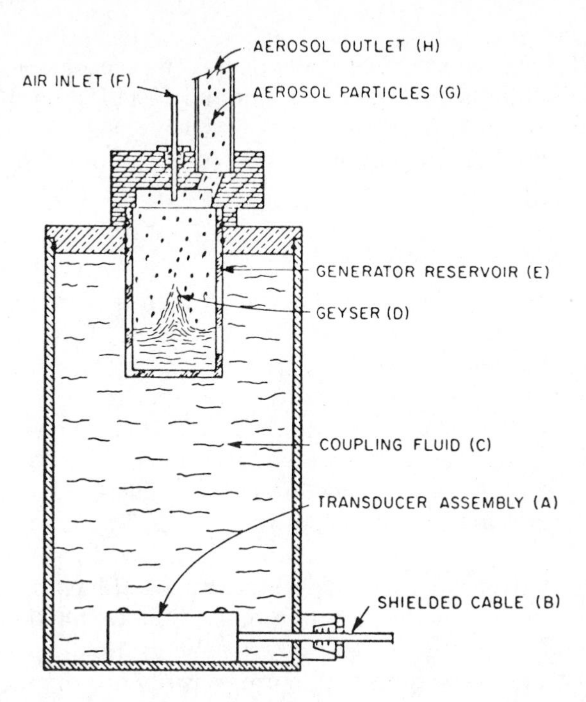

Figure 16.41. Sectional schematic view of an operating Ultrasonic Aerosol Generator: transducer assembly *A* receives power through shielded cable *B*, generates an acoustic field in the coupling fluid *C*, creating an ultrasonic geyser *D* in the generator reservoir *E*, and air entering at *F* carries away aerosol *G* through outlet *H*. Figure by G. J. Newton, reproduced from Raabe (20).

pension of the particles (latex, bacteria, etc.) in which the concentration is sufficiently dilute in the liquid phase that the probability of more than one particle being present in each droplet is acceptably small (66–68). This will result in a high vapor to particle ratio, thus limiting the mass concentration of aerosol produced. Another approach is to use a colloid as the feed liquid. In this case the diameter of the colloid particles can be orders of magnitude smaller than the particles in the resulting aerosol. Thus the volume of the droplet, and the size of the dried aggregate particles, is determined by the solids content of the sol.

An aerosol with chemical properties different from those of the feed material can be produced by utilizing suitable gas phase reactions such as polymerization or oxidation. Kanapilly et al. (69) describe the generation of spherical particles of insoluble oxides from aqueous solutions with heat treatment of the aerosols. This procedure involves (*a*) nebulizing a solution of metal ions in chelated form, (*b*) drying the droplets, (*c*) passing the aerosol through a high-temperature heating column to produce the spherical oxide particles, and (*d*) cooling the aerosol with the addition of diluting air. Another example of aerosol alteration is the production of spherical aluminosilicate particles with entrapped radionuclides by heat fusion of clay aerosols (70). This method involves (*a*) ion exchange of the desired radionuclide cation into clay in aqueous suspension and washing away of the unexchanged fraction, (*b*) nebulization of the suspension yielding a clay aerosol, and (*c*) heat fusion of clay aerosol, removing water and forming an aerosol of smooth solid spheres.

4.7.6 Characterizing Aerosols

4.7.6.1 Size Dispersion. The size dispersion of a test aerosol produced by a laboratory generator is determined by the characteristics of the generator and feed materials. The data on size included in the preceding discussion on generator characteristics, and in Table 16.7, indicate the approximate range obtainable in normal operation. The actual size distribution in a given case should always be measured with appropriate techniques and instrumentation. Sampling for particle size analysis has been discussed by Knutson (3). Sampling and analytical techniques have also been reviewed by Raabe (20) and Giever (71).

The distribution of droplets produced by nebulizers and some dry dust generators can usually be described by assuming that the logarithms of size are normally distributed. This log-normal distribution of sizes allows for simple mathematical transformation (72) and usually describes volume distributions satisfactorily (73). The characteristic parameters of a log-normal distribution are the median (or geometric mean) and the geometric standard deviation σ_g. The median of a distribution of diameters is called the count median diameter CMD; the median diameter based on the surface area is called the surface median diameter SMD; the median of the mass or volume distribution of the droplets or particles is called either the mass median diameter MMD or the volume median diameter VMD. These are related as follows.

$$\ln(\text{SMD}) = \ln(\text{CMD}) + 2\ln^2\sigma_g$$

$$\ln(\text{MMD}) = \ln(\text{CMD}) + 3\ln^2\sigma_g$$

in which ln designates the natural logarithm. A representative log-normal distribution appears in Figure 16.42 for a CMD equal to 1 μm and a σ_g equal to 2.

When particles are classified on the basis of their airborne behavior, a parameter called aerodynamic diameter is often used. It refers to the size of a unit density sphere having the same terminal settling velocity as the particle in question. For radioactive particles, an ICRP task group (74) has used the.parameter aerodynamic mass activity diameter (AMAD), which is the aerodynamic median size for airborne particulate activity.

7.7.6.2 Physical and Chemical Properties. An aerosol of a pure material having the desired physical and chemical characteristics can be prepared by dispersing that material into the air by any appropriate technique previously described. It is also possible to produce aerosols that differ in physical and/or chemical properties from the feed material. For example, particle size can be varied by dissolving or sus-

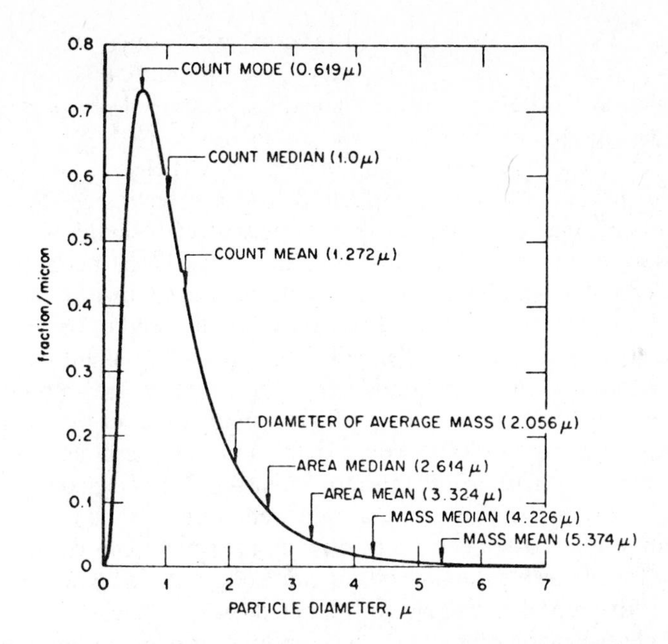

Figure 16.42. An example of the log-normal distribution function in normalized linear form for CMD = 1.0 μm and σ_g = 2.0 μm, showing the mode, median, and mean diameters, the mass distribution median and mean diameters, and the diameter of average mass. Graph courtesy Dr. Otto G. Raabe (20).

pending the material in a suitable volatile solvent that evaporates in the air to leave residue particles smaller than the nebulized droplets.

Solid aerosols resulting from droplet evaporation are generally spherical, but not always. Too rapid solvent evaporation, low pH and the presence of impurities may cause the dried particles to be wrinkled or to assume various shapes (75).

Aerosols produced from aqueous solutions (and some other methods) are charged by the random imbalance of ions in the droplets as they form (76). After evaporation, aerosol particles can be relatively highly charged; this may cause a small evaporating droplet to break up if the Rayleigh limit (77) is reached because of the repelling forces of the electrostatic charges overcoming the liquid surface tension (78). In some cases the net charge on a particle may be tens or even hundreds of electronic charge units, which will affect both the aerosol stability and behavior. Therefore a reduction in the net charges on aerosols produced by nebulization is desirable and in some experiments may be imperative. This can be accomplished either by mixing the aerosol with bipolar ions (79) or by passing it through a highly ionized volume near a radioactive source (80).

4.7.7 Detection of Aerosol Particles and Tagging Techniques

For many applications, such as efficiency testing of aerosol samplers or filters, it is often necessary to be able to measure concentrations that differ by several orders of magnitude. This type of testing can be done with untagged particles, such as polystyrene latex, using sensitive light-scattering photometers for concentration measurements. When particles other than the test aerosol are present, however, as in many field test situations,this method should not be used. Also, light-scattering techniques can be used over only a limited range of particle size, and the equipment is relatively expensive. Another approach to efficiency testing entails a microscopic count and/or size analyses of upstream and downstream samplers. However, this procedure is so tedious and time-consuming that it is seldom the method of choice.

Particle detection is often facilitated by incorporating dye or radiosotope tags in the particles in their production. Test aerosols composed of or containing fluorimetric dyes that can be analyzed in solutions containing as little as 10^{-10} g/cm^3 have been used for such applications (49). The particles are soluble in water or alcohol and can be quantitatively leached from many types of filters and collection surfaces for analysis. Colorimetric dyes such as methylene blue, which is used in the British Standard Test for Respirator Canisters (48), can be used in similar fashion when extremes of sensitivity are not required.

Radioisotope tags have been used in many forms and can usually be detected at extremely low concentrations. Spurny and Lodge (35) have discussed a variety of techniques for preparing radioactively labeled aerosols, including (1) preparation by means of neutron activation of aerosols in a nuclear pile or other neutron source, (2) labeling by means of decay products of radon and thoron, (3) preparation by means of radioactively labeled elements and compounds (condensation aerosols, disperse aerosols, and plasma aerosols), and (4) preparation by means of radioactively labeled condensation nuclei.

Method 2 refers to a process in which the particle surface is tagged while the particle is airborne. Procedures for surface tagging of polystyrene latex particles with isotopes in liquid suspensions by emulsion–polymerization reactions have been described by Black and Walsh (81), Bogen (82), and Singer et al. (83). Flachsbart and Stöber (84) have described a technique for growing uniform silica particles in suspension and incorporating various radioactive tags.

Other insoluble test aerosols containing nonleaching radioisotope tags have been produced by several techniques. The technique of heat fusion of ion-exchange clays (70) was discussed in the preceding section on insoluble aerosols. Techniques for producing insoluble spherical aggregate particles by nebulizing colloidal suspensions and plastics in solution have been described (57, 75). These aerosols made from colloids can be tagged with radioisotopes by mixing the nonradioactive colloid with a much lower mass concentration of an insoluble radioactive colloid before nebulization. The plastic particles can be tagged with radioisotopes in chelated form, dissolved in the plastic solution (75, 85–87).

4.8 Calibration of Sampler's Collection Efficiency

4.8.1 Use of Well-Characterized Test Atmospheres

To test the collection efficiency of a sampler for a given contaminant, it is necessary either (1) to conduct the test in the field using a proved reference instrument or

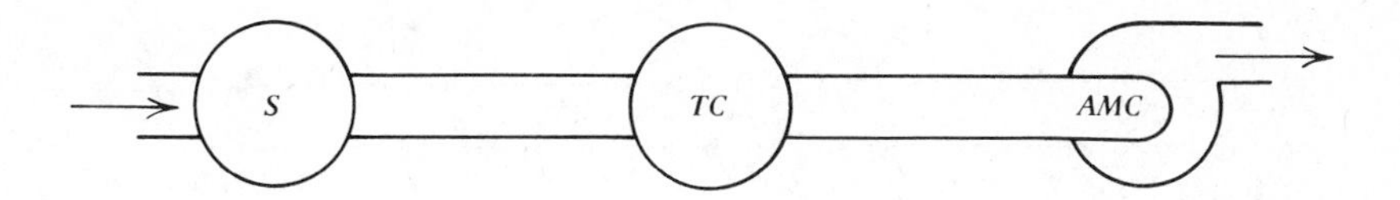

Figure 16.43. Sampler efficiency evaluation with downstream total collector; analysis of collections in sample under test *S* and total collector *TC; AMC*, air mover, flowmeter, and flow control.

technique as a reference standard or (2) to reproduce the atmosphere in a laboratory chamber or flow system. Techniques and equipment for producing such atmospheres are discussed in Sections 4.6 and 4.7.

4.8.2 Analysis of Sampler's Collection and Downstream Total Collector

The best approach to use in the analysis of a sampler's collection is to operate the sampler under test in series with downstream total collector, as illustrated in Figure 16.43. The sampler's efficiency is then determined by the ratio of the sampler's retention to the retention in the sampler and downstream collector combined. This approach is not always feasible, however. When the penetration is estimated from downstream samples there may be additional errors if the samples are not representative.

4.8.3 Analysis of Sampler's Collection and Downstream Samples

It is not always possible or feasible to collect quantitatively all the test material that penetrates the sampler being evaluated. For example, a total collector might add too much flow resistance to the system or be too bulky for efficient analysis. In this case, the degree of penetration can be estimated from an analysis of a sample of the downstream atmosphere, as illustrated in Figure 16.44. When this approach is used, it may be necessary to collect a series of samples across the flow profile, rather than a single sample, to obtain a true average concentration of the penetrating atmosphere.

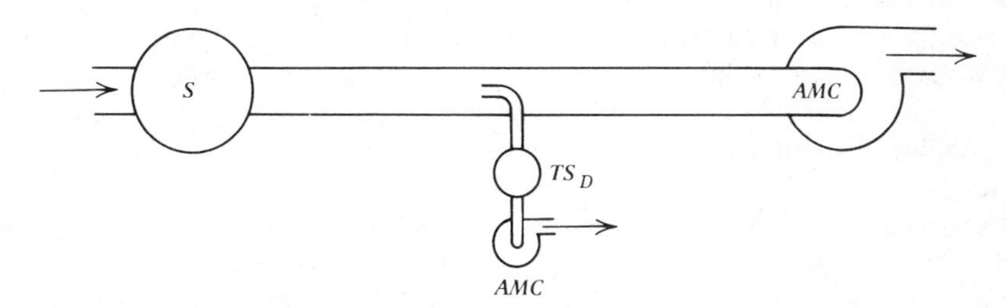

Figure 16.44. Sampler efficiency evaluation with downstream concentration sampler: analysis of collections in sample under test *S* and downstream sampler total collector TS_D.

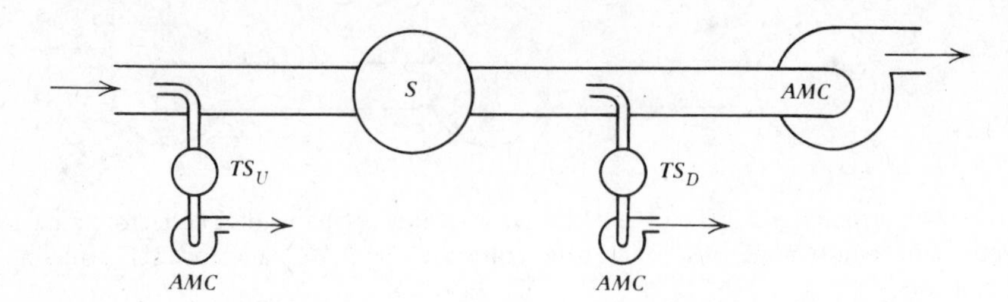

Figure 16.45. Sampler efficiency evaluation with upstream and downstream concentration samplers: analysis of collections in upstream and downstream samplers, total collection, TS_U and TS_D.

4.8.4 *Analysis of Upstream and Downstream Samples*

In some cases it is not possible to recover or otherwise measure the material trapped within elements of the sampling train such as sampling probes. The magnitude of such losses can be determined by comparing the concentrations upstream and downstream of the elements in question, as schematized in Figure 16.45.

4.9 Determination of Sample Stability and/or Recovery

The stability and the recovery of trace contaminants from sampling substrates are difficult to predict or control. Thus these factors are best explored by realistic calibration tests.

If the sample is divided into a number of aliquots that are analyzed individually at periodic intervals, it is possible to determine the long-term rate of sample degradation, or any tendency for reduced recovery efficiences with time. These analyses would not, however, provide any information on losses that may have occurred during or immediately after collection because they had different rate constants. Such losses should be investigated using spiked samples.

Analysis of Spiked Samples. If known amounts of the contaminants of interest are intentionally added to the sample substrate, subsequent analysis of sample aliquots will permit calculation of sample recovery efficiency and rate of deterioration. These results will be valid only insofar as the added material is equivalent in all respects to the material in the ambient air. There are two basic approaches to spiked sample analyses: (1) the addition of known quantities to blank samples, and (2) the addition of radioactive isotopes to either blank or actual field collected samples.

When the material being analyzed is available in tagged form, the tag can be added to the sample in negligible or at least known low concentrations. If there are losses in sample processing or analysis, the fractional recovery of the tagged molecules will provide a basis for estimating the comparable loss that took place in the untagged molecules of the same species.

4.10 Calibration or Sensor Response

Direct-reading instruments are generally delivered with a direct reading panel meter, a set of calibration curves, or both. The unwary and inexperienced user tends to believe the manufacturer's calibration, and this often leads to grief and error. Any instrument with calibration adjustment screws should of course be suspect, because such adjustments can easily be changed intentionally or accidentally—for example, in shipment.

All instruments should be checked against appropriate calibration standards and atmospheres immediately upon receipt and periodically thereafter. Procedures for establishing test atmospheres are discussed earlier in this chapter. Verification of the concentrations of such test atmospheres should be performed whenever possible using analytical techniques that are referee tested or otherwise known to be reliable.

With these techniques, calibration curves for direct-reading instruments can be tested or generated. When environmental factors such as temperature, ambient pressure, and radiant energy may be expected to influence the results, these effects should be explored with appropriate tests whenever possible. Similarly, the effects of cocontaminants and water vapor on instrument response should also be explored.

5 ESTIMATION OF ERRORS

5.1 Sources of Sampling and Analytical Errors

The difference between the air concentration reported for an air contaminant on the basis of a meter reading or laboratory analysis, and the true concentration at that time and place represents the error of the measurement. The overall error is often due to a number of smaller component errors rather than to a single cause. To minimize the overall error, it is usually necessary to analyze each of its potential components, concentrating one's efforts on reducing the component error that is largest. It would not be productive to reduce the uncertainty in the analytical procedure from 10 to 1.0 percent when the error associated with the sample volume measurement is ±15 percent.

Sampling problems are so varied in practice that it is possible only to generalize on the likely sources of error to be encountered in typical sampling situations. In analyzing a particular sampling problem, consideration should be given to each of the following:

1. Flow rate and sample volume.
2. Collection efficiency.
3. Sample stability under conditions anticipated for sampling, storage, and transport.
4. Efficiency of recovery from sampling substrate.
5. Analytical background and interferences introduced by sampling substrate.
6. Effect of atmospheric cocontaminants on samples during collection, storage, and analyses.

5.2 Cumulative Statistical Error

The most probable value of the cumulative error E_e can be calculated from the following equation:

$$E_e = [E_1^2 + E_2^2 + E_3^2 + \ldots + E_n^2]^{1/2}$$

For example, if accuracies of the flow rate measurement, sampling time, recovery, and analysis are ± 15, 2, 10, and 10 percent, respectively, and there are no other significant sources of error, the cumulative error would be

$$E_e = [15^2 + 2^2 + 10^2 + 10^2] = [429]^{1/2} = \pm 20.7\%$$

It should be remembered that this provides an estimate of the deviation of the measured concentration from the true concentration at the time and place the sample was collected. As an estimate of the average concentration to which a worker was exposed in performing a given operation, it would have additional uncertainty, depending on the variability of concentration with time and space at the work station.

6 SUMMARY AND CONCLUSIONS

Determinations of the concentrations of trace level contaminants in air and of heat stress, noise, and radiant energies are subject to numerous variables, many of them difficult to control. Thus it is prudent to perform frequent calibration checks on all industrial hygiene instruments. Such calibrations should be based on realistic simulations of the conditions encountered in the field.

The production of test atmospheres in the range of occupational threshold limits is often difficult. This chapter provides a review of available techniques for the production of test atmospheres of gases, vapors, and aerosols, with sketches of many of the more useful techniques.

Extreme care should be exercised in performing all calibration procedures. The following guidelines should be followed:

1. Use standard or reference atmospheres, instruments, and devices with care and attention to detail.
2. Check all standard materials and instruments and procedures periodically to determine their stability and/or operating condition.
3. Perform calibrations whenever a device has been changed, repaired, received from a manufacturer, subjected to use, mishandled, or damaged, and at any time when a question arises with respect to its accuracy.
4. Understand the operation of an instrument before attempting to calibrate it, and use a procedure or setup that will not change the characteristics of the instrument or standard within the operating range required.

5. When in doubt about procedures or data, make certain of their validity before proceeding to the next operation.
6. Keep all sampling and calibration train connections as short and free of constrictions and resistance as possible.
7. Exercise extreme care in reading scales, timing, adjusting, and leveling if needed, and in all other operations involved.
8. Allow sufficient time for equilibrium to be established, inertia to be overcome, and conditions to stabilize.
9. Obtain enough points or different rates of flow on a calibration curve to give confidence in the plot obtained. Each point should be made up of more than one reading whenever practical.
10. Maintain a complete permanent record of all procedures, data, and results. This should include trial runs, known faulty data with appropriate comments, instrument identification, connection sizes, barometric pressure, and temperature.
11. When a calibration differs from previous records, determine the cause of change before accepting the new data or repeating the procedure.
12. Identify calibration curves and factors properly with respect to conditions of calibration, device calibrated and what it was calibrated against, units involved, range and precision of calibration, date, and name of the person who performed the actual procedure. Often it is convenient to indicate where the original data are filed and to attach a tag to the instrument indicating the items just listed.

REFERENCES

1. Code of Federal Regulations, Title 40, Part 58. *Ambient Air Quality Surveillance, Appendix D-Network Design for State and Local Air Monitoring Stations (SLAMS) and National Air Monitoring Stations (NAMS),* U.S. Government Printing Office, Washington DC, 1981, pp. 149–159.
2. P. M. Eller, Ed., *NIOSH Manual of Analytical Methods, 3rd ed.,* US DHHS, CDC, NIOSH, Cincinnati, OH, 1984.
3. American Conference of Governmental Industrial Hygienists, *Air Sampling Instruments*, 7th ed., ACGIH, Cincinnati, OH, 1989.
4. N. A. Leidel, K. A. Busch, and J. R. Lynch, *Occupational Exposure Sampling Strategy Manual*, US DHEW, CDC, NIOSH, Cincinnati, OH, 1977.
5. D. Craig, *Health Phys.*, **21**, 328–332 (1971).
6. J. H. Perry et al., Eds., *Chemical Engineering Handbook*, 4th ed., McGraw-Hill, New York, 1963.
7. American Society of Mechanical Engineers, "Flow Measurement by Means of Standardized Nozzles and Orifice Plates," ASME Power Test Code (PTC 19.5.4–1959), ASME, New York, 1959.
8. B. E. Saltzman, *Anal. Chem.*, **33**, 1100–1112 (1961).

9. G. Hama, *Air Eng.*, **9**, 18 (1967).
10. H. D. Axelrod and J. P. Lodge, in *Air Pollution*, Vol. 3, 3rd ed., A. C. Stern, Ed., Academic Press, New York, 1976, pp. 145–182.
11. W. D. Conner and J. S. Nader, *Am. Ind. Hyg. Assoc. J.*, **25**, 291–297 (1964).
12. A. N. Setterlind, *Am. Ind. Hyg. Assoc. Quart.*, **14**, 113–120 (1953).
13. L. Silverman, in *Air Pollution Handbook*, P. L. Magill, F. R. Holden, and C. Ackley, Eds., McGraw-Hill, New York, 1956, pp. 12:1–12:48.
14. H. N. Cotabish, P. W. McConnaughey, and H. C. Messer, *Am. Ind. Hyg. Assoc. J.*, **22**, 392–402 (1961).
15. A. Goetz and T. Kallai, *J. Air Pollut. Control Assoc.*, **12**, 437–443 (1962).
16. G. O. Nelson and K. S. Griggs, *Rev. Sci. Instrum.*, **39**, 927–928 (1968).
17. G. O. Nelson, *Rev. Sci. Instrum.*, **41**, 776–777 (1960).
18. A. E. O'Keefe and G. O. Ortman, *Anal. Chem.*, **38**, 760–763 (1966).
19. T. T. Mercer, M. I. Tillery, and H. Y. Chow, *Am. Ind. Hyg. Assoc. J.*, **29**, 66–78 (1968).
20. O. G. Raabe, in *Inhalation Carcinogenesis*, M. G. Hanna, P. Nettesheim and J. R. Gilbert, Eds., CONF-691001, Clearinghouse for Federal Scientific and Technical Information, NBS, U.S. Department of Commerce, Springfield, VA, April 1970.
21. M. Kerker, *Adv. Colloid Interface Sci*, 5, 105–172 (1975).
22. D. A. Fraser, R. E. Bales, M. Lippmann, and H. E. Stokinger, "Exposure Chambers for Research in Animal Inhalation," Public Health Monograph No. 57, Public Health Service Publication No. 662, U.S. Government Printing Office, Washington, DC, 1959.
23. N. A. Fuchs and A. G. Sutugin, in *Aerosol Science*, C. N. Davies, Ed., Academic Press, London, 1966, pp. 1–30.
24. B. Wilson and V. K. LaMer, *J. Ind. Hyg. Toxicol.*, **30**, 265–280 (1948).
25. V. K. LaMer and D. Sinclair, *An Improved Homogeneous Aerosol Generator*, OSRD Report No. 1668, Department of Commerce, Washington, DC, 1943.
26. D. C. F. Muir, *Ann. Occup. Hyg.*, **8**, 233–238 (1965).
27. C. M. Huang, M. Kerker, E. Matijevic, and D. D. Cooke, *J. Colloid Interface Sci.*, **33**, 244 (1970).
28. E. Rapaport and S. G. Weinstock, *Experimentia*, **11**:9, 363 (1955).
29. B. Y. H. Liu and K. W. Lee, *Am. Industr. Hyg. Assoc. J.*, **36**, 861–865 (1975).
30. V. Prodi, in *Assessment of Airborne Particles*, T. T. Mercer, P. E. Morrow, and W. Stober, Eds., Charles C. Thomas, Springfield, IL, 1971, pp. 169–181.
31. K. Homma, *Ind. Health*, **4**, 129–137 (1966).
32. P. Movilliat, *Ann. Occup. Hyg.*, **4**, 275 (1962).
33. E. Matijevic, W. F. Espenscheid, and M. Kerker, *J. Colloid Interface Sci.*, **18**, 91–93 (1963).
34. S. Kitani and S. Ouchi, *J. Colloid Interface Sci.*, **23**, 200–202 (1967).
35. K. R. Spurny and J. P. Lodge, Jr., *Atmos. Environ.*, **2**, 429–440 (1968).
36. R. Ebens, *Staub*, **29**, 89–92 (1969).
37. B. M. Wright, *J. Sci. Instrum.*, **27**, 12–15 (1950).
38. S. Laskin, S. Posner, and R. Drew, paper presented at the annual meeting of the American Industrial Hygiene Association, St. Louis, May 1968.

39. R. L. Dimmock, *AMA Arch. Ind. Health*, **20**, 8–14 (July 1959).
40. J. R. Brown, J. Horwood, and E. Mastromatteo, *Ann. Occup. Hyg.*, **5**, 145–147 (1962).
41. V. Timbrell, A. W. Hyett, and J. W. Skidmore, *Ann. Occup. Hyg.*, **11**, 273–281 (1968).
42. P. F. Holt and D. K. Young, *Ann. Occup. Hyg.*, **2**, 249–256 (1960).
43. J. C. Couchman, *Metallic Microsphere Generation*, EG&G, Inc., Santa Barbara, CA, 1966.
44. M. Polydorova, *Staub* (Engl. transl.), **29**, 38 (1969).
45. F. G. Karioris and B. R. Fish, *J. Colloid Sci.*, **17**, 155–161 (1962).
46. M. Tomaides and K. T. Whitby, *Proceedings of the Seventh International Conference on Condensation and Ice Nuclei*, Academia, Prague, 1969.
47. J. D. Holmgren, J. O. Gibson, and C. Sheer, *J. Electrochem. Soc.*, **3**, 362–369 (1964).
48. British Standards Institute, "Methylene Blue Particulate Test for Respiratory Canister," B.S. No. 2577, British Standards Institute, London, 1955.
49. K. T. Whitby, D. A. Lundgren, and C. M. Peterson, *Int. J. Air Water Pollut.*, **9**, 263–277 (1965).
50. B. M. Wright, *Lancet*, 24–25 (1958).
51. L. Dautrebande, *Microaerosols*, Academic Press, New York, 1962.
52. V. K. LaMer and R. Gruen, *Trans. Faraday Soc.*, **48**, 410–415 (1952).
53. C. Orr, F. K. Hurd, and W. J. Corbett, *J. Colloid Sci*, **13**, 472–482 (1952).
54. M. J. Pilat and R. J. Charlson, *J. Rech. Atmos.*, **2**, 165–170 (1966).
55. C. C. Snead and J. T. Zung, *J. Colloid Interface Sci.*, **27**, 25–31 (1968).
56. W. H. Walton and W. C. Prewett, *Proc. Phys. Soc.*, (London), **62**, 341–350 (1949).
57. M. Lippmann and R. E. Albert, *Am. Ind. Hyg. Assoc. J.*, **28**, 501–506 (1967).
58. N. A. Dimmick, *Nature*, **166**, 686–687 (1950).
59. L. Strom, *Rev. Sci. Instrum.*, **40**, 778–782 (1969).
60. W. R. Wolf, *Rev. Sci. Instrum.*, **32**, 1124–1129 (1961).
61. R. N. Berglund and B. Y. H. Liu, *Environ. Sci. Technol.*, **7**, 147 (1973).
62. M. A. Nawab and S. G. Mason, *J. Colloid Sci.*, **12**, 179–187 (1958).
63. E. P. Yurkstas and C. J. Meisenzehl, "Solid Homogeneous Aerosol Production by Electrical Atomization," University of Rochester Atomic Energy Report No. UR-652, Rochester, NY, October 30, 1964.
64. V. A. Drozin, *J. Colloid Sci.*, **10**, 158 (1955).
65. B. Y. H. Liu and D. Y. H. Pui, *J. Colloid Interface Sci.*, **10**, 158, (1955).
66. P. C. Reist and W. A. Burgess, *J. Colloid Interface Sci.*, **24**, 271–273 (1967).
67. O. G. Raabe, *Am. Ind. Hyg. Assoc. J.*, **29**, 439–443 (1968).
68. S. C. Stern, J. S. Baumstark, A. I. Schekman, and R. K. Olson, *J. Appl. Phys.*, **30**, 952–953 (1959).
69. G. M. Kanapilly, O. G. Raabe, and G. J. Newton, *Am. Ind. Hyg. Assoc. J.*, **30**, 125 (1969) (abstract).
70. G. M. Kanipilly, O. G. Raabe, and G. J. Newton, *Aerosol Sci.*, **1**, 313 (1970).
71. P. M. Gieyer in *Air Pollution*, Vol. 3, 3rd Ed., A. C. Stern, Ed., Academic Press, New York, 1976, pp. 3–50.
72. T. Hatch and S. P. Choate, *J. Franklin Inst.*, **207**, 369–387 (1929).

73. T. T. Mercer, R. F. Goddard, and R. L. Flores, *Ann. Allergy*, **23**, 314–326 (1967).
74. P. E. Morrow, *Health Phys.*, **12**, 173–208 (1966).
75. R. E. Albert, H. G. Petrow, A. S. Salam, and J. R. Spiegelman, *Health Phys.*, **10**, 933–940 (1964).
76. T. T. Mercer, *Health Phys.*, **10**, 873–887 (1964).
77. L. Rayleigh, *Phil. Mag.*, **14**, 184–186 (1882).
78. K. T. Whitby and B. Y. H. Liu, in *Aerosol Science*, C. N. Davies, Ed., Academic Press, New York, 1966, pp. 59–86.
79. K. T. Whitby, *Rev. Sci. Instrum.*, **32**, 351–355 (1961).
80. S. L. Soong, M. S. thesis, University of Rochester, Rochester, NY, 1968.
81. A. Black and M. Walsh, *Ann. Occup. Hyg.*, **13**, 87–100 (1970).
82. D. C. Bogen, *Ann. Industr. Hyg. Assoc. J.*, **31**, 349–352 (May-June 1970).
83. M. Singer, C. J. Van Oss, and W. Wanderhoff, *J. Reticuloendothei. Soc.*, **6**, 281–286 (1969).
84. H. Flachsbart and W. Stöber, *J. Colloid Interface Sci.*, **30**, 568–573 (1969).
85. R. E. Albert, J. Spiegelman, M. Lippmann, and R. Bennett, *Arch. Environ. Health*, **17**, 50–58 (July 1968).
86. J. R. Spiegelman, G. D. Hanson, A. Lazarus, R. J. Bennett, M. Lippmann, and R. E. Albert, *Arch. Environ. Health*, **17**, 321–326 (1968).
87. D. V. Booker, A. C. Chamberlain, J. Rundo, D. C. F. Muir, and M. L. Thomson, *Nature*, **215**, 30–33 (1967).
88. T. T. Mercer, R. F. Goddard, and R. L. Flores, *Ann. Allergy*, **26**, 18–27 (1968).
89. K. E. Lauterbach, A. D. Hayes, and M. A. Coelho, *AMA Arch. Ind. Health*, **13**, 156–160 (1956).
90. K. R. May, *J. Aerosol Sci.*, **4**, 235–243 (1973).
91. O. G. Raabe, in *Fine Particles*, B. Y. H. Liu, Ed., Academic Press, New York, 1976, pp. 60–110.

CHAPTER SEVENTEEN

Indoor Air Quality in Nonindustrial Occupational Environments

Philip R. Morey, Ph.D., C.I.H., and Jaswant Singh, Ph.D., C.I.H.

1 INTRODUCTION

1.1 Historical

Concern over the quality of air in indoor environments has historically included viewpoints that outdoor ventilation air is required both to prevent *adverse health effects* and to provide for *comfort* of occupants. Thus over two centuries ago, Benjamin Franklin wrote that " . . . I am persuaded that no common air from without is so unwholesome as the air within a closed room that has been often breathed and not changed" (1). He further stated that outdoor and "cool air does good to persons in the smallpox and other fevers. It is hoped, that in another century or two we may find out that it is not bad even for people in health." Accordingly, ventilation codes by the late nineteenth century recommended the provision of large amounts of outdoor air to lower the risk of disease from certain infective agents such as that causing tuberculosis.

In the 1930s, studies by Yaglow et al. (2) showed that between 10 and 30 cubic feet per minute (cfm) of outdoor air was required in ventilation air to prevent annoyance in interior environments due to human body odors. An important objective of the current American Society of Heating, Refrigerating, and Air-con-

Patty's Industrial Hygiene and Toxicology, Fourth Edition, Volume 1, Part A, Edited by George D. Clayton and Florence E. Clayton
ISBN 0-471-50197-2

ditioning Engineers (ASHRAE) Standards 55 (3) and 62 (4) is to provide for comfortable environmental conditions for a majority (80 percent or more) of occupants in indoor environments.

LIST OF ABBREVIATIONS USED IN CHAPTER SEVENTEEN

ACGIH	American Conference of Governmental Industrial Hygienists
AHU	Air handling unit
ASHRAE	American Society of Heating, Refrigerating, and Air-Conditioning Engineers
Bq	Becquerel
BRI	Building-related illness
°C	Degrees Celsius
cfm	Cubic feet per minute
EPA	U.S. Environmental Protection Agency
ETS	Environmental tobacco smoke
HVAC	Heating, ventilating, and air-conditioning
IAQ	Indoor air quality
l/sec	Liters per second
NAAQS	National Ambient Air Quality Standards
NIOSH	National Institute for Occupational Safety and Health
NOPES	Non-occupational Pesticide Exposure Study
OSHA	Occupational Safety and Health Administration
pCi/L	Picocurie per liter
ppb	Parts per billion
ppm	Parts per million
SBS	Sick building syndrome
SVOC	Semivolatile organic compound
TLV	Threshold limit value
TWA	Time weighted average
VAV	Variable air volume
VOC	Volatile organic compound
VVOC	Very volatile organic compound
WL	Working level

1.2 Sick Building Syndrome

In a landmark paper entitled "The Sick Building Syndrome" (SBS), Jan Stolwijk (5) described a constellation of nonspecific complaints that occurs at higher prevalence rates in some problem buildings. The 14 complaints listed in his paper included the following:

1. Mucous membrane irritation
2. Eye irritation
3. Headache
4. Odor
5. Skin irritation/rash
6. Sinus congestion
7. Cough
8. Sore throat
9. Shortness of breath
10. Abnormal taste
11. Dizziness
12. Fatigue
13. Nausea
14. Wheezing and hypersensitivity

Although these complaints seem to be present at some level in all buildings, their prevalence appears highest in a subset of problem or "sick" buildings. Most SBS complaints are associated with occupant discomfort and annoyance and not documentable clinical disease. Stolwijk (5) estimated that although 500 to 5000 buildings had been evaluated (by 1984) the cause(s) of SBS in the vast majority of studies was unknown. It was clear at this time from studies being carried out at the U.S. National Institute for Occupational Safety and Health (NIOSH) that inadequate or ineffective ventilation was associated with the majority of occurrences of SBS. Thus in studies up to 1986 (6), NIOSH categorized indoor air quality (IAQ) complaints in nonindustrial commercial buildings as being brought about by the following causes: inadequate ventilation (50 percent), internal contaminants (23 percent), external contaminants (11 percent), and microorganisms (5 percent). In the remaining 11 percent of the buildings studied the cause of air quality problems could not be determined.

A number of factors collectively make IAQ an important issue in the 1990s, whereas 40 or 50 years ago it was not recognized as a problem.

Construction practices today are vastly different from those employed half a century ago. Windows in modern buildings generally do not open, whereas in the 1940s natural ventilation (opened windows) was common. In modern buildings, reliance is placed on the heating, ventilating, and air-conditioning (HVAC) system to transport outdoor air to the breathing zone. Energy conservation, however, is a major factor in the design and operation of most modern buildings. This is translated by architects, design engineers, and mechanical contractors into tighter construction of the building envelope (less infiltration) and restrictions on the intake of outdoor air during the winter and summer seasons to minimize heating and cooling costs. The net effect on occupants is often less outdoor air at the breathing zone.

Construction materials have markedly changed over the past 50 years. Stone, wood, and other "natural" construction/finishing materials have largely been replaced by synthetics. The ceiling tiles, wall and floor coverings, and even desks and chairs in modern buildings are primarily synthetic in nature. Volatile organic compounds (VOCs), many of which are toxic, are emitted into the indoor environment from modern finishing and construction materials. Volatile agents from cleaning and graphics materials, pesticides, hydraulic elevator fluid, and cosmetics are additional sources of indoor air pollution.

Heightened public awareness of IAQ probably began in North America in 1976 with the occurrence of Legionnaires' disease in a Philadelphia hotel where 29 fatalities occurred because of indoor exposure to *Legionella*. A 1984 telephone survey of 600 office workers estimated that 20 percent of the respondents felt that their productivity was hampered by poor IAQ (7). Public awareness of IAQ is heightened today by the realization that the concentration of air pollutants such as VOCs, combustion products, and radon is often considerably greater indoors than outdoors.

Other factors of importance in the perception of IAQ as an important issue include:

- The greater susceptibility of the elderly and the very young to effects of indoor pollutants.
- The expectation of a cleaner indoor environment; this is analogous to national efforts over the past 20 years to reduce ambient (outdoor) air pollutants.
- The increasing importance of comfort and productivity parameters to a work force in an office-type service-dominated economy.
- Advances in analytical techniques have made possible the detection of concentrations of air pollutants in the parts per billion range.

1.3 Building-Related Illness

In industrialized countries most people spend up to 90 percent of a 24-hr day in the indoor environment (office, transportation, residence). Concern has thus been raised over the potential effects of long-term exposure to indoor air contaminants. Indeed the risks associated with breathing indoor air are now thought generally to exceed the risks associated with pollutants in ambient air or in drinking water (8) at the currently regulated limits. Of all the types of air pollutants, radon, VOCs, environmental tobacco smoke (ETS), and microbiological agents probably present the greatest risk of premature death or morbidity arising from exposure in the indoor environment.

Among indoor air pollutants, exposure to radon may present the greatest risk of premature death (from lung cancer) (8). Risk due to exposure to VOCs in indoor air may be of a similar magnitude as that associated with exposure to chemicals in industrial environments (8). Microbial aerosols in indoor environments may cause significant adverse health effects, especially among the population with a predisposition to asthma or allergy or with an enhanced susceptibility to infectious agents.

Illness in the general population caused by respiratory infection affects almost all North Americans at least once per year. Recent epidemiologic studies (9) show that the incidence of respiratory illness is significantly higher in military barracks that are mechanically ventilated than in those in which outdoor air is at least partially provided through open windows. Thus occupants in mechanically ventilated buildings have a higher risk of developing respiratory infections owing to recirculation of microbial agents (most likely viruses) in the indoor air.

1.4 Definition of Sick Building Syndrome and Building-Related Illness

Over the past 10 years, two terms have been used to define the maladies that are associated with occupancy in nonindustrial indoor environments, namely, SBS and building-related illness (BRI).

Sick building syndrome (SBS, sometimes called tight building syndrome), as used in this chapter, is defined as a constellation of nonspecific, primarily sensory complaints including eye, nose, and throat irritation, headache, fatigue, skin irritation, mild neurotoxic symptoms, and odor annoyance. A complaint rate (during a formal epidemiologic survey) of 20 percent is probably needed for a health problem to qualify as SBS. Affected occupants generally report relief when they breath outdoor air for even a short period of time. Because SBS symptoms are nonspecific and can be potentially associated with many different types of contaminant sources, an exact cause for SBS is not readily apparent even in buildings that have been extensively studied.

Building-related illness is used to distinguish those instances where occupant health problems are closely recognizable as a disease upon medical examination and are putatively associated with indoor environmental exposure. Examples of BRI and the air contaminants that are involved in disease etiology include:

- Cancer; caused by gaseous and particulate components of ETS, some VOCs, and radon
- Legionnaires' disease; caused by *Legionella* species and serotypes
- Hypersensitivity pneumonitis, humidifier fever, allergic rhinitis, and asthma; caused by microbial allergens
- Dermatitis; caused by fibers from man-made insulation and some volatile organic compounds

It should be noted that the same agents may cause both SBS and BRI. For example, exposure to VOCs, ETS, or combustion products may cause SBS. Chronic exposures to the same air contaminants potentially may also elicit BRI.

It should be recognized that "sick buildings" represent a subset of problem buildings including those in which occupants may be exposed to ergonomic problems, safety hazards, and other concerns. Presumably the vast majority of buildings in North America are "nonproblem" or "healthy" buildings (see Section 5.2).

In any discussion of sick buildings, three basic approaches should be followed, namely, a consideration of the types of pollutants that may be present (see Section 2), the illnesses and discomfort phenomena experienced by occupants (see Sections 3–5), and the pollutant sources and HVAC systems in buildings where problems occur (see Sections 6–8).

2 INDOOR AIR QUALITY PARAMETERS

In the following discussion the pollutants that are important in IAQ evaluations are reviewed. Emphasis is placed on recent developments. The reader is also referred to past reviews (10–13) on IAQ.

2.1 Bioaerosols

The term bioaerosol is used to describe airborne microbiological particulate matter including that derived from viruses, bacteria, fungi, protozoa, and their cellular or cell wall components. Bioaerosols are ubiquitous in both indoor and outdoor environments.

In the past, much attention has been given to saprophytic fungi (which derive their nutrients from dead organic material), probably because of readily available sampling and analytical methods for this type of microorganism. Agents such as viruses, rickettsia, chlamydia, protozoa, and many pathogenic fungi and bacteria are more difficult to culture and air sampling methodology for these agents is just becoming available (14).

2.1.1 Sources

Microorganisms are essential components of the earth's terrestrial and aquatic ecosystems. Microorganisms break down the complex molecules found in dead organic materials from plants and animals and recycle minerals and carbon to simple substances such as carbon dioxide and nitrates, which are then used by the major producers of the planet, namely, green plants. The presence of microorganisms such as saprophytic bacteria and fungi in the soil and in the atmosphere is a normal occurrence. Thus *Cladosporium* spores are almost always found in outdoor air. *Cladosporium* spores are also found normally in the indoor environment depending on the amount of outdoor air that infiltrates into interior spaces and that is brought into the HVAC system. Bacteria that are saprophytic (for example, *Staphylococcus*) and viruses that are obligate parasites (for example, Influenza A) are shed by humans and are normally present in increased numbers in indoor environments.

Although microorganisms are normally present in indoor environments, the availability of moisture and substrate in interior niches has been associated with the amplification of some microbial agents to the extent that the interior environment is microbiologically atypical. Thus certain types of humidifiers, water spray systems, and porous man-made insulation can be reservoirs and amplification sites for fungi, bacteria (including actinomycetes), protozoa, and even nematodes in indoor environments (15–18). Excessive air moisture (19) and floods (20) have also been associated with the proliferation of microorganisms indoors. Turbulence associated with the start-up of HVAC systems and maintenance operations in mechanical systems may cause significant elevations in concentrations of bacteria and fungi in occupied spaces (21).

Cooling towers, evaporative condensers, and water service systems all provide water and nutrients for amplification of microorganisms such as *Legionella*. Amplification of microbial populations to concentrations that might be considered atypical or excessive is generally caused by poor preventive maintenance of these systems.

2.1.2 Health Effects

Microorganisms in indoor environments may cause BRI that is infective and/or allergic in nature (22, 23). Some microorganisms may produce volatile chemicals

(24) that are malodorous or irritative (25) and therefore may be important in the etiology of SBS.

Legionellosis is the collective name given to two distinct types of illnesses, Legionnaires' disease (see Section 4.4) and Pontiac fever. Legionnaires' disease is a true infection (the lung is the target organ) that can result in pneumonia requiring hospitalization and appropriate antibiotic therapy. *Legionella pneumophila*, a relatively recently discovered bacterium (the genus name *Legionella* is derived from the American Legion Convention in Philadelphia in 1976; the species name *pneumophila* means "lung loving"), is the microorganism that the public most closely associates with infectious disease in buildings. A second form of legionellosis, Pontiac fever, is milder and nonpneumonic. Recovery usually occurs without hospitalization or use of antibiotics.

At least 23 species and over 40 serogroups of Legionellae have been identified in the microbiology literature (26). These gram-negative bacteria require special media containing the amino acid L-cysteine for growth, and they are commonly found outdoors in natural (fresh) water. Because *Legionella* requires special nutrients for growth and does not produce endospores, this microorganism is very difficult to recover from the air. Collection and analysis of water specimens (not air samples) is thus usually relied upon for the detection of *Legionella*.

Outbreaks of infective illness in the indoor environment may be caused by other types of microorganisms such as viruses. In an airliner cabin the majority of the passengers developed influenza following exposure to one acutely ill person (27). The plane had been disabled on a runway for several hours during which time the ventilation system was inoperative. This outbreak was recognized only because the majority of affected passengers were treated by the same medical personnel.

Histoplasmosis, an infective illness caused by the fungus *Histoplasma capsulatum* may occur (rarely) as a BRI among individuals involved in the removal of bat droppings in attics of abandoned buildings (28). Presumably asexual spores (conidia) from this fungus are inhaled by workers without adequate respiratory protection.

Microorganisms may cause BRI in indoor environments by affecting the immune system. Thus allergic respiratory disease may develop in response to inhaled particulates containing microorganisms or their components such as spores, enzymes, and cell wall fragments. There have been numerous reports of allergic respiratory illness (often called humidifier fever in Europe and hypersensitivity pneumonitis in North America) in which affected people manifest acute symptoms such as malaise, fever, chills, shortness of breath, and cough (29–31). These illnesses usually occur as a response to microbiological contaminants originating from HVAC system components such as humidifiers and water spray systems or other components of buildings that have been damaged by chronic floods (20, 32). Unlike SBS, affected individuals usually experience relief only after having left the building for an extended period of time.

Microbiological contamination was important in about 5 percent of the IAQ evaluations carried out by NIOSH (6). This may be a substantial underestimate because many NIOSH evaluations were superficial in nature, and microorganisms may not have even been considered. In addition, almost all NIOSH evaluations did not consider the possible importance of microbial volatiles, endotoxins, mycotoxins, and nonviable particulate. Several recent studies discussed in Sections 3.2 and 4.5 suggest a far greater importance of microorganisms in indoor air quality issues than previously suspected.

2.1.3 Sampling

The principles of sampling and analysis for microorganisms are reviewed by Chatigny in the American Conference of Governmental Industrial Hygienists' (ACGIH) *Air Sampling Instruments* manual (33). The ACGIH Bioaerosols Committee has developed a generic protocol (34) for the collection of microbial particulate. Some of the salient points of this protocol plus our personal views are listed below.

2.1.3.1 Preassessment. Sampling for microorganisms should be undertaken especially when medical evidence suggests the occurrence of diseases such as humidifier fever, hypersensitivity pneumonitis, allergic asthma, and allergic rhinitis. Prior to any sampling, a walkthrough examination of the indoor environment is recommended for visual detection of possible microbial reservoirs and amplification sites. If a microbial reservoir/amplifier is identified, it is often useful as an aid to future laboratory analysis of subsequent air samples to obtain a bulk or source sample from the reservoir or amplifier. Also, actions carried out at this stage of the evaluation to remove clearly identified reservoirs and amplifiers are preferable to complicated and costly air sampling procedures.

2.1.3.2 Air Sampling. The same principles that affect the collection of any particulate aerosol also govern air sampling for microorganisms (34). In general, culture plate impactors, including multiple- and single-stage sieve devices, as well as slit-to-agar samplers are most useful in office environments where relatively low concentrations of microorganisms are expected. Filter cassette samplers are useful for microorganisms or components of microorganisms (for example, endotoxin) that are resistant to desiccation.

2.1.3.3 Data Interpretation. Rank order assessment is recommended as a means for interpreting air sampling data for saprophytic microorganisms (34). Individual taxa are listed in descending order of abundance for indoor sites and for a control (usually outdoor air) location. The predominance of one or more taxa indoors, but not in the control site, suggests the presence of an amplifier (for that taxa) in the building. In the example given in Table 17.1, *Tritirachium* and *Aspergillus* species (fungi) were the predominant taxa present indoors whereas *Cladosporium* and *Fusarium* dominated outdoor collections. Clearly, *Tritirachium* and *Aspergillus* were being amplified indoors. Further investigation in this building showed that *Tritirachium* and *Aspergillus* were amplifying in moist wall cavities in the building.

Table 17.1. Airborne Fungi Present in Indoor and in Outdoor Air

Location	Cfu/m^3[a]	Rank Order of Taxa
Outdoors	210	*Cladosporium* > *Fusarium* > *Epicoccum* > *Aspergillus*
Office 1	2500	*Tritirachium* > *Aspergillus* > *Cladosporium*
Office 2	700	*Tritirachium* > *Aspergillus* > *Cladosporium*

[a] Cfu/m^3 means colony-forming units per cubic meter of air. Culture media was malt extract agar.

2.1.4 *Guidelines*

At present there are no numerical guidelines for bioaerosol exposure in indoor or outdoor environments for the following reasons (35):

- Incomplete base-line data for concentrations and types of microbial particulate indoors, especially as affected by geographic, seasonal, and type-of-building parameters
- Absence of epidemiologic data relating bioaerosol exposure to BRI
- Enormous variability in microbial particulate including viable cells, dead spores, toxins, and antigens
- Enormous variation in human susceptibility to microbial agents making generalized estimates of health risk problematic

However, even in the absence of numerical guidelines, bioaerosol sampling data can and should be interpreted based on a number of considerations as follows:

- Rank order assessment of the types of microbial agents present in indoor and outdoor air (34)
- Medical or laboratory evidence that a BRI is caused by a type(s) of microorganism (34)
- Indoor/outdoor concentration ratios for various kinds of microbial agents (18, 35)

In order for a microbial agent to cause BRI, it must be transported in sufficient dose to the breathing zone of a susceptible occupant. The concepts of reservoir, amplifier, and disseminator (22, 34, 35) must be considered in data interpretation.

2.2 Volatile Organic Compounds

2.2.1 *Characterization of Volatile Organic Compounds in Indoor Air*

The VOCs are probably the one class of air pollutant most commonly associated with indoor air pollution and SBS. VOCs are characterized by boiling points ranging from about 50 to 100°C to 240 to 260°C (36) and include alcohols, aldehydes, olefins, terpenes, aliphatic and aromatic hydrocarbons, and chlorinated hydrocarbons. Those VOCs with boiling points less than 50 to 100°C are referred to as *very* volatile organic compounds (VVOC). Those with boiling points above 240–260°C are called *semivolatile* organic compounds (SVOC). SVOCs include pesticides, polynuclear aromatic compounds, and certain plasticizers such as phthalate esters (37).

Even in those geographic areas where outdoor air pollution from sources such as petroleum refineries is significant, indoor total VOC concentrations have been shown to be 2 to 10 times higher than those outdoors (38). In new office buildings the total VOC concentration at the time of initial occupancy is often 50 to 100 times that present in outdoor air. The variety of VOCs found in indoor air is almost always greater than that in outdoor air, largely because of the very large number of possible sources in indoor environments. Although VOCs in indoor nonindustrial environments are generally present at concentrations considered to be "elevated" with respect to the outdoor air, the absolute concentrations of specific VOCs is almost always two to three orders of magnitude *less* than that considered significant in industrial workplaces.

2.2.2 *Nature of VOCs and Their Sources*

Aliphatic and aromatic hydrocarbons are the VOCs most often present as contaminants in indoor air. Aliphatic compounds such as decane and undecane may be present indoors in new buildings at concentrations of 100 or even 1000 times that present outdoors. Liquid-process photocopies are important sources of branched aliphatic hydrocarbons (39). The concentration of aliphatic hydrocarbons in buildings with liquid-process photocopiers may rise to several times outdoor levels even under conditions of good outdoor air ventilation.

The presence of benzene indoors is often associated with ETS. Toluene, xylene, and ethylbenzene are other aromatic hydrocarbons commonly found in nonindustrial indoor environments. It has been estimated that the arithmetic mean concentration of toluene in residences is 20 $\mu g/m^3$, or about 10 times that present in the outdoor air (36).

A wide variety of chlorinated hydrocarbons are commonly found indoors. Chloroform at concentrations of 1 to 4 $\mu g/m^3$ occurs in residential environments (37), primarily as a result of release from hot water during showering and bathing. 1,1,1-Trichloroethane is often found in office air because solvents containing this compound are used to clean work station panels. Methylene chloride, a common solvent component, is released into the indoor environment often from paint strippers and cleaners.

Terpenes from fresheners, cleaners, polishes, and deodorants are found in measurable concentrations in indoor air (37). Limonene from lemon scent and pinene from pine scent are common examples of volatile terpenes present in some offices. Other VOCs found in indoor air include cyclohexenes and freons. 4-Phenylcyclohexene from styrene–butadiene carpet backing has been associated with allergic responses in some buildings. Freon 113 may be released in appreciable amounts from certain types of laser printers (40).

Among the VOCs, formaldehyde is most closely associated by the public with indoor air pollution. Although formaldehyde may be emitted from a number of sources such as from gas stoves and from smoking, its major source indoors is from construction materials such as particle board, fiberboard, and plywood. Concentrations of formaldehyde in residential buildings is higher than in office buildings because of the relatively large ratio of pressed wood products to air volume in the former as compared to the latter type of buildings (41). Thus although concentrations of formaldehyde in residences in one study often exceeded 35 ppb, this VOC is usually found at levels below 35 ppb in office buildings (Figure 2 in Reference 41).

Examples of SVOCs that may be present as significant pollutants in indoor air include polynuclear aromatic hydrocarbons, primarily from combustion products, and phthalate ester plasticizers found in paints, floor coverings, and many plastic-based materials. Benzo-*a*-pyrene may originate from sources such as ETS, wood smoke, and unvented kerosene heaters. Dibutyl phthalate is a plasticizer emitted from paints. This SVOC binds to particulate aerosols, settles on leaves of houseplants, and has been shown to cause necrosis in leaf tissue by interfering with chlorophyll synthesis (42, 43).

Studies on the sources of VOCs in nonindustrial indoor environments are confounded by the variable nature of emissions from potential sources (42). Emissions from some sources such as building materials are *continuous* and *regular*, whereas emissions from other sources (e.g., paints used in renovation work) are *continuous* but *irregular* in nature. Emission of VOCs from yet other sources are *intermittent* and *regular* (for example, VOCs in combustion products from gas stoves) or *intermittent* and *irregular* (for example, VOCs from carpet shampoos).

Major sources of VOCs in buildings include construction materials, maintenance products, and combustion and consumer materials (41). The VOCs in building structural and finishing materials are often unreacted, residual solvents. These VOCs are emitted from building surfaces that cover large areas (e.g., work station panels). Building maintenance products including alcohols, organic acids, alkanes, and aromatic and chlorinated hydrocarbons are used continuously but in an episodic (irregular) manner in buildings. In some buildings, significant indoor VOC pollution results from the entrainment of nearby industrial emissions. However, in most buildings, the outdoor VOCs present only a minor source of indoor VOCs; most investigators use outdoor VOCs as a base line against which indoor-generated VOCs are compared.

A major difficulty encountered when attempting to identify exact sources of emissions in buildings is that some porous indoor building materials act as sinks for VOCs from other sources. Subsequently, under other conditions, the same building materials can act as secondary VOC emission sources (41). Berglund et al. (42) studied the primary and secondary emission characteristics of floor, ceiling, and wall materials from a 7-year-old building. In chamber tests where ventilation with VOC-free air was provided, the emission of 27 VOCs declined to below detectable limits after 29 days. The 7-year-old materials, however, continued to emit constant low levels of 13 other VOCs. Berglund et al. (42) concluded that the 27 VOCs whose emissions declined to below detectable limits were initially *adsorbed* from room air and then desorbed and emitted when the materials were ventilated by clean air. They point out that secondary emission of adsorbed VOCs is a major impediment to tracing the origin of specific VOCs in the indoor air.

Chamber test techniques are available to measure the emission rates of individual VOCs from construction and finishing materials (44). The comparative emission rates of VOCs from adhesives was found to vary from product to product. For example, the emission of toluene was found to range from 140 $\mu g/m^2hr$ to 7.4 mg/m^2hr (44). As this type of emission rate data becomes available for more commercial products it may be possible to control VOCs in buildings by source substitution (see Section 8.2).

2.2.3 Health Effects of VOCs

Adverse health responses potentially caused by VOCs in nonindustrial indoor environments fall into three categories, namely, (*a*) irritant effects including the perception of unpleasant odors and mucous membrane irritation, (*b*) systemic effects such as fatigue and difficulty concentrating, and (*c*) toxic effects such as carcinogenicity (41).

Many of the VOCs emitted from new furnishings and building materials are mucous membrane irritants. Molhave et al. (45) exposed 62 people with a history of difficulty with poor air quality in a chamber setting to a mixture of 22 irritant VOCs (total concentration 5 to 25 mg/m^3). Subjective tests showed that the perception of poor air quality and odor intensity increased with increase in the total concentration of VOCs. Similarly the perception of mucous membrane irritation was elevated at both 5 and 25 mg/m^3 concentrations. It has been hypothesized that IAQ complaints of eye, nose, and throat irritation so common in SBS may be due to exposure to low concentrations of VOC mixtures that are often found in new or renovated buildings (46). Molhave's chamber exposure tests have been criticized because the 62 people studied may not be representative of the general population and the mixture of 22 VOCs used was not typical of that normally found in office buildings.

Adverse effects of VOCs may include systemic responses such as fatigue, short-term memory loss, and difficulty in concentrating. Indeed, Molhave and co-workers (45, 46) found that scores from digit span tests (a measure of short-term memory impairment) were decreased during exposure of test subjects in chambers to mixtures of VOCs. These systemic effects of VOCs are thought to be mediated by the central nervous system in a manner analogous to solvent neurotoxicity seen in occupational settings (47).

There is concern about the chronic adverse health effects due to VOC exposure because some VOCs commonly found in indoor air are human (benzene) or animal (chloroform, trichloroethylene, tetrachloroethylene, carbon tetrachloride, *p*-dichlorobenzene) carcinogens. Some VOCs are also genotoxic. Theoretical risk assessment studies indicate that chronic exposure risk due to VOCs in residential indoor air is greater than that associated with exposure to VOCs found in the outdoor air or to VOCs in drinking water (48, 49).

2.2.4 Sampling

Existing methods of sampling for VOCs in industrial workplaces are not readily adaptable to nonindustrial indoor studies. Methods developed for industrial workplaces are often bulky, noisy, and validated for about an order of magnitude below the applicable threshold limit value (TLV). Concentrations of VOCs found in nonindustrial indoor air are usually two or more orders of magnitude lower than industrial TLVs (37). More sensitive sampling and analytical methods are therefore required.

The use of sorbents such as Tenax, Ambersorb, and activated charcoal combined with analysis involving gas chromatography and mass spectrometry are extensively used for VOC sampling and analysis in IAQ studies (50). Tenax was the sorbent of choice in the most extensive indoor study of VOCs conducted to date (51, 52). Consecutive, 12-hr personal samples were collected in 600 homes in five cities. Personal exposures to a target list of 16 compounds were determined. Results showed that indoor concentrations of VOCs consistently exceeded outdoor levels by factors of 2 to 10. Because Tenax does not collect VOCs with boiling points below 70°C, multiple sorbents were used in some recent studies to collect with VVOCs (as well as SVOCs) (39, 50).

During the past few years passive sampling devices based on the principle of molecular diffusion of VOCs onto charcoal have been successfully used in IAQ studies (53). Sensitivity for a target list of VOCs including toluene, ethylbenzene, and *n*-tridecane are on the order of 0.1 $\mu g/m^3$. A recent study using a modification of this technique in residences showed concentrations of chlorinated hydrocarbon indoors over several weeks to be consistently higher than those outdoors (54).

2.2.5 Interpretation of VOC Data

Interpreting VOC sampling results for indoor nonindustrial environments is highly complex because of the absence of VOC dose–response data. Also the potential effects can include mucous membrane irritation, increased perception of odor intensity, systemic effects mediated by the central nervous system, and potential chronic effects such as carcinogenesis. Molhave et al. (45) have suggested that the threshold in the total VOC concentration above which irritation and odor annoyance occur ranges from 0.16 to 5 mg/m^3. This threshold, however, cannot be exactly defined because of varying human sensitivity to VOCs, presence of other air con-

Table 17.2. Volatile Organic Compounds in Two Office Buildings

Compound	VOC Concentrations In Outdoor Air ($\mu g/m^3$)	Indoor Air ($\mu g/m^3$)
Building 1		
Total VOCs	62	2,200
Tetrachloroethylene	6	700
Cyclohexanes	1	140
Dodecane	4	200
Chloroform	1	40
Building 2		
Total VOCs	542	448
Methylene chloride	400	40
Decane	6	200
Ethylbenzene	2	80

Sample collection on Tenax sorbent; analysis by gas chromatography/mass spectrometry.

taminants, and unpredictable irritative effects of different VOC mixtures on the human sensory system.

To determine whether or not a mixture of VOCs is typical or atypical, it is useful to determine both the concentration of all (total) VOCs as well as specific VOCs both indoors and outdoors. In Building 1 (Table 17.2) the total concentration of VOCs indoors is almost 30 times higher than that in outdoor air. The presence of elevated concentrations of four specific VOCs indoors (Table 17.2) suggests the presence of strong emission sources such as from cleaning chemicals and from unreacted solvents present in new construction materials.

The equivalent total concentrations of VOCs indoors and outdoors for Building 2 (Table 17.2) might at first suggest an absence of unusual VOC exposures. However, examination of the concentration data for three specific VOCs indicates that methylene chloride from the outdoor air is being entrained in the building. In addition, there is indication of indoor emission sources for aliphatic and aromatic hydrocarbons.

Variables such as building age and outdoor air ventilation must also be considered when interpreting VOC sampling results. In new office buildings, the indoor/outdoor concentration ratio of total VOCs may be 50 or 100 to 1 (55, 56). With adequate outdoor air ventilation, these ratios fall to less than 5 to 1 after 4 or 5 months of aging. In older buildings with continuous, regular and irregular emission sources, indoor/outdoor concentration ratios of total VOCs may vary from nearly 1 when maximum amounts of outdoor air are being used in HVAC systems to greater than 10 during winter and summer months when minimum amounts of outdoor air are being used (18).

2.3 Pesticides

Pesticides including insecticides, termiticides, and fungicides are often used in interior spaces to control a wide variety of organisms including wood-boring insects, moths, and fungi. Although pesticides are by definition poisons, their toxicity varies toward different types of organisms. Several million pounds of naphthalene and paradichlorobenzene are used annually in U.S. homes as moth repellents (57). Similar quantities of pentachlorophenol are used annually in U.S. homes to protect wood and paints from degradation by insects and fungi. Termiticides such as chlordane, heptachlor, aldrin, dieldrin, and chlorpyrifos have been used or are still used in the soil or under the foundations of buildings.

A large number of the general population is exposed to pesticides in their homes. Indoor concentrations of chlorpyrifos in the range of 0.2 to 2.0 $\mu g/m^3$ have been reported to occur for up to 30 days after its application (57). Chlordane at concentrations of 0.1 to 10 $\mu g/m^3$ may be present indoors for years after initial application (57). Many studies indicate that insecticides such as diazinon, heptachlor, lindane, and so on occur at elevated concentrations indoors versus those outdoors (58, 59). One recent study showed that the indoor to outdoor concentration ratio of 2-chlordane was as high as 60 in the living area and 1000 in the basement of one home (60).

Health effect data available for pesticides are based primarily on oral or dermal exposures from animal studies. Little information is available on adverse health effects of inhalation exposures. It is significant, however, that most pesticide-related injuries (including those from organophosphate insecticides) occur in the home (57). Of the accidents that were not related to ingestion, both inhalation and dermal exposures were found to be equally important (57).

A recent study showed organophosphate pesticides being involved in the development of BRI (organophosphate poisoning) in an office environment (61). Another study of over 600 employees who worked at veterinary clinics, pet shops, and facilities (in California) where flea control products were used showed that there was a significant elevation in adverse health problems ranging from headache and fatigue to respiratory problems, suggestive of systemic pesticide poisoning.

In 1985 the U.S. Environmental Protection Agency (EPA) initiated studies on pesticide exposures in the general population. The Non-Occupational Pesticide Exposure Study (NOPES) has facilitated the development of improved methodologies for sampling for pesticides in indoor environments (58, 62). Using techniques described in these studies, it is now possible to sample and analyze for over 50 specific organochlorine, organophosphate, organonitrogen, and pyrethroid pesticides using polyurethane foam sorbent with sensitivities as low as 0.01 $\mu g/m^3$. For example, among 50 residences monitored in the Jacksonville, Florida area, 46 contained detectable chlorpyrifos (mean concentration 0.47 $\mu g/m^3$). Chlorpyrifos was detectable in outdoor air around only 9 of the 50 residences, with a mean concentration of 0.059 μg/m³ (58). Concentrations of chlorpyrifos as high as 37 μg/m³ have been found in other residences (37).

Because of the potential chronic health effects from exposure to termiticides, the National Research Council has recommended that airborne concentrations in indoor air be limited as follows: chlorpyrifos (10 μg/m³), chlordane (5 μg/m³), heptachlor (2 μg/m³), and dieldrin and aldrin (1 μg/m³) (63). It should be noted that these exposure levels are more than an order of magnitude lower than concentrations considered acceptable in industrial workplaces.

2.4 Combustion Products

A variety of combustion sources contribute to indoor air pollution. In residential environments, these include gas ranges, unventilated kerosene or gas heaters, and wood- or coal-burning stoves. Combustion products from vehicles and other external sources may be entrained into the outdoor air inlets of HVAC systems of commercial buildings. Carbon dioxide (see also Section 6.3) and water vapor are the major products of combustion. However, other combustion by-products such as carbon monoxide (CO), nitrogen dioxide (NO_2), sulfur dioxide (SO_2), and respirable particulates are significant causes of indoor air pollution. The type and amount of combustion by-products in indoor air of commercial buildings depend upon the type of fuel consumed (for example, diesel oil used by trucks at a building loading dock contains significant amounts of sulfur, and SO_2 is an important combustion by-product) and the location of outdoor air inlets and emission sources. Combustion processes that are oxygen-starved and characterized by yellow-colored flames are characterized by elevated CO emissions (64). Combustion under oxygen-rich conditions results in higher flame temperatures thus emitting greater amounts of oxides of nitrogen. Considerable literature is available describing characteristics of combustion devices found in indoor environments (primarily residences) and factors affecting the emission rates of various combustion by-products (65, 66).

Indoor air pollutants such as CO and NO_2 originate from multiple sources, and concentrations indoors depend on parameters such as source emission rates, the volume of air indoors, outdoor air ventilation rates, and HVAC system characteristics. Pollutants such as CO and NO_2 are emitted intermittently and are usually concentrated only in certain areas of the building. Outdoor conditions, such as ambient concentrations of NO_2, have a strong influence on indoor pollutant concentrations (67).

Considerable literature exists on exposure to combustion by-products in residential environments. For example, use of an unvented gas stove adds about 25 ppb of NO_2 to the background concentration in indoor air (10). Peak concentrations of NO_2 in kitchens during cooking with a gas range may be elevated by 200 to 400 ppb (68). In a study carried out at the Newark International Airport in 1985, 24-hr time weighted average (TWA) indoor concentrations of NO_2 indoors in gate areas ranged from 19 to 116 ppb (69). Average outdoor concentrations were higher and varied from 41 to 233 ppb. During cooking with a gas range, CO concentrations in residences typically range from 2 to 6 ppm. CO levels in vehicles during urban commuting may become elevated by as much as five times that present in ambient air (10).

It is well known that CO combines with hemoglobin to form carboxyhemoglobin, thereby resulting in a decline in the oxygen-carrying capacity of the blood. Carboxyhemoglobin levels higher than 4 to 5 percent are known to exacerbate symptoms of individuals with preexisting cardiovascular disease (12). Limiting average CO exposures to 9 ppm (maximum) for 8 hr or to 35 ppm for 1 hr, as specified in the National Ambient Air Quality Standards (NAAQS) (70), is intended to provide a margin of safety with regard to carboxyhemoglobin buildup in individuals with cardiovascular disease. The effects of chronic exposure to lower levels of CO that may occur in buildings are currently controversial (10).

Most studies on the health effects of NO_2 at concentrations commonly found in nonindustrial indoor environments have emphasized respiratory symptoms and illnesses. Some (but not all) epidemiologic studies have shown a higher prevalence of respiratory symptoms in children exposed to NO_2 at levels below 1 ppm. Airways reactivity may be increased in some asthmatics by exposure levels that occur in homes (10). Experimental animal studies indicate an elevated incidence of respiratory infections from exposures to NO_2 concentrations above 1 ppm.

The monitoring of combustion by-products such as CO, NO_2, and SO_2 are reviewed elsewhere (62); only a few recent advances are described here. A passive sampling device for NO_2 using triethanolamine-impregnated filter with subsequent analysis by ion chromatography has recently become available (71). Concentrations as low as 12 ppb are detectable with sampling time as short as 1 hr. Good agreement between this passive sampling device and a reference chemiluminescence detector in the NO_2 concentration range of 10 to 60 ppb has been achieved (71). Other studies have shown that continuous chemiluminescence methods for measuring NO_2 concentrations are applicable to IAQ studies (72).

Establishing combustion by-products as the cause of air quality complaints in a building is a difficult task. In one large building where a portion of an office floor had been vacated because of suspected exposure to combustion products, round-the-clock sampling for NO_2 for 6 days was necessary before it could be demonstrated that combustion products from a loading dock were being entrained in the HVAC system outdoor air inlet serving the affected office (18). The concentrations of NO_2 in the outdoor air inlet, in the vacated office, and in the outdoor air on the roof far removed from emission sources were 2.0, 0.7, and 0.08 ppm, respectively.

Table 1 of ASHRAE Standard 62-1989 (4) lists maximum concentrations of certain combustion by-products that are considered acceptable for introduction with outdoor air into HVAC systems. For example, NO_2 concentrations in outdoor air used for ventilation should not exceed 55 ppb (on a yearly basis). An addendum to ASHRAE Standard 62-1989 (73) recommends that respirable particulate levels not exceed 50 $\mu g/m^3$. In spite of considerable research, there is an absence of consensus on what constitutes "safe" exposure levels for various combustion products (10). For example, although it appears that adverse pulmonary function response to NO_2 occurs at exposures above 300 ppb (74), there is considerable diversity of opinion on the range of the lowest-observed-effect level (80 ppb, see

Reference 75; 100 ppb, see Reference 76). Obtaining "safe" indoor target levels of combustion by-products such as NO_2 is difficult owing to the poor ambient air quality in some urban areas where NAAQS (70) are frequently exceeded (18).

2.5 Environmental Tobacco Smoke

Environmental tobacco smoke (ETS) is composed of a complex mixture of chemicals including combustion gases/and respirable particles. Because tobacco leaf does not burn completely, ETS contains more than 4700 chemical compounds (77) including nicotine, tars (containing a variety of polynuclear aromatic hydrocarbons), vinyl chloride, formaldehyde, benzene, styrene, ammonia, NO_2, SO_2, CO, hydrogen cyanide, and arsenic. Forty-three carcinogenic compounds are found in ETS.

ETS is derived from mainstream smoke that is drawn through the cigarette by the smoker and from sidestream smoke that is emitted from the cigarette itself directly into room air between puffs. Individual chemical components of ETS may be found in both the gaseous and the particulate phases. During smoking, particulate ETS may adsorb on surfaces in the occupied space and ventilation system. Gaseous components can then be reemitted into the indoor air from adsorbed particulate (78). Smoking characteristically results in a significant rise in the concentration of respirable particulate with a mass median diameter of about 0.2 to 0.4 μm (79). Although many chemical components of ETS can also be attributed to other sources, nicotine is considered a specific marker for this indoor pollutant.

Adverse effects of ETS can be divided into annoyance and discomfort complaints and chronic disease including increased frequency of respiratory illness, loss of lung function, and increased risk of lung cancer. Eye, nose, and throat irritation are frequent complaints in people exposed to ETS. Odor annoyance due to ETS is perceived at concentrations much lower than those that cause irritation (80); consequently much greater amounts of outdoor air are required to control odor annoyance as opposed to the acute irritation effects of ETS (81, 82). Both odor annoyance and irritational effects are thought to be caused by the gaseous phase of ETS. In recent IAQ studies in a series of Danish buildings, authors estimated that about 25 percent of the perception of stale, unacceptable air is due to smoking (83; see Section 3.4).

The major chronic health effect from ETS is lung cancer and is reviewed elsewhere (10). In 1985 three major study panels were convened, namely, by the U.S. Public Health Service, the National Research Council, and the Federal Interagency Task Force on Environmental Cancer, Heart and Lung Disease, to consider the risk associated with breathing ETS in indoor air (so called passive smoking). All groups arrived at the same conclusion, namely, that passive smoking significantly increased the risk of lung cancer in adults (77, 84).

The presence of ETS in indoor air is associated with increased concentrations of respirable particulate, nicotine, and a variety of other contaminants. One smoker consuming a pack of cigarettes daily in a residence contributes approximately 20 $\mu g/m^3$ of respirable particulate to the (24-hr) particle concentration (10, 85). Concentrations of respirable particulates less than 2.5 μm in aerodynamic diameter in indoor environments where smoking is permitted may rise up to 500 μg/m^3 (79).

Nicotine is the most specific marker for ETS. A sampler using a filter treated with sodium bisulfate followed by analysis by gas chromatography with a nitrogen–phosphorus sensitive detector has been successfully utilized in quantifying ETS (86). This method can detect nicotine concentrations below 0.5 μg/m^3.

Miesner et al. (79) measured nicotine concentrations using this passive sampler method in a number of office environments with varying smoking policies and found an increase in nicotine levels as respirable particulate concentrations increased. Nicotine and respirable particulate levels in a designated smoking room in one office were 26.5 μg/m^3 and 520.8 μg/m^3, respectively (79). In a nonsmoking office located directly above a floor where smoking was permitted, the concentration of nicotine was 2.0 μg/m^3.

A recent ASTM area/personal sampling method for nicotine, using XAD-4 sorbent with a limit of quantitation of 0.2 μg/m^3 (8-hr sample duration flow rate 1.7 l/min) has been developed and should prove useful in future studies on ETS (87).

In spite of considerable research, the interpretation of sampling data for ETS is still controversial, and for in-depth discussion the reader is referred to a recent publication of the National Research Council (88). An example provided by Repace (89) illustrates the interpretation methodology applied in one study of comparative ETS exposure in smoking and nonsmoking sections of the U.S. State Department Cafeteria in Washington, DC. Respirable particulate concentrations in the smoking section were about 95 μg/m^3 whereas those in an equivalent sized, nonsmoking section were 40 μg/m^3. Background respirable particulate in the outdoor air in Washington, DC at the same time was 10 μg/m^3. Using outdoor air as a reference point, it can be concluded that although separation of smokers from nonsmokers did not entirely protect the latter, it did significantly lessen their exposure to some tobacco combustion products.

2.6 Bioeffluents

Body odors (bioeffluents) are composed of a mixture of gases at very low concentrations emitted from breath, the skin (perspiration), and the digestive tract. Emissions from individuals vary greatly and are influenced by diet, activity level, and personal hygiene. Ventilation standards for large commercial buildings have historically been based on providing adequate amounts of outdoor air to dilute bioeffluents so that at least 80 percent of a panel of observers visiting a building find the indoor air "acceptable" (90, 91). The amount of outdoor air required to dilute bioeffluents to an acceptable level is directly proportional to the activity and density of the occupants in the space and is inversely proportional to the room volume available for each person (see Figure 7.2 in Reference 13).

Bioeffluents are perceived especially when people enter an occupied space for the first time. People who have been in a room for some time may fail to notice an odor because their olfactory sense becomes fatigued or becomes adapted to the odor (92). Thus a group of persons in an occupied space may not notice unpleasant body odors whereas visitors are acutely affected. For this reason, visitors and not building occupants should be used in the subjective evaluation of ventilation air (see Appendix C in Reference 4). Although not widely used in the definition of SBS in North America, persistent odors (resulting at least in part from bioeffluents) are considered to be an important cause of SBS symptoms (5, 93).

Because the gases that compose body odors are diverse and present in very low concentrations, objective measurement of exposure to these indoor air pollutants is difficult. In one experimental study, bioeffluent concentrations were measured in a large auditorium before and during human occupancy (94). The presence of a number of organic (acetone, butyric acid, ethanol, and methanol) and inorganic (carbon dioxide, ammonia, and hydrogen sulfide) bioeffluents were associated with human occupancy. For example, the methanol concentration increased from about 1 ppb before occupancy to 20 to 45 ppb during occupancy.

Techniques involving subjective evaluations of bioeffluents have been described (92). In a typical evaluation, a group of occupants is commonly seated in an auditorium for a certain period of time under a defined ventilation condition (for example, outdoor air ventilation at a certain rate per person). Panels of visitors are brought into the auditorium and asked to rate the acceptability of the air and to make a subjective judgment on the intensity (e.g., slight odor, strong odor) of any perceived odor. Visitors are asked to make their judgments immediately upon entering the auditorium. After making a subjective evaluation, visitors are taken to a well ventilated room or zone to readapt their olfactory senses to outdoor air.

In a study involving 106 occupants and panels consisting of a total of 79 visitors, it was shown that there was a close correlation between the percent of dissatisfied visitors and their subjective judgments of odor intensity (92). In addition, approximately 80 percent of the visitors found the air quality to be acceptable (20 percent were dissatisfied) when the subjective judgment of odor intensity was intermediate between "slight odor" and "moderate odor."

When the outdoor air ventilation rate in the auditorium was approximately 8 l/sec per person (16 cfm) 80 percent of visitors found the air quality acceptable (20 percent dissatisfied) (92). With a ventilation rate of only 2.5 l/sec per person, the number of occupants dissatisfied exceeded 30 percent. However, the occupants who were present all of the time in the auditorium failed to perceive an effect of the outdoor air ventilation on the acceptability of auditorium air.

2.7 Radon

Radon-222 (radon) is a noble gas and a decay product of radium-226, which in turn is a product of the decay of the uranium-238 series. Uranium-238 and radium-226 are found in very small amounts in most rocks and soil. Because the half-life of radium-226 is about 1600 years, radon gas is released at an almost constant rate into the soil atmosphere (13). Because it is not chemically bound or attached to other materials, radon migrates through soil porosities and is released into the ambient air or it diffuses through cracks and pores in building foundations and enters the indoor environment. Radon may also enter the indoor environment as a solution in well water or by diffusion from construction materials containing radium or uranium. Infiltration of soil gas into buildings is by far the largest contributor to indoor radon concentrations. The amount of radon that enters indoor air is influenced by a number of factors including the amount of radium in the soil or rock around the building foundation, the permeability of the soil around the foundation, the air pressure inside the building (for example, air pressure indoors may be lowered by the operation of exhaust fans), and the extent of cracks or openings in the building foundation. In general, radon is more of a problem in buildings with a high foundation surface area and low internal volume (residences) as opposed to those structures with a relatively low foundation surface area and large internal volume (large commercial office buildings).

Radon has a half-life of 3.8 days and its decay progeny (polonium-218, lead-214, bismuth-214, and polonium-214) have half-lives of less than 30 min. Unlike radon, which is almost chemically inert, the decay products tend to adhere to dust particles and surfaces including those in the lung. Thus, although an inhaled radon atom is likely to be exhaled before it decays, inhaled radon progeny will likely decay (emit alpha or beta particles or gamma rays) before they can be removed by normal lung clearance mechanisms. Polonium-218 and polonium-214 both emit alpha particles, and consequently are the radon progeny of major concern in terms of potential adverse health effects (62).

There may be a significant distinction in health risk from radon progeny that attach to dust particles (attached decay products) and those progeny that do not attach (unattached decay products). The latter agglomerate to form very small particles (ranging from 0.002 to 0.02 μm), which have a higher probability of being deposited deep in the lung. Radon decay products attached to dust particles are more likely deposited in the moist epithelial lining of the bronchi. Some mathematical models indicate that a higher lung dose of alpha emissions is associated with deposition of unattached radon decay products (62).

Amounts of radioactive materials are specified in terms of their activity or the rate at which atoms decay radioactivity. One picocurie (pCi), the term traditionally used in the United States, corresponds to 0.037 decays per second; the international unit for activity, the Becquerel (Bq), corresponds to one decay per second (95).

Radon decay product concentration has traditionally been measured in terms of potential alpha energy concentrations expressed in the working level (WL) unit. A radon level of 1.0 pCi/l in dynamic equilibrium with its decay products is equivalent to a decay product concentration of about 0.005 WL (62, 95). Extensive studies on radon exposure in U.S. residences indicate that the yearly average exposure is about 1.5 pCi/l with approximately 1 to 3 percent of homes exceeding 8 pCi/l (95). In commercial buildings and health care facilities, radon levels are generally less than 2 pCi/l (96).

Inhalation of radon decay products is a significant health risk (lung cancer). Estimates of health risk due to radon exposure are based on extrapolation from epidemiologic studies on underground miners whose exposures were higher than those characteristic of indoor environments. Although the extrapolation calculations are somewhat uncertain, it is estimated that aggregate health effects from exposure to radon decay products range from 5000 to 20,000 lung cancer deaths per year in the United States (62).

Short-term and long-term measurement methods are available for the quantification of indoor radon (95). Short-term screening measurements are generally carried out over a period of 1 to 7 days using canisters containing activated charcoal. Gamma radiation from the decay products collected in the activated charcoal is measured to quantify radon levels. Long-term radon measurements are commonly carried out over periods of up to a year using alpha track detectors. Ionization tracks in the plastic detector film produced by the emission of alpha particles from radon and its decay products are enlarged by etching and then visually counted. Other detection methods and extensive discussions on measurement protocols are reviewed elsewhere (62, 95, 97).

Measurements of indoor radon concentration can vary on a hourly, daily, and seasonal basis. Thus measurements made under conditions of maximum outdoor air ventilation (open windows) can underestimate the average annual radon level. Indoor measurements made when a building is subject to minimum outdoor air ventilation (for example, during the winter) tend to overestimate average concentrations. The most reliable estimates of radon concentration are those made over periods of 6 months to 1 year.

The EPA currently recommends a two-phase strategy for measuring indoor radon concentrations in residences. A short-term measurement of radon is carried out under closed-house conditions in a living area where elevated concentrations might be expected (ground level). If measured concentrations indoors are between 4 and 20 pCi/l, follow-up measurements made in the living space over a 12-month period are recommended. If short-term measurements exceed 20 pCi/l, intensive follow-up tests are recommended because these concentrations can significantly increase health risk (98). Follow-up measurement is probably unnecessary when screening measurements show that radon levels are less than 4 pCi/l. Interpretation of indoor radon measurements should take into consideration that naturally occurring radon concentrations in outdoor air range from 0.1 to 1 pCi/l.

2.8 Man-Made Fibrous Dusts

Man-made fibrous materials are used in many aspects of building construction including accoustic insulation systems, thermal insulation, ceiling tiles, and office partition panels. In HVAC systems, man-made fibrous insulation is often used as an accoustic/thermal liner on the internal surfaces of AHU plenums as well as in the air supply duct system. Several studies of fibrous glass-lined air supply systems have indicated that under normal maintenance and operational conditions, there is no or little evidence of erosion of fibrous materials from internal HVAC system surfaces (99, 100). Thus in a study in one building, the average concentration of airborne fibrous glass in filtered outdoor air entering the HVAC system, in the air supply ducts themselves, and in the occupied space was 0.05, 0.03, and 0.17 fibers/liter, respectively (see Table 1 in Reference 99). However, anecdotal comments by other investigators suggest that erosion of fibers from insulation, especially lining the internal surfaces of AHU plenums, can occur (17, 31, 101). Several residential studies do present qualitative data showing that man-made fibrous materials from air conditioning ducts or from filters can be sources of significant amounts of airborne fibers (102, 103).

Studies conducted in Danish schools do not support the hypothesis that airborne man-made fibers in the indoor environment cause SBS or BRI (104). However, these studies do show a positive correlation between skin irritation and the presence of settled fibers on horizontal surfaces.

A case study in a residence (105) showed that airborne fibrous glass caused intense itching without rash. Itching was characteristically relieved by showering. The source of airborne fibrous glass in this case study was a filter that had been improperly fitted into a forced air heating system.

Instances of irritation caused by exposure to man-made fibers in indoor environments are likely due to improper maintenance of insulation surfaces (for example, deterioration caused by facility personnel walking on insulation materials in AHU plenums) or by improper installation (for example, exposure of cut fiber glass surfaces to the airstream). Practices for proper installation of fiber glass materials in HVAC systems are described by the Sheet Metal and Air-conditioning Contractors' National Association (106).

3 NONSPECIFIC COMPLAINTS IN BUILDINGS

SBS is characterized by a number of nonspecific symptoms including mucous membrane irritation, eye irritation, headache, odor annoyance, sinus congestion, and fatigue. Ideally, in order to understand the etiology of SBS the prevalence of symptoms in occupants should be studied, preferably in problem and nonproblem buildings. Objectives of such case-control studies should include a determination if the percentage of dissatisfied occupants exceeds some unacceptable level, and if dissatisfaction can be related to building, air contaminant, or work-practice variables. A number of different approaches have been used in attempts to elucidate the etiology of the nonspecific symptoms characteristic of SBS.

3.1 SBS Studies in the United States

By the end of the 1970s, numerous annoyance and discomfort complaints were being reported to local and federal agencies by occupants in office environments. Common health complaints investigated by NIOSH included eye irritation, dry throat, headache, fatigue, and sinus congestion (6, 107). Complaints that were temporally correlated with occupancy in the building were termed SBS. The etiology of complaints in 52 percent of the problem buildings studied by NIOSH by the mid-1980s was ascribed to "inadequate ventilation." Elevated CO_2 levels were often used as a surrogate in these studies for inadequate ventilation. NIOSH studies also indicated that air quality problems in other buildings were associated with indoor pollutant sources (16 percent of buildings), entrainment of outdoor contaminants (10 percent), unknown causes (12 percent), microbial contaminants (5 percent), and building fabric contamination (4 percent) (107). It is characteristic of NIOSH studies that even in the most superficial studies emphasis was almost always placed on assigning a cause to building problems and then recommending remedial actions. Studies in control buildings were rare and almost no attempt was made epidemiologically in NIOSH studies to relate SBS symptoms such as eye irritation to specific building variables such as the type of HVAC system, the nature of construction and finishing materials, and indoor concentrations of ETS, VOCs, and CO_2.

It has been pointed out that the diagnosis of SBS in the United States in the 1980s was primarily made on the basis of "exclusion" (101). Thus alternative sources of annoyance and irritation were eliminated on the basis of review of occupant complaints, a walk-through evaluation of the HVAC system, and an evaluation of processes that occurred in the occupied space. The inability of industrial hygiene measurement techniques to identify specific chemical irritants led to the interpretation by many that inadequate ventilation was the cause of nonspecific symptoms characteristic of SBS. Thus SBS investigations in the United States have mostly been of a case study–problem solving nature. Little effort has been expended in quantifying the causes of specific SBS symptoms among occupants in representative nonindustrial buildings.

3.2 British Epidemiologic Studies

Several systematic studies on complaints related to building characteristics have been carried out in British office buildings (108, 109). In the largest study, complainant symptoms were recorded using a common protocol in 47 different groups of occupants in 42 buildings. Most of the buildings selected for study were chosen without any prior knowledge of whether or not occupant complaints existed. The questionnaire used in these studies elicited information on whether the following 10 symptoms occurred during the past year and whether symptoms disappeared during periods when occupants were away from the building: dryness of eyes, itching of eyes, stuffy nose, runny nose, dry throat, lethargy, headache, fever, breathing difficulty, and chest tightness. Repeated use of the questionnaire in 6 buildings over a 1 to 2 year period showed that the response of randomly selected populations in different buildings was repeatable. Certain buildings could be shown repeatedly to have a higher prevalence rate of "building sickness" than others (108).

Ventilation systems serving the 47 different occupant groups studied by Burge et al. (108) were categorized as *natural* (open windows, no forced air), *mechanical* (forced air system without cooling or humidification), *local induction units* (Sec. 6), *central induction/fan coil units*, and *variable or constant air volume system*. The latter three ventilation categories were characterized by cooling of air, and in some cases, also by humidification. Questionnaire results showed that the lowest prevalence of work-related symptoms was found in the naturally or mechanically ventilated categories. Although there were considerable variations between buildings of each ventilation type, the highest prevalence rates were found in ventilation systems with induction or fan coil units. A somewhat intermediate symptom prevalence occurred in variable/constant air systems. For example, prevalence of itchy eyes was 32 to 33 percent among occupants in zones served by induction unit/fan coil units, 29 percent in variable/constant volume systems, and only 20 to 22 percent in naturally or mechanically ventilated systems (see Table 5 in Reference 108). The overall most common work-related symptoms reported by occupants in the 42 buildings were lethargy (57 percent), stuffy nose (52 percent), dry throat (46 percent), headache (43 percent), and itchy eyes (28 percent).

These British studies show that naturally ventilated and nonair-conditioned, mechanically ventilated buildings are the healthiest workplaces (108). Because the other three ventilation types examined are characterized by air-conditioning and, in some cases, also by humidification, it has been hypothesized that microbiological air contaminants may be responsible for some of the higher prevalence rates of work-related symptoms found in these studies (101, 108).

3.3 Danish Epidemiologic Studies

Systematic questionnaire studies of work-related symptoms such as mucosal irritation, headache, and fatigue were carried out in 14 town halls and 14 affiliated buildings in the Copenhagen area (110). The buildings chosen in this study were not previously categorized as being sick or healthy (111). The prevalence of work-related symptoms such as eye, nose, and throat irritation as well as headache and fatigue and a number of other environmental complaints were determined for over 3000 office workers present in the buildings (110, 112). Environmental measurements such as the concentration of total VOCs and CO_2, the microbial content in floor dusts, and thermal/air moisture parameters were made in one representative office in each building. Two unique parameters, the "fleece" and "shelf" factors, were measured in each study office. The fleece factor is a measure of the surface area of all porous room furnishings (e.g., carpets, drapes, upholstery) divided by the room volume. The shelf factor is the length of open shelves in the study room divided by total volume of the room studied (110).

Work-related symptoms reported most frequently in the Danish Town Hall Studies included eye, nose, and throat irritation, fatigue, and headache (110). A great variation in the prevalence of work-related symptoms was found between buildings (see Table 4 in Reference 110). Symptoms were somewhat higher among occupants who worked frequently with photocopiers, carbonless paper, and visual display terminals. Higher, though not statistically significant, prevalence rates of work-related symptoms were found in mechanically as compared to naturally ventilated buildings (see Figure 2 in Reference 110). Workplace environmental factors such as total amount of floor dusts and the fleece and shelf factors of studied offices could be related to symptom prevalence (111). Certain indoor environmental factors such as CO_2 concentrations, however, could not be related to the prevalence of work-related symptoms (101).

The Danish Town Hall Study is an important milestone in IAQ research because it indicates that there is considerable variation in the prevalence rate of SBS symptoms between buildings. In addition, novel factors such as the amount of adsorptive and absorptive (fleece and shelf factors) areas for VOCs, combustion products, and microorganisms have been postulated to explain SBS causation (101, 113).

3.4 The Olf and Decipol Concepts

A prominent complaint in buildings is the perception of odors. Indeed, most ventilation rates for large commercial buildings are based primarily on the provision of adequate amounts of outdoor air in order to dilute human bioeffluents (see Section 2.6). An approach toward understanding SBS (83, 114) uses bioeffluents from humans as a standard for quantifying the perception of air quality in a building.

The bioeffluents emitted from a standard person (age 18 to 30 years; skin area 1.8 m^2, daily change of underwear; 0.7 baths per day; sedentary, metabolic rate of 58 W/m^2) is defined as one standard *olf*. The unacceptability of the indoor air caused by other pollutant sources such as building furnishings or the HVAC system itself are defined in terms of the number of standard olfs needed to cause the same degree of dissatisfaction. Thus the moist dirt in a HVAC system might be the source of enough odor to cause the same equivalent dissatisfaction as 14 standard persons or 14 olfs. The measurement of olf values for different sources requires that a panel of judges determine the acceptability of the air in a space ventilated by a given flow rate of unpolluted or outdoor air.

The *decipol* is a recent term introduced by Fanger (114) to quantify the perception (by the nose) of air pollutant concentrations. One decipol is equivalent to the perceived air pollution that would cause the same dissatisfaction as the bioeffluents from a standard person (one olf) diluted by 10 l/sec of unpolluted or outdoor air. In the panel studies carried out by Fanger and colleagues (83), 1 decipol is the perceived air pollution that results in causing about 15 percent of judges to find the air unacceptable when entering the occupied space. Three decipols is the perceived air pollution that causes slightly more than 30 percent of judges to be dissatisfied with the air quality.

The olf and decipol concepts were first used to quantify air pollution sources in 15 office buildings and five assembly halls in the Copenhagen area (83). Buildings were randomly chosen without knowledge of any IAQ problems, and air in all buildings was conditioned by HVAC systems. Panels of judges consisting of both smokers and nonsmokers visited each building and immediately assessed air quality during periods when the building was unoccupied and unventilated, unoccupied and ventilated, and occupied and ventilated. At the same time, environmental measurements including the concentrations of CO_2, particulates, and total VOCs were made in the occupied spaces. Fanger et al. (83) found that although the ventilation rate in offices was on average 25 l/sec per occupant (50 cfm/occupant), the percentage of dissatisfied judges was in excess of 30 percent. For each olf associated with human occupancy (bioeffluents), 6 or 7 olfs were estimated to be emitted from other sources in the buildings. The perception of poor air quality in these studies was assigned to the following pollutant sources: the HVAC system

(42 percent), ETS (25 percent), furnishings and construction materials (20 percent), and occupants themselves (13 percent). Concentrations of CO_2, particulates, and VOCs were shown to be poor predictors of perceived air quality.

A number of important conclusions can be made from Fanger's studies (83, 114). Provision of excessive outdoor ventilation air (for example, 25 l/sec per occupant) may not be adequate to reduce the percent of dissatisfied visitors to acceptable levels. The HVAC system itself can be the major source of air pollutants that cause the perception of poor air quality. Low-olf cleaning methods (better preventive maintenance) and low-olf construction materials must be utilized to control indoor air pollution sources (83).

The decipol concept (115) allows investigators to look at the problem of SBS from a different perspective. "Healthy" buildings are likely to be those with reduced pollutant sources and perceived air pollution of less than one decipol. Buildings where many are dissatisfied are those with decipol values considerably greater than one, and these structures are commonly identified as "sick buildings."

4 BUILDING-RELATED ILLNESS

Health effects caused by exposure to a variety of indoor pollutants including ETS, combustion products, VOCs, microorganisms, and radon have been reviewed (10, 11). BRI refers to those health problems that are related to exposure in the indoor environment and are recognizable by medical examination and/or laboratory tests. Because of the long latency period and relatively small populations involved, some BRI such as cancer that may be caused by exposure indoors to carcinogens such as asbestos, ETS, radon, and some VOCs are estimated through risk analysis (101; Section 7.4). Thus it is not practical to study the carcinogenic risk of VOC exposure in a building-by-building manner.

Other BRIs attributable to specific agents, however, may be studied on a building-by-building basis. Thus individual cases of organophosphate poisoning, carbon monoxide poisoning, or rash due to exposure to a fibrous dust can be successfully investigated in individual buildings even when only one or a few occupants are affected (101).

In the sections that follow the etiology of BRI caused by allergic and infective mechanisms is reviewed.

4.1 Hypersensitivity Pneumonitis

Hypersensitivity pneumonitis is an immunologic lung disease that occurs in some individuals after inhalation of organic dusts. Hypersensitivity pneumonitis is suspected when symptoms such as fever, cough, and chest tightness occur several hours after exposure. The diagnosis of the disease is made on the basis of a physician's review of patient symptomology plus a battery of tests (101) including restrictive pulmonary function measurements, precipitins (IgG antibodies) to extracts of microbial agents collected in the building, and in rare instances, inhalation exposure of the patient to suspect antigens.

A wide variety of specific microbial agents have been implicated in outbreaks of hypersensitivity pneumonitis. *Cladosporium* (116), thermophilic actinomycetes (117), and *Bacillus subtilis* (118) have each been implicated as etiologic agents in indoor residential outbreaks of this BRI. In small buildings or portions of buildings, *Penicillium* species have been shown to be causative agents of disease (119, 120).

In large buildings with complex HVAC systems and with many potential sites of microbial amplification, outbreaks of hypersensitivity pneumonitis often cannot be ascribed to a single agent (16, 20, 32). For example, an outbreak of acute hypersensitivity pneumonitis associated with cafeteria flooding in a large office building could not be etiologically related to potential agents such as *Thermoactinomyces* and *Acanthamoeba polyphaga* even though they were abundant in environmental samples (20). Changing patterns of microbial amplification and dissemination as well as loss of viability and antigenicity of disseminated organic dusts may account for the difficulty in establishing specific etiology in these instances of BRI.

4.2 Humidifier Fever

Episodes of fever, muscle aches, and malaise with only minor pulmonary function changes have been associated with inhalation of aerosols from humidifiers contaminated with gram-negative bacteria and protozoa (30, 121). Symptoms of this flu-like illness generally subside within a day after exposure without any long-term adverse effects.

An example of this type of BRI occurred in an office where three of the seven occupants reported attacks of fever and chills that started late during the workday and lasted well into the night (122). Illness was associated with the use of a humidifier on occasions when the air was considered too dry for comfort. Analysis of water from the humidifier reservoir showed that *Flavobacterium* was present at a concentration of about 8×10^4/ml. The concentration of airborne *Flavobacterium* increased from nondetectable when the humidifier was not running to about 3000/m^3 within 15 min of operation. Endotoxin from this gram-negative bacterium was thought to be the etiologic agent of this outbreak of humidifier fever (122). Subsequent work by Rylander and Haglind (123) showed that endotoxin from humidifiers at a concentration of about 130 to 390 ng/m^3 from a *Pseudomonas* species caused another outbreak of this BRI.

4.3 Asthma

Asthma is characterized by symptoms of wheezing, cough, chest tightness and shortness of breath, and reversible airflow limitation during pulmonary function tests (101). The development of occupational asthma may occur as the result of exposure to a sensitizing agent at work. Sensitizing agents include chemical agents such as those used to cure epoxy resins as well as pollen, dander, and urinary proteins from some animals and from microbial agents such as *Rhodotorula* (124).

The concept that asthma may be a BRI is a recent one. Finnegan and Pickering (125) described a case of occupational asthma in an office building where the conditioned air was humidified by a large water spray system that was cleaned and disinfected with a biocide on a monthly basis. Peak flow (lung function) measurements made at 2-hr intervals over 2 weeks showed a decline that correlated with periods of occupancy in the building. When the humidifier was turned off over a subsequent 10-day period, only minimal reductions in peak flow measurements occurred (see Figure 1 in Reference 126). Airborne antigens from the humidifier were suspected to be the etiologic agent of this BRI. Biocides and possibly endotoxins in humidifiers have been suggested as causative agents of other cases of building-associated asthma (126).

4.4 Legionellosis

Legionnaires' disease and Pontiac fever are both caused by *Legionella* species and both may be building-related (see Section 2.1.2). Sources of *Legionella* in indoor air include entrainment of aerosols from cooling towers and evaporative condensers into HVAC system outdoor air intakes and the generation of aerosols containing *Legionella* from indoor sources such as shower heads, humidifiers, whirlpools, and saunas (127–129). The hotel-associated outbreak of Legionnaires' disease described by Band et al. (130) illustrates the unexpected pathways that may lead to *Legionella* exposure in the indoor air. A cooling tower on the roof of the hotel was shown to be the reservoir where the bacterium was selectively amplifying. Aerosol from the tower was disseminated into a meeting room via a nearby open chimney that connected to a room fireplace. The operation at the same time of a large exhaust fan in the meeting room kept this indoor space under negative pressure relative to the ambient air. The proximity of an open building vent to a poorly maintained cooling tower and the operation of an exhaust fan collectively were associated with the high attack rate of Legionnaires' disease in the meeting room.

The risk of legionellosis outbreaks can be minimized by proper attention to the design, operation, and maintenance of building heat rejection and hot water service systems (131–133).

4.5 Acute Febrile Respiratory Illness

Respiratory tract infections may be caused by a number of pathogens including RNA (for example, influenza, coronavirus, and rhinovirus) and DNA (adenovirus) viruses. Infection by influenza virus and adenovirus readily occurs by aerosolization of droplets from the human respiratory tract. The outbreak of influenza among 72 percent of passengers in an airliner cabin exposed to a single index case (27; see Section 2.1) provides an example of how respiratory pathogens can be dispersed into a confined indoor environment.

The risk of elevated respiratory tract infections in modern buildings as a function of decreased outdoor air ventilation has been addressed in a 47-month surveillance study of U.S. Army recruits (9). Hospitalizations caused by acute respiratory disease were monitored in modern and old barrack buildings at four army training centers. The modern buildings were characterized by very minimal outdoor air ventilation (about 1.8 cfm/occupant; see Reference 134) especially when the outdoor temperature was more than 5 degrees warmer or cooler than that indoors. By contrast, the use of outdoor air for ventilation in old buildings was estimated to be about 14.4 cfm/occupant (134).

The acute respiratory disease rate in the modern buildings was about 50 percent higher than that in the old buildings (9). For unimmunized populations in modern barracks it was not uncommon to find elevated attack rates both during winter and summer months. By contrast, in old buildings respiratory illness had a typical seasonal pattern with highest incidence in the winter (see Figure 1 in Reference 9).

Dilution of indoor air with inadequate amounts of outdoor air likely results in the concentration of airborne infectious agents throughout the occupied space. The economic consequences of this type of BRI in terms of absenteeism and decreased productivity probably results in losses in excess of $10 billion annually (135).

5 OTHER BUILDING-RELATED PROBLEMS

Traditionally, the focus of most SBS and BRI evaluations in the United States has been on chemical, physical, and microbiological agents or on the HVAC systems. Several distinct types of problems including thermal discomfort, productivity, and psychosocial issues may also occur in buildings. These topics are described briefly and related to SBS and BRI where applicable.

5.1 Thermal Discomfort

Thermal environmental complaints are common in office buildings. Often IAQ problems that are initially perceived to be due to chemical or ventilation parameters are found to be caused by unsatisfactory air temperature and air movement parameters upon detailed evaluation.

In the 1970s, the imposition in buildings of temperature restrictions as a means of reducing energy costs led to widespread thermal environmental problems characterized by complaints that the air was too warm or cold, too humid, or "stuffy" (136). Human acceptance of a thermal environment is related to a number of variables such as metabolic heat production, the transfer of heat between the occupant and the environment, and physiological and body temperature adjustments. Heat transfer is influenced by variables such as dry bulb temperature, thermal radiation, moisture levels, air velocity, clothing insulation, and metabolic activity. The body is in thermal equilibrium when the net heat gain or loss is zero. Thermal comfort is presumed to be optimal when the body is close to or at thermal equilibrium with its indoor environment. A number of factors in offices may adversely affect the body's thermal equilibrium and lead to thermal dissatisfaction. Examples include asymmetric thermal radiation that affects people sitting near sun-facing windows, vertical temperature differences in a room, cold drafts, and cold floors and walls.

Performance criteria for environmental conditions that 80 percent or more sedentary occupants in indoor environments will find acceptable have been defined in ASHRAE Standard 55-1981 (3). This standard recommends ranges of temperature and air moisture levels as well as limits on thermal radiation, air velocity, and other variables that should collectively be acceptable to the majority (80 percent) of occupants.

A recent revision to ASHRAE Standard 55 (137) has been proposed. An important change in the draft of Standard 55 involves the definition of an "acceptable" thermal environment. Although recognizing that it is impossible to provide a thermal environment that is acceptable to everyone, an acceptable thermal environment is now defined as one that would be perceived as satisfactory by 90 percent of the occupants.

The public review draft of ASHRAE Standard 55 recognizes that dissatisfaction from warm or cool discomfort for the body as a whole may be minimized by provision of acceptable ranges of operative temperature and humidity. These acceptable comfort ranges are listed as follows and apply to people attired in typical summer or winter clothing and primarily carrying out light sedentary work.

- Winter: operative temperature range 68 to 76°F; relative humidity 30 to 60 percent
- Summer: operative temperature range 73 to 80°F; relative humidity 30 to 60 percent

An important change in ASHRAE Standard 55-1981r (137) is the recommendation that acceptable air moisture levels for comfort be defined by a 30 to 60 percent *relative humidity* range and not by the *dew point* temperature range of 35 to 62°F as recommended by ASHRAE Standard 55-1981 (3). This recommended change is based on consideration of dry skin, mucous membrane irritation, and microbial growth that occur when relative humidities fall outside the 30 to 60 percent range.

Other performance criteria enumerated in ASHRAE Standard 55-1981r (137) include the following, which are designed to prevent dissatisfaction caused by heating or cooling of one part of the body:

- Radiant asymmetry should be less than 9°F in the vertical direction and less than 18°F horizontally.
- The vertical air temperature gradient measured 4 in. and 76 in. above the floor should not exceed 5°F.
- The surface temperature of the floor shall be in the range from 65 to 84°F.

ASHRAE Standard 55-1981r (137) provides guidance and protocols with regard to the measuring range, accuracy, and response time of instruments used to record thermal parameters such as air and mean radiant temperatures, air motion, relative humidity, and radiant temperature asymmetry.

Carlton-Foss (136) describes several examples where an understanding of the etiology of thermal discomfort is based on both ventilation and building parameters. Thus in buildings where the heating system thermostat has been set back for the weekend, thermal dissatisfaction complaints are typically greatest on Monday, especially in interior occupied spaces. Because the interior and exterior zones may have different heating systems (often none exist in the interior zones in large buildings) and may be constructed with different materials, it takes a longer time for air temperatures in interior zones to reach the acceptable comfort range. In another example, temperature gradients of about 5°F existed between peripheral and interior zones because thermostats were not properly located; dissimilar cooling zones were controlled by one thermostat (136).

A recent symposium (138) contains several articles on human responses in buildings located in hot and humid areas. One study (139) found that people in Japanese buildings perceive indoor air as being thermally neutral at temperatures slightly higher than those perceived to be neutral by people in Denmark and the United States.

5.2 Economic Costs of Poor Indoor Air Quality

The exact number of occupants affected by SBS in nonresidential, nonindustrial buildings is unknown. Studies in Britain indicate prevalence rates of SBS symptoms in randomly selected buildings exceeds 40 percent (108). The Danish Town Hall Studies showed that irritational complaints were present at a prevalence rate of 28 percent in buildings without any previously known IAQ problems (110). A telephone survey of 600 United States office workers in 48 states showed that SBS complaints such as headache, irritational problems, and fatigue were prevalent among 39 to 56 percent of respondents (7). The same study showed that approximately 20 percent of respondents perceived that poor IAQ interfered with their productivity in the office. Current estimates suggest that about 20 to 30 percent of the nonindustrial building stock may be classified as "problem buildings"; that is, those with a significant number of occupants manifesting symptoms of SBS and impaired productivity because of poor IAQ (140).

The total number of nonindustrial commercial buildings in the United States is about 4,000,000 (141). Assuming that each building has an average area of 13,000 ft^2, and an occupancy of 3 to 5/1000 ft^2, and that approximately 20 to 30 percent of buildings are environmentally problem-type, Woods (140) has estimated that between 30 and 70 million people may be exposed to poor indoor air, and therefore approximately 20 to 50 million of these occupants may have SBS symptoms.

Some estimates suggest that as many as one-third of problem buildings are characterized by BRI (140). Although this estimate may be too high (the diagnosis of BRI was usually made without medical input), it is clear that the costs associated with each instance of BRI can be significant.

Several instances where BRI has led to the evacuation of a building or portion of a building and the renovation/reconstruction of the affected structure are found in the literature (16, 20, 32). The building studied by Arnow et al. (16) originally housed about 1000 employees and was vacated for about 3 years during renovation. Renovations in the buildings studied by Hodgson et al. (20, 32) required 1 to 3 years. Costs associated with renovations of these buildings varied from $150,000 to $10,000,000. Direct health care and litigation costs in these types of problem buildings were also significant.

The EPA (98) has estimated the direct health care costs of cancer for certain indoor air pollutants such as radon ($426 million/year), VOCs ($25 to $125 million/year), and ETS ($0.26 to $111 million/year). Direct health care costs due to visits to physicians by those affected by respiratory tract infections are estimated to be $15 billion (135). It is likely, based on the results of the study of the acute respiratory febrile illness by Brundage et al. (9), that a significant percentage of these physician visits are due to building-associated respiratory tract infection (see Section 4.5).

So far there is very little direct evidence of IAQ issues adversely affecting the marketability of nonindustrial commercial buildings (98). Indirect evidence, however, suggests that as building owners and tenants become more cognizant of IAQ, issues such as the adequacy of original HVAC system design and control strategies as well as ease of facilities maintenance will become an important aspect of real estate transactions (140). In addition, some indoor air contaminants such as combustion products, ETS, and fungi adversely affect a building's value by degrading construction materials or by increasing housekeeping expenses. The cost of failures in telephone switching equipment and electronics has been estimated by one source to be in the range of $10,000 to $380,000 per event (98).

The EPA (98) in a recent report to Congress has estimated that among the 64 million U.S. office workers poor IAQ leads to an excess of 0.24 doctor visits per person per year at cost of about $0.5 billion. The indirect employer costs associated with absenteeism and decreased productivity (3 percent productivity decline; extra 0.6 sick days per person per year) have been estimated by the EPA to be in the tens of billions of dollars per year. In a comprehensive review of the economic consequences of indoor air pollution, Woods (140) estimated that the indirect cost to U.S. employers for three sick days per year *or* for a loss of six minutes of concentration ability per day is about $10 billion annually.

Avoiding the adverse economic consequences of indoor air pollutants necessitates investment in better design, operation, and maintenance of buildings. In a recent study on the origin of microbial contamination in 18 buildings, the failure of maintenance programs was recognized in 13 instances (142). In 11 of the 18 buildings, the HVAC system equipment itself had no provision for access so as to facilitate the preventive maintenance needed to prevent microbial amplification. The transformation of problem buildings into healthy buildings almost always includes improving building performance (140).

Costs for improvement in operation and maintenance of existing buildings and better design and construction for new buildings to improve IAQ (Section 8) should be balanced against potential gains in occupant productivity. Various costs associated with the construction and operation of nonindustrial buildings are estimated as follows:

- Salary cost: \$100 to \$300/(ft^2) (year) (140, 143)
- Lease cost: \$15 to \$50/(ft^2) (year) (140)
- Building construction cost: \$50 to \$125/gross ft^2 (140)
- Capital assets cost (furnishings, equipment): \$20 to \$100/ft^2 of floor area (140)
- Maintenance and operation cost: \$2 to \$4/(ft^2) (year) (140)
- Total environmental control cost: \$2 to \$10/(ft^2) (year) (98, 143)
- Total utility cost: \$2 to \$4/(ft^2) (year) (140)
- Cost to increase outdoor air ventilation from 5 to 20 cfm/occupant: $\$<0.5$/(ft^2) (year) (144)
- Costs to improve operation and maintenance programs: \$0.25 to \$1.0/(ft^2) (year) (140)

The costs associated with improved building performance are potentially more than offset through improved productivity of building occupants (140). It thus makes good economic as well as environmental sense to transform a problem building into a healthy building.

5.3 Crisis Buildings and Mass Psychogenic Illness

Baker (145) has reviewed the psychological and social factors that are important in office environments and may be important as predisposing factors in building problems. He points out that cues from the environment such as unusual odors or stuffy air over a period of time may be interpreted by occupants as an indicator that a toxic agent exists in the indoor air. This phenomenon may lead to the development of a "crisis building." In a very different situation, an anxiety reaction known as mass psychogenic illness may occur over a matter of a few hours or days. Mass psychogenic illness is characterized by hyperventilation and often by visits to hospital emergency rooms.

Stress and workplace organizational factors are considered to be predisposing factors for the development of crisis buildings and mass psychogenic illness. Many stresses exist in the modern office, such as routine and boring work, low pay, low status, little chance for advancement, and lack of control over rigid work patterns.

Baker (145) has described how a toxic-hysteria crisis (crisis building) can develop in a stressful office environment. One or more occupants report nonspecific symptoms such as odor annoyance, headache, or irritational complaints. Over a period of weeks or months occupants come to believe that a toxic agent is present in their office environment. Management on the other hand believes that this is merely hysteria. Environmental investigations to detect toxic agents are inconclusive. Deep and sincere concern by occupants about their safety leads to heightened anxiety. An indoor environment in which occupants feel that it is no longer safe to work is a crisis building (145).

A number of intervention strategies have been suggested for dealing with crisis buildings (145). Most important is an understanding of the underlying stress factors and organizational characteristics of the office. The likelihood that the employee symptoms have a multifactorial basis should be considered. Effective communications between employees and management should be encouraged. This might involve a joint environmental committee to monitor symptoms and measure parameters such as thermal environmental conditions (145). Changes in rigid workplace organization and involvement of employees in the control of their workplace environment should be considered.

In contrast to the phenomenon of a crisis building, mass psychogenic illness is recognized by the acuteness of the anxiety reaction and the characteristic occurrence of hyperventilation, often followed by symptoms such as headache, dizziness, faintness, nausea, and weakness (29, 145).

Gender-specific attack rates may occur and, most importantly, illness is transmitted both along sight lines and/or verbally. The triggering event for mass psychogenic illness is often a strong index case(s) such as an individual who precipitates an epidemic by *announcing* to office colleagues that a toxic agent in the air caused illness (146) or such as several people who develop symptoms such as uncontrollable itching in view of their colleagues (147). Mass psychogenic illness has been reviewed elsewhere (148–150). It should be realized that although instances of mass psychogenic illness have occurred in nonindustrial buildings, the vast majority of IAQ complaints do have an objective basis (29, 145).

6 VENTILATION SYSTEMS IN COMMERCIAL BUILDINGS

A HVAC system should provide the occupied space with acceptable IAQ and acceptable thermal conditions. Thus the HVAC system should provide conditioned air so that the range of air temperatures and moisture levels in occupied spaces conforms to the guidelines of ASHRAE Standard 55 (3, 137) described in Section 5.1. The HVAC system should also provide to the occupied zone an appropriate amount of outdoor air (see Section 6.5) and recycled (return) air that is free from harmful air contaminants and to which a majority of people do not express dissatisfaction (4).

6.1 Description of a Typical HVAC System

A typical HVAC system serving a mechanically ventilated commercial building is described. More detailed discussion of HVAC system characteristics (and deficiencies) are found elsewhere (31, 151, 152).

The HVAC system of a large mechanically ventilated building contains one or more air handling units (AHU). In the mixed air plenum of an AHU, outdoor air is mixed with a portion of the HVAC system's return air (see diagrams and figures in References 31, 151). The amount of outdoor air that enters the mixing plenum must be sufficient to deliver a minimum of 15 cfm per person to the occupied zone.

The mixture of outdoor and return air enters a filter bank that contains filters of varying efficiency as measured by the dust spot efficiency method (153). Highly efficient bag filters found in some HVAC systems remove 60 to 90 percent of the fine airborne dusts that would otherwise visually soil interior surfaces in occupied spaces. Unfortunately, in many buildings only low efficiency (20 percent or less) filters are utilized to remove particulates from the ventilation air.

Special filters that are almost never found in commercial building HVAC systems are needed to remove gases and VOCs. ASHRAE Standard 62-1989 (4), however, requires that if contaminant concentrations in outdoor air exceed limits set by the EPA NAAQS (70), outdoor air entering HVAC systems must be appropriately filtered. Thus HVAC filter systems of buildings located in urban areas where ozone and NO_2 frequently exceed NAAQS limits should be equipped to remove these gases.

The air mixture after filtration enters the heat exchanger section where heat is either added to or removed from the airstream as required to maintain the thermal comfort of occupants in the building.

During the summer air-conditioning season, moisture is removed from the airstream as it passes over the cooling coils (151). This moisture collects in drain pans beneath the heat exchanger and should exit the AHU through drain lines with deep sealed traps. Water should not be allowed to stagnate in drain pans or in other portions of the AHU (4).

AHUs must be maintained in order to provide acceptable performance. Easy access into the mixed air, filter, heat exchanger, and fan plenums is required for routine maintenance.

After passing through the heat exchanger and the supply fan, conditioned air is distributed to the occupied spaces. The main air supply ductwork is usually constructed of sheet metal and it, as well as the plenums housing the fan and heat exchanger, may be internally insulated with fiber glass liner. Internal fiber glass liner in these plenums and ducts must be undamaged and have a structural integrity that does not allow for loose fibers to be entrained in the airstream.

Air from main supply ducts enters rigid branch ducts which in modern buildings (designed since the 1970s) often contain variable air volume (VAV) terminals. According to the requirements of zone thermostats, the VAV terminals modulate the flow of conditioned air to each zone. VAV terminals should always provide some continuous airflow to the occupied zone even when the thermostat is satisfied. Defective hardware or control systems for VAV terminals in buildings often result in HVAC system performance that does not meet design specifications.

In one building, average outdoor air flow to zones served by various VAV terminals ranged from 0 to 39.0 cfm/occupant (154). ASHRAE Standard 62-1989 (4) clearly requires that when the total supply of conditioned air to an occupied zone is reduced a minimum amount of outdoor air must still be provided (151, 155).

In a VAV system, the volume of air delivered to the zone is varied to maintain the interior temperature. In a constant air volume system that is often found in older buildings, the interior space temperature is maintained by varying the temperature of the conditioned air (151).

Air from rigid branch ducts usually enters a flexible duct and then passes into the occupied space through a ceiling-mounted diffuser. A portion of room air is then entrained into the supply airstream being discharged into the occupied space. If air supply inlets and return air outlets are properly sized and selected, outdoor air can be effectively distributed to the occupied zone (151).

The fraction of the outdoor air delivered to the occupied space (via the diffuser) that actually reaches the occupied zone (generally 3 to 72 in. above the floor) is defined as ventilation effectiveness. The minimum outdoor air ventilation rates recommended in ASHRAE Standard 62-1989 (4) assume a ventilation effectiveness that approaches 100 percent. Actual measurements of ventilation efficiency indicate that most HVAC systems perform at less than 100 percent efficiency in delivering outdoor air to the occupied zone (156, 157). One study reported ventilation effectiveness measurements for different office configurations in the range of 0.57 to 0.76 (158). It is implicit in ASHRAE Standard 62-1989 (4) that more than the minimum amount of outdoor air as required by the ventilation rate procedure is required to compensate for imperfect mixing of outdoor air in the occupied space.

In many modern buildings, the above-ceiling cavity is used for the passage of return air from the occupied space back to the AHU. The use of a ceiling plenum instead of return ducts is associated with a number of problems discussed elsewhere (151).

Some return air is discharged from the HVAC system through a relief louver. The amount of outdoor air brought into the HVAC system should be slightly more than the combined total of relief air plus air exhausted from the building by toilet and local exhaust fans. The building is thus maintained slightly positive with respect to the atmosphere so as to prevent the infiltration of unfiltered and unconditioned air through loose construction and open doors.

In many buildings, fan coil and induction units located along interior walls are used to condition the air in perimeter zones. Fan coil units contain small fans, low-efficiency filters, and small heat exchangers. These units condition and recirculate supply air (usually without any outdoor air supply) in peripheral zones. Induction units are generally supplied with conditioned outdoor air from a central AHU. Conditioned air from the unit passes through nozzles and mixes with a portion of the room air. Because a large building may contain several hundred fan coil or induction units, maintenance of this part of the HVAC system is often neglected.

6.2 Humidification and Dehumidification

In hospital critical care areas and in computer rooms, moisture is often added to ventilation air to maintain humidity levels usually between 40 and 50 percent. A relative humidity in occupied spaces of between 30 and 60 percent is recommended by ASHRAE Standard 55-1981r (137).

Moisture is preferably added to the ventilation air in commercial buildings by humidifiers that inject steam or water vapor into the AHU (often in fan plenum) or in the main supply air ductwork (151). Injection nozzles should not be located near porous interior insulation that can become wet and offer a niche for the amplification of microorganisms.

HVAC system humidifiers should preferably use steam as a moisture source. The supply of steam introduced into the HVAC system should be free of volatile amine corrosion inhibitors, especially morpholine (159).

Cold-water humidifiers that contain reservoirs of water can become contaminated with microorganisms that may cause BRIs such as hypersensitivity pneumonitis, humidifier fever, and asthma. The use of biocides in these types of humidifiers may be ineffective (160, 161) or may in itself cause asthmatic reactions (see Section 4.3). Preventive maintenance programs for cold-water humidifiers used in HVAC systems have been described in detail elsewhere (162, 163).

During the air-conditioning season, as air passes around cooling coils of fan coil units and AHUs, the dry bulb temperature approaches the dew point temperature. Consequently, air downstream of cooling coils has a humidity close to 100 percent. Organic dusts and debris not removed during filtration can pass through the heat exchanger section and become entrained in moist, porous, internal insulation in downstream plenums and air supply ductwork. Amplification of fungi and bacteria may thus occur on the nutrients trapped in these HVAC system locations (142).

During the air-conditioning season, the cooling coil's chilled water temperature may be raised in an effort to save energy costs. This results in an elevated relative humidity in occupied spaces. As the relative humidity in occupied spaces rises above about 70 percent, the equilibrium moisture content in organic dusts and materials approaches a level sufficient to support fungal spore germination and proliferation (19, 164). The upper relative humidity limit of 60 percent now proposed in ASHRAE Standard 55-1981r (137) provides a margin of safety that should prevent amplification of fungi in occupied spaces.

6.3 CO_2 Concentration

In several studies, CO_2 concentrations above 1000 ppm have been associated with an increase in occupant complaints in buildings (165, 166). The normal concentration of CO_2 in the outdoor air is about 325 ppm. In the indoor environment the CO_2 concentration is always somewhat elevated relative to the outdoor level because a CO_2 concentration of about 38,000 ppm is present in the air exhaled from the lung.

ASHRAE Standard 62-1989 recommends that steady-state levels of CO_2 in indoor air not exceed 1000 ppm, which is equivalent to the provision of about 15 cfm of outdoor air per occupant (134). Panel studies (92; see Section 2.6) have shown that approximately 15 cfm outdoor air per occupant is needed to dilute human body odor adequately to a concentration where 20 percent or less of visitors are dissatisfied (80 percent acceptability) with the IAQ. The maintenance of CO_2 at 1000 ppm or less is a convenient surrogate for assuring that outdoor air ventilation rates do not fall below 15 cfm per occupant.

Although CO_2 concentration remains a good surrogate for human-generated contaminants such as body odor (167) and a good predictor of overoccupancy conditions, the consensus of recent formal epidemiologic studies shows that prevalence of SBS symptoms is not related to CO_2 levels (101, 112). Studies summarized by Fanger (168) indicate that air contaminants derived from nonhuman sources such as HVAC systems and interior construction and finishing materials are major sources of perceived dissatisfaction with indoor air. The concentration of CO_2 in indoor air would, of course, have no relationship with these emission sources.

6.4 Evolution of ASHRAE Standard 62

The concept that ingress of clean outdoor air will promote the health and well being of occupants indoors is over 200 years old. Benjamin Franklin expressed this opinion when he wrote (1, 90):

> I considered (fresh air) an enemy, and closed with extreme care every crevice in the room I inhabited. Experience has convinced me of my error. I am now persuaded that no common air from without is so unwholesome as the air within a closed room that has often been breathed and not changed.

He further wrote:

> You physicians have of late happily discovered, after a contrary opinion had prevailed some ages, that fresh and cool air does good to persons in the small-pox and other fever. It is hoped, that in another century or two we may find out that it is not bad even for people in health.

Tredgold suggested in 1824 that a minimum of 3 cfm per person of outdoor air was required to dilute the CO_2 exhaled by miners (90). Medical opinion by the end of the nineteenth century suggested that 30 to 60 cfm of outdoor air per person was helpful in lowering the risk of tuberculosis and this led in 1895 to the recommendation of a minimum of 30 cfm of outdoor air per person in the first ASHRAE ventilation standard (90). In the 1930s, studies by Yaglow et al. (2) showed that from 10 to 30 cfm of outdoor air per person was required to provide acceptable dilution of ETS and bioeffluents.

The ASHRAE Standard for Natural and Mechanical Ventilation (91) subsequently prescribed a "minimum" outdoor air ventilation rate of 5 cfm per person to provide acceptable but not odorfree conditions and a "recommended" outdoor air ventilation rate (generally a minimum of 10 cfm outdoor air per occupant or more) to control odor annoyance. The minimum outdoor air ventilation rate recommended in ASHRAE Standard 62-1973 (91) (5 cfm per occupant) was intended to prevent indoor CO_2 levels from rising above 2500 ppm, or half the recognized occupational TLV for this gas.

As a result of the energy crisis of the mid 1970s, ASHRAE Standard 90 (169), published in 1975, recommended that only the "minimum" outdoor air ventilation rate of ASHRAE Standard 62-1973 (91) (5 cfm outdoor air per occupant) be used for ventilation (90). Concerns about providing enough outdoor air to dilute ETS and bioeffluents to acceptable levels were overlooked. Many of the HVAC systems designed for buildings constructed or retrofitted in the 1970s and 1980s have been designed on the basis of the "minimum" outdoor ventilation rate recommended in ASHRAE Standard 90. Thus it is not surprising that air quality complaints have occurred in these buildings.

ASHRAE Standard 62-1981 (170) partially rectified the error of Standard 90 by recommending a minimum outdoor air ventilation rate of 20 cfm per occupant for zones where smoking was permitted. The 5 cfm of outdoor air per occupant value recommended for nonsmoking zones was intended to prevent CO_2 concentrations from rising above 2500 ppm. In practice, many building designers throughout the 1980s continued to specify outdoor air ventilation rates between 5 and 10 cfm per occupant, by misinterpreting ASHRAE Standard 62-1981 (170) to imply that the nonsmoking and smoking ventilation rates could be averaged according to the anticipated number of smokers in the building.

6.5 ASHRAE Standard 62-1989

ASHRAE Standard 62-1989 (4) is probably the most important document in the IAQ literature. It reflects a consensus reached since 1983 by knowledgeable individuals from engineering, industrial, and academic groups. Janssen (134) points out that the ventilation rates recommended in Standard 62-1989 for the most part are similar to "recommended" rates in Standard 62-1973 and to the rates recommended for smoking environments in Standard 62-1981.

A key feature of Standard 62-1989 (4) and its *ventilation rate procedure* is the increase in the minimum outdoor ventilation rate from 5 to 15 cfm per person. Outdoor air requirements recommended by the ventilation rate procedure make no distinction between "smoking-allowed" and "smoking-prohibited" areas. A minimum of 15 cfm of outdoor air per person as specified in the ventilation rate procedure is recommended because new research indicated that this is the *minimum* amount of outdoor air needed to dilute body and tobacco smoke odors to acceptable levels (134). The outdoor air requirements specified in the ventilation rate procedure must be delivered to the occupied zone. Design assumptions with regard to ventilation rates and air distribution to the occupied zone are required by Standard 62-1989.

Standard 62-1989 also requires that the design documentation for a HVAC system state clearly which assumptions are used in design. This allows others to estimate the limits of the HVAC system in removing air contaminants prior to commissioning and prior to the introduction of new contaminant sources into the occupied space.

A key provision in Standard 62-1989 now requires that when the supply of air to the occupied zone is reduced (for example, in VAV systems), provision be made to maintain minimum flow rates of outdoor air throughout the occupied zone.

Building maintenance is recognized as an important factor in providing acceptable IAQ. Thus AHUs and fan coil units should be easily accessible for both inspection and maintenance. Specific mention is made of avoiding stagnant water in HVAC systems. Caution is urged in the use of recirculated water spray systems that are prone to microbial contamination. Special care is urged to prevent entrainment of moisture drift from cooling towers into outdoor air inlets. However, a minimum distance between makeup air inlets and external contaminant sources such as cooling towers, sanitary vents, and loading docks is not specified.

The outdoor air used in HVAC systems must meet NAAQS for priority pollutants before it is introduced into the occupied zone. According to this criterion, the average long-term concentration of NO_2 and SO_2 for example, in outdoor air introduced into HVAC systems, shall not exceed 0.055 and 0.03 ppm, respectively. Ozone shall not exceed a short-term concentration of 0.12 ppm. A proposed addendum to Standard 62-1989 recommends that in accordance with the 1987 revision of the NAAQS, the long-term respirable particulate concentration in outdoor air used in HVAC systems shall not exceed 50 $\mu g/m^3$ (73). When outdoor air is unacceptable, contaminant levels must be reduced by filtration to acceptable limits.

The *indoor air quality* procedure for achieving acceptable IAQ is retained in Standard 62-1989 (4). In this procedure it is required that the concentration of air contaminants in the occupied zone be held below acceptable limits. The outdoor air ventilation rate is left unspecified. Guidance in Standard 62-1989 as to what concentrations of indoor air contaminants are acceptable is furnished for outdoor contaminants covered in the NAAQS and for four additional air contaminants of indoor origin, namely, CO_2 (1000 ppm), chlordane (5 $\mu g/m^3$), ozone (0.05 ppm), and radon (0.027 WL). Because acceptable limits and health effects data are currently unavailable for the vast majority of other indoor air contaminants (see Section 2), very few designers will likely use the air quality procedure of the standard. The standard does, however, make provision for the subjective evaluation of indoor air contaminants much like that carried out by Fanger et al. (83) in Danish buildings as one means of implementing the indoor air quality procedure (also see Appendix C of Reference 4).

Some experimental work has been initiated with sensors having a capacity to detect thermal parameters as well as air contaminant concentrations (171). In theory, these devices would replace thermostats in the occupied space and control HVAC system operation and IAQ much like that envisioned in the air quality procedure of Standard 62-1989. Major difficulties with this approach at present include the limitations in terms of sensitivity of detector systems for concentrations of interest in indoor air and the absence of acceptable concentration guidelines for most indoor air pollutants.

The scientific basis of Standard 62-1989 has recently been discussed by two of its committee members (172). A major weakness of Standard 62-1989 is the absence of health effects data for long-term exposures to low concentrations of pollutants commonly found in indoor air. The ideal standard of the future would be one combining a performance path with an upper limit on the source strength for each type of pollutant *and* a prescriptive path for the designer specifying a lower limit on the ventilation rate that would maintain the pollutant concentration within acceptable health guidelines (172). The current ventilation rate procedure of Standard 62-1989 likely to be used by designers for buildings built in the 1990s largely ignores the concept of pollutant source strength. Setting source strength upper limits (limiting olfs) and specifying associated minimal ventilation rates will be major IAQ research issues during the 1990s (172).

7 IAQ EVALUATION PROTOCOLS AND GUIDELINES

Approaches to solving SBS problems using traditional industrial hygiene techniques are generally inadequate. The traditional industrial hygiene approach to solving occupational health problems involves recognition of the hazard (because input and output materials are known, the identity of air pollutants is generally preestablished), measuring the contaminant(s) by NIOSH- or Occupational Safety and Health Administration (OSHA)-approved procedures, interpretation of analytical data in terms of TWA, TLVs, and suggestion of corrective action such as containment of the contaminant at its source. A common result of this approach when applied to IAQ evaluations is that compliance with industrial workplace standards is demonstrated but building-associated complaints from occupants persist (90).

For IAQ evaluations, more sensitive protocols are needed because comfort, occupant well-being, and general population susceptibilities are parameters that are not addressed by industrial workplace TLVs (173). In IAQ evaluations, sources of contaminants such as bioeffluents and VOCs are diverse and diffuse. Concentrations of air contaminants are often one to four orders of magnitude less than the TLVs. Contaminant control strategy in mechanically ventilated buildings is based almost universally on dilution ventilation by a HVAC system that is designed primarily to provide acceptable thermal conditions in the occupied space.

The evaluation approach described here is basically that of building diagnostics, a process wherein an expert uses available knowledge, techniques, and instruments to predict the likely performance of a building over a period of time (174, 173). The net result of building diagnostics is the development of recommendations to improve both HVAC system and occupant responses. Diagnostics as used in the context of IAQ evaluations is usually divided into a qualitative and a quantitative phase.

An extensive literature is available on protocols for dealing with SBS and BRI, and the reader is referred to these publications for alternative approaches to those presented here (175–178).

7.1 Qualitative Evaluation

The qualitative IAQ evaluation begins with a consultation phase that may be accomplished through an extensive telephone interview or by a site visit. During this phase of the evaluation, objectives and scope are defined and a preliminary hypothesis as to the likely reason for occupant complaints is formulated.

At the building site, the nature of the health complaints is reviewed, environmental factors that may be responsible for complaints are visually evaluated, and an engineering assessment of the HVAC system is conducted. An objective of the qualitative evaluation is to provide recommendations for remedial actions in a manner that avoids the necessity of costly objective measurement of air contaminant concentrations or HVAC system performance parameters.

Instrumentation used during the qualitative IAQ evaluation is limited, and intended primarily to guide the investigator in verifying the hypothesis as to the nature of the building-associated problem. Direct-reading instruments to measure parameters such as relative humidity, operative temperature, and CO_2 and respirable particulate concentrations are often useful.

A key aspect of the qualitative evaluation is the analysis of a building's HVAC system. HVAC system mechanical components are visually examined for deficiencies with regard to design, operation, and maintenance parameters. The control strategies that govern HVAC system operation must be thoroughly understood. Essential items for the qualitative evaluation of the HVAC system include a smoke pencil for visualization of airflow patterns, a complete set of mechanical plans and specifications, and the on-site availability of the building facility engineer.

Care must be taken at this phase of the evaluation to exclude the possibility that occupant complaints are due to lighting, acoustic, ergonomic, or psychosocial problems. Thus the possibility that complaints are correlated with labor management difficulties or with occupant perceptions that a toxic agent is present in the air (see Section 5.3) should be explored. The widespread tendency of American investigators to ascribe SBS problems to inadequate outdoor air ventilation should be carefully reviewed in terms of the work of Fanger et al. (83) and Burge et al. (108). Thus occupant problems may not be resolved by increasing outdoor air ventilation if there are intense sources of contaminants (olfs) in the occupied spaces (for example, VOCs from new furnishings) or if contaminants originate from the HVAC system itself. Consideration (during the qualitative IAQ evaluation) must be given to the design and/or energy parameters of a building. Thus it will be difficult to rectify IAQ problems in a building designed according to ASHRAE Standard 90 (169; see Section 6.4).

When health complaints are reviewed, it is essential to determine if the perceived problem is one of occupant discomfort and annoyance (SBS, thermal discomfort) or if one or more occupants have a BRI. When a BRI is apparent, immediate medical attention for the affected occupants is required concomitant with the initiation of appropriate remedial actions (173).

Performance criteria for maintaining an indoor environment in which 80 percent or more of occupants do not express annoyance or dissatisfaction should be reviewed with building management/occupants during the qualitative evaluation (173). Some performance criteria for comfort are based on ASHRAE Standards 55-1981r (137) and 62-1989 (4), on the NAAQS, on odor recognition thresholds, and on professional judgment (see Table 1 in Reference 173). Exact performance criteria for the vast majority of contaminants found in indoor air are unknown. However, the prevention of discomfort and annoyance in nonindustrial environments, such as offices, implies control of air contaminants to concentrations several orders of magnitude below industrial TLVs. An in-depth discussion of the qualitative IAQ evaluation is found elsewhere (173).

7.2 Major Components of the Qualitative IAQ Evaluation

Review all telephone conversations and written information relative to existence of thermal environmental problems, SBS, BRI, ergonomic issues, and potential psychosocial issues. Define a hypothesis. What is the likely problem based on preassessment telephone interviews and review of available documentation? Be prepared to discuss your hypothesis and your plan for the qualitative evaluation. Also be prepared to conduct an exit interview (later in day) at which time major observations, conclusions, and preliminary recommendations should be discussed with building management and employee representatives.

Carry out interviews with office managers and employees or their representatives in order to form a conclusion as to the presence or absence of SBS. Determine by interview if higher attack rates of SBS occur in certain occupied zones. Elicit information on whether or not prevalence of SBS symptoms follow a daily, weekly, or seasonal pattern.

Obtain information on the possible existence of BRI among one or more occupants during interviews with management and employee representatives. What evidence suggests that occupant illnesses are building-related? For example, is there evidence to suggest that acute febrile respiratory illness among building occupants follows a nonseasonal pattern and occurs at a higher prevalence rate than in the general community population?

Determine by the interview process if occupant complaints have a thermal environmental basis. Is there a seasonal or a zonal aspect (for example, thermal complaints in interior zones only) to thermal problems? Finally, are complaints in the building multifactorial, suggesting that SBS, BRI, and possibly thermal environmental problems all coexist (140, 173)?

Observations on possible contaminant sources (as follows) made during the qualitative evaluation are important both in verifying the preliminary hypothesis of complaint etiology and in formulating recommendations for corrective actions. Because a building is dynamic in operation, observations should be made with regard to present and past environmental conditions.

Microbiological Contaminant Sources

________ Floods

________ Humidifiers (central HVAC system or portable; ultrasonic, cool mist, etc.)

________ Excessive relative humidity or dew point temperature in occupied zone

________ Moisture in fan coil units, induction units, AHUs

________ Condensed water or cold surfaces

_______ Bark chips, wicker, or other organic materials that may have been wet
_______ Porous moist insulation in HVAC system
_______ Stagnant water in HVAC system
_______ Bioaerosol sources at or near outdoor air inlets or in locations where air infiltrates into buildings
_______ Wet fibrous insulation in common return ceiling air plenum
_______ Inadequate provision for maintenance of HVAC system in design; inadequate housekeeping for properly designed equipment

Combustion Product Contaminant Sources

_______ Location of contaminant sources such as garages, parking lots, loading docks, boiler stacks, and furnace flues, versus location of HVAC system outdoor air intakes doors/operable windows
_______ Determine if building or building zone is negatively pressurized relative to source of contaminant
_______ Transmission of contaminant by stack effect

ETS Contaminants

_______ Building smoking policy per each AHU
_______ Location of smokers
_______ Separate HVAC system or local exhaust for ETS
_______ Concentrations of respirable particulate present in smoking zones, nonsmoking zones, and in outdoor air; ETS migration patterns
_______ Annoyance from ETS particulates
_______ Odor annoyance

VOC Sources

_______ Age of construction and finishing materials by zone
_______ Was building properly ventilated during commissioning?
_______ Use of pesticides and fertilizers
_______ Special processes such as printing, liquid-process photocopiers; graphics divisions
_______ Proper local exhaust ventilation for laboratories
_______ What types of cleaning chemicals are used? Are cleaning chemicals stored in a well ventilated area?
_______ Unusual outdoor sources such as roof tar
_______ Consider both sources and sinks for VOCs in occupied space and HVAC system

_______ Fleecing factor (110)
_______ Shelf factor (110)

Odors

_______ Bioeffluents; number of people/ft^2; volume of room
_______ Enough outdoor air ventilation to prevent odor dissatisfaction (92)
_______ Kitchen odors
_______ Odors from laboratories and cleaning chemicals
_______ Odors from sanitary vents or cooling towers near outdoor air intake
_______ Perception of the acceptability of indoor air within 2 min upon entering building
_______ Sewer gas from drains when traps dry out

Nuisance Particulates as Contaminant Source

_______ Presence of construction dusts
_______ Dust from poor housekeeping
_______ Particulate from copies or processes
_______ Particulate entrained in outdoor air intake
_______ Fiber fallout from fiber glass batts on ceiling tiles

Other Factors Relating to Contaminant Sources

_______ Elicit information on possible contaminants in relation to time sequence of occupant complaints
_______ Effect of HVAC system dynamics on perception of pollutant presence. For example, complaints occur when HVAC system is on or off, or when 100 percent outdoor air is used, or when minimal outdoor air is used.

Similar checklists have been developed for the qualitative evaluation of HVAC system design, maintenance, and operation (173, 176). For example, information should be elicited on the original HVAC system control strategy, on the control strategy as actually installed, and on existing control strategy (173).

The product of the qualitative evaluation is a report that summarizes visual observations and redefines the preliminary hypothesis on complaint etiology. The most important component of the qualitative evaluation report is recommendations for improved building performance generally involving a combination of source control measures and an upgrading of HVAC system design, operation, and maintenance strategies.

7.3 Quantitative Evaluation

In some building investigations, objective measurements of contaminant concentrations, subjective measurements of occupant perception of acceptability, and measurement of HVAC system operational parameters are required to document the hypothesis, to justify recommendations (173, 179), or for litigation purposes.

Based on observations made during the qualitative evaluation, air sampling should be carried out in at least five separate sites including (*a*) the zone with the most susceptible occupant(s) (for example, room with several SBS complainants or BRI case location), (*b*) the zone with the poorest apparent ventilation effectiveness, (*c*) the zone with the highest apparent contaminant sources (for example, location where new finishing materials are being installed), (*d*) an indoor location without occupant complaints and with acceptable ventilation effectiveness, and (*e*) as a reference point, the outdoor air. A vast literature is available on sampling and analytical methods applicable for indoor contaminants (180, 181) including microorganisms (34), VOCs (41, 182), respirable particulates (182), and combustion products and ETS (81, 183, 184). In order to make appropriate interpretations, sampling for indoor air pollutants must be carried out with a complete understanding of HVAC system dynamics. For example, concentrations of total and specific VOCs in an office building may vary by an order of magnitude depending on whether the occupied zone is being provided with maximum (economizer) or minimum amounts of outdoor air (18).

In some quantitative evaluations, subjective measures of physical and chemical parameters are made by questionnaires (185, 186) or by panels of judges (83, 92). Interpretation of subjective data in light of air sampling results requires that both types of objective measurements be carried out simultaneously.

A simple questionnaire that asks the occupant to evaluate the acceptability of the indoor environment in terms of air quality, thermal, acoustic, and lighting parameters is available (173, 187). A small amount of complaint data is elicited in the questionnaire so as to relate environmental acceptability parameters with prevalence of SBS.

The analysis and interpretation of the sampling results obtained during the quantitative evaluation are illustrated in the following example. Qualitative evaluation of a one-year-old office building suggested that the widespread prevalence of irritational complaints among occupants was due to VOCs emitted from new finishings and construction materials. Subsequent quantitative evaluation (Table 2, building 1) indicated that even a year after initial occupancy, total VOC concentrations were about 30 times higher indoors than outdoors. The prior qualitative evaluation had revealed that VAV terminals serving many occupied zones did not have minimal set points to assure continuous flow of outdoor air under partial thermal load conditions. The building had (unfortunately) been designed according to the criteria of ASHRAE Standard 90 (169) and during periods of maximum occupancy the HVAC system could supply only about 5 cfm of outdoor air to the occupied zone assuming perfect ventilation efficiency. The quantitative evaluation report recommended that, as an interim measure, VOC concentrations and SBS complaints could be reduced if the HVAC system is operated 24 hours a day, 7 days a week. Subsequent air sampling for VOCs was recommended to document the effectiveness of this recommendation. Because the maximum outdoor air ventilation rate was much less than that required to prevent odor dissatisfaction in a substantial number of occupants (92), a redesign of the HVAC system outdoor air intake capacity to comply with ASHRAE Standard 62-1989 (4) was also recommended. The quantitative evaluation report also described actions that could be taken to reduce VOC concentrations by control of housekeeping chemicals and by control of VOC emissions during anticipated renovations.

7.4 IAQ Regulations and Guidelines

No U.S. federal agency has authority to regulate SBS or BRI. The Clean Air Act of 1970 (U.S. Public Law 90-148) cannot be used to regulate IAQ (188). According to the EPA (98), Congress in 1970 conceived of air pollution as an outdoor phenomenon and contemplated regulation only for outdoor air. The only U.S. federal agency to implement comprehensive air quality regulations for indoor environments is OSHA (98), but these regulatory efforts are directed only to the industrial workplace. Although U.S. federal agencies do not have any comprehensive regulatory authority for IAQ, several agencies (for example, The Consumer Products Safety Commission and Housing and Urban Development) have been involved in efforts to limit emissions of pollutants from products or materials used indoors.

IAQ legislation at the federal level in the United States will likely be enacted in the near future (189). Although regulation will not likely be a primary focus of any foreseeable legislation, it is significant that federal efforts will lead to a comprehensive national program for IAQ that includes nonindustrial workplaces, commercial and government buildings, and residences (189). Armstrong et al. (190) and Berry (191) have recently reviewed the scope of Canadian and U.S. federal IAQ programs. The extensive complexities dealing with the development of regulation for just one indoor air contaminant, namely, formaldehyde, have been reviewed elsewhere (192).

A number of states including Massachusetts, Minnesota, Wisconsin, and California have developed IAQ programs in research, mitigation, and in some instances, the regulation of smoking in public buildings and the use of unvented heating appliances. The proposed Standard for IAQ in New Jersey (193) and the Report of the Interministerial Committee on IAQ in Ontario (194) are briefly reviewed to illustrate the trends on guideline development occurring at the state and provincial levels.

The scope of the proposed New Jersey IAQ legislation would cover only buildings occupied by public employees. The comprehensive draft legislation addresses records that must be kept by employers, procedures for dealing with IAQ complaints, methods to be used for control of air contaminants, and ventilation parameters (193). Some components of the proposed legislation are described as follows:

- Employers will be required to maintain written records of HVAC system operating and maintenance procedures. This would include the control system logic that maintains the required outdoor air supply rate in relation to changes in outdoor air temperature. The method and equipment that will be used to provide acceptable amounts of outdoor air in VAV systems operating at part load conditions shall be described.
- The employer shall designate a person responsible for maintaining IAQ and for operating and maintaining the building's HVAC system. This individual shall have a minimum of 40 hours training in ventilation system operation and maintenance by an institution approved by the New Jersey Department of Health.
- The proposed standard includes a number of actions that are intended to control contaminants at their sources. For example, internal fiber glass insulation cannot be used in ductwork if the fiber glass is directly exposed to the airstream. Exposure of occupants to VOC emissions shall be reduced. Building outdoor air intakes shall be protected from VOCs and SVOCs associated with roofing operations. Newly installed humidification systems shall be designed and operated so that microbial growth and standing water are avoided. Biocides shall not be present in humidifier moisture added to indoor air.

The Report of the Interministerial Committee on IAQ in Ontario (194) discusses IAQ issues, health effects, comfort factors, the derivation of recommended guidelines, an evaluation protocol, and recommendations. Some components of the Ontario documents are described as follows:

- "Health" in the indoor nonindustrial environment is defined according to the World Health Organization's definition, namely, "a state of complete physical, mental, and social well-being, and not merely the absence of disease or infirmity (195)."
- Recommendations for existing buildings include the provision of a minimum of 15 cfm of outdoor air per person to ensure that CO_2 concentrations do not exceed 1000 ppm.
- In new buildings, HVAC system equipment should be designed for ease of maintenance and inspection. Building design should provide for pollutant removal at the source. Materials low in pollutant emissions should preferentially be used in construction.
- Designated smoking areas should be ventilated at a minimum rate of 60 cfm of outdoor air per occupant and exhaust air from these areas should not be recirculated to other building zones.

In nonindustrial indoor environments, air contaminants are widely recognized as a threat to human health (196–199). Air contaminants are present over a wide concentration range in the indoor air. Total public health risk is estimated on the basis of the populations exposed to contaminant concentration ranges. Stolwijk (197) has estimated lifetime and annual risk per unit of contaminant concentration for a stated adverse health effect per million people. Risk estimates are presented for lung cancer due to radon, leukemia due to benzene, and lower lung-related illnesses due to NO_2 and respirable particulate exposures. This approach to determining the adverse health consequences of exposure to indoor pollutants provides a basis for future studies on performance criteria for IAQ guidelines.

In the United States there are approximately 5000 building codes (200). Most of these are prescriptive in that they have evolved by consensus opinion and are used in building design. ASHRAE consensus standards are used by many but not all code bodies. It is of interest to note that most building codes still use ASHRAE Standard 62-1973 and not Standard 62-1981, because the American National Standards Institute did not approve Standard 62-1981 (200). A recent paper by Ferahian (201) reviews the issue of building codes and IAQ.

Performance criteria that could form the basis of IAQ guidelines have been influenced by both health and comfort concerns. Concern for both health and comfort is perhaps best embodied in the preface of the ACGIH TLVs for chemical substances in the work environment (202):

> The basis on which the values are established may differ from substance to substance; protection against impairment of health may be a guiding factor for some, whereas reasonable freedom from irritation, narcosis, nuisance or other forms of stress may form the basis for others.

Although it is clear that the ACGIH TLVs deal with exposure to chemical substances in the industrial workplace, it is evident that the portion of the TLV rationale dealing with "reasonable freedom from irritation, narcosis, nuisance or other forms of stress" is similar to the World Health Organization definition of health, namely, "a state of complete physical, mental, and social well-being, and not merely the absence of disease or infirmity" (90, 195). Detailed performance criteria for comfort that are based on scientific data and that could be used as guidelines for evaluation and control of SBS are not yet available. However, Tables 1 and 3 in ASHRAE Standard 62-1989 (4) and Tables C-1 through C-4 in Appendix C of that standard provide some basis for the beginning of the development of IAQ performance criteria for many of the pollutants discussed above in Section 2.

Performance criteria especially for VOCs may also be derived from yet additional sources (18, 36, 181).

Performance criteria for exposures to microorganisms are not available (34, 35). However, guidelines for exposure to antigen from dust mites have recently been proposed (203, 204). Thus in order to prevent sensitization, a maximum allowable concentration of 0.6 mg guanine/g of dust (equal to about 100 mites/g) have been proposed. To prevent acute asthma attacks, a maximum allowable concentration of 500 mites/g (equivalent to 10 μg/g Group I *Dermatophagoides* allergen) is recommended (203).

8 CONTROL MEASURES

Approaches to IAQ control may be reactive or proactive. A major purpose of both the qualitative and quantitative IAQ evaluation is a report in which specific recommendations are made to remediate the causes of SBS or BRI. This reactive approach to problem solving is the focus of recent ASHRAE symposia (205–208) where ventilation strategies are usually emphasized. Source control is another major remediation approach (8, 98, 209). Proactive control measures are embodied in commissioning methods for buildings and their HVAC systems (210, 211). Some reactive and proactive approaches for IAQ control are reviewed here.

8.1 Remedial Actions in Problem Buildings

A very common method often recommended to control IAQ problems (6, 107) uses outdoor air to dilute indoor air pollutants. Relatively clean outdoor air is delivered to the occupied zone where concentrations of contaminants generated in the occupied spaces are lowered to acceptable levels. This method of IAQ control is embodied in the ventilation rate procedure of ASHRAE Standard 62-1989 (4) wherein acceptable air quality is achieved by providing the occupied zone with at least a minimum quantity of outdoor air based on the number of occupants in the space.

The work of Fanger et al. (83) clearly shows that IAQ problems are not simply ameliorated through the provision of more outdoor air. Significant occupant dissatisfaction still occurs in some buildings even when the outdoor air ventilation rate is 50 cfm per occupant or even higher. Fanger et al. (83) found the HVAC system itself was the major source of occupant perception of poor IAQ. Over the past few years, the focus of HVAC system remedial actions has been directed toward improvement of design, maintenance, and control strategies and not merely the improvement of dilution ventilation in the occupied space.

Inadequate maintenance of HVAC system is probably the single most important factor responsible for poor air quality in many buildings (31, 90). Thus plenums in AHUs and air supply ducts may be dirty, filters may be excessively loaded with dust, porous interior insulation adjacent to the airstream may become encrusted with dirt, damper linkages may be disconnected, control systems may be inoperative or abandoned, and fan coil and induction units may be seldom cleaned. Remedial actions based on upgrading maintenance are usually straightforward (151) but often overlooked by building operators because of perceived lack of return on investment (see Section 5.2).

Recommendations to improve IAQ are often based on rectifying inadequate HVAC system design. Thus outdoor air intakes may be located at ground level and sometimes easily contaminated by emissions from nearby vehicular traffic. Outdoor air intakes on roofs may be contaminated by discharges from nearby cooling towers, sanitary vents, and toilet and kitchen exhausts. Corrective actions associated with these design deficiencies are often neglected by building operators because of the costs associated with retrofitting the HVAC system.

Design deficiency is often manifest in poor air distribution (ventilation effectiveness) in the occupied zone. Thus the location and style of supply diffusers and return air grilles may be governed more by appearance than by functional requirements (90). The use of distributed heat pump systems especially in speculative buildings often results in zones characterized by poor air distribution and stale or stuffy air. A paper by Woods et al. (212) discusses corrective actions for these design problems.

Remedial measures involving specific types of pollutants often emphasize source control or control by filtration (98). Source control may be achieved by substitution or source modification. A recommendation to replace liquid-process photocopiers that emit copious quantities of VOCs with other types of nonpolluting copiers is an example of source control. A recommendation to reduce the concentration of combustion products being entrained in the indoor air from a loading dock area by replacement of gasoline-powered vehicles with electric-powered ones is an example of source modification.

Remediation of IAQ problems may include air cleaning strategies involving removal of contaminants by particulate filtration, electrostatic precipitation, and gaseous sorption. These strategies are reviewed elsewhere (213–215).

The difficulties of applying conventional air cleaning methods to one class of indoor pollutants (VOCs) is illustrated by the work of Ensor et al. (214). Laboratory experiments were carried out in which concentrations of specific VOCs such as benzene were predicted both upstream and downstream of an activated carbon filter bank. Calculations showed that a filter bed (100 cfm airflow for 6 months through a 4-ft^2 filter) of excessive depth was required to reduce the concentration of benzene from 500 to 200 $\mu g/m^3$ (214). Although the preceding example is not necessarily representative of difficulties accompanying remedial strategies based on air cleaning, it does serve to emphasize that source control is always the preferred approach for contaminant mitigation.

8.2 Proactive IAQ Actions for New Buildings

"Healthy" buildings can be constructed if IAQ issues are properly addressed through all major phases of a building project, namely, planning, design, construction, commissioning, and initial occupancy. Important proactive actions needed to ensure IAQ in new construction are enumerated here based on the review by Levin (216).

During the planning phase for a building, IAQ goals should be defined by both the architect and the owner. If occupant activities such as extensive photocopying or printing work can be clearly identified, provision for containment of potential contaminants can be made during planning. If the end user of a building (as in the case of speculative ventures) is unknown, the architect can recommend that flexible design be used so as to provide acceptable environmental quality to a variety of potential users. Because major financial decisions are made during project planning, this is the time for the architect to convince the owner to allocate funds for specialized IAQ studies and for flexibility in construction schedules to allow for preoccupancy testing of building and HVAC system performance (216). As a result of proper planning, the mechanical engineer responsible for the building should be given clear performance specifications on IAQ matters such as (*a*) general location of the outdoor air supply inlet, (*b*) the outdoor air quantity to be supplied to the occupied zone, and (*c*) the necessity based on ambient air pollution data of cleaning outdoor air to be used in the HVAC system. Major IAQ problems such as the entrainment of combustion products into an outdoor air inlet should be clearly avoided at the planning stage.

During the design phase of a building project, important decisions are made that affect future IAQ performance. These include (*a*) the calculation of thermal loads and the design of the mechanical systems that will provide thermal comfort to occupants, (*b*) the selection of major construction materials, (*c*) the exact location of outdoor air intakes, (*d*) the choice of a VAV or constant volume mechanical system, and (*e*) specification for the total supply and outdoor air volumes from diffusers taking into account anticipated deficiencies in ventilation effectiveness (156). Design specifications that include the selection of construction materials that are less prone to microbiological deterioration and that limit niches for insect pests (thus limiting future indoor pesticide use) can have profound effects on future building performance.

During construction planning, specifications are made for the selection of materials for carpet, flooring, wall covering, and insulation systems (216). Considerations are necessary at this stage for the selection of materials with the lowest possible VOC source strengths and for installation of these materials under temporary ventilation conditions so that adsorption (and later desorption) onto other porous materials is minimized (42).

The process of verifying the performance of HVAC and other building systems is known as commissioning (210). One of the most important aspects of commissioning and one that prevents many of the reactive IAQ problems discussed in Section 8.1 is the training of personnel in HVAC system control strategies, maintenance, and diagnosis of possible problems (210).

IAQ problems are often greatest during the initial occupancy phase of a building project. Tenants may demand to occupy a building before the HVAC system is fully functional or when construction activities are occurring in adjacent zones. Proper planning during the initial phase of a building project will prevent IAQ problems during initial occupancy. Occupancy of the building should occur only after the HVAC system is fully functional. HVAC system operational protocols that provide temporary maximum outdoor air ventilation during both occupied and unoccupied hours may be required in the first few weeks or months of building operation (216). The building commissioning process should assure that HVAC and other systems are operated and maintained properly during subsequent occupancy in the building.

Levin (217) has developed detailed proactive procedures to control indoor VOCs. His work is a model of how source control can be achieved for one potential indoor air pollutant during new construction. Considerable literature is currently available on VOC emissions from a wide variety of materials used in the construction phase of a building project (55, 56, 218, 219, 221). Levin (217) divides the evaluation of building materials for potential VOC emission into four sequential steps namely, (*a*) identifying these products or materials that are likely to emit toxic or irritating VOCs, (*b*) a screening of these products or materials with regard to information on the extent of their use and potential health effects, (*c*) review of applicable literature on emission testing for similar materials, and (*d*) determination of the source strength of VOC emissions from certain materials and recommendations for materials selection and handling in order to minimize potential indoor air contamination.

During the first step of evaluation, a list is made of all materials used in the project that might contain VOCs. This includes materials such as pesticides that may be used in foundation work and materials used in construction, insulation, finishing, and HVAC systems (217). During the next phase of the evaluation, those products that may emit VOCs that are irritating or toxic and that are used in large amounts (e.g., carpets, interior partitions, and adhesives) are identified.

Emission testing data available in the literature is reviewed relative to building materials likely to cause a problem. For example, the extensive use of polyurethane floor finishes in the building will likely be associated with the emission of benzene derivatives (220). A review of the literature may reveal if caulks or sealants proposed for the building are slow or fast drying and therefore likely to emit VOCs over long or short periods of time (217).

During the analysis and recommendation phase, potential suppliers are asked to submit data showing how their installed construction material or product is the least toxic or emits the "lowest" amount of VOCs (217). Thus a successful carpet vendor may be chosen on the basis that the carpet is made of the least toxic chemicals and is preconditioned to remove residual VOCs by ventilation in a warehouse before installation in the building.

During this final phase of building materials evaluation, the architect or designer will likely specify a number of control measures to prevent the dissemination of VOCs from newly installed materials such as adhesives and paints to "VOC sinks" in the building. Thus temporary ventilation during carpet installation using only the air supply portion of the HVAC system and direct exhaust of air to the outside through windows may be required to prevent the adsorption of strong VOC emissions into porous sinks such as those often found in common return plenums (217). Other porous (fleecy) materials may have to be protected from VOCs by temporary use of vapor barriers or plastic coverings (217). Round-the-clock operation of HVAC systems under conditions of maximum outdoor air supply are often required during the commissioning stage to lower VOC concentrations.

In applying a set of proactive recommendations to new construction it should be realized that each building is unique in that different contaminants and environmental variables may be expected. Thus in the new building studied by Hodgson and Daisey (222), the major source of VOCs in the indoor air was not the building itself, but rather processes such as liquid-process photocopiers and combustion emissions from a basement loading dock.

REFERENCES

1. P. R. Morey and J. E. Woods, *Occup. Med: State of the Art Rev.*, **2**, 547 (1987).
2. C. P. Yaglow, E. C. Riley, and D. I. Coggins, *ASHRAE Trans.*, **42**, 133 (1936).
3. American Society of Heating, Refrigerating and Air-Conditioning Engineers, *Thermal Environmental Conditions for Human Occupancy*, Standard 55-1981, Atlanta, 1981.
4. American Society of Heating, Refrigerating and Air-Conditioning Engineers, *Ventilation for Acceptable Indoor Air Quality*, Standard 62-1989, Atlanta, 1989.
5. J. A. J. Stolwijk, in *Proceedings of the 3rd International Conference on Indoor Air Quality and Climate*, Vol. 1, Stockholm, Sweden, 1984, pp. 23–29.
6. K. M. Wallingford and J. Carpenter, in *Proceedings IAQ '89 Managing Indoor Air for Health and Energy Conservation*, ASHRAE, Atlanta, 1986, pp. 448–453.
7. J. E. Woods, G. M. Drewry and P. R. Morey, in *Proceedings of the 4th International Conference on Indoor Air Quality and Climate*, Vol. 2, Berlin (West), 1987, pp. 464–468.
8. A. V. Nero, Jr., *Sci. Am.*, **258**, 42 (1988).
9. J. F. Brundage, R. M. Scott, W. M. Lednar, D. W. Smith, and R. N. Miller, *J. Am. Med. Assoc.*, **259**, 2108 (1988).
10. J. M. Samet, M. C. Marbury, and J. D. Spengler, *Am. Rev. Respir. Dis.*, **136**, 1486 (1987).
11. J. M. Samet, M. C. Marbury, and J. D. Spengler, *Am. Rev. Respir. Dis.*, **137**, 221 (1988).
12. American Society of Heating, Refrigerating and Air-Conditioning Engineers, *Indoor Air Quality Position Paper*, Atlanta, August 11, 1987.
13. R. A. Wadden and P. A. Scheff, *Indoor Air Pollution*, John Wiley, 1983.
14. American Society for Testing and Materials, Conference on Biological Contaminants in Indoor Environments, ASTM STP 1071, Philadelphia, In Press.
15. O. Strindehag, I. Josefsson, and E. Henningson, in *Healthy Buildings '88*, Vol. 3, Stockholm, Sweden, 1988, pp. 611–620.
16. P. M. Arnow, J. N. Fink, D. P. Schleuter, J. J. Barboriak, G. Mallison, S. I. Said, S. Martin, G. F. Unger, G. T. Scanlon, and V. P. Kurup, *Am. J. Med.*, **64**, 237 (1978).
17. P. R. Morey, M. J. Hodgson, W. G. Sorenson, G. J. Kullman, W. W. Rhodes, and G. S. Visvesvara, *ASHRAE Trans.*, **93**, 399 (1986).
18. P. R. Morey and B. A. Jenkins, in *Proceedings of IAQ '89, The Human Equation: Health and Comfort*, ASHRAE, Atlanta, 1989, pp. 67–71.
19. G. Brundrett and A. H. S. Onions, *J. Consumer Stud. Home Econ.*, **4**, 311 (1980).
20. M. J. Hodgson, P. R. Morey, M. Attfield, W. Sorenson, J. N. Fink, W. W. Rhodes, G. S. Visvesvara, *Arc. Environ. Health*, **40**, 96 (1985).
21. S. Yoshizawa, F. Sugawara, S. Ozawo, Y. Kohsaka, and A. Matsumae, in *Proceedings of the 4th International Conference on Indoor Air Quality and Climate*, Vol. 1, Berlin (West), 1987, pp. 627–631.
22. P. R. Morey and J. C. Feeley, Sr., *ASTM Standardization News*, **16**, 54 (1988).
23. H. Burge, *Occupational Medicine: State of the Art Reviews*, Vol. 4, J. E. Cone and M. J. Hodgson, Eds., Hanley and Belfus, Philadelphia, 1989, pp. 713–721.

24. A. Hyppel, in *Proceedings of 3rd International Conference on Indoor Air Quality and Climate*, Vol. 3, Stockholm, Sweden, 1984, pp. 443–447.
25. K. Holmberg, in *Proceedings of the 4th International Conference on Indoor Air Quality and Climate*, Vol. 1, Berlin (West), 1987, pp. 637–642.
26. J. C. Feeley, in *Architectural Design and Indoor Microbial Pollution*, R. B. Kundsin, Ed., Oxford University Press, New York, 1988, pp. 218–227.
27. M. R. Moser, T. R. Bender, H. S. Margolis, G. R. Noble, A. P. Kendal, and D. G. Ritter, *Am. J. Epidemiol.*, **110**, 1 (1979).
28. P. C. Bartlett, L. A. Vonbehren, R. P. Tewari, R. J. Martin, L. Eagleton, M. J. Isaac, and P. S. Kulkarni, *Am. J. Pub. Health*, **72**, 1369 (1982).
29. M. J. Hodgson and P. R. Morey, in *Immunology and Allergy Clinics of North America*, Vol. 9, W. R. Solomon, Ed., W. B. Saunders, 1989, pp. 399–412.
30. J. H. Edwards, *Br. J. Med.*, **37**, 55 (1980).
31. P. R. Morey, in *Architectural Design and Indoor Microbial Pollution*, R. B. Kundsin, Ed., Oxford University Press, New York, 1988, pp. 40–80.
32. M. J. Hodgson, P. R. Morey, J. S. Simon, T. D. Waters, and J. N. Fink, *Am. J. Epidemiol.*, **125**, 631 (1987).
33. M. Chatigny, in *Air Sampling Instruments*, 6th ed., American Conference of Governmental Industrial Hygienists, Cincinnati, 1983, pp. E2–E9.
34. American Conference of Governmental Industrial Hygienists, *Guidelines for the Assessment of Bioaerosols in the Indoor Environment*, Cincinnati, OH, 1989.
35. P. R. Morey, in *Indoor Air Quality*, International Symposium, American Industrial Hygiene Association, Akron, OH, in press.
36. World Health Organization, *Draft Final Report*, *Organic Pollutants*, Working Group on Indoor Air Quality, Berlin (West), August 23–28, 1987.
37. R. G. Lewis and L. A. Wallace, *ASTM Standardization News*, **16**, 40 (1988).
38. L. A. Wallace, T. D. Hartwell, K. Perritt, L. S. Sheldon, L. Michael, and E. D. Pellizzari, in *Proceedings of the 4th International Conference on Indoor Air Quality and Climate*, Vol. 1, Berlin (West), 1987, pp. 117–121.
39. D. S. Walkinshaw, Y. Tsuchiya, and I. Hoffmann, in *IAQ '87, Practical Control of Indoor Air Problems*, ASHRAE, 1987, pp. 139–149.
40. A. Sonnino and I. Pavan, in *Ergonomics and Health in Modern Offices*, E. Grandjean, Ed., Taylor and Francis, London, 1984, pp. 82–85.
41. J. R. Girman, in *Occupational Medicine: State of the Art Reviews*, Vol. 4, J. E. Cone and M. J. Hodgson, Eds., Hanley and Belfus, Philadelphia, 1989, pp. 695–712.
42. B. Berglund, I. Johansson, and T. Lindvall, in *Healthy Buildings 88*, Vol. 3, Stockholm, Sweden, 1988, pp. 299–309.
43. H. I. Virgin, in *Proceedings of the 3rd International Conference on Indoor Air Quality and Climate*, Vol. 3, Stockholm, Sweden, 1984, pp. 355–360.
44. J. R. Girman, A. T. Hodgson, and A. S. Newton, *Environ. Int.*, **12**, 317 (1986).
45. L. Molhave, B. Bach, and O. F. Pedersen, *Environ. Int.*, **12**, 167 (1986).
46. S. Kjaergaard, L. Molhave, and O. F. Pedersen, in *Proceedings of the 4th International Conference on Indoor Air Quality and Climate*, Vol. 1, Berlin (West), 1987, pp. 97–101.

47. M. J. Hodgson, in *Occupational Medicine: State of the Art Reviews*, Vol. 4, J. E. Cone and M. J. Hodgson, Eds., Hanley & Belfus, Philadelphia, 1989, pp. 593–606.
48. J. McCann, L. Horn, J. Girman, and A. V. Nero, in *Short-Term Bioassays in the Analysis of Complex Mixtures*, Vol. 5, S. S. Sandhu, D. M. DeMarini, M. J. Mass et al., Eds., Plenum Press, New York, 1987, pp. 325–354.
49. M. Tancrede, R. Wilson, L. Zeise, and E. A. C. Crouch, *Atm. Environ.*, **21**, 2187 (1987).
50. A. T. Hodgson and J. R. Girman, *Design and Protocol for Monitoring Indoor Air Quality ASTM STP*, 1002, N. L. Nagda and J. P. Harper, Eds., American Society for Testing and Materials, Philadelphia, 1989, pp. 244–256.
51. L. A. Wallace, *The Total Exposure Assessment Methodology (TEAM) Study: Summary and Analysis*, Vol. 1, Office of Research and Development, U.S. Environmental Protection Agency, 1987.
52. L. Wallace and C. A. Clayton, in *Proceedings of the 4th International Conference on Indoor Air Quality and Climate*, Vol. 1, Berlin (West), 1987, pp. 183–187.
53. H. C. Shields and C. J. Weschler, *J. Air Pollut. Control Assoc.*, **37**, 1039 (1987).
54. M. A. Cohen, P. B. Ryan, Y. Yanagisawa, J. D. Spengler, H. Ozkaynak, and P. S. Epstein, *J. Air Pollut. Control Assoc.*, **39**, 1086 (1989).
55. L. S. Sheldon, R. W. Handy, T. D. Hartwell, R. W. Whitmore, H. S. Zelon, and E. D. Pillizzari, *Indoor Air Quality in Public Buildings*, Vol. 1, Washington, DC, EPA/600/S6-88/009A (1988).
56. L. Sheldon, H. Zelon, J. Sickles, C. Easton, T. Hartwell, and L. Wallace, *Indoor Air Quality in Public Buildings*, Vol. II, Research Triangle Park, NC, EPA/600/S6-88/009b (1988).
57. J. C. Reinert, in *Proceedings of the 3rd International Conference on Indoor Air Quality and Climate*, Vol. 1, Stockholm, Sweden, 1984, pp. 233–238.
58. R. G. Lewis and A. E. Bond, in *Proceedings of the 4th International Conference on Indoor Air Quality and Climate*, Vol. 1, Berlin (West), 1987, pp. 195–199.
59. A. Zsolnay, I. Gebefugi, and F. Korte, in *Proceedings of the 4th International Conference on Indoor Air Quality and Climate*, Vol. 1, Berlin (West), 1987, pp. 262–264.
60. D. J. Anderson and R. A. Hites, *Environ. Sci. Technol.*, **22**, 717 (1988).
61. M. J. Hodgson, G. D. Block, and D. K. Parkinson, *J. Occup. Med.*, **18**, 434 (1986).
62. *EPA Indoor Air Quality Implementation Plan*, Appendix A, Preliminary Indoor Pollution Information Assessment, Washington, DC, EPA/600/8-87/014, 1987.
63. National Research Council, *An Assessment of the Health Risks of Seven Pesticides Used for Termite Control*, Committee on Toxicology, Washington, DC, 1982.
64. J. D. Spengler and M. A. Cohen, in R. B. Gammage and S. V. Kaye, Eds., *Indoor Air and Human Health*, Lewis Publishers, 1985, pp. 261–278.
65. U.S. Department of Energy, *Indoor Air Quality Environmental Information Handbook: Combustion Sources*, DOE/EV/10450-1, 1985.
66. J. L. Woodring, T. L. Duffy, J. T. Davis, and R. R. Bechtold, *Am. Ind. Hyg. Assoc. J.*, **46**, 350 (1985).
67. I. H. Billick, J. D. Spengler, P. B. Ryan, P. E. Baker, and S. D. Colome, *J. Air Pollut. Control Assoc.*, **39**, 1169 (1989).
68. J. D. Spengler and K. Sexton, *Science*, **221**, 9 (1983).

69. G. D. Thurston, in *Proceedings of the 4th International Conference on Indoor Air Quality and Climate*, Vol. 1, Berlin (West), 1987, pp. 451–455.
70. Code of Federal Regulations, Title 40, Part 50 (40CFR50), *National Primary and Secondary Ambient Air Quality Standards*, U.S. Environmental Protection Agency.
71. J. D. Mulik, R. G. Lewis, and W. A. McClenny, *Anal. Chem.*, **61**, 187 (1989).
72. A. L. Wilson, S. D. Colome, P. E. Baker, S. J. Cunningham, E. W. Becker, and W. G. Bope, in *Proceedings of the 4th International Conference on Indoor Air Quality and Climate*, Vol. 1, Berlin (West), 1987, pp. 456–459.
73. Amer. Soc. of Heating, Refrigerating, and Air-Conditioning Engineers, *Addendum to Ventilation for Acceptable Indoor Air Quality*, (ASHRAE Standard 62-1989), Public Review Draft, November, 1989.
74. J. F. Goldstein, L. R. Andrews, K. Lieber, G. Fourtrakis, F. Kazembe, P. Huang, and C. Hayes, in *Proceedings of the 4th International Conference on Indoor Air Quality and Climate*, Vol. 1, Berlin (West), 1987, pp. 293–297.
75. World Health Organization, *Air Quality Guidelines for Europe*, WHO Regional Publications, European Series No. 23, 1987, pp. 297–314.
76. W. J. Parkhurst, J. P. Harper, J. D. Spengler, L. P. Fraumeni, A. M. Majahad, and J. W. Cropp, Paper 88-109.4, Air Pollution Control Association, 1988, 12 pp.
77. U.S. Environmental Protection Agency, *Indoor Air Facts*, No. 5, ANR-445, 1989.
78. G. H. Clausen, in *Proc. IAQ '88, Engineering Solutions to Indoor Air Problems*, ASHRAE, Atlanta, 1988, pp. 267–274.
79. E. A. Miesner, S. N. Rudnich, F. C. Hu, J. D. Spengler, L. Preller, H. Ozkaynak, and W. Nelson, Paper 88-76.4, Air Pollution Control Association, 1988, 16pp.
80. G. H. Clausen, S. B. Muller, P. O. Fanger, B. P. Leaderer, and R. Dietz, in *Proc. IAQ '86, Managing Indoor Air for Health and Energy Conservation*, ASHRAE, Atlanta, 1986, pp. 119–125.
81. M. J. Hodgson, in *Occupational Medicine: State of the Art Reviews*, Vol. 4, J. E. Cone and M. J. Hodgson, Eds., Hanley & Belfus, Philadelphia, 1989, pp. 735–740.
82. W. S. Cain, B. P. Leaderer, R. Isseroff, L. G. Berglund, R. J. Huey, E. D. Lipsitt, and D. Perlman, *Atm. Environ.*, **17**, 1183 (1983).
83. P. O. Fanger, J. Lauridsen, P. Bluyssen, and G. Clausen, *Energy and Buildings*, **12**, 7 (1988).
84. U.S. Public Health Service, *Surgeon Generals' Report: The Health Consequences of Involuntary Smoking*, USDHHS, CDC, 87-8398, 1987.
85. J. D. Spengler, D. W. Dockery, W. A. Turner, J. M. Wolfson, and B. G. Ferris, in *Atmos. Environ.*, **15**, 23 (1981).
86. S. K. Hammond, J. Coghlin, and B. P. Leaderer, in *Proceedings of the 4th International Conference on Indoor Air Quality and Climate*, Vol. 2, Berlin (West), 1987, pp. 131–136.
87. American Society for Testing and Materials, *Standard Test Method for Nicotine in Indoor Air*, ASTM D22.05, draft #3, December 20, 1988.
88. National Research Council, *Environmental Tobacco Smoke: Measuring Exposures and Assessing Health Effects*, Committee on Passive Smoking, Washington, DC, National Academy Press, 1986.
89. J. L. Repace, *Bull. N.Y. Acad. Med.*, **57**, 936 (1981).

90. J. E. Woods, P. R. Morey, and J. A. J. Stolwijk, in *Chartered Institute of Building Services Engineers, Advances in Air-Conditioning Conference*, London, October 7, 1988, pp. 15–25.
91. American Society of Heating Refrigerating, and Air-Conditioning Engineers, *Standards for Natural and Mechanical Ventilation*, Standard 62-1973 (ANSI 3 194-1-1977), Atlanta, 1977.
92. B. Berg-Munch, G. Clausen, and P. O. Fanger, *Environ. Int.*, **12**, 195 (1986).
93. B. Berglund, paper presented at the *Indoor Air Quality Symposium at Georgia Institute of Technology*, September 20–22, 1988, 6 pp.
94. T. C. Wang, *ASHRAE Trans.*, **81**, 32–44 (1975).
95. R. C. Fortmann, *ASTM Standardization News*, **16**, 50 (1988).
96. L. S. Sheldon, H. Zelon, J. Sickles, C. Eaton, T. Hartwell, and R. Jungers, Paper 88-109.8, Air Pollution Control Association, 1988, 9 pp.
97. D. T. Harrje, L. M. Hubbard, and D. C. Sanchez, in *Healthy Buildings '88*, Vol. 2, Stockholm, Sweden, 1988, pp. 143–152.
98. U.S. Environmental Protection Agency, *Report to Congress on Indoor Air Quality; Volume II: Assessment and Control of Indoor Air Pollution*, 1989, EPA/400/1-89/001C.
99. J. L. R. Balzer, W. C. Cooper, and D. P. Fowler, *Am. Ind. Hyg. Assoc. J.*, **32**, 512 (1971).
100. R. R. Gamboa, B. P. Gallagher, and K. R. Mattews, in *Proc. IAQ '88, Engineering Solutions to Indoor Air Problems*, ASHRAE, Atlanta, 1988, pp. 25–33.
101. K. Kreiss, in *Occupational Medicine: State of the Art Reviews*, Vol. 4, J. E. Cone and M. J. Hodgson, Eds., Hanley & Belfus, Philadelphia, 1989, pp. 575–592.
102. J. D. Sacca, *Ann. Allergy*, **34**, 105 (1975).
103. H. H. Newball and S. A. Brahim, *Environ. Res.*, **12**, 201 (1976).
104. A. Rindely, C. Hugod, E. Bach, and N. O. Breum, in *Proceedings of the 4th International Conference on Indoor Air Quality and Climate*, Vol. 1, Berlin (West), 1987, pp. 590–594.
105. A. Bohroff, *J. Am. Med. Assoc.*, **186**, 30 (1963).
106. Sheet Metal and Air-Conditioning Contractors' National Association, *Fibrous Glass Duct Construction Standards*, 5th ed., 1979.
107. M. S. Crandall, Paper 88-110.3, Air Pollution Control Association, 1988, 11pp.
108. S. Burge, A. Hedge, S. Wilson, J. H. Bass, and A. Robertson, *Ann. Occup. Hyg.*, **31**, 493 (1987).
109. J. Harrison, A. C. Pickering, M. J. Finnegan, P. K. C. Austwick, in *Proceedings of the 4th International Conference on Indoor Air Quality and Climate*, Vol. 2, Berlin (West), 1987, pp. 487–491.
110. P. Skov, O. Valbjorn et al., *Environ. Int.*, **13**, 339 (1987).
111. O. Valbjorn and P. Skov, in *Proceedings of the 4th International Conference on Indoor Air Quality and Climate*, Vol. 2, Berlin (West), 1987, pp. 593–597.
112. P. Skov and O. Valbjorn, in *Proceedings of the 4th International Conference on Indoor Air Quality and Climate*, Vol. 2, Berlin (West), 1987, pp. 439–443.
113. P. H. Nielsen, in *Proceedings of the 4th International Conference on Indoor Air Quality and Climate*, Vol. 2, Berlin (West), 1987, pp. 598–602.
114. P. O. Fanger, *Energy and Buildings*, **12**, 1 (1988).

115. P. O. Fanger, in *Healthy Buildings '88*, Vol. I, Stockholm, Sweden, 1988, pp. 39–51.
116. R. L. Jacors, R. E. Thorner, J. R. Holcomb, L. A. Schwietz, and F. O. Jacobs, *Ann. Int. Med.*, **105**, 204 (1986).
117. J. N. Fink, E. F. Banaszak, W. H. Thiede, and J. J. Barboriak, *Ann. Internal Med.*, **74**, 80 (1971).
118. C. L. Johnson, J. L. Bernstein, J. S. Gallagher, P. F. Bonventre, and S. M. Brooks, *Am. Rev. Respir. Dis.*, **122**, 339 (1980).
119. G. O. Solley and R. E. Hyatt, *J. Allergy Clin. Immunol.*, **65**, 65 (1980).
120. R. S. Bernstein, W. G. Sorenson, D. Garabrant, C. Reaux, and R. Treitman, *Am. Ind. Hyg. Assoc. J.*, **44**, 161 (1983).
121. MRC Symposium, *Thorax*, **32**, 653 (1977).
122. R. Rylander, P. Haglind, M. Lundholm, J. Mattsby, and K. Stengvist, *Clinical Allergy*, **8**, 511 (1978).
123. R. Rylander and P. Haglind, *J. Clin. Allergy*, **14**, 109 (1984).
124. W. R. Soloman, *J. Allergy Clin. Immunol.*, **54**, 222 (1974).
125. M. J. Finnegan and A. C. Pickering, in *Proceedings of the 3rd International Conference on Indoor Air Quality and Climate*, Vol. I, Stockholm, Sweden, 1984, pp. 257–262.
126. M. J. Finnegan and A. C. Pickering, *Clin. Allergy*, **16**, 389 (1986).
127. P. L. Garbe, B. J. Davis, J. S. Weisfeld, L. Markowitz, P. Miner, F. Garrity, J. M. Barbaree, and A. L. Reingold, *J. Am. Med. Assoc.*, **254**, 521 (1985).
128. S. J. States, L. F. Conley, J. M. Kuchta, B. M. Oleck, M. J. Lipovich, R. S. Wolford, R. M. Wadowsky, A. M. McNamara, J. L. Sykora, G. Keleti, and R. B. Yee, *Appl. Environ. Microbiol.*, **53**, 979 (1987).
129. S. Friedman, K. Spitalny, J. Barbaree, Y. Faur, and R. McKinney, *Am. J. Publ. Health*, **77**, 568 (1987).
130. J. D. Band, M. La Venture, J. P. Davis, G. F. Mallison, P. Skaliy, P. S. Hayes, W. L. Schell, H. Weiss, D. J. Greenberg, and D. W. Fraser, *J. Am. Med. Assoc.*, **245**, 2404 (1981).
131. Chartered Institution of Building Services Engineers, *Minimising the Risk of Legionnaires' Disease*, TMB, 1987, 20 pp.
132. J. M. Sykes and A. M. Brazier, *Ann. Occup. Hyg.*, **32**, 63 (1988).
133. Wisconsin Division of Health, *Control of Legionella in Cooling Towers, Summary Guidelines*, 1987.
134. J. E. Janssen, *ASHRAE J.*, **31**, 40 (1989).
135. R. E. Dixon, *Am. J. Med.*, **78**(suppl. 6B), 45–51 (1985).
136. J. A. Carlton-Foss, *Ann. Am. Conf. Gov. Ind. Hyg.*, **10**, 93 (1984).
137. American Society of Heating, Refrigerating and Air-Conditioning Engineers, *Thermal Environmental Conditions for Human Occupancy*, Public Review Draft, 1989, BSR/ ASHRAE 55-81r.
138. American Society of Heating, Refrigerating, and Air-Conditioning Engineers, *Far East Conference on Air-Conditioning in Hot Climates*, Singapore, September 1987.
139. S. Tanabe and K. Kimura, *ASHRAE Far East Conference on Air-Conditioning in Hot Climates*, Singapore, 1987, pp. 3–21.
140. J. E. Woods, in *Occupational Medicine: State of the Art Reviews*, Vol. 4, J. E. Cone and M. J. Hodgson, Eds., Hanley & Belfus, Philadelphia, 1989, 753–770.

141. U.S. Department of Energy, *Nonresidential Building Energy Consumption Survey: Characteristics of Commercial Buildings*, Washington, DC, Energy Information, DOE/EIA-0246(83), 1985.
142. P. R. Morey, in *Proc. IAQ '88, Engineering Solutions to Indoor Air Problems*, ASHRAE, Atlanta, 1988, pp. 10–24.
143. C. E. Dorgan, in *Proc. IAQ '88 Engineering Solutions to Indoor Air Problems*, ASHRAE, Atlanta, 1988, pp. 77–83.
144. J. H. Eto and C. Mayer, *ASHRAE Trans.*, 94 (part 2) (1988).
145. D. B. Baker, in *Occupational Medicine: State of the Art Reviews*, Vol. 4, J. E. Cone and M. J. Hodgson, Eds., Hanley & Belfus, Philadelphia, 1989, pp. 607–624.
146. R. W. Alexander and J. J. Fedoruk, *J. Occup. Med.*, **28**, 42 (1986).
147. R. J. Levine, D. J. Sexton, F. J. Romm, B. T. Wood, and J. Kaiser, *Lancet*, 1500 (1974).
148. P. A. Boxer, *J. Occup. Med.*, **27**, 867 (1985).
149. G. J. Magarian, *Medicine*, **61**, 219 (1982).
150. M. J. Colligan and M. J. Smith, *J. Occup. Med.*, **20**, 401 (1978).
151. P. R. Morey and D. E. Shattuck, in *Occupational Medicine: State of the Art Reviews*, Vol. 4, J. E. Cone and M. J. Hodgson, Eds., Hanley & Belfus, Philadelphia, 1989, pp. 625–642.
152. P. E. McNall and A. K. Persily, *Ann. Am. Conf. Gov. Ind. Hyg.*, **10**, 49 (1984).
153. American Society of Heating, Refrigerating, and Air-Conditioning Engineers, *Gravimetric and Dust Spot Procedures for Testing Air Cleaning Devices Used in General Ventilation for Removing Particulate Matter*, Public Review Draft, ASHRAE Standard 52.1p, 1989.
154. H. Levin and T. J. Phillips, in *Design and Protocol for Monitoring Indoor Air Quality, ASTM STP* 1002, N. L. Nagda and J. P. Harper, Eds., American Society for Testing and Materials, Philadelphia, 1989, pp. 99–110.
155. A. E. Wheeler, in *Proceedings IAQ '88 Engineering Solutions to Indoor Air Problems*, ASHRAE, Atlanta, 1988, pp. 99–107.
156. H. Levin, in *Proceedings IAQ '88, Engineering Solutions to Indoor Air Problems*, ASHRAE, Atlanta, 1988, pp. 61–76.
157. A. K. Persilly, in *Proceedings IAQ '86, Managing Indoor Air for Health and Energy Conservation*, ASHRAE, Atlanta, 1986, pp. 548–558.
158. F. J. Offermann and D. Int-Hout, in *Proceedings of the 4th International Conference on Indoor Air Quality and Climate*, Vol. 3, Berlin (West), 1987, pp. 313–318.
159. National Research Council, *An Assessment of the Health Risks of Morpholine and Diethylaminoethanol*, National Academy Press, Washington, DC, 1983.
160. M. Ganier, P. Lieberman, J. Fink, and D. G. Lockwood, *Chest*, **77**, 183 (1980).
161. A. Cockcroft, J. Edwards, C. Bevan, I. Campbell, G. Collins, K. Houston, D. Jenkins, S. Lathum, M. Sanders, and D. Trotman, *Br. J. Ind. Med.*, **38**, 144 (1981).
162. B. Ager and J. Ticker, *Ann. Occup. Hyg.*, **27**, 341 (1983).
163. G. Brundrett, J. Collins, G. Da Costa, J. H. Edwards, W. P. Jones, D. J. Newson, and R. J. Stone, *J. Chart Inst. Bldg. Serv. Eng.*, **3**, 35 (1981).
164. S. Block, *Appl. Microbiol.*, **1**, 287 (1953).
165. G. S. Raijhans, *Occup. Health Ontario*, **4**, 160 (1983).

166. K. Ikeda, S. Yoshizawa, T. Irie, et al., *ASHRAE Trans.*, Paper No. 2955, 1986.
167. M. Narasak, in *Proceedings of the 4th International Conference on Indoor Air Quality and Climate*, Vol. 3, Berlin (West), 1987, pp. 277–282.
168. P. O. Fanger, *ASHRAE J.*, **31**, 33 (1989).
169. American Society of Heating Refrigerating, and Air-Conditioning Engineers, *Energy Conservation in New Building Design*, ASHRAE Standard 90-75, Atlanta, 1975.
170. American Society of Heating Refrigerating, and Air-Conditioning Engineers, *Ventilation for Acceptable Indoor Air Quality*, ASHRAE Standard 62-1981.
171. C. F. Klein, S. M. Relwani, and D. J. Moschandreas, in *Proceedings of the 4th International Conference on Indoor Air Quality and Climate*, Vol. 3, Berlin (West), 1987, pp. 249–252.
172. D. T. Grimsrud and K. Y. Teichman, *ASHRAE J.*, **31**, 51 (1989).
173. J. E. Woods, P. R. Morey, and D. R. Rask, in *Design and Protocol for Monitoring Indoor Air Quality*, N. L. Nagda and J. P. Harper, Eds., American Society for Testing and Materials, Philadelphia, 1989, pp. 80–98.
174. National Research Council, *Building Diagnostics, A Conceptional Framework*, Building Research Board, National Academy Press, Washington, DC, 1985.
175. Electric Power Research Institute, *Manual on Indoor Air Quality*, EM-3469, 1984.
176. P. Quinlan, J. M. Macher, L. E. Alevantis, and J. E. Cone, in *Occupational Medicine: State of the Art Reviews*, Vol. 4, J. E. Cone and M. J. Hodgson, Eds., Hanley & Belfus, Philadelphia, 1989, pp. 771–797.
177. C. Y. Shaw, Paper 88-110.4, Air Pollution Control Association, 1988, 19 pp.
178. D. A. Sterling, D. J. Moschandreas, and S. M. Relwani, in *Proceedings of the 4th International Conference on Indoor Air Quality and Climate*, Vol. 2, Berlin (West) 1987, pp. 444–448.
179. A. K. Persily, W. A. Turner, H. A. Burge, and R. A. Grot, in *Design and Protocol for Monitoring Indoor Air and Materials, ASTM STP* 1002, N. L. Nagda and J. P. Harper, Eds., American Society for Testing and Materials, Philadelphia, 1989, pp. 35–50.
180. H. Levin, Paper 88-94.5, Air Pollution Control Association, 1988, 15 pp.
181. R. B. Gammage and W. S. Kerbel, in *Proceedings of the 4th International Conference on Indoor Air Quality and Climate*, Vol. 3, Berlin (West), 1987, pp. 567–572.
182. C. J. Weschler, H. C. Shields, S. P. Keltey, L. A. Psota-Kelty, and J. D. Sinclair, in *Design and Protocol for Monitoring Indoor Air Quality, ASTM STP* 1002, N. L. Nagda and J. P. Harper, Eds., American Society for Testing and Materials, Philadelphia, 1989, pp. 9–34.
183. J. D. Spengler, J. Ware, F. Speizer, B. Ferris, D. Dockery, E. Lebret, and B. Brunnekreef, in *Proceedings of the 4th International Conference on Indoor Air Quality and Climate*, Vol. 2, Berlin (West), 1987, pp. 218–223.
184. W. E. Lambert and J. M. Samet, in *Occupational Medicine: State of the Art Reviews*, Vol. 4, J. E. Cone and M. J. Hodgson, Eds., Hanley & Belfus, Philadelphia, 1989, pp. 723–733.
185. M. D. Lebowitz, J. J. Quackenboss, M. L. Soczek, S. D. Colome, and P. J. Lioy, in *Design and Protocol for Monitoring Indoor Air Quality, ASTM STP* 1002, N. L. Nagda and J. P. Harper, Eds., American Society for Testing and Materials, Philadelphia, 1989, pp. 203–216.

186. M. D. Lebowitz, J. J. Quackenboss, M. L. Soczek, M. Kollander, and S. Colome, *J. Air Pollut. Control Assoc.*, **39**, 1411 (1989).
187. F. H. Rohles, J. E. Woods, and P. R. Morey, in *Proceedings of the 4th International Conference on Indoor Air Quality and Climate*, Vol. 2, Berlin (West), 1987, pp. 520–524.
188. U.S. Environmental Protection Agency, *Report to Congress on Indoor Air Quality, Executive Summary and Recommendations*, 1989, EPA/400/1-89/001A.
189. E. L. Besch, in *Occupational Medicine: State of the Art Reviews*, Vol. 4, J. E. Cone and M. J. Hodgson, Eds., Hanley & Belfus, Philadelphia, 1989, pp. 741–752.
190. V. C. Armstrong, M. E. Meek, and D. S. Walkinshaw, Paper 88-94.3, Air Pollution Control Association, 1988, 12 pp.
191. M. A. Berry, Paper 88-94.7, Air Pollution Control Association, 1988, 12 pp.
192. C. B. Meyer and P. M. Ayd, in *Healthy Buildings '88*, Vol. 3, Stockholm, Sweden, 1988, pp. 381–390.
193. New Jersey Department of Labor, *Safety and Health Standards for Public Employees, Standard for Indoor Air Quality*, NJAC, 12, 100-114, Draft, January 23, 1989.
194. Province of Ontario, *Draft Report of the Interministerial Committee on Indoor Air Quality*, Ministry of Labor, Toronto, 1988.
195. *Constitution of the World Health Organization*, 1946, Official Record of the World Health Organization, Vol. 2, p. 100.
196. T. Lindvall, in *Proceedings of the 4th International Conference on Indoor Air Quality and Climate*, Vol. 4, Berlin (West), 1987, pp. 117–133.
197. J. A. J. Stolwijk, in *Proceedings of the 4th International Conference on Indoor Air Quality and Climate*, Vol. 4, Berlin (West), 1987, pp. 3–8.
198. D. J. Moschandreas, in *Healthy Buildings '88*, Vol. I, Stockholm, Sweden, 1988, pp. 19–23.
199. P. F. Ricci and T. E. McKone, in *Healthy Buildings '88*, Vol. I, Stockholm, Sweden, 1988, pp. 183–187.
200. P. McNall and K. Teichman, in *Healthy Buildings '88*, Vol. I, Stockholm, Sweden, 1988, pp. 163–168.
201. R. H. Ferahian, in *Healthy Buildings '88*, Vol. 3, Stockholm, Sweden, 1988, pp. 671–678.
202. Amer. Conference of Governmental Industrial Hygienists, *Threshold Limit Values and Biological Exposure Indices for 1989–1990*, Cincinnati, OH.
203. T. A. E. Platts-Mills, M. D. Chapman, S. M. Pollart, P. W. Heymann, and C. M. Luczynska, Paper 88-110.1, Air Pollution Control Association, 1988, 11 pp.
204. J. E. M. H. Van Bronswijk, in *Proceedings of the 4th International Conference on Indoor Air Quality and Climate*, Vol. 2, Berlin (West), 1987, pp. 747–751.
205. American Society of Heating, Refrigerating, and Air-Conditioning Engineers, in *Proceedings of IAQ '86, Managing Indoor Air for Health and Energy Conservation*, Atlanta, 1986.
206. American Society of Heating, Refrigerating, and Air-Conditioning Engineers, *Proceedings of IAQ '87, Practical Control of Indoor Air Problem*, Atlanta, 1987.
207. American Society of Heating, Refrigerating, and Air-Conditioning Engineers, *Proceedings of IAQ '88, Engineering Solutions to Indoor Air Problems*, Atlanta, 1988.

208. American Society of Heating, Refrigerating, and Air-Conditioning Engineers, *Proceedings of IAQ '89, The Human Equation: Health and Comfort*, Atlanta, 1989.
209. W. J. Fisk, R. K. Spencer, D. T. Grimsrud, F. J. Offermann, B. Petersen, and R. Sextro, *Indoor Air Quality Control Techniques: Radon, Formaldehyde, Combustion Products*, Pollution Technology Review, No. 144, Noyes Data Corp., 1987, 245 pp.
210. C. N. Lawson, *ASHRAE J.*, **31**, 56 (1989).
211. V. Loftness and V. Hartkopf, in *Occupational Medicine: State of the Art Reviews*, Vol. 4, J. E. Cone and M. J. Hodgson, Eds., Hanley & Belfus, Philadelphia, 1989, pp. 643–665.
212. J. E. Woods, J. E. Janssen, P. R. Morey, and D. R. Rask, in *Proceedings IAQ '87, Practical Control of Indoor Air Problems*, ASHRAE, Atlanta, 1987, pp. 338–348.
213. P. E. McNall, in *Proceedings IAQ '88, Engineering Solutions to Indoor Air Problems*, ASHRAE, Atlanta, 1988, pp. 51–60.
214. D. S. Ensor, A. S. Viner, J. T. Hanley, P. A. Lawless, K. Ramanathan, M. K. Owen, T. Yamamoto, and L. E. Sparks, in *Proceedings IAQ '88, Engineering Solutions to Indoor Air Problems*, ASHRAE, Atlanta, 1988, pp. 111–129.
215. M. Meckler and J. E. Janssen, in *Proceedings IAQ '88, Engineering Solutions to Indoor Air Problems*, ASHRAE, Atlanta, pp. 131–147.
216. H. Levin, in *Proceedings IAQ '87, Practical Control of Indoor Air Problems*, ASHRAE, Atlanta, 1987, pp. 157–170.
217. H. Levin, in *Occupational Medicine: State of the Art Reviews*, Vol. 4, J. E. Cone and M. J. Hodgson, Eds., Hanley and Belfus, Philadelphia, 1989, pp. 667–693.
218. C. W. Bayer and M. S. Black, in *Design and Protocol for Monitoring Indoor Air Quality, ASTM STP* 1002, N. L. Nagda and J. P. Harper, Eds., American Society for Testing and Materials, Philadelphia, 1989, pp. 234–243.
219. D. Jennings, D. Eyre, and M. Small, *The Safety Categorization of Sealants According to Their Volatile Emissions*, Report No. M91-11/1-4, 1988E, Ottawa, Canada, Ministry of Supply and Services, 1988.
220. B. A. Tichenor, in *Proceedings of the 4th International Conference on Indoor Air Quality and Climate*, Vol. 1, Berlin (West), 1987, pp. 8–15.
221. M. C. Baechler, H. Ozkaynak, J. D. Spengler, L. A. Wallance, and W. C. Nelson, Paper 89-86.7, Air and Waste Management Association, 1989, 14 pp.
222. A. T. Hodgson and J. M. Daisey, Paper 89-80.7, Air and Waste Management Association, 1989, 15 pp.

CHAPTER EIGHTEEN

Potential Exposures in the Manufacturing Industry—Their Recognition and Control

William A. Burgess, C.I.H.

1 INTRODUCTION

Although employment in the United States is shifting from manufacturing to the service sector, manufacturing continues to employ in excess of 20 million workers in workplaces that present both traditional and new occupational health hazards. To understand the nature of these hazards, the occupational health professional must understand not only the toxicology of industrial materials but the manufacturing technology that defines how contaminants are released from the process, the physical form of the contaminants, and the route of exposure. Physical stresses including noise, vibration, heat, and ionizing and nonionizing radiation must also be attributed to the manufacturing processes. Fourteen specific operations representing both large employment and potential health hazard to the worker have been chosen for discussion in this chapter. This discussion is based on the previous Patty chapters on this subject and a more recent update by me (1).

2 ABRASIVE BLASTING

Abrasive blasting is practiced in a number of occupational settings including bridge and building construction, shipbuilding and repair, founding, and metal finishing.

Patty's Industrial Hygiene and Toxicology, Fourth Edition, Volume 1, Part A, Edited by George D. Clayton and Florence E. Clayton
ISBN 0-471-50197-2

The process is used in heavy industry as an initial cleaning step to remove surface coatings and scale, rust, or fused sand in preparation for subsequent finishing operations. No other method is as effective and economical for this purpose. Abrasive blasting is used in intermediate finishing operations to remove flashing, tooling marks, or burrs from cast, welded, or machined fabrications, and to provide a matte finish to enhance bonding of paint or other coatings. In addition, chilled-iron and steel shot abrasive can be used to peen the metal surface to improve resistance to fatigue stress, reduce surface porosity, and increase the wearability of parts.

Various abrasives are used in blast cleaning installations. Ground corn cobs, nut shells, and glass beads and spheres are used for light cleaning. A new system applicable to light and moderate duty cleaning utilizes carbon dioxide pellets. The common heavy-duty abrasives in order of use are silica sand, iron and steel shot and grit, aluminum oxide (Al_2O_3), silicon carbide (SiC), metallurgical slag, and aluminum shot.

Three major types of blasting equipment are used in industry and construction. In pressure blasting, a venturi driven by compressed air aspirates the abrasive from a storage "pot" to a nozzle where it is directed to the workpiece at high velocity by the operator. In a second system used in blasting enclosures, a high-speed centrifugal impeller projects the abrasive at the workpiece. Finally, a wet blast technique utilizes a suspension of abrasive in water, which is directed at the workpiece from a nozzle using air and water pressure.

2.1 Application and Hazards

In open air blasting in construction and shipyard applications, general area contamination occurs unless isolation of the work area can be achieved. As a result, the blasting crew, including the operator, "pot man," and cleanup personnel, all may be exposed to high dust concentrations depending on the existing wind conditions. Open air work is difficult to control and has resulted in silicosis in sandblasters and nearby workers.

Abrasive-vacuum blasting is a relatively new technique that has good application in outdoor abrasive blasting operations where one cannot achieve control by isolation or use of an exhausted enclosure. The abrasive nozzle is coupled with a low-volume, high-velocity exhaust system that provides pickup of spent abrasive and workpiece debris directly at the site of blasting. No data have been published in the open literature on the collection efficiency of this system. Anecdotal information suggests it may be useful only for small work.

A variety of exhausted enclosures are used in the metalworking industry, including cabinets and exhausted rooms, whereas in automatic barrel, rotating table, and tunnel units the operator controls the operation from outside an enclosure. In most cases, these industrial units have integral local exhaust ventilation systems.

The two obvious health hazards one must consider in reviewing abrasive blasting operations are airborne contaminants and noise. Dust exposure includes the base metal being blasted, the surface coating or contamination being removed, and the abrasive in use and any abrasive contaminant from previous operations. In the United States, the widespread use of sand containing high concentrations of crystalline quartz continues to present a major hazard to workers. In the United States, a prevalence rate of silicosis of 6.5 percent occurs in the abrasive blasting population of 100,000. A number of European nations have forbidden the use of silica sand for this application, but this effective control measure has not been adopted in the United States. Exposure to the metal shot abrasives and the two synthetic materials, silicon carbide and aluminum oxide, has not resulted in demonstrable lung diseases. The use of abrasives based on metallurgical slags warrants attention owing to the presence of heavy metal contamination.

In most cases, the base metal being blasted is iron or steel, and the resulting exposure to iron dust presents a limited hazard. If blasting is carried out on metal alloys containing such materials as nickel, manganese, lead, or chromium, the hazard should be evaluated by air sampling. The surface coating or contamination on the workpiece frequently presents a major inhalation hazard. In the foundry it may be fused silica sand; in ship repair, the surface coating may be a lead-based paint or an organic mercury or tin biocide. In addition to direct exposure to the dust from the surface coating, one must guard against concentrating the contaminant from this coating in the abrasive recycle system. Abrasives used to remove lead paint from a bridge structure may contain up to 1 percent lead by weight.

When blasting units are used in open air or in blasting rooms, the operators are exposed to noise levels in the range of 85 to 118 dBA. The noise exposure at the operator's location at cabinets or automatic rooms ranges from 83 to 110 dBA. In a survey of 22 abrasive blasting facilities, 17 exceeded the Occupational Safety and Health Administration (OSHA) Hearing Conservation Amendment Standard of 90 dBA.

2.2 Control

The ventilation requirements for abrasive blasting enclosures have evolved over several decades, and effective design criteria are now available. The minimum exhaust volumes, based on seals and curtains in new condition, include 20 air changes per hour for cabinets with a minimum of 500 fpm (feet per minute) through all baffled inlets; the tumble and swing table blast units also require 500 fpm through all openings with doors closed and 200 fpm through curtained openings; and the abrasive blasting rooms require 80 to 100 fpm downdraft velocity depending on the room geometry (2, 3). The nature of the operation necessitates rugged enclosures and easily maintained curtains and seals to eliminate outward dust leakage. Systems to be used for organic abrasives require special design features because of the fire and explosion hazard. Operators should not be permitted to work inside such enclosures.

Dust control on abrasive blasting equipment depends to a large degree on the integrity of the enclosure. All units should be inspected periodically, including baffle plates at air inlets, gaskets around doors and windows, gloves and sleeves on cabinets, star gaskets at hose inlets, and the major structural seams of the enclosure. The dust collection system should also be inspected to ensure that fines are removed and the necessary periodic maintenance is carried out. If possible, the operation in walk-in rooms should be set up so that the direct blast of the nozzle does not hit the door.

Operators directly exposed to the blasting operation, as in open air or blasting room operations, must be provided with a National Institute for Occupational Safety and Health (NIOSH)-approved type CE abrasive blasting hood or helmet. The intake for the air compressor providing the respirable air supply should be located in an area free from air contamination, and the quality of the air delivered to the respirator must be checked periodically. The respirator must be used in the context of a full respirator program, and rigorous maintenance of the equipment is essential. Serious health hazards may result from inadequate abrasive blasting respirator programs. To evaluate truly the protection from such equipment, air samples should be taken inside the helmet.

3 ACID AND ALKALI CLEANING OF METALS

After the removal of major soils, oils, and grease by degreasing, metal parts are often treated in acid and alkaline baths to condition the parts for electroplating or other finishes. The principal hazard in this series of operations is exposure to acid and alkaline mist released by heating, air agitation, gassing from electrolytic operation, or cross contamination between tanks.

3.1 Acid Pickling and Bright Dip

Pickling or descaling is a technique used to remove oxide scales formed from heat treatment, welding, and hot forming operations prior to surface finishing. On low- and high-carbon steels, the scale is iron oxide, whereas on stainless steels it is composed of oxides of iron, chromium, nickel, and other alloying metals (4). The term pickling is apparently derived from the early practice of cleaning metal parts by dipping them in vinegar.

Scale and rust are commonly removed from low and medium alloy steels using a nonelectrolytic immersion bath of 5 to 15 percent sulfuric acid at a temperature of 60 to 82°C (140 to 180°F) or a 10 to 25 percent hydrochloric acid bath at room temperature. Because sulfuric and hydrochloric acids frequently cause pitting, phosphoric acid may be used to remove hard scale and water scale. When steel parts are pickled, a residue or smut is frequently left on the surface. This smut can be removed by an anodic electropickle, alkaline electrocleaning, or ultrasonic cleaning with an alkaline cleaning solution.

Nitric acid is frequently used in pickling stainless steel. This strong oxidizer ensures that the thin protective oxide coating is not removed from the metal surface. In practice, nitric acid is used in conjunction with hydrofluoric, sulfuric, and hydrochloric acids. These acids are reducing agents and eliminate the scale by directly reducing the oxide. The most common descaling process for chromium–nickel based stainless steel uses nitric acid in the concentration range of 5 to 25 percent in conjunction with hydrochloric acid at 0.5 to 3 percent. For light scale removal, the concentrations are 12 to 15 percent nitric acid and 1 percent of hydrofluoric acid by volume at a bath temperature of 120 to 140°F; for heavy oxides, the concentration of hydrofluoric is increased to 2 to 3 percent. Frequently operators do not wish to use hydrofluoric acid. In this case, nitric acid is used with hydrochloric acid in a ratio of at least 10:1. The efficiency of these pickling operations for stainless steels can be improved by operating at higher temperatures, ultrasonically or electrolytically. After pickling of stainless steels, an operation called passivation is conducted in nitric acid to dissolve any residual iron on the surfaces and establish a protective oxide film to resist corrosion.

Pickling operations on nonferrous metals such as aluminum, magnesium, zinc, and lead can take several forms. Aluminum parts do not require acid pickle as do steels, but light scale stains can be removed by a bath of 25 percent nitric acid and 2 percent hydrofluoric acid, concentrated nitric–hydrofluoric acid, phosphoric acid, or various bright dips. Deoxidizing of magnesium is accomplished with baths of chromic acid or 30 percent hydrofluoric acid at room temperature. Dilute sulfuric acid or sulfuric–nitric acid baths are used on zinc. Titanium is treated with 50 percent nitric acid and 5 percent hydrofluoric acid.

Acid bright dips are usually mixtures of nitric and sulfuric acids employed to provide a mirrorlike surface on cadmium, magnesium, copper, copper alloys, silver, and in some cases, stainless steel.

The air contaminants released from pickling and bright dips include not only the mists of the acids used in the process but nitrogen oxides, if nitric acid is employed, and hydrogen chloride gas from processes using hydrochloric acid. The bath component and the physical form of the contaminant for all common pickling and bright dip baths are listed in the American Conference of Governmental Industrial Hygienists (ACGIH) Ventilation Manual (3). The extent of the problems depends on bath temperature, surface area of work, current density (if bath is electrolytic), and whether the bath contains inhibitors that produce a foam blanket on the bath or that lower the surface tension of the bath and thereby reduce misting. It is a general rule that local exhaust ventilation is required for pickling and acid dip tanks operating at elevated temperature and for electrolytic processes. The decision whether to exhaust a tank and, if so, what exhaust rate to use can be made using standard methods (2, 3, 5). Owing to carry-over of pickling acids in the rinse, the rinse tanks may also require local exhaust ventilation. Bright dip baths require more efficient hooding and exhaust than do most other acid treatment operations.

Minimum safe practices for pickling and bright dip operators have been proposed by Spring (4):

1. Hands and faces should be washed before eating, smoking, or leaving plant. Eating and smoking should not be permitted at the work location.
2. Only authorized employees should be permitted to make additions of chemicals to baths.

3. Face shields, chemical handlers' goggles, rubber gloves, rubber aprons, and rubber platers' boots should be worn when adding chemicals to baths and when cleaning or repairing tanks.
4. Chemicals contacting the body should be washed off immediately and medical assistance obtained.
5. Supervisor should be notified of any change in procedures or unusual occurrences.

3.2 Alkaline Treatment

3.2.1 Alkaline Immersion Cleaning

Acid and alkaline cleaning techniques are complementary in terms of the cleaning tasks that can be accomplished. Alkaline soak, spray, and electrolytic cleaning systems are superior to acid cleaning for removal of oil, gases, buffing compounds, certain soils, and paint. A range of alkaline cleansers including sodium hydroxide, potassium hydroxide, sodium carbonate, sodium meta- or orthosilicate, trisodium phosphate, borax, and tetrasodium pyrophosphate are used for both soak and electrolytic alkaline cleaning solutions.

The composition of the alkaline bath may be complex, with a number of additives to handle specific tasks (4). In nonelectrolytic cleaning of rust from steels, the bath may contain 50 to 80 percent caustic soda in addition to chelating and sequestering agents. The parts are immersed for 10 to 15 min and rinsed with a spray. This cycle is repeated until the parts are derusted. Although excellent for rust, this technique does not remove scale.

Electrolytic alkaline cleaning is an aggressive cleaning method and usually follows some other mechanical or chemical primary cleaning. The bath is an electrolytic cell powered by direct current with the workpiece conventionally the cathode, and an inert electrode as the anode. The water dissociates; oxygen is released at the anode and hydrogen at the cathode. An initial nonelectrolytic soak is used to loosen the soil; when the system is powered, the hydrogen gas generated at the workpiece causes agitation of the surface soils with excellent soil removal. Cleaning is enhanced by periodically reversing the current of the bath.

The gases released at the electrodes by the dissociation of water may result in the release of caustic mist and steam at the surface of the bath. Mist generation is greatest under cathodic cleaning (workpiece is the cathode) and varies with bath concentration, temperature, and current density.

Surfactants and additives that provide a foam blanket are important to the proper operation of the bath. Ideally, the foam blanket should be 5 to 8 cm thick to trap the released gas bubbles and thereby minimize misting. The additives are adjusted to the individual cleaning job because soil properties may either enhance or suppress foam generation. If the foam blanket is too thin, the gas may escape, causing a significant alkaline mist to become airborne; if too thick, the blanket may trap hydrogen and oxygen with resulting minor explosions ignited by sparking electrodes.

Aluminum may be cleaned and etched with a strong alkaline solution in a cathodic cleaning configuration. Surface removal can be controlled to etch the parts lightly, remove surface imperfections, and provide a matte finish. Alkaline baths can be used also to mill aluminum parts chemically to size with removal of large quantities of metal.

To reduce the etch rate for aluminum, triethanolamine and salts of strontium, barium, and calcium are added to alkaline dip baths for aluminum; stannous or cobalt salts are used to increase the etch rate. The surface appearance of the etched part can be modified by adding sodium nitrate to the bath.

3.2.2 Salt Baths

A bath of molten caustic at 370 to 540°C (700 to 1000°F) is used for initial cleaning and descaling of cast iron, copper, aluminum, and nickel with subsequent quenching and acid pickling. The advantages claimed for this type of cleaning are that pre-cleaning is not required and the process provides a good bond surface for a subsequent finish. Metal oxides of chromium, nickel, and iron are removed without attacking the base metal; oils are burned off, and graphite, carbon, and sand are removed. The oxides collect as sludge on the bottom on the bath. The other contaminants combine with the molten caustic, float on the surface of the bath, or are volatized as vapor or fume.

Molten sodium hydroxide is used at 430 to 540°C (800 to 1000°F) for general-purpose descaling and removal of sand on castings. A modification of this bath, the Virgo process, uses a few percent of sodium nitrate or a chlorate salt to enhance the performance of the bath. A reducing process utilizes sodium hydride in the bath at 370°C (700°F) to reduce oxides to their metallic state. The bath utilizes fused liquid anhydrous sodium hydroxide with up to 2 percent sodium hydride, which is generated in accessory equipment by reacting metallic sodium with hydrogen. All molten baths require subsequent quenching, pickling, and rinsing. The quenching operation dislodges the scale through steam generation and thermal shock.

These baths require well-defined operating procedures owing to the hazard from molten caustic. Local exhaust is necessary, and the tank must be equipped with a complete enclosure to protect the operator from violent splashing as the part is immersed in the bath. Quenching tanks and pickling tanks must also be provided with local exhaust ventilation. Ventilation standards have been proposed for these operations (3, 5).

4 DEGREASING

For many decades the principal application of degreasing technology has been in the metalworking industry for the removal of machining oils, grease, drawing oils, chips, and other soils from metal parts. The technology has expanded greatly prompted by advances in both degreasing equipment and solvents. At this time it is probably the most common industrial process extending across all industries including jewelry, electronics, special optics, electrical, machining, instrumentation, and even rubber and plastic goods. The three degreasing techniques used in industry are cold degreasing, soak cleaning, and vapor phase degreasing. The significant occupational health problems associated with these degreasing processes are described below.

4.1 Cold Degreasing

The term cold degreasing identifies the use of a solvent at room temperature in which parts are dipped, sprayed, brushed, wiped, or agitated for removal of oil and grease. It is the simplest of all degreasing processes, requiring only a simple container with the solvent of choice. It is widely used in small production shops, maintenance and repair shops, and automotive garages. The solvents used vary from low-volatility, high-flash petroleum distillates such as mineral spirits and Stoddard Solvent to solvents of high volatility including aromatic hydrocarbons, chlorinated hydrocarbons, and ketones.

Skin contact with cold degreasing materials should be avoided by work practices and the use of protective clothing. Glasses and a face shield should be used to protect the eyes and face from accidental splashing during dipping and spraying. Solvent tanks should be provided with a cover and, if volatile solvents are used, the dip tank and drain station should be provided with ventilation control.

Spraying with high-flash petroleum distillates such as Stoddard Solvent, mineral spirits, or kerosene is a widely used method of cleaning oils and grease from metals. This operation should be provided with suitable local exhaust ventilation. The hood may be a conventional spray booth type and may be fitted with a fire door and automatic extinguishers. The fire hazard in spraying a high-flash petroleum solvent is comparable to spraying many lacquers and paints.

4.2 Emulsion Cleaners

Emulsion cleaners containing petroleum and coal-tar solvents are commonly used in power washers and soak tanks. When used in soak tanks at room temperature, ventilation is not required. When the cleaner is sprayed or used hot, the operation should be confined and provided with local exhaust ventilation. Emulsion cleaners containing cresylic acid, phenols, or halogenated hydrocarbons should also be provided with local exhaust ventilation, and skin contact should be carefully avoided.

4.3 Vapor Degreasing

A vapor degreaser is a tank containing a quantity of solvent heated to its boiling point. The solvent vapor rises and fills the tank to an elevation determined by the location of a condenser. The vapor condenses and returns to the liquid sump. The tank has a freeboard that extends above the condenser to minimize air currents inside the tank (Figure 18.1). As the parts are lowered into the hot vapor, the

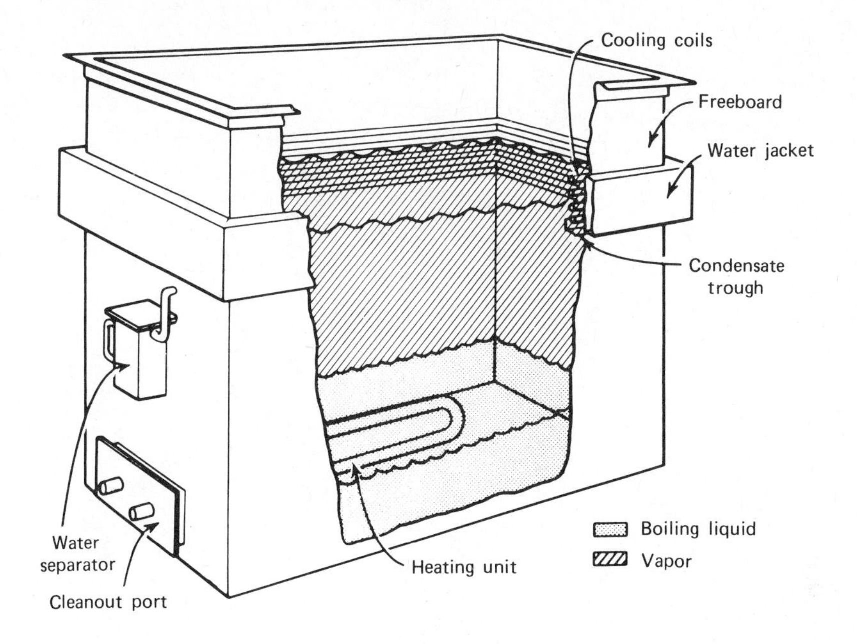

Figure 18.1. Major components of a vapor phase degreaser.

vapor condenses on the cold part and dissolves the surface oils and greases. This oily condensate drops back into the liquid solvent at the base of the tank. The solvent is continuously evaporated to form the vapor blanket. Because the oils are not vaporized, they remain to form a sludge in the bottom of the tank. The scrubbing action of the condensing vapor continues until the temperature of the part reaches the temperature of the vapor whereupon condensation stops, the part appears dry, and it is removed from the degreaser. The time required to reach this point depends on the particular solvent, the temperature of the vapor, the weight of the part, and its specific heat. The vapor phase degreaser does an excellent job of drying parts after aqueous cleaning and prior to plating; it is frequently used for this purpose in the jewelry industry.

4.3.1 Types of Vapor Phase Degreasers and Solvents

The simplest form of vapor phase degreaser, shown in Figure 18.1, utilizes only the vapor for cleaning. The straight vapor-cycle degreaser is not effective on small, light work because the part reaches the temperature of the vapor before the condensing action has cleaned the part. Also, the straight vapor cycle does not remove insoluble surface soils. For such applications, the vapor-spray-cycle degreaser is frequently used. The part to be cleaned is first placed in the vapor zone as in the straight vapor-cycle degreaser. A portion of the vapor is condensed by a cooling coil and fills a liquid solvent reservoir. This warm liquid solvent is pumped to a spray lance, which can be used to direct the solvent on the part, washing off surface

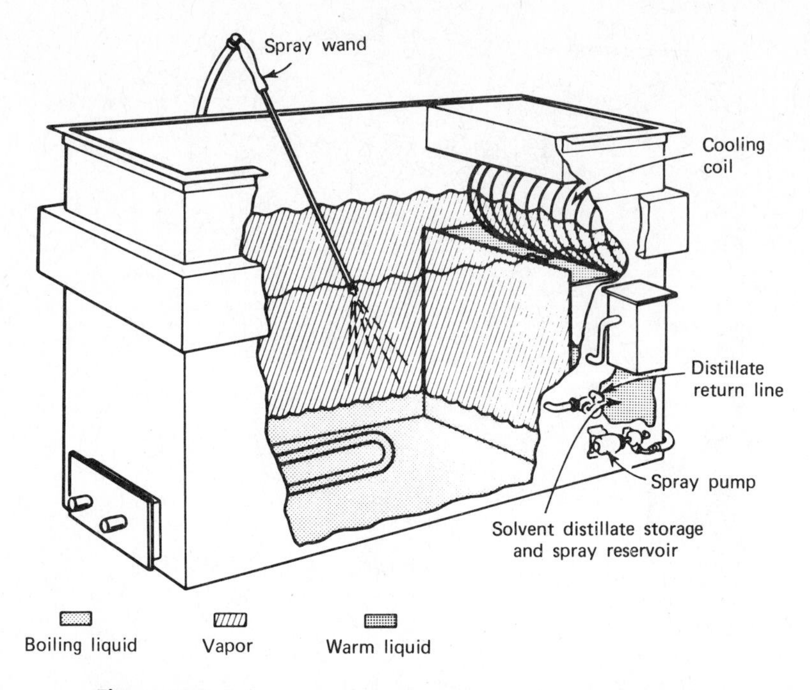

Figure 18.2. A vapor phase degreaser with a spray wand.

oils and cooling the part, thereby permitting final cleaning by vapor condensation (Figure 18.2).

A third degreaser design has two compartments, one with warm liquid solvent and a second compartment with a vapor zone. The work sequence is vapor, liquid, and vapor. This degreaser is used for heavily soiled parts with involved geometry or to clean a basket of small parts that nest together. Finally, a three-compartment degreaser has vapor, boiling and warm liquid compartments with a vapor, boiling liquid, warm liquid, and vapor work sequence. Other specialty degreasers encountered in industry include enclosed conveyorized units for continuous production cleaning.

Over the past decade, ultrasonic cleaning modules installed in vapor degreasers have found broad application for critical cleaning jobs. In an ultrasonic degreaser, a transducer operating in the range of 20 to 40 kHz is mounted at the base of a liquid immersion solvent tank. The transducer alternately compresses and expands the solvent forming small bubbles that cavitate or collapse at the surface of the workpiece. The cavitation phenomenon disrupts the adhering soils and cleans the part. Ultrasonic degreasers use chlorinated solvents at 32 to 49°C (90 to 120°F) and aqueous solutions at 43 to 71°C (110 to 160°F). These degreasers commonly employ refrigerated or water-chilled coils for control of solvent vapors; the manufacturers claim that local exhaust ventilation is not needed in this configuration.

The solvents commonly used with vapor phase degreasers include trichloroethylene, perchloroethylene, 1,1,1-trichloroethane, methylene chloride, and a series of Freon® solvents (Table 18.1) (6, 7). The degreaser must be designed for and used with a specific solvent. Most chlorinated degreasing solvents sold under trade names contain a stabilizer present in a concentration of less than 5 percent. The purpose of the stabilizer is to neutralize any free acid that might result from oxidation of the degreasing liquid in the presence of air, hydrolysis in the presence of water, or pyrolysis under the influence of high temperatures. The stabilizer is not a critical issue in establishing health risk to the worker owing to its low concentration; the solvent itself is usually the predictor of risk.

The heat input to the conventional vapor phase degreaser is commonly provided by steam although electricity and gas are also available. The recent emphasis on energy conservation has had an impact on the choice of solvents for vapor phase degreasers. To minimize energy, one must consider the boiling temperature, specific heat, and latent heat of vaporization of the candidate solvents. The chlorinated degreasing solvents (trichloroethylene, methylene chloride, 1,1,1-trichloroethane, and perchloroethylene) are still the major degreasing solvents in use.

The loss of degreaser solvent to the workplace obviously depends on a number of operating conditions including the type and properties of contaminants removed, cleaning cycle, volume of material processed, design of parts being cleaned, and most importantly, work practices. The loss of solvent from an idling open-top degreaser located in an area without significant drafts expressed as pounds per square foot per hour is 0.14 for 1,1,1-trichloroethane, 0.20 for trichloroethylene, 0.26 for methylene chloride, and 0.29 for perchloroethylene (7). Airborne concentrations at the operator's breathing zone measured by a direct-reading instrument during active operation of degreasers vary widely with the concentration increasing from lowering parts, holding parts in the degreaser, and during removal of parts. The time weighted average concentrations for a full 8-hr shift developed by a state agency based on personal air sampling ranged from 5 to 311 ppm for trichloroethylene, and 5 to 174 ppm for perchloroethylene.

4.3.2 Control

The vapor level in the degreaser is controlled by a nest of water-fed condensing coils located on the inside perimeter of the tank (Figure 18.1). In addition, a water jacket positioned on the outside of the tank keeps the freeboard cool. The vertical distance between the lowest point at which vapors can escape from the degreaser machine and the highest normal vapor level is called the freeboard. The freeboard should be at least 15 in. and not less than one-half to three-fifths the width of the machine. The effluent water from the coils and water jacket should be regulated to 32 to 49°C (90 to 120°F); a temperature indicator or control is desirable.

Properly designed vapor degreasers have a thermostat located a few inches above the normal vapor level to shut off the source of heat if the vapor rises above the condensing surface. A thermostat is also immersed in the boiling liquid; if overheating occurs, the heat source is turned off.

There is a difference of opinion on the need for local exhaust ventilation on vapor phase degreasers. Authorities frequently cite the room volume as a guide in

Table 18.1. Properties of Vapor Degreasing Solvents

	Trichloroethylene	Perchloroethylene	Methylene Chloride	Trichlorotrifluoroethane[a]	Methyl Chloroform (1,1,1-Trichloroethane)
Boiling point					
°C	87	121	40	48	74
°F	188	250	104	118	165
Flammability	Nonflammable under vapor degreasing conditions				
Latent heat of vaporization (b.p.), Btu/lb	103	90	142	63	105
Specific gravity					
Vapor (air = 1.00)	4.53	5.72	2.93	6.75	4.60
Liquid (water = 1.00)	1.464	1.623	1.326	1.514	1.327

[a]Binary azeotropes are also available with ethyl alcohol, isopropyl alcohol, acetone, and methylene chloride.

determining if ventilation is needed: Local exhaust ventilation is needed if there is less than 2000 ft^3 in the room for each square foot of solvent surface, or if the room is smaller than 25,000 ft^3. In fact, ventilation control requirements depend on the degreaser design, location, maintenance, and operating practices. Local exhaust increases solvent loss, and the installation may require solvent recovery before discharging to outdoors. To ensure effective use of local exhaust, the units should be installed away from drafts from open windows, spray booths, space heaters, supply air grilles, and fans. Equally important is the parts loading and unloading station. When baskets of small parts are degreased, it is not possible to eliminate drag out completely and the unloading station usually requires local exhaust.

When degreasers are installed in pits, mechanical exhaust ventilation should be provided at the lowest part of the pit. Open flames, electric heating elements, and welding operations should be divorced from the degreaser locations because the solvent is degraded by both direct flame and ultraviolet (UV) radiation, thereby producing toxic air contaminants.

Later in this chapter under the discussion of welding, reference is made to the decomposition of chlorinated solvent under thermal and UV stress with the formation of chlorine, hydrogen chloride, and phosgene. Because degreasers using such solvents are frequently located near welding operations, this problem warrants attention. In a laboratory study of the decomposition potential of methyl chloride, methylene chloride, carbon tetrachloride, ethylene dichloride, 1,1,1-trichloroethane, *o*-dichlorobenzene, trichloroethylene, and perchloroethylene, only the latter two solvents decomposed in the welding environment to form dangerous levels of phosgene, chlorine, and hydrogen chloride (8). All chlorinated materials thermally degrade if introduced to direct-fired combustion units commonly used in industry. If a highly corroded heater is noted in the degreaser area, it may indicate that toxic and corrosive air contaminants are being generated.

Installation instructions and operating precautions for the use of conventional vapor phase degreasers have been proposed by various authorities (9). The following minimum instructions should be observed at all installations:

1. If the unit is equipped with a water condenser, the water should be turned on before the solvent is heated.
2. Water temperature should be maintained between 27 and 43°C (81 and 110°F).
3. Work should not be placed in and removed from the vapor faster than 11 fpm (.055 m/sec). If a hoist is not available, a support should be positioned to hold the work in the vapor. This minimizes the time the operator must spend in the high exposure zone.
4. The part must be kept in the vapor until it reaches vapor temperature and is visually dry.
5. Parts should be loaded to minimize pullout. For example, cup-shaped parts should be inverted.

6. Overloading should be avoided because it will cause displacement of vapor into the workroom.
7. The work should be sprayed with the lance below the vapor level.
8. Proper heat input must be available to ensure vapor level recovery when large loads are placed in the degreaser.
9. A thermostat should be installed in the boiling solvent to prevent overheating of the solvent.
10. A thermostat vapor level control must be installed above the vapor level inside the degreaser and set for the particular solvent in use.
11. The degreaser tank should be covered when not in use.
12. Hot solvent should not be removed from the degreaser for other degreasing applications nor should garments be cleaned in the degreaser.
13. An emergency eyewash station would be located near the degreaser for prompt irrigation of the eye in case of an accidental splash.

To ensure efficient and safe operation, vapor phase degreasers should be cleaned when the contamination level reaches 25 percent. The solvent should be distilled off until the heating surface or element is $1\frac{1}{2}$ in. below the solvent level or until the solvent vapors fail to rise to the collecting trough. After cooling, the oil and solvent should be drained off and the sludge removed. It is important that the solvent be cooled prior to draining. In addition to placing the operators at risk, removing hot solvent causes serious air contamination and frequently requires the evacuation of plant personnel from the building. A fire hazard may exist during the cleaning of machines heated by gas or electricity because the flash point of the residual oil may be reached and because trichloroethylene itself is flammable at elevated temperatures. After sludge and solvent removal, the degreaser must be mechanically ventilated before any maintenance work is undertaken. A person should not be permitted to enter a degreaser or place his or her head in one until all controls for entry into a confined space have been put in place. Anyone entering a degreaser should wear a respirator suitable for conditions immediately hazardous to life, as well as a lifeline held by an attendant. Anesthetic concentrations of vapor may be encountered, and oxygen concentration may be insufficient. Such an atmosphere may cause unconsciousness with little or no warning. Deaths occur each year because of failure to observe these precautions.

The substitution of one degreaser solvent for another as a control technique must be done with caution. Such a decision should not be based solely on the relative exposure standards but must consider the type of toxic effect, the photochemical properties, the physical properties of the solvents including vapor pressure, and other parameters describing occupational and environmental risk.

Medical control is especially important for degreasing operators and should include both preplacement and periodic physical examinations. Specific medical screening is advisable for persons with cardiovascular disease.

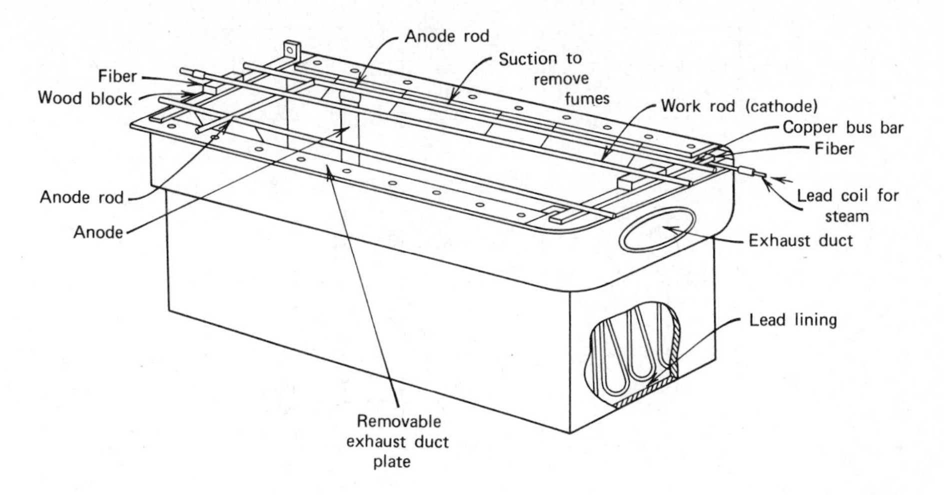

Figure 18.3. An electroplating tank.

5 ELECTROPLATING

Metal, plastic, and rubber parts are plated to prevent rusting and corrosion, for appearance, to reduce electrical contact resistance, to provide electrical insulation, as a base for soldering operations, and to improve wearability. The common plating metals include cadmium, chromium, copper, gold, nickel, silver, and zinc. Prior to electroplating, the parts must be cleaned and the surfaces treated as described in Sections 2 and 3.

There are approximately 160,000 electroplating workers in the United States; independent job shops average 10 workers, and the captive electroplating shops have twice that number. A number of epidemiologic studies have identified a series of health effects in electroplaters ranging from dermatitis to elevated mortality for a series of cancers.

5.1 Electroplating Techniques

The basic electroplating system is shown in Figure 18.3. The plating tank contains an electrolyte consisting of a metal salt of the metal to be applied dissolved in water. Two electrodes powered by a low-voltage dc power supply are immersed in the electrolyte. The cathode is the workpiece to be plated, and the anode is either an inert electrode or, most frequently, a slab or a basket of spheres of the metal to be deposited. When power is applied, the metal ions deposit out of the bath on the cathode or workpiece. Water is dissociated, releasing hydrogen at the cathode and oxygen at the anode. The anode may be designed to replenish the metallic ion concentration in the bath. Current density expressed in amperes per unit area of workpiece surface varies depending on the operation. In addition to the dissolved salt containing the metallic ion, the plating bath may contain additives

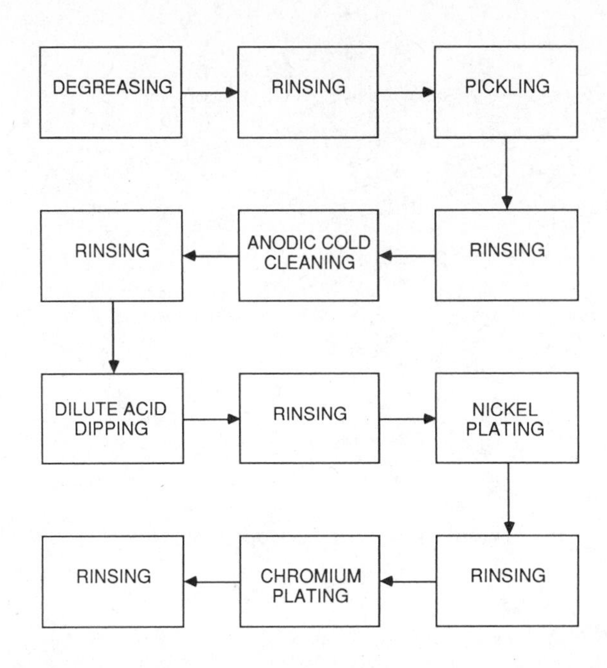

Figure 18.4. The steps involved in chromium plating of steel.

to adjust the electrical conductivity of the bath, define the type of plating deposit, and buffer the pH of the bath.

Anodizing, a common surface treatment for decoration, corrosion resistance, and electrical insulation on such metals as magnesium, aluminum, and titanium, operates in a different fashion. The workpiece is the anode, and the cathode is a lead bar. The oxygen formed at the workpiece causes a controlled surface oxidation. The process is conducted in a sulfuric or chromic acid bath with high current density, and because its efficiency is quite low, the amount of misting is high.

In conventional plating operations, individual parts on a hanger or a rack of small parts are manually hung from the cathode bar. If many small pieces are to be plated, the parts may be placed in a perforated plastic barrel in electrical contact with the cathode bar, and the barrel is immersed in the bath. The parts are tumbled to achieve a uniform plating.

In a small job-shop operation, the parts are transferred manually from tank to tank as dictated by the type of plating operation. The series of steps necessary in one plating operation (Figure 18.4) illustrates the complexity of the operation. After surface cleaning and preparation, the electroplate steps are completed with a water rinse tank isolating each tank from contamination. In high production shops, an automatic transfer unit is programmed to cycle the parts from tank to tank, and the worker is required only to load and unload the racks or baskets. Automatic plating operations may permit exhaust hood enclosures on the tanks and therefore more effective control of air contaminants. Exposure is also limited because the worker is stationed at one loading position and is not directly exposed to air contaminants released at the tanks.

5.2 Air Contaminants

The principal source of air contamination in electroplating operations is the release of the bath electrolyte to the air by the gassing of the bath. As mentioned above, the bath operates as an electrolytic cell, so water is dissociated and hydrogen is released at the cathode and oxygen at the anode. The gases released at the electrodes rise to the surface of the bath and burst, generating a respirable mist that becomes airborne. The presence of this mist can be detected by placing a clean piece of paper parallel to the bath an inch or so above the surface. The mist generation rate depends on the bath efficiency. In copper plating, the efficiency of the plating bath is nearly 100 percent; that is, essentially all the energy goes into the plating operation and little into the electrolytic dissociation of water (10). Nickel plating baths operate at 95 percent efficiency so only 5 percent of the energy is directed to dissociation of water, and misting is minimal. However, chromium plating operations are quite inefficient, and up to 90 percent of the total energy may be devoted to dissociation of the bath with resulting severe gassing, and resulting potential exposure of the operator to chromic acid mist. Although the contamination generation rate of the bath is governed principally by the efficiency of the bath, it also varies with the metallic ion concentration in the bath, the current density, the nature of bath additives, and bath temperature. Air or mechanical agitation of the bath used to improve plating quality may also release the bath as droplets.

The health significance of the mist generated by electroplating processes depends, of course, on the contents of the bath. The electrolyte mist released from the bath is alkaline or acidic depending on the specific electroplating process. A majority of the alkaline baths based on cyanide salts solutions are used for cadmium, copper, silver, brass, and bronze plating. Acidic solutions are used for chromium, copper, nickel, and tin. The exact composition of the baths can be obtained from an electroplaters' handbook or, if proprietary, from the supplier. An inventory of the nature of the chemicals in the common electroplating baths, the form in which they are released to the air, and the rate of gassing has evolved over the past several decades, drawing heavily on the experience of the state industrial hygiene programs in New York and Michigan (2). These data are useful in defining the nature of the contaminant and the air sampling procedure necessary to define the worker exposure. As noted below, these data are also valuable in defining the ventilation requirements for various plating operations.

5.3 Control

Proprietary bath additives are available to reduce the surface tension of the electrolyte and therefore reduce misting. The minimum effective viscosity is 35 dynes/cm, but one should operate at a viscosity of 25 dynes/cm. Another additive provides a thick foam that traps the mist released from the bath. This agent is best used for tanks that operate continuously. A layer of plastic chips, beads, or balls on the surface of the bath also trap the mist and permit it to drain back into the bath. Where possible, tanks should be provided with covers to reduce bath loss.

Although the use of the above mist suppressants are helpful, they will not alone control airborne contaminants from plating tanks at an acceptable level. Local exhaust ventilation in the form of lateral slot or upward plenum slotted hoods is the principal control measure.

Until the 1950s, it was considered good practice to exhaust all tanks containing nitric acid, chromic acid, hydrofluoric acid, hot cyanide, and alkaline solutions and hot water at 0.51 m^3/sec per square meter (100 to 150 cfm per square foot) of tank area. In some cases this was quite adequate, in some cases, not. The design approach described in the *Industrial Ventilation Manual* (3) provides a firm basis for the control of electroplating air contaminants. This procedure permits one to determine a minimum capture velocity based on the hazard potential of the bath and the rate of contaminant generation. The exhaust volume is based on the capture velocity and the tank measurements and geometry.

In addition to the proper design and installation of good local exhaust ventilation, one must provide adequate replacement air, backflow dampers on any combustion devices to prevent carbon monoxide contamination of the workplace, and suitable air cleaning.

As is the case in all industrial processes, the effectiveness of the exhaust ventilation may be evaluated by air sampling at the workplace and direct ventilation measurements at the tank. Owing to the severe corrosion of duct work, periodic checks of the exhaust systems in plating shops are necessary. One should determine the exhaust volumes from each tank by Pitot measurements and compare the observed values with the recommended exhaust rate. Qualitative assessment of the ventilation is possible using smoke tubes or other tracers. It is common practice to measure the hood slot velocity with a Velometer, average the readings, and calculate the exhaust volume from these measurements. This technique has limited value owing to the difficulty in defining average velocity in a narrow slot. Measurement of slot velocity is useful in determining the uniformity of the exhaust over the length of the tank, but that is all. The effects of room drafts on capture velocity should be identified. In many cases, the use of partitions to minimize the disruptive effects of drafts may greatly improve the installed ventilation.

Because a low-voltage dc power supply is used, an electrical hazard does not exist at the plating tanks. A fire and explosion risk may result from solvent degreasing and spray painting conducted in areas contiguous with the plating area. The major chemical safety hazards are due to handling concentrated acids and alkalies and the accidental mixing of acids with cyanides and sulfides during plating, bath preparation, and waste disposal with the formation of hydrogen cyanide and hydrogen sulfide.

The educated use of protective equipment by electroplaters is extremely important in preventing contact with the various sensitizers and corrosive materials encountered in the plating shop. The minimum protective clothing should include rubber gloves, aprons, boots, and chemical handler's goggles. Aprons should come below the top of the boots. All personnel should have a change of clothing available at the workplace. If solutions are splashed on the work clothing, they should be removed, the skin washed, and the worker should change to clean garments. A shower and eyewash station serviced with tempered water should be available at the workplace. The wide range of chemicals presents a major dermatitis hazard to the plater, and skin contact must be avoided. Nickel is a skin sensitizer and may cause nickel itch, developing into a rash with skin ulcerations.

A summary of the health hazards encountered in electroplating shops and the available controls is presented in Table 18.2. More information on controls can be found in References 11 and 12.

6 FORGING

The practice of forging, the compressive deformation of metal between dies to a given geometry, imparts certain desirable metallurgical properties to the workpiece that cannot be obtained in any other fashion. Industry could not function as it now does without forged parts. A wide range of metals and alloys are processed by cold and hot forging. Because hot forging presents the major occupational health hazards, it will be given attention in this discussion. More than one-half the metal worked is low-carbon steel, low-alloy metals, and aluminum; the balance of the metals worked include a range of nickel–chromium base alloys, magnesium, and titanium.

6.1 Forging Practice

Forging processes are classified by the type of equipment used to form the metal part in an open or closed impression die (13). In drop hammer forging, the bottom half of the die is positioned on an anvil isolated from the frame of the hammer. The top half of the die is attached to a vertical ram that is raised by an air or steam cylinder and then dropped by gravity or driven downward by air or steam (Figure 18.5). An operator controls the force of the impact and the frequency of the blows.

The second type of forging utilizes a forging press. Presses may be mechanical—that is, the energy is stored in a flywheel—or hydraulic with the top die driven downward by a large hydraulic piston. Presses provide a slow squeezing action rather than the rapid blows of the hammers.

The forging technique may be impression die, roll ring, open die, or upset forging. In hot forging, metal stock is cut to size, the part is heated to forging temperature, and the workpiece is forged between die blocks. The dies may be cold or heated to effect quality forging. A recent advance has been isothermal hot forging in which the die temperature approaches that of the workpiece. If single cavity dies are used, the part might be cycled between reheat furnaces and a

Table 18.2. Summary of Major Electroplating Health Hazards

Exposure	How Contamination Occurs
Inhalation	
Mist, gases, and vapors	
Hydrogen cyanide	Accidental mixing of cyanide solutions and acids
Chromic acid	Released as a mist during chrome plating and anodizing
Hydrogen sulfide	Accidental mixing of sulfide solutions and acids
Nitrogen oxides	Released from pickling baths containing nitric acid
Dust	Released during weighing and transferring of solid bath materials, including cyanides and cadmium salts
Fumes	Generated during on-site repair of lead-lined tanks using torch-burning techniques
Ingestion	
Workplace particles	Accidental ingestion during smoking and eating at workplace
Skin Contact	
Cyanide compounds	Absorption through the skin
Solvents	Defatting by solvents
Irritants	Primary irritants contacting the skin
Contact allergens	Sensitization
Control Technology	
Local exhaust ventilation	
Mist reduction	
Reduce surface tension	
Coat surface	
Tank covers	
Isolation of stored chemicals	

progression of dies mounted on a series of hammers. For small parts with simple geometry, the necessary impressions or cavities may be cut in a single die and the forging can be completed at one hammer or press without reheating.

Between hammer blows, lubricants are applied by swab or spray to the die face and to the workpiece positioned on the bottom die. The die lubricant is designed to prevent sticking or fusing of the part in the dies, improve metal flow, act as a parting agent, and reduce wear of the die. The composition of the lubricant and its impact on the workplace air quality are discussed later in this section.

In practice, forging of a complex part of moderate size may take a four-man crew consisting of the heater, helper, lubricator, and hammer operator. The forging stock is loaded into a gas- or oil-fired furnace by the heater and brought to a forging temperature ranging from 800°F (aluminum) to 2400°F (alloy steel). Box, slot, or rotary hearth furnaces are used to preheat the stock to these forging temperatures.

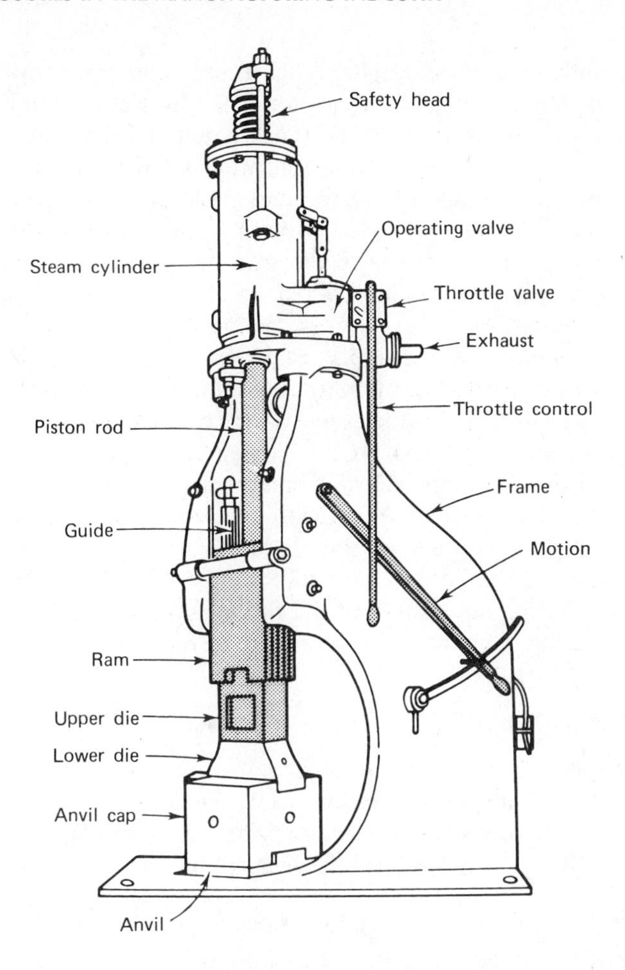

Figure 18.5. A forging hammer operated by steam.

At the breakdown hammer, the surfaces of both dies are blown off with compressed air, and forging compound is applied with spray or swab by the lubricator. The heater and a helper transfer the stock from the furnace to the breakdown hammer by forklift truck, by crane, or manually if the piece is small. The piece is positioned on the die by the hammer operator and the lubricator. As many as 8 to 10 blows are struck by the forge hammer operator. Between strikes, the die lubricant is applied to the top die by the lubricator. There is usually some overspray, and lubricant mist is a significant air contaminant. If the die lubricant is petroleum based, it will burn when it hits the hot workpiece and die, generating a cloud of oil mist, sooty particulates, and a range of gases and vapors of the type noted with poor combustion of a heavy oil. The forging area may be equipped with local exhaust hoods; however, this is rare. In most cases, the contaminants are released to the workroom and controlled by dilution ventilation.

When the rough form of the forging is achieved at the breakdown hammer, the forging is returned to the second furnace for reheat to forging temperature. After reheat, the forging is transferred to the finish hammer where the forging is completed using the same technique as on the breakdown hammer. Again, forging compound or lubricant is applied to the dies between strikes, and the same air contaminant cloud is generated as described on the breakdown hammer. After finish forging, the pieces are taken to the trim press where the metal flashing is removed by a trim die. The flashing is placed in a dumpster, and the forging is taken to the finishing room.

Forging press operations are similar to hammer operations. In a large press operation, the crew consists of an operator, barman, oiler, and helper. The operator usually works at a control console some distance from the press but the others work directly at the press and have greater exposure to the air contaminants formed during the operation. Again the stock or billet is first heated in a furnace and then transferred to the press by a forklift truck. The lubricator blows off the die with compressed air and then sprays on the die lubricant or forging compounds. The piece is positioned on the bottom die and is forged with a single stroke. There is exposure to mist during application of the forging compound and to the products of combustion during burn-off if it is an oil-based lubricant.

6.2 Air Contamination

Significant air contamination may occur from furnace operation, the application of die lubricants, the forging operation itself, and the heating of the dies. The various finishing operations also contribute particulate contamination, but because these operations have been discussed elsewhere, they are not treated in detail in this discussion.

The stock heating furnaces are positioned adjacent to the forging hammer or press they service. The furnaces are fired with No. 5 fuel oil or with gas. It is not uncommon for the products of combustion to be released from short stacks directly to the workplace and then be removed by a roof exhauster. If gas is used, contamination of the workplace with the products of combustion should be evaluated but are probably not of great significance. However, the release of combustion products from oil-fired units may present a significant contamination problem. In addition to the conventional products of combustion, including carbon monoxide, one may have sulfur dioxide formed from fuel oils that contain from 0.5 to 3 percent sulfur. In most cases, the furnaces used on the forging line are air atmosphere furnaces. Controlled atmosphere furnaces used on special operations either may burn natural gas to provide a reducing or oxidizing atmosphere or may possibly utilize an ammonia cracking unit to provide the inert atmosphere. The potential health implications of such operations are described in Section 9.

A second source of carbon monoxide and other products of combustion is associated with the heating of the dies installed on the forging hammers or presses. Initially, dies are brought to temperature in a die furnace and installed hot in the hammer or press. These dies receive supplementary heating to keep them at temperature with an open kerosene or gas burner flame impinging on the die. This open flame may contribute to the air contamination occurring at the forging operation.

The application of forging compounds (die lubricants) to hot dies has been a major contributor to forge shop exposures since the industry started. The early lubricants consisted of natural graphite and animal fat added to an oil base. Present die lubricants are a sophisticated blend of components designed for specific forging applications. When using petroleum-based lubricants, the oil burns off, resulting in a heavy particulate cloud containing oil mist, sooty particulates, trace metals such as vanadium, and polynuclear aromatic hydrocarbons. Sulfur dioxide may also be formed from the combustion of residual oils. In such an operation, the concentrations of respirable mass particulates and benzo-*a*-pyrene require that controls be installed. Usually the concentrations of sulfur dioxide, aldehydes, and nitrogen dioxide are not excessive (14).

Owing to air contamination from the oil-based die lubricants, water-based forging compounds were introduced in the industry as early as 1950. In the past decade, they have gained widespread acceptance when their application is acceptable from a production standpoint.

Few data are available on the concentrations of metal dust or fume generated from the workpiece during forging operations. The contamination generation rate depends on forging temperature, vapor pressure of the metal at forging temperature, metal flow characteristics, and the oxidizing potential of the metal. If conventional steels are worked, the concentration of iron in the air is probably not significant; however, the forging of alloys containing chromium and nickel may warrant specific air sampling. Specialty forging of highly toxic metals, such as beryllium, contribute significant air contamination unless specific controls are installed.

The machining of die blocks may be conducted as a part of the forge operation or done by an outside supplier. In most cases, the metal stock is a steel alloy that may contain chromium and molybdenum. Because most of the work is milling using coolants, it does not represent a significant source of air contamination. Specific comments on machining operations are presented in Section 10.

6.3 Heat Stress

In high production forge shops, heat stress may be a major health problem. The principal heat load is due to radiation from furnaces, dies, and the workpiece and to the worker metabolic load when material handling is done manually. In certain parts of the country, convective heat load may be significant. A survey must be conducted to determine the origin of the heat load to the worker so that controls may be installed. Controls may include reduction of the work load by improved materials handling and plant layout, shielding of furnaces to reduce the radiation load, and spot cooling. Impressive reduction in heat stress can be accomplished if reflective shields are installed on the stock heating furnaces, although one must face the challenge of keeping these shields clean. The opening height of furnace doors must be kept at a minimum to reduce exposure of the workers to direct furnace radiation. The industry frequently uses steam and airlines mounted at the bottom of the door opening surface which release steam or air when the door opens. The steam release may reduce the radiant load somewhat, although no data are available to support this claim. Lightweight movable plates in front of the furnace loading door may also be effective if used efficiently. Pedestal fans are commonly used for cooling of workers and to blow smoke away from the forging crew. For more information on heat stress, see Chapter 21.

6.4 Noise

Forging was one of the first industrial operations to be identified as causing noise-induced hearing loss. The noise hazard is due to hammer impact noise and, to a lesser degree, "pink" noise from the steam and air cylinders and various air solenoids. The hammer operation may be as slow as one blow per minute or as high as 60 blows per minute for a small air lift hammer. Impact noise may be as high as 140 db. The air release noise can be controlled by mufflers; however, reduction of the primary forging impact noise is more elusive. This forge shop noise problem is normally controlled by ear protection and a complete hearing conservation program. See Chapter 23 for additional information on noise.

6.5 Controls

A combination of local and dilution exhaust ventilation is used in most large forge shops to control air contaminants. The small job shop frequently operates without local exhaust ventilation control. Dilution exhaust ventilation is reasonably effective for carbon monoxide control from many fugitive sources but does not control the smoke cloud from forging operation. Hot forging hammer and presses utilizing oil-based die lubricants should be equipped with local exhaust ventilation. The industry frequently installs local hooding that also acts as a shield to control flying scale during the air blowoff of dies. To date, there are no design criteria published on local exhaust ventilation for forging operations.

6.6 Protective Equipment

Protective clothing requirements vary greatly with the type of forging. At a minimum, one should consider head protection and the mandatory use of protective goggles and shoes. On certain operations, gauntlet gloves and shoulder length fireproof sleeves are necessary. If there is heavy use of oil-based die lubricants, suitable oil and fireproof aprons and leggings should be worn by the forging crew. For the heaters, goggles with colored lenses and wire mesh face screen for infrared protection may be in order.

7 FOUNDRY OPERATIONS

The founding of metal is the pouring of molten metal into a cavity formed in some type of molding media. The principal media used is silica sand. The mold cavity may contain a refractory core to define a void in the casting. After cooling, the mold is taken to a shake-out facility where the molding media is removed from around the casting. The casting is cleaned, extraneous cast metal is removed, and the molding sand is recycled to be conditioned for reuse. These foundry processes are shown in Figure 18.6 for a gray iron foundry using a cupola furnace (15).

The foundry industry is the sixth largest industrial sector, based on sales, with annual production rates of 20 to 25 million tons of castings from 4000 foundries with a work population of 350,000 to 400,000. Approximately one-quarter of the foundries are captive foundries, and the balance are independent facilities (16).

The health and safety problems in foundries are myriad, including exposures to mineral dusts, predominantly silica, metal fumes, carbon monoxide, and resin systems used as bonding agents. The physical hazards include noise, vibration, and heat stress. Although the hazard from exposure to silica has been known for decades, the prevalence of silicosis in foundrymen is not known, although a series of recent studies indicate it may be high. A review of one set of OSHA inspection reports reveals that 40 percent of the airborne dust samples taken in foundries exceeded the OSHA PEL for silica and 53 percent exceeded the NIOSH REL (17). There is now definitive evidence for increased risk of lung cancer in foundrymen associated with exposure to polycyclic aromatic hydrocarbons. Another problem of major importance is the high prevalence of vibration white finger (VWF) in foundry finishing room personnel. In the 1970s, OSHA identified foundries as a high hazard industry and included it in the National Emphasis Program.

This section discusses the occupational health hazards in molding, coremaking, pouring, shake-out, and casting cleaning and finishing. Limited attention is given to the cleaning and finishing operations because these are covered in Sections 2 and 8. The coverage focuses on ferrous foundries, that is, those casting iron and steel. The nonferrous foundries, which include aluminum, magnesium, brass, and bronze, are cited when specific hazards are unique to the process.

7.1 Molding

During the 1980s, approximately one-half of both ferrous and nonferrous castings were produced in green sand molds in the United States. This molding sand is a mix of silica sand with clay as a plastic agent and cereal, sugar, and starch added to increase mold strength. Sea coal, pitch, lignite, or asphalt is added to the molding sand to improve casting quality. When the metal is poured, the sea coal pyrolyzes, forming carbon and gases that reduce metal penetration into the sand and improve the finish of the metal surface.

The silica sand is mixed with the other ingredients in a muller. Loading of the high silica sand is dusty, and the muller must be provided with local exhaust ventilation. After mixing, the sand is placed in storage bins or silos; when required,

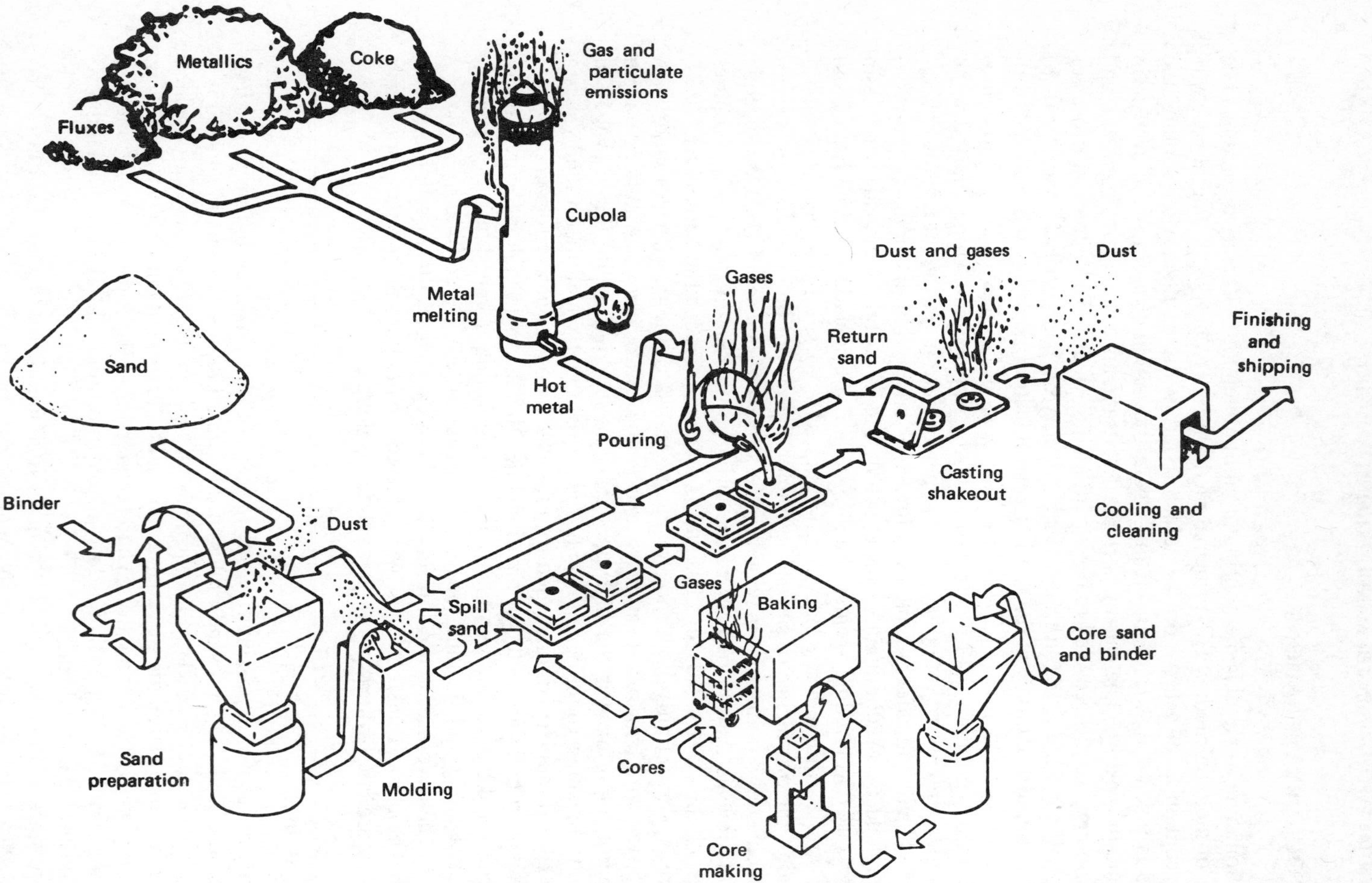

Figure 18.6. A gray iron cupola foundry.

it is mechanically transferred to delivery bins at the molding positions. At this time, the sand is plastic and cohesive and is not a dust hazard. After the casting and shake-out operations, the molding sand is a friable, dusty mass and must be conditioned for reuse. The dustiness of the molding sand is at a minimum after sand preparation and at a maximum directly after shake-out.

One-half of sand molds in the 1980s were made with the green sand process. Of the balance, the distribution is 10 percent by carbon dioxide molding, 8 percent by shell molding, and less than 10 percent by no-bake and hot and cold box systems. The no-bake molding procedures are gradually replacing the green sand process.

7.1.1 Green Sand Molding

The steps involved in making a simple mold from green sand are shown in Figure 18.7. A replica or pattern of the metal part to be cast is made from wood or metal. The pattern is designed so that it can be withdrawn after the sand has been packed around it. In the manual bench molding operation, a metal or wooden frame called a flask is used to hold the mold. The top half of the flask is called the cope, and the bottom is the drag. If additional height is needed, an extra section known as the cheek is mounted between the cope and the drag. The drag is inverted and placed on a molding board; the pattern is positioned on the board inside the flask. A fine, clean facing sand is then screened (riddled) over the pattern and board and packed in place. Next, heavy backing sand is rammed in place manually or by a pneumatic tamper. The excess sand is struck off, and the drag is inverted.

A parting compound is dusted on the face of the pattern and drag face so that the completed cope can be easily removed. The matching cope is positioned on the drag and pinned in place, and the cope pattern is assembled. The cope is then completed in the same fashion as the drag; wooden spacers are positioned to form the sprue and riser channels.

The sprue and associated gating convey the molten metal to the mold cavity. The riser acts as a vent and provides a reservoir of molten metal to handle shrinkage when the metal cools. Various vent holes are also introduced into the mold. The cope is then removed, and the cope and drag are positioned face up so the patterns can be removed. Sprue, riser, and runner forms are also removed at this time. If there are to be cavities in the final casting, cores must be positioned in the mold. The making of the sand cores is described below; however, these are refractory parts with mechanical strength and are positioned in core prints in the mold. With the cavity defined, the cope and drag are reassembled and clamped together ready for pouring. The finished sand mold is shown in Figure 18.7.

As indicated earlier, the new molding sand is not dusty and does not result in a significant silica exposure during the molding operation; however, the facing sand and the parting material may represent a dust exposure. In recent years, nonsiliceous materials such as chromite and zircon have replaced silica facing sand. Common parting compounds are frequently applied as spray suspensions in water, aliphatic hydrocarbons, 1,1,1-trichloroethane, or a Freon® solvent. If spray appli-

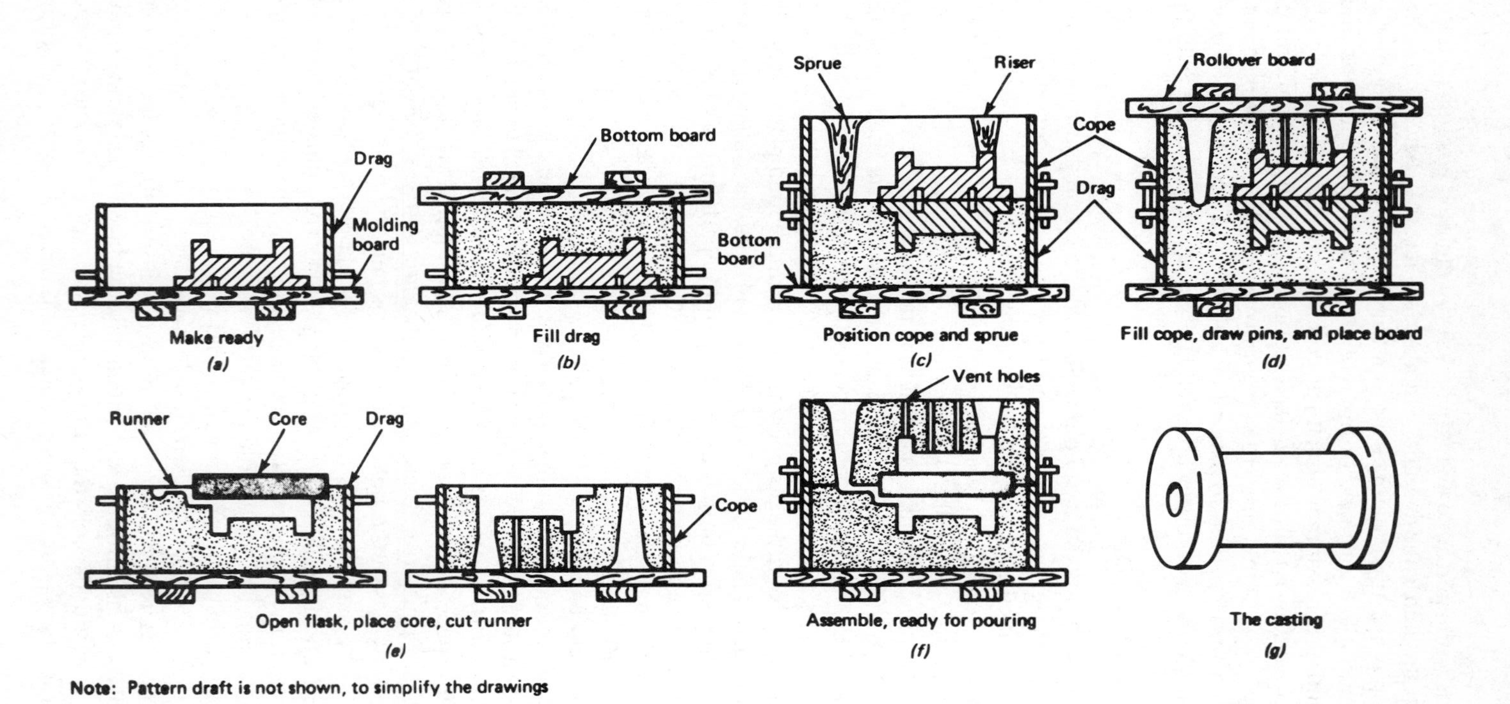

Figure 18.7. Preparation of a green sand mold.

cation of the suspensions is required, either airless or electrostatic spray techniques should be used to minimize overspray and rebound.

In a high production shop, the green sand mold described above would be made on a molding machine. This equipment is designed to pack the sand firmly in the flasks and manipulate the flasks, pattern, and completed molds semiautomatically, thereby minimizing the physical labor of the molder and improving mold quality. The molding machine tables are equipped with air cylinders that initially jolt the sand into place in the flask and then squeeze the sand to pack it around the pattern in a reproducible manner. If the work is too large for machine molding, it is done with large floor molds using flasks that are handled by crane. These flasks are filled with the molding sand by a mobile sand slinger, which is a single blade centrifugal blower located at the end of a boom and supplied with molding sand from a hopper. The operator can remotely direct the high-velocity sand stream to various parts of the flask and produce a compact mold. If the cast part is too large for available flasks, a pit is dug in the foundry floor, and the molding and pouring are conducted in place. The various molding techniques increase both noise and vibration problems to the industry.

The conventional green sand molding process continues to represent one-half of the molding done in the United States; however, many of the coremaking techniques are now used in mold making. In addition, there are a number of specialty molding processes used principally in steel and nonferrous foundries, which are described in the following sections.

7.1.2 Shell Molding

The shell molding process uses a resin bonded sand for high production of complex-shaped, small parts. Shell molding sand is coated with phenol or urea formaldehyde resin using a cold, warm, or hot coating technique. In the cold technique, the resin, hexamethylenetetramine, and calcium stearate are blended with dry sand in a mixer. In the warm coating process, the resin–sand mix is dried with a 149 to 177°C (300 to 350°F) airstream. Flaked resin is blended with the sand in the hot coating technique, quenched with a slurry of hexamethylenetetramine and calcium stearate, and then aerated at ambient temperature. The modern foundry obtains precoated shell molding sand, thereby eliminating these mixing processes at the foundry.

A manual, dump-box technique for shell molding is frequently used. In this process, a machined metal pattern reflects the geometry of the part to be cast. The pattern is sprayed with a parting agent such as a silicone and mounted on a resin–sand dump box. The pattern is heated by the oven, and the dump box is then inverted. The resin hits the hot pattern, and a skin of partially cured resin–sand is formed over the pattern. The dump box is turned upright, and an oven is positioned over the pattern to effect a complete cure of the resin at temperatures of 316 to 427°C (600 to 800°F). In another high production technique, the shell molding sand is blown into a heated die where the resin bond is formed. After the cured shell is stripped from the pattern or the die, it is assembled with a matching shell to form the mold. The cope and drag are either glued or clamped together, and frequently they are supported by sand or steel shot in a flask for casting.

The operator of a shell molding machine is exposed to phenol, formaldehyde, ammonia, carbon monoxide, hexamethylenetetramine, and carbon monoxide. Local exhaust ventilation is required on shell molding machines at the cope rack, mold press, and storage rack. A fresh air supply diffuser may be required to control heat stress on this operation.

7.1.3 Investment Casting (Lost Wax)

The lost wax or investment casting method is an ancient technique now used industrially for precision casting of a range of products including turbine vanes and nozzles. Expendable patterns made from hard wax or plastic such as polystyrene are cast in a metal die. The formed patterns are assembled in a tree-like cluster with a common sprue so that a large number of parts can be obtained from one pouring. The cluster may be dipped in a slurry to obtain a structural skin or placed in a metal flask and invested with a refractory mixture of plaster of paris, talc, silica, and water (or possibly an alcohol). After the investment is cured, the flask is inverted and placed in an oven at a temperature of approximately 2000°F. If wax is used, the patterns melt, and the wax drains out of the flask; if a plastic is used, it is vaporized. Molten metal is poured into the cavity, the metal is cooled, and the parts are retrieved from the flask.

The hazards in this operation vary depending on the exact materials in use. The preparation of the silica slurry involves an exposure to silica sand and alcohol. Low-temperature melting of the wax should not present a problem unless chlorinated waxes are used in the process. The vaporization of the plastic patterns may release a number of thermal degradation products depending on the specific plastic and the oven temperature. Removing the parts may involve an abrasive blasting operation with possible dust exposure.

7.1.4 Full Mold

In the full mold (lost foam, evaporative pattern, or expendable-pattern) process, an expendable pattern in the image of the part to be cast is made of polystyrene foam. In most cases, the pattern can be constructed so that cores are not required. Previously, the patterns were machined from polystyrene foam blocks on numerically controlled lathes. The latest technique involves casting the patterns in an aluminum die using polystyrene beads and a pentane blowing agent. The castings are coated with a water-soluble silica refractory (18). The pattern is positioned in a molding box with the necessary sprues and risers also made of polystyrene. Molding sand is placed around the pattern and gently rammed in place. When the molten metal is poured into the sprue, the polystyrene evaporates, and the gases diffuse into the sand mass. The entire polystyrene pattern evaporates in this manner, and the molten metal fills the space occupied by the pattern. An advantage of this process is the minimum cleaning required of the cast part.

The hazards from this operation include silica dust exposure and the gases vented from the vaporized polystyrene. Laboratory investigations of plastic thermal degradation suggest that one should be prepared to sample for styrene monomer, carbon monoxide, and possibly benzene.

7.2 Coremaking

The core is a refractory element placed in the mold to define a cavity in the final casting. Because the molten metal will flow around it, the core must be mechanically strong at that point and yet become friable after pouring and cooling to allow easy removal from the casting. Cores are made in a similar fashion to the molds described above. In the oldest system, core sand is prepared in a muller by mixing silica sand with an organic binder such as linseed oil and starch or dextrin. The sand is packed in a core box with a cavity defining the shape of the core. The fragile core is then removed and cured in a core oven at approximately 204°C (400°F). Oil-based core ovens are notorious air pollution sources, releasing acrolein and other aldehydes to the workplace and the neighborhood. For this reason and other production reasons, such ovens are rarely seen in a modern foundry.

In the past two decades, there has been a revolution in core manufacturing techniques. The new coremaking procedures involve a series of resin binder systems. Certain binder systems require oven heating; others require gassing to cure the system; and there are many no-bake systems. As described below, these systems may or may not release air contaminants during curing, but most do in pouring and shakeout operations (16, 19, 20, 21).

In high production foundries, cores are made on automatic core machines with multiple cavities. The core machines fill the cavities with the core sand–resin mix (core blowing) gas or heat the core, purge, and eject the core. After the cores are ejected, the operator cleans and repairs them and applies a coating. The cleaning process may involve either a simple dressing off with a knife and hand or machine sanding with some dust exposure. The repair of small defects is done with a putty knife and a filler compound. The cores may be dip or spray coated; a popular coating is a paraffin material in an isopropyl alcohol vehicle. The entire operation may be conducted by one person or, in a large foundry, the tasks may be divided between a muller operator, core machine operator, and a core finisher.

The specific resin systems presently in use and the worker exposures to airborne contaminants are outlined below.

7.2.1 Sodium Silicate System

The sodium silicate system has been in widespread use since its introduction in the mid-1950s and is now the most widely used gas/vapor cured system. Happily it represents a minimal health hazard to the workers. The core sand is prepared by mixing 2 to 6 percent sodium silicate (water glass) with silica sand and certain other additives including sugars, iron oxide, carbon, and polymers. After the corebox is filled and faced, the core is cured by passing carbon dioxide through the core sand. A reaction takes place forming sodium carbonate and a silicon dioxide gel and fixing the sand in place with a strong bond. Chemical hardeners now used in lieu of carbon dioxide include ferrosilicon, sodium silicon fluoride, dicalcium silicate, Portland cement, glycerol diacetate triacetate, and ethylene glycol diacetate.

Workers handling the concentrated water glass, a strongly alkaline solution, should wear personal protective equipment including gloves, aprons, and goggles. The chemical hardeners also require care in handling.

The carbon dioxide vented from the corebox does not normally present a problem in an open workplace; however, if the process is conducted in a molding pit or other enclosed space, significant carbon dioxide concentrations may exist. If carbon dioxide is used to cure the cores, the thermal degradation products released during the later pouring and shake-out operations are those common to any sand molding operation. The use of the organic additives results in the formation of carbon monoxide and other products during pouring. A complete description of the thermal degradation products of these activators is not known at this time.

7.2.2 Hot Box

In this process, thermosetting resins such as phenol formaldehyde, urea formaldehyde, furfuryl alcohol, or other combinations are mixed with sand and a catalyst to form a system that will cure to a solid mass in a heated corebox. A variety of catalysts and hardeners including ammonium chloride, ammonium nitrate, and magnesium oxide are in use. After being mixed in a muller, the sand, resin, and catalyst are injected into a mold heated to a temperature of 204 to 260°C (400 to 500°F). The compressed core sand mix is cured to a solid mass in a minute or two. The core is then ejected to a cooling station where the cure continues throughout the mass of the core. During this period, significant off-gassing occurs, and good local exhaust ventilation is required.

Precautions are required in handling the resin and catalyst in concentrated form while preparing the mix. Such precautions should include skin and eye protection for both the urea- and phenol-based resins. Ventilation control is required on the mixer, the coremaking machine including the core blowers, the cool-down location, and pouring, casting, cooling, and shake-out stations.

The air contaminants released during hot box core manufacture and shakeout operations include ammonia, formaldehyde, phenol, furfuryl alcohol, and carbon monoxide. The exact contaminants depend on the type of resin system in use. Air concentration data on coremaking operations with good control have been developed in a NIOSH study (16).

7.2.3 No Bake

Efforts to eliminate oven and corebox heating processes led to the development of a series of sand–resin–catalyst systems that cure at room temperature. These systems are now also used in mold making. As noted below, these systems do involve unique exposure patterns to contaminants not previously seen in the foundry environment.

The most common system is a phenol resin used with or without furan resins.

Acid catalysts including phosphoric, toluene sulfonic, benzene sulfonic, and sulfuric acids are used with these systems.

Urethane no-bake systems are formed when phenol resin is used in conjunction with isocyanates. A liquid cured no-bake system includes a phenol–formaldehyde resin binder, methylene dissocyanate (MDI) catalyst, and a pyridine compound. The resin is added to the sand and then the catalyst; the reaction takes place rapidly, and the core may be stripped in less than 15 min. A similar system based on gas curing uses the same sand–resin mix with an MDI catalyst, but the vapor of triethylamine or ethylamine is blown into the corebox for curing.

A range of hazards exist from no-bake resin operations. Skin and eye contact with the principal resins should be avoided. The strong acids used as catalysts in the phenolic–furan systems demand attention to safe handling procedures and personal protective equipment. Exposure to the isocyanates may cause pulmonary asthma at low concentrations; the gassing agents used in the gas-cured phenolic–urethane systems have known systemic toxicity. Local exhaust ventilation is required where the sand–resin systems are weighed out and mixed at the core machine, core storage, and during pouring and shake-out. Skin contact must be minimized and scrupulous housekeeping practices followed and encouraged by disposable paper covers on work surfaces. A set of tools should be assigned solely to the resin–catalyst work stations to prevent contaminating other work sites.

7.2.4 Shell Coremaking

This system is identical to the shell molding process described under molding and includes skin contact with hexamethylene tetramine and air contaminants such as ammonia, phenol, and formaldehyde. The thermal degradation products noted in later steps are similar to those for other phenol- and urea-based systems.

7.3 Metal Melting and Pouring

These processes include the preparation of the furnace charge materials, preheating of furnace and ladles, melting of the charge in the furnace, fluxing of the melt both in the furnace and at the ladle to remove silicates and oxides, inoculation of the charge with materials for improved metallurgical properties, tapping of the furnace, pouring from furnace to a receiving ladle and subsequent transfer to smaller pouring ladles, and pouring the melt into the prepared molds. The major health hazards in both ferrous and nonferrous operations include exposure to toxic metal fumes, carbon monoxide from the furnaces and molds, other toxic gases, and heat stress.

The potential exposures to metal fumes are listed in Table 18.3 and to thermal degradation products from the molds and cores in Table 18.4.

7.3.1 Arc Furnaces

The electric arc furnace, handling 2 to 200 tons per melt, has replaced the open hearth as the major furnace in large steel foundries. The furnace is charged with ingot, scrap, and the necessary alloying metals. An arc is drawn between the three

Table 18.3. Dust and Fume Exposures from Metal Melting and Pouring

Metal	Dust or Fume	Occurrence
Iron and steel	Iron oxide	Common
	Lead, leaded steel	Common
	Manganese	Common
	Silica	Common
	Carbon monoxide	Common
	Acrolein	Rare
Bronze and brass	Copper	Common
	Zinc	Common
	Lead	Common
	Manganese	Rare
	Phosphine	Rare
	Silica	Common
	Carbon monoxide	Common
Aluminum	Aluminum	Common
Magnesium	Magnesium	Common
	Fluorides	Common
	Sulfur dioxide	Common
Zinc	Zinc	Common
Cadmium	Cadmium	Common
Lead alloys	Lead	Common
	Antimony	Common
	Tin	Common
Beryllium	Beryllium	Common
Beryllium–copper	Beryllium	Common

Source: Reference 35.

carbon electrodes and the charge, heating the charge and quickly melting it. A slag cover is formed with various fluxing agents to reduce oxidation of surface metal, refine the metal, and protect the roof of the furnace from damage from excessive heat radiation. When the melt is ready for pouring, the electrodes are raised and the furnace is tilted to pour to a receiving ladle.

These furnaces produce tremendous quantities of metal fume, resulting in both a workplace and air pollution problem. Local exhaust ventilation systems based on side draft or enclosing hoods are available to handle fume generation during the melt cycle (3). A slag door hood and pour spout hood must be included in the ventilation system. Because metal fume escapes during the operation, high-volume roof exhausters are located over the furnace area to remove fugitive losses. Another approach to local exhaust ventilation on this equipment is the use of partitions dropped from the roof to form a modified canopy hood. However, this presents a difficult air cleaning problem, namely, a large air volume and a low fume concentration. A serious noise problem also occurs from the intermittent make and break of the arc on these furnaces.

Table 18.4. Thermal Decomposition Products of Molds and Cores

Air Contaminant	Binder Source
Carbon monoxide	All organic binders and additives
Carbon dioxide	All organic binders, silicate–CO_2
Hydrocarbons	All organic binders, some inorganics
Hydrogen cyanide	Nitrogen-containing binders
Ammonia	Nitrogen-containing binders
Formaldehyde	Urea–, phenol–, and furfuryl alcohol–formaldehyde resins, decomposition of other organic binders
Phenol	Phenolic resins, phenolic urethanes
Furfuryl alcohol	Furan binders
Sulfur dioxide and hydrogen sulfide	Sulfonic acid catalysts
Phosphines and phosphorus oxides	Phosphoric acid catalysts, phosphate binder

7.3.2 Induction Furnace

This furnace is used widely in both nonferrous and alloy steels foundries. The melting refractory in a crucible is surrounded by water-cooled copper coils powered by a high-frequency power supply. The outer winding induces current flow in the outer edge of the metal charge and, owing to the high resistance, the metal charge is heated and melting progresses from the edge of the charge to the center.

The metal fumes from this furnace are best controlled by enclosing hoods, although canopy hoods and dilution ventilation are frequently utilized. For low-alloy steels, the canopy hood or dilution ventilation may provide adequate control; however, an enclosing hood is required for nonferrous metals.

7.3.3 Crucible Furnaces

The crucible or ceramic pot containing the charge is heated directly by a gas or oil burner or occasionally coke. This furnace is widely used for nonferrous alloys. The principal hazards are carbon monoxide, metal fumes, burner noise, and heat stress. Ventilation is usually accomplished by a canopy hood at the furnace line, although perimeter slot hoods provide better control.

7.3.4 Cupola

Over 60 percent of the gray iron castings produced in the United States are produced by the cupola. It is the most economical way to convert scrap and pig iron to usable molten iron. As produced, the metal is gray iron; if inoculated with manganese or cerium at the ladle, a ductile iron is formed. The cupola furnace shell is a right cylinder lined with refractory brick. The base of the cupola contains a burner to ignite a bed of coke. Alternate layers of coke, limestone, and metal are periodically charged to the furnace, with the coke providing both the fuel and a reducing environment.

A major hazard from cupola operation is exposure to carbon monoxide. This is especially true if twin cupolas are in use and one is under repair. Iron oxide fume is not a direct problem in the workplace; however, scrap iron may contain small quantities of toxic metals, such as lead and cadmium, which may present significant exposures.

Personnel who operate and repair cupolas have a serious heat problem. Where a fixed pouring station is used, an air-conditioned work station may be installed for the ladle man. Reflective barriers are also effective in minimizing the radiant heat load.

7.3.5 Transfer, Pouring, Cooling

Fume release during pouring to the receiving ladle, subsequent transfer to the pouring ladle, and pouring at the mold line require local exhaust ventilation. Two approaches are possible for ventilation control of pouring operations. If the molds are conveyorized, the pouring may be done at a fixed location with an installed local exhaust hood. If the molds are poured on the floor, the ladle must be equipped with a mobile hood.

Inoculation of heats with special metals places the worker in close proximity to the furnace or ladle with resulting high exposures to heat, metal fume, noise, and accidental metal splashes. The most recent techniques of inoculation, including bell and sandwich inoculation, reduce these hazards although they still warrant attention. Lead, a common inoculant to improve machinability of steel, illustrates the potential hazard during inoculation. Depending on how it is added, from 20 to 90 percent of the lead may be released to the atmosphere.

Concentrations of lead up to 1.0 mg/m^3 may be noted at the operator's level during inoculation, and high levels may exist in the crane cab over the ladle area.

Silicate-forming materials and oxides must be removed from the metal before pouring. This requires the addition of fluxing agents and removal of the slag that is formed. To protect the worker from fume exposure and heat stress, a combination of shielding and distance is frequently used. On large operations with severe heat stress, mechanical slagging is conducted by remote control.

Operators directly involved in handling the molten metal may require personal protective equipment including aluminized clothing, tinted glasses and face shields, and head and foot protection. A combination of engineering and work practice controls must be installed to handle heat stress. The acclimatization of the worker requires at least a week, so care should be observed during the first days of the warm season and during the start-up week after vacation.

The crane operator has a special exposure that warrants attention. The metal fume lost from the floor operations is cleared by roof exhausters, and high air concentrations of fume may be noted at the crane level. Obviously, if remote radio-controlled cranes are used, the operator is not exposed. If an operator must be positioned in the crane, one should consider enclosure of the cab with a supply of filtered, conditioned air.

After pouring and before shake-out, molds are moved by a conveyor to a staging area for cooling. A critical exposure to carbon monoxide may occur in this area especially if sea coal or other carbonaceous material is added to the molding sand to improve the surface of the casting. The sea coal forms methane, which burns off with the production of carbon monoxide. I have noted concentrations of carbon monoxide of 200 to 300 ppm in the staging area of a small gray iron foundry. Control may be achieved by routing pallet molds through an exhausted cooling tunnel.

7.4 Shake-out

The casting is removed from the mold at the shake-out position. In the simplest system, small- to medium-sized castings are placed on a vibrating screen; the molding sand drops to a hopper through the screen for reconditioning; the flask is routed back to the molding line; and the casting is hooked free for cleaning. Small castings can be more efficiently removed from the flasks by a "punch-out" process. This method is superior to shake-out because it generates less dust. Shake-out is still required for removal of surface sand and cores. This is a hot, demanding job with serious exposure to particulates, gases, vapors, noise, and vibration.

Considerable data are available that demonstrate significant air concentrations of silica at foundry shake-out operations. The resin–sand systems used in molding and coremaking are the source of various gases and vapors reflecting the composition of the resin-catalyst system as shown in Table 18.4. If sea coal is used in the molding sand, there may be a serious exposure to carbon monoxide. Few published data are available on the exposure of shake-out operators to toxic metal particulates resulting from sand contamination, although this must be considered a potential problem in foundries using a variety of scrap in furnace charging.

The principal control of air contaminants at the shake-out is local exhaust ventilation utilizing enclosures and side draft and downdraft hoods. With the exception of well-designed enclosures for automatic shakeout of small castings, none of the systems is completely effective. A number of approaches have been considered to replace or augment the conventional shake-out. Rather than vibrating the casting free from the molding sand, hydro-blast units have been utilized to free the casting while generating little dust. Small molds may be placed in a special enclosed tumbler where the parts are separated from the sand. Failing good control, the shake-out is frequently conducted off-shift to reduce the number of persons exposed.

The removal of cores from the castings is accomplished either at the shake-out or at a special core knockout station. High-pressure streams have been successfully applied on this operation.

7.5 Cleaning and Finishing

After shake-out, the casting is processed in the finishing room. Complex cores require special core knockouts, which may be either manual or mechanized operations depending on the core geometry and production rate. The extraneous metal including sprue, risers, and gates on gray iron castings are removed by a sharp rap of a hammer; torch burning or cutoff wheels are used on steel and alloys castings. The parts are frequently cleaned by abrasive blasting or hydroblast, and rough finishing is done by chipping and grinding. If the castings are clean, there is limited exposure to silica sand during finishing; however, in steel foundries, the sand fuses on the surface, and some of the silica may be converted to tridymite or cristobalite. The occupational health problems in this area include dust exposure, noise, and vibration. Silica dust is the major health hazard in the cleaning room.

It is clear that a significant percentage of foundry workers in the United States continue to be exposed to excessive concentrations of silica dust. Silica concentrations in shake-out, chipping, grinding, and other cleaning operations are among the highest. The iron and steel foundries have higher exposures than nonferrous, apparently owing to the greater sand penetration of castings from high-temperature casting.

7.6 Control Technology

Although other air contaminants exist, silica dust and carbon monoxide continue to be the major health hazards in the foundry. Noise, heat, and vibration are important physical stresses in foundry work. The American National Standards Institute has published a series of standards that provide guidance on health and safety controls for all major foundry operations (22).

The hazard from silica varies with the physical state of the sand, its chemical composition, and the method of handling. Because the sand is initially dry, the mixing or mulling operation is dusty, and local exhaust ventilation is required even though the sand is coarse. After this point, the moisture content is high and little dusting occurs during the molding operation. After casting, the sand becomes friable and dusty; serious dust exposures may occur during shakeout and sand conditioning. Research conducted by the British Cast Iron Research Association (BCIRA) has demonstrated that foundry sand with a moisture content greater than approximately 30 percent does not present a dust problem during normal handling. This laboratory observation is borne out in plant observations where foundry operations right up to the pouring station present little dustiness potential. From that point on, the sand becomes friable, dry, and dusty. It is therefore imperative to add moisture immediately at the shake-out by the following techniques:

- Use a pan mill at the shake-out and immediately add water.
- Add water at the belt from the shake-out.
- Add prepared molding sand with high moisture content to the shake-out sand.
- Where possible, add sea coal and clay as a slurry.

Additional advice is presented by the BCIRA on the proper design of conveyor belts and hoists to minimize spillage. Foundry sand spilled from conveyors dries out and easily becomes resuspended, increasing workers' exposure to silica.

Historically, the approach to dust control in sand foundries has been the use of control ventilation coupled with housekeeping and wet methods. Low-volume, high-velocity capture systems are finding increased application in grinding and chipping operations in finishing rooms. In recent years, substitution of nonsiliceous parting compounds for silica flour has demonstrated the value of substitution as a control. Changes in procedure, such as the Schumacher method, which mixes some prepared moist molding sand with dry shake-out sand before it is conveyed back from the shake-out, result in impressive dust reduction. The introduction of permanent mold techniques will have a major impact on silica sand usage and resulting worker exposure.

Foundry noise is a major problem. The American Foundrymen's Society has published an excellent review of current knowledge of noise control on chipping and grinding, burners, electric arc furnaces, shake-out, molding operations, core and molding machines, and conveyors. This document treats the most difficult problems in the foundry and provides state-of-the-art technology solutions that have been evaluated in the industry (23).

Until the late 1970s, little attention was given to the occurrence of vibration-induced disease, such as vibration-induced white finger (VWF), in the foundry population. Studies in United States foundries have demonstrated that VWF is a problem of major proportion, with a minimum prevalence of VWF of 50 percent in cleaning and finishing room personnel and latency periods as short as 1 to 2 years. A guide published in the United Kingdom recommends exposure limits for hand-transmitted vibration from vibratory tools of the type used in the foundry (24). The minimum control approach to this problem proposed by Pelmear and Kitchener includes (25):

- Identification of high risk situations
- Redesign of tools to minimize vibration
- Use of vibration isolation pads where possible
- Introduction of alternative work methods
- Education of workers
- Preeployment and periodic examinations of vibration-exposed persons
- Transfer of affected workers to alternative work

GRINDING, POLISHING, BUFFING

These operations are grouped together for discussion because they all involve controlled use of bonded abrasives for metal finishing operations; in many cases, the operations are conducted in the sequence noted. This discussion covers the nonprecision applications of these techniques.

8.1 Processes and Materials

Nondimensional application of grinding techniques includes cutoff operations in foundries where gates, sprues, and risers are removed, rough grinding of forgings

Table 18.5. Grinding Wheel Specification Nomenclature

Abrasives (First Letter in Specification)	
A	Aluminum oxide
C	Silicon carbide
D or ND	Natural diamonds
SD	Synthetic diamonds
CB or CBN	Cubic boron nitride
Bond (Last Letter in Specification)	
V	Vitrified
B	Resinoid
R	Rubber
S	Silicate

and castings, facing-off weldments, and grinding out major surface imperfections in metal fabrications. Grinding is frequently done with wheels and disks of various geometries made up of selected abrasives in bonding structural matrices. The common abrasives are aluminum oxide and silicon carbide; less common are diamond and cubic boron nitride. A variety of bonding materials are available to provide mechanical strength and yet release the spent abrasive granules to renew the cutting surface. Vitrified glass is the most common bonding agent. The grinding wheel is made by mixing clay and feldspar with the abrasive, pressing it in shape and firing it at high temperature to form a glass coating and bond for the abrasive grains. Resinoid wheel bonds, based on thermosetting resins such as phenol–formaldehyde, are used for diamond and boron nitride wheels and are reinforced with metal or fiber glass for heavy-duty applications including cutoff wheels. Sodium silicate (water glass), the softest bonding agent, is used on grinding wheels that require high wheel wear while remaining cool. Rubber bonding agents are routinely used for final finishing, polishing applications, and cutoff wheels.

The abrasive industry utilizes a standard labeling nomenclature to identify the grinding wheel design (26). This information is invaluable in identifying the possible air contaminants released from the grinding wheel. Of specific interest to the occupational health specialist is the identification of the abrasives and the bonding agent (Table 18.5).

Little information is available on the generation rate of grinding wheel debris on various applications; however, the wheel components normally make up a small fraction of the total airborne particles released during grinding. The bulk of the particles are released from the workpiece. After use, the grinding wheel may load or plug, and the wheel must be "dressed" with a diamond tool or "crushed dressed" with a steel roller. During this brief period, a significant amount of the wheel is removed and a small quantity may become airborne.

Polishing techniques are used to remove workpiece surface imperfections such as tool marks. This technique may be used to remove as much as 0.1 mm of stock from the workpiece. The abrasive, again usually aluminum oxide or silicon dioxide, is bonded to the surface of a belt, disk, or wheel structure in a closely governed geometry, and the workpiece is commonly applied to the moving abrasive carrier by hand.

The buffing process differs from grinding and polishing in that little metal is removed from the workpiece. The process merely provides a high luster surface by smearing any surface roughness with a lightweight abrasive. Red rouge (ferric oxide) and green rouge (chromium oxide) are used for soft metals, aluminum oxide for harder metals. The abrasive is blended in a grease or wax carrier that is packaged in a bar or tube form. The buffing wheel is made from cotton or wool disks sewn together to form a wheel or "buff." The abrasive is applied to the perimeter of the wheel, and the workpiece is then pressed against the rotating wheel. The wheels are normally mounted on a buffing lathe that is similar to a grinding stand. During heavy-duty operations, the surface temperature of the workpiece may reach 150°C (300°F).

8.2 Exposures and Control

The hazard potential from grinding, polishing, and buffing operations depends on the specific operation, the workpiece metal and its surface coating, and the type of abrasive system in use. A NIOSH-sponsored study of the ventilation requirements for grinding, polishing, and buffing operations shows that the major source of airborne particles in grinding and polishing is the workpiece, whereas the abrasive and the wheel represent the principal sources of contamination in buffing (27). Few data are available on the influence of the metal type on airborne levels of contaminants, although limited tests on three metals in the aforementioned study showed that titanium generated the most dust, a 1018 steel alloy was the next dustiest, and aluminum was the least dusty. Certainly the type of metal worked and the construction of the abrasive system govern the generation rate and the size characteristics of the dust; however, the data necessary for a detailed assessment of this matter are not available.

A listing of the metal and alloys worked and information on the nature of the materials released from the abrasive system are needed in order to evaluate the exposure of the operator. In many cases, the exposure to dust can be evaluated by means of personal air sampling with gravimetric analysis. If dusts of toxic metals are released then specific analysis for these contaminants is necessary.

Is ventilation control necessary? The need for local exhaust ventilation on grinding operations has been addressed by British authorities (28), who state that control is required if one is grinding toxic metals and alloys, ferrous and nonferrous castings produced by sand molding, and metal surfaces coated with toxic material.

The same source also states that control is not needed if steel surfaces are ground prior to welding, forging, stamping and precision grinding. This statement may apply to mild steel; but if alloy steels are worked, ventilation may be required during these operations. Control of exposures from grinding operations should include the removal of fused sand or toxic metal coatings from the workpiece by such techniques as abrasive blasting under controlled conditions.

One can extend the application of the above guideline on grinding to polishing. The ventilation requirements for buffing, however, are principally based on the large amount of debris released from the wheel, which may present a housekeeping problem and a potential fire risk.

The hood designs for grinding, polishing, and buffing are based on a tight hood enclosure with minimum wheel-hood clearance to control dust at a minimum exhaust volume. An adjustable tongue on the hood also reduces the air induced by wheel rotation at high speed. On buffing operations, the hood is usually designed with a settling chamber to minimize plugging of the duct with coarse wheel debris.

The conventional ventilation control techniques for fixed location grinding, polishing, and buffing are well described in the ACGIH and American National Standards Institute (ANSI) publications (3, 29). It is difficult to provide effective ventilation controls for portable grinders used on large weldments, castings, and forgings. Flexible exterior hoods positioned by the worker may be effective; and for large, high production shops, the hood design may be customized for a specific workpiece. High-velocity, low-volume exhaust systems with the exhaust integral to the grinder also are suitable for some applications (30). Worker resistance to this equipment may be minimized by suspending the grinder from an overhead boom counterweighted suspension to minimize effort in maneuvering the grinder. On many operations, neither of these approaches will work and a grinding booth is the best alternative. Heinsohn et al. (31) has presented design information on such booths.

The performance of the conventional hoods on grinding, polishing, and buffing operations should be checked periodically. Because the hood entry loss factor for conventional hoods is known, the hood static suction method is an efficient way to evaluate the exhaust rates quickly.

The hazard from bursting wheels operated at high speed and the fire hazard from handling certain metals such as aluminum and magnesium are not covered here, but these problems affect the design of hoods and the design of wet dust collection systems.

The hazard from vibration from hand tools leading to VWF was introduced in the discussion of foundries. This hazard is generally acknowledged in the use of portable tools, but the same problem exists in pedestal mounted tools. In one study of pedestal grinders, the total work population in a finishing room developed VWF with an average latency period of 10.3 months (32).

9 HEAT TREATING

A range of heat treating methods for metal alloys is available to improve the strength, impact resistance, hardness, durability, and heat and corrosion resistance of the workpiece (33). In the most common procedures, metals are hardened by heating the workpiece to a high temperature with subsequent rapid cooling. Softening processes normally involve only heating, or heating with slow cooling.

9.1 Surface Hardening

Case hardening, the production of a hard surface or case to the workpiece, is normally accomplished by diffusing carbon or nitrogen into the metal surface to a given depth to achieve the hardening of the alloy. This process may be accomplished in air, in atmosphere furnaces, or in immersion baths by one of the following methods.

9.1.1 Carburizing

In this process, the workpiece is heated in a gaseous or liquid environment containing high concentrations of a carbon-bearing material that is the source of the diffused carbon. In gas carburizing, the parts are heated in a furnace containing hydrocarbon gases or carbon monoxide; in pack carburizing, the part is covered with carbonaceous material that burns to produce the carbon-bearing gas blanket. In liquid carburizing, the workpiece is immersed in a molten bath that is the source of carbon.

In gas carburizing, the furnace atmosphere is supplied by an atmosphere generator. In its simplest form, the generator burns or reacts a fuel such as natural gas under controlled conditions to produce the correct concentration of carbon monoxide, and this gas is supplied to the furnace. Because carbon monoxide concentrations up to 40 percent may be used, small leaks may result in significant workroom exposure. Workplace exposures to carbon monoxide greater than 100 ppm are common in gas carburizing installations.

To control emission from gas carburizing operations, the combustion processes should be closely controlled, furnaces maintained in tight condition, dilution ventilation installed to remove fugitive leaks, furnaces provided with flame curtains at doors to control escaping gases, and self-contained breathing apparatus available for escape and repair operations (34). In liquid carburizing, a molten bath of sodium cyanide and sodium carbonate provides a limited amount of nitrogen and the necessary carbon for surface hardening.

9.1.2 Cyaniding

The conventional method of liquid carbonitriding is immersion in a cyanide bath with a subsequent quench. The part is commonly held in a sodium cyanide bath at temperatures above 870°C (1600°F) for 30 to 60 min. The air contaminant released from this prccess is sodium carbonate; cyanide compounds apparently are not released, although there are no citations in the open literature demonstrating this. Local or dilution ventilation is frequently applied to this process, although standards have not been proposed. The handling of cyanide salts requires the same precautions as those noted under electroplating, that is, secure and dry storage, isolation from acids, and planned disposal of waste. Care must be taken in handling quench liquids because the cyanide salt residue on the part will in time contaminate the quench liquid.

9.1.3 Gas Nitriding

Gas nitriding is a common means of achieving hardening by the diffusion of nitrogen into the metal. This process utilizes a furnace atmosphere of ammonia operating at 510 to 570°C (950 to 1050°F). The handling of ammonia in this operation is hazardous in terms of fire, explosion, and toxicity.

If the furnace environment contains 10 percent natural gas and 5 percent ammonia, carbon monoxide and ammonia both are available to diffuse into the metal and carbonitriding occurs.

9.2 Annealing

Annealing is a heat treatment technique applied to soften the metal to improve its machinability, relieve it of stress after certain operations, or increase its softness for other reasons. The process varies depending on the alloy and the use of the part, but in all cases it involves heating at a given temperature for a specific time and then cooling at a desired rate.

The quench baths may be water, oil, molten salt, liquid air, or brine. The potential problems range from a nuisance problem due to release of steam from a water bath to acrolein or other thermal degradation products from oil. Local exhaust ventilation may be necessary on oil quench tanks.

9.3 Hazard Potential

The principal problems in heat treating operations are due to the special furnace environments, especially carbon monoxide, and the special hazards from handling bath materials. Although the hazard potential is significant from these operations, few data are available. In the only published data on air contaminants in metal annealing and hardening operations, Elkins noted hazardous concentrations for carbon monoxide in 19 out of 108 installations, and 59 out of 167 exceeded the current maximum allowable concentration for lead (35). Hazardous cyanide concentrations were not observed.

Salt bath temperature controls must be reliable, and the baths must be equipped with automatic shutdowns. Before these baths are brought down to room temperature, rods should be inserted into the bath. The rods provide vent holes to release gases when the bath is again brought up to temperature. If this is not done and gas is occluded in the bath, explosions or blowouts may occur. Parts must be clean and dry before immersion in baths for residual grease, paint, and oil may cause explosions. These baths may require local exhaust ventilation although no design data are available.

Lead baths held at temperatures between 540°C (1000°F) and 820°C (1500°F) require exhaust ventilation. Significant lead exposures may occur when removing the dross floating on this molten lead. Oil quench tanks are often ventilated to remove the irritating smoke that evolves during their use. Some tanks have cooling coils to control the temperature of the oil and to reduce both smoke production and the fire hazard. There are no ventilation standards published for these operations.

Where sprinkler systems are used, canopies should be erected above all oil, salt, and metal baths to prevent water from cascading into them. Any workman who happened to be adjacent to a hot bath when water struck it would be in grave danger.

10 METAL MACHINING

The fabrication of metal parts is done with a variety of machine tools, the most common of which are the lathe, drill press, miller, shaper, planer, and surface grinder. The occupational health hazards from these operations are similar so they are grouped together under conventional machining. Many of the same techniques are also used for machining plastics and composite materials. It is estimated that there are over one million machinists in the United States working in varied settings from small model shops to large production facilities. Two rapidly expanding techniques, electrochemical and electrical discharge machining, are also discussed in this section. This brief overview of machining does not include a number of advanced techniques such as laser and high-pressure water machining.

10.1 Conventional Tool Machining

The major machining operations of turning, milling, and drilling utilize cutting tools that shear metal from the workpiece as either the workpiece or the tool rotates (Figure 18.8). A thin running coil is formed that normally breaks at the tool to form small chips. Extremes of temperature and pressure occur at the interface between the cutting tool and the work. To cool this point, provide an interface lubricant, and help flush away the chips, a coolant or cutting oil is directed on the cutting tool in a solid stream (flood) or a mist.

The airborne particles generated by these machining operations depend on the type of base metal and cutting tool, the dust-forming characteristics of the metal, the machining technique, and the coolant and the manner in which it is applied. Each of these concerns is addressed briefly in this section.

The type of metal being machined is of course of paramount concern. The metals range from mild steel with no potential health hazard as a result of conventional machining to various high-temperature and stainless alloys incorporating known toxic metals including lead, chromium, nickel, and cobalt, which may present low airborne exposures to toxic metals depending on the machining technique. Finally, highly toxic metals, such as beryllium, do present significant exposures that require vigorous control in any machining operation. Under normal machining operations,

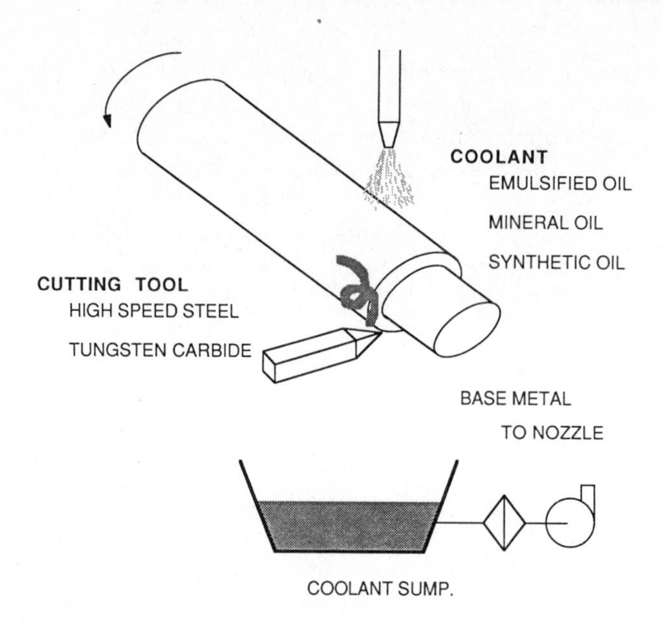

Figure 18.8. Major components in turning metals on a lathe.

excluding dimensional grinding, the airborne dust concentration from conventional metals and alloys is probably minimal.

A range of specialized alloys has been developed for use in the manufacture of cutting tools. These materials include (*a*) high carbon steels with alloying elements of vanadium, chromium, and manganese; (*b*) high-speed steels containing manganese and tungsten; (*c*) special cobalt steels; (*d*) cast alloys of tungsten, chromium, and cobalt; and (*e*) tungsten carbide. The loss of material from the cutting tool is insignificant during conventional machining and therefore airborne dust concentrations from the tool do not represent a potential hazard. Preparing the cutting tools may involve a significant exposure to toxic metal dusts during grinding and dressing operations, especially on tungsten carbide and other hard metals, and such operations should be provided with local exhaust ventilation.

Coolants and cutting fluids are designed to cool and lubricate the point of the cutting tool and flush away chips. These fluids are presently available in the form of (*a*) soluble (emulsified) cutting oils based on mineral oil emulsified in water with soaps or sulfonates; (*b*) straight cutting oils based on a complex mixture of paraffinic, naphthenic, and aromatic mineral oils with the addition of fatty acids; (*c*) synthetic oils of varying composition; and a mixture of *a* and *c*. A description of the composition of the three types of cutting fluids including special additives is shown in Table 18.6 (36).

Cutting fluids present two potential health problems: extensive skin contact with the cutting fluids and the inhalation of respirable oil mist. It has been estimated that over 400,000 cases of dermatitis occur in the United States each year from contact with coolants and cutting fluids. Soluble oils frequently cause eczematous

Table 18.6. Composition of Cutting Fluids

I. Mineral oil
   1. Base 60–100%, paraffinic or naphthenic
   2. Polar additives
      a. Animal and vegetable oils, fats, and waxes to wet and penetrate the chip/tool interface
      b. Synthetic boundary lubricants: esters, fatty oils and acids, poly or complex alcohols
   3. Extreme pressure (EP) lubricants
      a. Sulfur-free, or combined as sulfurized mineral oil or sulfurized fat
      b. Chlorine, as long-chain chlorinated wax or chlorinated ester
      c. Combination: sulfo-chlorinated mineral oil or sulfo-chlorinated fatty oil
      d. Phosphorus, as organic phosphate or metallic phosphate
   4. Germicides

II. Emulsified oil (soluble oil)—opaque, milky appearance
   1. Base: mineral oil, comprising 50–90% of the concentrate; in use the concentrate is diluted with water in ratios of 1 : 5 to 1 : 50
   2. Emulsifiers: petroleum sulfonates, amine soaps, rosin soaps, naphthenic acids
   3. Polar additives: sperm oil, lard oil, and esters
   4. Extreme pressure (EP) lubricants
   5. Corrosion inhibitors: polar organics, e.g., hydroxylamines
   6. Germicides
   7. Dyes

III. Synthetics (transparent)
   1. Base: water, comprising 50–80% of the concentrate; in use the concentrate is diluted with water in ratios of 1 : 10 to 1 : 200. True synthetics contain no oil. Semisynthetics are available that contain mineral oil present in amounts of 5–25% of the concentrate
   2. Corrosion inhibitors
      a. Inorganics: borates, nitrites, nitrates, phosphates
      b. Organics: amines, nitrites (amines and nitrites are typical and cheap)
   3. Surfactants
   4. Lubricants: esters
   5. Dyes
   6. Germicides

Source: Reference 36.

dermatitis whereas the straight oils (insoluble) cause folliculitis. One also notes occasional sensitization to coolants.

There continues to be a difference of opinion on the role of bacterial contamination of fluids in dermatitis; however, there is agreement that maintenance of coolants is of hygienic significance. A coolant sampling procedure is now available that permits evaluation of aerobic bacteria, yeasts, and fungi concentration in coolants using simple dip slides.

Cutting fluids may become contaminated with trace quantities of the base metal being machined. These dissolved metals in the fluid are then contacted by the worker; a sensitizing dermatitis may result when machining chromium, nickel, and cobalt alloys.

The application of the cutting fluid to hot, rotating parts releases an oil mist that causes the characteristic smell in the machine shop. The introduction of mist application of fluids in the 1950s also contributed to this problem. The health effects from extended exposure to airborne mists of mineral oil and synthetic coolants is not clear. Previous exposures to high concentrations of nonrefined mineral oils may have resulted in a slight excess in cancer mortality. Exposures since 1950 to low to moderate concentrations of high-quality mineral oil coolants present no significant increase in cancer risk (37). Decoulfe states the exposure to mineral oil-based coolants does not represent a health hazard from respiratory cancers, but it may be associated with gastrointestinal cancers (38). Ongoing epidemiology in the automotive industry may resolve the question of risk from mineral oils and identify the health risk from synthetic oil exposures.

An excellent pamphlet on lubricating and coolant oils prepared by Esso outlines procedures to minimize exposure to cutting fluids (39). Selected material is quoted below:

1. Avoid all unnecessary contact with mineral or synthetic oils. Minimize contact by using splash guards, protective gloves and protective aprons, etc. Use goggles or face visors when handling soluble oil concentrate. The golden rule is: don't wear oil-soaked clothing and never put oil rags into pockets.

2. Encourage workers to wear clean work clothes, since oil-soaked clothing may hold the oil in contact with the skin longer than would otherwise occur. This applies particularly to underclothes, which should be changed frequently and washed thoroughly before re-use. Consider the provision of one locker for work clothes and a separate one for street clothes.

3. Consider the use of short-sleeved overalls rather than long-sleeved garments for workers handling metalworking fluids when friction on the skin from cuffs saturated in oil can promote skin problems.

4. Removal of oil from the skin as soon as possible if contact does occur. This means the installation of easily accessible wash basins and the provision of mild soap and clean towels in adequate supply. Avoid strong soaps and detergents, and abrasive-type skin cleaners.

5. Encourage workers to take showers at the end of a day's work in order to remove all traces of oil from the skin.

6. Do not allow solvents to be used for cleansing the skin. Use only warm water, mild soap and a soft brush or in combination with a mild proprietary skin cleanser.

7. Encourage the use of a skin reconditioning cream at the end of the shift, after washing hands. These products help to replace the natural fats and oils removed from the skin by exposure to oils and by washing, and are a very important part of a skin conservation programme.

8. Encourage the use of barrier cream before starting work, and also after each time hands are washed. The barrier creams of choice vary with different oils, so ensure that the correct one is used.
9. Avoid unnecessary exposure of workers to oil mist or vapours. In any event, ensure that breathing zone levels of oil mist are well below the recommended permissible concentration of 5 milligrams/cubic metre air.
10. See that all cuts and scratches receive prompt medical attention.
11. Prevent contamination of all oils particularly the soluble oils, and minimize the use of biocides. Ensure that soluble oils are used only at the recommended dilution ratios.
12. Programme the regular cleaning of coolants systems.
13. Obey any special instructions on product labels. In common with most other industries, the petroleum industry is increasing its use of precautionary labelling.
14. Use correct work technique—particularly for soluble oil concentrates which may irritate skin and eyes. Handling concentrate and preparing dilutions requires careful precautionary measures: the use of goggles or a face-visor, impervious gloves, etc. Use warning notices, placards, etc., to draw attention to the need for good personal hygiene and good work practices.

In many cases, the machining operations on such metals as magnesium and titanium may generate explosive concentrations of dust. Frequently, these operations must be segregated, and the operations must be conducted with suitable ventilation control and air cleaning. High-density layout of machine tools frequently results in workplace exposures above 85 dBA, the level that triggers the OSHA Hearing Conservation Program.

The exhaust hoods commonly used to capture toxic metal dusts from machining include conventional exterior hoods, high-velocity, low-volume capture hoods, and enclosures. A number of special hood enclosures have been utilized in beryllium fabrication shops.

The discovery that cutting fluids containing both nitrates and amines additives that could result in the generations of various carcinogenic nitrosamines prompted a NIOSH field study of exposure to cutting fluid mists (36). Average concentrations of oil mist for a variety of machining operations ranged from 0.2 to 1.9 mg/m^3. The authors stated that concentrations of oil mist can be maintained below the present TLV of 5 mg/m^3 by the use of oil containing anti-mist additives, hood enclosures with suitable exhaust ventilation, and air cleaning. Because carcinogenic polyaromatic hydrocarbons were tentatively identified in certain bulk cutting oils, the authors recommended that appropriate chemical studies be conducted to evaluate this problem. Since this study, the industry has redesigned the coolant chemicals to minimize the formation of nitrosamines.

10.2 Electrochemical Machining

The electrochemical machining (ECM) process utilizes a dc electrolytic bath operating at low voltage and high current density. The workpiece is the anode; and the cutting tool, or cathode, is machined to reflect the geometry of the hole to be cut in the workpiece. As electrolyte is pumped through the space between the tool and the workpiece, metal ions are removed from the workpiece and are swept away by the electrolyte. The tool is fed into the workpiece to complete the cut. The electrolyte varies with the operation; one manufacturer states the electrolyte is 10 percent sulfuric acid.

Because the ECM method is fast, produces an excellent surface finish, does not produce burrs, and produces little tool wear, it is widely used for cutting irregular shaped holes in hard, tough metals.

In the operation, the electrolyte is dissociated and hydrogen is released at the cathode. A dense mist or smoke is released from the electrolyte bath. Local exhaust ventilation must be provided to remove this mist and ensure hydrogen concentrations do not approach the lower flammability limit.

10.3 Electrical Discharge Machining

A spark-gap technique is the basis for the electrical discharge machining (EDM) procedure, which is a popular machining technique for large precise work such as die sinking or the drilling of small holes in complex parts. In one system, a graphite tool is machined to the precise size and shape of the hole to be cut. The workpiece (anode) and the tool (cathode) are immersed in a dielectric oil bath and powered by a low-voltage dc power supply. The voltage across the gap increases until breakdown occurs and there is a spark discharge across the gap, which produces a high temperature at the discharge point. This spark erodes a small quantity of the metal from the workpiece. The cycle is repeated at a frequency of 200 to 500 Hz with rather slow, accurate cutting of the workpiece. In more advanced systems, the cutting element is a wire that is programmed to track the cutting profile. The wire systems can also be used to drill small holes.

The hazards from this process are minimal and are principally associated with the oil. In light cutting jobs, a petroleum distillate such as Stoddard Solvent is commonly used, whereas in large work a mineral oil is the dielectric. When heavy oil mists are encountered, local exhaust ventilation is needed. The oil gradually becomes contaminated with small hollow spheres of metal eroded from the part. As in the case of conventional machining, these metals may dissolve in the oil and present a dermatitis problem. An ultra-high efficiency filter should be placed in the oil recirculating line to remove metal particles.

11 NONDESTRUCTIVE TESTING

With the increase in manufacturing technology in the last decade of the twentieth century, a need has developed for in-plant inspection techniques. The most common procedures now in use in the metalworking industries are discussed in this section.

11.1 Industrial Radiography

Radiography is used principally in industry for the examination of metal fabrications such as weldments, castings, and forgings in a variety of settings. Specially designed shielded cabinets may be located in manufacturing areas for in-process examination of parts. Large components may be transported to shielded rooms for examination. Radiography may be performed in open shop areas, on construction sites, on board ships, and along pipelines.

The process of radiography consists of exposing the object to be examined to X-rays or gamma rays from one side and measuring the amount of radiation that emerges from the opposite side. This measurement is usually made with film or a fluoroscopic screen to provide a visual, two-dimensional display of the radiation distribution and any subsurface porosities.

The principal potential hazard in industrial radiography is exposure to ionizing radiation. This section deals with the minimum safety precautions designed to minimize worker exposure to radiation (X-rays and gamma rays) sources.

11.1.1 X-Ray Sources

X-Rays used in industrial radiography are produced electrically and therefore fall into the category of "electronic product radiation." For this reason the design and manufacture of industrial X-ray generators are regulated by the Food and Drug Administration, Center for Devices and Radiological Health. ANSI has developed a standard for the design and manufacture of these devices (40). These standards specify maximum allowable radiation intensities outside the useful beam. They require warning lights on both the control panel and the tube head to indicate when X-rays are being generated.

The use of industrial X-ray generators is regulated by OSHA. Additionally, many states regulate the use of industrial X-ray generators within their own jurisdictions. Areas in which radiography is performed must be posted with signs that bear the radiation caution symbol and a warning statement. Access to these areas must be secured against unauthorized entry. When radiography is being performed in open manufacturing areas, it is essential to instruct other workers in the identification and meaning of these warning signs in order to minimize these unnecessary exposures.

Radiographic operators are required to wear personal monitoring devices to measure the magnitude of their exposure to radiation. Typical devices are film badges, thermoluminescent dosimeters, and direct-reading pocket dosimeters. It is also advisable for radiographic operators to use audible alarm dosimeters or "chirpers." These devices emit an audible signal or "chirp" when exposed to radiation. The frequency of the signal is proportional to the radiation intensity. They are useful in warning operators who unknowingly enter a field of radiation (41).

The operators must be trained in the use of radiation survey instruments to monitor radiation levels to which they are exposed and to assure that the X-ray source is turned off at the conclusion of the operation. A wide variety of radiation survey instruments is available for use in industrial radiography. When using industrial X-ray generators, especially relatively low energy generators, it is imperative that the instrument has an appropriate energy range for measuring the energy of radiation used.

Radiation exposure is not the only potential hazard associated with industrial X-ray generators. These devices use high-voltage power supplies for the production of X-rays, and consideration must be given to the electrical hazards associated with using this equipment.

11.1.2 Gamma Ray Sources

Gamma rays used in industrial radiography are produced as a result of the decay of radioactive nuclei. The principal radioisotopes used in industrial radiography are iridium-192 and cobalt-60. Other radioisotopes, such as ytterbium-169 and thulium-170, are much less common but have some limited applicability. Radioisotope sources produce gamma rays with discrete energies as opposed to the continuous spectrum of energies produced by X-ray generators. Owing to radioactive decay, the activity of radioisotope sources decreases exponentially with time.

Unlike X-ray generators, radioisotope sources require no external source of energy, which makes their use attractive in performing radiography in remote locations such as pipelines. However, because they are not energized by an external power supply, these sources cannot be "turned off," and they continuously emit gamma rays. For this reason, certain additional safety precautions must be exercised.

In industrial radiographic sources, the radioisotope is sealed inside a source capsule that is usually fabricated from stainless steel. The radioisotope source capsule is stored inside a shielded container or "pig" when not in use, to reduce the radiation intensities in the surrounding areas. In practice, the film is positioned and the radioactive source is then moved from the pig to the desired exposure position through a flexible tube by means of a mechanical actuator. At the end of the exposure, the operator retracts the source to the storage position in the pig.

The design, manufacture, and use of radioisotope sources and exposure devices for industrial radiography are regulated by the U.S. Nuclear Regulatory Commission. However, the Nuclear Regulatory Commission has entered into an agreement with "agreement states" for the latter to regulate radioisotope radiography within their jurisdictions. Organizations that wish to perform radioisotope radiography must obtain a license from either the Nuclear Regulatory Commission or the agreement state. In order to obtain this license, they must describe their safety procedures and equipment to the licensing authority.

Radiographic operators must receive training as required by the regulatory bodies including instruction in their own organization's safety procedures, a formal radiation safety training course, and a period of on-the-job training where the trainees work under the direct personal supervision of a qualified radiographer. At the conclusion of this training, operators must demonstrate their knowledge and competence to the licensee's management.

Areas in which radiography is being performed must be posted with radiation warning signs, and access to these areas must be secured as described earlier. An operator performing radioisotope radiography must wear both a direct-reading pocket dosimeter and either a film badge or a thermoluminescent dosimeter. Additionally, the operator must use a calibrated radiation survey instrument during all radiographic operations. In order to reduce the radiation intensity in the area after an exposure, the operator must retract the source to a shielded position within the exposure device. The only method available to the operator for assuring that the source is shielded properly is a radiation survey of the area. The operator should survey the entire perimeter of the exposure device and the entire length of guide tube and source stop after each radiographic operation to assure that the source has been fully and properly shielded.

Occasional radiation exposure incidents have occurred in industrial radioisotope radiography. These incidents generally result because the operator fails to return the source properly to a shielded position in the exposure device and then approaches the exposure device or source stop without making a proper radiation survey. The importance of making a proper radiation survey at the conclusion of each radiographic exposure cannot be overemphasized.

The frequency of radiation exposure incidents may be reduced through the use of audible alarm dosimeters. As described earlier, these devices emit an audible signal when exposed to radiation. In cases where operators have not properly shielded sources and fail to make a proper survey, the audible alarm dosimeter might alert the operator to abnormal radiation levels. Some current types of radiographic exposure devices incorporate source position indicators that provide a visual signal if the source is not stored properly. Use of these devices may also reduce the frequency of radiation exposure incidents.

11.2 Magnetic Particle Inspection

This procedure is suitable for detecting surface discontinuities, especially cracks, in magnetic materials. The procedure is simple and of relatively low cost; as a result, it is widely applied in metal fabrication plants.

The parts to be inspected first undergo vigorous cleaning and then are magnetized. Magnetic particles are applied as a powder or a suspension of particles in a carrier liquid. The powders are available in color for daylight viewing or as fluorescent particles for more rigorous inspection with a UV lamp. Surface imperfections, such as cracks, result in leakage of the magnetic field with resulting adherence of the particles. The defined geometry of the imperfection stands out in a properly illuminated environment or with a hand-held UV lamp.

Worker exposures using this technique are minor, consisting of skin contact and minimal air contamination from the suspension field, usually an aliphatic hydrocarbon; exposure to the magnetic field during the inspection/process; and exposure to the lamp output. General exhaust ventilation usually is adequate for vapors from the liquid carrier, and gloves, apron, and eye protection should be used to minimize skin and eye contact. A proper filter on the UV lamp minimizes exposure to UV B radiation. The hazard from magnetic fields is now under investigation.

11.3 Liquid Penetrant

This procedure complements magnetic particle inspection for surface cracks and weldment failures because it can be used on nonmagnetic materials. A colored or fluorescent liquid used as the penetrant is flowed or sprayed on the surface, or the part is dipped. The excess penetrant is removed, in some systems by adding an emulsifier and removing with water, or by wiping off the excess. A developer is applied to intensify the definition of the crack, and the inspection is carried out by daylight or UV lamp depending on the penetrant system of choice. The penetrant is then removed from the workpiece by either water or solvent such as mineral spirits.

The inspectors' exposures are to the solvent or carrier used with the penetrant, emulsifier, developer, and the final cleaning agent. The solvent is water, mineral spirits, or in some cases, halogenated compounds.

11.4 Ultrasound

Pulse echo and transmission type ultrasonic inspection have a wide range of application for both surface and subsurface flaws. This procedure resolves voids much smaller than all other methods including radiographic procedures. The procedure commonly involves immersing the part in water to improve coupling between the ultrasound transmitter/receiver and the workpiece. The reported response on other applications of higher energy ultrasound systems include local heating and subjective effects. Health effects have not been reported on its application in nondestructive testing of metals.

12 THERMAL SPRAYING

A practical technique for spraying molten metal was invented in the 1920s and has found application for applying metals, ceramics, and plastic powder to workpieces for corrosion protection, to build up worn or corroded parts, to improve wear resistance, to reduce production costs, and as a decorative surface. The health hazards from thermal spraying of metals was first observed in 1922 in the United Kingdom when operators suffered from lead poisoning (42).

The application head for thermal spraying takes many forms and dictates the range of hazards one may encounter from the operation. The four principal techniques presently in use in industry are described below.

12.1 Spraying Methods

12.1.1 Flame

The coating material in wire form is fed to a gun operated with air/oxygen and a combustible gas such as acetylene, propane, or natural gas. The wire is melted in the oxygen–fuel flame, disintegrated with a coaxial stream of compressed air, and propelled from the torch at velocities up to 240 m/sec (48,000 fpm). The material bonds to the workpiece by a combination of mechanical interlocking of the molten, platelet-form particles and a cementation of partially oxidized material.

The material can also be sprayed in powder form. The fuel gases in this system are either acetylene or hydrogen used in combination with oxygen. The powder is aspirated by the oxygen/fuel-gas stream, and the molten particles are deposited on the workpiece at velocities below 30 m/sec (6,000 fpm). This technique may also be used to apply ceramics, carbides, metals, and alloys containing a flux that is fused at high temperature in a separate operation.

12.1.2 Detonation

In this unit, the gases are fed to a combustion chamber where they are ignited by a spark plug at the rate of 260 firings per minute. Metal powder is fed to the chamber, and the explosions drive the melted powder to the workpiece at velocities of approximately 760 m/sec (150,000 fpm). Tungsten carbide, chromium carbide, and aluminum oxide are applied with this technique.

12.1.3 Plasma

An electric arc is established in the controlled atmosphere of a special nozzle. Argon is passed through the arc, where it ionizes to form a plasma that continues through the nozzle and recombines to create temperatures as high as 16,700°C (30,000°F). Metal alloy, ceramic, and carbide powders are melted in the stream and are released from the gun at a velocity of 300 to 600 m/sec (60,000 to 120,000 fpm).

12.1.4 Wire

Two consumable metal wire electrodes in this process are made of the metal to be sprayed. As the wires feed into the gun, they establish an arc as in a conventional arc welding unit; the molten metal is disintegrated by compressed air, and the molten particles are projected to the workpiece at high velocity.

12.2 Hazards and Controls

A major hazard common to all techniques is the potential exposure to toxic metal fumes. If the vapor pressure of the metal at the application temperature is high, this will be reflected in high air concentrations of metal fume. If the metal is overheated because of malfunction or poor adjustment of the gun, this problem is accentuated. The deposition efficiency, that is, the percent of the metal sprayed deposited on the workpiece, varies between application techniques and the coating metal. For flame spraying with wire, the deposition efficiency varies from 90 percent for aluminum to 55 percent for lead; the plasma arc technique ranges from 50 percent for chromium carbide to as high as 90 percent for alumina–titania.

Deposition efficiency obviously has a major impact on air concentrations in the workplace. Because the airborne level varies with the type of metal sprayed, the application technique, the fuel in use, and the tool–workpiece geometry, it is impossible to estimate the level of contamination that may occur from a given operation. For this reason, if a toxic metal is sprayed, air samples should be taken to define the worker exposure.

The principal control of air contaminants generated during flame and arc spraying is local exhaust ventilation (43). Nontoxic metals may be exhausted with a freely suspended open hood with a face velocity of 1.02 m/sec (200 fpm) or an enclosing hood with a minimum face velocity of 0.64 m/sec (125 fpm). Toxic metal fumes should be exhausted with an enclosed hood with a face velocity of 1.02 m/sec (200 fpm). The operator should wear a positive pressure air-supplied respirator while metallizing with toxic metals. One must also consider the exposure of other workers in the area, because systems with 50 percent overspray obviously represent a significant source of general contamination.

All procedures result in significant noise exposures and require effective hearing conservation programs. Flame spraying techniques represent the lowest exposures whereas exposures greater than 100 dBA result from detonation and plasma thermal spraying. Appreciable noise reduction can be achieved by changing the operating characteristics of the spray equipment. On wire and powder guns, the noise level can be reduced 3 to 5 dBA by lowering the gas flow rate. Reduced amperage on plasma flow spray guns and electric arc guns is effective in reducing noise output from this equipment. However, there is a penalty for this noise reduction in lower spray rates and a poor quality surface coating. If one cannot accept these penalties, the manufacturers suggest the normal approach to noise control including isolation, the use of hearing protectors, and various work practices.

The wire arc and plasma procedures present additional hazards reflecting their welding ancestry including UV B exposure and the generation of ozone and nitrogen dioxide. If ventilation is effective in controlling metal fumes, it will probably control ozone and nitrogen dioxide, although this fact should be established by air sampling.

In addition to air-supplied respirators, operators handling toxic metals such as cadmium and lead should wear protective gloves and coveralls and be required to strip, bathe, and change to clean clothes prior to leaving the plant.

Two special problems are encountered in routine metallizing operations. Unless ventilation control is effective, particles will deposit on various plant structures and may become a potential fire hazard. If one uses a scrubber to collect the particles, hydrogen may form in the sludge, resulting in a potential fire and explosion hazard.

13 PAINTING

Paint products are used widely in industry to provide a surface coating for protection against corrosion, for appearance, as electrical insulation, for fire retardation, and for other special purposes. The widespread application of this technology from small job shops to highly automated painting of automobiles includes more than half a million workers. Although the health status of painters has not been well defined, the studies that have been completed suggests acute and chronic central nervous system effects, hematologic disorders, and excess mortality from cancer. In addition, respiratory sensitization may occur from two-part urethane systems, and the amine catalyst in spray paints may cause skin sensitization. The hazards associated with the industrial application of paint products are discussed here.

13.1 Types of Paints

The term paint is commonly used to identify a range of organic coatings including paints, varnishes, enamels, and lacquers. Conventional paint is an inorganic pigment dispersed in a vehicle consisting of a binder and a solvent with selected fillers and additives. The conventional varnish is a nonpigmented product based on oil and natural resin in a solvent that dries first by the evaporation of the solvent and then by the oxidation of the resin binder. Varnishes are also available based on synthetic resins. A pigmented varnish is called an enamel. Lacquers are coatings in which the solvent evaporates leaving a film that can be redissolved in the original solvent.

In the past decade, paint formulations have been influenced by environmental regulations limiting the release of volatile organic solvents. Although conventional solvent-based paints will continue to see wide application, the major thrust is to convert to formulations with low solvent content. Such systems include solvent-based paints with high solids content (>70 percent by weight), nonaqueous dispersions, powder, two-part catalyzed systems, and water-borne paints.

13.2 Composition of Paint

The conventional solvent-based paints consist of the vehicle, filler, and additives. The vehicle represents the total liquid content of the paint and includes the binder and solvent. The binder, which is the film-forming ingredient, may be a naturally occurring oil or resin including linseed oil and oleoresinous materials or synthetic material such as alkyd resins.

The fillers include pigments and extenders, which historically have presented a major hazard in painting. The common white pigments include bentonite and kaolin clay, talc, titanium dioxide, and zinc oxide. Mineral dust used as extenders to control viscosity, texture, and gloss include talc, clay, calcium carbonate, barite, and both crystalline and amorphous silica. The pigments and extenders do not represent an exposure during brush or roller application, but when sanded or removed preparatory to repainting, these materials are released to the air and may present significant exposures. A group of pigments including lead carbonate, cadmium red, and chrome green and yellow do represent a critical exposure, both during spray application and surface preparation.

Additives to hasten drying, reduce skin formation in the can, and control fungus are added to paints. The fungicides warrant special attention because the ingredients, including copper and zinc naphthenate, copper oxide, and tributyltin oxide, are biologically active.

The solvent systems are varied and complex. The most common organic solvents include aliphatic and aromatic hydrocarbons, ketones, alcohols, glycols, and glycol ether/esters. These solvents have high vapor pressure and represent the critical worker exposure component in most painting techniques. High-solids paint systems (low solvent) represent a major contribution to reduced solvent exposure.

The ultimate high-solids, low-solvent formulation is a dry powder paint. This coating technique provides a high-quality job while eliminating solvent exposure for the worker. The application technique, to be described later, utilizes a dry powder formulated to contain a resin, pigment, and additives. Thermosetting resins, used for decorative and protective coatings, are based on epoxy, polyester, and acrylic resins. Thermoplastic resins are applied in thick coatings for critical applications.

Water-borne paints now represent 15 to 20 percent of construction and industry paints, and expanded use is anticipated. The present systems use binders based on copolymers of several monomers including acrylic acid, butyl acrylate, ethyl hexyl acrylate, and styrene. The other major ingredients and their maximum concentrations include pigments (54 percent), coalescing solvents (15 percent), surfactants (5 percent), biocides (1.1 percent), plasticizers (2.2 percent), and driers in lesser amounts (44). The pigments include the conventional materials mentioned under solvent-based paints. Coalescing solvents, including hydrocarbons, alcohols, esters, glycol, and glycol ether/esters, although present in low concentrations, may present an inhalation hazard. The proprietary biocides in use frequently result in exposure to significant airborne formaldehyde concentrations. The surfactants in use include known skin irritants and sensitizers (44).

A single study of water-borne paints and worker exposure confirms that these systems represent an improvement over solvent-based systems. The authors of this study recommend that paint formulations be designed to eliminate formaldehyde, minimize the unreacted monomer content in the various polymers, eliminate ethylene glycol ethers because of their reproductive hazard, and minimize the ammonia content (44).

Special attention must be given two-component epoxy and urethane paint systems. The urethanes consist of a polyurethane prepolymer containing a reactive isocyanate; the second component is the polyester. Mixing of the two materials initiates a reaction with the formation of a chemically resistant coating. In the early formulations, the unreacted toluene diisocyanate (TDI) resulted in significant airborne exposures to TDI and resulting respiratory sensitization. Formulations have been changed to include TDI adducts and derivatives that are stated to have minimized the respiratory hazard from these systems. For ease of application, single package systems have been developed using a blocked isocyanate. Unblocking takes place when the finish is baked and the isocyanate is released to react with the polyester.

Epoxy paint systems offer excellent adhesive properties, resistance to abrasion and chemicals, and stability at high temperatures. The conventional epoxy system is a two-component system consisting of a resin based on the reaction products of bisphenol A and epichlorohydrin. The resin may be modified by reactive diluents such as glycidyl ethers. The second component, the hardener or curing agent, was initially based on low molecular weight, highly reactive amines.

Epoxy resins can be divided into three grades. The solid grades are felt to be innocuous; however, skin irritation may occur from solvents used to take up the resin. The liquid grades are mild to moderate skin irritants. The low-viscosity glycidyl ether modifiers are skin irritants and sensitizers and have systemic toxicity. The use of low molecular weight aliphatic amines such as diethylenetetramine and triethylenetetramine, both strong skin irritants and sensitizers, presented major health hazards during the early use of epoxy systems. This has been overcome to a degree by the use of low-volatility amine adducts and high molecular weight amine curing agents.

13.3 Operations and Exposures

In the industrial setting, paints can be applied to parts by a myriad of processes including brush, roller, dip, flow, curtain, tumbling, conventional air spray, airless spray, heated systems, disk spraying, and powder coating. Conventional air spraying is the most common method encountered in industry and presents the principal hazards owing to overspray. The use of airless and hot spray techniques minimizes mist and solvent exposure to the operator, as does electrostatic spraying. The electrostatic technique, now commonly used with many installations, places a charge on the paint mist particle so it is attracted to the part to be painted, thereby reducing rebound and overspray.

In powder coating, a powder, which represents the paint formulation including the resin, pigment, and additives, is conveyed from a powder reservoir to the spray gun. In the gun, a charge is imparted to the individual paint particles. This dry powder is sprayed on the electrically grounded workpiece, and the parts are baked to fuse the powder film into a continuous coating.

The operator is exposed to the solvent or thinner in processes in which the paint is flowed on, as in brushing and dipping, and during drying of the parts. However, during atomization techniques, the exposure is to both the solvent and the paint mist. The level of exposure reflects the overspray and rebound that occur during spraying.

A common exposure to organic vapors occurs when spray operators place the freshly sprayed parts on a rack directly behind them. The air movement to the spray booth sweeps over the drying parts and past the breathing zone of the operator, resulting in an exposure to solvent vapors. Drying stations and baking ovens must therefore be exhausted. The choice of exhaust control on tumbling and roll applications depends on the surface area of the parts and the nature of the solvent.

13.4 Controls

Most industrial flow and spray painting operations utilizing solvent-based paints require exhaust ventilation for control of solvent vapors at the point of application

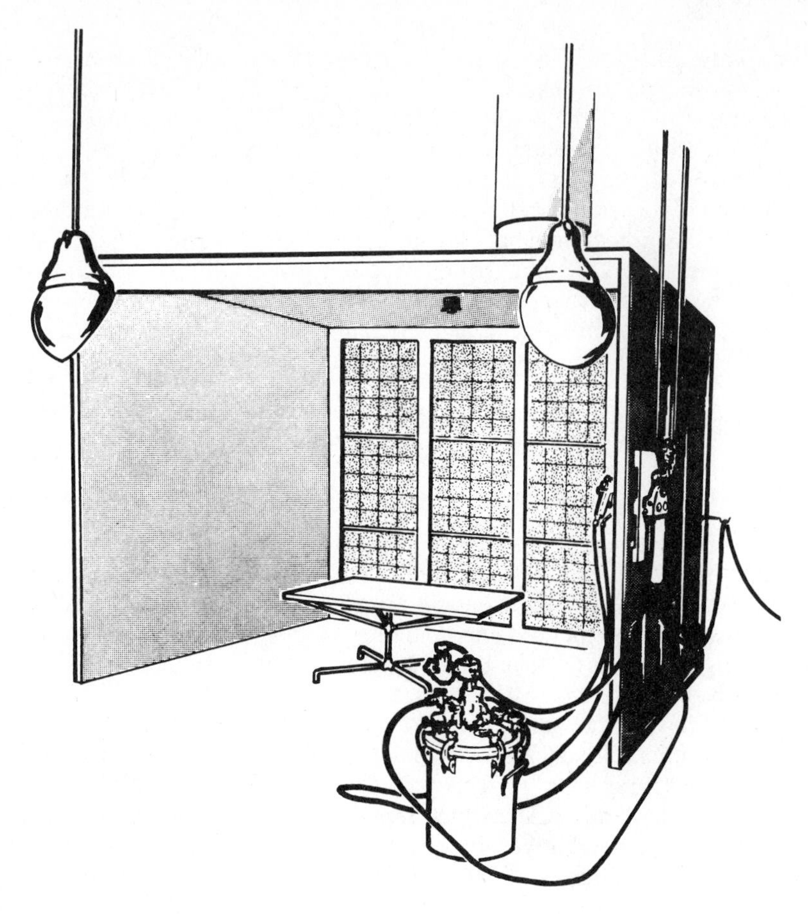

Figure 18.9. Paint spray booth.

and during drying and baking operations (45, 46). Water-based paints may require ventilation only when spray application is utilized. Flow application of solvent-based paints require local exhaust ventilation depending on the application technique. The usual ventilation control for spray application of solvent-based paints is a spray booth, room, or tunnel provided with some type of paint spray mist arrestor before the effluent is exhausted outdoors. The degree of control of paint mist and solvent varies with the application method, that is, whether it is air atomization, airless, or electrostatic painting. The latter two techniques call for somewhat lower exhaust volumes. The advent of robotic spray painting permits the design of exhausted enclosures and minimizes worker exposure.

The design of the conventional paint spray booth is shown in Figure 18.9. The booths are commonly equipped with a water curtain or a throwaway dry filter to provide paint mist removal. The efficiency of these devices against paint mist has not been evaluated. Neither of these systems, of course, removes solvent vapors from the air stream.

In industrial spray painting of parts, the simple instructions contained in several state codes on spray painting should be observed:

1. Do not spray toward a person.
2. Automate spray booth operations where possible to reduce exposure.
3. Maintain 2-ft clearance between the sides of the booth and large flat surfaces to be sprayed.
4. Keep the distance between the nozzle and the part to be sprayed to less than 12 in.
5. Do not position work so that the operator is between the exhaust and the spray gun or disk.
6. Locate the drying room so that air does not pass over drying objects to exhaust hood past the breathing zone of the operator.

Controls in the application of two-component urethane and epoxy paint systems must include excellent housekeeping, effective ventilation control, and protective clothing; moreover, in applications not effectively controlled by ventilation, the operators should wear air-supplied respirators. Adequate washing facilities should be available, and eating, drinking, and smoking should be prohibited in the work area.

Dermatitis due to primary irritation and defatting from solvents or thinners as well as sensitization from epoxy systems is not uncommon. Skin contact must be minimized, rigorous personal cleanliness encouraged, and suitable protective equipment used by the operator.

14 SOLDERING AND BRAZING

Soft soldering is the joining of metal by surface adhesion without melting the base metal. This technique uses a filler metal (solder) with a melting point less than 316°C (600°F); hard solder is used in the range of 316 to 427°C (600 to 800°F). These temperature ranges differentiate soldering from brazing, which utilizes a filler metal with a melting point greater than 427°C (800°F).

14.1 Soldering

To understand the potential health hazards from soldering operations, one must be familiar with the composition of the solder and fluxes in use and the applicable production techniques.

14.1.1 Flux

All metals including the noble metals have a film of tarnish that must be removed in order to wet the metal with solder effectively and accomplish a good mechanical bond. The tarnish takes the form of oxides, sulfides, carbonates, and other corrosion

Table 18.7. Common Flux Materials

Type	Typical fluxes	Vehicle
	Inorganic	
Acids	Hydrochloric, hydrofluoric, orthophosphoric	Water, petrolatum paste
Salts	Zinc chloride, ammonium chloride, tin chloride	Water, petrolatum paste, polyethylene glycol
Gases	Hydrogen-forming gas; dry HCI	None
	Organic, Nonrosin Base	
Acids	Lactic, oleic, stearic, glutamic, phthalic	Water, organic solvents, petrolatum paste, polyethylene glycol
Halogens	Aniline hydrochloride, glutamic acid hydrochloride, bromide derivatives of palmitic acid, hydrazine hydrochloride or hydrobromide	Water, organic solvents, polyethylene glycol
Amines and amides	Urea, ethylenediamine, mono- and triethanolamine	Water, organic solvents, petrolatum paste, polyethylene glycol
	Organic, Rosin Base	
Superactivated	Rosin or resin with strong activators	Alcohols, organic solvents, glycols
Activated (RA)	Rosin or resin with activator	Alcohols, organic solvents, glycols
Mildly activated (RMA)	Rosin with activator	Alcohols, organic solvents, glycols
Nonactivated (water-white rosin) (R)	Rosin only	Alcohols, organic solvents, glycols

Source: Reference 47.

products (47). The flux, which may be a solid, liquid, or gas, is designed to remove any adsorbed gases and tarnish from the surface of the base metal and keep it clean until the solder is applied. The molten solder displaces the residual flux and wets the base metal to accomplish the bond.

A range of organic and inorganic materials is used in the design of soldering fluxes as shown in Table 18.7. The flux is usually a corrosive cleaner frequently used with a volatile solvent or vehicle. Rosin, which is a common base for organic fluxes, contains abietic acid as the active material. Manko (47) has described the performance of this flux on a copper base metal which has an oxidized surface. When the flux is applied, the surface oxide reacts with the abietic acid to form copper-abiet. The copper-abiet mixes with the unreacted rosin on the now untarnished surface of the base metal. When the molten solder is applied, it displaces the copper-abiet and the unreacted rosin to reveal the clean metal surface, thereby ensuring a good bond. Activated rosin flux is formulated by adding an organic chloride compound to the base rosin–alcohol liquid flux to provide active corrosive cleaning.

The three major flux bases are inorganic, organic nonrosin, and organic rosin, as shown in Table 18.7. A proprietary flux may contain several of these materials. The inorganic fluxes may be based on any of the acids commonly used for etch baths in electroplating operations. The inorganic salts, such as zinc chloride, react with water to form hydrochloric acid, which converts the base metal oxide to a chloride. The metal chloride is water soluble and can be easily removed by a water rinse to provide an untarnished surface for soldering. A combination of salts is frequently used to achieve a flux with a melting point below that of the solder so that no flux residual is left on the part. The last inorganic technique utilizes hot hydrogen chloride gas, which sweeps over the base metal reducing the oxide.

The second class of materials, organic nonrosin, is less corrosive and thereby slower acting. In general, these fluxes do not present as severe a handling hazard as the inorganic acids. This is not true of the organic halides whose degradation products are very corrosive and warrant careful handling and ventilation control. The amines and amides in this second class of flux materials are common ingredients whose degradation products are very corrosive.

The rosin base fluxes are inactive at room temperature but at soldering temperature are activated to remove tarnish. This material continues to be a popular flux compound because the residual material is chemically and electrically benign and need not be removed as vigorously as the residual products of the corrosive fluxes.

The cleaning of the soldered parts may range from a simple hot water or detergent rinse to a degreasing technique using a range of organic solvents. These cleaning techniques may require local exhaust ventilation.

14.1.2 Solder

The most common solder contains 65 percent tin and 35 percent lead. Traces of other metals including cadmium, bismuth, copper, indium, iron, aluminum, nickel, zinc, and arsenic are present. A number of special solders contain antimony in concentrations up to 5 percent. The melting point of these solders is quite low; and at these temperatures, the vapor pressure of lead and antimony usually do not result in significant air concentrations of metal fume. When solder contains silver the technique is still properly called soldering, not, as frequently stated, brazing. The composition of the common solders encountered in industry can be obtained from the manufacturers.

14.1.3 Application Techniques

Soldering is a fastening technique used in a wide range of products from simple mechanical assemblies to complex electronic systems. The soldering process includes cleaning the base metal and other components for soldering, fluxing, the actual soldering, and post-soldering cleaning.

14.1.4 Initial Cleaning of Base Metals

Prior to fluxing and soldering, the base metal must be cleaned to remove oil, grease, wax, and other surface debris. Unless this is done, the flux will not be able to attack and remove the metal surface tarnish. The procedures include cold solvent degreasing, vapor degreasing, and ultrasonic degreasing. If the base metal has been heat treated, the resulting surface scale must be removed.

In many cases, mild abrasive blasting techniques are utilized to remove heavy tarnish before fluxing. In electrical soldering, the insulation on the wire must be stripped back to permit soldering. Stripping is accomplished by mechanical techniques such as cutters or wire brushes, chemical strippers, and thermal techniques. Mechanical stripping of asbestos-based insulation obviously presents a potential health hazard, and this operation must be controlled by local exhaust ventilation. The hazard from chemical stripping depends on the chemical used to strip the insulation; however, at a minimum, the stripper will be a very corrosive agent. Thermal stripping of wire at high production rates may present a problem owing to the thermal degradation products of the insulation. Hot wire stripping of fluorocarbon insulation such as Teflon may cause polymer fume fever if operations are not controlled by ventilation. Other plastic insulation such as polyvinyl chloride may produce irritating and toxic thermal degradation products.

14.1.5 Fluxing Operations

Proprietary fluxes are available in solid, paste, and liquid form for various applications and may be cut with volatile vehicles such as alcohols to vary their viscosity. The flux may be applied by one of 10 techniques shown in Table 18.8 depending on the workpiece and the production rate. Because flux is corrosive, skin contact must be minimized by specific work practices and good housekeeping. The two techniques that utilize spray application of the flux require local exhaust ventilation to prevent air contamination from the flux mist.

14.1.6 Soldering and Cleaning

In two of the fluxing techniques listed in Table 18.8, the flux and solder are applied together. One need merely apply the necessary heat to bring the system first to the melting point of the flux and then to the melting point of the solder. In most applications, however, the flux and solder are applied separately, although the two operations may be closely integrated and frequently are in a continuous operation.

The soldering techniques used for manual soldering operations on parts that have been fluxed are the soldering iron and solder pot. A number of variations of

Table 18.8. Soldering Flux Application Techniques

Method	Application Technique	Use
Brushing	Applied manually by paint, acid, or rotary brushes	Copper pipe, job shop printed circuit board, large structural parts
Rolling	Paint roller application	Precision soldering, printed circuit board, suitable for automation
Spraying	Spray painting equipment	Automatic soldering operations, not effective for selective application
Rotary screen	Liquid flux picked up by screen and air directs it to part	Printed circuit board application
Foaming	Work passes over air agitated foam at surface of flux tank	Selective fluxing, automatic printed circuit board lines
Dipping	Simple dip tank	Wide application for manual and automatic operations on all parts
Wave fluxing	Liquid flux pumped through trough forming wave through which work is dipped	High-speed automated operation
Floating	Solid flux on surface melts providing liquid layer	Tinning of wire and strips of material
Cored solder	Flux inside solder wire melts, flows to surface, and fluxes before solder melts	Wide range of manual operations
Solder paste	Solder blended with flux, applied manually	Component and hybrid microelectric soldering

Source: Reference 47.

the solder pot have been introduced to handle high production soldering of printed circuit boards (PCB) in the electronic industry. In the drag solder technique, the PCB is positioned horizontally and pulled along the surface of a shallow molten solder bath behind a skimmer plate that removes the dross. This process is usually integrated with a cleaning and fluxing station in a single automated unit. Another system designed for automation is wave soldering. In this technique, a standing wave is formed by pumping the solder through a spout. Again, the conveyorized PCBs are pulled through the flowing solder.

The potential health hazards to the soldering operators are minimal. A thermostatically controlled lead–tin solder pot operates at temperatures too low to generate significant concentrations of lead fume. The use of activated rosin fluxing agents may result in the release of thermal degradation products that may require control by local exhaust ventilation. The handling of solder dross during cleanup and maintenance may result in exposure to lead dust.

Furnace soldering is widely used for the assembly of semiconductors. The parts are positioned in a holding jig, and flux and solder pre-forms are positioned at the soldering locations. Parts are processed by batch or continuous furnace operations. Because small quantities of flux and solder are used at closely regulated furnace temperatures, there is no significant health hazard from this operation. In another operation, the parts are prefluxed and solder pre-forms are processed through a vapor phase unit operating with a chlorofluorocarbon (CFC) which brings the assembly up to a precise temperature in a gradual fashion without undue thermal stress to the part. As in the case of the vapor degreaser, the condenser and freeboard ensure minimal release of the CFC to the workplace, but this should be evaluated.

After soldering operations are carried out, some flux residue and its degradation products remain on the base metal. Both water-soluble and solvent-soluble materials normally exist so that it is necessary to clean with both systems using detergents and saponifiers in one case and common chlorinated and fluorinated hydrocarbons in the other. The CFCs are being replaced with cleaners such as limonene which are less threatening to the environment. The processing equipment may include ultrasonic cleaners and vapor degreasers.

14.1.7 Controls

The fluxes may represent the most significant hazard from soldering. The conventional pure rosin fluxes are not difficult to handle, but highly activated fluxes warrant special handling instruction. A range of alcohols, including methanol, ethanol, and isopropyl alcohol are used as volatile vehicles for fluxes. The special fluxes should be evaluated under conditions of use to determine worker exposure. Rosin is a common cause of allergic contact dermatitis, and the fume causes an allergic asthma. Initially, it was believed the abietic acid was the allergen; however, it is the oxidation products of the abietic acid and dehydroabietic acid that are the offending substances (48). It should be noted that lead solders have been banned for use in potable water supplies.

14.2 Brazing (49)

This technique of joining metals is discussed in this section because it is commonly identified as a soldering operation in industry. It is more properly identified as a welding process and is so classified by the American Welding Society. Brazing techniques are widely used in the manufacture of refrigerators, electronics, jewelry, and aerospace components to join both similar and dissimilar metals. Although the final joint looks similar to a soft solder bond, it is much stronger and the joint requires little finish. As mentioned earlier in this section, brazing is defined as a technique for joining metals that are heated above 430°C (800°F), whereas soldering is conducted below 430°C. The temperature of the operation is of major importance because it determines the vapor pressure of the metals that are heated and therefore the concentration of metal fumes to which the operator is exposed.

14.2.1 Flux and Filler Metals

Flux is frequently used; however, certain metals may be joined without flux. The flux is chosen to prevent oxidation of the base metal and not to prepare the surface as is the case with soldering. The common fluxes are based on fluorine, chlorine, and phosphorus compounds and present the same health hazards as fluxes used in soldering; that is, they are corrosive to the skin and may cause respiratory irritation. A range of filler metals used in brazing rods or wire may include phosphorus, silver, zinc, copper, cadmium, nickel, chromium, beryllium, magnesium, and lithium. The selection of the proper filler metal is the key to quality brazing.

14.2.2 Application Techniques

Brazing of small job lots that do not require close temperature control are routinely done with a torch. More critical, high production operations are accomplished by dip techniques in a molten bath, by brazing furnaces using either an ammonia or hydrogen atmosphere, or by induction heating.

The brazing temperatures define the relative hazard from the various operations. As an example, the melting point of cadmium is approximately 1400°C. The vapor pressure of cadmium and the resulting airborne fume concentrations increase dramatically with temperature. The filler metals with the higher brazing temperatures will therefore present the most severe exposure to cadmium. The exposure to fresh cadmium fume during brazing of low-alloy steels, stainless steels, and nickel alloys has resulted in documented cases of occupational disease and represents the major hazard from these operations. This is especially true of torch brazing, where temperature extremes may occur. On the other hand, the temperatures of furnace and induction heating operations may be controlled to ± 5°C.

One study of brazing in a pipe shop and on board ship included air sampling for cadmium while brazing with a filler rod containing from 10 to 24 percent cadmium. The mean concentrations during shipboard operation was 0.45 mg/m^3 with a maximum of 1.40 mg/m^3.

14.2.3 Controls

Controls on brazing operations must obviously be based on the identification of the composition of the filler rod. Local ventilation control is necessary in operations where toxic metal fumes may be generated from the brazing components or from parts plated with cadmium or other toxic metals. In the use of common fluxes, one should minimize skin contact owing to their corrosiveness and provide exhaust ventilation to control airborne thermal degradation products released during the brazing operation.

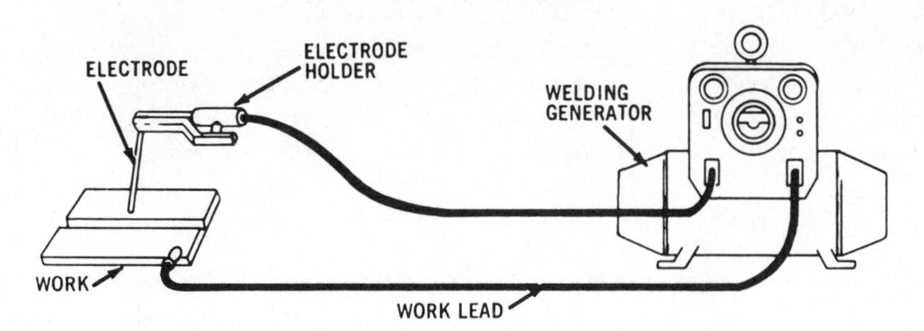

Figure 18.10. Shielded metal arc welding (SMAW).

15 WELDING

Welding is a process for joining metals in which coalescence is produced by heating the metals to a suitable temperature. Over a dozen welding procedures are commonly encountered in industry. Four welding techniques have been chosen for discussion because they represent 80 to 90 percent of all manufacturing and maintenance welding. These procedures are used by more than half a million welders and helpers in the United States.

In the nonpressure welding techniques to be discussed in this section, metal is vaporized and then condenses to form initially a fume in the 0.01 to 0.1 μm particle size range that rapidly agglomerates. The source of this respirable metal fume is the base metal, the metal coating on the workpiece, the electrode, and the fluxing agents associated with the particular welding system. A range of gases and vapors including carbon monoxide, ozone, and nitrogen dioxide may be generated depending on the welding process. The wavelength and intensity of the electromagnetic radiation emitted from the arc depends on the welding procedure, inerting gas, and the base metal.

The common welding techniques are reviewed assuming that the welding is done on unalloyed steel, or so-called mild steel. The nature of the base metal is important in evaluating the metal fume exposure, and occasionally it has impact in other areas. When such impact is important, it is discussed.

15.1 Shielded Metal Arc Welding

Shielded metal arc (SMA) welding (Figure 18.10) is commonly called stick or electrode welding. An electric arc is drawn between a welding rod and the workpiece, melting the metal along a seam or a surface. The molten metal from the workpiece and the electrode form a common puddle and cool to form the bead and its slag cover. Either dc or ac power is used in straight (electrode negative, work positive) or reverse polarity. The most common technique involves dc voltages of 10 to 50 and a wide range of current up to 2000 A. Although operating voltages are low, under certain conditions an electrical hazard may exist.

The welding rod or electrode may have significant occupational health implications. Initially a bare electrode was used to establish the arc and act as filler metal. Now the electrode covering may contain 20 to 30 organic and inorganic compounds and perform several functions. The principal function of the electrode coating is to release a shielding gas such as carbon dioxide to ensure that air does not enter the arc puddle and thereby cause failure of the weld. In addition, the covering stabilizes the arc, provides a flux and slag producer to remove oxygen from the weldment, adds alloying metal, and controls the viscosity of the metal. A complete occupational health survey of shielded metal arc welding requires the identification of the rod and its covering. The composition of the electrode can be obtained from the American Welding Society (AWS) classification number stamped on the electrode.

The electrode mass that appears as airborne fume may include iron oxides, manganese oxide, fluorides, silicon dioxide, and compounds of titanium, nickel, chromium, molybdenum, vanadium, tungsten, copper, cobalt, lead, and zinc. Silicon dioxide is routinely reported as present in welding fume in an amorphous form or as silicates, not in the highly toxic crystalline form.

15.1.1 Metal Fume Exposure

The potential health hazards from exposure to metal fume during shielded metal arc welding obviously depend on the metal being welded and the composition of the welding electrode. The principal component of the fume generated from mild steel is iron oxide. The hazard from exposure to iron oxide fume appears to be limited. The deposition of iron oxide particles in the lung does cause a benign pneumoconiosis known as siderosis. There is no functional impairment of the lung nor is there fibrous tissue proliferation. In a comprehensive review of conflicting data, Stokinger has shown that iron oxide is not carcinogenic to man (50).

The concentration of metal fume to which the welder is exposed depends not only on the alloy composition but on the welding conditions including the current density (amperes per unit area of electrode), wire feed rate, the arc time, which may vary from 10 to 30 percent, the power configuration, that is, dc or ac supply, and straight or reverse polarity. The work environment also defines the level of exposure to welding fume and includes the type and quality of exhaust ventilation and whether the welding is done in an open, enclosed, or confined space.

Many of the data on metal fume exposures have been generated from shipyard studies. The air concentration of welding fume in several studies ranged from less than 5 to over 100 mg/m^3 depending on the welding process, ventilation, and the degree of enclosure. In these early studies, the air samples were normally taken outside of the welding helmet. Recent studies have shown the concentration at the breathing zone inside the helmet ranges from one-half to one-tenth the outside concentrations. If SMA welding is conducted on stainless steel the chromium concentrations may exceed the TLV; in the case of alloys with greater than 50 percent nickel, the concentration of nickel fume may exceed 1 mg/m^3 (51).

15.1.2 Gases and Vapors

Shielded metal arc welding has the potential to produce nitrogen oxides; however, this is not normally a problem in open shop welding. In over 100 samples of SMA

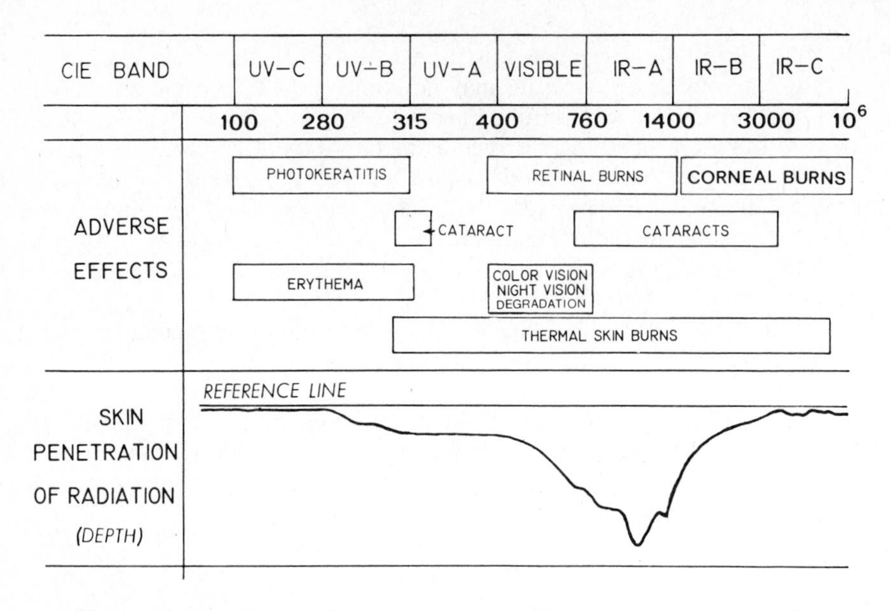

Figure 18.11. Effects of electromagnetic radiation on the eye and skin.

welding in a shipyard, I have not identified an exposure to nitrogen dioxide in excess of 0.5 ppm under a wide range of operating conditions. Ozone is also fixed by the arc, but again this is not a significant contaminant in SMA welding operations. Carbon monoxide and carbon dioxide are produced from the electrode cover, but air concentrations are usually minimal.

Low-hydrogen electrodes are used with conventional arc welding systems to maintain a hydrogen-free arc environment for critical welding tasks on certain steels. The electrode coating is a calcium carbonate–calcium fluoride system with various deoxidizers and alloying elements such as carbon, manganese, silicon, chromium, nickel, molybdenum, and vanadium. A large part of this coating and of all electrode coatings becomes airborne during welding (52). In addition to hydrogen fluoride, sodium, potassium, and calcium fluorides are present as particles in the fume cone.

Exposure to fumes from low hydrogen welding under conditions of poor ventilation may prompt complaints of nose and throat irritation and chronic nosebleeds. There has been no evidence of systemic fluorosis from this exposure. Monitoring can be accomplished by both air sampling and urinary fluoride measurements.

15.1.3 Radiation

The radiation generated by SMA welding covers the spectrum from the infrared C range of wavelengths to the UV C range. The acute condition known to the welder as "arc eye," "sand in the eye," or "flash burn" is due to exposures in the UV B range. (See Figure 18.11.) The radiation in this range is completely absorbed in the corneal epithelium of the eye and causes a severe photokeratitis. Severe

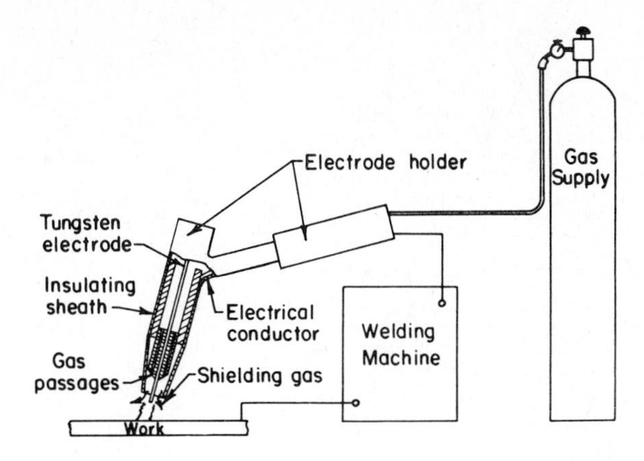

Figure 18.12. Gas tungsten arc welding (GTAW).

pain occurs 5 to 6 hr after exposure to the arc, and the condition clears within 24 hr. Welders experience this condition only once and then protect themselves against a recurrence by the use of a welding helmet with a proper filter, in addition to tinted safety goggles. Skin erythema or reddening may also be induced by exposure to UV C and UV B as shown in Figure 18.11 (53).

15.2 Gas Tungsten Arc Welding

Although shielded metal arc welding using coated electrodes is an effective way to weld many ferrous metals, it is not practical for welding aluminum, magnesium, and other reactive metals. The introduction of inert gas in the 1930s to blanket the arc environment and prevent the intrusion of oxygen and hydrogen into the weld provided a solution to this problem. In gas tungsten arc welding (GTA) (Figure 18.12), also known as tungsten inert gas and Heliarc welding, the arc is established between a nonconsumable tungsten electrode and the workpiece producing the heat to melt the abutting edges of the metal to be joined. Argon or helium is fed to the annular space around the electrode to maintain the inert environment. A manually fed filler rod is commonly used. The GTA technique is routinely used on low-hazard materials such as aluminum and magnesium in addition to a number of alloys including stainless steel, nickel alloys, copper–nickel, brasses, silver, bronze, and a variety of low-alloy steels that may have industrial hygiene significance.

The welding fume concentrations in GTA welding are lower than in manual stick welding and gas metal arc welding. High-energy GTA arc produces nitrogen dioxide concentrations at the welder's position; the maximum concentration noted by me is 3.0 ppm. Argon results in higher concentrations of nitrogen dioxide than helium (54).

The inert gas technique introduced a new dimension in the welder's exposure to electromagnetic radiation from the arc with energies an order of magnitude

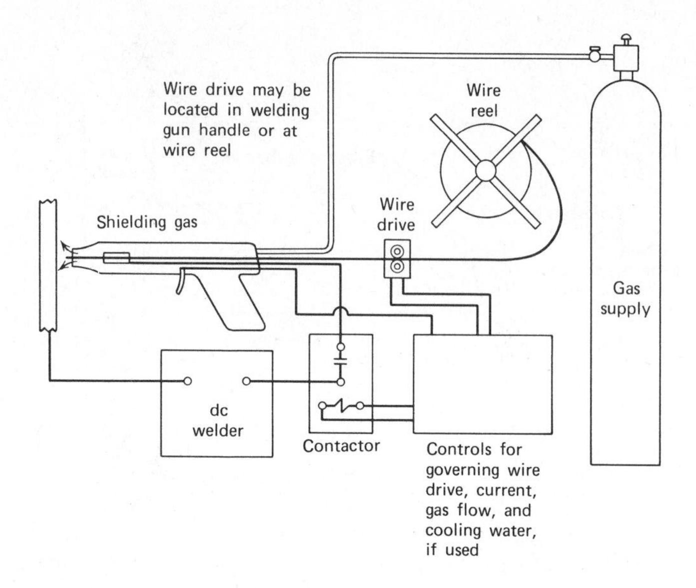

Figure 18.13. Gas metal arc welding (GMAW).

greater than SMA welding. The energy in the UV B range, especially in the region of 290 nm, is the most biologically effective radiation and will produce skin erythema and photokeratitis. The energy concentrated in the wavelengths below 200 nm (UV A) is most important in fixing oxygen as ozone. The GTA procedure produces a rich, broad spectral distribution with important energies in these wavelengths. The ozone concentration is higher when welding on aluminum than on steel, and argon produces higher concentrations of ozone than helium owing to its stronger spectral emission. The spectral energy depends on current density. Ozone concentrations may exceed the TLV under conditions of poor ventilation.

15.3 Gas Metal Arc Welding

In the 1940s, a consumable wire electrode was developed to replace the nonconsumable tungsten electrode used in the GTA system. Originally developed to weld thick, thermally conductive plate, the gas metal arc welding (GMA) process (also known as manual inert gas welding) now has widespread application for aluminum, copper, magnesium, nickel alloys, and titanium, as well as steel alloys.

In this system (Figure 18.13), the welding torch has a center consumable wire that maintains the arc as it melts into the weld puddle. Around this electrode is an annular passage for the flow of helium, argon, carbon dioxide, nitrogen, or a blend of these gases. The wire usually has a composition the same as or similar to the base metal with a flash coating of copper to ensure electrical contact in the gun and to prevent rusting.

An improvement in GMA welding is the use of a flux-cored consumable electrode. The electrode is a hollow wire with the core filled with various deoxifiers, fluxing agents, and metal powders. The arc may be shielded with carbon dioxide or the inert gas may be generated by the flux core.

Metal fume concentrations from GMA welding frequently exceed the TLV on mild steel. If stainless steel is welded the chromium concentrations may exceed the TLV and a significant percent may be in the form of hexavalent chromium. As in the case of SMA, welding on nickel alloys produces significant concentrations of nickel fume. Nitrogen dioxide concentrations are on the same order of magnitude as SMA welding; however, ozone concentrations are much higher with the GMA technique. The ozone generation rate increases with an increase in current density but plateaus rapidly. The arc length and the inert gas flow rate do not have a significant impact on the ozone generation rate.

Carbon dioxide is widely used in GMA welding because of its attractive price; argon and helium cost approximately 15 times as much as carbon dioxide. The carbon dioxide process is similar to other inert gas arc welding shielding techniques, and one encounters the usual problem of metal fume, ozone, oxides of nitrogen, decomposition of chlorinated hydrocarbons solvents, and UV radiation. In addition, the carbon dioxide gas is reduced to form carbon monoxide. The generation rate of carbon monoxide depends on current density, gas flow rate, and the base metal being welded. Although the concentrations of carbon monoxide may exceed 100 ppm in the fume cone, the concentration drops off rapidly with distance, and with reasonable ventilation, hazardous concentrations should not exist at the breathing zone.

The intensity of the radiation emitted from the arc is, as in the case of GTA welding, an order of magnitude greater than that noted with shielded metal arc welding. The impact of such a rich radiation source in the UV B and UV C wavelengths has been covered in the GTA discussion. With both GTA and GMA procedures trichloroethylene and other chlorinated hydrocarbon vapors are decomposed by UV from the arc-forming chlorine, hydrogen chloride, phosgene, and other compounds. The solvent is degraded in the UV field and not directly in the arc. Studies by Dahlberg showed that hazardous concentrations of phosgene could occur with trichloroethylene and 1,1,1-trichloroethane even though the solvent vapor concentrations were below the appropriate TLV for the solvent (55). The GMA technique produces higher concentrations of phosgene than GTA welding under comparable operating conditions. Dahlberg also identified dichloroacetyl chloride as a principal product of decomposition that could act as a warning agent because of its lacrimatory action. In a follow-up study, perchloroethylene was shown to be less stable than other chlorinated hydrocarbons. When this solvent was degraded, phosgene was formed rapidly, and dichloroacetyl chloride was also formed in the UV field.

In summation, perchloroethylene, trichloroethylene, and 1,1,1-trichloroethane vapors in the UV field of high-energy arcs such as those produced by GTA and GMA welding may generate hazardous concentrations of toxic air contaminants. Other chlorinated solvents may present a problem depending on the operating

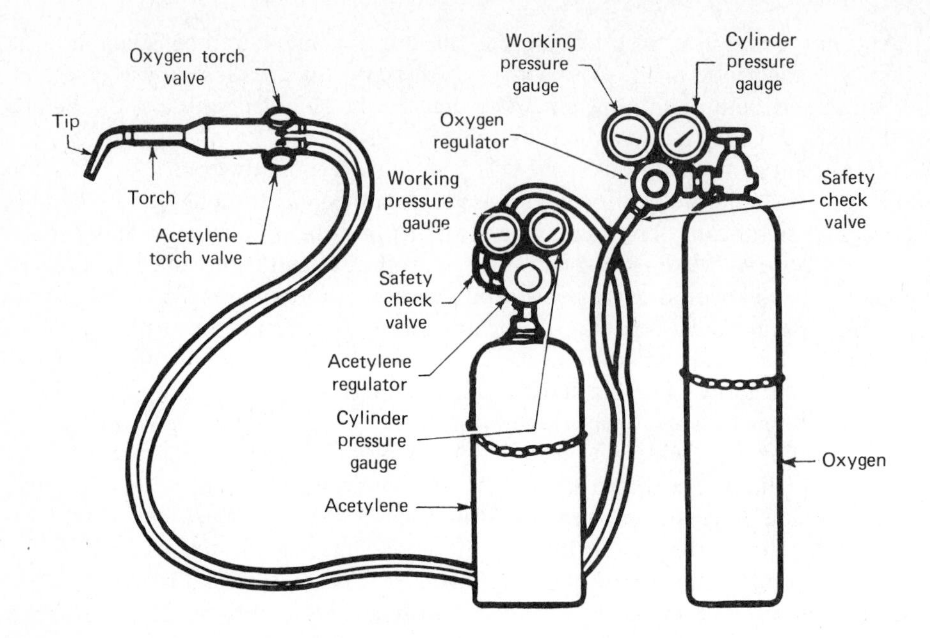

Figure 18.14. Gas welding.

conditions; therefore diagnostic air sampling should be performed. The principal effort should be the control of the vapors to ensure that they do not appear in the UV field. Delivery of parts directly from degreasing to the welding area has, in my experience, presented major problems owing to pullout and trapping of solvent in the geometry of the part.

15.4 Gas Welding

In the gas welding process (Figure 18.14), the heat of fusion is obtained from the combustion of oxygen and one of several gases including acetylene, methylacetylene propadiene (MAPP), propane, butane, and hydrogen. The flame melts the workpiece, and a filter rod is manually fed into the joint. Gas welding is used widely for light sheet metal and repair work. The hazards from gas welding are minimal compared to those from arc welding techniques.

The uncoated filler rod is usually of the same composition as the metal being welded except on iron where a bronze rod is used. Paste flux is applied by first dipping the rod into the flux. Fluxes are used on cast iron, some steel alloys, and nonferrous work to remove oxides or assist in fusion. Borax-based fluxes are used widely on nonferrous work whereas chlorine, fluorine, and bromine compounds of lithium, potassium, sodium, and magnesium are used on gas welding of aluminum and magnesium. The metal fume originates from the base metal, filler metal, and the flux. The fume concentration encountered in field welding operations depends principally on the degree of enclosure in the work area and the quality of ventilation.

Table 18.9. Contaminants From Welding Operations

Contaminant	Source
Metal fumes	
Iron	Parent iron or steel metal, electrode
Chromium	Stainless steel, electrode, plating, chrome-primed metal
Nickel	Stainless steel, nickel-clad steel
Zinc	Galvanized or zinc-primed steel
Copper	Coating on filler wire, sheaths on air-carbon arc gouging electrodes, nonferrous alloys
Vanadium, manganese, and molybdenum	Welding rod, alloys in steel
Tin	Tin-coated steel
Cadmium	Plating
Lead	Lead paint, electrode coating
Fluorides	Flux on electrodes
Gases and vapors	
Carbon monoxide	CO_2 shielded, GMA, carbon arc gouging, oxy-gas
Ozone	GTA, GMA, carbon arc gouging
Nitrogen dioxide	GMA, all flame processes

Source: Reference 1.

The principal hazard in gas welding in confined spaces is due to the formation of nitrogen dioxide. Higher concentrations occurred when the torch was burning without active welding. Striekevskia found concentrations of nitrogen dioxide of 150 ppm in a space without ventilation and of 26 ppm in a space with ventilation (56). This investigator cautions that phosphine may be present as a contaminant in acetylene and that carbon monoxide may be generated during heating of cold metal with gas burner.

The radiation from gas welding is quite different from arc welding. The principal emissions are in the visible and IR A, B, C wavelengths and require the use of light tinted goggles for work. Ultraviolet radiation from gas welding is negligible.

15.5 Control of Exposure

In addition to the fume exposures that occur while welding steel alloys, nonferrous metal, and copper alloys, metal coatings and the welding electrodes also contribute to the metal fume exposure. Table 18.9 shows the wide range of metal fumes encountered in welding. Recent interest in the possible carcinogenicity of chromium and nickel fume has prompted special attention to these exposures during welding. Stern has suggested that welders who work on stainless and alloy steels containing chromium and nickel may be at elevated risk from respiratory cancer (57). Lead has been used as an alloy in steels to improve its machinability, and welding on such material requires rigorous control. This is also the case with manganese used in steel alloys to improve metallurgical properties. Beryllium, probably the most toxic alloying metal, is alloyed with copper and warrants close control during welding and brazing operations.

Welding or cutting on workpieces that have metallic coatings may be especially hazardous. Lead-based paints have been used commonly to paint marine and structural members. Welding on these surfaces during repair and shipbreaking generates high concentrations of lead fume. In my experience, cutting and welding on structural steel covered with lead paint results in concentrations exceeding 1.0 mg/m^3 in well-ventilated conditions out of doors.

Steel is galvanized by dipping in molten zinc. Air concentrations of zinc during the welding of galvanized steel and steel painted with zinc silicate range from 3 to 12 times the zinc TLV under conditions of poor ventilation. Concentrations are lower with oxygen–acetylene torch work than with arc cutting methods. When local exhaust ventilation is established, the TLV for zinc is seldom exceeded.

The hazard from burning and welding of pipe coated with a zinc-rich silicate can be minimized in two ways. Because pipes are joined end to end, the principal recommendation is to mask the pipe ends with tape before painting (58). If it is necessary to weld pipe after it is painted, the first step before welding is to remove the paint by hand filing, power brush or grinding, scratching, or abrasive blasting. If the material cannot be removed, suitable respiratory protection is required; in many cases, air-supplied respirators may be required.

A variety of techniques to reduce ozone from GMA welding, including the use of magnesium wire, local exhaust ventilation, and addition of nitric oxide, have been attempted without success. A stainless steel mesh shroud positioned on the welding head controlled concentrations to below 0.1 ppm (59, 60).

Hazardous concentrations of nitrogen dioxide may be generated in enclosed spaces in short periods of time and therefore require effective exhaust ventilation.

15.5.1 Radiant Energy

Eye protection from exposure to UV B and UV C wavelengths is obtained with filter glasses in the welding helmet of the correct shade, as recommended by the AWS. The shade choice may be as low as 8 to 10 for light manual electrode welding, or as high as 14 for plasma welding. Eye protection must also be afforded other workers in the area. To minimize the hazard to nonwelders in the area, flash screens or barriers should be installed. Semitransparent welding curtains provide effective protection against the hazards of UV and infrared from welding operations. Sliney et al. have provided a format for choice of the most effective curtain for protection while permitting adequate visibility (61). Plano goggles or lightly tinted safety glasses may be adequate if one is some distance from the operation. If one is 30 to 40 ft away, eye protection is probably not needed for conventional welding; however, high current density GMA welding may require that the individual be 100 ft away before direct viewing is possible without eye injury. On gas welding and cutting operations, infrared wavelengths must be attenuated by proper eye protection for both worker health and comfort.

The attenuation afforded by tinted lens is available in National Bureau of Standards reports, and the quality of welding lens has been evaluated by NIOSH. Glass safety goggles provide some protection from UV; however, plastic safety glasses may not.

The UV radiation from inert gas-shielded operations causes skin erythema or reddening; therefore, welders must adequately protect their faces, necks, and arms. Heavy chrome leather vest armlets and gloves must be used with such high-energy arc operations.

15.5.2 Decomposition of Chlorinated Hydrocarbon Solvents

The decomposition of chlorinated hydrocarbon solvents occurs in the UV field around the arc and not in the arc itself. The most effective control is to prevent the solvent vapors from entering the welding area in detectable concentrations. Merely maintaining the concentration of solvent below the TLV is not satisfactory in itself. If vapors cannot be excluded from the workplace, the UV field should be reduced to a minimum by shielding the arc. Pyrex glass is an effective shield that permits the welder to view his or her work. Rigorous shielding of the arc is frequently possible at fixed station work locations, but it may not be feasible in field welding operations.

15.5.3 System Design

The advent of robotics has permitted remote operation of welding equipment, thereby minimizing direct operator exposure to metal fumes, toxic gases, and radiation.

REFERENCES

1. W. Burgess, *Recognition of Health Hazards in Industry: A Review of Materials and Processes*, John Wiley, New York, 1981.
2. W. A. Burgess, M. J. Ellenbecker, and R. D. Treitman, *Ventilation for Control of The Work Environment*, John Wiley, New York, 1989.
3. Committee on Industrial Ventilation, American Conference of Governmental Industrial Hygienists, *Industrial Ventilation: A Manual of Recommended Practice*, 20th ed., ACGIH, Lansing, MI, 1988.
4. S. Spring, *Industrial Cleaning*, Prism Press, Melbourne, 1974.
5. American National Standards Institute, "Practices for Ventilation and Operation of Open-Surface Tanks," ANSI Z9.1-1977, New York, 1977.
6. "Freon Solvent Data," Bulletin No. FST-1, DuPont, Wilmington, DE, 1987.
7. "How To Select A Vapor Degreasing Solvent," Bulletin Form 100-5321-78, Dow Chemical Co., Midland, MI (1978).
8. "Modern Vapor Degreasing and Dow Chlorinated Solvents," Dow Bulletin, Form No. 100-5185-77, Dow Chemical Co., Midland, MI (1977).
9. *Manual on Vapor Degreasing*, 3rd ed., American Society for Testing and Materials, Philadelphia, 1989.

10. D. R. Blair, *Principles of Metal Surface Treatment and Protection*, Pergamon Press, Oxford, 1972.
11. "Safety Manual for Electroplating and Finishing Shops," American Electroplaters and Surface Finishers Society, Orlando, FL, 1989.
12. "Control Technology Assessment: Metal Plating and Cleaning Operations," DHHS (NIOSH) Publication No. 85-102, Cincinnati, OH, 1985.
13. *Forging Handbook*, Forging Industry Association, Cleveland, OH, 1985.
14. A. H. Goldsmith, K. W. Vorpahl, K. A. French, P. T. Jordan, and N. B. Jurinski, *Am. Ind. Hyg. Assoc. J.*, **37**, 217–226, 1976.
15. B. S. Gutow, *Environ. Sci. Technol.*, **6**, 790 (1972).
16. "Recommendations for Control of Occupational Safety and Health Hazards—Foundries," DHHS (NIOSH) Publication No. 85-116, Cincinnati, OH, 1985.
17. J. Oudiz, J. A. Brown, H. A. Ayer, and S. Samuels, *Am. Ind. Hyg. Assoc. J.*, **44**, 374 (1983).
18. D. R. Dreger, *Machine Design*, 47 (Nov. 25, 1982).
19. International Molders and Allied Workers Union (AFL-CIO-CLC), *No-Bakes and Others: Common Chemical Binders and Their Hazards*, undated.
20. U. Knecht, H. J. Elliehausen, and H. J. Woitowitz, *Br. J. Ind. Med.*, **43**, 834 (1986).
21. R. H. Toeniskoetter and R. J. Schafer, "Industrial Hygiene Aspects of the Use of Sand Binders and Additives," in *Proceedings of the Working Environment in Iron Foundries*, March 22–24, 1977, British Cast Iron Research Association, Birmingham, England, 1977.
22. American National Standards Institute, ANSI Z241.1-1989—"Safety requirements for sand preparation, molding and coremaking in the sand foundry industry," ANSI Z241.2-1989—"Safety requirements for melting and pouring of metals in the foundry industry," ANSI Z241.3-1989—"Safety requirements for cleaning and finishing of castings," New York, 1989.
23. *Industrial Noise Control—An Engineering Guide*, American Foundrymen's Society, Inc., Des Plaines, IL, 1985.
24. D. K. Verma, D. C. F. Muir, S. Cunliffe, J. A. Julian, J. H. Vogt, and J. Rosenfeld, *Ann. Occup. Hyg.*, **25**, 17 (1982).
25. P. L. Pelmear and R. Kitchener, "The Effects and Measurement of Vibration," in *Proceedings of the Working Environment in Iron Foundries*, March 22–24, 1977, British Cast Iron Research Association, Birmingham, England, 1977.
26. American National Standards Institute, "Safety Requirements for the Use, Care, and Protection of Abrasive Wheels, B7.1-1978, ANSI, New York, 1978.
27. E. K. Bastress et al., "Ventilation Requirements for Grinding, Polishing, and Buffing Operations," Report No. 0213, IKOR Inc., Burlington, MA, June, 1973.
28. "Dust Control: The Low Volume–High Velocity System," Technical Data Note 1 (2nd rev.), Department of Employment, HM Factory Inspectorate, London, England.
29. American National Standards Institute, "Ventilation Control of Grinding, Polishing, Buffing Operations," Z 43.1-1966, ANSI, New York, 1966.
30. *Portable Grinding Machines: Control of Dust*, Health and Safety Series Booklet HS(g) 18, Health and Safety Executive, H. M. Stationary Office, London, England, 1982.
31. R. J. Heinsohn, D. Johnson, and J. W. Davis, *Am. Ind. Hyg. Assoc. J.*, **43**, 587 (1982).

32. J. Starek, M. Farkkila, S. Aatola, I. Pyykko, and O. Korhonen, *Br. J. Ind. Med.*, **40**, 426 (1983).
33. *Heat Treaters Guide*, American Society for Metals, New York, 1982.
34. J. Danielson, Ed., *Air Pollution Engineering Manual*, 2nd ed., Publication No. AP-40, U.S. Government Printing Office, Washington, DC, 1973.
35. H. B. Elkins, *The Chemistry of Industrial Toxicology*, John Wiley, New York, 1959.
36. D. O'Brien and J. C. Frede, "Guidelines for the Control of Exposure to Metalworking Fluids," Department of Health, Education and Welfare, NIOSH Publication No. 78-165, Cincinnati, OH, 1978.
37. E. Bingham, R. P. Trosset, and D. Warshawsky, *J. Environ. Pathol. Toxicol.*, **3**, 483 (1979).
38. P. Decoulfe, *J. Nat. Cancer Inst.*, **61**, 1025 (1978).
39. "This is About Health, Esso Lubricating Oils and Cutting Fluids," Esso Petroleum Company, Ltd., 1979.
40. American National Standards Institute, "Radiological Safety Standard for the Design of Radiographic and Fluoroscopic Industrial X-ray Equipment," ANSI/NBS 123-1976, ANSI, New York, 1976.
41. *Isotope Radiography Safety Handbook*, Technical Operations, Inc., Burlington, MA, 1981.
42. W. E. Ballard, *Ann. Occup. Hyg.*, **15**, 101 (1972).
43. J. H. Hogapian, and E. K. Bastress, "Recommended Industrial Ventilation Guidelines," Department of Health, Education and Welfare, NIOSH Publication No. 76-162, Cincinnati, OH, 1976.
44. M. K. Hansen, M. Larsen, and K.-H. Cohr, *Scand. J. Work Environ. Health*, **13**, 473 (1987).
45. D. M. O'Brien and D. E. Hurley, "An Evaluation of Engineering Control Technology for Spray Painting," DHHS (NIOSH) Publication No. 81-121, USDHHS, CDS, Cincinnati, OH (1981).
46. "Recommendations for Control of Occupational Safety and Health Hazards—Manufacture of Paint and Allied Coating Products," DHHS (NIOSH) Publication No. 84-115, CDC, NIOSH, 1984.
47. H. H. Manko, *Solders and Soldering*, McGraw-Hill, New York, 1979.
48. P. S. Burge, M. G. Harries, I. M. O'Brien et al., *Clin. Allergy*, **8**, 1 (1978).
49. "The Brazing Book," Handy & Harman, New York, 1985.
50. H. Stokinger, *Am. Ind. Hyg. Assoc. J.*, **45**, 127 (1984).
51. J. F. VanDerWal, *Ann. Occup. Hyg.*, **34**, 45 (1990).
52. M. Pantucek, *Am. Ind. Hyg. Assoc. J.*, **32**, 687 (1971).
53. D. Sliney and M. Wolbarsht, *Safety with Lasers and Other Optical Sources*, Plenum, New York, 1980.
54. J. Ferry and G. Ginter, *Welding J.*, **32**, 396 (1953).
55. J. Dahlberg, *Ann. Occup. Hyg.*, **14**, 259 (1971).
56. I. Striekevskia, *Welding Prod.*, **7**, 40 (1961).
57. R. M. Stern, A. Berlin, A. C. Fletcher, and J. Jarvisalo, *Health Hazards and Biological Effects of Welding Fumes and Gases*, Excerpta Medica ICS 676, Elsevier, Amsterdam, 1986.

58. F. Venable, *Esso Med. Bull.*, **39**, 129 (1979).
59. A. G. Faggetter, V. E. Freeman, and H. R. Hosein, *Am. Ind. Hyg. Assoc. Y.*, **44**, 316 (1983).
60. M. J. Tinkler, Measurement and Control of Ozone Evolution During Aluminum GMA Welding, *Colloquium on Welding and Health*, July, 1980, Brazil.
61. D. H. Sliney, C. E. Moss, C. G. Miller, and J. B. Stephens, *Appl. Optics*, **20**, 2352 (1981).

CHAPTER NINETEEN

Respiratory Protective Devices

Darrel D. Douglas, C.I.H.

1 INTRODUCTION

The *American Heritage Dictionary* defines respirator first as "an apparatus used in administering artificial respiration," and secondly as "a screenlike device worn over the mouth or nose, or both, to protect the respiratory tract." In the field of occupational health and safety, the second definition applies. A respirator protects the wearer from toxic agents in the air.

The use of respirators seems uncomplicated. Place a mask on the user, seal it tight to the face or head and either supply breathing air to the mask or let the user inhale through a filter that removes all harmful materials. Its apparent simplicity and low cost make it the first type of protection chosen. In practice it is neither cheap nor simple. Some of the problems that may arise are:

A. The mask to face seal is not satisfactory.
B. The user does not understand how or why the respirator is to be used.
C. An ineffective filter or sorbent is supplied.
D. The respirator is not maintained and can no longer function.
E. Filter or sorbent cartridge costs may quickly exceed the original cost of an air-purifying respirator.
F. The original reason for choosing the respirator has changed.

Respirators are mentioned in history as early as Roman times, and scattered mention of them occurs in reports of industrial processes during the Middle Ages.

Patty's Industrial Hygiene and Toxicology, Fourth Edition, Volume 1, Part A, Edited by George D. Clayton and Florence E. Clayton
ISBN 0-471-50197-2

Until the nineteenth century, all devices appear to be air-purifying devices intended to prevent the inhalation of aerosols. They were varied in design, ranging from animal bladders or rags wrapped around the nose and mouth to full face masks made of glass with air inlets covered by particulate filters. During the 1800s masks were produced that combined aerosol filters and vapor sorbing materials. These advances were made primarily for fire fighters.

Chemical warfare agents, introduced in World War I, focused attention on the need for adequate respirators. In the United States this work was successfully carried out for the army by the Bureau of Mines. After the war, misuse of surplus army gas masks by civilians highlighted the need for respiratory protection standards. The Bureau of Mines developed these standards, and in 1920 approved their first respirator, a self-contained breathing apparatus (SCBA) (1). Out of this effort grew the federal respirator approval system that is still in effect. Many of the respirator developments made during the decade following 1920 remain in use today (2).

The respirators available today offer good protection to the user, provided the standards for respirator use are followed. These standards influence both respirator use and design. Three major standards or standard-producing agencies are:

A. The American National Standards Institute (ANSI), "Practices for Respiratory Protection"
B. National Institute for Occupational Safety and Health (NIOSH) respirator approval regulations (30CFR11) and respirator use recommendations
C. Occupational Safety and Health Administration (OSHA) general (29CFR1910.134) and specific respirator use regulations

OSHA respirator regulations, 1910.134, which define how respirators will be used in the workplace, are adapted from the ANSI Z88.2-1969, "Practices for Respiratory Protection." OSHA also provides more detailed respirator information as regulations covering specific materials are published.

All OSHA regulations require that only NIOSH-approved respirators be used in the workplace. The NIOSH respirator approval regulations, which are the present-day Bureau of Mines approvals, assure the user that the approved respirator has met the minimum standards for its class. Understanding the limitations of the NIOSH approval is important prior to using the respirator.

2 RESPIRATOR CLASSIFICATION

There are two types of respirators, air purifying and air supplying. Each type may have tight-fitting facepieces and loose-fitting hoods that cover the head and may cover the body. An important aspect of respirator operation and classification is the air pressure within the facepiece. If facepiece pressure is lower than the outside air pressure, it is classified as negative; if above, it is positive. The concept of

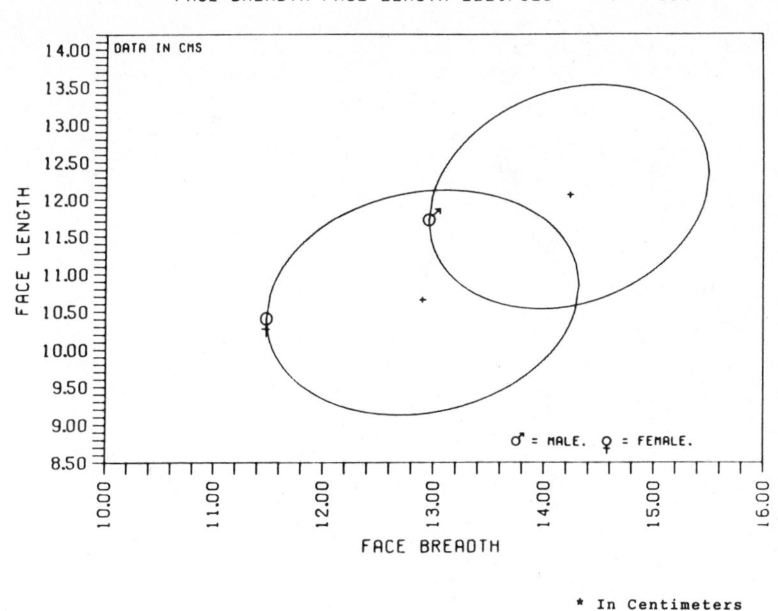

Figure 19.1. Facial size grouping, male and female (3).

negative and positive pressure operation is important when considering potential contaminant leakage into the respirator.

3 RESPIRATOR FACEPIECES

3.1 Anthropometric Considerations for Respirators Facepieces

In order to be effective the facepiece must fit the face. Until the 1970s respirator manufacturers usually produced their respirator facepiece in one size only. Although each company's facepiece was different, the respirators tended to fit males better than females. Figure 19.1 depicts the facial size groupings of U.S. Air Force Personnel using face width and face length (3). Within each oval are 95 percent of the facial sizes for males and females. Respirator sizes did not span the two ovals well, so that most females, and many small males, had difficulty finding a respirator facepiece that would fit well.

Interest in respirator fit led to a study of facial measurements from which three measurements were chosen for use in facial sizing (3). The principal dimensions chosen in relation to respirator face fit are face length (Menton–nasal root depression), face width (bizygomatic breadth), and lip length (Figure 19.2). The dimensions of face length and face width are used for full face masks, and face length

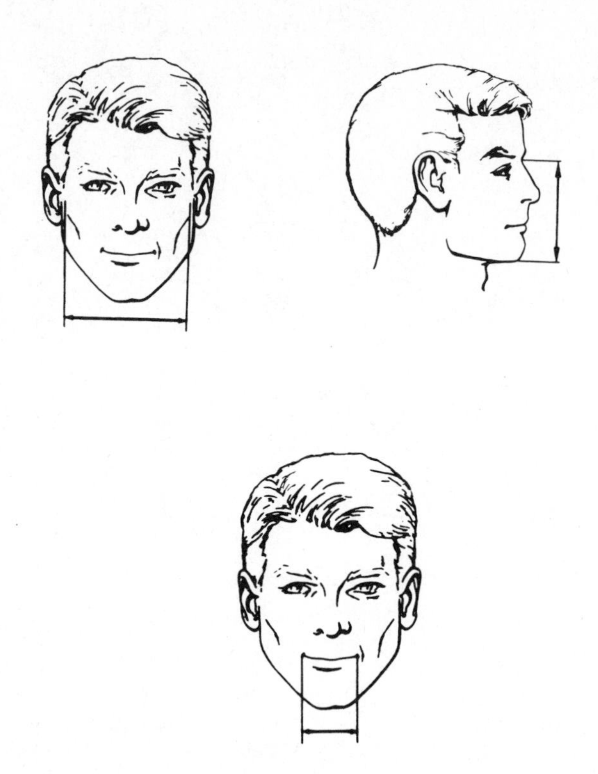

Figure 19.2. Facial measurements for respirator test panels (3).

and lip length are used for half and quarter masks. These dimensions are easily measured, and readily understood by the respirator user. A disadvantage of using these facial dimensions to relate respirator efficiency to facial size is that they are only two dimensional.

Utilizing U.S. Air Force anthropometric studies, facial sizes for two 25-person panels were established to represent the working population (4). The panels in Figure 19.3 for full and half face masks present the population distribution for facial sizes related to face length and face width and lip length and face width, respectively. These panels have been utilized to determine how well respirators fit the general population.

Today most manufacturers supply half masks in three sizes which cover the facial groupings, and at least two manufacturers supply their full facepiece in three sizes (see Figures 19.4 and 19.5). With the expansion in respirator sizes it is now possible for most women to find a respirator that fits.

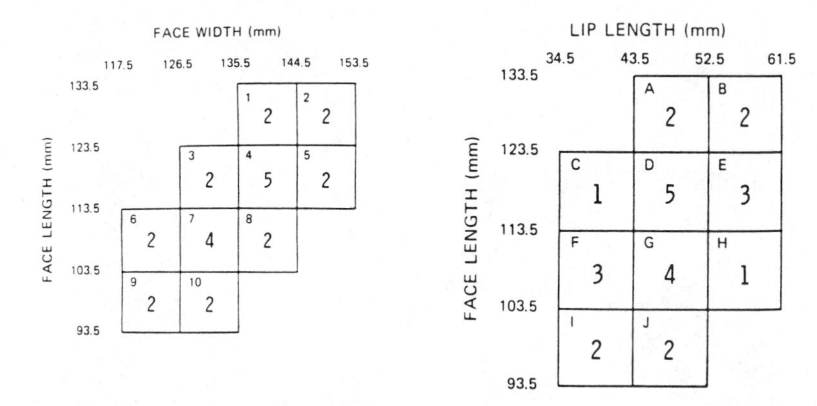

Figure 19.3. Twenty-five person respirator test panels. Left: full face mask panel. Right: half mask panel (4).

3.2 Tight-Fitting Facepieces

The tight-fitting facepiece is intended to adhere snugly to the skin of the wearer. It is available in three varieties: quarter mask, half mask, and full face mask. The quarter mask covers the nose and mouth; the half mask covers the nose, mouth,

Figure 19.4. Half mask respirator in three sizes, large, medium, and small. Model 1482 Twin Cartridge Respirator, Pro-Tech Respirators.

Figure 19.5. Full face respirator in three sizes, extra large, large, and small. Model 65 Twin Cartridge Respirator, Scott Aviation.

and chin; and the full face mask covers the entire face from chin to hairline and from ear to ear.

A device that does not fit easily into the foregoing categories is the mouthpiece respirator. A mouthpiece is held in the wearer's mouth, and a clamp is placed over the nostrils. The lips are pursed tightly around the mouthpiece, and all air comes through the filtering device or air supply attached to the mouthpiece. It has been used for mine rescue and is now used as an emergency escape device.

The facepiece is usually held in place by elastic or rubber straps. Quarter and half masks may be secured by one strap attached to each side of the facepiece, that is, two-point suspension, or two straps attached at two points on each side of the facepiece, that is, four-point suspension. The four straps, instead of being attached to tabs on the edge of the facepiece, may be part of a yoke that is fastened to the facepiece by one or two points in the front of the facepiece. Full face masks have a head harness attached to the facepiece at four, five, or six points. The large sealing surface of the full face mask and the distribution of the headband attachment assists in maintaining a stable facepiece with less slippage than is experienced with quarter or half masks.

3.3 Loose-Fitting Respirators

The best known loose-fitting respirator is the supplied-air hood (SAH) used by the abrasive blaster. The hood covers the head, neck, and upper torso, and usually includes a neck cuff. Air is provided through a hose leading into the hood. Because the hood is not tight fitting, it is important that sufficient air be provided to maintain an outward flow of air and prevent contaminants from entering the hood. See Figure 19.6.

Figure 19.6. Abrasive blasting supplied-air hood. 77 Series Supplied-Air Respirator, E. D. Bullard Company.

Not all SAHs are as rugged or durable as the abrasive blaster's hood, which meets NIOSH approval regulations requiring a durable covering to withstand the rigors of the abrasive blasting atmosphere. There are lightweight plastic and paper SAHs available that can be used in nonabrasive blasting environments.

3.3.1 Supplied-Air Helmets

A helmet with face shield but no cape or neck cuff is available. Air is supplied to the rear of the helmet and travels inside the top and comes down along the face shield.

3.3.2 Supplied-Air Suits

The hood can be extended to cover the entire body and provide whole body protection. Disposable impervious clothing is available for this purpose. There are no NIOSH approval regulations for supplied-air suits.

4 AIR-PURIFYING RESPIRATORS (APR), NONPOWERED

An APR must have a tight-fitting facepiece because it depends on negative pressure produced when the user inhales to induce airflow into the facepiece. Some supplied-air respirators also depend on the inhalation of the user to provide airflow into the facepiece. The facepieces used for negative pressure air purifying devices can be quarter, half, full face, or mouthpiece respirators.

4.1 Respirator Facepiece Valves

4.1.1 Inhalation Valves

Inhalation valves prevent moisture-laden exhaled air from coming in contact with sorbents and filters. Exhaled air tends to increase breathing resistance of the filters and cause them to be changed more frequently.

4.1.2 Exhalation Valves

With the exception of some disposable respirators, NIOSH approval regulations require the provision of exhalation valves, which provide a one-way exit for exhaled air. This important part, out of sight and often neglected, needs frequent checking to ensure that it is working (Figure 19.7). A malfunctioning exhalation valve can render useless both positive and negative pressure respirators. If an obstruction such as a hair or a dirt particle comes between the valve and the valve seat, a serious leak can occur.

Exhalation valves must also have a valve cover. This serves the dual purpose of protecting the valve from physical damage and retaining a small reservoir of exhaled air. A finite time elapses between the beginning of inhalation and the closure of the exhalation valve. During this period the clean, recently exhaled air prevents toxic material from entering the valve.

4.2 Removal of Aerosols, Gases, and Vapors

With the exception of disposable respirators, the air-purifying elements used with respirators are designed to be replaced. Air-purifying units can be divided into two functions: aerosol removal, and vapor and gas removal. In the former case the size of the aerosol influences the removal process used regardless of the aerosol composition. In the latter case the vapor or gas to be removed may require a specific chemical to capture it.

4.2.1 Aerosol Capture

Aerosols are removed from the air using a variety of filtration mechanisms. Regardless of the chemical composition it is the size of aerosol that determines the type of filter necessary for capture.

As air passes through a fibrous media, aerosols are removed by the mechanical

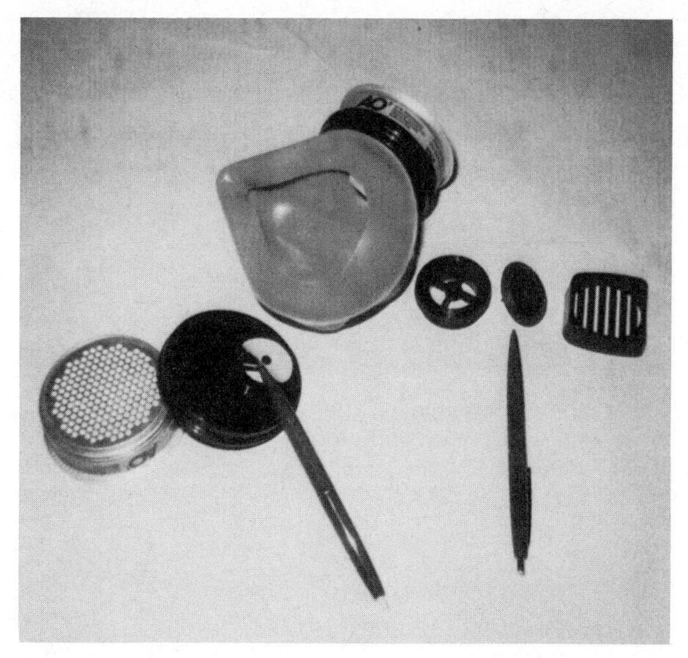

Figure 19.7. Inhalation valve and exhalation valve assembly, air-purifying respirator. AO 5 Star Dual Element Air Purifying Respirator, AO Safety Products.

phenomena of impaction, interception, and diffusion. The efficiency of aerosol removal is related to its size, with 0.5-μm-diameter aerosols being the hardest to capture, and the filter fiber medium used. The efficiency of all mechanisms is improved as the filter fiber diameter is decreased and the number of fibers increased. The resistance to airflow per unit area of filter usually increases as the filter fiber diameter decreases.

A filter with high collection efficiency for small aerosols usually has a higher resistance to airflow than a low-efficiency filter. The high-efficiency filter area must be larger to keep the inhalation flow resistance equal to that of the low-efficiency filter. During use the efficiency of filters improves, and breathing resistance increases, as the aerosols lodge on the filter surface.

LIST OF ABBREVIATIONS USED IN CHAPTER NINETEEN

ANSI	American National Standards Institute
APF	Assigned protection factor
APR	Air-purifying respirator
BM	Bureau of Mines
CB	Certification branch
CFM	Continuous flow mode
cfm	Cubic feet per minute

CNC	Condensation nucleus counter
DM	Demand mode
DOE	Department of Energy
DOP	Dioctylphthalate
IAA	Isoamyl acetate
IDLH	Immediately dangerous to life or health
l/m	Liters per minute
LANL	Los Alamos National Laboratory
MMAD	Mass median aerodynamic diameter
MWR	Maximum work rate
NIOSH	National Institute for Occupational Safety and Health
OSHA	Occupational Safety and Health Administration
PAPR	Powered air-purifying respirator
PDM	Pressure demand mode
PEL	Permissible exposure limit
PF	Protection factor
PIF	Peak inspiratory flow
QLFT	Qualitative fit test
QNFT	Quantitative fit test
SAH	Supplied-air hood
SAR	Supplied-air respirator
SAR-CF	Supplied-air respirator, continuous flow mode
SAR-DM	Supplied-air respirator, demand mode
SAR-PDM	Supplied-air respirator, pressure demand mode
SCBA	Self-contained breathing apparatus
SCBA-DM	Self-contained breathing apparatus, demand mode
SCBA-PDM	Self-contained breathing apparatus, pressure demand mode
WPF	Workplace protection factor

In addition to the mechanical methods of entrapment, an electrostatic charge can increase the filter efficiency. The resin added to resin felt dust filters is given a high electrostatic charge at the factory, which can remain unchanged for a period of years. Electrostatic filters have lower resistance to airflow without decreasing filter efficiency than filters of similar efficiency without an electrostatic charge.

4.2.1.1 Filter Classification. NIOSH approval regulations have four aerosol filter classifications: dust, dust and mist, fume, and high efficiency.

4.2.1.2 Dust Filters. Fiber glass and resin felt are materials used for dust and mist filters. Fiber glass has low resistance and good efficiency to dust particles above 1 μm. Resin felt, using an electrostatic charge to collect aerosols, has good efficiency for dust particles and low resistance to flow. The low resistance means the size of filter can be reduced.

Lowry and Ortiz and their co-workers found that resin felt filters exposed to high heat, high humidity, or liquid aerosols show a marked reduction in efficiency,

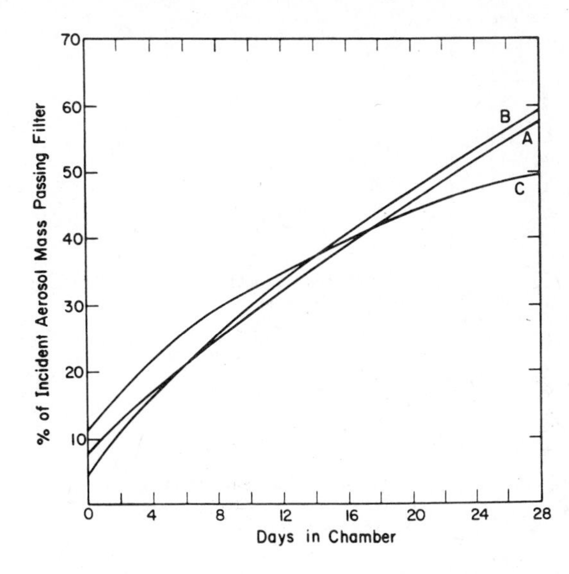

Figure 19.8. Effects of electrostatic filters stored at 32°C and 90 percent relative humidity (5).

presumably caused by a reduction in the electrostatic charge (5, 6). (See Figures 19.8 and 19.9.) Fiber glass filters are not affected. It is important that these filters be stored away from high heat and humidity, and not used for the collection of liquid aerosols.

Dust filters are tested using a silica aerosol with a 0.4 to 0.6 μm geometric mean diameter, with a maximum acceptable aerosol penetration of 1.04 percent of the challenge concentration.

4.2.1.3 Mist Filters. Liquid aerosols are captured in the same manner as solid aerosols with the exception that the electrostatic charge is not effective. Mist filters are tested with a silica mist generated using a 270-mesh screen with a maximum acceptable aerosol penetration of 1.25 percent of the challenge concentration (7). The actual particle size of the silica mist aerosol is not specified.

4.2.1.4 Fume Filters. The fume filter is satisfactory to remove the most common fume encountered, welding fumes, and is unaffected by liquid aerosols. The fume filter is tested using a freshly generated lead fume with a count median diameter

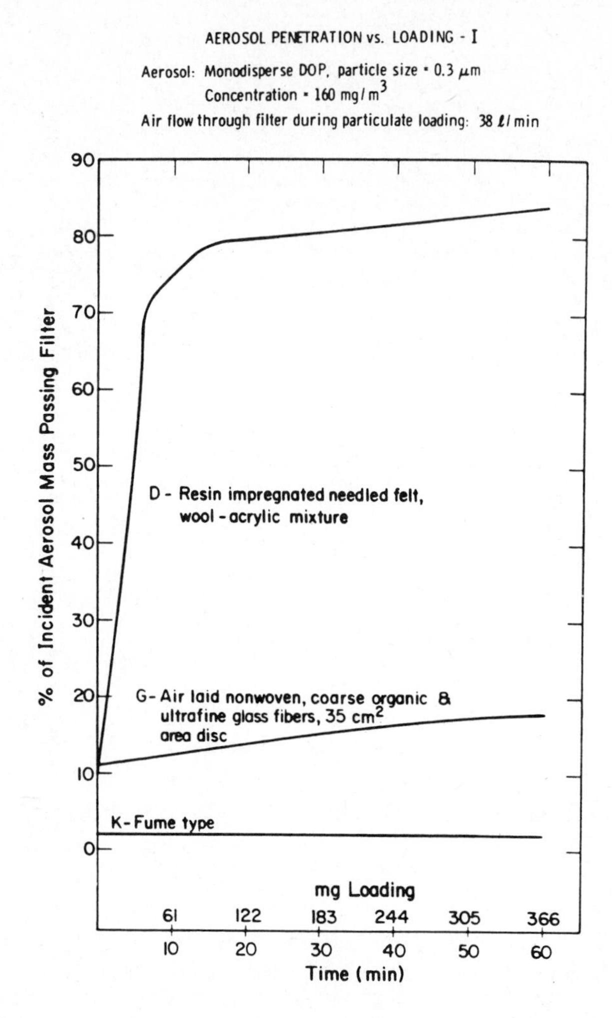

Figure 19.9. Liquid aerosol penetration of electrostatic filters (5).

of 0.23 μm with a maximum acceptable aerosol penetration of 1.0 percent of the challenge concentration (5, 8).

4.2.1.5 High-Efficiency Filters. These filters are thin sheets of filter material with small fiber diameter and high resistance to flow per unit area. The high-efficiency filter was originally designed for use with radioactive aerosols. It is now used for all highly toxic aerosols. High-efficiency filters are tested using a monodisperse 0.3 μm mass median aerodynamic diameter (MMAD) dioctylphthalate (DOP) aerosol with a maximum acceptable aerosol penetration of 0.03 percent of the challenge concentration (7).

4.2.1.6 Filter Efficiency Comparison. Comparison between filter classes is difficult because the tests are not done in the same manner. Dust, mist, and fume filters are tested by collecting the aerosols that penetrate the filter over a 90- to 312-min period. This method does not provide data on initial aerosol penetration which may be higher at the beginning when the filter is not loaded. The high efficiency filter test is dynamic, measuring penetration in a period of a few seconds.

4.2.1.7 Filter Selection. The NIOSH approval regulations consider toxicity as well as aerosol size. If the aerosol permissible exposure limit (PEL) is above 0.05 mg/m^3, then a dust, mist, or fume filter can be used depending on the aerosol size and physical state. If the aerosol PEL is below 0.05 mg/m^3, then a high-efficiency filter must be used regardless of aerosol size.

4.2.1.8 Asbestos Filters. If air-purifying respirators are used for asbestos exposure, OSHA accepts only high-efficiency filters. None of the test aerosols mentioned are fibers, and there is concern about the ability of the filters tested to trap fibers. Ortiz et al. studied the penetration of chrysotile asbestos fibers using a dust and mist filter; two disposable dust and mist masks; a dust, mist, and fume filter; and a high-efficiency filter (6). They found all filters as purchased had less than 0.1 percent penetration of chrysotile asbestos fibers except one disposable mask that had a 2.8 percent penetration.

4.2.2 Gas and Vapor Removal

Gas and vapor removal is more complex than aerosol capture, which is sensitive to aerosol size but not to chemical composition. Sorbents are porous and have a large surface area per unit weight—up to 1500 m^2/g. As the gas or vapor passes through the sorbent bed, the molecules are sorbed on the surface.

Gases and vapors are captured by adsorption, chemisorption, absorption, or catalysis when passing through a particulate (sorbent) bed of chemicals. Adsorption is the physical attraction of a vapor to an adsorbent surface. Chemisorption occurs when the sorbent surface and the vapor bond chemically. Absorption occurs when the vapor penetrates below the surface of the absorbent and chemically reacts with it. If the solid material causes the vapor to change chemically without affecting the solid then it is a catalytic reaction.

The sorbent most widely used is activated carbon or charcoal. Alone it is an excellent adsorbent of many organic vapors. Impregnating it with specific materials increases its retention efficiency for additional gases and vapors.

4.2.2.1 Carbon Monoxide. Carbon monoxide is not sorbed, but by catalytic action is oxidized into carbon dioxide. The carbon monoxide canister is filled with a mixture of manganese and copper oxides, known as Hopcalite. This mixture catalytically converts carbon monoxide to carbon dioxide without the presence of moisture. Moisture stops the reaction. To prevent the entry of moisture carbon monoxide canisters have layers of drying agents on both sides of the catalytic agent. Because carbon monoxide has no warning odor, an end of service life indicator shows when the drying agent is saturated and the life of the canister is ended.

4.2.2.2 Combination Cartridges and Canisters. Sorbents are packaged as cartridges for respirators, and chin and chest style canisters for gas masks. The smaller sized cartridge with a single sorbent functions the same as a larger canister, but the maximum use concentration or the useful life of the unit can be greater with the canister. There are cartridges and canisters containing more than one sorbent, and they may be combined with an aerosol filter (see Table 19.1).

The universal canister for gas masks packages four different sorbent agents and an aerosol filter in a single container. At one time it was considered to be satisfactory for use against unknown airborne hazards, now the province of the self-contained breathing apparatus. NIOSH approval procedures specify that it is tested with five gases and vapors, as shown in Table 19.1, and that the aerosol filter pass the high-efficiency filter penetration test. The test gas concentrations are the same as that used for a single sorbent, but the minimum service life is reduced by half. Universal canisters should be used with caution and replaced after each use. They should not be used if only one specific contaminant is expected.

The most common of the multiple sorbent and filter units are the paint spray and the pesticide cartridges. The paint spray unit combines an organic vapor cartridge and an aerosol filter. Besides passing the organic vapor challenge, the cartridge must pass two special aerosol tests for enamels and lacquers.

From 1950 to 1967 the U.S. Department of Agriculture performed tests on respirator chemical cartridges to determine their efficiency when used with pesticides (9). NIOSH now approves pesticide cartridges and canisters. Similar to the paint spray cartridge, the pesticide cartridge must provide protection from organic vapors and aerosols. Cartridge and canister must meet fume and high-efficiency filter requirements, respectively. Respirator manufacturers may have additional data on the efficiency of this cartridge with specific pesticides. This unit is not approved for use with fumigants.

4.2.2.3 Effectiveness of Gas and Vapor Cartridges and Canisters. The effectiveness of any sorption cartridge or canister should be established for the particular gas or vapor to be captured. Gas and vapor canisters are reviewed by the NIOSH CB to determine their effectiveness. This review system does not test the sorption efficiency for all gases and vapors found in the workplace. Presently there are 16 gases and vapors used to test sorbents regularly. There are two general sorbent categories, acid gases and organic vapors, with two acid gases and one organic vapor used as test indicators. Information on sorption effectiveness is not readily available for many chemicals. The technical literature should be reviewed and respirator manufacturers and NIOSH contacted for specific information.

Nelson and Harder did extensive work in determining the service life for organic vapor cartridges exposed to various organic vapors (10). They exposed commercial organic vapor cartridges to high concentrations of organic vapors and recorded the time interval until the organic vapor was detected coming through the cartridge. Using the NIOSH procedure for organic vapor cartridge testing, they tested the cartridges against 107 organic vapors. (See Table 19.2.) Nineteen of the 107 materials tested failed in less than 50 min, which is the minimum time the cartridge

Table 19.1. Test Agent Test Concentrations and Minimum Service Life 30CFR11, Subparts I, L, M, and N

Test Agent	Subpart I Gas Mask Chest Canister		Subpart I Gas Mask All-Purpose Canister		Subpart I Gas Mask Chin Style Canister		Subpart I Gas Mask Chin Style All-Purpose Canister		Subpart I Gas Mask Escape Canisters		Subpart L Chemical Cartridge Respirator		Subpart N Vinyl Chloride	
	TC (ppm)	MSL (min)	TC (ppm)	MSL (min)	TC (ppm)	MSL (min)	TC (ppm)	MSL (min)	TC (ppm)	MSL (min)	TC (ppm)	MSL (min)	MSL (min)	TC (ppm)
Acid gases														
Sulfur dioxide	20,000	12	20,000	6	5,000	12	5,000	6	5,000	12	500	30		
Chlorine	20,000	12	20,000	6	5,000	12	5,000	6	5,000	12	500	35		
Organic vapors														
Carbon tetrachloride	20,000	12	20,000	6	5,000	12	5,000	6	5,000	12	1,000	50		
Ammonia	30,000	12	30,000	6	5,000	12	5,000		5,000	12	1,000	50		
Carbon monoxide	20,000	60	20,000	60	20,000	60	20,000	60	10,000	60				
Ethylene oxide	50	240							5,000	12				
Formaldehyde	500	120									100	50		
Hydrogen fluoride	20	12							5,000	12				
Phosphine	1,500	12												
Vinyl chloride	25	360											10	90
Chlorine dioxide											500	30		
Hydrogen chloride											500	50		
Hydrogen cyanide									5,000	12				
Hydrogen sulfide									5,000	12	1,000	30		
Mercury											21.5[a]	480		
Methylamine											1,000	25		

TC = Test concentration, MSL = minimum service life.

[a]Milligrams per cubic meter.

Table 19.2. Effect of Solvent Vapor on Respiratory Cartridge Efficiency[a]

Solvent	Time to Reach 10 ppm (min)	Solvent	Time to Reach 10 ppm (min)	Solvent	Time to Reach 10 ppm (min)
Aromatics[b]					
Benzene	73	1,2-Dichloropropane	65	2-Heptanone	101
Toluene	94	1,4-Dichlorobutane	108	Cyclohexanone	126
Ethyl benzene	84	*o*-Dichlorobenzene	109	5-Methyl-3-heptanone	86
m-Xylene	99	Trichlorides		3-Methylcyclohexanone	101
Cumene	81	Chloroform	33	Diisobutyl ketone	71
Mesitylene	86	Methylchloroform	40	4-Methylcyclohexanone	111
Alcohols[b]		Trichloroethylene	55	Alkanes[c]	
Methanol	0.2	1,1,2-Trichloroethane	72	Pentane	61
Ethanol	28	1,2,3-Trichloropropane	111	Hexane	52
Isopropanol	54	Tetra- and pentachlorides[b]		Methylcyclopentane	62
Allyl alcohol	66	Carbon tetrachloride	77	Cyclohexane	69
n-Propanol	70	Perchloroethylene	107	Cyclohexene	86
sec-Butanol	96	1,1,2,2-Tetrachloroethane	104	2,2,4-Trimethylpentane	68
Butanol	115	Pentachloroethane	93	Heptane	78
2-Methoxyethanol	116	Acetates[b]		Methylcyclohexane	69
Isoamyl alcohol	97	Methyl acetate	33	5-Ethylidene-2-norbornene	87
4-Methyl-2-pentanol	75	Vinyl acetate	55	Nonane	76
2-Ethoxyethanol	77	Ethyl acetate	67	Decane	71

Amyl alcohol	102	Isopropyl acetate	65	Amines	
2-Ethyl-1-butanol	76.5	Isopropenyl acetate	83	Methylamine	12
Monochlorides[b]		Propyl acetate	79	Ethylamine	40
Methyl chloride	0.05	Allyl acetate	76	Isopropylamine	66
Vinyl chloride	3.8	*sec*-Butyl acetate	83	Propylamine	90
Ethyl chloride	5.6	Butyl acetate	77	Diethylamine	88
Allyl chloride	31	Isopentyl acetate	71	Butylamine	110
1-Chloropropane	25	2-Methoxyethyl acetate	93	Triethylamine	81
1-Chlorobutane	72	1,3-Dimethylbutyl acetate	61	Dipropylamine	93
Chlorocyclopentane	78	Amyl acetate	73	Diisopropylamine	77
Chlorobenzene	107	2-Ethoxyethyl acetate	80	Cyclohexylamine	112
1-Chlorohexane	77	Hexyl acetate	67	Dibutylamine	76
o-Chlorotoluene	102	Ketones[c]		Miscellaneous materials[c]	
1-Chloroheptane	82	Acetone	37	Acrylonitrile	49
3-(Chloromethyl)heptane	63	2-Butanone	82	Pyridine	119
Dichlorides[b]		2-Pentanone	104	1-Nitropropane	143
Dichloromethane	10	3-Pentanone	94	Methyl iodide	12
cis-1,2-Dichloroethylene	30	4-Methyl-2-Pentanone	96	Dibromomethane	82
trans-1,2-Dichloroethylene	33	Mesityl oxide	122	1,2-Dibromoethane	141
1,1-Dichloroethane	23	Cyclopentanone	141	Acetic anhydride	124
1,2-Dichloroethane	54	3-Heptanone	91	Bromobenzene	142

[a]Cartridge pairs tested at 1000 ppm and 53.3 l/m.
[b]Mine Safety Appliance Company cartridges.
[c]American Optical Corporation cartridges.

Source: Reference 10.

must last under NIOSH test conditions. Eleven of the 19 failures occurred in the chlorinated hydrocarbon group.

Considering the variations in the life-spans of different organic vapors and organic vapor cartridges, when should cartridges be replaced? The usual answer is when the user smells the vapor coming through the cartridge. This is why NIOSH does not approve the use of cartridges for materials with poor warning properties. The human nose, as a vapor detection device, has a wide variability among users.

NIOSH will and has approved sorption cartridges for materials with poor warning properties if they have an end of service life indicator on the cartridge. Such end of service life devices are available for carbon monoxide, mercury vapor, and vinyl chloride. Without an end of service life indicator, information concerning the effectiveness of the sorbent with the contaminant should be obtained, and a conservative time of service set. The nose can then be used as a secondary alarm.

5 SUPPLIED-AIR RESPIRATORS

Supplied-air respirators (SARs) do not have air-purifying filters and cartridges; instead, they depend on delivering acceptable quality breathing air to the user from an external source.

5.1 Hose Masks, Powered and Nonpowered

The oldest SAR unit is the type B hose mask. It consists of a large-diameter hose, anchored in an area with acceptable quality breathing air and connected to the wearer's facepiece. The diameter of the hose is large, which reduces inhalation resistance. Another version is the type A hose mask, which has a hand- or electric-powered air mover connected to the fresh air source to assist in supplying air. The Bureau of Mines has approved hose masks since 1929. Considering the supplied-air equipment available today, the hose masks are no longer needed.

5.2 Supplied-Air Respirators, Type C

The type C supplied-air respirator differs from the hose masks in that the air is delivered to the facepiece under pressure up to 125 psi. It can be obtained in demand mode (DM), pressure demand mode (PDM), and continuous flow mode (CFM) of operation.

5.2.1 Demand Mode

The SAR-DM, as the name implies, supplies air to the user on demand. When the user inhales, the respirator airflow regulation valve opens and air under pressure enters the respirator. The negative pressure in the facepiece may reach 50 mm H_2O before the air starts to flow. Airflow ceases when inhalation ceases. The minimum air flow rate is 115 liter/min (l/m). The facepiece pressure is comparable to the negative pressure present when using an air-purifying respirator.

The SAR may have a quarter, half, or full facepiece. In each case, because of the negative pressure in the facepiece during each inhalation, the degree of facepiece leakage is comparable to that of an air-purifying respirator with the same facepiece. Because SAR-DM respirators do not offer protection to the user that is superior to the air-purifying respirator, their use is not recommended.

5.2.2 Positive Pressure Respirators

The problem of leakage into negative pressure facepieces led to the positive pressure facepiece, which greatly reduces contaminant leakage into the respirator facepiece. Positive pressure can also be maintained by a continuous airflow, such as that used for SAHs and suits.

5.2.2.1 Pressure Demand Mode The SAR-PD is similar to the SAR-DM, already described, except for the exhalation valve and the airflow regulation unit. The exhalation valve is spring loaded so that a facepiece pressure of up to 38 mm H_2O can be maintained. When inhalation occurs the facepiece pressure drops and at some point above atmospheric the airflow regulation valve opens. Normally when inhalation occurs, the positive pressure inside the facepiece is reduced but remains positive. It is possible under high work rates for the user to demand a higher instantaneous airflow than the SAR can deliver. Under this condition, negative pressure may occur in the facepiece for a short period of time. Regardless, the PDM provides the greatest degree of protection for the user of SAR.

5.2.2.2 Continuous Flow Mode. There are tight-fitting and loose-fitting continuous flow supplied air respirators. The loose fitting supplied air respirator may consist of a supplied air hood, helmet, blouse, or full suit.

5.2.2.3 Tight-fitting Facepieces. The tight-fitting respirator is simply a facepiece with a continual flow of air through the mask. It requires a minimum continuous air flow of 115 l/m when in use.

5.2.2.4 Supplied Air Hoods. The SAH introduces the air supply through a manifold above the head of the user. The air flows downward and exits under the edges of the hood. Better designs have neck cuffs which tend to create a slight positive pressure in the hood. NIOSH regulations call for a minimum airflow of 170 l/m for SAHs.

The minimum air flow for SAHs was investigated in 1933 by Bloomfield and Greenburg (11), prior to the Bureau of Mines establishing the approval regulations for SAHs. A sandblasting operation was set up and samples taken inside and outside a SAH worn by the abrasive blaster while working. Bloomfield and Greenburg found that contamination in the SAH while sandblasting was minimal when the airflow to the SAH was maintained at 170 l/m or above. The sandblasting operation does not appear to involve heavy physical activity.

5.2.2.5 Supplied Air Suits. There are no official regulations governing the design, sale, and use of supplied air suits, but the Department of Energy (DOE) has a supplied-air suit protocol for use in reviewing suits used by DOE contractors. Much is unknown about the design and use of airline suits. The traditional figure of 170 l/m for SAHs may not provide sufficient airflow to prevent inward leakage for these devices. Airline suits should be tested to determine their efficiency prior to use in atmospheres of high toxicity.

5.2.3 Air Supply Hose and Pressure Regulator

The NIOSH-approved SAR includes the facepiece, breathing tube, belt-mounted air control valve, and air hose of 15 to 300 ft long, and for DM and PDM equipment an airflow regulator. This entire assembly must be purchased from the respirator manufacturer. The user is responsible for providing the air supply, any necessary airline filters, manifold with airflow control valve for the air hose connection, and the airline pressure gauge. The airflow is controlled by setting the airline pressure at the manifold according to the manufacturer's instructions. Problems are encountered in ensuring the minimum required flow is delivered. Changes such as an airline that is longer than design length, has a smaller diameter, or has been spliced can result in a drastic reduction in the air supply with the same setting of the pressure gauge. It will not be possible to know whether the airflow to the respirator meets the minimum airflow required if any non-NIOSH approved components are used. The airflow, being set by a valve at the supply manifold, can be changed only by manually adjusting the airflow control valve. This is in contrast with the open-circuit self-contained breathing apparatus, which has an airflow that varies automatically with the inhalation pressure of the user.

5.3 Self-Contained Breathing Apparatus

The SCBA respirator carries an air supply of compressed air, compressed or liquid oxygen, or oxygen-generating chemicals. No external connection to an air supply is necessary.

5.3.1 Open-Circuit Self-Contained Breathing Apparatus

The most widely used SCBA is an open-circuit device consisting of a compressed air tank, airflow regulator, airline, and facepiece, all of which weigh less than 35 lb (see Figure 19.10). In the United States SCBAs are available in units with air tanks pressured to 2200 or 4500 psi. The 4500-psi unit can have an air supply tank that is physically smaller than a 2200-psi tank of the same capacity. The airflow regulator functions in a similar manner as the SAR. It opens to admit air into the facepiece when the pressure in the facepiece is decreased. The airflow can vary depending on the inhalation pressure. In contrast to the SAR regulator, it may be a two-stage pressure-reducing regulator because the air supply tank is under high pressure. In addition, there is a manual bypass valve for emergency use by the wearer if the regulator malfunctions.

The air supply tanks may be steel, aluminum, fiber glass-wrapped aluminum,

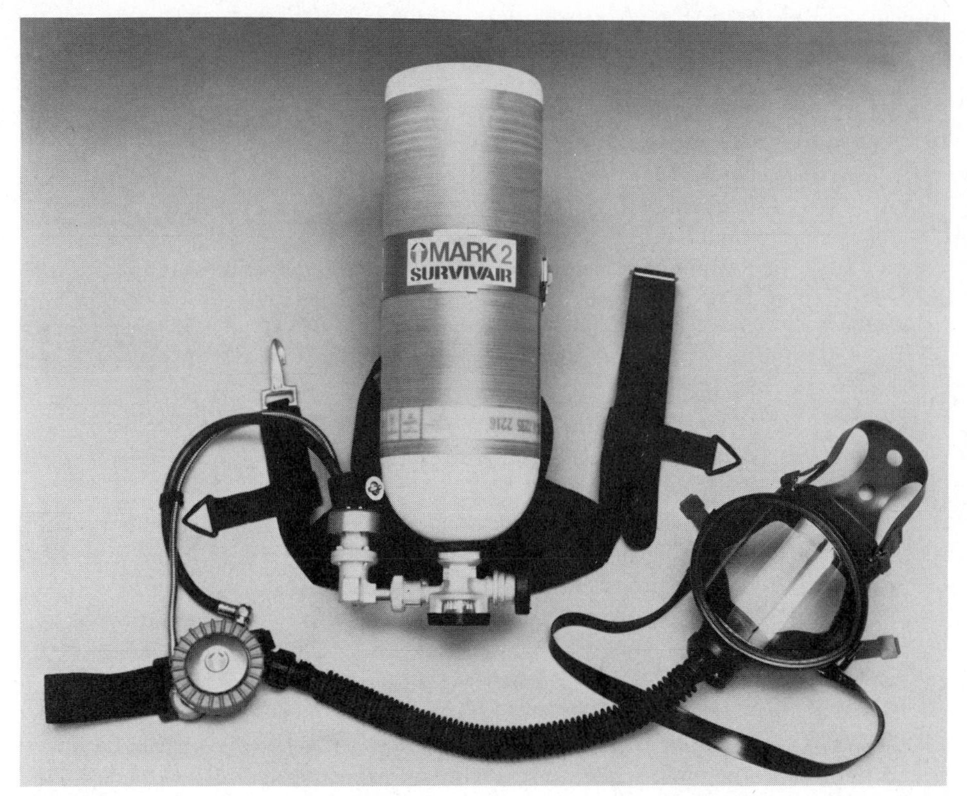

Figure 19.10. Self-contained breathing apparatus, open circuit. Model Mark 2, Survivair Company.

or a composite tank of an aluminum bladder inside a filament wound fiber glass bottle. The fiberglass bottles and the fiberglass wrapped aluminum tanks reduce the weight of the air tanks greatly.

The customary SCBA for entry and use in hazardous atmospheres is rated at 30 min; however, SCBA units for escape only can be rated at 3, 5, 10, and 15 min. Air supply ratings given with this equipment are nominal and should not be regarded as absolute. High work rates combined with the stress of emergency conditions will reduce the actual use time of the air tank.

The SCBA can be obtained in demand or pressure demand mode. Only the pressure demand units should be used because they provide the highest degree of protection of any respirator. Only SCBA-PD respirators are recommended for use in environments that may be immediately dangerous to life or health (IDLH).

5.3.2 Closed-Circuit Self-Contained Breathing Apparatus

Instead of exhausting the exhaled air, as is done with the open-circuit devices, the exhaled air from a closed-circuit SCBA is reused. Closed-circuit devices, available with up to a 4-hr service life, recirculate the exhaled air through a chemical to remove the carbon dioxide. The remaining exhaled air is sent to the breathing bag, the breathing air reservoir for the facepiece. Because the volume of air is decreased by the carbon dioxide removed, a small amount of oxygen is added to make up the volume. The oxygen can be in a liquid or gaseous state.

Oxygen may also be stored as a solid chemical compound for use with a closed-circuit SCBA. The chemical, usually a type of peroxide, reacts in the presence of moisture and carbon dioxide to produce pure oxygen. Because moisture and carbon dioxide are both components of exhaled breath, a usable breathing cycle is formed by recirculating exhaled breath through the oxygen-rich chemical. The carbon dioxide is removed as it becomes part of the reaction to release the oxygen. The air then enters a breathing bag for mixing before being drawn into the facepiece. A disadvantage of this unit is that once the reaction has been started in the canister, it cannot be terminated until the chemical reaction is complete. Canisters are available for up to 2-hr duration. This type of unit is used as an emergency escape device for mines.

Most SCBA-closed circuit are demand mode devices. The units have been used successfully for many years in mine rescue operations. The rescue teams are given extensive training in the use of the equipment to insure proper use.

6 POWERED AIR-PURIFYING RESPIRATORS

The powered air-purifying respirator (PAPR) is one with a battery-operated blower added to provide the energy necessary to force air through the filters or cartridges and supply it to the user. PAPRs are a mixture of SARs and APRs. NIOSH uses APR approval requirements for dust, fume, mist, and vapor filters and cartridges, and adds the SAR requirements that they supply 115 l/m of air for tight-fitting masks and 170 l/m for hoods and helmets. The PAPR is available in three configurations: tight-fitting facepieces, hoods, and helmets (see Figure 19.11).

PAPRs are advantageous for use in areas where it is difficult to provide air for SAR respirators or mobility requirements proscribe an air hose. Because these are air-purifying units only, they must not be used in an IDLH environment or in the presence of atmospheric contaminants for which the filters or sorbents are not designed.

The PAPR, in good operating condition, has been considered as providing protection equivalent to SARs. Workplace evaluations have shown this is not correct. The helmet PAPR appears to be susceptible to adverse air currents in the workplace that reduce the protection provided. Similarly, the conventional PAPR has been shown to provide a lower degree of protection than SARs. PAPR devices should be used with caution if a high degree of protection is provided.

The helmet model differs from the conventional SAH. It resembles an oversize hard hat with a face shield. Inside the helmet near the back there is a motor, blower, and filter unit. Air is drawn from the back of the helmet through the filter and blown out the front where it exits along the sides and bottom of the face shield.

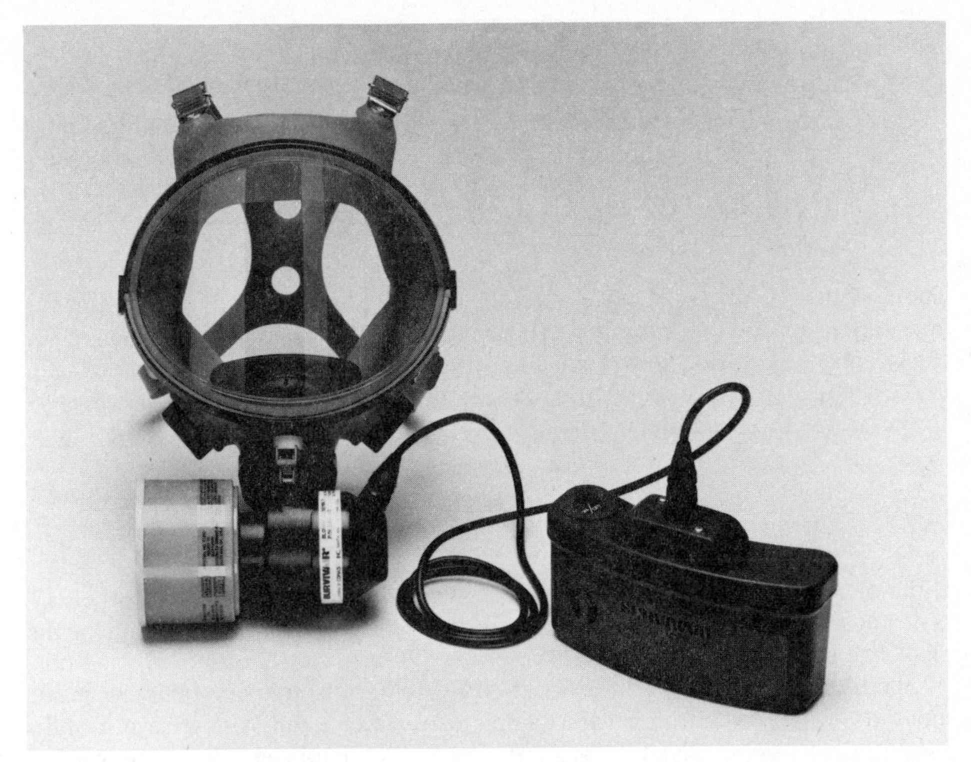

Figure 19.11. Tight-fitting PAPR with blower, motor, and filter element mounted on facepiece. Survivair Company.

7 RESPIRATOR EFFICIENCY

The efficiency of any respiratory protective system can be evaluated by measuring the amount of outside contaminant that enters the wearer's respiratory system compared with the total contaminant present. The efficiency is affected by the respirator cartridge, filter, cartridge holder, valve or regulator malfunction, face to facepiece seal, and air supply for SARs and SCBAs. Some of these problems can be found by visual inspection. Problems involving the respirator fit and the air supply require additional information.

7.1 Respirator Fitting Tests

It is important to determine that the respirator fits satisfactorily before entry into a hazardous atmosphere. There are three types of respirator fit tests:

A. Quantitative fit tests, which determine the degree of protection provided by the respirator

B. Qualitative fit tests, which use simpler equipment than that required for quantitative fit tests, and which depend on the subjective opinion of the user as to whether the respirator leaks
C. Fit checks, which are used whenever the respirator is donned and just prior to entering a toxic atmosphere, to determine if the facepiece seal is satisfactory

7.1.1 Quantitative Fit Tests

The quantitative fit test (QNFT) measures the contaminant concentration in the atmosphere around the respirator user, and compares it to the concentration inside the facepiece. The protection provided, expressed as the respirator protection factor (PF), is equal to the concentration outside the respirator divided by the concentration inside the respirator:

$$\text{PF} = \frac{\text{Concentration outside respirator}}{\text{Concentration inside respirator}}$$

If the outside concentration is considered the 100 percent concentration, the concentration inside the respirator can be expressed as percent penetration of the challenge concentration.

The ideal QNFT should measure the PF on the job, utilizing sampling inside and outside the respirator. Such a sample taken over a full shift or several shifts would provide good QNFT data for the individual employee. Presently this is a difficult process and not often done. It is done to evaluate how well a respirator works under actual field conditions and/or compare QNFT data obtained in the laboratory to that actually achieved in the workplace. PFs obtained in the workplace are known as workplace protection factors (WPF).

At this time there are no QNFT standards. ANSI Z88.2-1980, "Practices for Respiratory Protection," has recommendations for laboratory QNFT procedures including specifications for fit test agents and suggested exercises. OSHA has, as appendices to some regulations, provided guidance to QNFT procedures. Most QNFT procedures in the United States since 1970 have followed test procedures used by the Los Alamos National Laboratory (LANL) (5, 12). These procedures, using DOP or sodium chloride as a test atmosphere, place the respirator wearer in a confined space containing the test atmosphere. The air within and without the facepiece is sampled while the subject pursues mild exercises.

Although all QNFTs follow the procedure outlined above, there are differences in how the tests are carried out. The principal differences involve the exercises to be carried out, the degree of physical work required during the test, placement of the respirator sampling probe and the air sampling rate, and the method of calculation used to determine the amount of penetration. There are also differences in the QNFT agents used.

7.1.1.1 Di-2-ethylhexyl Phthalate. Di-2-ethylhexyl phthalate was the most widely used QNFT agent and is frequently referred to as dioctylphthalate (DOP). As a thermogenerated monodispersed liquid aerosol with a 0.3-μm MMAD, this aerosol is used to test high-efficiency filters. QNFT using thermogenerated DOP has been done (13), but the potential for the formation of decomposition products that may be toxic led to its abandonment as a fit test agent. Air-generated DOP, also known as polydispersed DOP, with a concentration of 15 to 20 mg/m^3 and a MMAD of 0.6 μm was adopted as a respirator fit test agent. The lower limit of detection is 0.005 percent of the challenge concentration. Continued questions about the toxicity of DOP led to its abandonment (14). Corn oil was found to have similar physical properties to DOP and replaced it in the operation of the commercially available QNFT equipment (15).

7.1.1.2 Sodium Chloride. Sodium chloride has been employed for respirator testing in England for many years. In the United States sodium chloride research for use in respirator testing was initiated at the Los Alamos National Laboratory (LANL) (5). The sodium chloride QNFT system developed by LANL is commercially produced with a test atmosphere of 15 mg/m^3 and a test aerosol of 0.6 μm MMAD. Some concern has been raised about the possible effect of the salt atmosphere on a hypertensive subject.

7.1.1.3 Dichlorodifluoromethane. Adley and Wisehart reported using dichlorodifluoromethane gas for QNFT in 1962 for both SARs and APRs (16). Da Rosa et al. reported using dichlorodifluoromethane for SCBA tests under high work rate conditions. The lower limit of detection was 0.0001 percent of the challenge concentration (17).

7.1.1.4 Sulfur Hexafluoride. Lowry et al. used sulfur hexafluoride (SF_6) in 1977 to compare gaseous and aerosol fit testing procedures. The lower limit of detection was 1.0 percent of the challenge concentration (18). Bentley et al. in 1984 used SF_6 as a fit test agent with SARs and SCBAs. The lower limit of detection was 0.00005 percent of the challenge concentration. They compared SF_6, argon, and sodium chloride and found good correlation between the three test agents (19).

7.1.1.5 Condensation Nucleus Counter. The condensation nucleus counter (CNC) operates differently from the QNFT agents described above. The unit, developed in a cooperative effort between the U.S. Army and private industry, is small, approximately 6 in. × 6 in. × 3 in. It does not require a separate test chamber, for it uses as a test atmosphere the dust particles in an office environment. Air drawn into the device by a small sampling pump passes through a heated chamber saturated with alcohol, and enters a cooled chamber that causes the alcohol vapor to condense on the dust particles. The particles grow to about 10 μm in size during the alcohol condensation and enter a forward light scattering chamber for counting. The programmed unit alternately draws air from the room and the respirator, records the dust concentration in each, compares them, and calculates a PF. The unit appears to perform comparably with the corn oil aerosol units now in use throughout the country (20).

The commercial model of the CNC designed for QNFT has a built-in QNFT program. When a subject with a probed respirator is connected with the unit, it starts a testing program that automatically samples from within and without the facepiece and automatically calculates the PF. The conventional QNFT units allow the operator control concerning the time necessary for each exercise and provide a visual record of the leakage pattern in each exercise. The CNC provides only a calculation of the PF for each set interval. Its portability and use of room air for a challenge atmosphere means it can be taken close to the work site and be available for QNFT quickly.

7.1.1.6 Quantitative Fit Test Use. QNFT use is not widespread outside the nuclear industry, although OSHA regulations requiring QNFT in industries that have asbestos or lead exposures have increased its usage. It has also been used as a research tool investigating the PFs provided by respirators.

Most of the PF research since 1970 has been done in the laboratory. Although many research laboratories are involved in these efforts today, E. C. Hyatt initiated these activities at LANL. The early QNFT research carried out at LANL used an exercise regime requiring light physical exertion to see if the face to facepiece seal would be broken during normal movements of the head, neck, and face. Respirator facepieces at the time these tests were run did not come in multiple sizes, but respirator testing represented a full range of facial sizes. Respirator studies found the half and full face masks available on the market at the time of the investigation were not able to fit adequately all the facial sizes present in the population (5). Tight-fitting SARs and SCBAs operated in the DM provided the user PFs similar to that provided by the equivalent air-purifying respirator (21–23); SAR-CF provided PFs lower than SAR-PD (23, 24); and SCBA-PD provided the best protection to the wearer (22, 25).

A report was issued with PFs assigned by respirator class known as assigned protection factors (APF), to be used when QNFT was not available (26). NIOSH and OSHA incorporated the PF data in their recommendations and regulations (27). As additional QNFT information became available both NIOSH and OSHA have edited their PF recommendations (28).

7.1.1.7 Quantitative Fit Test Problems. Continuing research in the laboratory and workplace has raised questions concerning the accuracy of APFs by respirator class. PF research in the workplace has shown that users obtain WPFs that range above and below the APFs. To date it has not been possible to establish a relationship between laboratory PFs obtained under conditions of light exercise and WPFs measured during workplace activities. Replicate tests on the same individual and the same respirator have shown the standard deviation of the PF varied widely between individuals (29).

Myers et al. have shown that the position of the facepiece sampling probe may influence the QNFT results (30). It has been assumed that a contaminant entering the facepiece through a leak in the facepiece seal will be thoroughly mixed with the air in the facepiece. Myers et al. has noted that streamlining can occur and the sample concentration removed from the facepiece is affected by the probe placement. The most widely used respirator sample probe location, near the mouth and nose just inside the respirator surface, appears to underestimate the actual contaminant concentration inhaled by the user.

The SCBA-PD is still considered to provide the best respiratory protection, but it has been found that pressure in the facepiece may be negative at high work rates.

7.1.1.8 Powered Air-Purifying Respirator: Correlation of Assigned Protection Factors and Workplace Protection Factors. PAPRs had been assigned as a class a PF of 1000. Three studies of helmet-type PAPR in the workplace found that the average WPFs were below 300 in one study and below 100 in the other two (31–33). In two laboratory studies that measured the PFs of two helmet PAPRs under strenuous work, PFs were above 900 for one study and below 100 for the other (34, 35).

Two studies of PFs for a tight-fitting PAPR at high work rates found the PF remained above 1000 at the high work rates (33, 34).

The design of the helmet respirator may make it susceptible to room air currents and points up the usefulness of the traditional neck cuff and cape of the SAH.

7.1.1.9 Protection Factors of Supplied Air Hoods, Supplied Air Respirators, and Self-Contained Breathing Apparatus in the Laboratory under Mild Work Conditions

Supplied-Air Hoods. Under laboratory conditions a study of 16 SAHs, using mild work exercises and an air supply of 170 l/m, found all achieved a PF of 500, 15 achieved a PF of 1000, and 14 achieved a PF of 2000. All hoods achieved a PF of 2000 at 400 l/m (23).

Skaggs et al. tested a SAH with an air supply of 170 l/m and found a PF of 20,000 under simulated work conditions. They noted reduced protection when a fan blew contaminant under an unsecured cape (35).

Supplied-Air Respirator. Of the SARs tested, fourteen tight-fitting SAR-CF, tested on a panel of 25 subjects representative of all facial sizes under laboratory conditions provided a PF of 50 to all subjects, a PF of 500 to 90 percent, and a PR of 5000 to 50 percent (22). If one poorly fitting respirator is omitted, then all had a PF of 500.

Seven tight-fitting SAR-PD, tested on a representative panel of 10 subjects, provided a PF of 2000, and six of the seven gave PFs of 10,000.

Six SCBA-PD tested on a representative panel of 10 provided a PF of 1000 to all, five of six provided a PF of 5000 to all and four of the six provided PFs of 10,000 to all. If the one poor fitting facepiece is omitted then all would receive a PF of 5000 (24).

The tight-fitting SAR-CF, SAR-PD, and SCBA-PD were not available in multiple size facepieces at the time of the tests.

7.1.1.10 Protection Factors at High Work Rates. SAR-PD and SCBA-PD rely on a positive pressure within the facepiece to provide a high PF. Research shows that at high work rates the facepiece pressure is not always positive. SARs have a set flow rate that does not vary unless the control valve is adjusted, but the SCBA flow varies depending on the pressure in the facepiece. The SCBA must be able to deliver up to 200 l/m upon demand, and some can deliver over 450 l/m (36). An air supply of 115 l/m, the minimum required for a SAR, is insufficient to maintain a positive pressure if the users' peak inspiratory flow (PIF) requirement exceeds 115 l/m and the pressure inside the facepiece will be negative for a brief period during the breathing cycle.

Self-Contained Breathing Apparatus. Stengel and Rodrigues bench-tested 14 SCBA-PD on a breathing machine set at a high work rate (minute volume of 100 l/m). Ten of the 14 SCBA-PD maintained a positive pressure in the facepiece when the tank pressure was full. At the alarm point, a signal to the user that the air supply is low and should be replaced soon, the positive pressure was maintained in only one of the six tested at this point (37).

Myhre et al. tested subjects wearing SCBA-PD at 65 and 80 percent MWR and found the pressure in the facepiece to be negative 8 percent and 36 percent of the time, respectively (38) Dahlback and Balldin, testing SCBA-PD, found negative pressure in the facepiece at 80 percent MWR (39).

Supplied-Air Respirators. Negative pressure has been measured within SAR-PD facepieces at work rates of 35 percent MWR, 65 percent MWR, 70 percent MWR, and 80 percent MWR (34, 40). At higher work rates the length of time the facepiece stays negative increases.

Supplied-Air Hoods. In the case of a SAH a negative pressure potential is created during the time of inhalation if the inspirational flow exceeds 170 l/m, the minimum approved flow. If the hood is rigid, than the intensity of the negative pressure within the hood will depend on the free volume in the hood, the airflow setting of the air supply, and the integrity of the neck cuff seal. The neck cuff should be tied as snugly as possible. If the SAH has a flexible covering, then an inhalation demand for more air than the air supply is delivering may not result in a negative pressure in the facepiece. Because the covering is flexible, part of the response to the negative pressure is a reduction in volume of the hood.

7.1.1.11 Work Rates. The quantity of air necessary for safe use of a SAR is that which will maintain a positive pressure inside the facepiece or hood. The exact amount of air is related to the work effort being performed—the harder the work, the more demand there is for air by the body. Wilson et al. recommend that SARs and SCBAs be designed to accommodate a PIF of 400 l/m because there are respirator users with such capabilities in the work force (41).

Although high work rates can produce negative pressure in the facepiece, how long can an individual sustain a high work rate? Raven estimates that the average individual can sustain a MWR for 3 to 5 min, a 70 percent MWR for 45 to 60 min, and a 50 percent MWR for an 8-hr period (42).

7.1.1.12 Assigned Protection Factors. Hyatt's "respirator protection factors" summarized early work on respirator efficiency and provided the first APF summary (26). The APFs should be reconsidered in light of WPF investigations and the potential of decreased PFs for SARs and PAPRs under high work rate conditions. There are PFs published by NIOSH, by OSHA, in specific regulations, and by respirator researchers, which can be consulted to choose an APF for a respirator class. The OSHA regulations should be followed as a minimum, but other information reviewed to see if the OSHA recommendation is current.

The APFs published by NIOSH are more conservative than OSHA (27). NIOSH believes a carcinogen exposure must have the best protection possible, an SCBA-PD. OSHA recognizes the problem of the carcinogen, but may allow an APR to be used.

7.2 Qualitative Fit Tests

The use of PFs presupposes that a QLFT program is carried out for personnel assigned to wear respirators. There is then reasonable assurance that the respirator fit is adequate to utilize the PF recommended.

QLFT is considered to be less accurate than QNFT, but successful passage of a QLFT has been used as evidence of some minimum level of protection. It is a test that needs a minimum of equipment and can be brought to the workplace. QLFTs have now been standardized so that their accuracy is improved, but the time necessary to administer the test has increased.

The QLFT methods expose the respirator user to an atmosphere with a contaminant, and the respirator is acceptable if the contaminant can not be detected by the wearer. These methods depend on the subjective opinion of the user to determine whether the respirator fits.

7.2.1 Qualitative Fit Test Agents

OSHA has accepted the three QLFT methods discussed below and published detailed instructions for their use (43). The instructions include the type of chamber, if any, to be used; the preparation of the test aerosol or vapor; the exercises to be carried out; and an item not stressed in the past, whether the user can detect the contaminant at low concentrations.

7.2.1.1 Isoamyl Acetate. Isoamyl acetate (IAA), also known as banana oil, is an organic vapor. It is used with an APR equipped with organic vapor cartridges. To use, the subject is first tested to see if the IAA can be detected at 2 ppm. If so, the respirator is donned and the subject enters a test hood that has a known amount of IAA in the air. The subject, as directed, carries out a series of head and breathing exercises. If the IAA is detected at any time, the test is terminated and declared an unsatisfactory fit.

There are disadvantages to this test. Care must be taken to prevent exposure of the subject to high concentrations of IAA before the test, for the olfactory bulbs are easily fatigued. The vapor is pleasant smelling, and test subjects may not signal immediately upon detecting the vapor. The cooperation and complete understanding of the test subject is vital.

7.2.1.2 Irritant Fume. Stannic chloride and titanium tetrachloride sublime upon exposure to moist air and produce an aerosol composed of intensely irritating fine particles (approximately 1 μm or less in diameter). These materials are available as smoke tubes that are marketed to check ventilation systems. Applied with caution, they can be used to conduct an excellent QLFT.

The respirator must be equipped with high-efficiency filters before starting the test. The subject adjusts the respirator and closes his eyes to begin the test. Smoke is blown toward the respirator from a distance of 2 ft. If no leakage is detected, the smoke tube is moved toward the subject. The subject, as directed, carries out a series of head and breathing exercises. If the irritant smoke is detected at any time, the test is terminated and declared an unsatisfactory fit.

A disadvantage of this test is the intensely irritating smoke. The test must be administered with care to prevent a high concentration inside the mask. The irritation is also an advantage because the subject reacts immediately on detecting the smoke.

7.2.1.3 Saccharin. Saccharin is used as a liquid aerosol produced by a hand nebulizer. It was developed for use with disposable respirators that did not have high-efficiency filters that could be used with the irritant fume test. The test depends on the subject being able to taste saccharin. To use, the subject is first tested to see if the saccharin can be detected by taste using a saccharin aerosol 1 percent as strong as the saccharin test solution. If the diluted saccharin solution is detected, the respirator is donned and the subject enters a test hood and a set quantity of saccharin aerosol is introduced. The subject, as directed, carries out a series of head and breathing exercises. If the saccharin is detected at any time, the test is terminated and declared an unsatisfactory fit.

7.3 Respirator Fit Checks

These are tests to be performed at the work site whenever the respirator is donned. The tests provide the user with knowledge that the respirator is seated properly on the face with no noticeable leaks.

Depending on the respirator in use, an assistant can introduce irritant smoke, an IAA-soaked swab, or saccharin aerosol close to the edge of the respirator to enable the user to check for leaks.

The following positive and negative pressure tests can be done by the user without assistance. Care must be taken not to disturb the respirator with pressure from the hands when making these checks.

The negative pressure test is performed on tight-fitting facepieces only, by covering the air inlet lightly, preferably with two pieces of light plastic, and inhaling slightly. If a leak exists, the slight vacuum can not be maintained. If leaks are found, adjust the respirator and try again.

The positive pressure test is performed by blocking the exhalation valve and exhaling lightly. If a leak is present, a slight positive pressure cannot be maintained. If leaks are found, adjust the respirator and try again.

8 PHYSIOLOGICAL CONSIDERATIONS IN THE USE OF RESPIRATORS

NIOSH (27) recommends, and in some cases OSHA requires, that a physician evaluate potential respirator users and determine whether they can use the respirator under consideration. OSHA places emphasis on the cardiopulmonary system and includes a pulmonary function examination. Before the examination the physician should be given information about the equipment to be used, its additional inspiratory and expiratory stress, the additional weight, whether it is used in conjunction with protective clothing that may cause an increase in the metabolic heat load, and what are the metabolic requirements of the task to be done.

Studies have been carried out to determine the effect of respirators on the cardiovascular and respiratory systems. Raven et al., testing groups with normal and abnormal pulmonary function on SAR-PD, noted some increase in systolic blood pressure at work rates above 50 percent (40). They noted no significant difference in response between the two groups at various work rates. They did find that at high work rates the facepiece pressure varied from -10 to 14 cm H_2O. Davis and Myhre and their co-workers found no significant cardiovascular response in using the SCBA at high work rates except the increased heart rate from carrying the 35-lb SCBA (44, 38).

Because the respirator represents a burden on the respiratory system, there has been concern about the degree of pulmonary impairment that would prevent the use of respirators. Raven et al. believe the workload of the task to be done must be matched with the pulmonary capacity of the employee while wearing the respirator (40). Wilson and Raven measured the pulmonary function of individuals with and without SAR-PD at various work rates. They outline a system that uses the maximum voluntary ventilation in 15 sec ($MVV_{.25}$) of the individual and the total energy cost of the task to be done to determine whether the individual can work in the respirator (45).

Some users experience breathing distress or claustrophobia while wearing a respirator (46). Recognition of such users should be made prior to entering life-threatening environments. Morgan and Raven, using an anxiety profile, are able to predict those who may experience problems wearing respirators in high work rate environments (47).

9 SELECTION OF THE PROPER RESPIRATOR

Respirator selection requires correctly matching the respirator with the hazard, the degree of hazard, and the user.

9.1 Preselection Information

The information required to choose a respirator is similar to that required for any industrial hygiene control method.

A. Material safety data sheet
B. Specific regulations or guidelines for the use of this material
C. Dust, mist, fume, vapor, or gas
D. Airborne concentration
E. IDLH concentration
F. PEL
G. Is the material detectable below the PEL?
H. Does material irritate the skin, nose, or eyes?
I. Does the material readily pass through the skin?
J. If the air contaminant is a gas or vapor, can the material be absorbed by a gas or vapor canister?
K. Carcinogen classification
L. Atmospheric oxygen concentration
M. PF required (contaminant concentration/PEL).

9.2 Respirator Selection

For assistance in the selection process there are three major resources:

1. OSHA General Respirator Regulations, 29CFR1910.134, and Respirator Regulations specific to a particular contaminant
2. NIOSH "Respirator Decision Logic" (28)
3. ANSI "Practices for Respiratory Protection" (48)

Compare the PF required with those of the selection guides. OSHA general rules do not provide assistance in choosing respirators by PF, but OSHA has some specific contaminant regulations that specify the respirators by PF levels. NIOSH has the most conservative PF recommendations. The ANSI Z88.2-1980 provides PF data based on older data and is in the process of revision. If the contaminant concentration is IDLH or there is an oxygen deficiency, the choice of respirators is limited to an SCBA.

Before deciding on an air-supplied or powered air-purifying respirator consider the estimated work rate for the task and any pulmonary function data on the respirator user. This can assist you in deciding on a minimum acceptable air flow to maintain positive pressure in the respirator.

10 TRAINING IN THE USE OF RESPIRATORS

The proper respirator is of no value without training for the user. The training should provide the user with the reason the respirator is needed, the effect of incorrect respirator use, and how to use the respirator correctly. The following is the minimum information which the respirator user must know. It is patterned after the training recommended by ANSI Z88.2-1980, "Practices for Respiratory Protection."

A. Explain why respiratory protection instead of engineering controls is being used. If engineering controls are being used, explain why they are not adequate, and what efforts are being made to reduce or eliminate the need for respirators.
B. Explain the health effects to the respirator user from the airborne contaminant.
C. Explain why the particular respirator class was selected.
D. Explain the operation and limitations of the respirators selected.
E. Explain how to inspect, don, check the fit, and wear the respirator selected by the user.
F. Ensure that each respirator wearer dons the respirator and wears it in a test atmosphere, and is qualitatively or quantitatively fit tested.
G. Explain how respirator maintenance and storage is done.
H. Explain how to recognize and cope with emergency situations.
I. Explain the respirator use regulations.
J. Explain the respirator use program.

11 CLEANING, MAINTENANCE, AND STORAGE OF RESPIRATORS

To ensure that the respirator remains serviceable a maintenance program must be in place the first day of respirator use. Respirators must be cleaned daily, and if used by different individuals, sanitized between uses. They must be inspected during cleaning, and all defective parts must be replaced before reuse. Finally, they must be stored with the goal of keeping them clean and in good condition until the next use.

There is no perfect way for these tasks to be accomplished. If only a few respirators are used, the user can clean and service the respirator. If there are many respirators used, it can be economical to set up a central respirator servicing program. This ensures that all the respirators receive the same service and reduces the surveillance tasks for management. Whether the task is carried out by the individual user or the central cleaning station, time is necessary to carry out this service and is a cost of the protection program. If this cost is not allocated, respirators will not be properly maintained and their usefulness will be questionable.

Instructions included with the respirator provide information on cleaning and maintenance. For half and full face masks, washing with soap and water, a disinfectant, and a water rinse not to exceed 120°F, along with a careful parts inspection, are usually all that is required. For complex apparatus, such as airline pressure reducing valves, check with the manufacturer about the users' maintenance responsibilities.

12 AIR SUPPLY FOR SELF-CONTAINED BREATHING APPARATUS AND AIRLINE RESPIRATORS

Breathing air for SARs and SCBAs can be supplied by air compressors or compressed air bottles. OSHA and NIOSH require breathing air meet the grade D standards of the Compressed Gas Association (49). Grade D standards are met if the following contaminant levels are not exceeded: 20 ppm carbon monoxide, 1000 ppm carbon dioxide, and 5 mg/m^3 oil mist.

There are two sources of contamination, the air intake and the air compression system.

12.1 The Air Intake

The air in the compressor system will contain the contaminants present at the air intake. Care should be taken to see that the air intake location is distant from any source of contamination such as automobile exhaust gases. In urban areas with serious air pollution problems it may be difficult to provide an air intake location free from contamination.

12.2 Air Compressors

The most common air compressor used for breathing air is the oil-lubricated compressor, which adds the oil mist and hydrocarbon vapors to the compressed air. If for some reason the compressor overheats, it is possible for partial combustion of the oil to occur and generate carbon monoxide. There are filters and sorbents available to remove the oil mist and hydrocarbon vapors from the compressed air, but carbon monoxide removal from the compressor airstream is more complex.

12.2.1 Carbon Monoxide

Carbon monoxide catalytic removal units contain Hopcalite to convert CO to CO_2. Because moisture ruins Hopcalite, drying units protect the Hopcalite. Dual drying towers are installed on some Hopcalite units so that while one is removing CO from the air supply, the other is being dried. When dry it will automatically cycle to replace the other unit.

Carbon monoxide catalysis units are not mandatory on oil-lubricated systems but periodic monitoring for CO is required. In addition, the compressor must be equipped with a high-temperature alarm to notify personnel that the compressor has overheated. Although not required, continuous-reading CO monitors should be installed to monitor the CO concentration in the breathing air.

12.2.2 Nonoil-lubricated Compressors

A solution to the potential problem of carbon monoxide is to use a nonoil-lubricated compressor. The diaphragm pump, oilless compressors with graphite or Teflon piston rings, and water-lubricated compressors are available.

12.3 Compressed Air Bottles

SAR-PD, which does not require a continuous flow of air, can use bottled air instead of an on-site air compressor. Breathing air, certified to meet grade D standards, can be purchased in cylinders holding more than 6000 liters. If the work requires only modest physical effort, it is possible that a bottle of air may last for 2 to 4 hr. For users needing small quantities of breathing air, purchasing bottled air means no need for a breathing air compressor or to install breathing air monitors and filters.

13 SPECIAL PROBLEMS WITH RESPIRATORS

13.1 Respirators and Hair

It is imperative that clean, smooth skin be in contact with the respirator sealing surface because even a mild growth of facial hair may interfere with this seal. Stobbe et al., in a review of facial hair and respirator fit studies, found that the face seal of negative pressure respirators was adversely affected by facial hair (50). The American Industrial Hygiene Association Board of Directors has approved and published a position paper of its Respirator Committee to the effect that facial hair in or along the face seal area of a tight-fitting respirator should not be permitted (51).

Restrictions on facial hair is an emotional subject to some, and company policies prohibiting facial hair, not carefully planned and carried out, have been successfully challenged (52). A policy statement prohibiting facial hair that interferes with wearing a respirator should be concise and include only respirator users. It must be part of a respirator program, but not the only part enforced.

13.2 Cold Temperatures

In cold temperatures problems of facepiece flexibility, visibility, and frozen valves must be considered. Those who work in cold environments should ask the manufacturer about respirators for cold weather. The facepiece will fog in cold weather when exposed to warm exhaled breath. A nose cup, which must be used at temperatures below 32°F, is available for all full face respirators to keep the warm air away from the vision port. Valves and regulators should be carefully checked before the respirator is used to ensure that they are functioning satisfactorily.

13.3 Corrective Lenses

Temple bars on eyeglasses interfere with the seal of full facepiece respirators and are not acceptable. Eyeglass kits are available from all manufacturers for use with full facepieces. If half and quarter masks interfere with proper seating of eyeglasses, the respirator fit is not satisfactory, and either another respirator or another style of eyeglass must be selected.

13.4 Communications

Communication through a facepiece is difficult, and trying to speak loudly may affect the face seal. There are several devices available to assist the respirator wearer. Some respirators have built-in speaking diaphragms. Although they are protected, care must be taken that the thin diaphragm is not pierced by thin wires or hot sparks. Wireless units are available that can be placed inside full face masks for radio communication. Throat and ear microphones are also available. None of these units require penetration of the facepiece that can affect the respirator approval. Use of any electrical device must also consider potential explosive problems if flammable vapors are present.

14 RESPIRATOR CERTIFICATION

14.1 Approved Respirators

The term "approved respirator" has become a part of the language of industrial hygiene. In 1920 the Bureau of Mines (BM) began to publish "Approval Schedules" for specific types of respirator, and approved their first respirator the same year (1). Manufacturers voluntarily submitted their devices for testing, which, if satisfactory, could be sold as BM approved.

In 1972 the testing and approval program was assumed by the Certification Branch (CB), Division of Safety Research, NIOSH, in Morgantown, West Virginia. The Mine Safety and Health Administration, U.S. Department of Labor, jointly approves respiratory equipment used in mining.

The approval categories and some of the approval requirements are given below. Table 19.3 summarizes fit test requirements in each category.

14.1.1 Subpart H: Self-Contained Breathing Apparatus

Devices can be open- or closed-circuit, in demand or pressure demand modes of operation. The maximum weight for open-circuit devices is 35 lb. The open-circuit SCBA must, at an inhalation resistance of 51 mm H_2O and an air pressure of 500 psi, deliver a minimum air flow of 200 l/m. The air supply can be a minimum of 3 min and a maximum of 240 min. Combination SAR and auxiliary SCBA units can be approved for use in IDLH atmospheres.

14.1.2 Subpart I: Gas Masks

Gas masks are defined as full facepieces attached to a canister that can be mounted on the front, back, or chin. An exception is the escape mask, which may have a half mask or a mouthpiece and nose clip. Table 19.1 lists the canister test agents. The regulations list only five test agents, but NIOSH may, on request, establish approval procedures for other materials. Multisorbent cartridges for respirators must meet the minimum service life for each sorbent, but multisorbent canisters for gas masks may reduce the minimum service life of each sorbent by half. The maximum inhalation and exhalation resistances are 85 and 20 mm H_2O, respectively.

14.1.3 Subpart J: Supplied Air Respirators

Supplied-air respirators are divided into three classes: classes A and B are hose masks, not widely used, and class C are the half masks, full face masks, and hoods or helmets supplied by a hose with air under pressure not to exceed 125 psi. Class C may be approved in demand, pressure demand, or continuous flow modes. Minimum air flows are 115 l/m for tight-fitting facepieces and 170 l/m for hood and helmets, and maximum air flow is 425 l/m. CE refers to devices for abrasive blasting that have outer coverings. Hoses for class C devices can be obtained in 15-, 25-, or 50-ft lengths up to a maximum of 300 ft. The maximum inhalation and exhalation resistances are both 50 mm H_2O.

14.1.4 Subpart K: Dust, Fume, and Mist Respirators

Approvals are granted for all types of facepieces under this section. In addition to half and full facepieces, hoods and helmets can be used with powered air-purifying units. The filters attached to the facepieces may be for protection against dust and/or fume and/or mist. The approval depends on the aerosol that is used for the test. Table 19.4 summarizes the test aerosols.

Single-use respirators are approved for use with dusts producing pneumoconiosis and fibrosis or other dusts of low toxicity. A fit test is not required for approval of single-use respirators, but they must pass the standard silica dust and silica mist test.

The maximum inhalation resistances range from 15 to 50 mm H_2O. An approval for asbestos fibers and radon daughters can be obtained if the filter resistance does not exceed 25 mm H_2O.

14.1.5 Subpart L: Chemical Cartridge Respirators

Chemical cartridge respirators are similar to gas masks in subpart I, but the quantity of available sorbent material is less. They are available as half masks, full face masks, and hoods or helmets. Although some of the approval test gases and vapors

Table 19.3. Respirator Fit Test Requirements for NIOSH Approval, 30CFR11

30CFR11 Subpart	Type of Facepiece	Test	Number of People Required for Test	Total Time of Test (min)	Exercises Required
Subpart H, SCBA	Half mask, full face mouthpiece	1000 ppm, isoamyl acetate QLFT	6	2	None required
Subpart I, gas masks	Half mask	100 ppm, isoamyl acetate QLFT	Not specified	8	Four 2-min exercises specified
	Full face mask and mouthpiece	1000 ppm, isoamyl acetate QLFT	Not specified	8	Four 2-min exercises specified
Subpart J, SAR					
Types A, B, C	Half mask, full face, hoods and helmets	1000 ppm, isoamyl acetate QLFT	Not specified	10	Two 5-min exercises specified
Type CE, approved for abrasive blasting	Full face, hoods and helmets	DOP, QNFT	Not specified	Not stated	Seven exercises, routine specified
Subpart K, dust, fume, and mist respirators					Not specified
If PEL > 0.05 mg/m^3					
Dust and mist	All facepieces, pow-	No fit test required	Not required	None	Not required

respirators	ered and nonpowered				
Fume respirators	All facepieces, powered and nonpowered	100 ppm, isoamyl acetate QLFT	Not specified	2	Not specified
If PEL is <0.05 mg/m^3	Half mask, powered and nonpowered	100 ppm, isoamyl acetate QLFT	Not specified	5	One 2-min and one 3-min exercise
If PEL is <0.05 mg/m^3	Full face, hoods, helmets and mouthpiece, powered and nonpowered	1000 ppm, isoamyl acetate QLFT	Not specified	5	One 2-min and one 3-min exercise
Subpart L, chemical cartridge respirators	Half mask	100 ppm, isoamyl acetate QLFT	Not specified	8	Four 2-min exercises specified
	Full face, hoods and helmets and mouthpiece, powered and nonpowered	1000 ppm, isoamyl acetate QLFT	Not specified	8	Four 2-min exercises specified
Subpart M, pesticide respirators—same as Subpart L.					
Subpart N, vinyl chloride respirators—same as Subpart L.					

Table 19.4. Respirator Filter Test Requirements, 30CFR11

	PEL above 0.05 mg/m^3			PEL below 0.05 mg/m^3— High-Efficiency Filter
	Dust Filter	Fume Filter	Mist Filter	
Aerosol test conditions	Silica dust with a geometric mean diameter of 0.4–0.6 μm; concn 50–60 mg/m^3	Molten lead with oxygen gas torch impinging on surface; concn 15–20 mg/m^3 as lead	Aqueous silica solution, passes 270 mesh; concn 20–25 mg/m^3 as silica	100 mg/m^3 of DOP at flow rates of 32 l/m and 85 l/m for 5–10 sec
Respirator filters				
Subpart K				
Respirator with filter	Test for 90 min at 32 l/m. Dust through filter must not exceed 0.52 mg/m^3	Test for 312 m at 32 l/m; lead through filter must not exceed 0.15 mg/m^3	Test for 312 m at 32 l/m; mist through filter must not exceed 0.25 mg/m^3	The DOP leakage shall not exceed 0.03% of the challenge concentration
Powered respirator, tight fitting	Test for 240 min at 115 l/m. Dust through filter must not exceed 0.52 mg/m^3	Test for 240 min at 115 l/m. Lead through filter must not exceed 0.15 mg/m^3	Test for 240 min at 115 l/m. Mist through filter must not exceed 0.25 mg/m^3	The DOP leakage shall not exceed 0.03% of the challenge concentration
Powered respirator, loose fitting	Test for 240 min at 170 l/m. Dust through filter must not exceed 0.52 mg/m^3	Test for 240 min at 170 l/m. Lead through filter must not exceed 0.15 mg/m^3	Test for 240 min at 170 l/m. Mist through filter must not exceed 0.25 mg/m^3	The DOP leakage shall not exceed 0.03% of the challenge concentration

Single-use respirator	Test for 90 min at 40 l/m. Dust through filter must not exceed 0.50 mg/m^3	No test scheduled	No test scheduled	No test scheduled
Subpart M				
Pesticide cartridge respirators	No test scheduled	Test for 90 min at 32 l/m; lead through filter must not exceed 0.15 mg/m^3	No test scheduled	No test scheduled
Respirator with canister	No test scheduled	Test for 90 min at 32 l/m; lead through filter must not exceed 0.15 mg/m^3	No test scheduled	The DOP leakage shall not exceed 0.03% of the challenge concentration
Powered respirator, tight fitting	No test scheduled	Test for 240 min at 115 l/m. Lead through filter must not exceed 0.15 mg/m^3	No test scheduled	The DOP leakage shall not exceed 0.03% of the challenge concentration
Powered respirator, loose fitting	No test scheduled	Test for 240 min at 170 l/m. Lead through filter must not exceed 0.15 mg/m^3	No test scheduled	The DOP leakage shall not exceed 0.03% of the challenge concentration

for chemical cartridge respirators are the same as those used to test gas masks, the test concentrations and the minimum sorbent life-span differ. Combination cartridges for paint spraying also are approved. The maximum inhalation and exhalation resistances are 70 and 20 mm H_2O, respectively.

14.1.6 Subpart M: Pesticide Respirators

A pesticide respirator consists of an organic vapor respirator approved in Subpart L and an aerosol filter. The aerosol filter for the cartridge must pass the lead fume test. The gas masks aerosol filter must pass the DOP aerosol test for high-efficiency filters. The maximum inhalation and exhalation resistances are 85 and 20 mm H_2O, respectively. Not approved for use with fumigants in gas or vapor form.

14.1.7 Subpart N: Special-Purpose Respirators

Only vinyl chloride respirators are classified as special-purpose devices at this time. An end of service life indicator is required, and the length of service life for the canister is specified by the manufacturer.

REFERENCES

1. B. J. Held, "History of Respiratory Protective Devices in the U. S., Pre World War I," Lawrence Livermore National Laboratory, Energy Research and Development Administration, contract W-7405-Eng-48.
2. W. P. Yant, "Bureau of Mines Approved Devices for Respiratory Protection," *J. Ind. Hyg.*, 473–480 (Nov. 1933).
3. J. T. McConville and E. Churchill, "Human Variability and Respirator Sizing," National Institute for Occupational Safety and Health, March 1976.
4. A. L. Hack and J. T. McConville, "Respirator Protection Factors: Part I—Development of an Anthropometric Test Panel," *Am. Ind. Hyg. J.*, **39**(12), 970–975 (1978).
5. D. D. Douglas, W. Revoir, J. A. Pritchard, A. L. Hack, L. A. Geoffrion, T. O. Davis, P. L. Lowry, C. P. Richards, L. D. Wheat, J. M. Bustos, and P. R. Hesch, "Respirators Studies for the National Institute for Occupational Safety and Health, July, 1974 through June 30, 1975," Los Alamos Scientific Laboratory Report No. LA-6386-PR, August 1976.
6. L. W. Ortiz, S. C. Soderholm, and F. O. Valdez, "Penetration of Respirator Filters by an Asbestos Aerosol," *Am. Ind. Hyg. Assoc. J.*, **49**(9), 451–460 (1988).
7. Code of Federal Regulations, 30/CFR/11.140-5, .140-6, .140-7, .140-11.
8. L. Reed, D. L. Smith, and E. S. Moyer, "Comparison of Respirator Particulate Filter Test Method," *Int. Soc. Respir. Protect. J.*, **4**(3), 43–60 (1986).
9. R. A. Fulton, F. F. Smith, and R. L. Busbey, "Respiratory Devices for Protection Against Certain Pesticides," ARS-33-76, U.S. Department of Agriculture, 1962.
10. G. O. Nelson and C. A. Harder, "Respirator Cartridge Efficiency Studies." *Am. Ind. Hyg. Assoc. J.*, **35**(7), 491–510 (1974).
11. J. J. Bloomfield and L. Greenburg, *J. Ind. Hyg.*, **15**(4) (July 1933).

12. E. C. Hyatt, J. A. Pritchard, A. L. Hack, B. J. Held, P. L. Lowry, D. A. Bevis, T. O. Moore, T. O. Davis, C. P. Richards, L. A. Geoffrion, and L. D. Wheat, "Respirators Studies for the National Institute for Occupational Safety and Health, July 1, 1972 through June 30, 1973," Los Alamos Scientific Laboratory Report No. LA-5620-PR, May 1974.
13. E. C. Hyatt and C. P. Richards, "A Study of Facepiece Leakage of Self-Contained Breathing Apparatus by DOP Man Tests Progress Report, July 1, 1971 through February 29, 1972," Los Alamos Scientific Laboratory Report No. LA-4927-PR, April 1972.
14. E. S. Kolesar, Jr., "Respirator Qualitative/Quantitative Fit Test Method Analysis," Aeromedical Review 2-80, USAF School of Aerospace Medicine, Aerospace Medical Division (AFSC), Brooks Air Force Base, Texas 78235, August 1980.
15. W. C. Hinds, J. M. Macher, and M. W. First, "Size Distribution of Aerosols Produced by the Laskin Aerosol Generator Using Substitute Materials for DOP," *Am. Ind. Hyg. Assoc. J.*, **44**(7), 495–500 (1983).
16. F. E. Adley and D. E. Wisehart, "Methods for Performance Testing of Respiratory Protective Equipment," *Am. Ind. Hyg. Assoc. J.*, **23**(4), 251–256 (1962).
17. R. A. da Roza, A. H. Biermann, and C. A. Cadena-Fix, "Simulated Workplace Protection Factor Measurement," Presented at the 1985 International Symposium on Respirator Test Technology, U.S. Army Chemical Research and Development Center, Baltimore, MD, October 16, 1985.
18. P. L. Lowry, C. P. Richards, L. A. Geoffrion, S. K. Yasuda, L. D. Wheat, J. M. Bustos, and D. D. Douglas, "Respirators Studies for the National Institute for Occupational Safety and Health," January 1–December 31, 1977, Los Alamos Scientific Laboratory Report No. LA-7317-PR, June 1978.
19. R. A. Bentley, G. J. Bostock, D. J. Longson, and M. W. Roff, "Determination of the Quantitative Fit Factors of Various Types of Respiratory Protective Equipment," *J. Int. Soc. Respir. Protect.*, **2**(4), 313–337, 1984.
20. R. Laye, "Evaluation of a Miniaturized Condensation Nucleus Counter for Measurement of Respirator Fit Factor," *J. Int. Soc. Respir. Protect.*, **5**(3), 1–8 (1987).
21. A. Hack, A. Trujillo, O. D. Bradley, and K. Carter, NUREG/CR-1586, LA-8432-MS, "Evaluation and Performance of Escape Type Self-Contained Breathing Apparatus," Los Alamos Scientific Laboratory, July 1980.
22. A. Hack, A. Trujillo, O. D. Bradley, and K. Carter, NUREG/CR-2652, LA-9266-MS, "Evaluation and Performance of Closed-Circuit Breathing Apparatus," Los Alamos Scientific Laboratory, April 1982.
23. A. L. Hack, O. D. Bradley, and A. Trujillo, "Respirator Protection Factors: Part II—Protection Factors of Supplied-Air Respirators," *Am. Ind. Hyg./Assoc. J.*, **41**(5), 376–380 (1980).
24. D. D. Douglas, P. R. Hesch, and P. L. Lowry, LA-NUREG-6612-MS, "Supplied-Air Hood Report," Los Alamos Scientific Laboratory, December 1976.
25. A. Hack, A. Trujillo, O. D. Bradley, and K. Carter, NUREG/CR-1235, LA-8188-MS, "Evaluation and Performance of Open-Circuit Breathing Apparatus," Los Alamos Scientific Laboratory, January 1980.
26. E. C. Hyatt, "Respirator Protection Factors," Los Alamos Scientific Laboratory Report No. LA-6084-MS, January 1976.
27. Standards Completion Program, "Respirator Decision Logic," National Institute for Occupational Safety and Health, Occupational Safety and Health Administration, 1975.

28.. W. A. Myers, N. J. Bollinger, T. K. Hodous, N. A. Leidel, S. H. Rabinovitz, and L. D. Reed, *Respirator Decision Logic*, National Institute for Occupational Safety and Health, Publication No. 87-108, May 1987.

29. R. A. da Roza, C. A. Dadena-Fix, G. J. Carlson, K. E. Hardis, and B. J. Held, "Reproducibility of Respirator Fit as Measured by Quantitative Fitting Tests," *Am. Ind. Hyg. Assoc. J.*, **44**(11), 788–794 (1983).

30. W. A. Myers, J. Allender, R. Plummer, and T. Stobbe, "Parameters that Bias the Measurement of Airborne Concentration within a Respirator," *Am. Ind. Hyg. Assoc. J.*, **47**(2), 106–114 (1986).

31. W. A. Myers, M. J. Peach III, K. Cutright, and W. Iskander, "Workplace Protection Factor Measurements on Powered Air-Purifying Respirators at a Secondary Lead Smelter: Results and Discussion," *Am. Ind. Hyg. Assoc. J.*, **45**(10), 681–684 (1984).

32. S. S. Que Hee and P. Lawrence, "Inhalation Exposure of Lead in Brass Foundry Workers: The Evaluation of the Effectiveness of Powered Air-Purifying Respirators and Engineering Controls," *Am. Ind. Hyg. Assoc. J.*, **44**(10), 746–751 (1983).

33. L. W. Grauvogel, Summary Report, "Effectiveness of a Positive Pressure Respirator for Controlling Lead Exposure in Acid Storage Battery Manufacturing," *Am. Ind. Hyg. Assoc. J.*, **47**(2), 144–146 (1986).

34. R. A. da Roza, C. A. Dadena-Fix, and J. E. Kramer, "Powered Air-Purifying Respirator Study, Final Report," UCRL-53757, Lawrence Livermore National Laboratory, July 1986.

35. B. J. Skaggs, J. M. Loibl, K. D. Carter, and E. C. Hyatt, "Effects of Temperature and Humidity on Respirator Fit Under Simulated Work Conditions," Los Alamos National Laboratory NUREG/CR-5090, LA-11136, July 1988.

36. T. O. Davis, O. Bradley, A. Trujillo, and D. D. Douglas, "Respirator Studies for the DOE Division of Operational and Environmental Safety," October 1, 1976– September 30, 1977, Los Alamos Scientific Laboratory Report No. LA-6969-PR, April 1978.

37. J. W. Stengel and R. Rodrigues, "Machine Testing of Self-Contained Breathing Apparatus at a High Work Rate Typical of Firefighting," *J. Int. Soc. Respir. Protect.*, **2**(4), 362–368 (1984).

38. L. G. Myhre, R. D. Holden, F. W. Baumgardner, and D. Tucker, "Physiological Limits of Firefighters," ESL-TR-79-06, Engineering and Services Laboratory, U.S. Air Force Engineering and Services Center, January 1979.

39. G. O. Dahlback and U. I. Balldin, "Physiological Effects of Pressure Demand Masks During Heavy Exercise," *Am. Ind. Hyg. Assoc. J.*, **45**(3), 177–181 (1984).

40. P. B. Raven, O. Bradley, D. Rohm-Young, F. L. McClure, and B. Skaggs, "Physiological Response To 'Pressure-Demand' Respirator Wear," *Am. Ind. Hyg. Assoc. J.*, **43**(10), 773–781 (1982).

41. J. B. Wilson, P. B. Raven, W. P. Morgan, S. A. Zinkgraf, R. G. Garmon, and A. W. Jackson, "Effects of Pressure-Demand Respirator Wear on Physiological and Perceptual Variables during Progressive Exercise to Maximal Levels," *Am. Ind. Hyg. Assoc. J.*, **50**(2), 85–94 (1989).

42. Peter B. Raven, Department of Physiology, Texas College of Osteopathic Medicine, personal communication, 1989.

43. 29CFR1910.1028(n), *Federal Register*, 56/176, 34562-34578, September 11, 1988.

44. T. O. Davis, P. B. Raven, C. L. Shafer, A. C. Linnebur, J. M. Bustos, L. D. Wheat,

and D. D. Douglas, "Respirator Studies for the ERDA Division of Safety, Standards, and Compliance, July 1, 1975–June 30, 1976," Los Alamos Scientific Laboratory Report No. LA-6733-PR, March 1977.

45. J. B. Wilson and P. B. Raven, "Clinical Pulmonary Function Tests as Predictors of Work Performance during Respirator Wear," *Am. Ind. Hyg. Assoc. J.*, **50**(1), 51–57 (1989).
46. W. P. Morgan, "Psychological Problems Associated with the Wearing of Industrial Respirators," *Am. Ind. Hyg. Assoc. J.*, **44**(9), 671–675 (1983).
47. W. P. Morgan and P. B. Raven, "Prediction of Distress for Individuals Wearing Industrial Respirators," *Am. Ind. Hyg. Assoc. J.*, **46**(7), 363–368 (1985).
48. American National Standard Institute, ANSI Z88.2-1969, Practices for Respiratory Protection.
49. CGA Specification G-7.1, (ANSI Z86.1-1972), "Commodity Specification for Air," Compressed Gas Association.
50. T. J. Stobbe, R. A. da Roza, and M. A. Watkins, "Facial Hair and Respirator Fit: A Review of the Literature," *Am. Ind. Hyg. Assoc. J.*, **49**(4), 199–203 (1988).
51. American Industrial Hygiene Association Respiratory Protection Committee, "Facial Hair and Tight Fitting Respiratory Protection," *Am. Ind. Hyg. Assoc. J.*, **49**(4), A-276 (1988).
52. G. L. Holt, "Employee Facial Hair Versus Employer Respirator Policies," *J. Appl. Ind. Hyg.*, **2**(5), 200–203 (1987).

CHAPTER TWENTY

Agricultural Hygiene

William Popendorf, Ph.D., C.I.H., and Kelley J. Donham, D.V.M.

AGRICULTURE

Modern agriculture, as practiced by Western cultures, is not the bucolic, healthful working environment fantasized in media classics such as Laura Ingalls Wilder's *Farmer Boy* (1). It certainly was, and most places still is, a disciplined life style that has bred strong character and moral instincts in virtually every culture. Agriculture in some cultures (and indeed some crops within all cultures) has yet to make the transition from human to animal to mechanized power, and from mechanical to chemical to genetically engineered tools. Where these transitions have occurred, production per farmer has increased (2); unfortunately so have health and safety stresses upon farmers. For various reasons the traditional industrial hygiene phases of recognition, evaluation, and control of health (and safety) hazards in agriculture have lagged behind those in general industry. In contrast to past media depictions, a series of articles entitled "A Harvest of Harm" describing the hazards of modern farm life recently earned Tom Knudson a Pulitzer Prize for national reporting (3). Table 20.1 gives a topical overview of injurious and physical agents, biological and chemical agents, and diseases of concern in agriculture.

This chapter first describes the dilemma of agriculture as an industry and as a way of life. An array of agents and health hazards characteristic of, if not unique to, agriculture are then reviewed. The chapter concludes with a discussion of the applicability of traditional industrial hygiene approaches to agriculture, potential strategic policies to stimulate industrial hygiene and safety services on the farm and ranch, and approaches employed in other countries to deliver such services.

Patty's Industrial Hygiene and Toxicology, Fourth Edition, Volume 1, Part A, Edited by George D. Clayton and Florence E. Clayton
ISBN 0-471-50197-2

Table 20.1. Overview of Safety and Physical Agents, Biological and Chemical Agents, and Diseases of Concern in Agricultural Hygiene[a]

Safety and Physical Agents	Biological and Chemical Agents	Diseases
Anhydrous ammonia	Asphyxiation/suffocation	Dermatoses
Commodity storage and transfer	Confined space	Chemical related
Electricity	Fumigation	Infectious related
Ergonomics	Silos	Insect related
Back injury	Carcinogens—cancer agents	Livestock related
Lifting	Dusts (inorganic aerosols)	Plant related
Repetitive trauma	Hydrogen sulfide (manure gas)	Immunologic diseases
Farm machinery	Microbiologic organisms	Allergic rhinitis
Balers	Infectious—communicable	Asthma
Chainsaws	Noninfectious bioaerosols	Skin
Combines	Parasites	Noninfectious diseases
Power take-off (PTO)	Nitrogen dioxide (e.g., silo filling)	Cancer
Rollover protection (ROPS)	Organic dusts	Hypertension and heart
Safety guards	Cotton dust	Respiratory system
Tractors	Endotoxin	Asthma (also immunologic diseases)
Fire	Grain dust	Bagassosis
Fuel storage (leaks and fires)	Mold spores (e.g., farmer's lung)	Bronchitis
Illumination (lighting)	Silo unloader's disease	Byssinosis
Lightning	Sugarcane (bagassosis)	Hypersensitivity pneumonitis
Liquified propane (LP) gas	Wood dust	Organic dust toxic syndrome (ODTS)
Livestock handling injuries	Pesticide activities	Pneumoconiosis
Physical environmental hazards	Applicator exposure hazards	Silo filler's disease
Noise	Dermal exposure—absorption	Organophosphate poisoning
Thermal (heat and cold)	Harvester reentry hazards	Zoonotic diseases
Vibration	Nontarget exposure hazards	
Psychological stress	Environmental hazards	
Sanitation (field)	Pesticides	
Transportation (highway)	Carbamates and thiocarbamates	
Welding	Chlorinated insecticides	
	Organophosphates	
	Phenoxy-aliphatic acid herbicides	
	Triazine herbicides	
	Waste handling (see hydrogen sulfide)	

[a]Adapted from ACGIH, 1984 (4).

	Intensity During Harvest	
	Low	High
Low	1. Mechanized crops	2. Bush and tree crops
High	3. Livestock	4. Row fruits and vegetables

Figure 20.1. Labor intensity during growing and harvest phases of agricultural production.

1.1 Agriculture as an Industry

The fact that the number of farms and full-time farmers has been consistently decreasing following World War I does not belittle the fact that agriculture still represents the largest occupational group in the United States, with some 10 to 20 million people, depending upon the criteria for "agriculture" (5). Consider the following:

- 2 million persons are solely or primarily self-employed in agriculture.
- 2.7 million are hired farm workers, of which approximately half are migrant or temporary farm workers.
- 3.1 million are part-time self-employed in agriculture.
- 6 million are family members living, and often also working, on these farms.

In addition, there are an estimated 0.5 to 1.0 million undocumented migrant workers, and an estimated 8 million workers in agribusiness, many of whom share similar exposures to production agriculture workers. Twenty years ago agriculture was the third most hazardous occupation (6–7). Today agriculture is the most hazardous occupational group in the United States, and yet it is virtually devoid of occupational health and safety services (3, 5, 7–10).

The intrinsically seasonal nature of agriculture not only causes the size of this work force to vary temporally and often geographically via migrant work groups (11), but it usually also has major effects on the nature and intensity of the work itself. Figure 20.1 diagrams the wide temporal diversity of manual labor inputs among various crops. The least labor-intensive category is the mechanized commodities (such as cotton, grain, processed tomatoes, and nut crops), which require minimum manual tending while growing and are also highly automated during harvest. The second category is bush and tree crops (such as coffee, tobacco, stone fruit, and citrus) which are not labor intensive during their growing phases but require high manual labor during harvest. The third is livestock, which tends to be broadly labor intensive throughout the production phases but has no harvest peak. The fourth category is row crops (such as strawberries, lettuce, and flowers), which are labor intensive throughout most of their growing and harvest phases. Of course, labor intensities of crops within these categories vary, and production practices vary around the world or even within regions for various reasons, but these categories demonstrate the distinctively wide diversity in the intensity and

temporal frequency of exposure to hazards that makes it difficult to generalize about agriculture as a whole.
cm

1.2 Agriculture as a Way of Life

Like general industry, modern agriculture is not a static industry. The hazards facing modern farmers are different from those facing farmers one or two generations ago. Agriculture is also unlike general industry socially, economically, psychologically, and geographically. This hinders the recognition of and response to sometimes hidden health and safety hazards of new technologies. Examples of major differences include the following (12):

- The workplace and the residence are co-located. Thus the hours of work are as long as necessary, and many of the hazards that affect the producer also affect the family, including the children.
- As self-employers or very small businesses, there is little or no distinction between management and labor, few legal or pre-selection barriers to entry such as age, sex, or even ability (versus desire), and no employee benefits such as sick leave, medical insurance, or workers' compensation.
- The intrinsically "risky" nature of agriculture as a business coupled with the inability of the farmer to change prices to reflect costs provides limited incentives to purchase, install, or maintain preventive safety and health controls. The agricultural economy forces farming to be highly self-selective for the physically fit and psychologically motivated.
- Similar economic pressures cause most of the U.S. agricultural work force to be medically underinsured, often uninsured, and almost universally underserved in terms of their proximity to medical and professional preventive services.
- Agriculture is a geographically dispersed industry with many small "factories" spread over a broad region. Thus most incidents of disease and injury occur in solitary, often isolated settings, not only aggravating epidemiologic "recognition" but also delaying discovery of the victim, access to immediate medical service, and later the provision of rehabilitation services.
- The psychological stoic self-image of farming as an independent life style rather than a business further heightens the disinterest of farmers in preventive health services and of governmental or organized efforts to provide such services. Only in the case of highly labor-intensive row crops (the fourth category above) have organized (union) activities contributed toward implementing effective preventive health and safety policies.

Coupled with these characteristics is the rate of structural transition within American agriculture. A number of feature articles (2, 11, 13) and a recent government report (14) describe economic-political, institutional, and technological forces contributing to consolidation and losses of roughly half of all individual and family farms every 20 years. These forces include inflation, foreign exchange rate, tax policies, subsidy programs, and economies of scale; farm programs, farm organizations, vertical integration; the availability of human capital; and genetic engineering, respectively.

2 AGRICULTURAL HEALTH AND SAFETY PROBLEMS

2.1 General Health Status

The occupational health status of individual farmers is linked to their general health status, and that in turn to their living environment and socioeconomic status. The socioeconomic status of the U.S. farm population is extremely varied. On the low end (annual sales of <$100,000), there are the migrant farm workers and small farms of Appalachia and the Deep South. In the middle (annual sales of $100,000–$250,000) are the medium-sized family farms of the greater Midwest and Northeast sections of the United States. At the highest end (>$250,000 annually) are the large corporate farms scattered across the Midwest, South, and Far West. Therefore control programs must also consider a farmer's socioeconomic status.

As a group, migrant workers have an increased incidence of infectious and parasitic diseases, and their living conditions are often marked by substandard housing and poor environmental sanitation (15, 16). As summarized by West (17) in her review of 1962 agricultural occupational disease reports in California, "Because of migrant status, seasonal work, language barriers, substandard education, marginal health, and poor hygiene, they [migrant workers] are the least able of any group to protect themselves against occupational hazards, particularly agricultural chemicals."

Middle-class self-employed farmers actually have a more favorable outcome statistic for certain major health conditions compared to the general population. They have lower overall rates of cancer (largely due to lower smoking prevalence), cardiovascular disease, and strokes (18–20). However, as pointed out by Knapp (12), it is not necessarily farming that makes a person healthy; the farmer, rather, must be a basically healthy person to farm.

2.2 Acute Trauma

The rates of accidental injury and death in agriculture, like accidents in general industry, are the best documented occupational hazard and therefore most easily recognized problem. Death rates in U.S. agriculture have been consistently near 50 per 100,000 workers for several decades; temporarily disabling accident rates are estimated at over 5 percent (21); and the overall injury and illness rate is at nearly 13 percent (22). These rates have been relatively constant over the past 20 to 30 years. But in comparison to the reduced death rates in other industrial sectors, the farm is now the most hazardous workplace in America.

The distribution of these accidents by causative agent is typified by an analysis

Table 20.2. The Distribution of 542 Farm Deaths in 1986 by Causative Agent[a]

Fraction of Deaths (%)	Causative Agent	Comments (Breakdown within Agent)
51	Farm tractors	(57% rollovers; 9% fall-offs/run-overs)
11	Buildings and structures	(33% grain suffocation; 13% silo gas)
6	Farm trucks	Mostly non-highway accidents
4	Barnyard equipment	Augers, skid loaders, etc.
4	PTO/power takeoff	Drives attachable implements
3	Electrocutions	Overhead wires only
3	Haying equipment	Entanglement, especially in round balers
3	Farm animals	Mostly bulls and horses
15	Other	

[a]Adapted from Reference 23.

of 542 deaths in 1986 reported by farm safety specialists in 25 states (23) as shown in Table 20.2. It is a discouraging indicator of the nature of the problem that tractor rollover protective structures (ROPS) are known to be an effective control that would virtually eliminate the most common cause of accidental death, but are not required to be installed in the United States and have until recently been a rarely purchased optional accessory. It is also noteworthy that in that same year the deaths of 66 farm children under age 11 were caused by a very similar distribution of agents, although the children were nominally not doing farm work (see also Reference 24).

2.3 Cumulative Musculoskeletal Trauma

Although there are limited data on the subject, it is still quite apparent that chronic musculoskeletal injuries from cumulative trauma are very common among agricultural workers. One survey reported farmers experience twice the prevalence of arthritis compared to all other workers (25). A Minnesota study demonstrated that farmers have a much higher hospitalization rate for musculoskeletal injuries compared to nonfarmers (26). Also, farmers are more frequently disabled from osteoarthritis than are other occupations (27). Impairment of the back has been found to be extremely common among agricultural workers (28).

In addition to these general musculoskeletal injuries, two specific conditions are apparent among farmers in certain regions. Researchers in central Wisconsin have described degenerative osteoarthritis of the knee as a common condition among dairy farmers, for which they coined the term "milker's knee" (29). Swedish and French workers have separately described an osteoarthritis of the hip joint linked frequently to farmers (30--31). Low back pain and osteoarthritis of the hip are thought to be related to low-frequency vibration and sudden jolts transmitted through the tractor seat (31--33).

Until recently there has been little research to evaluate and control ergonomic hazards in agriculture; therefore, few specific recommendations are available. However, there has been a good deal of research in the design of tractor seats in recent years, particularly in West Germany and the United States. Certainly the newer designs of tractor seats, which include back support and pneumatic or hydraulic suspension, are much more "worker-friendly" than seats on older tractors (34). Milking parlors or strap-on spring-loaded milking stools, knee pads, and extended handles on milking machines may reduce the trauma to dairy farmers' knees (34). Long-handled hoes would help spare the back of row-crop workers. Education to include back-hardening schools could help, if an effective dissemination system were in place.

2.4 Acute Respiratory Hazards

A wide range of acute respiratory hazards exists in agriculture: some in confined spaces, others in only enclosed spaces, a few in open spaces. Many are not unique to agriculture; others are associated with commodities and/or processes unique to agriculture (35). A wide range of morbidity and mortality findings suggests that respiratory hazards may represent the greatest health hazard to farmers.

It is not surprising that acute respiratory hazards exist in fully confined spaces (such as the asphyxiation hazard from oxygen depletion in airtight silos), but modern semi-enclosed animal production buildings also have a wide range of acute contaminants. Mulhausen et al. (36) found that air quality in poultry barns frequently exceeded exposure limits of 25 ppm for ammonia during fall and winter and sometimes even exceeded its STEL of 35 ppm; dust levels (about 18 percent respirable) ranged broadly from 1 to 14 mg/m^3 and up to 28 mg/m^3 during "loadout" (when the birds are gathered and taken to market); H_2S was undetected. Donham et al. (37–39) surveyed swine barns and found 50 percent exceeded the TLV for ammonia, and many exceeded those for CO_2, H_2S, and CO (from unvented space heaters) as well. Dust levels were similar to poultry, ranging from 2.4 to 16 mg/m^3. In both settings it is important to remember that these largely organic aerosols are more biologically active than nuisance dusts; a 2.4 mg/m^3 guideline has been recommended for swine dust (40).

Mercaptans and organic acids of many sorts have been identified in the gases emanating from the anaerobic decay of manure typically stored in a pit under most hog and some dairy barns (41–43). Under normal conditions, H_2S is not at levels of great health concern (37–38); however, when the manure is agitated prior to pump-out to be returned to the fields as fertilizer, H_2S can upon rare occasions rapidly reach fatal levels (44–46). During agitation we have measured levels of H_2S as high as 300 ppm in the work room and 1500 ppm in the pit. Manure gas deaths often involve multiple victims during futile rescue attempts (45, 47).

Although the existence of acute hazards to farmers and rural residents has been documented for each agent, information concerning the magnitude and frequency of these acute hazards is limited by poor reporting and a lack of systematic surveillance. The following history of silo gas is representative of the fragmented information available. Occupational hazards associated with silo gas were first reported in 1914 via case studies of four fatalities attributed to carbon dioxide (48).

It was not until the 1950s that investigations revealed the presence and importance of nitrogen dioxide (49–51). The major portion of toxic NO_2 (and its dimer N_2O_4) is believed to be produced from organic nitrates, aggravated by the addition of heavy nitrate fertilizer and/or drought conditions (51). The process of NO_2 production takes several hours to begin and peaks in 2 to 5 days. Measurements of NO_2 concentrations have been sporadic; Peterson et al. (51) reported NO_2 as high as 150 ppm two days after filling. The hazard is primarily associated with vertical silos using corn as the silage (probably the most common but not exclusive combination). Recurrent reviews of the literature discuss the difficulty in diagnosing nonfatal cases of the disease owing to the multiple and usually latent phases of its clinical manifestations (52–54). It is likely that only very severe and fatal cases associated with silo filling are reported with any consistency. Thus not only is the true incidence of clinical cases unknown, but no systematic surveys have been conducted of typical concentrations, from which the potential for more frequent (including less severe) respiratory injury might be extrapolated.

2.5 Delayed and Chronic Respiratory Hazards

Soil-derived dust can contain many respiratory hazards. Although the presence of quartz or other forms of crystalline free silica in agricultural soils has been recognized (55), there is some evidence that pneumoconiosis can also develop among farm workers from more common silicates (56). Asbestos can be a component of some soils (57). Coccidioidomycosis is associated with arthrospores in arid soils of the Southwest (58). Histoplasmosis and blastomycosis are additional soil-borne fungi which may result in chronic lung disease (59).

Inorganic soil-derived aerosols can be generated from both mechanized and manual farming operations as a function of activity and soil moisture (55, 57). Casterton (60) summarized data indicating dust levels within the plume of a variety of field implements ranging from 100 to 200 mg/m^3; this is of course reduced by distance and wind, to 10 to 20 mg/m^3 near the driver; an enclosed cab could be expected to decrease levels to less than 2.5 mg/m^3. Popendorf et al. (55) tracked the buildup of dust on foliage in the arid climate of central California for up to 6 months without rain. The actions of manual harvesters disturbing this foliage created median total aerosol concentrations of approximately 15, 20, and 30 mg/m^3 for peach, grape, and citrus harvesters, respectively. These aerosols ranged from 2 to 10 percent respirable by mass and were only partly correlated with foliar dust concentrations at the time of harvest. The crystalline free quartz content of these dusts ranged between 10 and 20 percent in the soil fines, 5 and 20 percent in the total and 1 and 10 percent in the respirable aerosols, but was sufficient to exceed health guidelines 20 to 50 percent of the time for grape and citrus harvesters, respectively (55). Clearly, there is a significant potential for classical pneumoconiosis from long-term exposure to inorganic dust from both mechanized and manual agricultural operations in dry-to-arid climates.

In moist climates, organic dust is a more pervasive and increasingly recognized respiratory hazard (61–63). The organic component is common in many types of agricultural production including grain, swine, poultry, dairy, and sugarcane. Organic dust from agricultural operations (referred to here as agricultural dust) is a complex mixture of biologically active materials (62). There are several respiratory conditions that are known to be caused by agricultural dust exposure, including (*a*) atopic asthma, (*b*) occupational asthma, (*c*) bronchitis, (*d*) organic dust toxic syndrome (ODTS), and (*e*) hypersensitivity pneumonitis (HP) or farmer's lung. Bagassosis is a well known HP caused by sugarcane processing. Byssinosis could be a combination of occupational asthma, bronchitis, and organic dust toxic syndrome (ODTS) caused by cotton dust, which is most noticeable on Monday mornings (64). It is quite difficult to identify which of the agents in the dust are responsible for the given condition(s). Some research has shown that grain mites or animal dander are important relative to atopic asthma (65–67), and endotoxin is probably related to bronchitis and ODTS (61, 68).

Some diseases such as atopic asthma, occupational asthma, and bronchitis are not specific but appear to be prevalent in agricultural populations (69). ODTS and HP, on the other hand, are largely specific to farm workers. The onset of ODTS symptoms are delayed 4 to 6 hr following exposure to very high concentrations of agricultural dusts that often (but not always) have a high mold or other microbe content (61, 70). ODTS is an acute influenza-like illness with headache, muscle aches and pains, fever, and malaise believed to be a direct effect of endotoxin and/or other microbial toxic component. The individual often recovers from these exposures in 24 to 72 hr, with no known residual effects except possible bronchitis and increased sensitivity to subsequent exposures. Based on case reports and symptoms survey, one of this chapter's authors (Donham) hypothesizes that a chronic form of ODTS may occur. Chronic ODTS may explain the group of symptoms seen in many farmers that include symptoms of fatigue, muscle aches and pains, and difficult breathing. It is felt these symptoms result from prolonged exposure to lower concentrations of microbial and total dust, with or without occasional high exposure peaks. Table 20.3 lists those settings where the acute and chronic exposures may originate.

The agricultural workers' version of HP is called Farmer's Lung (FL). FL was first recognized back in the early 1930s (71). Although its clinical symptoms are similar to ODTS, FL is thought to be a specific, delayed hypersensitivity reaction to certain thermophilic bacteria and fungal spores, such as from moldy hay. FL is generally more severe in the acute stage than ODTS and may lead to chronic lung scarring and interstitial fibrosis (72). Despite changing agricultural practices that may have made FL much less common today, it is still important to the long-term health of the farmer to make the differential diagnosis between FL and ODTS, as outlined elsewhere (73).

Control of dust-related agricultural respiratory diseases should rely first on reduction of the dust source, second on ventilation, and third on personal protection. Reduction of acute exposure may involve applying moisture to the top of the material to reduce its aerosolization when disturbed; to apply this principle to some farm operations (e.g., silo unloading) may require special techniques (74). High ventilation of farm shops or animal confinement buildings is often resisted by

Table 20.3. Sources of Agricultural Dust Exposure Resulting in Organic Dust Toxic Syndrome, Bronchitis, and/or Hypersensitivity Pneumonitis

	Expected Exposure Levels		
Typical Activities	Total Aerosols (mg/m³)	Endotoxin (μg/m³)	Microbes cfu/m³
Acute Exposures:	10–100	0.5–5	10^7–10^{10}
Opening up silos (silo unloader's disease)			
Wood chip handling			
Cleaning grain bins			
Handling moldy grain			
Handling compost			
Chronic Exposures:	1–10	0.1–0.5	10^4–10^7
Swine confinement building			
Poultry confinement building			
Dairy barn			
Mushroom production building			
Grain handling			

operators who prefer to conserve heat in cold winter climates. The use of respirators should be considered a temporary and supplemental protection. The principle of respirator use in agriculture is similar to that in any other industry, except that there are no trained persons available to supervise the respiratory program on an individual farm. The availability of assistance in selection and fit of respirators is also a problem in agricultural communities.

2.6 Agricultural Chemicals

Agricultural chemicals are the traditional whipping boy of environmental and occupational health concern for farmers, especially farm workers. Although chemicals do indeed present hazards when misused, it should be apparent herein that pesticides and fertilizers represent only a narrow spectrum of the occupational risks within agriculture.

The Federal Insecticide, Fungicide and Rodenticide Act (FIFRA) refers to pesticides as "economic poisons" intended to prevent, destroy, repel, or mitigate "any insects, rodents, nematodes, fungi, or weeds or any other form of life declared to be pests, . . . and any substance or mixture of substances intended for use as a plant regulator, defoliant, or desiccant." Toxicologically, the major field-use agricultural pesticides can be broken down into six chemical groups of organophosphate, carbamate and thiocarbamate, and chlorinated insecticides and phenoxyaliphatic acids, triazine, and bipyridyl herbicides. Additionally fumigants such as phosphine and a decreasing range of volatile organics are used in produce storage areas (which are often also confined spaces) (75), and disinfectants are used increasingly in indoor animal production facilities. Although reviews of their toxicities are readily available (e.g., Reference 76), the industrial hygiene aspects of their use practices, levels of exposure, and the efficacy of exposure controls are less accessible.

Pesticides can present a hazard to applicators, to harvesters reentering a sprayed field, and to rural residents via air, water, and even food contamination. Methods to assess exposure include direct methods via dermal patches (77, 78), skin washes (77–80), dietary surveillance (81), and fluorescent tracers (82); indirect methods such as biochemical response, e.g. change in cholinesterase activity (78, 83–84), urinary excretion (78, 85–92), and chromosomal aberrations (93); and epidemiologic-response methods, for example, morbidity (94–97) and mortality (96, 98–100). Archetypical of the difficulty of investigating health effects among diversely exposed and dispersed populations was the discovery of testicular atrophy and sperm count depression among applicators of the nematocide dibromochloropropane (95) following its initial discovery by and among pesticide formulators (101).

Differences in the above methods of assessment complicate comparisons among the multiple routes of exposure contributing to farmers' total doses. Dermal, inhalation, and ingestion are all possible to varying degrees during application (87, 102–106), harvest (78, 107), and local or general environmental contamination of residences via spray drift and volatilization (108–111), food, or groundwater (112–115). Indoor agricultural uses of pesticides (e.g., grain fumigation and especially greenhouses) represent a specialized environment often with a higher airborne exposure (116–117). Some studies have found favorable comparisons between direct and indirect methods of assessment (78, 87, 90); others have found differences or a lack of correlation (e.g., 88, 89). It is often forgotten that correlations should not be expected between direct measures of exposure (chemicals that were prevented from actually reaching the skin by the collection media) and indirect measures (which reflect that which reaches and is absorbed) when conducted on the same subjects. These differences should not cause the industrial hygienist to lose sight of the goal not only to assess risk but also to create better control.

Reentry hazard (going into a field after a pesticide application) represents a long unrecognized hazard often attributed to poor sanitation, water, or food poisoning (112). The classic study by Milby et al. (118) typifies the evolutionary impact of new analytic technologies upon the investigation and understanding of pesticide hazards, in this case using the then-new process of gas chromatography to find for the first time that a more toxic "oxon" analogue of the applied thio-phosphate insecticide sometimes forms in field residues. Since that time, the frequency of oxon production in leaf residues (119) and its importance to harvester acute poisoning has been clarified (78, 107, 120).

Not surprisingly, levels of exposure vary by task. It may be somewhat surprising that the highest exposure can be to flaggers, who mark succeeding passes by an aerial applicator (103), and if not properly supervised, they may also be the least protected by clothing. Broadly speaking, variations in exposures are unrelated to the particular chemical being used but can be expected to range as shown in Table

Table 20.4. Expected Ranges for Dermal Exposure to Pesticides

Task	Range (mg/hr)
Flaggers	3 –300
Mixer-loaders	10 –100
Applicators	2 –10
Harvesters	0.5–30

20.4 as a function of pesticide formulation and concentration, application process and equipment, clothing and personal techniques amenable to education, and uncontrolled conditions (like weather and foliage) (121).

Exposure controls include personal protection, particularly clothing and gloves rather than respirators (80, 82, 121), and engineering/mechanical controls (60, 102, 122–123). One of the unanswered questions is the effectiveness of home laundering to remove pesticides and the likelihood of clothing residuals continuing to cause exposure (82, 124–126). Where surveillance is in place and protection breaks down, preplanned medical management is essential. Because of the acute danger when organophosphate (OP) pesticides are in use (and to a lesser degree carbamates), the existence of a cholinesterase monitoring program is important. Fortunately, guidelines for such monitoring and diagnosis of OP poisoning are well established (78, 83, 127–128).

Acute pesticide poisoning accounted for 10 percent of all hospital admittances of farmers and agricultural workers in Colorado, Iowa, and South Carolina during 1971–1973; this rate extrapolated to 9.1 per 100,000, for the 3-year study period. Organophosphates were responsible for 64 percent of these observed cases (129). Studies in both the United States and abroad have shown that only about 25 percent of the acute pesticide poisoning fatalities are of occupational origin. Of the remaining 75 percent in California, nearly 60 percent were children, frequently due to improperly stored insecticides (130–131); in the Third World, 75 percent were suicides (132). Elevated frequencies of suicides and accidents (even among U.S. farmers (143), indicates that rural life in general and farming in particular is stressful; in less developed countries, pesticides are merely an available, convenient, and perhaps economic vehicle for suicide.

Chronic mortality studies among rural residents and/or farmers in several states and countries (Table 20.5) have revealed statistically significant and/or recurring associations between farming and leukemia, lymphoma (particularly non-Hodgkins lymphoma or NHL), multiple myeloma, and lip, prostate, and skin cancers. An interesting exception was the study in North Carolina in which all farmers had significantly elevated PMRs for tuberculosis and for diseases of the skin and subcutaneous tissue (mostly infections or chronic ulcers). Suspect cancer causative agent(s) include exposures to pesticides (especially herbicides), zoonotic viruses, and ultraviolet light (sun), but no clear etiology has been established (148–149). A weak link in most of these studies has been the accurate estimates of exposure.

Table 20.5. Summary of Epidemiologic Mortality Studies Among Rural Residents

Location	Year	Reference	Finding[a]
Washington	1976	133	Multiple myeloma; lymphocytic leukemia; lip, liver, and connective tissue cancers
All states	1977	134	Bladder and prostate cancer and leukemia not associated with ag chemical use
California	1980	136	Skin, lymphocytic leukemia, and Hodgkin's disease
Iowa	1981	137	Leukemia; NHL; multiple myeloma; Hodgkin's disease; lip, stomach, and prostate cancers
Wisconsin	1981	138	Leukemia
Wisconsin	1982	139	NHL
Illinois	1984	140	NHL and prostate cancer
Wisconsin	1984	141	Multiple myeloma; some associated with poultry-borne virus; weak connection with fertilizer and insecticide use
California	1984	142	Stomach cancer and lymphoma (and cervical cancer)
North Carolina	1985	143	Tuberculosis, diseases of the skin and subcutaneous tissue (infections and chronic ulcers), accidents, and suicide; not ag chemical related
Utah	1985	144	NHL
Sweden	1986	145	Malignant lymphoma and soft tissue sarcoma
Japan	1986	146	Biliary tract cancer mortality
Kansas	1986	147	NHL

[a]NHL = non-Hodgkin's lymphoma.

The historical interest of control (and the focus of label instructions to users) is the respiratory route of exposure, but the repeated finding by direct measurement is that the dermal route tends to be 100× larger (78, 104, 105, 121). This finding, coupled with the fact that most insecticides are by design readily absorbed via the intact skin, indicates that the dermal route is usually the most important. Herbicides are not as well dermally absorbed, but even for compounds like paraquat, the skin can be an important route of entry if proper use practices are not followed (35).

A wide variety of disinfectants are used in livestock operations, especially dairy farms and a growing number of large hog buildings. They include chlorine, quaternary ammonia compounds, organic iodines, cresol-based compounds, and formaldehyde emitters, and often one of a variety of detergents. Certain individuals may develop contact dermatitis or an allergic contact dermatitis from these chemicals (150). Prevention is based on selection and use of chemicals that are not known as irritants or sensitizers. Rubber gloves should be worn as a rule during operations that require a great deal of contact with the chemicals, such as cleaning milking equipment. An alternative or supplement to gloves is the use of protective hand creams (150).

The other category of agricultural chemicals is the fertilizers. Anhydrous ammonia is the most heavily used fertilizer in production agriculture. It is stored and sold in liquid form under pressure by farm supply firms scattered in agricultural areas. It is transferred into small portable nurse tanks, transported to the field, and applied by either the farmer or an employee of the farm supply firm. Because of this pattern, farmers as well as farm supply employees and local residents around the supply firm are at risk of exposure.

Anhydrous ammonia is hazardous because it is highly hygroscopic, highly caustic, and extremely cold (−28°F under pressure). When this material contacts the skin it desiccates, penetrates, and freezes tissue, with the severity depending on the extent of contamination. Anhydrous ammonia is particularly hazardous to the eyes, because almost any eye contact with this chemical results in permanent blindness (151). Inhaling the material can result in severe damage to the upper respiratory tract, resulting in bronchiectasis as a possible sequela (152).

Most of the occupational exposures occur during transfer of the chemical from the bulk tank to the nurse tank, and secondarily during field application. Faulty couplings, bleeder valves, shutoff valves, worn hoses, and plugged applicator tips are common roots of an injury. Control should revolve around establishing a routine inspection and maintenance of bulk storage and nurse tanks at the farm supply firm. Hazard communication to employees and farmers is also very important to establish consistent wearing of eye protection and having clean water available to flush eyes and skin in case of contact.

Ammonia fertilizers have been used so extensively in modern agriculture, that the groundwater supplies in many intensive farming areas have become contaminated with nitrates (153), increasing the potential for nitrate poisoning in infants (the blue baby syndrome) (154). There is also a growing question as to the cancer risk because of carcinogenic nitrosamines that form in the drinking water secondary to nitrate contamination (155–158, see also Section 2.11). As a result, there are efforts in many states to establish a policy to reduce the total use of ammonia fertilizers and to develop techniques that keep the chemical out of the groundwater (159).

2.7 Veterinary Biological and Antibiotics

Biologicals are made from living products to enhance the immunity of an animal to a specific infectious disease or diseases. They may be live attenuated microbes, killed viruses (vaccines), killed bacteria (bacterins), or inactivated bacterial toxins (toxoids). All of the above products are intended to enhance the active immunity of the host. These products may also contain adjuvants which enhance the immunogenicity of the products. Another group of biologicals enhances the passive immunity of the host by injecting antibodies produced in another animal. These products may be crude blood sera from a hyperimmunized animal (antiserum), more refined globulin fractions of the sera, or genetically engineered products.

The main risk groups are those involved in livestock production and related veterinary care who administer these products to animals. Besides veterinarians and their assistants, farmers, ranchers, their family members, and employees all may be at risk. Operations involving swine, poultry, beef, dairy cattle, and sheep all may have an inherent risk for exposure. A government-regulated disease control program in effect for certain diseases (e.g., brucellosis, pseudorabies) requires that a veterinarian administer the biological. Otherwise the producer, as well as the veterinarian, may administer any of these biologicals.

The hazard is associated with either accidental inoculation or splashing the product into the eyes or mucous membrane or contamination of the broken skin. The result may be an infection (certain live products), inflammation, or an allergic reaction. Inflammation or allergic reactions may occur from inoculating either live or killed products, the adjuvant, or the foreign protein in the product. Inoculation with a dirty needle also has the extra risk of causing infections of environmental origin.

The primary products that have been associated with occupational illnesses include brucellosis strain 19, *Escherichia coli* bacterins, Jhone's disease bacterin, erysipelas vaccines, contagious ecthyma vaccine, and Newcastle disease vaccine. The most frequent reports of occupational illnesses associated with biologicals involve veterinarians using brucellosis strain 19, which is a live product containing an adjuvant. Veterinarians have become ill either by splashing the material in their eyes or by accidental needle sticks. The results may be infection, inflammation, and allergic reaction. The infection mimics the acute infection seen from acquisition of the disease directly from either cattle or swine, and treatment should be no different. If the person had a previous exposure to brucellosis (many veterinarians practicing before the mid 1960s had previous exposures), they may develop severe inflammatory and allergic reactions in addition to an infection (160). The reaction is characterized by severe localized swelling and pain extending from the site of the inoculation. The swelling and allergic reaction must be treated in addition to the infection in these cases. Disability may last for days to weeks in the worst cases.

Newcastle disease and contagious ecthyma (orf) vaccines are live products used in chickens and sheep, respectively. Newcastle vaccine is applied inside poultry buildings via a nebulizer. Workers who contaminate their eyes with this vaccine may acquire a moderate conjunctivitis with influenza-like systemic symptoms. Orf vaccine can cause the same pox-like lesions at the site of inoculation as a naturally acquired infection. Both of these diseases are self-limited and disability will only last for a few days, unless the orf lesions are numerous (161, 162).

Jhone's, *E. coli*, and most erysipelas biologicals are bacterins, and therefore injuries induced by these products are limited to the inflammatory response induced by the adjuvants.

Control of injuries associated with biologicals revolves around good animal handling techniques and facilities, because most of the accidental needle punctures are secondary to uncontrolled and untimely movements of stressed animals. The proper construction of animal handling facilities has been reviewed by Grandin (163). The use of pneumatic syringes, lock-on needle hubs, and multiple-dose syringes will also help reduce injuries. Eye protection is indicated in many instances, and a full face respirator is necessary for aerosolized vaccines such as Newcastle.

Antibiotics are products derived from (or synthesized) from living organisms, mainly mold species of the genus *Streptomyces*. Antibacterials are chemical compounds not from living organisms, but used in the same manner to treat infectious diseases therapeutically. They are also used widely in livestock production for improvement of rate of weight gain and feed efficiency in cattle, swine, and poultry. Livestock producers, veterinarians, and feed manufacturers and formulators are commonly exposed to these agents by direct contact with antibiotic-containing feeds, or via aerosol exposure within livestock buildings or within feed preparation areas on the farm or in feed manufacturing plants. There are two main occupational hazards: (*a*) allergic reactions and (*b*) the development of antibiotic-resistant infections. There are many different products used as feed additives, but the main ones include penicillin, tetracycline, sulfamethazine, erythromycin, and virginiamycin. These same products plus many more are used therapeutically. Penicillin is the primary agent that may induce an allergic reaction manifest in the form of a skin reaction from direct contact, or possibly a systemic reaction from inhalation or inoculation.

A variety of these agents may induce development of resistant organisms in the gut flora of exposed individuals. The resulting health impact of this is not clear-cut. However, there have been some cases of severe resistant salmonellosis traced to direct animal contact (164) and in people who were treated with antibiotics for a condition unrelated to salmonella. The latter case is a result of an overgrowth of the resistant organisms secondary to the antibiotic treatment.

Although the full importance of antibiotics as an agricultural health hazard is unknown, it is prudent to take some control measures. Feed formulation, grinding, mixing, and storing operations should be closed systems. General dust control procedures should be utilized in both feed preparation areas and in animal feeding operations. Until dust control procedures are proven effective, dust masks should be worn in conjunction with other engineering and work practice procedures. In addition, an emphasis should be placed on removal of those antibiotics used in human health from feed additives and a rotation of the particular type of antibiotic used should be considered.

2.8 Zoonoses

Zoonoses are infectious diseases common to animals and man. At least 24 of the over 150 such diseases known worldwide are occupational hazards for agricultural workers in North America (59, 165). Some of these diseases may be contracted directly from animals, whereas many are contracted from the natural environment that is part of the farmer's workplace. A list of recognized agricultural zoonoses was prepared by Donham and Horvath elsewhere (20).

The agricultural worker's risk of acquiring a zoonotic infection varies with the type and species of animal and the geographic location (166, 167). For example, dairy farmers in North America are at risk to acquire ringworm, milker's nodules, or brucellosis. Beef cattle producers are more prone to acquire rabies, anthrax, or leptospirosis. Besides livestock producers, those doing related service work (e.g., veterinarians) or animal processing are also at risk for certain zoonotic infections. Turkey processing workers are known to be at risk particularly for ornithosis, red meat processing workers for brucellosis and leptospirosis, and hair and hide processors for anthrax (5, 59, 168).

Control of these infections in the production phase depends largely on an awareness of the specific hazards, good preventive veterinary care, hazard communication, and medical backup, especially in cases where serological monitoring of animals or people may be indicated. For livestock producers, close animal health monitoring and veterinary preventive practices are best. In processing, early identification of infected animals as they come into the plant and appropriate handling of them is important. In some cases sanitation and personal protection are important. The key is developing both an understanding of certain generic features of this group of diseases and an awareness of conditions and agricultural activities that increase infection risks within specific locations (as reviewed elsewhere 59, 165, 169). Such an awareness is essential to enable the hygienist to anticipate, recognize, evaluate, and design a control program for zoonotic infections.

2.9 Skin Diseases

Diseases of the skin are very common in agriculture (170, 171). Compared to other occupational groups, farmers have a proportionately higher prevalence of skin diseases (172), and in some regions skin diseases are the most common condition reported by agricultural workers (173). Common agricultural skin diseases, causative agents, and suggested methods of control are listed in Table 20.6. (See also Chapter 10.)

Irritant contact dermatitis is perhaps the most common type of agricultural dermatoses (171–178). There is no particular subgroup of agricultural workers that is free from contacting a substance that may cause an inflammatory response to the skin. Irritant substances are ubiquitous and include ammonia fertilizers, several pesticides, soaps, petroleum products, and solvents. Avoidance schemes must include work practices to eliminate or reduce exposure to the most irritative substances and/or the use of personal protection equipment.

Allergic contact dermatitis is typified by poison ivy or poison oak reactions. These are exquisite sensitizers, as are certain herbicides and pesticides (170). These reactions are more difficult to control, because just a small amount of the offending material may produce a reaction.

Table 20.6. Skin Conditions of Agricultural Workers: The Principal Agents, Symptoms, and Prevention

Classification of Skin Condition	Agent	Description of Condition	Control
Contact dermatitis Irritant contact dermatitis	Ammonia fertilizers Animal feed additives (e.g., ethoxquin, cobalt) Insecticides (e.g., inorganic sulfur, petroleum, and coaltar derivatives) Plants: bulbs of tulips, hyacinths, onion, and garlic; vegetable crops (e.g., carrots, asparagus, celery, parsnips, lettuce) Herbicides (e.g., trichloroacetic acid, paraquat) Fumigants (e.g., ethylene oxide and methylbromide)	Dermatitis usually on hands and arms and other points of contact	Assure proper dilution of chemicals Wear protective clothing Wash hands, arms, and other contact areas frequently
Allergic contact dermatitis	Herbicides (e.g., propachlor, thiram, maleic hydrazide, randox, barban, nitrofen, dazomet, lasso) Insecticides (pyrethrum, rotenone, malathion, phenothiazine, naled, ditalimfos, omite, dazomet, dinobuton) Animal feed additives: antibiotics (e.g., penicillin, spiromycin phenothiazine) Plants: poison ivy, poison oak, poison sumac ragweed	Acute inflammatory response with swelling, possibly reddish, elevated eruptions, blisters, pruritus, usually on hands and arms	Same as above, plus: Wash clothes that contact offending substances Any work practice change that will limit contact with offending substance
Photocontact dermatitis (includes both photoirri-	Creosote Feed additive (e.g., phenothiazine) Plants containing furocoumarins	Combination of sunlight and skin exposure to offending substance induces variable dermatitis de-	Wash hands and contact areas of skin frequently Protective clothing (e.g., gloves and

tant and photoallergic contact dermatitis)	(e.g., carrots, celery, parsley, parsnips, limes, lemons); ragweed, oleoresins	pending on amount and type of exposure Exposure to furocoumarin-containing plants; blisters followed by hyperpigmentation in bizarre, streaked pattern	long-sleeved shirt)
Sun-induced dermatoses	Sunlight (ultraviolet radiation)	Sunburn Wrinkling of skin Actinic keratoses Squamous cell carcinoma Basal cell carcinoma	Protective clothing (e.g., wide-brimmed hat, long-sleeved shirt) Sunscreen (e.g., para-aminobenzoic acid)
Infectious dermatoses	Cattle, swine, rodent animal ringworm (*Tichophyton verrucosum*, *Microsporum nanum*, *T. metagophytes*, respectively) Sheep pox virus (orf or contagious ecthyma) Cattle pseudocowpox virus (milker's nodules)	Animal ringworm in human: highly inflamed scaly lesions on hands, arms, face, and head Orf: lesions on hands and arms, develop as red papules, progress to an ulcerative lesion Milker's nodules: multiple solitary, wart-like lesions on hands and arms	Appropriate veterinary treatment and prevention Wear protective clothing when handling infected animals Practice good sanitation of the animal environment
Heat-induced dermatoses	Moist, hot environments	Miliaria rubra (prickly heat): an exanthematous eruption of the skin, mainly under the arms and around the belt line; caused by inflammation of eccrine sweat glands	Wear loose-fitting, well-ventilated clothing Ventilate the work environment Daily bathing with a good soap
Arthropod-induced dermatoses	Chiggers Animal mites Grain mites Hymenoptera (e.g., bees, wasps, hornets, yellow jackets, fire ants)	Red macules, papules, pruritic lesions, possibly vesicles Sensitivity may vary with repeated exposure Anaphylactic reaction possible	Wear light-colored, nonflowery clothing Avoid perfumes Use insect repellant (e.g., diethyltoluamide)

[a]Adapted from References 5, 172, and 178.

Sun-induced dermatoses include sunburn and skin cancers (170, 179). Acute sunburn may be prevented by the use of sunscreens and protective clothing. More important is the cumulative effect of sun exposure, which may produce a variety of lesions about the face and arms. Actinic keratoses are common in older farm workers. Twenty-five percent of the pre-neoplastic lesions may develop into squamous cell carcinomas, the most common skin cancer. These do not tend to be malignant unless they occur on the lip but usually require surgical removal. Basal cell carcinomas are less common but have a greater tendency to become malignant. Melanomas are highly malignant, but fortunately the least common of these skin tumors.

Heat-induced dermatoses are not generally very serious, but they can be quite uncomfortable and may take several days to recover. The primary problem here is an inflammation of the eccrine sweat ducts, resulting in a pruritic eruption called prickly heat or miliaria rubra (170).

Infections are primarily a result of viruses and fungal agents of animal origin. Ringworm of cattle would perhaps be the most infectious, followed by contagious ecthyma (a viral disease of sheep) and milker's nodules (a viral disease of cattle) (170, 180–181).

Chiggers, grain mites, animal mites, bees, and wasps all can cause significant hazard to the skin of agricultural workers (182). The lesions vary from a mild skin rash to an anaphylactic reaction from stings of bees and wasps.

2.10 Physical Agents

It should come as no surprise that mechanization has had a major impact upon noise-induced hearing loss among farmers. In fact, surveys show that farmers today suffer a higher incidence of hearing loss compared to other occupational groups (e.g., 31, 183, 184). Even some non-mechanized modern farming practices can result in high noise exposure levels, as can be seen in Table 20.7. Sullivan et al. (187) conducted a year-long study of the noise environment of agricultural workers on six Nebraska farms and 67 farm workers. Thirty-eight percent of their machines produced sound levels in excess of 90 dB. To cope with farming's temporal variability, they time-weight averaged over monthly intervals and found 39 percent of farm workers exceeded 90 dB OSHA limits for a total of 15 percent of the months (187).

Traditional methods to prevent noise-induced hearing loss among general industrial workers (as described in Chapter 23) are broadly applicable to farmers, with the obstacles of long-term capital investments characteristic of large mechanized pieces of equipment and a generalized lack of knowledge regarding short-term personal protective equipment (188).

Heat, vibration, and ergonomic hazards are all prevalent in agriculture. Heat (and cold in many regions) is a seasonal stressor for outdoor workers generally (see also Chapter 21). Incidence of heat-induced illness is rarely reported, for example, heatstroke reported by West (17) and elevated PMRs from exposure to heat or cold by Une et al. (189), reflecting poor-to-no epidemiologic surveillance

Table 20.7. Typical Noise Levels during Selected Farming Operations

Activity	dBA
Chainsaws	105–112
Vane-axial grain drying fan	100–110
Combine, full throttle	102–107
Corn grinder	94–103
Squealing sows	95–102
Bed chopper	94–102
Hay choppers and balers	95–100
Tractor, full throttle	
Next to tractor	102
On seat, no cab	93
In cab	82–85
Harvestore unloader/conveyer	85
Milking parlor	76–84

Adapted from (184–185).

of this population, a large measure of self-selection within the work force, and a certain measure of self-pacing and flexibility in their work hours. Whole body vibration was very common on tractors; pathological radiological changes in the spine and complaints of low-back pain were found to be associated with both total years and hours per year of tractor driving (32), but vibration has been greatly reduced in newer designs (34, 190). Segmental vibration among farmers is most common from chainsaws, although many hand tools also contribute; exposure to farmers is usually limited to short periods (191). Again actual incidence of vibratory white fingers among farmers has not been reported. As discussed in Section 2.3, ergonomics is only beginning to have a major impact on agriculture (34), notably on the agricultural tractor cab, mechanized milking equipment, and the banning of the short hoe. Systematic study of additional hazards for disabled farmers returning to work is also in its infancy (192).

2.11 Cancer

Compared to the general population, farmers have lower overall cancer rates (138, 148); that is, they have lower rates for the most common cancers, principally those related to smoking [lung, esophageal, and mouth (192–193)]. It is a fact that farmers do smoke less; approximately 17 percent of farmers smoke compared to 34 percent of the general population (18). In spite of this lower overall rate as mentioned in Section 2.6, there are several cancers for which farmers are at increased risk. These include leukemia, non-Hodgkin's lymphoma, Hodgkin's disease, multiple myeloma, and cancers of the lip, prostate, stomach, kidney, and brain (134, 138, 149, 155–157, 193–196).

Determining the risk factors for these various cancers has been a very difficult problem because of long latency periods and difficulties in obtaining accurate exposure-classification data. In summarizing the available information, the strongest evidence for a risk factor is for lip and skin cancer, which is quite clearly related to sun exposure (5, 148). Risk factors most extensively studied are for the reticuloendothelial cancers. Exposure to dairy cattle, poultry, corn production, fertilizers, and pesticides, have all been shown to be risk factors for leukemia. Some studies suggest exposure to bovine leukemia virus via cattle is a risk factor; however, later studies point out that pesticide exposure may be more important (197–199). Risk factor data for the other agricultural cancers are quite sparse. Table 20.5 summarizes the available information reported in more detail elsewhere.

There is a great deal of research to be done regarding risk factors for agricultural cancers. Until further information is available, about the only thing the hygienist can tell farmers with certainty is that they can reduce their risk of skin cancer by wearing protective clothing, sun screen, and installing shade devices on their tractors and other pertinent equipment.

2.12 Mental Stress

Farmers die of suicide at a greater frequency (133) and suffer more frequent mental disability relative to other occupations (28). A recent study regarding social concerns within farm families suggest dysfunctional families, divorce, alcohol abuse, and children having problems are all more common within the farm community (200). The Iowa Farm Family Survey of 1988 indicated that farmers rated stress as one of their major concerns (159). Compounding inherent, endemic stressors in farming are episodic events such as the farm economic crisis of 1982–1987 and the drought of 1988–1989. Mental stress not only should be considered an important occupational health issue for farm families but may also contribute to more frequent injuries (201).

Most types of farming include a series of work-cycle peaks (e.g., Figure 20.1) that can be complicated by adverse weather conditions and machinery breakdowns (202). The stoic and independent nature of many farmers makes them reluctant to talk to anybody about these problems, let alone seek professional help (203). Also most rural communities lack the organized support systems that many urban centers often have, and in both areas, the extended family is not the support structure that it once was. To make ends meet economically in today's farm families, it is very common for one or both spouses to work full or part time off the farm (see Section 1.1). This increases family stress and creates a child-care problem. All too often children are found in the workplace, where it is difficult to supervise them and they frequently become accident victims (23, 24).

Control of this problem is certainly difficult. A few innovative, largely pro-active programs are being tried in communities scattered around the country but there needs to be greater activity in this area (203, 204). One such program in Iowa is called "Farmer-to-Farmer." Seeking out farm families in trouble and getting them together for discussion and mutual support with other families with similar problems seems to be a successful way to get help to the stoic independent farmer. The agricultural hygienist needs to be aware of "farm psychology" and find ways to deal with the problem.

2.13 Emerging Hazards

The ability to deliver effective prevention programs to the farm community should include the ability to anticipate developing occupational hazards. New genetically engineered crops, livestock, pesticides, and hormones will substantially increase productivity, forcing less efficient farmers out of business and concentrating agriculture even further into larger and fewer operations. Although consolidation will increase the need for hired labor, farm mechanization will eliminate many of the menial labor-intensive operations with which hired farm labor is primarily involved today. The farm employee will necessarily become more technically skilled and the need for temporary migrant labor will probably diminish. For example, the operation of specialized methods of livestock production requires rather sophisticated year-round labor, but the longer daily exposure to organic dust has created health hazards with which farmers have never before had to deal. The farm manager must become aware of these hazards, and of the opportunities and responsibilities to control this working environment, not only for the enhanced production and profitability of the farm, but also for the provision of preventive health and safety services.

Among the more specific future health hazards one can anticipate is that continued ozone depletion will result in increased ultraviolet light exposure to the farm population and a greater risk for skin cancer. On the biological side, Lyme disease, a tick-transmitted disease first recognized in the northeastern states, is now recognized in much of the upper Midwest (205, 206). This generalized illness can result in prolonged arthritic disability. Its latency and dynamic environmental prevalence precludes an accurate assessment of its potentially large impact at this time.

Aflatoxin is known to be an extremely toxic carcinogen in at least eight species of test animals but has long thought to represent only an oral risk to man. A recent follow-up of a small (60–70 persons) cohort exposed to roughly 5 pg/m^3 aflatoxin on airborne organic dust from peanut meal showed a 2.5 to 4.4 increased risk of cancer of all types for different exposed time periods (207). This diverse pattern of human toxic responses to aflatoxin is not inconsistent with the cancer findings in Table 20.5 (207, 208). The presence of aflatoxin in corn (209) and corn dust (210, 211), its extrapolation to exposures of 1 to 20 pg/m^3 on the farm (35), and the increased risk of *Aspergillus flavus* infestation during draught conditions (211) suggest that airborne agricultural exposures could be of considerable and growing concern.

Pesticides and nitrate fertilizers are known to contaminate rural water supplies (113–115, 155–157), raising a hypothetical concern for increased toxic effects at some point in the future. This concern and others are likely to lead to decreasing use of pesticides and high-volume fertilizers, to be replaced by integrated pest management, genetically engineered tools, and increased soil tillage, exposure to machinery, and new agents for which not all health effects will be known at the outset. Whatever the potential health hazards of these new products, agricultural workers are likely to receive the highest exposures and exhibit the first effects. Suffice it to say that not only should agricultural hygienists keep informed generally about new technologies and specific products that become available, but they are among the best qualified to anticipate their hazards and feasible controls.

3 SUMMATION

3.1 Lessons from General Industry

The industrial hygiene paradigm of anticipation, recognition, evaluation, and control can, in principle, be applied to agriculture with the following translations.

- *Anticipating* health and safety hazards is the preventive application of a dose–response knowledge data base. Response data for hazards unique to agriculture must be generated by either mandatory or funded research surveillance systems. Transferring experience from other industries requires either a knowledge of dose (exposures generated by a given act), the ability to assess dose in real time, or an assumption of worst case. In widely diverse settings (characteristic of agriculture), the worst case is much worse than the average case. To protect against the worst case requires perceptually overly restrictive controls for everyone else, which is contrary to the intrinsically risk-taking philosophy of farming. Thus a great deal of probability salesmanship would have to go into preventive programs based on anticipation, unless a solid understanding of exposure mechanisms is established.
- To apply the tools of *recognition* requires interest. Although fragmented information about injury, disease, and death risks in agriculture is available, complete information is lacking. Decades ago Knapp complained about the lack of good scientific epidemiologic studies in agriculture (12). Much of the fragmented data generated since is not current and (especially for health hazards) does not reflect current and evolving technologies. Even if current risks were known, the dispersed and locally innovative nature of agriculture would bolster a natural bias toward the use of new unevaluated technologies (212).
- If the desire were present, the technology exists to *evaluate* essentially all the agricultural hazards noted above. Although usually rewarding in the long run, agriculture is an inherently risky venture with a slow economic rate of return. Even in general industry, a person with less wealth will be more willing to accept job risks (e.g., the average blue collar worker would accept $900/yr more for perceived risky jobs) (212). And to the degree that most individual producers recognize agriculture as being hazardous, the value of evaluation is dimmed by the psychological perception that risks are intrinsic to farming, the cost of voluntary prevention is not competitive, and not to farm would entail a break with tradition, a loss of prestige and self-image, trying to find

Table 20.8. Distribution of 210 M$ Total Federal Expenditures for Protective Labor Services in Fiscal Year 1986

Area of Expenditure	$/Worker	$/ Death	$/Disabling Injury
Mining	182.00	363,400	4540.00
General industry	4.34	39,770	231.00
Agriculture	.30	606	5.71

Source: Reference 217.

an often scarce local alternative job with limited job skills, and/or the economic costs of retraining and even relocating.

- Implementing a *control* requires resources. Among the resources readily accessible to the producer, other than time, are land, water (usually), a wide variety of equipment, and the innovative skills to use them. The farmers' view of time and money have been inexorably bound up in the argument of whether farming is a way of life or a business (as one of the characters in Wilder's book said about new technology in the 1800s, "All it saves is time, son. And what good is time with nothing to do") (1). Because income to producers is largely limited to those commodities producible and marketable via the existing regional infrastructure, a farmer's access to the universally limiting resource of money is achievable only by producing more (compatible with a strong work ethic typical of highly agricultural communities), producing better (a weakly marketable option), or producing more cheaply (an option conducive to operating without "optional" safety features).

3.2 Control Policies and Strategies

The provision of industrial hygiene (and expanded safety) services to agriculture could be initiated via governmental requirements, private economic incentives, organized producer ("grass-roots") demands, or a combination of these. The lack of U.S. governmental interest is dramatized by the comparison of federal expenditures in Table 20.8 and a long-standing fragmentation of agricultural health and safety among multiple agencies (3, 8, 213). Recent governmental interest subsequent to a major policy conference and report (159) may remain or disinterest may return, in which case one must look to private forces to initiate interest in or actually provide preventive occupational health and safety services to the agricultural industry. Generic options and approaches for any agency to implement such services are outlined below. The model most likely to be successful in any culture is not at all established.

3.2.1 Research

Research on any or all of the traditional elements of anticipation, recognition, evaluation, or control can contribute to society's knowledge, but because of the technical and often interdisciplinary and specialized nature of health research and the equally segmented and organizationally flat nature of agriculture, such knowledge is not either available or used by policy administrators and is often unusable by individual farmers who constitute agriculture. It could also be added that a communication gap exists between agricultural, general industry, environmental, and medical researchers who often publish in diverse literature. Thus collectively a great deal more is known by some than by any or all.

3.2.2 Education Policies

Education is the least restrictive but most passive preventive measure. In the best of circumstances, education is the cornerstone of interest to recognize, of desire to evaluate, and of allocating resources to control health and safety hazards. Agricultural health and safety education has been federally supported, albeit weakly for over 50 years. Its inability by itself to reduce injury and accidental death in pace with other industries may be attributed to a lack of good, relevant research, the specialized nature of its dissemination, ineffective marketing, or reliance only on voluntary compliance.

3.2.3 Certification

Certification of chemicals, equipment, implements, or structures via a voluntary standards process offers the next least restrictive option to prevent unsafe practices or environments. FIEI (the Farm Implement and Equipment Institute) is one organization that has acted to adopt consensus safety and health production standards for agriculture. In certain markets, labor-management agreements have become de facto certification requirements. Many insurers, but not all banks or lenders, actively enforce rudimentary on-farm certification requirements in their dealings with farmers. A variation on voluntary compliance is the delegation of Pesticide Certification Training (required by the Environmental Protection Agency, EPA, for the purchase of certain commercial pesticides) to many land grant universities; the responsibility for similar certification programs could be (but is not) shared with producer organizations. Although the threat of liability litigation can hinder expansion of voluntary certification standards, farm consolidation and incorporation can encourage them.

3.2.4 Cost Reduction

Deductions in the cost of doing business are a potential incentive to encourage safer production practices. One model for such incentives is a government tax subsidy (e.g., the 1970s energy conservation tax deduction in the United States and a 1980s workers' compensation insurance safety equipment rebate program in Ontario). Another model would be insurance discounts for meeting certain safety/health criteria. Related approaches could include the involvement of financial lenders (e.g., by affecting a farmer's credit rating) or health care providers (e.g., preventive health and safety services to farmers). The potential benefits of this latter approach are currently being tested in rural Iowa (214).

3.2.5 Taxation

Taxation can take various forms to create a financial disincentive for producers to choose or continue to use relatively unsafe practices. In one form this option is currently being used via governmental taxation to fund the Occupational Safety and Health Administration (OSHA) and other preventive health and safety services (note that farmers too are paying for OSHA but are not receiving its benefits). Voluntary tax (sometimes referred to as a mill tax) is an alternative program with control and benefits vested in farmer organizations such as commodity groups, the Farm Bureau, coop businesses, or the integrated system of on-farm and clinical health services funded in Sweden via Lantbrukshälsan (215). An intermediate example can be found in the funding of the "Farmsafe" educational services through worker's compensation fees in Ontario.

3.2.6 Regulation

Regulation in the United States implies either specification or performance standards, a system of inspectors, and (usually) financial penalties. Governmental control is characteristically political, bureaucratic, and restrictive. Although the passage of occupational health and safety legislation for general industry required broad political support, agriculture is the only sector to have purposefully precluded itself from most OSHA requirements (perhaps to its detriment). The U.S. EPA as authorized by FIFRA legislation) has completely eliminated exposure to a small number of hazardous agents by cancellation of pesticide registrations, and has implemented wider but more limited controls via applicator certification and label use restrictions (largely without enforcement). Where OSHA has attempted to impose controls, the temporal and geographic diversity of agriculture have combined to present a bureaucratic dilemma: the attempt to protect employees in one crop or region requires overly restrictive protection in other settings. The justification of the 1988 Field Sanitation standard (29 CFR 1928.110) varies from essential to largely redundant, depending upon the setting. Although the OSHA Act states a preference toward performance standards, the level of exposure and compliance during an agricultural operation is difficult to assess or inspect because of its seasonal or otherwise transient nature. To the degree that agricultural regulations expand, specification standards assuring at least a minimum level of protection are perhaps not only more palatable but also more effective in developing an awareness that just as in general industry, so too in agriculture, safe work practices pay.

3.3 Model Programs

The development of preventive occupational health and safety services to U.S. agriculture suffers from a lack of both a clear governmental policy at the top and local leadership to express an interest at the bottom. Available services are fragmented (8), such as the various programs for migrant workers (which primarily stress medical services and surveillance, part-time but not hygiene), the cooperative extension service (which is largely limited to one person per state to disseminate educational materials), and the Farm Bureau (which functions similarly to the extension service for their insurees). These activities are very limited in scope and magnitude versus the breadth and size of the industry's hazards.

In the late 1970s both Finland and Sweden initiated model programs to deliver comprehensive occupational health services to their farm families (33, 215). Sweden's Lantbrukshälsan clinics provide medical surveillance, medical treatment, preventive physiotherapy, education, and on-the-farm industrial hygiene and safety services. Via these voluntary but subsidized programs, the majority of farmers in these countries now have access to occupational services similar to those in general industry. These countries have been the example for Norway, Denmark, The Netherlands, and other countries who are trying to establish similar programs. France and Germany also have farm programs but they are not nearly so comprehensive; their programs are primarily through their insurance systems and concentrate on medical issues in France and equipment safety features in Germany. Australia is initiating a new program modeled after the Scandinavian approach. Ontario in Canada has a well-developed program based primarily on education but includes some on-the-farm hygiene and safety services.

Suffice it to say that the small independent programs in the United States are quite behind all of these countries in providing services to farmers. The University of Iowa has had an active research and teaching activity at the Institute of Agricultural Medicine since 1955 (6). However, until recently service activities have not been a large part of their activity. In 1987, two state-funded model projects were initiated to deliver comprehensive services through community hospitals with consultation, training of their medical staff, farm educational program development, and referral services provided by core staff at the university. This program (the Iowa Agricultural Health and Safety Service Program) has now been expanded to a total of six community sites (214). Evaluations will follow to determine if this community hospital mechanism is a feasible option to provide needed services.

The New York Center for Agricultural Medicine at Cooperstown, New York is a recently initiated activity that will attempt to provide services out of their hospital and network several other regional hospitals. The Marshfield Medical Clinic at Marshfield, Wisconsin, has been active in treating farmers with occupational illnesses and doing research in agricultural lung illnesses. They have more recently expanded their activities in farmer education. South Carolina has initiated a program that unites the land grant university extension safety specialists with the medical school to deliver health and safety education programs; this Agro-medicine Program is the first of its kind in the United States, and certainly expands the traditional extension approach, but not yet to the degree of those mentioned above.

A greater emphasis must be placed on model programs offering comprehensive, interdisciplinary services rather than the piecemeal programs of the past. Policy strategies to implement services from research to education via consultation, certification, taxation, or regulation need coordination and strong leadership. Such calls have been made for decades (213). The fact that most new agribusiness and food employees are not agricultural school graduates (216) suggests that such leadership is likely to come from other backgrounds, perhaps even industrial hygiene.

4 CONCLUSION

The future of agricultural hygiene will continue to be affected by the economic and technologic forces that have promoted the progressive consolidation of farms into larger, more capital intensive operations (2, 11–14). New technologies will require more training, and from consolidation will evolve the stratification of agricultural producers into managers and hired employees. These economic and social forces should stimulate a growing interest in product safety, in occupational safety and health, and in more complete management services to both the traditional and the consolidated farm. Evidence for such interest is already seen internationally (218) and in the United States at the national level (by congressional funding beginning in 1990 for increased occupational health and safety research by agencies such as NIH and CDC), at the state level (e.g. new more comprehensive, legislatively funded programs in California, Iowa, Minnesota, and New York), and at local levels (centered around rural community hospitals (214)). A host of new and diverse professionals are becoming interested in the field, as noted by the recently formed National Coalition for Occupational Safety and Health (159). These forces and disciplines may advantageously blend with the traditional extension services characteristic of U.S. land grant colleges and universities. It is hoped that this transition to more specialized and comprehensive services will include agricultural hygiene as a growing opportunity.

REFERENCES

1. L. I. Wilder, *Farmer Boy*, Harper and Row, New York, 1933/1971.
2. S. S. Batie and R. G. Healy, "The Future of American Agriculture," *Sci. Am.*, **248**(2), 45–53 (1983).
3. T. A. Knudson, "A Harvest of Harm," A series of six feature articles in the *Des Moines Register*, Des Moines, IA, September 16–30, 1984.
4. American Conference of Governmental Industrial Hygienists, *Agricultural Health and Safety Resource Directory*, ACGIH, Cincinnati, OH, 1984.
5. C. F. Mutel and K. J. Donham, *Medical Practices in Rural Communities*, Springer, New York, pp. 77–78, 1983.
6. L. Lawhorne, "The Health of Farmers," *J. Iowa Med. Soc.*, **66**(10), 409–418 (1976).
7. D. J. Murphy, "A Perspective on Problems in Agricultural Safety," *Prof. Saf.*, **26**(12), 11–15 (1981).
8. ACGIH Agricultural Health and Safety Committee, Guest Editorial: "Solomon's Baby and the Farmer," *Am. Ind. Hyg. Assoc. J.*, **41**(1), A-4 (1980).
9. L. J. Elliott, "Agriculture: The Most Hazardous and Underserved Occupation," *Appl. Ind. Hyg.*, **4**, F-8 (1989).
10. K. J. Donham, "Agricultural Occupational and Environmental Health: Policy Strategies for the Future—Prologue," *Am. J. Ind. Med.*, **18**(2), 107–119 (1990).
11. P. L. Martin and A. L. Olmstead, "The Agricultural Mechanization Controversy," *Science*, **227**, 601–606 (1985).

12. L. W. Knapp, "Agricultural Injury Prevention," *J. Occup. Med.*, **7**(11), 545–553 (1965).
13. L. Tweeten, "The Economics of Small Farms," *Science*, **219**, 1037–1041 (1983).
14. Office of Technology Assessment (1986), "Technology, Public Policy, and the Changing Structure of American Agriculture," OTA-F-285, U.S. Government Printing Office, March, 1986.
15. P. H. Poma, "Impact of Culture on Health Care: Hispanos," *Iowa Med. J.*, **126**, 451–458 (1979).
16. K. J. Donham and C. F. Mutel, "Agricultural Medicine: The Missing Component in the Rural Health Movement," *J. Family Pract.*, **14**, 511–520 (1982).
17. I. West, "Occupational Disease of Farm Workers," *Arch. Environ. Health*, **9**, 92–98 (1964).
18. P. R. Pomrehn, R. B. Wallace, L. F. Burmeister, "Ischemic Heart Disease Mortality in Iowa Farmers," *J. Am. Med. Assoc.*, **248**, 1073–1076 (1982).
19. A. D. Stark, H.-G. Chang, E. F. Fitzgerald, K. Riccardi, and R. R. Stone, "A Retrospective Cohort Study of Mortality among New York State Farm Bureau Members," *Arch. Environ. Health* **42**(4), 204–212 (1987).
20. K. J. Donham and E. Horvath, "Agricultural Occupational Medicine," Chap. 59 in *Occupational Medicine*, 2nd ed., C. Zenz, (1988), pp. 933–957.
21. National Safety Council, "Accident Facts", NSC, Chicago, IL, (1987).
22. D. H. Cordes and D. Foster, "Health Hazards of Farming," *Am. Family Pract.*, **38**, 233–244 (1988).
23. A. B. Skromme, "The First Annual US Farm Accident Report—1986," Moline, IL, 1988.
24. F. P. Rivara, "Fatal and Nonfatal Farm Injuries to Children and Adolescents in the United States," *Pediatrics*, **76**(4), 567–573 (1985).
25. T. D. Woolsey, "Prevalence of Arthritis and Rheumatism in the United States," *Public Health Rep.*, **67**, 505–512 (1952).
26. J. A. Burkhart, C. F. Egleston, and R. J. Voss, "The Rural Health Study: A Comparison of Hospital Experience Between Farmers and NonFarmers in a Rural Area of Minnesota," DHEW (NIOSH), pp. 78–184 (1978).
27. J. Kennedy and T. J. Fishback, "*Occupational Characteristics of Disabled Workers:* Social Security Disability Benefits Awards to Workers during 1969–1972," U.S. DHHS Publ. No. (NIOSH), 80–145, U.S. Government Printing Office, Washington, DC, 1980.
28. L. D. Haber, "Disabling Effects of Chronic Disease and Impairment," *J. Chronic Dis.*, **24**, 482–483 (1971).
29. C. L. Anderson, P. S. Treuhaft, W. E. Pierce, and E. P. Horvath, "Degenerative Knee Disease among Dairy Farmers," in *Principles of Health and Safety in Agriculture*, J. A. Dosman and D. W. Cockcroft, Eds., CRC Press, Boca Raton, FL, 1989 pp. 367–379.
30. P. Louyot and R. Savin, "La coxarthrose chez l'agriculteur," *Rev. Rhum. Mal. Osteoartic.*, **33**, 625–632 (1966).
31. A. Thelin, "Work and Health among Farmers," *Scand. J. Soc. Med.*, **8**(Suppl. 22), 5–125 (1980).
32. C. Hulshof and B. V. vanZanten, "Whole-body Vibration and Low Back Pain: A

Review of Epidemiologic Studies," *Int. Arch. Occup. Environ. Health*, **59**, 205–220 (1987).

33. K. Husman, V. Notkola, R. Virolainen, K. Tupi, J. Nuutinen, J. Penttinen, and J. Heikkonen, "Farmers' Occupational Health Program in Finland, 1979–1988, From Research to Practice," *Scand. J. Work. Env. Health*, **14**(Supp. 1), 118–120 (1988).
34. J. Matthews, "Ergonomics And Farm Machinery," *J. Soc. Occup. Med.* **33**(3), 126–136 (1983).
35. W. Popendorf, K. J. Donham, D. N. Easton, and J. Silk, "A Synopsis of Agricultural Respiratory Hazards." *Am. Ind. Hyg. Assoc. J.*, **46**(3), 154–161 (1985).
36. J. R. Mulhausen, C. E. McJilton, P. T. Redig, and K. A. Janni, "Aspergillus and Other Human Respiratory Disease Agents in Turkey Confinement Houses," *Am. Ind. Hyg. Assoc. J.*, **48**(11), 894–899 (1987).
37. K. J. Donham, M. Rubino, T. D. Thedell, and J. Kammermeyer, "Potential Health Hazards to Agricultural Workers in Swine Confinement Buildings," *J. Occup. Med.*, **19**(6), 383–387 (1977).
38. K. J. Donham and W. Popendorf, "Ambient Levels of Selected Gases Inside Swine Confinement Buildings," *Am. Ind. Hyg. Assoc. J.*, **46**, 658–661 (1985).
39. K. J. Donham, L. J. Scallon, W. Popendorf, M. W. Treuhaft, and R. C. Roberts, "Characterization of Dusts Collected from Swine Confinement Buildings." *Am. Ind. Hyg. Assoc. J.*, **47**(7), 404–410 (1986).
40. K. J. Donham, P. Hagland, Y. Peterson, R. Rylander, and L. Belin, "Environmental and Health Studies of Farm Workers in Swedish Swine Confinement Buildings," *Br. J. Ind. Med.*, **46**, 31–37 (1989).
41. W. E. Burnett, "Air Pollution From Animal Wastes. Determination Of Malodors By Gas Chromatographic And Organoleptic Techniques," *Environ. Sci. Technol.*, **3**(8), 744–749 (1969).
42. J. A. Merkel, T. E. Hazen, and J. R. Miner, "Identification of Gases in a Confinement Swine Building Environment," *Trans. ASAE*, **12**, 310–315 (1969).
43. W. C. Banwart and J. M. Brenner, "Identification of Sulfur Gases Evolved from Animal Manures," *J. Environ. Qual.*, **4**(3), 363–366 (1975).
44. D. L. Morese and M. A. Woodbury, "Death Caused by Fermenting Manure," *J. Am. Med. Assoc.*, **245**(1), 63–64 (1981).
45. K. J. Donham, L. W. Knapp, R. Monson, and K. Gustafson, "Acute Toxic Exposure To Gases From Liquid Manure," *J. Occup. Med.*, **24**(2), 142–145 (1982).
46. S. R. Hagley and D. L. South, "Fatal Inhalation Of Liquid Manure Gas," *Med. J. Australia*, **2**, 459–460 (1983).
47. Anonymous, "Fatalities Attributed to Methane Asphyxia in Manure Waste Pits—Ohio, Michigan, 1989," *MMWR*, **38**(33), 583–586 (1989).
48. E. R. Hayhusrt and E. Scott, "Four Cases of Sudden Death in a Silo," *J. Am. Med. Assoc.*, **63**, 1570–1572 (1914).
49. R. R. Grayson, "Silage Gas Poisoning: Nitrogen Dioxide Pneumonia, a New Disease in Agricultural Workers," *Ann. Int. Med.*, **45**, 393–408 (1956).
50. T. Lowry and L. M. Schuman, "Silo-filler's Disease: A Syndrome Caused by Nitrogen Dioxide." *J. Am. Med. Assoc.*, **162**, 153–160 (1956).
51. W. H. Peterson, R. H. Burris, S. Rameshchandra, and H. N. Little, "Production of

Toxic Gas (Nitrogen Dioxides) in Silage Making," *Agric. Food Chem*, **6**, 121–126 (1958).

52. R. J. Ramirez and A. R. Dowell, "Silo Filler's Disease: Nitrogen Dioxide-Induced Lung Injury. Long-term Follow-up and Review of the Literature," *Ann. Intern. Med.*, **74**, 569–576 (1971).
53. E. G. Scott and W. B. Hunt, "Silo-filler's Disease," *Chest*, **63**, 701–706 (1973).
54. E. D. Horvath, G. A. do Pico, and R. A. Barbee et al., "Nitrogen Dioxide-Induced Pulmonary Disease," *J. Occup. Med.*, **20**, 103–110 (1978).
55. W. Popendorf, A. Pryor, and H. R. Wenk, "Mineral Dust in Manual Harvest Operations," *Ann. Am. Conf. Gov. Ind. Hyg.*, **2**, 101–115 (1982).
56. R. P. Sherman, M. L. Barman, and J. L. Abrahams, "Silicate Pneumoconiosis of Farm Workers," *Lab. Invest.*, **40**(5), 576–582 (1979).
57. W. Popendorf and H. R. Wenk, "Chrysotile Asbestos in a Vehicular Recreation Area: A Case Study," in *Environmental Effects of Off-Road Vehicles—Impacts and Management in Arid Regions*, H. G. Wilshire and R. H. Webb, Eds. Springer, New York, 1983, pp. 375–396.
58. W. M. Johnson, "Occupational Factors in Coccidioidomycosis," *J. Occup. Med.*, **23**(5), 367–374 (1981).
59. K. J. Donham, "Zoonotic Diseases of Occupational Significance in Agriculture: A Review," *Int. J. Zoonoses*, **12**, 163–191 (1985).
60. R. H. Casterton, "Enclosed Environments on Agricultural Tractors," *Ann. Am. Conf. Gov. Ind. Hyg.*, **2**, 121–127 (1982).
61. R. Rylander: Lung Diseases Caused by Organic Dusts in the Farm Environment, *Am. J. Ind. Med.*, **10**, 221–227 (1986).
62. R. Rylander, Y. Peterson, K. J. Donham, Eds., "Health Effects of Organic Dusts in the Farm Environment," Proceedings of an International Workshop held in Skokolster Sweden, April 23–25, 1985, *Am. J. Ind. Med.*, **10**, 193–340 (1986).
63. R. Rylander and Y. Peterson, Eds., "Proceedings of an International Workshop Held in Skokolster Sweden, October 24–27, 1988," *Am. J. Ind. Med.*, **17**(1), 1–148 (1990).
64. J. A. Merchant, "Agricultural Respiratory Diseases," *Sem. Respir. Med.*, **7**, 211–224 (1986).
65. I. I. Lutsky, G. L. Baum, H. Teichtahl et al. "Respiratory Diseases in Animal House Workers," *Eur. J. Respir. Dis.*, **69**, 29–35 (1986).
66. A. D. Blainey, M. D. Topping, S. Ollier, and R. J. Davies, "Respiratory Symptoms in Arable Farm Workers: Role of Storage Mites," *Thorax*, **43**, 697–702 (1988).
67. M. Van Hage-Hamsten, E. Ihre, O. Zetterström et al., "Bronchial Provocation Studies in Farmers with Positive RAST to the Storage Mite —*Lepidoglyphus destructor*," *Allergy*, **43**, 545–551 (1988).
68. K. J. Donham, "Hazardous Agents in Agricultural Dusts and Methods of Evaluation," *Am. J. Ind. Med.*, **10**, 205–220 (1986).
69. K. J. Donham, J. A. Merchant, W. Popendorf, and L. Burmeister, "Preventing Respiratory Disease in Swine Confinement Workers: Intervention through Applied Epidemiology, Education, and Consultation," *Am. J. Ind. Med.* **18**(3), 241–261 (1990).
70. A. Rask-Anderson, "Organic Dust Toxic Syndrome among Farmers," *Br. J. Ind. Med.*, **46**, 233–238 (1989).

71. J. M. Campbell, "Acute Symptoms Following Work with Hay," *Br. Med. J.*, **2**, 1143–1144 (1932).
72. M. Arden-Jones, "Farmer's Lung: An Overview and Prospects," *Ann. Am. Conf. Gov. Ind. Hyg.*, **2**, 172–182 (1982).
73. E. O. Terho, "Diagnostic Criteria for Farmer's Lung Disease," *Am. J. Ind. Med.*, **10**, 329 (1986).
74. D. S. Pratt, L. Stallones, D. Darrow, and J. J. May, "Acute Respiratory Illness Associated with Silo Unloading," *Am. J. Ind. Med.*, **10**, 328–329 (1986).
75. D. P. Morgan, C. F. Mutel, C. Cain, and K. J. Donham, "Pesticide Poisoning and Injuries: Where, When and How," Slide tape presentation No. AO-1941, General Services Administration, National Audiovisual Center, 1979.
76. M. Moses, "Pesticides," in W. H. Rom, Ed., *Environmental and Occupational Medicine,* Little, Brown, Boston, MA, 1983, pp. 547–571.
77. W. F. Durham and H. R. Wolfe, "Measurement of the Exposure of Workers to Pesticides," *Bull. World Health Org.*, **26**, 75–91 (1962).
78. W. Popendorf and J. T. Leffingwell, "Regulating OP Pesticide Residues for Farmworker Protection," *Residues Rev.*, **82**, 125–201 (1982).
79. R. R. Keenan and S. B. Cole, "A Sampling and Analytical Procedure for Skin-contamination Evaluation," *Am. Ind. Hyg. Assoc. J.*, **43**(7):473–476 (1982).
80. J. E. Davies, V. H. Freed, H. F. Enos, R. C. Duncan, A. Barquet, C. Morgade, L. J. Peters, and J. X. Danauskas, "Reduction of Pesticide Exposure with Protective Clothing for Applicators and Mixers," *J. Occup. Med.*, **24**(6), 464–468 (1982).
81. W. J. Hayes, "Monitoring Food and People for Pesticide Content," *Scientific Aspects of Pest Control*, National Research Council, National Academy of Sciences, Washington, D.C., Publication No. 1402, 1966, pp. 314–342.
82. R. A. Fenske, "Use of Fluorescent Tracers and Video Imaging to Evaluate Chemical Protective Clothing during Pesticide Applications," in *Performance of Protective Clothing: Second Symposium*, S. Z. Mansdorf, R. Sager, and A. P. Nielsen, Eds., American Society for Testing and Materials, Philadelphia, 1988, pp. 630–639.
83. M. A. Quinones, J. D. Bogen, D. B. Louria, A. el Nakah, and C. Hansen, Depressed Cholinesterase Activities among Farm Workers in New Jersey, *Sci. Total Environ.*, **6**, 155–159 (1976).
84. J. H. Wills and K. P. DuBois, "The Measurement and Significance of Changes in the Cholinesterase Activities of Erythrocytes and Plasma in Man and Animals," *CRC Crit. Rev. Toxicol.*, **1**, 153–202 (1972).
85. W. F. Durham, H. R. Wolfe, and J. W. Elliott, "Absorption and Excretion of Parathion by Spraymen," *Arch. Environ. Health*, **24**, 381–387 (1972).
86. D. E. Bradway, T. M. Shafik, and E. M. Lores, "Comparison of Cholinesterase Activity, Residue Levels, and Urinary Metabolite Excretion of Rats Exposed to Organophosphorus Pesticides," *J. Agric. Food Chem.*, **25**(6), 1353–1358 (1977).
87. D. P. Morgan, H. L. Hetzler, E. F. Slach, and L. I. Lin, "Urinary Excretion of Paranitrophenol and Alkyl Phosphates Following Ingestion of Methyl or Ethyl Parathion by Humans," *Arch. Environ. Contam. Toxicol.*, **6**, 159–173 (1977).
88. G. A. Wojeck, H. N. Nigg, J. H. Stamper, and D. E. Bradway, "Worker Exposure to Ethion in Florida Citrus," *Arch. Environ. Contam. Toxicol.*, **10**, 725–735 (1981).
89. C. A. Franklin, R. A. Fenske, R. Greenhalgh, L. Mathieu, H. V. Denley, J. T.

Leffingwell, and R. C. Spear, "Correlation of Urinary Pesticide Metabolite Excretion with Estimated Dermal Contact in the Course of Occupational Exposure to Guthion," *J. Toxicol. Environ. Health*, **7**(5), 715–731 (1981).

90. T. L. Lavy, J. S. Shepard, J. D. Mattice, "Exposure Measurements Of Applicators Spraying (2,4,5–Trichlorophenoxy)acetic Acid in the Forest," *J. Agric. Food Chem.*, **28**(3), 626–630 (1980).
91. H. N. Nigg and J. H. Stamper, "Field Studies: Methods Overview," in R. C. Honeycutt, G. Zweig, and N. N. Ragsdale, Eds., *Dermal Exposure Related to Pesticide Use*, American Chemical Society, Washington, DC, 1985, pp. 95–108.
92. J. Osterloh, R. Wang, F. Schneider, and K Maddy, "Biological Monitoring of Dichloropropene: Air Concentrations, Urinary Metabolite, and Renal Enzyme Excretion," *Arch. Environ. Health*, **44**(4), 207–213 (1989).
93. J. Yoder, M. Watson, and W. W. Benson, "Lymphocyte Chromosome Analysis of Agricultural Workers during Extensive Occupational Exposure to Pesticides," *Mut. Res.*, **21**, 335–340 (1973).
94. R. J. Korsak and M. M. Sato, "Effects of Chronic Organophosphate Pesticide Exposure on the Central Nervous System," *Clin. Toxicol.*, **11**(1), 83–95 (1977).
95. R. I. Glass, R. N. Lyness, D. C. Mengle, K. E. Powell, and E. Kahn, "Sperm Count Depression in Pesticide Applicators Exposed to Dibromochloropropane," *Am. J. Epidemiol.*, **109**(3), 346–351 (1979).
96. D. P. Morgan, L. I. Lin, and H. H. Saikaly, "Morbidity and Mortality in Workers Occupationally Exposed to Pesticides," *Arch. Environ. Contam. Toxicol.*, **9**(3), 349–382 (1980).
97. D. S. Sharp, B. Eskenazi, R. Harrison, P. Callas, and A. H. Smith, "Delayed Health Hazards of Pesticides Exposure," *Ann. Rev. Public Health*, **7**:441–471 (1986).
98. H. H. Wang and B. MacMahon, "Mortality of Pesticide Applicators," *J. Occup. Med.*, **21**(11), 741–744 (1979).
99. A. Blair, D. J. Grauman, J. H. Lubin, and J. F. Fraumeni, "Lung Cancer And Other Causes Of Death Among Licensed Pesticide Applicators," *J. Nat. Cancer Inst.*, **71**(1), 31–37 (1983).
100. B. MacMahon, R. R. Monson, H. H. Wang, and T. Zheng, "A Second Follow-Up of Mortality in a Cohort of Pesticide Applicators," *J. Occup. Med.*, **30**(5), 429–432 (1988).
101. D. Whorton, R. M. Krauss, S. Marshall, and T. H. Mibly, "Infertility in Male Pesticide Workers," *Lancet*, **7**, 1259–1261 (1977).
102. G. E. Carman, Y. Iwata, J. L. Pappas, J. R. O'Neal, and F. A. Gunther, "Pesticide Applicator Exposure To Insecticides During Treatment Of Citrus Trees With Oscillating Boom And Airblast Units," *Arch. Environ. Contam. Toxicol.*, **11**(6), 651–659 (1982).
103. Y. H. Atallah, W. P. Cahill, and D. M. Whitacre, "Exposure of Pesticide Applicators and Support Personnel to *o*-Ethyl-*o*-(4–nitrophenyl)phenylphosphonothioate (EPN)," *Arch. Environ. Contam. Toxicol.*, **11**(2), 219–225 (1982).
104. H. N. Nigg and J. H. Stamper, "Exposure of Spray Applicators and Mixer-Loaders to Chlorobenzilate Miticide in Florida Citrus Groves." *Arch. Environ. Contam. Toxicol.*, **12**, 477–482, (1983).
105. J. M. Devine, G. B. Kinoshita, R. P. Peterson, and G. L. Picard, "Farm Worker

Exposure to Terbufos [phosphorodithioic acid] during Planting Operations of Corn," *Arch. Environ. Contam. Toxicol.*, **15**, 113–119 (1986).

106. W. Popendorf, "Mechanisms of Clothing Exposure and Dermal Dosing during Spray Application," in *Performance of Protective Clothing: Second Symposium*, S. Z. Mansdorf, R. Sager, and A. P. Nielsen, Eds., American Society for Testing and Materials, Philadelphia, 1988, pp. 611–624.
107. H. N. Nigg, J. H. Stamper, and R. M. Queen, "The Development And Use Of A Universal Model To Predict Tree Crop Harvester Pesticide Exposure," *Am. Ind. Hyg. Assoc. J.*, **45**(3), 182–186 (1984).
108. W. F. Spencer, W. J. Farmer, and M. M. Cliath, "Pesticide Volatilization," *Residue Rev.*, **49**, 1–47 (1973).
109. J. Maybank, K. Yoshida, and R. Grover, "Spray Drift from Agricultural Pesticide Applications," *J. Air Pollut. Control Assoc.*, **28**, 1009–1014 (1978).
110. A. W. Taylor, "Post-application Volatilization of Pesticides under Field Conditions," *J. Air Pollut. Control Assoc.*, **28**, 922–927 (1978).
111. W. M. Draper, R. D. Gibson, and J. C. Street, "Drift from and Transport Subsequent to a Commercial Aerial Application of Carbofuran: an Estimation of Potential Human Exposure," *Bull. Environ. Contam. Toxicol.*, **26**, 537–543 (1981).
112. J. F. Armstrong, H. R. Wolfe, S. W. Comer, D. C. Staiff, "Oral Exposure of Workers to Parathion through Contamination of Food Items," *Bull. Environ. Contam. Toxicol.*, **10**(6), 321–327 (1973).
113. National Research Council, *Regulating Pesticides in Food*, National Academy Press, Washington, DC, 1987.
114. F. M. D'Itri and L. Wolfson, *Rural Groundwater Contamination*, Lewis Publishers, Chelsea, MI, 1987.
115. D. Fairchild, *Ground Water Quality and Agricultural Practices*, Lewis Publishers, Chelsea, MI, 1987.
116. J. H. Stamper, H. N. Nigg, W. D. Mahon, A. P. Nielsen, and M. D. Royer, "Pesticide Exposure to Greenhouse Foggers," *Chemosphere*, **17**(5), 1007–1023 (1988).
117. J. Liesivuori, S. Liukkonen, and P. Pirhonen, "Reentry Intervals after Pesticide Application in Greenhouses," *Scand. J. Work, Environ. Health*, **14**(Suppl. 1), 35–36 (1988).
118. T. H. Milby, F. Ottoboni, H. W. Mitchell, "Parathion Residue Poisoning Among Orchard Workers," *J. Am. Med. Assoc.*, **189**(5), 351–356 (1964).
119. W. J. Popendorf and J. T. Leffingwell, "Natural Variations in the Decay and Oxidation of Parathion Foliar Residues," *J. Agric. Food. Chem.*, **26**(2), 437–441 (1978).
120. R. C. Spear, W. Popendorf, W. F. Spencer, and T. H. Milby, "Worker Poisonings due to Paraoxon Residues," *J. Occup. Med.*, **19**(6), 411–414 (1977).
121. W. Popendorf, "Mechanisms of Clothing Exposure and Dermal Dosing during Spray Application," in *Performance of Protective Clothing: Second Symposium*, S. Z. Mansdorf, R. Sager, and A. P. Nielsen, Eds., American Society for Testing and Materials, Philadelphia, 1988, pp. 611–624.
122. E. F. Taschenberg, J. B. Bourke, D. F. Minnick, "An Air Filter-Pressurization Unit to Protect the Tractor Operator Applying Pesticides," *Bull. Environ. Contam. Toxicol.*, **13**(3), 263–268 (1975).
123. C. Lunchick, A. P. Nielsen, and J. C. Reinert, "Engineering Controls and Protective

Clothing in the Reduction of Pesticide Exposure to Tractor Drivers." in *Performance of Protective Clothing: Second Symposium*, S. Z. Mansdorf, R. Sager, and A. P. Nielsen, Eds., American Society for Testing and Materials, Philadelphia, 1988, pp. 605–610.

124. T. H. Lillie, J. M. Livingston, and M. A. Hamilton, "Recommendations for Selecting and Decontaminating Pesticide Applicator Clothing," *Bull. Environ. Contam. Toxicol.*, **27**(5), 716–723 (1981).
125. T. H. Lillie, R. E. Hampson, Y. Hishioka, and M. A. Hamilton, "Effectiveness of Detergent and Detergent Plus Bleach for Decontaminating Pesticide Applicator Clothing," *Bull. Environ. Contam. Toxicol.*, **29**(1), 89–94 (1982).
126. J. F. Stone and H. M. Stahr, "Pesticide Residues in Clothing," *J. Environ. Health*, **51**(5), 273–276 (1989).
127. J. E. Midtling, P. G. Barnett, M. J. Coye, A. R. Velasco, P. Romero, C. L. Clements, M. A. O'Malley, M. W. Tobin, T. G. Rose, and I. H. Monosson, "Clinical Management of Field Worker Organophosphate Poisoning," *West. J. Med.*, **142**(4), 514–518 (1985).
128. D. P. Morgan, *Recognition and Management of Pesticide Poisonings, Fourth Edition* Health Effects Division, U.S. Environmental Protection Agency, U.S. Government Printing Office, Washington, DC, 1989.
129. E. P. Savage, "Acute Pesticide Poisonings," *Pesticide Residue Hazards to Farm Workers*, Proceedings of a Workshop Held February 1976, HEW Publ. No. (NIOSH) 76-191, 1976, pp. 63–65.
130. G. D. Kleinman, "Occupational Disease in California Attributed To Pesticides And Other Agricultural Chemicals," Bureau of Occupational Health, Department of Public Health, State of California, 1965.
131. K. T. Maddy and S. Edmiston, "Selected Incidents of Illnesses and Injuries Related to Exposure to Pesticides Reported by Physicians in California in 1986," *Vet. Human Toxicol.*, **30**(3), 246–254 (1988).
132. L. B. L. De-Alwis and M. S. L. Salgado, "Agrochemical Poisoning in Sri Lanka," *Foren. Sci. Int.*, **36**(1/2), 81–89 (1988).
133. S. Milham, "Occupational Mortality in Washington State, 1950–1971, Vol. I–III," U.S. DHEW Publ, Nos. (NIOSH) 76-175 A,B,C, U.S. Government Printing Office, 1976.
134. P. Decoufle, K. Stanislawczyk, K. Houten, I. D. Bross, E. Viadana, "A Retrospective Survey of Cancer in Relation to Occupation," U.S. Department of Health Education and Welfare, DHEW (NIOSH) Publ. No. 77-178, 1977.
135. A. Blair and T. L. Thomas, "Leukemia Among Nebraska Farmers: A Death Certificate Study," *Am. J. Epidemiol.*, **110**(3), 264–273 (1979).
136. G. R. Petersen, "Occupational Mortality in the State of California, 1959–1961," U.S. DHEW (NIOSH) Publ. No. 80-104, U.S. Government Printing Office, Washington DC, 1980.
137. L. F. Burmeister, "Cancer Mortality in Iowa Farmers, 1971–1978," *J. Nat. Cancer Inst.*, **66**, 461–464 (1981).
138. A. Blair and D. W. White, "A Death Certificate Study of Leukemia among Farmers from Wisconsin," *J. Nat. Cancer Inst.*, **66**, 1027–1030 (1981).

139. K. P. Cantor, "Farming and Mortality from Non-Hodgkin's Lymphoma: A Case-Control Study," *Int. J. Cancer*, **29**(3), 239–247 (1982).
140. D. P. Buesching and L. Wollstadt, "Cancer Mortality Among Farmers," *J. Nat. Cancer Inst.*, **72**(3), 503–504 (1984).
141. K. P. Cantor and A. Blair, "Farming And Mortality From Multiple Myeloma: A Case-Control Study With The Use of Death Certificates," *J. Nat. Cancer Inst.*, **72**(2), 251–255 (1984).
142. H. A. Stubbs, J. Harris, and R. C. Spear, "A Proportionate Mortality Analysis of California Agricultural Workers, 1978–1979," *Am. J. Ind. Med.*, **6**, 305–320 (1984).
143. E. Delzell and S. Grufferman, "Mortality Among White And Nonwhite Farmers In North Carolina, 1976–1978," *Am. J. Epidemiol.*, **121**(3), 391–402 (1985).
144. M. C. Schumacher, "Farming Occupations and Mortality from Non-Hodgkin's Lymphoma in Utah—a Case-control Study," *J. Occup. Med.*, **27**, 580–584 (1985).
145. L. Hardell and O. Axelson, "Phenoxyherbicides and Other Pesticides in the Etiology of Cancer: Some Comments on Swedish Experiences," in *Cancer Prevention: Strategies in the Workplace*, C. E. Becker and M. J. Coye, Eds., Hemisphere Publishing, Washington, DC, 1986, pp. 107–119.
146. M. Yamamoto, K. Endoh, S. Toyama, H. Sakai, N. Shibuya, S. Takagi, J. Magara, and K. Fujiguchi, "Biliary Tract Cancers in Japan: A Study from the Point of View of Environmental Epidemiology," *Acta Med. Biol.*, **34**(2), 65–76 (1986).
147. S. K. Hoar, A. Blair, F. F. Holmes, C. D. Boysen, R. J. Robel, R. Hoover, and J. F. Fraumeni, "Agricultural Herbicide Use and Risk of Lymphoma and Soft-Tissue Sarcoma," *J. Am. Med. Assoc.*, **256**(9), 1141–1147 (1986).
148. A. Blair, H. Malker, K. P. Cantor, L. Burmeister, and K. Wiklund, "Cancer among Farmers: A Review," *Scand. J. Work Environ. Health*, **11**, 397–407 (1985).
149. M. Schenker and S. McCurdy, "Pesticides, Viruses, and Sunlight in the Etiology of Cancer Among Agricultural Workers," in *Cancer Prevention: Strategies in the Workplace*, C. E. Becker and M. J. Coye, Eds., Hemisphere Publishing, Washington, DC, 1986, pp. 29–37.
150. R. L. Zuehlke, "Common Cutaneous Problems in Agricultural Work," in *Proceedings of Conference on Agricultural Health and Safety*, Society of Occupational and Environmental Health, Environmental Sciences Laboratory, New York, 1975, pp. 54–67.
151. S. Helmers, F. H. Top, and L. W. Knapp, "Ammonia Injuries in Agriculture," *J. Iowa Med. Soc.*, **61**(5), 271–280 (1971).
152. I. Kass, N. Zamel, C. A. Dobry, and M. Holzer, "Bronchiectasis Following Ammonia Burns of the Respiratory Tract: A Review of Two Cases," *Chest*, **62**, 282–285 (1972).
153. J. A. Davies, Ed., *Ground Water Protection SW 8–86*, U.S. Environmental Protection Agency, 1980, pp. 1–19.
154. J. White, Jr., "Relative Significance of Dietary Sources of Nitrate and Nitrite," *J. Agric. Food Chem.*, **23**, 886–891 (1975).
155. R. Zaldivar and H. Robinson, "Epidemiological Investigations on Stomach Cancer Mortality in Chilans: Association with Nitrate Fertilizer," *Z. Kerbsforsch.*, **80**, 289–295 (1973).
156. R. Zaldivar and W. H. Wetterstrand, "Further Evidence of a Positive Correlation between Exposure to Nitrate Fertilizer (NaNO3) and Gastric Cancer Death Rates: Nitrates and Nitrosamines," *Experientia*, **31**, 1354–1355 (1975).

157. R. Zaldivar, "Nitrate Fertilizers as Environmental Pollutants: Positive Correlation between Nitrates (NaNO3 and KNO3) Use per Unit Area and Stomach Cancer Mortality," *Experientia*, **33**, 264–265 (1977).

158. R. C. Shank, "Toxicology of *N*-nitroso-compounds," *Toxicol. Appl. Pharmacology*, **31**, 361–368 (1975).

159. J. A. Merchant, B. C. Kross, K. J. Donham, and D. S. Pratt, "Agriculture at Risk: A Report to the Nation," National Coalition for Agricultural Safety and Health, Iowa City, IA, 1989.

160. W. W. Spink, "The Significance of Bacterial Hypersensitivity in Human Brucellosis; Studies on Infection due to Strain 19," *Brucella Abortus*, **47**, 861–873 (1957).

161. A. H. Keeney and M. C. Hunter, "Human Infection with Newcastle Virus of Fowls," *Arch. Opthalmol.*, **44**, 573–580 (1950).

162. U. W. Leavell, Jr., M. S. McNamara, R. Muelling, et al., "Orf—Report of 19 Human Cases with Clinical and Pathological Observations," *J. Am. Med. Assoc.*, **204**, 109–116 (1968).

163. T. Grandin, "Animal Handling and Farm Animal Behavior," *Vet. Clin. North Am. Food Anim. Pract.*, **3**, 324–336 (1987).

164. R. W. Lyons, C. L. Samples, H. N. DeSilva et al., "An Epidemic of Resistant Salmonella in a Nursery—Animal-to-Human Spread," *J. Am. Med. Assoc.*, **243**, 546–547 (1980).

165. P. N. Acha and B. Szyfres, "Zoonoses and Communicable Diseases Common to Man and Animals," Scientific Publication No. 354, Pan American Health Organization, Washington DC, 1980.

166. K. J. Donham, "Infectious Diseases Common to Animals and Man of Occupational Significance to Agricultural Workers," In *Proceedings of Conference on Agricultural Health and Safety*, New York Society for Occupational and Environmental Health, Environmental Sciences Laboratory, New York, 1975, pp. 160–175.

167. K. J. Donham and C. Mutel, "Zoonoses: An Overview," slide tape presentation No. WC-950, General Services Administration, National Audiovisual Center, 1978.

168. K. Hedberg, K. White, J. Forfang, et al., "An Outbreak of Psitacosis in Minnesota Turkey Industry Workers: Implications for Modes of Transmission and Control," *Am. J. Epidemiol.*, **130**, 569–577 (1989).

169. P. R. Schnurrenberger and W. T. Hubbert, *An Outline of Zoonoses*, Iowa State University Press, Ames, IA, 1981.

170. R. L. Zuehlke, C. F. Mutel, and K. J. Donham, "Skin Diseases of Agricultural Workers," Slide presentation no. AO-7249, General Services Administration, National Audiovisual Center, 1983.

171. D. Hogan and P. Lane, "Dermatologic Disorders in Agriculture," *Occup. Med. State Art Rev.*, **1**, 285–300 (1986).

172. C. L. Wand, "The Problem of Skin Diseases in Industry," Office of Occupational Safety and Health Statistics, U.S. Department of Labor, U.S. Government Printing Office, Washington, DC, 1978.

173. W. B. Whiting, "Occupational Illness and Injuries of California Agriculture Workers," *J. Occup. Med.*, **17**(3), 177–181 (1975).

174. R. M. Caplan, "Cutaneous Hazards Posed by Agricultural Chemicals," *J. Iowa Med Soc.*, **59**(4), 295–299 (1969).

175. J. C. TeLintum and J. P. Nater, "Allergic Contact Dermatitis Caused by Rubber Chemicals in Dairy Workers," *Dermatologica*, **148**, 42–44 (1974).
176. D. Burrows, "Contact Dermatitis in Animal Feed Mill Workers," *Br. J. Dermatol.*, **92**, 167–170 (1975).
177. R. D. Peachey, "Skin Hazards in Farming," *Br. J. Dermatol.*, **105**(Suppl. 21), 45–50 (1981).
178. J. S. Pasricha and R. Gupta, "Contact Dermatitis Due to Calcium Ammonium Nitrate," *Contact Dermatitis*, **9**, 149 (1983).
179. J. C. Whitakar, W. R. Lee, and J. E. Downes, "Squamous Cell Cancer in Northwest of England: 1967–1969 and its Relation to Occupation," *Br. J. Ind. Med.*, **36**, 43–51 (1979).
180. U. W. Leavell and J. A. Phillips, "Milker's Nodules," *Arch. Dermatol.*, **111**, 1307–1311 (1975).
181. L. Chmel, J. Buchvald, and M. Valentova, "Ringworm Infection Among Agricultural Workers," *Int. J. Epidemiol.*, **5**(3), 291–295 (1976).
182. W. L. Krinsky, "Dermatoses Associated with the Bites of Mites and Tics," *Int. J. Dermatol.*, **22**(2), 75–91 (1983).
183. D. M. Lierle and S. H. Reger, "The Effect of Tractor Noise on the Auditory Sensitivity of Tractor Operators," *Ann. Otol. Rhin. Laryng.*, **67**, 372–388 (1958).
184. M. Feldman and C. D. E. Downing, "Tractor Noise Pollution on the Farm—Problems and Recommendations," *Can. Agric. Eng.*, **14**(1), 2–5 (1972).
185. H. H. Jones and J. L. Oser, "Farm Equipment Noise Exposure Levels," *Am. Ind. Hyg. Assoc. J.*, **29**(2), 146–151 (1968).
186. E. W. Simpson and I. L. Deshayes, "Tractors Produce Ear Damaging Noise," *J. Environ. Health*, **31**(4), 347–350 (1969).
187. N. W. Sullivan, R. D. Schneider, and K. Von-Bargen, "Noise Exposure Patterns of Agricultural Employees," *Prof. Saf.*, **26**(12), 16–21 (1981).
188. C. E. McJilton and R. A. Aherin, "Getting the Message to the Farmer," *Am. Ind. Hyg. Assoc. J.*, **43**(6), 469–471 (1982).
189. H. Une, S. H. Schuman, S. T. Caldwell, and N. H. Whitlock, "Agricultural Life-Style: A Mortality Study among Male Farmers in South Carolina, 1983–1984" *South. Med. J.*, **80**(9), 1137–1140 (1987).
190. C. W. Suggs, L. F. Stikeleather, and C. F. Abrams, "Field Tests of an Active-Seat Suspension for Off-Road Vehicles," *Am. Soc. Agric. Eng.*, **13**(5), 608–611 (1970).
191. M. Takamatsu, T. Sakurai, and C. P. Chang, "Vibration Disease among Farmers Induced by Vibration Tools," Proceedings of the VII International Congress of Rural Medicine, Salt Lake City, Utah, Sept. 17–21, 1978, *Int. Assoc. Agric. Med.*, 188–190 (1978).
192. W. E. Field and R. L. Tormoehlen, "Impact of Physical Handicaps on Operators of Agricultural Equipment," *Appl. Ergonomics*, **16**(3), 179–182 (1985).
193. L. F. Burmeister, S. F. Van Lier, and P. Isacson, "Leukemia in Farm Practices in Iowa," *Am. J. Epidemiol.*, **115**, 720–728 (1982).
194. L. F. Burmeister, G. D. Everett, S. Van Lier, and P. Isacson, "Selected Cancer Mortality in Farm Practices in Iowa," *Am. J. Epidemiol.*, **118**, 72–77 (1983).
195. S. Milham, "Leukemia and Multiple Myeloma in Farmers," *Am. J. Epidemiol.*, **94**, 307–310 (1971).

196. A. Blair, G. Everett, K. Cantor, R. Gibson, L. Schuman, P. Isacson, W. Blattner, and S. Van Lier, "Leukemia and Farm Practices," *Am. J. Epidemiol.*, **122**, 535 (1985).
197. K. J. Donham, M. J. Van Der Maaten, J. M. Miller, B. C. Kruse, N. J. Rubino, "Serioepidemiologic Studies of the Possible Relationships of Human and Bovine Leukemia: Brief Communication," *J. Nat. Cancer Inst.*, **59**, 851–853 (1977).
198. K. J. Donham, J. Berg, and B. Sawin, "Epidemiologic Relationships of the Bovine Population and Human Leukemia in Iowa," *Am. J. Epidemiol.*, **112**, 80–92 (1980).
199. K. J. Donham, L. Burmeister, S. vanLier, and T. Greiner, "Relationships of Bovine Leukemia Virus Prevalence in Dairy Herds and Density of Dairy Cattle, to Human Lymphocytic Leukemia," *Am. J. Vet. Res.*, **48**, 235–238 (1987).
200. E. Elam and R. Rauncy, *Rural Health Crisis (A Project Report)*. Northwest Services, Inc., Mound City, MO, 1986.
201. W. E. Field, "Effects of Stress on the Performance of Agricultural Equipment Operators," *SAE Tech. Paper Series, 800932*, Warrendale, PA, 1980.
202. L. M. Haverstock, "Farm Stress: Research Considerations," in J. A. Dosman and D. W. Cockcroft, Eds., in *Principles of Health and Safety in Agriculture*, CRC Press, Boca Raton, FL, 1989, pp. 381–384.
203. C. N. Larson, S. Kuperman, and R. E. Smith, "Rural Psychiatry: A Definition of the Field," in *Principles of Health and Safety in Agriculture*, J. A. Dosman and D. W. Cockcroft, Eds., CRC Press, Boca Raton, FL, 1989, pp. 385–388.
204. W. G. Hollister, "Innovations in Mental Health Service Delivery in Rural Areas," in *Principles of Health and Safety in Agriculture*, J. A. Dosman and D. W. Cockcroft, Eds., CRC Press, Boca Raton, FL, 1989, pp. 399–401.
205. R. Reotutar, "The Ominous Spread of Borrelia Burgdorferi Infection," *J. Am. Vet. Med. Assoc.*, **194**, 1387–1391 (1989).
206. B. S. Schwartz and M. D. Goldstein, "Lyme Disease: A Review for the Occupational Physician," *J. Occup. Med.*, **31**, 735–742 (1989).
207. R. B. Hayes, J. P. Van Nieuwenhuize, J. W. Raatgever, and F. J. W. Ten Kate, "Aflatoxin Exposures in the Industrial Setting: an Epidemiological Study of Mortality," *Fd. Chem. Toxicol.*, **22**(1), 39–43 (1984).
208. C. S. Baxter, H. E. Wey, and W. R. Burg, "A Prospective Analysis of the Potential Risk Associated with Inhalation of Aflatoxin-Contaminated Grain Dusts," *Fd. Cosmet. Toxicol.*, **19**, 765–769 (1981).
209. S. G. Schmitt and C. R. Hurburgh, "Measurement and Distribution of Aflatoxin in 1983 Iowa Corn," ASAE paper No. 85–3533, *Am. Soc. Agric. Engineers*, St. Joseph, MI, 1985.
210. W. G. Sorensen, J. P. Simpson, M. J. Peach, T. D. Thedell, and S. A. Olenchock, "Aflatoxin in Respirable Corn Dust Particles," *J. Toxicol. Environ. Health*, **7**, 669–672 (1981).
211. O. Shotwell and W. Burg, "Aflatoxin in Corn: Potential Hazard to Agricultural Workers," *Ann. Am. Conf. Gov. Ind. Hyg.*, **2**, 69–86 (1982).
212. W. K. Viscusi, *Risk By Choice: Regulating Health and Safety in the Workplace*, Harvard University Press, MA, 1983.
213. C. M. Berry, "Organized Research in Agricultural Health and Safety," *Am. J. Public Health*, **55**(3), 424–428 (1965).

214. J. Gay, K. Donham, S. Leonard, "Iowa Agricultural Health and Safety Service Project," *Am. J. Ind. Med.* **18**(4), 385–389 (1990).

215. S. Höguland, "Farmer's Occupational Health in Sweden—10 years Experience," 10th International Congress of International Medicine and Rural Health, Pecs, Hungary, 1989.

216. C. Tevis, "Ag Schools Out of Step" *Successful Farming*, **87**(12), 17 (1989).

217. M. A. Purschwitz and W. E. Field, "Federal Funding for Farm Safety Relative to Other Safety Programs," *Proc. of the 1987 National Institute for Farm Safety Summer Meeting*, National Institute for Farm Safety (NIFS), Columbia, MO, 1987.

218. S. Höglund, "Farmers' Occupational Health Care—Worldwide," *Am. J. Ind. Med.*, **18**(4), 365–370 (1990).

CHAPTER TWENTY-ONE

Heat Stress: Its Effects, Measurement, and Control

John E. Mutchler, C.I.H.

1 INTRODUCTION

This chapter covers the health effects, environmental aspects, engineering control, and management of heat stress in industry. Heat represents one of the classical physical stresses in the occupational environment. Overexposure to heat has been a problem for centuries. Only in modern times, however, have the physiological and environmental characteristics of heat exposure been related sufficiently to develop a good understanding of the safe limits of exposure and suitable control methods for work in hot environments.

1.1 Significance of Heat Stress in Industry

As industry has developed, through the Industrial Revolution to our present highly technological society, on-the-job potential for injury and illness from acute exposure to heat has increased far beyond that known earlier to home-centered craftsmen. Among the more dangerous original industrial vocations were those using molten materials such as glass and metal. In these first "hot industries," the ever-present danger of burns, explosions, and spills of molten material were well-known and accepted, as were potential illness and death from very hard physical work in excessively hot environments.

Historically, except for slave laborers in conquered ancient lands, heat stress

Patty's Industrial Hygiene and Toxicology, Fourth Edition, Volume 1, Part A, Edited by George D. Clayton and Florence E. Clayton
ISBN 0-471-50197-2

likely manifested itself first as a serious occupational hazard among armies operating in warm climates. Interestingly, our present-day standards of good practice and emerging regulatory requirements for control of heat stress in the workplace stem largely from experience in the armed forces, including several studies at military training centers in the United States (1–4).

The mining industry has also provided much information on heat disorders and tolerance limits, not only because of traditionally substantial mortality and morbidity from overexposure, but also because of concern for reduced productivity that accompanies excessive thermal exposure in the workplace (5–9).

LIST OF ABBREVIATIONS USED IN CHAPTER TWENTY-ONE

AET	Allowable exposure time as calculated from elements of the heat stress index (HSI)
C	Rate of heat exchange (net) by convection between an individual and the environment
C	Ceiling limits (WBGT) recommended by NIOSH for all workers in hot jobs
E	Rate of heat loss by evaporation of water from the skin
EDZ	Environment-drive zone—the range of heat load ($M + R + C$) beyond the prescriptive zone in which physiological response is affected drastically by the thermal environment
E_{max}	Maximum evaporative heat loss by water vapor uptake in the air at prevailing meteorologic conditions
E_{req}	Heat loss required solely by evaporation of sweat to maintain body heat balance
ET	Effective temperature—an index used to estimate the effect of temperature, humidity, and air movement on the subjective sensation of warmth
ETCR	Effective temperature corrected for radiation—an index for estimating the effect of temperature, humidity, and air movement on the subjective sensation of warmth using globe temperature rather than air temperature
GT	The temperature inside a blackened, hollow, thin copper globe measured by a thermometer whose sensing element is at the center of the sphere
HSI	An index of heat stress derived from the ratio of E_{req} to E_{max}
M	Rate of transformation of chemical energy into energy used for performing work and producing heat
MRT	The mean radiant (surface) temperature of the material and objects totally surrounding an individual
NWB	The wet-bulb temperature measured under conditions of the prevailing (natural) air movement
PZ	Prescriptive zone—that range of environmental heat load ($M + R + C$) at which the physiologic strain (heart rate and core body temperature) is independent of the thermal environment

R	Rate of heat exchange by radiation between two radiant surfaces of different temperatures
RAL	Recommended alert limits (WBGT) specified by NIOSH for unacclimated, healthy workers
REL	Recommended exposure limits (WBGT) specified by NIOSH for acclimated, healthy workers
RH	Relative humidity—the ratio of the water vapor pressure in the ambient air (VP_A) to the water vapor pressure in saturated air at the same temperature
ΔS	Change in heat content of the body
T_A	The temperature of the air surrounding a body (also dry-bulb temperature)
TLV	Threshold limit values specified by the American Conference of Governmental Industrial Hygienists
T_S	The mean of skin temperatures taken at several locations and weighted for skin area
ULPZ	The level of heat stress at the interface of the PZ (upper limit) and the EDZ (lower limit)
V	Air velocity
VP_A	The partial pressure exerted by water vapor in the air
VP_S	Vapor pressure exerted by water on the skin
WBGT	Wet bulb globe temperature—an empirical index of heat stress obtained by weighting NWB, GT, and T_A (outdoors with solar load)
WGT	Wet globe temperature—an empirical index of heat stress as measured within a 3-in. copper sphere covered by a black cloth kept at 100 percent wettedness

Historically, the workers in most hot jobs have been a highly select population. Those who cannot cope with the prevailing hot conditions seek less demanding employment. As a result of this selection process, the majority of workers in hot jobs perform at a high level and are highly adaptable to work in heat (10).

In addition to ever-increasing levels of mechanization, there are widespread, often undocumented work practices in industry that help relieve workers of heat strain on excessively hot days. Such practices include the following:

1. Performing only unavoidable operations and postponing other less important tasks.
2. Reassigning workers in auxiliary jobs to help those who work in the hottest areas.
3. Substituting younger and more fit workers to relieve the older and less fit.

In the "hot-dry" industries, such as steel mills, forge shops, and glass manufacturing plants, the thermal load on a worker increases from the sensible heat that escapes from process equipment and operations into the surrounding work space. The major factor is radiant heat from the surfaces of tanks, hot metals, and the like; the heat load contributed by the surrounding hot air is secondary. In this type of hot industry, little or no moisture is added to the air, except by changing weather conditions. Therefore little or no decrement occurs in the evaporative cooling capacity of the work environment. Nevertheless, the requirements for sweating and evaporation are considerable because of the radiant and convective thermal loads.

In the "warm-moist" industries, such as laundries, papermaking, and mines, in which large amounts of water are used for dust control, workplace humidity rises from wet processes or escaping steam. Although relatively little sensible heat transfers from the process to the work environment, high relative humidities greatly reduce the evaporative cooling capacity. Consequently, workers may be unable to dispose adequately of metabolic heat as well as any additional heat gained from the environment.

Today we would expect deaths and heat-related illnesses from involuntary overwork to be rare in the United States. One of the characteristic features of industrial exposure to heat today is the relatively infrequent occurrence of frank illness. This is due largely to the body's capability to maintain a thermal balance with the environment, including voluntary reduction in the metabolic generation of heat when necessary.

Universal as heat may be, heat stress in modern industry tends to be either a potentially serious issue, especially in the so-called hot industries, or a problem that only occasionally requires industrial hygiene attention. Although definitively accurate data are not available on the frequency of heat casualties (11), and although heat stress is widely recognized as a serious potential hazard, most industrial hygienists give heat exposure relatively low priority in their day-to-day resolution of worker health hazards. A review of the *Statistical Abstracts of the United States* for the number of workers in industries where heat stress is a potential health hazard shows that a conservative estimate would be 5 to 10 million workers (12).

As our technology advances, we are also discovering that the interaction of heat and other chemical and physical stresses imposes a concern above and beyond exposure to heat alone. In part, these combined or interactive effects are attributable to certain peculiar characteristics of heat on physiological mechanisms. Such effects underscore a more general, growing concern among environmental health investigators that the combined action of two or more physical and/or chemical stresses can lead to adverse effects that are unacceptable in view of today's standards of comfort and societal prohibitions against health impairment. Studies of such interactions not only indicate the complexity of multiple stresses, but also raise fundamental questions about the validity of exposure standards for physical stresses and toxic substances based on single-stressor investigations.

1.2 Development of Heat Stress Standards

The first American standard for work in hot environments dates back to 1947, when the Committee on Atmospheric Comfort published its report, "Thermal Standards in Industry" (13). The criteria and permissible exposure limits in this standard were intended only as a guide. The committee recommended that each industry develop its own standard because of the complexity of industrial work and the variation in tolerance to heat among individual workers. As general criteria, the committee cited the comfort and health of individuals, their work output, and their physiological and psychological reaction to work. It used an index known as "effective temperature" (ET) as the basis of its guidelines, but the guidelines included exposure limits for only two levels of work. The higher level was 432 kcal/hr, and the lower was given only in the qualitative terminology, "light sedentary activities."

1.2.1 Current Status—OSHA

The only federal standard in the United States for occupational exposure to heat stress is that implicit in the "general duty clause" of the Occupational Safety and Health Act of 1970 (14). This legislation stipulates that each employer has a general duty to provide each employee "employment free from recognized hazards causing or likely to cause physical harm." The Occupational Safety and Health Administration (OSHA) has received recommendations from both the National Institute for Occupational Safety and Health (NIOSH) and a statutory Standards Advisory Committee on Heat Stress, but a specific OSHA standard covering exposure to heat has not been issued.

Any heat stress standard proposed by OSHA would likely be a "work practices standard." This means that when the value of an environmental index, such as wet bulb globe temperature (WBGT), reaches a threshold level based on various levels of work load and possibly other environmental criteria, certain actions and procedures would be required to minimize the effect of exposure. These work practices would undoubtedly include the following:

1. Modifying the bodily heat production (metabolic rate) for the task being performed.
2. Limiting the number and duration of exposures.
3. Reducing heat exchange with the thermal environment by feasible engineering controls and, if necessary and appropriate, by protective clothing.

1.2.2 Threshold Limit Values

In 1971, the American conference of Governmental Industrial Hygienists (ACGIH) published a notice of intent to establish threshold limit values (TLVs) for heat stress (15). These TLVs rest on the assumption that acclimatized, fully clothed workers whose deep body temperatures are 38°C (100.4°F) or less are not subjected to excessive heat stress. The ACGIH adopted the TLVs in 1974, and they have remained essentially unchanged (16).

Technology continues to advance in physiological monitoring, including indications of heat strain, but it is not practical to monitor workers' deep body temperatures. Therefore industrial hygienists must rely primarily on the measurement of certain environmental factors that correlate with physiological responses to heat. Of the several environmental indices of heat stress in common use, the WBGT,

Table 21.1. Permissible Heat Exposure TLVs (16)

	Permissible TLV [°C (°F) WBGT]		
Work–Rest Regimen	Light Work Load, ≤ 200 kcal/hr	Moderate Work Load, 200–350 kcal/hr	Heavy Work Load, 350–500 kcal/hr
Continuous work	30.0 (86)	26.7 (80)	25.0 (77)
75% work, 25% rest each hour	30.6 (87)	28.0 (82)	25.9 (79)
50% work, 50% rest each hour	31.4 (89)	29.4 (85)	27.9 (82)
25% work, 75% rest each hour	32.2 (90)	31.1 (88)	30.0 (86)

in the opinion of ACGIH, remains the most suitable index for assessing heat stress for exposed workers.

Table 21.1 shows the TLVs for heat stress. These TLVs emphasize the significance of work load level and the percentage of time spent working. The specific WBGT values in Table 21.1 were derived from data in the literature correlating deep body temperatures with work load and environmental conditions. In describing the TLVs, the Physical Agents Committee of ACGIH points out the following considerations (15, 16):

1. The permissible exposure limits in the TLVs assume that the WBGT value of the resting place is the same or very close to that of the workplace. If the resting place is air-conditioned, a time-weighted average value should be used for both environmental and metabolic heat.
2. The TLVs also assume use of light summer clothing worn customarily by workers when employed under hot environmental conditions. If special clothing is required which impedes sweat evaporation or has higher insulation value, the worker's heat tolerance is reduced, and the TLVs do not apply.

1.2.3 NIOSH Recommendations

The National Institute for Occupational Safety and Health (NIOSH) first published its *Criteria for a Recommended Standard . . . . Occupational Exposure to Hot Environments* in 1972 (10). Again, NIOSH specified WGBT as the index to be used for invoking work practices to reduce the risk of adverse effects from heat exposure, including interactions between excessive heat and toxic chemical and physical agents.

The NIOSH Criteria Document differed from the ACGIH TLV in several respects. A threshold WBGT was specified at 26°C (70°F) for continuous heavy work over 1 hr. NIOSH recommended that when exposure exceeds this level, any one or a combination of work practices should be initiated to ensure that body core temperature does not exceed 38.0°C (100.4°F). In addition, the Criteria Document outlined other requirements for management when exposures exceed 26°C (79°F)

Table 21.2. Threshold WBGT Values Recommended by the Standards Advisory Committee (18)

	Threshold WBGT Values in °C (°F)	
Work Load	Low Air Velocity, ≤1.5 m/sec (≤300 fpm)	High Air Velocity, >1.5 m/sec (>300 fpm)
Light (level 2), ≤200 kcal/hr (≤795 Btu/hr)	30 (86)	32 (90)
Moderate (level 3), 200–300 kcal/hr (800–1190 Btu/hr)	28 (82)	31 (87)
Heavy (level 4), >300 kcal/hr (1190 Btu/hr)	26 (79)	29 (84)

WBGT. These include preplacement and periodic medical examinations, WBGT profiles for each workplace, and detailed record keeping.

The NIOSH Criteria Document received widespread negative response from the hot industries, especially in the southern states, where outdoor WBGT values exceed 79°F nearly every day during the warm seasons (17). In 1973, the U.S. Secretary of Labor appointed a Standards Advisory Committee on Heat Stress to help resolve some of the discrepancies in and objections to the NIOSH Criteria Document relative to industrial experience. This panel reported its "Recommended Standard for Work in Hot Environments" to the Secretary of Labor in January, 1974 (18).

The Standards Advisory Committee modified the recommendations in the 1972 NIOSH Criteria Document by differentiating between threshold WBGT values on the basis of both work load and air speed. Table 21.2 summarizes the threshold WBGT values recommended by the Standards Advisory Committee.

Perhaps the most important aspect of the Standards Advisory Committee's work was the status of the final draft report; it was not a consensus recommendation. The Committee voted 10 to 5 that the draft be the basis of OSHA rule making, although there was substantial support for the idea that a heat stress standard was not appropriate at that time (17).

In 1986, NIOSH updated its Criteria Document to reflect continuing research and experience on assessing, evaluating, and controlling occupational heat stress (19). In its revised recommendations, NIOSH relies on the use of recommended alert limits (RALs) and recommended exposure limits (RELs). It asserts that nearly all healthy workers who are not acclimatized to working in hot environments and who are exposed to combinations of environmental and metabolic heat less than the NIOSH RALs should be able to work without increased risk of acute adverse health effects. Healthy workers who are heat-acclimatized to working in hot environments and who are exposed to combinations of environmental and metabolic heat less than the NIOSH RELs should be capable of working without adverse effects. The estimates of both environmental and metabolic heat are expressed as 1-hr time-weighted averages (TWAs) (19).

NIOSH also recommends that no worker be exposed at combinations of environmental and metabolic heat exceeding its "ceiling limits" without adequate heat-protective clothing and equipment.

In the revised Criteria Document, "healthy workers" refers to those not excluded from placement in hot-environment jobs by any of several criteria. These exclusionary criteria are qualitative in that the epidemiologic parameters of sensitivity, specificity, and predictive power of the evaluation methods are not documented fully. The recommended exclusionary criteria represent the best judgment of NIOSH based on the available data and comments of peer reviewers. These include both absolute and relative indicators related to age, gender, percent body fat, medical and occupational history, specific chronic diseases or therapeutic regimens, and the results of clinical tests such as maximum aerobic capacity, electrocardiography, pulmonary function, and chest X-rays.

A brief summary of key recommendations in the revised NIOSH Criteria Document follows (19):

A. *Unacclimatized Workers*: Total heat exposure to workers should be controlled so that unprotected healthy workers not yet acclimatized to working in hot environments are exposed to combinations of metabolic and environmental heat no greater than the applicable RALs shown in Figure 21.1.

B. *Acclimatized Workers*: Total heat exposure to workers should be controlled so that unprotected healthy workers who are acclimatized to working in hot environments are exposed to combinations of metabolic and environmental heat no greater than the applicable RELs shown in Figure 21.2.

C. *Ceiling Limits*: No worker should be exposed to combinations of metabolic and environmental heat exceeding the applicable ceiling limits (CL) in Figures 21.1 and 21.2 unless provided with and properly using appropriate and adequate heat-protective clothing and equipment.

D. *Environmental Heat Exposures*: These are assessed by the WBGT method or an equivalent technique with conversion to WBGT values. When air- and vapor-impermeable protective clothing is worn, the dry-bulb temperature or the adjusted dry-bulb temperature is a more appropriate measurement than WBGT. Environmental heat measurements should be taken at or as close as possible to the work area where exposure takes place. When a worker in a hot job moves between two or more areas with variable levels of environmental heat, or when the environmental heat varies substantially in the single hot area, the environmental heat exposures must be measured at each area and during each period of constant heat levels where employees are exposed. Hourly TWA WBGTs must be calculated for the combination of jobs (tasks), including all scheduled and unscheduled rest periods.

E. *Modifications of Work Conditions*: Environmental heat measurements must be taken at least hourly during the hottest portion of each work shift, during

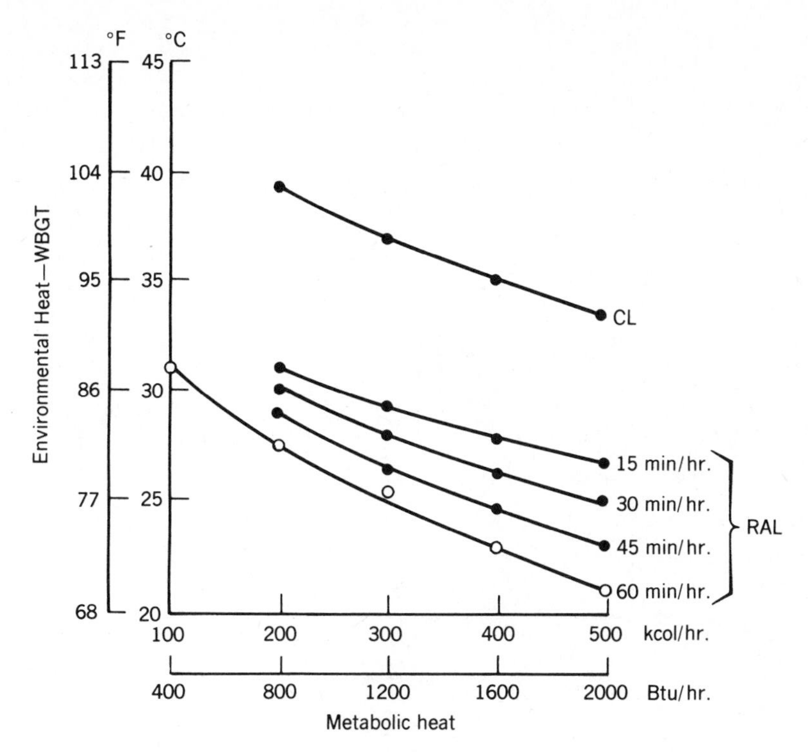

Figure 21.1. Recommended heat-stress alert limits (RALs) for heat-unacclimatized workers (19). CL = ceiling limit.

the hottest months of the year, and when a heat wave occurs or is predicted. If two such sequential measurements exceed the applicable RAL or REL, work conditions must then be modified by use of appropriate engineering controls, work practices, or other measures until two sequential measures conform with the exposure limits of the NIOSH-recommended standard.

1.2.4 International Standards and Recommendations

Several countries have developed and published standards, recommendations, and guidelines for limiting the exposure of workers to potentially harmful levels of occupational heat stress. These proposals range from official national regulations to unofficially suggested practices and guidelines proposed by institutions or agencies concerned with the health and safety of workers under conditions of high heat load. Most of these documents have in common the use of (1) WBGT as the index for expressing the environmental heat load, and (2) some method for estimating metabolic heat production. The permissible total heat load typically is expressed as a WBGT value for all levels of physical work ranging from resting to very heavy work (19).

In 1982 the International Organization for Standardization (ISO) adopted and

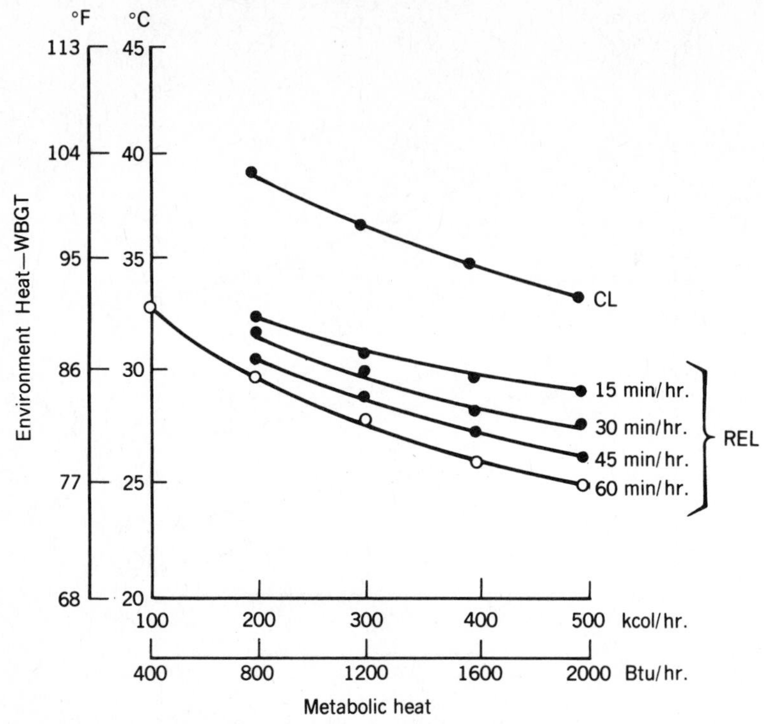

Figure 21.2. Recommended heat-stress exposure limits (RELs) for heat-acclimatized workers (19). CL = ceiling limit.

published an international standard on "Hot Environments—Estimation of the Heat Stress on Working Man Based on the WBGT Index (Wet Bulb Globe Temperature)" (20). The ISO standard closely resembles the ACGIH TLV for heat stress. The basic premise of both is that no worker be exposed to any combination of environmental heat and physical activity that would cause body core temperature to exceed 38°C (100.4°F). In addition, the ISO standard references tables for estimating the metabolic heat load. The ISO standard includes the classification of metabolic heat by examples of activities representative of five metabolic rates (rest, low metabolic rate, moderate metabolic rate, high metabolic rate, and very high metabolic rate). It presents upper-permissible WBGT values for each of these work load categories for both heat-acclimatized and heat-unacclimatized workers.

2 WORKER–ENVIRONMENT THERMAL EXCHANGE

An essential condition for health and normal body function is that body core temperature remain within the acceptable range of about 37°C (98.5°F) ± 1°C (1.8°F). This steady-state requirement depends on the controlled exchange of heat between the body and the environment. The rate and amount of necessary heat exchange are governed by the fundamental laws of thermodynamics. The heat that must be exchanged is a function of (1) the total heat produced by the body (metabolic heat) and (2) the heat gained, if any, from the environment. The rate of heat exchange with the environment depends essentially on air temperature and humidity, skin temperature, air velocity, evaporation of sweat, radiant temperature, and the type, amount, and characteristics of the clothing worn. Respiratory heat loss is generally of minor consequence except during vigorous activity in very dry environments.

2.1 Heat Balance Equation

The basic thermodynamic processes affecting heat exchange between the body and the environment can be described by the equation (21)

$$\Delta S = M + C + R - E$$

where ΔS = rate of change in intrinsic body heat content
M = rate of metabolism or heat produced within the body
C = rate of convective heat exchange
R = rate of radiative heat exchange
E = rate of evaporative heat loss

As outlined below, use of this equation requires measurements or reliable estimates of metabolic heat production, air temperature, air water-vapor pressure, air velocity, and mean radiant temperature.

2.2 Metabolic Heat Production

The energy required to sustain life and all bodily functions comes from the body by the exothermic oxidation, under enzymatic control, of food-derived fuel: carbohydrates, fats, and proteins (22). Although such oxidation occurs at rates and temperatures much different from those during equivalent combustion in the external environment, the thermodynamics of the oxidation is the same in both instances. Thus the same quantities of food substrates yield essentially the same thermal energy when oxidized at low temperature inside the body or at high temperature outside the body. This equivalency provides a basis for "indirect calorimetry," allowing metabolic heat production to be measured indirectly by the rate of oxygen uptake of the body. One liter of oxygen is about equivalent to a metabolic heat output of 5 kcal. An average man at rest consumes about 0.3 liter of oxygen per minute, equivalent to a metabolic heat production of 1.5 kcal/min or about 90 kcal/hr (22).

As indicated above, the energy cost of an activity (metabolic heat) is a major element in the heat exchange between the human body and the environment. The metabolic heat value (M) can be either measured or estimated. Estimation, with reference to accepted tables of energy expenditure, represents the prevailing op-

Table 21.3. Estimating Energy Cost of Work by Task Analysis (19)

A. Body Position and Movement		kcal/min[a]
Sitting		0.3
Standing		0.6
Walking		2.0–3.0
Walking uphill		0.8 per meter rise
B. Type of Work	**Average Cost (kcal/min)**	**Range (kcal/min)**
Hand work		
Light	0.4	0.2–1.2
Heavy	0.9	
Work with one arm		
Light	1.0	0.7–2.5
Heavy	1.8	
Work with both arms		
Light	1.5	1.0–3.5
Heavy	2.5	
Work with whole body		
Light	3.5	2.5–9.0
Moderate	5.0	
Heavy	7.0	
Very heavy	9.0	
C. Basal Metabolism	1.0	
D. Sample Calculation		**Average Cost (kcal/min)**
Assembling work with heavy hand tools		
1. Standing		0.6
2. Two-arm work		3.5
3. Basal metabolism		1.0
Total metabolic rate (M)		5.1

[a]For standard worker of 70-kg body weight (154 lb) and 1.8 m^3 body surface (19.4 ft^2).

erational procedure for practicing industrial hygienists. Individual variation in metabolic heat production is relatively small for resting, but wide differences exist for maximal physical effort, depending largely on body size, muscular development, physical fitness, and age (22). Table 21.3 shows the latest NIOSH-recommended task analysis method for estimating metabolic heat production (19).

2.3 Environmental Heat Exchange

The major mechanisms of heat exchange between a worker and the environment are convection, radiation, and evaporation. Other than for brief periods of body contact with hot tools, equipment, floors, and so on, conduction plays a minor role in occupational heat stress.

2.3.1 Convection

The rate of convective heat exchange between the skin of a person and the ambient air immediately surrounding the skin is a function of the temperature difference between the ambient air (T_A) and the mean skin temperature (T_S) as well as air velocity over the skin (V). This relationship is stated algebraically for the "standard worker" wearing the customary one-layer work clothing ensemble as follows (19, 21):

$$C = 7.0V^{0.6}(T_A - T_S)$$

where C = convective heat exchange (kcal/hr)
V = air velocity in meters per second (m/sec)
T_A = air temperature (°C)
T_S = mean weighted skin temperature (usually assumed to be 35°C)

Note that when $T_A > 35$°C, there is a gain in body heat from the ambient air by convection; when $T_A < 35$°C, heat is lost from the body to the ambient air by convection.

The empirically obtained exponent, 0.6, applies to forced convection over vertical cylinders, the geometric configuration that corresponds to a standing worker. The equation accounts for the surface area for an "average" person (1.8 m^2). For an individual whose surface area differs from average, the value of C may be adjusted by multiplying by the ratio of actual area to 1.8 (11).

2.3.2 Radiation

Radiative exchange (R) is a function primarily of the difference between the mean radiant temperature of the solid surroundings and skin temperature. Technically, radiative heat exchange is a function of the fourth power of absolute temperature, but a first-order approximation is sufficiently accurate for estimating R in the applications outlined here (19, 21):

$$R = 6.6(MRT - T_S)$$

where R = radiant heat exchange (kcal/hr)
MRT = mean radiant temperature of the solid surroundings (°C)
T_S = skin surface temperature (°C)

It is helpful to examine this overall expression of thermal balance in the context of both "hot-dry" and "warm-moist" environments. If the solid objects surrounding a worker in a hot-dry environment are hotter than skin temperature, the radiant heat gain may exceed the capacity of the sweating mechanism to provide sufficient cooling, and body temperature will rise. In a warm-moist environment, heat load from radiation may not be great, but the humid environment severely limits the sweating-convection mechanism. Again, when the maximum evaporative capacity is not enough to permit dissipation of the body's heat load, body temperature must rise.

2.3.3 Evaporation

The evaporation of water (sweat) from the skin surface results in significant heat loss from the body. The maximum evaporative capacity (and heat loss) is a function of air motion (V) and the water vapor pressure difference between the ambient air (VP_A) and the wetted skin at skin temperature (VP_S). The equation defining this relationship for the customary one-layer clothed worker follows (19, 21):

$$E = 14V^{0.6}(VP_S - VP_A)$$

where E = evaporative heat loss (kcal/hr)
V = air velocity (m/sec)
VP_S = vapor pressure of water on skin (assumed to be 42 mm Hg at 35°C skin temperature)
VP_A = water vapor pressure of ambient air (mm Hg)

In hot-dry environments, E may be limited by the amount of perspiration that can be produced by the worker. The maximum sweat production that can be maintained by the average adult male throughout a 8-hr shift is slightly above 1 l/hr, equivalent to an evaporative heat loss of about 600 kcal/hr (2400 Btu/hr) (11).

2.4 Effects of Clothing

Clothing serves not only as an insulative barrier between the skin and the environment but can also protect against hazardous chemical, physical, and biological agents. Clothing alters the rate of heat exchange between the skin and the ambient air by convection, radiation, and evaporation, and this effect can be substantial. In general, the thicker and greater the air and vapor impermeability of the clothing, the greater is its interference with convective, radiative, and evaporative heat exchange. When estimating worker–environment heat exchange, correction factors must be used to reflect the type, amount, and characteristics of the clothing being worn. This is especially important when the clothing differs substantially (i.e., more than one layer and/or greater air and vapor impermeability) from the customary one-layer ensemble. This clothing correction for dry heat exchange is nondimensional (23–25).

This "effect of clothing" is an active area in heat stress research. The revised NIOSH Criteria Document includes some basic guidelines for handling this effect (19). Corrections of the REL and RAL can be made for a variety of environmental and metabolic heat loads. When a two-layer clothing system is worn, the REL and RAL should be lowered by 2°C (3.6°F). When a partially air- and/or vapor-impermeable ensemble or heat-reflective or protective aprons, leggings, gauntlets, and so on are worn, the REL and RAL should be lowered by 4°C (7.2°F). These suggested corrections of the REL or RAL are judgments and have not been substantiated by controlled laboratory studies or long-term industrial experience (19).

3 PHYSIOLOGY OF HEAT STRESS

Our current understanding of the physiological effects of working under hot conditions has developed and emerged from several significant laboratory and field studies. The modern era of such investigations began in the 1920s with the work of the American Society of Heating and Ventilating Engineers. Since then extremely important contributions have been made by such investigators as Bedford et al. (5, 6), Yaglou et al. (1, 26), Leithead (27, 28), Lind (27, 29–31), Minard et al. (2–4, 32, 33), Hatch (34–36), Haines (35), Hertig et al. (37, 38), Belding et al. (21, 32, 36, 38), and Wyndham et al. (8, 9, 39, 40), among others.

Regardless of the thermal environment, humans attempt to maintain steady heat content and body temperature through a set of involuntary compensating mechanisms. In that context, several physiological systems influence body temperature regulation, thermal exchange with the environment, manifestations of heat strain, heat disorders, and tolerance factors.

3.1 Body Temperature Regulation

As warm-blooded homeotherms, humans must maintain an internal body temperature within a narrow range (near 37.0°C) (41). Hypothermia results if the core temperature of the body falls below 35.0°C, and death is likely at core temperatures below 27°C. Hyperthermia results when body core temperature exceeds 40°C in the absence of sweating; death will likely occur when body core temperature exceeds about 42°C.

Our inherent thermal homeostasis provides temperature regulation within suitable limits by the involuntary control of blood flow from sites of metabolic heat production (muscles and deep tissues) to the cooler body surface (skin). Here, heat loss takes place through the mechanisms of radiation, convection, and evaporation (see Section 2.3). Thus a worker reacts dynamically to changes in the thermal environment, always striving to maintain a core body temperature within the critical range.

3.1.1 Hypothalamic Regulation of Temperature

The temperature-regulating center for humans lies in a region at the base of the brain known as the hypothalamus. The hypothalamus responds to increases in its own temperature as well as to incoming nerve impulses from temperature receptors in the skin (22). It activates heat loss through increased blood flow to the skin and from sweating. Such physiological reactivity to elevated body temperature leads involuntarily to interaction with the thermal environment so as to offset the increase, thus maintaining body temperature within an acceptable range.

3.1.2 Muscular Activity

Muscles represent the largest single group of tissues in the body, typically accounting for about 45 percent of body weight. The bony skeleton, on which the muscles operate to generate their forces, represents about 15 percent of body weight. Even at rest, however, the muscles account for about 20 to 25 percent of the total metabolic heat production. Metabolic heat produced at rest is similar for all individuals when expressed per unit of surface area or of lean or fat-free body weight. On the other hand, the heat produced by the muscles during exercise or work can be much higher, most of which must be dissipated to maintain a heat balance with the environment (19).

The capacity for prolonged work of moderate intensity in hot environments is affected adversely by dehydration—often associated with a reduction in skin blood flow and sweat rate—and a concomitant rise in body temperature and heart rate. If the total heat load is high and the sweat rate is high, it is increasingly more difficult to replace water lost by sweating (750 to 1000 ml/hr). The thirst mechanism is usually not strong enough to drive one to drink the large quantities of fluid needed to replace water volume lost in sweat. Considerable evidence supports the concept that as the body temperature increases in a hot work environment, endurance for physical work is decreased. Similarly, as the environmental heat stress increases, many of the psychomotor, vigilance, and other measurable psychological tasks show decrements in performance (42–46).

3.1.3 Circulatory Regulation

The circulatory system provides the transport mechanism needed to deliver oxygen and foodstuffs to all tissues and to convey unwanted metabolites and heat away from the tissues. Nevertheless, the heart may not provide enough cardiac output to meet the peak needs of all of the body's organ systems as well as the need for dissipation of body heat. The autonomic nervous system and endocrine system control the allocation of blood flow among competing organ systems.

During exercise or physical work, there is widespread sympathetic vasoconstriction throughout the body, even in the cutaneous bed. The increase in blood supply to the active muscles is assured by the action of locally produced vasodilator substances and by neural input. In inactive vascular beds, there is progressive vasoconstriction with intensity of the muscle activity (19).

3.1.4 Sweating Mechanism

Sweat glands exist in abundance in the outer layers of the skin. They are stimulated by cholinergic sympathetic nerves and secrete a hypotonic watery solution onto the surface of the skin. Sweat production at rates of about 1 l/hr has been recorded frequently in industrial work. This represents a large potential source of cooling if the sweat can be evaporated, because each liter of sweat evaporated from the skin surface represents a loss of about 580 kcal of heat to the environment. Large losses of water by sweating can also pose a potential threat to successful thermoregulation because a progressive depletion of body water occurs if the fluid lost is not replaced. Dehydration affects thermoregulation significantly and contributes to a rise in core temperature (19).

The difference in water vapor pressure on sweat-wetted skin surface and the air layer next to the skin controls the rate of sweat evaporation, as does the speed of air movement over the skin (see Section 2.3.3). As a consequence, hot environments with high humidity limit the amount of sweat that can evaporate. Sweat not evaporated drips or flows from the skin and does not result in any heat loss from the body. This can be deleterious, because it still represents a significant loss of water and salt from the body.

Sodium chloride and potassium are very important constituents of sweat. In most circumstances, a sodium chloride deficit does not occur readily, because normal dietary intake provides about 10 to 15 g/day. Nevertheless, the salt content of sweat in unacclimatized individuals may be as high as 4 g/l. Acclimatized individuals will show lower concentrations, 1 g/l or less. Thus it is possible for a heat-unacclimatized individual with restricted salt intake to develop a negative salt balance. In view of the high incidence of elevated blood pressure in the U.S. population and the relatively high salt content of the typical U.S. diet, recommending increased salt intake is probably not warranted (19). Rehydration will occur more rapidly when sodium, the major electrolyte lost in sweat, is taken in with replacement fluids.

Rehydration is imperative to replace the water lost by sweating. It is not uncommon for workers in hot environments to lose 6 to 8 quarts of sweat (about 1 l/hr) during a work shift. If this fluid is not replaced, there is a progressive decrease of body water with a shrinkage not only of the extracellular space and interstitial and plasma volumes but also of water in the cells. There is clear evidence that the amount of sweat production depends on the state of hydration (27, 47, 48). Progressive dehydration results in lower sweat production and a corresponding increase in body temperature, a potentially dangerous situation.

Sweat lost in large quantities is often difficult to replace completely as the work shift proceeds, and it is not uncommon for individuals to register a water deficit of 2 to 3 percent of body weight. Because the normal thirst mechanism is not sensitive enough to ensure a sufficient water intake, every effort should be made to encourage individuals to drink water or a low-sodium noncarbonated beverage. One pint (two cups) should be consumed for every pound of weight lost by sweating. The fluid should be as palatable as possible at 10 to 15°C (50 to 60°F). Taking small quantities at frequent intervals, about 150 to 200 ml every 15 to 20 min, is a more effective regimen for practical fluid replacement than taking 750 ml or more once an hour (19).

3.1.5 Acclimatization to Heat

Acclimatization refers to a set of adaptive physiological and psychological adjustments that occur when an individual accustomed to working in a temperate environment undertakes work in a hot environment. These progressive adjustments occur over periods of increasing duration and reduce the strain experienced on initial exposure to heat. This enhanced tolerance allows a person to work effectively under conditions that might have been unendurable before acclimatization.

When workers are initially exposed to hot work environments, they can show signs of distress and discomfort; develop increased core temperatures and heart rates; complain of headache, giddiness, or nausea; and present other symptoms of incipient heat exhaustion. Yet on repeated exposure, there is a marked adaptation in which the principal physiological benefit appears to result from an increased sweating efficiency (earlier onset, greater sweat production, and lower electrolyte concentration) and a concomitant stabilization of the circulation. After heat exposure on several successive days, the individuals perform the same work with a much lower core temperature and heart rate and higher sweat rate (reduced thermoregulatory strain), and with none of the distressing symptoms that may be experienced initially. Acclimatization to heat is a remarkable example of physiological adaptation that is well demonstrated in laboratory experiments and field experience (49, 50). Full heat acclimatization occurs with relatively brief daily exposures to working in the heat. The minimum exposure time for achieving heat acclimatization is a continuous exposure of about 100 min daily (27, 50).

Heat acclimatization represents a dynamic state of conditioning rather than a long-term change in innate physiology. The level of acclimatization is relative to the initial level of physical fitness and the total heat stress experienced by an individual. Thus a worker who does only light work indoors in a hot climate will not achieve the level of acclimatization needed to work outdoors (with the additional heat load from the sun) or to do harder physical work in the same hot environment indoors.

Failure to replace the water lost in sweat will retard or even prevent the development of the physiological adaptations characteristic of acclimatization. Although acclimatization will be reasonably well maintained for a few days of nonheat exposure, absence from work in the heat for a week or more results in a significant loss in the beneficial adaptations. Nevertheless, heat acclimatization usually can be regained in two to three days upon return to a hot job (50).

3.2 Indexes of Heat Strain

Heat strain is a reactive physiological manifestation of environmental heat stress. The burden of balancing bodily heat gain and heat loss—a balance needed to prevent elevation in body temperature—falls primarily on the sweating mechanism and the circulatory system. Several indexes of heat strain have been identified and studied in the evaluation of the relationships between heat stress and strain with respect to the thermal environment. The most common indexes of heat strain include sweat rate, sweat evaporation rate (skin temperature), heart rate, and core temperature.

3.2.1 Sweat Rate

Sweating occurs over a wide range of thermal exposure in amounts sufficient to achieve enough evaporative cooling (E) to offset the total heat load, represented by $M + R + C$. The rate of secretion of individual sweat glands as well as the number of active glands determines the sweat rate. Under maximum thermal insult, some 2.5 million sweat glands secrete at peak rates of more than 3 kg/hr for up to an hour; highly acclimatized men can maintain rates of 1 to 1.5 kg/hr for several hours (21). When sweat evaporates freely from the skin, sweat rate equals the rate of evaporation. Evaporative cooling (E) is then regulated under steady-state conditions of work and heat exposure to balance the heat load ($M + R + C$). The central drive for sweating is determined by metabolic work rate (M); however, actual sweat output is modulated by skin temperature to meet the evaporative requirements up to the limits for sweating capacity.

3.2.2 Sweat Evaporation Rate (Skin Temperature)

The thermodynamics of water vaporization suggest that 1 g of sweat can eliminate 0.58 kcal of body heat upon evaporation. The efficiency of cooling by sweat depends on the rate of evaporation E, which is determined by the gradient between the vapor pressure of the wetted skin (VP_S) and the partial pressure (vapor pressure) of water vapor in the ambient air (VP_A), multiplied by a root function of effective air velocity at the skin surface ($V^{0.6}$) and the fraction of body surface that is wetted (see Section 2.3.3).

When the evaporation of sweat is restricted, skin temperature (T_S) rises above that observed under less humid conditions at the same air temperature. When this occurs, more sweat glands are activated, thereby increasing the fraction of wetted body surface. If such a response achieves the degree of cooling needed to balance the heat load ($M + R + C$), the body core temperature remains essentially unchanged (22).

At higher levels of VP_A or lower air velocities, the fraction of the body surface that is wetted increases until the body is completely wetted. Any further increase in sweat production does not contribute to cooling because the liquid perspiration drips off the body and is wasted as a coolant. Under higher levels of VP_A with further restrictions on evaporation, body heat is not dissipated, thus raising the temperature of the skin T_S as well as the deep body temperature. The response is a greater central drive for sweating, with VP_S and evaporative rate E increasing as well. As a result, a new thermal balance may be established, but at a higher body temperature and at the cost of increased thermoregulatory strain. A modest rise is physiologically acceptable, but any substantial increase in body temperature is accompanied by serious heat strain (22).

Sweat rate in the zone of free evaporation changes in proportion to heat load. In the zone of restricted evaporation, when the wetted body surface area approaches total body area, the sweat rate exceeds the evaporative capacity and is proportional to the increase in the core temperature and the skin temperature.

Sweat rate thus represents an index of heat stress over the entire range of compensation. Furthermore, it is also an index of heat strain in the zone of time-limited compensation where its rise parallels deep body temperature and heart rate. Sweat rate serves as a time-weighted average of heat stress and is measured by the change in body weight over a given time period, corrected for weight gained by water and food intake and weight lost in urine and feces (22).

3.2.3 Heart Rate (Circulatory Strain)

Maintaining a thermal balance without rise in body core temperature requires an effective physiological response to increase the transport of metabolic heat to the skin. This occurs by augmented blood flow through the dermal vessels, but under extreme conditions, this can tax the capacity for cardiac output.

Heart rate responds both to the demand for increased cardiac output required by working muscles and to the added circulatory strain imposed by heat exposure; therefore it is a useful index of total heat load. Heart rate or pulse rate is one of the most reliable indexes of heat strain. The heart rate of a subject reflects the combined demands of environmental heat, work level, elevation of body temperature, and individual cardiovascular fitness. As Figure 21.3 shows, heart rate increases disproportionately with heat load (51). This pattern reflects the body's increasingly futile effort to avoid a rise in core temperature as heat load rises. As the external heat load increases, skin temperature T_S also rises, decreasing the thermal gradient between body core and skin temperature and thereby demanding an increased volume of dermal blood flow.

Because heart rate reflects the combined influence of environmental heat, work level, elevation of body temperature, and individual cardiovascular fitness, a given heart rate has a relatively consistent physiological meaning among workers. For example, a rate of 180 to 200 beats per minute represents the maximum capacity for most adults and is sustainable for only a few minutes. In the same manner, a given heart rate signifies approximately the same level of strain regardless of the degree of an individual's fitness. Yet a person with high cardiovascular fitness will achieve a given heart rate at a higher level of thermal stress than an unfit person.

The detrimental effect of heat stress on work performance is indicated by an increase in heart rate both at work and in recovery. Brouha observed that progressive deterioration of pulse rate does not occur when the pulse rate at 0.5 to 1 min after the end of a work period does not exceed 110 beats/min and when the rate falls at least 10 beats/min within the next 2 min (52). Brouha proposed a simple guide to ensure that persons performing intermittent work in heat will remain in thermal balance for a full work shift without cumulative effects of strain. Using his guidelines, pulse rate is counted for the last 30 sec of each of the first 3 min after rest begins. If the first recovery pulse (i.e., from 30 to 60 sec) is maintained at 110 beats/min or below and deceleration between the first and third minute is at least 10 beats/min, no increasing strain occurs as the work shift progresses (52).

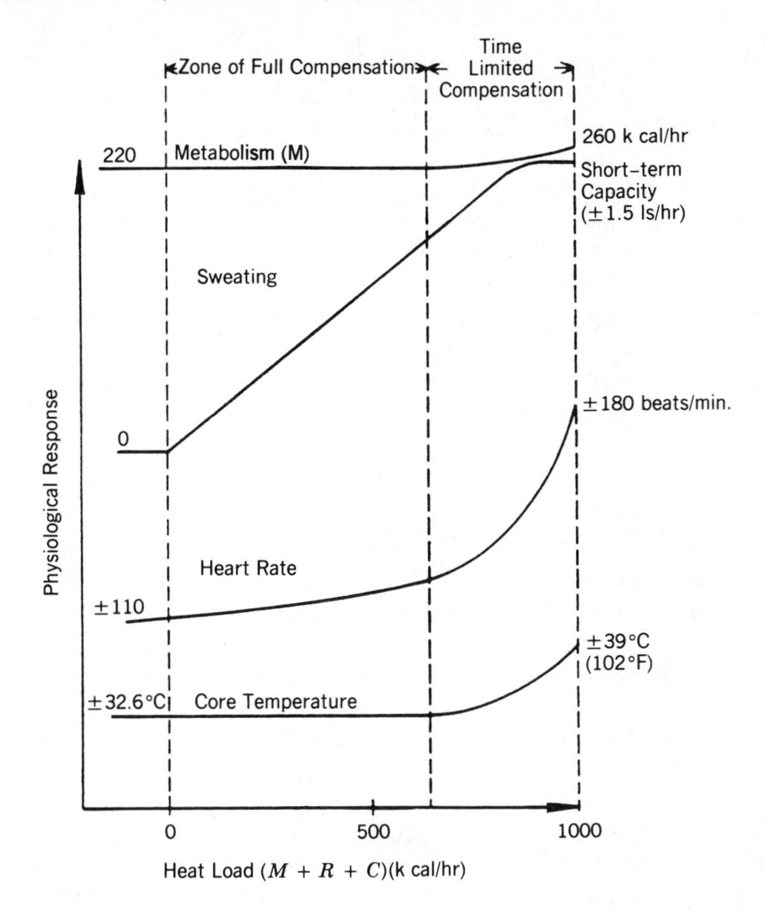

Figure 21.3. Relationship between heat load and physiological response (51).

An integrated or time-weighted heart rate level observed during an entire work shift reflects sustained elevations in peak rates as well as recovery and resting rates and thus can serve as a guide in assessing circulatory strain. Electronic devices for integrating total heart rate are now generally available and provide a valuable tool for refining investigations of the relation between heat stress and heat strain.

3.2.4 Body Core Temperature

The temperature control center in the brain (hypothalamus) reacts to adjust internal temperature to a level determined by an individual's metabolic rate M. As the total heat load $(M + R + C)$ increases, sweat rate and blood flow rate increase proportionally, with body core temperature maintained at a uniform level determined only by metabolic rate. Lind (29) termed this level of thermal equilibrium the "prescriptive zone," to indicate the range of thermal environments without strain on homeostatic control of body core temperature.

The data of Robinson (53) and others (54–57) show that the upper limit of the prescriptive zone in highly acclimatized men working at 300 kcal/hr is 31 to 32°C effective temperature (ET). Lind (58) recommends an ET of 27.5°C as a realistic limit for nonacclimatized men of varying physical fitness working at this level. The wide difference in these two limits symbolizes the perplexing nature of the problem of establishing rational standards for industrial heat stress (22).

In the prescriptive zone, the mechanisms of circulatory control and sweating are so effective that an acceptable body core temperature is maintained over a wide range of environmental temperatures. When the core temperature exceeds a critical level, injury to the temperature regulatory center may result, causing reduced sweating and a dangerous further temperature rise. Thus an increase in body core temperature is a very significant index of heat strain and one that has received much attention in efforts to define acceptable levels of heat stress.

3.3 Heat Disorders and Tolerance Factors

Environmental heat and the inability to remove metabolic heat lead to well-known reactions in humans, including increased cardiovascular activity, sweating, and increased body core temperature. Viewed from a wider perspective, three overlapping responses to heat can occur: psychological, psychophysiological, and physiological.

In addition to the physiological effects of heat, subclinical effects may modify performance or behavior or potentiate response to other stresses imposed simultaneously. There is no demonstrable effect on health at low levels of thermal stress. As heat load increases, however, there appears to be a higher order of psychophysiological disturbance, an increase in the frequency of errors, a higher frequency of accidents, and a reduction in efficiency in the performance of skilled physical tasks. At even higher levels of exposure increasingly well-defined disturbance of physiological well-being occurs, with strain on the heart and circulatory system, and overloading of the mechanisms of electrolyte and water balance in the body (10). Nevertheless, one of the most striking features of heat stress and tolerance to such stress is the high degree of individual susceptibility to heat. This section examines the various heat disorders in more detail and focuses on several factors that can influence heat tolerance.

3.3.1 Psychophysiological Effects

Under conditions of heat stress, increased demands are made for blood flow to the periphery of the body, diverting some of the cardiac output and rendering it unavailable to active muscles. Accordingly, as upper limits of tolerance are reached, work output must be reduced (59). Wyndham et al. demonstrated that real limits of endurance exist (39).

Effects of heat stress at levels lower than limiting levels include lack of efficiency in performing heavy tasks (60) and interference with skilled manipulations or psychological tasks (59). Pepler has reviewed the effects of heat on skilled tasks and mental tasks; qualitatively, there is no doubt that heat interferes with these types of activity (61, 62). Furthermore, it is common experience that heat exposure accelerates the onset of fatigue. One study showed that the production of delicately assembled items (based on incentive pay) included higher scrap rates when the subjects were working in heat, and although production of good pieces was maintained, lower efficiency resulted (63). In another study, Pepler (64) showed that relatively slight increases in environmental temperature affect classroom learning adversely. Increased bodily temperature and discomfort also increase irritation, anger, and other untoward emotional reactions (27, 65).

3.3.2 Interactive Effects

Much of our insight into the combined effects of heat and other stresses, such as noise and vibration, is expected to accumulate from the space program and related aerospace medical research. Crew members in aircraft and space vehicles are exposed to combinations of environmental stresses that traditionally have been studied one-by-one in laboratory research.

The combination of heat and carbon monoxide (CO) has been shown to have a deleterious effect greater than that due to either stress alone (37). The subjects reported persistent headaches, anorexia, irritability, depression, and general malaise; the effects were more pronounced in women than in men. These symptoms were markedly more severe after exposure to heat and CO than after exposure to either alone. It is interesting to note that these physiological disturbances were more severe in the hours following the exposures.

Renshaw (66) investigated the effects of noise and heat on performance on a five-choice serial reaction task. The effect of heat was found to be statistically significant. Performance was poorer at 90°F effective temperature than at 72°F effective temperature when the noise level was held constant.

3.3.3 Acute Heat Disorders

A variety of heat disorders can occur when individuals are exposed to excessive heat. These illnesses range from simple postural heat syncope (fainting) to the life-threatening complexities of heatstroke. The heat disorders are interrelated and seldom occur as discrete entities. A common feature of most heat-related disorders is some degree of elevated body temperature which may then be complicated by dehydration. The prognosis depends on the absolute level of the elevated body temperature, the promptness of treatment to lower the body temperature, and the extent of deficiency or imbalance of fluids and electrolytes. A summary of the classification, clinical features, prevention, and treatment of the acute heat illnesses is shown in Table 21.4 (22, 27). Some minor disorders not discussed in detail in this section are also included.

3.3.3.1 Heatstroke. Heatstroke is a serious disorder that is linked to failure of the thermoregulatory center. The classical description of heatstroke includes (1) a major disruption of central nervous function (unconsciousness or convulsions),

Table 21.4. Classification, Medical Aspects, and Prevention of Heat Illness[a]

	Heatstroke
Clinical features	(1) Hot dry skin: red, mottled, or cyanotic; (2) high and rising core temperature, 40.5°C and above; (3) mental confusion, loss of consciousness, convulsions, or coma, as core temperature continues to rise. Fatal if treatment delayed
Underlying physiological disturbance	Failure of the central drive for sweating leading to loss of evaporative cooling and an uncontrolled, accelerating rise in core temperature. Failure of sweating may be partial rather than complete
Predisposing factors	(1) Sustained exertion in heat by unacclimatized workers, (2) obesity and lack of physical fitness, (3) recent alcohol intake, (4) dehydration, (5) individual susceptibility, (6) chronic cardiovascular disease
Treatment	Immediate and rapid cooling by immersion in chilled water with massage, or by wrapping in wet sheet with vigorous fanning with cool, dry air. Avoid overcooling. Treat shock if present
Prevention	Medical screening of workers. Placement based on health and physical fitness. Acclimatization for 5–7 days by graded work and heat exposure. Monitoring workers during sustained work in severe heat

	Heat Syncope
Clinical features	Fainting while standing erect and immobile in heat
Underlying physiological disturbance	Pooling of blood in dilated vessels of skin and lower parts of the body
Predisposing factors	Lack of acclimatization
Treatment	Remove to cooler area. Recovery typically prompt and complete
Prevention	Acclimatization. Intermittent activity to assist venous return to heart

(2) a lack of sweating, and (3) a rectal temperature in excess of 41°C (105.8°F) (27, 67–69). The local circumstances of metabolic and environmental heat loads that give rise to the disorder are highly variable and are difficult to reconstruct with accuracy (19).

Heatstroke calls for emergency medical action. Any procedure from the onset

Table 21.4. (continued)

	Heat Exhaustion
Clinical features	(1) Fatigue, nausea, headache, giddiness; (2) skin clammy and moist, complexion pale, muddy, or with hectic flush; (3) may faint on standing with rapid, thready pulse and low blood pressure; (4) oral temperature normal or low but rectal temperature usually elevated (37.5 to 38.5°C). Water-restriction type: urine volume small, highly concentrated. Salt-restriction type: urine less concentrated; chlorides less than 3 g/l.
Underlying physiological disturbance	(1) Dehydration from deficiency of water and/or salt intake, (2) depletion of circulating blood volume, (3) circulatory strain from competing demands for blood flow to skin and to active muscles
Predisposing factors	(1) Sustained exertion in heat, (2) lack of acclimatization, (3) failure to replace water and/or salt lost in sweat
Treatment	Remove to cooler environment. Administer fluids by mouth or give intravenous infusions of normal saline (0.9%) if patient is unconscious or vomiting. Keep at rest until urine volume and content indicate water and electrolyte balances have been restored
Prevention	Acclimatize workers using a breaking-in schedule for 5 to 7 days. Supplement dietary salt only during acclimatization. Ample drinking water to be available at all times and to be taken frequently during work shift
	Heat Cramps
Clinical features	Painful spasms of muscles used during work (arms, legs, or abdominal). Onset during or after work hours
Underlying physiological disturbance	Loss of body salt in sweat. Water intake dilutes electrolytes. Water enters muscles, causing spasm
Predisposing factors	(1) Heavy sweating during hot work, (2) drinking large volumes of water without replacing electrolytes
Treatment	Salted liquids by mouth, or more prompt relief by intravenous infusion
Prevention	Adequate salt intake with meals. In unacclimatized workers, provide salted (0.1%) drinking water or equivalent electrolyte replacement beverage

Table 21.4. (continued)

Heat Rash	
Clinical features	Profuse tiny raised red vesicles (blisterlike) on affected areas. Pricking sensations during heat exposure
Underlying physiological disturbance	Plugging of sweat gland ducts, with retention of sweat and inflammatory reaction
Predisposing factors	Unrelieved exposure to humid heat and skin continuously wet with unevaporated sweat
Treatment	Mild drying lotions. Skin cleanliness to prevent infection
Prevention	Cool sleeping quarters or convection to allow skin to dry between heat exposures
Heat Fatigue—Transient	
Clinical features	Impaired performance of skilled sensorimotor, mental, or vigilance tasks, in heat
Underlying physiological disturbance	Discomfort and physiological strain
Predisposing factors	Performance decrement greater in unacclimatized and unskilled workers
Treatment	Not indicated unless accompanied by other heat illness
Prevention	Acclimatization and training for work in the heat

[a]Adapted from References 22 and 27.

that will cool the patient improves the prognosis. Placing the patient in a shady area, removing outer clothing and wetting the skin, and increasing air movement to enhance evaporative cooling are all needed and appropriate until professional methods of cooling and assessing the degree of the disorder are available. By the time a patient is admitted to a hospital, the heatstroke may progress to a multisystem lesion affecting virtually all tissues and organs (70). In the typical clinical presentation, the central nervous system is disorganized, and there is commonly evidence of fragility of small blood vessels, possibly coupled with the loss of integrity of cellular membranes in many tissues. The blood-clotting mechanism is often severely disturbed, as are liver and kidney functions. It is not clear, however, whether these conditions are present at the onset of the disorder, or whether their development requires a combination of elevated body temperature and a certain period for tissue or cellular damage to occur. Postmortem evaluation indicates that few tissues escape pathological involvement. Early recognition of the disorder or its impending onset, together with appropriate treatment, reduces the death rate considerably as well as the extent of organ and tissue involvement (19).

3.3.3.2 ***Heat Exhaustion.*** Heat exhaustion is a milder form of heat disorder linked to depletion of body fluids and electrolytes. The critical event is low arterial blood pressure caused partly from inadequate cardiac output and partly from widespread vasodilation. The chief factors leading to heat exhaustion include:

1. Increased vascular dilation with decreased capacity of circulation to meet the demands for heat loss to the environment, exercise, and digestive activities
2. Decreased blood volume due to dehydration, gravitational edema, vasodilatation, or lack of salt (electrolyte imbalance)

Heat exhaustion usually yields readily to prompt treatment. This disorder has been encountered frequently in experimental assessment of heat tolerance. Characteristically, it is accompanied by a small increase in body temperature (38 to 39°C or 100.4 to 102.2°F). The symptoms of headache, nausea, vertigo, weakness, thirst, and giddiness are common both to heat exhaustion and the early stage of heatstroke. There is wide individual variation in the ability to tolerate increased body temperature; some individuals cannot tolerate rectal temperatures of 38 to 39°C, and others continue to perform well at even higher rectal temperatures (71).

There are many variants in the development of heat disorders. Failure to replace water may predispose the individual to one or more of the heat disorders and may complicate an already complex situation. Therefore cases of hyperpyrexia can be precipitated by dehydration. It is unlikely there is only one cause of hyperpyrexia without some influence from another. Recent data suggest that cases of heat exhaustion can be expected to occur about 10 times more frequently than cases of heatstroke (67).

3.3.3.3 ***Heat Cramps.*** Heat cramps are fairly common among individuals who work vigorously in the heat. Heat cramps are spasms in the voluntary muscles following a reduction in the concentration of sodium chloride in blood below a certain critical level. They are attributable to a continued loss of salt in the sweat, accompanied by copious intake of water without appropriate replacement of salt. Other electrolytes such as Mg^{2+}, Ca^{2+}, and K^{+} may also be involved. Cramps often occur in the muscles principally used during work and can be readily alleviated by rest, ingestion of water, and correction of electrolyte imbalance in body fluids (19).

3.3.3.4 ***Heat Rashes*** The most common heat rash is prickly heat (miliaria rubra), which appears as red papules, usually in areas where the clothing is restrictive, and gives rise to a prickling sensation, particularly as sweating increases. It occurs in skin that is persistently wetted by unevaporated sweat, apparently because the keratinous layers of the skin absorb water, swell, and mechanically obstruct the sweat ducts (72–74). The papules may become infected unless they are treated.

Another skin disorder (miliaria crystallina) appears with the onset of sweating in skin previously injured at the surface, commonly in sunburned areas. The damage prevents the escape of sweat with the formation of small to large watery vesicles which rapidly subside once sweating stops, and the problem ceases to exist after the damaged skin is sloughed.

Miliaria profunda occurs when the blockage of sweat ducts occurs below the skin surface. This rash may also follow sunburn injury, but has been reported to occur without clear evidence of previous skin injury. Discrete and pale elevations of the skin, resembling gooseflesh, are present.

In most cases, the rashes disappear when the individuals are returned to cool environments. It seems likely that none of the rashes occur (or if they do, with greatly reduced frequency) when a substantial part of the day is spent in cool and/or dry areas so that the skin surface can dry.

3.3.4 Age

Older, healthy workers perform well in hot jobs if allowed to proceed at a self-regulated pace. Under demands for sustained work output in heat, however, an older worker is at a distinct disadvantage compared with younger colleagues. First, the maximum oxygen uptake at maximum work capacity declines 20 to 30 percent between ages 30 and 65, leaving an older worker with less cardiocirculatory reserve.

Second, under levels of heat stress above the prescriptive zone, an older worker compensates for heat loads less effectively than do younger persons, as indicated by higher core temperature and peripheral blood flow for the same work output (31). This occurs because of a delay in the onset of sweating and a lower sweat-rate capacity with age, thus resulting in greater heat storage during work and longer time required for heat recovery.

Aging also leads to a more sluggish response of the sweat glands, which leads to less effective control of body temperature. Aging also results in a greater level of skin blood flow associated with exposure to heat (19). When two groups of male coal miners of average age 47 and 27 years worked in several comfortable or cool environments, they showed little difference in their responses to heat near the REL with light work. In hotter environments, however, the older men showed substantially greater thermoregulatory strain than their younger counterparts. The older men also had lower aerobic work capacities (75). In analyzing the distribution of 5 years' accumulation of data on heatstroke in South African gold mines, Strydom (76) found a marked increase in heatstroke with increasing age of the workers. Thus men over 40 years of age represented less than 10 percent of the mining population, but they accounted for 50 percent of the fatal and 25 percent of the nonfatal cases of heatstroke. The incidence of cases per 100,000 workers was 10 or more times greater for men over 40 years than for men under 25 years of age. In all the experimental and epidemiologic studies described above, the workers had been medically examined and were considered free of disease. Total body water decreases with age, which may be a factor in the observed higher incidence of fatal and nonfatal heatstroke in the older group (19).

3.3.5 Gender

Because of a lower aerobic capacity, the average woman, similar to a small man, is at a disadvantage when she has to perform the same hot job as the average-sized man. Although all aspects of heat tolerance in women have not been fully examined, gender-related differences in thermoregulatory capacities have been reviewed extensively. Lower sweat rates for females are widely reported in the literature, and related differences in physiological response by males and females are well established (77). Nevertheless, when working at similar proportions of their maximum aerobic capacity, women perform similarly or only slightly less well than men (78–80).

3.3.6 Body Fat

It is well established that obesity predisposes individuals to heat disorders (27). The acquisition of fat means that additional weight must be carried, thereby calling for a greater expenditure of energy to perform a given task and use of a greater proportion of the aerobic capacity. In addition, the body surface to body weight ratio (m^2 to kg) becomes less favorable for heat dissipation. Probably more important is the lower physical fitness and decreased maximum work capacity and cardiovascular capacity frequently associated with obesity. The increased layer of subcutaneous fat provides an insulative barrier between the skin and the deep-lying tissues. The fat layer theoretically reduces the direct transfer of heat from the muscles to the skin (81).

3.3.7 Water and Electrolyte Balance

Sustained, effective work performance in heat requires, among other things, a replacement of body water and electrolytes lost through sweating. A fully acclimatized worker weighing 70 kg secretes up to 6 to 8 kg of sweat per 8-hr shift. If this water is not replaced by drinking, continued sweating will draw on water reserves from both tissues and body cells, leading to dehydration (22).

Water lost from sweating 1 kg (1.4 percent of body weight in a 70-kg person) can be tolerated without serious effects. Water deficits of 1.5 kg or more during work in the heat reduce the volume of circulating blood, resulting in signs and symptoms of increasing heat strain, including elevated heart rate and body temperature, thirst, and severe heat discomfort. At water deficits of 2 to 4 kg (3 to 6 percent of body weight), work performance is impaired; continued work under such conditions leads to heat exhaustion (19, 22). Therefore, sweating workers should drink at frequent intervals—at least two or three times per hour—to assure adequate fluid replacement.

Sodium chloride lost in sweat must be replaced by ingestion. A typical American diet (10 to 15 g/day of salt) supplies the needs of an acclimatized worker producing 6 to 8 kg of sweat during a single shift, for 1 kg of sweat contains 1 to 2 g of salt (24). For the period of acclimatization, however, workers with no previous heat exposure require additional salt. Although maximal sweating rates in unacclimatized persons are lower, salt concentrations are higher than after acclimatization. Thus an unacclimatized person may lose 18 to 30 g of salt per day. Salt supplements in the form of tablets (preferably impregnated, to avoid gastric irritation) may be ingested if ample water is available. A preferable practice is to use salted water (one tablespoon per gallon) or ingest a rehydration beverage containing sodium.

Depletion of body salt may occur in unacclimatized workers exposed to heat who replace water losses without adequate salt intake. This leads to progressive dehydration because thermoregulatory controls in the human body are geared to maintain a balance between electrolyte concentration in tissue fluids and body cells. Deficient salt intake with continued intake of water dilutes the tissue fluid, which in turn suppresses the release of antidiuretic hormone (ADH) by the pituitary gland. The kidney then fails to reabsorb water and excretes dilute urine containing little salt (36).

Under these conditions, homeostasis maintains the electrolyte concentration of body fluids but at the cost of depleting body fluids and ensuing dehydration. Under continued heat stress, the symptoms of heat exhaustion (elevated heart rate and body temperature plus severe discomfort) develop similarly to those resulting from water restriction, but signs of circulatory insufficiency are more severe and there is notably little thirst. An excellent diagnostic tool for salt deficiency is the presence of a very low level of chloride (less than 3 g/l) in the urine (22).

On a short-term basis, sweating workers drinking large volumes of unsalted water may develop heat cramps, extremely painful spasms of the muscles used while working, such as arms, legs, and abdomen. The dilution of tissue fluid around the working muscle results in transfer of water into muscle fibers, causing the spasms. Treatments of the clinical symptoms of water and salt depletion are similar, namely, the replacement of water and/or salt by oral ingestion of salted liquids in mild cases or the intravenous infusion of saline in more serious cases. Any excess of salt or water over actual needs is readily controlled by urinary excretion, assuming a healthy heart and kidneys.

3.3.8 Alcohol and Drugs

3.3.8.1 Alcohol. Alcohol has been associated as a contributing factor in the occurrence of heatstroke (27). This drug interferes with central and peripheral nervous function and is associated with dehydration by suppressing ADH production. Notwithstanding the potential hazards from central nervous system (CNS) depression, the ingestion of alcohol before or during work in the heat should not be permitted, because it reduces heat tolerance and increases the risk of heat illnesses.

3.3.8.2 Therapeutic Drugs. Many drugs—including diuretics and antihypertensives—prescribed for therapeutic purposes can interfere with thermoregulation (67). Some of these drugs are anticholinergic in nature or involve inhibition of monoamine oxidative reactions. Nevertheless, almost any drug that affects CNS activity, cardiovascular reserve, or body hydration could potentially affect heat tolerance. Any worker subject to heat stress who requires drug therapy should be under the supervision of a physician who understands the potential ramifications of medications on heat tolerance. A worker taking therapeutic medications who is exposed even intermittently or occasionally to a hot environment should seek the guidance of a physician.

3.3.8.3 Social Drugs. It is difficult to separate the heat-disorder implications of drugs used therapeutically from those which are used socially. Nevertheless, there are many drugs other than alcohol that are used on social occasions. Some of these have been implicated in cases of heat disorder, sometimes leading to death (67).

3.3.9 Physical Fitness

Physical conditioning alone does not assure heat acclimatization, but persons with a high level of physical fitness have a distinct advantage (22). Physical training, without heat exposure, improves heat tolerance as indicated by somewhat lower heart rates and core temperatures when compared with persons who are exposed to heat before conditioning. Sweat rates do not increase, and skin temperature remains high. Physical conditioning enhances heat tolerance by increasing the functional capacity of the cardiocirculatory system. Increased heat tolerance occurs by two important changes. First, an increase occurs in the number of capillary blood vessels relative to muscle mass, providing a larger interface between blood and muscle for the exchange of oxygen and waste products. Second, the increased tone of small veins from nonmuscle tissue reduces their volumetric capacity during exercise and thus increases pressure on large, central veins returning blood to the heart. This increase in venous return causes cardiac output to increase during work with less need to accelerate the heart. Therefore a physically conditioned person, by virtue of having a higher maximum ventilatory capacity, has a wider margin of safety in coping with the added circulatory strain of working under heat stress.

4 MEASUREMENT OF THE THERMAL ENVIRONMENT

Heat stress refers to the aggregate of environmental and physical factors that constitute the total heat load imposed on the body. Assessment of heat stress always includes some theoretical or empirical combination of environmental variables to describe the severity of the thermal environment and the capacity of that environment to facilitate heat exchange with a worker. As discussed in Section 2, both the degree of discomfort and the level of stress caused by environmental heat depend on air temperature, relative humidity, velocity of air movement, and temperature of the surrounding surfaces with which the body exchanges heat by radiation.

4.1 Indexes of Heat Stress

Investigators in this field have devised several indexes for assessing heat stress, ranging from as simple a parameter as dry-bulb temperature to algebraic combinations of multiple environmental variables, some with correction factors and modifications. Several indexes of thermal stress have survived or emerged as those most commonly encountered in the evaluation and control of industrial heat stress. Beyond dry-bulb and wet-bulb temperatures, these include the effective temperature (ET), defined by Houghton and Yaglou (82); equivalent effective temperature corrected for radiation (ETCR), a modification of ET to make allowance for radiant heat (83); the heat stress index (HSI) (36); the wet bulb globe temperature (WBGT) (1); and the wet globe temperature (84).

Investigators who have tried to develop optimal heat stress indexes by combining environmental measurements have approached the problem in three general ways. The first approach is based on the thermometric scale, the second on the rate of sweating, and the third on calculations of heat load and of the evaporative capacity of the environment.

4.1.1 Dry-Bulb and Wet-Bulb Temperatures

The dry-bulb air temperature (T_A) is the temperature of the ambient air as measured with a thermometer or equivalent instrument (see Section 4.2). It is the simplest to measure of climatic factors. ISO temperature units are degrees Celsius (°C) and degrees Kelvin (°K = °C + 273).

The natural wet-bulb temperature (NWB) is the temperature measured by a thermometer or equivalent sensor, which is covered by a wetted cotton wick and which is exposed only to the naturally prevailing air movement. Accurate measurement of NWB requires use of a clean wick, distilled water, and shielding to prevent radiant heat gain.

The psychrometric wet-bulb temperature (WB) is obtained when the wetted wick covering the sensor is exposed to a high forced air movement. The WB is commonly measured with a psychrometer which consists of two mercury-in-glass thermometers mounted alongside each other on the frame of the psychrometer. One thermometer is used to measure the WB by covering its bulb with a clean cotton wick wetter with water, and the second measures the dry-bulb temperature (T_A). The air movement is obtained manually with a sling psychrometer or mechanically with a motor-driven psychrometer.

Although these simple temperature parameters are convenient to obtain, they provide less-than-complete information on the thermal exchange between a worker and the occupational environment. Sole use of dry-bulb temperature when the temperature exceeds the comfort zone is justified only for situations where the worker is wearing completely vapor- and air-impermeable encapsulating protective clothing. Wet-bulb temperature is most applicable for assessing heat stress and predicting heat strain in hot, humid situations where WB approaches skin temperature, the radiant heat load is minimal, and air velocity is low.

4.1.2 Effective Temperature

Effective temperature (ET) has a long history and wide application as a useful index of thermal comfort. Effective temperature emerged in the search for design criteria for thermal comfort in occupied space and was introduced in 1923 by Houghton and Yaglou (82). Effective temperature is an empirical sensory index combining dry- and wet-bulb temperatures and air speed to yield a thermal sensation equivalent to that at a given temperature of still, saturated air. To develop the effective temperature scale, subjects were exposed to one combination and then another of the various parameters of air temperature, air motion, and humidity. On the basis of a large number of trials, nomograms were developed that characterized equivalent environments, expressed in terms of the temperature of a still, saturated environment.

Wyndham, in studying gold miners, concluded that below 26°C ET the risk of heatstroke was negligible and that the risk of fatal heatstroke began at 28°C ET (9). Up to 32°C ET there was a slow, steady rise in the risk of both fatal and nonfatal cases of heatstroke, but above 33°C ET the risk increased sharply. Similar values were observed by Brief and Confer in a comparison of heat stress indices (85).

4.1.3 Equivalent Effective Temperature Corrected for Radiation

Bedford showed that the ET scale could be corrected to account for radiation from hot surroundings by using the black globe temperature (see Section 4.2.4) in place of the dry-bulb temperature (83). Present use of the equivalent effective temperature corrected for radiation (ETCR) scale incorporates such modifications into a single chart.

In spite of widespread use, both the ET and ETCR have serious limitations as indexes of heat stress (27):

1. ET was developed on the basis of transient thermal sensations. This emphasis tended to neglect the importance of the sorption and desorption of moisture from the subject's clothing.
2. The ET scale was developed using clothed subjects in the dress of that era.
3. The subjects were sedentary. Later modifications were made to include the effective metabolic rate. Nevertheless, the scale was designed primarily for environments reasonably near the comfort zone, and extrapolation to thermally stressful environments is tenuous.

4.1.4 Heat Stress Index

The heat stress index (HSI), developed by Belding and Hatch at the University of Pittsburgh in the mid-1950s (36), combines the environmental heat-exchange components of radiation R and convection C with metabolic heat M in an expression of stress in terms of the required sweat evaporation E_{req}. Stated algebraically,

$$E_{req} = M \pm R \pm C$$

The resulting physiological strain is determined by the ratio (times 100) of the

Table 21.5. Evaluation of Values in Belding and Hatch HSI (36)

Index of Heat Stress (HSI)	Physiological and Hygienic Implications of 8-hr Exposures to Various Heat Stresses
−20 −10	*Mild cold strain.* This condition frequently exists in areas wherer persons recover from exposure to heat.
0	*No thermal strain*
+10 20 30	*Mild to moderate heat strain.* For a job that involves higher intellectual functions, dexterity, or alertness, subtle to substantial decrements in performance may be expected. In performance of heavy physical work, little decrement is expected unless ability of individuals to perform such work under no thermal stress is marginal
40 50 60	*Severe heat strain*, involving a threat to health unless persons are physically fit. Break-in period required for those not previously acclimatized. Some decrement in performance of physical work is to be expected. Medical selection of personnel desirable because these conditions are unsuitable for those with cardiovascular or respiratory impairment or with chronic dermatitis. These working conditions are also unsuitable for activities requiring sustained mental effort
70 80 90	*Very severe heat strain.* Only a small percentage of the population may be expected to qualify for this work. Personnel should be selected (*a*) by medical examination and (*b*) by trial on the job (after acclimatization). Special measures are needed to assure adequate water and salt intake. Amelioration of working conditions by any feasible means is highly desirable and may be expected to decrease the health hazard, while increasing efficiency on the job. Slight "indisposition" that in most jobs would be insufficient to affect performance may render employees unfit for this exposure
100	*Maximum strain* tolerated daily by fit, acclimatized young men

stress (E_{req}) to the maximum evaporative capacity of the environment (E_{max}), and HSI is defined as follows:

$$\text{HSI} = 100\left(\frac{E_{req}}{E_{max}}\right)$$

When HSI exceeds 100, body heating occurs; when HSI is less than 100, body cooling occurs. Values of E_{req} and E_{max} may be computed by means of appropriate equations or by a convenient nomogram method developed by McKarns and Brief (86).

Table 21.5 shows the physiological and hygienic implications of 8-hr exposures to various levels of the HSI (11).

The heat stress index finds application in the engineering analysis of occupational heat exposure and also as a predictor of the "allowable exposure time" (AET) (85). For an average man, AET is given by the equation:

$$\text{AET} = \frac{250 - 60}{E_{\text{req}} - E_{\text{max}}}$$

The HSI loses some applicability at very high heat stress conditions. It also does not identify correctly the heat stress differences resulting from hot, dry and hot, and humid conditions. The strain resulting from metabolic versus environmental heat may not be differentiated because $E_{\text{req}}/E_{\text{max}}$ is a ratio that may disguise the absolute values of the two factors (19).

Nevertheless, HSI and its components offer an excellent starting point for specifying corrective measures when heat exposure is excessive. The relative values of the convective exchange C, radiant exchange R, and evaporative capacity E_{max} not only provide a rigorous way to estimate heat stress, but also offer a basis for a rational approach to corrective engineering measures. This subject is developed further in Section 6.

4.1.5 Wet Bulb Globe Temperature

The WBGT index was intended originally as a simple expression of heat stress for use in military training where men were exercising outdoors in conditions of high solar radiation (1). It proved very successful in monitoring heat stress and minimizing heat casualties in the United States and has been widely adopted. The WBGT index provides a convenient method to assess quickly, with a minimum of operator skills, conditions that pose threats of thermal strain. Because of its simplicity and close correlation with ETCR, it was adopted in 1971 (and has prevailed) as the principal index for the TLV for heat stress established by the ACGIH (14).

Fundamentally, the WBGT index is an algebraic approximation of the effective temperature concept. As such, it has the built-in limitations of the effective temperature, but it also has the advantage that air or wind velocity does not have to be measured directly to calculate the intensity of WBGT. For outdoor use with solar load, the index is derived from the formula:

$$\text{WBGT} = 0.7NWB + 0.2GT + 0.1T_A$$

where NWB = natural wet-bulb temperature
GT = globe temperature
T_A = dry-bulb (air) temperature

For indoor use, the weighted expression becomes:

$$\text{WBGT} = 0.7NWB + 0.3GT$$

Although direct measurement of air movement is not required for the computation of WBGT, allowances are made for this factor by the use of the naturally convected wet-bulb sensor.

NIOSH endorses WBGT as the preferred measure of severity of occupational exposures to heat stress (10, 19). The main criteria used by NIOSH for the selection of a suitable index were that (1) the measurements and calculations must be simple and (2) index values must be predictive of the physiological strain of heat exposure.

There is little dispute that the WBGT index is convenient and simple to use, but WBGT is not a perfect predictor of physiological strain. Ramanathan and Belding, among others, have shown clearly that environmental combinations yielding the same WBGT levels result in different physiological strains in individuals working at a moderate level (87). This suggests that WBGT has limitations as a heat stress index, especially at high levels of severity. In addition, when impermeable clothing is worn, the WBGT will not be a relevant index, because evaporative cooling (wet-bulb temperature) will be limited. Nevertheless, WBGT has become the index most frequently used and recommended for use throughout the world.

4.1.6 Wet Globe Temperature (WGT)

The wet globe thermometer includes a hollow, 3-in. copper sphere covered by a black cloth that is kept at 100-percent wettedness from a water reservoir. The sensing element of a thermometer is located at the inside center of the copper sphere, and the temperature inside the sphere is read on a dial on the end of the stem. The wet sphere exchanges heat with the environment by the same mechanism that a person with a totally wetted skin would use in the same environment; that is, heat exchange by convection, radiation, and evaporation is integrated into a single instrument reading (84). The stabilization time of the instrument ranges from about 5 to 15 min depending on the magnitude of the heat-load differential (19).

During the past several years, the WGT has been used in many laboratory studies and field situations where it has been compared with the WBGT (88–92). In general, the correlation between the two is high. Nevertheless, the relationship between the two is not constant for all combinations of environmental variables. Correction factors ranging between 1°C (1.8°F) and 7°C (12.6°F) have been suggested. A simple approximation of the relationship is WBGT = WGT + 2°C for conditions of moderate radiant heat and humidity. These approximations are likely adequate for general monitoring in industry (19).

4.2 Sensing Instruments

The four primary environmental factors that determine the rate of heat transfer between a worker and the surroundings are air temperature, humidity, air velocity, and radiation (solar and infrared). This section discusses the instruments used and techniques involved in measuring these four variables. Technology is advancing steadily in this aspect, and coverage here is limited to instruments that are widely available and easily adapted to field use.

4.2.1 Thermometry

Air temperature can be measured by a variety of instruments, and each has advantages under specific circumstances.

Liquid-in-Glass Thermometers. The most widely used instrument for measuring air temperature is a glass thermometer in which mercury or alcohol is the expanding liquid. This type of instrument is relatively inexpensive and is available in various temperature ranges and with varying degrees of accuracy. The response time of a thermometer depends primarily on the bulb size and the air velocity at the bulb. Transient temperature readings should not be recorded until the thermometer reaches steady state.

Bimetallic Thermometers. A bimetallic thermometer consists of thin, bonded strips of two different metals, each having a unique coefficient of expansion. Unequal expansion occurs when there is a change in temperature, causing the elements to bend and producing a displacement of the free end. This displacement is transmitted through a suitable linkage to a needle indicator that moves across a scale to indicate the temperature. The bimetallic thermometer element is widely used in the common dial thermometer and in many inexpensive, self-contained temperature recorders.

Thermocouples. A thermocouple is a junction of wires of two different metals. Such thermojunctions release an electromotive force (emf) that varies with the temperature of the junction. In a circuit containing two thermocouples, the emf in the unit depends on the temperature difference between the two junctions. If one of the junctions is held at a constant reference temperature, the temperature of the other junction can be determined by the measured emf on the circuit. The emf in a thermocouple circuit may be measured accurately with a potentiometer or a high resistance millivolt meter. Thermocouples of copper and constantan are commonly used for environmental temperatures ranging to 70°F.

Although use of thermocouples entails the initial cost of a potentiometer, together with its bulk, thermocouples can give remote readings. In such applications, simultaneous readings may be taken at one place for several work sites. In addition, the equilibration time required with varying temperatures is much less than that of mercury-in-glass thermometers.

Thermistors. A thermistor or resistance thermometer is a semiconductor (typically a metal wire) that exhibits a substantial change in resistance with a small change in temperature. Because the resistance of a thermistor is on the order of thousands of ohms, incremental resistance added by lead wires up to about 25 m is immaterial. Therefore, like thermocouples, thermistors offer the advantage of remote monitoring.

Difficulties with Air Temperature Measurements. Thermal radiation can cause serious errors in the measurement of air temperature. When the surrounding surfaces are warmer or cooler than the air in a space, radiation effects will cause a sensor used to measure air temperature to become warmer or cooler than the air. Because of their smaller size, thermocouples and thermistors are less affected by radiation than mercury-in-glass thermometers. The temperature distortion due to radiation can be reduced by shielding, by increasing the velocity of air movement over the sensor, or by a combination of the two. Heavy aluminum foil positioned loosely around the sensor provides a simple, effective shield. It is important, however, that a shield not restrict the free flow of air around the sensor element.

4.2.2 Humidity

Humidity refers to water vapor in the atmosphere. The level of humidity can be expressed quantitatively in several different ways, each expression being applicable to certain processes or problems. The terms "specific humidity" or "dew-point" temperature are useful to an air-conditioning engineer. In heat stress calculations, "vapor pressure" and "partial pressure" are convenient terms. Only "dew-point temperature" and "relative humidity" can be measured directly by ordinary instrumentation. The other expressions of humidity must be derived from other humidity-related measurements.

Psychrometers. All thermodynamic properties of mixtures of air and water vapor can be determined readily from knowledge of the dry- and wet-bulb temperatures. These temperatures are measured with a psychrometer, an instrument with two sensors, one of which is enclosed in a wetted wick or sock. The sensor in the wetted wick is cooled by evaporation, depressing its temperature below that of the dry-bulb sensor. The air velocity over the wetted sensor must be at least 900 ft/min to reach the true (psychrometric) wet-bulb temperature (93). The psychrometric wet-bulb temperature (WB) is sometimes referred to as the "vented wet-bulb."

A sling psychrometer uses a pair of liquid-filled thermometers, one being covered by a wetted cotton wick. The operator holds the instrument by a handle and whirls the pair of thermometers to effect the required air velocity over the wetted bulb. After repeatedly whirling the sling psychrometer until no further depression of wet-bulb temperature is noted, the final readings of the two thermometers define dry-bulb and psychrometric wet-bulb temperatures.

Aspirated psychrometers include two stationary temperature-sensing elements, one dry and one covered with a wetted wick. Airflow over them is produced by a small fan or a squeeze bulb. The sensors of an aspirated psychrometer are usually shielded to prevent radiation-related errors in the indicated temperatures.

Natural wet-bulb temperature (*NWB*) is used in the determination of the wet bulb globe temperature (WBGT) index. The *NWB* is obtained by exposing a thermometer with a wetted wick to natural air movement without regard to the minimum air velocity. Therefore *NWB* is always less than or equal to psychrometric wet-bulb temperature.

Hygrometers. Hygrometers are classified as one of two types, "organic" and "dew point." An organic hygrometer is a direct-reading device for measuring relative humidity. Relative humidity refers to the percentage saturation of water vapor in air at a given air temperature. The sensing element of a hygrometer can be made of an organic material that undergoes dimensional changes as a function of the relative humidity. This type of sensor is widely used in self-contained humidity recorders and in control instruments.

Dew-point hygrometers provide a direct measure of dew-point temperature. This is accomplished by noting the temperature of a highly polished surface at the time condensation starts to form. The condensing plate can be cooled thermoelectrically or by other means, such as the evaporation of a refrigerant at the back of the plate. The onset of condensation can be determined by visual inspection or by use of a photoelectric cell and light source. The temperature at the surface is usually measured by thermocouples affixed to it. The dew-point hygrometer is not a widely used field instrument for measuring humidity in the industrial setting. Nevertheless, it is useful in the laboratory and finds specific application in estimating the moisture content of air or other gases at elevated temperatures.

4.2.3 Air Velocity

Any instrument used to measure air movement in the ambient work environment must be capable of measuring very low air velocities and must be nondirectional. These two requirements can be fulfilled only by some type of thermal anemometer. Higher air velocities can be measured adequately by vane anemometers.

4.2.3.1 Thermal Anemometers. A thermal anemometer has an electrically heated temperature-sensing element that is cooled by air passing over it. Because a thermal anemometer responds to mass flow over the sensor, a correction is required when the gas density differs greatly from standard air density.

Heated Thermocouple Anemometer. A heated thermocouple anemometer consists of heated and unheated thermocouples connected in series and exposed to the airstream. The resultant emf (voltage) is a function of temperature difference between the heated sensor and the air. The response of the instrument is calibrated to indicate air velocity.

Hot-Wire Anemometer. The sensor in a hot-wire anemometer is a resistance thermometer element of fine wire that is heated. Its temperature and electrical resistance vary with the velocity of air passing over it. The air velocity is calibrated to and determined by the measurement of resistance.

Heated-Bulb Thermometer Anemometer. The heated bulb-thermometer anemometer includes two matched mercury-in-glass thermometers. The temperature of one thermometer is raised by passing a known current through a fine resistance wire wound around the bulb. The two thermometers are exposed to the air simultaneously, and their temperatures are read. The air velocity is calibrated to and determined by the temperature difference.

4.2.3.2 Vane Anemometers. Vane anemometers include rotating vane and swinging vane anemometers. A rotating vane anemometer consists of a light, rotating wind-driven wheel enclosed in a ring. Using recording dials, it indicates the number of revolutions of the wheel or the linear distance in meters or feet. In order to determine air or wind velocity, a stopwatch must be used to record the elapsed time, although newer models have a digital readout. The swinging anemometer consists of a vane enclosed in a case that has inlet and outlet air openings. The vane is placed in the pathway of the air, and the movement of the air causes the vane to deflect. This deflection can be translated to a direct reading of the wind velocity by means of a gear train. Rotating vane anemometers typically are more accurate than swinging vane anemometers (19).

Another type of rotating anemometer consists of three or four hemispherical cups mounted radially from a vertical shaft. Wind from any direction causes the cups to rotate the shaft, and wind speed is then determined from the shaft speed (94).

4.2.4 Radiation

Radiant heat sources in the workplace include infrared sources indoors and solar radiation outdoors. The radiant heat exchange between a worker and the surrounding surfaces of the work space can be computed, at least theoretically, if the absolute temperatures and emissivities of all surfaces are known. In practical applications, however, such an analytical solution would be extremely tedious. Therefore an empirical approximation of radiant heat exchange has been developed and is used widely for indoor sources of radiation.

The black globe thermometer is used widely for assessing thermal radiation in the occupational environment (95, 96). A temperature sensor is located at the center of a thin-walled copper sphere. The outer surface of the globe is painted flat black. When the globe thermometer is used in a space in which surface temperatures are higher than the air temperature, the globe temperature (GT) rises above the air temperature because of the radiant heat adsorbed by its black surface. As this occurs, the globe begins to lose heat by convection. Its temperature stabilizes when the rate of heat gained by radiation equals the rate of heat lost by convection. Usually a minimum of 15 min is required for the globe temperature to reach steady state. The Vernon globe thermometer, sometimes called the standard 6-in. black globe, is recommended by NIOSH to measure black globe temperature (19).

Mean radiant temperature (MRT) represents the uniform black body temperature of an imaginary enclosure, equivalent to the actual environment, with which a worker exchanges radiative heat. The MRT can be calculated from the globe thermometer temperature, air velocity, and air temperature using the following equation (19):

$$MRT = GT + 1.8V^{0.5}(GT - T_A)$$

where MRT = mean radiant temperature (°C)
GT = black globe temperature (°C)
V = air velocity (m/sec)
T_A = dry-bulb (air) temperature (°C)

MRT is a very localized index and refers only to the specific point at which the measurements (GT, V, T_A) are taken.

Solar or outdoor radiation can be classified as direct, diffuse, or reflected. Direct radiation comes from the solid angle of the sun's disk. Diffuse radiation (sky radiation) refers to the scattered and reflected radiation coming from the whole hemisphere after shading the solid angle of the sun's disk. Reflected solar radiation is that reflected from the ground or water surface(s). The total solar radiant load is the sum of the direct, diffuse, and reflected solar radiation as modified by clothing worn and position of the body relative to the solar radiation (97).

Direct solar radiation can be measured with a pyrheliometer. A pyrheliometer includes a tube directed at the sun's disk as well as a thermal sensor. Generally, a pyrheliometer with a thermopile as sensor and a view angle of 5.7° is recommended (98, 99). Two varieties of pyrheliometers are widely used: the Ångstrom compensation pyrheliometer and the Smithsonian silver disk pyrheliometer, each of which uses a slightly different scale factor.

Diffuse and total solar radiation can be measured with a pyranometer. Pyranometers can be used for measuring solar or other radiation between 0.35 and 2.5 μm, which includes the ultraviolet, visible, and infrared range (19). A pyranometer for measuring diffuse radiation is fitted with a disk or a shading ring to prevent direct solar radiation from reaching the sensor. The receiver usually resembles a hemispherical dome to provide a 180° view angle for total sun and sky radiation. The receiver is used in an inverted position to measure reflected radiation. The thermal sensor could be a thermopile, a silicon cell, or a bimetallic strip.

4.3 Integrating Instruments

Many attempts have been made to develop a single-reading heat stress instrument for assessing and integrating the environmental factors of air temperature, air speed, humidity, and radiant temperature. Among the instruments that have been proposed are the globe thermometer (100), the heated globe thermometer (101), the wet Kata thermometer (102), the eupathescope (103), the thermointegrator (104), and the wet globe thermometer (84). Several of these instruments have been largely abandoned, and the search continues for a suitable single-reading instrument.

4.3.1 Requirements and Limitations

In an excellent review of the design requirements and limitations of a single-reading heat stress meter, Professor Hatch showed that despite the mathematical equivalence of human and instrument, the capacity of heat exchange by convection and evaporation is inherently greater for a physical instrument of simple geometric shape and small size than for a much larger person (105).

Single-reading, integrating instruments have the appeal of convenience that surely will continue, but their use in the context of today's state of the art in heat stress evaluation may offer convenience at the cost of knowing the levels of the key stress components of the thermal environment. Often there is a dual purpose in assessing heat stress: (1) to determine the level of stress and (2) to determine in what manner and to what extent control procedures can be employed to reduce the stress to an acceptable level. There is a parallel in industrial hygiene with respect to exposure levels estimated on the basis of integrated samples, such as full-shift personal samples, and exposure estimates constructed from discrete exposure intervals based on the temporal and spatial variability of contaminant levels in the work environment. Although a single, time-weighted exposure level obtained from an integrated sample is convenient, the convenience comes at the expense of the diagnostic information needed to assess the causes for exposure and the basis for a rational, cost-effective approach to control strategy.

4.3.2 Wet Globe Temperature

As reviewed earlier, Botsford has developed a wet globe instrument consisting of a small copper sphere fitted with a black cotton wick and a water reservoir (84). This device not only integrates the effects of air temperature, mean radiant temperature, and air motion, it also includes the effect of fractional wetness. The objective of the "Botsball" is to provide an index and model that simulates a worker's exchange of heat with the environment by convection, radiation, and evaporation. Although the relative values of the three components of heat transfer from the wet globe and the human body may differ considerably, a growing number of investigators have reported good correlation between wet globe temperature and other well-known heat stress indexes (88–92, 106, 107).

4.3.3 WBGT Index

The TLVs for heat stress as well as the recommendations to OSHA from both NIOSH (1972 and 1986) and the Standards Advisory Committee (1974) specify the use of the WBGT index. Based on the equations for calculating WBGT (Section 4.1.5), one must know the natural wet-bulb temperature (NWB), the globe temperature (GT), and, in the case of a solar load, the dry-bulb air temperature (T_A). The instruments and techniques for measuring these parameters have been reviewed in Section 4.2. A suggested arrangement for manual measurement of WBGT appears in Figure 21.4.

The U.S. Army has developed a portable, hand-held, mechanical WBGT device to replace the standard WBGT setup for use in the field. The Weksler miniaturized kit (WMK) uses a small shielded dry-bulb and natural wet-bulb thermometer along with a modified black globe. It includes a slide rule to calculate the WBGT index after reading the thermometers. Although the WMK is easier to use than the standard WBGT setup, it still uses relatively fragile thermometers (107).

Several WBGT instruments have been developed to sense and indicate dry-bulb,

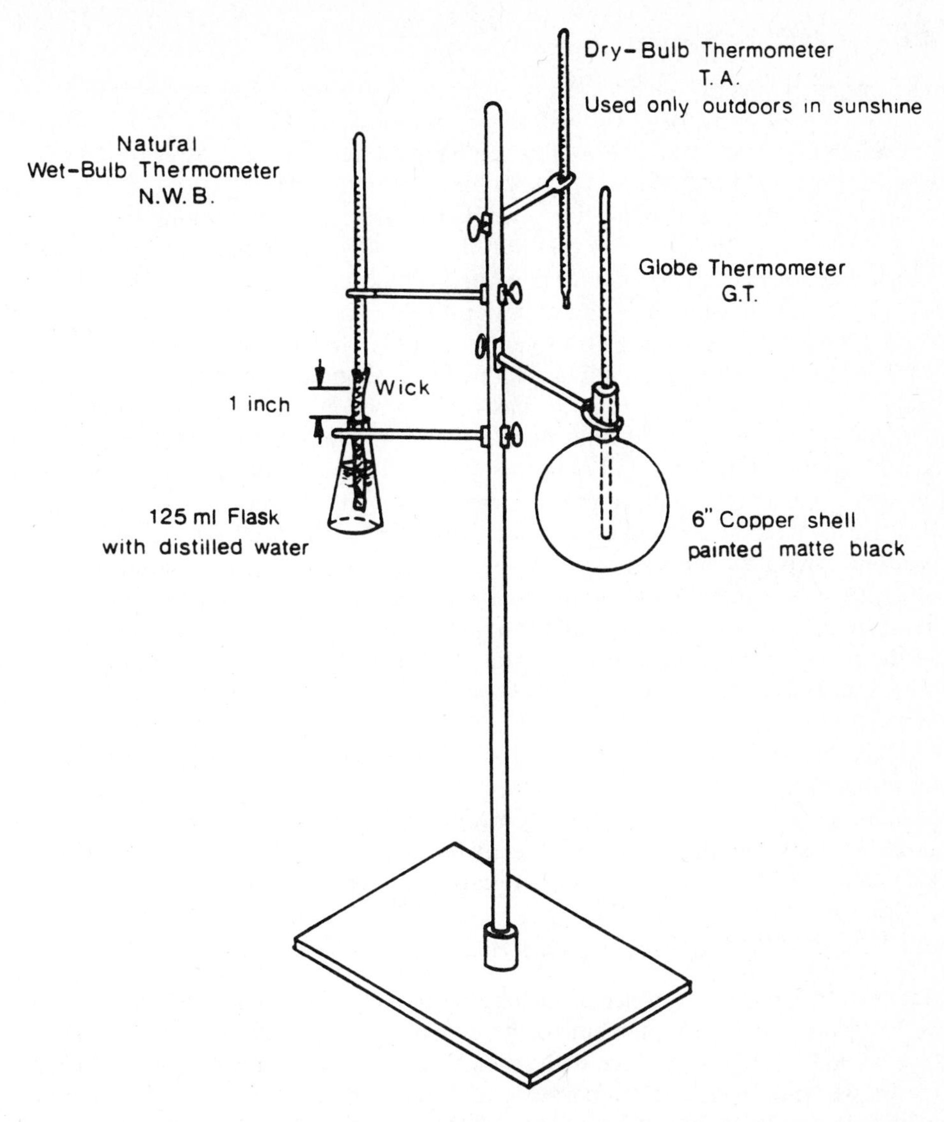

Figure 21.4. Arrangement for manual measurement of WBGT.

natural wet-bulb, and globe temperatures, as well as to integrate the values to yield WBGT. Electronic, direct-reading instruments are now available commercially to measure WBGT. These electronic WBGT instruments are easy to use but require careful handling and cannot be exposed to the elements (rain, sandstorms, etc.). The temperature sensors are vulnerable to rough handling, and the direct-reading instruments are relatively expensive (107).

4.4 The Effect of Weather

The potential severity of heat stress is determined largely by weather conditions even in the hot industries. The prevailing meteorological conditions, the onset of hot weather, and heat wave episodes can have a very significant effect on the heat stress experienced by workers.

Attempts have been made by several industries and institutions to establish qualitative or semiquantitative relationships between outdoor weather conditions and the degree of environmental heat stress imposed on workers. Most of these studies have been limited to the consideration of specific locations and, until recently, have resulted mainly in subjective correlations.

A 1966 study (108) offers a limited functional relationship between outdoor and indoor climates developed on the basis of afternoon measurements of climatic variables taken on four days during the summer. Curves of the temperature differences between indoors and outdoors were developed for individual work areas in a shop, forming a family of hyperbolas. A "temperature parameter," independent of outside climatic conditions, was developed based on the indoor and outdoor dry-bulb temperatures. This parameter is determined as the mean value of temperature differential of two pairs of measurements and depends on the heating value of the work area and the rate of ventilation in the shop. Limitations of the model include insensitivity to rapid changes in outside temperature and a narrow applicable time span (1 to 4 P.M.) during which deterministic measurements are taken.

Other investigators (109, 110) used a thermal circuit concept to estimate indoor air temperature and relative humidity, for these are affected by outdoor air temperature and solar radiation. Because no indoor heat source was considered, however, inside climate was estimated directly from outdoor meteorological conditions, using only building structure and materials as the critical variables.

4.4.1 National Weather Service Information

The National Weather Service provides a set of daily environmental measurements that can be a useful supplement to the climatic factors measured at a work site. The National Weather Service data include daily observations at 3-hr intervals for air temperature, wet-bulb temperature, dew point temperature, relative humidity, wind velocity, sky cover, ceiling, and visibility. A summary of daily environmental measurements includes maximum, minimum, and average temperatures, as well as wind velocity (direction and speed), extent of sunshine, and sky cover. These data, where available, can be used for approximate assessment of the worksite environmental heat load for outdoor jobs or for some indoor jobs where air conditioning is not in use.

4.4.2 Predictions of WBGT

In the mid 1970s, NIOSH commissioned a major study to develop predictive models than can be used to calculate WBGT from estimates of a minimum number of conventional meteorological variables (111). The resulting models were based on data acquired during 27 week-long surveys conducted over a 2-year period in 15 representative hot industries in three climatic regions of the United States.

Using the recommended methodology and three sets of relatively simple equations, WBGT can be calculated directly from independent variables that quantify outdoor weather conditions. The first set of models is predictive in nature and is designed to estimate future levels of WBGT from weather forecasts. The second series can be used to estimate current levels of WBGT at a work site from current values of meteorological variables reported routinely by a nearby office of the National Weather Service. The third set also estimates current levels of WBGT, but uses outdoor meteorological variables measured locally in the immediate vicinity of the workplace of interest.

For each of the three sets of models, one of seven equations is selected as the most appropriate for any given work site, depending on the range of differences between inside and outside dry-bulb temperature. The appropriate model is then used to estimate WBGT from either direct knowledge of, or forecasts on, prevailing weather conditions (111).

5 ASSESSMENT OF HEAT STRESS AND STRAIN

The specification of workplace standards or guidelines for limiting heat stress requires an understanding of the relationship between environmental heat stress and physiological strain. The objective of a heat stress standard is to limit the level of health risk (level of strain and the danger of incurring heat-related illnesses) associated with the total heat load (environmental and metabolic) imposed on a worker in a hot environment. The methodology of risk estimation has improved during the past few years but still lacks a high level of accuracy (19). Although several studies show generally good correlation between some index of environmental heat stress and one or more indexes of physiological strain, no single stress index and no single strain index have proved sufficient to characterize fully the complex stress–strain relationship over the entire range of conditions expected in industry.

5.1 Correlation of Stress and Strain

As research continues on the environmental and physiological aspects of heat stress, a primary goal is to define direct and indirect physiological limitations in the context of the industrial environment, and to translate these acceptable physiological limits to corresponding levels of environmental heat stress. Many investigators share credit for advancing our level of understanding on these issues.

The basic approaches used to correlate heat stress and physiological strain fall into three types: theoretical or semitheoretical calculations, empirical evaluations under controlled conditions, and epidemiologic-type studies conducted on workers exposed in occupational settings.

5.1.1 Calculated Limits of Heat Exposure

Gagge et al. (112) defined an upper limit of tolerable heat exposure at which the body heat balance is just maintained, at a given work rate, without a significant rise in skin temperature. Starting with the fundamental heat balance, Haines and Hatch (35) derived a thermal balance line for skin temperature on the basis of the ratio of E_{req}, the rate of evaporative cooling required to maintain thermal balance, to E_{max}, the maximum rate of evaporative cooling possible in a given environmental situation, using either the "operative temperature" (113) or the Vernon globe temperature (114).

Belding and Hatch (36) combined the two limiting requirements to define the heat stress index, (HSI), the percentage ratio of E_{req} to E_{max}. They related several levels of HSI to expected physiological or psychological effects. Unfortunately, the HSI does not provide the complete solution to the problem of measuring and predicting heat stress and strain (51). Experience showed that the effects did not always occur as predicted. Therefore modifications were attempted to increase the accuracy of HSI and to use the heat balance concept to derive better correlative indices of heat stress and strain.

5.1.2 Empirical Studies in Controlled Conditions

Although several new or modified heat stress indexes have appeared in the literature, they usually have been tested in experimental programs designed to evaluate their relation to selected physiological responses. As Peterson (115) has indicated, many physiological parameters have been related to limited measures of heat stress but not necessarily to any of the heat stress indexes.

In a comprehensive project typical of the empirical approach identified earlier, Peterson designed and executed a study to relate several heat stress indexes to several measures of physiological response. His objective was to identify an optimal indicator of strain as well as the physiological responses that are sensitive to change as a function of heat stress. Although WBGT was not included among the heat stress indexes considered, the results demonstrated that at least three indexes of strain were necessary to describe the probable response of man to the thermal environment. Sweat rate was the biothermal strain best correlated with all stress indexes except the equivalent operative ambient temperature. This suggests that sweat rate is a suitable parameter in terms of total strain. Neither the actual nor predicted sweat rate correlated well, however, with ear, rectal, or skin temperatures, suggesting that the total-strain concept probably has limited utility (115).

Another study of three fit subjects under controlled environmental conditions evaluated the relationship among environmental parameters, physiological strain parameters, and heat stress indexes (87); WBGT was included as a heat stress index at levels of 68, 85, and 89°F WBGT. The authors concluded that although a given level of WBGT has meaning dependent on environmental conditions, higher levels of the WBGT scale do not always signify greater strain, especially when the environment is relatively dry. Furthermore, the WBGT values failed to parallel the increased physiological strain of varying humidity, air speed, or radiant heat.

A more serious inconsistency was that a higher WBGT level did not produce greater strain than a lower WBGT level in dry environments. Therefore WBGT would seem to have limited value as a predictor of physiological strain at the higher heat stress levels that may be encountered in industry (87).

The same study showed that the heat stress index differentiates between environmental conditions in the correct order of physiological strain, although the resulting strain seems to be overemphasized. The wet globe temperature, however, did not differentiate between two widely different levels of physiological strain in the participants (87).

5.1.3 Workers Exposed in Occupational Settings

Since the 1920s researchers have focused on the relationships between stress and strain for workers in hot industries (5, 13, 116–119). Nevertheless, few epidemiologic studies have been reported that relate the length and intensity of heat exposure to the long-term health experience of workers. Health experience statistics, although appearing in several studies, tend to relate qualitative differences in some dependent variable representing morbidity or mortality to some single categorical or independent environmental variable (6, 7, 27, 120–123).

Much information relating heat stress and strain has accumulated over the past 40 years in the South African deep mines. From data derived from laboratory studies, a series of curves has been prepared to predict the probability of a worker's body temperature reaching dangerous levels when working under various levels of heat stress (124, 125). Based on these data and on epidemiologic data on heatstroke from miners, estimates of the probabilities of reaching dangerously high rectal temperatures have been made. If a body temperature of 40°C (104°F) is accepted as the threshold temperature at which a worker is in imminent danger of irreversible heatstroke, the estimated probability of reaching this body temperature is 10^{-6} for workers exposed to an effective temperature (ET) of 34.6°C (94.3°F), 10^{-4} at 35.3°C (95.5°F), 10^{-2} at 35.8°C (96.4°F), and $10^{-0.5}$ at 36.6°C (97.9°F). These ET correlates were established for conditions with relative humidity near 100%; whether they are equally valid for these same ET values for low humidities has not been proven. Probabilities of body temperature reaching designated levels at various ET values are also available for unacclimatized men (124, 125). Although these estimates have proven to be useful in preventing heat casualties under the conditions of work and heat found in the South African mines, their direct application to general industrial environments has not been validated (19).

Our present understanding of the effects of heat on health and safety suggests that standards of exposure should be based on the acute effects. It is clear, however, that the physiological consequences of exposure to heat are not directly proportional to the intensity of heat stress over the entire tolerable range. Physiological strain increases exponentially in the upper range of heat stress. Thus a small incremental increase at high levels of heat stress can result in a large increase in strain. Furthermore, many factors mitigate the relationship between exposure and effects, as discussed in Section 3.3. Any exposure limits stipulated for heat stress must reflect a recognition that the specific environment–worker–job situation defines a total stress that must remain within acceptable levels by adequate control of one or a combination of factors.

5.2 Rationale for a Heat Stress Standard

Many persons and agencies have proposed several expressions of permissible exposure limits for work in hot environments. These recommendations or standards typically include some heat stress index, and most account for the four crucial climatic factors: air temperature, humidity, radiant heat, and air velocity. Several indexes consider the work load as well. A World Health Organization (WHO) panel of experts has attested to the shortcomings in virtually all the indexes as well as the proposed upper limits set forth for each of these parameters (49).

Many of the recommended standards for thermal exposure have been derived from laboratory experience rather than from studies on industrial workers. This means the experimental subjects are not necessarily representative in age and fitness of an employed work force. In addition, the severity of physiological strain is known to correlate poorly with various levels of each of the heat stress indexes. In a study previously referenced, the use of at least three indexes simultaneously was recommended to obtain an adequate evaluation of the heat strain of an exposed worker (115).

The WHO panel of experts recommended that a deep body temperature of 38°C be considered the limit of permissible exposure for work in heat (49). This is consistent with the observation that body core temperature in excess of 38°C increases the likelihood of heat disorder or illness (29, 126).

5.2.1 Criteria for a Standard Heat Stress Index

The NIOSH Criteria Documents set forth several reasonable criteria for any heat stress index to be used in industry (10, 19):

1. Feasibility and accuracy must be proven through industrial use.
2. All important factors (environmental, metabolic, clothing, physical condition, etc.) must be considered.
3. Required measurements and calculations must be simple.
4. Index exposure limits must be supported by corresponding physiological and/or psychological responses which reflect an increased risk to safety and health.
5. The index must be applicable for setting limits under a wide range of environmental and metabolic conditions.

Several indexes satisfy the first criterion, including ET, WBGT, WGT, and HSI. An estimate of the work load is included in HSI, and this confers an advantage over WBGT, WGT, and ET with respect to the second criterion. The measurements and calculations are much simpler for ET, WGT, and WBGT than for HSI, however, and even WBGT is somewhat simpler to use than ET. With respect to the fourth criterion, all indexes have shortcomings. Considering the applicability for use in regulatory limits, HSI has the advantage of being used to calculate an allowable exposure time and minimum recovery time for a given heat stress condition. Nevertheless, WBGT and WGT are the most convenient indexes with which to monitor levels of exposure because they can be read directly from an integrating instrument or, in the worst case, WBGT can be derived easily from sets of two or three environmental measurements.

Even though HSI offers several distinct supplementary advantages with respect to the engineering analysis of a given heat stress exposure, WBGT remains the preferred, prevailing environmental index for heat stress.

5.2.2 Prescriptive and Environment-Driven Zones

The "prescriptive zone" (PZ) refers to the range of environmental conditions in which deep body temperature is determined by work intensity only. The prescriptive zone has been defined empirically in terms of effective temperature by a series of experiments (rectal vs. effective temperature) at three levels of metabolic rate: 180, 300, and 420 kcal/hr (29). Such studies show clearly that steady-state rectal temperature is not a function of effective temperature until critical levels of effective temperature (dependent somewhat on M) are reached, at which point rectal temperature increases sharply. The range of environmental heat stress within which core temperature rises sharply with elevations in climatic conditions is called the "environment-driven zone" (EDZ). The value of an environmental heat stress index at the interface of the PZ and the EDZ is called the "upper limit of prescriptive zone" (ULPZ). The value of the ULPZ varies among individuals. It is higher for workers who are acclimatized to heat, but it is lower when more clothing is worn (10).

To be certain that 95 percent of a heat-acclimatized population wearing work clothing will not have core temperatures in excess of 38°C, the level of environmental heat stress at which a 5-percentile person will reach the ULPZ must be established (10). This value must then be corrected for the level of acclimatization and clothing. Lind and Liddell found the ULPZ to be 80.5°F ET for a group of 128 men of average physical fitness (127).

Permissible exposure limits for heat stress cannot be based on 8-hr average values because exposures in excess of 1 hr may cause a worker to accumulate enough heat that body temperature will rise sufficiently to induce an acute heat disorder. Thus in continuous heat exposure, hourly averages are necessary. If the exposure is intermittent, the accumulation of heat will be slowed, allowing the use of a 2-hr average exposure (10).

Another study by Lind justifies the use of time-weighted average hourly work load values for intermittent work (30). The results show that for ULPZ, a certain hourly amount of work can be performed either continuously at a lower rate or at a higher rate interrupted with rest periods. In another study, the ULPZ was found to be the same for men of different ages; thus no correction for age is required (128). When older persons are exposed to a strenuous heat load, increased caution is advisable because of lower physiological capacities and higher overall susceptibility to disease.

5.2.3 *Work Practices Versus Environmental Standard*

Lind's work on the prescriptive zone represents a sound basis for the development of an environmental heat stress standard. It combines both the climatic and work load conditions that are imposed on the worker, and it can be monitored with a convenient index such as WBGT. Nevertheless, a number of practical shortcomings and unresolved questions remain (10, 19).

It is important that the upper limit of prescriptive zone concept be confirmed with data from a representative industrial work force. More data are required on the age and sex distribution of the work force. Also to be clarified is the effect of the self-selection process that normally occurs in an industrial situation when a worker self-determines the ability to endure high levels of heat stress. This consideration alone may have resulted in past heat stress standards that were unrealistic for an industrial work force. The lack of data regarding intermittent exposures to heat is another major unresolved question of the effect of heat stress on the workforce. In addition, differences in sweat loss under a wide variety of industrial conditions still have not been thoroughly studied. The highly variable work loads and the intermittency of strenuous work loads that are normal in industrial operations may also have a major effect on heat stress (10, 19).

The answers to these questions require additional research to validate the techniques presently proposed for the evaluation of heat stress conditions. Nevertheless, our current level of understanding justifies a "work practices standard" rather than an environmental standard for heat stress. Clearly the available environmental indexes are sufficient to estimate threshold limits of heat stress that can be used to initiate work practices to protect the industrial worker adequately against heat.

5.3 Options for Control

Before selecting control measures, it is crucial to identify the components of heat stress to which workers are exposed currently or expected to be exposed in new operations. Only then, by examining the alternatives, can the most effective means of control be selected. A review of the heat balance equation described in Section 2.1 suggests that total heat stress can be reduced only by modifying one or more of the following factors: metabolic heat production, heat exchange by convection, heat exchange by radiation, or heat exchange by evaporation. Environmental heat load (C, R, and E) can be reduced by engineering controls (e.g., ventilation, air conditioning, screening, insulation, process or operational modification), as well as protective clothing and equipment. Metabolic heat production can be modified by work practices and application of labor-reducing devices.

Clearly, heat stress control for an individual worker depends on several elements of appropriate behavior and environmental control. The most critical of these are listed below:

1. Bodily heat production
2. Number and duration of exposures
3. Heat exchange components as affected by environmental factors (MRT, E_{max}, V, T_A, and VP_A)
4. Thermal conditions of the rest area
5. Clothing and protective equipment

The third element is covered in greater detail in Section 6. It is important, however, to be aware of the full range of options for control in a specific situation. Table 21.6 provides a summary or checklist of actions for controlling heat stress and strain. In addition to engineering controls, these include work practices and administrative measures which are useful in reducing or eliminating heat stress (19).

6 ENGINEERING CONTROL OF HEAT STRESS

The control of heat in industry to ensure that exposures fall within acceptable limits requires the application of feasible engineering procedures. Cost-effective engineering control of heat stress calls for an understanding of the physiological response of man in a hot environment. The underlying worker-environment thermal balance (Section 2) suggest that when a worker is exposed to elevated temperatures, the rate of heat loss decreases. The role of engineering controls is to help sustain a rate of bodily heat loss equal to the imposition of heat from the environment. In general this can be accomplished by increasing the velocity of air across the body, a technique useful within certain limits of temperature and humidity. Above such limits it may be necessary to reduce the surrounding temperature. In practice, it is often necessary to combine these two approaches to achieve an acceptable thermal work environment.

The important role of engineering methods in regulating heat stress is underscored by the recommendations of the Department of Labor Standards Advisory Committee on Heat Stress (18). In its report, the committee sets forth special work practices needed to bring the 2-hr average heat exposure level within the limits of its recommended threshold WBGT values. The first such practice is that "the employer should adopt engineering controls which are appropriate for reducing and controlling the level of heat exposure" (18).

An engineering approach to the reduction of heat stress generally parallels the same control strategy applied for other workplace environmental stresses such as airborne gases, vapors, and dusts. Among the alternative control methods are substitution, control at the source, local controls, and general ventilation. Another approach to an analysis of engineering control for heat stress focuses on the important environmental components of the thermal balance: convection, evaporative capacity, and radiation. This section combines both viewpoints and presents control methods that not only reflect a traditional approach to engineering control alternatives, but add appropriate emphasis to the aspects of heat control that characterize this specific physical stress.

Table 21.6. Checklist for Controlling Heat Stress and Strain (19)

Item	Actions for Consideration
Heat Components	
M, body heat production	Reduce physical demands of the work; use powered assistance for heavy tasks
R, radiative load	Interpose line-of-sight barrier; furnace wall insulation, metallic reflecting screen, or heat-reflecting screen; cover exposed parts of body
C, connective load	If air temperature is above 35°C (95°F), reduce air temperature, reduce air speed across skin, and wear clothing
	If air temperature is below 35°C (95°F), increase air speed across skin and reduce clothing
E_{max}, maximum evaporative cooling by sweating	Increase by decreasing humidity and increasing air speed
	Decrease clothing
Work practices	Shorten duration of each exposure; use more frequent short exposures rather than fewer long exposures
	Schedule very hot jobs in cooler part of day when possible
Exposure limit	Self-limiting, based on formal indoctrination of workers and supervisors on signs and symptoms of heat strain
Recovery	Air-conditioned space nearby
Personal protection	Cooled air, cooled fluid, or ice-cooled conditioning clothing
	Reflective clothing or aprons
Other considerations	Determine medical status by evaluation, primarily of cardiovascular system
	Careful breaking in of unacclimatized workers
	Fluid intake at frequent intervals to prevent dehydration
	Fatigue or mild illness not related to the job may temporarily contraindicate exposure (e.g., low-grade infection, diarrhea, sleeplessness, alcohol ingestion)
Heat wave	Introduce heat alert program (HAP)

6.1 Control at the Source

The first and most fundamental approach to engineering control of heat in the workplace is to examine options for eliminating heat at its point of generation. Occasionally, one feasible alternative is to change the operation or substitute a process component of lower temperature for one of higher temperature. One well-known example is the use of induction heating rather than direct-fired furnaces for certain forging operations.

As with most environmental stresses, heat is controlled most effectively if it is regulated at the source, to prevent "contamination" of the space occupied by workers. In general, the options for control of heat at the source are isolation, reduction in emissivity, insulation, radiation shielding, and local exhaust ventilation.

6.1.1 Isolation

The most practical method for limiting heat exposure from hot processing operations that are difficult to control or for operations that are extremely hot is to isolate the heat source. Such operations might be partitioned and separated from the rest of the facility, located in a separate building, or relocated outdoors with minimal shelter. A typical example is an industrial boiler, invariably segregated from the other operations in the same facility.

6.1.2 Reduced Emissivity of Hot Surfaces

The rate at which heat is radiated from the surface of a hot source can often be lowered if the emissivity of the source is reduced through surface treatment. When an oven, boiler, or other hot surface is covered with aluminum paint, the reduced emissivity of its surface offers two advantages. Less heat radiates to workers nearby; and heat is conserved inside the unit—representing a substantial savings in energy costs (129).

The emissivity of a hot source can also be lowered by sheathing or by covering the source with sheet aluminum. Other metals, such as galvanized steel, have relatively low emissivities but are more expensive to use than aluminum. In addition, the emissivity of galvanized steel increases faster with aging than with aluminum sheet (129).

Because of reduced emissivity, structural steel members will radiate less heat if they are painted with aluminum paint. Even though aluminum-painted surfaces have a higher emissivity than aluminum sheet, they do not radiate as much heat as oil-based painted steel at a given temperature. Table 21.7 shows the emissivity and reflectivity (Section 6.1.4) of various common materials (129).

6.1.3 Insulation

"Insulation" is not mutually exclusive of "isolation" in the context of engineering control of heat stress. Insulation also prevents the escape of sensible and radiant heat into the work environment. A well-known example of insulation that also has implications for energy conservation is that of pipe-covering insulation on steam lines. By reducing the escape of heat into the environment, such insulation clearly helps conserve energy and fuel resources.

In addition to reducing radiative exchange, insulation reduces the convective

Table 21.7. Relative Efficiencies of Common Shielding Materials (129)

Surface	Reflectivity of Radiant Heat Incident Upon Surface (%)	Emissivity of Radiant Heat from Surface (%)
Aluminum, bright	95	5
Polished aluminum	92	8
Polished tin	92	8
Zinc, bright	90	10
Aluminum, oxidized	84	16
Varnished aluminum	80	20
Varnished tin	80	20
Zinc, oxidized	73	27
Aluminum paint, new, clean	65	35
Iron, sheet, smooth	45	55
Aluminum paint, dull, dirty	40	60
Iron, sheet, oxidized	35	65
Steel and iron	10–20	80–90
Brick	4–20	80–96
Lacquer, black	10	90
Wood	4–8	92–96
Asbestos board	4–8	92–96
Oxide paints, all colors	6	94
Lacquer, flat black	3	97

heat transfer from hot equipment to the work environment by minimizing local convective currents that form when air that contacts very hot surfaces is heated.

6.1.4 Radiation Shielding

Shielding against radiant heat represents an extremely important control measure. The characteristics of radiant heat are quite different from those of high air temperatures, and the difference must be understood if engineering controls are to be effective. Radiant heat passes through air without heating the air; it heats only the objects in its path that are capable of absorbing it. Shielding of radiant heat sources means putting a barrier between the worker and the source to protect the worker from being a targeted receptor of the radiant energy. Radiation shielding can be classified into reflecting, absorbing, transparent, and flexible shields.

Reflective Shields. Reflective shields are constructed from sheets of aluminum, stainless steel, or other bright-surface metallic materials. Aluminum offers the advantage of 85 to 95 percent reflectivity. It is also used as shielding in the form of foil with insulative backing, and in aluminized paint, with reduced effectiveness. Reflectivity of other shielding materials is shown in Table 21.7.

Successful use of aluminum as shielding requires an understanding of certain principles (129):

1. There must be an aluminum-to-air surface; the shield cannot be embedded in other materials.
2. The shield should not be painted or enameled.
3. The shield should be kept free of oil, grease, and dirt to maximize reflectivity.
4. When used to enclose a hot source, the shield should be separated from the source by several inches.
5. Corrugated sheeting should be arranged so that the corrugations run vertically rather than horizontally, to help maintain a surface free of foreign matter.

Absorptive Shielding. Absorption shielding absorbs infrared radiation readily. This type of shielding, preferably flat black, is constructed typically of two or three sheets separated by air spaces. Heat can then be removed by water flowing between two metal plates in the shield, transferring heat from the shield by conduction (129). The surface(s) of absorptive shielding exposed to work areas should be constructed of aluminum or aluminized to reduce emissivity.

Transparent Shielding. Transparent shielding consists of two general types: special glass and metallic mesh. Special glass reduces transmission of infrared radiation because it is either "heat absorptive" or "infrared reflecting." Infrared reflecting glass is used commonly in the windows of control rooms amid excessive heat sources. Metallic mesh shielding uses chains and wire mesh to provide partial reflectance and to help reduce the radiant heat reaching an operator (129). Such use is warranted where manual operations prevent use of a solid barrier.

Flexible Shielding. Flexible shielding uses fabrics treated with aluminum. When worn as aprons or other items of clothing, they protect against radiant heat by reflecting up to 90 percent. Reflective garments are useful for protection against very localized and directional radiant sources).

6.1.5 Local Exhaust Ventilation

Canopy hoods with natural draft or mechanical exhaust ventilation are used commonly over furnaces and similar hot equipment. Although the benefits of such an application are obvious, because heated air has a natural tendency to rise, it must be remembered that local ventilation removes only convective heat. Radiant energy losses, whose magnitude often overrides convective losses, are not controlled by local exhaust hooding. Radiation shielding must be used as well to control what is likely to be the larger fraction of the total heat escaping from the hot process (130).

6.2 Localized Cooling at Work Stations

When practical considerations limit the feasibility of heat control at the source or throughout the general work area, relief can be provided at localized areas. In such instances, cool air is introduced in sufficient quantity to surround the worker with an "independent" atmospheric environment. This local relief, often termed "spot cooling," can serve two functions, depending on the relative magnitudes of the radiant and convective components in the total heat load.

If the overall thermal load is primarily convective in the form of hot air surrounding the worker, the local relief system displaces the hot air immediately around the individual with cooler air having a higher velocity. If such air is available at a suitable temperature and is introduced without mixing with the hot ambient air, no further cooling of the worker is necessary.

When there is a significant radiation load, however, the local relief system must provide some actual cooling to offset the radiant energy that penetrates the mass of air surrounding the worker. Here the temperature of the supplied air must be low enough to make the convection component C negative in the heat exchange model to offset R, the radiation component.

Air movement within the local relief zone is desirable to maintain adequate evaporative cooling. Nevertheless, the use of high air velocities (increased evaporative cooling) to offset radiation may be counterproductive if the convective load is increased (where air temperature exceeds skin temperature), causing a greater sweating demand. This defines a rationale for not using simple human-cooling fans, which are often used inappropriately throughout industry.

A major problem in the design of localized cooling systems is the introduction of supplied air in such a way that minimal mixing takes place with the surrounding hot air. High-velocity jets encourage mixing, yet are necessary to "throw" the air into the work zone from an off-site supply duct.

In extreme heat, workers should be stationed inside an insulated, locally cooled observation booth or relief room to which they can return after brief periods of high heat exposure. The practical utility of such a protected room will vary from industry to industry. Air-conditioned crane cabs represent one application of this concept now in common use.

6.3 General Ventilation

A common method for heat removal in the hot industries is general ventilation, making use of wall openings for the entrance of cool outside air and roof openings, commonly of the gravity type, for the discharge of heated air.

Although a higher fraction of heat loss from hot sources occurs by radiation, such radiant heat is absorbed by walls and other solid structures. These heated surfaces then become secondary sources of convective heat and also act as secondary radiators. Eventually, most of the heat must be removed from the enclosed space by ventilation; the balance is lost by conduction through walls. Therefore general ventilation is an essential part of an overall heat control strategy. It cannot offset direct radiant heat exposures, however, and even in a secondary role, often fails to function as needed because of inadequate supply of make-up air. Insufficient area of openings and poor location of inlets often result in improper distribution or too little air within the building (131).

6.3.1 *Fresh Air Supply*

The use of uncooled outside air is sometimes effective in controlling heat, but a more practical approach for a new facility is to design a general ventilation system with the capability of cooling outside air before distributing it throughout the plant.

Evaporative Cooling. Evaporative cooling is a well established method for lowering the dry-bulb (air) temperature. An evaporative cooler provides intimate contact between the incoming air and water by using sprays or wetted filters, and upon such contact, the air is cooled adiabatically. This means that no appreciable heat is transferred to or from the air. By converting sensible heat in the airstream to latent heat, dry-bulb temperature is caused to drop as the relative humidity increases. Even though additional humidity is added to the incoming air, the effect is not necessarily negative because the new dry-bulb temperature may be sufficiently below skin temperature to allow the worker to be cooled by convection.

Chilled Coil Systems. In a chilled coil system, air passes over coils containing a medium whose temperature is sufficiently below that of the air to result in satisfactory cooling. These systems are usually one of the following types (131):

1. *Water-Cooled Coil.* In this system—the simplest—water is circulated directly through the coil while air from the conditioned space is passed over the coil to remove heat and moisture. As a general rule, water must enter the coil at not more than 11°C, and it is wasted after leaving the coil unless it is used for process work.
2. *Cooled-Water System.* When water is not available from a sufficient supply or at a satisfactory temperature, it can be cooled artificially and recirculated. This system is often used when the coolant must be piped long distances and a variety of loads must be overcome. Usually the heat is removed from the water by mechanical refrigeration, but in some intermittent operations it is removed by the melting of ice.
3. *Refrigerated Coils.* As the name implies, the coils are cooled by the direct vaporization and expansion of a liquid refrigerant. This method is widely used in small units as well as large central systems.

6.3.2 *Distribution of Make-up Air*

Convective flow around hot bodies is naturally upward. For an ideal system of general ventilation, combined with radiation shielding, the inlet air should enter near floor level, be directed toward the workers, and flow toward the hot equipment. In this way the coolest air available is received by the workers before its temperature is increased by mixing with warm building air or circulation over hot processes. This air then flows toward the hot equipment and, as its temperature increases, rises and escapes through vent openings in the roof. The combination of the motion and the cool temperature of the air yields the maximum comfort that can be obtained for the worker without artificially cooling the air. Contrary to general practice, provisions for proper distribution of the air supply should receive the same careful consideration that is given to the selection of exhaust equipment (131).

6.3.3 Removal of Heated Air

Increasingly, buildings are constructed to provide for the natural removal of heated air. For example, glass manufacturing facilities usually have large gravity roof ventilators for exhaust purposes. Unfortunately many hot industries are characterized by large, flat building configurations. Exhaust fans or gravity ventilators without a forced air supply may not provide satisfactory ventilation patterns in the building's workspace.

The basic strategy when general ventilation is used to remove heated air positions the exhaust openings, either natural draft or mechanically operated, above the sources of heat and as close to them as practical (129).

6.4 Moisture Control

The importance of moisture control is underscored in the warm-moist industries where high temperatures and high relative humidities prevail. In some industries relative humidity is deliberately maintained at a high level to maintain product quality or to prevent static electricity. Examples are textile mills, munitions plants, coal mines, and flour mills. Whenever feasible, dehumidification can help to offset heat stress because it reduces the partial pressure of water vapor in air, which increases the evaporative capacity (Section 2.3).

In a general sense, moisture control includes both prevention of increased humidity and the use of dehumidification procedures. Such unsophisticated yet effective controls as enclosing hot water tanks, covering drains carrying hot water, and repairing leaky joints and valves in steam piping offer direct measures to help alleviate heat stress in warm-moist industries.

Dehumidification, aimed at reducing workplace humidity rather than preventing its increase, can be accomplished by refrigeration, absorption, or adsorption. In the context of occupational heat exposures, refrigeration is the most widely used technique to condition the air in relief areas, operating booths, or other local or regional portions of an industrial facility. Although it is clearly impractical to air condition some operations, such as a hot rolling mill in a steel plant, the use of air conditioning is useful as a heat control method, especially in combination with a work-rest regimen for a given job function. The concept of mechanical refrigeration to reduce the temperature (and humidity) of supply air is mentioned explicitly in the list of engineering controls of the Standards Advisory Committee on Heat Stress (18).

6.5 Typical Examples

The first step in the engineering control of heat stress is to identify the magnitude of the various components of stress to which the workers are exposed in existing

Table 21.8. Use of Ventilation to Offset High Humidity (132)

Data	Before Ventilation Added: Semi-clothed	Before Ventilation Added: Fully Clothed	After Ventilation Increased: Semi-clothed
GT (°C)	35.0	35.0	35.0
T_A (°C)	35.0	35.0	35.0
NWB (°C)	31.4	31.4	31.1
VP_A (mm Hg)	30	30	30
V (m/min)	15	15	45
WBGT (°C)	32.5	32.5	32.3
MRT (°C)	35.0	35.0	35.0
R (kcal/hr)	0	0	0
C (kcal/hr)	0	0	0
M (kcal/hr)	200	200	200
E_{req} (kcal/hr)	200	200	200
E_{max} (kcal/hr)	130	80	250
Sweat rate (l/hr)	0.33[a]	0.33[a]	0.33

[a]Dripping.

or expected conditions. Only by knowing the relative magnitude and the factors contributing to the heat stress components, can strategies be selected to help solve the problem with a cost-effective program.

6.5.1 Increasing Ventilation to Offset High Humidity (129, 132)

Table 21.8 gives environmental data typical for a hypothetical laundry, where a small wall fan moves air at a relatively low velocity. Figures for both semi-clothed and clothed workers indicate the stress added from clothing in high humidity environments. Here WBGT is at 32.5°C, showing that some control measures are necessary. Note that *R* and *C* are both zero because of the simplified thermal conditions assumed, namely, air and skin temperature are both 35°C.

In this case, increasing air speed from 15 to 45 m/min produces an E_{max} greater than E_{req}, a condition necessary to avoid bodily accumulation of heat and its consequent strain. In the "before" situation, the workers are sweating at a near-maximum rate but cannot achieve thermal balance because a deficit of 70 kcal/hr exists between E_{req} and E_{max}. With clothing, the deficit increases to 120 kcal/hr. At 120 kcal/hr, the limit of tolerance would be reached within 1 hr.

Tripling the ventilation rate effectively doubles the evaporative cooling and, under such conditions, E_{max} exceeds E_{req}. It should be noted that even under these conditions WBGT is at 32.3°C, an unacceptable level. Nevertheless, an acclimatized person could continue to work in this environment, and the 0.33 l/hr sweat rate would not create undue strain. The increased ventilation has created a thermal condition that allows the sweat to be evaporated efficiently enough that the skin is no longer dripping wet. At higher air velocities, additional cooling would occur,

Table 21.9. Use of Evaporative Cooling in a Hot-Dry Environment (132)

Data	Before Cooling Installed: Fully Clothed	Before Cooling Installed: Semi-clothed	After Evaporative Cooling Installed: Fully Clothed
GT (°C)	46.1	46.1	35.0
T_A (°C)	42.8	42.8	29.4
NWB (°C)	24.2	24.2	24.4
VP_A (mm Hg)	10	10	18.5
V (m/min)	110	110	110
WBGT (°C)	30.8	30.8	27.6
MRT (°C)	54.4	54.4	41.7
R (kcal/hr)	130	220	50
C (kcal/hr)	80	130	−60
M (kcal/hr)	200	200	200
E_{req} (kcal/hr)	410	550	190
E_{max} (kcal/hr)	650	1090	480
Sweat rate (l/hr)	0.68	0.91	0.32

but there may be an upper limit of air speed that could interfere with laundry operations.

A more effective, permanent control method would be to reduce the moisture content of the unknown air. This should be done as close as possible to the source of water vapor when the source is within the workplace. If the source of humidity were outside the workroom, mechanical air conditioning would likely be required to reduce the temperature and moisture level of the work space (132).

6.5.2 Use of Evaporative Cooling in a Hot-Dry Environment (129, 132)

Table 21.9 shows data for a hot, dry environment where the presence of high air speed becomes a liability, and the wearing of clothing becomes advantageous. Because the air speed is already at 110 m/min, incremental increases in air velocity would not be expected to make major improvements in evaporative capacity. If the workers were semi-nude rather than clothed, the increased thermal load from both R and C created by a greater exposure of skin would account for an increase in sweat rate from 0.68 to 0.91 l/hr. The benefit of greater evaporative capacity (E_{max}) from less clothing does not improve the situation. It follows that when MRT and T_A are above 35°C and humidity is low, the wearing of full clothing can reduce heat stress. In addition, the type and weight of clothing can be optimized for different hot work situations.

With low humidity (VP_A = 10 mm Hg), a better solution to the problem would be the installation of an evaporative cooler. Assume that in the situation described, the inside temperature is usually 5°C hotter than out-of-doors owing to process heat and building insulation. If the outside temperature does not exceed 40°C and VP_A is about 10 mm Hg, outside air could be drawn through a water spray and the temperature reduced to that of the prevailing outdoor wet-bulb temperature

Table 21.10. Use of Radiant Heat Shielding (132)

Data	Before Shielding Installed: Clothed	After Shielding Installed: Clothed
GT (°C)	71.7	43.3
T_A (°C)	47.8	43.3
NWB (°C)	36.3	29.8
VP_A (mm Hg)	24.5	24.5
V (m/min)	240	240
WBGT (°C)	46.9	33.8
MRT (°C)	159	43.3
R (kcal/hr)	850	60
C (kcal/hr)	210	140
M (kcal/hr)	200	200
E_{req} (kcal/hr)	1260	400
E_{max} (kcal/hr)	630	630
Sweat rate (l/hr)	2.1[a]	0.65

[a]Dripping.

(23°C). VP_A would then be increased to 18.5 mm Hg, and the temperature of the workplace could be reduced to 29°C. The "after" calculation of the heat components shows a negative value for the convective heat component with an E_{req} of only 190 kcal/hr. Because E_{max} exceeds E_{req}, the sweat rate of 0.32 l/hr could be accomplished easily without undue heat strain.

6.5.3 Use of Radiant Heat Shielding (129, 132)

When the main component of heat stress comes from a radiant source, such as a furnace wall, the first control priority should be shielding of the radiant energy. Increased air velocity does little to improve such situations. Table 21.10 represents another "before and after" situation taken from the work of Leihnard et al. (133). Here the task requires skimming dross from molten bars of aluminum. Manipulation of a ladle at a fixed station, with moderate use of shoulder and arm muscles, suggests a metabolic rate of about 200 kcal/hr. Air is directed at the worker from an overhead duct at a velocity of 240 m/min. The humidity is quite high, and WBGT is 46.9°C, an extremely stressful situation. Calculation of the heat components confirms this, and the worker would have to take frequent work breaks to maintain thermal equilibrium.

Despite use of clothing and a face shield, the workers were able to perform this task only a few minutes at a time, and heat exhaustion was not uncommon. Control of the high heat exposure was achieved by placing a finished aluminum sheet between the heat source and the worker. Infrared reflecting glass at face level enabled the workers to see the work. Spaces were left in the aluminum sheet to permit workers to operate the ladle. After shielding was installed, GT and T_A were both reduced to 43.3°C, and WBGT was lowered to 33.8°C. E_{max} was then greater than E_{req}, and it was possible for a worker to evaporate about 0.65 liter of sweat per hour, thus maintaining a thermal balance. The previous sweat rate was an impossible-to-sustain level of 2.1 l/hr. Although there may be large errors in the estimate of R at extremely high globe temperatures, the maximum relief that could be expected from shielding was actually achieved in the case cited (132).

It should be noted that a polished metallic surface will not maintain its shielding effectiveness (reflectivity) if allowed to become dirty. This is true even for fabrics coated with very fine metallic particles. Even a thin layer of grease or oil can change the emissivity of a polished surface from 0.1 to 0.9. To provide shielding that does not interfere with performance of the work task, a curtain of metal chains can be installed; the chains reduce radiation but can be parted as required. Another approach is the use of a mechanically activated door, opened only during ejection or manipulation of a product. Also, remote-operated tools can be used. Partial barriers can also be used effectively because radiant heating from an open portal is limited to the line of sight (129, 132).

7 MANAGEMENT OF EMPLOYEE HEAT EXPOSURE

Many situations exist in industry where the complete control of heat stress by engineering methods alone may be impossible or impractical, where the level of environmental heat stress may be unpredictable and variable, or where the exposure time may vary with the task and with unforeseen critical events. Typical examples include asbestos workers, hazardous materials workers, and welders/burners. When engineering control of heat stress is neither practical nor complete, other solutions must be used to keep the level of total heat stress on the worker within limits which will not be accompanied by an increased risk of heat illness.

Accordingly, the overall management of heat stress in industry could, and often should, include each of the following elements:

1. Employee and supervisory training
2. Medical supervision
3. Acclimatization
4. Work-rest regimen
5. Water and electrolyte provisions
6. Environmental monitoring
7. Auxiliary body cooling and protection
8. Forecast of episodes and heat alert programs

7.1 Employee and Supervisory Training

Exposure to excessively hot environmental conditions clearly can lead to primary heat illnesses, unsafe acts, or increased susceptibility to toxic chemicals and physical substances. Individual employees and supervisors can reduce the likelihood of ill effects from a hot work environment through application of basic health and safety procedures. Each employee who may be exposed to heat, as well as each supervisor, should receive a safety training program to become aware of the following points, as a minimum (10, 18, 19):

1. Instruction on the causes and recognition of symptoms of heat disorders
2. Information concerning heat acclimatization
3. Information concerning water and salt replacement
4. Instruction on the possible combined effects of heat and (*a*) alcoholic beverages, (*b*) prescription and nonprescription drugs, including blood pressure drugs and diuretics (*c*) toxic agents, and (*d*) other physical agents
5. Information concerning the use of appropriate protective clothing and equipment

7.2 Medical Supervision

Medical supervision in the management of employee heat exposure includes both the process of selection and the periodic examination of workers. The work practices specified by NIOSH (10, 19) and the Standards Advisory Committee (18) include a preplacement medical evaluation to determine a worker's fitness for work in heat, with special emphasis on the cardiovascular system, and a review of medical history with reference to heat disorders and all related physiologic systems. Periodic medical examinations are recommended for all employees working in conditions of extreme heat exposure, with an examination at least annually of all workers exposed to heat stress at levels exceeding the NIOSH-proposed RAL's.

7.2.1 Medical Examinations

A preplacement examination for a worker applying for a hot job serves to determine mental, physical, and emotional qualifications to perform the job assignment with reasonable efficiency and without risk to one's health and safety or to that of fellow employees (134).

The NIOSH Criteria Documents suggest that the examining physician should seek to discern possible evidence of intolerance to heat, occupational or nonoccupational. A history of successful adaptation to heat exposure on previous jobs can be used for predicting the effectiveness of a worker's future performance under heat stress, assuming that levels of work load and heat exposure are equivalent and that no significant alteration has occurred in health status since the time of the previous employment (10, 19).

After the age of 45, or as specified by the responsible health professional, periodic physical and laboratory examinations should be designed to detect the onset of chronic impairments of the cardiocirculatory and cardiorespiratory systems and to detect metabolic, skin, and renal disease for employees in hot jobs. During these periodic examinations, any incidents of acute illness or injury, either occupational or nonoccupational, during the interval between examinations should be evaluated carefully. Repeated accidental injuries on the job or frequent absence due to illness should alert the physician to possible heat intolerance or the presence of an aggravating stress such as carbon monoxide in combination with the exposure to heat. Nutritional status should be noted, and advice to correct overweight should be offered (10, 19).

7.2.2 Emergency Medical Care

The NIOSH Criteria Documents and prevailing practice suggest that during working hours a person trained in first aid and in the recognition of the signs and symptoms of any heat disorder be available (10, 19). The Standards Advisory Committee recommended among its work practices for extreme heat exposure that workers be under the observation of a trained supervisor or co-worker who can note any early signs of heat effects (18).

Supervisors and selected personnel should be trained to recognize the signs and symptoms of heat disorders and to administer first aid. The most serious emergency is heatstroke, signaled by dry, hot, red, or mottled skin, mental confusion, delirium, convulsions or coma, and a high and rising rectal temperature, usually 41°C, but occasionally as low as 40°C (see Section 3.3).

7.3 Acclimatization

As indicated earlier, acclimatization refers to the adaptive process that results in a decrement in the physiological response produced by the application of a constant environmental stress. Upon initial exposure to a hot environment, there is a noticeable impairment in the ability to work, with physiological strain. If the exposure is repeated on several successive days, however, a gradual improvement takes place in the ability to work with a corresponding decrease in strain. After 5 to 7 days of the acclimatization process, subjective discomfort almost disappears, body temperature and heart rate are lower, sweat is more profuse and dilute; and substantial improvement takes place in the ability to perform vigorous work.

Being physically fit for the job enhances (but does not replace heat acclimatization) heat tolerance for both heat-acclimatized and unacclimatized workers. The time required to develop heat acclimatization in unfit individuals is about 50 percent greater than in the physically fit (19).

NIOSH recommends that for workers who have had previous experience with the hot job, the acclimatization regimen should be exposure for 50 percent on day one, 60 percent on day two, 80 percent on day three, and 100 percent on day four. New workers should begin with 20 percent exposure on day one and a 20 percent increase on each successive day (19).

7.4 Work–Rest Regimen

The overall management of heat stress must include work practices that acknowledge how the duration of work periods, the frequency and length of rest pauses, and the pace and tempo of work can be adjusted to avoid heat strain. The appropriateness and length of rest periods obviously depend on the severity and duration of heat stress. Rest periods should be taken before excessive fatigue develops and should be long enough to reduce the pulse rate to below 100 beats/min (129).

Brouha (52) showed that the heart rates of men who rest in an air-conditioned room after working in hot environments not only drop much faster but also fall to a lower level than is recorded in hot rest areas. Air-conditioned rest areas and low humidity encourage recovery markedly.

The NIOSH Criteria Documents recommend that a work and rest regimen be implemented to reduce the peaks of physiological strain and to improve recovery during rest periods (10, 19). There are several ways to control the daily length of time and temperature to which a worker is exposed in heat stress conditions: (19)

- When possible, schedule hot jobs for a cooler part of the day (early morning, late afternoon, or evening).
- Schedule routine maintenance and repair work in hot areas for the cooler seasons of the year.
- Alter the work–rest regimen to permit more rest time.
- Provide cool areas for rest and recovery.
- Use extra personnel to reduce the exposure time of each member of the work team.
- Permit freedom to interrupt work when a worker feels extreme heat discomfort.

7.5 Water and Electrolyte Provisions

To ensure that fluid lost in the sweat and urine is replaced (at least hourly) during the work shift, an adequate fluid supply and intake are essential for heat tolerance and prevention of heat-induced illnesses.

Fluid intake during a work period should ideally approximate the amount of sweat produced. Work in hot environments could result in a sweat production of 1 to 3 gal per shift. If this loss is not replaced, severe dehydration results. Unfortunately, thirst is not an adequate drive to stimulate total replenishment. An ample supply of cool fluid should be readily available, and workers should be encouraged to drink every 15 to 20 min.

Rehydration will occur more rapidly when beverages containing sodium—the major electrolyte lost in sweat—are consumed. Ingesting a beverage containing sodium allows the plasma sodium to remain elevated during the rehydration period, helps maintain thirst, and delays stimulation of urine production. The rehydration beverage could also contain glucose or sucrose because these carbohydrates provide a source of energy for working muscles, stimulate fluid absorption in the gut, and improve beverage taste.

Electrolyte balance in body fluids must be maintained to minimize the risk of heat-induced illnesses. For heat-unacclimatized workers who may be on a restricted salt diet, additional salting of the food, with a physician's concurrence, during the first few days of heat exposure may be appropriate to replace the salt lost in the sweat. Acclimatized workers lose relatively little salt in the sweat; therefore salt supplementation of the normal U.S. diet is usually not required (19).

7.6 Environmental Monitoring

Knowledge of the levels of thermal stress is critical for the control of heat stress in industry. Management in a hot industry should have, as a minimum, a WBGT profile for the hot jobs and the hot areas throughout each facility. In addition, an engineering assessment of heat stress by measurement of the values of the various environmental components should be made using the concepts and measurements required for the heat stress index (HSI).

The use of appropriate work practices requires some "action level" based on a measurement or estimate of an environmental heat stress index such as WBGT. The NIOSH Criteria Documents suggest that a WBGT profile be established for each workplace for winter and summer seasons to serve as a guide for deciding when work practices should be initiated. After such initial profiles have been established, monitoring should be conducted during the warmest part of each succeeding year (10, 19).

The Standards Advisory Committee specified in its recommendation that a worker's thermal exposure be based on the hottest 2-hr period of the work shift in which regular work is performed (18). Accordingly, maximum WBGT values during the hottest period would be used to characterize the temperature level at that workplace. If no single work station accounts for more than half of this 2-hr period, the WBGT value that is compared with the recommended threshold levels (Table 21.1) must be determined as a time-weighted average as described below:

$$\text{WBGT} = \frac{1}{120} \sum_{i=1}^{h} (\text{WBGT})_i \times T_i$$

where WBGT = 2-hr, time-weighted wet-bulb globe temperature
$(\text{WBGT})_i$ = WBGT during interval i
i = a discrete exposure interval
T_i = duration of a discrete exposure interval (minutes)

The Standards Advisory Committee recommendation is to estimate work load during the hottest two hours of a work shift to classify it as light, moderate, or heavy. As with WBGT, this work load level can be time weighted (18).

In addition, air speed must be estimated or measured to implement the recommendation of the Standards Advisory Committee because the threshold levels of WBGT are based on a division of air velocity at 300 ft/min (18). It should be noted that air velocity is not required to assess WBGT, which depends only on natural wet-bulb temperature and black globe temperature for indoor exposures.

7.7 Auxiliary Body Cooling and Protection

As indicated earlier, there are four basic approaches to resolving an unacceptable heat stress situation: (1) modify the worker by heat acclimatization; (2) modify the clothing or equipment; (3) modify the work; or (4) modify the environment. Heat tolerance can be enhanced if individuals are fully heat-acclimated, have received good training in the use of and practice of wearing protective clothing, are in good physical condition, and are encouraged to drink rehydration beverage(s) as necessary to compensate for sweat water loss.

If heat acclimatization and physical fitness are not sufficient to alleviate the heat stress and reduce the risk of heat illnesses, only the latter three solutions are left to deal with the problem. When air temperature is above 35°C (95°F) with a relative humidity of 75 to 85 percent or when there are intense radiant heat sources, a suitable, sometimes more functional approach is to modify the clothing to include some form of auxiliary body cooling for limited periods of time. A properly designed system will reduce heat stress, preclude large amounts of drinking water, and allow unimpaired performance across a wide range of climatic factors (19). A seated individual will rarely require more than 85 kcal/hr or 345 Btu/hr of auxiliary cooling, and the most active individuals no more than 345 kcal/hr or 1380 Btu/hr unless working at a level where physical exhaustion per se would limit the duration of work. Some form of heat-protective clothing or equipment should be provided for exposures at heat-stress levels that exceed the NIOSH ceiling limit in Figures 21.1 and 21.2 (19).

Auxiliary cooling systems can range from such simple approaches as an ice vest, prefrozen and worn under the clothing, to more complex systems; however, costs and maintenance are considerations of varying magnitude in all of these systems. At least four auxiliary cooling approaches have been evaluated: (1) water-cooled garments, (2) an air-cooled vest, (3) an ice packet vest, and (4) a wettable cover. Each of these cooling approaches might be applied in alleviating risk of severe heat stress in a specific industrial setting (135, 136).

7.7.1 Water-cooled Garments

Water-cooled garments include (1) a water-cooled hood that provides cooling to the head, (2) a water-cooled vest that provides cooling to the head and torso, (3) a short, water-cooled undergarment that provides cooling to the torso, arms, and legs, and (4) a long water-cooled undergarment that provides cooling to the head, torso, arms, and legs. None of these water-cooled systems provide cooling to the hands and feet.

Water-cooled garments and headgear require a battery-driven circulating pump and container where the circulating fluid is cooled by the ice. The weight of the batteries, container, and pump limit the amount of ice that can be carried. The amount of ice available determines the effective time of the water-cooled garment.

7.7.2 Air-Cooled Garments

Air-cooled suits and/or hoods that distribute cooling air next to the skin are available. The total heat exchange from a completely sweat-wetted skin when cooling air is supplied to the air-cooled suit is a function of cooling-air temperature and cooling-airflow rate. Both the total heat exchanges and the cooling power increase with cooling airflow rate and decrease with increasing cooling air inlet temperature.

The use of a vortex tube as a source of cooled air for body cooling is applicable in many hot industrial situations. The vortex tube, which is attached to the worker, requires a constant source of compressed air supplied through an air hose. The hose connecting the vortex tube to the compressed air source limits the area in which the worker can operate. Unless mobility of the worker is required, the vortex tube, even though noisy, should be considered as a simple source of cooled air.

7.7.3 *Ice Packet Vest*

The available ice packet vests may contain as many as 72 ice packets; each packet has a surface area of approximately 64 cm^2 and contains about 46 g of water. These ice packets are generally secured to the vest by tape. The cooling provided by each individual ice packet will vary with time and with its contact pressure with the body surface, plus any heating effect of the clothing and hot environment; thus the environmental conditions have an effect on both the cooling provided and the duration of time this cooling is provided. Solid carbon dioxide in plastic packets can be used instead of ice packets in some models (19).

In environments of 29.4°C (84.9°F) at 85 percent relative humidity and 35.0°C (95°F) at 62 percent relative humidity, an ice packet vest can still provide some cooling up to 4 hr of operation (about 2 to 3 hr of effective cooling is usually the case). In an environment of 51.7°C (125.1°F) at 25 percent relative humidity, any benefit is negligible after about 3 hr of operation. With 60 percent of the ice packets in place in the vest, the cooling provided may be negligible after 2 hr of operation. Because the ice packet vest does not provide continuous and regulated cooling over an indefinite time period, exposure to a hot environment would require replacement of the frozen vests every 3 to 4 hr. Nevertheless, the cooling is supplied noise-free and independent of any energy source or umbilical cord that limits a worker's mobility. The greatest potential for the ice packet vest appears to be for work where other conditions limit the length of exposure, for example, short duration tasks and emergency repairs. The ice packet vest is also relatively cheaper than other cooling approaches (19).

7.7.4 *Wetted Overgarments*

A wetted cotton terry cloth coverall or a two-piece cotton cover that extends from just above the boots and from the wrists to a V-neck when used with impermeable protective clothing can be a simple and effective auxiliary cooling garment.

Predicted values of supplementary cooling and of the minimal water requirements to maintain the cover wet in various combinations of air temperature, relative humidity, and wind speed can be calculated. Under environmental conditions of low humidity and high temperatures where evaporation of moisture from the wet cover garment is not restricted, this approach to auxiliary cooling can be effective, relatively simple, and inexpensive to use.

7.8 Forecast of Episodes and Heat Alert Programs

In many hot industries, heat stress tends to be a seasonal phenomenon. Sudden heat waves early in the summer can create acute situations. For other facilities, heat stress may be a significant problem only during the hottest episodes of the summer season (10, 18, 19).

Accordingly, management should be interested in any episodes that are forecast by the National Weather Service or other forecasting agencies. Depending on the severity of the measured or expected levels of heat stress, additional work practices such as the distribution of total work load, scheduling of hottest jobs for the coolest part of the work shift, or other administrative measures may be appropriate. Although many of the hot industries have subjective correlations to relate weather with the intensity of heat stress in the occupational environment, there now exists a systematic modeling technique to relate WBGT at any worksite to external weather conditions, current or predicted (111). However, such modeling is more convenient, and certainly more reliable, if WBGT profiles have been developed at the work stations where significant heat stress can be anticipated.

In some plants where heat illnesses and disorders occur mainly during hot spells in the summer, a Heat Alert Program (HAP) can be established for preventive purposes. Although such programs differ in detail from one facility to another, the underlying concept is identical, that is, to take advantage of the weather forecast of the National Weather Service. If a hot spell is predicted for the next day or days, a declaration of "Heat Alert" can help prevent heat casualties. One such HAP is described below (19, 70):

1. A "Heat Alert Committee" consisting of an occupational physician or nurse, industrial hygienist, safety engineer, operations engineer, and a key manager is defined and activated. This committee provides for the following action plan:
   a. Establish criteria for the declaration of a Heat Alert. For example, a Heat Alert could be declared if the area weather forecast for the next day predicts a maximum air temperature of 35°C (95°F) or above or a maximum of 32°C (90°F) if the predicted maximum is 5°C (9°F) above the maximum reached in any of the preceding 3 days.
   b. Assume an effective training program for all involved in the HAP, including procedures to follow should a Heat Alert be declared. This training must give special emphasis to the prevention and early recognition of heat illnesses and first-aid procedures when a heat illness occurs.
   c. Instruct the supervisors to:

      (1) Reverse winterization of the plant, that is, open windows, doors, skylights, and vents according to instructions for greatest ventilating efficiency at places where high air movement is needed.

      (2) Check drinking fountains, fans, and air conditioners to make sure they are functional, necessary maintenance and repair is performed, these facilities are rechecked regularly, and workers know how to use them.

   d. Ascertain that in the medical department, as well as at the job sites, all resources and facilities required to give first aid in case of a heat illness are in a state of readiness.
2. During the state of Heat Alert these procedures are to be followed:
   a. Postpone tasks that are not urgent (e.g., preventive maintenance involving high activity or heat exposure) until the hot spell is over.
   b. Increase the number of workers in each team in order to reduce each worker's heat exposure. Introduce new workers gradually to allow acclimatization (follow heat-acclimatization procedure).
   c. Increase rest allowances. Allow workers to recover in air-conditioned rest places.
   d. Turn off all heat sources that are not absolutely necessary.
   e. Remind workers to drink water or other suitable fluids frequently, in small amounts, to prevent excessive dehydration, to weigh themselves before and after the work shift, and to be sure to maintain body weight through fluid replacement.
   f. Monitor the environmental heat at the job sites and resting places.
   g. Check the oral temperature of workers during their most severe heat-exposure period.
   h. Exercise additional caution on the first day of a shift change to make sure that workers are not overexposed to heat, because they may have lost some acclimatization over the weekend or during days off.
   i. Send any worker who shows signs of a heat disorder, even minor, for medical evaluation. A physician's permission to return to work must be given in writing.
   j. Limit overtime work.

Table 21.6 shows a typical checklist of the variety and number of control methods available for managing occupational heat stress (19).

REFERENCES

1. C. P. Yaglou and D. Minard, *Arch. Ind. Health*, **16**, 302–316 (1957).
2. D. Minard, Research Report No. 4, Contract No. MR 005.01-0001.01, Naval Medical Research Institute, Bethesda, MD, 1961.
3. D. Minard and R. L. O'Brien, Research Report No. 7, Contract No. MR 005.01-0001.01, Naval Medical Research Institute, Bethesda, MD, 1964.
4. D. Minard, *Mil. Med.*, **126**, 261–272 (1961).
5. T. Bedford and C. G. Warner, *J. Ind. Hyg. Toxicol.*, **13**, 252 (1931).
6. H. M. Vernon, T. Bedford, and C. G. Warner, Industrial Fatigue Research Board Report No. 62, Her Majesty's Stationery Office, London, 1931.

7. C. P. Yaglou, *J. Ind. Hyg. Toxicol.*, **19**, 12–43 (1937).
8. C. H. Wyndham, *Am. Ind. Hyg. Assoc. J.*, **35**, 113–136 (1974).
9. C. H. Wyndham, *Ergonomics*, **5**, 434–444 (1962).
10. *Criteria for a Recommended Standard . . . Occupational Exposure to Hot Environments*, U.S. Department of Health, Education and Welfare, NIOSH, HSM-72-10269, 1972.
11. B. A. Hertig, "Thermal Standards and Measurement Techniques," in *The Industrial Environment—Its Evaluation and Control*, U.S. Department of Health, Education and Welfare, NIOSH, 1973.
12. *Statistical Abstracts of the United States*, 105th ed. National Data Book and Guide Sources, U.S. Department of Commerce, Bureau of Census, 1985.
13. C. P. Yaglou, "Thermal Standards in Industry," in *Report of the Committee on Atmospheric Comfort*, Year Book Publishers, Chicago, 1947.
14. Occupational Safety and Health Act PL91-596, 91st Congress, S.2193, U.S. Department of Labor, Washington, DC, 1970.
15. American Conference of Governmental Industrial Hygienists, Threshold Limit Values for Chemical Substances and Physical Agents in the Workroom Environment with Intended Changes for 1971, ACGIH, Cincinnati, Ohio, 1971.
16. American Conference of Governmental Industrial Hygienists, Threshold Limit Values and Biological Exposure Indeces for 1988–1989, ACGIH, Cincinnati, Ohio, 1988.
17. B. A. Hertig, "Work in Hot Environments: Threshold Limit Values and Proposed Standards," in *Industrial Environmental Health—The Worker and the Community*, Academic Press, New York, 1975.
18. "Recommendations for a Standard for Work in Hot Environments," Draft No. 5, Standards Advisory Committee on Heat Stress, Department of Labor, Washington, DC, 1974.
19. *Criteria for a Recommended Standard . . . Occupational Exposure to Hot Environments*, Revised Criteria 1986, U.S. Department of Health and Human Services, NIOSH, April, 1986.
20. "Hot Environments—Estimation of Heat Stress on Working Man Based on the WBGT Index (Wet Bulb Globe Temperature)," ISO 7243-1982, 1982.
21. H. S. Belding, Evaluation of Stresses of Exposure to Heat, Grant No. EC 00202-16, University of Pittsburgh, 1971.
22. D. Minard, "Physiology of Heat Stress," in *The Industrial Environment—Its Evaluation and Control*, U.S. Department of Health, Education and Welfare, NIOSH, 1973.
23. R. F. Goldman, "Prediction of Human Heat Tolerance," in *Environmental Stress*, J. L. Folinsbee, J. A. Wanger, J. A. Boria, B. L. Drinkwater, J. A. Gliner, and J. F. Bedi, Eds., Academic Press, New York, 1978, pp. 53–69.
24. J. J. Vogt, V. Candas, and J. P. Libert, *Ergonomics*, **25**, 285–294 (1982).
25. E. A. McCullough, E. J. Arpin, B. Jones, S. A. Konz, and R. H. Rohles, *ASHRAE Trans.*, **88**, 1077–1094 (1982).
26. C. P. Yaglou, *J. Ind. Hyg.*, **9**, 297 (1927).
27. C. S. Leithead and A. R. Lind, *Heat Stress and Heat Disorders*, F. A. Davis, Philadelphia, 1964.
28. C. S. Leithead, *Bull. WHO*, **38**, 649–657 (1968).
29. A. R. Lind, *J. Appl. Physiol.*, **18**, 51–56 (1963).

30. A. R. Lind, *J. Appl. Physiol.*, **18**, 57–60 (1963).
31. A. R. Lind, *J. Appl. Physiol.*, **28**, 50 (1970).
32. D. Minard, H. S. Belding, and J. R. Kingston, *J. Am. Med. Assoc.*, **165**, 1813–1818 (1967).
33. D. Minard, Naval Medical Research Report No. 6, Bethesda, MD, 1964.
34. T. F. Hatch, *Heat. Pip. Air Cond.*, **23**, 140 (1951).
35. G. F. Haines, Jr. and T. F. Hatch, *Heating Venti.*, **49**, 93–104 (1952).
36. H. S. Belding and T. F. Hatch, *Heat. Pip. Air Cond.*, **27**, 129–135 (1955).
37. B. A. Hertig, D. W. Badger, P. J. Schmitz, and L. D. Siler, *Am. Ind. Hyg. Assoc. J.*, **32**, 4 (1971).
38. H. S. Belding, B. A. Hertig, and M. L. Reidesel, *Am. Ind. Hyg. Assoc. J.*, **21**, 25 (1960).
39. C. H. Wyndham, N. B. Strydom, H. M. Cooke, and J. S. Martiz, "Studies on the Effects of Heat on Performance of Work," Applied Physiology Laboratory Reports, Transvaal and Orange Free State Chamber of Mines, South Africa, 1959.
40. C. H. Wyndham, *Ergonomics*, **5**, 115 (1962).
41. J. D. Walters, *Ann. Occup. Hyg.*, **17**, 255–264 (1975).
42. P. A. Hancock, *Aviat. Space Environ. Med.*, **52**, 177–180 (1981).
43. J. D. Ramsey and S. J. Morrissey, *Appl. Ergonomics*, **9**, 66–72 (1978).
44. P. A. Hancock, *Aviat. Space Environ. Med.*, **53**, 778–784 (1982).
45. B. Givoni and Y. Rim, *Ergonomics*, **5**, 99–119 (1962).
46. K. Marg, *Plant Eng.*, 73–74 (1983).
47. J. E. Greenleaf and M. H. Harrison, *Exercise, Nutrition and Health*, American Chemical Society, Washington, DC, 1985.
48. A. Henschel, "The Environment and Performance," in *Physiology of Work Capacity and Fatigue*, E. Simonson, Ed., Charles C Thomas, Springfield, IL, 1971, pp. 325–347.
49. World Health Organization, "Health Factors Involved in Working Under Conditions of Heat Strerss," WHO, Geneva, 1969.
50. A. R. Lind and D. E. Bass, *Fed. Proc.*, **22**, 704–708 (1963).
51. H. S. Belding, "Work in Hot Environments," in *Industrial Hygiene Highlights*, Vol. 1, Industrial Hygiene Foundation of America, Pittsburgh, 1968.
52. L. A. Brouha, *Physiology in Industry-Evaluation of Industrial Stresses by the Physiological Reactions of the Worker*, 2nd ed., Pergamon Press, Elmsford, NY, 1967.
53. S. Robinson, "Physiological Adjustments to Heat," in *Physiology of Heat Regulation and the Science of Clothing*, Saunders, Philadelphia, 1949.
54. L. W. Eichna, W. F. Ashe, W. B. Bean, and W. B. Shelley, *J. Ind. Hyg. Toxicol.* **27**, 59 (1945).
55. L. W. Eichna, C. R. Park, N. Nelson, S. Horvath, and E. D. Palmes, *Am. J. Physiol.*, **5**, 299 (1952).
56. C. H. Wyndham, W. M. Bouwer, M. G. Devine, H. E. Paterson, and D. K. C. MacDonald, *J. Appl. Physiol.*, **5**, 299 (1952).
57. E. Kamon and H. S. Belding, *J. Appl. Physiol.*, **31**, 472 (1971).
58. A. R. Lind, *J. Appl. Physiol.*, **28**, 57 (1970).

59. H. S. Belding, "Resistance to Heat in Man and Other Homeothermic Animals," in *Thermobiology*, A. H. Rose, Ed., Academic Press, London, 1967.
60. C. W. Suggs and W. E. Splinter, *J. Appl. Physiol.*, **16**, 413 (1961).
61. R. D. Pepler, "Performance and Well-Being in Heat," in *Temperature—Its Measurement and Control in Science and Industry*, Vol. 3, Part 3, J. D. Hardy, Ed., Reinhold, New York, 1963.
62. R. D. Pepler, "Psychological Effects of Heat," in *Heat Stress and Heat Disorders*, F. A. Davis, Philadelphia, 1964.
63. B. C. Duggar, "Trial of an Assembly Job for Evaluating the Effects of Heat on Human Performance," M.S. Thesis, University of Pittsburgh, 1956.
64. R. D. Pepler, "Variations in Students' Test Performance and in Classroom Temperatures in Climate Controlled and Non-Climate Controlled Schools," *ASHRAE Trans.*, **47** Part II, 35–42 (1971).
65. J. B. Moses, *Ind. Med. Surg.*, **22**, 20 (1960).
66. F. Renshaw, *Am. Ind. Hyg. Assoc. J., Abstr. Suppl.*, **38** (1971).
67. M. Khagali and J. S. R. Hayes, *Heatstroke and Temperature Regulation*, Academic Press, Sydney, 1983.
68. D. Minard and L. Copman, "Elevation of Body Temperature in Disease," in *Temperature: Its Measurement and Control in Science and Industry*, J. O. Hardy, Ed., Reinhold, New York, 1963.
69. S. Shibolet, M. C. Lancaster, and Y. Danon, *Aviat. Space Environ. Med.*, **47**, 280–301 (1976).
70. F. N. Dukes-Dobos, *Scand. J. Work Environ. Health*, **7**, 73–83 (1981).
71. R. J. T. Joy and R. F. Goldman, *Mil. Med.*, **133**, 458–470 (1968).
72. J. P. DeBenedetto and S. M. Worobec, *Occup. Health Safety*, **54**, 35–38 (1985).
73. K. B. Pandolf, T. B. Griffin, E. H. Munro, and R. F. Goldman, *Am. J. Physiol.*, **239**, 226–232 (1980).
74. K. B. Pandolf, T. B. Griffin, E. H. Munro, and R. F. Goldman, *Am. J. Physiol.*, **239**, 233–240 (1980).
75. A. R. Lind, P. W. Humphreys, K. J. Collins, K. Foster, and K. F. Sweetland, *J. Appl. Physiol.*, **28**, 50–56 (1970).
76. N. B. Strydom, *J. S. Afr. Inst. Mining Metall.*, **72**, 112–114 (1971).
77. J. D. Ramsey, *Am. Ind. Hyg. Assoc. J.*, **39**, 491–495 (1978).
78. B. A. Avellini, E. Kamon, and J. T. Krajewski, *J. Appl. Physiol.*, **49**, 254–261 (1980).
79. B. L. Drinkwater, J. E. Denton, I. C. Kupprat, S. Talag, and S. M. Horvath, *J. Appl. Physiol.*, **41**, 815–821 (1976).
80. A. J. Frye and E. Kamon, *J. Appl. Physiol.*, **50**, 65–71 (1981).
81. C. L. Wells and E. R. Buskirk, *J. Appl. Physiol.*, **31**, 858–868 (1971).
82. F. C. Houghton and C. P. Yaglou, *J. Am. Soc. Heat. Vent. Eng.*, **29**, 165–176 (1923).
83. T. Bedford, Medical Research Council War Memo No. 17, London, 1946.
84. J. H. Botsford, *Am. Ind. Hyg. Assoc. J.*, **32**, 1–10 (1971).
85. R. S. Brief and R. G. Confer, *Am. Ind. Hyg. Assoc. J.*, **32**, 11–16 (1971).
86. J. S. McKarns and R. S. Brief, *Heat. Pip. Air Cond.*, **38**, 113 (1966).
87. N. L. Ramanathan and H. S. Belding, *Am. Ind. Hyg. Assoc. J.*, **34**, 375–383 (1973).

88. V. M. Ciricello and S. H. Snook, *Am. Ind. Hyg. Assoc. J.*, **38**, 264–271 (1971).
89. A. T. Johnson and G. D. Kirk, *Am. Ind. Hyg. Assoc. J.*, **41**, 361–366 (1980).
90. M. Y. Beshir, J. D. Ramsey and C. L. Burford, *Ergonomics*, **25**, 247–254 (1982).
91. M. Y. Beshir, *Am. Ind. Hyg. Assoc. J.*, **42**, 81–87 (1981).
92. R. D. Parker and F. D. Pierce, *Am. Ind. Hyg. Assoc. J.*, **45**, 405–415 (1984).
93. World Meteorological Organization, Guide to Meteorological Instruments and Observing Practices, WMO, Geneva, 1971.
94. *ASHRAE Handbook*, 1981 Fundamentals, The American Society for Heating, Refrigeration and Air Conditioning, Atlanta, 1981.
95. T. Bedford, *Basic Principles of Ventilation and Heating*, K. J. Lewis, London, 1948.
96. T. Bedford and C. G. Warner, *J. Hyg.*, **34**, 458 (1934).
97. W. L. Roller and R. F. Goldman, *Int. J. Biomeoterol.*, **11**, 329–336 (1967).
98. A. P. Garg, Treatise on Solar Energy, in: *Fundamentals of Solar Energy*, Vol. 1, John Wiley, New York, 1982.
99. R. W. Allen, M. D. Ellis, and A. W. Hart, *Industrial Hygiene*, Prentice-Hall, Englewood, NJ, 1976, pp. 87–114.
100. H. M. Vernon, *J. Physiol.*, **70**, 15 (1970).
101. C. P. Yaglou, *J. Ind. Hyg.*, September 1935.
102. L. Hill, Medical Research Council Special Report No. 32, Part I, London, 1919.
103. A. F. Dufton, *Physiol. Mag.*, **9**, 858 (1930).
104. C. E. A. Winslow and L. Greenburg, *Am. Soc. Heat. Vent. Eng. Trans.*, **41**, 149 (1935).
105. T. F. Hatch, *Am. Ind. Hyg. Assoc. J.*, **34**, 66–72 (1973).
106. J. E. Mutchler and J. L. Vecchio, *Am. Ind. Hyg. Assoc. J.*, **38**, 253–263 (1977).
107. J. W. Jabara, *Appl. Ind. Hyg.*, **3**, 303–309 (1988).
108. V. Basus, "A Temperature Parameter Useful in Evaluation of Summer Microclimatic Conditions in Spacious Manufacturing Shops and Smaller Individual Working Places," STS Translation, from *Zdrav. Tech. Vzduchotech. (Prague)*, **10**(1), 29–36 (1967).
109. A. G. Loudon, *J. Inst. Heat Vent. Eng. (London)*, **37**, 280–292 (1970).
110. G. V. Parmelee, *Bull. Am. Meteorol. Soc.*, **36**, 256–264 (1955).
111. J. E. Mutchler, D. D. Malzahn, J. L. Vecchio, and R. D. Soule, *Am. Ind. Hyg. Assoc. J.*, **37**, 151–164 (1976).
112. A. P. Gagge, L. P. Herrington, and C. E. A. Winslow, *Am. J. Hyg.*, **26**, 84 (1937).
113. C. E. A. Winslow and L. P. Herrington, *Temperature and Human Life*, Princeton University Press, Princeton, NJ, 1949.
114. H. M. Vernon, *J. Ind. Hyg. Toxicol.*, **19**, 498 (1937).
115. J. E. Peterson, *Am. Ind. Hyg. Assoc. J.*, **31**, 305–317 (1970).
116. H. C. Bazett and J. B. S. Haldane, *J. Physiol.*, **4**, 252 (1921).
117. D. B. Dill, *Life, Heat, and Altitude*, Harvard University Press, Cambridge, MA, 1938.
118. C. K. Drinker, *J. Ind. Hyg. Toxicol.*, **18**, 471 (1936).
119. J. H. Talbott, *Medicine*, **14**, 323 (1935).
120. H. M. Vernon, T. Bedford, and C. G. Warner, Industrial Fatigue Research Board Report No. 51, Her Majesty's Stationery Office, London, 1928.

121. H. M. Vernon and E. A. Rusher, Industrial Fatigue Research Board Report No. 5, Her Majesty's Stationery Office, London, 1920.
122. R. H. Britten and L. R. Thompson, U.S. Public Health Service Bulletin No. 163, 1926.
123. C. R. Bell, in "Hot Environments and Performance," in the *Effects of Abnormal Physical Conditions at Work*, E. M. Davis, P. R. Davis, and F. H. Tyrer, Eds., Livingston, London, 1967.
124. J. M. Stewart, "The Use of Heat Transfer and Limiting Physiological Criteria as a Basis for Setting Heat Stress Limits," Second International Mine Ventilation Congress, Reno, Nevada, 1979.
125. C. H. Wyndham and A. J. Heyns, *Arch. Sci. Physiol.*, **27**, A545–A562 (1973).
126. F. H. Fuller and L. Brouha, *ASHRAE J.*, **39** (1966).
127. A. R. Lind and F. D. K. Liddell, "The Influence of Individual Variation on the Prescriptive Zone of Climates," National Coal Board Memo, London, 1963.
128. A. R. Lind, P. W. Humphreys, K. Collins, and K. Foster, "The Influence of Aging on the Prescriptive Zone of Climates," National Coal Board Memo, London, 1963.
129. American Industrial Hygiene Association, *Heating and Cooling for Man in Industry*, 2nd ed., AIHA, Akron, OH, 1975.
130. B. A. Hertig, "Control of the Thermal Environment," in *Ergonomics and Physical Environmental Factors*, International Labor Office, Occupational Safety and Health Series No. 21, Geneva, 1970.
131. T. F. Hatch, "Heat Control in the Hot Industries," in *Industrial Hygiene and Toxicology*, Vol. 1, F. A. Patty, Ed., Wiley-Interscience, New York, 1958.
132. R. S. Brief and R. G. Confer, *Med. Bull.*, **33**, 229–253 (1973).
133. W. F. Leihnard, R. S. McClintock, and J. P. Hughes, "Appraisal of Heat Exposures in an Aluminum Plant," Paper No. 8-3, presented at the 13th International Congress on Occupational Health, New York, 1960.
134. Scope, Objectives, and Functions of Occupational Health Programs, American Medical Association Council on Occupational Health, Chicago, revised December, 1971.
135. R. F. Goldman, "Evaluating the Effects of Clothing of the Wearer," in *Bioengineering, Thermal Physiology and Comfort*, Elsevier, New York, 1981.
136. R. F. Goldman, *Arch. Sci. Physiol.*, **27**, A137–147 (1973).

CHAPTER TWENTY-TWO

Air Pollution Controls

John E. Mutchler, C.I.H. and Mark A. Golembiewski, C.I.H.

This chapter focuses on the requirement for air pollution control, principles of air cleaning, selection of suitable control methods, and economic and energy resources needed to install and operate emission control systems of various types.

1 RELATION OF ATMOSPHERIC EMISSIONS TO WORKPLACE AIR QUALITY

Although the close relation between air pollution controls and industrial hygiene should be obvious to a practicing industrial hygienist, it is worthwhile to review the basis for this important concept. In this day of specialization, some tend to think of air pollution engineering as a field of endeavor somewhat removed from industrial hygiene. This unfortunate compartmentalization, found all too often in governmental and industrial organizations, can detract significantly from a full understanding of air pollution and industrial hygiene, fostering less than a total approach to effective problem solving in either air cleaning or occupational health. Only when an industrial hygienist applies expertise to achieve the optimal overall control strategy will due regard be given to controlling occupational exposures and maintaining suitable ambient air quality.

The most basic conceptual entity common to air pollution control and industrial hygiene is the "source-process." The source-process represents a common denominator between workroom air quality and atmospheric emissions. With this concept as a basis, the interrelating effect of the source-process falls into two general categories: operations with and without direct atmospheric exhaust.

Patty's Industrial Hygiene and Toxicology, Fourth Edition, Volume 1, Part A, Edited by George D. Clayton and Florence E. Clayton
ISBN 0-471-50197-2

1.1 Operations Exhausted Directly to Atmosphere

When an industrial operation or process is exhausted directly to the atmosphere, as with a hooded metallurgical furnace, workroom air quality is affected directly by the design and performance of the exhaust system. An improperly designed hood or a hood evacuated with a less-than-sufficient volumetric rate of air will contaminate the occupational environment and affect workers in the vicinity of the furnace. This is a simple yet clear illustration of one form of the close relation between atmospheric emissions and occupational exposure.

Another example of the close relation between atmospheric emissions and workplace air quality is the use of coke-side sheds adjacent to coke-oven batteries. These shed structures, when properly designed and evacuated, can provide effective capture of virtually all particulate and gaseous emissions emanating from the coke side of coke-oven batteries: door leaks, pushing emissions, and emissions from the coke car in transit to the quenching station. On the other hand, the same structures present a semiconfined space for containing coke-side emissions, and under conditions of inadequate exhaust, they can restrict the dilution and dispersion of emissions from the coke ovens, thereby jeopardizing the quality of the work environment for persons within the shedded structure. In this instance, the shed serves as the first of four components of the typical "local" exhaust system: hood, ductwork, air cleaner, and air-moving device (1).

1.2 Operations Exhausted Indirectly to Atmosphere

In some situations, the first step in the eventual, ultimate outdoor emission of materials generated from a source-process is the dispersion of the contaminant throughout an enclosed workplace, followed by significant release to the atmosphere throughout the general ventilation system, natural or mechanical. This delayed, decentralized mechanism for atmospheric release has been a reality in industry for decades, but it draws increasing attention as our regulatory standards impinge on the generation and emission of toxic materials, including those deemed "hazardous air pollutants" by the Administrator of the Environmental Protection Agency (EPA) (2). Materials such as beryllium, asbestos, and vinyl chloride are officially labeled "air pollutants" without regard to the location or mechanism of atmospheric release—dispersion and subsequent release through general ventilation, including natural exfiltration through doors, windows, and other openings in a building structure.

1.3 Recirculation of Exhaust Air

The recirculation of exhaust air into the workroom provides a clear example of the close relationship between atmospheric emissions and workplace air quality. Industrial hygienists have long been concerned about the undesirable recirculation of exhaust air, particularly in the case of atmospheric emissions that are captured in the ambient airflow wake of a building. This phenomenon can occur when an exhaust discharge point is close to the building roof line and the contaminated air is drawn back into the building by virtue of natural or mechanical ventilation.

There has been a long-standing tradition in industrial hygiene engineering to disallow the intentional recirculation of any exhaust air, even after cleaning, if the air contains toxic materials. As the costs of heating and cooling make-up air have risen significantly, however, this maxim has been modified. In some cases, recirculation may be the only really feasible method of providing suitable general ventilation. In other cases, the economic and energy conservation considerations may be important enough to warrant the additional capital and operating expenses required to safely provide the desired air recirculation.

It is now understood that the feasibility of recirculation and the final design of a suitable system will depend on a number of industrial hygiene and engineering factors. The following criteria should be used to establish the feasibility of recirculation and to assure adequate worker protection (3):

1. The chemical, physical and toxicological characteristics of the contaminants in the airstream to be recirculated must be identified and evaluated. Exhaust air containing contaminants whose toxicity is unknown, or for which there is no established safe exposure level, should not be recirculated.

2. All federal, state, and local regulations regarding recirculation must be reviewed to determine if the recirculation system under review may be restricted or prohibited.

3. The effect of a malfunction of the recirculation system must be addressed. Recirculation should not be attempted if a malfunction could result in exposure levels that would cause serious worker health problems. Substances that can cause permanent damage or significant physiological harm from a short overexposure should not be recirculated.

4. The availability of a suitable air cleaning device must be determined. The air cleaner must be capable of yielding a contaminant concentration in the effluent airstream that is sufficiently low to achieve acceptable workplace concentrations.

5. The recirculation system must incorporate a monitoring system that provides an accurate warning signal if a malfunction should occur, and that is capable of initiating corrective action or process shutdown before harmful contaminant concentrations can build up in the workplace. Monitoring can be accomplished by a number of methods, depending on the type and hazardous nature of the substance. Examples include area monitoring for nuisance-type substances and on-line monitors or filter pressure drop indicators for more hazardous materials.

The permissible concentration of a contaminant in recirculated air may be calculated by the following equation if the system is operating under steady-state conditions (3):

$$C_r = 0.5\ (\mathrm{TLV} - C_0)\ \frac{Q_t}{Q_r}\ \frac{1}{K}$$

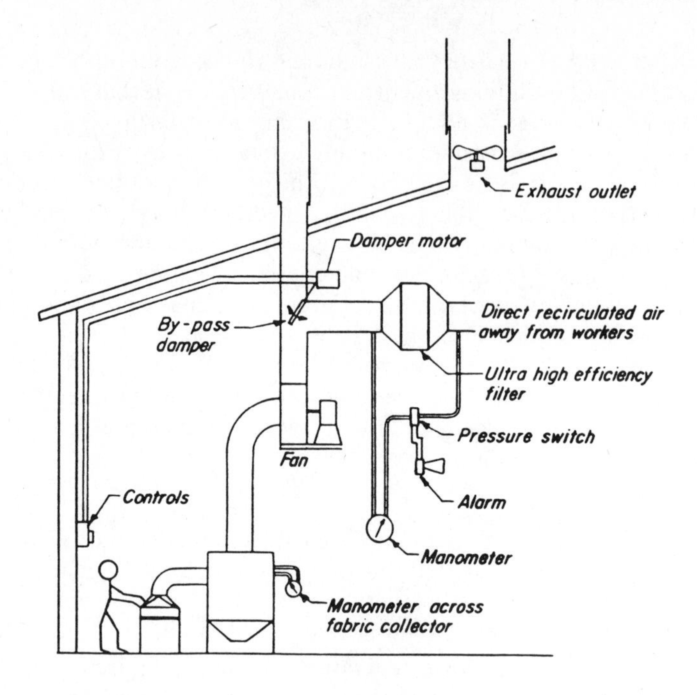

Figure 22.1. Example of a recirculation system from a particulate control device.

where C_r = concentration of the contaminant in the recirculated air stream from the air cleaning system before any mixing
TLV = threshold limit value of the contaminant (4)
C_0 = concentration of the contaminant in the worker's breathing zone without recirculation
Q_t = total ventilation flow rate through the affected work space
Q_r = recirculated air flow rate
K = "effectiveness of mixing" factor, usually between 3 ("good mixing") and 10 ("poor mixing")

Figure 22.1 gives an example of a suitable system for the recirculation of air from a particulate air cleaning device.

2 EMISSION CONTROL REQUIREMENTS

Requirements for emission controls generally find their basis in two broad types of overlapping expectations, regulatory standards and engineering performance specifications. Most air pollution regulations in the United States relate to, or at a minimum conform with, federal regulations based on the Clean Air Act as amended (2). Engineering specifications for air pollution controls have developed into increasingly precise stipulations that often reference the regulatory constraints.

Table 22.1. Significant Deterioration Limitations (5)

Pollutant	Deterioration Limitation ($\mu g/m^3$) Area Class I	Area Class II	Area Class III
Particulate matter			
Annual geometric mean	5	19	37
24-hr maximum	10	37	75
Sulfur dioxide			
Annual arithmetic mean	2	20	40
24-hr maximum	5	91	182
3-hr maximum	25	512	700
Nitrogen dioxide			
Annual arithmetic mean	2.5	25	50

2.1 Federal Clean Air Act

The Clean Air Act (CAA) of 1970 provided the EPA with broad powers to adopt and enforce air pollution emission regulations. The agency subsequently promulgated primary (designed to protect public health) and secondary (designed to protect public welfare) National Ambient Air Quality Standards (NAAQS). These standards have set maximum ambient concentrations for sulfur dioxide, particulate matter, carbon monoxide, nitrogen dioxide, ozone, and lead. The CAA authorizes the EPA Administrator to designate air quality-control regions (AQCRs) for regulatory purposes and states are required to list the status of NAAQS attainment for each criteria pollutant in each AQCR.

The principal method for attaining and maintaining the NAAQS is the state implementation plan, which is prepared and approved by each state and subject to final approval by EPA (5). These plans are directed at source control and establish timetables for attaining the NAAQS. They are intended to provide the framework for determining whether individual facilities operate in a manner consistent with the achievement of air quality goals. States may adopt requirements that are more stringent than the federal regulations.

Amendments to the CAA have also produced requirements that address attainment areas and the prevention of significant deterioration of air quality (PSD) within those areas, as well as nonattainment areas (2). The PSD regulations apply to areas where ambient air quality is superior to the NAAQS and contain detailed permit provisions governing construction and modification of major stationary sources. The PSD provisions establish three classes of clean air areas and maximum allowable increases in pollution levels over base-line measurements. Accordingly, permission for the construction or expansion of industrial plants with major emissions is contingent on increasing the ambient pollutant levels by no more than the levels shown in Table 22.1. These facilities must use the best available control technology (BACT). The PSD provisions also establish requirements to protect visibility in PSD areas.

An area where the air quality is measured or estimated to be worse than the NAAQS is designated as a "nonattainment area." States must implement a program to ensure attainment of the NAAQS in these areas, including a permit program similar to that established under the PSD regulations. New or modified emission sources in a nonattainment area, however, are subject to an emission offset policy. That is, emissions from the existing sources in the area must be reduced by an amount more than adequate to offset the new plant's emissions. In addition, the new source must use the lowest achievable emission rate.

The most direct and explicit requirements for emission controls set in motion by the CAA are those for a list of major sources and for highly toxic substances.

New Source Performance Standards (NSPS) are specific limitations for stationary source categories that might contribute significantly to pollution that could endanger public health or welfare. This provision of the law was intended to ensure that new stationary sources be designed, built, equipped, and maintained to reduce emissions to a minimum, regardless of whether the sources are located in a "clean" area or in an area that requires strict controls. The NSPS are designed to reflect the degree of emission reduction that EPA determines to be achievable for each source category by applying the "best technological system of continuous emission reduction" that has been adequately demonstrated for the source.

National Emission Standards for Hazardous Air Pollutants have been established under Section 112 of the CAA. These regulations apply to specific substances for which no ambient air quality standards have been set, and that cause or contribute to air pollution that may result in an increase in mortality or cause serious irreversible or incapacitating illness. The list of such hazardous substances now includes asbestos, beryllium, mercury, vinyl chloride, benzene, radionuclides, and inorganic arsenic.

The Clean Air Act amendments enacted in November 1990 touch on virtually every aspect of air pollution law. They regulate mobile and stationary sources, large and small businesses, routine and toxic emissions, and consumer products.

The essence of the bill's air toxics section is a list of 189 Hazardous Air Pollutants (HAPs) that EPA must regulate. The Agency's task is to determine "maximum achievable control technology" (MACT) for these substances. EPA may add or delete chemicals from the list, and private parties may petition for additions and deletions. The Agency has 10 years to issue MACT standards for all sources of the 189 chemicals. Sources that emit more than 10 tons per year of any substance or 25 tons per year of any combination of listed chemicals are covered by this title. EPA must establish categories for these sources for the purpose of promulgating standards.

The MACT standards will differ for new and existing sources. MACT for new sources refers to the most stringent level currently achieved with a similar source. In most cases, this limitation will be equivalent to the BACT as defined by current PSD rules. MACT for existing sources would be the average control in place for the best 12 percent of similar sources. For any category with fewer than thirty sources, MACT would be the average of the five best-performing sources. In addition, any source making a 90-percent reduction of a listed chemical will have a 6-year extension in the deadline for achieving the MACT standard.

Six years from enactment, EPA must complete a study and report to Congress on the level of risk remaining after the MACT reductions are in place. If Congress does not act on these recommendations, EPA must develop health-based emission standards that would limit the cancer risk to exposed individuals to no more than one case in one million.

2.2 Typical Regulatory Requirements

Within today's multiplicity of regulatory constraints for abating air pollution from stationary sources, three basic requirements emerge to affect the operator or owner of emission sources. These are quantitative emission standards, subjective prohibitions, and installation and operating permits (air use approval).

2.2.1 Quantitative Emission Standards

Local, state, and federal regulations provide complex spectrum of emission standards, ranging from general specifications of engineering standards to very precise emission limitations in specific concentration or mass-effluent units. Increasingly, these emission limitations are related to specific sampling and analytical methods. Typical of this class of regulations are those referenced to process-weight, effluent grain loading, and efflux weight limitations.

2.2.2 Subjective Prohibitions

2.2.2.1 Plume Visibility. Regulations covering plume appearance are usually expressed in terms of an allowable equivalent opacity or Ringelmann number, a subjective measure of visual obscuration. The original use of plume visibility as a regulatory tool covered the evaluation of the density of black smoke. Today most visual plumes are neither black nor soot containing; their visibility is due to fly ash, dust, fumes, and condensed water vapor, and they often are white or light colored.

Although plume visibility is widely used by enforcement agencies as a convenient index for monitoring emission sources, certain inconsistencies in the equivalent opacity concept are obvious to the objective mind. These include the dependence of plume opacity on particle size, shape, and refractivity, the typical lack of correlation between opacity and the amount of particulate matter in a plume, and the dependence of the opacity on background and relative position of the observer with respect to the sun.

To decrease the subjective element in plume appearance reading, the EPA and other agencies have developed "smoke schools," where observers are certified on their ability to read "calibrated" dark and light plumes accurately and consistently.

2.2.2.2 Nuisance and Trespass. Typical of the nuisance and trespass class of restriction are the portions of air pollution regulations prohibiting the emission of an air contaminant that causes or "will cause" detriment to the safety, health, welfare, or comfort of any person, or that causes or "will cause" damage to property or business. Such sections of regulations are typically known as "nuisance clauses," and they supplement the quantitative or objective restrictions, as well as providing for administrative and legal relief to receptors when emissions jeopardize aesthetic values, comfort, or well-being.

2.2.3 Installation and Operating Permits

The 1990 Clean Air Act Amendments establish a new federal permit system in which stationary sources will need permits that contain all applicable emission control requirements. EPA has 1 year to promulgate permit regulations, and the program is slated to go into effect 4 years from enactment. Emitters defined as major sources, those subject to acid rain controls, those subject to new source performance requirements, and plants emitting hazardous air pollutants are required to obtain these federally sanctioned permits.

States must also meet minimum criteria for permit enforcement programs. EPA has authority to object to a state-issued permit if it conflicts with Clean Air Act requirements. The permitting authority has 90 days to revise the permit; otherwise, EPA must issue a final ruling on the permit.

It is very common today, especially at the state and local agency levels, for air pollution regulations to require at least one and usually two permits before a newly operating emission source or emission control system is certified to be in compliance with regulatory requirements. An "installation permit" (or a "permit to construct") is a sanction granted by the regulatory agency before construction begins and after consideration by the agency of the emission sources and the nature and engineering characteristics of the control system. It is at this stage that predictive dispersion modeling, even if rudimentary, can help assure maintenance of acceptable air quality downwind from the proposed installation.

2.2.3.1 Predictive Air Quality Modeling. The well-known technique of air quality dispersion modeling has become a widespread requirement for predicting the impact of new or modified stationary emission sources and emission control systems in terms of expected downwind concentrations of the contaminant. Preconstruction dispersion modeling examines the proposed emission or control system configuration and any alternatives being considered, to estimate the maximum downwind concentrations as they relate to the national ambient air quality standards and, where appropriate, to the EPA PSD regulations for sulfur dioxide, nitrogen dioxide, and suspended particulate matter. Such modeling is mandated by the EPA for the various state agencies as they administer the EPA-approved implementation plans designed to achieve the ambient concentrations specified in the national air quality standards and to restrict significant increases in the atmospheric burden in regions of superior air quality. Increasingly, predictive modeling is also being required by state and local enforcement agencies concerned with limiting the release of "air toxics"; that is, air contaminants with significant toxic properties.

2.2.3.2 Certification. After installation of the source or control system is complete and the system begins to operate, a second permit must be obtained in many jurisdictions, attesting that the source operates in a manner satisfactory to the control agency. After an operating permit is granted after inspection and observation of the operating facility by a representative from the air pollution control agency. Nevertheless, in many instances compliance source testing is required to document the level of effluent with reference to the applicable emission limitations.

2.3 Engineering Performance Specifications

Today's requirements for an emission control system usually include engineering performance specifications. It is technically sound, in general, that the performance specifications of air or gas cleaning equipment for a given application include any or all of the following factors:

1. Range of air-cleaning capacity in terms of volumetric flow rate
2. The exit concentration of material in the system exhaust gases
3. Mass emission rate of discharged materials
4. Efficiency of control system based on mass flow rate
5. Acceptance test methods

The value of specifying these parameters among such traditional specifications as materials of construction, fabrication, and inspection is obvious, yet these factors are important enough to warrant review.

2.3.1 Range of Air-Cleaning Capacity

An emission control system must be designed to operate at sufficient capacity to exhaust all the gases that come from the contaminant-generating source. The application of even the latest technology to a given source is rather meaningless if the control unit is undersized and unable to capture and exhaust all the contaminants. Examples of inadequate system design are commonplace in industry, especially in primary and secondary metallurgical industries.

2.3.2 Exit Concentration of Material

If the object of an air-cleaning system is the recovery of suspended materials, it is important to know the percentage or amount of material collected or retained in the equipment. Usually, however, even though the recovery of material is an important consideration, the percentage of material collected is only incidental. If the object is to produce air of better quality, either for supply to occupied spaces or for discharge outdoors, performance should be specified in terms of the concentration of contaminant in the outgoing airstream.

2.3.3 Emission Rate of Discharged Material

The emission rate of material, mass per unit time, is defined by the concentration of material in the exit gas stream and the volumetric flow rate of the gas stream.

Specification of only one or the other of these two factors is insufficient to characterize the expected performance of an air pollution control system. In many applications in industry poor maintenance or other factors result in an exit flow rate from a control system that is greater than the inlet flow rate; specification of the concentration alone could be misleading because of a dilution effect caused by the increased outlet flow rate.

Furthermore, the real impact of air pollution must be measured in terms of the emission rate to the atmosphere. Only this parameter provides a meaningful index of mass contribution to the atmospheric burden.

Finally, it is important to specify the emission rate of an air pollution control system either directly or indirectly (by flow rate and by concentration) because emission regulations are invariably expressed in terms of mass emission rate or emission factors that depend on both the mass emission rate and the process rate. When combined, mass emission rate and process rate yield such dimensional emission terms as pounds per ton of product produced, or pounds per ton of feed material.

2.3.4 Collector Efficiency

The term "efficiency" in air pollution control does not refer to the proportion of energy used effectively in the cleaning process. Rather, "efficiency" is used almost exclusively to indicate the proportion of material removed from an airstream, without regard to the amount of power required. Energy consumption, an important consideration in air pollution control, is treated separately in Section 3.

If a material were completely removed from an airstream by an air-cleaning device, the efficiency would be rated as 100 percent. The general definition of "efficiency" with respect to the degree of separation affected by an air pollution control system is the "emission rate entering the collector minus the emission rate leaving the collector divided by the emission rate entering the collector." For example, if the mass flow rate of material in a contaminated airstream entering a collector is 100 kg/hr and the emission rate is 5 kg/hr, the "efficiency" of the control device is

$$100\left(\frac{100 - 5}{100}\right), \text{ or 95 percent}$$

2.3.5 Acceptance Test Methods

Measurement of the control efficiency of a collector is accomplished with source sampling equipment operating as a miniature-scale separating and collection system with very high inherent effectiveness. The measured emissions, however, depend very much on the method of sampling. Therefore the performance of an air or gas cleaning system must be specified with reference to the sampling method to be used to determine the nature, amount, and mass flux of contaminants in the inlet and outlet of the system. Accordingly, the specific method of testing should be named in a statement of expected performance. If the specified method is not a standard method, such as an ASTM or EPA method, the sampling device or method should be described in enough detail to permit replication.

3 ECONOMIC IMPACT OF AIR POLLUTION CONTROLS

Air pollution represents a problem for which the goals of public policy are "noneconomic," yet have important economic aspects. Federal, state, and local governments have major programs aimed at the prevention and abatement of atmospheric pollution. Nevertheless, many economic issues related to these programs are inadequately understood or documented. For example, would it be cheaper and more effective to disperse activities that pollute the atmosphere, to burn higher quality fuel, or to remove pollutants from process materials before effluents are discharged to the atmosphere? Current answers to these and similar questions are generally inadequate.

A macroeconomic analysis of air pollution control issues, although a very interesting and relevant topic, is not the subject of this section. Rather, the goal is to present some macroeconomic considerations in air pollution control in terms of the cost–benefit relationships, the resources necessary to apply certain types of control, and the minimum requirements of an economic feasibility analysis needed to specify the most "economic" control methods for a given stationary source of air pollution from among suitable technical alternatives.

Many air pollution control problems can be solved by more than one alternative measure, but to identify the optimal method for controlling emissions, each technically adequate solution must be evaluated carefully before implementation. Sometimes basic measures such as the substitution of fuels or raw materials and the modification or replacement of processes provide the most cost-effective solution, and these avenues should never be overlooked (see Section 4).

3.1 Cost-Effectiveness Relationships

A cost-effectiveness "variable" measures all costs associated with a given project as a function of achievable reduction in pollutant emissions. For example, when estimating the total cost of an emission control system, we must consider raw materials and fuel used in the control process, needed alterations in process equipment, control hardware, auxiliary equipment, disposal or reuse of collected materials, and similar factors.

Figure 22.2 plots a typical cost-effectiveness relationship in stationary source emission control. The cost of control is represented on the vertical axis, and the quantity of material discharged is represented on the horizontal axis. Point *P* indicates the uncontrolled state at which there are no control costs. As control efficiency improves, the quantity of emissions is reduced and the cost of control increases. In most cases, the marginal cost of control is smaller at lower levels of efficiency (higher emissions) near point *P* of the curve. This cost-effectiveness curve

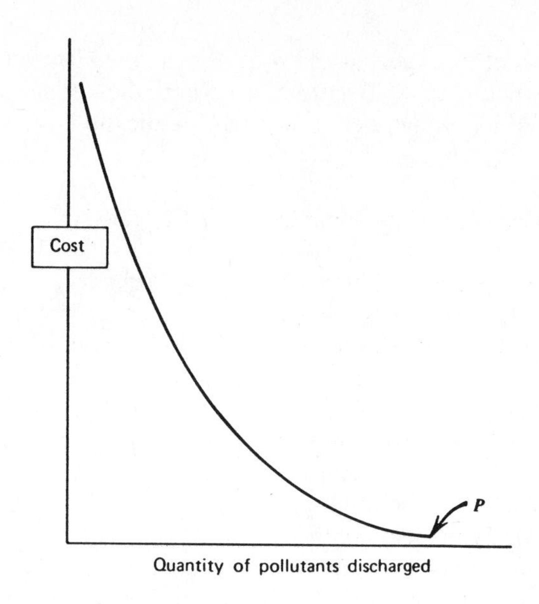

Figure 22.2. Cost-effective relationship in stationary source emission control (6).

demonstrates an important reality: as the degree of control increases, greater increments of costs are usually required for corresponding increments in emission abatement.

Obviously, cost-effectiveness information is needed for sound decision making in emission control. Although several technically feasible measures may be available for controlling an emission source, in most cases the least-cost solution for a source can be calculated at various levels of control. After each alternative is evaluated, and future process expansions and the likelihood of more rigid control restrictions are considered, sufficient information should be available on which to base an intelligent control decision.

3.2 Control Cost Elements

The cost of installing and operating an air pollution control system depends on many direct and indirect cost factors. An accurate analysis of control costs for a specific emission source must include an evaluation of all relevant factors. The definable control costs are those directly associated with the installation and operation of control systems. These cost elements have been compiled into a logical organization for accounting purposes and are presented in Table 22.2. The control cost elements that receive attention in this discussion are capital investment, maintenance and operating costs (including energy consumption), and conservation of material and energy in the design and operation of air pollution control systems.

Table 22.2. Control Cost Elements (6)

Control Cost Elements
Capital investment
Engineering studies
Land
Control hardware
Auxiliary equipment
Operating supply inventory
Installation
Startup
Structure modification
Maintenance and operation
Utilities
Labor
Supplies and materials
Treatment and disposal of collected material
Capital charges
Taxes
Insurance
Interest

3.3 Capital Investment

The *installed cost* quoted by a manufacturer of air pollution control equipment should result from an analysis of the specific emission source. This cost usually includes three of the eight capital investment items listed in Table 22.2: control hardware costs, auxiliary equipment costs, and costs for field installation. Basic control hardware usually includes built-in instrumentation and pumps.

Purchase costs are the amounts charged by a manufacturer for equipment constructed of standard materials. Purchase costs depend to a large degree on the size and collection efficiency of the control device. In addition, equipment fabricated with special materials for extremely high temperatures or corrosive applications generally costs much more than equipment constructed from standard materials.

3.4 Maintenance and Operating Costs

The costs of operating and maintaining air pollution controls depend not only on the inherent characteristics of the different control methods, but also on a wide range of quality and suitability of control equipment, the user's understanding of its operation, and his commitment to maintaining it in reliable operation.

Maintenance costs represent the expenditures required to sustain the operation of a control device at its designed efficiency, using a scheduled maintenance program and necessary replacement of any defective component parts. Simple control devices of low efficiency have low maintenance costs, and complex, highly efficient control systems generally have high maintenance costs.

Annual operating costs refer to the yearly expense of operating an emission control system at its designed collection efficiency. These costs depend directly on the following factors:

1. Volumetric rate of gas cleaned
2. Pressure drop across the control system
3. Annual operating time
4. Consumption rate and cost of electricity
5. Mechanical efficiency of the air-moving device and, where applicable, two additional items:
6. Scrubbing liquor consumption and replacement costs
7. Consumption rate and cost of fuel

The energy requirement for air pollution control equipment is an important and significant component of operating costs. The increased cost of fuel points to a growing emphasis on the role of energy conservation in industry generally, and on pollution control equipment specifically. Traditionally, the energy requirements for emission control equipment have been included in an overall consideration and expression of operating costs, but the continuing emphasis on energy costs leads logically to the isolation of this cost element for at least two purposes: (1) to relate operating costs to energy costs, and (2) to locate potential areas for energy conservation.

Because air pollution control equipment ranges from natural-draft stacks to combinations of collectors (cyclones, filters, electrostatic precipitators, scrubbers, and stacks with induced or forced-draft fan units), the energy consumption requirements of the various types of pollution control equipment must be examined separately. Table 22.3 gives the most commonly used air pollution control equipment and the associated power requirements. Power requirements are listed in kilowatts, because most pollution control equipment operates on electricity. Pressure drop is expressed in inches of water. The heating requirements for afterburners and regeneration of adsorption beds are given in British thermal units per hour. The typical requirements in Table 22.3 can be added directly to obtain an estimate of total energy, pressure drop, and heating requirements for a complex chain of combined equipment. The total pressure drop can then be used to estimate the total fan power requirement.

As an example, consider a 120-MW coal-burning power plant (7). The boiler exhaust emission controls could include an electrostatic precipitator for removing fly ash particulates, an alkali scrubber system for removing sulfur dioxide, and a fan and a tall stack to help achieve acceptable air quality by dispersion. The equipment power requirements for this system could then be estimated as follows:

Item	Electrical Energy (kW/1000 scfm)
Electrostatic precipitator	0.3
Limestone scrubber	12.1
Tall stack	0
Fan	0.1
Total exhaust system power	12.5

For a 120-MW power plant with a stack flow rate of 300,000 standard cubic feet per minute (scfm) the electrical power required would be $300 \times 12.5 = 3750$ kW.

3.5 Material Conservation

Historically, except for precious metal applications, most air pollution control systems have collected material of little economic worth. Nevertheless, as resource recovery becomes more feasible because of current and expected shortages of many material resources, the economic worth of material collected by emission controls may gain in significance in the economic analysis of air pollution controls.

The alternatives for handling collected particulate remain as follows:

1. Recycle material to the process.
2. Sell material as collected.
3. Convert material to salable products.
4. Discard material in the most economical (and environmentally acceptable) manner.

In some process operations, collected material is sufficiently valuable to warrant its return to the process, and the value of the recovered material can partially or wholly pay for the collection equipment. In many applications, however, the cost of the high-efficiency control systems necessary to achieve desired ambient air quality would be greater than the revenue returned for recovery of the material collected.

The cement industry provides an example of collected material being routinely returned to the process. Not only does recovered dust, in such situations, have value as a raw material, its recovery also reduces disposal costs, as well as costs related to the preparation of raw materials used in the process.

Although material collected by air pollution control equipment may be unsuitable for return to a process within that plant, it may be suitable for another manufacturing activity. Hence it may be treated and sold to another firm that can use the material. Untreated, pulverized fly ash, for example, cannot be reused in a furnace, but it can be sold as a raw material to a cement manufacturer. It can also be used as a soil conditioner, as an asphalt filler, or as landfill material. When

Table 22.3. Power Requirements for Air Pollution Equipment (7)

Item	Pressure Drop (in. H_2O)	Power[a] Required (kW per 1000 scfm), Typical Range	Remarks
Exhaust stacks	0	0	—[b]
Filters and separators			
Baghouses, cloth filters	1–30	0.1–1.0	Power function of rapping rate, particle size, cloth and dust-layer pressure drop
Impingement and gravitational separators	0.1–1.5	None	
Cyclones			
Single	0.1–2	None	ΔP = 1 to 20 inlet velocity heads
High efficiency	2–10	None	
Dry centrifugal	2–4	None	Combined separator-fan effect
Packed bed	1–10	None	
Electrostatic precipitators			
High voltage	0.1–1	0.2–0.6	
Low voltage	0.1–1	0.01–0.04	
Scrubbers		1–12	Power requirements for fan and pump
Centrifugal, mechanical venturi	2–8	None	
High pressure drop	10–60	None	3 gal/min per 1000 cfm typical liquor rate
Low pressure drop	0.5–100	None	3 gal/min per 1000 cfm typical liquor rate

Spray tower	1–2	None	
Impingement and entrainment	4–20	None	
Packed-bed absorption	0.5 per foot of thickness	None	
Plate absorption plus liquor circulation pump[c]	1–3 per plate 1–60 ft	None (see manufacturer's pump curve)	1–20 gal/min liquor, some sprays to 600 psi
Cyclone	0.1–10	None	
Adsorption beds	1–30	100–200 Btu/lb adsorbent to heat to 600°F without heat exchange	Regeneration rate dependent on adsorbent and pollutant concentration; fan power may also be required for hot gas regeneration 1500°F, with 80% heat exchanger (2×10^5 Btu/hr); 1500°F, no regeneration, heat exchanger (14×10^5 Btu/hr)
Acoustic fume afterburners	0.1–2 (0.05–1 per in. of bed)	2 to 14×10^5 Btu/hr	
Catalytic oxidizers	0.5–2 (0.05–1 per in. of bed)	0.1 to 1.0×10^6 Btu/hr	300–900°F, 80% heat exchanger to 900°F, without heat exchanger depends on inlet temperature

[a]"None" indicates that no direct power is required; the only power requirement is indirect from fans (due to pressure drop in equipment).
[b]The rise of hot gas in the stack creates a pressure head of about 0.5 in. H_2O, but this is counteracted by friction losses in the stack plus the desired velocity head leaving the stack for dispersions.
[c]Add scrubber pump power to scrubber power requirements.

treated, pulverized fly ash can yield an even more valuable product. Some utilities, for example, sinter pulverized fly ash to produce a light-weight aggregate that can be used to manufacture bricks and light-weight building blocks.

3.6 Energy Recovery

Although there has been much comment about the need for industrial energy conservation, the fact remains that most exhaust stacks in industry are very hot, suggesting that much energy is being wasted.

The most economical approach to energy conservation and environmental control is a combined approach whenever possible. Real savings can be gained by combining heat recovery with air pollution control. Whereas in the past, low fuel costs made it uneconomical to install heat recovery equipment in many industrial applications such as on stacks of direct-fired furnaces, the higher cost of fuel today has turned the spotlight on fuel economy now and in the foreseeable future.

A major advantage in the combined approach to energy and air pollution controls is the drop in temperature, caused by the recovery of sensible heat energy from an airstream, that can lead to a dramatic reduction in the volumetric exhaust rate and the requisite cost and size of air pollution control equipment. Thus it is possible to save not only on capital expenses, but also on operating costs, as well as in the consumption of increasingly expensive fuel. Recovered heat can be used to reduce fuel requirements by preheating primary combustion air, preheating air going to another process, or preheating other fuel; or it can be used in heating buildings.

3.7 Economic Analysis Models

Cost estimates are presented for the installation and operation of five general types of control equipment:

1. Gravitational and inertial collectors
2. Fabric filters
3. Wet scrubbers
4. Electrostatic precipitators
5. Afterburners

3.7.1 Gravitational and Inertial Collectors

In general, the only significant cost for operating mechanical collectors is the electric power cost, which varies with the unit size and the pressure drop. Because pressure deep in gravitational collectors is low, operational costs associated with these units are considered to be insignificant. Maintenance cost includes the costs of servicing the fan motor, replacing any lining worn by abrasion, and, for multiclone collectors, flushing the clogged small diameter tubes.

The theoretical annual cost G of operation and maintenance for centrifugal collectors can be expressed as follows (6):

$$G = S\left(\frac{0.7457PHK}{6356E} + M\right) \tag{2}$$

where
S = design capacity of the collectors (actual cubic feet per minute, acfm)
P = pressure drop (in. H_2O)
E = fan efficiency, assumed to be 60 percent (expressed as 0.60)
0.7457 = a constant (1 hp = 0.7457 kW)
H = annual operating time (assumed 8760 hr)
K = power cost ($/kW-hr)
M = maintenance cost ($/acfm)

For computational purposes, the cost formula can be simplified as follows (6):

$$G = S(195.5 \times 10^{-6}\ PHK + M)$$

3.7.2 Fabric Filters

Operating costs for fabric filters include power costs for operating the fan and the bag cleaning device. These costs vary directly with the sizes of the equipment and the pressure drop. Maintenance costs include costs for servicing the fan and shaking mechanism, emptying the hoppers, and replacing the worn bags.

The theoretical annual cost G for operation and maintenance of fabric filters is as follows (6):

$$G = S\left(\frac{0.7457PHK}{6356E} + M\right) \tag{3}$$

with the units as defined as in Equation 2.

For computational purposes, the cost formula can be simplified as follows (6):

$$G = S(195.5 \times 10^{-6}PHK + M)$$

3.7.3 Wet Collectors

The operating costs for a wet collector include power and scrubbing liquor costs. Power costs vary with equipment size, liquor circulation rate, and pressure drop. Liquor consumption varies with equipment size and stack gas temperature. Maintenance includes servicing the fan or compressor motor, servicing the pump, replacing worn linings, cleaning piping, and any necessary chemical treatment of the liquor in the circulation system.

The theoretical annual cost G of operation and maintenance for wet collectors can be expressed as follows (6):

$$G = S\left[0.7457HK\left(\frac{P}{6356E} + \frac{Qg}{1722F} + \frac{Qh}{3960F}\right) + WHL + M\right] \tag{4}$$

Table 22.4. Contact Power Requirements for Wet Scrubbers (6)

Scrubber Efficiency	"Scrubbing" (Contact) Power (hp/acfm)
Low	0.0013
Medium	0.0035
High	0.015

where S = design capacity of the wet collector (acfm)
0.7457 = a constant (1 hp = 0.7457 kW)
H = annual operating time (assumed 8760 hr)
K = power costs (dollars per kilowatt-hour)
P = pressure drop across fan (in. H_2O)
Q = liquor circulation (gal/acfm)
g = liquor pressure at the collector (psig)
h = physical height liquor is pumped in circulation system (ft)
W = makeup liquor consumption (gal/acfm)
L = liquor cost ($/gal)
M = maintenance cost ($/acfm)
E = fan efficiency, assumed to be 60 percent (expressed as 0.60)
F = pump efficiency, assumed to be 50 percent (expressed as 0.50)

This equation can be simplified according to Semrau's total "contacting power" concept (8). Semrau shows that efficiency is proportional to the total energy input to meet fan and nozzle power requirements. The scrubbing (contact) power factors in Table 22.4 were calculated from typical performance data listed in manufacturers' brochures. These factors are in general agreement with data reported by Semrau. Using Semrau's concept, the equation for operating cost can be simplified as follows (6):

$$G = S\left[0.7457HK\left(Z + \frac{Qh}{1980}\right) + WHL + M\right] \tag{5}$$

where Z = contact power, that is, total power input required for collection efficiency (hp/acfm; see Table 22.4). It is a combination of:

1. Fan horsepower per acfm

$$= \frac{P}{6356E} \tag{6}$$

2. Pump horsepower per acfm

$$= \frac{Qg}{1722F} \text{ the power to atomize water through a nozzle} \tag{7}$$

The pump horsepower $Qh/1980$ required to provide pressure head is not included in the contact power requirements.

3.7.4 Electrostatic Precipitators

The only operating cost considered in running electrostatic precipitators is the power cost for ionizing the gas and operating the fan. Depending on the length and design of the upstream and downstream ductwork, the fan energy requirements may be relatively small. Because the pressure drop across the equipment is usually less than 0.5 H_2O, the cost of operating the fan is assumed to be negligible if minimal ductwork is used. The power cost varies with the efficiency and the size of the equipment.

Maintenance usually requires the services of an engineer or highly trained operator, in addition to regular maintenance personnel. Maintenance includes servicing fans and replacing damaged wires and rectifiers.

The theoretical annual cost G for operation and maintenance of electrostatic precipitators is as follows (6):

$$G = S(JHK + M) \tag{8}$$

where S = design capacity of the electrostatic precipitator (acfm)
J = power requirements (kW/acfm)
H = annual operating time (assumed 8760 hr)
K = power cost ($/kW-hr)
M = maintenance cost ($/acfm)

3.7.5 Afterburners

The major operating cost item for afterburners is fuel. Fuel requirements are a direct function of the gas volume, the enthalpy of the gas, and the difference between inlet and outlet gas temperatures. For most applications, the inlet gas temperature at the source ranges from 50 to 400°F. Outlet temperatures may vary from 1200 to 1500°F for direct flame afterburners and from 730 to 1200°F for catalytic afterburners. The use of heat exchangers may bring about a 50 percent reduction in the temperature difference.

The equation for calculating the operation and maintenance costs G is as follows (6):

$$G = S\left(\frac{0.7457PHK}{6356E} + HF + M\right) \tag{9}$$

where S = design capacity of the afterburner (acfm)
P = pressure drop (in. H_2O)

E = fan efficiency, assumed to be 60 percent (expressed as 0.60)
0.7457 = a constant (1 hp = 0.7457 kW)
H = annual operating time (assumed 8760 hr)
K = power cost (\$/kW-hr)
F = fuel cost (\$/acfm-hr)
M = maintenance cost (\$/acfm)

For computational purposes, the cost formula is simplified as follows (6):

$$G = S(195.5 \times 10^{-6}PHK + HF + M)$$

4 PROCESS AND SYSTEM CONTROL

Process and system control implies a careful review of a production unit operation within the context of air pollution control to examine whether the manufacturing process is optimal, considering the emission rate or necessary control and treatment of the process effluents.

A fundamental notion in air pollution control asserts, "The problem can be solved best if it is solved at the source of emissions." This simple rule, if applied consistently, could greatly reduce the number and complexity of emission problems in industry. Methods for total or partial "control at the source" include elimination of emissions, minimization of emissions, and concentration of contaminants before discharge.

4.1 Elimination of Emissions

An excellent example of the potentially beneficial effect of material substitution is conversion from a "dirty" fuel to a "clean" fuel, such as a switch from coal to gas in utility boilers. In the late 1960s and early 1970s many state and local pollution control agencies enacted regulations to curb sulfur dioxide emissions from coal- and oil-burning power plants. The effect of such regulations was a trend toward the use of low sulfur oil and coal and, where possible, to the use of the cleanest fossil fuel, natural gas. Because of the shortages and higher costs of natural gas and low sulfur coal and oil, however, the achievement and maintenance of air quality goals cannot be accomplished solely by switching to naturally available low sulfur fuels. It is apparent that coal with medium to high sulfur content will have to be burned in increasing quantities by power plants to meet growth requirements and to safeguard an adequate supply of clean fuels for residential, commercial, and industrial use. However, the recent identification of acid rain as a serious air pollution problem in certain regions of North America will likely result in regulatory mandates for stricter limitations on sulfur dioxide and nitrogen dioxide emissions from power plants.

4.1.2 Process Changes

Process changes can be as effective as material substitution in eliminating air polluting emissions. It has often been possible and even profitable in the chemical and petroleum industries to control the loss of volatile organic materials to the atmosphere by condensation, leading to the "reuse" of otherwise fugitive vapors. The compressors, absorbers, and condenser units on petroleum product process vessels provide a good example of such process modification.

In the case of burning fossil fuels, process change can also be construed to encompass the addition of plants auxiliary to the main process that generate substitute fuels. One familiar example is the recycling of distilled volatiles in coke production. When high sulfur coals are distilled destructively, they produce gas streams that when treated, are useful energy sources, producing low sulfur oxides emissions on burning. Recycling a portion of this fuel to the coke ovens supplies energy for further fuel gas production from the next charge of coal in the ovens while also providing product coke.

The addition of a step to reduce atmospheric emissions has been commonplace in brass foundry practice when indirect fired furnaces are employed. A fluxing material applied to the surface of the molten grass serves as an evaporation barrier and reduces the evolution of brass fumes; this additional step was developed strictly as an air pollution control measure.

4.1.3 Equipment Substitution

An example of the benefit to be derived from equipment substitution was a trend in the polyvinyl chloride (PVC) industry in the 1970s to change to large polymerization reactors, which not only boosted productivity but helped to stem fugitive emissions of vinyl chloride monomer. A large PVC reaction vessel has only half the possible fugitive emissions leak points of two smaller units of the same total capacity (9). As late as 1972, the PVC industry relied on reactors of capacity smaller than 7500 gallons for 85 percent of its output. By the end of the decade, 20,000-gal vessels and even some massive 55,000-gal reactors were employed in the PVC industry to control vinyl chloride emissions.

4.2 Reduction of Emissions

One way to reduce air pollution control costs is to minimize, if not eliminate, release of the contaminant. Such action is cost effective both for fugitive emissions and process emissions.

4.2.1 Fugitive Emissions

Leaking conveyor systems can be modified to eliminate spilling and dusting. Airtight enclosures can be built around conveyors, or cover housings can be equipped with flanges and soft gaskets. Shaft bearings can be redesigned and relocated inside the conveyor housing to prevent dust leakage around the shaft.

Tanks and bins should have airtight covers and joints sealed with cemented strips of plastic or rubber sheeting. Each bin or tank normally should have only one vent where temperature difference can create a natural draft from the vent; other openings should be kept closed.

Where possible, interconnecting a series of vents from several tanks should be considered. While one tank is being filled, another can be emptied, thus reducing the need for exhausting air to the atmosphere. It may also be possible to have one small dust collector serve a number of units.

4.2.2 Process Emissions

When contaminants are a by-product of a production operation, generation can often be minimized by changes in operating conditions. Lower combustion temperatures will reduce nitrogen oxides formation. Levels of fluorine-containing compounds, released when some ores are heated, frequently can be decreased if water vapor is eliminated from the atmosphere.

Reducing excess air in coal- or oil-fired boiler systems can result in a substantial reduction in the conversion of sulfur dioxide to sulfur trioxide, thus reducing sulfuric acid emissions.

4.3 Concentrating Pollutants at the Source

One major phase of process and system control involves centralizing and decreasing the number of emission points and reducing the effluent volume of the exhaust gas to be treated.

4.3.1 Reducing the Number of Emission Points

Many industrial operations, such as machining in metalwork industries, are characterized by the replication of identical or similar operations. Examples include grinding, sawing, and cutting. Control of worker exposures and effective, economical emission control can often be best achieved by clustering the operations together and exhausting them locally and simultaneously to a common collection system, rather than exhausting each operation to an individual collector. The concept of centralized collection and abatement offers the advantage of economy of scale for collection systems; this tends to decrease control costs per unit weight of material collected. This advantage is due in part to the relatively high concentration of pollutants at a combined source compared to the more dilute concentrations, which would be experienced from collecting and exhausting materials from each similar operation independently.

4.3.2 Volume Reduction by Cooling

Many dust collectors such as bag filters and precipitators are sized more on a volumetric flow basis than on a mass flow basis. If the effluent is a hot, dusty gas, cooling can reduce the volume appreciably. Cooling by the addition of cold air is the poorest method from the standpoint of reducing cost. Radiation panels, fin surfaces, waste heat boilers, forced convection, heat interchange, and direct spray cooling with water are all possible means of effecting volume reduction.

4.3.3 Optimal Local Exhaust Ventilation

Well-designed and properly installed local exhaust systems are an important aspect of control measures to minimize exhaust flow rate and maximize concentrations in exhaust gases to be cleared before discharge to the atmosphere. By careful design, local exhaust hoods can be applied to assure complete capture of all contaminants. Close-fitting hoods not only ensure better control of materials that may escape into the work environment, they also provide minimal exhaust ventilation of processors that produce contaminants.

5 TECHNICAL CRITERIA FOR SELECTING AIR-CLEANING METHODS

The cost effectiveness of an air pollution control system is closely related to its degree of control, its capacity, and the type of control system. Often more than one type of control could be used to solve a specific emission problem. Nevertheless, there is generally only one type that is optimal from both technical and economic standpoints.

Some of the most basic technical factors affecting selection of equipment are as follows:

1. System volumetric flow rate
2. Concentration (loading) of contaminant
3. Size distribution of particulate contaminants
4. Degree of cleaning required
5. Conditions of air or gas stream with reference to temperature, moisture content, and chemical composition
6. Characteristics of the contaminant, such as corrosiveness, solubility, reactivity, adhesion or packing tendencies, specific gravity, surface, and shape

This section focuses on these and other technical criteria that are helpful in selecting emission controls.

5.1 Performance Objectives

The prime factor in the selection of control equipment is the maximum amount or rate of contaminant to be discharged to the atmosphere. Knowledge of this amount, together with knowledge of the amount of contaminant entering a proposed collection system, defines the required collection efficiency; this begins the technical selection of air-cleaning methods.

If the material to be collected is particulate, it is essential to understand that collectors have different efficiencies for different-sized particles. Therefore the particle size distribution of the emission must be known before the collector efficiency required for each particle size range can be determined. Collector efficiency also varies with gas flow rate and with properties of the carrier gas, which may fluctuate with flow rate or time. Such variations must be considered carefully in determining collector efficiency.

When the material to be collected is a gas or vapor, it is necessary to know to what extent the material is soluble in the scrubbing liquid or retainable on the absorbing surfaces. The determinations must be made for the concentrations expected in the collector inlet and outlet streams and for the conditions of temperature, pressure, and flow rate expected.

5.2 Contaminant Properties

Contaminant properties, as distinguished from carrier gas properties, comprise both chemical and physical characteristics of the material to be removed from an exhaust airstream.

5.2.1 Loading

Contaminant loading from many processes varies over a wide range for an operating cycle. Variations of an order of magnitude in concentration are not uncommon. Well-known examples of such variation include the basic oxygen furnace in steel making and soot blowing in a steam boiler. Contaminant loading may also vary with the carrier gas flow rate. Particularly in the case of gases, concentration is all-important in predicting removal efficiency and specifying system design parameters.

5.2.2 Composition

The composition of a contaminant affects both physical and chemical properties. Chemical properties, in turn, affect physical properties. For example, if collected material is to be used in a process or shipped in a dry state, a dry collector is indicated. If the collected material has a very high intrinsic value, a very efficient collector is called for.

Because chemical and physical properties vary with composition, a collector must be able to cope with both expected and unexpected composition changes. For example, in the secondary aluminum industry a collector must be able to deal with the evolution of aluminum chloride during chlorine "demaging." The aluminum chloride levels vary widely throughout the cycle; peak levels last for only a few minutes but continue to develop in decreasing amounts throughout a full cycle of 16 hr or more.

Further examples include absorption, in which solubility may be important to the ease with which the absorbent may be regenerated, and scrubbing to remove particulate matter, in which wettability of the dust by the scrubbing liquor aids the collector mechanisms and the basic separating forces that determine scrubber performance.

5.2.3 *Combustibility*

Generally, it is not desirable to use a collection system that permits accumulation of "pockets" of contaminant when the contaminant collected is explosive or combustible. Systems handling such materials must be protected against accumulation of static charges. Electrostatic precipitators are not suitable for such contaminants because of their tendency to spark. Wet collection by scrubbing or absorption methods is especially appropriate. Some dusts, however, such as magnesium, are pyrophoric in the presence of small amounts of water. In combustion (with or without a catalyst), explosibility must always be considered.

5.2.4 *Reactivity*

Certain obvious precautions must be taken in the selection of equipment for the collection of reactive contaminants. In filtration, selection of the filtering media can present a special problem. In absorption, because certain applications require that the absorbed contaminant react with the absorbent, the degree of reactivity is important. Where scrubbers are used to remove corrosive gases or particulates, the potential corrosiveness must be balanced against the potential savings when using corrosion-resistant construction. The decision is thus whether to use more expensive materials or to incur higher maintenance costs.

5.2.5 *Electrical and Sonic Properties*

The electrical properties of the contaminant can influence the performance of several types of collector. Electrical properties are considered to be a factor influencing the buildup of solids in inertial collectors. In electrostatic precipitators, such electrical properties of the contaminant as the resistivity are of paramount importance in determining collection efficiency and precipitator size, and they influence the ease with which collected particulate is removed by periodic cleaning of the collection surfaces. In fabric filtration, electrostatic phenomena may have direct and observable effects on the process of cake formation and the subsequent ease of cake removal. In spray towers, where liquid droplets are formed and contact between these droplets and contaminant particles is required for particle collection, the electrical charge on both particles and droplets is an important parameter in determining collection efficiency. The process is most efficient when the charges on the droplet attract rather than repel those on the particle. Sonic properties are significant where sonic agglomeration is employed.

5.2.6 *Toxicity*

The degree of contaminant toxicity influences collector efficiency requirements and may necessitate the use of equipment that will provide ultrahigh efficiency. Toxicity also affects the means for removal of collected contaminant from the collector and its disposal, as well as the means of servicing and maintaining the collector. However, toxicity of the contaminant does not influence the removal mechanisms of any collection technique.

5.2.7 *Particle Size, Shape, and Density*

Size, shape, and density are three properties of particulate matter that determine the magnitude of forces resisting movement of a particle through a gas stream. These are the major factors determining the effectiveness of removal in inertial collectors, gravity collectors, venturi scrubbers, and electrostatic precipitators. In these collections, resisting forces are balanced against some removal force (e.g., centrifugal force in cyclones) that is applied in the control device, and the magnitude of the net force tending to remove the particle determines the effectiveness of the equipment.

Size, shape, and density of a particle can be related to terminal settling velocity, which is a useful parameter in the selection of equipment for particulate control. Settling velocity is derived from Stokes' law, which equates the velocity at which a particle will fall at constant speed (because of a balance of the frictional drag force and the downward force of gravity) to the properties of the particle and the viscosity of the gas stream through which it is settling.

Terminal settling velocities can be determined by any of a number of standard techniques and used in the quantitative evaluation of the difficulties to be anticipated in designing particulate removal equipment.

Because particle size is related to the ease with which individual particles are removed from a gas stream, it is apparent that size distribution largely determines the overall efficiency of a particular piece of control equipment. Generally, the smaller the particle size to be removed, the greater the expenditure required for power or equipment or both. To increase the efficiencies obtainable with scrubbers, it is necessary to expend additional power either to produce high gas stream velocities, as in the venturi scrubber, or to produce finely divided spray water. Cyclones call for the use of a larger number of small units for higher efficiency in a given situation. Both the power cost and equipment cost are increased. Achieving higher efficiencies for electrostatic precipitators necessitates the use of a number of units or fields in series because there is an approximately inverse logarithmic relationship between outlet concentration and the size of collection equipment. A precipitator giving 90 percent efficiency must be approximately doubled in size to give 99 percent efficiency and approximately tripled to give 99.9 percent efficiency.

5.2.8 *Hygroscopicity*

Although not specifically related to any removal mechanism, hygroscopicity may be a measure of how readily particulates will cake or tend to accumulate in equipment if moisture is present. If such accumulation occurs on a fabric filter, it may completely blind it and prevent gas flow.

5.2.9 *Agglomerating Characteristics*

Collectors are sometimes used in series, with the first collector acting as an agglomerator and the second collecting the particles agglomerated in the first one. In carbon black collection, for example, extremely fine particles are first agglomerated; then they can be collected practicably.

5.2.10 Flow Properties

Flow properties of the material are mainly related to the ease with which the collected dust may be discharged from the collector. Extreme stickiness may eliminate the possibility of using equipment such as fabric filters. Hopper size and shape depend in part on the packing characteristics or bulk density of the collected material and its flow properties. Hygroscopic materials, or collected dust tending to cake, exhibit flow properties that change with time as the dust remains in the hopper. Hopper heaters or rappers may be required.

5.3 Carrier Gas Properties

Carrier gas properties are important insofar as they affect the selection of control equipment, especially with reference to composition, reactivity, conditions of temperature, pressure, moisture, solubility, condensability, combustibility, and toxicity.

5.3.1 Composition

Gas composition affects physical and chemical properties, which are important to the extent that there may be chemical reactions between the contaminants and the collector, either in its structure or in its contents. One common reaction between gaseous components and equipment is the corrosion of metallic parts of collectors when gases contain sulfur oxides and water vapor.

Composition, concentration, and chemical reaction properties of the inlet stream determine the collection efficiency in packed-tower scrubbers removing gaseous or vapor phase contaminants.

5.3.2 Temperature

The temperature of the carrier gas principally influences the volume of the carrier gas and the materials of construction of the collector. The volume of the carrier gas influences the size and cost of the collector and the concentration of the contaminant per unit volume; concentration in turn is the driving force for removal. In addition, viscosity, density, and other gas properties are temperature dependent. Temperature also affects the vapor–liquid equilibria in gaseous contaminant scrubbing such that scrubbing efficiency for partially soluble gases decreases with increasing temperature.

Adsorption processes are generally exothermic and are impracticable at higher temperature, the absorbability being inversely proportional to the temperature (when the reaction is primarily physical and is not influenced by an accompanying chemical reaction). Similarly, in absorption (where gas solubility depends on the temperature of the solvent), temperature effects may have significance if the concentration of the soluble material is such that an appreciable temperature rise results. When combustion is used as a means for contaminant removal, the gas temperature affects the heat balance, which is the vital factor in the economics of the process. In electrostatic precipitation, both dust resistivity and the dielectric strength of the gas are temperature dependent.

Wet processes cannot be used at a temperature that would cause the liquid to freeze, boil, or evaporate too rapidly. Filter media can be used only in the temperature range within which they are stable. The filter structure must retain structural integrity at the operating temperature.

Finally, cool gases flowing from a stack downstream of control equipment disperse in the atmosphere less effectively than do hot gases. Consequently, benefits derived from partial cleaning accompanied by cooling may be offset if the cooler exhaust gas cannot be well dispersed. This is a factor of importance in wet cleaning processes for hot gases, where the advantage gained by cleaning is sometimes offset by downwash from the stack near the plant because the exhaust gas is cooled. In the case of wet collection devices, the effluent gases may present a visually objectionable steam plume or even "rain out" in the stack vicinity. Raising the discharge gas temperature may eliminate or reduce these problems.

5.3.3 Pressure

In general, carrier gas pressure much higher or lower than atmospheric pressure requires that the control equipment be designed as a pressure vessel. Some types of equipment are much more amenable to being designed into pressure vessels than others. For example, catalytic converters are incorporated in pressure processes for the production of nitric acid and provide an economical process for the reduction of nitrogen oxides of nitrogen before release to the atmosphere.

Pressure of the carrier gas is not of prime importance in particulate collection except for its influence on gas density, viscosity, and electrical properties. It may, however, have importance in certain special situations, for example, when the choice is between high-efficiency scrubbers and other devices for collection of particulate matter. The available source pressure can be used to overcome the high pressure drop across the scrubber, reducing the high power requirement that often limits the utilization of scrubbers. In adsorption, high pressure favors removal and may be required in some situations.

5.3.4 Viscosity

Viscosity is important to collection techniques in two respects. First, it is important to the removal mechanisms in many situations (inertial collection, gravity collection, and electrostatic precipitation). Particulate removal technique involves migration of the particles through the gas stream under the influence of some removal force. Ease of migration decreases with increasing viscosity of the gas stream. Second, viscosity influences the pressure drop across the collector, thereby becoming a major parameter in power requirement computation.

5.3.5 Density

Density of the carrier gas, for the most part, has no significant effect in most real gas cleaning processes, although the difference between particle density and gas density appears as a factor in the theoretical analysis of all gravitational and centrifugal collection devices. Particle density is so much greater than gas density that the usual changes in gas density have negligible effects. Carrier gas density does influence fan power requirements and therefore is important to fan selection and operating cost. Furthermore, special precautions must be taken in "cold start-up" of a fan designed to operate at high temperature, to ensure that the motor capacity is not exceeded.

5.3.6 Humidity

Humidity of the carrier gas stream may affect the selection and performance of control equipment in any of several basically different ways. High humidity may contribute to accumulations of solids and lead to the caking and blocking of inertial collectors as well as the caking of filter media. It can also result in cold spot condensation and aggravation of corrosion problems. In addition, the water vapor may act on the basic mechanism of removal in electrostatic precipitation and greatly influence resistivity. In catalytic combustion it may be an important consideration in the heat balance that must be maintained. In adsorption it may tend to limit the capacity of the bed of water preferentially or concurrently adsorbed with the contaminant. Even in filtration it may influence agglomeration and produce subtle effects.

The above-mentioned considerations are the main limitations on the utilization of evaporative cooling in spite of its obvious power advantage. When humidity is a serious problem for one of the foregoing reasons, scrubbers, or adsorption towers may be particularly appropriate devices. Humidity also affects the appearance of the stack exhaust gas discharged from wet collector devices. Because steam plumes are visually objectionable, sometimes it is necessary to heat exhaust gases to raise their dew point before discharge. This can add considerable operating expense to the total system.

5.3.7 Combustibility

The handling of a carrier gas that is flammable or explosive requires certain precautions. The most important is making sure that the carrier gas is either above the upper explosive limit or below the lower explosive limit for any air admixture that may exist or occur. The use of water scrubbing or adsorption may be an effective means of minimizing the hazards in some instances. Electrostatic precipitators are often impractical, for they tend to spark and may ignite the gas.

5.3.8 Reactivity

A reactive carrier gas presents special problems. In filtration, for example, the presence of gaseous fluorides may eliminate the possibility of high-temperature filtration using glass fiber fabrics. In adsorption, carrier gas must not react preferentially with the adsorbents. For example, silica gel is not appropriate for adsorption of contaminants when water vapor is a component of the carrier gas stream.

Also, the magnitude of this problem may be greater when a high-temperature process is involved. On the one hand, devices that use water may be eliminated from consideration if the carrier gas reacts with water. On the other hand, scrubbers may be especially appropriate in that they tend to be relatively small and require small amounts of construction material, permitting the use of corrosion-resistant components, with lower relative increase in cost.

5.3.9 Toxicity

When the carrier gas is toxic or is an irritant, special precautions are needed in the construction of the collector, the ductwork, and the means of discharge to the atmosphere. The entire system, including the stack, should be under negative pressure and the stack must be of tight construction. Because the collector is often under "suction," special means such as "airlocks" must be provided for removing the contaminant from the hoppers if collection is by a dry technique. Special precautions may also be required for service and maintenance operations on the equipment.

5.3.10 Electrical and Sonic Properties

Electrical properties are important to electrostatic precipitation because the rate or ease of ionization will influence removal mechanisms.

Generally speaking, intensity of Brownian motion and gas viscosity increase with gas temperature. These factors are important gas stream characteristics that relate to the "sonic properties" of the stream. Increases in either property tend to increase the effectiveness with which sonic energy can be used to produce particle agglomeration.

5.4 Flow Characteristics of Carrier Gas

5.4.1 Volumetric Flow Rate

The rate of evolution from the process, the temperature of the effluent, and the degree and the means by which it is cooled, if cooling is used, fix the rate at which carrier gases must be treated, and therefore the size of removal equipment and the rate at which gas passes through it. For economic reasons, it is desirable to minimize the size of the equipment. Optimizing the size and velocity relationship involves consideration of two effects: (1) reduction in size results in increased power requirements for handling a given amount of gas because of increased pressure loss within the control device, and (2) velocity exerts an effect on the removal mechanisms. For example, higher velocities favor removal in inertial equipment up to the point of turbulence, but beyond this, increased velocity results in decreased efficiency. In gravity settling chambers, flow velocity determines the smallest size that will be removed. In venturi scrubbers, efficiency is proportional to velocity through the system. In absorption, velocity affects film resistance to mass transfer. In filtration, the resistance of the medium often varies with velocity because of changes in dust cake permeability with flow. In adsorption, velocity across the bed should not exceed the maximum that permits effective removal. Optimal velocities have not generally been established with certainty for any of the control processes because they are highly influenced by the properties of the contaminant and carrier gas as well as by the design of the equipment.

5.4.2 Variations in Flow Rate

Rate variations result in velocity changes, thereby influencing equipment efficiency and pressure drop. Various control techniques differ in their abilities to adjust to flow changes. When rate variations are inescapable, it is necessary to (1) design for extreme conditions, (2) employ devices that will correct for flow changes, or (3) use a collector that is inherently positive in its operation. Filtration is most easily adapted to extreme rate variations because it presents a positive barrier for particulate removal. This process is subject to pressure drop variations, however, and generally the air-moving equipment will not deliver at a constant rate when pressure drop increases. In most other control techniques, variations in flow produce a change in the effectiveness of removal.

One means of coping with rate variation is the use of two collectors in series, one that improves performance with increasing flow (e.g., multicyclone) and one whose performance decreases with increasing flow (e.g., electrostatic precipitator). Some venturi scrubbers are equipped with automatically controlled, variable size throats. Changes in gas flow rate are automatically sensed, and the throat's cross-sectional area is changed correspondingly to maintain a relatively constant pressure drop and efficiency over a relatively wide range of conditions.

6 CLASSIFICATION OF AIR POLLUTION CONTROLS

In a broad sense, air pollution controls for stationary source emissions include process and system control, air-cleaning methods, and the use of tall stacks. Process and system control were covered in Section 4. This section focuses on the other types of air pollution control and offers an overview of the role of each method. More detailed information on air-cleaning methods follows in Sections 7 through 12.

6.1 Tall Stacks

In the context of air pollution control, a tall stack has the simple function of discharging exhaust gases at an elevation high enough to put the maximum concentration of contaminants experienced at ground level within acceptable limits. However, it is unwise and often illegal to rely only on tall stacks for solving air pollution problems. Furthermore, the use of a tall stack may serve to satisfy ground level concentration requirements, but does nothing to reduce the quantity of pollutants being emitted into the atmosphere and may contribute to air quality problems at distances far downwind from the emission source.

Through the Clean Air Act, the EPA now regulates permissible stack heights for new emission sources (2). The agency has defined a "good engineering practice (GEP) stack height" as the greater of either 65 m or the height computed from a formula that takes into account the dimensions of nearby structures. This provision is intended to ensure that emissions from a stack do not result in excessive concentrations of pollutants as a result of atmospheric downwash, wakes, or eddy current effects.

Ground level concentrations depend on the strength of the emission source, the physical and chemical nature of materials discharged, atmospheric conditions, stack and exhaust gas parameters, topography, and the aerodynamic characteristics of the physical surroundings. The state of the art in atmospheric dispersion modeling continues to advance, and air quality modeling is embodied in many state and local air pollution regulations and is sanctioned as well by the EPA (see Section 2.2).

The computational details of atmospheric dispersion modeling are outside the scope of this discussion. Several excellent publications (10–12) illustrate the rationale and methods for estimating ground level concentrations downwind from stationary sources.

6.2 Gravitational and Inertial Separation

Gravitational and inertial separation are the simpler forms of particulate collectors, but several different types and configurations are included. Each type is built to incorporate a hopper into which the collected material will eventually settle. The performance of this type of collector depends either on velocity reduction to permit settling or on the application of centrifugal force that increases the effective mass of particles. The common types of mechanical collectors include settling chambers, cyclones, and multiple cyclones.

6.2.1 Settling Chambers

A settling chamber may be nothing more than a long, straight, bottomless duct over a hopper, or it may consist of a group of horizontal passages formed by shelves in a chamber. A baffle chamber consists simply of a short chamber with horizontal entry and exit, using single or multiple vertical baffles. The inertia of the entrained particles causes them to strike the plates, enabling the dust to fall into the collection hopper. Settling chambers require a large space and are effective only on particles with diameters greater than 50 to 100 μm. However, their resistance to airflow, as measured by the pressure drop, is low. Collectors of this class are frequently used as precleaners for coarse particles, preceding other, more efficient types of collector.

6.2.2 Cyclones

In a cyclone, a vortex created within the collector propels particles to locations from which they may be removed from the collector; the devices can be operated either wet or dry. Cyclones may either deposit the collected particulate matter in a hopper or concentrate it into a stream of carrier gas that flows to another separator, usually of a different, more efficient type, for ultimate collection. As long as the interior of a cyclone remains clean, pressure drop does not increase with time. Up to a certain limit, both collection efficiency and pressure drop (usually less than 2 in. H_2O, gauge) increase with flow rate through a cyclone. This type of collector is best applied in the removal of coarse dusts.

Cyclones are frequently used in parallel but seldom in series. When they are used in series it is to accomplish a special objective, such as to provide a backup in case the dust discharge of the primary cyclone fails to function. Small-diameter cyclones are more effective than larger ones; centrifugal force for a given tangential velocity varies inversely with the radius. Accordingly, multiple cyclones, banks of small cyclones arranged in parallel, are used commonly to maximize particulate control efficiency within the inertial collector concept. Whereas single cyclones can be several feet in diameter, multiple cyclones are often less than 12 in. in diameter; pressure drop across this configuration is typically 3 to 8 in. H_2O, gauge.

6.2.3 Other Devices

Other types of inertial separators include impingement collectors, consisting of a series of nozzles, orifices, or slots each followed by a baffle or plate surface, and dynamic collectors, consisting of a power-driven centrifugal fan with the provisions and housing for skimming off a layer of gas in which dust has been concentrated.

6.3 Filtration

Filters are devices used for removal of particulate matter from gas streams by retention of particles in and around a porous structure through which the gas flows. The porous structure is most commonly a woven or felted fabric, but pierced, woven, or sintered metal can be used, as well as beds of a large variety of substances (fibers, metal turnings, coke, slag roll, sand, etc.).

Unless operated wet to keep the interstices clean, filters in general improve in retention efficiency as the interstices in the porous structure begin to be filled by collected material. These collected particles form a pore structure of their own, supported by the filter, and because of the increased surface area, they have the ability to intercept and retain other particles. This increase in retention efficiency is accompanied by an increase in pressure drop through the filter. To prevent decrease in gas flow, therefore, the filter must be cleaned continuously or periodically, or else replaced after a certain length of time.

For controlling air pollution, the fabric filter collector, commonly known as a "baghouse," together with its fan or air mover, can be likened to a giant vacuum cleaner. Baghouses utilize fabrics, woven or felted, from natural or synthetic fibers. In its most usual configuration, a fabric filter consists of a series of cylindrical bags or tubes, vertically mounted, and dust is deposited on either the inner or outer surfaces of the fabric. Fabric filters are generally useful where:

1. Very high particulate control efficiencies are desired.
2. Temperatures are below 550°F.
3. Gas temperatures are above the dew point.
4. Valuable material is to be collected dry.

6.4 Electrostatic Precipitation

Electrostatic precipitators are devices in which one or more high-intensity electrical fields cause particles to acquire an electrical charge and migrate to a collecting surface. The collecting surface may be either dry or wet. Because the collecting force is applied only to the particles and not to the gas, the pressure drop of the gas is only that due to flow through a duct having the configuration of the collector. Hence pressure drop is very low and does not tend to increase with time. In general, collection efficiency increases with the length of passage through an electrostatic precipitator. Therefore replicate precipitator sections are employed in series to obtain higher collection efficiency.

Electrostatic precipitators may incorporate pipes or flat plates as grounded electrodes onto which electrically charged dust particles are deposited for subsequent removal. In any electrostatic precipitator, three functional units exist by virtue of three specific operations tending to occur simultaneously.

1. Charging of particulate
2. Collection of particulate
3. Removal and transport of collected particulate

Electrostatic precipitators find their greatest use in situations where:

1. Gas volumes are relatively large.
2. High efficiencies are required on fine particulate.
3. Valuable materials are to be collected dry.
4. Relatively high-temperature gas streams must be cleaned.

6.5 Liquid Scrubbing

The prime means of collection in wet scrubbers is a liquid introduced into the collector for contact with the contaminant aerosol. Scrubbers are used both to remove gases and vapor phase contaminants and to separate particulate from the carrier gas. The scrubber liquid may be wet, or it may dissolve or react chemically with the contaminant collected.

Methods of effecting contact between scrubbing liquid and carrier gas include the following:

1. Spraying the liquid into open chambers containing various forms of baffles, grilles, or packing

2. Flowing the liquid into these structures over weirs
3. Bubbling the gas through tanks or troughs of liquids
4. Using gas flow to create droplets from liquid introduced at a location of high gas velocity

Scrubber liquid frequently can be recirculated after the collected contaminant is partially or completely removed. In other cases, all or part of the liquid must be discarded. In general, as long as the interior elements of the scrubber remain clean, the pressure drop does not increase with time. Collection efficiency tends to increase with increasing gas flow rate, provided the liquid feed keeps pace with gas flow and carryout or entrainment of liquid with the effluent gases is prevented effectively.

Scrubbers are generally of either the low-energy or high-energy type. Examples of low-energy scrubbers are spray chambers, centrifugal units, impingement scrubbers, most packed beds, and the submerged nozzle type of scrubber. High-energy scrubbers, especially the venturi scrubber, find wide application in situations calling for the removal of fine particulate matter at high efficiency levels. In the most common design, a venturi scrubber consists of a venturi-shaped air passage with radially directed water jets located just before or at the high-velocity throat. The water is broken into fine droplets by the action of the high velocity of the gas stream. The particulate matter is deposited on the water droplet by impaction, diffusion, and condensation. Coarse droplets of water, together with the entrained dust, are readily separated by a comparatively simple demister section on the venturi discharge.

Wet scrubbers can handle high temperatures and are often used simultaneously as particulate collectors and heat transfer devices. Wet scrubbers generally are applicable and should be considered in a feasibility analysis if any one of the following situation descriptions applies:

1. The gaseous vapor phase contaminant is soluble in water (particulate need not be soluble).
2. The exhaust gas stream contains both gaseous and particulate contaminants.
3. Exhaust gases are combustible.
4. Cooling is desirable with an increase in moisture content satisfactory.

6.6 Gas–Solid Adsorption

Absorbers are devices in which contaminant gases or vapors are retained on the surface of a porous medium through which the carrier gas flows. The medium most commonly used for adsorption is activated carbon. The design of an adsorber parallels that of a filtration device for particulate matter in that the gas flows through a porous bed. In the case of an adsorber, however, the adsorption bed should be preceded by a filter, to protect it from plugging due to particulate matter.

In true adsorption, there is no irreversible chemical reaction between the adsorbent and the adsorbed gas or vapor. Therefore the adsorbed gas or vapor can be driven off the adsorbent by heat, vacuum, steam, or other means. In some adsorbers the adsorbent is regenerated in this manner for reuse. In other applications the spent adsorbent is discarded and replaced with fresh adsorbent. Pressure drop through an adsorber that does not handle gas contaminated by particulate matter does not increase with time but does increase with gas flow rate. The relationship between adsorption efficiency and gas flow rate depends entirely on adsorber design and on the characteristics of the material being collected.

Carbon adsorption is generally carried out in large horizontal fixed beds with depths of 3 to 25 ft. Such units can handle from 2000 to about 50,000 cfm and are often equipped with blowers, condensers, separators, and controls. Typically, an installation includes two carbon beds; one is onstream while the other is being regenerated.

Although molecular sieves have reached the commercial stage in several new applications, activated carbon remains the most important dry adsorbent in gaseous emission control. Molecular sieves, like activated carbon, usually require fixed-bed units with sequence valves for switching the beds from adsorption to regeneration. The key element in these systems is molecular sieve–adsorbent blends: synthetic crystalline metallic-alumina silicates that are highly porous adsorbents for some liquids and gases.

6.7 Combustion

Many processes produce gas streams that bear organic materials having little recovery value or containing toxic or odorous materials that can be oxidized to less harmful combustion products. In such cases thermal oxidation may be the optimal control route, especially if the gas streams are combustible. There are three methods of combustion in common use today: thermal oxidation, direct flame incineration, and catalytic oxidation.

Thermal oxidizers or afterburners can be used when the contaminant is combustible. The contaminated airstream is introduced to an open flame or heating device, then it goes to a residence chamber where combustibles are oxidized to carbon dioxide and water vapor. Most combustible contaminants can be oxidized at temperatures between 100 and 1500°F. The residence chamber must provide sufficient dwell time and turbulence to allow complete oxidation.

Direct combustors differ from thermal oxidizers by introducing the contaminated gases and auxiliary air into the burner itself as fuel. Auxiliary fuel, usually natural gas or oil, is generally necessary for ignition and may or may not be required to sustain burning.

Catalytic oxidizers may be used when the contaminant is combustible. The contaminated gas stream is preheated, then passed through a catalyst bed that promotes oxidation of the combustibles to carbon dioxide and water vapor. Metals of the platinum family are commonly used catalysts, and they promote oxidation at temperatures between 700 and 900°F.

To use either thermal or catalytic oxidation, the combustible contaminant concentration must be below the lower explosive limit. Equipment specifically designed for control of gaseous or vapor contaminants should be applied with caution when the airstream also contains solid particles. Solid particulate matter can plug catalysts and, if noncombustible, it will not be converted in thermal oxidizers and direct combustors.

Thermal and direct combustors usually have lower capital cost requirements but higher operating costs because an auxiliary fuel is often required for burning. This cost is offset when heat recovery is employed. The catalytic approach has high capital cost because of the relatively expensive catalyst, but it needs less fuel. Either method provides a clean, odorless effluent if the exit gas temperature is sufficiently high.

6.8 Combination Systems

Within the wide spectrum of stationary emission sources, there are several applications for coupling complementary control devices into an effective combination system. Selection of the best combination and sequence of control methods depends to some degree on past experience, but excellent results can be achieved by careful analysis of the contaminated air or gas to be treated.

A good example of a combination system is the inertial collector–electrostatic precipitator system, applicable when a wide range of particle sizes occurs in the gas, or where dust loadings are high. Examples are blast furnaces and coal-fired boilers. Another innovative combination uses the carbon absorber and incinerator to minimize operating costs. In this control system, organic vapors are adsorbed for several hours on activated carbon; during this time, the incinerator is turned off. When the carbon bed is regenerated, the desorbent vapors, including the steam used to heat the bed, can be sent directly to the incinerator for destruction. Not only is auxiliary fuel saved by this intermittent operation, but the desorbent may be self-sustaining during incineration. Furthermore, incinerator size is reduced dramatically compared with direct-fired equipment. Capital costs may sometimes be offsetting, however.

Two principal arrangements of multiple-unit air-cleaning systems are recognized: parallel combination and serial combination. Parallel combination is simply a means of providing a wide range of flexibility in adjusting the capacity of a system to the airflow rate. It permits small, highly effective elements, such as centrifugal tubes (multiple cyclones), to be assembled in parallel arrangement for handling equal shares of the total gas flow. Serial combination presents the problem of determining the best sequence, if different types of cleaning units are to be used, or the best combination of preparation and separation units to meet the range of contaminants that must be removed from the airstream.

6.8.1 Two-Stage Cleaning Systems

Two-stage systems comprise a preparation stage and a separation stage. Examples of this arrangement include the following:

1. Two-stage electrostatic precipitators, with particle charging preceding precipitation
2. Venturi scrubbers, with simultaneous agitation and liquid injection preceding cyclone separation
3. Sonic agglomeration systems, with simultaneous agglomeration and gravitational precipitation followed by inertial or centrifugal separation

6.8.2 Sequence of Separating Stages

The selectivity of particulate separators, or their fractionating tendencies, makes it advisable to combine stages of cleaning that are complementary. The sequence of treatment is generally, but not invariably, from the types of separator least able to remove submicrometer particles to those at the final stages most likely to succeed at this removal.

The agglomerating action of certain types of separator results in large aggregates of flocs blowing intermittently from dry collecting surfaces into the airstream. When this condition occurs, because of excessive air velocities or poor retentivity at the contact surfaces, it is good practice to interpose coarse particle separators to reduce the load on subsequent stages of air cleaning.

Some of the variation in airborne matter is gradually suppressed or damped as the air progresses through a multistage system of cleaning. The process fluctuations may not be eliminated, but they frequently are modified so greatly in the initial stages of air treatment that the final, and most critical, stages are protected against disturbing variations in the character of materials traveling with the air.

6.8.3 Concentration and Subdivision of Airstreams

As the air passes through a cleaning system, it may be handled as a single stream from inlet to outlet. There are numerous installations, however, in which the airstream is subdivided one or more times to specialize the task of cleaning and to make each stage more effective. A common example is conveyance of the small-volume, high-concentration effluent from the apex of the periphery of a cyclone to a secondary separator generating greater centrifugal force.

7 GRAVITATIONAL AND INERTIAL SEPARATION

The simplest type of particulate collector is commonly known as a "mechanical collector." Mechanical collectors utilize either gravity or inertia to remove relatively large particles from suspension in a moving gas stream. In these collectors the gas stream is made to flow in a path that either enhances the gravitational separation or changes direction such that particles cannot easily follow because of their inertia. Most mechanical collectors operate in a dry condition, although water is sometimes used in conjunction with a mechanical collector to aid in continuous cleaning of the control device.

7.1 Range of Performance

Gravitational and inertial separation devices range in capture efficiency from less than 50 to about 90 percent, depending on particle size and type of collector. Generally, as the velocity through the collector increases, control efficiency increases, except for gravitational settling chambers where the opposite effect occurs.

Pressure drop across an inertial collector of this type is one indicator of relative efficiency. Although a direct relationship between drop and efficiency cannot always be calculated easily, there is usually a significant correlation between the two; control efficiency usually increases with pressure drop. One important aspect of this phenomenon is that pressure drop increases require greater amounts of energy to operate the collection system.

Table 22.5 compares characteristics of various kinds of gravitational and inertial separators.

7.2 Settling Chambers

Settling chambers represent the oldest form of air pollution control device. Because of their simplicity of design and relatively low operating costs, they are most commonly used today as precleaners for more efficient particulate control methods that consume higher amounts of energy.

Settling chambers use the force of gravity to separate dust and mist from the gas stream by slowing down the gas stream so that particles will settle out into a hopper (Figure 22.3) or onto shelves from which they can be removed (Figure 22.4).

From a practical standpoint, settling chambers are not employed for the removal of particles less than 50 μm in diameter because of the excessive size of the equipment required for collection below this size limit. To prevent reentrainment of the separated particles, the velocity of the gas stream entering the settling chamber should not exceed 600 fpm.

The efficiency of a settling chamber can be calculated by the following equation (14):

$$E = \frac{100 U_t WL}{Q} \tag{10}$$

where E = efficiency, weight percent of particles of settling velocity U_t (dimensionless)

U_t = terminal settling velocity of dust (ft/sec)

L = chamber length (ft)

W = chamber width (ft)

Q = gas flowrate (ft/sec)

This efficiency model assumes good distribution at the inlet and outlet of the chamber. Combining this equation with Stokes' law, the minimum particle size that

Table 22.5. Characteristics of Mechanical Collection Equipment (13)

Collector Type	Space Requirements	Volume Range	Efficiency by Weight	Pressure Loss[a] (in. H_2O)	Temperature Limitations	Power[b] (hp per 1000 cfm gas)	Application Areas
Settling chamber	Large	Space available only limitation	Good above 500 μm	0.2–0.5	700–1000°F, limited only by materials of construction	0.04–0.12	Precollector for fly ash metallurgical dust, can be used for any large dust particles above 50 μm
Conventional cyclone	Large	Normal range up to 50,000 cfm	Approximately 50% on 20 μm	1–3	700–1000°F, limited only by materials of construction	0.24–0.73	Woodworking, paper, buffing, fibers, etc.; well suited for dry dust particles 20 μm and above
High-efficiency cyclone	Medium	Normal range up to 12,000 cfm	Approximately 80% to 10 μm	3–5	700–1000°F, limited only by materials of construction	0.73–1.2	Woodworking, material conveying, product recovery, etc.; well suited for dry dust particles 10 μm and above
Multitube cyclone	Small	Normal range up to 100,000 cfm	90% on 7.5 μm	4.5	700–1000°F	1.1	Precollector for electrostatic precipitator on fly ash, product

								recovery, etc.; well suited for dry dust particles 5 μm and above
Dynamic precipitator	Small	17,000 cfm	80% on 15 μm	No loss (true fan)	700°F		Power consumption will depend on selection point, mechanical efficiency in usual selection range from 40 to 50%	Woodworking, nonproduction buffing, metal working, etc.; well suited for dry dust particles 10 μm and above
Impingement separator	Small	Space available only limitation	90% on 10 μm	1–5	700°F		0.24–1.2	Certain types used for collecting coarse particles in boiler fly ash and cement clinker cooler; recent designs used for cleaning atmospheric air to diesel engines and gas turbines

[a]Pressure drop is based on standard conditions.
[b]Power consumption figured from: horsepower = cfm × total pressure/6356 × ME; mechanical efficiency (ME) assumed to be 65%.

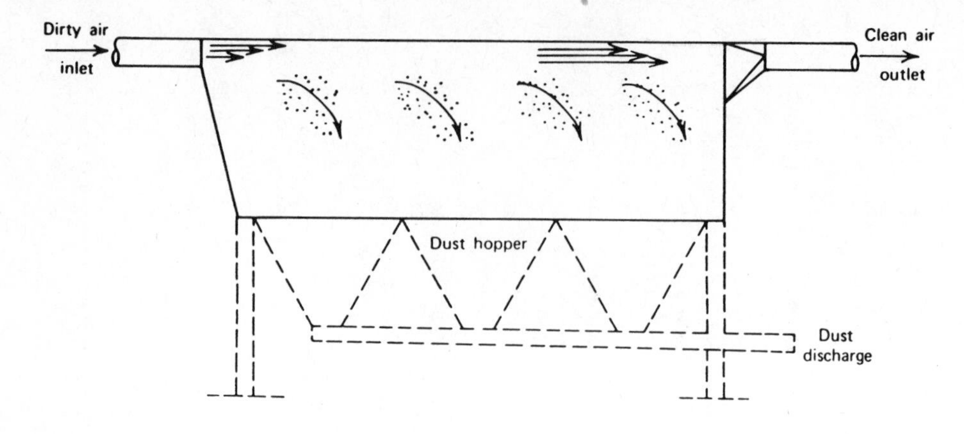

Figure 22.3. Gravity dust-settling chamber (13).

can be completely separated from the gas stream can be calculated by the following equation (15):

$$d_p = \left[\frac{18\mu HV}{gL(\rho_p - \rho_g)}\right]^{1/2} \tag{11}$$

where d_p = minimum size particle collected completely (ft)
μ = gas viscosity (lb/ft-sec)
H = chamber height (ft)
V = gas velocity (ft/sec)
g = gravitational constant (32.2 ft/sec^2)
ρ_p = particle density (lb/ft^3)
ρ_g = gas density (lb/ft^3)

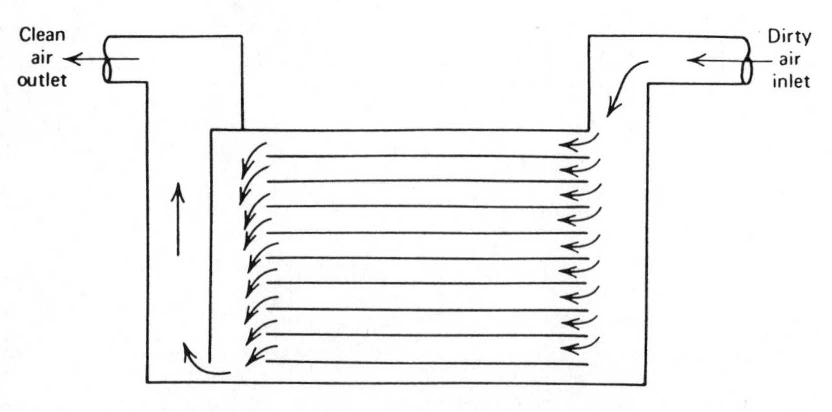

Figure 22.4. Multiple-tray dust collector (6).

If there are horizontal plates or trays in the chamber, the effective settling distance is reduced, and the efficiency and minimum particle size that can be completely separated are given by Equations 12 and 13, respectively (15):

$$E = \frac{NU_tWL}{Q} \tag{12}$$

$$d_p = \left[\frac{18\mu Q}{gNWL(\rho_p - \rho_g)}\right]^{1/2} \tag{13}$$

where N is the number of plates and other units are as in Equations 10 and 11.

The chief advantages of settling chambers are their low operating cost, relatively low initial cost, simple construction, low maintenance costs, and low pressure drop in operation.

Their main disadvantages are large space requirements and low collection efficiency for smaller particles. When using shelves to increase efficiency, this characteristic also causes a problem with cleaning the settled particulate off the shelves; this is often done by rinsing the plates with water.

7.3 Inertial Separators

Inertial separators represent one of the simplest pollution control devices. They take up less room than settling chambers and are more efficient. Inertial separators are relatively inexpensive to acquire and operate, but they are not very efficient, and like settling chambers, they are used basically for precleaning of gas streams. Inertial separators include all dry-type collectors that utilize the relatively great inertia of the particles to effect particulate–gas separation. Two basic types of inertia separation equipment are simple impaction separators, which employ incremental changes of the carrier gas stream direction to exert the greater inertial effects of the particles, and cyclonic separators, which produce continuous centrifugal force as a means of exerting the greater inertial effects. The cyclones are discussed in a subsequent section.

Impingement or impaction separation occurs when the gas stream suddenly changes its direction because of the presence of an obstructing body. The impingement efficiency, defined as the fraction of particles in the gas volume swept by the obstructing body which will impinge on that body, is given as follows:

$$E = \frac{U_tV}{D_B} \tag{14}$$

where E = separation efficiency (dimensionless)
U_t = terminal settling velocity of particle (ft/sec)
V = gas stream velocity (ft/sec)
D_B = equivalent diameter of obstructing body (ft)

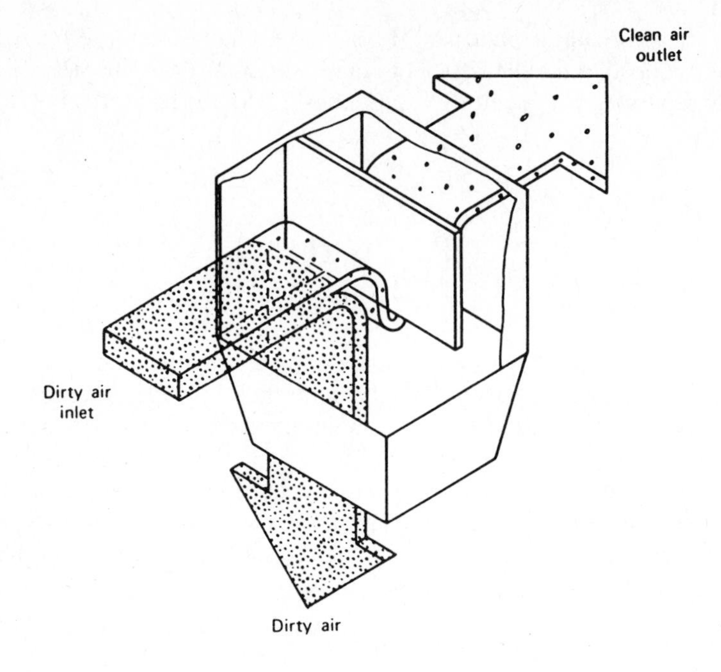

Figure 22.5. Baffled chamber.

Impingement separators vary with the configuration of the obstructing bodies and include baffle type, orifice impaction type, high-velocity gas reversal type, and louver type.

The baffled chamber (Figure 22.5) is the simplest of the impingement separators. It uses one or two plates as impingement sites to stop larger particles and cause them to fall into a dust-collecting bin. This equipment can remove particles larger than 20 μm in diameter with pressure drops varying from 0.5 to 1.5 in H_2O.

The orifice impaction collector (Figure 22.6) may consist of many nozzles, slots, or orifices followed by a plate or baffle for an impingement surface from which the particles can fall into the dust bin. This device can remove particles greater than 2 μm in diameter. It is more efficient than the simple baffle type but more expensive. Normal velocities through the orifices are about 50 to 100 ft/sec and the pressure drop is approximately 2.5 orifice velocity heads.

The louvered impingement separator uses many impingement surfaces on an angle, to rebound the particles into a secondary airstream and allow the cleaner air to pass through the louver. The particulate is thus concentrated into a secondary airstream whose flowrate is 5 to 10 percent that of the primary airstream. This concentrated airstream is usually passed through another more efficient cleaner for discharge into the atmosphere. The efficiency of this type of collector is basically a function of the louver spacing, closer spacings producing higher efficiencies. The

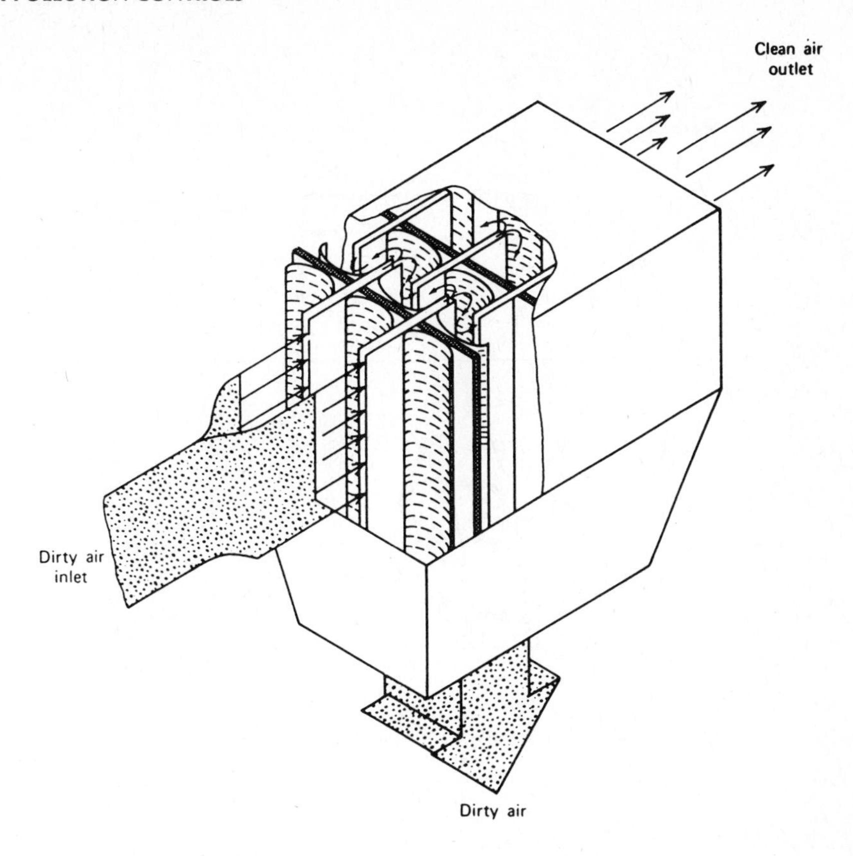

Figure 22.6. Orifice impaction collector.

two principal arrangements of louvers are the flat louver impingement separator (Figure 22.7) and the conical louver impingement separator (Figure 22.8).

7.4 Dynamic Separators

Dynamic separators, sometimes termed rotary centrifugal separators, are the newest gravitational and inertial devices used for collecting particulate. Dynamic precipitators use centrifugal force to separate particulate from an airstream. The precipitator concentrates the dust around the impellers, and it is dropped into a hopper. These devices are about 80 percent efficient on particles larger than 15 μm in diameter.

The major advantage of the dynamic separator is its small size, which is most helpful when a facility needs many independently operated precipitators but has only limited space. Because this separator is a true fan, there is no pressure drop across the device. Its major limitation is its tendency toward plugging and imbalance as well as wearing of the blades.

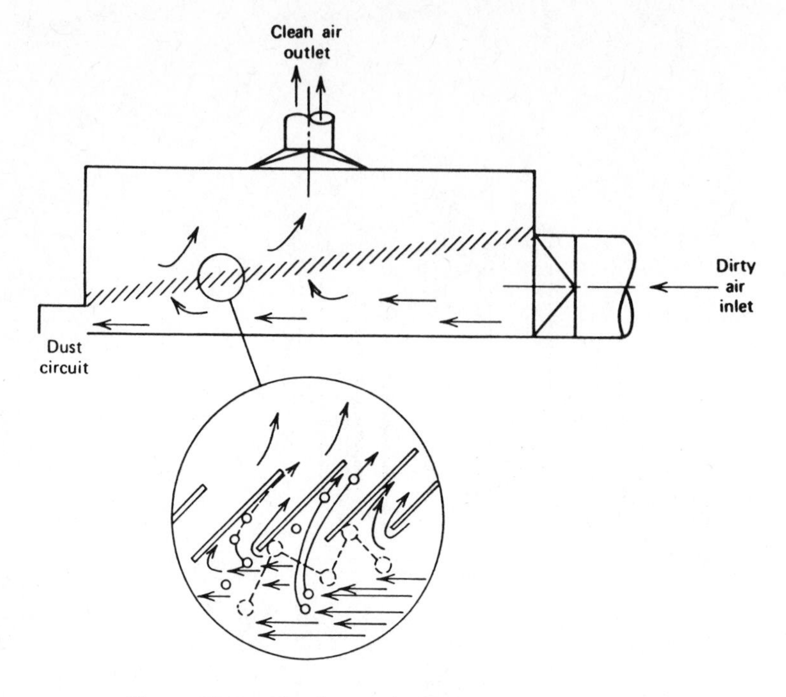

Figure 22.7. Flat louver impingement separator (16).

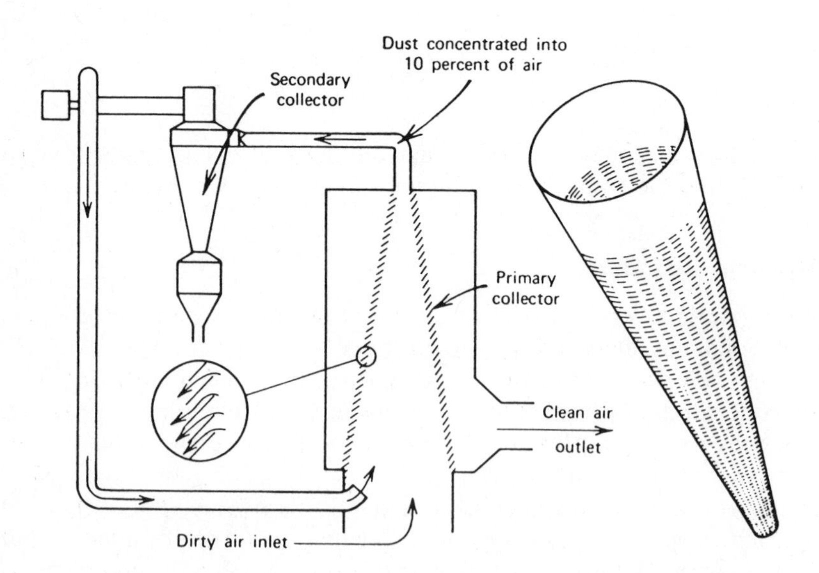

Figure 22.8. Conical louver impingement separator (16).

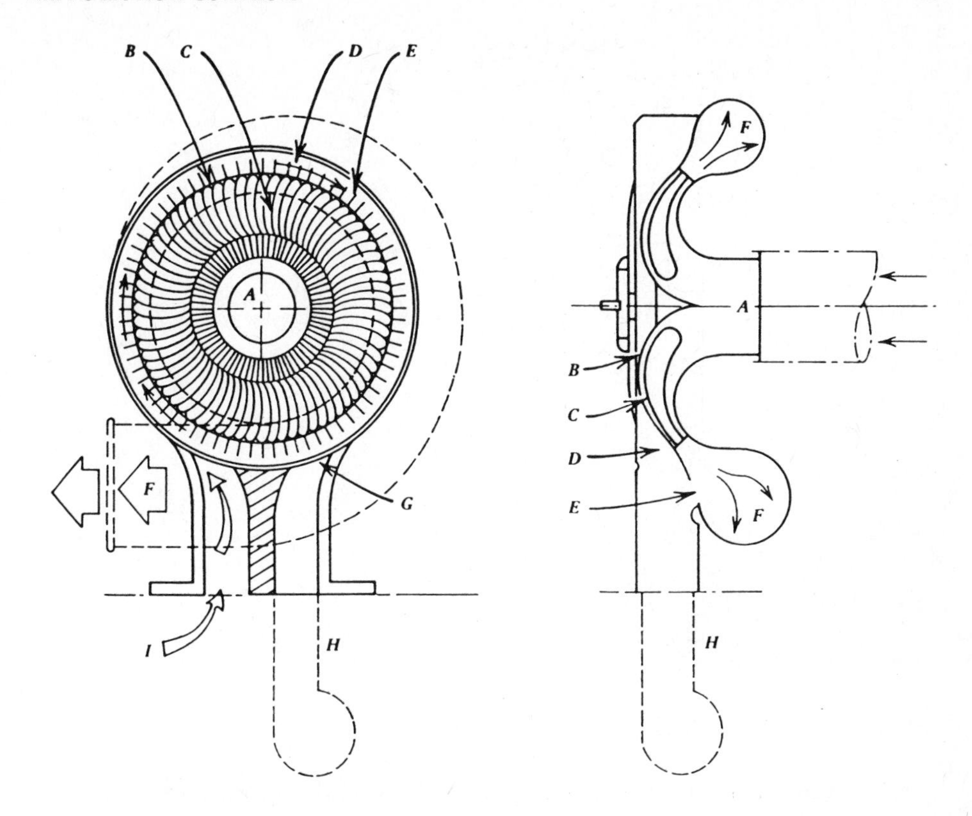

Figure 22.9. Dynamic separator: *A*, center inlet; *B*, interception disk; *C*, impeller blades; *D*, opening for dust escape; *E*, chamber; *F*, scroll-shaped discharge chamber; *G*, point of deflection into hopper *H*; *I*, secondary air return (16).

A dynamic precipitator (see Figure 22.9) separates particles by first drawing the gas stream into it; then as the particulate impacts the impeller, centrifugal force throws the heavier particles to the periphery of the housing. The lighter particles are impacted on the blades and glide along the blade surface to the outside edge where they are also thrown to the periphery of the housing. The particles are then discharged out of the annular slot to the dust collection bin. The cleaned air is discharged into a scroll-shaped discharger.

7.5 Cyclones

Cyclones have long been regarded as one of the simplest and most economical mechanical collectors. They can be used as precleaners or as final collectors. The primary elements of a cyclone are a gas inlet that produces the vortex, an axial outlet for cleaned gas, and a dust discharge opening. Several different types of cyclones (Figures 22.10 through 22.13) are as follows:

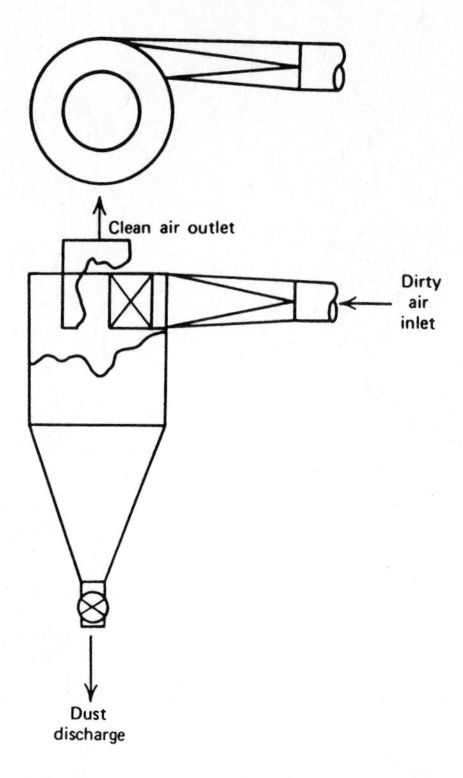

Figure 22.10. Cyclone with tangential inlet and axial discharge (15).

1. Tangential inlet with axial dust discharge (most common)
2. Tangential inlet with peripheral dust discharge
3. Axial inlet through swirl vanes with axial dust discharge
4. Axial inlet through swirl vanes with peripheral dust discharge

Centrifugal force is the precipitating force for particulate and droplets. The air to be cleaned usually enters at the top of a cyclone, either by tangential flow or by curved vanes in an axial inlet. As the air turns, it flows down the cyclone; near the bottom, it reverses direction and flows up the center of the cyclone through the axial exhaust port. Because air is always spinning in the same direction, suspended particles are being forced to the outside wall, where they slide down to the discharge chute (Figure 22.14). Standard cyclone design dimensions are given in Table 22.6.

Efficiency of a cyclone is defined as the fractional weight of particles collected. The major parameter in the prediction of collection efficiency is particle size. The particle size that can be removed from the inlet gas stream at an efficiency of 50 percent in a cyclone is defined as the particle cut size d and is represented in the following relationship (18):

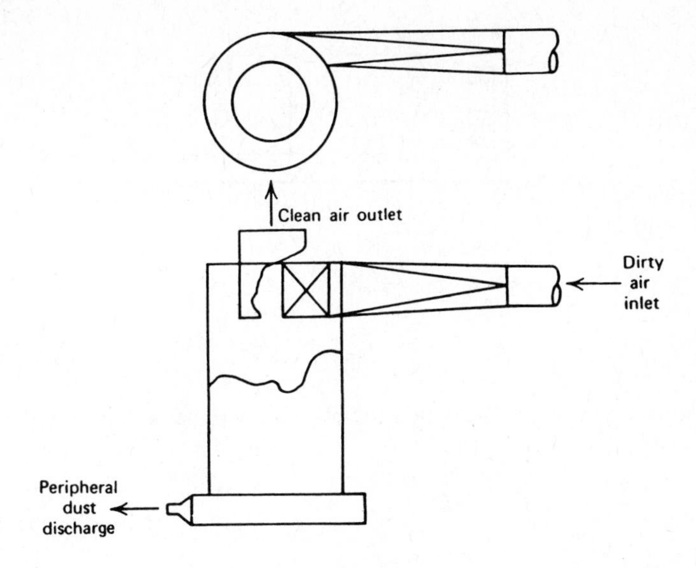

Figure 22.11. Cyclone with tangential inlet and peripheral discharge (15).

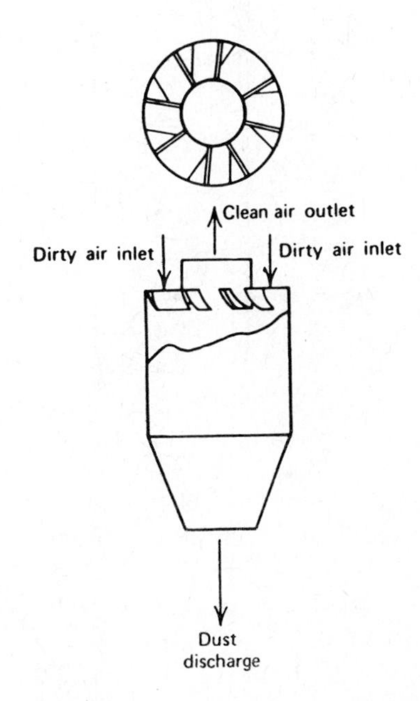

Figure 22.12. Cyclone with axial inlet and axial discharge (15).

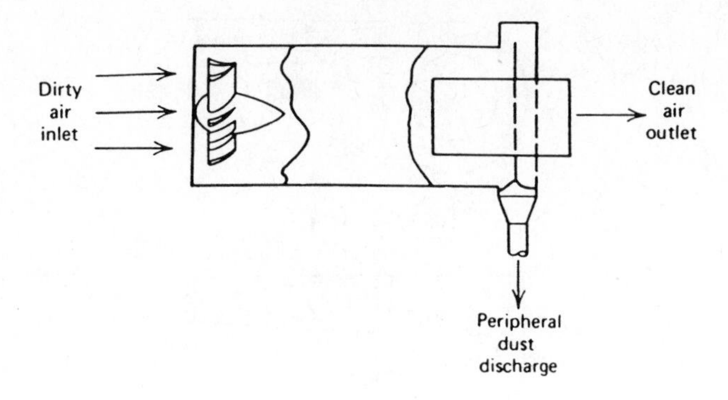

Figure 22.13. Cyclone with axial inlet and peripheral discharge (15).

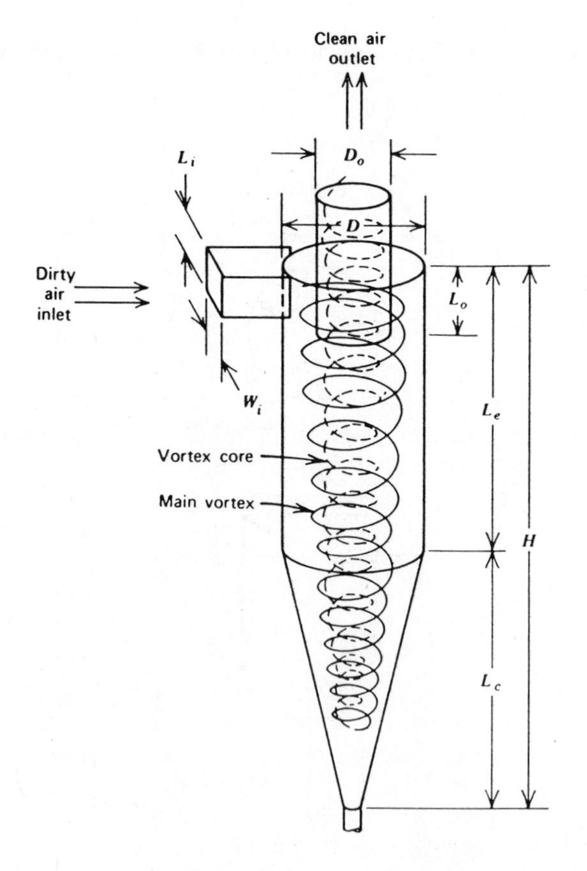

Figure 22.14. Typical cyclone; for abbreviations and standard dimensions, see Table 22.6 (17).

Table 22.6. Standard Basic Cyclone Design Dimensions

Parameter	Conventional Cyclone	High-Throughput Cyclone	High-Efficiency Cyclone
Cyclone diameter, D	D	D	D
Cyclone length, L_e	$2D$	$1.5D$	$1.5D$
Cone length, L_c	$2D$	$2.5D$	$2.5D$
Total height, H	$4D$	$4D$	$4D$
Outlet length, L_o	$0.675D$	$0.875D$	$0.5D$
Inlet height, L_i	$0.5D$	$0.75D$	$0.5D$
Inlet width, W_i	$0.25D$	$0.375D$	$0.2D$
Outlet diameter, D_o	$0.5D$	$0.75D$	$0.5D$

$$d_{pc} = \left[\frac{9\mu W}{2NV(\rho_p - \rho_g)\pi}\right]^{1/2} \quad (15)$$

where d_{pc} = diameter cut size particle collected at 50 percent efficiency (ft)
μ = gas viscosity (lb mass/sec-ft = centipoise $\times$ 0.672×10^{-3})
W = cyclone inset width (ft)
N = effective number of turns within cyclone
V = inlet gas velocity (ft/sec)
ρ_p = true particle density (lb/ft^3)
ρ_g = gas density (lb/ft^3)
π = constant, 3.1416

This equation, together with Figure 22.15, permits the accurate prediction of the collection efficiency of a cyclone when the particle size is known. The design factor having the greatest effect on collection efficiency is cyclone diameter. For a given pressure drop, the smaller the diameter of the unit, the higher the collection efficiency obtained. The efficiency of a cyclone also increases with increased gas inlet velocity, cyclone body length, and ratio of body diameter to gas outlet diameter. Conversely, efficiency decreases with increased gas temperature, gas outlet diameter, and inlet area.

The pressure drop across cyclones is another factor affecting collection efficiency. It varies between approximately 2 and 8 in. H_2O, gauge; pressure drop increases with the square of the inlet velocity. Collection efficiency also increases with the square of the velocity, but not as rapidly as does the pressure drop.

The inlet velocity can be increased only to the point (70 ft/sec) at which the turbulence is not causing reentrainment of the particles being separated. After this point, the efficiency decreases with increased inlet velocity. Efficiency also increases as the dust loading increases, as long as the dust does not plug the cyclone or cause it to be eroded severely. In addition, the size of the particles affects the efficiency: the larger the particle, the higher the collection efficiency.

One of the greatest advantages of cyclones is that they can be made of many

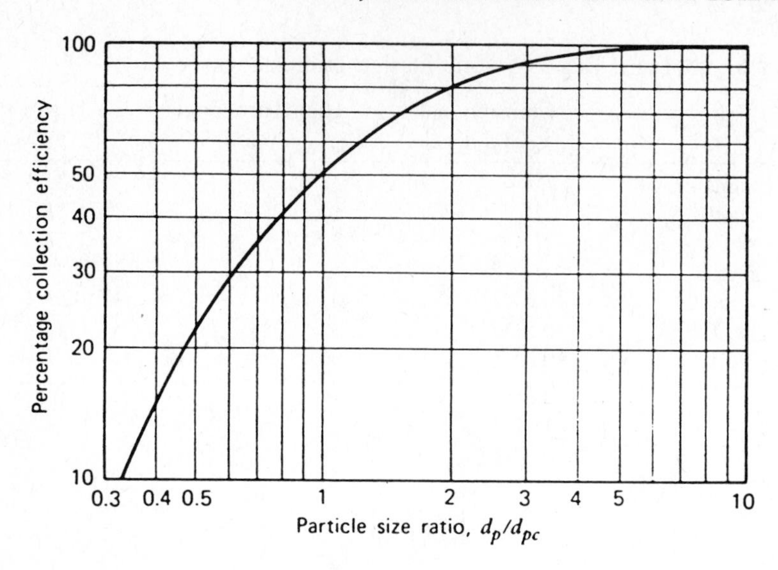

Figure 22.15. Cyclone collection efficiency as a function of particle size ratio (18).

materials and thus are able to handle almost any type of contaminant. Other advantages of cyclones include their ability to handle high dust loads, relatively small size, relatively low initial and maintenance costs, and, excluding mechanical centrifugals, temperature and process limitations imposed only by materials and construction. Because cyclones have larger pressure drops than settling chambers or inertial separators, their cost of operation is higher, but still lower than that of most other cleaners. A disadvantage of cyclones is that efficiency drops off with decreased dust loading. Plugging can also occur, but if the particle characteristics are taken into consideration, this problem can be avoided. Low collection efficiencies for particles below 10 μm in diameter is another disadvantage.

7.6 Wet Cyclones

A wet cyclone is nothing more than a dry cyclone with an inlet for water or some other fluid to be impacted with the incoming particles, rinsing the particles from the cyclone. It costs more to operate than a dry cyclone because of the increased cost for the fluid introduced into the cyclone and for requisite cleaning of the discharge fluid to remove suspended or dissolved particulate.

The forces acting to remove the particles are the same as for dry cyclones. The smaller particles are made larger by impaction with fluid droplets, to increase the efficiency for smaller particles. Because the cyclone walls bear a liquid film, the chance for particulate reentrainment is reduced; thus efficiencies are greater.

Wet cyclones offer the ability to take the liquid from any of several points other than the axial part of the vortex cone, which means that the vortex itself is not distributed. A wet cyclone usually has fewer erosion problems than a dry cyclone,

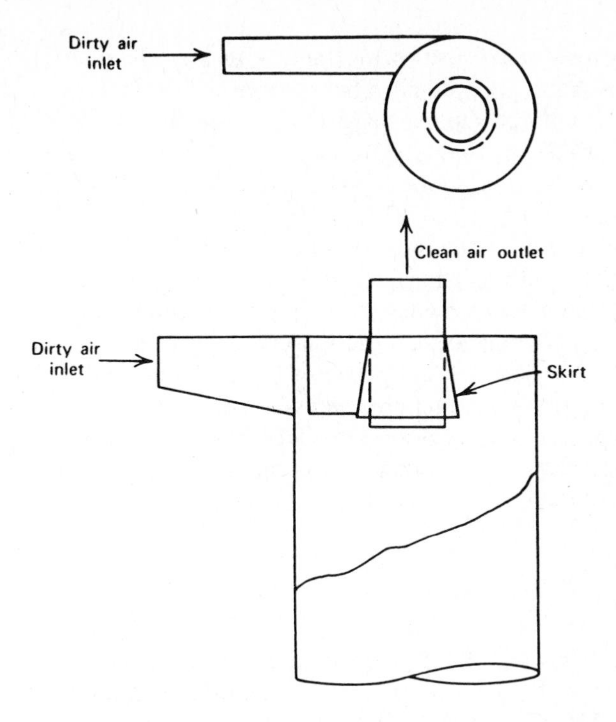

Figure 22.16. Gas outlet skirt for wet cyclone (15).

but if corrosive dust or gas is present, water may activate a serious corrosion problem. Disposal or cleaning of the contaminated water or liquid is another problem to be solved in wet operations.

Wet cyclones entail some design problems that are not present with dry cyclones. First, the liquid tends to creep along the walls of the cyclone to the outlet, where droplets are sheared off into the outlet stream. This can be corrected by installing a skirt on the exit (Figure 22.16) so that the droplets go into the vortex rather than the exit gas stream. If the velocity is too high in the cyclone, very small droplets cannot be recollected. The recommended inlet velocity is not to exceed 150 ft/sec at atmospheric pressure and for injected water; this estimate varies with different gases and liquids but is a general rule.

7.7 Multiple Cyclones

Multiple cyclones are simply many cyclones put in parallel or series. Cyclones in series are not used very often. Cyclones in parallel constitute the most common configuration for multiple cyclones. Space requirements are moderate for multiple cyclones, but these devices are more expensive than regular single cyclones to purchase and design. Operational costs for multiple cyclones are about the same as those for single cyclones. The devices are usually used as a final control but also serve as precleaners in some applications.

Multiple cyclones make use of the same forces to separate the particles from the airstream as the single cyclone. The multiple cyclones usually consist of smaller cyclones that have more "separation force" for a given overall energy input. When many small (3- to 12-in. diameter) cyclones are used in parallel, they are more efficient than one larger cyclone of the same capacity. Their pressure drops are usually between 3 and 6 in. H_2O.

Multiple cyclones offer the most significant advantage of ability to handle large airflows with heavy dust loadings and still exhibit good efficiency. When properly designed inlet ductwork provides even flow to all the individual cyclones, the multiple cyclone performs at peak design efficiency. This also ensures that there will not be any backflow through some of the cyclones, causing particles to flow from the dust bin to the exit plenum. Many multiple cyclones have several gas inlet and exit plenums and several dust exit valves; this allows any part of the multiple cyclone to be shut off from another part in the event that a cyclone plugs or needs repairs.

8 FILTRATION

Filtration represents the oldest, and inherently the most reliable, of the many methods by which particulate matter (dusts, mists, and fumes) can be removed from gases. Filters are especially desirable for extracting particulate matter from gases produced by industrial operations because they generally offer very high collection efficiencies with only moderate power consumption. Initial investment costs and maintenance expenditures can range from relatively low to comparatively high, depending on the size and density of the particulate matter being collected, as well as the quantity and temperature of the dusty gas to be cleaned.

Filters are most readily classified according to filtering media. For particulate emissions control, the different medial can be broadly categorized as (1) woven fibrous mats and aggregate beds, (2) paper filters, and (3) fabric filters.

8.1 Range of Performance

Filtration is highly effective for small particulates, even down to less than 0.05 μm in diameter, if the flow through the filter is kept low enough. Filtration can be used on almost any process emitting small particles of dust. Limiting factors, however, are the characteristics of the gas stream. If the unfiltered stream is too hot, it may be necessary to precool the gas stream. If the gas stream is too heavily loaded, a precleaner may be necessary to rid the stream of larger particles. The cost of auxiliary equipment required for such pretreatment may equal or even exceed that of the filter itself. There are also several causes for filter failure, including blinding, caking, burning, abrasion, chemical attack, and aging.

Designs of filters and filter enclosures are many. Filters can be configured as mats, panels, tubes, or envelopes. When tubular fabric filters are enclosed, the structure and its contents are often called a "baghouse."

8.2 Fibrous Mats and Aggregate Beds

Fibrous mats and aggregate-bed filters are characterized by high porosity; both are composed largely (97 to 99 percent) of void spaces. Such void space is usually much larger than the particles being collected; thus the mechanisms of sieving and straining of particles are of no significance in these filters. The predominant forces functioning in the cleaning of gas streams using large void filters include impaction, impingement, and surface attraction. Impaction and impingement are effective when the particles are made to change direction quickly and impact or impinge on the surface of the filter medium. Surface-attractive forces are mostly electrostatic, and before they contribute significantly to total collection, the dust particles must come within several particle diameters.

Efficiencies can be extremely high even with very low dust loadings in the input airstream. Early applications using sand as the filtering medium reported collection efficiencies of 99.7 percent for submicrometer particles (19). More recently, glass fiber beds have been used and efficiencies of 99.99 percent have been reported for dust loadings of 0.00002 to 0.00004 grains/ft^3 (20).

Advantages of aggregate-bed filters include a longer life without frequent cleaning, high dust storage capacity with a modest increase in airflow resistance, and application to high-temperature emissions. One disadvantage is difficulty in cleaning; many fibrous mats are simply discarded. Large space requirements also pose a problem when these filters are used.

These filters range from very inexpensive mats that are changed and discarded to very expensive beds that are almost never replaced. Deep beds and fibrous mats are relatively inexpensive to operate because of low pressure losses (between 0.1 and 1 in. H_2O, gauge) in most operations.

Designs for deep-bed filters are numerous. Almost any material can be chosen to make up the bed. Sand has been used most often, but glass fiber beds have been used recently. Sulfuric and phosphoric acid plants use what are known as "coke boxes" for collection of acids. A lead or ceramic-lined box is filled with several feet of $\frac{1}{4}$- to $\frac{1}{2}$-in. diameter coke to collect the mechanically produced mist at an efficiency of 80 to 90 percent. Condensed mist is generally too small to be collected by the coke box.

The mats are of many materials. Glass fibers, stainless steel, brass, and aluminum are all used in mats for different types of airstreams. Fibrous mats must be cleaned when their pressure loss becomes too high or the flow rate becomes too low. Cleaning can be accomplished by removing a portion of the filter bed continually and replacing it with new or cleaned portions. Some cleaning is accomplished by reversing the airflow and vibrating the bed or using shock waves. Some mats are self-cleaning and some are continuous cleaning.

Designs for self-cleaning beds are numerous. For example, an automatic viscous filter (Figure 22.17) uses an oil film to ensure that the particles are not reentrained in the airstream. The airstream must pass through the filter twice. The plates on the conveyer belt open and close at the top and bottom of the filter. At the bottom,

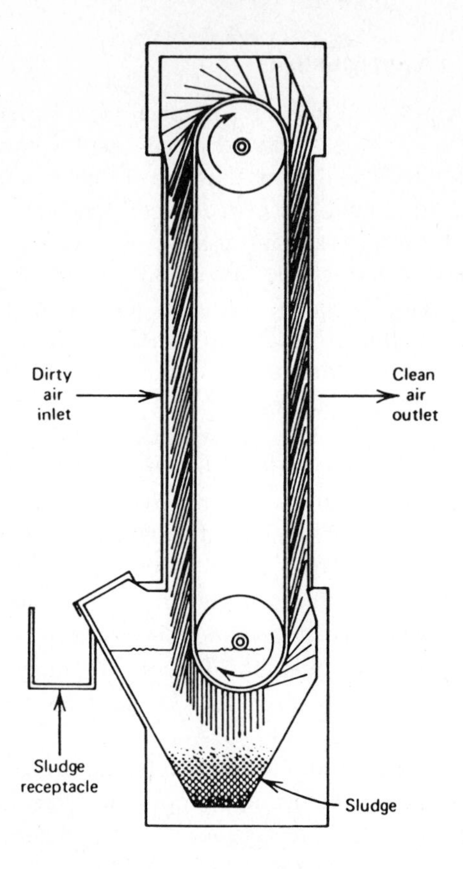

Figure 22.17. Self-cleaning automatic viscous filter (20).

the plates are plunged into an oil reservoir and cleaned by agitation. Upon emerging from the oil bath, they are cleaned further and reoiled.

Another type of self-cleaning mat is the water spray or wet filter (Figure 22.18). The mats are sprayed continuously while the airstream flows through the filter. The particles are dislodged by the water and flow with it to the sump.

8.3 Paper Filters

Paper filters are of relatively recent design for air pollution control and find service chiefly where ultrahigh efficiencies are needed. These filters have come into wide use where very clean air is essential, as in "white rooms" of hospitals, data processing centers, the aerospace industry, food processing plants, and semiconductor manufacturing.

Paper filters can be made of minerals, asbestos, or glass microfibers, with or without binders that add strength, formability, or water resistance. Glass microfilters are the most popular because of their fire resistance and availability. The

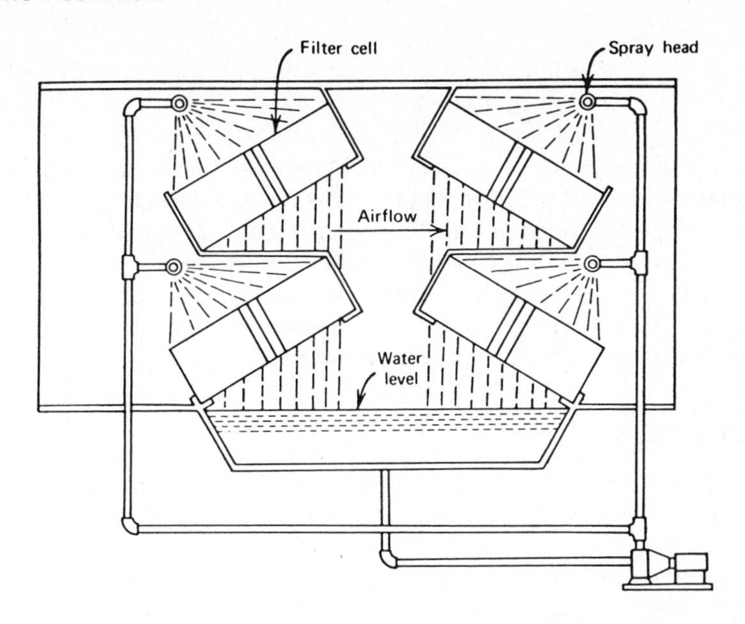

Figure 22.18. Self-cleaning wet filter (20).

frames that contain a set of filters known as "packs" or "plugs" can be made of steel, aluminum, hardboard, or plywood, depending on the application. These frames are normally fitted with gaskets on both sides to ensure a good seal; the seals are made of appropriate materials for the application (e.g., neoprene, mineral wool, asbestos, or rubber).

Paper filters may be flat, cylindrical, or any shape that is practical for the application and space available. Fluting of the paper medium increases the surface area of a filter compared to that of a flat fiber; this can save appreciably in equipment size used to house the filtering elements.

Paper filter systems require moderate initial capital investment, and operating costs are relatively low. Variations in design, however, can make capital investment costs very high, depending on the application.

Paper filters use all mechanisms of particle capture and retention. The most significant is diffusion. Paper filters are often used as final filters to remove very small particles or dusts in very low concentrations. Any larger particles still present in the gas stream may be captured, but the life of a filter is shortened measurably if too many large particles are captured on a continuous paper filter. Many times high-efficiency, inexpensive precleaners are used before the paper filter to make the paper medium last longer. Because the filtration is mostly a surface action phenomenon, the dust storage capacity is a function of the surface area.

The efficiency of paper filter media is very high, 99.97 percent by weight for the best commercial paper filters. The dioctyl phthalate (DOP) method for calculating the efficiency of a filter is employed for paper filters. The test method is the U.S. Army Chemical Corps. DOP Smoke Penetration and Air Resistance Test No. MIL-STD-282, Method 102.9.1 (21).

To ensure good collection efficiencies throughout their lifetime, paper filters must have larger pressure losses when new compared with other filter types. This means that the ratio of pressure loss in a new filter to a spent filter is larger. Flow velocity through the paper is usually around 5 ft/min.

Paper filters are advantageous in that they have a long life expectancy, usually 1 to 2 years, and are relatively inexpensive to replace. Paper filters cannot be cleaned and reused. Other disadvantages include the necessity to provide low flow rates and low dust loadings to the paper filtering element.

8.4 Fabric Filters

One of the most positive methods for removing solid particulate contaminants from gas streams is filtration through fabric media. A fabric filter is capable of providing a high collection efficiency for particles as small as 0.5 μm and will remove a substantial quantity of particles as small as 0.01 μm.

Fabric filters are usually tubular (bags) or flat and made of woven or felted synthetic fabric. The dirty gas stream passes through the fabric, and the particles are collected on the upstream side by the filtration of the fabric. The dust retained on the fabric is periodically shaken off and falls into a collecting hopper.

The structure in which the bags hang is known as a baghouse. The bags have an average life of 18 to 36 months. The number of bags may vary from one to several thousand. The baghouse may have one compartment or many, making it possible to clean one while others are still in service.

Removal of the particles from the gas stream is not a simple filtration or sieving process, because the pores of the fabric employed in fabric filters are normally many times the size of the particles collected, sometimes 100 times larger or more. The collection of particles takes place through interception and impaction of the particles on the fabric filters, and Brownian diffusion, electrostatic attraction, and gravitational setting within the pores. Once a mat or a cake of dust is accumulated, further collection is accomplished by the mechanism of mat or cake sieving, as well as by the foregoing mechanisms. Periodically the accumulated dust is removed, but some residual dust remains and serves as an aid to further filtering.

Direct interception occurs whenever the gas streamline, along with a particle, approaches a filter element. Inertial impaction occurs when a particle unable to follow the streamline curving around an obstacle comes closer to the filter element than it would have come if it had approached along the streamline. Small particles, usually less than 0.2 μm, do not follow the streamline because collision with gas molecules occurs, resulting in a random Brownian motion that increases the chance of contact between the particles and the filter element. Electrostatic attraction results from electrostatic forces drawing particles and filter elements together whenever either or both possess a static charge.

The major particulate removal mechanisms as they apply to a single fiber in a fabric filter are shown in the following tabulation:

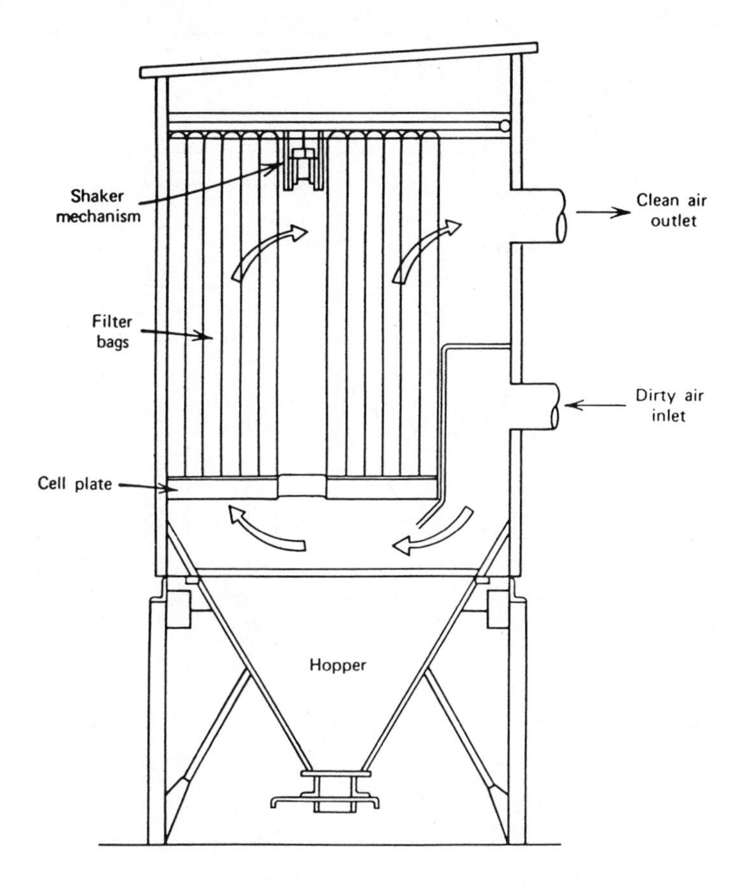

Figure 22.19. Shaker-type baghouse (13).

Primary Collection Mechanism	Diameter of Particle (μm)
Direct interception	>1
Impingement	>1
Diffusion	<0.01–0.2
Electrostatic attraction	<0.01
Gravity	>1

Fabric filter systems (i.e., baghouses) are characterized and identified according to the method used to remove collected dust from the filters. This is accomplished in a variety of types, including shaker (Figure 22.19), reverse flow (Figure 22.20), reverse jet (Figure 22.21), and reverse pulse (Figure 22.22).

The fundamental criterion in applying any baghouse to any application is the air to cloth ratio, defined as

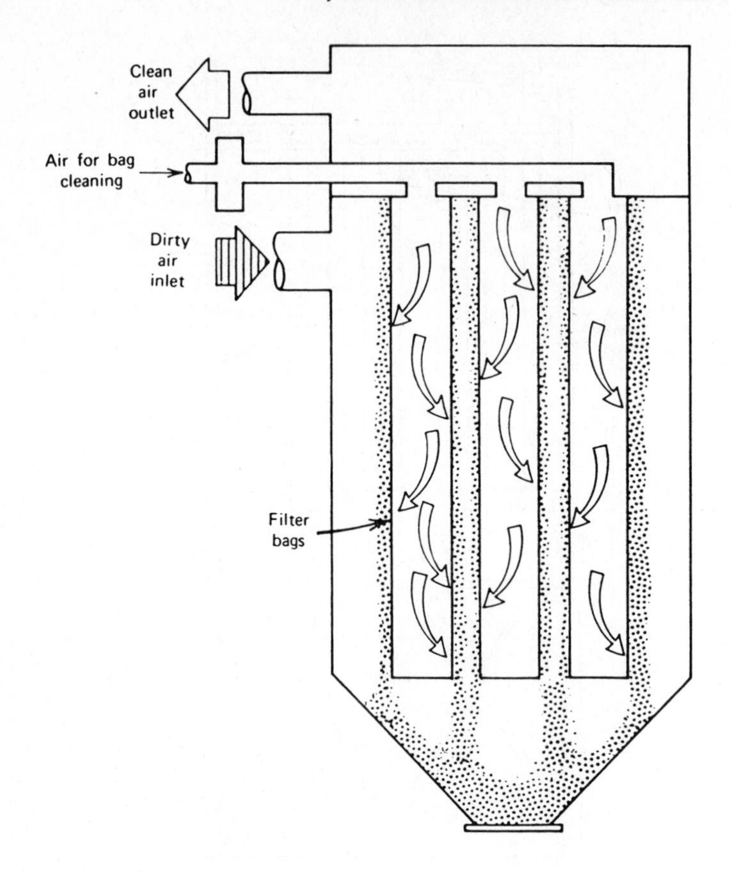

Figure 22.20. Reverse airflow baghouse.

$$\frac{A}{C} = \frac{Q}{A} \tag{16}$$

where A/C = air-to-cloth ratio (ft/min)
Q = volumetric gas flow rate (acfm)
A = net cloth area (ft^2)

Air-to-cloth ratio is equal to the superficial face velocity of the air as it passes the cloth. Shaker and reverse airflow baghouses usually operate at an air-to-cloth ratio of from 1 to 3, whereas a reverse pulse baghouse operates at about three to six times this range.

A second important factor in baghouse application is the type of filter material. Woven cloth is used in shaker and reverse airflow baghouses; felted cloth is selected for reverse pulse baghouses. One of the characteristics of filter fabrics is the permeability, which is expressed as the air volume, in actual cubic feet per minute, passing through a square foot of clean cloth with a pressure differential of 0.5 in. H_2O,

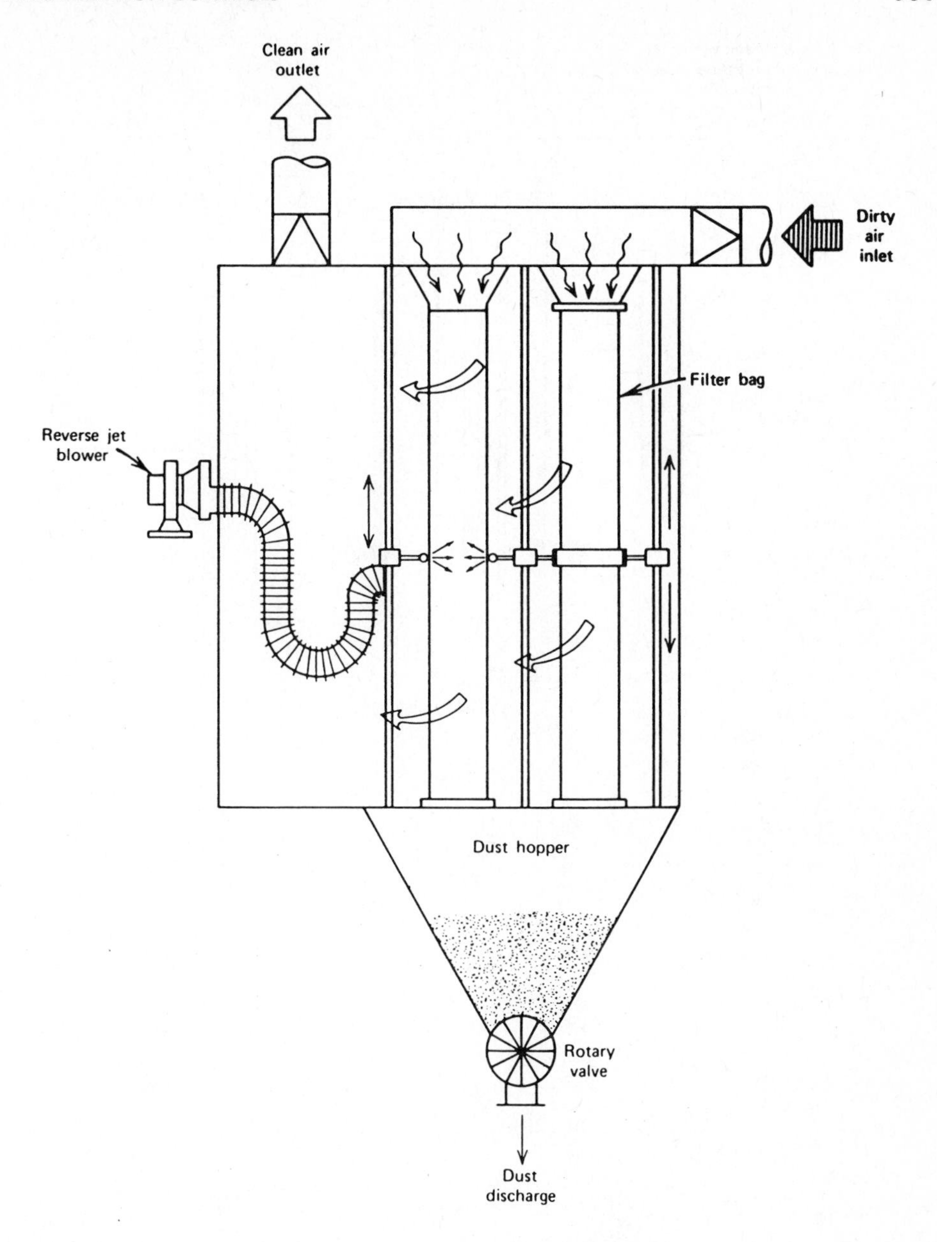

Figure 22.21. Reverse jet baghouse (20).

gauge. The overall range of permeability is from 10 to 110 acfm/ft^2, but it usually ranges from 10 to 30 acfm/ft^2 for all types of baghouses.

Another important operating characteristic is the filter drag, given as

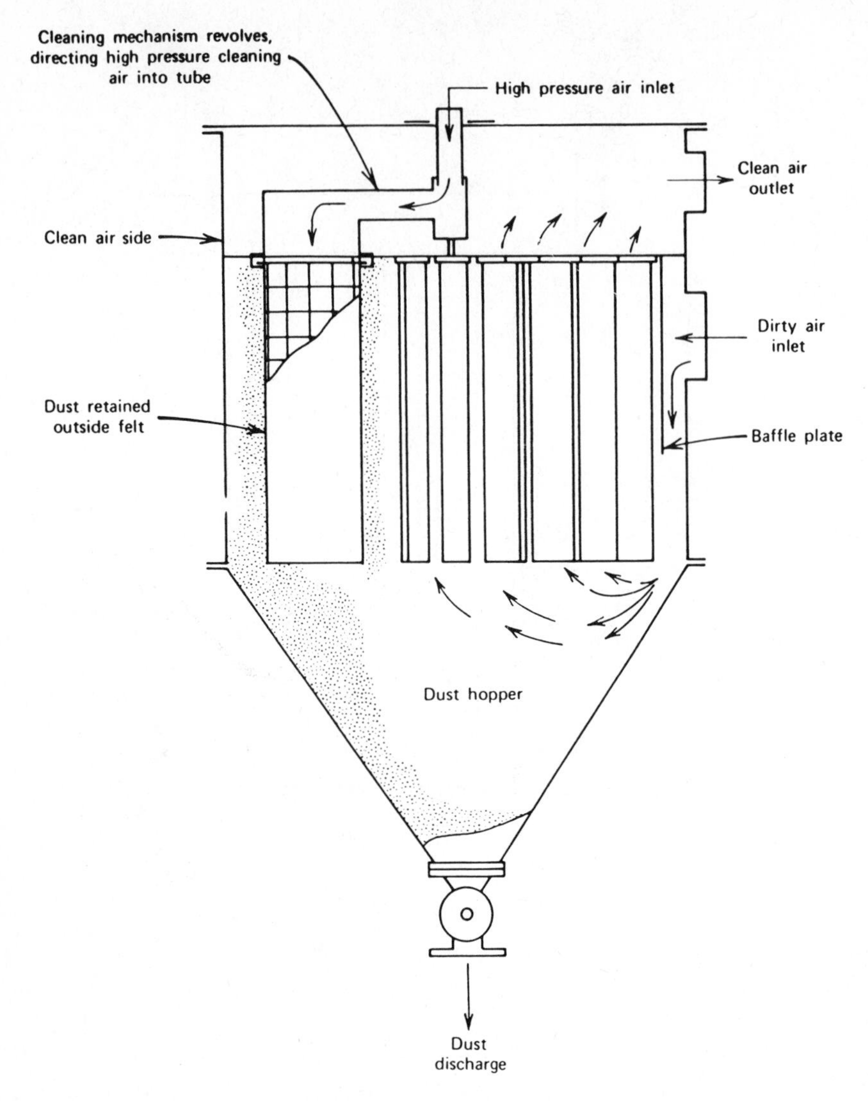

Figure 22.22. Reverse pulse baghouse (13).

$$S = \frac{\Delta PA}{Q} \tag{17}$$

where S = filter drag (in. H_2O/per fpm)
ΔP = pressure drop across filter (usually 2 to 6 in. H_2O)
A = net cloth area (ft^2)
Q = volumetric gas flow rate (acfm)

Figure 22.23 shows the effect of filter drag on outlet concentration (22).

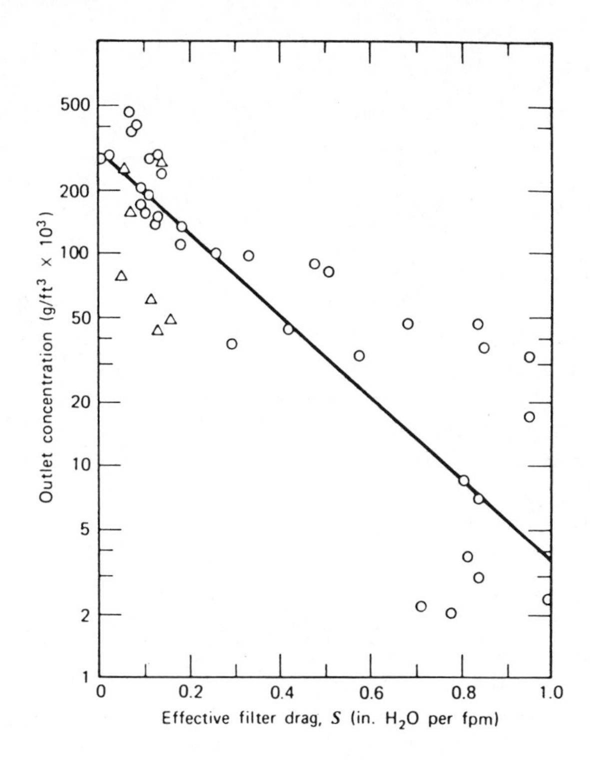

Figure 22.23. Baghouse outlet concentration as a function of filter drag (22).

9 LIQUID SCRUBBING

"Liquid scrubbing" denotes a process whereby soluble gases or particulate contaminants are removed from a carrier gas stream by contacting the contaminated gas stream with a suitable liquid to decrease the concentration of the contaminant. Scrubber geometry, contacting media, and the scrubbing liquor are design variables that have been the subject of years of investigation to optimize scrubber performance in a variety of applications.

Although the liquid used in liquid scrubbing is generally recirculated, the need to discharge some portion of the scrubbing liquid can create water pollution problems complex enough to render liquid scrubbing infeasible. Nonetheless, the application of liquid scrubbers to air pollution abatement strategies finds optimal use when soluble gaseous contaminants of fine particulate contaminants must be removed to high efficiency levels, when the gases involved are combustible, when cooling is desired, and when increased moisture content can be tolerated. When one or more of these prerequisites is met, liquid scrubbing may very well provide the only applicable air pollution abatement strategy within the contexts of economic and technical feasibility.

9.1 Range of Performance

The type of scrubber used in a particular application depends mostly on the characteristics of the contaminants being scrubbed and the degree of control required. Almost any device in which good contact is promoted between the scrubbing liquor and a contaminated gas stream will absorb to some degree both gaseous and particulate contaminants. The questions of what type of scrubber to specify in an individual case hinges on the efficiency desired, the properties of the contaminant, and the merging of these two factors within the context of minimum operating costs and energy consumption. For example, although a high-energy venturi scrubber using water as the scrubbing liquor will efficiently absorb some water-soluble gases, the energy requirement of the venturi scrubber can be an order of magnitude greater than that of a packed tower to promote the same degree of mass transfer.

The design parameters and established correlations useful in specifying and sizing liquid scrubbers can be subdivided into two general classes: (1) mass transfer dynamics for gases dissolving and/or reacting with the scrubbing liquor, and (2) momentum transfer dynamics for particulate matter colliding with and being entrained in the liquid scrubbing medium. In the first class, either gas–liquid equilibrium data and mass transfer coefficients must be established by experience with the actual species of chemical constituents being encountered, or existing data for chemically similar scrubbing system applications may be extrapolated, in the hope that useful design information can be extracted and the new system design parameters thereby established. In the second class, semitheoretical considerations provide useful design criteria in specifying scrubbers that utilize inertial impaction or Brownian motion to extract particulate matter. Even in this case, however, empirical design data supply the soundest base on which to specify the type and size of liquid scrubber needed to ensure efficient collection of particulate matter at minimum feasible capital and operating cost.

Packed towers using water as a scrubbing medium have found wide applications in the chemical industry to absorb water-soluble gases and vapors, both for pollution abatement and for product recovery. In the power industry, scrubbers are frequently used in the absorption of sulfur dioxide, and they function efficiently as long as the partial pressure of the sulfur oxides in the gas stream is greater than the vapor pressure of the sulfur oxides existing at the surface of the scrubbing liquor. Such a concentration difference provides a driving force to promote mass transfer of sulfur oxides from the gas into the liquid stream. Because the solubility of most water-soluble gases is limited by equilibrium relationships, the amount of scrubbing liquor specified per unit volume of contaminated gas is determined using mass balance calculations and equilibrium relationships and giving due consideration to any chemicals added to the scrubbing liquor to increase absorption efficiency.

A wide variety of scrubbers, operating in pressure drop ranges of 0.5 to more than 80 in. H_2O have been used successfully to absorb particulate matter ranging in size from tens of micrometers down to material of submicrometer size. In any particulate scrubber, the design must provide as large a constant area as possible for scrubbing liquor and particulate matter to interface. In a venturi scrubber, for example, the submicrometer dust impacts on the surfaces of dispersed liquid droplets and penetrates into and becomes part of the liquid droplets, which are subsequently agglomerated, coalesced, and removed from the gas stream.

Although contact area is a prerequisite to particulate absorption in liquid scrubbing, the nature of the dust and its physical properties in relation to the scrubbing liquor determine whether the dust can be wetted by the scrubbing liquid or whether it can even penetrate the surface of the scrubbing liquor. For example, in the scrubbing of iron oxide dust using water, the relatively high surface tension of water does not permit iron oxide to be "wetted" well enough to assure permanent retention in the scrubbing water. Addition of "surfactants" to the scrubbing water decreases the surface tension, thereby permitting the particles to penetrate the "skin" of the droplets at the gas–liquid interface.

9.2 Spray Chambers

Spray-type scrubbers are useful in collecting particulate matter and cases in liquid droplets that are dispersed in a chamber using spray nozzles to atomize the liquid. The geometry of the spray chamber can vary from a simple straight cylinder to configurations designed to provide maximum contact area and contact time between droplets and gases passing through the atomized sprayer.

In the simplest type of spray chamber, a vertical tower is the container in which droplets and gas meet. In vertical spray chambers, the droplet velocity eventually becomes the terminal settling velocity for a given droplet. The droplet velocity relative to the particulate collision velocity determines efficiency. Characteristically, spray chambers exhibit low pressure drop, utilize high relative rates of scrubbing liquid per unit volume of gas, and are generally very low in cost because of their simple design. Generally, spray chambers exhibit high collection efficiency for particles that are larger than 10 μm in diameter.

Usually the gases are introduced to the lower section of a spray tower, and the sprays, positioned at the top of the tower, inject scrubbing liquid at some velocity determined by the nozzle orifice (Figure 22.24). The residence time and the average relative velocity difference between the upward traveling gas and the downward traveling droplets determine the amount of contaminant removal that can be expected for a given tower size and ratio of scrubbing liquid to gas flow rate.

Some spray towers introduce the gases at the base of the tower in tangential fashion, thereby imparting a spiraling motion to the gases. Not only is longer contact time obtained in this manner, but spray droplets are driven to the walls of the spray tower, where they impinge on and drain down the walls as a liquid film.

In spray chambers, collection efficiency of particulate matter increases as droplet size decreases. Therefore the use of the high-pressure atomizing nozzles to produce fine droplets enhances collection efficiency near the nozzle. These small droplets quickly decelerate, however, because of aerodynamic drag. Therefore the relative velocity difference between the spray droplets and the particulate traveling with the gas falls off so rapidly that the collection efficiency is almost zero in the rest

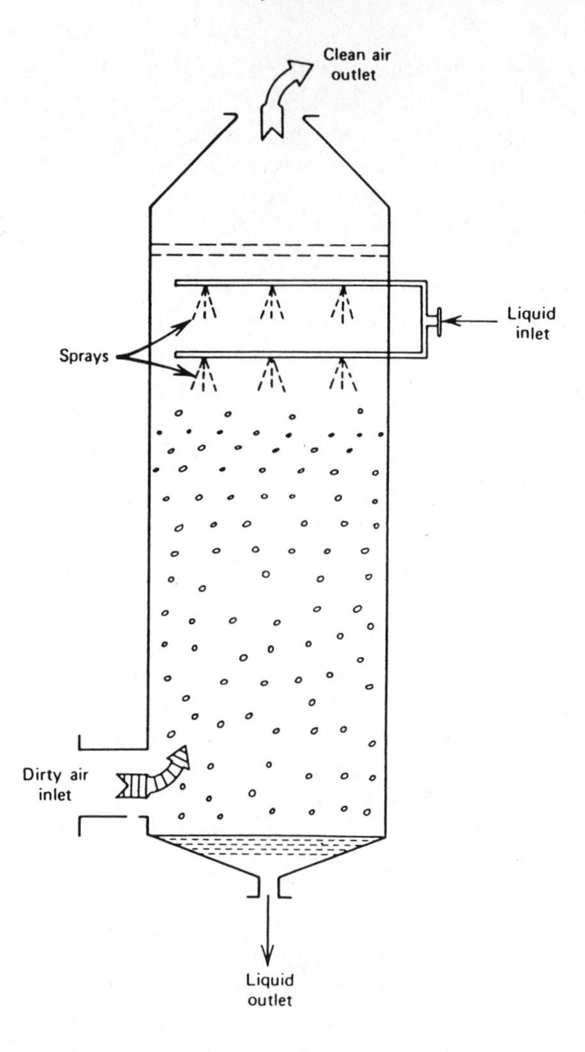

Figure 22.24. Spray tower (23).

of the tower a short distance from the nozzle orifices. An optimal droplet size is 500 to 1000 μm. These droplets in spray chambers scrub relatively efficiently over long enough path lengths throughout the spray tower to provide an optimal collection efficiency.

Because of the relatively simple design, spray towers also serve a useful function as precoolers or quench chambers for relatively quickly dropping the temperature of hot gases as they are emitted from a wide variety of processes. Very often these hot gases contain a relatively high solids content, because the spray chamber is most likely to be the first piece of gas conditioning equipment the stream encounters before final discharge to the atmosphere. The spray chamber can handle recycled scrubbing liquid with relatively high solids concentrations suspended in the scrubbing liquid because of its simple design and because the clearance in the spray nozzle orifice is relatively large. Therefore the spray chamber is ideally suited both as an initial solid particulate remover and as a gas conditioner useful in preparing the gas for subsequent downstream processing equipment, which is probably more complex in design and more efficient in particulate extraction.

Specification of pertinent design parameters for a particular application depends to a large extent on the type of contaminant being scrubbed in the spray chamber. A rule of thumb, however, would indicate that in simple spray towers from 5 to 20 gal of scrubbing liquid per 1000 ft^3 of contaminated gas provides optimal liquid gas contact.

In applications featuring the scrubbing of gases, equilibrium considerations come into play and computation of the number of mass transfer units is necessary. These calculations are theoretical and contain proportionality constants that depend on the physical and chemical properties of the contaminants being scrubbed; therefore empirical data are required to provide reasonably accurate specifications. The calculations are similar to those made in sizing packed-tower scrubbers.

9.3 Packed Towers

Scrubbers containing packing material such as Pall rings, Berl or Interlox saddles, or Raschig rings are termed packed-bed scrubbers or packed towers. The packing provides high surface area of liquid films ideal for mass transfer. In general, packed-bed scrubbers can be classified as countercurrent, cocurrent, and cross flow. In the countercurrent packed-bed scrubber (Figure 22.25) scrubbing liquid, injected at the top of the packing, trickles down through the packing material as gases pass up. Countercurrent scrubbers generally cannot be operated at liquid or gas flow rates as high as those used in cocurrent packed-bed scrubbers, where the gas and liquid phases flow in the same direction. Cocurrent scrubbers, in general, provide lower mass transfer cubic foot of packing than countercurrent units.

Packed-bed scrubbers are normally used to remove gaseous contaminants from airstreams. In general, when high loading of particulate is a possibility in such applications, the packing media may be fouled by deposited particulate matter, resulting in eventual blockage of the unit. Because gaseous contaminant removal is limited by equilibrium considerations, the efficiency of the unit, as well as the absolute concentration of contaminant in the gas stream leaving the packed-bed scrubber, is affected by whether a cocurrent or countercurrent scrubber is specified.

The cross-flow scrubber is the most capable of the three types of packed-bed scrubber of coping with deposited solids on the tower packing. In these units, scrubbing liquid trickles through the packing from the top of the packing bed, while the dirty gas moves horizontally through the bed. This arrangement provides good washing of the media and simultaneously is a very stable configuration, difficult to flood.

Packed-bed scrubbers have been used historically in chemical operations in a wide variety of applications. By adding various chemical substances to the scrubbing liquor, pollutant gases can be removed to very low outlet concentrations if sufficient

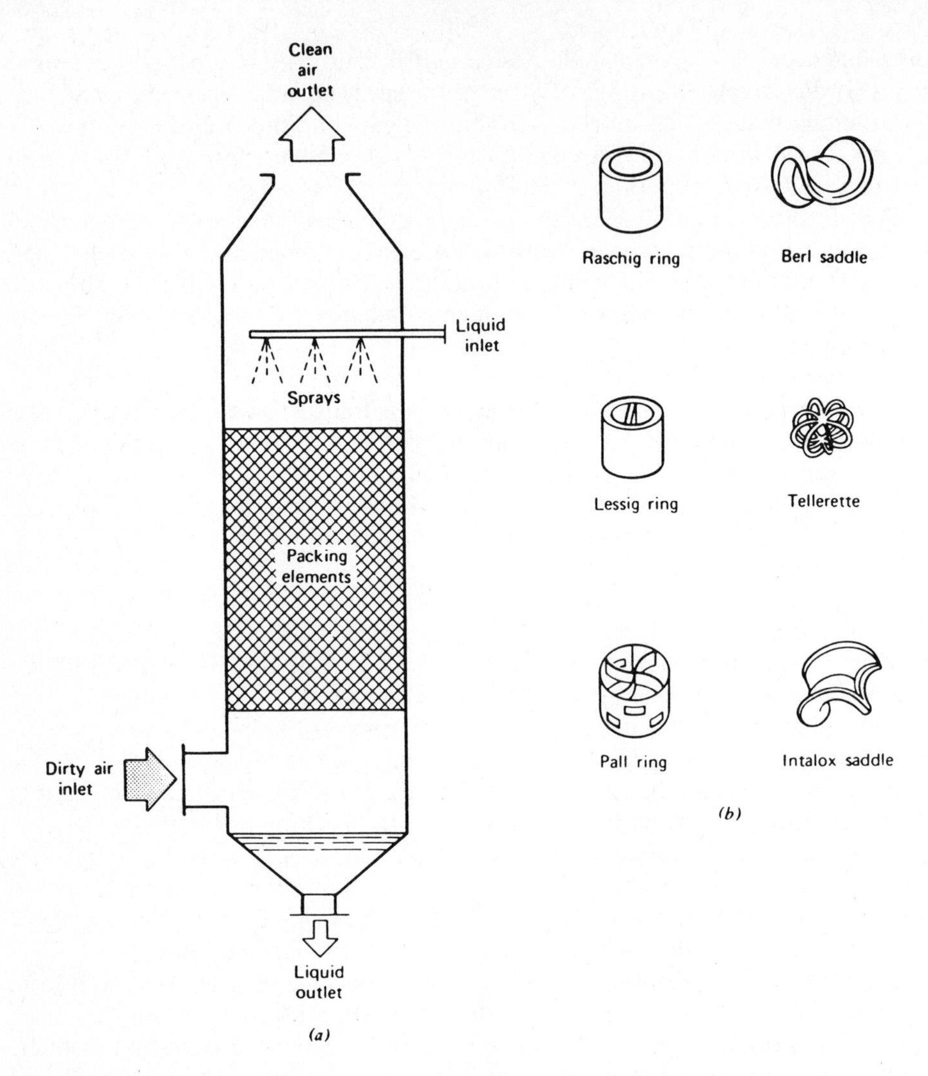

Figure 22.25. (*a*) Countercurrent packed-bed scrubber and (*b*) packings (23).

fan power is provided to overcome the bed pressure drop and the absolute volume of packing and tower height are sufficiently large. Typically, empirical design data are necessary to determine mass transfer coefficients useful in predicting efficiencies for particular combinations of contaminant and scrubbing liquid. Pressure drops through packed-bed towers can be quite low. Although the pressure drop depends on the flow rate through the tower and the liquid–gas ratio, as well as the type of packing, typically 0.5 in. H_2O per foot of packing is the pressure loss in packed-bed towers.

Packed beds offer the added advantage of being intrinsic mist eliminators. Because flow rate is characteristically low, entrainment of liquid into the gas leaving the tower is much less than in scrubbers of other types where the liquid phase is dispersed in fine droplets. Furthermore, because tower height is somewhat greater than packing height, scrubbing efficiency of the packed tower can be increased by simply adding more packing, at minimal cost to the overall system. Thus packed-bed scrubbers are relatively flexible to process changes that may alter the concentration or nature of gaseous contaminants in their inlet streams.

Section 9.6 discusses the pertinent design parameters and applicable equations necessary to size a packed-bed scrubber for a variety of conditions. In general, pertinent parameters include the concentration of the contaminant in the gas stream admitted to the scrubber, the required efficiency of the unit, the ratio of scrubbing liquor flow rate to inlet gas flow rate, an estimate of the mass transfer area per cubic foot of packing in the packed bed, and equilibrium data specifying the concentration of contaminant in the scrubbing liquor.

9.4 Orifice Scrubbers

Orifice scrubbers use the carrier gas velocity to promote dispersal of the scrubbing liquid and turbulent contact of the contaminated gas stream with the dispersed liquid (Figure 22.26). Both the efficiency and the pressure drop depend on the gas flow rate through the scrubber and the ratio of the scrubbing liquor injected per unit volume of gas. Orifice scrubbers make use of internal geometric designs that attempt to supply scrubbing liquid uniformly across the cross section of gas flowing through the unit. The kinetic energy of the gas supplies the energy needed to disperse the liquid phase. In the turbulent contacting zone, usually a short distance downstream of the orifice, the intensely turbulent motions provide violent contact between particulate matter and the larger scrubbing liquid droplets, which possess high inertia.

Downstream of the contact zone, mist eliminators agglomerate and coalesce the droplets of scrubbing liquor containing the absorbed contaminant. Depending on the energy supplied to the unit, scrubbing efficiencies can range from 85 percent to more than 99.5 percent, depending on the nature and physical properties of the contaminant being removed. The optimum use of input energy to effect a given degree of contaminant removal is the subject of many innovative designs applied to a wide variety of industrial process emissions.

There are numerous geometries and internal designs for orifice scrubbers. This general classification includes all scrubbers that depend on contacting by passing the gas and liquid phases through some type of opening. In the more sophisticated designs, the opening may be varied to accommodate changes in requisite removal efficiency or gas flow rate. The basic principle of an orifice scrubber having a fixed orifice size dictates that scrubbing efficiency will fall off markedly if gas velocity through the orifice is decreased. By providing pressure drop sensors to the variable pressure drop orifice type scrubber, automatic operation can compensate for wide variations in gas flow rate, thus ensuring relatively constant collection efficiency.

Another special type of orifice scrubber is the venturi scrubber (Figure 22.27). Relatively high velocities between the gas and liquid droplets promote high col-

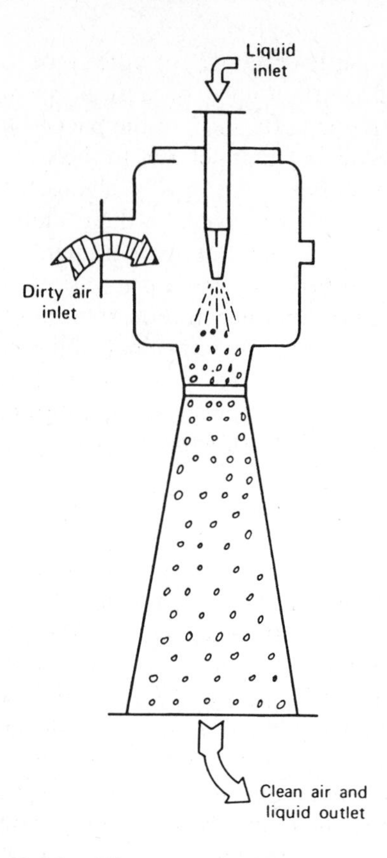

Figure 22.26. Ejector-type orifice scrubber (23).

lection efficiency in this unit. Such high velocities require a large energy input, raising operating costs to high levels, especially in applications processing large gas volumes. Venturi scrubbers typically use gas velocities of 200 to 400 ft/sec. At the venturi throat, the intensely turbulent action and shearing forces produce extremely fine water droplets that collect particulate matter very efficiently. At the instant of formation, these droplets move relatively slowly, and the fast-moving dust particles collide with nearly 100 percent efficiency with any droplets they may en-

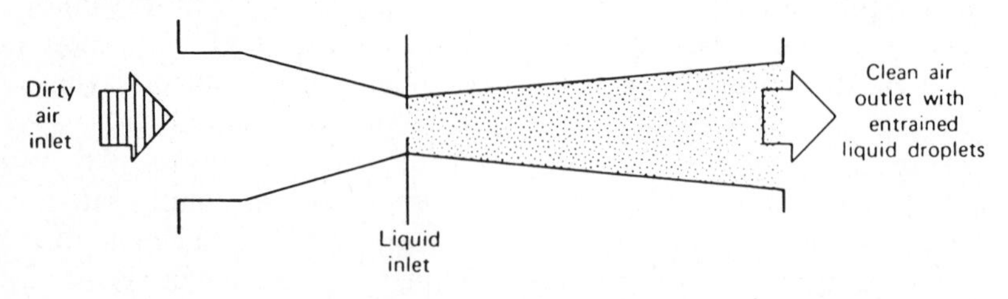

Figure 22.27. Venturi scrubber (23).

counter. In the diverging section of venturi scrubbers, the droplets are slowed in preparation for removal in downstream mist eliminators.

Ejector venturis use spray nozzles and high-pressure liquid to collect both particulate and gaseous contaminants and also to move the gas through the unit. These scrubbers have no fan to provide requisite gas flow. Thus fan power costs are eliminated, but relatively high energy consumption is still necessary because of the relatively high scrubber liquid pumping costs. Modifications to this type of device have featured sonic nozzles in the ejector venturi to give high collection efficiency while conserving somewhat on high liquid pumping costs.

The collection efficiency of the venturi scrubber increases with increased energy input. Usually a venturi scrubber operating in a pressure drop range of 30 to 40 in. H_2O is capable of an almost total collection of particles ranging from 0.2 to 1.0 μm.

9.5 Mist Eliminators

Depending on the type of the scrubber used, the number and size of water droplets entrained in the gas stream emerging from the scrubber may present objectionable stack gas discharge. If adequate mist elimination is not practiced, "raining" of contaminated scrubbing liquid droplets near the discharge stack can mean noncompliance with both mass emission and opacity regulations.

Normally the droplets emerging from a scrubber with inadequate mist elimination contain dissolved gases and/or captured particulate matter. Depending on the size of these droplets, fallout may occur a considerable distance beyond the point of discharge. In some cases, depending on ambient conditions and on the degree of saturation of the plume emerging from the scrubber, the droplets either grow because of further condensation or evaporate, thus liberating the captured contaminant.

Generally droplets formed from tearing of liquid sheets of water (e.g., in orifice-type scrubbers) are 200 μm in diameter and larger. In extremely high-energy applications, droplets as small as 40 μm may be produced. In low-energy systems using coarse spray nozzles, the droplets may be as large as 1000 to 1500 μm. Clearly the type of device that must be used to ensure adequate mist elimination is partly a function of the size of the droplets.

Mist eliminators are devices that are designed to use physical separation processes, which depend on gravitational and inertial forces to "knock out" the droplets from the emerging gas stream. One such device appears in Figure 22-28. Devices used include packed beds, cyclone collectors, gravitational settling chambers, and chambers containing baffles that make a tortuous "zigzag" path for the gases to follow, while more inert water droplets impinge on the solid surfaces. No matter what type of device is used, once the scrubbing liquid droplets are captured, if the mist eliminator is operated at proper velocities, the droplets coalesce and drain down to some quiescent zone in the mist eliminator for subsequent treatment.

One common problem with some mist eliminator designs is reentrainment. If the liquid load is too great for the mist eliminator, the onset of reentrainment

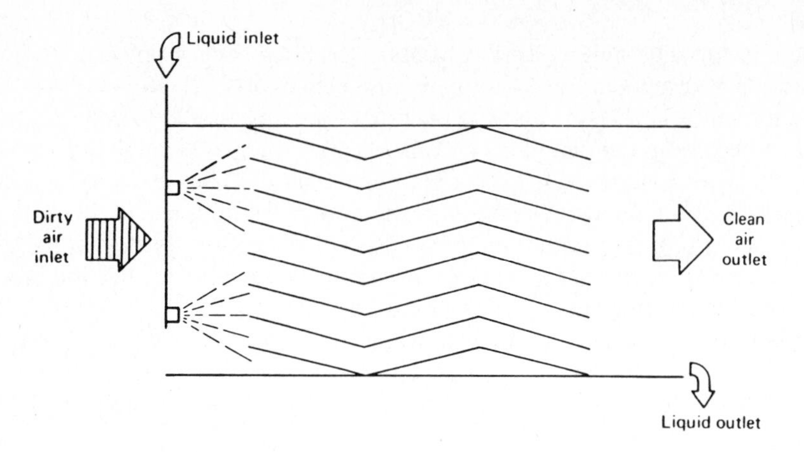

Figure 22.28. Baffle-type mist eliminator (23).

begins at lower than normal gas velocities. That is, for a given gas velocity, as the amount of liquid removed from the mist eliminator increases, the efficiency remains relatively constant up until the capacity of the mist eliminator is exceeded; then efficiently drops markedly.

Cyclone-type mist eliminators function basically on the same principles delineated in Sections 7.5 and 7.6. Some departure from theory occurs, depending on reentrainment or shearing effects at the cyclone walls, causing the liquid film to be sheared and reconverted to atomized droplets if cyclone operation is improper.

The fiber-type mist eliminator is a relatively new device. The fibrous bed provides a tortuous path for the carrier gas laden with entrained droplets. In one type of application using fibrous mist eliminators, an annular cylinder is used within a concentric shell. The entering wet gases pass readily through the annulus, and the liquid drains down the inner walls of the annular cylinder. Gases pass up through the center and exit through the top of the mist eliminator. The water is collected at the bottom of the outer shell and is drained off.

Specification of any mist eliminator must involve consideration of the pressure drop that the mist eliminator will add to the total control system. Frequently manufacturers' specifications are the best source of such empirical data. When the emerging gas stream is saturated and still hot, liquid entrainment may have a problem if the gases are discharged to a relatively cool ambient environment. In such cases, it may be necessary to heat the exhaust gases to minimize production of steam plumes or condensation before the gases have a chance to mix and disperse in the atmosphere.

9.6 Design Parameters

The preceding sections discussed some design parameters for the particular types of scrubber delineated. Because of the largely empirical correlations, frequently rules of thumb are necessary to specify scrubbers either for unique applications or when the state of the art and data correlation for the particular unit are not yet sufficiently refined to permit accurate prediction of collection efficiencies.

There do exist, however, some concepts fundamental to nearly any type of wet scrubber designed to remove the general gaseous and/or particulate contaminant from the carrier gas stream. In the case of gaseous contaminant removal, the concepts of "transfer units" and "theoretically equivalent height of packing" form the basis of these correlations. In the case of particulate removal, inertial impaction parameters form the basis of predicting the size of droplets required to collect a dust of known size distribution.

9.6.1 Particulate Scrubbing

Particulate removal by wet scrubbing is accomplished by liquid droplet impaction and by mechanisms that depend on inertial forces. Statistical probability and impaction correlations are the basis of mathematical modeling in spray collector design. The impaction parameter is the basis of nearly all collection mechanisms that involve inertial impaction of particles on droplets. It also takes into account the departure of solid particles from gas streamlines in passing around, or colliding with, liquid droplets. A general rule of thumb, according to impaction correlations, is the higher the impaction parameter, the higher the collection efficiency.

9.6.2 Gaseous Scrubbing

In gaseous contaminant scrubbing, the design equations are based not on particle dynamics but on mass transfer characteristics. In its most general form, the design equation describing the height of the packed column required to perform a given gaseous separation is complex and involves mass transfer coefficients, vapor phase concentrations of the gaseous contaminant along the length of the column, and the tower cross-sectional area. The required height is essentially a function of the height of a "transfer unit" multiplied by the "number of transfer units" required. The relationships involved can be simplified so that for a given scrubbing efficiency, inlet gas flow rate, and entrained acid content, the volume of packing material required is completely determined if the scrubbing liquor flow rate is known and the mass transfer coefficient is defined.

10 ELECTROSTATIC PRECIPITATION

Electrostatic precipitation is a process by which particulate matter is separated from a carrier gas stream using electrostatic forces and deposited on solid surfaces for subsequent removal. Electrostatic precipitators find wide application in many segments of pollution abatement systems. One of the largest users of electrostatic precipitation is the electric power industry. The need to remove particulate matter from large volumes of combustion-exhaust gases produced by coal-fired boilers calls for a collector that can extract particulate matter efficiently with small consumption of energy. This all-important advantage renders electrostatic precipitation an extremely attractive control alternative from the viewpoint of low operating cost. Unfortunately the relatively large initial capital investment cost needed to construct and install these massive and complex steel structures may render the application somewhat less attractive from the point of view of annual cost.

10.1 Range of Performance

The high-voltage electrostatic precipitator has been applied at more installations emitting large gas volumes than any other class of high-efficiency particulate collecting device (6). Although the device can efficiently extract even submicrometer material, the precipitator commonly treats dusts whose particle size distributions indicate a maximum diameter of approximately 20 μm. In those cases, more coarse dusts are also included in the contaminated gas stream and the precipitator is often preceded by mechanical collectors.

Precipitators applied to the electric power industry have been designed to handle gas volumes ranging from 50,000 to 2×10^{-6} cfm (6). Both "cold-side" and "hot-side" electrostatic precipitators have been installed that efficiently remove particulate matter from gas streams ranging from ambient temperatures to 1650°F, respectively (24). Depending on the location of the precipitator in the control system (i.e., whether gas flows through the precipitator by way of an induced or forced-draft fan), the unit may function under negative or positive pressures ranging from several inches of water, gauge, to 150 psig (6). Characteristically, the pressure drop through electrostatic precipitators, mostly open chambers with nonrestrictive gas passages, is on the order of 0.5 to 2 in. H_2O, gauge.

Typically, one stage of an electrostatic precipitator provides from 70 to 90 percent collection efficiency. By providing sequential stages, the overall precipitator efficiency can be increased to very high values. For example, in typical installations abating fly ash of moderate resistivity, one electrostatic precipitator stage is capable of producing nominally 80 percent collection efficiency. The remaining 20 percent of the entering particulate matter can be removed to an efficiency of 80 percent, approximately, in a second stage. The removal, then, of 80 percent plus 16 percent in the second stage (80 percent of the dust penetrating the first stage) provides an overall collection efficiency of 96 percent for the two-stage precipitator. Additional stages can be installed. Using this estimate of collection efficiency, a six-stage precipitator would provide a precipitator collection efficiency in excess of 99 percent.

10.2 Mechanisms of Precipitation

The electrostatic precipitator relies on three basic mechanisms to extract particulate matter from entering carrier gases. First, the entrained particulate matter must be electrically charged in the initial stages of the unit. Second, the charged particulate matter is propelled by a voltage gradient (induced between high voltage electrodes and the grounded collecting surfaces), causing the particulate matter to collect on the grounded surface. Third, the particulate matter must be removed from the collection surfaces to some external container such as a hopper or collecting bin.

The high-voltage potential across the discharge and collecting electrodes causes a corona discharge to be established in the region of the discharge electrode, and this forms a powerful ionizing field. When the particles in the gas stream pass through the field, they become charged and begin to migrate toward the collecting surfaces under the influence of the potential gradient existing between the electrodes.

Once the dust is deposited on the grounded surface, easy removal from the collecting surface depends on many factors. Dust resistivity is one prime factor dictating the collection efficiency and degree of reentrainment that can be expected under a given set of design conditions. If dust resistivity is too low, particles once deposited on the grounded surface may lose their charge and reenter the gas stream from the electrode collecting surface. If resistivity is too high, particles once deposited can cause "corona quenching" or "back corona." In this condition, the voltage gradient between dust deposited on the plates and the high-voltage electrode begins to diminish because of the large voltage drop in the dust layer accumulated on the plate. This decreases collection efficiency, because the voltage gradient is all-important in determining the rate or migration of particulate matter toward the collecting surface.

In many electrostatic precipitators the dust is "rapped" from the collection surface by impairing a sharp blow to some area of the surface. The dust then cascades in a "sheeting action" down the collecting surface into a hopper. The extent to which this dust is removed in large pieces determines the degree of reentrainment upon rapping. Plate design, rapper design, and use of modular sections of precipitators are among the widely varying approaches employed by precipitator manufacturers.

10.3 Plate-Type Precipitators

Two basic types of electrostatic precipitator have been used: plate type and pipe type. In plate-type devices the collecting surface consists of a series of parallel, vertical grounded steel plates between which the gas flows (Figure 22.29). Alterations to the geometry of the plates are implemented by various manufacturers in an attempt to provide better collection and more efficient removal of collected dust upon rapping. Some plate-type precipitators are equipped with shielding-type structures installed vertically at incremental distances along the plate surface. These structures represent an attempt to minimize reentrainment as the dust sheet falls behind the structure. The design of these protuberances affects the geometry and intensity of the electrostatic field set up between the high-voltage electrodes and the plates. Thus the spacing, size, and shape have been the subject of years of developmental work and research sponsored by precipitator manufacturers.

Most commonly, the plate-type precipitator finds application in the collection of solid particulate matter or dust. Plate-type precipitators consume approximately

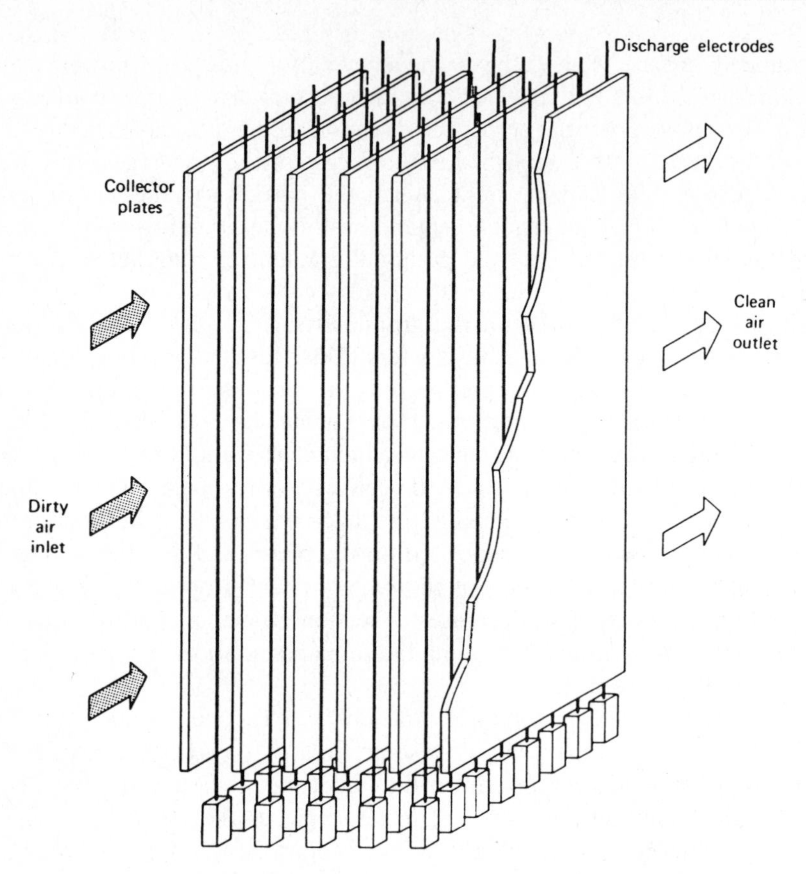

Figure 22.29. Plate-type electrostatic precipitator.

200 W per 1000 cfm. This extremely low power requirement is affected also, however, by the resistivity of the dust and the geometry of the precipitator.

Sometimes the dust is removed from plate-type precipitators by washing. Strategically placed nozzles outside the electrostatic field are timed for actuation at various intervals, supplying irrigation to flush away dust such as sticky or oily particulate matter that is especially difficult to remove.

One disadvantage of plate-type precipitators with smooth plates (i.e., not equipped with a certain type of protuberance) is that they permit the gradual creepage of deposited particulate toward the precipitator outlet. Collected dust may be gradually moved by aerodynamic drag from the traveling gas stream, or particles may "jump" along the plate, depending on dust resistivity. In general, capital investment costs of the plate-type precipitator may be lower because of the simplicity of plate design as compared with pipe-type precipitators. The exact design, however, dictates actual assembly and construction costs.

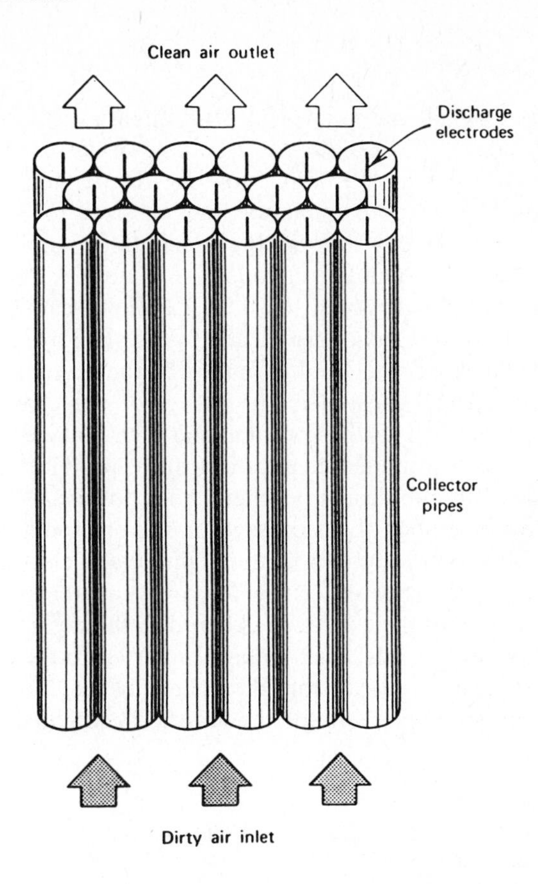

Figure 22.30. Pipe-type electrostatic precipitator.

10.4 Pipe-Type Precipitators

Pipe-type precipitators generally feature high-voltage electrodes positioned at the center line of a grounded pipe (Figure 22.30). The gas flows through a bank of these pipes parallel to the high-voltage wire. Dust deposited on the inner walls of the grounded pipe is removed in much the same manner as in the plate-type precipitator, that is, by rapping.

Some studies indicate that the pipe-type precipitators should have superior electrical characteristics (16). The pipe-type precipitator is prone, however, to creepage of collected dust toward the clean gas exit. In general, the design parameters for the specification of pipe-type precipitators may vary in the actual numerical values used to specify the number of pipes. The theoretical models on which sizing correlations are based, however, still include consideration for dust resistivity, total plate collection area, and gas velocity through the pipe.

10.5 Design Parameters

Precipitator size is normally based on a predetermined estimation of the resistivity of the dust, knowledge of the volumetric flow rate of gas to be processed by the precipitator, and an estimate of the dust loading. Depending on dust resistivity, the cross-sectional area of the precipitator may be sized for the known gas flow rate to provided a line velocity through the precipitator parallel to the collection surface of 2 to 6 ft/sec. The number of collecting plates (i.e., plates that form gas passages or channels through which the dirty gas flows parallel to the plates) is fixed by this calculated cross-sectional area and by the plate spacing. In conventional precipitators, the plate spacing is predetermined by the maximum voltage to which the high-voltage electrode, located in the center of the channels, is subjected. Typically, 45,000 V applied to these center-line high-voltage electrodes, located typically $4\frac{1}{2}$ in. from the grounded electrode surface, produces 10 k/V·in. voltage gradient. This rule-of-thumb design perimeter then forms the basis for the determination for the total number of gas passages in the precipitator and all necessary hardware needed to support and power the high-voltage electrodes.

Manufacturers claim higher efficiency for a given area of collecting surface depending on the design of the collecting electrodes, the design of the high-voltage electrodes, the wave form of the high voltage imparted to the electrodes, and the mechanism of particulate removal from the collecting surfaces. The collection efficiency of an electrostatic precipitator, however, is estimated using the Deutsch equation (31):

$$E = 100\left(1 - \exp\frac{-AW}{Q}\right) \tag{18}$$

where E = collection efficiency (percent)
A = area of collecting electrodes (ft^2)
Q = gas flow rate (acfm)
W = particle migration velocity (ft/min)

Generally, the migration velocity is an empirically determined sizing factor for a given application, grain loading, and dust resistivity. This experience factor is often a closely guarded number among precipitator designers and manufacturers.

The design parameter of dust resistivity is a prime component in determining the migration velocity W. Dust resistivity measurements are frequently made on site. Resistivities vary from 10^9 to 10^{15} ohm-centimeters and are a function of the nature of the dust and the temperature. Some precipitator designers analyze collected dust in the laboratory for dust resistivity, yet dust resistivity measurements directly from the gas stream are indispensable in providing accurate design information. For example, it is not uncommon to find one to three orders of magnitude difference in measured dust resistivity when comparing field measurements to laboratory measurements of collected dust.

Rules of thumb indicate that the migration velocity W varies from 2 to 14 cm/

sec, depending on the aforementioned design parameters. Typical values for coal-fired boilers range from 6 to 10 cm/sec. Clearly, from the foregoing equation, the required collecting plate area for a given gas volume and collection efficiency is inversely proportional to the migration velocity. Therefore, choosing a conservative migration velocity of 6 as opposed to 10 cm/sec requires approximately 1.7 times the plate area.

11 GAS–SOLID ADSORPTION

Adsorption denotes a diffusion process whereby certain gases are retained selectively on the surface, or in the pores of interstices, of specially prepared solids. The principle has been used successfully for many years in the separation of mixtures and in product improvement through removal of odorous contaminants or through decolorizing.

In many applications, the primary purpose of adsorption is to render an operation economical by recovery of valuable materials for reuse or resale. A good illustration of this is the long-standing practice in several industries of recovering organic solvent vapors by adsorption. In such applications, the adsorbed materials may be "desorbed" by a variety of methods and concentrated and purified for reuse. The adsorbing solid is regenerated and can be resumed for many such cycles.

11.1 Range of Performance

In atmospheric pollution control as practiced today, the principle of adsorption is employed primarily to prevent odorous and offensive organic vapors from escaping into populated areas. A typical example is the removal of vapors generated by printing plants in large cities. Occasionally, however, adsorption is used to remove trace concentrations of highly toxic gaseous materials, such as radioactive iodine vapors.

Fixed-bed adsorbers are used when trace concentrations are encountered, whereas for heavier concentrations, regenerative adsorbers are needed. Product recovery from regenerative adsorbers (e.g., solvent recovery in the dry-cleaning field) may pay for the emission control system and its operation, or at least offset part of the annual cost.

Recently, carbon bed adsorption techniques have been applied to the goal of lessening the impact of energy use on processes that rely on incineration techniques to abate combustible process emissions. In earlier applications of incinerators, airstreams laden with vapors containing organic material have been directly incinerated; some of the incinerators use heat recovery to minimize the amount of auxiliary fuel required to support combustion and provide efficient abatement. In practice, either the organic vapor concentration in the airstream's incinerator was low due to process dynamics, or the concentration of organic vapors was necessarily limited to comply with safety code requirements (e.g., maintaining the concentration of combustible vapors at 25 percent of the lower explosive limit).

In recent applications, carbon bed adsorbers have been used to adsorb organic vapors over prolonged periods (e.g., 2 to 8 hr). Upon saturation, the bed can be desorbed using saturated or superheated steam. The effluent from the bed, consisting mostly of water vapor and concentrated organic compounds, can be sent to an intermittently fired incinerator, which is designed to operate only during bed desorption. In most cases, the concentration of organic material has been raised sufficiently so that by mixing auxiliary air at the point of combustion, the combustion process sustains itself with little or no auxiliary fuel. Furthermore, because the desorbed stream consists of steam and organic compounds, the danger of explosion is eliminated in most cases (because the amount of air in the desorbed stream is very small, the concentration of organics is therefore above the upper explosive limit).

The nuisance aspect of industrial odorous emissions is becoming a more significant factor in the network of developing emission regulations. Characteristically, odors can be produced by organic compounds at very low concentrations and emission rates, which means that the odorous emission is not limited by organic compound emission regulations. Such low emissions are ideally suited to adsorption processes, especially those utilizing activated carbon. Furthermore, because the absolute amount of odorous substances is usually so low, the adsorbent can be used to extended periods and discarded with no significant impact on system operating cost. Thus adsorption systems have found wide use in rendering plants and industrial processes emitting high molecular weight hydrocarbons.

In general, the range of performance of gas–solid adsorption systems is related to many factors controlling the effectiveness of physical adsorption in dynamic systems. These factors include surface area of the adsorbent, the affinity of the adsorbent for the adsorbate, the density and vapor pressure of the adsorbate, the concentration of the adsorbate, the system temperature and pressure, the dwell time, the bed packing geometry, the thermodynamics of adsorption of solvent mixtures, and the mechanism of removal of heat of adsorption (13). All these factors are interrelated, requiring careful analysis and specification before installing or even considering a carbon adsorption system as a viable pollution abatement system in a particular application.

Carbon adsorption has been used in many areas of the food processing industry: the handling and blending of spices, canning, cooking, frying, and fermentation processes, for example. In the manufacture and use of chemicals, process discharges have been abated by carbon adsorption to minimize objectionable emissions in the manufacture of odorous substances such as pesticides, glue, fertilizers, and pharmaceutical products, in paint and varnish production, and in the release of odorous vapors emitted to the atmosphere from tanks during filing operations (25).

Aside from activated carbon, gas–solid adsorbers use other materials for reversible or irreversible removal of gaseous constituents. Among these are alumina, bauxite, and silica gel (26). Activated carbon, however, constitutes one of the most popular types of solid adsorbent, mostly because of its low cost and its ability to be regenerated.

11.2 Physical Adsorption

Physical adsorption denotes a type of adsorption process wherein multiple layers of adsorbed molecules accumulate on the surface of the adsorbing solid. In this process, the number of layers that can be made to accumulate on the activated carbon surface is related to the concentration of the adsorbate in the gas stream (27).

Physical adsorption processes generally entail reversible processes. In the case of adsorption of organic solvents, for example, on the surface and interstices of activated carbon, the rate at which adsorbate molecules are "driven" from the relatively homogeneous gas phase into the pores of the activated carbon depends to some extent on the concentration of the gas. At the surface of the adsorbing solid, molecules are attracted and deposited, providing a concentration gradient useful in causing migration of solvent molecules into the activated carbon pores. From the standpoint of residence time and the ease of saturation of activated carbon, therefore, solvent concentration in the gas phase partly determines the amount of activated carbon required and the optimal placement or geometry of the carbon relative to the flowing gas stream. The more molecular layers that can be accumulated on the activated carbon interstitial area, the less carbon is required per pound of solvent to be adsorbed. Conversely, for a given residence time of contaminated gas in an activated carbon bed, the higher the concentration, the higher will be the extraction efficiency. These considerations apply in physical adsorption processes, whether the adsorbing solid is activated carbon, a siliceous adsorbent, or a metallic oxide adsorbent.

In the case of activated carbon adsorption, the micropore structure of the carbon surface provides an extremely large surface area for a given weight of carbon; typically, several hundred thousand square feet of available surface is distributed throughout the interstices of 1 lb of activated carbon (25). Generally, the adsorptive capacity of a particular activated carbon is measured by its ability to retain a given chemical entity at various concentrations of that chemical species. These empirical data are normally obtained by laboratory investigations performed by the suppliers of activated carbon. The data are useful for specifying the equilibrium concentrations theoretically obtainable at a given set of conditions, assuming that sufficient residence time is available in the bed (i.e., that sufficient carbon is present for a specified gas flow rate and velocity). From adsorption "isotherms" (supplied by activated carbon manufacturers), the minimum amount of carbon required can be determined for a given adsorption cycle time. The amount of solvent adsorbed on the carbon at this "saturation point" can be theoretically achieved, but only at the expense of low collection efficiency near the end of the adsorption cycle. Adsorption isotherm data do not necessarily specify adsorption efficiency, for efficiency may vary with time since last regeneration or composition of mixtures being absorbed, as well as with bed geometry and solvent-laden gas flow rate through the bed.

As a rule, the particular geometric design and arrangement of activated carbon in an activated carbon bed adsorber is best specified by well-informed engineers who specialize in adsorption equipment. Even then, pilot plant work may be required before a full-scale carbon bed absorber can be specified or designed. Generally, bed geometrics can be arranged in layers, beds, cylindrical canisters, or a variety of other shapes to provide optimal contact time for a given concentration of material and a given gas flow rate.

11.3 Polar Adsorption

Polar adsorption denotes a gas–solid adsorption process whereby molecules of a given adsorbate are attracted to and deposited on the surface of an adsorbing solid by virtue of the polarity of the adsorbate. Some siliceous adsorbents depend on the polarity of gas molecules for selective removal from a gas stream. For example, synthetic zeolites are adsorbents that can be produced with specific uniform pore diameters, giving them the ability to segregate molecules in the liquid or gas stage on the basis of the shape of the adsorbent molecule (25). Adsorbents of this type will not even adsorb organic molecules of the same size as their pores from a moist airstream, for the water molecules are differentially adsorbed because of the chemical structure of the adsorbing solid (25).

11.4 Chemical Adsorption

Chemical adsorption takes place when a chemical reaction occurs between the adsorbed molecule and the solid adsorbent, resulting in the formation of a chemical compound (27). Chemical adsorption may be contrasted with physical adsorption, where the forces holding the adsorbate to the solid surface are weak. In chemical adsorption, usually only one molecular layer is formed, and the chemical reaction is irreversible except in cases when the energy of reaction can be applied to the solid adsorbent to reverse the chemical reaction.

Many of the same factors useful in sizing the specifying physical adsorption equipment apply in chemical adsorption. As in applications featuring the removal of low-concentration, high molecular weight, odorous substances, the contaminated adsorption bed is usually discarded or shipped for external regeneration after its efficiency has degraded. The cost of operating chemical adsorption equipment may be prohibitive in applications involving either high flow rate or a concentration of adsorbate in the incoming gas stream that causes the bed to deteriorate quickly. Alternatively, chemical adsorption systems can be applied to streams where the contaminant is present in high concentrations but is emitted only intermittently.

11.5 Design Parameters

One of the first steps in specifying the quantity of activated carbon required for a given cycle time in an activated carbon absorber is to establish what chemical species and concentrations are present in the incoming gas stream. Adsorption isotherms are available from activated carbon suppliers either for individual chemical constituents being adsorbed, or for chemical species that are similar enough in chemical structure or physical properties to enable reasonable approximations regarding adsorbability and retention. Such empirical data, established by the suppliers, apply for the single chemical species being adsorbed on a given type of activated carbon. When mixtures of organic species are being processed, it can be expected that the higher molecular weight substances are preferentially adsorbed, and lower molecular weight species are adsorbed initially to a lesser efficiency than would be predicted if the lower molecular weight species were adsorbed alone. As the adsorption cycle proceeds, the effluent can be expected to be rich in low molecular weight compounds from the mixture, and high molecular weight compounds will be adsorbed at the highest efficiency.

Once the minimum bed size based on a given cycle time and equilibrium retention relation is calculated, the bed geometry and actual amount of carbon must be determined by consideration of the retention time in the absorber for a given adsorption efficiency, sufficient capacity to ensure an economical service life, low resistance to airflow to minimize overall system pressure drop, uniformity of airflow distribution over the bed to ensure full utilization, pretreatment of the air to remove particulate matter that would gradually impair and poison the bed, and provision for some manual or automatic mode for bed regeneration (25).

The geometry and depth of the absorbing bed depend primarily on the superficial gas velocity through the bed, the concentration of contaminants in the incoming airstream, and the cycle time. As adsorption continues, the portion of the activated carbon bed first encountered by the incoming gas becomes rich and eventually fully saturated in adsorbate. Deeper areas of the bed are progressively less and less saturated with adsorbate until some small but finite level of penetration of adsorbate can be observed at the bed exit. As the adsorption cycle continues, this "adsorption wave" proceeds through the bed until the concentration of the exit gas reaches a finite value in excess of allowable emissions. At this point "breakthrough" is said to have occurred, and it is necessary to regenerate the bed. By arranging the given necessary amount of activated carbon in an optimal geometry, breakthrough can be postponed to the point at which minimum utilities are required for bed desorption and the maximum practical level of adsorbate is present in the desorption stream.

It is not possible to set down rules of design encompassing all applications of carbon bed absorption to pollution abatement processes. Because of the wide variation in possible solvent mixtures and concentrations, the bed geometry is usually specified from previous experience or extrapolated from performance data obtained in a similar type of installation.

12 COMBUSTION

Combustion, sometimes termed "incineration," oxidizes combustible matter entrained in exhaust gases, thus converting undesirable organic matter to less objectionable products, principally carbon dioxide and water.

In many instances combustors are the most practical means for bringing equipment into compliance with air pollution control regulations. These devices have been employed to reduce or eliminate smoke, odors, and particulate matter in cases when such emissions might exceed the limits set by law.

Equipment in operation today falls essentially into two classifications, direct flame combustion and catalytic combustion. Direct flame incineration has been more widely used and has, in general, been more successfully applied. However, properly designed catalytic incineration units that perform satisfactorily have been constructed.

Catalytic burners differ from direct flame burners in that they allow organic vapors to be oxidized at temperatures considerably below their autoignition point and without direct flame contact upon passage through certain catalysts.

With the increased cost of gas and oil, efficient heat recovery systems must be employed with either catalytic or direct flame burners to make them economically sound. The arrangement of these devices is limited only by the type of operation and by other plant facilities. Effluent gases from these units often are used in dryoff ovens, water heaters, space heaters, and many other types of heat exchanger, to make use of their content.

12.1 Range of Performance

Prior to the energy crisis in the early 1970s, any organic emissions not economically recoverable using the control techniques outlined in the preceding sections were often subjected to direct incineration. With the advent of higher energy costs and the unavailability of auxiliary fuel to support the combustion, the range of performance of combustion equipment has become the subject of innovative designs.

The myriad of combustible organic compounds emitted to the atmosphere from our numerous manufacturing processes calls for the application of carefully designed combustion equipment. The design must be based on the particular contaminants being combusted, the concentration of these contaminants in the carrier gas stream, and the flow rate of the gas stream, which affects the physical size of the combustion equipment.

The specifically designed and engineered systems used to control the process emissions are as diverse as the industrial operations emitting combustible compounds. Typical examples of processes requiring combustion as a means for the control of hydrocarbon emissions are industrial drying processes, baking of paints, application of enamels and printing ink, application of coatings and impregnates to paper, manufacturing of fabric and plastic, and manufacturing of paints, varnishes, and organic chemicals, as well as synthetic fibers and natural rubber (28). In all these processes, flares, furnaces, and catalytic combustors have found suitable application, and the materials emitted may be oxidized to odorless, colorless, and innocuous carbon dioxide and water vapor (28).

The incineration of gases and vapors has found wide application in the control of both gaseous and liquid wastes. In the area of gaseous emissions, use of incineration to minimize odorous pollutants is often the only available control alternative. Because most odorous compounds are characteristically low in concentration, the choice of carbon bed adsorption, for example, may prove to be economically or even technologically infeasible because of low control efficiencies for certain compounds.

The opacity of plumes emitted from various processes has historically been reduced by passing the otherwise highly visible emission through a combustion apparatus. Examples include application to coffee roasters, smokehouses, and some enamel baking ovens. In addition, afterburners are sometimes used to heat otherwise wet and highly visible stack gases that would emit a harmless but visually objectionable steam plume.

In some applications the organic gases and vapors being emitted from a process are classified as "reactive hydrocarbons." These compounds, when discharged to the atmosphere, are known to contribute to the formation of smog and otherwise irritating compounds. Afterburners can be used to convert most of these reactive hydrocarbons into carbon dioxide and water vapor.

Combustion equipment has been used extensively in refineries and chemical plants as a method of disposal for unusable waste and as a method of reducing explosion hazards. In these applications, ultimate destruction using a combustion-based process must be carefully designed to avoid creating a hazard to nearby, highly flammable storage tanks or to the process itself.

12.2 Flares

A flare is a method for the efficient oxidation of combustible gases when these are present in a stream that is within or about the limits of flammability (28). Flares usually find their widest application in petrochemical plants, especially in the disposal of waste gases that are often mixed with other inert gases such as nitrogen or carbon dioxide. Additionally, many of the chemical processing plants that produce, use, or otherwise discharge highly dangerous or toxic gases are often equipped with flares designed to be activated under emergency conditions, in the event that such toxic gases should require immediate discharge.

Generally, flares are designed after thorough analysis of the gas stream being incinerated. Flare heights are established by taking into account the heat and light emitted from the flare, and they are supposed to ensure sufficient mixing to prevent unburned organic compounds that may penetrate the flare from presenting a hazard or a nuisance to the surrounding area.

Often when flares are applied to installations where the concentration of combustible gases is not sufficient to ensure or maintain the persistence of the flare, auxiliary fuel must be added to provide for efficient combustion and sustained burning.

In some applications, where the materials being combusted are difficult to burn, auxiliary air must be introduced at the point of combustion at the top of the flare, using steam jets or other satisfactory means to ensure good fuel–air mixing.

One design parameter in the sizing and specification of flares is the ratio of hydrogen to carbon in the materials being combusted. As an example, in low molecular weight, aliphatic hydrocarbons (e.g., methane) the high ratio of hydrogen to carbon guarantees burning with very little, if any, soot or smoke. In contrast, double-bonded or triple-bonded hydrocarbon molecules (e.g., acetylene) burn with a very sooty flame because the ratio of hydrogen to carbon is low. Normally the discharge from a flare should be limited by design to carbon dioxide and water vapor and should certainly exclude the emissions of visually objectionable products of combustion such as carbonaceous soot.

When condensable mists are present in the gases to be flared, inertial separators should be provided at the base of the flare to permit automatic separation and automatic drainage of these liquefiable compounds, which may otherwise flow back through the flare and down its walls, causing flashback and/or an explosion.

12.3 Direct Flame Combustion

Direct combustors represent a broad classification of combustion equipment in which auxiliary fuel is added and heat recovery may or may not be practiced to ensure efficient combustion and minimization of operating costs. The designs of direct combustors are as varied as the manufacturer's experience. Generally, direct combustors rely on three factors to provide efficient conversion of the organic compounds to carbon dioxide and water:

1. Sufficient time in the incinerator to allow complete conversion to carbon dioxide and water vapor, all other conditions being specified
2. Sufficient fuel to ensure combustion temperatures high enough to permit complete combustion within the allowable residence time in the incinerator
3. Sufficient turbulence in the incinerator to provide for good mixing and contacting of the combustible materials with the active flame front

Direct combustors equipped with heat-exchange units are often used to preheat the incoming gases to a temperature high enough to ensure more complete and efficient combustion and to save on total energy consumption needed to preheat the incoming gases to the combustion temperature. This heat exchange can be derived in a variety of ways, but it essentially causes the removal from the incinerator discharge of waste heat that would otherwise pass into the atmosphere, and recycles this heat to the incoming gases by way of some physical medium. Heat exchange efficiencies range from 20 percent to more than 90 percent depending on the design of the heat exchanger, the pressure drop through the device, and the materials of construction; these variables in design also affect, of course, the initial capital investment cost and the operating costs.

Direct combustors are also useful in the removal of combustible particulate matter, particularly that which is odorous. When such particulate matter is of submicrometer size, this method of control is often the only available or most technologically and economically feasible technique.

Incinerators are also quite widely used because they occupy relatively little space, have simple construction, and generate low maintenance requirements (6).

Multiple-chamber direct combustors provide a useful means for increasing combustion efficiency and overall conversion to carbon dioxide and water for a given quantity of fuel input. The multiple chambers provide additional residence time and therefore additional combustion efficiency, which may be required to convert difficult-to-burn organic compounds to carbon dioxide and water.

12.4 Catalytic Combustion

In catalytic combustors, reactions occur that convert compounds into carbon dioxide and water on the surface of a "catalyst," usually composed of platinum or palladium, with the end result that auxiliary fuel costs are minimized.

Catalytic combustion has come into wide use in installations where concentrations of hydrocarbons are relatively low and large amounts of auxiliary fuel would be required to sustain combustion. The systems do not find wide applicability in processes involving an inlet gas stream that can contain materials capable of "poisoning" the catalyst, thus rendering it ineffective. Additionally, systems containing high amounts of particulate matter often cannot be incinerated using catalytic systems, or else they must be equipped with high-efficiency precleaners to prevent the particulate matter from coating the surface of the catalyst, making it ineffective.

Basically, a catalytic system includes a preheat burner, exhaust fan, and catalyst elements, as well as control and safety equipment (31). The preheat burner usually raises the incoming gas stream to a temperature of 700 to 900°F. As the heated gases pass through the catalyst bed, the heat of reaction raises the temperature further to levels comparable to those found in direct combustors. The obvious advantage of minimal fuel input renders catalytic incineration an attractive alternative where applicable.

Typically, for a 10,000-scfm system, capital investment costs for catalytic incinerators may be approximately the same as direct flame incinerators; however, annual fuel costs may be reduced to about 2 percent of flame-type incinerators depending on hydrocarbon concentration and incoming gas temperature (31).

Catalytic incinerators may also be equipped with heat recovery devices, which may be used to recapture otherwise wasted heat and inject it into the incoming gas stream, thereby minimizing the already low preheat fuel requirements.

Even when well-controlled with precleaners, catalytic incinerators require periodic washing and maintenance to remove particulate matter that accumulates after long operating periods. Additionally, catalysts often require periodic reactivation because materials such as phosphorus, silica, and lead, even when present in trace amounts, shorten the active life of the catalysts (29).

12.5 Design Parameters

Depending on the type of incinerator selected, the sizing and specification of the unit rely primarily on thermodynamic calculations of auxiliary fuel requirements and system residence times. In general, the calculations are based on the simple concept that all incoming materials must be preheated to the combustion temperature, generally in excess of 1200°F and usually 1500°F, so that conversion to carbon dioxide and water vapor occurs if sufficient residence time is provided for the heated constituents to interact with the provided oxygen.

When concentrations of hydrocarbons in the gas stream are low, the specific heat of the gas stream may be regarded as equal to that of air. In general, the rate of heat input Q required to raise the temperature of the incoming gas from inlet conditions to 1200°F is expressed simply by

$$Q = SCFM\ \rho_g C_p\ (1200 - T) \tag{19}$$

where Q = required rate of heat input (Btu/min)
$SCFM$ = gas flow rate at standard conditions (cfm)
ρ_g = gas density at standard conditions (lb/ft^3)
C_p = average specific heat for air over temperature range (Btu/(lb)(°F))
T = incoming gas temperature (°F)

To provide the requisite heat input, the total amount of auxiliary fuel may be determined by considering the heating value of the fuel being used as well as the heating value of any organic compounds present in the incoming stream. Assuming that the incoming stream has an average concentration C_a of hydrocarbons (lb/ft^3), and that this stream of hydrocarbons provides Q_{hc}, the quantity of heat per pound of hydrocarbons, the available heat from the organic content of the stream Q_{oc} is calculated by the following equation:

$$Q_{oc} = SCFM\ C_a Q_{hc} \tag{20}$$

Now, subtracting this available energy from the preheating requirements, the total amount of heat required Q_{req} may be computed if the heating value of the available fuel Q is known:

$$Q_{req} = Q - Q_{oc} = AF \times Q_f \tag{21}$$

where AF = auxiliary fuel rate required (lb/min)
Q_f = heat content of auxiliary fuel (Btu/lb)

If heat recovery equipment is available, the quantity of available heat may be subtracted directly from the heat computed in Equation 21 and the fuel requirements thereby proportionally reduced. In practice, even though the heating requirements are small or even negative according to these simplistic energy balances, it is likely that auxiliary fuel will still be required because no account has been made in this analysis for heat losses in the total system.

Thermodynamic calculations therefore are useful in arriving at fuel requirements; however, the specification of total residence time or degree of turbulence involves an experience factor available from the supplier of the combustion equipment. As a rule, though, a residence time of 0.5 to 1 sec in the primary incinerator with at least 0.2 sec in its secondary chambers for direct flame combustion methods is considered to be adequate reaction time to convert oxidizable substances to carbon dioxide and water vapor, presuming sufficient or excess oxygen. The residence time required for efficient combustion in catalytic incinerators is reduced by at least one-fifth to one-tenth of the time required for direct flame combustion (30). This residence time must be computed knowing the cross-sectional area of the incinerator and the volume of gases in actual cubic feet per minute as determined from the combustion temperature.

To provide adequate turbulence, the internal design of the incinerator often features baffles and/or relatively tortuous turns. This allows good mixing while minimizing system pressure drop. The gas velocity required for good mixing is considered to be about 2100 ft/min (30).

13 SELECTING AND APPLYING CONTROL METHODS

The preceding sections of this chapter have presented and summarized the regulatory, economic, and technical considerations involved in the selection of the optimal method of emission control for a specific stationary source. The application of these basic factors in an orderly sequence of decisions may require not only a carefully conceived and applied methodology, but a perspective broader than that of this chapter.

The selection of the "best" control method for a specific emission source can be a very simple and obvious choice, or it can entail a complex decision-making procedure, especially important when the expected cost of control—usually related to the magnitude and strength of the emission—is large relative to the financial resources of the owner or operator of the source.

Because of technical uncertainties and ever-present alternatives, it would be very desirable to have an inclusive, systematic approach to the selection of a control system for a given application. Such a strategy should combine the type of information and methods presented in this chapter with an even more basic understanding of certain very crucial steps in a whole sequence of decisions necessary to ensure success of any emission cleanup effort. The ultimate performance of an emission control system rests very heavily on the avoidance of oversights during one of the planning steps, as well as the proper use and understanding of the basic fundamentals for selecting optimal control methods. Sound and accurate emissions data are essential, and skillful interpretation of all available data can spell the difference between a control system that functions efficiently and one that performs only marginally.

The overall strategy that should be incorporated when addressing the need to control a stationary emission entails (31) defining the emission limit, identifying all related emission sources, investigating process modifications, defining the technical aspects of control problems, and selecting the optimal control system.

13.1 Defining the Emission Limit

The emission limit that applies for any given pollution control problem is the most basic information and, indeed, a building block with which to develop and specify the optimal control system. It is important to understand that emissions are not always covered by an established legal or regulatory limitation. At this writing, the EPA has established emission standards for hazardous air pollutants and also has specified maximum permissible emission levels of the "criteria pollutants" (particulate matter, carbon monoxide, sulfur dioxide, nitrogen oxides, and volatile organic compounds) for about 58 categories of new sources. Although these and related standards will continue to be promulgated, it is highly unlikely that federal standards will cover most of the industrial situations for which emission "controls" are needed or being planned. State and local standards also may not apply specifically to a given plant emission. Nevertheless, virtually all emissions require control under widely applicable and generally phrased "nuisance," "opacity," or "odor" regulations. Others logically demand control if they constitute a substance capable of being a potential toxic hazard.

In these cases, the responsible design engineer will have to determine an emission level without simple reference to a regulatory action. Sources of emissions that are likely in violation of nuisance, opacity, or odor limitations call for expert analysis, for all three areas could require the application of the most recent technology to assess the magnitude of the problem and to specify a viable abatement strategy. In some cases the engineer should seek the advice and counsel of a toxicologist or industrial hygienist. Such professional specialists can sometimes give a reasonable exposure level for the general population that reflects the stability of the material and the possible chronic effects of environmental exposure. A note of caution should be injected here with respect to threshold limit values published by the American Conference of Governmental Industrial Hygienists (4). These values were derived and are intended for the control of worker exposures; they cannot be used reliably to determine relative toxicities, to evaluate air pollution problems, or to assess the effects of continuous exposures. The relation between industrial hygiene and air pollution is indeed a close one, but it does not include an indiscriminate cross-referencing of TLVs with respect to ambient levels or emission limits for stationary source controls.

13.2 Identifying All Related Emission Sources

Many times when emission controls are specified, it is later found that all emission sources have not been included in the "process envelope" of the control system because of insufficient care in the planning stage.

Normally, it is not enough to attach a well-designed control device to the main exhaust vent emitting the pollutant. With the very low emission levels required for the more toxic materials, emissions from sources other than the main process vent can, in total, overshadow even those from the main exhaust. It is good practice, particularly when dealing with toxic pollutants, to study the entire process and identify all points of emission and all possible solutions to the control problem. Some frequently overlooked emission points that have been found to contribute heavily to process emissions are as follows (33):

1. Accidental releases
   a. Spills
   b. Relief valve operation
2. Uncollected emissions
   a. Tank breathing
   b. Packing gland or rotary seal leakage
   c. Vacuum pump discharges
   d. Sampling station emissions
   e. Flange leaks
   f. Manufacturing area ventilation systems
3. Reemission of collected materials
   a. Vaporization from water wastes in ditches or canals
   b. Vaporization from aeration basins
   c. Reentrainment or vaporization from landfills
   d. Losses during transfer operations

13.3 Process Modification

This discussion of various air-cleaning methods began, as it always must, with a suggestion that process and system control represents a sound first step in making decisions in the selection and specification of emission controls. If the problem can be solved at its source, this approach offers several advantages, including a typical economic advantage, over the installation of some added or incremental type of control system.

Process modification is usually a most economical way to reduce emissions because little or no capital is needed to purchase control equipment. In addition, there can be improvements in operating efficiency that will reduce material or energy losses, and the cost of a terminal control system, if one is ever required, can also be cut because it has to handle less material. It is important to retain the option of process modification from the very beginning to avoid costly repetition of an engineering analysis once the add-on controls have been specified and the cost has been estimated to be exorbitantly expensive. Some of the techniques (Section 4) that can be helpful include the following (31):

1. Substitution of a less toxic or less volatile solvent for the one being used
2. Replacement of a raw material with a purer grade, to reduce the amount of inerts vented from the process or the formation of undesirable impurities and by-products
3. Changing the process operating conditions to reduce the amount of undesirable by-products formed

4. Recycling process streams to recover waste products, conserve materials, or diminish the formation of an undesirable by-product by the law of mass action
5. Enclosing certain process steps to lessen contact of volatile materials with air

13.4 Defining the Technical Aspects of the Control Problem

An important final step before selecting a control method is to define the properties of the exhaust gas stream; the basic data needed were reviewed in Section 5. Such data can come from several sources, however, and the information obtained must be viewed with objectivity if not skepticism. Laboratory data, for example, are never quite the same as results reported under actual plant conditions. Empirical prediction of emission factors or rates cannot replace actual source testing of stack gases. Particle size measurement is more reliable than literature data citing "typical" size distributions from various processes. In addition, the method of sampling where source testing is employed is a most important consideration that can yield differences both in reported results and the interpretation of those results.

Some common pitfalls to avoid in the technical definition of a control problem include failure to recognize the presence of another phase, failure to recognize the presence of fine particulates, and failure to recognize variations in the characteristics of the emission (31).

If particulates, especially fines, are present when only a gaseous type of effluent is expected—or vice versa—serious control efficiency problems can follow, because most devices designed for removing gaseous pollutants are far less efficient for particulates, and most particulate control devices are inefficient at removing gases. Only a few devices do both jobs well. Some packed-tower scrubbers or carbon adsorption units can be rendered inefficient by particulate matter, and baghouse filters can be corroded by even trace amounts of gases or vapors.

The problem is particularly complex when the particulate is an aerosol resulting from the condensation of a relatively nonvolatile material. In this case, a sample at the only available sampling point may give one answer, whereas a sample further upstream might show more gaseous material and one downstream might show more particulate. Often sulfur trioxide (SO_3) presents this type of difficulty (33). Dry SO_3 is gaseous and can be absorbed in 98 percent sulfuric acid (H_2SO_4) or 80 percent isopropanol. When SO_3 contacts water, however, it hydrates to H_2SO_4 and condenses to a submicrometer aerosol that is not collected well by either reagent. In this form, it should be collected with a dry filter. The EPA Sampling Method 8 (34) for this substance employs both an absorber and a filter. In the plant, however, dual control devices are usually prohibitively expensive.

The failure to recognize the presence of fine particles in a control system exhaust stream can lead to great difficulties. Until relatively recently, this was a minor problem. But now fine particles are known as a serious environmental hazard. EPA has recognized this fact from a regulatory standpoint and the NAAQS have been amended so that the attainment of the primary and secondary standards for particulate matter is based on PM_{10}. PM_{10} is defined as particles with an aerodynamic diameter less than or equal to a nominal 10 μm. An ambient air sampling

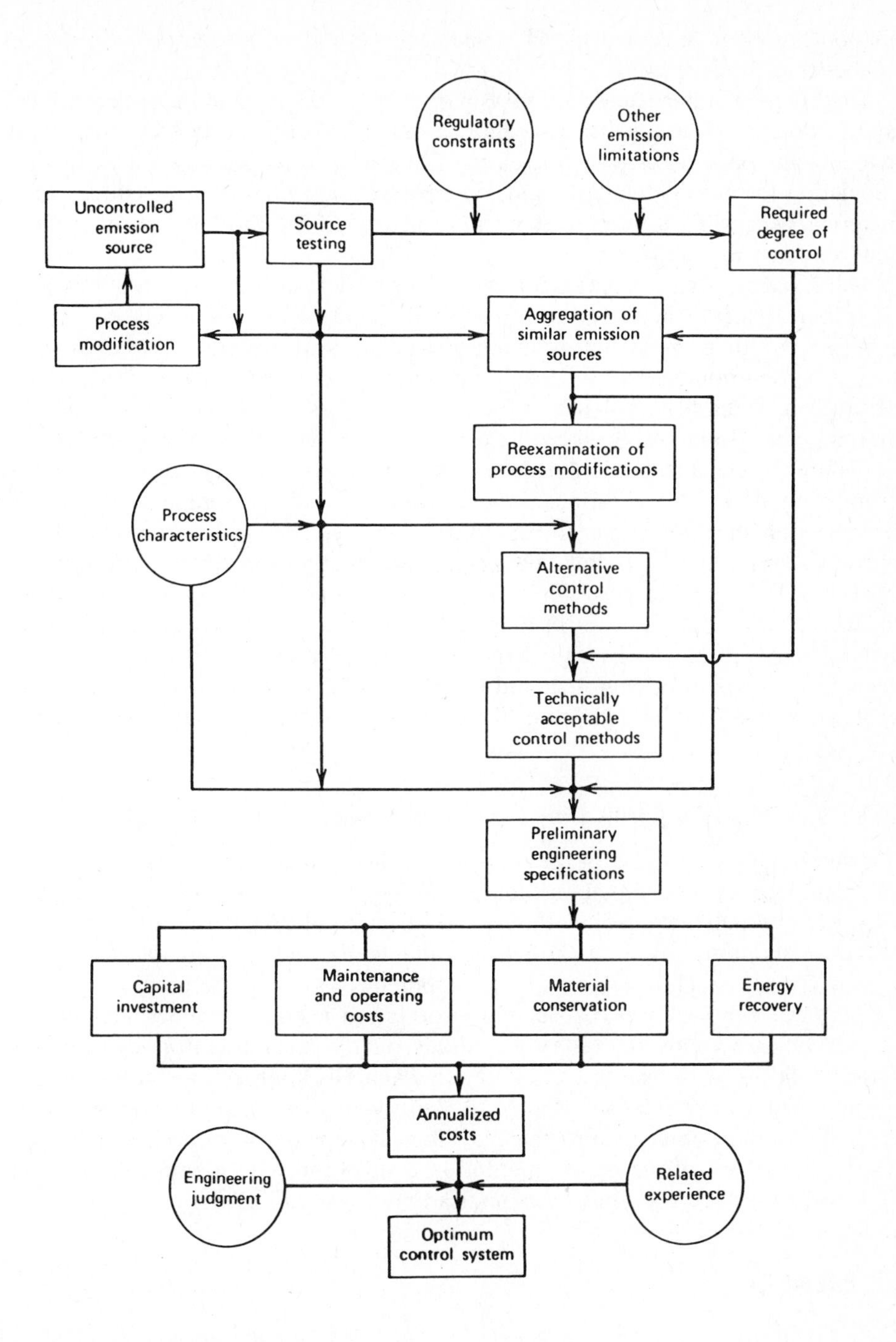

Figure 22.31. Technical-economic decision model for selecting emission controls; circled items denote required inputs of information data.

method has been specified by EPA that collects only those particles in the air smaller than 10 μm.

The fineness of particulate matter is particularly important in the treatment and control of toxic pollutants for two reasons (31). First, fine particles stay airborne longer than coarse ones, thus increasing the chance of exposing the surrounding population to the pollutant. Second, the low emission levels permissible for toxic materials generally require the removal of nearly all the fine, difficult-to-remove material from the emissions.

The presence of fine particles can sometimes be detected by the hazy appearance of a plume, except when coarse particles or steam are also present. Unless one has experience with a similar emission, the particle size distribution of the emission should be determined with a cascade impactor or an optical device. Care should be taken to extend the test into sufficiently fine sizes, to ensure that all particles that may have to be removed by the control device will be accounted for.

Failure to recognize variations in the characteristics of the emission is the third problem that can lead to serious design errors. Virtually all emissions vary as the temperatures of both the air and the process change. Emissions from continuous processes vary from day to day, and some, particularly from batch processes, vary from hour to hour. The ranges of such variations should be determined to evaluate the effect on a potential control device. If the control device is designed for the worst possible case, it will usually handle less severe cases. Nevertheless, problems peculiar to the control device should be recognized, such as the sensitivity of the venturi scrubber to flow rate, or of the electrostatic precipitator to dust resistivity (31).

13.5 Selection of Controls with a Technical-Economic Decision Model

Clearly, the basic strategy for the selection of the optimal control method must be an integrated approach that includes, at a minimum, identification of suitable technical alternative control methods, and selection of the alternative that is most attractive economically from among the suitable alternatives available.

It is not too much to expect the development of a computer program utilizing all relevant technical and cost-related information needed to select not only the most technically appropriate control methods, but also the most attractive economic choice. The need to use an overall systems analysis approach to the problem is obvious; indeed it is relatively simple to identify and interrelate the various major types of information needed to achieve a logical stepwise set of decisions that will lead to the choice of the "best" method of control for a given emission problem. One such approach is outlined schematically in Figure 22.31.

REFERENCES

1. J. E. Mutchler, "Principles of Ventilation," in *The Industrial Environment—Its Evaluation and Control*, U.S. Department of Health, Education, and Welfare, NIOSH, Cincinnati, OH, 1973.

2. Clean Air Act, as amended through December 8, 1983, 42 U.S.C. 7401 et seq., *Environ. Rep.*, S-693, **71**, 1101 (1984).
3. American Conference of Governmental Industrial Hygienists, *Industrial Ventilation—A Manual of Recommended Practice*, 19th Ed., ACGIH, Lansing, MI, 1986.
4. American Conference of Governmental Industrial Hygienists, *Threshold Limit Values and Biological Exposure Indicies for 1988–1989*, ACGIH, Cincinnati, OH, 1988.
5. Environmental Protection Agency Regulations on Approval and Promulgation of Implementation Plans, as amended through October 17, 1988, 40 CFR 52, *Environ. Rep.*, S-820, **125**, 0201 (1988).
6. *Control Techniques for Particulate Air Pollutants*, Department of Health, Education and Welfare, National Air Pollution Control Administration, Washington, DC, 1969.
7. F. I. Honea, *Chem. Eng. Deskbook*, 81-55-60 (1974).
8. K. T. Semrau, *J. Air Pollut. Control Assoc.*, **13**, 587–594 (1963).
9. "PVC Makers Move to Mop Up Monomer Emissions," *Chem. Eng.*, **82**, 25–27 (1975).
10. D. B. Turner, *Workbook of Atmospheric Dispersion Estimates*, U.S. Environmental Protection Agency, Office of Air Programs, Publication No. AP-26 (revised), 1970.
11. U.S. Environmental Protection Agency, *Guideline on Air Quality Models* (*Revised*), Office of Air Quality Planning and Standards, USEPA, Research Triangle Park, NC, 1986.
12. U.S. Environmental Protection Agency, *Guidelines for Air Quality Maintenance Planning and Analysis—Volume 10* (*Revised*): *Procedures for Evaluating Air Quality Impact of New Stationary Sources*, Office of Air and Waste Management, Office of Air Quality Planning and Standards, USEPA, Research Triangle Park, NC, 1977.
13. American Industrial Hygiene Association, *Air Pollution Manual*, Part II, AIHA, Detroit, 1968.
14. A. H. Rose, D. G. Stephan, and R. L. Stenburg, "Control by Process Changes or Equipment," in *Air Pollution*, Columbia University Press, New York, 1961.
15. K. J. Caplan, "Source Control by Centrifugal Force and Gravity," in *Air Pollution*, Vol. 3, 2nd ed., A. C. Stern, Ed., Academic Press, New York, 1968.
16. C. J. Stairmand, *J. Inst. Fuel*, **29**, 58–81 (1956).
17. H. E. Hesketh, *Understanding and Controlling Air Pollution*, 2nd ed., Ann Arbor Publishers, Ann Arbor, MI, 1974.
18. C. E. Lapple, *Chem. Eng.*, **58**, 145–151 (1951).
19. U.S. Atomic Energy Commission, Hanford, WA, 1948.
20. K. Iinoya and C. Orr, Jr., "Source Control by Filtration," in *Air Pollution*, Vol. 3, 2nd ed., A. C. Stern, Ed., Academic Press, New York, 1968.
21. B. Goyer, R. Gruen, and V. K. Lamer, *J. Phys. Chem.*, **58**, 137 (1954).
22. D. C. Drehmel, "Relationship between Fabric Structure and Filtration Performance in Dust Filtration," U.S. Environmental Protection Agency, Control Systems Laboratory, Research Triangle Park, NC, 1973.
23. S. Calvert, J. Goldshmid, D. Leith, and D. Mehta, *Wet Scrubber System Study*, Vol. 1, U.S. Environmental Protection Agency, Control Systems Division, Office of Air Programs, Research Triangle Park, NC, 1972.
24. A. B. Walker, *Pollut. Eng.*, **2**, 20–22 (1970).

25. A. Turk, "Source Control by Gas-Solid Adsorption," in *Air Pollution*, Vol. 3, 2nd ed., A. C. Stern, Ed., Academic Press, New York, 1968.
26. J. A. Danielson, Ed., *Air Pollution Engineering Manual*, 2nd ed., U.S. Environmental Protection Agency, Office of Air and Water Programs, Research Triangle Park, NC, 1973.
27. P. N. Cheremisinoff and A. C. Moressi, *Pollut. Eng.*, **6**, 66–68 (1974).
28. H. J. Paulus, "Nuisance Abatement by Combustion," in *Air Pollution*, Vol. 3, 2nd ed., A. C. Stern, Ed., Academic Press, New York, 1968.
29. G. L. Brewer, *Chem. Eng.*, **75**, 160–165 (1968).
30. L. Thomaides, *Pollut. Eng.*, **3**, 32–33, 1971.
31. W. L. O'Connell, *Chem. Eng. Deskbook*, **83**, 97–106 (1976).
32. Environmental Protection Agency Regulations on Standards of Performance for New Stationary Sources, as amended through June 15, 1989, 40 CFR 60, *Environ. Rep.*, S-838, **120**, 301 (1989).

CHAPTER TWENTY-THREE

Industrial Noise and Conservation of Hearing

Paul L. Michael, Ph.D.

1 INTRODUCTION

A number of noise problems continue to challenge occupational health professionals even though most have been with us for many years. Current noise problems include:

1. Temporary and/or permanent hearing impairment (1–64, 109)
2. Interference with speech, and warning signals (65–74)
3. Interference with hearing machine sounds that may have significant effects on safety and work performance (75–82)
4. Stresses from noise, at times interacting with other stressors, to produce unwanted effects on the health and safety of workers (13, 21–26)
5. Physiological and psychological effects of infrasound (250–251) and ultrasound (163–249)

The effectiveness of hearing conservation programs continues to vary widely from one industrial plant to another and, in some cases, from one work area to another within the same plant. A primary reason for this lack of progress is that a disproportionate amount of time and effort has been spent in (1) redefining the problem, (2) confirming that there is a problem without doing anything about it, (3) using shortcut and misleading performance ratings of hearing protectors, and

Patty's Industrial Hygiene and Toxicology, Fourth Edition, Volume 1, Part A, Edited by George D. Clayton and Florence E. Clayton
ISBN 0-471-50197-2

(4) using the same safety factor for all noise exposure situations, which causes attention to be drawn away from the real problems.

The fact that workers, hearing conservationists, and workplace situations may be very different from one place to another must be recognized. Some problems are much more difficult than others. *Different solutions and safety factors* must be developed and used in order to focus on real problems. Only by placing the focus on the inadequacy or lack of hearing conservation programs (not on protector ratings), are we going to develop *the best combination of effective hearing conservation programs, work safety, work effectiveness, and necessary communication.*

2 PHYSICS OF SOUND

Technically, the sensation of sound results from oscillations in pressure, stress, particle displacement, and particle velocity, in any elastic medium that connects the sound source with the ear. When sound is transmitted through air, it is usually described in terms of propagated changes in pressure that alternate above and below atmospheric pressure. These pressure changes are produced when vibrating objects (sound sources) cause regions of high and low pressure that propagate from the sound sources.

The characteristics of the sound received by an ear depend on the rate at which the sound source vibrates, the amplitude of the vibration, and the characteristics of the conducting medium. A sound may have a single rate, or frequency, of pressure alternations but most sounds have many frequency components. Each of these frequency components, or bands of sound, may have a different amplitude.

2.1 Terminology

Absorption Coefficient: The sound absorption coefficient of a given surface is the ratio of the sound energy absorbed by the surface to the sound energy incident upon the surface.

Acoustic Intensity: See Sound Intensity.

Acoustic Power: See Sound Power.

Acoustic Pressure: See Sound Pressure.

Ambient Noise: The overall composite of sounds in an environment.

Amplitude: The quantity of sound produced at a given location away from the source, *or* the overall ability of the source to emit sound. The amount of sound at a location away from the source is generally described by the sound pressure or sound intensity, whereas the ability of the source to produce noise is described by the sound power of the source.

Anechoic Room: A room that has essentially no boundaries to reflect sound energy generated therein. Thus any sound generated within is referred to as being in a free field (see definition).

Audiogram: A recording of hearing levels referenced to a statistically normal sound pressure as a function of frequency.

Audiometer: An instrument for measuring hearing thresholds.

Continuous Spectrum: A spectrum of sound that has components distributed over a known frequency range.

Cycle: A cycle of a periodic function, such as a single-frequency sound, is the complete sequence of values that occur in a period.

Cycles per Second: See Frequency.

Decibel: A convenient means for describing the logarithmic level of sound intensity, sound power, or sound pressure above arbitrarily chosen reference values (see text).

Diffuse Sound Field: A sound field that has sound pressure levels that are essentially the same throughout, with the directional incidence of energy flux randomly distributed.

Effective Sound Pressure: The sound pressure at a given location, derived by calculating the root mean square (rms) value of the instantaneous sound pressures measured over a period of time at that location.

Free Field: A field that exists in a homogeneous isotropic medium, free from boundaries. In a free field, sound radiated from a source can be measured accurately without influence from the test space. True free-field conditions are rarely found, except in expensive anechoic test chambers. However, approximate free-field conditions may be found in any homogeneous space where the distance from the reflecting surfaces to the measuring location is much greater than the wavelengths of the sound being measured.

Frequency: The rate at which complete cycles of high- and low-pressure regions are produced by the sound source. The unit of frequency is the cycle per second (cps) or, preferably, the hertz (Hz). The frequency range of the human ear is highly dependent upon the individual and on the sound level, but an ear with normal hearing has a frequency range of approximately 20 to 20,000 Hz at moderate to high sound levels. The frequency of a sound wave heard by a listener is the same as the vibrating source if the distance between the source and the listener remains constant. The frequency heard by the listener increases or decreases as the distance from the source decreases or increases—the Doppler effect (American National Standards Institute, 1978).

Hertz: See Frequency.

Infrasonic Frequency: Sounds having major frequency components well below audibile range (below 20 Hz).

Intensity: See Sound Intensity.

Level: The level of any quantity, when described in decibels, is the logarithm of the ratio of that quantity to a reference value in the same units as the specified quantity.

Loudness: An observer's impression of the sound's amplitude, which includes the response characteristics of the ear.

Natural Frequency: See Resonance.

Noise: The terms noise and sound are often used interchangeably, but generally, sound describes useful communication or pleasant sounds, such as music, whereas noise is dissonance or unwanted sound.

Noise Reduction Coefficient (NRC): The arithmetic average of the sound absorption coefficients of a material at 250, 500, 1000, and 2000 Hz.

Octave Band: A frequency bandwidth that has an upper band-edge frequency equal to twice its lower band-edge frequency.

Peak Sound Pressure Level: The maximum instantaneous level that occurs over any specified period of time.

Period (T): The time (in seconds) required for one cycle of pressure change to take place; hence it is the reciprocal of the frequency.

Pitch: A measure of auditory sensation that depends primarily on frequency but also on the pressure and wave form of the sound stimulus.

Power: See Sound Power.

Pure Tone: A sound wave with only one sinusoidal change of pressure with time.

Random Incidence Sound Field: See Diffuse Sound Field.

Random Noise: A noise made up of many frequency components whose instantaneous amplitudes occur randomly as a function of time.

Resonance: A system is in resonance when any change in the frequency of forced oscillation causes a decrease in the response of the system.

Reverberation: Reverberation occurs when sound persists after direct reception of the sound has stopped. The reverberation of a space is specified by the reverberation time, which is the time required, after the source has stopped radiating sound, for the rms pressure to decrease 60 dB from its steady-state level.

Root Mean Square Sound Pressure: The root-mean-square (rms) value of a changing quantity, such as sound pressure, is the square root of the mean of the squares of the instantaneous values of the quantity.

Sound: See Noise.

Sound Intensity (I): The average rate at which sound energy is transmitted through a unit area normal to the direction of propagation. The units used for sound intensity are joules per square meter (J/m^2 sec). Sound intensity is also expressed in terms of a sound intensity level (L_I) in decibels referred to 10^{-12} W/m^2.

Sound Power (P): The total sound energy radiated by the source per unit time. Sound power is normally expressed in terms of a sound power level (L_p) decibels referenced to 10^{-12} W.

Sound Pressure (p): The rms value of the pressure above and below atmospheric pressure of steady-state noise. Short-term or impulse-type noises are described by peak pressure values. The units used to describe sound pressures are pascals (Pa), newtons per square meter (N/m^2), dynes per square centimeter (dyn/cm^2), or microbars (μbar).

Sound Pressure Level (L_p): Sound pressure is also described in terms of sound pressure level (L_p) in decibels referenced to 20 μ Pa.

Standing Waves: Periodic waves that have a fixed distribution in the propagation medium.

Transmission Loss (TL): Ten times the logarithm (to the base 10) of the ratio of the incident acoustic energy to the acoustic energy transmitted through a sound barrier.

Ultrasonic: The frequency of ultrasonic sound is higher than that of audible sound.

Velocity: The speed at which the regions of sound-producing pressure changes move away from the sound source is called the velocity of propagation. Sound velocity (c) varies directly with the square root of the density and inversely with the compressibility of the transmitting medium; however, the velocity of sound is usually considered constant under normal conditions. For example, the velocity of sound is approximately 344 m/sec (1130 ft/sec) in air, 1433 m/sec (4700 ft/sec) in water, 3962 m/sec (13,000 ft/sec) in wood, and 5029 m/sec (16,500 ft/sec) in steel.

Wavelength (λ): The distance required to complete one pressure cycle. The wavelength, a very useful tool in noise control, is calculated from known values of frequency (f) and velocity (c): $\lambda = c/f$.

White Noise: White noise has an essentially random spectrum with equal energy per unit frequency bandwidth over a specified frequency band.

2.2 Units for Noise Measurements

2.2.1 Sound Pressure and Sound Pressure Level

The range of root-mean-square (rms) sound pressures is usually very wide. Sound pressures well above the pain threshold, about 20 N/m^2, or 20 Pascals (Pa), are found in some work areas, whereas sound pressures down to the threshold of hearing (about 0.00002 Pa) must be used for hearing measurements. The wide range of sound exposures of more than 10^6 Pa cannot be scaled linearly with a practical instrument because to obtain the desired accuracy at low levels, such a scale would have to be many miles long. To cover this wide range of sound pressure with a reasonable number of scale divisions and to provide a scale that responds more closely to the response of the human ear, the logarithmic decibel (dB) scale is used.

By definition, the decibel is a unit without dimensions; it is the logarithm of the *ratio* of a *measured quantity* to a *reference quantity*. The decibel is difficult to use and to understand because it is often used with different reference quantities. Acoustic intensity, acoustic power, hearing thresholds, electric voltage, electric current, electric power, and sound pressure level may all be expressed in decibels, each having a different reference. Obviously the decibel has no meaning unless a specific reference quantity is specified, or understood.

Most sound-measuring instruments are calibrated to provide a reading of rms

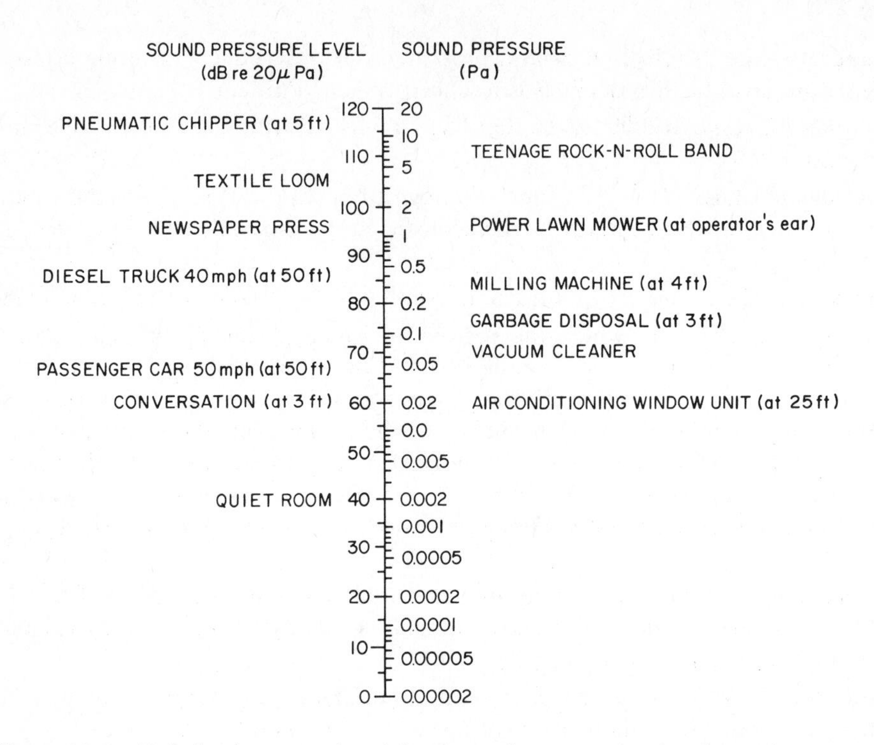

Figure 23.1. Relation between A-weighted sound pressure level and sound pressure.

sound pressures on a logarithmic scale in decibels. The decibel reading taken from such an instrument is called the sound pressure level L_p. The term level is used because the measured pressure is at a particular level above a given pressure reference. For sound measurements in air 0.00002 Pa, or 20 μPa, commonly serves as the reference sound pressure. This reference is an arbitrary pressure chosen many years ago because it was thought to approximate the normal threshold of human hearing at 1000 Hz. Mathematically, L_p is written as follows:

$$L_p = 20 \log \frac{p}{p_0} \text{ dB} \tag{1}$$

where p is the measured rms sound pressure, p_0 is the reference sound pressure (generally in N/m^2 or Pa), and the logarithm is to the base 10. Thus for technical purposes, L_p should be written in terms of decibels referenced to a specified pressure level. For example, in air the notation for L_p is commonly abbreviated as "dB re 20 μPa," where μ is 10^{-6}. Most instruments measure sound pressure levels in dB re 20 μPa so calculations are usually not necessary.

Figure 23.1 shows the relationship between sound pressure (in pascals) and sound pressure level (in dB re 20 μPa). This figure illustrates the advantage of using the decibel scale rather than the wide range of direct pressure measurements. It is of interest to note that any doubling of a pressure is equivalent to a 6-dB change in level. For example, a range of 20 to 40 μPa, which might be found in hearing measurements, or a range of 10 to 20 Pa, which might be found in hearing conservation programs, are both ranges of 6 dB. This allows reasonable accuracy for both low- and high-sound pressure levels.

The L_p referenced to 20 μPa may be written in any of the following six forms:

$$\begin{aligned} L_p &= 20 \log(p/0.00002) \\ &= 20(\log p - \log 0.00002) \\ &= 20[\log p - (\log 2 - \log 10^5)] \\ &= 20[\log p - (0.3 - 5)] \\ &= 20(\log p + 4.7) \\ &= 20 \log p + 94 \text{ re } 20\ \mu\text{Pa} \end{aligned}$$

2.2.2 *Sound Intensity and Sound Intensity Level*

Sound intensity at any specified location may be defined as the average acoustic energy per unit time passing through a unit area that is normal to the direction of propagation. For a spherical or free-progressive sound wave, the intensity may be expressed by

$$I = \frac{p^2}{\rho c} \tag{2}$$

where p is the rms sound pressure, ρ (rho) is the density of the medium, and c is the speed of sound in the medium. Sound intensity units, like sound pressure units, cover a wide range, and it is often desirable to use decibel levels to compress the measuring scale. To be consistent with Equations 1 and 2, intensity level is defined as

$$L_I = 10 \log \frac{I}{I_0} \text{ dB} \tag{3}$$

where I is the measured intensity at some given distance from the source and I_0 is a reference intensity. The reference intensity commonly used to 10^{-12} W/m^2. In air, this reference closely corresponds to the reference pressure 20 μPa used for sound pressure levels.

2.2.3 *Sound Power and Sound Power Level*

Sound power P is used to describe the sound source in terms of the amount of acoustic energy that is produced per unit time (joules per second or watts). Sound power is related to the average sound intensity produced in free-field conditions at a distance r from a point source by

$$P = I_{avg} 4\pi r^2 \tag{4}$$

where I_{avg} is the average intensity at a distance r from a sound source whose acoustic power is P. The quantity $4\pi r^2$ is the area of a sphere surrounding the source over which the intensity is averaged. It is obvious from Equation 4 that the intensity decreases with the square of the distance from the source, hence the well-known inverse square law.

Power units, as pressure and intensity units, are often described in terms of decibel levels because of the wide range of powers covered in practical applications. Power level L_P is defined by

$$L_p = 20 \log \frac{P}{P_0} \text{ dB} \tag{5}$$

where P is the power of the source in watts (W) and P_0 is the reference power. The reference power consistent with the references used for pressure and intensity is 10^{-12} W. Figure 23.2 shows the relation between sound power in watts and sound power level in dB re 10^{-12} W.

2.2.4 *Sound Power versus Sound Intensity versus Sound Pressure*

Noise control problems may require a practical knowledge of the relationship between pressure, intensity, and power. For example, consider the prediction of sound pressure levels that would be produced around a proposed machine location from the sound power level provided for the machine.

Example: The manufacturer of a noisy machine states that this machine has an acoustic power output of 1 W. Predict the sound pressure level at a location 100 ft from the machine. From Equations 2 and 4, *in a free field and for an omnidirectional source*, sound power is

$$P = I_{avg}\, 4\pi r^2 = \frac{(p_{avg}^2\, 4\pi r^2)}{\rho c} \tag{6}$$

where

$$p_{avg} = \left(\frac{P\rho c}{4\pi r^2}\right)^{1/2} \tag{7}$$

If P is given in watts, r in feet, p in pascals, and ρ in kg/m^3, then with standard conditions, Equation 7 may be rewritten as

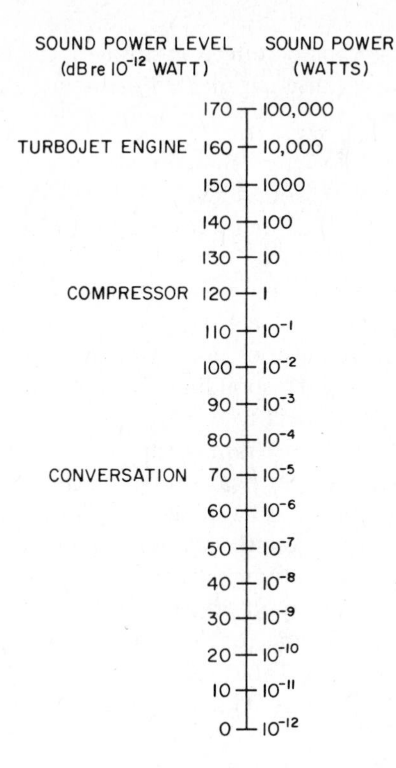

Figure 23.2. Relation between sound power level and sound power.

$$p_{avg} = \left(\frac{3.5P \times 10^2}{r^2}\right)^{1/2}$$

and, for this example,

$$p_{avg} = \left(\frac{3.5P \times 1.0 \times 10^2}{100^2}\right)^{1/2} = 0.187 \text{ Pa}$$

The sound pressure level may be determined from Equation 1:

$$L_p = 20 \log \left(\frac{0.187}{0.00002}\right) \text{dB} = 79.4 \text{ dB re } 20 \ \mu\text{Pa}$$

Admittedly, there are few truly free-field situations, and few omnidirectional sources, so the absolute value of the sound pressure level is in question. However, comparison or rank ordering of different machines can be made and at least a rough estimate of the sound pressure is available. Noise levels in locations that are reverberant, or where there are many reflecting surfaces, can be expected to be higher than that predicted because noise is reflected back to the point of measurement.

2.2.5 Combining Sound Levels

It is often necessary to combine sound levels—for example, to combine frequency band levels to obtain the overall or total sound pressure level from band levels within the overall noise. Another example is the estimation of total sound pressure level resulting from adding a machine of known noise spectrum to a noise environment of known characteristics. Simple addition of individual sound pressure levels, which are logarithmic quantities, constitutes multiplication of pressure ratios; therefore the sound pressure corresponding to each sound pressure level must be determined and added with respect to existing phase relationships.

For the most part, industrial noise is made up of sources having many frequencies (broad band), with nearly random phase relationships. Sound pressure levels of different random noise sources can be added by (1) converting the levels of each to pressure units, (2) converting to intensity units that may be added arithmetically, (3) then reconverting the resultant intensity to pressure, and finally, (4) converting the resultant pressure value to sound pressure levels in decibels. Equations 1 and 2 can be used in free-field conditions for this purpose. Acceptable accuracy is found in most cases, however, by using Table 23.1 for adding sound pressure levels from two separate random noise sources. The use of this table may be illustrated by the following example.

Example: If the octave band sound pressure levels measured for a random noise source are 85, 88, 91, 94, 95, 100, 97, 90, and 88 dB, respectively, they can be added as follows: Begin by adding the highest levels so that calculations may be discontinued when there is no significant difference when lower values are added to the total. That is, the difference between 100 and 97 dB is 3dB; therefore, opposite the range 2.8 to 3.0 in the left-hand column of Table 23.1, read a value of 1.8 in the right-hand column and add this value to the higher of the two levels: $100 + 1.8 = 101.8$ dB. This resultant is now added to the next highest level by repeating the process: $101.8 - 95 = 6.8$; from the table read an amount to be added of 0.8. Thus $101.8 + 0.8 = 102.6$ dB. This procedure is continued with each reading to arrive at an overall sound pressure level of 104 dB.

It may be convenient to commit to memory a simplified set of values from Table 23.1 (see Table 23.2) that will enable an accuracy within about 1 dB.

The overall sound pressure level calculated in the example above corresponds to the value that would be found by reading a sound level meter at this location with the frequency weighting set so that the levels at each frequency in the spectrum are weighted equally. Common names given to this frequency weighting are flat, linear, 20 kHz, and overall (see Section 5.1).

The corresponding A-weighted sound pressure level (dBA) found in many noise regulations may also be calculated from octave band values such as those in the

Table 23.1. Table for Combining Decibel Levels of Noises with Random Frequency Characteristics

Numerical Difference between Levels (dB)	Amount to be Added to the Higher Level (dB)
0.0–0.1	3.0
0.2–0.3	2.9
0.4–0.5	2.8
0.6–0.7	2.7
0.8–0.9	2.6
1.0–1.2	2.5
1.3–1.4	2.4
1.5–1.6	2.3
1.7–1.9	2.2
2.0–2.1	2.1
2.2–2.4	2.0
2.5–2.7	1.9
2.8–3.0	1.8
3.1–3.3	1.7
3.4–3.6	1.6
3.7–4.0	1.5
4.1–4.3	1.4
4.4–4.7	1.3
4.8–5.1	1.2
5.2–5.6	1.1
5.7–6.1	1.0
6.2–6.6	0.9
6.7–7.2	0.8
7.3–7.9	0.7
8.0–8.6	0.6
8.7–9.6	0.5
9.7–10.7	0.4
10.8–12.2	0.3
12.3–14.5	0.2
14.6–19.3	0.1
19.4–∞	0.0

foregoing example if the adjustments given in Table 23.3 are first applied. For example, the octave band (OB) levels with A-weighting (see Section 5.1) corresponding to the example would be:

OB center frequency (Hz)	31.5	63	125	250	500	1000	2000	4000	8000
L_p (A-Wtd)	45.8	61.9	77.8	85.4	91.7	100	98.2	91.0	86.9

Table 23.2. Simplified Table for Combining Decibel Levels of Noise with Random Frequency Characteristics

Numerical Differences between Levels	Amount To Be Added to the Higher Level
0–1	3
2–4	2
5–9	1
>10	0

These octave band levels with A-frequency weighting can be added by this procedure to obtain the resultant A-weighted level, which is about 103 dBA.

Most industrial noises have random frequency characteristics, and can be combined as described in the preceding paragraphs. In the few cases when noises have significant pitched or major pure tone components, these calculations are not accurate, and phase relationships must be considered. In areas where significant pitched noises are present, standing waves often can be recognized by the presence of rapidly varying sound pressure levels over short distances. It is not practical to try to predict levels in areas where standing waves are present.

When the sound pressure levels of two pitched sources are added, it might be assumed that the resultant sound pressure level L_P(R) will be less as often as it is greater than the level of a single source. In almost all cases, however, the resultant L_P(R) is greater than either single source. The reason for this may be seen if two pure tone sources are added at several specified phase differences (Figure 23.3). At zero phase difference, the resultant of two identical pure tone sources is 6 dB greater than either single level. At a phase difference of 90°, the resultant is 3 dB greater than either level. Between 90° and 0°, the resultant is somewhere between 3 and 6 dB greater than either level. At a phase difference of 120°, the resultant is equal to the individual levels; and between 120° and 90°, the resultant is between 0 and 3 dB greater than either level. At 180° there is complete cancellation of sound. Obviously, the resultant L_P(R) is greater than the individual levels for all phase differences from 0° and 120°, but less than individual levels for phase differences from 120° and 180°. Also, most pitched tones are not single tones but combinations thereof; thus almost all points in the noise fields have pressure levels that exceed the individual levels.

The most common frequency bandwidth used for industrial noise measurements is the octave band. A frequency band is said to be an octave in width when its upper band-edge frequency f_2 is twice the lower band-edge frequency f_1:

$$f_2 = 2f_1 \tag{8}$$

Octave bands are commonly used for measurements directly related to the effects

Table 23.3. Sound Level Meter Random Incidence Relative Response Level as a Function of Frequency for Various Weightings

Frequency (Hz)	A-Weighting Relative Response (dB)	B-Weighting Relative Response (dB)	C-Weighting Relative Response (dB)
10	−70.4	−38.2	−14.3
12.5	−63.4	−33.2	−11.2
16	−56.7	−28.5	− 8.5
20	−50.5	−24.2	− 6.2
25	−44.7	−20.4	− 4.4
31.5	−39.4	−17.1	− 3.0
40	−34.6	−14.2	− 2.0
50	−30.2	−11.6	− 1.3
63	−26.2	− 9.3	− 0.8
80	−22.5	− 7.4	− 0.5
100	−19.1	− 5.6	− 0.3
125	−16.1	− 4.2	− 0.2
160	−13.4	− 3.0	− 0.1
200	−10.9	− 2.0	0
250	− 8.6	− 1.3	0
315	− 6.6	− 0.8	0
400	− 4.8	− 0.5	0
500	− 3.2	− 0.3	0
630	− 1.9	− 0.1	0
800	− 0.8	0	0
1,000	0	0	0
1,250	+ 0.6	0	0
1,600	+ 1.0	0	− 0.1
2,000	+ 1.2	− 0.1	− 0.2
2,500	+ 1.3	− 0.2	− 0.3
3,150	+ 1.2	− 0.4	− 0.5
4,000	+ 1.0	− 0.7	− 0.8
5,000	+ 0.5	− 1.2	− 1.3
6,300	− 0.1	− 1.9	− 2.0
8,000	− 1.1	− 2.9	− 3.0
10,000	− 2.5	− 4.3	− 4.4
12,500	− 4.3	− 6.1	− 6.2
16,000	− 6.6	− 8.4	− 8.5
20,000	− 9.3	−11.1	−11.2

of noise on the ear, and for some noise control purposes, because they provide the maximum amount of information in a reasonable number of measurements.

When more specific characteristics of a noise source are required (e.g., for pinpointing a particular noise source in a background of other sources), it is necessary to use frequency bandwidths that are narrower than octave bands. Half-

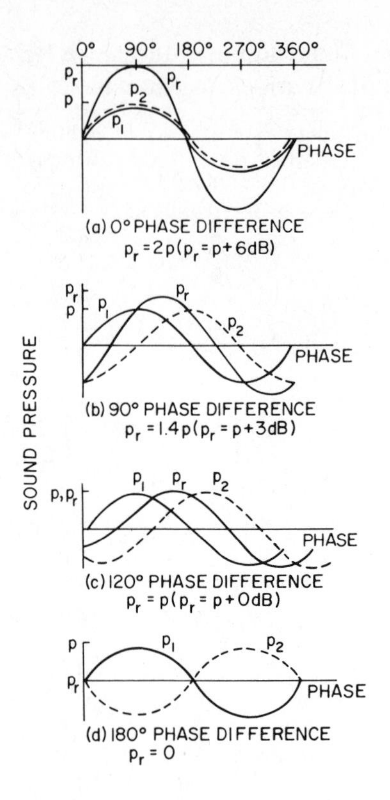

Figure 23.3. Combinations of two pure tone noises, (p_1 and p_2) phase differences.

octave, third-octave, and narrower bands are used for these purposes. A half-octave bandwidth is defined as a band whose upper band-edge frequency f_2 is the square root of twice the lower band-edge frequency f_1:

$$f_2 = (2)^{1/2} f_1 \tag{9}$$

A third-octave bandwidth is a band whose upper band-edge frequency f_2 is the cube root of twice the lower band-edge frequency

$$f_2 = (2)^{1/3} f_1 \tag{10}$$

The center frequency f_m of any of these bands is the square root of the product of the high and low band-edge frequencies (geometric mean):

$$f_m = (f_2 f_1)^{1/2} \tag{11}$$

It should be noted that the upper and lower band-edge frequencies describing a frequency band do not imply abrupt cutoffs at these frequencies. These band-edge frequencies are conventionally used as the 3-dB-down points of gradually sloping curves that meet the American National Standard Specification for Octave-Band and Fractional-Octave Band Analog and Digital Filters (S1.11-1986) (106).

Noise measurement data (rms) taken with analyzers of a given bandwidth may be converted to another given bandwidth if the frequency range covered has a continuous spectrum with no prominent changes in level. The conversion may be made in terms of sound pressure levels by

$$L_p(\mathrm{A}) = L_p(\mathrm{B}) - 10 \log \left(\frac{\delta f(\mathrm{B})}{\delta f(\mathrm{A})} \right) \tag{12}$$

where $L_p(\mathrm{A})$ is the sound pressure level (dB) of the band having a width $\delta(\mathrm{A})$ Hz, and $L_p(\mathrm{B})$ is the sound pressure level (dB) of the band having a width $\delta(\mathrm{B})$ Hz. Sound pressure levels for different bandwidths of flat, continuous spectrum noises may also be converted to spectrum levels. The spectrum level, which is used at times by noise control engineers as a general noise level rating, describes a continuous spectrum, wide-band noise in terms of its energy equivalent in a band 1 Hz wide, assuming there are no prominent peaks. The spectrum level $L_p(\mathrm{S})$ may be determined by

$$L_p(\mathrm{S}) = L_p(f) - 10 \log \delta(f) \tag{13}$$

where $L_p(f)$ is the sound pressure level of the band having a width of $\delta(f)$ Hz.

It should be emphasized that accurate conversion of sound pressure levels from one bandwidth to another by the method just described can be accomplished only when the frequency bands have flat continuous spectra.

3 THE EAR

The normal human ear has a remarkable frequency range that covers a range from about 20 to 20,000 Hz at common loudness levels (6, 16, 18, 21, 93). The characteristics of any individual ear over this wide frequency range are extremely complex, and understanding the ear's responses is made even more difficult because of large differences among individuals' responses. An ear's response characteristics may change as a result of physical or mental conditions, sound level, medications, environmental stresses, diseases, and other factors.

A normal healthy human ear also effectively transduces a remarkable range of sound levels. It is sensitive to very low sound pressures that produce a displacement of the eardrum no greater than the diameter of a hydrogen molecule. At the other extreme, it can transduce sounds whose sound pressures are more than 10^6 times greater than the ear's lower threshold value; however, exposure to high level sounds may cause temporary or permanent damage to the ear.

The ear may be divided into three sections (Figure 23.4). Sound incident upon

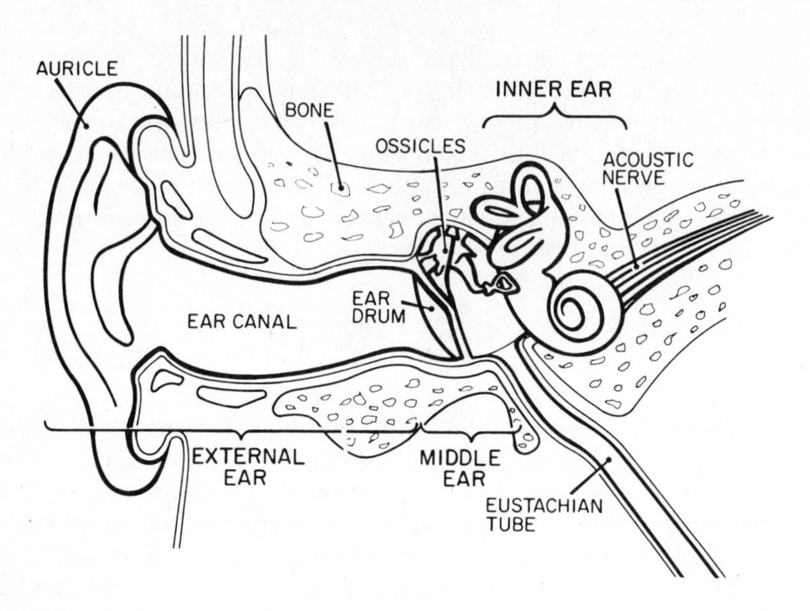

Figure 23.4. Cross section of the ear showing the external, middle, and inner ear configurations.

the ear travels through the ear canal to the eardrum, which separates the external and middle ear sections (Figure 23.5). The combined alternating sound pressures that are incident upon the eardrum cause the eardrum to vibrate with the same relative characteristics as the sound source(s). The mechanical vibration of the eardrum is then coupled through the three bones of the middle ear to the oval window of the inner ear. The vibration of the oval window is then coupled to the fluid contained in the inner ear (Figure 23.6). During its vibration, when the oval window moves inward, it pushes the fluid through the cochlea, which causes the round window to bulge outward (Figure 23.7). During the outward motion of the oval window, the round window bulges inward to compensate for the fluid flowing toward the oval window. The movement of the inner ear fluid is detected by thousands of transducers located within the cochlea. This complex information is then transmitted to the brain, through the eighth nerve, for further analysis and readout.

3.1 External Ear

The auricle, sometimes called the pinna (Figure 23.5), plays a significant part in the hearing process only at very high audible frequencies, where its size is large compared with that of a wavelength (for these frequencies). The auricle helps direct these high-frequency sounds into the ear canal, and it assists the overall hearing system in determining the direction from which the sound comes.

Ear canals have many sizes and shapes, even for the same individual. They are

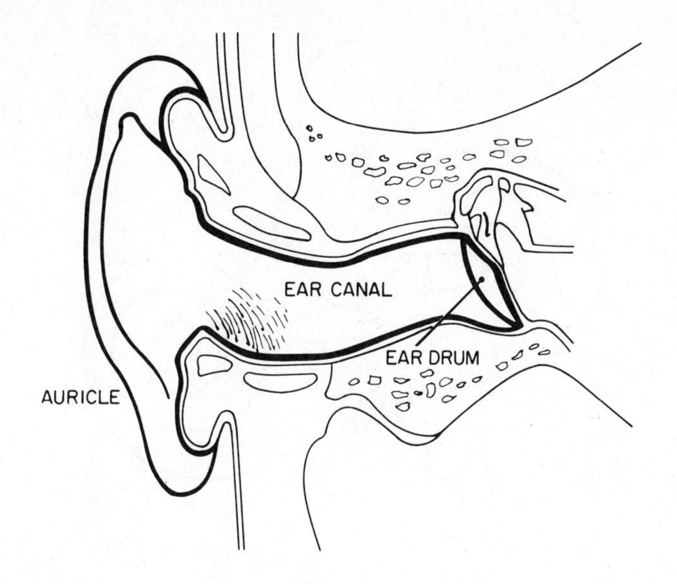

Figure 23.5. The external ear.

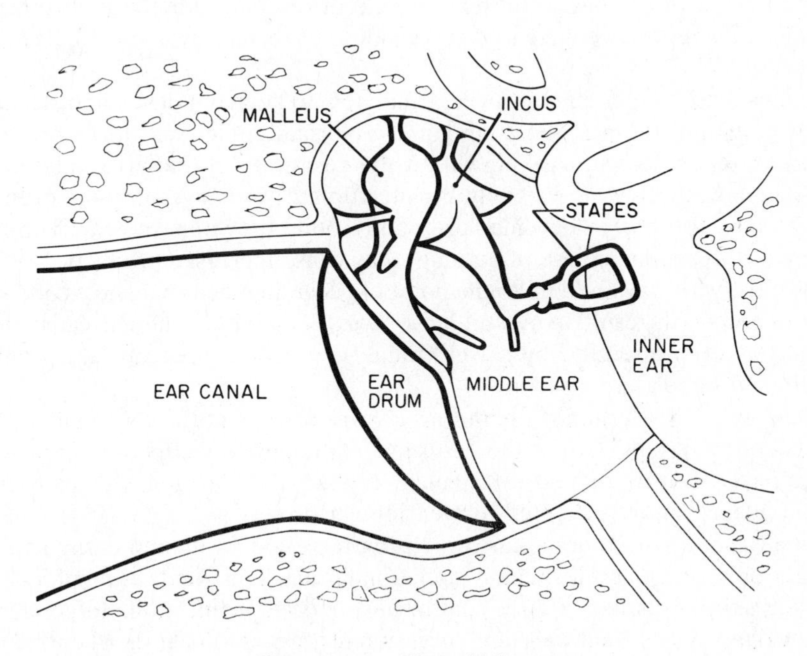

Figure 23.6. The middle ear.

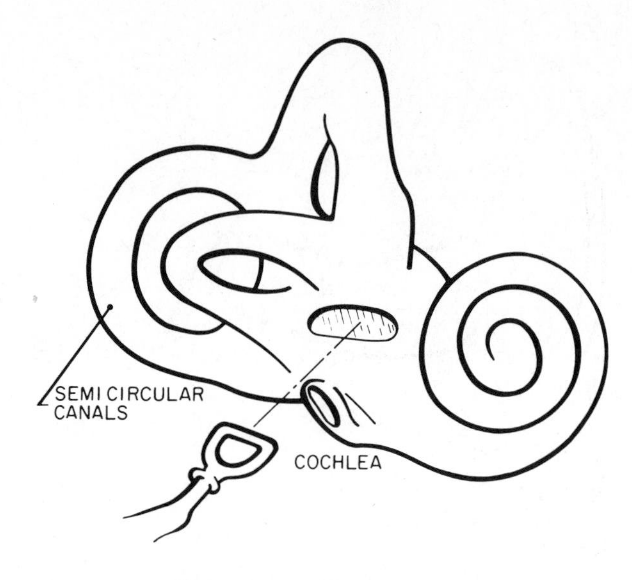

Figure 23.7. The inner ear.

seldom as straight as indicated in Figures 23.4 and 23.5, and the shape and size of ear canals differ significantly among individuals and even between ears of the same individual. The average length of the ear canal is about 1 in. When closed at one end by the eardrum, it has a quarter-wavelength resonance of about 3000 Hz in a free-field. This resonance increases the response of the ear by about 10 dB at 3000 Hz.

The hairs at the outer end of the ear canal help to keep out dust and dirt; further into the canal are the wax-secreting glands. Normally, ear wax flows toward the entrance of the ear canal, carrying with it the dust and dirt that accumulate in the canal. The normal flow of wax may be interrupted by changes in body chemistry that can cause the wax to become hard and to build up within the ear. Too much cleaning or the prolonged use of earplugs may cause increased production of wax. At times, the wax may build up to the point of occluding the canal and a conductive loss of hearing will result. Any buildup of wax deep within the ear canal should be removed very carefully, by a well-trained person, to prevent damage to the eardrum and middle ear structures.

During welding or grinding operations a spark may enter the ear canal and burn the canal or a large portion of the eardrum. Although very effective surgical procedures have been developed to repair or replace the eardrum, this painful and costly accident can be prevented by wearing ear protectors.

The surface of the external ear canal is extremely delicate and easily irritated. Cleaning or scratching with match sticks, nails, hairpins, and other objects can break the skin and cause a very painful and persistent infection. Infections can cause swelling of the canal walls and, occasionally, a loss of hearing when the canal swells shut. An infected ear should be given prompt attention by a physician.

3.2 Eardrum

The eardrum is a very thin and delicate membrane that responds to the very low sound pressures at the lower threshold of normal hearing; yet it is seldom damaged by common continuous high level noises. Although an eardrum may be damaged by an explosion or a rapid change in ambient pressure, the often repeated statement "the noise was so loud it almost burst my eardrums" is rarely true for common steady-state noise exposures.

When an eardrum is ruptured, the attached middle-ear ossicle bones may be dislocated; thus the eardrum should be carefully examined immediately after the injury occurs to determine whether realignment of the ossicle bones is necessary. In a high percentage of cases, surgical procedures are successful in realigning dislocated ossicles, so that little or no significant loss in hearing acuity results from this injury.

3.3 Middle Ear

The air-filled space between the eardrum and the inner ear is called the middle ear (Figure 23.6). The middle ear contains three small bones—the malleus (hammer), the incus (anvil), and the stapes (stirrups)—that mechanically connect the eardrum to the oval window of the inner ear.

The eardrum has an area about 20 times that of the oval window, thereby providing a mechanical advantage of about 20. The ossicles provide an additional mechanical advantage of about 3. Hence the overall mechanical advantage of the middle-ear structure is $3 \times 20 = 60$ at the natural resonant frequency of the system. This complex system also acts as an ear protector by mismatching impedances through the involuntary relaxation of coupling efficiency between the ossicle bones. The reaction time for the middle-ear system is approximately 10 msec.

The most common problem encountered in the middle ear is infection. This dark, damp, air-filled space is completely enclosed except for the small eustachian tube that connects the space to the back of the throat; thus it is very susceptible to infection, particularly in teenagers. If the eustachian tube is closed as a result of an infection or an allergy (Figure 23.6), the pressure inside the middle ear cannot be equalized with that of the surrounding atmosphere. In such an event, a significant change in atmospheric pressure, such as that encountered in an airplane or when driving in mountainous territory, may produce a loss of hearing sensitivity and extreme discomfort as a result of the displacement of the eardrum toward the low-pressure side. Even a healthy ear may suffer a temporary loss of hearing sensitivity if the eustachian tube becomes blocked, but this loss of hearing can often be restored simply by swallowing or chewing gum to open the eustachian tube momentarily.

Another middle-ear problem may result from an abnormal bone growth (otosclerosis) around the ossicle bones, restricting their normal movement. The cause of otosclerosis is not totally understood, but heredity is considered to be an important factor. The conductive type of hearing loss from otosclerosis is generally observed first at low frequencies, it then extends to higher frequencies, and eventually may result in a severe loss in hearing sensitivity over a wide frequency range. Hearing aids may often restore hearing sensitivity lost as a result of otosclerosis, but effective surgical procedures have been refined to such a point that they are often recommended. An important side benefit of an effective hearing conservation program is the early detection of such hearing impairments as otosclerosis.

3.4 Inner Ear

The inner ear is completely surrounded by bone (Figure 23.7). One end of the space inside the bony shell of the inner ear is shaped like a snail shell; it contains the cochlea. The other end of the inner ear has the shape of three semicircular loops. The fluid-filled cochlea serves to detect and analyze incoming sound signals and to translate them into nerve impulses that are transmitted to the brain. The semicircular canals contain sensors for balance and orientation.

In operation, sound energy is coupled into the inner ear by the stapes, whose base is coupled into the oval window of the cochlea. The oval window and the round window located below it are covered by thin, elastic membranes to contain the few drops of fluid within the cochlea. As the stapes forces the oval window in and out with the dynamic characteristics of the incident sound, the fluid of the cochlea is moved with the same characteristic motions. Thousands of hair cells located along the two and one-half turns of the cochlea detect and analyze these motions and translate them into nerve impulses. The nerve impluses, in turn, are transmitted to the brain for further analysis and interpretation.

The hair cells within the cochlea may be damaged by old age, disease, certain drugs, blows to the head, and exposure to high levels of noise (109). Unfortunately the characteristics of the hearing losses resulting from these various causes are often very similar, and it is impossible to determine the etiology of a particular case from an audiogram.

3.5 How Noise Damages Hearing

Noise-induced hearing loss may be temporary or permanent depending on the *level and frequency characteristics of the noise*, the *duration of exposures*, and the *susceptibility of the individual*. Usually temporary losses of hearing sensitivity diminish, and the originial sensitivities are restored within about 16 hr (1–3, 109); however, temporary losses may last for weeks in some cases. When temporary losses do not recover completely before other significant exposures, permanent losses may be produced. Permanent losses are irreversible and cannot be corrected by conventional surgical or therapeutic procedures.

Noise-induced damage generally occurs in hair cells located within the cochlea, a sensorineural loss. For common broad-band noise exposures, hearing acuity is generally affected first in the frequency range from 2000 to 6000 Hz, with most affected persons showing a loss or "dip" at 4000 Hz. For noise exposures having significant components concentrated in narrow frequency bands below 4000 Hz, impairments usually are found about one-half to one octave *above* the predominant exposure frequencies. For noise exposures having significant components concentrated in narrow frequency bands *above* 4000 Hz, impairments are often found about one-half of one octave below the predominant exposure frequencies. If high-level exposures are continued, the loss of hearing generally increases around 4000 Hz and spreads to lower frequencies (109).

Noise-induced hearing loss is an insidious problem because a person does not necessarily experience pain before learning that severe hearing damage has taken place. The damage may occur instantaneously, or over a long period of time, depending upon the noise characteristics. Generally, impulsive or impact noises are most likely to produce significant losses with short exposure periods, and steady-state continuous noises are responsible for impairments that develop over a long period of time.

Even after a significant amount of damage, a person with noise-induced hearing loss is able to hear common, low-frequency (vowel) sounds very well, but the high frequencies (consonants) in speech are not heard. Loudness levels may be nearly normal, but intelligibility will be poor in some situations. A noise-induced hearing loss becomes particularly noticeable when speech communication is attempted in noisy reverberant areas. Speech is masked most effectively by background noises having major frequency components in the speech frequencies, such as those found at a cocktail party.

3.5.1 Harmful Noise Exposures and Damage-Risk Criteria

Four major factors contribute to the development of noise-induced hearing impairment: (1) the noise level (loudness), (2) the frequency content (tone), (3) the time of exposure, and (4) the susceptibility of the individual. There is a very large number of different combinations of noise levels, frequency component distributions, and times of exposure. In addition, there are wide differences in the susceptibility of individuals to noise-induced hearing impairment. Studies leading to the development of comprehensive damage-risk criteria therefore require complex statistical studies of large groups over long periods of time.

There are also many factors that compromise the accuracy of most studies. For example:

1. Hearing changes found in a study may be due to causes other than noise exposures at work. Other common sensorineural hearing impairments caused by nontypical aging, childhood diseases, medications, blows to the head, and heredity factors may cause similar high-frequency losses and may produce similar audiograms for individuals.
2. Hearing changes found in a study may be due in part to noise exposures away from work that are not considered.

Thus it is impossible to set the particular exposure level as a dividing line between safe and unsafe conditions that would apply for all individuals. Damage-risk criteria are also compromised by practical limits. Usually exposure level limits are established as a compromise between (1) the amount of hearing impairment that may result from a specified exposure dose and (2) the economic or other impact that may result from noise control expenditures (Section 6).

3.6 The Problem: At Work

Comprehensive data are not available for an accurate determination of the number of people who have small degrees of noise-induced hearing impairment (1–3, 6, 109). The best estimates of the number of persons who have significant hearing impairment as a result of overexposure to noise are based on a comparison of the number of those with hearing impairments found in high-noise work areas with members of the general population, who have relatively low noise exposures (25). These studies show that significant hearing impairments for industrial populations are 10 to 30 percent greater for all ages than for general populations that have relatively low level noise exposures. At age 55, for example, 22 percent of a group that has had low noise exposures may show significant hearing impairment, whereas in an industrial high noise exposure group, the figure is 46 percent. Significant hearing loss is defined in all cases to be greater than 25-dB hearing level (referenced to the American National Standards Institute (ANSI) Specification S3.6-1969) averaged at 500, 1000, and 2000 Hz.

Unfortunately, there are very few persons in our society who are not exposed to potentially hazardous noise exposures away from work. Also, noise levels vary significantly from one location to another, and from one time to another, away from work as well as at work. Therefore very large samples must be used over very long periods of time to be meaningful. Most studies of this kind are necessarily limited to the extent that estimates for situations not covered are rough at best.

Another way of estimating the magnitude of the problem is to consider compensation costs. The Veterans Administration sets the cost of compensation for hearing loss compensation (primary disability) for all veterans at $187,546,536 for 1988, and at $2,111,453,508 for the period 1968–1988.

3.7 The Problem: Away from Work

High-level noise exposures away from work can be just as harmful as those at work, so all exposures must be considered in order to get an accurate assessment of a daily noise exposure. Potentially hazardous noises away from work may result from exposures to noises from chainsaws, airplanes, lawn mowers, motorboats, motorcycles, automobile or motorcycle races, farm equipment, loud music, shooting hobbies (skeet, targets, etc.), and shop tools. Even riding in a car at legal speeds with the window down or in subways may be harmful to some noise-sensitive individuals.

A hearing conservation program cannot be effective unless each individual observes the rules limiting noise exposure both at work and away from work. All hearing conservation programs must emphasize the need for constant awareness of noise hazards through a continuing program of education and enforcement.

4 HEARING MEASUREMENT

The only way to monitor the overall effectiveness of a hearing conservation program is to check periodically the hearing of all persons exposed to potentially hazardous noises. Air conduction hearing thresholds must be checked at 500, 1000, 2000, 3000, 4000, and 6000 Hz (Figure 23.8). It takes only a little more time to measure the threshold at 8000 Hz, however, so this is strongly recommended.

To assure the accuracy of air conduction thresholds, audiometers must be calibrated periodically and a quiet test environment must be maintained (5, 93). A well-trained audiometric technician or hearing conservationist must be used to perform these measurements (13). Requirements for audiometer performance and for the background noise limits in test rooms have been specified in standards published by the American National Standards Institute (ANSI) (5, 93, 94). Guidelines for training hearing conservationists have been established by the Council for Accreditation in Occupational Hearing Conservation (5, 13). These guidelines may be obtained from the Council for Accreditation in Occupational Hearing Conservation Manual, Fischler's Printing, Cherry Hill, NJ.

4.1 Audiometers

The instrument used for measuring pure tone, air conduction hearing thresholds, an audiometer, may be designed for manual or self-recording (sometimes called automatic) operation. Both the manual and self-recording instruments must be operated by properly trained persons (5, 13).

The ANSI Standard for Audiometers (93) provides the specifications that the instruments must meet to provide accurate information. Instrumentation inaccuracies are seldom obvious, and there is a strong tendency to accept dial readings as being accurate. Thus calibration is often neglected. Audiometers may lose their specified accuracy very quickly if they are handled roughly. Earphones are particularly susceptible to damage from rough handling; a janitor may drop them on the floor while cleaning and not inform the person in charge of testing. Rough handling during shipment could cause unseen damage so that even a new instrument may be out of calibration when received. The normal aging of components and heat may cause changes in audiometer accuracy. A common cause of overheating is storing an audiometer in a closed car on a warm day.

Dust or dirt inside the audiometer can cause switches to produce electrical noise or to wear excessively. Poor electrical contacts may produce intermittent operation or poor accuracy. Dust covers should always be used to protect the instrument when not in use, and the exterior of the instrument should be cleaned periodically to prevent dust and dirt from getting inside the case.

High humidity, salt air, and acid fumes may corrode electrical contacts in switches within an audiometer. The increased resistance that results from corroded contacts may cause electrical noise or affect the accuracy of the instrument.

If the operating characteristics of the audiometer change suddenly and the change is significant, the instrument should be serviced. Many changes in the instrument

ENVIRONMENTAL ACOUSTICS LABORATORY

AUDIOGRAM RECORD

ENVIRONMENTAL ACOUSTICS LABORATORY
E.A.L.
THE PENNSYLVANIA STATE UNIVERSITY

Name____________________ Title____________________

Soc. Sec. No.____________________ Birth date__________ Sex______

Date	Tester	Right Ear							Left Ear							Comments
		1000	2000	3000	4000	6000	8000	500	1000	2000	3000	4000	6000	8000	500	

Job Assignment	Date	C/T	Job Assignment	Date	C/T

Figure 23.8. Serial-type audiogram form.

occur slowly, however, and may not be noticed, so instruments must be calibrated periodically. OSHA requires that biological calibrations (audiograms taken on persons with known stable hearing thresholds) be performed before each day's use, and that acoustic calibrations be made at least annually (5, p. 9778). Accuracy checks should also be made any time there are any reasons to suspect a problem.

More than one biological calibration check per day may be desirable if a large number of thresholds are being measured. Biological checks made both before and after each workday would assure that no more than one day of testing would have to be repeated. With just one biological calibration each day, it may be necessary to repeat two full days of testing if the instrument is found to be out of calibration.

Hearing thresholds may drop as much as 20 dB temporarily because of allergies, colds, or other causes; therefore, it is strongly recommended that at least two normal-hearing persons be made available for biological tests. The hearing conservationist may serve as one if he or she has normal hearing.

The technician *should* also check the audiometer periodically using the following additional procedures:

1. Check all control knobs on the audiometer to be sure that they are tight on their shafts and not misaligned.

2. Straighten the earphone cords so that there are no sharp bends or knots. Worn or cracked cord covers should be replaced.

3. With the dials set at 2000 Hz and 60 dB, test the earphone cords for electrical continuity by listening while bending the cords along their length. Any scratching noise, intermittency, or change in test tone indicates a need for new cords. A new acoustical calibration is not required when replacing earphone cords.

4. Clean the earphone cushions regularly with a damp cloth. Cushions should be replaced if they are not resilient or if cracks, bubbles, or crevices develop. A new acoustic calibration is not required when replacing earphone cushions.

5. With the tone control set on 2000 Hz, check the linearity of the hearing level control by listening to the earphone while slowly increasing the hearing level from threshold. Each 5-dB step should produce a noticeable increase in loudness without changes in tone quality or any other audible extraneous noise.

6. With dials set at 2000 Hz and 60 dB, test the operation of the tone interrupter by listening to the earphones and operating the interrupter several times. No audible noises (such as clicks or scratches) or changes in tone quality should be heard when the interrupter switch is used.

7. With the hearing level control set at 60 dB and the test earphone jack disconnected from the audiometer, listen for extraneous noises from the case and other earphone. No noise should be heard when the tone control is switched to each test tone while wearing the earphones.

When an audiometer is sent to a laboratory for calibration, an understanding should be reached with the laboratory with regard to the kind of calibration to be performed. Some laboratories may calibrate audiometers by checking only a single hearing level for the various test tones, but this is not enough to assure accurate hearing threshold measurements. Hearing level accuracy should be checked for each test tone at each 5-dB interval throughout the operating range. For industrial applications, the range should cover at least 10 to 70 dB hearing levels (93). In addition, many other specifications listed in the ANSI Standard for Audiometers (93) should be checked including those on the attenuator linearity, interrupter operation, distortion, and test tone accuracy and purity.

Many audiometers in use today may not have been calibrated for several years, and many may not meet pertinent ANSI Standard specifications. Furthermore, differences of several decibels may be found in threshold measurement data taken with two audiometers that meet opposite extremes of the allowed limits. Thus the need for dependable calibration services that will produce accurate adjustments and correction data is emphasized. The pertinent performance specifications may not be well understood by some of the laboratories offering calibration services. Therefore a statement that the audiometer meets specifications should not be accepted without an explanation of the specific calibration procedures used and a copy of the calibration data.

4.2 Test Rooms

The sound pressure level of the background noise in rooms used for measuring hearing thresholds must be limited to prevent masking effects that cause misleading, elevated threshold values. The maximum allowable sound pressure levels for test rooms specified by the Occupational Safety and Health Administration (OSHA) (5) may mask some thresholds that are better (lower hearing levels) than above 10 dB hearing level, so quieter test areas are recommended.

More specifically, the maximum allowable sound pressure levels specified in ANSI S3.1-1977 (R 1986) are recommended in order to measure 0 dB hearing levels with no more than 1-dB threshold elevation for one-ear listening. These background noise levels (Table 23.4) are difficult to obtain in noisy areas, and may be below the practical limit for room noise in many industrial locations. Thus it has been the accepted practice for some industries to use the 10-dB hearing level as the lowest hearing threshold measurement level (5, 93, 94). If the 10-dB hearing level is to be used as the lowest threshold level measured, the background noise limits shown in Table 23.4 should be adjusted upward by 10 dB.

In addition to the noise level requirements of Table 23.4, subjective tests should be made inside the closed test booth on location to determine that no noises (e.g., talking and heel clicking) are audible. Any audible noise may distract the subject and interfere with the hearing threshold measurements. These noises must be eliminated, or the tests must be delayed until no noise can be heard. Short impulse-type noises, in particular, may be heard even though the measured sound pressure levels are below the limits in Table 23.4.

4.2.1 Portable or Prefabricated Rooms

The noise reduction provided by a good prefabricated audiometric test booth should be adequate to permit limited range threshold measurements (at and above 10-dB

Table 23.4. Maximum Allowable Sound Pressure Levels for One-Ear Listening for No More than a One-Decibel Threshold Elevation re the Reference Threshold Levels Given in ANSI S36-1969 (Rounded to Nearest 0.5 dB)

Test Tone Frequency (Hz)	Octave Band Levels (dB re 20 μPa)
125	34.5
250	23.0
500	21.5
750	22.5
1000	29.5
1500	29.0
2000	34.5
3000	39.0
4000	42.0
6000	41.0
8000	45.0

hearing level) in many reasonably quiet areas selected for hearing tests. However, it is recommended that the burden of on-location performance be placed on the supplier, to be sure that the booth is erected properly. The supplier should be willing to guarantee in writing the performance of the test room or booth when installed at the specified test site if the supplier does the installation. Assembly of the test room is an important part of the room's performance, and misunderstandings are common unless a clear agreement with a performance guarantee on their installation is included. This should also take care of unforeseen factors such as poor vibration isolation.

In addition to the attenuation characteristics and the cost of prefabricated test rooms, several other features should be considered when making a selection:

1. Size and appearance are important.
2. Will opening and closing the door result in wearing of contact material (seals), necessitating frequent replacement?
3. Are the interior surfaces durable and easily cleaned?
4. Is the door easily opened from the inside, so that subjects will not feel "locked in"?
5. The observation window and seating arrangement must provide an easy view of the subject, but the subject must not be able to see the technician or audiologist.

6. If the booth is equipped with a ventilation system, the noise should be below the limits specified for the room.
7. Portability of the room is seldom an important factor, because test rooms are rarely moved.

4.2.2 Test Rooms in a Quiet Building

A prefabricated hearing test room is not always necessary. Modifications in an existing quiet room may prove to be satisfactory. Generally, this room must be located some distance away from high-noise areas, but this depends on building structures and other factors.

Buildings with heavy masonry walls, floors, and ceilings usually provide good noise isolation if care is taken to prevent leakage paths such as small cracks that might be found around windows, doors, electrical fixtures, and pipes. A 1.5×1.5 in. hole in a wall transmits about the same amount of acoustic energy as 100 ft^2 of wall area that has a transmission loss of 40 dB. Wherever possible, all holes and cracks should be sealed permanently. Frequently used doors or windows should be sealed with flexible gaskets.

Leakage and reradiation of noise can occur through thin or light sections such as single-pane windows and doors. Additional noise reduction can be obtained by installing double doors or double-pane windows.

Structure-borne vibration can be transmitted through heavy walls and reradiated into the air of an enclosed space. If noise levels from structure-borne sources are high, "room within a room" construction may be required, or perhaps the conducting path can be broken by relatively simple means. Professional help is usually advisable when this solution is selected.

Although noise-absorbing materials within the room have little effect on the amount of noise leaking into the room from the outside, they will prevent noise levels from building up. For rooms of moderate size, adequate sound absorption is provided by a carpet on the floor and full drapes on two walls. If sound-absorbing treatment is used, it should be distributed on the ceiling and two adjoining walls for maximum effectiveness.

Two general principles are normally applied in the design of ventilation systems for hearing test rooms: one is to use long inlet and outlet ducts that are heavily lined with noise-absorbing material; the other is to use low air velocities that will minimize noise caused by air turbulence. The amount of noise reduction resulting from increased duct size, increased thickness of adsorption materials, and reduced air velocity may be found in the literature (157–161).

4.3 Hearing Threshold Measurements

Normally, the purpose of hearing testing in industrial hearing conservation programs is to monitor the effectiveness of hearing conservation procedures. If this is the case, there is no need for diagnostic information that would require the use of sophisticated audiometric techniques. Pure tone, air conduction hearing thresholds are usually adequate for industrial monitoring purposes (13).

Properly calibrated manual or self-recording audiometers may be used to monitor hearing thresholds in industry. The recommended procedure for manual threshold testing is as follows:

1. Check the audiometer to be sure that the earphone cords and electrical power connections are properly made.
2. Before the earphones are placed on the subject, make the following adjustments on the audiometer.
   a. Place the power switch in the "on" position.
   b. Set the interrupter control switch to the "normally off" position.
   c. Set the "hearing level" control to 0 dB.
   d. Set the "frequency" of "tone" control to 1000 Hz.
   e. Set the earphone selector to the "right ear" position.
3. Seat the subject facing away from the audiometer so that he or she cannot see the instrument or the operator. Give concise but complete instructions regarding the responses expected during the test procedures. For example, instruct the subject to raise a finger or hand immediately when he hears a test tone, to hold this response until he can no longer hear the tone, and then immediately lower the finger or hand.

   It is advisable to memorize (or read) a set of instructions to ensure that no aspect of the instructions will be omitted. During the instructions it is helpful to raise and lower the hearing level control slowly while the earphones are held about 1 ft from the subject's ear to illustrate the nature of a test tone. The subject should be advised that the pitch and level of the test tone will be changed during the test procedures.
4. Carefully place the earphones on the subject; they should be comfortably positioned directly over the entrance to the external ear canal. Be sure that the right earphone (red) is over the right hear and the left earphone (blue) is over the left ear. Earrings, hearing aids, glasses, and other obstacles must be removed from between the ears and the earphones.
5. Reset the hearing level control to 0 dB.
6. Press and hold the interrupter switch (to prevent the tone to the subject) and slowly increase the test tone level (by turning the hearing level control) until the subject signals that the tone is heard.
7. Release the tone interrupter switch (tone off). The subject should signal that he no longer hears the tone.
8. Press the interrupter switch and present the tone again at the same level for about 3 sec. The subject should signal that he hears the tone; if not, return to step 5.

   In the following steps all adjustments of tone level must be made with the tone off. Also, all presentations of the tone shall be for a duration of about 3 sec.
9. After the initial level of response has been established by two responses

from the subject, lower the tone level about 10 dB to the nearest multiple of 5 on the hearing level scale (e.g., if the initial level is 32 dB, lower the hearing level control to 20 dB).

10. Present the test tone at the lower level found in step 9.
11. If the subject does not respond to the tone at this level, raise the level by 5 dB and present the tone again. If the subject responds to the tone at the level established in step 9, lower the tone level by 5 dB and present the tone again. Presenting the tone once at each of these levels is sufficient. Additional attempts to improve the threshold values should not be made.
12. Continue the bracketing procedure in 5-dB steps as described in step 11 until two positive and two negative responses are found between consecutive 5-dB test levels (e.g., two positive responses are found at a 15-dB hearing level and two negative responses are found at a 10-dB hearing level). The threshold level for the 1000-Hz test tone is taken as the level corresponding to the positive responses (i.e., 15 dB).
13. Record the threshold value at 1000 Hz on the audiogram record (Figure 23.8). Conventionally, a red "O" is used for recording thresholds for the right ear and a blue "X" for the left ear if a graphical presentation is to be used. If the values are to be recorded in tabular form, the right-ear thresholds are conventionally recorded in red and the left-ear values in blue.
14. Repeat steps 5 through 13 to determine and record thresholds for the right ear at test tones of 2000, 3000, 4000, 6000, 8000, 500, and 1000 Hz, in that order.
15. Set the earphone selector to the "left-ear" position and repeat steps 5 through 14.
16. Remove the earphones and reset the audiometer as indicated in step 2.

The recommended procedure for using the self-recording audiometer is as follows:

1. Check the audiometer to be sure that the earphone cords and the electrical power connections are properly made.
2. Before the earphones are placed on the subject, make the following adjustments on the audiometer.
   a. Place the power switch in the "on" position.
   b. Place the audiogram recording chart on the recording audiometer (Figure 23.9).
   c. Place the recording pen at the start position at the upper left-hand corner of the chart.
3. Seat the subject so that he can see neither the audiometer nor the operator.
4. Give concise but complete instructions regarding the responses expected during the test procedures. For example, instruct the subject to push the control button when he hears a test tone and to hold it until he can no longer hear

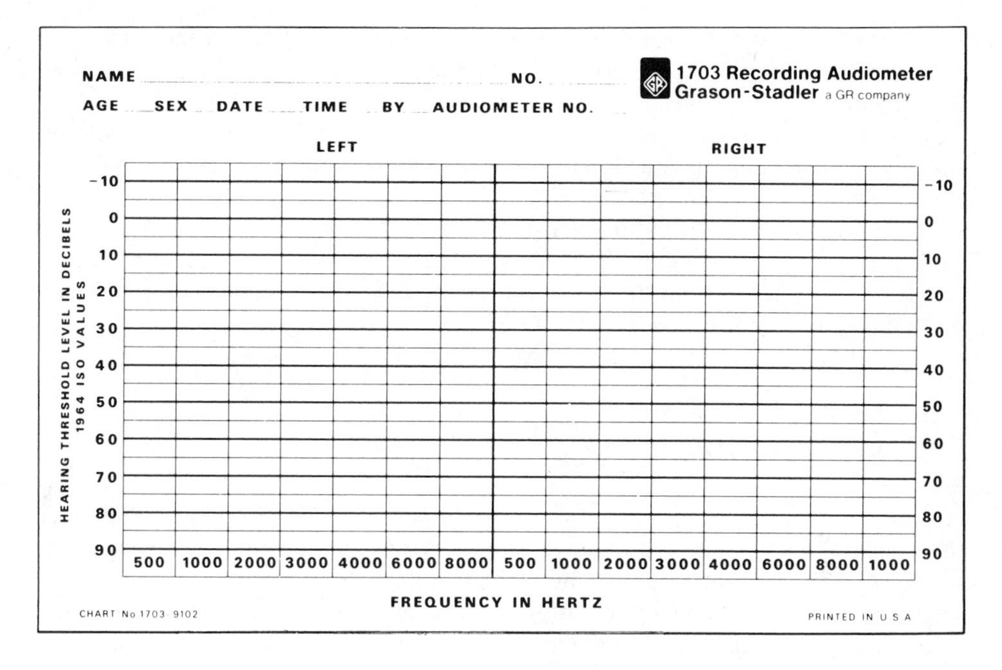

Figure 23.9. Recording-type audiogram form.

the tone, then immediately remove his finger from the button. It is advisable to memorize a set of instructions to ensure that no aspect of the instructions is left out. During the instructions it is helpful to raise and lower the hearing level control slowly while the earphones are held about 1 ft from the subject's ear to illustrate the nature of a test tone. The subject should be advised that the pitch and level of the test tone will be changed during the test procedures and that the test tone will automatically be transferred to the other ear during the test.

4. Carefully place the earphones on the subject; they should be comfortably positioned directly over the entrance to the external ear canal. Be sure that the right earphone (red) is over the right ear and the left earphone (blue) is over the left ear. Earrings, hearing aids, glasses, and other obstacles must be removed from between the ears and the earphones.
5. Reset the recording pen to the start position.
6. Press the start button.
7. Observe the test continuously to be sure that the instructions are being followed.
8. When the test is completed, remove the earphones and recording chart, then reset the recording pen to start position.

4.4 Records

The hearing conservationist's records must be complete and accurate if they are to have medicolegal significance. Records should be kept in ink, without erasures.

If a mistake is made in recording, a line should be drawn through the erroneous entry, and the initials of the person making the recording should be placed above the line along with the date. The entry must not be obliterated.

The model, serial number, and calibration dates of the instruments should be recorded on each audiogram. Records should also be kept of periodic noise level measurements in the test space.

In addition to the threshold levels, an audiogram should have space provided for the recording of pertinent medical and noise exposure information. Some typical questions are listed on the form given in Figure 23.10. Additional materials may be found in References 13, 16, 30, and 57.

5 NOISE MEASUREMENT

A wide variety of new sound measuring equipment has been made available over the past few years. There are sound level meters that provide only the basic overall weighted measurements required by OSHA and others that provide a very wide range of functions including integration for dose and impulse noise measurements. The sophisticated measuring equipment can be used to obtain an enormous amount of data in a relatively short period of time. The weak link in sound measuring equipment, the microphone, has also been much improved in both performance and durability in recent years.

Even with this much improved equipment, careful consideration of the objectives of the measurement must be made before the equipment is selected. The microphone must function reliably over the range of sound frequencies and levels that are to be measured, and it must be readily adaptable for use with any instruments selected (105–113). The microphone must also be capable of operating in the temperature and humidity conditions that may be encountered, and its operation must not be adversely affected by other environmental factors such as electromagnetic energy or vibration.

If sound levels change significantly in a short time, conventional sound level meters may not respond fast enough to afford accurate readings (105); therefore special instrumentation—such as sound level meters with integrating circuits, impulse meters, or oscilloscopes—is required (108). In all cases, accurate sound level measurements require a well-trained operator and calibration equipment. A calibrator specifically designed for the microphone must be used before and after the measurements to assure the required accuracy (112).

5.1 The Sound Level Meter

The basic sound level meter consists of a microphone, an amplifier–attenuator circuit, and an indicating meter. The microphone transforms airborne acoustic pressure variations into electrical signals and feeds them to a carefully calibrated amplifier–attenuator circuit. The electrical signals are then directed through a

Name__________ Date________ Recorded by________ Employee initial________

HISTORY	YES	NO	Comments
Have you had a previous hearing test?			
Have you ever had hearing trouble?			
Do you now have any trouble hearing?			
Have you ever worked in a noisy industry?			
Do you think you can hear better in your Right ear?			
or Left ear?			
Have you ever had noises in your ears?			
Have you ever had dizziness?			
Have you ever had a head injury?			
Has anyone in your family lost his hearing before age 50?			
Have you ever had measles, mumps. or scarlet fever?			
Do you have any allergies?			
Are you now taking or have you regularly taken drugs, antibiotics, or medication?			
Have you ever had an earache?			
Have your ears ever run? Right ear?			
Left ear?			
Have you been in the Military service? Describe			
Have you been exposed to any sort of gunfire? Describe			
Do you have a second job? Explain			
What hobbies do you have?			

Figure 23.10. Medical history form.

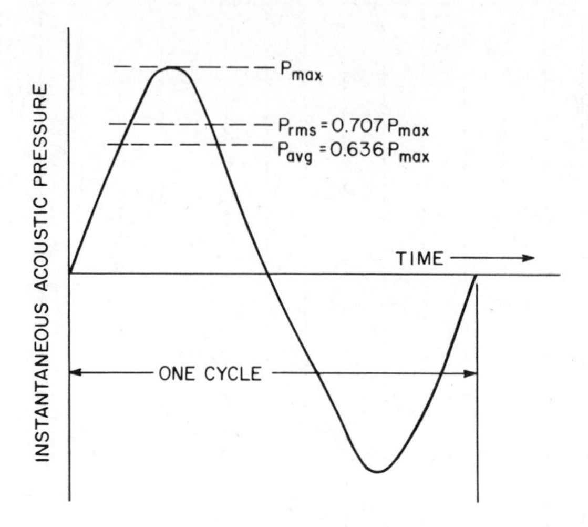

Figure 23.11. A comparison of maximum p_{max} or peak, root-mean-square p_{rms}, and rectified-average p_{avg} values of acoustic pressure.

logarithmic weighting network to an indicating meter, where the sound pressure is displayed in dB re 0.00002 Pa.

Most sound-measuring instruments present the sound pressure levels in terms of its rms value, which is defined as the square root of the mean-squared displacements during one period. The rms value is useful for hearing conservation purposes because it is related to acoustic power and it correlates well with human response. Also, the rms value of a random noise is directly proportional to the bandwidth; hence the rms value of any bandwidth is the logarithmic sum of the rms values of its component narrow bands. For example, octave band levels may be added logarithmically to find the overall level for the frequency range covered. The rms value of pure tones or sine waves is equal to 0.707 times the maximum value (Figure 23.11).

The rms values cannot be used to describe prominent peak pressures of noise that extend several decibels above a relatively constant background noise; maximum, or peak, values are used for this purpose. On the other hand, peak readings are of relatively little value for measuring sustained noises unless the wave form is known to be sinusoidal, because the relationship between the peak reading and the acoustic power changes with the complexity of the wave. As the wave form becomes more complex, the peak value can be as much as 25 dB above the measured rms value.

In addition to the rms and peak values, a rectified average value of the acoustic pressure is sometimes used for noise measurements. A rectified average value is an average taken over a period of time without regard to whether the instantaneous signal values are positive or negative. The rectified average value of a sine wave is equal to 0.636 times the peak value. For complex wave forms, the rectified average value may fall as much as 2 dB below the rms value. In some cases rectified average characteristics have been used in sound level meters by adjusting the output to read 1 dB above the rms level for sine wave signals, to ensure the average reading in always within 1 dB of the true rms value. Figure 23.11 compares rms, maximum, and rectified average values of a sinusoidal wave.

5.1.1 *Meter Indication and Response Speed*

The indicating meter of a sound level meter may have ballistic characteristics that are not constant over its entire dynamic range, or scale, and these will result in different readings, depending on the attenuator setting and the portion of the meter scale used. When a difference in readings is noted, the reading should be taken using the higher part of the meter scale (the lowest attenuator setting), because the ballistics are generally more carefully controlled in this portion of the scale.

A general-purpose (Type 2) level meter, still included in OSHA regulations, has fast and slow meter response characteristics that may be used for measuring sustained noise (5, 105). The fast response enables the meter to reach within 4 dB of its calibrated reading for a 0.2-sec pulse of 1000 Hz; thus it can be used to measure, with reasonable accuracy, noise levels that do not change substantially in periods less than 0.2 sec. The slow response is intended to provide an averaging effect that will make widely fluctuating sound levels easier to read; however, this setting will not provide accurate readings if the sound levels change significantly in less than 0.5 sec. Newer sound level meters meet tighter specifications than the Type 2 (105).

5.1.2 *Frequency-Weighting Networks*

General-purpose sound level meters are normally equipped with three frequency-weighting networks (A, B, and C) that can be used to approximate the frequency distribution of noise over the audible spectrum (43). These three frequency weightings, given in Table 23.3, were chosen because (1) they approximate the response characteristics of the ear at various sound levels, and (2) they can be easily produced with a few common electronic components. A linear, flat, or overall response, also included on some expensive sound level meters, weights all frequencies equally.

The A-frequency weighting approximates the response characteristics of the ear for low-level sound (below about 55 dB re 0.00002 Pa). The B-frequency weighting is intended to approximate the response of the ear for levels between 55 and 85 dB, and the C-frequency weighting corresponds to the response of the ear for levels above 85 dB.

In use, the frequency distribution of noise energy can be approximated by comparing the levels measured with each of the frequency weightings. For example, if the noise levels measured using the A and C networks are approximately equal, it can be reasoned that most of the noise energy is above 1000 Hz, because this is the only portion of the spectrum in which the weightings are similar. On the other hand, a large difference between these readings indicates that most of the energy will be found below 1000 Hz.

Many specific uses have been made of the individual weightings in addition to the frequency distribution of noise. In particular, the A network has been given prominence in recent years as a means for estimating annoyance caused by noise (255) and for estimating the risk of noise-induced hearing damage (1, 5).

5.2 Microphones

The three basic types of microphones used for noise measurements are piezoelectric, dynamic, and condenser. Each type has advantages and disadvantages that depend on the specific measurement situation; all three types can be made to meet the American National Standard Specification for Sound Meters (S1.4.1971) (44, 275). The dynamic microphone has not been used as much as the other two types in recent years, but older ones are still suitable for some applications.

New piezoelectric and condenser microphones (1) are much less expensive than earlier ones, (2) have excellent frequency responses from a few hertz to 10,000 Hz (and up to 100 kHz if needed), and (3) have good reliability records. It is often said that it is difficult to find a bad microphone these days. Piezoelectric and condenser microphones are now provided as original equipment with most sound measuring instruments with tighter tolerance limits as included in the ANSI Specification for Sound Level Meters (105).

Most new sound measuring instruments cover a dynamic range of at least 40 to 140 dB re 0.00002 Pa. More expensive equipment permits measurements in octave bands from about 0 dB to well above 160 dB, depending on the microphones selected. Most modern noise-measuring instruments are not damaged by exposures to normal ranges of temperature and humidity; however, temporary erroneous readings may result from condensation when they are moved from cold locations to warm, humid areas.

5.2.1 Microphone Directional Characteristics

Most noises encountered in industry are produced from many different noise sources combined with their reflected energies. Noise usually comes from many directions and may be considered to be randomly incident upon a microphone diaphragm. For this reason, microphones are sometimes calibrated for randomly incident sound.

Depending on the design and purpose of a microphone, it may be calibrated for grazing, perpendicular, or random incidence, or it may be calibrated for use in couplers (pressure calibration). Thus care must be taken to use the microphone in the manner specified by the manufacturer.

A microphone calibrated with randomly incident sound should be pointed at an angle to the major noise source that is specified by the manufacturer. An angle of about 70° from the axis of the microphone is often used to produce characteristics similar to randomly incident waves, but the angle for each microphone should be supplied by the manufacturer.

A free-field microphone is calibrated to measure sounds perpendicularly incident to the microphone diaphragm; thus it should be pointed directly at the source to be measured. A pressure-type microphone is designed for use in a coupler such as

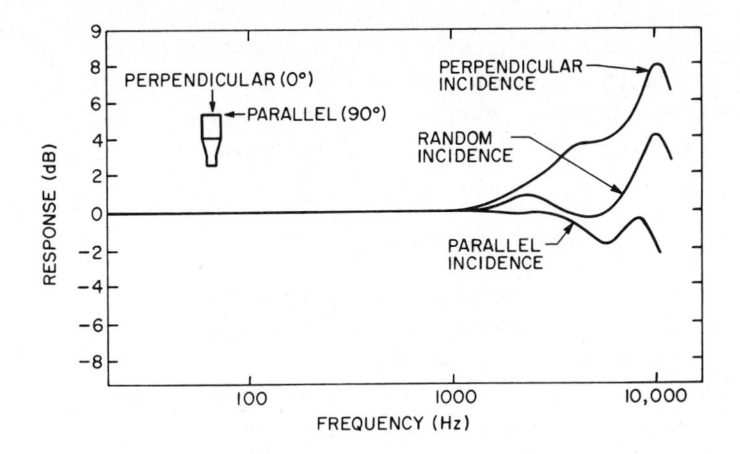

Figure 23.12. Directional characteristics of a piezoelectric microphone.

those used fōr calibrating audiometers; however, this microphone can be used to measure noise over most of the audible spectrum if the noise propagation is at grazing incidence to the diaphragm and the microphone calibration curve is used.

Microphones commonly used with sound measuring equipment are often said to be nearly omnidirectional but, strictly speaking, this is generally true only for frequencies below 1000 Hz (Figure 23.12). When measurments are to be made of high-frequency noise produced by a directional noise source (i.e., where a high percentage of the noise energy is coming from one direction), the orientation of these microphones becomes very important.

Directional characteristics of microphones may be used to advantage at times. For example, an improved signal-to-noise ratio may be obtained for sound pressure level measurements of a given source by using 0° incidence when high background levels are being produced by sources at other locations. Erroneous readings caused by reflected high-frequency sound emitted by other sources but coming from the same direction may be checked with directional microphones by rotating the microphone about an axis coincident with the direction of incident sound. Reflected energy is evidenced by a variation in level as the microphone is rotated. The microphone orientation that corresponds to the lowest reading should be chosen, because the reflection error would be minimized at this position.

Special-purpose microphones with sharp directional characteristics may also be used to advantage in some locations. These microphones are particularly useful for locating specific high-frequency noise sources in the presence of other noise sources.

5.2.2 Microphone Cables

Cables may be used with many microphones, but instructions must be followed for each microphone to avoid errors. Cable corrections may or may not be required for modern microphones, depending on the preamplifier design and the calibration data that are provided. Some microphones are calibrated when mounted directly on sound measuring equipment; others are calibrated with cables attached, and others have built-in preamplifiers that permit the use of certain cables (certain lengths and capacitance). Instructions with the microphones usually provide this information.

5.2.3 *Temperature and Humidity Effects*

Corrections may also be necessary for temperature effects, particularly if a high level of accuracy is required. These corrections should be included in instruction manuals supplied by the instrument manufacturers.

There are also cases where high humidity may significantly affect the performance of a condenser microphone, and *warnings may or may not be given in the instructions*. If an unusually high erratic reading occurs when used in high humidity it may be necessary to store a condenser microphone in a container with a desiccant for several hours before use.

5.3 Frequency Analyzers

The rough estimate of frequency-response characteristics provided by the weighting networks of a sound level meter is not always adequate. In such cases the output of the sound level meter can be fed into a suitable analyzer that will provide more specific frequency distribution characteristics of the sound pressure (45). The linear network of the sound level meter should be used when the output is to be fed to an analyzer. If the sound level meter does not have a linear network, the C-network may be used for analyses over the major portion of the audible spectrum (see Table 23.3). Alternatively, a combination sound level meter and analyzer can be used.

5.3.1 *Octave Band Analyzers*

The octave band (OB) analyzer is the most common type of noise filter used for hearing conservation purposes because the OB normally provides adequate spectral information with a minimum number of measurements. An OB is defined as any bandwidth having an upper band-edge frequency f_2 equal to twice the lower band-edge frequency f_1 (Equation 14).

Octave band analyzers generally have their bands centered at 31.5, 63, 125, 250, 500, . . ., 8000 Hz (S1.6-1984) (83). The center frequency (geometric mean) of an OB, or other bandwidths (f_c), is found from the square root of the product of the upper (f_2) and lower (f_1) band-edge frequencies (Equation 15). Octave band-edge frequencies corresponding to the preferred center frequencies can be calculated using two equations (14 and 15) with two unknowns. For example, the band-edge frequencies corresponding to a center frequency (f_c) of 1000 Hz can be calculated as follows:

$$f_2 = 2f_1 \tag{14}$$

and

$$f_c = (f_1 f_2)^{1/2} \tag{15}$$

Substituting for $f_2 = 2f_1$ from Equation 14 into Equation 15,

$$f_c = 1000 = (f_1 \times 2f_1)^{1/2} = f_1(2)^{1/2}$$

Therefore $f_1 = 1000/1.414 = 707$ Hz and, from Equation 14, $f_2 = 2 \times 707 = 1414$ Hz.

5.3.2 *Half-Octave and Third-Octave Analyzers*

When more specific spectral information is required than provided by octave bands, narrower bands must be used. The number of measurements necessary to cover the overall frequency range increases as the bandwidth of the analyzer is made smaller; thus a compromise must be reached between the resolution required and the number of measurements required.

Half-octave and one-third octave filters are the next steps in resolution beyond octave band analyzers. A half-octave is a bandwidth with an upper band-edge frequency equal to $(2)^{1/2}$ times its lower band-edge frequency. A third-octave has an upper band-edge frequency that is $(2)^{1/3}$ times its lower band-edge frequency.

5.3.3 *Adjustable Bandwidth Broad Band Analyzers*

Some analyzers are designed with independently adjustable upper and lower band-edge frequencies. This design permits a wide selection of bandwidths in octaves, multiples of octaves, or fractions of an octave. The smallest fraction of an octave usually available on these adjustable bandwidth analyzers is about one-tenth, and the largest extends up to the overall reading.

In addition to the obvious advantage of being able to choose the proper bandwidth for a particular job, these analyzers permit the selection of any octave, rather than a preselected series of octaves. The disadvantage of these instruments is their relatively large size.

5.3.4 *Narrow Band Analyzers*

Analyzers with bandwidths narrower than tenth-octaves are normally referred to as narrow-band analyzers. These analyzers may have either constant-percentage bandwidths, or constant bandwidths, and they are often continuously adjustable.

The constant-percentage narrow band analyzer is similar to the octave band analyzer in that in both cases the bandwidths vary with frequency. As its name indicates, a constant-percentage analyzer has a bandwidth that is a constant percentage of the center frequency to which it is tuned. Typically, a bandwidth of about $\frac{1}{30}$-octave might be selected with these analyzers.

The bandwidth of a constant bandwidth analyzer remains the same (constant) for all center frequencies. Provision may be made on some instruments to change the bandwidths, but typically a constant bandwidth analyzer remains fixed at a few hertz.

The constant bandwidth analyzer normally provides a narrower bandwidth and better discrimination outside the passband than the constant percentage analyzer; therefore it is often the best choice when discrete frequency components are to be measured. Also, it usually covers the entire spectrum with a single dial sweep, thus facilitating coupling to recorders for automatic analysis.

Most constant percentage analyzers require band switching to sweep the audible spectrum. On the other hand, caution must be exercised to avoid serious errors when constant bandwidth analyzers are used to analyze noises that have frequency modulation, or warbling, of components. Frequency-modulated noises are commonly produced by reciprocating-type noise sources in some machinery. Frequency-modulated noise is not a major problem if constant percentage analyzers are used.

5.4 Measurement of Impulse or Impact Noise

The inertia of the indicating meters of general-purpose sound level meters prevents accurate, direct measurements of single-impulse noises that change level significantly in less than 0.2 sec. Typical noises with short time constants are those produced by drop hammers, explosives, and other noises with short, sharp, clanging characteristics.

If much detailed information is required, impulse noise characteristics may be measured directly from a calibrated oscilloscope with a long persistence screen. Photographic accessories may also be used to obtain permanent records.

The oscilloscope is usually connected to the output of a sound level meter having a wide frequency response and calibrated with a known sound level of sinusoidal characteristics. The screen of the oscilloscope can be calibrated directly in decibels (rms) by comparing the oscilloscope deflection produced by a sinusoidal signal with the reading of the second level meter. Several calibration points may be fixed on the oscilloscope screen by providing various signal levels into the sound level meter, or the scale may be determined from a single calibration level by using linear equivalents to decibels.

For example, half of a given deflection for a sine wave signal will be equivalent to a 6-dB drop in level on an oscilloscope, and 0.316 times the deflection will be the equivalent to a 10-dB drop in level. These equivalent values may be calculated from the equation

$$\text{drop in level (dB)} = 20 \log_{10}\left(\frac{d_1}{d_2}\right)$$

where d_1 and d_2 are the small and large linear screen deflectors being compared. It should be noted that this calibration, using a sine wave, is for convenience, and that a constant factor of 3 dB must be added to the rms calibration to obtain the true instantaneous peak values for sine waves. The relationship of rms to peak values is more complex for nonsinusoidal waves (108).

Care must be taken while using an oscilloscope driven by a sound level meter to prevent errors resulting from overloading. If the oscilloscope deflections show a sharp clipping action at a given amplitude (flat top on a sine wave), the attenuator settings on one or both instruments may require adjustment upward. Also, a check should be made to determine whether the indicating meter of the sound level meter affects the wave form produced on the oscilloscope. This may be done by switching the meter out of the circuit, to a battery-check position, and observing the wave form. If the oscilloscope wave form is changed in any way by the indicating meter, it should be removed from the circuit each time a deflection is measured on the oscilloscope.

The oscilloscope is inconvenient to use in many field applications because it is relatively large and complex. Also, many oscilloscopes require ac power, and the supply voltage may vary in the field, causing changes in calibration.

For field applications it is often convenient to use peak-reading impact noise analyzers. These battery-driven instruments do not provide as much information as the oscilloscope, but they are often adequate. The electrical energy produced by an impulse noise is stored for a short time by these instruments in capacitor-type circuits, permitting the acquisition of information on the maximum peak level, on the average level over a period of time, and on the duration of the impact noise. As with the oscilloscope, care must be taken not to overload the sound level meter driving these impact noise meters.

5.5 Magnetic Field and Vibration Effects

The response of sound level meters and analyzers may be affected by strong alternating magnetic fields found around some electrical equipment and furnaces. Dynamic microphones, coils, and transformers are particularly susceptible to hum pickup from these fields, but other types of microphone are also susceptible, and some condenser microphones may also be affected by strong magnetic fields. It is good practice to check the operation of sound level meters any time there is any suspicion that a strong magnetic field exists. A check can be made conveniently by replacing the microphone cartridge with a dummy microphone. The equipment manufacturer can supply information on building a dummy microphone.

Vibration of the microphone or measuring instrument may also cause erroneous readings, and in some cases strong vibrations may permanently damage the equipment. It is always good practice to isolate sound measuring equipment mechanically from any vibrating surface. Holding the equipment in the hands or placing it on a foam rubber pad is satisfactory in most cases. To determine whether the equipment is being affected by vibration, observe the meter reading while the noise source is shut off, if this can be done without changing the vibration. If the meter reading drops by more than 10 dB when the noise of shut off, the effects of vibration are not significant. If the noise cannot be shut off without changing the vibration, the same result can be obtained by replacing the microphone cartridge with a dummy microphone. The equipment manufacturer should be able to supply information on building a dummy microphone.

5.6 Tape Recording of Noise

It is sometimes convenient to record a noise so that it can be analyzed at a later date. This is particularly helpful when lengthy narrow band analyses are to be made, or when very short, transient-type noises are to be analyzed. To avoid errors, however, extreme care must be taken in the calibration and use of the recorder. Also, direct sound pressure measurement and analysis should be made during the recording procedure so that the operator will know when additional measurements or data are necessary.

Many professional or broadcast-quality tape recorders are satisfactory for noise-recording applications; however, the microphone must have the proper characteristics. Frequently the specifications given for a tape recorder do not include the microphone characteristics, and the microphone may be of very poor quality for some purposes. When the tape recorder is not specifically built for measuring noise, it is usually good practice to connect the input of the recorder to the output of a properly calibrated sound level meter. As is the case when attaching any accessory equipment to sound level meters, it is important that the impedances be properly matched. The bridging input of a tape recorder is satisfactory for the output circuits of most sound level meters.

When a tape recorder is used to record noises that have no high prominent peaks, the recording level should be set so that the recorder meter (VU meter) reads between −6 and 0 dB. This setting assumes that a sinusoidal signal reading of +10 dB on the VU meter will correspond to about 2 or 3 percent distortion according to standard recording practice.

If the recorded noise has prominent peaks, it is good practice to make at least two additional recordings with the input attenuator set, giving recording levels between −6 and 0, −16 and −10, and −26 and −20 VU. If there is less than 10 dB between any two of these adjacent 10-dB steps, overloading has occurred at the higher recorded level, and the lower of the two steps should be used.

It is important to calibrate the combination of tape recorder and sound level meter at known level and tone control settings throughout the frequency range before the recordings are made. Before each series of measurements, a pressure level calibration should be made by recording the overall sound pressure level reading by stating the levels orally, along with the tape recorder dial settings. It is also good practice to state orally on each recording the type and serial numbers of the microphone and sound level meter, the location and orientation of the microphone, the description of the noise source and surroundings, and other pertinent information. This practice of noting information orally on the tape often prevents information from being lost or confused with other tapes.

5.7 Graphic Level Recording

A graphic level recorder may be coupled to the output of a sound level meter or analyzer to provide a continuous written record of the output level. Many graphic level recorders provide records in the conventional rms logarithmic form used by sound level meters; thus the data may be read directly in decibels. Some older recorders use rectified average response characteristics, and corrections must be made to convert these recordings to true rms values. As with sound level meters, these recorders are intended primarily for the recording of sustained noises without short or prominent impact-type peak levels. The equipment manufacturer or instruction manuals should be consulted to determine the limitations of each graphic level recorder.

5.8 Instrument Calibration

If valid data are to be obtained, it is essential that all equipment for the measurement and analysis of sound be in calibration. When equipment is purchased from the manufacturer, it should have been calibrated to the pertinent ANSI specifications (105–112). However, it is the responsibility of the equipment user to keep the instrument in calibration by periodic checks.

Most general-purpose sound measuring instruments have built-in calibration circuits that may be used for checking electrical gain. Acoustic calibrators are available for checking the overall acoustical and electrical performance at one or more frequencies. These electrical and acoustic calibrations should be made according to the manufacturer's instructions at the beginning and at the end of each day's measurements. A battery check should also be made at these times. These calibration procedures cannot be considered to be of high absolute accuracy, nor will they allow the operator to detect changes in performance at frequencies other than that used for calibration. They do serve as a warning of most common instrument failures, thus avoiding many invalid measurements.

Periodically sound measuring instruments should be sent back to the manufacturer or to a competent acoustic laboratory for calibration at several frequencies throughout the instrument range. These calibrations require technical competence and the use of expensive equipment. How frequently these complete calibrations should be made depends on the purpose of the measurements and how roughly the instruments are handled. In most cases it is good practice to have a complete calibration performed at least once a year, and at any time an unusual reading is found.

6 NOISE EXPOSURE LIMITS AND OSHA

6.1 Background

More than 20 years have passed since the Safety and Health Standards for Federal Supply Contracts (Walsh–Healy Public Contracts Act), U.S. Department of Labor, was revised on May 20, 1969 (2) to include the first national noise exposure regulations in the United States. These regulations were based on information provided by an Intersociety Committee on Noise Exposure Control (268) and the National

Institute for Occupational Safety and Health (NIOSH) (1), and on the Threshold Limit Value (TLV) of Noise developed by the American Conference of Governmental Industrial Hygienists (ACGIH) in 1968 (109).

The rules and regulations promulgated under the Walsh–Healy Public Contracts Act were then adopted by OSHA. The Occupational Safety and Health Act of 1970 (2), which was established by U.S. Public Law 91-596, uses essentially the same noise exposure criteria as the Walsh–Healy regulations but encompasses a much wider general workplace coverage.

The noise exposure limits set forth in the 1970 standards have separate criteria for continuous and impulsive noises. A continuous noise exposure limit of 90 dB, measured with an A-frequency weighting, is set for exposure of 8 hr/day, with higher levels being permitted over less time at the rate of 5 dB for halving of exposure time (i.e., 95 dBA for 4 hr, 100 dBA for 2 hr, 105 dBA for 1 hr, 110 dBA for 30 min, and 115 dBA for 15 min). Exposures to continuous noise levels greater than 115 dBA are not allowed under any circumstances. When the daily noise exposure is composed of two or more periods of exposure at different (non-impulsive) levels, their combined effect is determined by adding the individual contributions as follows:

$$\frac{C_1}{T_1} + \frac{C_2}{T_2} + \frac{C_3}{T_3} + \ldots + \frac{C_N}{T_N}$$

The values C_1 to C_N indicate the times of exposure to specified levels of noise, and the corresponding values of T indicate the total time of exposure permitted at each of these levels. If the sum of the individual contributions exceeds 1.0, the mixed exposures are considered to exceed the overall limit value. For example, if a person were exposed to 90 dBA for 5 hr, 100 dBA for 1 hr, and 75 dBA for 3 hr during an 8-hr working day, the times of exposure are $C_1 = 5$ hr, $C_2 = 1$ hr; and $C_3 = 3$ hr; the corresponding OSHA time limits for these exposures are $T_1 = 8$ hr, $T_2 = 2$ hr, and $T_3 =$ infinity. Therefore, because 3 divided by infinity is zero, there is no contribution from the 75-dBA exposure. Hence the combined exposure dose for this person is $\frac{5}{8} + \frac{1}{2} = 1.125$ (about 113 percent), which slightly exceeds the specified limit of 1.0 (100 percent).

The limit for impulse noise exposures is simply 140-dB peak sound pressure level with no limit for the number of discrete impulses allowed per day at levels below 140 dB. Formulas having other limiting numbers and levels were proposed by OSHA in 1974 but these algorithms have never been adopted.

A significant hearing conservation amendment was made to the OSHA occupational noise standard in 1983 (5). Some of the amended sections are discussed in the following sections.

6.2 Base-Line Audiograms

A base-line audiogram must be established within 6 months of an employee's first exposure at or above an 8-hr time-weighted average (TWA) of 85 dB A-weighted (dBA) or equivalently, a dose of fifty (50) percent. All subsequent audiograms are to be compared to the base-line audiogram to determine if changes have been made in hearing thresholds.

New base-lines may be established if the audiologist or physician in charge of the hearing conservation program determines that old audiograms are not valid, if a significant threshold shift (STS) is established, or if the annual audiogram indicates significant improvement over the baseline audiogram (4, page 9777).

NOTE: Not all audiologists or physicians are qualified industrial hearing conservationists. Employers should carefully investigate the professional qualifications and experience of any person employed for industrial hearing conservation. Generally, expertise only in clinical practice or certification in specialty fields other than hearing conservation is not sufficient.

6.3 Annual Audiograms

Annual audiograms must be taken on all employees who are exposed to an 8-hr time-weighted average (TWA) noise exposure of 85 dBA or greater. If a STS, "a change in hearing threshold relative to the baseline audiogram of an average of 10 dB or more at 2000, 3000, and 4000 Hz in either ear," is established by an annual audiogram, it must be reported by letter to the affected employee within 21 days. No details are given on what is to be done pending the results of a retest, but if the retest supports the finding of a STS, it might be assumed that another 21 days would be allowed to report the results to the employee. If no STS is found on the retest, it might be assumed that no letter is required. Exceptions of longer time periods are given when thresholds are measured by outside firms using testing vans.

In practice, a retest is justified in most cases when a STS is found because there are many common reasons why a 10-dB STS may be temporary. All retests should be completed within 21 days to avoid unnecessarily upsetting an employee with a letter advising him or her of a STS *if* there is none.

When a STS is established, a letter must be written to the employee advising him or her of this fact within the 21-day period. This person's annual audiogram should be used in the future as the new base-line audiogram. Old audiograms must not be discarded when new base-lines are established for an employee (4, page 9777). The letter used to inform an employee of a STS should also be used to educate and motivate the employee to take better care of his or her hearing both at and away from work. A sample letter is presented in Appendix 23.1.

6.3.1 Chronic Hearing Problems

Chronic hearing problems may cause temporary but significant changes in hearing levels over relatively short periods of time, so annual audiograms on these persons will have doubtful value during these times. Even though these annual audiograms may cause confusion, and arguments may be given that audiograms are being taken by the employee's physician, annual audiograms must be taken by the employer in any case.

NOTE: Care should be taken to make detailed notes, in ink, on *all* of these audiograms explaining that the observed hearing levels may have been caused by this chronic problem. Not only are the annual audiograms required by OSHA, but the employer should monitor all employees' hearing levels in any case because the employee's physician may not be qualified to take audiograms, and legal claims may be based on the physician's audiograms.

6.3.2 Record Retention

The 1983 amendment requires that audiograms be retained for the length of employment for the affected employee. Questions have arisen regarding whether or not a new base-line audiogram can be established each time an employee is terminated and rehired. Obviously, OSHA does not intend that an employee can be deliberately fired and rehired within a short time period so that a new base-line audiogram can be established. Such actions would, in fact, be undesirable for both the employer and the employee. Both should profit by having complete medical records retained after the employee is terminated for several reasons, including compensation claims and general health references.

In most cases, records should be retained at least 2 years, and in many cases longer, after an employee is terminated. These records could be valuable if there is a question of legal responsibility for an employee's hearing impairment, or if the employee is to return to this job. Termination audiograms are not required by OSHA, but they should be seriously considered by employers because of potential liability problems (4, p. 9778).

6.4 Inclusion of 85 dBA in the *C/T* Calculation

The limit of 85 dBA for 8 hr is added as an action level in the 1983 amendment. The 1983 OSHA amendment states that the "employer shall administer a continuing, effective hearing conservation program, as described in paragraphs (c) through (o) of this section, whenever employee noise exposures equal or exceed an 8-hour time-weighted average sound level (TWA) of 85 dB measured on the A-scale (slow response), or equivalently, a dose of fifty percent." This is to say that all employees with an exposure dose of 50 percent or higher shall have their hearing thresholds measured. If any of these persons develop a STS they must wear hearing protectors (4, pp. 9776–9779).

6.5 Integration of Steady-State and Impulsive Noise

The 1983 OSHA amendment states that "all continuous, intermittent, *and impulsive* sound levels from 80 decibels to 130 decibels shall be integrated into the noise measurements." The earlier limit for impulsive noise exposures (140-dB peak level) is still in effect, so there are two procedures for limiting impulsive sound exposures (4, pp. 9776–9780).

The integration of continuous and impulsive noises is done automatically if exposure levels are measured using a dosimeter. Sound level meters including the old ANSI Type 2 instruments are still allowed for measuring continuous noise exposures, using the limits discussed in the preceding sections. Therefore impulse noises must then be measured separately with an impulse measuring instrument, or a calibrated oscilloscope, using the separate limit of 140 dB peak sound pressure level. There is no formula for calculating an integrated dose from separate measurements of continuous and impulsive noise exposures.

The 1974 proposed graduated scale for impulsive noise exposure limits (140 dB for the first 100 impulses, 130 dB for between 100 and 1000 impulses, 120 dB for between 1000 and 10000 impulses, etc., in an 8-hr period) has not been adopted by OSHA. However, this conservative guideline is recommended.

6.6 Impulse Noise limits

When constant or steady-state noise is measured with sound level meters, the impulse noise exposure limit is 140 dB. There are no adjustments for the number of impulses.

6.7 Practical Considerations

The limits specified in the OSHA noise exposure regulations are the most restrictive limits deemed feasible with due consideration given to other important factors, such as economic impact. These limits are not intended to provide complete protection for all persons. It is estimated that 85 percent of persons exposed to the OSHA limits of 90 dBA (TWA) for 8 hr/day, 5 days a week, for about 10 years, will not develop a significant hearing impairment. The other 15 percent of persons exposed to these limits would probably have various levels of hearing impairment depending upon their susceptibility. Exposures outside the workplace may contribute significantly to the hearing impairment for some persons.

OSHA exposure limits are intended only as *minimum* action levels. Wherever feasible, hearing conservation measures should be extended to reduce exposure levels *below the limits specified* both at and away from the workplace.

7 HEARING CONSERVATION PROGRAMS

7.1 General

The objective of hearing conservation programs is to prevent noise-induced hearing loss. This obvious fact is often forgotten, or ignored, when pressures are applied for compliance with local, state, or federal rules and regulations on noise exposures. Because compliance with rules and regulations will not always prevent noise-induced hearing impairment in susceptible individuals, every effort should be made to use limits lower than OSHA's wherever possible.

Compromises of noise exposure limits have been made depending upon economic feasibility in most rules and regulations. The OSHA rules are no exception. Most of these rules and regulations have been based on exposure levels from all work areas or industries, regardless of the level or difficulty of the noise problems. That is, the relative economic impact of any given limiting noise dose on typical "heavy industries" having relatively high noise exposure levels can be expected to be higher than on typical "light industries" having significantly lower noise exposure levels. This approach was chosen by OSHA to provide simpler enforcement conditions. Hence it is economically feasible in many work areas having relatively low noise exposures to select exposure limits that are significantly lower than OSHA's.

The lowest feasible noise exposure levels are obviously desirable for the health, safety, and well-being of workers (75–82). In addition, these lower limits are of significant value to employers because morale and work productivity of workers should be maximized (76–82), and the number of compensation claims for noise-induced hearing impairment should be minimized.

7.2 Specific Requirements

An effective hearing conservation program should provide for:

1. The identification of noise hazard areas, and the performance of noise exposure measurements. (see Section 5)
2. The reduction of the noise exposure to safe levels with hearing protectors and/or engineering noise control procedures (see Sections 8 and 9)
3. The measurement of exposed persons' hearing thresholds to monitor the effectiveness of the program (see Section 4)
4. The education and motivation of employees and management about the need for hearing conservation, and the instruction of employees in the use and care of personal hearing protectors (see Section 7)
5. The maintenance of accurate and reliable records of hearing and noise exposure measurements (see Sections 4–6, 8, and 9)
6. The referral of employees who have abnormal hearing thresholds for examination and diagnosis (see Section 6)

7.2.1 Identification of Noise Hazard Areas

Noise hazard areas having continuous noise characteristics are identified by means of sound level meter measurements with A-frequency weighting or with dosimeters (see Section 5.1 and 5.2). When significant impulse-type noises are present, hazard areas may be established with noise dosimeters or impulse measuring devices (see Section 5.4).

Action noise exposure doses must be established that are at least as low as those specified by the OSHA Rules and Regulations (3–5). A noise hazard dose for continuing steady-state noise must be described in terms of both A-weighted sound pressure level and time of exposure. In order to select the best hearing protectors for particular noise spectra, and for hearing warning signals, machine sounds, or speech communication, octave band sound pressure levels must be used (see Section 8.2).

7.2.2 *Reduction of Noise Exposure Levels*

As soon as a noise hazard area has been determined, hearing protective devices should be provided to reduce the exposures to safe levels (see Section 8). Engineering control means (see Section 9) should then be employed where they are feasible. This will often require the use of experts in noise control to work with those at the plant, and/or others, who understand the machines and their operation (see Section 8). If it is not feasible to reduce exposure levels to the limits selected by engineering control means, the use of hearing protectors must be continued.

Continued monitoring of the effectiveness of hearing protector devices (hearing threshold measurements) must be maintained (see Sections 7.3–7.10) until engineering control procedures (see Section 9) have reduced the noise exposures to safe levels. It is strongly advisable to continue monitoring the effectiveness of the hearing conservation *program* even after engineering control measures have been successfully installed if there is any reason to suspect that noise exposures at, or away from, work are significant.

7.2.3 *Hearing Measurement and Record Keeping*

Periodic hearing threshold measurements (see Section 4) are necessary to monitor the effectiveness of personal hearing protectors. High-level noise exposures away from work may be partially responsible for any STS (significant threshold shift); hence it is important to make a significant effort to determine the cause of any threshold shift. Affected persons should be interviewed and their audiological historical data sheets should be updated periodically.

Hearing conservation measures away from work can be encouraged or assisted by lending or giving hearing protectors to employees for use with noisy activities away from work. The cost of supplying hearing protectors for this use may be far less than hearing compensation costs, and it often directs more attention to the hearing conservation program at work.

If a STS is found in an annual audiogram, a letter must be sent to the employee informing him or her of the results of the audiogram. See Appendix 23.1 for a sample letter for this purpose. Even though follow-up letters to employees are not required after the base-line audiogram, letters should be used to advise new employees of hearing impairments greater than would be expected for their age. These employees may be more susceptible to noise-induced hearing impairment, so this early warning should result in lower exposures at and away from work during the following year.

Management sometimes resists contacting employees when it is not required. Supporting argument may be needed, therefore, even though a letter after the base-line audiogram should prevent rather than cause problems. For example, if these letters are written, employees are much more likely to pay attention to the hearing conservation procedures, at and away from work, during the following year so they are less likely to have a STS on their first annual audiogram. Also, employees have responded positively to this letter by assuming greater responsibility for their own health and safety. Some have apologized when a STS developed, saying that they had forgotten their protectors when they used guns or chainsaws. If the warning letter is not written, these persons are not only more likely to get a STS, but they are apt to blame employers for preemployment impairment.

Even though not an OSHA requirement, records should be retained at least 2 years, and in many cases longer, after an employee is terminated. These records could be valuable if there is a question of legal responsibility for an employee's hearing impairment, or if the employee is to return to this job. Also, *termination audiograms* are not required by OSHA, but they should be taken because of potential liability for hearing impairments after termination.

7.3 Setting up a Hearing Conservation Program

Setting up a hearing conservation program entails the development and maintenance of all of the factors mentioned above. Ideally, a team made up of management, industrial hygiene, preventive medicine, safety, and production personnel will work together toward building an effective hearing conservation program. The first objective for this group would be to become aware of *their* problems and to consider possible ways to limit noise exposure levels as much as possible, at least to levels set by OSHA.

In practice, these committees may not work well except as advisors on general items such as company policy. Many successful hearing conservation programs have been highly dependent upon the competence and dedication of one person *who is given the time and support required to develop and maintain the program.* This person must be motivated and knowledgable, and respected by both management and workers. Extensive formal training in any specific field is not usually necessary, but he or she must have a good practical knowledge of noise-induced hearing impairment, hearing threshold measurement, and personal hearing protectors.

If an outstanding person is not available for this job, it may be necessary to use a hearing conservation committee to establish general policies and to appoint different individuals to be responsible for carrying out special assignments. Generally, it is highly desirable to have well-defined tasks given to each individual.

The development of effective and practical hearing conservation programs has proven to be difficult. Many ineffective programs remain even after years of effort. Two major reasons for poor programs are:

1. The responsibility for the program is given to a person within the company who is already overburdened with other jobs, and essentially no support is allotted for a hearing conservation program.

2. The responsibility for the hearing conservation program is contracted to an outside firm that claims to have a "hearing conservationist." One of the apparent reasons for hiring an outside firm is to transfer responsibility for the hearing conservation program to the contractor. Unfortunately, there is often a communication

problem because the employer and/or the contractor does not understand the requirements of a hearing conservation program. The employer often believes that the contractor is handling the complete hearing conservation program, but only hearing or noise measurements are made, and the contractor leaves without taking any further action. Money is spent inefficiently and employees continue to have their hearing impaired. In addition, evidence is gathered showing that hearing impairment is occurring, and there is negligence when no action is taken.

Generally, the responsibility for developing an overall hearing conservation program must be accepted by the employer. At least one person within the company who has overall knowledge of essential components of the company's operations must work with any outside contractors that are used. The company representative must have knowledge of production processes, company politics, individual interactions and characteristics, and so forth. In many cases, this knowledge must be used to establish an overall effective program. An outside firm or individual cannot be expected to know these things, nor is an outside party likely to have the deep interest in the employees' or the company's well-being that an employee should have. Also, an employee is usually the only logical person to maintain the program continuously after it is once established.

Outside firms may be useful for specific jobs such as hearing measurements, noise measurements, or noise control projects, but it is usually to the company's benefit to have a knowledgeable employee work with the specialist. Information on individuals and detailed information on machine operation (access areas needed for operation and production flow, etc.) are often needed for the best results. Otherwise, costly mistakes can be made that may require repeating expensive projects.

Personnel and work conditions may vary widely from one plant, or work area, to another. Monitoring safety procedures is much easier where workers are concentrated in areas where they are easily visible than in situations where they are widely scattered. Significant differences are found in management/employee relationships, and in motivation, education, and communication skills of hearing conservationists from one location to another, etc. As a result, a program that works well in one place may fail in others. It is generally necessary, therefore, to develop a program for each situation if it is to be effective.

Obviously, top management must support the program, and others in middle management should be enthusiastic and knowledgeable. Perhaps even more important than middle management are the floor supervisors and the plant nurses. These key persons often have the respect of workers, and they may be the only persons at the plant with whom some workers will communicate freely. The floor supervisor must be genuinely concerned about the health and safety of the employee and be able to answer questions about the effects of noise exposures, the use of safety equipment, and the overall importance of the program.

7.4 Selection of a Hearing Protector

Reasons for ineffective hearing protector programs often may be pinpointed by complaints from persons wearing protectors. The most common complaints are related to comfort and communication but there may be other complaints based on not understanding the necessity for wearing protectors. Ideally, a hearing protector should be selected with the following objectives in mind:

1. The ear must be protected with an adequate margin of safety, but without unnecessarily reducing important communication (hearing warning signals, machines, speech, etc.). The protector's attenuation characteristics therefore should be selected to match those of the noise exposure spectra as closely as possible.
2. Environmental conditions and the kind of activity required by the job may significantly influence the comfort and general acceptability of a hearing protector. For example, insert-type protectors may be a better choice than muff types for use in high temperatures, vigorous activities, or close quarters. In very dirty areas, for very sensitive ear canals, or for ease of monitoring muff types may be preferred to insert types.
3. Some wearers cannot be fitted properly by certain types of protectors. Performance, comfort, and/or effectiveness may be significantly enhanced by the proper choice of protector for some individuals. Wearing time is often an important consideration in making this decision.
4. It is difficult to predict or understand the acceptance of protector types at times. Important factors may include (*a*) the wearer's appearance or hair style while wearing the protector, and (*b*) something said about particular models, or just a desire to be different from others.

The effectiveness of a hearing conservation program, job performance, health and safety considerations, and morale may be adversely affected if attention is not given to all these factors. In particular, attention should be given to requests for changes in protector types. Obviously, there is no single protector, or protector type, that is best for all individuals or all situations. A choice of several protectors should be offered. This action also reinforces the importance of the program.

The acceptance of protectors can be helped, in some cases, by issuing selected protectors to management a few days before they are issued to workers. This procedure may cause some employees to react with demands that they be given these protectors too, thereby creating a more positive attitude when protectors are issued. In all cases supervisory personnel and visitors must be required to obey *all* safety and health rules if safety equipment is to be used effectively. Exceptions cannot be tolerated.

7.5 Safety Factors

Safety factors must be developed and used with the attenuation values supplied with hearing protectors (see Section 8). The laboratory data supplied with protectors are measured using "best-fit" conditions (122–151), which may be acceptable for the best programs. However, the effectiveness of hearing conservation programs vary significantly so that widely different safety factors may be required to assure adequate protection in many work areas. For example, it is much easier to develop an effective hearing conservation program for a small well-defined work area (model shops), than it is for large open spaces where workers move around over large areas (forests, farms, mines, etc.).

The use of a single overall safety factor for all work areas is, at times, considered attractive because of its simplicity. However, this simplicity comes at the price of very large safety factors in order to protect workers in the worst programs. It follows that communication will be restricted by unnecessary protection for large numbers of workers who have more effective hearing conservation programs (see Section 8.4). In addition, the real problem of poor hearing conservation programs will remain hidden or ignored, and the high percentage of ineffective programs will continue.

7.5.1 Noise Reduction Ratings

Hearing protectors are often selected with consideration being given only to the magnitude of a single-number performance rating; generally, the Environmental Protection Agency (EPA) Noise Reduction Rating (NRR)(132). Although a single-number rating is useful in initiating a program when more precise octave or one-third octave noise measurement data are not available, its use also requires large safety factors because of its lack of precision. Even with the large safety factors now being used with the NRR, there are situations where there are underprotected individuals. OSHA regulations specify that a safety factor of 7 dB is to be subtracted from any A-weighted exposure level when the NRR is used because of this inaccuracy (3–5). The NRR misleads the user of muff-type hearing protectors in typical noise exposure spectra in the steel industry by more than ±8 dB as compared to calculations based on the more accurate NIOSH Method 1 (276). The calculation of the NRR includes additional safety factors of two times the standard deviation plus 3 dB for spectral uncertainty, in addition to the 7 dB specified by OSHA. Therefore the total safety factor applied to laboratory data because of the poor accuracy of the NRR can easily exceed 15 dB, and this does not include the safety factors intended for poor hearing conservation programs discussed in the above paragraphs.

When the large safety factors required for use of the single-number NRR rating are added to the single-number safety factor sometimes proposed for poor hearing conservation programs, it is obvious that a very high price is being paid for simplicity. Many workers will be significantly overprotected, which may cause a variety of problems ranging from serious injuries when warning signals are not heard to reduce work efficiency when machine sounds are masked. In addition, unnecessarily masked communications may cause significant annoyance and stress-related effects that can in turn cause wearers deliberately to disable their protectors so that they can communicate.

Another weakness of single-number ratings, such as the NRR, is that they are often based on just one or two of the nine third-octave test signals used in laboratory measurements. Generally, the controlling test signals are below 1000 Hz and performance levels for other test signals may have little or no effect on the final NRR. For example, two protectors may have the same EPA NRR because of their limiting attenuation values at test signals centered at 250 or 500 Hz, but one of these protectors may provide more than 15 dB greater protection than the other for higher frequencies, above 1000 Hz, without this information being indicated. The NRR values in this example would be a reasonably accurate assessment of protector performance if the highest exposures were centered at 250 or 500 Hz, but if the noise exposures contain prominent high-frequency components the NRR values may be very misleading.

The more precise long method for calculating hearing protector performance ratings can be used to obtain a more accurate estimate of protection, while at the same time affording a more accurate means of maximizing communication. None of the safety factors discussed above for the NRR rating are required when octave band sound pressure levels are used to determine exposure levels under hearing protectors according to the more accurate NIOSH Method 1 (276).

7.6 Communication Without Hearing Protectors

Performance and safety aspects of a job often depend on the workers' ability to hear warning signals, machine sounds, and speech in the presence of high noise levels. The effect of noise on communication depends to a large extent on the spectrum of the noise, the hearing characteristics of the worker, and the attenuation characteristics of hearing protectors, if they are used.

For a normal-hearing person, speech communication is affected most when the noise has high-level components in the speech frequency range from about 400 to 3000 Hz. Speech interference studies (65–74) show that conversational speech begins to be difficult for a speaker and a normal-hearing listener, separated by about 2 ft, when broad-band noise levels approach about 88 dBA. Hearing-impaired persons have much more difficulty communicating in noise than do persons having normal hearing; the degree of difficulty depends upon the amount and kind of impairment.

7.7 Communication With Hearing Protectors

Few hearing protectors are selected and purchased with any thought being given to communication of any kind. In fact, meaningful consideration of communication cannot be given based on a single-number hearing protector rating such as the NRR. The lack of precision of single-number hearing protector ratings (NRR) also must be accompanied by large safety factors based on worst-case conditions, and these safety factors may lead to overprotection in some cases of more than 15 dB.

Overprotection may be considered as acceptable, or even desirable against many health and safety hazards, but overprotection against noise exposure may cause significant communication problems. Maximizing communication while wearing hearing protectors often improves safety and work efficiency conditions, and can prolong the working lifetime of skilled workers having high-frequency hearing impairment. Unnecessary protection from hearing protectors may also lead to the deliberate misuse or rejection of hearing protectors where the wearer wishes to communicate with fellow workers.

7.7.1 Normal-Hearing Persons

Hearing protectors interfere with speech communication in quiet environments for most persons, regardless of their hearing characteristics. Normal-hearing persons can often raise their voice levels to provide satisfactory communication in noise.

When wearing hearing protectors in noise levels between 88 and about 97 dBA normal-hearing persons often complain that the protectors impede communication. Above about 97 dBA background noise levels, normal-hearing persons are often able to communicate about as well with as without wearing protectors (46, 66, 69). In fact, protectors may improve speech communication for some normal-hearing persons when background noise levels are higher than 97 dBA because speech-to-noise ratios are held relatively constant, and distortion is reduced. Optimal communication is usually provided when the protector's attenuation characteristics are matched to those of the noise spectra (46, 66, 69).

7.7.2 Persons with Sensorineural Impairments

Most work areas have a significant percentage of persons with high-frequency hearing impairment. These persons often experience a greater reduction in intelligibility in noisy areas than persons having normal hearing, and they seldom find any advantage to wearing protectors, regardless of background noise levels. It is very important, therefore, to match hearing protector attenuation characteristics to those of the noise exposure spectra as closely as possible for these persons in order to provide the best possible communication conditions. This can be accomplished best by using octave band sound pressure levels and the long method of calculation of attenuation characteristics for hearing protectors.

7.7.3 Filter-Type Protectors

Employees and hearing conservationists are often attracted to protectors advertised as providing a "filter" that allows speech to be heard but blocks harmful noise. These filter-type devices generally provide less protection than conventional protectors, particularly at frequencies below 1000 Hz. These protectors may afford adequate protection in modest noise exposure levels (see Section 7.6), and they may provide better communication than conventional protectors. This type of protector is particularly useful for intermittent noise exposures where there are relatively quiet periods between short-term noise exposures. As with other types of protectors, the amount of protection required for each noise exposure spectra must be considered carefully before the protectors are used (see Section 8.4).

In most cases, filter-type protectors are more likely to be acceptable for use in noises having major components above 1000 Hz. The "filters" of many of these protectors are simply holes through the protectors which permit low frequencies (below about 1000 Hz) to pass without significant attenuation; higher frequencies are attenuated as if there were no hole.

7.8 Communication Through Radio Headsets

Wearing insert-type protectors under electronic communication headsets may improve the clarity of speech when used in high background noise. The improved perception of the speaker's own voice may also allow better control over modulation of his or her own voice which, in turn, may reduce distortion that often accompanies loud speech and shouting. Also, listeners can adjust their electronic gain or volume control to obtain the best reception levels.

Speech perception is not always improved by the use of earplugs under communication headsets when noise levels are above about 130 dB overall. In these rare cases it is often desirable to improve the signal-to-noise ratio at the microphone as well as at the receiver. One common method for increasing the signal-to-noise ratio at the microphone is to use noise cancellation principles.

A noise-canceling microphone picks up ambient noise through two apertures, one on either side of its sensing element, so that a portion of the low-frequency noise is canceled. The sensing element of the microphone is oriented so that the speech signal enters mainly through one aperture when the microphone is held close to the mouth; thus speech is not canceled.

In noise levels above about 120 dB re 20 μPa overall, the signal-to-noise ratio may also be improved at times by reducing the background noise at the microphone by noise shields that encase the sensing element. Efficient noise-attenuating shields tightly held around the mouth may be used with microphone systems to transmit intelligible communication in wide-band noise levels exceeding 140 dB.

An electronic noise-canceling technique has also been used with muff-type hearing protectors to increase intelligibility in noisy environments (66–69, 269–270). The incident noise has its phase changed 180° electronically, and the phase-adjusted signal is then fed into the headset where noise canceling is used to improve the speech-to-noise ratio, thereby improving communication. This use of noise cancellation is mainly effective in low frequencies (at and below 1500 Hz).

7.9 Rank-Ordering of Communication in Selected Noise Spectra

Protection should be provided only where required in order to optimize the hearing of warning signals, machine sounds, and required speech in the work area. The selection of the best protectors for providing adequate protection along with the best possible communication in a specified noise exposure spectrum is complicated because persons with widely different hearing characteristics must be able to receive the required communication. Hence there is a need for a sophisticated selection procedure that will provide information that is simple to use.

One such procedure for rank-ordering communication scores of selected hearing protectors, while at the same time restricting the choice to those affording acceptable noise exposure levels, has been developed with the support of the American Iron and Steel Institute (AISI) (276). This procedure is based on the American National Standard Methods for the Calculation of the Articulation Index (AI), ANSI S3.5-1969 (R1986) (271). Articulation indexes are calculated for each of 18 typical work areas with distances between the talker and the listener of 1, 1.5, or 3 ft with adjustments in the AI made for:

1. Typical octave band noise exposures spectra (18) selected for the steel industry
2. Attenuation characteristics of the 104 different hearing protectors that were selected by the users
3. Three classes of hearing characteristics (normal hearing, moderately impaired, and severely impaired)
4. Expected voice levels for each noise exposure
5. Nonlinear growth of masking

The hearing protectors selected for this study were the ones used by workers in each industrial plant. Both the NIOSH "long method" (1, 142) and the single-number NRR were used to calculate the noise exposures for covered and uncovered ears. Adjustments were made to account for the natural raising of the voice when background noise levels increase, and for the nonlinear growth of masking.

In practice, an employer must provide the following information:

1. A limiting exposure level (risk factor) at or below that specified by OSHA
2. The normal communication distance (or distances)
3. A listing of hearing protectors to be included
4. Octave band sound pressure levels must be measured or supplied for each work area

Articulation index (AI) values are calculated so the *relative* AI scores are established for each hearing protector when they are used in each noise spectrum (area) for normal, moderately impaired, and severely impaired hearing conditions, and for specified communication distances. The use of limited familiar words and phrases is assumed for these calculations. Each of these tables is likely to rank-order the protectors differently for communication.

A sample sheet showing the rank-ordering of 20 selected hearing protectors for simple speech communication in 12 different work areas is shown in Table 23.5. This table, calculated for persons having moderate hearing impairment, is based on an exposure limit of 80 dBA time-weighted average (2–5). This is an arbitrarily chosen exposure limit, more restrictive than OSHA's, that may be selected as being economically feasible for many industrial plants. Any other exposure limit lower than that specified by OSHA can be used. The hearing protector mean attenuation values are reduced by two standard deviations as in the NIOSH long method (145).

The user of these tables need not have knowledge of acoustics nor audiology. Protectors having higher numbers for given work areas (where the protector row

Table 23.5. Relative Communication Scores at 1-Foot Distance in 12 Selected Industry Work Areas Under 20 Different Hearing Protectors[a]

Hearing	Work Areas											
Protector	1	2	3	4	5	6	7	8	9	10	11	12
1	7	5	3	5	8	7	5	5	6	4	4	5
2	7	5	4	5	8		5	5		4	5	6
3		5	4	5						4	4	
4	7	5	3	4	8	7	5	5	6	3	4	5
5	6	4	4	5			5	5	6	4	4	5
6	7	5	3	4	8	7	5	5	6	4	4	5
7	8	6	3	4	9	7	5	6	7	4	6	6
8	7	5	3	4	8	7	5	6	7	4	5	6
9	7	5	3	4	8	6	4	5	6	3	5	6
10	7	5	3	5	8	7	5	5	7	4	5	6
11	7	5	3	5	8	7	5	5	6	4	4	5
12	7	5	3	4	8	7	5	5	7	4	5	6
13	7	5	3	5	8		5	5		4	5	6
14	6	5	3	4	7	6	4	5	6	3	4	5
15	7	5	3	4	8	7	4	5	6	3	5	6
16	6	4	4	5			5			4	4	5
17	6	4	4	5	7		5	4	6	4	4	5
18	7	5	3	4	8	7	5	5	6	4	4	5
19	6	4	3	4	7	7	4	4	5	3	3	4
20	7	5	3	4	8	7	4	5	7	4	5	6

[a]Entries are made only in cases where noise exposures are less than 80 dBA TWA. These data are based on persons having moderate hearing impairment.

intersects with the work area column) have higher communication scores. Hearing protectors having no number at this intersection do not provide enough protection in this area to meet the exposure limit of 80 dBA TWA (or any other that may be chosen).

The tables developed so far are intended for simple speech communication (relatively few familiar words). It should also be possible to develop relative communication scores for specific spectra from warning signals, machines, and so on.

To summarize, this procedure should simplify the selection and use of hearing protectors, and it adds the important communications aspect that has been neglected in most hearing conservation programs. It should also afford much more precision in rating relative communication levels under hearing protectors than has been possible using a rule-of-thumb procedure.

8 HEARING PROTECTORS

8.1 General

Personal hearing protectors can provide adequate protection against noise-induced hearing impairment in a high percentage of industrial work areas (128, 144, 148) *if* the protectors are properly selected, fitted, and worn. In many cases, however, protectors are not worn effectively, if at all (see Section 7).

Even though the use of hearing protectors is one of the most important parts of a hearing conservation program, only a fraction of the total money is spent for this purpose. Ironically, a much larger percentage of hearing conservation effort and money is often spent over long time periods on hearing threshold measurements that are used for monitoring the effectiveness of poor or nonexistent hearing protector programs. Not only is this a waste of money but, when it is demonstrated that hearing protectors are used ineffectively, and nothing is done to rectify the problem, threshold measurement data are evidence of negligence that may be used against the employer (see Section 9).

Hearing threshold measurements are important when action is taken based on the results. In fact, the only practical means for evaluating the effectiveness of personal protectors is to monitor thresholds periodically. If no hearing losses are observed—other than those due to the aging process—the program may be considered to be successful. Noise-induced hearing impairments usually develop slowly, however, so it may take years for the results from a hearing monitoring program to become meaningful. The careful selection of protectors and a continuing hearing conservation program, including close supervision by floor supervisors, are therefore very important.

An effective hearing protector program seldom develops automatically simply by making hearing threshold or noise measurements. Noise does not have to be painful to be potentially harmful, so many employees do not understand the need for wearing protectors. A significant effort must be made to develop and to maintain an effective program.

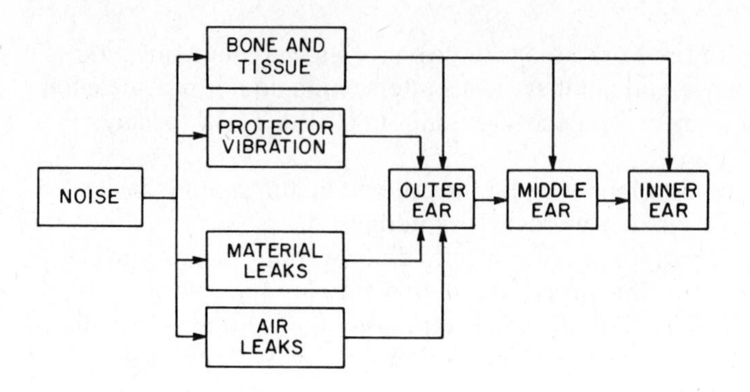

Figure 23.13. Noise pathways to the inner ear.

Unfortunately, there are few convenient and reliable sources of information available to guide employers on how to set up effective hearing conservation programs. It often follows that employers are at times too willing to take shortcuts, or to allow persons selling a product or a service make decisions for them. This is particularly true for small industries.

8.2 Performance Limitations

A primary limitation of protection afforded by a hearing protector is the way it is fitted and worn. Other important limitations of a hearing protector depend upon its construction and on the physiological and anatomical characteristics of the wearer. Sound energy may reach the inner ears of persons wearing protectors by four different pathways: (1) by passing through bone and tissue around the protector; (2) by causing vibration of the protector, which in turn generates sound into the external ear canal; (3) by passing through leaks in the protector; and (4) by passing through leaks around the protector. These pathways are illustrated in Figure 23.13.

Even if there are no acoustic leaks through or around a hearing protector, some noise reaches the inner ear by bone and tissue conduction or protector vibration *if* noise levels are sufficiently high. The practical limits set by the bone- and tissue-conduction threshold vary significantly among individuals, and among protector types, generally from about 40 to 55 dB. Limits set by protector vibration also vary widely, generally from about 25 to 40 dB, depending upon the protector type and design and on the materials used. Contact surface area and compliance of materials are major contributing factors. The results of studies on these limitations are influenced significantly by procedures and techniques used and by the choice of subjects. These very wide ranges of limits are therefore of limited value for use for individuals (144, 266, 267). If hearing protectors are to provide noise reduction values approaching practical limits, acoustic leaks through and around the protectors must be minimized by proper fitting and wearing.

8.3 Types of Hearing Protectors

Hearing protectors usually take the form of either insert types that seal against the ear canal walls or muff types that seal against the head around the ear. There are also concha-seated devices that provide an acoustic seal at the entrance of the external ear canal. There is no "best" type for all situations (see Section 7.4). However, some types are better than others for use in specific noise exposures, for some work activities, or for some environmental conditions.

8.3.1 Insert Types

Ear canals differ widely in size, shape, and position among individuals and even within the same individual. Several different types of earplugs may be required therefore to fit the wide differences in ear canal configurations. No single insert-type protector is best for all individuals, nor for all situations.

Ear canals vary in cross-sectional diameter from about 3 to 14 mm, but most are between 5 and 11 mm. Most ear canals are elliptically shaped, but some are round, and many have only a small slit-like opening that may open into a larger diameter. Some canals are directed in a straight line toward the center of the head, but most are directed toward the front of the head and they all may bend in various ways.

In many cases, there is only a small space available to accommodate an earplug, but it must fit snugly to be effective. Most ear canals can be opened and straightened by pulling the external ear back, or directly away from the head, making it possible to seat many earplugs securely. For comfort and plug retention, ear canals must return to their approximate normal configuration once the protector is seated.

8.3.1.1 Sized Ear Plugs. Few single-sized molded earplugs provide consistently high levels of protection for a large range of ear canal sizes and shapes. Hence most of the more widely accepted molded ear plugs come in two or more sizes. The best molded ear plugs are made of soft and flexible materials that will conform readily to the shape of the ear canal for a snug, airtight, and comfortable fit.

Earplugs must be nontoxic and have smooth surfaces that are easily cleaned with soap and water. Earplugs are usually made of materials that will retain their size and flexibility over long periods of time; however, some persons have ear wax that may cause significant changes in size and flexibility of some protectors after relatively short periods of use.

The earplug size distribution for a large group of males is approximately as follows: 5 percent extra small, 15 percent small, 30 percent medium, 30 percent large, 15 percent extra large, and 5 percent larger than those supplied by many earplug manufacturers. The equal percentage of wearers for medium and large sizes indicates that many persons are fitted with earplugs that are too small.

If individuals are permitted to fit themselves, they often select a size on the basis of comfort rather than on the amount of protection provided. Some ear canals increase in size slightly with the regular use of earplugs, and earplugs may shrink in size. Thus if a given ear size falls between two plug sizes, it is advisable to choose the larger rather than the smaller of the two. It follows that the fit of earplugs should be checked periodically.

A common (and often valid) complaint is that the protector case costs more than the earplugs; however, a good case keeps earplugs clean, in good condition, and readily available when needed. Also, the case can be used repeatedly for replacement plugs that may be bought in bulk. An earplug container of good quality should outlast several pairs of plugs.

8.3.1.2 Malleable Earplugs. Malleable earplugs are made of materials such as cotton, paper, plastic, wax, glass wool, and mixtures of these materials. Typically, a small cone or cylinder of the material is hand-formed and inserted into the ear with sufficient force so that the material conforms to the shape of the canal and holds itself in position. Manufacturer's instructions should be followed regarding the depth of insertion of the plug into the canal. In any case, care should always be taken to avoid deep insertions that may cause the material to touch the eardrum.

Malleable earplugs should be formed and inserted with clean hands, because dirt or foreign objects placed in the ear canal may cause irritation or infection. Malleable plugs should be carefully inserted at the beginning of a work shift and, if removed, they should not be reinserted until the hands are cleaned. For this reason, malleable plugs (and to a somewhat lesser extent, all earplugs) may be a poor choice if the work area is dirty or subjected to intermittent high level noises, where it may be desirable to remove and reinsert protective devices during the work period.

On the other hand, malleable plugs have the obvious advantage of fitting almost any ear canal, eliminating the need to keep a stock of various sizes, as is necessary for most molded earplugs. The cost of any throw-away protectors may be relatively high, depending upon how they are used.

8.3.1.3 Foam-type Plugs. Earplugs made of plastic foam cylinders have been popular during the last decade. Most of these protectors are comfortable to wear and they provide very good protection at all test frequencies *if* they are worn properly. Generally, these plugs must be inserted about three-fourths of their length into the canal and held for 1 min while they expand to fill the canal in order to provide the rated protection level. If these plugs are not inserted properly, their performance may drop significantly. It may not be possible to insert foam plugs properly in some small, rapid bending, or slit-shaped canals.

8.3.2 Concha-Seated Ear Protectors

Protectors that provide an acoustic seal in the concha and/or at the entrance to the ear canal are often referred to as concha-seated devices. Hearing-aid type molds used as hearing protectors are considered in this category, although some of these molds extend far into the canal. Another protector design in this class makes use of various plug shapes attached to a light-weight headband that holds the plugs in the entrance of the external ear canal. The performance of this kind of protector varies significantly among different models. The level of comfort varies among the many different models.

Because the materials used for hearing aid molds do not usually expand or conform after insertion, the molds must be made to fit very well in order to get the tight fit required for good attenuation. Hence significant differences in attenuation and comfort ratings may be found among these protectors made by different manufacturers.

If individually molded protectors are well made, protection levels are typically very good. Also, when made properly, these protectors can be expected to provide among the most consistent protection levels in daily use of all protector types (128, 129). The initial cost of individually molded protectors is high, but most should last a long period of time.

8.3.3 Earmuffs

Manufacturers of muff-type hearing protectors usually offer a choice of two or more models having different performance levels. Sealing materials that contact the skin are generally selected to be nontoxic, and the fit, comfort, and general performance depend upon the designs used.

The performance levels of muff-type protectors depend upon many individual factors and how well these factors blend together. Generally, larger and heavier protectors provide greater protection, but the wearing comfort may not be as good for long wearing periods. Hence workers often accept light-weight protectors more readily than the larger ones. Other important factors include:

1. The suspension must distribute the force around the seal, with slightly more force at the bottom to prevent leaks in the hollow behind the ear.
2. The suspension mounting should be such that the proper seal is made against the head automatically without adjustments by the wearer.
3. The earcups must be formed of a rigid, dense, imperforate material to prevent leaks and significant resonances.
4. The size of the enclosed volume within the muff shell (particularly for low frequencies) and the mass of the protectors are directly related to the attenuation provided.
5. The inside of each earcup should be partially filled with an open-cell material to absorb high-frequency resonant noises and to dampen movement of the shell. The material placed inside the cup should not contact the external ear; otherwise discomfort to the wearer and soiling of the lining may result.

Ear seals should have a small circumference so that the acoustic seal takes place over the smallest possible irregularities in head contour, and leaks caused by jaw and neck movements are minimized. However, a compromise between the small seal circumference (performance) and the number of persons that can be fitted properly must be made.

Earmuff cushions are generally made of a smooth, plastic envelope filled with a foam or fluid material. Because skin oil and perspiration may have adverse effects on cushion materials, the soft and pliant cushions may tend to become stiff and to shrink after extended use. Fluid-filled cushions generally provide superior performance, but they occasionally have a leakage problem. Foam-filled seals should have small holes to allow air to escape within the muff when mounted. Seals filled with trapped air may vibrate causing a loss of attenuation in low frequencies. Most earmuffs are equipped with easily replaceable seals.

Earmuffs normally provide maximum protection when placed on flat and smooth surfaces; thus less protection should be expected when muffs are worn over long hair, glasses, or other uneven surfaces. Glasses with plastic temples may cause losses in attenuation of from 1 to 8 dB (144, 267). In some cases, this loss of protection can be reduced substantially if small, close-fitting, wire or elastic-band temples are used. Acoustic seal covers provided to absorb perspiration may also reduce the amount of attentuation by several decibels, because noise may leak through porous materials.

Because the loss of protection is directly proportional to the size of the uneven obstructions under the seal, every effort should be made to minimize these obstructions. If long coarse hair or other significant obstructions cannot be avoided, it may be advisable to use earplugs.

The force applied by muff suspensions is often directly related to the level of noise attenuation provided by some protectors. On the other hand, the wearing comfort of a muff-type protector is generally inversely related to the suspension forces, so a compromise must be made between performance and comfort. Muff suspensions should never be deliberately sprung to reduce the applied force. Not only may a loss of attenuation be expected, but the distribution of force around the seal may be changed, which may cause an additional reduction in performance. To assure the expected performance, the applied force should be measured periodically, and the muff should be visually inspected.

Muff-type protectors are sometimes chosen because their use can be monitored from greater distances than insert-type protectors. They are also easier to fit than insert types because one size fits most persons. Comfort may also be better for muff-type protectors for *some persons in some work areas*. However, muff-type protectors may be very uncomfortable to wear in hot work areas, particularly when the work involves a vigorous activity. Muff-type protectors are also a poor choice when work is performed in areas having limited head space.

8.3.4 Summary of Advantages and Disadvantages of Protector Types

Both insert- and muff-type hearing protectors have distinct advantages and disadvantages. Some *advantages of insert-type protectors* are:

1. They are small and easily carried.
2. They can be worn conveniently and effectively without interference from glasses, headgear, earrings, or hair.
3. They are normally comfortable to wear in hot environments.

4. They do not restrict head movement in close quarters.
5. The cost of sized earplugs (except for some throw-away types and molded protectors) is significantly less than muffs.

Some *disadvantages of insert-type protectors* are:

1. Sized and molded insert protectors require more time and effort for fitting than do muffs.
2. The amount of protection provided by a good earplug may be less and more variable between wearers than that provided by a good muff-type protector.
3. Dirt may be inserted into the ear canal if earplugs are inserted with dirty hands.
4. The wearing of earplugs is difficult to monitor because they cannot be seen at a significant distance.
5. Earplugs should be worn only in healthy ear canals, and even then, acceptance may take some time for some individuals.

Some *advantages of muff-type protectors* are:

1. A good muff-type protector generally provides more consistent attenuation among wearers than good earplugs.
2. One size fits most heads.
3. It is easy to monitor groups wearing muffs because they are easily seen at a distance.
4. At the beginning of a hearing conservation program, muffs are usually accepted more readily than are earplugs.
5. Muffs can be worn despite minor ear infections.
6. Muffs are not easily misplaced or lost.

Some *disadvantages of muff-type protectors* are:

1. They may be uncomfortable in hot environments.
2. They are not easily carried or stored.
3. They are not convenient to wear without interference from glasses, headgear, earrings, or hair.
4. Usage or deliberate bending of suspension band may reduce protection significantly.
5. They may restrict head movement in close quarters.
6. They are more expensive than most insert-type protectors.

8.4 Protection Levels and Ratings

Performance requirements for hearing protectors can be determined considering (1) the noise exposure spectra (2), the hearing conservation noise exposure limits

(OSHA limits or lower), and (3) the hearing protector's mean attenuation and standard deviations values.

Noise exposures should be determined using octave band analyzers that meet or exceed ANSI specifications (106), or sound level meters that meet or exceed ANSI specifications (105). Use of the octave band measurements is preferred over the shortcut method that uses A-weighted sound pressure levels (dBA) in order to maximize communication. Also, the accuracy of dBA noise exposure data is poor for some spectra, hence requiring large safety factors (5).

The goal for the hearing conservation exposure limits should be to prevent noise-induced hearing impairment, and not simply to meet the OSHA limits. Management must be made aware of the risk they are taking if they simply select the OSHA limits (see Section 6).

8.4.1 Calculations Using the Long and the Single-Number (NRR) Methods

Attenuation mean and standard deviation values are available from the suppliers of protectors that can be used in calculations to obtain estimates of exposures while wearing the protector (139). By subtracting two standard deviations (SDs) from each mean attenuation value measured in the laboratory (mean − 2SD), you are assured that the resultant values were exceeded about 98 percent of the time during tests on subjects in the laboratory. The ANSI standard requires that laboratory data are measured using "best-fit" conditions. That is, subjects are given careful wearing instructions, and they are allowed time to adjust the protectors for best fit before being tested. For many protectors, this procedure is more practical than it appears because subjects used for laboratory measurements have only a short period of time to become familiar with each protector, whereas the users at work should become more skilled in using the protector effectively, if they want to do so, over a longer period of time. This procedure has been used for more than 30 years in ANSI specifications because this is the only way reliable rank-ordering of hearing protector performance can be accomplished.

Protector attenuation values for each noise spectrum should be calculated using the most accurate NIOSH Method 1 (122, 134). An example using the NIOSH Method 1 "long method" (122, 134, 139, 142, 145) for determining the amount of protection provided by a hearing protector is shown in Table 23.6. Steps taken in this calculation are as follows:

1. Measure and record the octave band exposure levels as shown in line 1 of Table 23.6.
2. Adjust these octave band values with A-frequency weightings for the center frequency in each octave band (line 2) and record the adjusted values in line 3. The overall A-weighted sound exposure level for unprotected ears is equal to the logarithmic sum of the A-weighted exposure levels in line 3 (95.2 dBA).
3. Record the mean hearing protector attenuation values supplied by the pro-

Table 23.6. Sample Calculations Using the Long Method to Determine A-Weighted Noise Exposure Levels Under a Hearing Protector When Used in Noise with Known Octave Band Levels

	Octave Band Center Frequency (Hz)						
	125	250	500	1000	2000	4000[a]	8000[b]
1. Measured octave band exposures in dB	93	91	91	89	88	86	87
2. A-weighted adjustments in dB	16.1	8.6	3.2	0.0	−1.2	−1.0	1.1
3. Unprotected ear A-weighted levels step 1–step 4 in dB	76.5	82.7	88.0	89.1	89.0	86.7	85.9
			logarithmic sum = 95.2 dBA				
4. Mean attenuation in dB at frequency (from laboratory)	18.0	20.0	30.0	39.0	34.0	32.0	36.0
5. Standard deviations in dB times 2	7.0	5.0	7.0	6.0	6.0	6.0	7.0
6. A-weighted levels in protected ear in dB	65.5	67.7	65.0	56.1	61.0	60.7	56.9
Overall A-weighted level under protector = 72.0 dBA							
Calculated protection = 23 dB; NRR = 23 dB							

[a] Average of 3150 and 4000 Hz.
[b] Average of 6300 and 8000 Hz.

tector distributor or manufacturer in step 4, and the corresponding standard deviation values times two in line 5.

4. Subtract the mean attenuation values in line 4 from the corresponding weighted levels in line 3 for each center frequency and add the two standard deviations (line 5) to the result to obtain the A-weighted exposure levels under the protector shown in line 6.
5. Add the A-weighted exposure levels in line 6 logarithmically to obtain an estimate of the highest sound pressure level (72 dBA) to which wearers of this protector would be exposed 97.5 percent of the time *if the protectors are fitted and worn properly*.

The A-weighted protection afforded in this example can be determined by subtracting the sound pressure level under the protector (line 6) from the sound pressure level incident on the uncovered ears (line 3), which is about 23 dB. This value is the same as the single-number rating (NRR) (132), which is calculated as shown in Table 23.7. However, the more accurate long method rating may be as much as ±8 dB different from the NRR, depending upon the noise exposure spectrum and the hearing protector used.

Two other examples of long method calculations for this protector *with different common noise exposure spectra* are shown in Tables 23.8 and 23.9. In these examples, the long method of calculation shows this same protector to provide as much as 28 dB attenuation in one area, but only 17 dB in another as compared to the NRR of 23 dB.

No adjustments have been made in these examples to account for improper use of protectors. A reasonably good hearing conservation program should have protection levels approaching the long method estimates based on laboratory data. Unfortunately, the results of a poor program cannot be predicted; exposure levels may be as low as 72 dBA, or they may be 95 dBA where protectors are not worn.

For a variety of reasons, the effectiveness of hearing conservation programs may differ widely (see Sections 7.4 and 7.5). Because of these significant differences in effectiveness among programs, it is strongly recommended that safety factors be chosen for specific work areas, rather than using the very large safety factors that may be necessary if only a single safety factor is assigned for all areas. The resulting sacrifice of communication for warning signals, machine sounds, and speech is a high price to pay just to avoid a small amount of work that may cost less than hearing threshold measurements for one year (see Section 7.5).

In summary, the proper use of a good hearing protector can provide adequate protection in most work environments where it is not feasible to use engineering control measures. Special care should be taken to obtain the best protectors for a given purpose to assure adequate protection and communication. A continuing effort must also be made to ensure that protectors are used properly and to monitor hearing thresholds regularly to be sure that workers are not being exposed to harmful noises at or away from work, or to ascertain that they may have other problems with their hearing.

An employer may be held responsible for any high-frequency hearing impairment

Table 23.7. Sample Calculation of the Environmental Protection Agency Noise Reduction Rating

	Octave Band Center Frequency (Hz)						
	125	250	500	1000	2000	4000[a]	8000[b]
1. Measured octave band exposures in dB	100	100	100	100	100	100	100
2. C-weighted adjustments in dB	0.2	0.0	0.0	0.0	−0.2	−0.8	−3.0
3. Unprotected ear C-weighted levels in dB (line 1 − line 2)	99.8	100.0	100.0	100.0	99.8	99.2	97.0
			logarithmic sum = 107.9				
4. A-weighted adjustments in dB	16.1	8.6	3.2	0.0	−1.2	−1.0	1.1
5. Unprotected ear A-weighted levels in dB (line 1 − line 4)	83.9	91.4	96.8	100.0	101.2	101.0	98.9
			logarithmic sum = 107.0 dB				
6. Mean attenuation in dB at frequency (from laboratory)	18.0	20.0	30.0	39.0	34.0	32.0	36.0
7. Standard deviations in dB times 2	7.0	5.0	7.0	6.0	6.0	6.0	7.0
8. A-weighted levels in protected ear in dB (line 5 − line 6 + line 7)	72.9	76.4	73.8	67.0	73.2	75.0	69.9
			logarithmic sum = 78.0 dB				
9. NRR = sum line 3 − sum line 8 − 3 dB = 108.0 − sum line 8 − 3 =	23dB						

[a]Average of 3150 and 4000 Hz.
[b]Average of 6300 and 8000 Hz.

Table 23.8. Sample Calculations Using the Long Method to Determine A-Weighted Noise Exposure Levels Under a Hearing Protector When Used in Noise with Known Octave Band Levels

	Octave Band Center Frequency (Hz)						
	125	250	500	1000	2000	4000[a]	8000[b]
1. Measured octave band exposures in dB	85	96	86	80	75	68	67
2. A-weighted adjustments in dB	16.1	8.6	3.2	0.0	−1.2	−1.0	1.1
3. Unprotected ear A-weighted levels step 1 − step 4 in dB	68.9	87.4	82.5	79.5	76.5	69.2	65.7
			logarithmic sum = 89.4 dBA				
4. Mean attenuation in dB at frequency (from laboratory)	18.0	20.0	30.0	39.0	34.0	32.0	36.0
5. Standard deviations in dB times 2	7.0	5.0	7.0	6.0	6.0	6.0	7.0
6. A-weighted levels in protected ear in dB	57.9	72.4	59.5	46.5	48.5	43.2	36.7
Overall A-weighted level under protector = 72.8 dBA							
Calculated protection = 17 dB; NRR = 23 dB							

[a]Average of 3150 and 4000 Hz.
[b]Average of 6300 and 8000 Hz.

Table 23.9. Sample Calculations Using the Long Method to Determine A-Weighted Noise Exposure Levels Under a Hearing Protector When Used in Noise with Known Octave Band Levels

	Octave Band Center Frequency (Hz)						
	125	250	500	1000	2000	4000[a]	8000[b]
1. Measured octave band exposures in dB	85	96	100	109	111	110	104
2. A-weighted adjustments in dB	16.1	8.6	3.2	0.0	−1.2	−1.0	1.1
3. Unprotected ear A-weighted levels step 1 − step 4 in dB	68.7	87.7	96.9	109.2	112.2	110.8	102.9
			logarithmic sum = 116.0 dBA				
4. Mean attenuation in dB at frequency (from laboratory)	18.0	20.0	30.0	39.0	34.0	32.0	36.0
5. Standard deviations in dB times 2	7.0	5.0	7.0	6.0	6.0	6.0	7.0
6. A-weighted levels in protected ear in dB	57.7	72.7	73.9	76.2	84.2	84.8	73.9
Overall A-weighted level under protector = 88.3 dBA							
Calculated protection = 28 dB; NRR = 23 dB							

[a]Average of 3150 and 4000 Hz.
[b]Average of 6300 and 8000 Hz.

regardless of the cause, because it is often impossible to apportion responsibility for a hearing impairment. If a STS is found on an annual audiogram (see Section 6), the cause must be investigated. If it is determined that noise exposures may be the problem, it may be necessary either for this person to use different protectors or a combination of insert- and muff-type protectors, or to limit the times of exposure by administrative means.

8.4.2 Protection Provided by Using Both Insert and Muff Types

The combined attenuation from wearing both a muff-type and an insert-type protector cannot be predicted accurately because of complex coupling factors. If the attenuation values of the muff and insert protectors are about the same in any specified frequency band being added, the resultant should be 3 to 6 dB greater than the higher of the two individual values. If one of these two protectors has an attenuation value that is significantly higher in this band, the increased attenuation by wearing both will be only slightly greater than the higher of the two.

Generally, it is necessary to measure the attenuation provided by the two protectors on a group of 10 subjects to obtain a good estimate. The A-weighted attenuation provided by any pair of protectors should be calculated using the long method described in Tables 23.6, 23.8, and 23.9.

9 NOISE CONTROL PROCEDURES

Ideally, engineering noise control procedures should be initiated in the design stages of building noisy equipment. Unfortunately, there are several economic reasons why more effort has not been made to build quiet equipment. The process of making equipment quiet generally increases its weight and size, making it clumsy to use, and it is often significantly more expensive. Another very important reason is that *most purchase orders are based only on price and performance; very few purchase orders specify quiet equipment*. In particular, there was almost no attention given to quiet equipment prior to 1969–1970, when the OSHA noise regulations were promulgated. To summarize, older manufacturers would probably not be in business today if they had built quiet equipment that would not have been competitive over the past several years.

Efforts have been made on several occasions by such groups as the American Industrial Hygiene Association (AIHA) noise committee to get users of noisy equipment to specify acceptable noise levels in their orders, but these efforts have met with only limited success. In many cases, buyers of noisy equipment do not specify quiet equipment because they are not aware of noise problems until the equipment is in place. Also, purchasing departments do not always welcome "help" in writing orders.

Even after a noisy machine is installed, engineering procedures are desirable, because they deal with predictable inanimate objects rather than relying on the uncertain cooperation of workers. Unfortunately, engineering procedures are not always practical, and it is necessary to use personal hearing protectors and/or administrative procedures to reduce exposures to the desired limits. The possibility of using a combination of all noise control methods should always be considered in order to reach the best possible solution.

9.1 Engineering Procedures

The first step in providing the quiet equipment is to make a strong effort to have purchase orders include noise limits. The desired quiet equipment may not be available, but at least these specifications will provide an incentive for the design of quiet products. Advantage has not been taken of available technology in many cases to produce quiet equipment.

The use of engineering noise control procedures on equipment already in operation is usually difficult and, in many cases, ineffective. Also, it may not be economically feasible to replace many long-lived, noisy machines with new quiet units. It is therefore imperative that the most efficient use be made of existing techniques to quiet equipment that is already in place. These techniques include the use of noise-absorbing materials, enclosure, or barriers, mechanical isolation, damping, reduced driving force, driving system modifications, and muffling (272, 273).

9.1.1 Absorption

Machines that contain cams, gears, reciprocating components, and metal stops are often located in large, acoustically reverberant areas that reflect and build up noise levels in the room. Frequently the noise levels in adjoining areas can be reduced significantly by using sound-absorbing materials on walls and ceilings. However, the amount of reduction close to the machines may be slight because most of the noise exposure energy is coming directly from the machines and not from the reflecting surfaces. The type, amount, configuration, and placement of absorption materials depend on the specific application; however, the choice of absorbing materials can be guided by the absorption coefficients listed in Table 23.10 (157–162).

9.1.2 Noise Barriers and Enclosures

The amount of noise reduction that can be attained with barriers depends on the characteristics of the noise source, the configuration and materials used for the barrier, and the acoustic environment on each side of the barrier. It is necessary to consider all these complex factors to determine the overall benefit of a barrier. The characteristics of the barrier itself can be determined using the transmission loss (TL) values of the material and the design used for the barriers shown in Table 23.11.

The noise reduction of barriers or enclosures may vary significantly. Single-wall barriers with no openings may provide as little as 2 to 5 dB reduction in the low frequencies and a 10 to 15 dB reduction in the high frequencies. Higher reduction

Table 23.10. Sound Absorption Coefficients of Materials (48)

Materials	Frequency (Hz)					
	125	250	500	1000	2000	4000
Brick	0.01	0.01	0.01	0.01	0.02	0.02
Glazed	0.03	0.03	0.03	0.04	0.05	0.07
Unglazed	0.01	0.01	0.02	0.02	0.02	0.03
Unglazed, painted						
Carpet						
Heavy, on concrete	0.02	0.06	0.14	0.37	0.60	0.65
On 40 oz hairfelt or foam rubber (carpet has coarse backing)	0.08	0.24	0.57	0.69	0.71	0.73
With impermeable latex backing on 40 oz hairfelt or foam rubber	0.08	0.27	0.39	0.34	0.48	0.63
Concrete block						
Coarse	0.36	0.27	0.39	0.34	0.48	0.63
Painted	0.10	0.05	0.06	0.07	0.09	0.08
Poured	0.01	0.01	0.02	0.02	0.02	0.03
Fabrics						
Light velour: 10 oz/yd² hung straight, in contact with wall	0.03	0.04	0.11	0.17	0.24	0.35
Medium velour: 14 oz/yd² draped to half-area	0.07	0.31	0.49	0.75	0.70	0.60
Heavy velour: 18 oz/yd² draped to half-area	0.14	0.35	0.55	0.72	0.70	0.65
Floors						
Concrete or terrazzo	0.01	0.01	0.015	0.02	0.02	0.02
Linoleum, asphalt, rubber, or cork tile on concrete	0.02	0.03	0.03	0.03	0.03	0.02
Wood	0.15	0.11	0.10	0.07	0.06	0.07

Wood parquet in asphalt on concrete	0.04	0.04	0.07	0.06	0.07	—
Glass						
Large panes of heavy plate glass	0.18	0.06	0.04	0.03	0.02	0.02
Ordinary window glass	0.35	0.25	0.18	0.12	0.07	0.04
Glass fiber						
Mounted with impervious backing: 3 lb/ft^3, 1 in. thick	0.14	0.55	0.67	0.97	0.90	0.85
Mounted with impervious backing: 3 lb/ft^3, 2 in. thick	0.39	0.78	0.94	0.96	0.85	0.84
Mounted with impervious backing: 3 lb/ft^3, 3 in. thick	0.43	0.91	0.99	0.98	0.95	0.93
Gypsum board: ½ in. nailed to 2 × 4's, 16 in. o.c.	0.29	0.10	0.05	0.04	0.07	0.09
Marble	0.01	0.01	0.01	0.01	0.02	0.02
Openings						
Stage, depending on furnishings			0.25–0.75			
Deep balcony, upholstered seats			0.50–1.00			
Grills, ventilating			0.15–0.50			
Grills, ventilating to outside			1.00			
Plaster						
Gypsum or lime, smooth finish on tile or brick	0.013	0.015	0.02	0.03	0.04	0.05
Gypsum or lime, rough finish on lath	0.14	0.10	0.06	0.05	0.04	0.03
With smooth finish	0.14	0.10	0.06	0.04	0.04	0.03
Plywood paneling: ⅜ in. thick	0.28	0.22	0.17	0.09	0.10	0.11
Sand						
Dry, 4 in. thick	0.15	0.35	0.40	0.50	0.55	0.80
Dry, 12 in. thick	0.20	0.30	0.40	0.50	0.60	0.75
14 lb H_2O/ft^3, 4 in. thick	0.05	0.05	0.05	0.05	0.05	0.15
Water	0.01	0.01	0.01	0.01	0.02	0.02

Table 23.11. Sound Transmission Loss[a] of General Building Materials and Structures (48)

Material or Structure	Frequency (Hz)								
	125	175	250	350	500	700	1000	2000	4000
Doors									
Heavy wooden door: special hardware; rubber gasket at top, sides, and bottom; 2.5 in. thick; 12.5 lb/ft²	30	30	30	29	24	25	26	37	36
Steel-clad door: well-sealed at door casing and threshold	42	47	51	48	48	45	46	48	45
Flush: hollow core; well-sealed at door casing and threshold	14	21	27	24	25	25	26	29	31
Solid oak: with cracks as ordinarily hung; 1.75 in. thick	12		15		20		22	16	
Wooden door (30 × 84 in.), special soundproof construction: well sealed at door casing and threshold; 3 in. thick; 7 lb/ft²	31	27	32	30	33	31	29	37	41
Glass									
0.125 in. thick; 1.5 lb/ft²	27	29	30	31	33	34	34	34	42
0.25 in. thick; 3 lb/ft²	27	29	31	32	33	34	34	34	42
0.5 in. thick; 6 lb/ft²	17	20	22	23	24	27	29	34	24
1 in. thick; 12 lb/ft²	27	31	32	33	35	36	32	37	44
Walls, homogeneous									
Steel sheet: fluted; 18 gauge stiffened at edges by 2 × 4 wood strips; joints sealed; 4.4 lb/ft²	30	20	20	21	22	17	30	29	31

Asbestos board: corrugated, stiffened horizontally by 2 × 8 in. wood beam; joints sealed; 7.0 lb/ft²	33	29	31	34	33	33	33	42	39
Sheet steel									
30 gauge; 0.012 in. thick; 0.5 lb/ft²	3	6		11		16		21	26
16 gauge; 0.598 in. thick; 2.5 lb/ft²	13	18		23		28		33	38
10 gauge; 0.1345 in. thick; 5.625 lb/ft²	18	23		28		33		38	43
0.25 in. thick; 10 lb/ft²	23	28	38	33	41	38	46	43	48
0.375 in. thick; 15/ft²	26	31	39	36	42	41	47	41	51
0.5 in. thick; 20 lb/ft²	28	33		38		43		48	53
Sheet aluminum									
16 gauge; 0.051 in. thick; 0.734 lb/ft²	5	8		13		18		23	28
10 gauge; 0.102 in. thick; 1.47 lb/ft²	8	14		19		24		29	34
Plywood									
0.25 in. thick; 0.73 lb/ft²		20		19		24		27	22
0.5 in. thick; 1.5 lb/ft²	8	14		19		24		29	34
0.75 in. thick; 2.25 lb/ft²	12	17		22		27		32	37
Sheet lead									
0.0625 in. thick; 3.9 lb/ft²			32		33		32	32	32
0.125 in. thick; 8.2 lb/ft²			31		27		37	44	33
Glass fiber board: 6 lb/ft³; 1 in. thick; 0.5 lb/ft²	5	5	5	5	5	4	4	4	3
Laminated glass fiber (FRP): 0.375 in. thick			26		31		38	37	38

(Continued on next page)

Table 23.11. (Continued)

Material or Structure	Frequency (Hz) 125	175	250	350	500	700	1000	2000	4000
Walls, nonhomogeneous									
Gypsum wallboard									
Two 0.5 in. sheets cemented together; joints wood battened; 1 in. thick; 4.5 lb/ft^2	24	25	29	32	31	33	32	30	34
Four 0.5 in. sheets cemented together; fastened together with sheet metal screws; dovetail-type joints paper taped; 2 in. thick; 8/9 lb/ft^2	28	35	32	37	34	36	40	38	49
0.25 in. plywood glued to both sides of 1 × 3 studs 16 in. o.c.; 3 in. thick; 2.5 lb/ft^2	16	16	18	20	26	27	28	37	33
Same as above, but 0.5 in. gypsum wallboard nailed to each face; 4 in. thick; 6/6 lb/ft^2	26	34	33	40	39	44	46	50	50
0.25 in. dense fiberboard on both sides of 2 × 4 wood studs, 16 in. o.c.; fiberboard joints at studs; 0.80 in. thick; 3.8 lb/ft^2	16	19	22	32	28	33	38	50	52
Soft-type fiberboard (0.75 in.) on both sides of 2 × 4 wood studs, 16 in. o.c.; fiberboard joints at studs; 5 in. thick; 4.3 lb/ft^2	21	18	21	27	31	32	38	49	53
0.5 in. gypsum wallboard on both sides of 2 × 4 wood studs, 16 in. o.c.; 4.5 in. thick; 5.9 lb/ft^2	20	22	27	35	37	39	43	48	43

Two 0.375 in. gypsum wallboard sheets glued together and applied to each side of 2 × 4 wood studs, 16 in. o.c.; 5 in. thick; 8.2 lb/ft^2	27	24	31	35	40	42	46	53	48
2 in. glass fiber (3 lb/ft^3) + lead vinyl composite; 0.87 lb/ft^2			4		4		13	26	31
0.375 in. steel + 2.375 in. polyurethane foam (2 lb/ft^2) + 0.0625 in. steel			38		52		55	64	77
Same as above, but 2.5 in. glass fiber (3 lb/ft^3) instead of foam			37		51		56	65	76
0.25 in. steel + 1 in. polyurethane foam (2 lb/ft^2) + 0.055 in. lead vinyl composite; 1.0 lb/ft^2			38		45		57	56	67
Concrete block, dense aggregate									
4 in. hollow, no surface treatment	30	36	39	41	43	44	47	54	50
4 in. hollow, one coat cement base paint on face	30	36	39	41	43	44	47	54	49
6 in. hollow, no surface treatment	37	46	50	50	50	53	56	56	46
6 in. hollow, one coat resin-emulsion paint each face	37	50	54	52	53	55	57	56	46
8 in. hollow, no surface treatment	40	47	53	54	54	56	58	58	50
8 in. hollow, two coats resin-emulsion paint each face	38	50	54	54	55	58	60	38	49

[a]The sound attenuation provided by a barrier to airborne diffuse sound energy may be described in terms of its sound transmission loss TL; TL is defined (in dB) as 10 times the logarithm to the base 10 of the ratio of the acoustic energy transmitted through a barrier to the acoustic energy incident upon its opposite side. It is a physical property of the barrier material, not of the construction techniques used.

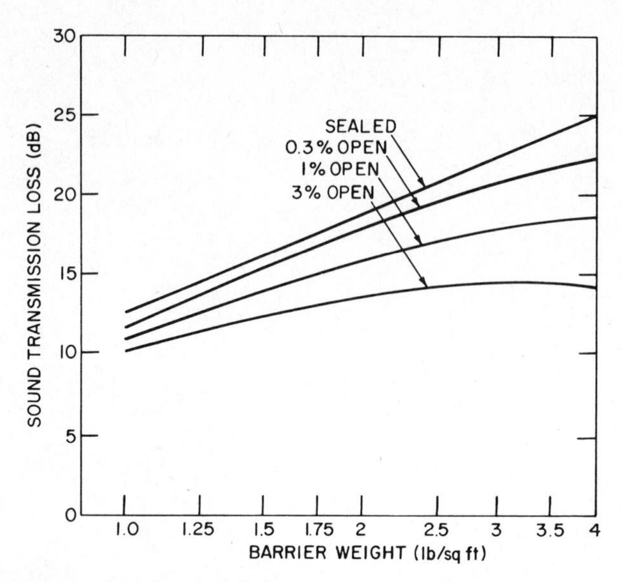

Figure 23.14. Average sound transmission loss of a single sound barrier as a function of barrier mass and percentage of open area.

values are possible with heavier barriers with greater surface areas. Higher values may also be expected when the source and/or the persons exposed are close to the barrier. The effects of two- or three-sided barriers are difficult to predict on a general basis; however, well-designed partial enclosures may provide noise reduction values of more than twice as much as single wall barriers. Complete enclosures from simple practical designs may provide noise reduction values in excess of 10 to 15 dB in the low frequencies and in excess of 30 dB in the high frequencies.

Precautions must be taken with any barrier or enclosure to be sure that there are no unnecessary openings. Figure 23.14 shows the average TL of a single barrier as a function of barrier mass and percentage of open area.

Example: An operator positioned close to a punch press that uses jets of compressed air to blow foreign particles from the die was exposed to excessive noise levels. A $\frac{1}{4}$-in. safety glass provided good visibility and access to the work position. The following noise reduction (NR) was provided at the operator's head position:

OB center frequency (Hz)	31.5	63	125	250	500	1000	2000	4000	8000
NR (dB)	—	—	1	2	3	9	14	20	22

Example: A sheet metal guard installed around a high-speed, rubber-toothed drive belt achieved the following noise reduction.

OB center frequency (Hz)	31.5	63	125	250	500	1000	2000	4000	8000
NR (dB)	—	—	—	—	—	—	7	9	19

Example: An electric motor-gear drive assembly was enclosed with $\frac{1}{8}$-in. thick steel with welded joints and lined with Fiberglass (No. 615) board 1 in. thick. Silencers for intake and exhaust ventilation air were constructed of 12-in. parallel plates 1 in. apart. The noise reduction achieved is as follows:

OB center frequency (Hz)	31.5	63	125	250	500	1000	2000	4000	8000
NR (dB)	—	5	6	12	14	25	35	24	23

9.1.3 Impact Noise, Noise Radiation, and Vibration Control

Example: A high-speed film-rewind machine (15 hp) produced excessive noise from the metal-to-metal impacts between gear teeth. Fiber gears were substituted, and the gears were flooded in oil. The noise reduction is as follows:

OB center frequency (Hz)	31.5	63	125	250	500	1000	2000	4000	8000
NR (dB)	—	10	6	5	5	8	20	16	14

Example: An 8-ft diameter hopper with an electric solenoid-type vibrator coupled solidly to a bottom bin was causing excessive noise. A live-bottom bin made by Vibra Screw was installed that required less vibratory power since only the cone is vibrated. Also, the new system had less radiation area and there were no metal-to-metal impacts. The noise reduction is as follows:

OB center frequency (Hz)	31.5	63	125	250	500	1000	2000	4000	8000
NR (dB)	—	7	6	20	22	16	12	12	9

Example: Screw machine stock tubes constructed of steel usually make excessive noise because there is nearly continuous impact between the tube and the screw stock. New tubes, such as the Corlett Turner silent Stock tube, constructed as a sandwich with an absorbent material between the outer steel tube and an inner helically wound liner, provide significantly lower noise levels. At 4000 rpm with $\frac{1}{2}$-in. hexagonal stock, the following noise reduction is achieved with the new tube:

OB center frequency (Hz)	31.5	63	125	250	500	1000	2000	4000	8000
NR (dB)	—	12	15	15	14	20	29	34	30

9.1.4 Acoustic Damping

Example: A metal enclosure around a rubber compounding mill vibrated freely, thus amplifying the motor, gear, and roll noises of the mill. An application to the inner surface of the metal enclosure of vibration-damping material ($\frac{1}{4}$-in. Aquaplas F 102A) reduced the noise as follows:

OB center frequency (Hz)	31.5	63	125	250	500	1000	2000	4000	8000
NR (dB)	10	9	9	13	9	7	8	10	11

Example: The guards and exhaust hoods of a 10-blade gang ripsaw was coated with 3M Underseal (EC-244). The following noise reduction was attained while the saw was idling:

OB center frequency (Hz)	31.5	63	125	250	500	1000	2000	4000	8000
NR (dB)	6	7	10	7	5	3	3	5	6

Example: The $\frac{3}{8}$-in. thick casing and the 1-in. thick steel base of a 2000-hp extruder gear were causing unwanted noise. Accelerometer measurements showed the casing and the base to be vibrating at about the same level. The casing was damped with a $\frac{1}{4}$-in. felt (No. 11, Anchor Packing Co.) plus an outer covering of $\frac{1}{4}$-in. steel. The felt-steel sandwich was bolted together on 8-in. centers. Because the steel base had ribs 9 in. deep that made the felt-steel damping impractical, the base was damped by a thick cover of sand. The following noise reduction was attained:

OB center frequency (Hz)	31.5	63	125	250	500	1000	2000	4000	8000
NR (dB)	—	—	—	—	4	17	26	24	18

9.1.5 Reduced Driving Force

The noise produced by an eccentric or imbalanced rotating member increases as the rotational speed is increased. One obvious noise control procedure is to balance dynamically all rotating parts. Also, these pieces should rotate concentrically. Proper maintenance of all bearing and other rotating contact surfaces is essential to keep equipment running quietly.

No machine should be operated at an unnecessarily high speed. In many instances a significant reduction in noise can be achieved by using a larger machine that can do the same job while operating at lower speeds.

The reduction of driving force in almost any form is an effective noise control procedure. In many instances, a reduction of driving force will provide the additional advantage of reduced radiation area.

Example: A blower exhaust system running at 705 rpm, 6-in. static pressure, and 13,800 cfm was badly out of balance, and bearings needed replacing. After new bearings were installed and the system balanced, the following noise reduction was found:

OB center frequency (Hz)	31.5	63	125	250	500	1000	2000	4000	8000
NR (dB)	—	3	3	11	12	11	10	8	10

Example: An oversized (36-in.) propeller-type fan mounted in the wall of a large reverberant room produced excessive noise when operated at 870 rpm. By reducing the fan speed from 870 to 690 rpm, it was possible to get the significant noise reduction shown below and still provide sufficient ventilation:

OB center frequency (Hz)	31.5	63	125	250	500	1000	2000	4000	8000
NR (dB)	—	3	7	8	12	9	8	6	44

Example: Small metal parts were dropped several inches into a metal chute, where they were moved by gravity onto another operation. The dropping distance and weight of the pieces could not be changed, but the chute surface was covered with a layer of paperboard $\frac{1}{16}$-in. thick, and this layer was covered by 18-gauge steel. The noise reduction produced by this sandwich covering was as follows:

OB center frequency (Hz)	31.5	63	125	250	500	1000	2000	4000	8000
NR (dB)	4	4	4	2	7	9	12	14	16

9.1.6 Muffler and Air Noise Generation Control

Example: The noise produced by an air-driven impact gun can be easily reduced by piping (with rubber hose) to a remote location. An internal muffler can also be used. The following noise reductions were achieved with an air gun running free:

OB center frequency (Hz)	31.5	63	125	250	500	1000	2000	4000	8000
NR (dB)									
Muffler	—	—	2	2	4	15	9	6	7
Rubber hose	—	—	19	17	30	42	29	28	28

Example: The discharge of a Gast Air Motor (models 4 AM and 6 AM) created excessive noise. A Burgess Manning Delta P—CA type muffler installed on the discharge outlet produced the following noise reduction:

OB center frequency (Hz)	31.5	63	125	250	500	1000	2000	4000	8000
NR (dB)	—	2	7	7	9	10	23	29	23

Example: Excessive blower noise was being produced by a pneumatic conveying system handling synthetic fiber fluff. An absorbing-type muffler was not desired because of the possibility of snagging and plugging. A resonant-type muffler supplied by Universal Silencer Corporation gave the following noise reduction:

OB center frequency (Hz)	31.5	63	125	250	500	1000	2000	4000	8000
NR (dB)	—	12	23	13	11	10	—	—	—

Example: The air intake of a gas turbine operating at 5800 rpm and 6200 hp created excessive noise. A parallel baffle muffler consisting of six plates, each $3\frac{1}{2}$-in. wide, filled with Fiberglass and faced with 18-gauge perforated sheet steel, was attached to the intake; then the baffle was connected to a duct of unlined steel plate $\frac{1}{4}$-in. thick. The cross section of the duct was 7 × 8 ft. The following noise reduction was achieved:

OB center frequency (Hz)	31.5	63	125	250	500	1000	2000	4000	8000
NR (dB)	—	—	10	16	22	33	35	27	26

Example: The noise produced by a tube reamer was reduced by the following values by mounting a Wilson 8500 muffler on the exhaust:

OB center frequency (Hz)	31.5	63	125	250	500	1000	2000	4000	8000
NR (dB)	2	2	3	10	23	26	28	16	18

9.1.7 Drive System Modification

Example: A rubber-toothed belt used to drive a pump was replaced by a V-belt drive, yielding the following noise reduction:

OB center frequency (Hz)	31.5	63	125	250	500	1000	2000	4000	8000
NR (dB)	—	5	5	4	2	—	8	17	18

10 COMMUNITY NOISE

Most community noise exposures, by themselves, do not cause noise-induced hearing impairment. On the other hand, any extended exposure to high-level noise from communities, at work, or elsewhere, may combine to become potentially harmful. Noise problems from communities do commonly entail communication and stress-related problems such as annoyance, and can become harmful to the general well-being of humans.

The effects of noise on communication can be measured with a reasonable degree of accuracy (65–74); however, other psychological and physiological responses to noise are extremely complicated and difficult to measure in a meaningful way. Attempts to correlate noise exposure levels and annoyance, or the general well-being of humans, are complicated by many factors such as:

1. The attitude of the listener toward the noise source
2. The history of individual noise exposures
3. The activities of the listeners and stresses on the listeners during the noise exposures
4. The hearing sensitivities of the listeners and other differences in individual responses

10.1 Nuisance-Type Regulations

For the most part, old rules and regulations were based on vaguely defined nuisance factors (252, 261–263). Nuisance-type regulations often take the form, "There shall be no unnecessary nor disturbing noise . . . ." An obvious weakness of this form of regulation is the failure to specify the conditions of how, when, and to whom noise is unnecessary or disturbing. Innumerable arguments may result in the interpretation of these laws when an attempt is made to enforce them. On the other hand, this kind of law may be useful in some cases where it is not possible to make reasonable decisions based on exposure (sound pressure) levels.

10.2 Performance-Type Regulations

Most recent zoning codes have specified maximum noise limits for various zones. These so-called performance zoning codes are more objective and easier to enforce than the nuisance laws. Unfortunately, many problems must be considered individually on a nuisance basis because of the many possible noise exposure situations. For example, a noise limit of 55 dBA may be generally acceptable for a given neighborhood. However, it is reasonable to assume that lawn mowers should be permitted and they may produce a noise level of 88 dBA at the property line. The lawn mower noise should be acceptable at certain hours when other persons' activities are not unreasonably affected. On the other hand, this same noise may not be acceptable if neighbors have guests on the adjoining lawn, if they are trying to sleep, and so on.

Obviously performance laws may be used to establish guidelines for reasonable noise exposure levels, but nuisance laws may also be needed in some instances. No law can replace the fact that all individuals must show consideration for their neighbors.

Communication among neighbors is essential if problems are to be avoided. Generally, complaints about a noisy party can be minimized by advising neighbors (tactfully) of the planned time for the party well in advance. In this way neighbors can minimize interruptions by adjusting their schedules. Face-to-face communication beforehand is usually a very helpful way to establish good will, and to avoid problems.

Most performance-type noise ordinances establish limiting levels that are well below the level where physiological damage may occur. Usually the limits are based on the number and level of adverse responses in the community rather than trying to measure individual annoyance reactions to noise. This decision is made in most cases because the group statistics involved in describing "community responses" to noise avoid many of the variables that are extremely difficult to account for in individual annoyance reactions.

Most performance-type noise codes specify limits of sound pressure level that are based on (1) a selected frequency weighting or noise analysis procedure, (2) the pattern of exposure times for various noise levels, (3) the ambient noise levels that would be expected in that particular kind of community without the offending noise source(s), and (4) a land-use zoning of the area. Regulations based on specific exposures—such as those from aircraft or from ground transportation—often work well at other locations having the same kinds of noise. Caution must be used, however, in applying these results to other kinds of noise exposure.

The document "Model Community Noise-Control Ordinance," which was published in 1975 by the EPA (262), will provide guidance in the development of performance-type noise control ordinances. This EPA document does not constitute a standard, specification, or regulation, but it does present available technical knowledge that may be used in a uniform and practical way by communities of various sizes to tailor ordinances to their specific conditions and goals.

10.3 Frequency Weighting or Analysis of Community Noise

Many frequency-weighting and multiple-band analysis procedures have been used in estimating community reaction to noise. A-weighted sound pressure level measurements are often used for setting noise exposure limits, and more complete analyses using octave, one-third octave, and narrower bands have been used in noise reaction measurements (252–263) and to pinpoint specific noise sources.

The single readings of sound pressure level have the obvious advantages of short measurement times and overall simplicity in data manipulation. The A-frequency weighting, which is provided on most sound level meters, has also been given considerable prominence in hearing conservation criteria (1–4) and for speech interference measurements (65–74). The readily available A-frequency weighting is considered by many investigators to be one of the most practical means of measuring noise exposures available at this time.

10.3.1 Aircraft Noise

Exposures to aircraft noise are often unique based on spectral and level characteristics, and on listeners' attitudes toward the source. Some of the more sophisticated noise analysis techniques for aircraft noise include the following:

1. Perceived noise level (PNL), measured in perceived noise decibels (PN dB)
2. Effective perceived noise levels (EPNL), measured in effective perceived noise decibels (EPN dB)
3. Noise exposure forecast (NEF), using EPN dB as the basic unit

PNL is an instantaneous measure of the noise produced at a given location, based on octave or third-octave sound pressure levels (264). EPNL values are derived from PNL instantaneous levels with added adjustments for high level, pure tone content, and flyover duration. NEF values are derived from EPNL levels with adjustments added for aircraft type, mix of aircraft, number of operations, runway utilization, flight path, operating procedures, and time of day.

Although the PNL, EPNL, and NEF procedures have been shown to be more accurate than single-frequency weightings for the prediction of community reaction to aircraft noise, their relatively high degree of complexity cannot be justified for many other uses. Differences between aircraft and nonaircraft noise exposure factors—such as noise characteristics, attitudes of listeners toward the source, and community composition—are often significant. Thus the complex data manipulation procedures of the PNL, EPNL, and NEF methods are not covered in this chapter. Complete details on the use of these aircraft noise measures can be found in EPA documents (261–264) and other sources (252–260).

10.3.2 Nonaircraft Noise

A-weighted sound levels are the most commonly used and the most practical means for measuring reactions to nonaircraft noise. Instrumentation with A-weightings is more readily available, and data manipulation is much simpler using this simple procedure than with other methods. There are also a few instances where other more complicated means have been shown to be superior for measurement of nonaircraft noises.

10.4 Noise Measurement Equipment

A wide range of instruments are available for community noise measurements. In many cases a sound level meter with an A-weighting is satisfactory when used with a timer. Any instrument used should meet the pertinent ANSI specifications (105–117), to assure accurate and legally admissible data.

Accessory equipment such as tape recorders or graphic level records may also be used with sound level meters; however, they must also conform with the pertinent sound level meter specifications. There are also instruments that are, in effect, complex combinations of sound level meters and accessories that can perform almost any recording task required. In addition, the output of some of these instruments can be connected into a computer to print out lengthy reports of time versus A-weighted sound pressure level.

10.5 Measurement Locations and Procedures

Sound pressure level measurements or recordings must be made at the place and time of annoyance. Outdoor measurements should be made at positions about 4 to 5 ft above the ground and as far as possible away from any solid vertical structure that may reflect sound. If the reflecting object is large (a house or a billboard), the distance away from the object should be about 50 ft. Complete descriptions of all measurement positions should be recorded, and particular attention must be given to conditions (barriers and uneven land) that might influence measurements (see Section 4).

The measuring equipment should be operated in the manner specified by the manufacturer, and care should be taken to prevent the measurements from being influenced by such factors as wind blowing over the microphone, vibration, and electromagnetic energy. Extremes in climate conditions may cause significant differences in measured levels; therefore measurements should be made under normal conditions.

10.5.1 Measuring Steady and Fluctuating Noises

Several different noise measurement procedures may be required in making community noise measurements. If the noise is continuous, has little variation in level, and is composed mostly of low-frequency sounds, almost any measuring procedure will be satisfactory. If noise levels vary widely in a complicated manner, however, it may be necessary to record the noise over a long period of time and to perform statistical analyses on the data.

Procedures for determining average values are not justified in some cases because community reactions often vary widely. One relatively simple procedure that is often adequate for sampling community noise is as follows:

1. Observe the A-weighted sound pressure level reading on a sound level meter with slow meter damping for 5 sec and record the best estimate of central tendency, and the range of the meter deflections in decibels.
2. Repeat the sampling described in step 1 until the number of central tendency readings equals or exceeds the total range (in decibels) of all the readings.
3. Find the arithmetic average of all the central tendency readings in steps 1 and 2, and call this estimate the community noise level for this particular measuring time and location.

10.5.2 Multiple Noise Sources

On occasion, several noise sources contribute to the measured A-weighted levels, and it may be desirable to determine the contributions of each source for purposes of noise control or for assessing the responsibility for the noise. The simplest way to obtain the contribution of each source is to measure the weighted levels produced by each individual source when the other sources are not operating. If it is not possible to obtain measures of the individual source contributions directly, it may be necessary to analyze the overall noise levels in narrow frequency bands, generally one-third octave band or narrower, and correlate the characteristics of the individual noise sources with the overall spectrum at the point in question.

Because analysis procedures are time-consuming and not always very meaningful, the individual sources should be measured directly if at all possible. These measurements may be taken at night, between work shifts, between process changes, or by special arrangements with supervisory personnel in charge of the individual operations.

10.5.3 Indoor Measurements

Indoor measurements should be made at locations at least 3 ft from walls or other reflecting surfaces. If regions of high and low levels are found close together (standing waves), three or four measurements about a foot apart should be averaged arithmetically.

10.5.4 Impulse Noise Measurements

Noise exposure criteria based on A-weighted sound pressure levels usually hold only for steady-state noises. The inertia of the sensing elements of conventional sound level meters prevent accurate readings of levels that change significantly within a short time and these instruments are calibrated to produce levels in root mean square (rms) rather than peak values. To obtain accurate peak factor readings for single pressure pulses, a peak-reading instrument or an oscillosope must be used. When noise impulses occur at a rate of more than 10 per second, and the variation in sound pressure level between pulses is less than 6 dB, a sound level meter can be used to estimate the rms sound pressure level.

Before effort is expended on impulse measurements, it is important to know how the information is to be used. Few rules and regulations have specific provisions for impulsive noise, and most of these documents simply specify an adjustment that is to be made to the reading of a conventional sound level meter for "impulsive or hammering noises." For such cases, accurate assessments of impulsive noise characteristics are not useful.

10.6 A Guide for Community Noise Criteria

Acceptable noise levels vary from one community to another depending on the history of noise exposures and other variables. In most cases, noise codes are tailored to the character and requirements of the particular community.

A typical noise code will have a steady-state noise limit of about 40 to 55 dBA for outdoor, daytime exposures in rural residential, hospital, or other normally quiet areas. The particular value of this base limit of sound pressure level should be selected with regard to the living habits, present and past, of the people in the area. For example, in a young neighborhood where there are many children, more noise may be tolerated than in an older neighborhood made up of older people.

Table 23.12. Estimated Community Response to Noise as Compared to the Noise Code Criteria

Difference between Measured Noise and Adjusted Noise Criterion (dBA)	Estimated Community Response
+0	No observed reaction
+5	Occasional complaints
+10	Widespread complaints
+15	Threats of community action
+20	Vigorous community action

When the base level is properly selected, there will be no observable reaction of the community to noises at or below this level.

The level selected for the base criterion may be completely impractical for other locations and conditions; therefore special adjustments must be made for each situation. It is impossible to provide the necessary adjustments for all situations in a document of reasonable size, but many situations can be covered if adjustments for the following factors are included:

1. The kind of community, that is, rural, suburban, urban (with some workplaces, businesses, or main roads), city (with heavy business and traffic, and heavy-industry noise)
2. The duration and time pattern of exposures
   a. Steady-state constant
   b. Intermittent steady state
   c. Impulsive noise
   d. Pure tone or whine characteristics
3. Time of day (i.e., daytime, evening, or night)
4. Indoors or outdoors
5. Climatic conditions
   a. Wind
   b. Precipitation
   c. Temperature
6. Special conditions for the district

The estimated community response to noise, compared to criteria prepared for a noise code as just described, might be summarized as in Table 23.12.

If the community response to noise with respect to the selected criteria does not follow the pattern given in Table 23.12, the criteria should be adjusted until this pattern is reached. Additional reference materials on community noise criteria may be found in References 252 to 263.

APPENDIX

Sample Letter to Persons Showing a Standard (Significant) Threshold Shift on a Serial Audiogram*

Dear ——————:

Your hearing thresholds measured on ————— indicate a change of more than ——— dB from (worse than) your base-line audiogram. This change may be temporary, or it may be permanent. Temporary changes may be the result of colds, allergies, medications, recent exposures to high-level noises, blows to the head, etc. Depending upon the severity of these and other factors, permanent impairments may also result.

If your hearing impairment is temporary, you should regain most of your hearing soon. To be safe, however, we recommend strongly that you have an audiological examination at the ——————————— Speech and Hearing Clinic (tel. xxx-xxx-xxxx). We have arranged for payment for this examination. Please continue to obey our hearing conservation rules at and away from work.

Sincerely,

*Required by OSHA if permanent significant (standard) threshold shift is determined.

REFERENCES

1. NIOSH, Criteria for a Recommended Standard: Occupational Exposure to Noise, NIOSH 73-11001.
2. Safety and Health Standards for Federal Supply Contracts (Walsh–Healy Public Contracts Act), U.S. Department of Labor, *Fed. Regist.*, **34**, 7948 (1969).
3. Occupational Safety and Health Standards (Williams–Steiger Occupational Safety and Health Act of 1970), U.S. Department of Labor, *Fed. Regist.*, **36**, 10518 (1971).
4. Occupational Noise Exposure; Hearing Conservation Amendment, Occupational Safety and Health Administration, 29 CFR Part 1910, *Fed. Regist.*, **46**(162) (August 21, 1981).
5. Occupational Noise Exposure; Hearing Conservation Amendment, Occupational Safety and Health Administration, 29 CFR Part 1910, *Fed. Regist.*, **48**(42) (March 3, 1983).
6. J. R. Anticaglia, "Physiology of Hearing," in *The Industrial Environment—Its Evaluation and Control*, U.S. DHEW, PHS, CDC, National Institute for Occupational Safety and Health, 1973.

7. A. Bell, "Noise—An Occupational Hazard and Public Nuisance," Public Health Paper, No. 30, Geneva, Switzerland, 1966.
8. G. R. Bienvenue and P. L. Michael, *Am. Ind. Hyg. Assoc., J.*, **41**, 535–541 (1980).
9. G. R. Bienvenue, P. L. Michael, and J. Violon-Singer, *Am. Ind. Hyg. Assoc. J.*, **37**, 628–634 (1976).
10. G. R. Bienvenue, J. Violon-Singer, and P. L. Michael, *Am. Ind. Hyg. Assoc. J.*, 133–337 (1977).
11. J. H. Botsford, "Noise Measurement and Acceptability Criteria," in *The Industrial Environment—Its Evaluation and Control*, U.S. DHEW, PHS, CDC, National Institute for Occupational Safety and Health, 1973.
12. W. Burns, *Noise and Man*, J. B. Lippincott, Philadelphia, 1968.
13. Council for Accreditation in Occupational Hearing Conservation Manual, Fischler's Printing, Cherry Hill, NJ, 1978.
14. Employer's Insurance of Wausau, "Guide for Industrial Audiometric Technicians," Employer's Insurance of Wausau, Wisconsin, 1963.
15. Environmental Acoustics Laboratory, The Pennsylvania State University, *Early Detection of Noise-Induced Hearing Impairment*, A Final Report to the Environmental Protection Agency, Contract No. 68-01-4498, 1978.
16. A. Feldmen and C. Grimes, *Hearing Conservation in Industry*, Williams and Wilkins, Baltimore, MD, 1985, pp. 77–88.
17. J. A. Gillies, Ed., *A Textbook of Aviation Physiology*, Pergamon Press, New York, 1965.
18. A. Glorig, *Noise and Your Ear*, Grune & Stratton, New York, 1958.
19. A. Glorig, *Sound Vib.*, **5**, 28–29 (1971).
20. "Hearing Conservation Program Manual for Federal Agencies," Standing Committee on Occupational Noise of the Federal Advisory Council on Occupational Safety and Health, 1983.
21. *Industrial Noise Manual*, 2nd, 3rd, and 4th eds., American Industrial Hygiene Association, Akron, OH, 1966, 1975, 1987.
22. *Industrial Noise Control Manual*. rev. ed., National Institute of Occupational Safety and Health, Cincinnati, OH, 1978.
23. Intersociety Committee on Industrial Audiometric Technician Training, "Guide for Training of Industrial Audiometric Technicians," *Am. Ind. Hyg. Assoc. J.*, **27**, 303 (1966).
24. K. Kryter, *The Effects of Noise on Man*, Academic Press, New York, 1970.
25. A. Lawther and D. W. Robinson, "Further Investigation of Tests for Susceptibility to Noise-Induced Hearing Loss," *ISVR Technical Report 149*, 1987.
26. A. Lawther and D. W. Robinson, "An Investigation of Tests for Susceptibility to Noise-Induced Hearing Loss," *ISVR Technical Report 138*, University of South-Hampton, 1986.
27. A. Lawther and D. W. Robinson, "Potential Factors in Resistance to Noise-Induced Hearing Loss," *Proc. Inst. Acoust.*, **9**, 217–219 (1987).
28. D. M. Lipscomb, *Hearing Conservation in Industry, Schools, and the Military*, College Hill, Boston, 1988.
29. P. L. Michael, "Noise," in *Industrial Environmental Health, the Worker and the Community*, Academic Press, New York, 1972, pp. 145–170.

30. P. L. Michael, "Training of Personnel Responsible for Enforcement of Noise Regulations in and Around Work Places, Internoise 74 Proceedings," Institute of Noise Control Engineering, Library of Congress Catalog No. 72-91606, 1974.
31. P. L. Michael and G. R. Bienvenue, *J. Acoust. Soc. Am.*, **60**(4), 944–952 (1976).
32. P. L. Michael and G. R. Bienvenue, *J. Acoust. Soc. Am.*, **67**(5), 1812–1815 (1980).
33. P. L. Michael and G. R. Bienvenue, *J. Acoust. Soc. Am.*, **67**(2), 693–698 (1980).
34. P. L. Michael, *Am. Ind. Hyg. Assoc. J.*, 1965.
35. P. L. Michael, *A. Ind. Nurses J.*, 7–10 (1965).
36. P. L. Michael, *J. Occup. Med.*, **10**, 67–71 (1968).
37. P. L. Michael, "Summary-Efforts of Noise on Man," Proceedings of the Conference on Noise as a Public Health Hazard, American Speech and Hearing Association, Report No. 4, 1969, pp. 157–161.
38. P. L. Michael and G. R. Bienvenue, *Am. Ind. Hyg. Assoc. J.*, **30**, 52–55 (1976).
39. P. L. Michael and G. R. Bienvenue, *J. Acoust. Soc. Am.*, **60**(6), 1640–1642 (1977).
40. P. L. Michael, *The Industrial Environment—Its Evaluation and Control*, In U.S. DHEW-PHS-CDC-NIOSH, 299–309, 1973.
41. P. L. Michael, "Industrial Noise and Conservation of Hearing," in Patty's *Industrial Hygiene and Toxicology*; 3rd rev. ed., Vol. 1, *General Principles*, John Wiley, New York, 1978.
42. P. L. Michael, "Noise," *Ind. Hyg. Found. Trans.*, **29**, 31–33 (1955).
43. P. L. Michael, "Noise Measurement and Personal Protection," Industrial Hygiene Foundation Transactions of the 20th Meeting held in Pittsburgh, PA, 1955, pp. 1–10.
44. P. L. Michael, "Noise at the Source," *National Safety Congress Transactions*, National Safety Council, Chicago, 1971.
45. P. L. Michael, "Noise," Mine Safety Appliances Company, 1963.
46. P. L. Michael, *Arch. Environ. Health*, **10**, 612–618 (1965).
47. P. L. Michael, *Min. Congr. J.*, 74–82 (1972).
48. P. L. Michael and J. Prout," Hearing Conservation for Miners," A tape-programmed slide lecture, U.S. Bureau of Mines, 1976.
49. P. L. Michael, *Health Saf.*, 31–32 (1975).
50. *Noise Control—A Guide for Workers and Employers*, Occupational Safety and Health Administration, Office of Information, No. OSHA-3048, 1980.
51. National Standards Institute, Draft American National Standard Methods for the Evaluation of the Potential Effect on Human Hearing of Sounds with Peak A-Weighted Sound Pressure Levels Above 120 Decibels and Peak C-Weighted Sound Pressure Levels Below 140 Decibels Draft ANSI S3.28-1986), Published for Trial Use, Comment, and Criticism for a Period of 5 Years, 1986.
52. J. B. Olishfski and E. Harford, Eds., *Industrial Noise and Hearing Conservation*, National Safety Council, Chicago, 1975.
53. H. O. Parrack, "Physiological Effects of Intense Sound," Eng. Div. Memo. Report No. MCREXD-695-71B, AD 732 935 Air Material Command, Wright-Patterson Air Force Base, OH, 1948.
54. D. W. Robinson, "The Concept of Noise Pollution Level, NPL Aero Report Ac 38," National Physical Laboratory, Aerodynamics Division, 1969.

55. D. W. Robinson, "Noise Exposure and Hearing: A New Look at the Experimental Data," *Health and Safety* Executive Research Report No. 1, 1987.
56. D. W. Robinson, *Br. J. Audiol.*, **22**(1), 5–20 (1987).
57. J. Sataloff and P. L. Michael, *Hearing Conservation*, Charles C. Thomas, Springfield, IL 1973, pp. 291–318.
58. P. E. Smith, *Am. Ind. Hyg. Assoc. J.*, **24**, 297 (1967).
59. University of Dayton Research Institute, "A Basis for Limiting Noise Exposure for Hearing Conservation," Final Technical Report UDRI-TR-73- 29, AD 767 274, 1973.
60. U.S. Congress, Occupational Safety and Health Act of 1970, Public Law 91-596, 91st Congress, Washington, DC, 1970.
61. U.S. Congress, Walsh–Healy Public Contracts Act, *Fed. Regist.*, **34**(96), 1969.
62. U.S. Department of Health, Education and Welfare, National Institute for Occupational Safety and Health: Occupational Exposure to Noise, Washington, DC, 1972.
63. U.S. Department of Labor, Williams–Steiger Occupational Safety and Health Act of 1970: Occupational Safety and Health Standards, *Fed. Regist.*, **36**, 10518 (1971).
64. Unpublished work by Paul L. Michael and David F. Bolka at the Pennsylvania State University, 1973.
65. American National Standards Institute (ANSI), American National Standard Procedure for Rating Noise with Respect to Speech Interference, ANSI S3. 14-1977, 1977.
66. G. R. Bienvenue and P. L. Michael, "Digital Processing Techniques in Speech Discrimination Testing," in *Rehabilitation Strategies for Sensorineural Hearing Loss*, Grune & Stratton, New York, 1979.
67. A. S. House, C. E. Williams, M. H. L. Hecker, and K. D. Kryter, *J. Acoust. Soc. Am.*, **37**, 158–166 (1965).
68. K. D. Kryter, "Speech Communication in Noise," AFCRCOTRO54-52, Washington, D.C.: Air Force Cambridge Research Center, Air Research and Development Command, 1955.
69. P. L. Michael et al., "Intelligibility Test on a Family of Electroacoustic Devices (FEADS)—Final Report," Contract No. N00953-70-M-2620, U.S. Navy, 1970.
70. P. L. Michael and J. Prout, "An Evaluation of Experimental and Developmental Communication Systems for Underground Coal Mines," A Final Report to the U.S. Bureau of Mines, Grant Agreement G01661, 1977.
71. H. Fletcher and W. A. Munson, *J. Acoust. Soc. Am.*, **5** (1933).
72. J. C. Webster, "Effects of Noise on the Hearing of Speech, Proceedings of the Second International Congress on Noise as a Public Health Problem," *Am. Speech Hear. Assoc. Rep.*, **5** (1969).
73. J. C. Webster, "Effects of Noise on Speech Intelligibility," *Am. Speech Hear. Assoc. Rep.*, **4**, 49–73 (1969).
74. J. C. Webster, *J. Audio Eng.* (1970).
75. A. Carpenter, "How Does Noise Affect the Individual?", *Impulse* **24** (1964).
76. A. Cohen, *Nat. Saf. News*, **108**(2), 93–99 (1973).
77. A. Cohen, *Nat. Saf. News*, **108**(3), 68–76 (1973).
78. A. Cohen, "Effects of Noise on Performance," *Proceedings of the International Congress of Occupational Health, Vienna, Austria*, 1966, pp. 157–160.
79. A. Cohen, "Noise and Psychologic State," *Proceedings of National Conference on*

Noise as a Public Health Hazard, American Speech and Hearing Association, 1969, pp. 89–98.

80. A. Cohen, *Trans. N.Y. Acad. Sci.*, **30**, 910–918 (1968).
81. W. F. Grether, "Noise and Human Performance," Report AMRL-TR-70-29, AD 729 213, Aerospace Medical Research Laboratory, Wright-Patterson Air Force Base, OH, 1971.
82. P. L. Michael and G. R. Bienvenue, "Industrial Noise and Man," in *Physiology and Productivity at Work—The Physical Environment*, John Wiley, Sussex, England, 1983.
83. American National Standards Institute, ANSI S1.6-1984 (ASA 53) American Standards Preferred Frequencies, Frequency Levels, and Band Numbers for Acoustical Measurements, 1984.
84. American National Standards Institute, ANSI S3.20-1973 (R 1986) Revision of S1.1-1960, American Standard Psychoacoustical Terminology, 1986.
85. American National Standards Institute, ANSI S1.8-1969 (R 1974) American Standard Preferred Reference Quantities for Acoustical Levels, 1969.
86. American National Standards Institute, Acoustical Terminology, S1.1-1960 (Revised 1971), 1971.
87. L. L. Beranek, *Acoustic Measurements*, John Wiley, New York, 1949.
88. L. L. Beranek, *Music, Acoustics and Architecture*, John Wiley, New York, 1962.
89. L. E. Kinsler and A. R. Frey, *Fundamentals of Acoustics*, John Wiley, New York, 1962.
90. L. E. Kinsler, A. R. Frey, A. B. Coppens, and J. V. Sanders, *Fundamentals of Acoustics*, American National Standard Specifications for Octave, Half-Octave and Third-Octave Band Filter Sets, ANSI S1.11 (R 1976), John Wiley, New York, 1982.
91. P. L. Michael, "Physics of Sound," in *The Industrial Environment—Its Evaluation and Control*, U.S. DHEW, PHS, CDC, National Institute for Occupational Safety and Health, 1973, 1983, 1989.
92. R. P. Hammernik, R. J. Salvi, and D. Henderson, "Blast Trauma: The Effect on Hearing," AD-A172 467/3/GAR, No. 1, University of Texas at Dallas, Richardson, TX, 1983, p. 339.
93. American National Standards Institute, American National Standard Specifications for Audiometers, ANSI S3.6-1989, New York, 1989.
94. American National Standards Institute, American National Standard Criteria for Permissible Ambient Noise during Audiometric Testing, ANSI S3.1-1977 (R 1986), 1986.
95. R. Carhart and J. F. Jerger, *J. Speech Hear. Disord.*, **24**, 330–345 (1959).
96. E. L. Eagles and L. G. Doerfler, *J. Speech Hear. Res.*, **4**, 149 (1961).
97. E. Harford, "Clinical Application and Significance of the SISI Test," in *Sensorineural Hearing Processes and Disorders*, A. B. Graham, Ed., Little, Brown, 1974, p. 223.
98. P. L. Michael and G. R. Bienvenue, "Noise Attenuation Characteristics of the MX-41/AR and the Telephonics Circumaural Audiometric Headsets," *J. Am. Aud. Soc.* **6**(4), 1–5 (1978).
99. P. L. Michael and G. R. Bienvenue, "Noise Attenuation Characteristics for Supra-Aural Audiometric Headsets using the Models MX-41/AR and 51 Earphone Cushions," *J. Acoust. Soc. Am.*, **70**, 5 (1981).
100. S. Rosen, D. Plester, A. El-Mofty, and H. V. Rosen, *Arch. Otolaryngol.*, **79** (1964).

101. J. Tonndorf and B. Kurman, *Ann. Oto. Rhino. Laryngol.*, **93**, 576–582 (1984).
102. E. Villchur, *J. Acoust. Soc. Am.*, **48**(6) Part 2, 1387–1396 (1970).
103. I. S. Whittle and M. E. Delaney, *J. Acoust. Soc. Am.*, **38**, 1187 (1966).
104. E. Zwicker, *Facts Models Hear.*, **30**, 132–141 (1974).
105. American National Standards Institute, 1983–1985. ANSI S1.4A-1985 (ASA 47) Amendment to S1.4-1983. American National Standard Specification for Sound Level Meters.
106. American National Standards Institute, ANSI S1.11-1986 (ASA 65) American Standard Specifications for Octave-Band and Fractional-Octave-Band Analog and Digital Filters, 1986.
107. American National Standards Institute, ANSI S1.42-1986 (ASA 64) American National Standard Design Response of Weighting Networks for Acoustical Measurements, 1986.
108. American National Standards Institute, ANSI S12.7-1986 (ASA 62) American Standards Methods for Measurement of Impulse Noise, 1986.
109. American National Standards Institute, DRAFT ANSI S3.28-1986, Draft American National Standard Methods for the Evaluation of the Potential Effect on Human Hearing of Sounds with Peak A-Weighted Sound Pressure Levels and Peak C-Weighted Sound Pressure Levels below 140 Decibels, 1986.
110. American National Standards Institute, ANSI S1.25-1978 (ASA 25), American National Standard Specification for Personal Noise Dosimeters, 1978.
111. American National Standards Institute, ANSI S1.13-1971 (R 1983) American Standard Methods for the Measurement of Sound Pressure Levels, 1983.
112. American National Specification for Acoustical Calibrators, ANSI S1.40-1984, Standards Secretariat, Acoustical Society of America, New York, 1984.
113. "Guidelines for Developing a Training Program in Noise Survey Techniques," Committee on Hearing and Bioacoustics (CHABA) ONRC Contract No. N00014-67-A-0244-0021. US EPA ONAC, 1975.
114. *J. Acoust. Soc. Am.*, 1959–1968, 1964–1988, Instrumentation and Techniques for Noise Measurement Cumulative Indices 43.85.F and 43.85.G (1959–1968), and *JASA* References to Contemporary Papers on Acoustics 43.85.F and 43.85.G (1964–1988).
115. *J. Acoust. Soc. Am.*, 1959–1968, 1964–1988, Acoustic Signal Processing Cumulative Indices 43.60 (1959–1968), and *JASA* References to Contemporary Papers on Acoustics 43.60 (1964–1988).
116. P. L. Michael, "Guidelines for Developing a Training Program in Noise Survey Techniques," ONRC Contract No. N00014-67-A-0244-0021, U.S. EPA ONAC, 1975.
117. P. L. Michael et al., "Development of Measurement Methodologies for Stationary Sound Sources," Report to U.S. Environmental Protection Agency, Office of Noise Abatement and Control, 1976.
118. NIOSH, Industrial Sound Level Meter Acoustical Calibrator Accuracy Test, un-numbered.
119. NIOSH, Industrial Sound Level Meter Attenuator Tolerance Test, un-numbered.
120. NIOSH, Industrial Sound Level Meter Magnetic Field Sensitivity Test, un-numbered.
121. NIOSH, Industrial Sound Level Meter Output Connection Test, un-numbered.
122. National Institute for Occupational Safety and Health, "A Report on the Performance of Personal Noise Dosimeters," pp. 78–186, 1982.

123. American National Standards Institute, ANSI S12.6-1984 (ASA 55) American Standards Method for the Measurement of the Real-Ear Attenuation of Hearing Protectors, 1984.
124. American National Standards Institute, ANSI S3.19-1974 (ASA 1) Method for the Measurement of Real-Ear Protection of Hearing Protectors and Physical Attenuation of Earmuffs, 1974.
125. American National Standard Method for the Measurement of the Real-Ear Attenuation of Hearing Protectors and Physical Attenuation of Earmuffs, ANSI S3.19-1974, American National Standards Institute, New York, 1974.
126. American National Standard Method for the Measurement of the Real-Ear Attenuation of Hearing Protectors, ANSI S12.6-1984, American National Standards Institute, New York, 1984.
127. American National Standard Institute, U.S.A. Standard Method for the Measurement of the Real-Ear Attenuation of Ear Protectors at Threshold, USAS Z24.22-1974, 1957.
128. R. G. Edwards, A. Broderson, W. W. Green, and B. L. Lampert, *Noise Control Eng. J.*, **20**, 6–15 (1983).
129. R. G. Edwards, W. P. Hauser, N. A. Moiseev, and A. B. Broderson, *Sound Vib.*, **12**, 12–22 (1978).
130. Environmental Acoustics Laboratory, "An Objective Method for Evaluating Ear Protectors," Final Report, Grant 1 R01 OH 00341 01 DHEW, PHS, 1972.
131. Environmental Acoustics Laboratory, "A Study of Roof Warning Signals and the Use of Personal Hearing Protection in Underground Coal Mines," USBOM Final Report Grant G0133026 DOI BOM, 1973.
132. Environmental Protection Agency, Noise Labeling Requirements for Hearing Protectors, *Fed. Regist.*, 40CFR Part 211, Subpart B, 56139-56147, No. 190, 1979.
133. Environmental Acoustics Laboratory, Aspects of Noise Generation and Hearing Protection in Underground Coal Mines, Final Report, Grant G 0122004, U.S. Bureau of Mines, 1972.
134. B. Lempert, *Sound Vib.*, **18**, 26–39 (1984).
135. P. L. Michael, R. Kerlin, G. R. Bienvenue, and J. Prout, "Aspects of Noise Generation and Hearing Protection in Underground Coal Mines," Final Report—Grant G 0122004, U.S. Bureau of Mines, 1972.
136. P. L. Michael, R. Kerlin, G. R. Bienvenue, and J. Prout, "A Real-Ear Field Method for the Measurement of the Noise Attenuation of Insert-type Hearing Protectors," Report for DHEW-NIOSH 76-181, 1975.
137. P. L. Michael, R. Kerlin, G. R. Bienvenue, and J. Prout, "A Study of Roof Warning Signals and the Use of Personal Hearing Protection in Underground Coal Mines," Final Report Grant G 0133026 DOI Bureau of Mines (BOM), 1973.
138. P. L. Michael, R. Kerlin, G. R. Bienvenue, and J. Prout, "An Objective Method for Evaluating Ear Protectors," Final Report Grant 1 R01 OH 00341 01 DHEW, PHS, 1972.
139. P. L. Michael, "Single-number Performance Factors for Hearing Protectors," in *Personal Hearing Protection in Industry*, P. W. Alberti, Ed., 1982, pp. 221–235.
140. P. L. Michael et al., "Evaluation of Speech Processing Systems, Evaluation of Electronic-Active Hearing Protectors for Use in Underground Coal Mines," Report to U.S. Department of the Interior, Bureau of Mines, 1976.

141. P. L. Michael, Personal Hearing Protection Evaluation—Present and Future, A List of Personal Hearing Protectors and Attenuation Data, DHEW-NIOSH, 76-120, 1971.
142. P. L. Michael and G. R. Bienvenue, "Hearing Protector Performance—an Update," *J. Am. Ind. Hyg. Assoc.*, **41**, 542–546 (1980).
143. P. L. Michael, "Evaluation of Principles Applicable to the Design of Ear Protectors and Communication Headsets for Use in Noise Environments," Ph.D. thesis, published by the University of Chicago Press, 1955.
144. C. W. Nixon and W. C. Knoblach, "Hearing Protection of Ear Muffs Worn over Eyeglasses," AMRL-TR-74-61, Aerospace Medical Research Laboratory, Wright-Patterson AFB, OH, June 1974.
145. "List of Personal Hearing Protectors and Attenuation Data," National Institute of Occupational Safety and Health, Cincinnati, OH, No. 76-120, 1975.
146. D. Ohlin, P. L. Michael, G. R. Bienvenue, and D. M. Rosenberg, *Sound Vib.*, **15**, 22–25 (1981).
147. D. Ohlin, Personal Hearing Protective Devices: Fitting and Use, USAEHA Technical Guide (Med), 1975.
148. J. F. Savell and E. H. Toothman, *Am. Ind. Hyg. Assoc. J.*, **48**, 23–27 (1987).
149. E. A. G. Shaw, "Hearing Protector Design Concepts and Performance Limitations," Integrated Symposium on Personal Hearing Protection Held in Toronto, May 14–16, 1980.
150. E. A. G. Shaw and G. J. Thiessen, *J. Acoust. Soc. Am.*, **30**, 24–36 (1958).
151. U.S. Department of the Interior, "Evaluation of Electronic-Active Hearing Protectors for Use in Underground Coal Mines," Report to U.S. Department of the Interior, Bureau of Mines, 1976.
152. American National Standards Institute, American National Standard Method for Rating the Sound Power Spectra of Small Stationary Noise Sources, ANSI S3.17-1975, 1975.
153. American National Standards Institute, American National Standard Method for the Designation of Sound Power Emitted by Machinery and Equipment, ANSI S1.23-1976, 1976.
154. American National Standards Institute, American National Standard Guidelines for the Use of Preparation Procedures for the Determination of Noise Emission from Sources, ANSI S12.1-1983, 1983.
155. American National Standards Institute, American National Standard Statistical Methods for Determining and Verifying Stated Noise Emission Values of Machinery and Equipment, ANSI S12.3-1985, 1985.
156. American National Standards Institute, American National Standard Guidelines for the Use of Sound Power Standards and for the Preparation of Noise Test Codes, ANSI S1.30-1979, 1985.
157. R. D. Berendt, E. L. R. Corliss, and M. S. Ojalvo, *NBS Handbook 119*, U.S. Department of Commerce, 1976.
158. M. J. Crocker and J. A. Price, *Noise and Noise Control*, Vol. I, CRC Press, Cleveland, OH, 1975.
159. C. M. Harris, Ed., *Handbook of Noise Control*, 2nd ed., McGraw-Hill, New York, 1979.
160. V. H. Hill, "Control of Noise Exposure," in *The Industrial Environment—Its Eval-*

uation and Control, U.S. DHEW, PHS, CDC, National Institute for Occupational Safety and Health, 1973.

161. P. H. Parkin et al., *Acoustics, Noise and Buildings*, Faber and Faber, London, 1979.
162. W. A. Rosenblith, K. N. Stevens, et al., 1953. *Handbook of Acoustic Noise Control*, Vol. 2, *Noise and Man*, WADC TR-52-204, Wright Air Development Center, Wright-Patterson Air Force Base.
163. E. Ackerman, *J. Appl. Phys.*, **24**, 1371–1373 (1953).
164. *J. Acoust. Soc. Am.*, 1964–1988, Chemical Effects of Ultrasound Cumulative Indices 43.35.V (1959–1968), and *JASA* References to Contemporary Papers on Acoustics 43.35.V.
165. *J. Acoust. Soc. Am.*, 1964–1988, Use of Ultrasonics in Nondestructive Testing of Materials Cumulative Indices 43.35.Z (1959–1968), and *JASA* References to Contemporary Papers on Acoustics 43.35.Z.
166. *J. Acoust. Soc. Am.*, 1964–1988, Biological Effects of Ultrasound, Ultrasonic Tomography Cumulative Indices 43.35.W (1959–1968), and *JASA* References to Contemporary Papers on Acoustics 43.35.W.
168. W. I. Acton, *J. Soc. Occup. Med.*, **33**, 107–113 (1983).
169. W. I. Acton, *Protection*, 14–10, 12–17 (1977).
170. W. I. Acton and M. B. Carson, *Br. J. Ind. Med.*, **24**, 297–304 (1967).
171. W. I. Acton, *Ann. Occup. Hyg.*, **11**, 227–234 (1968).
172. W. I. Acton, *Industrial Noise Manual*, 1st ed., American Industrial Hygiene Association, 1974.
173. W. I. Acton, *Ann. Occup. Hyg.*, **18**, 267–268 (1977).
174. V. M. Albers, Suggested Experiments for Laboratory Courses in Acoustics and Vibrations, The Pennsylvania State University Press, University Park, PA, 1972.
175. C. H. Allen, H. Frings, and I. Rudnick, *J. Acoust. Soc. Am.*, **20**, 62–65 (1948).
176. American Conference of Governmental Industrial Hygienists, *Threshold Limits of Physical Agents*, ACGIH, Cincinnati, 1969.
177. American Conference of Governmental Industrial Hygienists, *Documentation of the Threshold Limit Value*, 4th ed., ACGIH, Cincinnati, 1980.
178. American Conference of Governmental Industrial Hygienists, *Threshold Limit Values and Biological Exposure Indicies for 1988–89 (1988–1989 Adoption)*, p. 107.
179. American Conference of Governmental Industrial Hygienists (ACGIH), *Threshold Limit Values for Chemical Substances and Physical Agents in the Workroom Environment*, ACGIH, Cincinnati, 1988–1989, p. 107.
180. Z. D. Angeluscheff, *J. Acoust. Soc. Am.*, **26**(142), 942 (1954).
181. Z. Z. Asbel, *Gigena truda i professional 'Nye Zabolevanija* (Moscow), **9**, 29; Abstract in *Occup. Safety Health Abstr.*, **4**, 104 (1966).
182. S. B. Barnett, *Acta Otolarynol*, **89** 424–432 (1980).
183. B. Bauer and A. L. DiMattia, *J. Acoust. Soc. Am.*, **51**, 1388 (1972).
184. D. A. Benwell and M. H. Repacholi, "Guidelines for the Safe Use of Ultrasound: Part II—Industrial and Commercial Applications," Bureau of Radiation and Medical Devices (BRMD), Non-Ionizing Radiation Section, Ottawa, Canada, 1980.
185. S. H. P. Bly and D. A. Benwell, "Guidelines for the Safe Use of Ultrasound: Part

II—Industrial and Commercial Applications," Bureau of Radiation and Medical Devices (BRMD), Non-Ionizing Radiation Section, Ottawa, Canada, 1989.

186. N. Byalco et al., "Certain Biochemical Abnormalities in Workers Exposed to High-Frequency Noise," *Dokl. Vses. Nauch.-Prakt. Soveshch. Izuchen. Deistriya Shuma Org.* (Moscow); abstract in *Excerpta Med.*, **79**, 570 (1963).
187. Canadian National Guidelines for the Safe Use of Ultrasound, Draft Revision of 80-EHD-60—Safety Code 24 Part II—Industrial and Commercial Applications.
188. K. I. Carnes and F. Dunn, *Radiat. Environ. Biophys.*, **25**, 235–240 (1986).
189. J. L. Campanile, "Effects of Air-Borne Ultrasound on Humans with Implications for Development of Exposure Limits," Master's thesis, The Pennsylvania State University, University Park, PA, 1973.
190. J. Cordell, "Physiological Effects of Airborne Ultrasound: A Bibliography with Abstracts," Library Report No. 4, Commonwealth Acoustics Laboratories, Sydney, Australia, 1968.
191. R. B. Crabtree and S. E. Forshaw, "Exposure for Ultrasonic Cleaner Noise in the Canadian Forces," DCIEM Technical Report 77X45, Downsview, Ont., 1977.
192. A. E. Crawford, *Ultrasound Engineering*, Butterworths, London, 1955.
193. P. A. Danner, E. Ackerman, and H. W. Frings, *J. Acoust. Soc. Am.*, **26**, 731 (1954).
194. H. Davis, *J. Acoust. Soc. Am.*, **20**, 605–607 (1948).
195. H. Davis, H. O. Parrack, and D. H. Eldredge, *Ann. Otol.*, **58**, 732 (1949).
196. E. DeSeta, G. A. Bertoli, and R. Filipo, *Audiology*, **24**, 254–259 (1985).
197. E. D. D. Dickson and D. L. Chadwick, *J. Laryng. Otol.*, **65**, 154 (1951).
198. E. D. D. Dickson and N. P. Watson, *J. Laryng. Otol.*, **69**, 276 (1949).
199. V. K. Dobroserdov, *Gig. Sanit.* (Moscow), **32**, 17 (1967).
200. J. S. Felton and C. Spencer, *Am. Ind. Hyg. Assoc. J.*, **22**, 136–147 (1961).
201. C. Glickstein, *Basic Ultrasonics*, John F. Rider, New York, 1960.
202. A. Gootbier, *Western Electric Eng.*, **13** (1969).
203. E. Grandjean, "Physiologische and Psychophysiologische Wirkingen des Larms," *Menschen Umwelt* **4** (1960).
204. V. M. Grigor'eva, "Effect of Ultrasonic Vibrations on Personnel Working with Ultrasonic Equipment," *Sov. Phys. Acoust.*, **11**, 426–427 (1966).
205. J. Grzesik and E. Pluta, *Int. Arch. Occup. Environ. Health*, **57**, 137 (1986).
206. J. Grzesik and E. Pluta, *Int. Arch. Occup. Environ. Health*, **53**, 77 (1983).
207. A. V. Haeff and C. Knox, *Science*, **139**, 590–592 (1963).
208. E. N. Harvey, *Biol. Bull.*, **59**, 306 (1930).
209. Hazards Control Information Exchange Bulletin I, University of California Radiation Laboratory, 1961.
210. B. A. Herman and D. Powell, "Airborne Ultrasound: Measurement and Possible Adverse Effects, HHS Publication (FDA) 81-8163, 1981.
211. C. R. Hill and G. ter Haar, "Ultrasound," in *Non-Ionizing Radiation Protection*, M. J. Suess Ed., World Health Organization (WHO) Regional Publications, Copenhagen, Denmark, 1988, pp. 199–228.
212. A. P. Hulst, "Ultrasonic Welding of Metals," *Ultrasonics*, 1972.
213. International Non-Ionizing Radiation Committee of the International Radiation Pro-

tection Association (IRPA), "Interim Guidelines on Limits of Human Exposure to Airborne Ultrasound," *Health Phys.*, **46**, 969 (1984).

214. M. Janousek, J. Gruberova, and K. Svitek, *Prac. Lek.*, **37**, 369–373 (1985) (*Ergon. Abstr.*, **1987**, 104691–104692).
215. E. Kelly, Ed., *Ultrasonic Energy, Biological Investigations and Medical Applications*, University of Illinois Press, Urbana, IL, 1965.
216. E. Kelly, Ed., *Ultrasound in Biology and Medicine*, Publication No. 3, American Institute of Biological Sciences, Washington, DC, 1957.
217. J. J. Knight, *Ultrasonics*, **6**, 39–42 (1968).
218. J. J. Knight, *J. Acoust. Soc. Am.*, **39**, 1184 (1966).
219. R. LeGrand, "Ultrasonics—Get Three New Jobs," *Am. Mach.*, 1971.
220. P. P. Lele, *Symposium on Biological Effects and Characterizations of Ultrasound Sources*, HEW Publication (FDA) 78-8048, 224-238, 1978.
221. Z. S. Lisichkina, "An Investigation of the Hygienic Characteristics of Ultrasonic Waves in Industry," Report Number 15487, 1982, Joint Publication, Research Service.
222. R. E. Litke, *Arch. Otolaryng.*, **94**, 255–257 (1971).
223. P. Lofstedt and I. Limnell, "Infrasonic Noise in Milling Plants," Invest. Rep. No. 1984.23, National Board of Occupational Safety and Health, Sola, Sweden, 1986 (*Ergon. Abstr.* 100351).
224. F. Massa, "Ultrasonic Transducers for Use in Air," *Proc. IEEE*, **53**, 1363–1371 (1965).
225. L. L. Miasnikov and E. N. Miasnikov, *Acoustica*, **21**, 118–120 (1969).
226. P. L. Michael, "Effects of Airborne Upper Sonic and Ultrasonic Acoustic Radiation and Proposed Threshold Limit Values for Human Exposure," invited paper presented at the Acoustical Society of America Meeting held in Cleveland May 15, 1986.
227. P. L. Michael, R. Kerlin, G. R. Bienvenue, and J. Prout, "An Evaluation of Industrial Acoustic Radiation Above 10 kHz," Final Report on Contract No. HSM-99-72-125, National Institute of Occupational Safety and Health, 1974.
228. J. D. Miller, *J. Acoust. Soc. Am.*, **56**, 729 (1974).
229. G. M. Naimark, J. Klair, and W. A. Mosher, *J Franklin Inst.*, **251**, 279–299, 402–408 (1951).
230. National Council on Radiation Protection and Measurements (NCRP), "Biological Effects of Ultrasound: Mechanisms and Clinical Implications," NCRP Report # 74, Bethesda, MD, 1983.
231. V. F. Nordreva, Ed., *Ultrasound in Industrial Processing and Control, Soviet Progress in Applied Ultrasonics*, Vol. 1, Authorized translation from Russian, Consultants Bureau, New York, 1964.
232. V. A. Nosov, *Ultrasonics in the Chemical Industry, Soviet Progress in Applied Ultrasonics*, Vol. 2, Authorized translation from Russian, Consultants Bureau, New York, 1965.
233. H. O. Parrack, *Int. Audio.*, **5**, 294–308 (1966).
234. H. O. Parrack, *Ind. Med. Surg.*, **2**, 156 (1952).
235. G. Portman, M. Portman, and L. Barbe, "Etude Experimentale (Fonctionelle et Histologique) de L'action des Ultra-Sons sur L'audition." *Acta Oto-Laryngol.* (Stockholm), Suppl. 100, 119–132 (1952) (English translation in *Transl. Beltone Inst. Hear. Res.*, No. 3, April 1956).

236. G. Portman, M. Portman, and L. J. Barbe, "Experimental Study (Functional and Histological) of the Effect of Ultrasonics on Hearing" (in French), *Acta Oto-Laryngol.* (Stockholm), Suppl. 100, 119–133 (1952).

237. G. Portman, M. Portman, and L. J. Barbe, "Experimental Study (Functional and Histological) of the Effect of Ultrasonics on Hearing," *Transl. Beltone Inst. Hear. Res.*, No. 3, 3–18 (1956).

238. J. D. Pye, "Ultrasonic Bioacoustics," Final Scientific Report, AFOSR 70-2570TR, AD 714 632, U of London King's College, London, 1970.

239. J. M. Reid and M. R. Sikov, Eds., *Interaction of Ultrasound and Biological Tissues, Proceedings of a Workshop Held at Battelle Seattle Research Center, Nov. 8-11, 1971*, DHEW Publication (FDA) 73-8008 (BRH/DBE 73-1) 1972.

240. J. S. Saby and H. A. Thorpe, *J. Acoust. Soc. Am.*, **18**, 271–273 (1946).

241. A. Shoh, "Industrial Applications of Ultrasound—A Review: High-Power Ultrasound," *IEEE, Trans. Sonics Ultrasonics* (SU-22), No. 2, 60–70 (1975).

242. E. B. Steinberg, "Ultrasonics in Industry," *Proc. IEEE*, **53**(10), 1292–1304 (1965).

243. H. F. Stewart and M. E. Stratmeyer, "An Overview of Ultrasound: Theory, Measurement, Medical Applications and Biological Effects," HHS Publication (FDA) 82-8190, 1978.

244. N. N. Shetalov, A. D. Sartausv, and K. V. Glotov, *Labor Hyg. Occup. Dis.*, **6**, 10-14 (1962).

245. United Nations (UN) Environment Programme, World Health Organization (WHO), International Radiation Protection Association, Ultrasound, Environmental Health Criteria No. 22, Geneva, 1982.

246. United States Air Force (USAF), Hazardous Noise Exposure, AF Regulation 161-35, 1973.

247. H. E. von Gierke, *J. Acoust. Soc. Am.*, **22**, 675 (1950).

248. C. Wiernicki and W. J. Karoly, *Am. Ind. Hyg. Assoc. J.*, **46**, 488–496 (1985).

249. R. W. Wood and A. L. Loomis, *Phil. Mag.*, **4** (1927).

250. U. Landstrum, *J. Low Freq. Noise Vib.*, **1**, 29–33 (1987) (*Phys. Abstr.* 126919).

251. D. L. Johnson, *J. Low Freq. Noise Vib.*, **2**, 60–67 (1987) (*Phys. Abstr.* 78964).

252. Environmental Acoustics Laboratory, The Pennsylvania State University, "Community Noise Fundamentals: A Training Manual," A Final Report to the U.S. Environmental Protection Agency, Contract No. 68-01-4892, 1979.

253. Environmental Acoustics Laboratory, The Pennsylvania State University, "Community Noise Fundamentals—Independent Study by Correspondence," U.S. Environmental Protection Agency, Office of Noise Abatement and Control, Final Report on Contract No. 68-01-3895, 1977.

254. ISO Recommendation R 1996, *Acoustics, Assessment of Noise with Respect to Community Response*, 1st ed., International Organization for Standardization, ISO/R 1996, 1971 (E), 1971.

255. P. L. Michael, R. Kerlin, G. R. Bienvenue, and J. Prout, "Community Noise Fundamentals: A Training Manual," Final Report to the U.S. Environmental Protection Agency, Contract No. 68-01-4892, 1977.

256. P. L. Michael, "Noise Propagation from NASA Langley Research Center Facilities, Hampton, VA," Final Report, Contract No. NASI-11926, NASA, 1973.

257. P. L. Michael, "Noise Survey of the NASA Langley Research Center, Hampton, VA," Final Report, Contract No. NASI-9688, NASA, 1970.
258. *J. Acoust. Soc. Am.*, Listing of Related Abstracts and Papers, **52**, 499–500 (1972).
259. American Industrial Hygiene Association, *Industrial Noise Manual*, AIHA, Akron, OH, 1966, 1975, 1986.
260. ISO Recommendation R 1996, Acoustics, "Assessment of Noise with Respect to Community Response," 1st ed., International Organization for Standardization, ISO/R 1996, 1971 (E), May 1971.
261. U.S. Environmental Protection Agency, "Laws and Regulatory Schemes for Noise Abatement," NTID 300.4, Washington, DC, December 1971.
262. U.S. Environmental Protection Agency, "Model Community Noise Control Ordinance," EPA Document 550/9-76-003, Washington, DC, September, 1975.
263. U.S. Environmental Protection Agency, "Information of Levels of Environmental Noise Requisite to Protect Public Health and Welfare with an Adequate Margin of Safety, Washington, DC, 1974.
264. K. Kryter, "Perceived Noisiness (Annoyance)," in *The Effects of Noise on Man*, Academic Press, New York, 1970, p. 269.
265. American Industrial Hygiene Association, *Industrial Noise Manual*, 2nd ed., AIHA, Akron, OH, 1964.
266. C. W. Nixon and H. E. von Gierke, "Experiments on Boone Conduction Threshold in a Free Sound Field," *J. Acoust. Soc. Am.*, **31**, 1121–1125 (1969).
267. C. W. Nixon, "Hearing Protective Devices: Ear Protectors," in *Handbook of Noise Control* 2nd ed., McGraw-Hill, New York, 1979.
268. Guidelines on Noise Exposure Control, Intersociety Committee on Noise Exposure Control, *J. Am. Ind. Hyg. Assoc.*, **28**, 418–424 (Sept.–Oct. 1967).
269. W. F. Meeker, "Active Ear Defender Systems: Component Considerations and Theory," Wright Air Development Center Technical Report No. WADC TR 57-368 (I), Wright-Patterson Air Force Base, Ohio, 1958.
270. W. F. Meeker, "Active Ear Defender Systems: Development of a Laboratory Model," Wright Air Development Center Technical Report No. WADC TR 57-368 (II), Wright-Patterson Air Force Base, Ohio, 1959.
271. American National Standards Institute, American National Standard Methods for the Calculation of the Articulation Index, ANSI S3.5-1969 (R 1986).
272. National Institute for Occupational Safety and Health, "Industrial Noise Control Manual," Technical Report 79-117.
273. National Institute for Occupational Safety and Health, "Compendium of Materials for Noise Control," Technical Report 80-116.
274. U.S. Department of Commerce/National Bureau of Standards, "Quieting: A Practical Guide to Noise Control," NBS Handbook 119, July, 1976.
275. American National Standards Institute, American National Standard Specification for Sound Level Meters, ANSI S1.4-1971.
276. American Iron and Steel Institute, "Steel Industry Hearing Protector Study," *Vol. 1, Applications Handbook*, Vol. 2 References, AISI, Washington, DC, 1983.

Index

Abrasive blasting, manufacturing hazards, 595–598
Abrasives, manufacturing hazards, 633–636
Absorption, *see also* Toxic material modes
  noise control procedures, 1009
  percutaneous, 256–257, 272–273
Academic programs, *see also* Education; Training programs
  consultants, individual and, 94–96
  consultant training role, 101
  in industrial hygiene, 8–9
Acanthamoeba polyphaga, hypersensitivity pneumonitis, 558
Acceptance test methods, engineering performance specifications, 848–849
Access, laboratory design and, 388
Accidents, *see also* Fire and explosion hazards
  Bhopal, India disaster, 75, 124, 131
  laboratory design and, 384
Acclimatization to heat:
  body temperature regulation, 780
  heat stress managment control, 826
Accreditation, *see also* Professional certification
  laboratories, (AIHA), 111, 442
  organizational consultants, (AIHA), 111
ACGIH, *see* American Conference of Governmental Industrial Hygienists (ACGIH)
Acid and alkali cleaning of metals, manufacturing hazards, 598–601
Acid pickling, manufacturing hazards, 598–600
Acid waste, laboratory plumbing design and, 406. *See also* Hazardous waste management and disposal
Acne, 256, 268
Acoustic damping, noise control procedures, 1018
Acquired immune deficiency syndrome (AIDS), 371, 378
Acroosteolysis, dermatoses, clinical appearance of, 272
Active transport, cell membrane transport of contaminants, 216–217
Acute eczematous contact dermatitis:
  clinical appearance of, 267
  treatment of, 277
Acute febrile respiratory illness, 559–560
Adjustable bandwidth broad band analyzer, 975
Administrative Procedures Act, 22–23
Adsorption, *see* Gas–solid adsorption
Advanced Notice of Proposed Rulemaking (ANPR), standard-setting process and, 24
Advisory role, of consultants, 106–107
Aerosol calibration:
  aerosol characterization, 520–521
  aerosol particle detection and tagging techniques, 521–522
  calibration instruments and techniques, 502–522
  condensation aerosol generation, monodisperse, 507, 511
  dispersion aerosol generation:
    dry, 511–513
    wet, 513–518
  overview of, 502, 507, 511
  solid insoluble aerosol generation with wet dispersion generators, 518–520
Aerosol capture, air-purifying respirators (nonpowered), 682–687
Aerosols, *see also* Air contaminants
  agricultural hygiene, 727–730
  bioaerosols, 536–539
    guidelines for, 539
    health effects of, 536–538

Aerosols (*Continued*)
sampling for, 538
sources of, 536
lung transit of, 220
particle size and respiratory system reaction to, 295
particulate matter contaminant, classification, 207
Aesthetic considerations, laboratory design, 385
Aflatoxin, agricultural hygiene, 743
Afterburners, 859–860
Age level:
dermatoses, indirect causes, 258
heat stress and, 790
of work force, 74
Agglomerating characteristics, air-cleaning method selection, 866
Aggregate beds, air pollution filtration control, 895–896
Agricola, Georgius, 2
Agricultural hygiene, 721–761
certification, 746
control policies and strategies, 745
cost reduction, 746
education policies, 746
health and safety problems, 725–744
cancer, 741–742
chemical hazards, 730–734
emerging hazards, 743–744
general health status, 725
mental stress, 742–743
physical agents, 740–741
respiratory hazards (acute), 727–728
respiratory hazards (delayed and chronic), 728–730
skin diseases, 737–740
trauma (acute), 725–726
trauma (cumulative musculoskeletal), 726–727
veterinary biologicals and antibiotics, 734–736
zoonoses, 736–737
industrial hygiene lessons applied to, 10, 744–745
industrial perspective on, 723–724
lifestyle perspective on, 724–725
model programs, 747–748
overview of, 721, 722
regulation, 747
research, 745–746
taxation, 747
AIDS, 371, 378
AIHA, *see* American Industrial Hygiene Association (AIHA)
Air, route of transmission, 194
Air compressor, supplied-air respirators, 708–709
Air conditioning, laboratory design, and air conditioning (HVAC) system, ventilation, 399–400. *See also* Heating, ventilation, and air conditioning (HVAC) system
Air contaminants, *see also* Aerosols
electroplating, 611
forging, 616–617
hazard communication program and, 149
industrial hygiene/medical surveys combined, 85
welding hazards, 667–668
Air-cooled garments, 829–830
Aircraft noise, 1022–1023
Air noise generation control, noise control procedures, 1019–1020
Air pollution:
contamination levels, public opinion and, 11
historical perspective on standards for, 15
indoor air quality, 26. *See also* Indoor air quality
organizational consultant services and, 116
workplace air quality and, 839–842
Air pollution controls, 839–936
classification of, 871–878
combination systems, 877–878
combustion, 876–877
electrostatic precipitation, 874
filtration, 873–874
gas–solid adsorption, 875–876
gravitational and inertial separation, 872–873
liquid scrubbing, 874–875
tall stacks, 871–872
combustion, 923–929
catalytic combustion, 927
design parameters of, 927–929
direct flame combustion, 926–927
flares, 925–926
overview of, 923–924
performance range of, 924–925
economic impacts, 849–860
capital investment, 851
control cost elements, 850
cost-effectiveness relationships, 849–850
economic analysis models for, 856–860
economic analysis models for (afterburners), 859–860
economic analysis models for (electrostatic precipitators), 859
economic analysis models for (fabric filters), 857
economic analysis models for (gravitational and inertial collectors), 856–857

Air pollution controls (*Continued*)
economic analysis models for (wet collectors), 857–859
energy recovery, 856
maintenance and operating costs, 851–853
material conservation, 853, 856
power requirements table, 854–855
electrostatic precipitation, 913–919
design parameters, 918–919
overview of, 913–914
performance range of, 914
pipe-type precipitators, 917
plate-type precipitators, 915–917
precipitation mechanisms, 914–915
emission control requirements, 842–849
Clean Air Act of 1970 (CCA), 843–845
typical regulatory requirements, 845–847
filtration, 894–903
fabric filters, 898–902
fibrous mats and aggregate beds, 895–896
paper filters, 896–898
performance range of, 894
gas–solid adsorption, 919–923
chemical adsorption, 922
design parameters of, 922–923
overview of, 919
performance range of, 919–920
physical adsorption, 921–922
polar adsorption, 922
gravitational and inertial separation, 878–894
characteristics table, 880–881
cyclones, 887–892
cyclones (multiple), 893–894
cyclones (wet), 892–893
dynamic separators, 885–887
inertial separators, 883–885
overview of, 878
performance range of, 879
settling chambers, 879, 882–883
industrial hygiene and, 839
liquid scrubbing, 903–913
design parameters, 912–913
mist eliminators, 911–912
orifice scrubbers, 909–911
overview of, 903
packed towers, 907–909
performance range of, 904–905
spray chambers, 905–907
process and system controls:
emission elimination, 860–861
emission reduction, 861–862
source concentration of pollutants, 862–863
selection and application, 929–934
emission limit definition, 930
emission source identification, 930–931
process modification, 931–932
technical aspects definition, 932–934
technical-economic decision model use, 934
selection criteria for, 863–871
carrier gas flow characteristics, 870–871
carrier gas properties, 867–870
contaminant properties, 864–867
performance objectives, 863–864
workplace air quality and atmospheric emissions, 839–842
Air pressure measurement, ventilation system measurement calibration, 465
Air-purifying respirator, *see* Respiratory protective devices
Air quality, *see* Air pollution; Indoor air quality
Air quality modeling, described, 846
Air samples and sampling:
biological agents monitoring, 200
instruments calibration for, 462–465
investigational survey, 82
records and reports, 83–84
Air supply hose, supplied-air respirators, 694
Air velocity measurement:
heat stress index, 801–802
ventilation system measurement calibration, 465
Air velocity meters (anemometers):
air velocity measurement, heat stress index, 801–802
calibration instruments and techniques, 486–488
Alarm systems, laboratory design, safety considerations, 420–421
Alchemy, 3
Alcohol, heat stress and, 792–793
Alkaline treatment of metals, manufacturing hazards, 600–601
Allowable concentrations, 15
Altered barometric pressure effects, *see* Barometric pressure effects
Alveolar wall, penetration of, 223
AMA Archives of Industrial Health, 6
American Academy of Industrial Hygiene (AAIH):
Code of Ethics of, 103, 104, 107, 181
membership in, 7–8
American Association for Labor Legislation, 6, 14
American Association of Industrial Physicians and Surgeons, 6, 14
American Board of Industrial Hygiene, 32
industrial hygienist, definition of, 180
professional certification and, 7–8
American Board of Medical Microbiology, 198
American Conference of Governmental Industrial Hygienists (ACGIH):
air pollution standards, 930

American Conference of Governmental Industrial Hygienists (*Continued*)
manufacturing hazards, grinding, polishing, and buffing operations, 636
membership of, 15
organization of, 5
professional certification and, 7
publications of, 7
standards established by, 6, 15
threshold limit values:
heat stress, 767–768
noise exposure limits, 980
Ventilation Manual:
acid and alkali cleaning of metals, 599
electroplating, 612
American Foundrymen's Society, manufacturing hazards, foundry, 633
American Industrial Hygiene Association (AIHA):
consultants listing of, 109–110
current membership of, 92
formation of, 5, 15
founding membership of, 92
future objectives of, 10
noise levels, 1008
organizational consultant accreditation program, 111
organizational consultants and, 111
professional certification and, 7
publications and, 6–7
quality control program, program audit, 456
standards and, 15
training programs offered by, 102
American Industrial Hygiene Association (AIHA) Journal, 6–7
organizational consultant activities, 116–117
organizational consultant listing, 119
American Industrial Hygiene Association (AIHA) Proficiency Analytical Testing (PAT) program:
laboratory selection, 120
quality control, external testing for accreditation, 442
American Iron and Steel Institute (AISI), noise exposure levels, 992–993
American Medical Association, 6
American Museum of Safety, *see* Safety Institute of America
American National Standard Methods for the Calculation of the Articulation Index (AI), noise exposure levels, 992–993
American National Standards Institute (ANSI):
audiometers, 962
hearing loss and, 958
hearing measurement standards, 959
hearing protector requirements, 1001–1002
hearing test rooms, 962
labeling standards of, 165
manufacturing hazards, grinding, polishing, and buffing operations, 636
noise exposure levels, 992–993
respiratory protective devices, 676
selection considerations, 706
training in use of, 707
standards of, 15
X-ray generator regulations, 645
American National Standard Specification for Octave-Band and Fractional-Octave Band Analog and Digital Filters, 951
American National Standard Specification for Sound Meters, 972
American Public Health Association, 6, 14
American Society of Heating, Refrigerating, and Air-conditioning Engineers (ASHRAE):
carbon dioxide standards, 568
evolution of, 569–570
humidification standards, 567
indoor air quality, evaluation protocols and guidelines, 573, 577
standards, indoor air quality, 531–532, 547
thermal discomfort, 561–562
ventilation system standards, 565, 566, 570–572
American Society of Safety Engineers, 9
American Standards Association (Z-37 Committee), standards of, 15
American Welding Society (AWS), 663
Analytical laboratories, *see* Laboratory services
Analytical survey, industrial hygiene survey, investigational, 80–82
Anemic anoxia, toxic material classification, pathological, 208–209
Anemometers, *see* Air velocity meters (anemometers)
Anesthetics, toxic material classification, pathological, 209
Anhydrous ammonia, agricultural hygiene, 734
Animal bites, dermatoses, direct causal factors, 267
Animal studies:
asbestos dust, 319, 321
dust effect on lung, 303–304
fibrous glass, 323–324
hyperoxia, 342
Annealing, 638
Annual audiogram, noise exposure limits, 981
Anoxia-producing agents, toxic material classification, pathological, 208–209
Anoxic anoxia, toxic material classification, pathological, 208

ANSI, *see* American National Standards Institute (ANSI)
Anthrax:
- agricultural hygiene, 737
- described, 195

Anthropometric considerations, facepieces, 677–678
Antibiotics, agricultural hygiene, 736
Antigen-antibody reaction, toxic material classification, pathological, 210–211
Applied Industrial Hygiene, 7
Applied Occupational and Environmental Hygiene, 7
Arc furnace, described, 627–628
Architecture, industrial hygiene's influence on, 10. *See also* Laboratory design
Arterial gas embolism, 350–351
Arthralgias, 338
Artists, historical perspective, 3
Asbestos dust:
- air-purifying respirators (nonpowered), 687
- described, 318–322
- filters for, air-purifying respirators (nonpowered), 687
- management of, organizational consultant services, 114–115
- organizational consultants and, 112
- removal of, 26
- risks of, industrial hygiene survey and, 75

Asbestos Hazard Emergency Response Act of 1986 (AHERA), 21, 75
Aseptic necrosis of bone, described, 351–352
ASHRAE, *see* American Society of Heating, Refrigerating, and Air-conditioning Engineers (ASHRAE)
ASME, hyperbaric chamber standards, 336
Asphyxiants, toxic material classification, pathological, 208–209
Assistant Secretary of Labor for Occupational Safety and Health, creation of office, 20
Associate industrial hygienist, training and qualifications of, 89–90
Asthma, building-related illness, 558–559
Atomic Energy Act, 16
Audiograms:
- letter indicating hearing problem, 1027
- noise exposure limits, 980–981

Audiometers:
- hearing measurement, 959–962
- self-recording audiometer, 966

Auditing, *see* Program auditing
Australia, 29–32
Australian Institute of Occupational Hygiene (AIOH), 32
Autopsy, dust effect on lung, 304
Auxiliary cooling and protection measures, heat stress managment control, 829–830

Bacillus subtilis, hypersensitivity pneumonitis, 558
Bacteria, *see* Biological agents determination
Baker, Sir George, 3
Balanced flow system, velocity calibration procedures, reciprocal, 492–493
Barometric pressure effects, 329–357
- compression and pressurized gas effects, 338–347
  - barotrauma of descent, 338–339
  - compression arthralgias, 338
  - diving, thermal problems and energy balance in, 344–347
  - inert gas narcosis and high-pressure nervous syndrome, 338–340
  - oxygen toxicity, 342–344
  - pulmonary function and gas exchange, 340–342
- decompression effects, 347–352
  - aseptic necrosis of bone, 351–352
  - decompression principles, 347–348
  - decompression sickness, 349–350
  - pulmonary barotrauma and arterial gas embolism, 350–351
- overview of, 329–330
- pressurized environments, 330–338
  - compressed air and mixed gas diving, 334–335
  - general principles of, 330–332
  - hyperbaric chambers, 335–336
  - immersion and breath-hold diving, 332–334
  - saturation diving, 335
  - submarines and submarine escape, 337–338
  - tunnels and caissons, 336–337
- recompression therapy and hyperbaric oxygen, 352–354

Barotrauma of descent, described, 338–339
Basal cell carcinoma, agricultural hygiene, 740
Basal cell epithelioma, dermatoses, clinical appearance of, 270
Base-line audiogram, noise exposure limits, 980–981
Beddoes, Thomas, 3
Benches and bench layout (laboratory design), 389–390, 396–397
Benzene:
- bone marrow accumulation of, 237
- leukemia and, 580
- manufacturing hazards, molding, 625
- standards for, 23

Bhopal, India disaster, 75, 124, 131

Biliary tract, elimination and excretion by, 245–246
Bimetallic thermometer, 799
Bingham, Eula, 126
Bioaerosols, 536–539, *See also* Aerosols; Biological agents determination
guidelines for, 539
health effects of, 536–538
sampling for, 538
sources of, 536
Bioeffluents, indoor air quality, 549–550
Biological agents determination, 193–203
agents listed, 195–197
case histories, 200–202
causal determination, 194–195
monitoring and, 199–200
overview of, 193
prevention, 198–199
problem evaluation in, 197–198
routes of transmission, 194
Biological infection, *see* Infection
Biotransformation, toxic material modes, 239–240
Birth defects:
health care field and, 372
placental barrier and, 234–235
Bismuth, liver accumulation of, 235
Black lung disease, *see* Coal worker's pneumoconioses (black lung disease)
Blood, *see also* Cardiovascular system
lung portal of contaminants and, 218–220
respiration and, 290–291
transport and distribution of contaminants by, 228–232
Blood–brain barrier, contaminant transfer and, 234
Blood organic acids, transport and distribution of contaminants, 231–232
Bloomfield, Jack, 5
Body core temperature, heat strain index, 783–784
Body fat:
heat stress and, 791
lipid-rich organs and tissues, contaminant deposition and accumulation in, 236–237
Body odor, indoor air quality, 549–550
Body temperature regulation, 777–780, *See also* Heat stress; Temperature
acclimatization to heat, 780
circulatory regulation, 778
diving, thermal problems, 344–347
hypothalamus, 777–778
muscular activity and, 778
skin function of, 256
sweating mechanism, 778–779
Bone:
aseptic necrosis of bone, described, 351–352
contaminant deposition and accumulation in, 237–239
Bone marrow, benzene accumulation in, 237
Brazing hazards, 660–661. *See also* Soldering
Bright dip, manufacturing hazards, 598–600
British Cast Iron Research Association (BCIRA), foundry hazards, 632
British Examination and Registration Board in Occupational Hygiene, 32
Bronchitis, chronic, dust and, 317–318
Bronchogenic cancer, dust effect on lung, 309
Brucellosis:
agricultural hygiene, 737
described, 195
Buffing operations, manufacturing hazards, 633–636
Building codes, indoor air pollution, 580
Building-related illness, 557–560. *See also* Sick building syndrome
acute febrile respiratory illness, 559–560
asthma, 558–559
defined, 535
humidifier fever, 558
hypersensitivity pneumonitis, 557–558
Legionnaires' disease, 559
overview of, 557
Bureau, *see entries under U.S. Bureau*
Bureau of Labor Statistics, Massachusetts, 14
Burns, dermatoses, direct causal factors, 265
Bush, George, 11, 25
Business practices, consultants, 108–109
Bypass flow indicator, volumetric flow rate, calibration instruments and techniques, 484

Cadmium, liver accumulation of, 235
Caissons and tunnels:
aseptic necrosis of bone, 351–352
described, 336–337
historical perspective on, 330
Calibration, 461–530
audiometers, 961–962
error estimation, 525–526
microphones, 972–973
noise measurement instruments, 979
overview of, 461–462
quality control, equipment, 433–436
sound level meter, 968
standards in, 469–470
types of, 462–469
air sampling instruments, 462–465
electromagnetic radiation measurement, 467–468

Calibration (*Continued*)
heat stress measurement, 465
noise-measuring instruments, 468–469
ventilation system measurement, 465
Calibration instruments and techniques, 470–525
aerosol calibration production, 502–522
aerosol characterization, 520–521
aerosol particle detection and tagging techniques, 521–522
condensation aerosol generation, monodisperse, 507, 511
dispersion aerosol generation (dry), 511–513
dispersion aerosol generation (wet), 513–518
overview of, 502, 507, 511
solid insoluble aerosol generation with wet dispersion generators, 518–520
air velocity meters (anemometers), 486–488
cumulative air volume, 470–476
dry gas meter, 475
frictionless piston meter, 472–473
positive displacement meter, 476
spirometer or gasometer, 471–472
water displacement, 471
wet test meter, 474–475
known vapor concentrations production, 494–502
dynamic systems (continuous flow), 498–502
overview of, 494–495
static systems (batch mixtures), 495–497
mass flow and tracer techniques, 485–486
sampler's collection efficiency calibration, 522–524
sample stability and/or recovery determination, 524–525
velocity calibration procedures, 488–494
volumetric flow rate, 476–484
overview of, 476–477
variable area meters (rotameters), 477–479
variable head meters, 479–484
Campylobacteriosis, described, 195
Canada, 32–36
Canadian Centre for Occupational Health and Safety, 34
Cancer and carcinogens:
agricultural hygiene, 732, 740, 741–742
asbestos dust and, 322
building-related illness, 534
coal worker's pneumoconioses and, 317
dermatoses, clinical appearance of, 269–270
dust effect on lung, 300–301, 309
labeling requirements, hazard communication program, 164
laboratory design, safety considerations, 411–412
manufacturing hazards, electroplating, 609
melanoma, dermatoses, clinical appearance of, 270
radon and, 550–552, 580
scrotal cancer, 3
tobacco smoke, indoor air quality, 548
Candida infection, dermatoses, direct causal factors, 266
Canisters, air-purifying respirators (nonpowered), 687–688, 692
Capital investment, air pollution controls, 851
Caplan's syndrome, coal worker's pneumoconioses and, 316
Carbonaceous dust, dust effect on lung, 315–317
Carbon dioxide, ventilation systems, 568–569
Carbon monoxide:
air-purifying respirators (nonpowered), 687
combustion products, indoor air quality, 546–548
manufacturing hazards:
heat treating, 638
molding, 625
supplied-air respirators, 708–709
welding hazards, 667
Carburizing, heat treating hazards, 637
Carcinogens, *see* Cancer and carcinogens
Cardiopulmonary resuscitation, laboratory safety considerations, 420
Cardiovascular system, *see also* Blood
body temperature regulation, 778
circulatory strain, heat strain index, 782–783
heart rate, heat strain index, 782–783
respiratory protective devices, 705
Career opportunities, *see* Employment (of hygienists)
Carrier gases:
flow characteristics of, air-cleaning method selection, 870–871
properties of, air-cleaning method selection, 867–870
Carter, Jimmy, 25, 128
Cartridges:
air-purifying respirators (nonpowered), 687–688, 692
certification of, 711, 716
Causal determination, biological agents determination, 194–195
Cell membrane transit (of contaminants), 211–217
carrier-mediated modes of transit, 216–217
cell membrane structure, 211–214
passive transport, 214–216
special transport processes, pinocytosis and phagocytosis, 217

Celsus, 254
Centers for Disease Control (CDC), biological agents prevention, 198–199
Central nervous system:
barometric pressure effects on, 339–340
contaminant deposition and accumulation in, 236
oxygen toxicity and, 342–344
CERCLA, *see* Comprehensive Environmental Response, Compensation, and Liability Act of 1980 (CERCLA)
Certification, *see* Professional certification
Certified Industrial Hygienist (CIH), consultants, individual, 93
Chain of custody, quality control program management, 427–430
Chalmydia, described, 195
Chemical adsorption, gas-solid adsorption, 922
Chemical hazards:
agricultural hygiene, 730–734
classification, toxic materials, 207
dermatoses, direct causal factors, 259–261
hazard communication standards, 127–128
storage, laboratory design, safety considerations, 410–413
Chemical industry:
hazard communication program responsiblities of, 139–144
hazard communication standards and, 128
labeling responsibility of, 141, 144, 163–164
material safety data sheets and, 140–141
trade secrets and, 135
Chemical treatment, laboratory design, hazardous waste diposal, 418
Chemical waste disposal, sources of information on, 418–419. *See also* Hazardous waste management and disposal
Chloracne, dermatoses, clinical appearance of, 268
Chlorinated biphenyls (PCBs):
milk concentrations of, 247
pharmacokinetic modeling of, 248–252
placental barrier and, 234
Chlorinated hydrocarbon solvents, welding hazards, 671
Chlorine, welding operations and, 607
Chlorofluorocarbon (CFC), soldering, 660
Chronic bronchitis, dust and, 317–318
Chronic eczematous contact dermatitis:
clinical appearance of, 268
treatment of, 277
Cigarette smoke, indoor air quality, 548–549
Circulatory system, *see* Blood; Cardiovascular system; Heart rate
Cladosporium, hypersensitivity pneumonitis, 558
Clean Air Act of 1970 (CAA), 21
criminal sanctions, 188
emission control requirements, 843–845
indoor air quality, evaluation protocols and guidelines, 578
tall stack requirements, 872
Cleaning and finishing, manufacturing hazards, 631–632
Clean Water Act of 1972:
criminal sanctions, 188
OSHAct and, 21
Clientage:
consultant's client relationship, 96–97
ethical considerations and, 107
fee-for-service basis, 108–109
Closed-circuit self-contained breathing apparatus (supplied-air respirators), 695–696
Clothing:
dermatoses prevention, 279–280
heat stress, worker-environment thermal exchange, 776–777
heat stress management, cooling garments, 829–830
Coal, carbonaceous dust and, 315–317
Coal Mine Health and Safety Act of 1969, 21
Coal worker's pneumoconioses (black lung disease), *see also* Pneumoconioses
described, 315–317
legislation and, 16
Coccidioidomycosis, described, 196
Cochlea, 956
Code of Business Practices, consultants, 108–109
Code of Ethics, *see also* Ethics
consultant litigation/expert witness role, 102, 103
organizational consultants and, 119
Cold, dermatoses, direct causal factors, 265. *See also* Temperature
Cold degreasing, described, 602
Collector efficiency, engineering performance specifications, 848
Combustibility:
carrier gas properties, 869
contaminant properties, 865
Combustion method, 923–929
air pollution controls classification, 876–877
catalytic combustion, 927
design parameters of, 927–929
direct flame combustion, 926–927
flares, 925–926
overview of, 923–924
performance range of, 924–925
Combustion products, indoor air quality, 546–548

Communication, respiratory protective device problems, 710
Community noise, 1020–1026. *See also* Noise
  frequency weighting or analysis of, 1022
    aircraft noise, 1022–1023
    nonaircraft noise, 1023
  guide for criteria in, 1025–1026
  measurement equipment, 1023
  measurement location and procedures, 1024–1025
  overview of, 1020–1021
  regulation:
    nuisance-type, 1021
    performance-type, 1021–1022
Community responsibility, historical perspective on, 4
Community right-to-know laws, *see also* Worker right-to-know laws
  hazard communication program and, 149–150
  historical perspective on, 124–125
  organizational consultants and, 113
Compensation claims, industrial hygiene survey, investigational, 80
Comprehensive Environmental Response, Compensation, and Liability Act of 1980 (CERCLA), 21
  criminal sanctions, 188
  industrial hygiene survey and, 75
Compressed air bottles, supplied air respirators, 709
Compressed air and mixed gas diving, pressure effects, 334–335
Compressed gases, chemical storage, laboratory design, safety considerations, 412–413
Compressed gas system, laboratory design, 406–407
Compression and pressurized gases, *see* Barometric pressure effects
Compression arthralgias, described, 338
Computer:
  consultant training role, 101–102
  industrial hygiene survey and, 74, 76, 84
  laboratory design, electrical power needs, 405
Concha-seated ear protector, described, 998–999
Condensation nucleus counter, respirator efficiency tests, 699–700
Confidentiality, consultants and, 107–108
Construction, laboratory design, 407–410
Consultants, 91–121, *See also* Individual consultants; Organizational consultants
  activities of, 98–107
    advisory role, 106–107
    legislative support role, 105–106
    litigation/expert witness roles, 102–103
    problem solving, 98–99
    program auditing, 100–101
    scientific study role, 103–105
    surveys, 99–100
    training role, 101–102
  American Industrial Hygiene Association, listing of, 109–110
  backgrounds of individuals, 92–95
  business aspects and, 108–109
  client relationships of, 96–97
  confidentiality and, 107–108
  ethical considerations and, 107–108
  historical perspective on, 91–92
  individual, advantages and disadvantages of, 97–98
  organizational support for, 110
  qualifications of, 95–96
Consulting firms, *see* Private consulting firms
Consulting organization, *see* Organizational consultant
Consumer Product Safety Act of 1972, 21
Consumer Products Safety Commission:
  industrial hygiene survey and, 74
  uniform standards and, 128
Contact dermatitis, *see also* Dermatitis; Dermatoses
  agricultural hygiene, 734, 737
  clinical appearance of, 267–268
  treatment of, 277
Contact urticaria, dermatoses, clinical appearance of, 271
Contagious ecthyma, agricultural hygiene, 735
Contaminants, *see also* Toxic material modes
  air-cleaning method selection, 864–867
  classification of:
    chemical, 207
    pathological, 207–211
    physical, 206–207
Continuous flow mode, supplied-air respirators, 693
Contracts, law and, 185–186
Control charts (statistical quality control), 443–447
  control limits calculation, 444–447
  control limits interpretation, 447
  parameters to be controlled, 443–444
  purposes of, 443
  trial control charts, 444
  types of, 444
Control limits (statistical control charts), 444–447
  calculation of, 444–447
  interpretation of, 447
Control procedure development, industrial hygiene survey, investigational, 82–83

Convection, heat stress, worker-environment thermal exchange, 775
Conventional tool machining, described, 639–643
Coremaking, 625–627
  hot box process, 626
  no bake process, 626–627
  overview of, 625
  shell coremaking, 627
  sodium silicate system, 625–626
Corporate policy, consultant role limited by, 100. *See also* Management
Corrective lenses, respiratory protective device problems, 710
Cost-benefit analysis, quality control programs and, 423
Cost-effectiveness relationships, air pollution controls, 849–850
Costs:
  agricultural hygiene, 746
  air pollution controls, 849–856
  economic analysis models (air pollution control), 856–860
  hazard communication program, 156–159
  hearing conservation program, 987
  hearing loss compensation, 958
  indoor air quality, 562–564
  quality control program management, 432–433
Court orders, *see* Judiciary; Law
Creams, dermatoses prevention, 280–281
Criminal sanctions, law, 187–189
Critical flow orific, volumetric flow rate, calibration instruments and techniques, 483–484
Crucible furnace, described, 629
Cumulative air volume, calibration instruments and techniques, 470–476
Cumulative musculoskeletal trauma, agricultural hygiene, 726–727
Cupola, described, 629–630
Custody, *see* Chain of custody
Cutaneous sensitizers, dermatoses, direct causal factors, 262
Cutting fluids, metal machining hazards, 640–643
Cyaniding, heat treating hazards, 637–638
Cyclones (gravitational and inertial separation), 887–892
  described, 872–873
  multiple, 893–894
  wet, 892–893
Cytomegalovirus:
  described, 195–196
  health care field, 372
Dalton's law, 218, 331–332
Data anlaysis techniques, quality control statistics, 447
Data validation, quality control statistics, 448
Davy, Sir Humphrey, 3
DDT:
  milk concentrations of, 247
  organ deposition of, 233
Decibel, sound measurement, 941
Decipol and olf concepts, sick building syndrome, 556–557
Decompression effects, 347–352
  aseptic necrosis of bone, 351–352
  decompression principles, 347–348
  decompression sickness, 349–350
  pulmonary barotrauma and arterial gas embolism, 350–351
Decompression sickness, 349–350
Deep Submergence Rescue Vessel (DSRV), 338
Degreasing technology, 601–608
  cold degreasing, 602
  control measures, 605, 607–608
  emulsion cleaners, 602
  vapor degreasing, 602–605
Demand mode, supplied-air respirators, 692–693
Density, air-cleaning method selection, carrier gas properties, 868–869
Deregulation, OSHA and, 75
Dermatitis, *see also* Contact dermatitis; Dermatoses
  agricultural hygiene, 734, 737
  electroplating, 609
  paints, 655
Dermatoses, 253–287
  agricultural hygiene, 737–740
  causal factors (direct), 259–267
    chemicals, 259–261
    cold, 265
    cutaneous sensitizers, 262
    electricity, 265
    heat, 265
    infection, 266–267
    ionizing radiation, 266
    laser radiation, 266
    mechanical injury, 264–265
    microwave radiation, 266
    photosensitivity, 263–264
    plants and woods, 262–263
    ultraviolet radiation, 265
  causal factors (indirect), 257–259
  clinical appearance of, 267–272
    acroosteolysis, 272
    acute eczematous contact dermatitis, 267
    chronic eczematous contact dermatitis, 268
    contact urticaria, 271

Dermatoses (*Continued*)
facial flush, 271–272
folliculitis, acne, and chloracne, 268
granulomas, 271
nail discoloration and dystrophy, 271
neoplasms, 269–270
pigmentary abnormalities, 269
sweat-induced reactions, 269
ulcerations, 270–271
diagnosis of, 273–276
historical perspective on, 253–254
incidence of, 254–255
medical controls, 282
overview of, 253
percutaneous absorption, systemic intoxication signs following, 272–273
prevention of, 278–282
skin's defensive function and, 255–257
treatment of, 276–277
welding hazards, 671
Detonation spraying, 649
Devonshire colic, 3
Dichlorodifluoromethane, respirator efficiency tests, 699
Di-2-ethylhexyl phthalate, respirator efficiency tests, 698–699
Dilution calibration, velocity calibration procedures, 493–494
1,1-Dimethylhydrazine, historical standards, 15
Directional microphones, noise measurement, 972–974
Director of industrial hygiene, training and qualifications of, 88–89
Directory of Academic Programs in Occupational Safety and Health, 8–9
Directory of College and University Safety Courses, 9
Discrimination, hiring practices, health care field, 370
Disease routes of transmission, *see* Routes of transmission
Distribution, *see* Toxic material modes
Distribution schemes, laboratory design, 399
Diving hazards:
barotrauma of descent, 338–339
compressed air and mixed gas diving, 334–335
compression arthralgias, 338
decompression sickness, 349–350
fitness to dive issue, 354–357
immersion and breath-hold diving, 332–334
inert gas narcosis and high-pressure nervous syndrome, 339–340
necrosis of bone, 351–352
pulmonary barotrauma and arterial gas embolism, 350–351
saturation diving, 335–336
thermal problems and energy balance in, 344–347
Downstream samples, sampler's collection efficiency calibration, 522–524
Drive system modification, noise control procedures, 1020
Drugs, heat stress and, 792–793
Dry-bulb temperatures, described, 794
Dry gas meter, cumulative air volume, calibration instruments and techniques, 475
Dufy, Raoul, 3
Dust, 289–327
agricultural hygiene, 727–730
altered lung structure and, 301–302
asbestos dust, 318–322
biological effects of, 308–318
fibrogenic dust, 309–317
nonfibrogenic (inert) dust, 308–309
biological protection against, 289–290, 297–298
bronchitis, chronic, and emphysema, 317–318
effects of, critique of studies on, 302–307
fibrous glass, 322–324
lung anatomy and, 290–294
lung reaction to, 299–301
man-made fibrous dust, indoor air quality, 552–553
manufacturing hazards:
abrasive blasting, 597–598
grinding, polishing, and buffing operations, 635
particle size and respiratory system reaction to, 295–298
particulate matter contaminant, classification, 207
pneumoconiosis, 324
Dust filters, air-purifying respirators (nonpowered), 684–685
Dynamic collectors, described, 873
Dynamic separators, described, 885–887
Dynamic systems (continuous flow), known vapor concentrations production, calibration instruments and techniques, 498–502

Ear, 951–958. *See also* Hearing; Noise
eardrum, 955
external ear, 952–954
frequency range of hearing, 951
inner ear, 956
middle ear, 955–956
noise damage to, 956–958
Eardrum:
described, 955
injury to, 954

Earmuffs, described, 999–1000
Earplugs, 997–998. *See also* Hearing protectors
Earth Day, 10
Ecology movement, industrial hygiene profession and, 10, 28
Economic analysis models (air pollution control), 856–860
  afterburners, 859–860
  electrostatic precipitators, 859
  fabric filters, 857
  gravitational and inertial collectors, 856–857
  wet collectors, 857–859
Eczema contagiosum, dermatoses, direct causal factors, 267
Eczematous contact dermatitis:
  clinical appearance of, 267–268
  treatment of, 277
Edgar, Robert, 131
Education, *see also* Academic programs; Training programs
  academic programs in industrial hygiene, 8–9
  agricultural hygiene, 746
  Australia, 31–32
  Canada, 35
  consultants, individual and, 94–95
  dermatoses prevention, 281
  Egypt, 37–38
  Finland, 40–41
  industrial hygiene survey personnel, 88–90
  Italy, 45–46
  Netherlands, 48–49
  qualifications of individual consultants, 95–96
  United Kingdom, 62
Education Opportunities in Industrial Hygiene, 9
Effective temperature concept, described, 794–795
Egregious violation, origin of term, 26
Egress:
  emergency evacuation and, 388
  laboratory design and, 388
Egypt, 36–38
85 dBA for 8 hr exposure, noise exposure limits, 982
Electrical burns, dermatoses, direct causal factors, 265
Electrical discharge machining, described, 644
Electrical power needs, laboratory design, 405
Electrochemical machining, described, 643–644
Electrolyte balance:
  heat stress and, 791–792
  heat stress management control and, 827–828
Electromagnetic radiation measurement, calibration of, 467–468
Electron microscope, laboratory plumbing design and, 406
Electroplating, 609–613
  air contaminants, 611
  control measures, 611–613
  techniques in, 609–611
Electrostatic precipitation, 913–919
  air pollution controls classification, 874
  design parameters, 918–919
  economic analysis models, 859
  overview of, 913–914
  performance range of, 914
  pipe-type precipitators, 917
  plate-type precipitators, 915–917
  precipitation mechanisms, 914–915
Elimination and excretion, 240–247. *See also* Toxic material modes
  biotransformation or metabolism and, 239–240
  kidney, 243–245
  kinetics of, 240–243
  overview of, 240
Ellenbog, Ulrich, 2
Emergency evacuation, laboratory design, safety considerations, 420–421
Emergency Exposure Limits, 15
Emergency facilities, laboratory design and, 388
Emergency Planning and Community Right-to-Know Act of 1986, *see* Superfund Amendments and Reauthorization Act of 1986 (SARA)
Emergency room, laboratory design, safety considerations, 420
Emission elimination, air pollution control, 860–861
Emission rate of discharged material, engineering performance specifications, 847–848
Emission reduction, air pollution control, 861–862
Emphysema, dust and, 317–318
Employees, *see also* Labor unions
  evaluation of, health care field, 365–369
  exposure and medical records of, hazard communication program, 149
  hazard communication program involvement of, 148
  information and training of, employer responsibility for, 145–146
Employers:
  hazard communication program responsibilities of, 144–146
  hearing protection responsibility of, 1004, 1008

Employers (*Continued*)
labeling responsibility of, 164
liability, individual liability, 186
Employment (of hygienists):
Canada, 35–36
industrial hygiene survey personnel, 88–90
Netherlands, 49
organizational consultants, 117–118
Emulsion cleaners, described, 602
Enclosures:
dermatoses prevention, 279
noise control procedures, 1009–1017
Endocytosis, cell membrane transport of contaminants, 217
Energy recovery, air pollution control costs, 856
Energy requirements, air pollution controls, power requirements table, 854–855
Enforcement:
criminal sanctions, 187–189
OSHA and, 87
Engineering:
heat stress controls by, 813–824
industrial hygiene's influence on, 10
organizational consultant services and, 116
Engineering performance specifications, air pollution controls, 847–849
England, *see* United Kingdom
Environmental considerations, laboratory design, 385
Environmental engineering firms, organizational consultant services and, 116
Environmental heat exchange, heat stress, worker-environment thermal exchange, 774–775
Environmental monitoring, *see* Monitoring
Environmental Pollution Control Act of 1972, 21
Eomites, route of transmission, 194
Epidemiology:
consultant role in, 104–105
dust effect on lung, 304–306
heat stress, 807–810
manufacturing hazards, electroplating, 609
sick building syndrome, 534, 569
Danish studies, 555–556
U.K. studies, 554–555
U.S. studies, 553–554
Epoxy paints, 652–653
Equipment, *see also* Calibration instruments and techniques
protective equipment:
dermatoses prevention, 279–280
manufacturing hazards, forging, 618
quality control program, 433–437
analytical instruments, 435–436
direct-reading instruments, 434–435
instrumental, 439–440
sampling equipment, 433–434
Equipment maintenance, *see also* Maintenance
laboratory design, safety considerations, 421–422
quality control program, 436–437
Equivalent effective temperature corrected for radiation, described, 795
Ergonomic hazards, agricultural hygiene, 741
Error estimation, calibration, 525–526
Erysipelothrix infections, described, 196
Erythrocytes, *see* Red blood cells
Escherichia coli, agricultural hygiene, 735–736
Ethics, *see also* Code of Ethics
consultants and, 107–108
law and, 181
organizational consultants and, 118
European Community (EC):
exposure limits, 67
Italy and, 45
member nations of, 67
Netherlands and, 50
Evacuation, laboratory design, safety considerations, 420–421
Evaporation, heat stress, worker-environment thermal exchange, 776
Evaporative cooling, heat stress control measures, 822–823
Exchange diffusion, cell membrane transport of contaminants, 216–217
Excretion, *see* Elimination and excretion
Executive Orders, OSHAct and, 22
Exhalation valves, facepieces (respiratory protective devices), 682
Exit concentration of material, engineering performance specifications, 847
Exit signs, laboratory design and, 389
Expansion planning, laboratory design and, 394–395
Expert witness, 102–103, 115, 189–191
Exposure levels:
industrial hygiene/medical surveys combined, 85
industrial hygiene survey, investigational, 79
Exposure limits:
Canada, 36
European Community (EC), 67
Finland, 42–43
historical perspective on development of, 13, 15
Netherlands, 49
Sweden, 55
United Kingdom, 63

Exposure and medical records (employee), hazard communication program and, 149
External ear, described, 952–954
Eye protection, welding hazards, 670–671
Eyewash stations and safety showers, laboratory design, safety considerations, 419

Fabric filters:
air pollution filtration control, 898–902
economic analysis models (air pollution control), 857
Facepieces (respiratory protective devices), 677–681
anthropometric considerations, 677–678
loose-fitting respirator, 680–681
supplied-air helmets, 681
supplied-air suits, 681
tight-fitting, 679–680, 693
valves, 682
Facial flush, dermatoses, clinical appearance of, 271–272
Facilitated diffusion, cell membrane transport of contaminants, 216–217
Facilities maintenance, *see* Maintenance
Factory Acts of 1833 (U.K.), 4, 13–14
Factory system, historical perspective on, 4, 13–14
Farmer's lung, 729
Fast vital capacity (FVC), dust effect on lung, 307
Febrile respiratory illness, acute, described, 559–560
Fecal-oral transmission, 194
Federal government, *see also entries under names of specific legislation*; Legislation; State and local governments
consultant problem-solving role, 98–99
hazard communication standards and, 128–130
industrial hygiene activities and, 87
Federal Hazardous Substances Act of 1966, 21
Federal Insecticide, Fungicide and Rodenticide Act (FIFRA), agricultural hygiene, 730
Federal Interagency Task Force on Environmental Cancer, Heart and Lung Disease, tobacco smoke, indoor air quality, 548
Fee-for-service basis, consultants, 108–109
Fertilizers, agricultural hygiene, 730, 734, 743
Fibrogenic dust, 309–317
carbonaceous dust, 315–317
free crystalline silica, 309–315
Fibrosis, *see* Pulmonary fibrosis
Fibrous glass, described, 322–324
Fibrous mats and aggregate beds, air pollution filtration systems, 895–896
Filters, air-purifying respirators (nonpowered), 684–687
Filter-type hearing protectors, communication with, 991–992
Filtration, 894–903
air pollution controls classification, 873–874
cell membrane transport of contaminants, 216
fabric filters, 898–902
fibrous mats and aggregate beds, 895–896
paper filters, 896–898
performance range of, 894
Finland, 38–43
Fire and explosion hazards, *see also* Accidents
air-cleaning method selection, 865, 869
laboratory design, 384
manufacturing hazards:
acid and alkali cleaning of metals, 600
degreasing technology, 608
thermal spraying, 650
Fire extinguishers and fire suppression systems, laboratory design, 419–420
First aid planning, laboratory design, safety considerations, 420
Fitness to dive issue, described, 354–357
Flame spraying, 649
Flammable liquids, laboratory design, safety considerations, 411
Flavobacterium, humidifier fever, 558
Fleas, dermatoses, direct causal factors, 266
Florio, James J., 131
Flow properties, air-cleaning method selection, 867
Flux:
brazing, 661
soldering, 655–658
Foam-type earplug, described, 998
Fog, particulate matter contaminant, classification, 207
Folliculitis, dermatoses, clinical appearance of, 268
Forecast and heat alert programs, heat stress managment control, 831–832
Forging, 613–618
air contamination, 616–617
controls, 618
heat stress, 617–618
noise, 618
process described, 613–616
protective equipment, 618
Formaldehyde, volatile organic compounds, 541
Forssman, Sven, 56

Foundry operations, 619–633
cleaning and finishing, 631–632
control technology, 632–633
coremaking, 625–627
metal melting and pouring, 627–631
molding, 621–625
shake-out, 631
France, 66
Franklin, Sir John, 3–4
Fraudulent misrepresentation, law, 184–185
Free crystalline silica, dust effect on lung, 309–315
Freon solvents, vapor degreasing, 605
Frequency analyzers, noise measurement, 974–976
Frequency-weighting networks, sound level meter, 971
Friberg, Lars T., 56
Frictionless piston meter, cumulative air volume, calibration instruments and techniques, 472–473
Fugitive emissions, air pollution control, 861–862
Full mold, described, 624–625
Fume filters, air-purifying respirators (nonpowered), 685–686
Fume hoods, laboratory design:
chemical storage under, safety considerations, 412–413
clean bench hoods, 405
generally, 400–403
glove boxes, 404
HEPA filters, 404–405
perchloric acid hoods, 403–404
Fumes:
particulate matter contaminant, classification, 207
welding hazards, 663
Fungus, *see* Biological agents determination
Furnishings, laboratory design and, 388–389

Galen, 2
Gamma rays, nondestructive testing, 645–647
Gas cylinders, laboratory design and, 389
Gas dilution systems, known vapor concentrations production, dynamic systems (continuous flow), 498–500
Gaseous scrubbing, liquid scrubbing design parameters, 913
Gases and vapors, *see also* Air pollution controls; Carrier gases
contaminant classification, 206–207
lung transit of, 218–220
removal, air-purifying respirators (nonpowered), 687–688, 692
welding hazards, 663–664
Gas exchange, respiratory system, barometric pressure effects and, 340–342
Gas masks, certification of, 711
Gas metal arc welding, described, 666–668
Gas nitriding, heat treating hazards, 638
Gasometer or spirometer, cumulative air volume, calibration instruments and techniques, 471–472
Gas–solid adsorption, 919–923
air pollution controls classification, 875–876
chemical adsorption, 922
design parameters of, 922–923
overview of, 919
performance range of, 919–920
physical adsorption, 921–922
polar adsorption, 922
Gastrointestinal tract:
biliary tract elimination and excretion, 245–246
contaminants transfer, 225–228
elimination and excretion by, 246
Gas tungsten arc welding, described, 665–666
Gas welding, described, 668–669
Gender differences, heat stress and, 791
Germany:
historical perspective, 13, 14
industrial hygiene in, 66
Giardiasis, described, 196
Glass, fibrous glass, described, 322–324
Gloves, dermatoses prevention, 280
Good practice elements, hazard communication program, 146–147
Granulomas, dermatoses, clinical appearance of, 271
Graphic level recording, noise measurement, 978–979
Gravitational and inertial separation, 878–894
air pollution controls classification, 872–873
characteristics table, 880–881
collectors, economic analysis models (air pollution control), 856–857
cyclones, 887–892
multiple, 893–894
wet, 892–893
dynamic separators, 885–887
inertial separators, 883–885
overview of, 878
performance range of, 879
settling chambers, 879, 882–883
Great Britain, *see* United Kingdom
Green, Henry L., 66

Green sand molding, described, 621–623
Grinding operations, manufacturing hazards, 633–636

Half-octave band analyzer, described, 975
Hamilton, Alice, 4
Hand cleansers, dermatoses prevention, 280
Handicapped person access, laboratory design and, 388
Harvard School of Public Health, 8, 14
Hazard communication program:
  application of, 137–139
  chemical manufacturer, supplier, or importer responsibilities under, 139–144
  employee input, 148
  employer responsibilities under, 144–146
  good practice elements in, 146–147
  hazardous materials information system for, 160–162
  labeling:
    employer's responsibility, 164
    manufacturer responsibility, 163–164
  management involvement in, 148
  management of, 151–156
    communication flow, 154–156
    policies and procedures, 151
    responsibilities, 152–154
  overview of, 123–124
  planning in, 156–159
  professionals involved in, 147–148
  program audit, 165–167
  record keeping for, 160
  related regulations, 149–150
  training tools in, 162–163
Hazard communication standards:
  controversial issues in:
    excluded articles, 134–135
    intrinsic hazard *vs.* risk, 134
    mixtures, 134
    scope, 133
    trade secrets, 135–137
  creation of, 123, 124
  events leading to, table, 126–127
  expanded rule, 131–133
  federal government and, 128–130
  legal challenges to, 130–131
  performance-oriented standard, 167
  uniform regulation for, 126–128
Hazardous air pollutants (HAPs), 844
Hazardous chemical, defined, 137
Hazardous exposures, health care field, evaluation factors, 370–372
Hazardous materials information system, hazard communication program, 160–162
Hazardous material tracking, hazardous material information system, 161–162
Hazardous Substance Act, criminal sanctions, 188
Hazardous waste management and disposal:
  laboratory design, safety considerations, 413–419
  laboratory plumbing design and, 406
  legislation on, 26
  organizational consultant services and, 116
  sources of information on, 418–419
Hazards, *see entries under specific hazards and hazardous processes*
Health and Safety Committee, health care field, 371–372
Health and safety considerations, laboratory design, 386–389
Health care field, 361–381. *See also entries under* Medical
  controls in, 377–380
  described, 362–363
  early detection problems in, 364
  employee evaluation, 365–369
  goal achievement in, 380
  hazardous exposure evaluation factors, 370–372
  hiring practices, 369–370
  historical perspective on, 361–362
  hospital and closed building problems, 373–376
  monitoring in, 376–377
  psychosocial factors in, 363–364
Hearing conservation programs, 983–995
  audiograms and, 981–982
  away from work exposures, 958
  effectiveness of, 937–938, 988
  hearing protector:
    communication with, 990–992
    communication without, 990
    radio headset communication, 992
    selection of, 987–988
  organization and management of, 986–987
  overview of, 983–984
  rank-ordering of communication, 992–995
  requirements of, 984–986
    hearing measurement and record keeping, 985–986
    noise exposure level reduction, 985
    noise hazard identification, 984–985
  safety factors, 988–990
Hearing loss:
  costs of compensation for, 958
  letter indicating hearing problem, 1027
  noise exposure limits, 981–982
  noise-induced, 956–958

Hearing measurement, 959–968
audiometers, 959–962
hearing conservation program, 985–986
overview of, 959
record keeping, 967–968
test rooms, 962–964
threshold measurements, 964–967
Hearing protectors, 995–1008
communication with, 990–991
communication without, 990
noise reduction ratings, 989–990
overview of, 995–996
performance limitations of, 996
protection levels and ratings, 1001–1008
radio headset communication with, 992
selection of, 987–988
types of, 996–1001
advantages/disadvantages summarized, 1000–1001
combined insert and muff types, 1008
concha-seated ear protector, 998–999
earmuffs, 999–1000
foam-type earplugs, 998
insert types, 997
malleable earplugs, 998
sized earplug, 997–998
Hearing threshold measurements, described, 964–967
Heart, *see* Blood; Cardiovascular system
Heart rate, heat strain index, 782–783
Heat, dermatoses, direct causal factors, 265. *See also* Heat stress
Heat Alert Program (HAP), heat stress management control, 831–832
Heat balance equation, heat stress, worker-environment thermal exchange, 773
Heat control, *see* Temperature control
Heat cramps, described, 789
Heated-bulb thermometer anemometer, heat stress measurement, 801
Heated element anemometer, air velocity meters (anemometers), calibration instruments and techniques, 488
Heated thermocouple anemometer, heat stress measurement, 801
Heat exhaustion, described, 789
Heating, ventilation, and air conditioning (HVAC) system, *see also* Ventilation and ventilation systems
combustion products, indoor air quality, 546
dust, man-made fibrous dust, 552
indoor air quality, evaluation protocols and guidelines, 572–573
laboratory design, 399–400
sick building syndrome, 556
sick building syndrome studies, 554
Heat loss, *see* Body temperature regulation
Heat rashes, described, 789–790
Heat stress, 763–837
assessment of, 807–813
control options, 812–813
standards rationale, 810–812
stress and psychological strain correlation, 807–810
disorders from, 784–790
acute disorders, 785–790
acute disorders (heat cramps), 789
acute disorders (heat exhaustion), 789
acute disorders (heat rashes), 789–790
acute disorders (heatstroke), 785–789
interactive effects, 785
psychophysiological effects, 784–785
tolerance factors and, 784
engineering control for, 813–824
examples of, 820–824
local cooling stations, 817–818
moisture control, 820
source of heat control, 814–817
ventilation (general), 818–820
management control, 824–832
acclimatization regimes, 826
auxiliary cooling and protection measures, 829–830
environmental monitoring, 828
forecast and heat alert programs, 831–832
medical supervision, 825–826
training programs, 824–825
water and electrolyte provisions, 827–828
work-rest regimen, 826–827
measurement of, calibration of instruments, 465
physiology of, 777–793
age level and, 790
alcohol and drugs and, 792–793
body fat and, 791
body temperature regulation, 777–780
gender and, 791
physical fitness and, 793
strain indexes (body core temperature), 782–783
strain indexes (heart rate), 782–783
strain indexes (sweat evaporation rate), 781–782
strain indexes (sweat rate), 781
water/electrolyte balance and, 791–792
significance of, in industry, 763–766
standards, 766–772
international, 771–772, 810

Heat stress (*Continued*)
NIOSH, 768–771
origin of, 766–767
OSHA, 767, 804
rationale for, 810
threshold limit values, 767–768, 804
thermal environment measurement, 793–807
dry-bulb and wet-bulb temperatures, 794
effective temperature concept, 794–795
equivalent effective temperature corrected for radiation, 795
heat stress index (HSI), 795–797
sensing instruments (air temperature measurement problems), 799–800
sensing instruments (air velocity measurement), 801–802
sensing instruments (bimetallic thermometer), 799
sensing instruments (humidity measurement), 800–801
sensing instruments (integrated instrument approach), 803–805
sensing instruments (liquid-in-glass themometer), 799
sensing instruments (radiation measurement), 802–803
sensing instruments (thermocouple), 799
sensing instruments (thermometry), 799
sensing instruments (thermosistor), 799
sensing instruments (weather effects), 806–807
wet globe temperature (WGT), 798, 804
worker-environment thermal exchange, 772–777
clothing effects, 776–777
convection, 775
environmental heat exchange, 774–775
evaporation, 776
heat balance equation, 773
metabolic heat production, 773–774
radiation, 775–776
Heat stress index (HSI), described, 795–797
Heat stroke:
agricultural hygiene, 741
described, 785–789
Heat treating, 636–639
annealing, 638
surface hardening, 637–638
Heavy metals, paints, 3
Heliox diving, pressure effects, 335
Helium diving, pressure effects, 335
Henry's law, 218
Hepatitis, health care field, controls, 378
Herbicides, agricultural hygiene, 733
Herpes infection, dermatoses, direct causal factors, 267
High-efficiency filters, air-purifying respirators (nonpowered), 686, 687
High-pressure nervous syndrome, described, 339–340
Hippocrates, 2
Hiring practices, health care field, 369–370
Histoplasmosis, described, 196
Histotoxic anoxia, toxic material classification, pathological, 209
Hodgkin's disease, agricultural hygiene, 741
Hoods, *see* Fume hoods; Supplied-air hoods
Hospitals, health care field, sick building syndrome, 373–376
Hot box process, described, 626
Hot-wire anemometer, heat stress measurement, 801
Housekeeping:
dermatoses prevention, 279
laboratory design, safety considerations, 421
Humidification and dehumidification, ventilation systems, 567–568
Humidifier fever, described, 558
Humidity:
air-cleaning method selection, carrier gas properties, 869
heat stress control measures, 820, 821–822
measurement of, heat stress, 800–801
microphone effects of, noise measurement, 974
HVAC system, *see* Heating, ventilation, and air conditioning (HVAC) system; Ventilation
Hydrogen chloride, welding operations and, 607
Hygrometer, 800–801
Hygroscopicity, air-cleaning method selection, 866
Hyperbaric chambers, described, 335–336
Hyperbaric environment, expansion of, 330. *See also* Barometric pressure effects
Hyperbaric oxygen and recompression therapy, 352–354
Hyperoxia, toxicty of, 342–344
Hypersensitivity pneumonitis, described, 557–558
Hypothalamus, body temperature regulation by, 777–778
Hypoxia, barometric pressure effects, 337

Ice packet vest, 830
Immersion and breath-hold diving, pressure effects, 332–334

Immune system, toxic material classification, pathological, 210–211
Impact and impulse noise:
community noise, measurement procedures, 1025
measurement of, 976–977
noise control procedures, 1017–1018
noise exposure limits, 982–983, 983
noise measurement, 976–977
Impingement collectors, described, 873
Impulse noise, *see* Impact and impulse noise
Incineration, *see* Combustion method
Individual consultants, *see also* Consultants; Organizational consultants
advantages and disadvantages of, compared to organizational consultants, 97–98
backgrounds of, 92–95
organizational consultant and, 91
qualifications of, 95–96
Individual liability, law, 186. *See also* Liability
Indoor air quality, 26, 531–594
building-related illness, 534–535, 557–560
acute febrile respiratory illness, 559–560
asthma, 558–559
humidifier fever, 558
hypersensitivity pneumonitis, 557–558
Legionnaires' disease, 559
overview of, 557
control measures for, 581–584
proactive measures, 582–584
remedial action, 581–582
costs of, 562–564
evaluation protocols and guidelines, 572–581
components of, 574–576
qualitative evaluation, 572–574
quantitative evaluation, 577–578
regulations and, 578–580
historical perspective on, 531–532
nonspecific complaints, 553–557. *See also* Sick building syndrome
parameters, 535–553
bioaerosols, 536–539
bioeffluents, 549–550
combustion products, 546–548
environmental tobacco smoke, 548–549
man-made fibrous dust, 552–553
pesticides, 545–546
radon, 550–552
volatile organic compounds, 540–545
sick building syndrome, 532–535
thermal discomfort, 560–562
ventilation systems and, 565–572
Indoor noise, measurement procedures for, 1025
Induction furnace, described, 629
Industrial diseases, national conference on, 6. *See also* Manufacturing hazards
Industrial hazards, *see* Manufacturing hazards
Industrial hygiene laboratory, *see entries under* Laboratory
Industrial hygiene survey, 73–90
administrative organization and, 87–88
changes in perspectives, 73–74
consultant role in, 99–100
deregulation era and, 75
future of, 75–76
indoor air quality, 572–581
medical survey combined with, 84–85
organizational consultant services, 114
OSHA and, 75, 86
private industry and, 86
sick building syndrome, hospitals and tight buildings, 374–376
training and qualifications of personnel for, 88–90
types of, 76–84
investigational survey, 79–84
preliminary survey, 77–79
U.S. Public Health Service and, 85
Industrial hygienist, *see also* Industrial and occupational hygiene
employment of:
Canada, 35–36
industrial hygiene survey personnel, 88–90
Netherlands, 49
organizational consultants, 117–118
expert witness role of, 190–191
future role of, 76
legal definition of, 180
training and qualification of personnel, 88–90
Industrial noise, *see* Noise
Industrial and occupational hygiene, *see also* Industrial hygienist
abroad, *see* International perspective
academic programs in, 8–9
agricultural hygiene and, 744–745
air pollution controls and, 839
defined, 1
Finland, 40
Netherlands, 48
Sweden, 52
future prospects for, 9–12
historical perspective on:
early, 2–4, 14
U.S., 4–7, 14–16
objectives of, 10
problems facing, 10–11
professional certification in, 7–8
professional responsibility of, 12

Industrial and occupational
hygiene (*Continued*)
profession and, 180
publications in, 6–7
quality control and, *see* Quality control
program
support personnel for, 90
Industrial and occupational hygiene consultant,
see Consultants; Organizational consultants;
Private consulting firms
Industrial radiography, nondestructive testing,
645–647
Industrial Revolution, 13, 254
Inert gas narcosis, described, 339–340
Inertial separators, described, 883–885. *See also*
Gravitational and inertial separation
Inert (nonfibrogenic) dust, biological effects of,
308–309. *See also* Dust
Infection:
agricultural hygiene, 740
building-related illness, 534
dermatoses, direct causal factors, 266
health care field:
controls, 378
preemployment evaluation, 366–368
teratogens, 372
routes of transmission of, 194. *See also* Routes
of transmission
viral, dermatoses, direct causal factors, 266–
267
zoonoses, agricultural hygiene, 736–737
Inhalation valves, facepieces (respiratory
protective devices), 682
Injury, *see* Mechanical injury; Trauma
Inner ear, described, 956
Inorganic dust, *see* Dust
Insects:
agricultural hygiene, 740
dermatoses, direct causal factors, 266
Insert-type hearing protector, described, 997
Installation permits, air pollution regulations,
846
Instrument calibration, *see* Calibration
Instrument layout, laboratory design and, 397–
398
Instruments, *see* Calibration instruments and
techniques; Equipment
Insulation, heat stress controls, 815–816
Insurance carriers, industrial hygienist
employment by, 86–87
Intentional misrepresentation, law, 184–185
Interagency Regulatory Liaison Group (IRLG),
uniform standards and, 128
Internal Revenue Service, consultants, business
practices, 109
International Occupational Hygiene Association
(IOHA), activities of, 68–69
International perspective, 29–71
Australia, 29–32
Canada, 32–36
Egypt, 36–38
European Community (EC), 45, 50, 67
Finland, 38–43
France, 66
Germany, 66
Italy, 43–46
Netherlands, 46–50
Sweden, 50–56
United Kingdom, 56–66
United Nations Organization, 67–68
International standards, heat stress, 771–772,
810
Intrinsic hazard *vs*. risk, hazard communication
standards, 134
Investigational industrial hygiene survey, *see
also* Industrial hygiene survey
analytical survey, 80–82
control procedure development, 82–83
plant survey, 79–80
records and reports, 83–84
sick building syndrome, hospitals and tight
buildings, 375–376
Investment casting (lost wax), described, 624
Ion-flow meters, mass flow and tracer
techniques, calibration instruments and
techniques, 485–486
Ionizing radiation, dermatoses, direct causal
factors, 266
Iron oxide, welding hazards, 663
Irritants:
irritant fume, respiratory fit test agents, 704
toxic material classification, pathological, 208
Isoamyl acetate, respiratory fit test agents, 703–
704
Isolation, dermatoses prevention, 279
Italy, 43–46

Joint Commission on Accreditation of
Healthcare Organizations (JCAHO), 362,
378
Journal of Industrial Hygiene, 6, 14
Judiciary:
consultant litigation/expert witness role, 102–
103
hazard communication standards and, 130–
133
OSHAct and, 23–24
Jurisdiction, OSHA, 26–27

Kidney:
biotransformation or metabolism of contaminants in, 239
contaminant deposition and accumulation in, 235–236
elimination and excretion by, 243–245
Kinetics, *see* Pharmacokinetic modeling
Klee, Paul, 3
Known vapor concentrations production, calibration instruments and techniques, 494–502

Labeling, *see also* Hazard communication standards
American National Standards Institute (ANSI), 165
chemical industry responsiblity for, 141, 144
employer responsibility for, 145
exempted chemicals from hazard communication requirements, 137
hazard communication program, 163–164
mixtures and, 134
Occupational Safety and Health Act of 1970 and, 125
uniform regulation for, 126–128
Laboratory analysis control (quality control), 438–442
external testing for accreditation, 442
instruments, 439–440
method adaption and validation, 439
sample identification and control, 438
samples (routine), 440–441
samples (single-time), 441–442
Laboratory design, 383–422
accidents and, 384
aesthetic and environmental considerations, 385
construction, 407–410
facility design, 389–407
laboratory planning, 396–398
overview of, 389–391
space planning, 391–395
health and safety considerations, 386–389
egress and handicapped person access, 388
emergency facilities, 388
features, 388
furnishings, 388
regulations, 386–387
ventilation, 387–388
laboratory functions and, 383–384
location of laboratory, 385
overview of, 383
quality control, facilities design and, 438
safety considerations, 410–422
chemical storage, 410–413
compressed gas storage, 413
emergency evacuation design, 420–421
fire extinguishers and fire suppression systems, 419–420
first aid facilities, 420
maintenance, 421–422
security, 422
showers and eyewash stations, 419
waste disposal, 413–419
security considerations, 422
services, 398–407
compressed gas system, 406–407
distribution schemes, 399
electrical power needs, 405
fume hoods, 400–405
heating, ventilation, and air conditioning, 399–400
plumbing needs, 406
Laboratory services:
accreditation, organizational consultants, AIHA, 111
hazard communication standards exemption, 138–139
organizational consultant services and, 116
quality control:
chain of custody and, 429–430
facilities design and, 438
purchase control, 437–438
reagent and reference standards, 437
sample handling, storage, and delivery, 442
selection of, 120
Labor force, *see* Work force
Labor unions, *see also* Employees
consultant advisory role, 106–107
consultant problem-solving role, 98–99
hazard communication standards and, 129, 130, 132, 133
industrial hygiene development and, 4, 5
organizational consultants and, 112
politics and, 18
Laser radiation, dermatoses, direct causal factors, 266
Law, 179–192
criminal sanctions, 187–189
historical perspective, 187–188
statutes, 188
U.S. v. Park, 188–189
ethical responsibilities, 181
expert witness and, 189–191
industrial hygienist definition, 180
legal documentation, industrial hygiene survey, investigational, 79
overview of, 179–180
potential liability theories (torts), 181–186
contracts, 185–186

Law (*Continued*)
generally, 181–182
intentional or fraudulent misrepresentation, 184–185
negligence, 182–184
profession and, 180
Workers' Compensation, 186
Lead:
manufacturing hazards, heat treating, 638–639
paints, 3
toxicity of, historical perspective on, 2, 3–4
Legal documentation, *see* Law
Legionnaires' disease and legionellosis:
bioaerosols, indoor air quality, 537
described, 196, 559
sick building syndrome and, 534
Legislation, *see also entries under names of specific legislation*; Federal government
Congressional hearings on occupational safety and health, 16–20
hazardous waste disposal, 26
historical perspective, 4–5, 14–16
Occupational Safety and Health Act of 1970, 21–24
future prospects for, 25–28
provision of, described, 20–21
United Kingdom, Factory Acts, 4, 13–14
Legislative support role, of consultants:
individual, 105–106
organizational, 115
Leptospirosis:
agricultural hygiene, 737
described, 196
Leukemia:
agricultural hygiene, 741
benzene and, 580
Liability:
individual liability, 186
organizational consultants and, 113, 119
potential liability theories, 181–186
professional liability, law, 181
Lift-slab construction, 25–26
Lighting, laboratory design and, 389
Lipid-rich organs and tissues, contaminant deposition and accumulation in, 236–237
Liquid dilution systems, known vapor concentrations production, dynamic systems (continuous flow), 500–502
Liquid-in-glass thermometer, 799
Liquid penetrant, nondestructive testing, 648
Liquid scrubbing, 903–913. *See also* Air pollution controls
air pollution controls classification, 874–875
design parameters, 912–913
mist eliminators, 911–912
orifice scrubbers, 909–911
overview of, 903
packed towers, 907–909
performance range of, 904–905
spray chambers, 905–907
Listeriosis, described, 196
Litigation/expert witness roles, of consultants, 102–103
Liver, contaminant deposition and accumulation in, 235–236
Local government, *see* State and local governments
Long and single-number (NRR) method, hearing protector requirements, 1002–1008
Loose-fitting respirator facepieces (respiratory protective devices), 680–681
Lost wax method (investment casting), described, 624
Lung, *see also* Respiratory system
anatomy of, and dust, 290–294
arterial gas embolism, described, 350–351
asbestos dust and, 318–322
barometric pressure effects on, 340–342
diving, barotrauma of descent, 338–339
dust effects on, critique of studies of, 302–307
dust particle size and, 295–298
elimination of contaminants by, 239, 246
oxygen toxicity and, 344
pulmonary barotrauma, described, 350–351
reaction to dust of, 299–301
respiratory function of, 298–299
self-cleaning function of, 297–298
Lung cancer, *see also* Cancer and carcinogens
asbestos dust and, 322
building-related illness, 534
dust effect on lung, 300–301, 309
radon, indoor air quality, 550–552
radon and, 580
tobacco smoke, indoor air quality, 548
Lung transit (of contaminants), 218–225
aerosol absorption, 220
gases and vapors absorption, 218–220
lung structure and function, 218
lymphatic system and, 224–225
particulate penetration and absorption, 220–223
respiratory tract clearance, 223–224
Luxon, S. G., 66
Lyme disease:
agricultural hygiene, 743
dermatoses, direct causal factors, 266
Lymphatic system:
lung portal of contaminants and, 224–225
transport and distribution of contaminants by, 232

Macrophage, dust reaction of lung, 297, 299–300
Magnetic field and vibration effects, noise measurement, 977–978
Magnetic particle inspection, nondestructive testing, 647–648
Maintenance:
  air pollution control costs, 851–853
  laboratory design, safety considerations, 421–422
  quality control program, equipment for, 436–437
  respiratory protective devices, 707–708
Malaria, dermatoses, direct causal factors, 266
Malleable earplug, described, 998
Management:
  chemical industry, hazard communication standards, 127–128
  consultant problem-solving role, 98–99
  corporate policy, consultant role limited by, 100
  ethical considerations and, 107
  hazard communication program involvement of, 148
  of hearing conservation program, 986–987
  heat stress control measures, 824–832
  industrial hygiene survey:
    control procedures, 83
    investigational, 81
    records and reports, 83, 84
  of quality control program, 425–433
Manufacturing hazards, 595–674
  abrasive blasting, 595–598
  acid and alkali cleaning of metals, 598–601
  brazing, 660–661
  degreasing technology, 601–608
    cold degreasing, 602
    control measures, 605, 607–608
    emulsion cleaners, 602
    vapor degreasing, 602–605
  electroplating, 609–613
    air contaminants, 611
    control measures, 611–613
    techniques in, 609–611
  forging, 613–618
    air contamination, 616–617
    controls, 618
    heat stress, 617–618
    noise, 618
    process described, 613–616
    protective equipment, 618
  foundry operations, 619–633
    cleaning and finishing, 631–632
    control technology, 632–633
    coremaking, 625–627
    metal melting and pouring, 627–631
    molding, 621–625
    overview of, 619–621
    shake-out, 631
  grinding, polishing, and buffing operations, 633–636
  heat treating, 636–639
    annealing, 638
    hazard potential, 638–639
    overview of, 636–637
    surface hardening, 637–638
  metal machining, 639–644
    conventional tool machining, 639–643
    electrical discharge machining, 644
    electrochemical machining, 643–644
  nondestructive testing, 644–648
    industrial radiography, 645–647
    liquid penetrant, 648
    magnetic particle inspection, 647–648
    ultrasound, 648
  overview of, 595
  painting, 650–655
    controls, 653–655
    operations and exposures, 653
    overview of, 650–651
    paint composition, 651–653
    paint types, 651
  soldering, 655–660
    application techniques, 658
    cleaning techniques, 658–660
    controls, 660
    flux composition, 655–657
    fluxing operations, 658
    initial cleaning procedures, 658
    overview of, 655
    solder composition, 657
  thermal spraying, 648–650
  welding, 662–674
    controls, 669–671
    gas metal arc welding, 666–668
    gas tungsten arc welding, 665–666
    gas welding, 668–669
    overview of, 662
    shielded metal arc welding, 662–665
Massachusetts, State Bureau of Labor Statistics of, 14
Mass flow and tracer techniques, calibration instruments and techniques, 485–486
Mass psychogenic illness concept, sick building syndrome, 564–565
Mastromatteo, Ernest, 33
Material conservation, air pollution control costs, 853, 856
Material safety data sheets (MSDS):
  chemical industry responsibility for, 140–141

Material safety data sheets (*Continued*)
employer responsibility for, 144
example of, 142–143
hazard communication program and, 138–139
hazard communication program audit, 165–167
hazard communication program record keeping, 160
hazard communication standards and, 125
hazardous material information system, 160–161
trade secrets and, 136
Maximal acceptable concentrations, 15
Maximum achievable control technology (MACT), described, 844–845
Maximum allowable concentrations of contaminants, 11, 15. *See also* Threshold limit values (TLVs)
Mechanical collectors, *see* Gravitational and inertial separation
Mechanical injury, dermatoses, direct causal factors, 264–265
Media, worker awareness of hazards, 80
Medical departments, industrial hygiene survey, preliminary, 78
Medical evaluation, diving, fitness to dive issue, 354–357
Medical field, *see* Health care field
Medical profession, historical perspective on, 3, 4
Medical records (employee), hazard communication program and, 149
Medical supervision, heat stress managment control, 825–826
Medical survey, industrial hygiene survey combined with, 84–85
Medical treatment:
dermatoses, 282
laboratory design, safety considerations, 420
recompression therapy and hyperbaric oxygen, 352–354
Melanoma:
agricultural hygiene, 740
dermatoses, clinical appearance of, 270
Mental stress, *see* Heat stress; Psychological considerations; Stress-related illness
Mercury:
liver accumulation of, 235
paints, 3
Mesothelial tumor, asbestos dust and, 322
Metabolic heat production, heat stress, worker-environment thermal exchange, 773–774
Metabolism, toxic material modes, 239–240
Metal and Nonmetallic Mine Safety and Health Act of 1966, 16
Metal machining, 639–644
conventional tool machining, 639–643
electrical discharge machining, 644
electrochemical machining, 643–644
Metal melting and pouring, 627–631
arc furnace, 627–628
crucible furnace, 629
cupola, 629–630
induction furnace, 629
overview of, 627
transfer, pouring, cooling, 630–631
Methylene chloride, vapor degreasing, 604, 606
Michaelis-Menton kinetics, 241
Microorganisms, *see* Bioaerosols; Biological agents determination
Microphones, noise measurement, 972–974
Microwave radiation, dermatoses, direct causal factors, 266
Middle Ages, 2
Middle ear, described, 955–956
Miliaria (prickly heat):
agricultural hygiene, 740
dermatoses:
clinical appearance of, 269
direct causal factors, 265
Milk, elimination and excretion by, 247
Milker's modules, agricultural hygiene, 737
Mining industry:
carbonaceous dust and, 315
historical perspective on, 2–3
Metal and Nonmetallic Mine Safety and Health Act of 1966, 16
Misrepresentation, law, 184–185
Mist, particulate matter contaminant, classification, 207
Mist eliminators, liquid scrubbing, 911–912
Mist filters, air-purifying respirators (nonpowered), 685
Mites, dermatoses, direct causal factors, 266
Mixture metering, mass flow and tracer techniques, calibration instruments and techniques, 485
Mixures, hazard communication standards and, 134
Model programs, agricultural hygiene, 747–748
Modes of toxins, *see* Toxic material modes
Moisture control, heat stress control, 820
Molding, 621–625
full mold, 624–625
green sand molding, 621–623
investment casting (lost wax), 624
shell molding, 623–624

Momsen lung, 337–338
Monitoring:
biological agents, 199–200
dermatoses prevention, 281
health care field, 376–377
hearing conservation programs, 987
heat stress managment control, 828
quality control programs and, 424
radon, 552
Mucus, dust reaction of lung, 299
Muffler and air noise generation control, noise control procedures, 1019–1020
Multi-industry approach, 17
Multiple cyclones, described, 893–894
Muscular activity, body temperature regulation and, 778

Nail discoloration and dystrophy, dermatoses, clinical appearance of, 271
Narcosis, inert gas narcosis, described, 339–340
Narcotics, toxic material classification, pathological, 209
Narrow band analyzer, described, 975–976
Nasopharynx, lung transit of contaminants and, 223
National Advisory Committee on Occupational Safety and Health (NACOSH), creation of, 21
National Ambient Air Quality Standards (NAAQS):
amendments to, 932
indoor air quality, 547
provisions of, 843
National Bureau of Standards, safety glasses, welding hazards, 670–671
National Environmental Policy Act of 1969, 21
National Institute for Occupational Safety and Health (NIOSH), 74
academic training and, 8–9
bioaerosols, indoor air quality, 537–538
creation of, 20
future prospects for, 25
hazard communication program trends, 150
hazard communication standards, 125
hearing protector requirements, 1002
heat stress, 774
wet bulb globe temperature predictive value, 806–807
heat stress management control, 825, 826, 827, 828, 829
heat stress standards, 768–771, 810–811
indoor air quality, evaluation protocols and guidelines, 572
industrial hygiene research activities of, 87
industrial hygiene survey and, 86
manufacturing hazards:
abrasive blasting, 598
grinding, polishing, and buffing operations, 635
metal machining hazards, 643
noise exposure regulations, 979–980
noise reduction ratings, 989, 990
organizational consultants and, 112
protective equipment, dermatoses prevention, 280
respirator efficiency tests, 703
respiratory protective devices, 676, 681, 682, 684, 688, 692
psychological considerations, 705
selection considerations, 706
safety glasses, welding hazards, 671
sick building syndrome and, 533
sick building syndrome studies, 553–554
standard-setting process and, 24
training programs offered by, 102
National Institute of Standards and Technology (NIST):
equipment calibration standards, 469
quality control, equipment calibration standards, 436
National Occupational Safety and Health Board, proposal of, 17–18
National Research Council:
indoor air quality, pesticides, 546
tobacco smoke, indoor air quality, 548
National Safety Council, 6, 9
National Weather Service, 806, 831
Negligence, law, 182–184. 182–184. *See also* Law; Liability
Neoplasms, dermatoses, clinical appearance of, 269–270
Netherlands, 46–50
Newcastle disease:
agricultural hygiene, 735
dermatoses, direct causal factors, 266–267
New Source Performance Standards (NSPS), described, 844
Nitrogen dioxide:
combustion products, indoor air quality, 546–548
historical standards, 15
lung illness due to, 580
predictive air quality modeling, 846
Nixon, Richard M,, 19
No bake process, described, 626–627
Nodular silicosis, dust effect on lung, 313–315
Noise, 937–1039
agricultural hygiene, 740

Noise (*Continued*)
community noise, 1020–1026
frequency weighting or analysis of, 1022
frequency weighting or analysis of (aircraft noise), 1022–1023
frequency weighting or analysis of (nonaircraft noise), 1023
guide for criteria in, 1025–1026
measurement equipment, 1023
measurement location and procedures, 1024–1025
overview of, 1020–1021
regulation (nuisance-type), 1021
regulation (performance-type), 1021–1022
current problems of, 937
ear and, 951–958. *See also* Ear
ear damage mechanism, 956–958
hearing conservation programs, 983–995
area identification, 984–985
hearing protector (communication with), 990–992
hearing protector (communication without), 990
hearing protector (selection), 987–989
organization and management of, 986–987
overview of, 983–984
radio headset communication, 992
rank-ordering of communication, 992–994
requirements of, 984–986
requirements of (hearing measurement and record keeping), 985–986
requirements of (noise exposure level reduction), 985
requirements of (noise hazard area identification), 984–985
safety factors, 988–990
hearing measurement, 959–968. *See also* Hearing measurement
hearing protectors, 995–1008
overview of, 995–996
performance limitations of, 996
protection levels and ratings, 1001–1008
types of, 996–1001
types of (advantages/disadvantages summarized), 1000–1001
types of (combined insert and muff types), 1008
types of (concha-seated ear protector), 998–999
types of (earmuffs), 999–1000
types of (foam-type earplugs), 998
types of (insert types), 997
types of (malleable earplugs), 998
types of (sized ear plugs), 997–998
manufacturing hazards, foundry, 633
OSHA exposure limits, 979–983
sound characteristics, 938–951. *See also* Sound measurement
terminology in, 938–941
Noise barriers and enclosures, noise control procedures, 1009–1017
Noise control procedures, 1008–1020
absorption, 1009
acoustic damping, 1018
drive system modification, 1020
impact noise, noise radiation, and vibration control, 1017–1018
muffler and air noise generation control, 1019–1020
noise barriers and enclosures, 1009–1017
overview of, 1008–1009
reduced driving force, 1018–1019
Noise exposure limits and levels, 979–983
annual audiogram, 981
base-line audiogram, 980–981
chronic hearing problems, 981–982
85 dBA in C/T calculation, 982
hearing conservation program, reductions in, 985
historical perspective on, 979–980
impulse noise limits, 983
practical considerations, 983
record retention, 982
steady-state and impulsive noise integration, 982–983
Noise measurement, 968–979
calibration of instruments for, 468–469, 979
frequency analyzers, 974–976
graphic level recording, 978–979
impulse or impact noise, 976–977
magnetic field and vibration effects, 977–978
microphones, 972–974
overview of, 968
sound level meter, 968–972
tape recording of noise, 978
Noise radiation, noise control procedures, 1017–1018
Noise reduction ratings, hearing protectors, 989–990
Nondestructive testing, 644–648
industrial radiography, 645–647
liquid penetrant, 648
magnetic particle inspection, 647–648
ultrasound, 648
Nonfibrogenic (inert) dust, biological effects of, 308–309
Noro, Leo, 39
Notice of Proposed Rulemaking (NPRM), standard-setting process and, 24

Noweir, M. H., 37
Nuclear hazards:
quantitative fit tests (respirator efficiency tests), 700
U.S. Department of Energy (DOE), 27
Nuisance and trespass restriction, air pollution controls, 845–846

Occupational acne, causes of, 256
Occupational bionomic system, concept of, 10
Occupational dermatoses, *see* Dermatoses
Occupational disease:
assumptions about, 1–2
historical perspective on, 2–4
Occupational disease prevention, *see* Prevention programs
Occupational Exposures to Hazardous Chemicals in Laboratories (OSHA), laboratory design and, 386–387
Occupational Health and Safety Act of Ontario, Canada, 34
Occupational health standards, *see* Standards
Occupational safety and health, legislative hearings on, 16–20
Occupational Safety and Health Act of 1970 (OSHAct), 5, 73
academic training and, 8
criminal sanctions, 188
dermatoses and, 255
hazard communication standards, 125
passage of, 19–20
provision of, described, 20–21
quality control programs and, 423
related governmental action to, 21–24
Occupational Safety and Health Administration (OSHA):
abrasive blasting standards, 597
agricultural hygiene, 740, 747
consultant problem-solving role, 99
deregulation and, 75
enforcement activities of, 87
establishment of, 6
future of, 25–28
hazard communication program audit, 165–167, 168–171
hazard communication standards and, 123, 128, 129–133, 149
health care field, 361–362, 363
controls, 379
preemployment evaluation, 367–368
hearing protection and, 961, 983–984, 986, 1001–1002
heat stress standards of, 767
indoor air quality, evaluation protocols and guidelines, 572, 578
industrial hygiene survey and, 74, 75, 80, 86
industrial hygienist and, 180
intrinsic hazard *vs.* risk, 134
labeling and, 163–164
laboratory design and, 386–387
metal machining hazards, 643
noise regulations, 968, 971, 979–983, 989, 1008
organizational consultants and, 111, 112
quality control:
data validation method of, 448
protocols of, 439
single-time samples, stop-gap methods, 441
respiratory protective devices, 676
efficiency tests, 703
psychological considerations, 705
selection considerations, 706
standards established by, 16
uniform standards and, 126–128
X-ray generator regulations, 645
Octave band:
industrial noise measurement, 948–951
sound pressure levels, 946
Octave band analyzer, described, 974–975
Odors, sick building syndrome, 556–557
Office of Management and Budget (OMB), powers of, 25
Ohman, Harry G., 56
Olf and decipol concepts, sick building syndrome, 556–557
Open-circuit self-contained breathing apparatus (supplied-air respirators), 694–695
Operating costs, air pollution control, 851–853
Operating permits, air pollution regulations, 846
Operating procedures, quality control program management, 426–427
Oral–fecal transmission, routes of transmission, 194
Organic acids, transport and distribution of contaminants, 231–232
Organization, quality control program management, 426
Organizational consultants, 111–120. *See also* Consultants; Individual consultants
career opportunities as, 117–118
emergence of, 111
ethical concerns and, 118
growth of, 111–113
individual consultant and, 91
individual consultant support relationship with, 110

Organizational consultants (*Continued*)
liability and, 119
needs served by, 113
selection of, 119
services offered by, 113–117
asbestos management, 114–115
field surveys, 114
indoor air quality, 115
industrial hygiene engineering, 116
laboratory services, 116, 120
litigation and expert witness, 115
miscellaneous services, 116–117
program development and auditing, 114
Orifice meter, volumetric flow rate, calibration instruments and techniques, 480
Orifice scrubbers, liquid scrubbing, 909–911
Ornithosis:
agricultural hygiene, 737
dermatoses, direct causal factors, 266–267
Oscilloscope, noise measurement, 976–977
Ossicle bones, 955
Oxygen toxicity, described, 342–344

Packed towers, liquid scrubbing, 907–909
Paints and painting, 650–655
controls, 653–655
operations and exposures, 653
overview of, 650–651
paint composition, 651–653
paint types, 651
toxins in, 3
Paint spray booth, 654–655
Paper filters, air pollution filtration control, 896–898
Paperwork Reduction Act of 1980, 22
Paracelsus, 3
Parasitic microorganisms, *see* Biological agents determination
Parasitic mites, dermatoses, direct causal factors, 266
Particle size (dust), respiratory system reaction and, 295–298
Particulates:
air-cleaning method selection, 866
contaminant classification, 207
lung transit of, 220–223
respirable, combustion products, indoor air quality, 546
Particulate scrubbing, liquid scrubbing design parameters, 913
Passive transport, cell membrane transport of contaminants, 214–216
Patch test, dermatoses diagnosis, 274–276
Pathological classification, toxic materials, 207–211
PBB:
milk concentrations of, 247
organ deposition of, 233
PCBs, *see* Chlorinated biphenyls (PCBs)
Perchloroethylene:
vapor degreasing, 604, 606
welding operations and, 607
Percutaneous absorption, systemic intoxication signs following, 272–273. *See also* Skin
Performance-oriented standard, hazard communication standards, 167
Permissible exposure limits, *see* Exposure limits
Permits, air pollution regulations, 846
Personal hygiene, dermatoses, indirect causes, 259
Personnel, future of, 76
Perspiration, *see* Sweat
Pesticide respirators, certification of, 716
Pesticides:
agricultural hygiene, 730–734, 743
indoor air quality, 545–546
Phagocytosis:
cell membrane transport of contaminants, 217
lung transit of contaminants and, 224
Pharmacokinetic modeling:
chlorinated biphenyls, 248–252
elimination and excretion, 240–243
heat stress and, 792–793
introduction to, 247–248
Phosgene, welding operations and, 607
Photosensitivity, dermatoses, direct causal factors, 263–264
Phthisis, *see* Tuberculosis
Physical adsorption, gas-solid adsorption, 921–922
Physical classification, toxic materials, 206–207
Physical fitness, heat stress and, 793
Pigmentary abnormalities, dermatoses, clinical appearance of, 269
Pinocytosis, cell membrane transport of contaminants, 217
Pipe-type precipitators, electrostatic precipitation, 917
Pitot tube, air velocity meters (anemometers), calibration instruments and techniques, 486–488
Placental barrier, contaminant transfer and, 234–235
Planning:
hazard communication program, 156–159
quality control program management, 426

Plants and woods, dermatoses:
  direct causal factors, 262–263
  photosensitizing plants, 264
Plant survey, industrial hygiene survey, investigational, 79–80
Plasma, transport and distribution of contaminants by, 230–232
Plasma spraying, 649
Plate-type precipitators, electrostatic precipitation, 915–917
Pliny the Elder, 2
Plumbing needs, laboratory design, 406
Plume visibility, air pollution controls, 845
Pneumoconioses:
  coal worker's pneumoconioses, 16, 315–317
  defined, 324
  legislation and, 16
  welding hazards, 663
Pneumonitis, hypersensitivity pneumonitis, described, 557–558
Poison ivy and poison oak, dematoses, 263
Poisons (systemic), toxic material classification, pathological, 209
Polar adsorption, gas-solid adsorption, 922
Policy, quality control program management, 426
Polishing operations, manufacturing hazards, 633–636
Politics:
  legislative hearings and, 16–20
  Occupational Safety and Health Act of 1970:
    future prospects for, 25–28
    provision of, described, 20–21
  threshold limit values (TLVs) and, 11–12
Polychlorinated biphenyls, *see* Chlorinated biphenyls (PCBs)
Polystyrene, manufacturing hazards, molding, 625
Polyurethane paints, 652–653
Pontiac fever:
  bioaerosols, indoor air quality, 537
  described, 196, 559
Positive displacement meter, cumulative air volume, calibration instruments and techniques, 476
Positive-pressure respirators, supplied-air respirators, 693
Potential liability theories, law, 181–186
Pott, Percival, 3
Powered air-purifying respirators:
  described, 696
  testing, 701
Precipitation mechanisms, electrostatic precipitation, 914–915
Precipitation method, *see* Electrostatic precipitation
Predictive air quality modeling, described, 846
Preemployment evaluation, health care field, 365–370
Preliminary industrial hygiene survey, described, 77–79
Pressure, air-cleaning method selection, carrier gas properties, 868
Pressure demand mode, supplied-air respirators, 693
Pressure effects, *see* Barometric pressure effects
Pressure measurement, ventilation system measurement calibration, 465
Pressure regulator, supplied-air respirators, 694
Pressure transducer, volumetric flow rate, calibration instruments and techniques, 483
Pressurized environments, 330–338
  compressed air and mixed gas diving, 334–335
  general principles of, 330–332
  hyperbaric chambers, 335–336
  immersion and breath-hold diving, 332–334
  saturation diving, 335
  submarines and submarine escape, 337–338
  tunnels and caissons, 336–337
Prevention programs:
  biological agents, 198–199
  historical perspective on, U.S., 13–15
  risk management and prevention programs (RMPPs), industrial hygiene survey and, 75
Prickly heat (miliaria):
  agricultural hygiene, 740
  dermatoses:
    clinical appearance of, 269
    direct causal factors, 265
Private consulting firms, industrial hygienist employment by, 87. *See also* Organizational consultants
Private industry:
  consultant advisory role, 106–107
  consultant's client relationship with, 96–97
  industrial hygienist employment by, 86
Problem-solving role, consultants, 98–99
Profession, industrial hygiene and, 180
Professional certification, *see also* Accreditation
  agricultural hygiene, 746
  air pollution controls, 847
  consultants, individual, 93
  director of industrial hygiene, 88
  in industrial hygiene, 7–8
  respiratory protective devices, 710–716
  United Kingdom, 63–65
Professional liability, law and, 181. *See also* Liability

Professional organizations, *see also entries under names of organizations*
  Australia, 32
  Canada, 36
  consultant advisory role, 106–107
  Finland, 39–40, 42
  Italy, 46
  Netherlands, 49–50
  organizational consultants and, 112
  Sweden, 54–55
  United Kingdom, 63
Professional training, *see* Academic programs; Education; Training programs
Proficiency Analytical Testing (PAT) program, *see* American Industrial Hygiene Association (AIHA) Proficiency Analytical Testing (PAT) program
Program auditing:
  consultant role in, 100–101
  hazard communication program, 165–167, 168–171
  organizational consultant services, 114
  quality control program, 448–456
    external evaluation, 456
    internal evaluation, 448–449
    self-appraisal check list for, 449–456
Program development, organizational consultant services, 114
Protective creams, dermatoses prevention, 280–281
Protective equipment:
  dermatoses prevention, 279–280
  manufacturing hazards, forging, 618
Protein binding, transport and distribution of contaminants, 230–231
Pseudomonas, humidifier fever, 558
Psychogenic illness concept, sick building syndrome, 564–565
Psychological considerations, respiratory protective devices, 705. *See also* Heat stress; Stress-related illness
Psychophysiological effects, heat stress, 784–785, 807–810
Psychrometer, 800
Publications:
  academic programs, 8–9
  in industrial hygiene, 6–7, 14
Public education:
  early goal of, 1
  worker awareness of hazards, 80
Public opinion, air pollution contamination levels and, 11
Pulmonary barotrauma, described, 350–351
Pulmonary fibrosis:
  asbestos and, 319–320
  dust reaction of lung and, 300
Pulmonary system, *see* Lung; Respiratory system
Purchase control, quality control, 437–438

Q fever:
  dermatoses, direct causal factors, 266–267
  described, 197
Qualitative fit tests, respirator efficiency, 703–704
Quality control program, 423–460
  equipment for, 433–437
    analytical instruments, 435–436
    direct-reading instruments, 434–435
    maintenance program, 436–437
    sampling equipment, 433–434
  laboratory analysis control, 438–442
    external testing for accreditation, 442
    instruments, 439–440
    method adaption and validation, 439
    sample identification and control, 438
    samples (routine), 440–441
    samples (single-time), 441–442
  management of, 425–433
    chain of custody, 427–430
    corrective actions, 431
    costs, 432–433
    objectives, 425–426
    operating procedures, 426–427
    organization, 426
    planning, 426
    policies, 426
    record keeping, 430–431
    training, 432
  overview of, 423–424
  program audit for, 448–456
    external evaluation, 456
    internal evaluation, 448–449
    self-appraisal check list for, 449–456
  purchase control, 437–438
  samples, handling, storage, and delivery of, 442
  scope of, 425
  standards, reagent and reference standards, 437
  statistical, 442–448
    control charts, 443–447
    data analysis techniques, 447
    data validation, 448
    sampling plans, 448
Quantitative emission standards, air pollution controls, 845
Quantitative fit tests (respirator efficiency tests), 698–703
  listing of, 698–700
  powered air-purifying respirator, 701

Quantitative fit tests (*Continued*)
problems of, 700–701
self-contained breathing apparatus, 701
supplied-air hoods, 701
supplied air respirators, 701
use of, 700
Québec, Canada, 34

Rabies, agricultural hygiene, 737
Race differences, dermatoses, indirect causes, 257
Radiation:
dermatoses, direct causal factors, 265, 265–266
electromagnetic radiation measurement, calibration of, 467–468
radiography, industrial, nondestructive testing, 645–647
thermal radiation:
effective temperature and, 795
heat stress, worker-environment thermal exchange, 775–776
heat stress control, 815–817, 823–824
heat stress measurement, 802–803
ultraviolet radiation:
dermatoses, direct causal factors, 265
manufacturing hazards, thermal spraying, 650
nondestructive testing:
liquid penetrant, 647–648
magnetic particle inspection, 647–648
welding hazards, 664–665, 666, 667, 669
Radiography, industrial, nondestructive testing, 645–647
Radio headsets, communication with, 992
Radiology:
coal worker's pneumoconioses and, 317
dust effect on lung, 306
Radon:
building-related illness, 534
indoor air quality, 550–552
lung cancer and, 580
measurement of, 552
OSHAct and, 26
Ramazzini, Bernardo, 3
Range of air-cleaning capacity, engineering performance specifications, 847
Reactivity, air-cleaning method selection, carrier gas properties, 869–870
Reagan, Ronald, 25, 128
Reagent and reference standards, quality control, 437
Recompression therapy and hyperbaric oxygen, 352–354
Record keeping:
hazard communication program, 160
hearing conservation program, 985–986
hearing measurement, 967–968
noise exposure limits, 982
quality control program management and, 430–431
Records and reports:
industrial hygiene survey, investigational, 82–83
sick building syndrome, hospitals and tight buildings, 375–376
Recycling, laboratory design, hazardous waste disposal, 417
Red blood cells, transport and distribution of contaminants by, 228–230
Reduced driving force, noise control procedures, 1018–1019
Refrigeration of chemicals, laboratory design, safety considerations, 412
Regulation, *see also entries under specific regulatory agencies and laws*
agricultural hygiene, 747
air pollution controls, 842–849
consultant legislative role in, 105–106
hazard communication program trends, 150
indoor air quality evaluation, 578–580
laboratory design, waste disposal, 413, 416
noise, community noise, 1021–1022
noise exposure regulations, 979–980
X-ray generator regulations, 645
Regulatory Flexibility Act of 1980, 22
Rehydration, sweating mechanism and, 779
Renoir, Pierre Auguste, 3
Reports and records, *see* Record keeping; Records and reports
Research, NIOSH and, 87
Research Triangle Institute, 9
Resource Conservation and Recovery Act (RCRA):
criminal sanctions, 188
hazard communication program and, 150
laboratory design, hazardous waste diposal regulations, 413, 416–417
Respirable particulates, combustion products, indoor air quality, 546
Respiration, dust and, *see* Dust
Respirator, *see also* Respiratory protective devices
defined, 675
historical perspective on, 675–676
Respiratory hazards, agricultural hygiene:
acute hazards, 727–728
delayed and chronic hazards, 728–730
Respiratory infection, building-related illness, 534

Respiratory protective devices, 675–719
air contaminants and, 82
air-purifying respirators (nonpowered), 682–692
aerosol capture, 682–687
facepiece valves, 682
gas and vapor removal, 687–688, 692
air-purifying respirators (powered), 696
testing of, 701
certification of, 710–716
classification of, 676–677
facepieces, 677–681
maintenance and cleaning of, 707–708
overview of, 675–676
problems with, 709–710
psychological considerations in use of, 705
respirator efficiency, 697–705
fit checks, 704–705
fitting tests, 697–703
qualitative fit tests, 703–704
selection considerations, 705–706
supplied-air respirators, 692–696
air supply considerations, 708–709
hose masks (powered and nonpowered), 692
self-contained breathing apparatus, 694–696
type C, 692–694
training in use of, 707
Respiratory system, *see also* Lung
barometric pressure effects on, 340–342
dust particle size and, 295–298
lung function in, 298–299
lung transit of contaminants and, 223–224
respiratory protective devices, 705
Rheumatoid arthritis, 3
Rhodotorula, asthma, 559
Rickettsial microorganisms, *see* Biological agents determination
Right-to-know laws, *see* Community right-to-know laws; Worker right-to-know laws
Ringworm, agricultural hygiene, 737, 740
Risk assessment, industrial hygiene survey and, 76
Risk management and prevention programs (RMPPs), industrial hygiene survey and, 75
Room size, laboratory design and, 392–394
Rotameters, *see* Variable area meters (rotameters)
Routes of transmission, 211–228. *See also* Toxic material modes
biological agents determination, 194
biotransformation or metabolism, 239–240
distribution to and deposition in organs and tissues, 232–239
bone deposition, 237–239
general factors in, 233–234
lipid-rich organs and tissues, 236–237
liver and kidney accumulation, 235–236
overview, 232–233
redistribution over time, 235
structures limiting distribution, 234–235
elimination and excretion of contaminants, 240–247
biliary excretion, 245–246
gastrointestinal tract excretion, 246
kidneys, 243–245
kinetics in, 240–243
lung elimination, 246
milk excretion, 247
overview, 240
perspiration and saliva excretion, 246–247
percutaneous absorption, systemic intoxication signs following, 272–273
pharmacokinetic modeling and, 248–252
portals of entry:
cell membrane, 211–217
gastrointestinal tract, 225–228
lungs, 218–225
skin, 228
skin, 256–257
transport and distribution, 228–232
blood elements in, 228–232
lymphatic vascular system in, 232
Rubella virus, health care field, 372
Rubens, Peter Paul, 3

Saccharin, respiratory fit test agents, 704
Safe Drinking Water Act of 1974, 21
Safety considerations, laboratory design, 410–422. *See also* Laboratory design, safety considerations
Safety glasses, welding hazards, 670–671
Safety Institute of America, establishment of, 6
Safety showers and eyewash stations, laboratory design, safety considerations, 419
Saliva, elimination and excretion by, 246–247
Salt baths, manufacturing hazards, 601
Sampler's collection efficiency calibration, calibration instruments and techniques, 522–524
Samples and sampling:
bioaerosols, indoor air quality, 538
error estimation, 525–526
industrial hygiene survey:
investigational, 82
records and reports, 83–84

Samples and sampling (*Continued*)
quality control:
chain of custody transfer and, 429
handling, storage, and delivery of, 442
routine samples, 440–441
sample identification, 438
single-time samples, 441–442
sampling plans, quality control statistics, 447
stability and/or recovery determination of, calibration instruments and techniques, 524–525
volatile organic compounds, indoor air quality, 543
SARA, *see* Superfund Amendments and Reauthorization Act of 1986 (SARA)
Saturation diving, pressure effects, 335–336
Scientific study, role of consultants, 103–105
Sclerosis, 3
Scrotal cancer, 3
SCUBA diving, *see also* Diving hazards
barotrauma of descent, 339
increase in hazards, 330
pressure effects, 334–335
Seasonal variation, dermatoses, indirect causes, 258
Security considerations, laboratory design, 422
Self-appraisal check list, quality control program, program auditing, 449–456
Self-contained breathing apparatus (supplied-air respirators), 694–696
certification of, 710–711
closed-circuit, 695–696
open-circuit, 694–695
testing, 701, 702
Sensitizers, toxic material classification, pathological, 210–211
Settling chambers, described, 872, 879, 882–883
Sewage systems, laboratory design, hazardous waste diposal, 418
Sex differences, heat stress and, 791
Shake-out, manufacturing hazards, 631
Shell coremaking, described, 627
Shell molding, described, 623–624
Sherwood, R. J., 66
Shielded metal arc welding, described, 662–665
Shifts, industrial hygiene survey, investigational, 81–82
Shigellosis, described, 197
Shipment, quality control:
chain of custody and, 429
sample handling, storage, and delivery, 442
Sick building syndrome, *see also* Building-related illness; Indoor air quality
carbon dioxide levels and, 569
concept of, 10
control measures for, 581–584
proactive measures, 582–584
remedial action, 581–582
costs of, 562–564
defined, 535
described, 532–532
epidemiology
Danish studies, 555–556
U.K. studies, 554–555
U.S. studies, 553–554
evaluation protocols and guidelines, 572–581
components of, 574–576
qualitative evaluation, 572–574
quantitative evaluation, 577–578
regulations and, 578–580
hospitals, 373–376
mass psychogenic illness concept, 564–565
olf and decipol concepts, 556–557
symptoms and characteristics of, 553
Siderosis:
dust effect on lung, 309
welding hazards, 663
Significant risk concept, OSHA and, 23–24
Silica dust, manufacturing hazards:
foundry, 632
molding, 624–625
Silicosis:
abrasive blasting hazards, 597
historical perspective on, 3
nodular silicosis, dust effect on lung, 313–315
Single-industry approach, 17
Sized ear plug, described, 997–998
Skin:
contaminants transfer, 228
defensive function of, 255–257
percutaneous absorption, systemic intoxication signs following, 272–273
Skin cancer, *see also* Cancer and carcinogens
agricultural hygiene, 732, 740
dermatoses and, 270
Skin disease, *see* Dermatoses
Skin temperature (sweat evaporation rate), heat strain index, 781–782
Skin ulcers:
dermatoses, clinical appearance of, 270–271
historical reporting of, 254
Slavery, 2, 13
Smog, particulate matter contaminant, classification, 207
Smoke, particulate matter contaminant, classification, 207
Social Security Act, job safety concept and, 14

Socioeconomic class:
  agricultural hygiene, 725
  coal worker's pneumoconioses (black lung disease), 315
  labor and, 2
Sodium chloride, respirator efficiency tests, 699
Sodium silicate system, described, 625–626
Soldering, 655–660. *See also* Brazing hazards
  application techniques, 658
  cleaning techniques, 658–660
  controls, 660
  flux composition, 655–657
  fluxing operations, 658
  initial cleaning procedures, 658
  overview of, 655
  solder composition, 657
Solvents:
  air-purifying respirators (nonpowered), canister effects of, 690–691
  chlorinated hydrocarbon solvents, welding hazards, 671
  degreasing technology, 601–608
  paints, 651, 652
  soldering, 658
Soot, cancer and, 3
Sound, terminology in, 938–941. *See also* Noise
Sound level meter, described, 968–972
Sound measurement, 938–951
  combining levels of, 946–951
  intensity and intensity level measurement, 943
  power and power level measurement, 943–944
  power/intensity/pressure interrelationships, 944–946
  pressure and pressure level measurement, 941–943
Spectrometer, laboratory plumbing design and, 406
Spiked samples, stability and/or recovery determination of, calibration instruments and techniques, 524
Spirometer or gasometer, cumulative air volume, calibration instruments and techniques, 471–472
Sporotrichosis, described, 197
Spray chambers, liquid scrubbing, 905–907
Spraying, *see* Thermal spraying
Squamous cell carcinoma, *see also* Cancer and carcinogens
  agricultural hygiene, 740
  dermatoses, clinical appearance of, 270
Standards:
  air pollution standards, U.S. Environmental Protection Agency (EPA), 930
  American National Standards Institute (ANSI):
    hearing measurement standards, 959
    hearing protector requirements, 1001–1002
    noise exposure levels, 992–993
  ASHRAE:
    carbon dioxide standards, 568
    evolution of, 569–570
    humidification standards, 567
    indoor air quality, evaluation protocols and guidelines, 577
    thermal discomfort, 561–562
    ventilation system standards, 565, 566
  benzene, 23
  calibration, 469–470
  enforcement of, historical perspective on, 15–16
  heat stress, 766–772
  historical perspective on establishment of, 6, 14, 15
  laboratory design and, 386–387
  National Ambient Air Quality Standards, indoor air quality, 547
  National Ambient Air Quality Standards (NAAQS), 843
  noise exposure regulations, 979–980
  OSHA authority in, Supreme Court decision, 23–24
  OSHAct, 21–22
  OSHA processes for setting, 24
  quality control, reagent and reference standards, 437
  respiratory protective devices, 676
  state laws, 15
  velocity calibration procedures, comparison of primary and secondary standards, 488–494
Standards Advisory Committee of Hazardous Materials Labeling, OSHA and, 125
Stapes, 956
State and local governments:
  air pollution regulations, 846
  enforcement of standards by, 15–16
  federal hazard communication standards and, 130–131
  hazardous material information system and, 161–162
  indoor air quality, evaluation protocols and guidelines, 578
  industrial hygiene survey and, 85
  organizational consultants and, 113
  shift to federal government from, 16
  solid and hazardous waste agencies of, 417
  variable standards among, 15
  worker right-to-know laws, 130

Static systems (batch mixtures), known vapor concentrations production, calibration instruments and techniques, 495–497
Statistical error, sampling errors, 526
Statistical quality control program, 442–448. *See also* Quality control program
  control charts, 443–447
  data analysis techniques, 447
  data validation, 448
  sampling plans, 448
Steady-state and impulsive noise integration, noise exposure limits, 982–983
Steiger, William, 19
Stop-gap methods, quality control, single-time samples, 441
Storage, laboratory design and, 389
Stress-related illness, *see also* Heat stress; Psychological considerations
  agricultural hygiene, 742–743
  industrial hygiene survey and, 74
Study, *see* Scientific study role
Styrene monomer, manufacturing hazards, molding, 625
Submarines and submarine escape:
  described, 337–338
  SCUBA diving and, 335
Substitution procedures, dermatoses prevention, 278–279
Sulfur dioxide:
  combustion products, indoor air quality, 546
  predictive air quality modeling, 846
Sulfur hexafluoride, respirator efficiency tests, 699
Sulfur trioxide, air pollution standards, 932
Sunlight, melanoma, dermatoses, clinical appearance of, 270
Superfund Amendments and Reauthorization Act of 1986 (SARA), 21
  hazard communication program and, 149–150
  hazardous material information system and, 162
  industrial hygiene survey and, 75
  right-to-know and, 124
Supplied-air helmets, facepieces (respiratory protective devices), 681
Supplied-air hoods:
  supplied-air respirators, 693
  testing, 701, 702
Supplied-air respirators, 692–696
  air supply considerations, 708–709
  certification of, 711
  hose masks (powered and nonpowered), 692
  self-contained breathing apparatus, 694–696
  testing, 701, 702
  type C, 692–694
Supplied-air suits:
  facepieces (respiratory protective devices), 681
  supplied-air respirators, 693–694
Supreme Court (U.S.), *see* United States Supreme Court
Survey, *see* Industrial hygiene survey
Suspended particulate matter, predictive air quality modeling, 846
Sweat:
  body temperature regulation, 778–779
  dermatoses, indirect causes, 258
  elimination and excretion by, 246–247
  rate of, heat strain index, 781
  skin heat control by, 256
Sweat evaporation rate (skin temperature), heat strain index, 781–782
Sweat-induced reactions, dermatoses, clinical appearance of, 269
Sweden, 50–56
Systemic intoxication, signs of, following percutaneous absorption, 272–273
Systemic poisons, toxic material classification, pathological, 209

Tall stacks, air pollution controls classification, 871–872
Tape recorder, noise measurement, 978
Taxation, agricultural hygiene, 747
Temperature, *see also* Body temperature regulation
  air-cleaning method selection, carrier gas properties, 867–868
  dermatoses, direct causal factors, 265
  diving, thermal problems, 344–347
  microphone effects of, noise measurement, 974
  respiratory protective device problems, 709–710
  thermal discomfort, indoor air quality, 560–562, 561–562
Teratogens:
  health care field and, 372
  placental barrier and, 234–235
Testing, *see* Nondestructive testing
Test rooms, hearing measurement, 962–964
Thackrah, Charles, 3
Thermal anemometer, heat stress measurement, 801
Thermal discomfort, indoor air quality, 560–562
Thermal meters, mass flow and tracer techniques, calibration instruments and techniques, 485

Thermal radiation:
effective tempertaure and, 795
heat stress, worker-environment thermal exchange, 775–776
heat stress control, 815–817, 823–824
heat stress measurement, 802–803
Thermal spraying, manufacturing hazards, 648–650
Thermal treatment, laboratory design, hazardous waste diposal, 417–418
Thermistors, 799
Thermoactinomyces, hypersensitivity pneumonitis, 558
Thermocouples, 799
Thermometry, described, 799–800
Third-octave band analyzer, described, 975
Threshold limit values (TLVs):
air pollution standards, 930
consultant study role in, 105
enforcement responsiblity for, 16
establishment of, 6, 15
future objectives and, 11
heat stress standards, 767–768, 804
noise exposure limits, 980
politics and, 11–12
Ticks, dermatoses, direct causal factors, 266
Tight buildings, health care field, sick building syndrome, 373–376
Tight-fitting facepieces:
respiratory protective devices, 679–680
supplied-air respirators, 693
Timeliness, industrial hygiene survey, records and reports, 84
Tobacco smoke, indoor air quality, 548–549
Torts, 181–186
contracts, 185–186
generally, 181–182
intentional or fraudulent misrepresentation, 184–185
negligence, 182–184
Toxicity:
carrier gas properties, air-cleaning method selection, 865
contaminant properties, air-cleaning method selection, 865
Toxic material modes, 205–252. *See also* Contaminants; Routes of transmission
biotransformation or metabolism, 239–240
classification of:
chemical, 207
pathological, 207–211
physical, 206–207
distribution to and deposition in organs and tissues, 232–239
bone deposition, 237–239
general factors in, 233–234
lipid-rich organs and tissues, 236–237
liver and kidney accumulation, 235–236
overview, 232–233
redistribution over time, 235
structures limiting distribution, 234–235
elimination and excretion of contaminants, 240–247
biliary excretion, 245–246
gastrointestinal tract excretion, 246
kidneys, 243–245
kinetics in, 240–243
lung elimination, 246
milk excretion, 247
overview, 240
perspiration and saliva excretion, 246–247
overview of, 205–206
percutaneous absorption, systemic intoxication signs following, 272–273
pharmacokinetic modeling and, 248–252
portals of entry, 211–228
cell membrane transit, 211–217
gastrointestinal tract transit, 225–228
lung transit, 218–225
skin transit, 225–228
skin, 256–257
transport and distribution, 227–232, 228–232
blood elements in, 228–232
lymphatic vascular system in, 232
Toxicokinetics, *see* Pharmacokinetic modeling
Toxicology Committee (AIHA), 15
Toxic Substances Control Act of 1976 (TSCA), 21
dermatoses and, 255
industrial hygiene survey and, 75
Toxins:
historical perspective on, 2–3, 4–5
laboratory design, safety considerations, 411–412
Tracheobronchial tract, lung transit of contaminants and, 223–224
Trade organizations, organizational consultants and, 112
Trade secrets, hazard communication standards and, 135–137
Training programs, *see also* Academic programs; Education
consultant role in, 101–102
hazard communication program, 162–163
heat stress managment control, 824–825
quality control program management, 432
respiratory protective devices use, 707
Transmission, *see* Routes of transmission
Trauma, *see also* Mechanical injury
acute trauma, 725–726
cumulative musculoskeletal trauma, 726–727

Trench foot, historical perspective on, 2
Trichloroethane:
  vapor degreasing, 604, 606
  welding hazards, 667–668
  welding operations and, 607
Trichloroethylene:
  historical standards, 15
  vapor degreasing, 604, 606
Tubercle bacillus, coal worker's pneumoconioses and, 316–317
Tuberculosis:
  health care field, controls, 378
  historical perspective on, 3
Tunnels and caissons:
  aseptic necrosis of bone, 351–352
  described, 336–337
  historical perspective on, 330
Two-stage cleaning systems, air pollution controls classification, 877–878

Ulcers of the skin, *see* Skin ulcers
Ultrasound, nondestructive testing, 648
Ultraviolet radiation:
  dermatoses, direct causal factors, 265
  manufacturing hazards, thermal spraying, 650
  nondestructive testing:
    liquid penetrant, 647–648
    magnetic particle inspection, 647–648
  welding hazards, 664–665, 666, 667, 669
Uniform regulation, hazard communication standards, 126–128
Unions, *see* Labor unions
United Kingdom:
  Factory Acts, 4, 13–14
  industrial hygiene in, 56–66
United Nations Organization, activities of, 67–68
United States:
  historical perspective, 4–7, 14–16
  legislative hearings in, 16–20
  Occupational Safety and Health Act of 1970
    future prospects for, 25–28
    provision of, described, 20–21
    related governmental actions to, 21–24
United States Airforce, facepieces (respiratory protective devices), 677–678
United States Atomic Energy Commission (AEC), 16, 17
United States Bureau of Labor, creation of, 6
United States Bureau of Labor Standards, enforcement by, 16
United States Bureau of Labor Statistics, dermatoses reported to, 254
United States Bureau of Mines:
  creation of, 6
  enforcement by, 17
  respirators, 676
United States Bureau of Occupational Health (BOSH), 73–74
United States Department of Energy (DOE), jurisdiction, 27
United States Department of Health, Education, and Welfare, politics and, 19
United States Department of Health and Human Services, industrial hygiene survey and, 86
United States Department of Justice, criminal sanctions, 187
United States Department of Labor:
  Assistant Secretary office created, 17, 19
  creation of, 6
  dermatoses reported to, 254
  heat stress standards, 813
  industrial hygiene survey and, 86
  noise exposure limits, 979
  politics and, 17, 19
United States Department of Transportation, laboratory design, hazardous waste diposal regulations, 416
United States Environmental Protection Agency (EPA):
  air pollution and, 11, 840, 930, 932, 934
  Clean Air Act of 1970 and, 843
  criminal sanctions, 187
  hazardous waste dispoal, 26
  indoor air quality, pesticides, 545
  industrial hygiene survey and, 74
  laboratory design, hazardous waste disposal regulations,
  noise reduction ratings, 989
  OSHA and, 26–27
  predictive air quality modeling, 846
  radon measurement, 552
  regulations of, 17
  standard-setting process and, 24
  tall stack requirements, 872
  uniform standards and, 126, 128
United States Food and Drug Administration, X-ray generator regulations, 645
United States Navy:
  hyperbaric chamber standards, 336
  submarines and submarine escape, 337–338
United States Nuclear Regulatory Commission, radiography, industrial, nondestructive testing, 646
United States Public Health Service, 5, 6, 73
  industrial hygiene/medical surveys combined, 85
  industrial hygiene survey and, 85
  tobacco smoke, indoor air quality, 548

United States Secretary of Labor:
enforcement and, 17
powers of, under OSHAct, 21–22
United States Supreme Court:
criminal sanctions, 188
OSHA standard-setting authority and, 23–24
United States v. Park, criminal sanctions, 188–189
Universities, consultant training role, 101. *See also* Academic programs
Upstream samples, sampler's collection efficiency calibration, 524
Urethane paints, 652–653
Urticaria, dermatoses, clinical appearance of, 271

Vane anemometer, heat stress measurement, 802
Vapor degreasing, described, 602–605
Vapors and gases, *see* Gases and vapors
Variable area meters (rotameters), volumetric flow rate, calibration instruments and techniques, 477–479
Variable head meters, volumetric flow rate, calibration instruments and techniques, 479–484
Varicella zoster virus, health care field, 372
VDT, *see* Video display terminal (VDT)
Velocity calibration procedures, calibration instruments and techniques, 488–494
Velometer, industrial hygiene survey, control procedures, 82–83
Ventilation and ventilation systems, 565–572. *See also* Heating, ventilation, and air conditioning (HVAC) system; Sick building syndrome
air pollution control, 863
ASHRAE and, 569–570, 570–572
bioeffluents, indoor air quality, 549–550
carbon dioxide concentrations, 568–569
dermatoses prevention, 279
described, 565–567
heat stress control:
general ventilation, 818–820
humidity and, 821–822
local exhaust ventilation, 817
humidification and dehumidification, 567–568
industrial hygiene survey, control procedures, 82–83
laboratory design, 399–400
laboratory design and, 387–388
manufacturing hazards:
abrasive blasting, 597–598
acid and alkali cleaning of metals, 599, 601
degreasing technology, 605, 607
grinding, polishing, and buffing operations, 635
thermal spraying, 650
measurement, calibration of, 465
overview of, 565
sick building syndrome and, 533
Venturi meter, volumetric flow rate, calibration instruments and techniques, 480–481
Veterinary biologicals and antibiotics, agricultural hygiene, 734–736
Vibration:
agricultural hygiene, 741
disease induced by, manufacturing hazards, foundry, 633
magnetic field effects and, noise measurement, 977–978
noise control procedures, 1017–1018
Video display terminal (VDT), laboratory design and, 389–390
Vigliani, Enrico, 44
Vinyl chloride respirators, certification of, 716
Viral infection, *see also* Bioaerosols; Biological agents determination
dermatoses, direct causal factors, 266–267
health care field, 372
Viscosity, air-cleaning method selection, carrier gas properties, 868
Volatile organic compounds (VOCs), 540–545
building-related illness, 534, 557
characterization of, indoor air quality, 540
data interpretation, indoor air quality, 543–545
health effects of, indoor air quality, 542–543
sampling for, indoor air quality, 543
sources and nature of, indoor air quality, 540–542
Volumetric flow rate, calibration instruments and techniques, 476–484

Walls, laboratory design and, 389
Walsh–Healy Public Contracts Act of 1936, 14–15, 16–17
noise exposure limits, 979, 980
Walton, W. H., 66
Warner, C. G., 66
Warts, dermatoses, clinical appearance of, 269–270
Waste disposal, laboratory design, safety considerations, 413–419. *See also* Hazardous waste management and disposal
Water, route of transmission, 194
heat stress managment control, 827–828
Water-cooled garments, 829

Water displacement, cumulative air volume, calibration instruments and techniques, 471
Water and electrolyte balance:
heat stress and, 791–792
Water pollution, organizational consultant services and, 116
Water Pollution Control Act, *see* Clean Water Act of 1972
Water-sensitive chemicals, laboratory design, safety considerations, 412
Water system, laboratory plumbing design and, 406
Weather and climate, heat stress and, 806–807
Welding, 662–674
degreasing technology and, 607
gas metal arc welding, 666–668
gas tungsten arc welding, 665–666
gas welding, 668–669
overview of, 662
shielded metal arc welding, 662–665
Wet bulb globe temperature (WBGT):
described, 797–798
heat stress empirical studies, 808–809
heat stress management control and, 828, 831
integrated heat stress indexes, 804–805
predictive value of, 806–807
Wet-bulb temperatures, described, 794
Wet collectors, economic analysis models (air pollution controls), 857–859
Wet cyclones, described, 892–893
Wet globe temperature (WGT):
described, 798
integrated heat stress indexes, 804
Wetted overgarments, 830
Wet test meter, cumulative air volume, calibration instruments and techniques, 474–475
Williams, Harrison, 19
Williams–Steiger Occupational Safety and Health Act of 1970 (OSHAct), *see* Occupational Safety and Health Act of 1970 (OSHAct)
Wire spraying, 649
Woods, *see* Plants and woods
Worker-environment thermal exchange, heat stress, 772–777
Worker right-to-know laws, *see also* Community right-to-know laws
dermatoses and, 255
historical perspective on, 124–125
legal challenges to, 130–131
organizational consultants and, 113
state and local governments, 130
Work force:
age level of, 74
agricultural, 723
manufacturing sector, 595
Workmen's compensation laws, 4, 186
Workplace biological agents determination, *see* Biological agents determination
Work-rest regimen, heat stress management control, 826–827
World Health Organization (WHO):
heat stress standards, 810
indoor air quality, evaluation protocols and guidelines, 580
Written hazard communication program, employer responsibility for, 144

Xenobiotics, *see* Contaminants; Toxic material modes
X-ray defraction instruments, laboratory plumbing design and, 406
X-ray hazards, health care field, 371
X-ray studies (industrial), nondestructive testing, 645–647
X-ray studies (medical):
asbestos dust, 321
coal worker's pneumoconioses and, 317
dust effect on lung, 306

Yant Memorial Award, 33, 37, 39, 44, 56, 66
Yeast infection, dermatoses, direct causal factors, 266
Yellow fever, dermatoses, direct causal factors, 266

Z-37 Committee (American Standards Association), standards of, 15
Zoonoses, agricultural hygiene, 736–737